T0231034

CRC Handbook
of
Materials Science

Volume I: General Properties

Volume II: Metals, Composites, and Refractory Materials

Volume III: Nonmetalic Materials and Applications

Editor

Charles T. Lynch, Ph.D.
Senior Scientist for Environmental Effects
Metals Behavior Branch, USAF
Air Force Wright Aeronautical Materials Laboratory
Wright-Patterson Air Force Base, Ohio

Volume IV: Wood

Editors

Robert Summitt, Ph.D.
Professor, Department of Metallurgy, Mechanics,
and Materials Science
Michigan State University
East Lansing, Michigan

Alan Sliker, Ph.D.
Professor of Wood Science
Forestry Department
Michigan State University
East Lansing, Michigan

CRC Handbook of Materials Science

Volume I:
General Properties

Editor

Charles T. Lynch, Ph.D.

Senior Scientist for Environmental Effects,
Metals Behavior Branch, USAF
Air Force Wright Aeronautical Materials Laboratory
Wright-Patterson Air Force Base, Ohio

CRC Press
Taylor & Francis Group
Boca Raton London New York

CRC Press is an imprint of the
Taylor & Francis Group, an **informa** business

CRC Press
Taylor & Francis Group
6000 Broken Sound Parkway NW, Suite 300
Boca Raton, FL 33487-2742

Reissued 2019 by CRC Press

© 1974 by Taylor & Francis Group, LLC
CRC Press is an imprint of Taylor & Francis Group, an Informa business

No claim to original U.S. Government works

A Library of Congress record exists under LC control number:

Publisher's Note
The publisher has gone to great lengths to ensure the quality of this reprint but points out that some imperfections in the original copies may be apparent.

Disclaimer
The publisher has made every effort to trace copyright holders and welcomes correspondence from those they have been unable to contact.

ISBN 13: 978-0-367-21163-9 (hbk)
ISBN 13: 978-0-367-21164-6 (pbk)
ISBN 13: 978-0-429-26579-2 (ebk)

Visit the Taylor & Francis Web site at http://www.taylorandfrancis.com and the
CRC Press Web site at http://www.crcpress.com

PREFACE

It has been the goal of the *CRC Handbook of Materials Science* to provide a current and readily accessible guide to the physical properties of solid state and structural materials. Interdisciplinary in approach and content, it covers the broadest variety of types of materials consistent with a reasonable size for the volumes, including materials of present commercial importance plus new biomedical, composite, and laser materials. This volume, General Properties, is the first of the three-volume *Handbook*; Metals, Composites, and Refractory Materials is the second. Volume III, Nonmetallic Materials and Applications, also contains a section on materials information and referral publications, data banks, and general handbooks.

During the approximately four years that it has taken to formulate and compile this *Handbook*, the importance of materials science has taken on a new dimension. The term "materials limited" has come into new prominence, enlarged from the narrower consideration of technical performance of given materials in given conditions of stress, environment, and so on, to encompass the availability of materials in commerce at a reasonable price. Our highly industrialized society, with its immense per capita consumption of raw materials, today finds itself facing the long-prophesized shortages of materials in many diverse areas of our economy. Those future shortages have become today's problems. As we find ourselves "materials limited" with respect to availability and price, with a growing concern for where our raw materials come from and how supplies may be manipulated to our national disadvantage, increased economic utilization of all our resources, and particularly our materials resources, becomes an American necessity. With this changing background the purpose for this type of compilation has broadened beyond a collection of data on physical properties to one of concern for comparative properties and alternative employment of materials. Therefore, at this time it seems particularly appropriate to enter this new addition to the CRC Handbook Series.

Most of the information presented in this *Handbook* is in tabular format for easy reference and comparability of various properties. In some cases it has seemed more advisable to retain written sections, but these have been kept to a minimum. The importance of having critically evaluated property data available on materials to solve modern problems is well understood. In this *Handbook* we seek to bridge the gap between uncritical data collections carrying all the published information for a single material class and general reference works with only limited property and classification data on materials. On the basis of advice from many and varied sources, numerous limitations and omissions have been necessary to retain a reasonable size. This reference is particularly aimed at the nonexperts, or those who are experts in one field but seek information on materials in another field. The expert normally has his own specific original sources available to guide him in his own area of expertise. He often needs assistance, however, to get started on something new. There is also considerable general information of interest to almost all scientists, engineers, and many administrators in the field of materials and materials applications. Comments and suggestions, and the calling to our attention of typographical errors, will be welcomed and are encouraged.

My sincere thanks is extended to all who have advised on the formulation, content, and coverage of this *Handbook*. I am grateful to many colleagues in industry, academic circles, and government for countless suggestions and specific contributions, and am particularly indebted to the Advisory Board and Contributors who have put so much of their time, effort, and talent into this compilation. Special appreciation is extended to the editorial staff of CRC Press, to Karen A. Gajewski, the Administrative Editor, and to Gerald A. Becker, Director of Editorial Operations.

I want to pay special tribute to my wife, Betty Ann, for her magnificent patience, encouragement, and assistance, and to our children, Karen, Ted Jr., Richard, and Thomas, for giving their Dad some space, quiet, and assistance in the compilation of a considerable amount of data.

Charles T. Lynch
Fairborn, Ohio
September 1974

THE EDITOR

Charles T. Lynch, Ph.D., is Senior Scientist for Environmental Effects in the Metals Behavior Branch of the Air Force Wright Aeronautical Materials Laboratory, Wright-Patterson Air Force Base, Ohio.

Dr. Lynch graduated from The George Washington University in 1955 with a B.S. degree in chemistry. He received his M.S. and Ph.D. degrees in analytical chemistry in 1957 and 1960, respectively, from the University of Illinois, Urbana.

Dr. Lynch served in the Air Force for several years before joining the Air Force Materials Laboratory as a civilian employee in 1962. Prior to his current position, he served as a research engineer, group leader for ceramic research, and Chief of the Advanced Metallurgical Studies Branch.

Dr. Lynch is a member of the American Chemical Society, American Ceramic Society, American Association for the Advancement of Science, Ohio Academy of Science, New York Academy of Science, Sigma Xi-RESA, and the Metallurgical Society of the AIME. He holds 13 patents and has published more than 60 research papers, over 70 national and international presentations, and one book, *Metal Matrix Composites,* written with J. P. Kershaw and published by The Chemical Rubber Company (now CRC Press) in 1972.

CONTRIBUTORS

C. Howard Adams
Manager of Product Engineering
Monsanto Company
Air Park Technical Center
Chesterfield, Missouri 63017

Allen M. Alper
Director of Research and Engineering
Chemical and Metallurgical Division
GTE Sylvania, Incorporated
Towanda, Pennsylvania 18848

Ray E. Bolz
Vice President and Dean of the Faculty
Worcester Polytechnic Institute
Worcester, Massachusetts 01609

Allen Brodsky
Health Physicist
U.S. Nuclear Regulatory
  Commission
Washington, D.C. 20555

D. F. Bunch
Atomics International
Canoga Park, California 91304

Donald E. Campbell
Senior Research Associate – Chemistry
Research and Development Division, Technical
  Staffs Services Laboratories
Corning Glass Works
Sullivan Park
Corning, New York 14830

William B. Cottrell
Director, Nuclear Safety Program
Oak Ridge National Laboratory
Oak Ridge, Tennessee 37830

Joseph E. Davison
Assistant Professor of Materials Engineering
University of Dayton
300 College Park
Dayton, Ohio 45409

Ed B. Fernsler
Technical Service Manager
Huntington Alloy Products Division
The International Nickel Company, Incorporated
  rated
Huntington, West Virginia 25720

Francis S. Galasso
Chief, Materials Science
United Aircraft Research Laboratories
East Hartford, Connecticut 06108

Henry E. Hagy
Senior Research Associate – Physics
Research and Development Division,
Technical Staffs Services Laboratories
Corning Glass Works
Sullivan Park
Corning, New York 14830

C. R. Hammond
Emhart Corporation
P.O. Box 1620
Hartford, Connecticut 06102

Michael Hoch
Professor, Department of Materials Science
and Metallurgical Engineering
University of Cincinnati
Clifton Avenue
Cincinnati, Ohio 45221

Bernard Jaffe
Vernitron Piezoelectric Division
232 Forbes Road
Bedford, Ohio 44146

Richard N. Kleiner
Section Head, Ceramics Department
Precision Materials Group
Chemical and Metallurgical Division
GTE Sylvania, Incorporated
Towanda, Pennsylvania 18848

George W. Latimer, Jr.
Group Leader, Analytical Methods
Mead Johnson Company
Evansville, Indiana 47721

Robert I. Leininger
Project Director, Biomaterials
Department of Biology, Environment, and
Chemistry
Battelle/Columbus Laboratories
505 King Avenue
Columbus, Ohio 43201

Robert S. Marvin
Office of Standard Reference Data
National Bureau of Standards
U.S. Department of Commerce
Washington, D.C. 20234

Eugene F. Murphy
Chief, Research and Development Division
Prosthetic and Sensory Aids Service
U.S. Veterans Administration
252 Seventh Avenue
New York, New York 10001

A. Pigeaud
Research Associate
Department of Metallurgy and Materials Sci-
ence
University of Cincinnati
Clifton Avenue
Cincinnati, Ohio 45221

B. W. Roberts
Director, Superconductive Materials Data
Center
General Electric Corporate Research and Devel-
opment
Box 8
Schenectady, New York 12301

Gail D. Schmidt
Chief, Radioactive Materials Branch
Division of Medical Radiation Exposure
Bureau of Radiological Health
U.S. Public Health Service
Rockville, Maryland 20852

James E. Selle
Senior Research Specialist
Mound Laboratory
Monsanto Research Corporation
Miamisburg, Ohio 45342

Gertrude B. Sherwood
Office of Standard Reference Data
National Bureau of Standards
U.S. Department of Commerce
Washington, D.C. 20234

Ward F. Simmons
Associate Director, Defense Metals Information
Center
Battelle/Columbus Laboratories
505 King Avenue
Columbus, Ohio 43201

George L. Tuve
2625 Exeter Road
Cleveland Heights, Ohio 44118

A. Bennett Wilson, Jr.
Temple University Health Science
Center
Krusen Research Center
Moss Rehabilitation Hospital
Philadelphia, Pennsylvania 19141

TABLE OF CONTENTS

Section 1

The Elements

THE ELEMENTS

One of the most striking facts about the elements is their unequal distribution and occurrence in nature. Present knowledge of the chemical composition of the universe, obtained from the study of the spectra of stars and nebulae, indicates that hydrogen is by far the most abundant element and may account for more than 90% of the atoms or about 75% of the mass of the universe. Helium atoms make up most of the remainder. All of the other elements together contribute only slightly to the total mass.

The chemical composition of the universe is undergoing continuous change. Hydrogen is being converted into helium, and helium is being changed into heavier elements. As time goes on, the ratio of heavier elements increases relative to hydrogen. Presumably, the process is not reversible.

Studies of the solar spectrum have led to the identification of 67 elements in the sun's atmosphere; however, all elements cannot be identified with the same degree of certainty. Other elements may be present in the sun, although they have not yet been detected spectroscopically. The element helium was discovered on the sun before it was found on earth. Some elements, such as scandium, are relatively more plentiful in the sun and stars than here on earth.

Minerals in lunar rocks brought back from the moon on the Apollo 11, 12, and 14 Missions consist predominantly of *plagioclase* [(Ca, Na)(Al, Si)$O_4 O_8$] and *pyroxene* [(Ca, Mg, Fe)$_2 Si_2 O_6$] — two minerals common in terrestrial volcanic rock. No new elements have been found on the moon that cannot be accounted for on earth; however, two minerals *armalcolite* [(F3, Mg)$Ti_2 O_5$] and *pyroxferroite* [CaFe$_6$(SiO$_3$)$_7$] are new. The oldest known terrestrial rocks are about 3.5 billion years old. One rock, known as the "Genesis Rock," brought back from the Apollo 15 Mission, is about 4.15 billion years old. This is only about one half billion years younger than the supposed age of the moon and solar system. Lunar rocks appear to be relatively enriched in refractory elements, such as chromium, titanium, zirconium, and the rare earths, and impoverished in volatile elements, such as the alkali metals, in chlorine, and in noble metals, such as nickel, platinum, and gold.

F. W. Clarke and others have carefully studied the composition of rocks making up the crust of the earth. Oxygen accounts for about 47% of the crust, by weight, while silicon comprises about 28%, and aluminum about 8%. These elements, plus iron, calcium, sodium, potassium, and magnesium, account for about 99% of the composition of the crust.

Many elements, such as tin, copper, zinc, lead, mercury, silver, platinum, antimony, arsenic and gold, which are so essential to our needs and civilization, are among some of the rarest elements in the earth's crust. These are made available to us only by the processes of concentration in ore bodies. Some of the so-called *rare earth* elements have been found to be much more plentiful than originally thought and are about as abundant as uranium, mercury, lead, or bismuth. The least abundant rare-earth or *lanthanide* element, thulium, is believed to be more plentiful on earth than silver, cadmium, gold, or iodine. Rubidium is the 16th most abundant element and is more plentiful than chlorine, while its compounds are little known in chemistry and commerce.

Ninety-one elements occur naturally on earth. Minute traces of plutonium-244 were recently discovered in rocks mined in So. Calif. This discovery supports the theory that heavy elements were produced during creation of the solar system. While technetium and promethium have not yet been found naturally on earth, they have been found to be present in stars. Technetium has been identified in the spectra of certain "late" type stars, and promethium lines have been identified in the spectra of a faintly visible star HR465 in Andromeda. Promethium must have been made very recently near the star's surface for no known isotope of this element has a half-life longer than 17.7 yr.

It has been suggested that californium is present in certain stellar explosions known as supernovae; however, this has not been proved. At present no elements are found elsewhere in the universe that cannot be accounted for here on earth.

All atomic mass numbers from 1 to 23 are found naturally on earth except for masses 5 and 8. About 280 stable and 67 naturally radioactive isotopes occur on earth, totaling 346. In addition, the neutron, technetium, promethium, and the transuranic elements (lying beyond uranium) up to Element 105 have been produced artificially. Laboratory processes have now extended the radioactive mass numbers beyond 238 to about 260. Each element from atomic number 1 to 104 is known to have at least one radioactive isotope. About 1700 different nuclides (the name given to different kinds of nuclei, whether they are of the same or different elements) are now recognized. Many stable and radioactive isotopes are now produced and distributed by the Oak Ridge National Laboratory, Oak Ridge, Tenn., U.S.A., to customers licensed by the U.S. Atomic Energy Commission.

Elements 89 (actinium) through 103 (lawrencium) are chemically similar to the rare earth or lanthanide elements (elements 57 to 71, inclusive). These elements therefore have been named *actinides* after the first member of this series. Those elements beyond uranium that have been produced artificially have the following names and symbols: neptunium, 93, (Np); plutonium, 94, (Pu); americium, 95, (Am); curium, 96, (Cm); berkelium, 97, (Bk); californium, 98, (Cf); einsteinium, 99, (Es); fermium, 100, (Fm); mendelevium, 101, (Md); nobelium, 102, (No); and lawrencium, 103, (Lr). It is now claimed that Elements 104 and 105 have been produced and positively identified. Element 104 is expected to have chemical properties similar to those of hafnium and would not be a member of the actinide series. Element 105 should have properties similar to those of tantalum.

There is presently some reason for optimism in producing elements beyond Element 105 either by bombardment of heavy isotopic targets with heavy ions, or by the irradiation of uranium or transuranic elements with the instantaneous high flux of neutrons produced by underground nuclear explosions. The limit will be set by the yields of the nuclear reactions and by the half-lives of radioactive decay. It has been suggested that Elements 102 and 103 have abnormally short lives only because they are in a pocket of instability, and that this region of instability might "heal" around Element 105. If so, it may be possible to produce heavier isotopes with longer half-lives. It has also been suggested that Element 114, with a mass number of 298, and Element 126, with a mass number of 310, may be sufficiently stable to make discovery and identification possible.

Calculations indicate that Element 110, a homolog of platinum, may have a half-life of as long as 100 million years. Searches have already been made by workers in a number of laboratories for Element 110 and its neighboring elements in naturally occurring platinum.

There are many claims in the literature of the existence of various allotropic modifications of the elements, some of which are based on incomplete evidence. Also the physical properties of an element may change drastically by the presence of small amounts of impurities. With new methods of purification, which are now able to produce elements with 99.9999 + % purity, it has been necessary to restudy the properties of the elements. For example, the melting point of thorium changes by several hundred degrees by the presence of a small percentage of ThO_2 as an impurity. Ordinary commercial tungsten is brittle and can be worked only with difficulty. Pure tungsten, however, is ductile enough to be drawn as a wire. In general, the value of physical properties given here applies to the pure element, when it is known.

Actinium (Gr. *aktis, aktinos,* beam or ray) — Ac; at. wt (227); at. no. 89; mp 1050°C; bp 3200°C ± 300°C (est.); sp gr 10.07 (calc.). Discovered by Andre Debierne in 1899 and independently by F. Giesel in 1902. Occurs naturally in association with uranium minerals. Actinium-227, a decay product of uranium-235, is a beta emitter with a 21.6-year half-life. Its principal decay products are thorium-227 (18.5-day half-life), radium-223 (11.4-day half-life), and a number of short-lived products including radon, bismuth, polonium, and lead isotopes. In equilibrium with its decay products, it is a powerful source of alpha rays. Actinium metal has been prepared by the reduction of actinium fluoride with lithium vapor at about 1100 to 1300°C. The chemical behavior of actinium is similar to that of the rare earths, particularly lanthanum. Purified actinium comes into equilibrium with its decay products at the end of 185 days, and then decays according to its 21.6-year half-life. It is about 150 times as active as radium, making it of value in the production of neutrons.

Aluminum (L. *alumen, alum*) — Al; at. wt 26.98154; at. no. 13; mp 660.37°C; bp 2467°C; sp gr 2.6989 (20°C); valence 3. The ancient Greeks and Romans used *alum* in medicine as an astringent, and as a mordant in dyeing. In 1761 de Morveau proposed the name *alumine* for the base in alum, and Lavoisier, in 1787, thought this to be the oxide of a still undiscovered metal. Wöhler is generally credited with having isolated the metal in 1827, although an impure form was prepared by Oersted 2 years earlier. In 1807 Davy proposed the name *alumium* for the metal, undiscovered at that time, and later agreeed to change it to *aluminum.* Shortly thereafter, the name *aluminium* was adopted to conform with the "ium" ending of most elements, and this spelling is now in use elsewhere in the world. *Aluminium* was also the accepted spelling in the U.S. until 1925 at which time the American Chemical Society officially decided to use the name *aluminum* thereafter in their publications. The method of obtaining aluminum metal by the electrolysis of alumina dissolved in *cryolite* was discovered in 1886 by Hall in the U.S. and about the same time by Heroult in France. Cryolite, a natural ore found in Greenland, is no longer widely used in commercial production, but has been replaced by an artificial mixture of sodium, aluminum, and calcium fluorides. *Bauxite,* an impure hydrated oxide ore, is found in large deposits in Arkansas, Jamaica, and elsewhere. The Bayer process is most commonly used today to refine bauxite so it can be accommodated in the Hall-Heroult refining process, used to produce most aluminum. Two new processes based on chlorination as a first step show promise to replace the Bayer-Hall process. In the Alcoa smelting process, conventionally made alumina is chlorinated and the aluminum chloride electrolyzed to yield metal and recyclable chlorine. In the Toth process low grade clay minerals can be used. The clay is calcined, chlorinated in the presence of coke, and then reacted with manganese metal to produce aluminum metal and manganese chloride. The latter is then processed to recycle manganese and chlorine. Aluminum is the most abundant metal to be found in the earth's crust (8.1%), but is never found free in nature. In addition to the minerals mentioned above, it is found in feldspars, granite, and in many other common minerals. Pure aluminum, a silvery white metal, possesses many desirable characteristics. It is light, nontoxic, has a pleasing appearance, can easily be formed, machined, or cast, has a high thermal conductivity, and has excellent corrosion resistance. It is non-magnetic and nonsparking, stands second among

metals in the scale of malleability, and sixth in ductility. It is extensively used for kitchen utensils, outside building decoration, and in thousands of industrial applications where a strong, light, easily constructed material is needed. Although its electrical conductivity is only about 60% that of copper per area of cross section, it is used in electrical transmission lines because of its light weight. Pure aluminum is soft and lacks strength, but it can be alloyed with small amounts of copper, magnesium, silicon, manganese, and other elements to impart a variety of useful properties. These alloys are of vital importance in the construction of modern aircraft and rockets. Aluminum, evaporated in a vacuum, forms a highly reflective coating for both visible light and radiant heat. These coatings soon form a thin layer of the protective oxide and do not deteriorate as do silver coatings. They have found application in coatings for telescope mirrors, in making decorative paper, packages, toys, and in many other uses. The compounds of greatest importance are aluminum oxide, the sulfate, and the soluble sulfate with potassium (alum). The oxide, alumina, occurs naturally as ruby, sapphire, corundum, and emery, and is used in glassmaking. Synthetic ruby and sapphire have found application in the construction of lasers for producing coherent light. In 1852, the price of aluminum was about $545/lb, and just before Hall's discovery in 1886, about $11.00. The price rapidly dropped to 30¢ and has been as low as 15¢/lb.

Americium (the Americas) − Am; at. wt (243); at. no. 95; mp 994 ± 4°C; bp 2607°C; sp gr 13.67 (20°C); valence 2, 3, 4, 5, or 6. Americium was the fourth transuranium element to be discovered; the isotope Am^{241} was identified by Seaborg, James, Morgan and Ghiorso late in 1944 at the wartime Metallurgical Laboratory (now the Argonne National Laboratory) of the University of Chicago as the result of successive neutron capture reactions by plutonium isotopes in a nuclear reactor:

$$Pu^{239}(n, \gamma)Pu^{240}(n, \gamma)Pu^{241} \xrightarrow{\beta^-} Am^{241}.$$

Since the isotope Am^{241} can be prepared in relatively pure form by extraction as a decay product over a period of years from strongly neutron-bombarded plutonium Pu^{241}, this isotope is used for much of the chemical investigation of this element. Better suited is the isotope Am^{243}

due to its longer half-life (8.8×10^3 years as compared to 470 years for Am^{241}). A mixture of the isotopes Am^{241}, Am^{242}, and Am^{243} can be prepared by intense neutron irradiation of Am^{241} according to the reactions Am^{241} (n,γ) Am^{242} (n,γ) Am^{243}. Nearly isotopically pure Am^{243} can be prepared by a sequence of neutron bombardments and chemical separations as follows: neutron-bombardment of Am^{241} yields Pu^{242} by the reactions Am^{241} (n,γ) $Am^{242} \xrightarrow{EC} Pu^{242}$; after chemical separation the Pu^{242} can be transformed to Am^{243} via the reactions Pu^{242} (n,γ) $Pu^{243} \xrightarrow{\beta^-} Am^{243}$, and the Am^{243} can be chemically separated. Fairly pure Pu^{242} can be prepared more simply by very intense neutron irradiation of Pu^{239} as the result of successive neutron-capture reactions. Americium metal has been prepared by reducing the trifluoride with barium vapor at $1000°$ to $1200°C$ or the dioxide by lanthanum metal. The luster of freshly prepared americium metal is whiter and more silvery than plutonium or neptunium prepared in the same manner. It appears to be more malleable than uranium or neptunium and tarnishes slowly in dry air at room temperature. Americium is thought to exist in two forms: an alpha form which has a double hexagonal close-packed structure, and a loose-packed cubic beta form. Americium must be handled with great care to avoid personal contamination. As little as 0.02 μg of Am^{241} is the allowable body burden (bone). The alpha activity from Am^{241} is about three times that of radium. When gram quantities of Am^{241} are handled, the intense gamma activity makes exposure a serious problem. Americium dioxide, AmO_2, is the most important oxide. AmF_3, AmF_4, $AmCl_3$, $AmBr_3$, and AmI_3, and other compounds have been prepared. The isotope Am^{241} has been used as a portable source for gamma radiography. It has also been used as a radioactive glass thickness gage for the flat glass industry. Americium-241 is available from the A.E.C. at a cost of $150/g and Americium-243 at a cost of $100/mg.

Antimony (L. *antimonium*) — Sb, (L. *stibium*, mark); at. wt 121.75; at. no. 51; mp 630.74°C; bp 1750°C; sp gr 6.691 (20°C); valence 0, -3, +3, or +5. Antimony was recognized in compounds by the ancients and was known as a metal at the beginning of the 17th century and possibly much earlier. It is not abundant, but is found in over 100 mineral species. It is sometimes found native, but more frequently as the sulfide, *stibnite* (Sb_2S_3); it is found as antimonides of the heavy metals, and as oxides. It is extracted from the sulfide by roasting to the oxide, which is reduced by salt and scrap iron; from its oxides it is also prepared by reduction with carbon. Two allotropic forms of antimony exist: the normal stable, metallic form, and the amorphous gray form. The so-called explosive antimony is an ill-defined material always containing an appreciable amount of halogen; therefore, it no longer warrants consideration as a separate allotrope. The yellow form, obtained by oxidation of *stibine*, SbH_3, is probably impure, and is not a distinct form. Metallic antimony is an extremely brittle metal of a flaky, crystalline texture. It is bluish-white and has a metallic luster. It is not acted on by air at room temperature, but burns brilliantly when heated with the formation of white fumes of Sb_2O_3. It is a poor conductor of heat and electricity, and has a hardness of 3 to 3.5. Antimony, available commercially with a purity of 99.999 + %, is finding use in semiconductor technology, for making infrared detectors, diodes, and Hall-effect devices. Commercial-grade antimony is widely used in alloys with percentages ranging from 1 to 20. It greatly increases the hardness and mechanical strength of lead. Batteries, antifriction alloys, type metal, cable sheathing, and minor products use about half the metal produced. Compounds taking up the other half are oxides, sulfides, sodium antimonate, and antimony trichloride. These are used in manufacturing flame-proofing compounds, paints, ceramic enamels, glass and pottery. Tartar emetic (hydrated potassium antimonyltartate) is used as a medicine. Antimony and many of its compounds are toxic. The maximum allowable concentration of antimony dust in the air is recommended to be 0.5 mg/cm³. Atmospheric concentrations of *stibine* should not exceed 0.1 ppm.

Argon (Gr. *argos*, inactive) — Ar; at. wt 39.948; at. no. 18; fp -189.2°C; bp -185.7°C; density 1.7837 g/l. Its presence in air was suspected by Cavendish in 1785; discovered by Lord Rayleigh and Sir William Ramsay in 1894. The gas is prepared by fractionation of liquid air, the atmosphere containing 0.94% argon. It is 2½ times as soluble in water as nitrogen, having about the same solubility as oxygen; best recognized by the characteristic lines in the red end of the spectrum. It is used in electric light bulbs and in fluorescent tubes at a pressure of about 3 mm, and in filling photo tubes, glow tubes, etc., and as a laser light source. Argon is also used as an inert gas shield for arc welding and cutting, as a blanket for the production of titanium and other reactive elements, and as a protective atmosphere for growing silicon and germanium crystals. Argon is colorless and odorless, both as a gas and liquid. It is available in high-purity form. Commercial argon is available at a cost of about $10\cancel{c}/ft^2$. Argon is considered to be a very inert gas and is not now known to form true chemical compounds, as do krypton, xenon, and radon. However, it does form a hydrate having a dissociation pressure of 105 atm at $0°C$. Ion molecules, such as $(ArKr)^+$, $(ArXe)^+$, $(NeAr)^+$, have been observed spectroscopically. Argon also forms a clathrate with β hydroquinone. This clathrate is stable and can be stored for a considerable time, but a true chemical bond does not exist. Van der Waals' forces act to hold the argon. Naturally occurring argon is a mixture of three isotopes. Five other radioactive isotopes are now known to exist.

Arsenic (L. *arsenicum*, Gr. *arsenikon*, yellow orpiment — identified with *arsenikos*, male, from the belief that metals were different sexes — Arab. *az-zernikh*, the orpiment from Persian *zerni-zar.* gold) — As; at. wt 74.9216; at. no. 33; valence −3, 0, +3 or +5. Elemental arsenic occurs in two solid modifications: yellow, and gray or metallic, with specific gravities of 1.97, and 5.73, respectively. Gray arsenic, the ordinary stable form, has a mp of $817°C$ (28 atm) and sublimes at $613°C$. Several other allotropic forms of arsenic are reported in the literature. It is believed that Albertus Magnus obtained the element in 1250 A.D. In 1649 Schroeder published two methods of preparing the element. It is found native, in the sulfides *realgar* and *orpiment*, as arsenides and sulfarsenides of heavy metals, as the oxide, and as arsenates. *Mispickel* or arsenopyrite (FeSAs) is the most common mineral, from which on heating the arsenic sublimes leaving ferrous sulfide. The element is a steel gray, very brittle, crystalline, semi-metallic solid; it tarnishes in air, and when heated is rapidly oxidized to arsenous oxide (As_2O_3) with the odor of garlic. Arsenic and its compounds are poisonous. The maximum allowable concentration of arsenic is recommended to be 0.5 mg/m^3 of air. Arsenic is also used in bronzing, pyrotechny, and for hardening and improving the sphericity of shot. The most important compounds are white arsenic (As_2O_3),

the sulfide, Paris green $3Cu(AsO_2)_2 \cdot Cu(C_2H_3O_2)_2$, calcium arsenate, and lead arsenate, the last three being used as agricultural insecticides and poisons. Marsh's test makes use of the formation and ready decomposition of arsine (AsH_3). Arsenic is available in high-purity form. It is finding increasing uses as a doping agent in solid-state devices, such as transistors. Gallium arsenide is used as a laser material to convert electricity directly into coherent light.

Astatine (Gr. *astatos*, unstable) — At; at. wt ~210; at. no. 85; mp $302°C$; bp $337°C$; valence probably 1, 3, 5, or 7. Synthesized in 1940 by D. R. Corson, K. R. MacKenzie, and E. Segre at the University of California by bombarding bismuth with alpha particles. The longest-lived isotope, At^{210}, has a half-life of only 8.3 hr. Twenty isotopes are known. Minute quantities of At^{215}, At^{218}, and At^{219} exist in equilibrium in nature with naturally occurring uranium and thorium isotopes, and traces of At^{217} are in equilibrium with U^{233} and Np^{239} resulting from interaction of thorium and uranium with naturally produced neutrons. The total amount of astatine present in the earth's crust, however, totals less than 1 oz. Asatine can be produced by bombarding bismuth with energetic alpha particles to obtain the relatively long-lived $At^{209-211}$, which can be distilled from the target by heating it in air. Only about 0.05 μg of astatine have been prepared to date. The "time of flight" mass spectrometer has been used to confirm that this highly radioactive halogen behaves chemically very much like other halogens, particularly iodine. The inter-halogen compounds AtI, AtBr, and AtCl are known to form, but it is not yet known if astatine forms diatomic astatine molecules. HAt and CH_3At (methyl astatide) have been detected. Astatine is said to be more metallic than iodine, and like iodine, it probably accumulates in the thyroid gland.

Barium (Gr. *barys*, heavy) — Ba; at. wt 137.34; at. no. 56; mp $725°C$; bp $1640°C$; sp gr 3.5 ($20°C$); valence 2. Baryta was distinguished from lime by Scheele in 1774; the element was discovered by Sir Humphry Davy in 1808. It is found only in combination with other elements chiefly in *barite* or *heavy spar* (sulfate) and *witherite* (carbonate) and is prepared by electrolysis of the chloride. Barium is a metallic element, soft, and when pure is silvery-white like lead; it belongs to the alkaline earth group, resembling calcium chemically. The metal oxidizes very easily and should be kept under petroleum or other suitable oxygen-free liquids to exclude air. It is decomposed by water or alcohol. The metal is used as a "getter" in vacuum tubes. The most important compounds are the peroxide (BaO_2), chloride, sulfate, carbonate, nitrate and chlorate. Lithopone, a pigment containing barium sulfate and zinc sulfide, has good covering power, and does not darken in the presence of sulfides. The sulfate, as permanent white or *blanc fixe*, is also used in paint, in x-ray diagnostic works, and in glassmaking. *Barite* is extensively used as a wetting agent in oilwell drilling fluids, and also in making rubber. The carbonate is used as a rat poison, while the nitrate and chlorate give colors in pyrotechny. The impure sulfide phosphoresces after exposure to the light. The compounds and the metal are not expensive. Barium metal (99.5 + % pure) costs about $20.00/lb. All barium compounds that are water or acid soluble are poisonous. Natural-occurring barium is a mixture of seven stable isotopes. Thirteen other radioactive isotopes are known to exist.

Berkelium (*Berkeley,* home of Univ. of Calif.) — Bk; at. wt (247); at. no. 97; valence 3 or 4; sp gr 14 (est.). Berkelium, the eighth member of the actinide transition series, was discovered in December 1949 by Thompson, Ghiorso, and Seaborg, and was the fifth transuranium element synthesized. It was produced by cyclotron bombardment of mg amounts of Am^{241} with helium ions at Berkeley, California. The first isotope produced had a mass number of 243 and decayed with a half-life of 4.6 hr. Eight isotopes are now known and have been synthesized. The existence of Bk^{249}, with a half-life of 314 days, makes it feasible to isolate berkelium in weighable amounts so that its properties can be investigated with macroscopic quantities. One of the first visible amounts of a pure berkelium compound — berkelium chloride — was produced in 1962. It weighed 3 billionth of a gram. Berkelium has not yet been prepared in elemental form, but it is expected to be a silvery metal, easily soluble in diluted mineral acids, and readily oxidized by air or oxygen at elevated temperatures to form the oxide. X-ray diffraction methods have been used to identify the following compounds: BkO_2, Bk_2O_3, BkF_3, $BkCl_3$, and $BkOCl$. As with other actinide elements, berkelium tends to accumulate in the skeletal system. The maximum permissible

body burden of Bk^{249} in the human skeleton is about 0.0004 µg. Because of its rarity, berkelium presently has no commercial or technological uses.

Beryllium (Gr. *berryllos, beryl;* **also called Glucinium or Glucinum,** Gr. *glykys,* sweet) — Be; at. wt. 9.01218; at. no. 4; mp 1278 ± 5°C; bp 2970°C; sp gr 1.848 (20°C); valence 2. Discovered as the oxide by Vauquelin in beryl and in emeralds in 1798. The metal was isolated in 1828 by Wöhler and by Bussy independently by the action of potassium on beryllium chloride. Beryllium is found in some 30 mineral species, the most important of which are *beryl, chrysoberyl,* and *phenacite. Aquamarine* and *emerald* are precious forms of *beryl. Beryl* ($3BeO \cdot Al_2O_3 \cdot 6SiO_2$) is the most important commercial source of the element and its compounds. Most of the metal is not prepared by reducing beryllium fluoride with magnesium metal. Beryllium metal did not become readily available to industry until 1957. The metal, steel gray in color, has many desirable properties. It is one of the lightest of all metals, and has one of the highest melting points of the light metals. Its modulus of elasticity is about one third greater than that of steel. It resists attack by concentrated nitric acid, has excellent thermal conductivity, and is nonmagnetic. It has a high permeability to x-rays, and when bombarded by alpha particles, as from radium or polonium, neutrons are produced in the ratio of about 30 neutrons/million alpha particles. At ordinary temperatures beryllium resists oxidation in air, although its ability to scratch glass is probably due to the formation of a thin layer of the oxide. Beryllium is used as an alloying agent in producing beryllium copper, which is extensively used for springs, electrical contacts, spot-welding electrodes, and nonsparking tools. It is finding application as a possible aerospace structural material. It is used in nuclear reactors as a reflector or moderator for it has a low thermal neutron absorption cross-section. It is used in gyroscopes, computer parts, and inertial guidance instruments where lightness, stiffness, and dimensional stability are required. The oxide has a very high melting point and is also used in nuclear work and ceramic applications. Beryllium and its salts are toxic and should be handled with the greatest of care. Beryllium and its compounds should not be tasted to verify the sweetish nature of beryllium (as did early experimenters). The metal, its alloys, and its salts can be handled safely if certain work codes are observed, but no attempt should be made to work with beryllium before becoming familiar with proper safeguards. The maximum allowable concentration of beryllium dust in an 8-hr day is recommended to be about 2 $\mu g/m^3$ in working areas. The average monthly concentration should not exceed 0.01 $\mu g/m^3$ in non-working areas. Beryllium metal in vacuum cast billet form is priced roughly at $70/lb. Fabricated forms are more expensive.

Bismuth (Ger. *Weisse Masse,* **white mass; later** *Wismuth* **and** *Bisemutum*) – Bi; at. wt. 208.9808. at no. 83; mp 271.3°C; bp 1560 ± 5°C; sp gr 9.747 (20°C); valence 3 or 5. In early times bismuth was confused with tin and lead. Claude Geoffroy the Younger showed it to be distinct from lead in 1753. It is a white, crystalline, brittle metal with a pinkish tinge. It occurs native. The most important ores are *bismuthinite* or *bismuth glance* (Bi_2S_3) and *bismite* (Bi_2O_3). Peru, Japan, Mexico, Bolivia, and Canada are major bismuth producers. Much of the bismuth produced in the U.S. is obtained as a by-product in refining lead, copper, tin, silver, and gold ores. Bismuth is the most diamagnetic of all metals, and the thermal conductivity is lower than any metal, except mercury. It has a high electrical resistance, and has the highest Hall effect of any metal (i.e., greatest increase in electrical resistance when placed in a magnetic field). "Bismanol" is a permanent magnet of high coercive force, made of MnBi, by the U.S. Naval Ordnance Laboratory. Bismuth expands 3.32% on solidification. This property makes bismuth alloys particularly suited to the making of sharp castings of objects subject to damage by high temperatures. With other metals, such as tin, cadmium, etc., bismuth forms low-melting alloys which are extensively used for safety devices used in fire detection and extinguishing systems. Bismuth is used in producing malleable irons and is finding use as a catalyst for making acrylic fibers. When bismuth is heated in air it burns with a blue flame forming yellow fumes of the oxide. The metal is also used as a thermocouple material (has highest negativity known), and has found application as a carrier for U^{235} or U^{233} fuel in atomic reactors. Its soluble salts are characterized by forming insoluble basic salts on the addition of water – a property sometimes used in detection work. Bismuth oxychloride is used extensively in cosmetics. Bismuth subnitrate and subcarbonate are used in medicine. High-purity bismuth metal costs about $4/lb.

Boron (Ar. *Buraq,* **Pers.** *Burah*) – B; at. wt. 10.81; at. no. 5; mp 2300°C; bp sublimes 2550°C; sp gr of crystals 2.34, of amorphous variety 2.37; valence 3. Boron compounds have been known for thousands of years, but the element was not discovered until 1808 by Sir Humphry Davy and by Gay-Lussac and Thenard. The element is not found free in nature, but occurs as orthoboric acid usually in certain volcanic spring waters and as borates in *borax* and *colemanite. Ulexite,* another boron mineral, is interesting as it is nature's own version of "fiber optics." By far the most important source of boron is the mineral *rasorite,* also known as kernite, found in the Mojave desert of California. Extensive *borax* deposits are also found in Turkey. Boron exists naturally at 19.78% $_5B^{10}$ isotope and 80.22% $_5B^{11}$ isotope. High-purity crystalline boron may be prepared by the vapor phase reduction of boron trichloride or tribromide with hydrogen on electrically heated filaments. The impure, or amorphous boron, a brownish-black powder, can be obtained by heating the trioxide with magnesium powder. Boron of 99.9999% purity has been produced and is available commercially. Elemental boron has an energy band gap of 1.50 to 1.56 electron volts, which is higher than that of either silicon or germanium. It has interesting optical characteristics, transmitting portions of the infrared, and is a poor conductor of electricity at room temperature, but a good conductor at high temperature. Amorphous boron is used in pyrotechnic flares to provide a distinctive green color, and in rockets as an igniter. The more important compounds of boron are boric, or boracic acid widely used as a mild antiseptic, and borax ($Na_2B_4O_7 \cdot 10H_2O$), which serves as a cleansing flux in welding and as a water softener in washing powders. Boron compounds are used in production of enamels for covering steel of refrigerators, washing machines, and like products. Boron compounds are also extensively used in the manufacture of borosilicate glasses. Boron coated on a thin tungsten wire substrate is the principal reinforcement for both resin and metal matrix composites. The isotope boron 10 is used as a control for nuclear reactors, as a shield for nuclear radiation, and in instruments used for detecting neutrons. Boron nitride has remarkable properties and can be used to make a material as hard as diamond. The nitride also behaves like an electrical insulator but conducts heat like a metal. It also has lubricating properties similar to graphite. The hydrides are easily oxidized with considerable energy liberation, and are being studied for use as rocket fuels. Demand is increasing for boron filaments, a high-strength, light-weight material chiefly employed for advanced aero-space structures. Amorphous boron (90% to 92% grade) costs about $12 to $30/lb depending on quantity ordered. Elemental boron is not considered to be a poison, but assimilation of its compounds has a cumulative poisonous effect.

Bromine (Gr. *bromos,* **stench)** — Br; at. wt. 79.904; at. no. 35; mp $-7.2°C$; bp $58.78°C$; density of gas 7.59 g/l, liquid 3.12 ($20°C$); valence 1, 3, 5, or 7. Discovered by Balard in 1826, but not prepared in quantity until 1860. A member of the halogen group of elements, it is obtained from natural brines from wells in Michigan and West Virginia and from sea water by displacement with chlorine; electrolysis might be used. Bromine is the only liquid nonmetallic element. It is a heavy, mobile, reddish-brown liquid, volatilizing readily at room temperature to a red vapor with a strong disagreeable odor, resembling chlorine, and having a very irritating effect on the eyes and throat; it is readily soluble in water or carbon disulfide, forming a red solution; it is less active than chlorine but more so than iodine; it unites readily with many elements and has a bleaching action; when spilled on the skin it produces painful sores. It presents a serious health hazard, and maximum safety precautions should be taken when handling it. About 80% of the bromine output in the U.S. is used in the production of ethylene dibromide, a lead scavenger used in making gasoline antiknock compounds. Bromine is also used in making fumigants, flameproofing agents, water purification compounds, dyes, medicinals, sanitizers, inorganic bromides for photography, etc. Organic bromides are also important.

Cadmium (L. *cadmia;* **Gr.** *kadmeia* — **ancient name for calamine, zinc carbonate)** — Cd; at. wt 112.40; at. no. 48; mp $320.9°C$; bp $765°C$; sp gr 8.65 ($20°C$); valence 2. Discovered by Stromeyer in 1817 from an impurity in zinc carbonate. Cadmium most often occurs in small quantities associated with zinc ores, such as *sphalerite* (ZnS). *Greenockite* (CdS) is the only mineral of any consequence bearing cadmium. Almost all cadmium is obtained as a by-product in the treatment of zinc, copper, and lead ores. It is a soft, bluish-white metal which is easily cut with a knife. It is similar in many respects to zinc. It is a component of some of the lowest melting alloys; it is used in bearing alloys with low coefficients of friction and great resistance to fatigue; it is used extensively in electroplating, which accounts for about 60% of its use. It is also used in many types of solder, for standard E.M.F. cells, for batteries, and as a barrier to control atomic fission. Cadmium compounds are used in black and white television phosphors and in blue and green phosphors for color TV tubes. It forms a number of salts of which the sulfate is the most common; the sulfide is used as a yellow pigment. Cadmium and solutions of its compounds are toxic. Failure to appreciate the toxic properties of cadmium may cause workers to be unwittingly exposed to dangerous fumes. Silver solder, for example, which contains cadmium, should be handled with care. Serious toxicity problems have been found from long-term exposure and work with cadmium plating baths. The recommended maximum allowable (8-hr day) concentration of cadmium vapor in air is 0.1 mg/m^3. The current price of cadmium is about $2.50/lb. It is available in high-purity form.

Calcium (L. *calx,* **lime)** — Ca; at. wt 40.08; at. no. 20; mp $839°C$; bp $1484°C$; sp gr 1.55 ($20°C$); valence 2. Though lime was prepared by the Romans in the first century under the name calx, not until 1808 was the metal discovered. After learning that Berzelius and Pontin prepared calcium amalgam by electrolyzing lime in mercury, Davy was able to isolate the impure metal. Calcium is a metallic element, fifth in abundance in the earth's crust, of which it forms more than three percent. It is an essential constituent of leaves, bones, teeth and shells. Never found in nature uncombined, it occurs abundantly as *limestone* (CaCO$_3$), *gypsum* (CaSO$_4$·2H$_2$O) and *fluorite* (CaF$_2$); *apatite* is the fluophosphate or chlorophosphate of calcium. The metal has a silvery color, is rather hard, and is prepared by electrolysis of the fused chloride to which calcium fluoride is added to lower the melting point. Chemically it is one of the alkaline earth elements; it readily forms a white coating of nitride in air, reacts with water, burns with a yellow red flame, forming largely the nitride. The metal is used as a reducing agent in preparing other metals, such as thorium, uranium, zirconium, etc., and is used as a deoxidizer, desulfurizer, or decarburizer for various ferrous and nonferrous alloys. It is also used as an alloying agent for aluminum, beryllium, copper, lead and magnesium alloys, and serves as a "getter" for residual gases in vacuum tubes, etc. Its natural and prepared compounds are widely used. Quicklime (CaO), made by heating limestone and changed into slaked lime by the careful addition of water, is the great cheap base of the chemical industry with countless uses. Mixed with sand it hardens as mortar and plaster by taking up carbon dioxide from the air. Calcium from limestone is an important element in Portland cement. The solubility of the carbonate in water containing carbon dioxide causes the formation of caves with stalactites and stalagmites and hardness in water. Other important compounds are the carbide (CaC_2), chloride ($CaCl_2$), cyanamide ($Ca(CN_2)$), hypochlorite ($Ca(OCl)_2$), nitrate ($Ca(NO_3)_2$), and sulfide (CaS).

Californium (State and University of California; Cf; at. wt (251); at. no. 98) — Californium, the sixth transuranium element to be discovered, was produced by Thompson, Street, Ghiorso, and Seaborg in 1950 by bombarding μg quantities of Cm^{242} with 35 MeV helium ions in the Berkeley 60-in. cyclotron. Californium (III) is the only ion stable in aqueous solutions, all attempts to reduce or oxidize californium (III) having failed. The isotope Cf^{249} results from the beta decay of Bk^{249} while the heavier isotopes are produced by intense neutron irradiation by the reactions:

$$Bk^{249}(n, \gamma)Bk^{250} \xrightarrow{\beta^-} Cf^{250} \text{ and } Cf^{249}(n, \gamma)Cf^{250}$$

followed by

$$Cf^{250}(n, \gamma)Cf^{251}(n, \gamma)Cf^{252}.$$

The existence of the isotopes Cf^{249}, Cf^{250}, Cf^{251} and Cf^{252} makes it feasible to isolate californium in weighable amounts so that its properties can be investigated with macroscopic quantities. Californium-252 is a very strong neutron emitter. One microgram releases 170 million neutrons/min which presents biological hazards. In 1960 a few tenths of a μg of californium trichloride $CfCl_3$, californium oxychloride, CfOCl, and californium oxide, Cf_2O_3, were first prepared. Reduction of californium to its metallic state has not yet been accomplished. Because californium is a very efficient source of neutrons, many new uses are expected for it. It has already found use in neutron moisture gages and in well-logging (the

determination of water and oil-bearing layers). It is also being used as a portable neutron source for discovery of metals, such as gold or silver, by on-the-spot activation analysis. Cf^{252} is now being offered for sale by the A.E.C. at a cost of $100 per 0.1 μg. It has been suggested that californium may be produced in certain stellar explosions, called *supernovae*, for the radioactive decay of Cf^{254} (55-day half-life) agrees with the characteristics of the light curves of such explosions observed through telescopes. This suggestion, however, is questioned.

Carbon (L. *carbo*, charcoal) — C; at. wt 12 exactly (C^{12}); at. wt (natural carbon) 12.011; at. no. 6; mp ~3550°C, graphite sublimes at 3367 ± 25°C; bp 4827°C; sp gr amorphous 1.8 to 2.1, graphite 1.9 to 2.3, diamond 3.15 to 3.53 (depending on variety); gem diamond 3.513 (25°C); valence 2, 3, or 4. Carbon, an element of prehistoric discovery, is very widely distributed in nature. It is found in abundance in the sun, stars, comets, and atmospheres of most planets. Carbon in the form of microscopic diamonds is found in some meteorites. Natural diamonds are found in *Kimberlite* of ancient volcanic "pipes," such as found in South Africa, Arkansas, and elsewhere. Diamonds are now also being recovered from the ocean floor off the Cape of Good Hope. About 30% of all industrial diamonds used in the U.S. are now made synthetically. Artificial diamonds are produced by a high temperature-high pressure process utilizing small amounts of catalysts. The method has the potential to manufacture high quality diamonds to rival natural diamonds. The energy of the sun and stars can be attributed at least in part to the well-known carbon-nitrogen cycle. Carbon is found free in nature in three allotropic forms: amorphous, graphite, and diamond. A fourth form, known as "white" carbon, is now thought to exist. Graphite is one of the softest known materials while diamond is one of the hardest. Graphite exists in two forms: alpha and beta. These have identical physical properties, except for their crystal structure. Naturally occurring graphites are reported to contain as much as 30% of the rhombohedral (beta) form, whereas synthetic materials contain only the alpha form. The hexagonal alpha type can be converted to the beta by mechanical treatment, and the beta form reverts to the alpha on heating it above 1000°C. In 1969 a new allotropic form of carbon was produced during the sublimation of pyrolytic graphite at low pressures. Under free-vaporization conditions above ~2550°K, "white" carbon forms as small transparent crystals on the edges of the basal planes of graphite. The interplanar spacings of "white" carbon are identical to those of a carbon form noted in the graphitic gneiss from the Ries (meteoritic) Crater of Germany. "White" carbon is a transparent birefringent material. Little information is presently available about this allotrope. In combination, carbon is found as carbon dioxide in the atmosphere of the earth and dissolved in all natural waters. It is a component of great rock masses in the form of carbonates of calcium (limestone), magnesium, and iron. Coal, petroleum, and natural gas are chiefly hydrocarbons. Carbon is unique among the elements in the vast number and variety of compounds it can form. With hydrogen, oxygen, and nitrogen, and other elements, it forms an infinite number of compounds, carbon atom often being linked to carbon atom. There are upwards of a million or more known carbon compounds, many thousands of which are vital to organic and life processes. Without carbon, the basis for life would be impossible. While it has been thought that silicon might take the place of carbon in forming a host of similar compounds, it is now not possible to form stable compounds with very long chains of silicon atoms. Some of the most important compounds of carbon are carbon dioxide (CO_2), carbon monoxide (CO), carbon disulfide (CS_2), chloroform ($CHCl_3$), carbon tetrachloride (CCl_4), methane (CH_4), ethylene (C_2H_4), acetylene (C_2H_2), benzene (C_6H_6), ethyl alcohol (C_2H_5OH), acetic acid (CH_3COOH), and their derivatives. Carbon has seven isotopes. In 1961 the International Union of Pure and Applied Chemistry adopted the isotope carbon-12 as the basis for atomic weights. Carbon-14, an isotope with a half-life of 5730 years, has been widely used to date such materials as wood, archeological specimens, etc. Carbon-13 is now commercially available at a cost of $700/g.

Cerium – (named for the asteroid *Ceres*, which was discovered in 1801 only 2 years before the element), Ce; at. wt 140.12; at. no. 58; mp 799°C; bp 3426°C; sp gr 6.657 (25°C); valence 3 or 4. Discovered in 1803 by Klaproth and by Berzelius and Hisinger; metal prepared by Hillebrand and Norton in 1875. Cerium is the most abundant of the metals of the so-called rare earths; it is found in a number of minerals including *allanite* (also

known as *orthite*), *monazite, bastnasite, cerite,* and *samarskite*. Monazite and bastnasite are presently the two most important sources of cerium. Large deposits of monazite found on the beaches of Travancore, India, in river sands in Brazil, and deposits of *allanite* in the Western United States, and *bastnasite* in Southern California will supply cerium, thorium and the other rare-earth metals for many years to come. Metallic cerium is prepared by metallothermic reduction techniques, such as by reducing cerous fluoride with calcium, or by electrolysis of molten cerous chloride or other cerous halides. The metallothermic technique is used to produce high-purity cerium. Cerium is especially interesting because of its variable electronic structure. The energy of the inner 4f level is nearly the same as that of the outer or valence electrons, and only small amounts of energy are required to change the relative occupancy of these electronic levels. This gives rise to dual valency states. For example, a volume change of about 10% occurs when cerium is subjected to high pressures or low temperatures. It appears that the valence changes from about 3 to 4 when it is cooled or compressed. The low temperature behavior of cerium is complex. Four allotropic modifications are thought to exist: cerium at room temperature and at atmospheric pressure is known as γ cerium. Upon cooling to -23°C, γ cerium changes to β cerium. The remaining γ cerium starts to change to α cerium when cooled to -158°C, and the transformation is complete at -196°C. α cerium has a density of 8.24. δ cerium exists above 726°C. At atmospheric pressure, liquid cerium is more dense than its solid form at the melting point. Cerium is an iron-gray lustrous metal. It is malleable, and oxidizes very readily at room temperature, especially in moist air. Except for europium, cerium is the most reactive of the "rare-earth" metals. It slowly decomposes in cold water, and rapidly in hot water. Alkali solutions and dilute and concentrated acids attack the metal rapidly. The pure metal is likely to ignite if scratched with a knife. Ceric salts are orange-red or yellowish; cerous salts are usually white. Cerium is a component of misch metal, which is extensively used in the manufacture of pyrophoric alloys for cigarette lighters, etc. While cerium is not radioactive, the impure commerical grade may contain traces of thorium, which is radioactive. The oxide is an important constituent of incandescent gas mantles and it is emerging as a hydrocarbon catalyst in "self-cleaning" ovens. In this application it can be incorporated into oven walls to prevent the collection of cooking residues. As ceric sulfate it finds extensive use as a volumetric oxidizing agent in quantitative analysis. Cerium compounds are used in the manufacture of glass, both as a component and a decolorizer. The oxide is finding increased use as a glass polishing agent instead of rouge, for it is much faster than rouge in polishing glass surfaces. Cerium, with other rare earths, is used in carbon-arc lighting, especially in the motion picture industry. It is also finding use as an important catalyst in petroleum refining and in ·metallurgical and nuclear applications. Commercial cerium metal costs about $75/lb. In small lots, 99.9% cerium costs about 65¢/g.

Cesium (L. *caesius*, sky blue) – Cs; at. wt 132.9054; at. no. 55; mp 28.40°C; bp 678.4°C; sp gr 1.873 (20°C); valence 1. Cesium was discovered spectroscopically by Bunsen and Kirchhoff in 1860 in mineral water from Dürkheim. Cesium, an alkali metal, occurs in *lepidolite, pollucite* (a hydrated silicate of aluminum and cesium) and in other sources. One of the world's richest sources of cesium is located at Bernie Lake, Manitoba. The deposits are estimated to contain 300,000 tons of pollucite, averaging 20% cesium. It can be isolated by electrolysis of the fused cyanide and by a number of other methods. Very pure, gas-free cesium can be prepared by thermal decomposition of cesium azide. The metal is characterized by a spectrum containing two bright lines in the blue along with several others in the red, yellow, and green. It is silvery-white, soft, and ductile. It is the most electro-positive and most alkaline element. Cesium, gallium and mercury are the only three metals that are liquid at room temperature. Cesium reacts explosively with cold water, and reacts with ice at temperatures above −116°C. Cesium hydroxide, the strongest base known, attacks glass. Because of its great affinity for oxygen the metal is used as a "getter" in radio tubes. It is also used in photo-electric cells, as well as a catalyst in the hydrogenation of certain organic compounds. The metal has recently found application in ion propulsion systems. Although these are not usable in the earth's atmosphere, 1 lb of cesium in outer space theoretically will propel a vehicle 140 times as far as the burning of the same amount of any known liquid or solid. Cesium is used in atomic clocks, which are accurate to 5 sec in 300 years. Its chief compounds are the chloride and the nitrate. The present price of cesium is about $100 to $150/lb, depending on quantity and purity.

Chlorine (Gr. *chloros*, greenish-yellow) – Cl; at. wt 35.453; at. no. 17; fp −100.98°C; bp −34.6°C; density 3.214 g/l; sp gr 1.56 (−33.6°C); valence 1, 3, 5, or 7. Discovered in 1744 by Scheele, who thought it contained oxygen; named in 1810 by Davy, who insisted it was an element. In nature it is found in the combined state only, chiefly with sodium as common salt (NaCl), *carnallite* ($KMgCl_3 \cdot 6H_2O$), and *sylvite* (KCl). It is a member of the halogen (salt forming) group of elements and is obtained from chlorides by the action of oxidizing agents and more often by electrolysis; it is a greenish-yellow gas, combining directly with nearly all elements. At 10°C one volume of water dissolves 3.10 vol of chlorine, at 30°C only 1.77 vol. Chlorine is widely used in making many everyday products. It is used for producing safe drinking water the world over. Even the smallest water supplies are now usually chlorinated. It is also extensively used in the production of paper products, dyestuffs, textiles, petroleum products, medicines, antiseptics, insecticides, foodstuffs, solvents, paints, plastics, and many other consumer products. Most of the chlorine produced is used in the manufacture of chlorinated compounds for sanitation, pulp bleaching, disinfectants, and textile processing. Further use is in the manufacture of chlorates, chloroform, carbon tetrachloride and in the extraction of bromine. Organic chemistry demands much from chlorine, both as an oxidizing agent and in substitution, since it often brings desired properties in an organic compound when substituted for hydrogen, as in one form of synthetic rubber. Chlorine is a respiratory irritant. The gas irritates the mucous membranes and the liquid burns the skin. As little as 3.5 ppm can be detected as an odor, and 1000 ppm is likely to be fatal after a few deep breaths. It was used as a war gas in 1915. The recommended maximum allowable concentration in air is about 1 ppm for prolonged exposure.

Chromium (Gr. *chroma*, color) – Cr; at. wt 51.996; at. no. 24; mp 1857°C; pb 2672°C; sp gr 7.18 to 7.20 (20°C); valence chiefly 2, 3, or 6. Discovered in 1797 by Vauquelin, who prepared the metal the next year. Chromium is a steel gray, lustrous, hard metal that takes a high polish. The principal ore is *chromite* (FeO · Cr_2O_3), which is found in Southern Rhodesia, U.S.S.R., Transvaal, Turkey, Iran, Albania, Finland, Malagasy, and the Philippines. The metal is usually produced by reducing the oxide with aluminum. Chromium is used to harden steel, to manufacture stainless steel, and to form many very useful alloys. Much is used in plating to produce a hard, beautiful surface and to prevent corrosion. Chromium is used to give glass an emerald green color. It finds wide use as a catalyst. All compounds of chromium are colored; the most important are the chromates of sodium and potassium (K_2CrO_4) and the dichromates ($K_2Cr_2O_7$) and the potassium and ammonium chrome alums, as $KCr(SO_4)_2$ · $12H_2O$. The dichromates are used as oxidizing agents in quantitative analysis, also in tanning leather. Other compounds are of industrial value; lead chromate is chrome yellow, a valued pigment. Chromium compounds are used in the textile industry as mordants, and by the aircraft and other industries for anodizing aluminum. The refractory industry has found chromite useful for forming bricks and shapes, as it has a high melting point, moderate thermal expansion, and stability of crystalline structure. Hexavalent chromium compounds are toxic. The recommended maximum allowable concentration of dusts and mists in air, measured as CrO_3, is 0.1 mg/m^3 for daily 8-hr exposure. Chromium is available in high-purity form.

Cobalt (*Kobold*, **from the German, goblin or evil spirit,** *cobalos*, **Greek, mine**) — Co; at. wt 58.9332; at. no. 27; mp 1495°C; bp 2870°C; sp gr 8.9 (20°C); valence 2 or 3. Discovered by Brandt about 1735. Cobalt occurs in the minerals *cobaltite, smaltite,* and *erythrite,* and is often associated with nickel, silver, lead, copper, and iron ores, from which it is most frequently obtained as a by-product. It is also present in meteorites. Important ore deposits are found in the Congo, Morocco, and Canada. Cobalt is a brittle, hard metal, closely resembling iron and nickel in appearance. It has a magnetic permeability of about two thirds that of iron. Cobalt tends to exist as a mixture of two allotropes over a wide temperature range. The β-form predominates below 400°C and the α above that temperature. The transformation is sluggish and accounts in part for the wide variation in reported data on physical properties of cobalt. It is alloyed with iron, nickel, and other metals to make Alnico, an alloy of

unusual magnetic strength with many important uses. Stellite alloys, containing cobalt, chromium, and tungsten, are used for high-speed, heavy-duty, high-temperature cutting tools, and for dies. Cobalt is also used in other magnet steels and stainless steels, and in alloys used in jet turbines and gas turbine generators. The metal is used in electroplating because of its appearance, hardness, and resistance to oxidation. The salts have been used for centuries for the production of brilliant and permanent blue colors in porcelain, glass, pottery, tiles and enamels. It is the principal ingredient in *Serves* and *Thenard's blue.* A solution of the chloride ($CoCl_2$ · $6H_2O$) is used as sympathetic ink. The cobalt ammines are of interest; the oxide and the nitrate are important. Cobalt carefully used in the form of the chloride, sulfate, acetate, or nitrate has been found effective in correcting a certain mineral deficiency disease in animals. Soils should contain 0.13 to 0.30 ppm of cobalt for proper animal nutrition. Cobalt-60, an artificial isotope, is an important gamma ray source, and is extensively used as a tracer and a radio-therapeutic agent. Single compact sources of Cobalt-60 are readily available and have a equivalent gamma ray output equal to thousands of grams of radium. The cost of Cobalt-60 varies from about 40¢ to $7.00/curie, depending on quantity and specific activity.

Columbium — (see Niobium).

Copper (L. *cuprum,* **from the island of Cyprus**) — Cu; at. wt 63.546; at. no. 29; mp 1083.4 ± 0.2°C; bp 2567°C; sp gr 8.96 (20°C); valence 1 or 2. The discovery of copper dates from prehistoric times; it is said to have been mined for more than 5000 years. It is one of man's most important metals. Copper is reddish-colored, takes on a bright metallic luster and is malleabie, ductile, and a good conductor of heat and electricity (second only to silver in electrical conductivity). The electrical industry is one of the greatest users of copper. Copper occasionally occurs native, and is found in many minerals, such as *cuprite, malachite, azurite, chalocopyrite,* and *bornite.* Large copper ore deposits are found in the U.S., Chile, Zambia, Zaire, Peru, and Rhodesia. The most important copper ores are the sulfides, oxides, and carbonates. From these copper is obtained by smelting, leaching and by electrolysis. Its alloys, brass and bronze, long used, are still very important; all American coins are now copper alloys; monel and gun metals also contain copper. The most important compounds are the oxide and the sulfate, blue vitriol; the latter has wide use as an agricultural poison and as an algicide in water purification. Copper compounds are widely used in analytical chemistry, such as Fehling's solution in tests for sugar. High-purity copper (99.999 + %) is available commercially.

Curium (Pierre and Marie Curie) — Cm; at. wt. 247; at. no. 96; mp 1340 ± 40°C; sp gr 13.51 (calc.); valence 3 and 4. Although curium follows americium in the periodic system, it was actually known before americium and was the third transuranium element to be discovered. It was identified by Seaborg, James, and Ghiorso in 1944 at the wartime Metallurgical Laboratory in Chicago as a result of helium-ion bombardment of Pu^{239} in the Berkeley, California 60-in. cyclotron. Visible amounts (30 μg) of Cm^{242}, in the form of the hydroxide, were first isolated by Werner and Perlman of the University of California in 1947. In 1950, Crane, Wallmann, and Cunningham found that the magnetic susceptibility of μg samples of CmF_3 was of the same magnitude as that of GdF_3. This provided direct experimental evidence for assigning a $5f^7$ electronic configuration to Cm^{+3}. In 1951, the same workers prepared curium in its elemental form for the first time. Thirteen isotopes of curium are now known. The most stable, Cm^{247}, with a half-life of 16 million years, is so short compared to the earth's age that any primordial curium must have disappeared long ago from the natural scene. Minute amounts of curium probably exist in natural deposits of uranium, as a result of a sequence of neutron captures and β^- decays sustained by the very low flux of neutrons naturally present in uranium ores. The presence of natural curium, however, has never been detected. Cm^{242} and Cm^{244} are available in multigram quantities. Cm^{248} has been produced only in mg amounts. Curium is similar in some regards to gadolinium, its rare earth homolog, but it has a more complex crystal structure. Curium is silver in color, is chemically reactive, and is more electropositive than aluminum. CmO_2, Cm_2O_3, CmF_3, CmF_4, $CmCl_3$, $CmBr_3$, and CmI_3 have been prepared. Most compounds of trivalent cerium are faintly yellow in color. The A.E.C. is attempting to produce several kilograms of Cm^{244}, an isotope with a 17.6-year half-life, by neutron irradiation of plutonium in a nuclear reactor. Ultimately, it is possible that it may be produced in ton quantities by converting a number of plutonium production reactors to the manufacture of Cm^{244}. Cm^{242} generates about three thermal watts of energy/g. This compares to one half thermal watt/g of Pu^{238}. This suggests use for curium as an isotopic power source. Cm^{244} is now offered for sale by the A.E.C. at $100/mg. Curium absorbed into the body accumulates in the bones, and is therefore very toxic as its radiation destroys the red-cell forming mechanism. The maximum permissible body burden of Cm^{244} in a human being is 0.002 μg.

Deuterium — an isotope of hydrogen (see Hydrogen).

Dysprosium (Gr. *dysprositos*, hard to get at) — Dy; at. wt 162.50; at. no. 66; mp 1412°C; bp 2562°C; sp gr 8.550 (25°C); valence 3. Dysprosium was discovered in 1886 by Lecoq de Boisbaudran, but not isolated. Neither the oxide nor the metal was available in relatively pure form until the development of ion-exchange separation and metallographic reduction techniques by Spedding and associates about 1950. Dysprosium occurs along with other so-called rare-earth or lanthanide elements in a variety of minerals, such as *xenotime, fergusonite, gadolinite, euxenite, polycrase,* and *blomstrandine.* The most important sources, however, are from *monazite* and *bastnasite.* Dysprosium can be prepared by reduction of the trifluoride with calcium. The element has a metallic, bright silver luster. It is relatively stable in air at room temperature, and is readily attacked and dissolved, with the evolution of hydrogen, by dilute and concentrated mineral acids. The metal is soft enough to be cut with a knife and can be machined without sparking if overheating is avoided. Small amounts of impurities can greatly affect its physical properties. While dysprosium has not yet found many applications, its thermal neutron absorption cross-section and high melting point suggest metallurgical uses in nuclear control applications for alloying with special stainless steels. A dysprosium oxide-nickel cermet has found use in cooling nuclear reactor control rods. This cermet absorbs neutrons readily without swelling or contracting under prolonged neutron bombardment. In combination with vanadium and other rare earths, dysprosium has been used in making laser materials. Dysprosium-cadmium calcogenides, as sources of infrared radiation, have been used for studying chemical reactions. The cost of dysprosium metal has dropped in recent years since the development of ion-exchange and solvent extraction techniques, and the discovery of large ore bodies. The metal is still expensive, however, and costs about 70¢/g or $190/lb in purities of 99 + %.

Einsteinium (Albert Einstein) − Es; at. wt. 254; at. no. 99. Einsteinium, the seventh transuranic element of the actinide series to be discovered, was identified by Ghiorso and co-workers at Berkeley in Dec. 1952 in debris from the first large thermonuclear or "hydrogen" bomb explosion, which took place in the Pacific in Nov. 1952. The isotope produced was the 20-day Es^{253} isotope. In 1961 a sufficient amount of einsteinium was produced to permit separation of a macroscopic amount of Es^{253}. This sample weighed about 0.01 μg. A special magnetic-type balance was used in making this determination. Es^{253} so produced was used to produce mendelevium (Element 101). Recently, about 3 μg of einsteinium were produced at Oak Ridge National Laboratories by irradiating for several years kg quantities of Pu^{239} in a reactor to produce Pu^{242}. This was then fabricated into pellets of plutonium oxide and aluminum powder, and loaded into target rods for an initial one-year irradiation at the A.E.C.'s Savannah River Plant, followed by irradiation in a HFIR (High Flux Isotopic Reactor). After 4 months in the HFIR the targets were removed for chemical separation of the einsteinium from californium. Twelve isotopes of einsteinium are now recognized. Es^{254} has the longest half-life (276 days). Tracer studies using Es^{253} show that einsteinium has chemical properties typical of a heavy trivalent, actinide element.

Element 104 − In 1964, workers of the Joint Nuclear Research Institute at Dubna (U.S.S.R.) bombarded plutonium with accelerated 113−115 MeV neon ions. By measuring fission tracks in a special glass with a microscope, they detected an isotope that decays by spontaneous fission. They suggested that this isotope, which had a half-life of 0.3 ± 0.1 sec might be 104^{260}, produced by the following reaction:

$$_{94}Pu^{242} + _{10}Ne^{22} \rightarrow 104^{260} + 4n.$$

Element 104, the first *transactinide* element, is expected to have chemical properties similar to those of hafnium. It would, for example, form a relatively volatile compound with chlorine (a tetrachloride). The Soviet scientists have performed experiments aimed at chemical identifica-

tion, and have attempted to show that the 0.3-sec activity is more volatile than that of the relatively nonvolatile actinide trichlorides. This experiment does not fulfill the test of chemically separating the new element from all others, but it provides important evidence for evaluation. New data, reportedly issued by Soviet scientists, have reduced the half-life of the isotope they worked with from 0.3 sec to 0.15 sec. The Dubna scientists suggest the name *kurchatovium* and symbol *Ku* for Element 104, in honor of Igor Vasilevich Kurchatov (1903−1960), late Head of Soviet Nuclear Research. In 1969, Ghiorso, Nurmia, Harris, K. A. Y. Eskola and P. L. Eskola, of the Univ. of Calif. at Berkeley, reported they had positively identified two, and possibly three, isotopes of Element 104. The Group also indicated that after repeated attempts, so far they have been unable to produce isotope 104^{260} reported by the Dubna Group in 1964. The discoveries at Berkeley were made by bombarding a target of Cf^{249} with C^{12} nuclei of 71 MeV, and C^{13} nuclei of 69 MeV. The combination of C^{12} with Cf^{249}, followed by instant emission of four neutrons, produced Element 104^{257}. This isotope has a half-life of 4 to 5 sec., decaying by emitting an alpha particle into No^{253}, with a half-life of 105 sec. The same reaction, except with the emission of three neutrons, was thought to have produced 104^{258}, with a half-life of about 1/100 sec. Element 104^{259} is formed by the merging of C^{13} nuclei with Cf^{249}, followed by emission of three neutrons. This isotope has a half-life of 3 to 4 sec, and decays by emitting an alpha particle into No^{255}, which has a half-life of 185 sec. Thousands of atoms of 104^{257} and 104^{259} have been detected. The Berkeley Group believe their identification of 104^{258} is correct, but they do not attach the same degree of confidence to this work as to their work on 104^{257} and 104^{259}. The Berkeley Group propose for the new element the name *rutherfordium* (Symbol R*f*), in honor of Ernest R. Rutherford, New Zealand physicist. The claims for discovery and the naming of Element 104 are still in question.

Element 105 − In 1967 G. N. Flerov reported that a Soviet team working at the Jt. Inst. for Nuclear Research at Dubna may have produced a few atoms of 105^{260} and 105^{261} by bombarding Am^{243} with Ne^{22}. Their evidence was based on time-coincidence measurements of alpha energies. More recently it was reported that early in 1970

Dubna scientists synthesized Element 105 and that by the end of April 1970 "had investigated all the types of decay of the new element and had determined its chemical properties." The Soviet Group have not proposed a name for Element 105. In late April 1970, it was announced that Ghiorso, Nurmia, Harris, K. A. Y. Eskola, and P. L. Eskola, working at the Univ. of Calif. at Berkeley, had positively identified Element 105. The discovery was made by bombarding a target of Cf^{249} with a beam of 84 MeV nitrogen nuclei in the Heavy Ion Linear Accelerator (HILAC). When a N^{15} nucleus is absorbed by a Cf^{249} nucleus, four neutrons are emitted and a new atom of 105^{260}, with a half-life of 1.6 sec., is formed. While the first atoms of Element 105 are said to have been detected conclusively on March 5, 1970, there is evidence that Element 105 had been formed in Berkeley experiments a year earlier by the method described. Ghiorso and his associates have attempted to confirm Soviet findings by more sophisticated methods without success. The Berkeley Group propose the name *hahnium*, after the late German scientist Otto Hahn (1879–1968), and *Ha*, for the chemical symbol.

More recently, in October 1971, it was announced that two new isotopes of Element 105 were synthesized with the heavy ion linear accelerator by A. Ghiorso and co-workers at Berkeley. Element 105^{261} was produced both by bombarding Cf^{250} with N^{15} and by bombarding Bk^{249} with O^{16}. The isotope emits 8.93-MeV α particles and decays to Lr^{257} with a half-life of about 1.8 sec. Element 105^{262} was produced by bombarding Bk^{249} with O^{18}. It emits 8.45 MeV α particles and decays to Lr^{258} with a half-life of about 40 sec.

Erbium *(Ytterby,* a town in Sweden) – Er; at. wt 167.26; at. no 68; mp 1529°C; bp 2863°C; sp gr 9.066 (25°C); valence 3. Erbium, one of the so-called rare-earth elements of the lanthanide series, is found in the mineral mentioned under dysprosium. In 1842 Mosander separated "yttria," found in the mineral *gadolinite*, into three fractions which he called *yttria, erbia,* and *terbia.* The names *erbia* and *terbia* became confused in this early period. After 1860, Mosander's *terbia* was known as *erbia*, and after 1877, the earlier known *erbia* became *terbia.* The *erbia* of this period was later shown to consist of five oxides, now known as *erbia, scandia, holmia, thulia,* and *ytterbia.* By 1905 Urbain and James independently succeeded

in isolating fairly pure Er_2O_3. Kelmm and Bommer first produced reasonably pure erbium metal in 1934 by reducing the anhydrous chloride with potassium vapor. The pure metal is soft and malleable and has a bright, silvery, metallic luster. As with other rare-earth metals, its properties depend to a certain extent on the impurities present. The metal is fairly stable in air and does not oxidize as rapidly as some of the other rare-earth metals. Natural-occurring erbium is a mixture of six isotopes, all of which are stable. Ten radioactive isotopes of erbium are also recognized. Recent production techniques, using ion-exchange reactions, have resulted in much lower prices of the rare-earth metals and their compounds in recent years. The cost of 99 + % erbium metal is about $1.00 per g, in small quantities. Erbium is finding nuclear and metallurgical uses. Added to vanadium, for example, erbium lowers the hardness and improves workability. Most of the rare-earth oxides have sharp absorption bands in the visible, ultraviolet, and near infrared. This property, associated with the electronic structure, gives beautiful pastel colors to many of the rare-earth salts. Erbium oxide gives a pink color and has been used as a colorant in glasses and porcelain enamel glazes.

Europium (Europe) – Eu; at. wt 151.96; at. no. 63; mp 822°C; bp 1597°C; sp gr 5.253 (25°C); valence 2 or 3. In 1890 Boisbaudran obtained basic fractions from samarium-gadolinium concentrates which had spark spectral lines not accounted for by samarium or gadolinium. These lines subsequently have been shown to belong to europium. The discovery of europium is generally credited to Demarcay, who separated the earth in reasonably pure form in 1901. The pure metal was not isolated until recent years. Europium is now prepared by mixing Eu_2O_3 with a 10% excess of lanthanum metal and heating the mixture in a tantalum crucible under high vacuum. The element is collected as a silvery-white metallic deposit on the walls of the crucible. As with other rare-earth metals, except for lanthanum, europium ignites in air at about 150 to 180°C. Europium is about as hard as lead and is quite ductile. It is the most reactive of the rare-earth metals, quickly oxidizing in air. It resembles calcium in its reaction with water. *Bastnasite* and *monazite* are the principal ores containing europium. Europium has been identified spectroscopically in the sun and certain stars. Seventeen isotopes are now recognized.

Europium isotopes are good neutron absorbers and are being studied for use in nuclear control applications. Europium oxide is now widely used as a phosphor activator, and europium activated yttrium vanadate is in commercial use as the red phosphor in color TV tubes. Europium-doped plastic has been used as a laser material. With the development of ion-exchange techniques and special processes, the cost of the metal has been greatly reduced in recent years. Europium is one of the rarest and most costly of the rare-earth metals. It is priced at about $11 to $15/g, or $4000/lb.

Fermium (Enrico Fermi) – Fm; at wt 257; at. no. 100. Fermium, the eighth transuranium element of the actinide series to be discovered, was identified by Ghiorso and co-workers in 1952 in the debris from a thermonuclear explosion in the Pacific in work involving the University of California Radiation Laboratory, the Argonne National Laboratory, and the Los Alamos Scientific Laboratory. The isotope produced was the 20-hr Fm^{255}. During 1953 and early 1954, while discovery of elements 99 and 100 was withheld from publication for security reasons, a group from the Nobel Institute of Physics in Stockholm bombarded U^{238} with O^{16} ions, and isolated a 30-min α-emitter, which they ascribed to 100^{250}, without claiming discovery of the element. This isotope has since been identified positively, and the 30-min half-life confirmed. The chemical properties of fermium have been studied solely with tracer amounts, and in normal aqueous media only the (III) oxidation state appears to exist. The isotope Fm^{254} and heavier isotopes can be produced by intense neutron irradiation of lower elements, such as plutonium, by a process of successive neutron capture interspersed with beta decays until these mass numbers and atomic numbers are reached. Ten isotopes of fermium are known to exist. Fm^{257}, with a half-life of about 80 days, is the longest lived. Fm^{250}, with a half-life of 30 min, has been shown to be a product of decay of Element 102^{254}. It was by chemical identification of Fm^{250} that it was certain that Element 102 (Nobelium) has been produced.

Fluorine (L. and F. *fluere*, flow, or flux) — F; at. wt 18.99840; at. no. 9; mp - 219.62°C (1 atm); bp -188.14°C (1 atm); density 1.696 g/l (0°C, 1 atm); sp gr of liquid 1.108 at bp; valence 1. In 1529, Georgius Agricola described the use of fluorspar as a flux, and as early at 1670 Schwandhard found that glass was etched when exposed to fluorspar treated with acid. Scheele and many later investigators, including Davy, Gay-Lussac, Lavoisier, and Thenard, experimented with hydrofluoric acid, some experiments ending in tragedy. The element was finally isolated in 1886 by Moisson after nearly 74 years of continuous effort. Fluorine occurs chiefly in *fluorspar* (CaF_2) and *cryolite* (Na_2AlF_6), but is rather widely distributed in other minerals. It is a member of the halogen family of elements, and obtained by electrolyzing a solution of potassium hydrogen fluoride in anhydrous hydrogen fluoride in a vessel of metal or transparent fluorspar. Modern commercial production methods are essentially variations on the procedures first used by Moisson. Fluorine is the most electronegative and reactive of all elements. It is a pale yellow, corrosive gas, which reacts with practically all organic and inorganic substances. Finely divided metals, glass, ceramics, carbon, and even water burn in fluorine with a bright flame. Until World War II, there was no commercial production of elemental fluorine. The atom-bomb project and nuclear energy applications, however, made it necessary to produce large quantities. Safe handling techniques have now been developed and it is possible at present to transport liquid fluorine by the ton. Fluorine and its compounds are used in producing uranium (from the hexafluoride) and more than 100 commercial fluorochemicals, including many well-known high-temperature plastics. Hydrofluoric acid is extensively used for etching the glass of light bulbs, etc. Fluorochloro hydrocarbons are extensively used in air conditioning and refrigeration. It has been suggested that fluorine can be substituted for hydrogen wherever it occurs in organic compounds, which could lead to an astronomical number of new fluorine compounds. The presence of fluorine as a soluble fluoride in drinking water to the extent of 2 ppm may cause mottled enamel in teeth, when used by children acquiring permanent teeth. After this was found to prevent tooth decay, smaller amounts of fluoride salts were added to the public drinking water of many cities and also to several commercial toothpastes as an effective dental prophylaxis. Elemental fluorine is being studied as a rocket propellant as it has an exceptionally high specific impulse value. Compounds of fluorine with rare gases have now been confirmed.

Fluorides of xenon, radon, and krypton are among those reported. Elemental fluorine and the fluoride ion are highly toxic. The free element has a characteristic pungent odor, detectable in concentrations as low at 20 parts per billion, which is below the safe working level. The recommended maximum allowable concentration for daily 8-hr exposure is 0.1 ppm.

Francium (France) — Fr; at. no. 87; at. wt 223; mp (27°C); bp (677°C); valence 1. Discovered in 1939 by Mlle. Marguerite Perey of the Curie Institute, Paris. Francium, the heaviest known member of the alkali metal series, occurs as a result of an alpha disintegration of actinium. It can also be made artifically by bombarding thorium with protons. While it occurs naturally in uranium minerals, there is probably less than an ounce of francium at any time in the total crust of the earth. It has the highest equivalent weight of any element, and is the most unstable of the first 101 elements of the periodic system. Twenty isotopes of francium are recognized. The longest lived, Fr^{223} (AcK), a daughter of Ac^{227}, has a half-life of 22 min. This is the only isotope of francium occurring in nature. Because all known isotopes of francium are highly unstable, knowledge of the chemical properties of this element comes from radiochemical techniques. No weighable quantity of the element has been prepared or isolated. The chemical properties of francium most closely resemble cesium.

Gadolinium (*gladolinite* — a mineral named for Gadolin, a Finnish chemist) — Gd; at. wt 157.25; at. no. 64; mp 1313 ± 1°C; sp gr 7.9004 (25°C); valence 3. Gadolinia, the oxide of gadolinium, was separated by Marignac in 1808 and Lecoq de Boisbaudran independently isolated the element from Mosander's "yttria" in 1886. The element was named for the mineral *gadolinite* from which this rare-earth was originally obtained. Gadolinium is found in several other minerals, including *monazite* and *bastnasite*, which are of commercial importance. The element has been isolated only in recent years. With the development of ion-exchange and solvent extraction techniques, the availability and price of gadolinium and the other rare-earth metals have greatly improved. Seventeen isotopes of gadolinium are now recognized. Seven occur naturally. The metal can be prepared by the reduction of the anhydrous fluoride with metallic calcium. As with other related rare-earth metals, it is silvery-white, has a metallic luster, and is malleable and ductile. At room temperature, gadolinium crystallizes in the hexagonal, close-packed α form. Upon heating to 1262°C α gadolinium transforms into the β form, which has a body-centered cubic structure. The metal is relatively stable in dry air, but in moist air, it tarnishes with the formation of a loosely adhering oxide form which spalls off and exposes more surface to oxidation. The metal reacts slowly with water and is soluble in dilute acid. Gadolinium has the highest thermal neutron capture cross-section of any known element (49,000 barns). Natural gadolinium is a mixture of seven isotopes. Two of these, Gd^{155} and Gd^{157}, have excellent capture characteristics, but they are present naturally in low concentrations. As a result, gadolinium has a very fast burnout rate and has limited use as a nuclear control rod material. It has been used in making gadolinium yttrium garnets, which have microwave applications. Compounds of gadolinium are used in making phosphors for color TV tubes. The metal has unusual superconductive properties. As little as 1% gadolinium has been found to improve the workability and resistance of iron, chromium, and related alloys to high temperatures and oxidation. Gadolinium ethyl sulfate has extremely low noise characteristics and may find use in duplicating the performance of h.f. amplifiers, such as the maser. The metal is ferromagnetic. Gadolinium is unique for its high magnetic moment and for its special Curie temperature (above which ferromagnetism vanishes) lying just at room temperature. This suggests uses as a magnetic component that senses hot and cold. The price of the metal is $1.25/g or $250/lb.

Gallium (L. *Gallia*, France) — Ga; at. wt 69.72; at. no. 31; mp 29.78°C; bp 2403°C; sp gr 5.904 (29.6°C) solid; sp gr 6.095 (29.8°C) liquid; valence 2 or 3. Predicted and described by Mendeleev as ekaaluminum, and discovered spectroscopically by Lecoq de Boisbaudran in 1875, who in the same year obtained the free metal by electrolysis of a solution of the hydroxide in KOH. Gallium is often found as a trace element in *diaspore*, *sphalerite*, *germanite*, *bauxite*, and *coal*. Some flue dusts from burning coal have been shown to contain as much as 1.5% gallium. It is the only metal, except for mercury, cesium, and rubidium, which can be liquid near room temperatures; this makes possible its use in high-temperature thermometers. It has one of the longest liquid ranges of any metal and has a low vapor pressure even at high temperatures. There is a strong tendency for gallium to supercool below its freezing point. Therefore, seeding may be necessary to initiate solidification. Ultra-pure gallium has a beautiful, silvery appearance and the solid metal exhibits a conchoidal fracture similar to glass. The metal expands 3.1% on solidifying; therefore, it should not be stored in glass or metal containers as they may break as the metal solidifies. Gallium wets glass or porcelain, and forms a brilliant mirror when it is painted on glass. It has found recent use in doping semiconductors and producing solid-state devices, such as transistors. High-purity gallium is attacked only slowly by mineral acids. Magnesium gallate containing divalent impurities, such as Mn^{+2}, is finding use in commercial ultra-violet activated powder phosphors. Gallium arsenide is capable of converting electricity directly into coherent light. Gallium readily alloys with most metals, and has been used as a component in low-melting alloys. Alloying with gallium is surprisingly easy despite a low melting point. This is because of a long liquid range and a low vapor pressure at temperatures generally employed for alloying and heat treatment. Its toxicity appears to be of a low order, but should be handled with care until more data are forthcoming. The metal can be supplied in ultra-pure form (99.99999 + %).

Germanium (L. *Germania*, Germany) — Ge; at. wt 72.59; at. no. 32; mp 937.4°C; bp 2830°C; sp gr 5.323 (25°C); valence 2 and 4. Predicted by Mendeleev in 1871 as ekasilicon, and discovered by Winkler in 1886. The metal is found in *argyrodite*, a sulfide of germanium and silver; in *germanite*, which contains 8% of the element; in zinc ores; in coal; and in other materials. The element is frequently obtained commercially from flue dusts of smelters processing zinc ores, and has been recovered from the by-products of combustion of certain coals. Its presence in coal insures a large reserve of the element in the years to come. Germanium can be separated from other metals by fractional distillation of its volatile tetrachloride. The tetrachloride may then be hydrolyzed to give GeO_2; then the dioxide can be reduced with hydrogen to give the metal. Recently developed zone-refining techniques permit the production of germanium of ultra-high purity. The element is a gray-white metalloid, and in its pure state is crystalline and brittle, retaining its luster in air at room temperature. It is a very important semi-conductor material. Zone refining techniques have led to production of crystalline germanium for semiconductor use with an inpurity of only 1 part in 10^{10}. Doped with arsenic, gallium or other elements, it is used as a transistor element in thousands of electronic applications. Transistors now provide the largest use for the element, but germanium is also finding many other applications including use as an alloying agent, as a phosphor in fluorescent lamps, and as a catalyst. Germanium oxide is transparent to the infrared and is used in infrared spectroscopes and other optical equipment, including extremely sensitive infrared detectors. Its high index of refraction and dispersion has made germanium useful as a component of glasses used in wide angle camera lenses and microscope objectives. The field of organo-germanium chemistry is becoming increasingly important. Certain germanium compounds have a low mammalian toxicity, but a marked activity against certain bacteria, which makes them of interest as chemotherapeutic agents. The cost of germanium is about 50¢/g for 99.9 + % purity in small lots.

Gold (Sanskrit Jval; Anglo-Saxon *gold*) — Au (L. *aurum*, shining dawn); at. wt 196.9665; at. no. 79; mp 1064.43°C; bp 2807°C; sp gr 19.32 (20°C); valence 1 or 3. Known and highly valued from earliest times. Gold is found in nature as the free metal and in tellurides; it is very widely distributed and is almost always associated with quartz or pyrite. It occurs in veins and alluvial deposits, and is often separated from rocks and other minerals by sluicing or panning operations. About two thirds of the world's gold output now comes from South Africa, and about two thirds of the total U.S. production comes from South Dakota and Nevada. The metal is recovered from its ores by cyaniding, amalgamating, and smelting processes. Refining is also frequently done by electrolysis. Gold occurs in sea water to the extent of 0.1 to 2 mg/ton, depending on the location where the sample is taken. As yet no method has been found for recovering gold from sea water profitably. It is estimated that all the gold in the world, so far refined, could be placed in a single cube fifty feet on a side. Of all the elements, gold in its pure state is undoubtedly the most beautiful. It is metallic, having a yellow color when in a mass, but when finely divided it may be black, ruby, or purple. The Purple of Cassius is a delicate test for auric gold. It is the most malleable and ductile metal; one ounce of gold can be beaten out to 300 ft^2. It is a soft metal and is usually alloyed to give it more strength. It is a good conductor of heat and electricity, and is unaffected by air and most reagents. It is used in coinage and is a standard for monetary systems in many countries. It is also extensively used for jewelry, decoration, dental work, and for plating. It is used for coating certain space satellites, as it is a good reflector of infrared, and is inert. Gold, like other precious metals, is measured in troy weight; when alloyed with other metals, the term *carat* is used to express the amount of gold present — 24 carats being pure gold. For many years the value of gold was set by the United States at \$20.67/troy oz; since 1934 this value was fixed by law at \$35.00/troy oz, 9/10th fine. On March 17, 1968, because of a gold crisis, a two-tiered pricing system was established whereby gold was still used to settle international accounts at the old \$35.00/troy oz price while the price of gold on the private market would be allowed to fluctuate. Since this time, the price of gold on the free market fluctuated between \$35 and \$170/troy oz. On March 19, 1968, President Johnson signed into law a bill removing the last statutory requirement for a gold backing against U.S. currency. On August 15, 1971, President Nixon announced an embargo on U.S. gold to settle international accounts, and on May 18, 1972, U.S. monetary gold was revalued at \$38/troy oz. The commonest gold compounds are auric chloride ($AuCl_3$) and chlorauic acid ($HAuCl_4$), the latter being used in photography for toning the silver image. Gold has 21 isotopes; Au^{198}, with a half-life of 2.7 days, is used for treating cancer and other diseases. Disodium aurothiomalate is administered intramuscularly as a treatment for arthritis. A mixture of one part nitric acid with three of hydrochloric acid is called *aqua regia,* because it dissolved Gold, the King of Metals. Gold is available commercially with a purity of 99.999 + %. For many years the temperature assigned to the freezing point of gold has been 1063.0°C. This has served as a calibration point for the International Temperature Scales (ITS-27 and ITS-48) and the International Practical Temperature Scale (IPTS-48). In 1968 a new International Practical Temperature Scale (IPTS-68) was adopted, which demands that the freezing point of gold be changed to 1064.43°C. Many of the scale changes are of minor significance to the routine user. IPTS-68 has defined several other fixed tempera-ture points, among which are the boiling points of hydrogen, neon, oxygen, and sulfur, and the freezing points of zinc, silver, tin, lead, antimony, and aluminum.

Hafnium *(Hafnia,* **Latin name for Copenhagen)** — Hf; at. wt 178.49; at. no. 72; mp 2227 ± 20°C; bp 4602°C; sp gr 13.31 (20°C); valence 4. Hafnium was thought to be present in various minerals and concentrations many years prior to its discovery, in 1923, credited to D. Coster and G. von Hevesey. On the basis of the Bohr theory, the new element was expected to be associated with zirconium. It was finally identified in *zircon* from Norway, by means of x-ray spectroscopic analysis. It was named in honor of the city in which the discovery was made. Most zirconium minerals contain 1 to 5% hafnium. It was originally separated from zirconium by repeated recrystallization of the double ammonium or potassium fluorides by von Hevesey and Jantzen. Metallic hafnium was first prepared by van Arkel and deBoer by passing the vapor of the tetraiodide over a heated tungsten filament. Almost all hafnium metal now produced is made by reducing the tetrachloride with magnesium or with sodium (Kroll Process). Hafnium is a ductile metal with a brilliant silver luster. Its properties are considerably influenced by the impurities of zirconium present. Of all the elements, zirconium and hafnium are two of the most difficult to separate. Their chemistry is almost identical; however, the density of zirconium is about half that of hafnium. Very pure hafnium has been produced with zirconium being the major impurity. Because hafnium has a good absorption cross-section for thermal neutrons (almost 600 times that of zirconium), has excellent mechanical properties, and is extremely corrosion resistant, it is used for reactor control rods. Such rods are used in nuclear submarines. Hafnium has been successfully alloyed with iron, titanium, niobium, tantalum, and other metals. Hafnium carbide is the most refractory binary composition known, and the nitride is the most refractory of all known metal nitrides (mp 3310°C). Hafnium is used in gas-filled and incandescent lamps, and is an efficient "getter" for scavenging oxygen and nitrogen. Finely divided hafnium is pyrophoric and can ignite spontaneously in air. Care should be taken when machining the metal or when handling hot sponge hafnium. At 700°C hafnium rapidly absorbs hydrogen to form the composition $HfH_{1.86}$.

Hafnium is resistant to concentrated alkalis, but at elevated temperatures reacts with oxygen, nitrogen, carbon, boron, sulfur, and silicon. Halogens react directly to form tetrahalides. The price of the metal is in the broad range of $30 to $175/lb, depending on purity and quantity. The yearly demand for hafnium in the U.S. is now in excess of 60,000 lb.

Hahnium (see Element 105)

Helium (Gr. *helios*, **the sun**) — He; at. wt 4.00260; at. no. 2; mp below −272.2°C (26 atm); bp −268.934°C; density 0.1785 g/l (0°C, 1 atm); liquid density 7.62 lb/ft^3 at. bp; valence usually 0. Evidence of the existence of helium was first obtained by Janssen during the solar eclipse of 1868 when he detected a new line in the solar spectrum; Lockyer and Frankland suggested the name *helium* for the new element; in 1895 Ramsay discovered helium in the uranium mineral *clevite*, and it was independently discovered in clevite by the Swedish chemists Cleve and Langlet about the same time. Rutherford and Royds in 1907 demonstrated that α particles are helium nuclei. Except for hydrogens, helium is the most abundant element found throughout the universe. It has been detected spectroscopically in great abundance, especially in the hotter stars, and it is an important component in both the proton-proton reaction and the carbon cycle, which accounts for the energy of the sun and stars. The fusion of hydrogen into helium provides the energy of the hydrogen bomb. The helium content of the atmosphere is about 1 part in 200,000. While it is present in various radioactive minerals as a decay product, the bulk of the Free World's supply is obtained from wells in Texas, Oklahoma, and Kansas. The only helium plant in the Free World outside the U.S. is near Swift River, Saskatchewan. The cost of helium fell from $2500/ft^3 in 1915 to 1.5 cent/ft^3 in 1940. The U.S. Bureau of Mines has set the price of Grade A helium at $35/1000 ft^3. Helium has the lowest mp of any element and has found wide use in cryogenic research as its bp is close to absolute zero. Its use in the study of superconductivity is vital. Using liquid helium, Kurti and co-workers, and others, have succeeded in obtaining temperatures of a few microdegrees Kelvin, by the adiabatic demagnetization of copper nuclei, starting from about 0.01°K. Five isotopes of helium are known. Liquid helium (He^4) exists in two forms: He^4 I and He^4 II, with a sharp transition point at 2.174°K (383 cm Hg). He^4 I (above this temperature) is a normal liquid, but He^4 II (below it) is unlike any other known substance. It expands on cooling; its conductivity for heat is enormous; and neither its heat conduction nor viscosity obeys normal rules. It has other peculiar properties. Helium is the only liquid that cannot be solidified by lowering the temperature. It remains liquid down to absolute zero at ordinary pressures, but it can readily be solidified by increasing the pressure. Solid He^3 and He^4 are unusual in that both can readily be changed in volume by more than 30% by application of pressure. The specific heat of helium gas is unusually high. The density of helium vapor at the normal boiling point is also very high, with the vapor expanding greatly when heated to room temperature. Containers filled with helium gas at 5 to 10°K should be treated as though they contained liquid helium due to the large increase in pressure resulting from warming the gas to room temperature. While helium normally has a 0 valence, it seems to have a weak tendency to combine with certain other elements. Means of preparing helium difluoride are being studied, and species, such as HeNe and the molecular ions He$_2^+$ and He$_2^{++}$ have been investigated. Helium is widely used as an inert gas shield for arc welding; as a protective gas in growing silicon and germanium crystals, and in titanium and zirconium production; as a cooling medium for nuclear reactors; and as a gas for supersonic wind tunnels. A mixture of 80% helium and 20% oxygen is used as an artificial atmosphere for divers and others working under pressure. Helium is extensively used for filling balloons as it is a much safer gas than hydrogen. While its density is almost twice that of hydrogen, it has about 98% of the lifting power of hydrogen. At sea level, 1000 ft^3 of helium lifts 68.5 lb. One of the recent largest uses for helium has been for pressuring liquid fuel rockets. A Saturn booster, such as used on the Apollo lunar missions, requires about 13 million ft^3 of helium for a firing, plus more for checkouts.

Holmium (L. *Holmia*, **for Stockholm**) — Ho; at. wt 164.9304; at. no. 67; mp 1474°C; bp 2695°C; sp gr 8.795 (25°C); valence +3. The spectral absorption bands of holmium were noticed in 1878 by the Swiss chemists Delafontaine and Soret, who announced the existence of an "Element X." Cleve, of Sweden, later independently discovered the element while working on erbia earth. The element is named after Cleve's native city. Pure holmia, the yellow oxide, was prepared by Homberg in 1911. Holmium occurs in *gadolinite, monazite,* and in other rare-earth minerals. It is commercially obtained from *monazite,* occurring in that mineral to the extent of about 0.05%. It has been isolated in pure form only in recent years. It can be separated from other rare earths by ion-exchange and solvent extraction techniques, and isolated by the reduction of its anhydrous chloride or fluoride with calcium metal. Pure homium has a metallic to bright silver luster. It is relatively soft and malleable, and is stable in dry air at room temperature, but rapidly oxidizes in moist air and at elevated temperatures. The metal has unusual magnetic properties. Few uses have yet been found for the element. The element, as with other rare earths, seems to have a low acute toxic rating. The price of 99 + % holmium metal is about $1.50/g or $500/lb.

Hydrogen (Gr. *hydro*, **water and** *genes*, **forming)** — H; at. wt (natural) 1.0080; at. wt (H^1) 1.0079; at. no. 1; mp $-259.14°C$; bp $-252.87°C$; density 0.08988 g/l; density (liquid) 70.8 g/l; ($-253°C$); density (solid) 70.6 g/l ($-262°C$); valence 1. Hydrogen was prepared many years before it was recognized as a distinct substance by Cavendish in 1766. It was named by Lavoisier. Hydrogen is the most abundant of all elements in the universe, and it is thought that the heavier elements were, and still are being built from hydrogen and helium. It has been estimated that hydrogen makes up more than 90% of all the atoms or three-quarters of the mass of the universe. It is found in the sun and most stars, and plays an important part in the proton-proton reaction and carbon-nitrogen cycle, which accounts for the energy of the sun and stars. It is thought that hydrogen is a major component of the planet Jupiter and that at some depth in the planet's interior the pressure is so great that solid molecular hydrogen is converted into solid metallic hydrogen. This transition is thought to take place at a pressure of one megabar. Ultimately it may be possible to produce metallic hydrogen in the laboratory. On earth, hydrogen occurs chiefly in combination with oxygen in water, but is also present in organic matter, such as living plants, petroleum, coal, etc. It is present as the free element in the atmosphere, but only to the extent of less than 1 part/million, by volume. It is the lightest of all gases, and combines with other elements, sometimes explosively, to form compounds. Great quantities of hydrogen are required commercially for the fixation of nitrogen from the air in the Haber ammonia process and for the hydrogenation of fats and oils. It is also used in large quantities in methanol production, in hydrodealkylation, hydrocracking, and hydrodesulfurization. It is also used as a rocket fuel, for welding, for production of hydrochloric acid, for the reduction of metallic ores, and for filling balloons. The lifting power of 1 ft^3 of hydrogen gas is about 0.076 lb at 0°C, 760 mm pressure. Production of hydrogen in the U.S. alone now amounts to hundreds of millions of ft^3/day. It is prepared by the action of steam on heated carbon, by decomposition of certain hydrocarbons with heat, by the electrolysis of water, or by the displacement from acids by certain metals. It is also produced by the action of sodium or potassium hydroxide on aluminum. Liquid hydrogen is important in cryogenics and in the study of superconductivity as its mp is only a few degrees above absolute zero. The ordinary isotope of hydrogen, $_1H^1$ is known as *protium.* In 1932 Urey announced the preparation of a stable isotope, deuterium, with an atomic weight of $2(_1H^2$ or D). Two years later an unstable isotope, tritium ($_1H^3$), with an atomic weight of 3 was discovered. Tritium has a half-life of about 12.5 years. One part deuterium is found to about 6000 ordinary hydrogen atoms. Tritium atoms are also present but in much smaller proportion. Tritium is readily produced in nuclear reactors, and is used in the production of the hydrogen bomb. It is also used as a radioactive agent making luminous paints, and as a tracer. The current price of tritium, to authorized personnel, is about $2/curie; deuterium gas is readily available, without permit, at about $1/liter. Heavy water, deuterium oxide (D_2O), which is used as a moderator to slow down neutrons, is available without permit at a cost of 6¢ to $1/g, depending on quantity and purity. Quite apart from isotopes, it has been shown that hydrogen gas under ordinary conditions is a mixture of two kinds of molecules, known as *ortho-* and *para*-hydrogen, which differ from one another by the spins of their electrons and nuclei. Normal hydrogen at room temperature contains 25% of the *para* form and 75% of the *ortho* form. The *ortho* form cannot be prepared in the pure state. Since the two forms differ in energy, the physical properties also differ. The melting and boiling points of *para*-hydrogen are about 0.1°C lower than those of normal hydrogen. Of current interest is a substance known as *polywater* or *anomalous water*, which some observers believe to be a polymer of water.

Indium (from the brilliant indigo line in its spectrum) − In; at. wt 114.82; at. no. 49; mp 156.61°C; bp 2080°C; sp gr 7.31 (20°C); valence 1, 2, or 3. Discovered by Reich and Richter, who later isolated the metal. Indium is most frequently associated with zinc minerals, and it is from these that most commercial indium is now obtained; however, it is also found in iron, lead, and copper ores. Until 1924, a gram or so constituted the world's supply of this element in isolated form. It is probably about as abundant as silver. One to 1½ million troy oz of indium are now produced annually in the Free World. Japan is presently producing more than 250,000 troy oz annually. The present cost of indium is about $1.50 to $5.00/troy oz, depending on quantity and purity. It is available in ultrapure form. Indium is a very soft, silvery-white metal with a brilliant luster. The pure metal gives a high-pitched "cry" when bent. It wets glass, as does gallium. It has found application in making low-melting alloys; an alloy of 24% indium-76% gallium is liquid at room temperature. It is used in making bearing alloys, germanium transistors, rectifiers, thermistors, and photoconductors. It can be plated onto metal and evaporated onto glass forming a mirror as good as that made with silver, but with more resistance to atmospheric corrosion. There is evidence that indium has a low order of toxicity. Normal hygienic precautions should provide adequate protection.

Iodine (Gr. *iodes*, violet) − I; at. wt 126.9045; at. no. 53; mp 113.5°C; bp 184.35°C; density of the gas 11.27 g/l; sp gr solid 4.93 (20°C); valence 1, 3, 5, or 7. Discovered by Courtois in 1811. Iodine, a halogen, occurs sparingly in the form of iodides in sea water from which it is assimilated by seaweeds, in Chilean saltpeter and nitrate-bearing earth, known as *caliche,* in brines from old sea deposits, and in brackish waters from oil and salt wells. Ultrapure iodine can be obtained from the reaction of potassium iodide with copper sulfate. Several other methods of isolating the element are known. Iodine is a bluish-black, lustrous solid, volatilizing at ordinary temperatures into a blue-violet gas with an irritating odor; it forms compounds with many elements, but is less active than

the other halogens, which displace it from iodides. Iodine exhibits some metallic-like properties. It dissolves readily in chloroform, carbon tetrachloride, or carbon disulfide to form beautiful purple solutions. It is only slightly soluble in water. Iodine compounds are important in organic chemistry and very useful in medicine. Twenty-four isotopes are recognized. Only one stable isotope, I^{127}, is found in nature. The artificial radioscope I^{131}, with a half-life of 8 days, has been used in treating the thyroid gland. The most common compounds are the iodides of sodium and potassium (KI) and the iodates (KIO_3). Lack of iodine is the cause of goiter. The iodide and thyroxin, which contain iodine, are used internally in medicine, and a solution of KI and iodine in alcohol is used for external wounds. Potassium iodide finds use in photography. The deep blue color with starch solution is characteristic of the free element. Care should be taken in handling and using iodine as contact with the skin can cause lesions; iodine vapor is intensely irritating to the eyes and mucous membranes. The recommended maximum allowable concentration in air is 1 mg/m^3.

Iridium (L. *iris*, rainbow) − Ir; at. wt 192.22; at. no. 77; mp 2410°C; bp 4130°C; sp gr 22.42 (17°C); valence 3 or 4. Discovered in 1803 by Tennant in the residue left when crude platinum is dissolved by aqua regia. The name iridium is appropriate, for its salts are highly colored. Iridium, a metal of the platinum family, is white, similar to platinum, but with a slight yellowish cast. It is very hard and brittle, making it very hard to machine, form or work. It is the most corrosion-resistant metal known, and was used in making the standard meter bar of Paris, which is a 90% platinum−10% iridium alloy. This meter bar has since been replaced as a fundamental unit of length (see under Krypton). Iridium occurs uncombined in nature with platinum and other metals of this family in alluvial deposits. It is recovered as a by-product from the nickel mining industry. Iridium has found use in making crucibles and apparatus for use at high temperatures. It is also used for electrical contacts. Iridium is widely used in thermocouples and resistor wires, alloyed with other precious metals such as 40% Ir to 60% Rh and 60% Ir to 40% Rh. Its principal use is as a hardening agent for platinum. With osmium, it forms an alloy which is used for tipping pens and compass bearings. The specific gravity of iridium is only very slightly lower than that of osmium, which has been generally credited as being the heaviest known element. Calculations of the densities of iridium and osmium from the space lattices give values of 22.65 and 22.61 g/cm^3, respectively. These values may be more reliable than actual physical measurements. At present, therefore, we know that either iridium or osmium is the densest known element, but the data do no yet allow selection between the two. Iridium costs about $160 to $190/troy oz.

Iron (Anglo-Saxon, *iron*) Fe, (L. *ferrum*) — at. wt 55.847; at. no. 26; mp 1535°C; bp 2750°C; sp gr 7.874 (20°C); valence 2, 3, 4, or 6. The use of iron is prehistoric. Genesis mentions that Tubal-Cain, seven generations from Adam, was "an instructer of every artificer in brass and iron." Homer mentions a ball of iron as the prize awarded to Achilles for athletic prowess. The metallurgy of iron began widespread development around 1400 to 1000 B.C., ushering in the Iron Age, with evidence of the ability to use steel-working and hardening to make weapons and agricultural implements. Today iron is the most useful metal and most widely used in modern civilization. The 1971 world output of iron ore is estimated at 785.8 million metric tons, pig iron at 430.12 million metric tons, and raw steel at 582.12 million metric tons. For the second year in a row the output of iron ore from the U.S.S.R. exceeded that of the U.S. at 203 to 82.1 million metric tons, and that of pig iron at 88.3 million metric tons compared to U.S. at 203 to 82.1 million million metric tons. The most significant change in production of iron ore has been the emergence of Australia as the number three producer of iron ore with France, fourth. For the first time in 1971 the production of steel in the U.S.S.R. exceeded that of the U.S. by 121 to 109.26 million metric tons, with Japan at 88.56 and West Germany at 40.31 following in tonnage. A remarkable iron pillar, dating to about A.D. 400, remains standing today in Delhi, India. This solid shaft of wrought iron is about 7¼ meters high by 40 cm in diameter. Corrosion to the pillar has been minimal although it has been exposed to the weather since its erection. Iron is a relatively abundant element in the universe. It is found in the sun and many types of stars in considerable quantity. Its nuclei are very stable. Iron is found native as a principal component of a class of meteorites known as *siderites,* and is a minor constituent of the other two classes. The core of the earth, 2150 miles in radius, is thought to be largely composed of iron with about 10% occluded hydrogen. The metal is the fourth most abundant element, by weight, making up the crust of the earth. The most common ore is *hematite* (Fe$_2$O$_3$), from which the metal is obtained by reduction with carbon. Iron is found in other widely distributed minerals, such as *magnetite* which is frequently seen as *black sands* along beaches and banks of streams. In addition to magnetite (Fe$_3$O$_4$), limonite (hydrated iron oxide), siderite (FeCO$_3$), and a magnetite bearing ore called taconite are important. *Taconite* is becoming increasingly important as a commercial ore. Common iron is a mixture of four isotopes, 56, 54, 57, and 58. Six other isotopes are known to exist. Iron is a vital constituent of plant and animal life, and appears in hemoglobin. The pure metal is not often encountered in commerce, but is usually alloyed with carbon or other metals. The pure metal is very reactive chemically, and rapidly corrodes especially in moist air or at elevated temperatures. It has four allotropic forms, or ferrites, known as α, β, γ, and δ, with transition points at 770°, 928°, and 1530°C. The α form is magnetic but when transformed into the β form, the magnetism disappears although the lattice remains unchanged. The relations of these forms are peculiar. Iron is smelted from the ore from a mixture with limestone and coke at high temperatures up to 1300°C, to produce cast iron or pig iron. Pig iron is an alloy of pure iron or ferrite containing about 3% of a glassy slag which is a complex high iron silicate containing S, Mn, and P, and other elements. It is hard, brittle, fairly fusible, and is used to produce other alloys. Wrought iron is obtained by oxidizing and fluxing the impurities from pig iron. It contains only a few tenths of % carbon, is tough, malleable, less fusible, and usually has a "fibrous" structure. Steel was originally made by long term heat treatment of wrought iron with charcoal to increase the carbon content to nearly 1%. In the Bessemer process introduced in 1856, an air blast was blown directly throught the melted pig iron to produce a high quality, relatively cheap product in large quantities. It revolutionized the industry and was followed by other large scale methods such as the open-hearth process. More recently the basic-oxygen process which uses 99.5% pure oxygen has been utilized by a growing number of firms.

Carbon steel is an alloy of iron with carbon. Carbon steels are divided into dead-soft steel (0.08 to 0.18% carbon), structural grade or mild steel (0.15 to 0.25% carbon), medium grade steel (0.25 to 0.35% carbon), medium-hard steel (0.35 to 0.65% carbon), and hard steel (0.65 to 0.85% carbon), the spring steels (0.85 to 1.05% carbon), and the high carbon tool steels (1.05 to 1.20% carbon). The addition of carbon to α-iron decreases the transition temperature to γ-iron to a minimum of 723°C at 0.8% carbon. Above this temperature the bcc structure of ferrite transforms to an fcc form called austenite which holds much greater amounts of carbon in solid solution than α-iron which forms ferrite containing a maximum of 0.025% carbon in solid solution at 723°C. When a steel containing more than 0.025% carbon is cooled below 723°C, the excess carbon forms a hard, brittle compound called cementite, Fe_3C. At the eutectoid point at 0.8% carbon, 723°C, the austenite decomposes on cooling to ferrite and cementite in a lamellar structure called pearlite. Controlled heat treating of austenite containing varying amounts of carbon, followed by various cooling rates, produces a wide variety of metallurgical structures. Very rapid quenching produces a ferrite supersaturated with carbon called martensite. Steels produced with a mixture of cementite and martensite give a maximum hardness. Alloy steels are carbon steels with major alloying elements added. These are principally nickel, chromium, manganese, vanadium, molybdenum, boron, tungsten, and cobalt. Stainless steels contain greater than 10% chromium. The most important of these is a steel containing 18% chromium and 8% nickel.

Krypton (Gr. *kryptos,* hidden) — Kr; at. wt 83.80; at. no. 36; mp -156.6°C, bp -152.30 ± 0.10°C; density 3.733 g/l (0°C); valence usually 0. Discovered in 1898 by Ramsay and Travers in the residue left after liquid air had nearly boiled away. Krypton is present in the air to the extent of about 1 part per million. It is one of the "noble" gases. It is characterized by its brilliant green and orange spectral lines. Naturally occurring krypton contains six stable isotopes. Fifteen other unstable isotopes are now recognized. The spectral lines of krypton are easily produced and some are very sharp. In 1960 it was internationally agreed that the fundamental unit of length, the meter, should be defined in terms of the orange-red spectral line of Kr^{86}, corresponding to the transition $5p[O_{1/2}]_1$

$- 6d[O_{1/2}]_1$ as follows: *1 meter = 1,650,763.73 wavelengths (in vacuo) of the orange-red line of Kr^{86}*. This replaces the standard meter of Paris, which was defined in terms of a bar made of a platinum-iridium alloy. Solid krypton is a white crystalline substance with a face-centered cubic structure which is common to all the "rare gases." While krypton is generally thought of as a rare gas that normally does not combine with other elements to form compounds, it now appears that the existence of some krypton compounds is established. Krypton difluoride has been prepared in gram quantities and can be made by several methods. A higher fluoride of krypton and a salt of an oxyacid of krypton also have been reported. Molecule ions of $ArKr^+$ and KrH^+ have been identified and investigated, and evidence is provided for the formation of KrXe or $KrXe^+$. Krypton clathrates have been prepared with hydroquinone and phenol. Kr^{85} has found recent application in chemical analysis. By imbedding the isotope in various solids, *kryptonates* are formed. The activity of these kryptonates is sensitive to chemical reactions at the surface. Estimates of the concentration of reactants are therefore made possible. Krypton is used commercially with argon as a low-pressure filling gas for fluorescent lights. It is used in certain photographic flash lamps for high-speed photography. Uses thus far have been limited because of its high cost. Krypton gas presently costs about $20/1.

Kurchatovium — (see Element 104).

Lanthanum (Gr. *lanthanein,* to lie hidden) — La; at. wt 138.9055; at. no. 57; mp 921°C; bp 3457°C; sp. gr 6.145 (25°C); valence 3. Mosander in 1839 extracted a new earth, *lanthana*, from impure cerium nitrate, and recognized the new element. Lanthanum is found in rare-earth minerals, such as *cerite, monazite, allanite,* and *bastnasite*. Monazite and bastnasite are principal ores in which lanthanum occurs in percentages up to 25% and 38%, respectively. Misch metal, used in making lighter flints, contains about 25% lanthanum. Lanthanum was isolated in relatively pure form in 1923. Ion exchange and solvent extraction techniques have led to much easier isolation of the so-called "rare-earth" elements. The availability of lanthanum and other rare earths has improved greatly in recent years. The metal can be produced by reducing the anhydrous fluoride with calcium. Lanthanum is silvery white, malleable, ductile, and soft enough to be cut with a knife. It is one of the most reactive of the rare-earth metals. It oxidizes rapidly when exposed to air. Cold water attacks lanthanum slowly, and hot water attacks it much more rapidly. The metal reacts directly with elemental carbon, nitrogen, boron, selenium, silicon, phosphorus, sulfur, and with halogens. At 310°C, lanthanum changes from a hexagonal to a face-centered cubic structure, and at 865°C it again transforms into a body-centered cubic structure. Natural lanthanum is a mixture of two stable isotopes, La^{138} and La^{139}. Seventeen other radioactive isotopes are recognized. Rare-earth compounds containing lanthanum are extensively used in carbon lighting applications, especially by the motion picture industry for studio lighting and projection. This application consumes about 25% of the rare-earth compounds produced. La_2O_3 improved the alkali resistance of glass, and is used in making special optical glasses. Small amounts of lanthanum, as an additive, can be used to produce nodular cast iron. Lanthanum and its compounds have a low to moderate acute toxicity rating; therefore, care should be taken in handling them. The price of 99.9% La_2O_3 is about $8/lb. The metal costs about $70/lb.

Lawrencium (Ernest O. Lawrence, inventor of the cyclotron) — Lr; at. no. 103; at. mass no. 257; valence +3(?). The last member of the 5f transition elements (actinide series) was discovered March 1961 by A. Ghiorso, T. Sikkeland, A. E. Larsh, and R. M. Latimer. A three microgram californium target, consisting of a mixture of isotopes of mass number 249, 250, 251, and 252 was bombarded with either B^{10} or B^{11}. The electrically charged transmutation nuclei recoiled with an atmosphere of helium and were collected on a thin copper conveyor tape which was then moved to place collected atoms in front of a series of solid-state detectors. The isotope of element 103 produced in this way decayed by emitting an 8.6 MeV alpha particle with a half-life of 8 sec. Flerov and associates of the Dubna Laboratory, in 1967, reported their inability to detect an alpha emitter with a half-life of 8 sec which was assigned by the Berkeley Group to 103^{257}. This assignment has been changed to Lr^{258} or Lr^{259}. In 1965 the Dubna workers found a longer-lived lawrencium isotope, Lr^{256}, with a half-life of 35 sec. In 1968 Ghiorso and associates at Berkeley were able to use a few atoms of this isotope to study the oxidation behavior of lawrencium. Using solvent extraction techniques, and working very rapidly, they extracted lawrencium ions from a buffered aqueous solution into an organic solvent, completing each extraction in about 30 sec. It was found that lawrencium behaves differently from dipositive nobelium and more like the tripositive elements earlier in the actinide series.

Lead (Anglo-Saxon, *lead*) — Pb (L. *plumbum*); at. wt 207.2; at. no. 82; mp 327.502°C; bp 1740°C; sp gr 11.35 (20°C); valence 2 or 4. Long known; mentioned in Exodus. The alchemists believed lead to be the oldest metal and associated it with the planet Saturn. Native lead occurs in nature, but it is rare. Lead is obtained chiefly from *galena* (PbS) by a roasting process. *Anglesite* ($PbSO_4$), *Cerrusite* ($PbCO_3$), and *Minim* (Pb_3O_4) are other common lead minerals. Lead is a bluish-white metal of bright luster, very soft, highly malleable, is ductile and a poor conductor of electricity. It is very resistant to corrosion; lead pipes bearing the insignia of Roman emperors, used as drains from the baths, are still in service. It is used as containers for corrosive liquids, such as in sulfuric acid chambers, and it may be toughened by the addition of a small percentage of antimony or other metals. Natural lead is a mixture of four stable isotopes: Pb^{204} (1.48%), Pb^{206} (23.6%), Pb^{207} (22.6%), and Pb^{208} (52.3%). Lead isotopes are the end products of each of the three series of naturally occurring radioactive elements: Pb^{206} for the uranium series, Pb^{207} for the actinium series, and Pb^{208} for the thorium series. Seventeen other isotopes of lead, all of which are radioactive, are recognized. Its alloys include solder, type metal, and various antifriction metals. Great quantities of lead, both as the metal and as the dioxide, are used in storage batteries. Much metal also goes into cable covering, plumbing, ammunition, and in the manufacture of lead tetraethyl, used as an anti-knock compound in gasoline. The metal is very effective as a sound absorber, is used as a radiation shield around x-ray equipment and nuclear reactors, and is used to absorb vibration. White lead, the basic lead carbonate, sublimed white lead ($PbSO_4$), chrome yellow ($PbCrO_4$), red lead (Pb_3O_4) and other lead compounds are extensively used in paints. Lead oxide is used in producing fine "crystal glass" and "flint glass" of a high index of refraction for achromatic lenses. The nitrate and the acetate are soluble salts. Lead salts, such as lead arsenate, have been used as insecticides, but their use in recent years has been practically eliminated and replaced with less harmful organic compounds. Care must be used in handling lead as it is a cumulative poison. Environmental concern with lead poisoning has resulted in a naturally programmed reduction in the concentration of lead in gasoline. Specific limits are to be set by the Environmental Protection Agency with the goal of a two thirds reduction in lead content within approximately 5 yr.

Lithium (Gr. *lithos***,** **stone)** – Li; at. wt 6.941; at. no. 3; mp 180.54°C; bp 1347°C; sp gr 0.534 (20°C); valence 1. Discovered by Arfvedson in 1817. Lithium is the lightest of all metals, with a density only about half that of water. It does not occur free in nature; combined it is found in small amounts in nearly all igneous rocks and in the waters of many mineral springs. *Lepidolite, spodumene, petalite*, and *amblygonite* are the more important minerals containing it. Lithium is presently being recovered from brines of Searles Lake, in California, and from the Great Salt Lake, Utah. Large deposits are found in Nevada and North Carolina. The metal is produced electrolytically from the fused chloride. Lithium is silvery in appearance much like Na and K, other members of the alkali metal series. It reacts with water, but not as vigorously as sodium. Lithium imparts a beautiful crimson color to a flame, but when the metal burns strongly the flame is a dazzling white. Since World War II, the production of lithium metal and its compounds has increased greatly. Because the metal has the highest specific heat of any solid element, it has found use in heat transfer applications; however, it is corrosive and requires special handling. The metal has been used as an alloying agent, is of interest in synthesis of organic compounds, and it has nuclear applications. It ranks as a leading contender as a battery anode material as it has a high electrochemical potential. Lithium is used in special glasses and ceramics. The glass for the 200-in. telescope at Mt. Palomar contains lithium as a minor ingredient. Lithium chloride is one of the most hygroscopic materials known, and it, as well as lithium bromide, is used in air conditioning and industrial drying systems. Lithium stearate is used as an all-purpose and high-temperature lubricant. Other lithium compounds are used in dry cells and storage batteries. The metal is priced at about $8/lb.

Lutetium (*Lutetia***, ancient name for Paris –** **sometimes called** *cassiopeium* **by the Germans)** – Lu; at. wt 174.97; at. no. 71; mp 1663°C; bp 3395°C; sp gr 9.840 (25°C); valence 3. In 1907

Urbain described a process by which Marignac's ytterbium (1879) could be separated into the two elements, ytterbium (neoytterbium) and lutetium. These elements were identical with "aldebaranium" and "cassiopeium" independently discovered by von Welsbach about the same time. Charles James of the University of New Hampshire also independently prepared the very pure oxide, *lutecia*, at this time. The spelling of the element was changed from *lutecium* to *leutetium* in 1949. Lutetium occurs in very small amounts in nearly all minerals contaning yttrium, and is present in *monazite* to the extent of about 0.003%, which is a commercial source. The pure metal has been isolated in recent years and is one of the most difficult to prepare. It can be prepared by the reduction of anhydrous $LuCl_3$ or LuF_3 by an alkali or alkaline earth metal. The metal is silvery white and relatively stable in air. While new techniques, including ion exchange reactions, have been developed to separate the various rare earth elements, lutetium is still the most costly of all naturally occurring rare earths. It is slightly more abundant than thulium. It is now priced at about $26/g or $8000/lb. Lu^{176} occurs naturally (2.6%) with Lu^{175} (97.4%). It is radioactive with a half-life of about 3×10^{10} years. Stable lutetium nuclides, which emit pure beta radiation after thermal neutron activation, can be used as a catalyst in cracking, alkylation, hydrogenation, and polymerization. Virtually no other commercial uses have been found yet for lutetium, as it is still one of the most costly natural elements. Lutetium, like other rare earth metals, has a low toxicity rating. It should be handled with care.

Magnesium (*Magnesia***, district in Thessaly)** – Mg; at. wt 24.305; at. no. 12; mp 648 ± 0.5°C; bp 1090°C; sp gr 1.738 (20°C); valence 2. Compounds of magnesium have long been known. Black recognized magnesium as an element in 1755. It was isolated by Davy in 1808, and prepared in coherent form by Bussy in 1831. Magnesium is the eighth most abundant element in the earth's crust. It does not occur uncombined, but is found in large deposits in the form of *magnesite, dolomite,* and other minerals. The metal is now principally obtained in the U.S. by electrolysis of fused magnesium chloride derived from brines, from wells and from sea water. Magnesium is a light, silvery white, and fairly tough metal. It tarnishes slightly in air, and finely divided magnesium readily ignites upon heating in air and burns with a dazzling white flame. It is used in flash-light photography, flares, and pyrotechnics, including incendiary bombs. It is one third lighter than aluminum, and in alloys is essential for airplane and missile construction. The metal improves the mechanical, fabrication, and welding characteristics of aluminum, when used as an alloying agent. Magnesium is used in producing nodular graphite in cast iron, and is used as an additive to conventional propellants. It is also used as a reducing agent in the production of pure uranium and other metals from their salts. The hydroxide (*Milk of Magnesia*), chloride, sulfate (Epsom salts), and citrate are used in medicine. Dead-burned magnesite is employed for refractory purposes, such as brick and liners in furnaces and converters. Organic magnesium compounds (Grignard's reaction) are important. Magnesium is an important element in both plant and animal life. Chlorophylls are magnesium-centered porphyrins. The adult daily requirement of magnesium is about 300 μg/day, but this is affected by various factors. Great care should be taken in handling magnesium metal, especially in the finely divided state, as serious fires can occur. Water should not be used on burning magnesium or on magnesium fires.

Maganese (L. *magnes*, magnet — from magnetic properties of pyrolusite; It. *manganese*, corrupt form of *magnesia*) — Mn; at. wt 54.9380; at. no. 25; mp 1244 ± 3°C; bp 1962°C; sp gr 7.21 to 7.44, depending on allotropic form; valence 1, 2, 3, 4, 6, or 7. Recognized by Scheele, Bergman, and others as an element and isolated by Gahn in 1774 by reduction of the dioxide with carbon. Manganese minerals are widely distributed; oxides, silicates, and carbonates are the most common. The recent discovery of large quantities of manganese nodules on the floor of the oceans appears promising as a new source of manganese. These nodules contain about 24% manganese together with many other elements in lesser abundance. Large deposits of nodules, extending over many square miles, have been found in Lake Michigan and Lake Superior. Pyrolusite (MnO_2) and rhodochrosite ($MnCO_3$) are common ores. The metal is obtained by reduction of the oxide with sodium, magnesium, aluminum, or by electrolysis. It is gray-white, resembling iron, but is harder and very brittle. The metal is reactive chemically, and decomposes cold water slowly. Manganese is used to form many important alloys. In steel, manganese improves the rolling and forging qualities,

strength, toughness, stiffness, wear resistance, hardness, and hardenability. With aluminum and antimony, especially with small amounts of copper, it forms highly ferromagnetic alloys. Manganese metal is ferromagnetic only after special treatment. The pure metal exists in four allotropic forms. The alpha form is stable at ordinary temperature; gamma manganese, which changes to alpha at ordinary temperatures, is said to be flexible, soft, easily cut, and capable of being bent. The dioxide (pyrolusite) is used as a depolarizer in dry cells, and is used to "decolorize" glass that is colored green by impurities of iron. Manganese by itself colors glass an amethyst color, and is responsible for the color of true amethyst. The dioxide is also used in the preparation of oxygen, chlorine, and in drying black paints. The permanganate is a powerful oxidizing agent and is used in quantitative analysis and in medicine. Manganese is widely distributed throughout the animal kingdom. It is an important trace element and may be essential for utilization of vitamin B_1. The world's first manganese ore pelletizing plant has recently been constructed in Brazil.

Mendelevium (Dmitri Mendeleev) — Md; at. wt 256; at. no. 101; valence +2, +3. Mendelevium, the ninth transuranium element of the actinide series to be discovered, was first identified by Ghiorso, Harvey, Choppin, Thompson, and Seaborg early in 1955 as a result of the bombardment of the isotope Es^{253} with helium ions in the Berkeley 60-in. cyclotron. The isotope produced was Md^{256}, which has a half-life of 77 min. This first identification was notable in that Md^{256} was synthesized on a one-atom-at-a-time basis. Four isotopes are now recognized. Md^{258} has a half-life of 2 months. This isotope has been produced by the bombardment of an isotope of einsteinium with ions of helium. It now appears possible that eventually enough Md^{258} can be made so that some of its physical properties can be determined. Md^{256} has been used to elucidate some of the chemical properties of mendelevium in aqueous solution. Experiments seem to show that the element possesses a moderately stable dipositive (II) oxidation state in addition to the tripositive (III) oxidation state, which is characteristic of actinide elements.

Mercury (Plant, *Mercury*) — Hg (*hydragyrum*, liquid silver); at. wt 200.59; at. no. 80; mp -38.87°C; bp 356.58°C; sp gr 13.546 (20°C); valence 1 or 2. Known to ancient Chinese and

Hindus; found in Egyptian tombs of 1500 B.C. Mercury is the only common metal liquid at ordinary temperatures. It only rarely occurs free in nature. The chief ore is *cinnabar* (HgS). Spain, Italy, and the USSR, together produce about 50% of the world's supply of the metal. The commercial unit for handling mercury is the "flask," which weighs 76 lb. In the past few years, the price has fluctuated widely between $150 and $270 per flask. The metal is obtained by heating cinnabar in a current of air and by condensing the vapor. It is a heavy, silvery white metal; a rather poor conductor of heat, as compared with other metals; and a fair conductor of electricity. The metal is widely used in laboratory work for making thermometers, barometers, diffusion pumps, and many other instruments. It is used in making mercury-vapor lamps and advertising signs, etc., and is used in mercury switches and other electrical apparatus. Other uses are for making pesticides, mercury cells for caustic-chlorine production, dental preparations, anti-fouling paint, batteries, and catalysts. The most important salts are mercuric chloride ($HgCl_2$, corrosive sublimate — a violent poison), mercurous chloride Hg_2Cl_2 (calomel — occasionally still used in medicine), mercury fulminate ($Hg(ONC)_2$, a detonator widely used in explosives), and mercuric sulfide (HgS, vermillion, a high-grade paint pigment). Organic mercury compounds are important. It has been found than an electrical discharge causes mercury vapor to combine with neon, argon, krypton, and xenon. These products, held together with van der Waals' forces, correspond to HgNe, HgAr, HgKr, and HgXe. It is not generally appreciated that mercury is a virulent poison and is readily absorbed through the respiratory tract, the gastrointestinal tract, or through unbroken skin. It acts as a cumulative poison since only small amounts of the element can be eliminated at a time by the human organism. The present adopted threshold limit value for mercury in air (except for alkyl compounds) is 0.05 mg/cù. meter. Since mercury is a very volatile element, dangerous levels are readily attained in air. Air saturated with mercury vapor at 20°C contains a concentration which exceeds the toxic limit by more than 100 times. The danger increases at higher temperatures. *It is therefore important that mercury be handled with utmost care.* Containers of mercury should be securely covered and spillage should be avoided. If it is necessary to heat mercury or mercury compounds, it should be done in a well-ventilated hood. Methyl mercury is a dangerous pollutant and is now widely found in water and streams.

Molybdenum (Gr. *molybdos*, lead) — Mo; at. wt 95.94; at. no. 42; mp 2617°C; bp 4612°C; sp gr 10.22 (20°C); valence 2, 3, 4?, 5?, or 6. Before Scheele recognized molybdenite as a distinct ore of a new element in 1778, it was confused with graphite and lead ore. The metal was prepared in an impure form in 1792 by Hjelm. Molybdenum does not occur native, but is obtained principally from *molybdenite* (MoS_2). *Wulfenite* ($PbMoO_4$) and *powellite* ($Ca(MoW)O_4$) are also minor commercial ores. Molybdenum is also recovered as a byproduct of copper and tungsten mining operations. The metal is prepared from the powder made by the hydrogen reduction of purified molybdic trioxide or ammonium molybdate. The metal is silvery white, very hard, but is softer and more ductile than tungsten. It has a high elastic modulus, and only tungsten and tantalum, of the more readily available metals, have higher melting points. It is a valuable alloying agent, as it contributes to the hardenability and toughness of quenched and tempered steels. It also improves the strength of steel at high temperatures. It is used in certain nickel-based alloys, such as the "Hastelloys," which are heat-resistant and corrosion-resistant to chemical solutions. Molybdenum oxidizes at elevated temperatures. The metal has found recent application as electrodes for electrically heated glass furnaces and forehearths. The metal is also used in nuclear energy applications and for missile and aircraft parts. Molybdenum wire is valuable for use as a filament material for metal evaporation work, and as a filament, grid, and screen material for electronic tubes. Molybdenum is an essential trace element in plant nutrition. Some lands are barren for lack of this element in the soil. Molybdenum sulfide is useful as a lubricant, especially at high temperatures where oils would decompose. Almost all ultra-high strength steels with minimum yield points up to 300,000 psi ($lb/in.^2$) contain molybdenum in amounts from 0.25% to 8%. Molybdenum powder is priced at about $4/lb, and bars rolled from arc-cast ingots cost about $15/lb.

Neodymium (Gr. *neos*, new, and *didymos*, twin) — Nd; at. wt 144.24; at. no. 60; mp 1020°C; bp 3168°C; sp gr 6.80 and 7.007, depending on allotropic form; valence 3. In 1841 Mosander extracted from *cerite* a new rose-colored oxide, which he believed to contain a new element. He named the element *didymium*, as it was "an inseparable twin brother of lanthanum." In 1885 von Welsbach separated didymium into two new elemental components, *neodymia* and *praseodymia*, by repeated fractionation of ammonium didymium nitrate. While the free metal is in *misch metal*, long known and used as a pyrophoric alloy for lighter flints, the element was not isolated in relatively pure form until 1925. Neodymium is present in misch metal to the extent of about 18%. It is present in the minerals *monazite* and *bastnasite*, which are principal sources of rare-earth metals. The element may be obtained by separating neodymium salts from other rare earths by ion exchange or solvent extraction techniques, and by reducing anhydrous halides, such as NdF_3, with calcium metal. Other separation techniques are possible. The metal has a bright silvery metallic luster. Neodymium is one of the more reactive rare-earth metals and quickly tarnishes in air, forming an oxide that spalls off and exposes metal to oxidation. The metal, therefore, should be kept under light mineral oil or sealed in a plastic material. Neodymium exists in two allotropic forms, with a transformation from a double hexagonal to a body-centered cubic structure taking place at 860°C. Natural neodymium is a mixture of seven stable isotopes. Nine other radioactive isotopes are recognized. Didymium, of which neodymium is a component, is used for coloring glass to make welder's goggles. By itself, neodymium colors glass delicate shades ranging from pure violet through wine-red and warm gray. Light transmitted through such glass shows unusually sharp absorption bands. The glass is useful in astronomical work to produce sharp bands by which spectral lines may be calibrated. Glass containing neodymium can be used as a laser material in place of ruby to produce coherent light. Neodymium slats are also used as a colorant for enamels. The price of the metal is about 50¢/g or $120/lb. Neodymium has a low-to-moderate acute toxic rating. As with other rare earths, neodymium should be handled with care.

Neon (Gr. *neos*, new) – Ne; at. wt 20.179; at. no. 10; mp -248.67°C; bp -246.048°C (1 atm); density of gas 0.89990 g/l (1 atm 0°C); density of liquid at pb $\overline{1}$.207 g/cm³; valence 0. Discovered by Ramsay and Travers in 1898. Neon is a rare gaseous element present in the atmosphere to the extent of 1 part in 65,000 of air. It is obtained by liquefaction of air and separated from the other gases by fractional distillation. Natural neon is a mixture of three isotopes. Five other unstable isotopes are known. It is a very inert element; however, it is said to form a compound with fluorine. It is still questionable if true compounds of neon exist, but evidence is mounting in favor of their existence. The following ions are known from optical and mass spectrometric studies: Ne_2^+, $(NeAr)^+$, $(NeH)^+$, and $(HeNe)^+$. Neon also forms an unstable hydrate. In a vacuum discharge tube, neon glows reddish orange. Of all the rare gases, the discharge of neon is the most intense at ordinary voltages and currents. Neon is used in making the common neon advertising signs, which accounts for its largest use. It is also used to make high-voltage indicators, lightning arrestors, wave meter tubes, and TV tubes. Neon and helium are used in making gas lasers. Liquid neon is now commercially available and is finding important application as an economical cryogenic refrigerant. It has over 40 times more refrigerating capacity per unit volume than liquid helium and more than three times that of liquid hydrogen. It is compact, inert, and is less expensive than helium when it meets refrigeration requirements. Neon gas costs about $1.50/l.

Neptunium (Planet *Neptune*) – Np; at. wt 237.0482; at. no. 93; mp 640 ± 1°C; bp 3902°C (est.); sp gr 20.25 (20°C); valence 3, 4, 5, and 6. Neptunium was the first of the synthetic transuranium elements of the actinide series to be discovered; the isotope Np^{239} was produced by McMillan and Abelson in 1940 at Berkeley, California, as the result of bombarding uranium with cyclotron-produced neutrons. The isotope Np^{237} (half-life of 2.14 ± 10^6 years) is currently obtained in gram quantities as a byproduct from nuclear reactors in the production of plutonium. Trace quantities of the element are actually found in nature due to transmutation reactions in uranium ores produced by the neutrons which are present. Neptunium is prepared by the reduction of NpF_3 with barium or lithium vapor at about 1200°C. Neptunium metal has a silvery appearance, is chemically reactive, and exists in at least three structural modifications: α-neptunium, orthorhombic, density – 20.25 g/cm³; β-neptunium (above 280°C), tetragonal, density (313°C) – 19.36 g/cm³; γ-neptunium (above 577°C), cubic, density (600°C) 18.0 g/cm³. Neptunium has four ionic oxidation states in solution: Np^{+3} (pale purple), analogous to the rare earth ion Pm^{+3}, Np^{+4} (yellow green), NpO_2^+ (green blue), and NpO_2^{++} (pale pink). These latter oxygenated species are in contrast to the rare earths which exhibit only simple ions of the (II), (III), and (IV) oxidation states in aqueous solution. The element forms tri- and tetrahalides such as NpF_3, NpF_4, $NpCl_4$, $NpBr_3$, NpI_3, and oxides of various compositions such as are found in the uranium-oxygen system, including Np_3O_8 and NpO_2. Fifteen isotopes of neptunium are now recognized. The A.E.C. has Np^{237} available for sale to its licensees and for export. This isotope can be used as a component in neutron detection instruments. It is offered at a price of \$225/g or \$0.25/mg for quantities less than 1 gram.

Nickel (Ger. *Nickel*, Satan or "Old Nick," and from *kupfernickel*, Old Nick's copper) — Ni; at. wt 58.71; at. no. 28; mp 1453°C; bp 2732°C; sp gr 8.902 (25°C); valence 0, 1, 2, 3. Discovered by Cronstedt in 1751 in kupfernickel (*niccolite*). Nickel is found as a constituent in most meteorites and often serves as one of the criteria for distinguishing a meteorite from other minerals. Iron meteorites, or *siderites*, may contain iron alloyed with from 5% to nearly 20% nickel. Nickel is obtained commercially from *pentlandite* and *pyrrhotite* of the Sudbury region of Ontario. This district produces about 70% of the nickel for the free world. Small deposits are found in Norway, New Caledonia, Cuba, Japan, and elsewhere. Nickel is silvery-white and takes on a high polish. It is hard, malleable, ductile, somewhat ferromagnetic, and a fair conductor of heat and electricity. It belongs to the iron-cobalt group of metals and is chiefly valuable for the alloys it forms. It is extensively used for making stainless steel and other corrosion-resistant alloys, such as Invar, Monel, Inconel, and the Hastelloys. Tubing made of a copper-nickel alloy is extensively used in making desalination plants for converting sea water into fresh water. Nickel is also now used extensively in coinage, in making nickel steel for armor plate and burglar-proof vaults, and is a component in Nichrome, Permalloy, and Constantan. Nickel added to glass gives a green color. Nickel plating is often used to provide a protective coating for other metals, and finely divided nickel is a catalyst for hydrogenating vegetable oils. It is also used in ceramics, in the manufacture of Alnico magnets, and in the Edison storage battery. The sulfate and the oxides are important compounds. Natural nickel is a mixture of five stable isotopes. Seven other unstable isotopes are known.

Niobium (*Niobe*, daughter of Tantalus), Nb; or **Columbium** (*Columbia*, name for America) — Cb; at. wt 92.9064; at. no. 41; mp 2468 ± 10°C; bp 4742°C; sp gr 8.57 (20°C); valence 2, 3, 4?, 5. Discovered in 1801 by Hatchett in an ore sent to England more than a century before by John Winthrop the Younger, first governor of Connecticut. The metal was first prepared in 1864 by Blomstrand, who reduced the chloride by heating it in a hydrogen atmosphere. The name "niobium" was adopted by the International Union of Pure and Applied Chemistry in 1950 after 100 years of controversy. Many leading chemical societies and government organizations refer to it by this name. Most metallurgists, leading metal societies, and all but one of the leading U.S. commercial producers, however, still refer to the metal as "columbium." The element is found in *niobite* (or *columbite*), *niobite-tantalite, pyrochlore*, and *euxenite*. Large deposits of niobium have been found associated with *carbonatites* (carbon-silicate rocks), as a constituent of *pyrochlore*. Extensive ore reserves are found in Canada, Brazil, Nigeria, the Congo, and the U.S. The metal can be isolated from tantalum, and prepared in several ways. It is a shiny, white, soft, and ductile metal, and takes on a bluish cast when exposed to air at room temperatures for a long time. The metal starts to oxidize in air at 200°C, and when processed at even moderate temperatures must be placed in a protective atmosphere. It is used as an alloying agent in carbon and alloy steels and in nonferrous metals. These alloys have improved strength and other desirable properties. The metal has a low capture cross-section for thermal neutrons. It is used in arc-welding rods for stabilized grades of stainless steel. Thousands of pounds of niobium have been used in advanced air frame systems, such as were used in the Gemini space program. The element has superconductive properties; superconductive magnets have been made with Nb-Zr wire, which retains its superconductivity in strong magnetic fields. This type of application offers hope of direct large-scale generation of electric power. Sixteen isotopes of niobium are known. Niobium metal (99.5% pure) is priced at about \$15/lb.

Nitrogen (L. *nitrum*, Gr. *nitron*, native soda; *genes, forming*) — N; at. wt 14.0067; at. no. 7; mp

−209.86°C; bp −195.8°C; density 1.2506 g/l; sp gr liquid 0.808 (−195.8°C), solid 1.026 (−252°C); valence 3 or 5. Discovered by Daniel Rutherford in 1772, but Scheele, Cavendish, Priestley, and others about the same time studied "burnt or dephlogisticated air," as air without oxygen was then called. Nitrogen makes up 78% of the air, by volume. The estimated amount of this element in the atmosphere is more than 4000 billion tons. From this inexhaustible source it can be obtained by liquefaction and fractional distillation. Nitrogen molecules give the orange-red, blue-green, blue-violet, and deep violet shades to the aurora. The element is so inert that Lavoisier named it *azote*, meaning without life, yet its compounds are so active as to be most important in foods, poisons, fertilizers, and explosives. Nitrogen also can be easily prepared by heating a water solution of ammonium nitrite. Nitrogen, as a gas, is colorless, odorless, and a generally inert element. As a liquid it is also colorless and odorless, and is similar in appearance to water. Two allotropic forms of solid nitrogen exist, with the transition from the α to the β form taking place at −237°C. When nitrogen is heated, it combines directly with magnesium, lithium, or calcium; when mixed with oxygen and subjected to electric sparks, it forms first nitric oxide (NO) and then the dioxide (NO_2); when heated under pressure with a catalyst with hydrogen, ammonia is formed (Haber process). The ammonia thus formed is of the utmost importance as it is used in fertilizers, and it can be oxidized to nitric acid (Ostwald process). The ammonia industry is the largest consumer of nitrogen. Large amounts of the gas are also used by the electronics industry, which uses the gas as a blanketing medium during production of such components as transistors, diodes, etc. The drug industry also uses large quantities. Nitrogen is used as a refrigerant both for the immersion freezing of food products and for transportation of foods. Liquid nitrogen is also used in missile work as a purge for components, insulators for space chambers, etc., and by the oil industry to build up great pressures in wells to force crude oil upward. Sodium and potassium nitrates are formed by the decomposition of organic matter with compounds of the metals present. In certain dry areas of the world these saltpeters are found in quantity. Ammonia, nitric acid, the nitrates, the five oxides (N_2O, NO, N_2O_3, NO_2, and N_2O_5), TNT, the cyanides, etc. are but a few of the important compounds. Nitrogen gas prices vary from 2¢ to $2.75 per 100 cu. ft, depending on purity, etc. Production of nitrogen in the U.S. is more than 160 billion cu. ft/year.

Nobelium (Alfred Nobel, discoverer of dynamite) − − No; at. no. 102; valence +2, +3; Nobelium was unambiguously discovered and identified in April 1958 at Berkeley by A. Ghiorso, T. Sikkeland, J. R. Walton, and G. T. Seaborg, who used a new double-recoil technique. A heavy-ion linear accelerator (HILAC) was used to bombard a thin target of curium (95% Cm^{244} and 4.5% Cm^{246}) with C^{12} ions to produce 102^{254} according to the Cm^{246} (C^{12}, 4n) reaction. Earlier in 1957 workers of the U.S., Britain, and Sweden announced the discovery of an isotope of element 102 with a 10-min half-life at 8.5 MeV, as a result of bombarding Cm^{244} with C^{13} nuclei. On the basis of this experiment the name *nobelium* was assigned and accepted by the Commission on Atomic Weights of the International Union of Pure and Applied Chemistry. The acceptance of the name was premature, for both Russian and American efforts now completely rule out the possibility of any isotope of element 102 having a half-life of 10 min in the vicinity of 8.5 MeV. Early work in 1957 on the search for this element, in Russia at the Kurchatov Institute, was marred by the assignment of 8.9 ± 0.4 MeV alpha radiation with a half-life of 2 to 40 sec, which was too indefinite to support claim to discovery. Confirmatory experiments at Berkeley in 1966 have shown the existence of 102^{254} with a 55-sec half-life, 102^{252} with a 2.3-sec half-life, and 102^{257} with a 23-sec half-life. Four other isotopes are now recognized, one of which — 102^{255} — has a half-life of 3 min. In view of the discoverer's traditional right to name an element, the Berkeley Group, in 1967, suggested that the hastily given name *nobelium*, along with the symbol No, be retained.

Osmium (Gr. *osme*, a smell) − Os; at. wt 190.2; at. no. 76; mp 3045 ± 30°C; bp 5027 ± 100°C; sp gr 22.57; valence 0 to +8, more usually +3, +4, +6, and +8. Discovered in 1803 by Tennant in the residue left when crude platinum is dissolved by aqua regia. Osmium occurs in *iridosime* and in platinum-bearing river sands of the Urals, North America, and South America. It is also found in the nickel-bearing ores of the Sudbury, Ontario region along with other platinum metals. While the quantity of platinum metals in these ores is very small, the large tonnages of nickel ores processed make commercial recovery possible. The metal is lustrous, bluish-white, extremely hard, and brittle even at high temperatures. It has the highest melting point and lowest vapor pressure of the platinum group. The metal is very difficult to fabricate, but the powder can be sintered in a hydrogen atmosphere at a temperature of 2000°C. The solid metal is not affected by air at room temperature, but the powdered or spongy metal slowly gives off osmium tetroxide, which is a powerful oxidizing agent and has a strong smell. The tetroxide is highly toxic, and boils at 130°C (760 mm). Concentrations in air as low as 10^{-7} g/m^2 can cause lung congestion, skin damage, or eye damage. The tetroxide has been used to detect fingerprints and to stain fatty tissue for microscope slides. The metal is almost entirely used to produce very hard alloys, with other metals of the platinum group, for fountain pen tips, instrument pivots, phonograph needles, and electrical contacts. The price of 99% pure osmium powder – the form usually supplied commercially – is about $5/g or $300 to $450/troy oz, depending on quantity and supplier. The measured densities of iridium and osmium seem to indicate that osmium is slightly more dense than iridium, and osmium has generally been credited with being the heaviest known element. Calculations of the density from the space lattice, which may be more reliable for these elements than actual measurements, however, give a density of 22.65 for iridium compared to 22.61 for osmium. At present, therefore, we know either iridium or osmium is the heaviest element, but the data do not allow selection between the two.

Oxygen (Gr. *oxys*, sharp, acid, and *genes*, forming; acid former) – O; at. wt (natural) 15.9994; at. no. 8; mp –218.4°C; bp 182.962°C; density 1.429 g/l (0°C); sp gr liquid 1.14 (–182.96°C); valence 2. For many centuries, workers from time-to-time realized air was composed of more than one component. The behavior of oxygen and nitrogen as components of air led to the advancement of the phlogiston theory of combustion, which captured the minds of chemists for a century. Oxygen was prepared by several workers, including Bayen and Borch, but they did not know how to collect it, did not study its properties, not did they recognize it as an elementary substance. Priestley is generally credited with its discovery, although Scheele also

discovered it independently. Oxygen is the third most abundant element found in the sun, and it plays a part in the carbon-nitrogen cycle – one process thought to give the sun and stars their energy. Oxygen under excited conditions is responsible for the bright-red and yellow-green colors of the aurora. Oxygen, as a gaseous element, forms 21% of the atmosphere by volume from which it can be obtained by liquefaction and fractional distillation. The element and its compounds made up 49.2%, by weight, of the earth's crust. About two thirds of the human body, and nine tenths of water are oxygen. In the laboratory it can be prepared by the electrolysis of water or by heating potassium chlorate with manganese dioxide as a catalyst. The gas is colorless, odorless, and tasteless. The liquid and solid forms are a pale blue color and are strongly paramagnetic. Ozone (O_3), a highly active allotropic form of oxygen, is formed by the action of an electrical discharge or ultra-violet light on oxygen. Ozone's presence in the atmosphere (amounting to the equivalent of a layer 3 mm thick of ordinary pressures and temperatures) is of vital importance in preventing ultra-violet rays from the sun from reaching the earth's surface and destroying life on earth. Undiluted ozone has a bluish color. Liquid ozone is bluish-black, and solid ozone is violet-black. Oxygen is very reactive and capable of combining with most elements. It is a component of hundreds of thousands of organic compounds. It is essential for respiration of all plants and animals and for practically all combustion. In hospitals it is frequently used to aid respiration of patients. Its atomic weight was used as a standard of comparison for each of the other elements until 1961 when the International Union of Pure and Applied Chemistry adopted carbon 12 as the new basis. Oxygen has eight isotopes. Natural oxygen is a mixture of three isotopes. Oxygen 18 occurs naturally, is stable, and is available commercially. Water (H_2O with 1.5% O^{18}) is also available. Commercial oxygen consumption in the U.S. is estimated at more than 350 billion ft^3/year, and the demand is expected to increase substantially in the next few years. Oxygen enrichment of steel blast furances accounts for the greatest use of the gas. Large quantities are also used in making synthesis gas for ammonia and methanol, ethylene oxide, and for oxy-acetylene welding. Air separation plants produce about 99% of the gas; electrolysis plants about 1%. The gas costs 2.5¢/ft^3 in small quantities, and from $6 to $12/ton in large quantities.

Palladium (named after the asteroid _Pallas_, discovered about the same time; Gr _Pallas_ goddess of wisdom) — Pd; at. wt 106.4; at. no. 46; mp 1552°C; bp 3140°C; sp gr 12.02 (20°C); valence 2, 3, or 4. Discovered in 1803 by Wollaston, palladium is found along with platinum and other metals of the platinum group in placer deposits of the U.S.S.R., South and North America, Ethiopia, and Australia. It is also found associated with the nickel-copper deposits of South Africa and Ontario. Its separation from the platinum metals depends upon the type of ore in which it is found. It is a steel-white metal, does not tarnish in air, and is the least dense and lowest melting of the platinum group of metals. When annealed, it is soft and ductile; cold working greatly increases its strength and hardness. Palladium is attacked by nitric and sulfuric acid. At room temperatures the metal has the unusual property of absorbing up to 900 times its own volume of hydrogen, possibly forming Pd$_2$H. It is not yet clear if this is a true compound. Hydrogen readily diffuses through heated palladium and this provides a means of purifying the gas. Finely divided palladium is a good catalyst and is used for hydrogenation and dehydrogenation reactions. It is alloyed and used in jewelry trades. White gold is an alloy of gold decolorized by the addition of palladium. Like gold, palladium can be beaten into leaf as thin as 1/250,000 in. The metal is used in dentistry, watchmaking, and in making surgical instruments and electrical contacts. The metal sells for about $40/troy oz.

Phosphorus (Gr. _phosphoros_, light-bearing; ancient name for the planet Venus when appearing before sunrise) — P; at. wt 30.97376; at. no. 15; mp (white) 44.1°C; bp (white) 280°C; sp gr (white) 1.82, (red) 2.20, (black) 2.25 to 2.69; valence 3 or 5. Discovered in 1669 by Brand, who prepared it from urine. Phosphorus exists in four or more allotropic forms: white (or yellow), red, and black (or violet). White phosphorus has two modifications: α and β with a transition temperature at −3.8°C. Never found free in nature, it is widely distributed in combination with minerals. _Phosphate_ rock, which contains the mineral _apatite_ — an impure tri-calcium phosphate — is an important source of the element. Large deposits are found in the U.S.S.R., in Morocco, and in Florida, Tennessee, Utah, Idaho, and elsewhere. Phosphorus is an essential ingredient of all cell protoplasm, nervous tissue, and bones. Ordinary phosphorus is a waxy white solid; when pure it is colorless and transparent. It is insoluble in water, but soluble in carbon disulfide. It takes fire spontaneously in air, burning to the pentoxide. It is very poisonous — 50 mg constituting an approximate fatal dose. The maximum recommended allowable concentration in air is 0.1 mg/m^3. White phosphorus should be kept under water as it is dangerously reactive in air, and it should be handled with forceps, as contact with the skin may cause severe burns. When exposed to sunlight or when heated in its own vapor to 250°C, it is converted to the red variety, which does not phosphoresce in air as does the white variety. This form does not ignite spontaneously and it is not as dangerous as white phosphorus. It should, however, be handled with care as it does convert to the white form at some temperatures and it emits highly toxic fumes of the oxides of phosphorus when heated. The red modification is fairly stable, sublimes with a vapor pressure of 1 atm at 417°C, and is used in the manufacture of safety matches, pyrotechnics, pesticides, incendiary shells, smoke bombs, tracer bullets, etc. White phosphorus may be made by several methods. By one process, tri-calcium phosphate, the essential ingredient of phosphate rock, is heated in the presence of carbon and silica in an electric furnace or fuel-fired blast furnace. Elementary phosphorus is liberated as vapor and may be collected under water. If desired, the phosphorus vapor and carbon monoxide produced by the reaction can be oxidized at once in the presence of moisture or water to produce phosphoric acid — an important compound in making super-phosphate fertilizers. In recent years, concentrated phosphoric acids, which may contain as much as 70 to 75% P$_2$O$_5$ content, have become of great importance to agriculture and farm production. World-wide demand for fertilizers has caused record phosphate production in recent years. Phosphates are used in the production of special glasses, such as those used for sodium lamps. Bone-ash, calcium phosphate, is also used to produce fine china-ware and to produce mono-calcium phosphate used in baking powder. Phosphorus is also important in the production of steels, phosphor bronze, and many other products. Trisodium phosphate is important as a cleaning agent, as a water-softener, and for preventing boiler scale and corrosion of pipes and boiler tubes. Organic compounds of phosphorus are important.

Platinum (Sp. *platina*, silver) — Pt; at. wt 195.09; at. no. 78; mp 1772°C; bp 3827 ± 100°C; sp gr 21.45 (20°C); valence 1?, 2, 3, or 4. Discovered in South America by Ulloa in 1735 and by Wood in 1741. The metal was used by pre-Colombian Indians. Platinum occurs native, accompanied by small quantities of iridium, osmium, palladium, ruthenium, and rhodium, all belonging to the same group of metals. These are found in the alluvial deposits of the Ural mountains, of Colombia, and of certain western American states. *Sperrylite* ($PtAs_2$) occurring with the nickel-bearing deposits of Sudbury, Ontario, is the source of a considerable amount of the metal. The large production of nickel offsets the fact that there is only one part of the platinum metals in two million parts of ore. Platinum is a beautiful silvery-white metal, when pure, and is malleable and ductile. It has a coefficient of expansion almost equal to that of soda-lime-silica glass, and is therefore used to make sealed electrodes in glass systems. The metal does not oxidize in air at any temperature, but is corroded by halogens, cyanides, sulfur, and caustic alkalis. It is insoluble in hydrochloric and nitric acid, but dissolves when they are mixed as *aqua regia,* forming chloroplatinic acid (H_2PtCl_6), an important compound. The metal is extensively used in jewelry, in wire and vessels from laboratory use, and in many valuable instruments, including thermocouple elements. It is also used for electrical contacts, corrosion-resistant apparatus, and in dentistry. Platinum-cobalt alloys have magnetic properties. One such alloy made of 76.7% Pt and 23.3% Co, by weight, is an extremely powerful magnet that offers a B-H (max) almost twice that of Alnico V. Platinum resistance wires are used for constructing high-temperature electric furnaces. The metal is used for coating missile nose cones, jet engine fuel nozzles, etc., which must perform reliably for long periods of time at high temperatures. The metal, like palladium, absorbs large volumes of hydrogen, retaining it at ordinary temperatures but giving it up at red heat. In the finely divided state platinum is an excellent catalyst, having long been used in the contact process for producing sulfuric acid. It is also used as a catalyst in cracking petroleum products. Fine platinum wire will glow red hot when placed in the vapor of methyl alcohol. It acts here as a catalyst, converting the alcohol to formaldehyde. This phenomenon has been used commercially to produce cigarette lighters and hand warmers. Hydrogen and oxygen explode in the presence of platinum. The price of platinum has varied widely; more than a century ago it was used to adulterate gold. It was nearly eight times as valuable as gold in 1920; the present price is about $140/troy oz. There is tremendous current interest in platinum for catalysts in antipollution devices for automobiles. Such a catalyst system would exert a strong, long-term demand for platinum.

Plutonium (Planet *pluto*) — Pu; at. no. 94; isotopic mass Pu^{239} 239.13 (physical scale); sp gr (α modification) 19.84 (25°C); mp 641°C; bp 3232°C; valence 3, 4, 5, or 6. Plutonium was the second transuranium element of the actinide series to be discovered. The isotope Pu^{238} was produced in 1940 by Seaborg, McMillan, Kennedy, and Wahl by deuteron bombardment of uranium in the 60-in. cyclotron at Berkeley, California. Plutonium also exists in trace quantities in naturally occurring uranium ores. It is formed in much the same manner as neptunium, by irradiation of natural uranium with the neutrons which are present. By far of greatest importance is the isotope Pu^{239}, with a half-life of 24,360 years, produced in extensive quantities in nuclear reactors from natural uranium:

$$U^{238}(n, \gamma)U^{239} \xrightarrow{\beta-} Np^{239} \xrightarrow{\beta-} Pu^{239}.$$

Fifteen isotopes of plutonium are known. Plutonium has assumed the position of dominant importance among the transuranium elements because of its successful use as an explosive ingredient in nuclear weapons and the place which it holds as a key material in the development of industrial use of nuclear power. One pound is equivalent to about 10 million kWh of heat energy. Its importance depends on the nuclear property of being readily fissionable with neutrons and its availability in quantity. The various nuclear applications of plutonium are well known. Pu^{238} has been used in the Apollo lunar missions to power seismic and other equipment on the lunar surface. As with neptunium and uranium, plutonium metal can be prepared by reduction of the trifluoride and alkaline-earth metals. The metal has a silvery appearance and takes on a yellow tarnish when slightly oxidized. It is chemically reactive. A relatively large piece of plutonium is warm to the touch because of the energy given off in alpha decay. Larger pieces will produce enough heat to boil water. The metal readily dissolves in concentrated hydrochloric acid, hydroiodic acid, or perchloric acid with formulation of the Pu^{+3} ion. The metal exhibits six allotropic modifications having various crystalline structures. The densities of these vary from 16.00 to 19.86 g/cm^3. Plutonium also exhibits four ionic valence states in aqueous solutions: Pu^{+3} (blue lavender), Pu^{+4} (yellow brown), PuO_2^{+2} (pink?), and PuO_2^{+2} (pink orange). The ion PuO_2^{+} is unstable in aqueous solutions, disproportionating into Pu^{+4} and PuO_2^{+}; the Pu^{+4} thus formed, however, oxidizes the PuO_2^{+2} into PuO_2^{+2}, itself being reduced to Pu^{+3}, giving finally Pu^{+3} and PuO_2^{+2}. Plutonium forms binary compounds with oxygen: PuO, PuO_2 and intermediate oxides of variable composition; with the halides: PuF_3, PuF_4, $PuCl_3$, $PuBr_3$, PuI_3; with carbon, nitrogen, and silicon: PuC, PuN, $PuSi_2$. Oxyhalides are also well known: $PuOCl$, $PuOBr$, $PuOi$. Because of the high rate of emission of alpha particles and the fact that the element is specifically absorbed by bone marrow, plutonium, as well as all of the other transuranium elements except neptunium, is a radiological poison and must be handled with special equipment and precautions. The maximum permissible body burden, or the amount that can be maintained indefinitely in an adult without producing significant body injury, for Pu^{239} is now set at 0.04 microcurie (0.6 μg). It is recommended that the concentration of plutonium in air not exceed 0.00003 μg/m^3. Plutonium, therefore, is one of the most dangerous poisons known. Precautions must also be taken to prevent the unintentional formation of a critical mass. Plutonium in liquid solution is more likely to become critical than solid plutonium. The shape of the mass must also be considered where criticality is concerned. Plutonium-238 is available from the A.E.C. at a cost of about $700/g (80 to 89% enriched).

Polonium (Poland, a native country of Mme. Curie) – Po; at. mass (∼ 210); at. no. 84; mp 254°C, bp 962°C; sp gr (alpha modification) 9.32; valence –2, 0, +2, +3(?), +4, and +6. Polonium was the first element discovered by Mme. Curie, in 1898, while seeking the cause of radioactivity of pitchblende from Joachimsthal, Bohemia. The electroscope showed it separating with bismuth. Polonium is also called Radium F. Polonium is a very rare natural element. Uranium ores contain only about 100 μg of the element per ton. Its abundance is only about 0.2% of that of radium. In 1934 it was found that when natural bismuth (Bi^{209}) was bombarded by neutrons, Bi^{210}, the parent of polonium, was obtained. Milligram amounts of polonium may now be prepared this way, by using the high neutron fluxes of nuclear reactors. Polonium-210 is a low-melting, fairly volatile metal, 50% of which is vaporized in air in 45 hr at 55°C. It is an alpha-emitter with a half-life of 138.39 days. A milligram emits as many alpha particles as 5 g of radium. The energy released by its decay is so large (27.5 cal per curie per day or 140 watts/g) that a capsule containing about half a gram reaches a temperature above 500°C. The capsule also presents a contact gamma-ray dose rate of 1.2 roentgens/hr. A few curies of polonium exhibit a blue glow, caused by excitation of the surrounding gas. Because almost all alpha radiation is stopped with the solid source and its container, giving up its energy, polonium has attracted attention for uses as a light-weight heat source for thermoelectric power in space satellites. Polonium has more isotopes than any other element. Twenty-seven isotopes of polonium are known, with atomic masses ranging from 192 to 218. Polonium-210 is the most readily available. Isotopes of mass 209 (half-life of 103 years) and mass 208 (half-life 2.9 years) can be prepared by alpha, proton, or deuteron bombardment of lead or bismuth in a cyclotron, but these are expensive to produce. Metallic polonium has been prepared from polonium hydroxide and some other polonium compounds in the presence of concentrated aqueous or anhydrous liquid ammonia. Two allotropic modifications are known to exist. Polonium is readily dissolved in dilute acids, but is only slightly soluble in alkalis. Polonium salts of organic acids char rapidly; halide ammines are reduced to the metal. Polonium can be mixed or alloyed with beryllium to provide a source of neutrons. It has been used in devices for eliminating static charges in textile mills, etc.; however, beta sources are more commonly used and are less dangerous. It is also used on brushes for removing dust from photographic films. The polonium for these is carefully sealed and controlled, minimizing hazards to the user. Polonium-210 is very dangerous to handle in even milligram or microgram amounts and special equipment and strict control are necessary. Damage arises from the complete adsorption of the energy of the alpha particle into tissue. The maximum permissible body-burden for ingested polonium is only 0.03 mCi, which represents a particle weighing only 6.8×10^{-12} g. Weight-for-weight it is about 2.5×10^{11} times as toxic as hydrocyanic acid. The maximum allowable concentration for soluble polonium compounds in air is 2×10^{-11} mCi/cc. Polonium is available commercially on special order with an A.E.C. permit from the Oak Ridge National Laboratory.

Potassium (English, *potash* — **pot ashes; L.** *kalium*; **Arab.** *quali,* **alkali)** — K; at. wt 39.09_8; at. no. 19; mp 63.65°C; bp 774°C; sp gr 0.862 (20°C); valence 1. Discovered in 1807 by Davy, who obtained it from caustic potash (KOH); this was the first metal isolated by electrolysis. The metal is the seventh most abundant and makes up about 2.4% by weight of the earth's crust. Most potassium minerals are insoluble and the metal is obtained from them only with great difficulty. Certain minerals, however, such as *sylvite, carnallite, langbeinite,* and *polyhalite* are found in ancient lake and sea beds and form rather extensive deposits from which potassium and its salts can readily be obtained. Potash is mined in Germany, New Mexico, California, Utah, and elsewhere. Large deposits of potash, found at a depth of some 3000 ft in Saskatchewan, promise to be important in coming years. Potassium is also found in the ocean, but is present only in relatively small amounts, compared to sodium. The greatest demand for potash has been in its use for fertilizers. Potassium is an essential constituent for plant growth and it is found in most soils. Potassium is never found free in nature, but is obtained by electrolysis of the hydroxide, much in the same manner as prepared by Davy. Thermal methods also are commonly used to produce potassium (such as by reduction of potassium compounds with CaC_2, C, Si, or Na). It is one of the most reactive and electropositive of metals; except for lithium, it is the lightest known metal. It is soft, easily cut with a knife, and is silvery in appearance immediately after a fresh surface is exposed. It rapidly oxidizes in air and must be preserved in a mineral oil, such as kerosene. As with other metals of the alkali group, it decomposes in water with the evolution of hydrogen. It catches fire spontaneously on water. Potassium and its salts impart a violet color to flames. Nine isotopes of potassium are known. Ordinary potassium is composed of three isotopes, one of which is K^{40} (0.00118%), a radioactive isotope with a half-life of 1.28×10^9 years. The radioactivity presents no appreciable hazard. An alloy of sodium and potassium (NaK) is used as a heat-transfer medium. Many potassium salts are of utmost importance. These include the hydroxide, nitrate, carbonate, chloride, chlorate, bromide, iodide, cyanide, sulfate, chromate, and dichromate. Metallic potassium is available commercially for about $2/lb in quantity.

Praseodymium (Gr. *prasios,* **green, and** *didymos,* **twin)** — Pr; at. wt 140.9077; at. no. 59; mp 931°C; bp 3512°C; sp gr (α) 6.773, (β) 6.64; valence 3 or 4. In 1841 Mosander extracted the rare earth *didymia* from *lanthana*; in 1879 Lecoq de Boisbaudran isolated a new earth, *samaria,* from didymia obtained from the mineral *samarskite.* Six years later, in 1885, von Welsbach separated didymia into two other earths, *praseodymia* and *neodymia,* which gave salts of different colors. As with other rare earths, compounds of these elements in solution have distinctive sharp spectral absorption bands or lines, some of which are only a few Ångstroms wide. The element occurs along the other rare-earth elements in a variety of minerals. *Monazite* and *bastnasite* are the two principal commercial sources of the rare-earth metals. Ion-exchange and solvent extraction techniques have led to much easier isolation of the rare earths and the cost has dropped greatly in the past few years. Praseodymiun can be prepared by several methods, such as by calcium reduction of anhydrous chloride or fluoride. Misch metal, used in making cigarette lighters, contains about 5% praseodymium metal. Praseodymium is soft, silvery, malleable, and ductile. It was prepared in relatively pure form in 1931. It is somewhat more resistant to corrosion in air than europium, lanthanum, cerium or neodymium, but it does develop a green oxide coating that spalls off, when exposed to air. As with other rare-earth metals it should be kept under a light mineral oil, or sealed in plastic. The rare-earth oxides, including Pr_2O_3, are among the most refractory substances known. Along with other rare earths, it is widely used as a core material for carbon arcs used by the motion picture industry for studio lighting and projection. Salts of praseodymium are used to color glasses and enamels; when mixed with certain other materials, praseodymium produces an intense and unusually clean yellow color in glass. Didymium glass, of which praseodymium is a component, is a colorant for welder's goggles. The metal (99 + % pure) is priced at about $2.00/g.

Promethium (*Prometheus*, who, according to mythology, stole fire from heaven) – Pm; at. no. 61; mp ~1080°C; bp 2460°C(?); sp gr . . .; valence 3. In 1902 Branner predicted the existence of an element between neodymium and samarium, and this was confirmed by Moseley in 1914. In 1941, workers at Ohio State University irradiated neodymium and praseodymium with neutrons, deuterons, and alpha particles, resp., and produced several new radioactivities, which most likely were those of Element 61. Wu and Segré, and Bethe, in 1942, confirmed the formation; however, chemical proof of the production of Element 61 was lacking because of the difficulty in separating the rare earths from each other at that time. In 1945, Marinsky, Glendenin, and Coryell made the first chemical identification by use of ion-exchange chromatography. Their work was done by fission of uranium and by neutron bombardment of neodymium. Searches for the element on earth have been fruitless, and it now appears that promethium is completely missing from the earth's crust. Promethium, however, has been identified in the spectrum of the star HR^{465} in Andromeda. This element is being formed recently near the star's surface, for no known isotope of promethium has a half-life longer than 17.7 yrs. Thirteen isotopes of promethium, with atomic masses from 141 to 154, are now known. Promethium–147, with a half-life of 2.5 yrs., is the most generally useful. Promethium–145 is the longest lived, and has a specific activity of 940 curies per gram. It is a soft beta emitter; although no gamma rays are emitted, x-radiation can be generated when beta particles impinge on elements of a high atomic number, and great care must be taken in handling it. Promethium salts luminesce in the dark with a pale blue or greenish glow, due to their high radioactivity. Ion-exchange methods led to the preparation of about 10 g of promethium from atomic reactor fuel processing wastes in early 1963. Little is yet generally known about the properties of metallic promethium. Two or more allotropic modifications are thought to exist. The element has applications as a beta source for thickness gages, and it can be absorbed by a phosphor to produce light. Light produced in this manner can be used for signs or signals that require dependable operation; it can be used as a nuclear-powered battery by capturing light in photocells which convert it into electric current. Such a battery, using Pm^{147}, would have a useful life of about 5 years. Promethium shows promise as a portable x-ray unit, and it may become useful as a heat source to provide auxiliary power for space probes and satellites. More than thirty promethium compounds have been prepared. Most are colored. Promethium–147 is available to A.E.C. licensees at a cost of about 50¢ per curie.

Protactinium (Gr. *protos*, first) – Pa; at. wt (231.0359); at. no. 91; mp < 1600°C; bp . . . ; sp gr 15.37 (calc.); valence 4 or 5. The first isotope of Element 91 to be discovered was Pa^{234m}, also known as UX_2, a short-lived member of the naturally occurring U^{238} decay series. It was identified by K. Fajans and O. H. Göhring in 1913 and they named the new element *brevium*. When the longer-lived isotope Pa^{231} was identified by Hahn and Meitner in 1918, the name protoactinium was adopted as being more consistent with the characteristics of the most abundant isotope. Soddy, Cranston, and Fleck were also active in this work. The name *protoactinium* was shortened to *protactinium* in 1949. In 1927 Grosse prepared 2 mg of a white powder, which was shown to be Pa_2O_5. Later, in 1934, from 0.1 g of pure Pa_2O_5 he isolated the element by two methods, one of which was by converting the oxide to an iodide and "cracking" it in a high vacuum by an electrically heated filament by the reaction

$$2PaI_5 \rightarrow 2Pa + 5I_2.$$

Protactinium has a bright metallic luster which it retains for some time in air. The element occurs in *pitchblende* to the extent of about 1 part Pa^{231} to 10 million of ore. Ores from the Congo have about 3 ppm. Protactinium has thirteen isotopes, the most common of which is Pa^{231} with a half-life of 32,500 years. A number of protactinium compounds are known, some of which are colored. The element is superconductive below 1.4°K. An indirect measurement indicates that protactinium has a vapor pressure of 5.1×10^{-5} at 1927°C. The element is a dangerous toxic material and requires precautions similar to those used when handling plutonium. In 1959 and 1961, it was announced that the Great Britian Atomic Energy Authority extracted by a 12-stage process 125 g of 99.9% protactinium, the world's only stock of the metal for many years to come. The extraction was made from 60 tons of waste material at a cost of about $500,000. Protactinium is one of the rarest and most expensive natural-occurring elements. It was reported that this stock was being distributed to laboratories around the world at a cost of about $2800/g. The element is an alpha emitter (5.0 MeV), and is a radiological hazard similar to polonium.

Radium (L. *radius*, ray) — Ra; at. wt (226.0254); at. no. 88; mp 700°C; bp 1140°C; sp gr 5?; valence 2. Radium was discovered in 1898 by M. and Mme. Curie in the *pitchblende* or *uraninite* of North Bohemia, in which it occurs. There is about 1 g of radium in 7 tons of pitchblende. The element was isolated in 1911 by Mme. Curie and Debierne by the electrolysis of a solution of pure radium chloride, employing a mercury cathode; on distillation in an atmosphere of hydrogen this amalgam yielded the pure metal. Originally, radium was obtained from the rich pitchblende ore found at Joachimsthal, Bohemia. The *carnotite* sands of Colorado furnish some radium, but richer ores are found in the Democratic Republic of Congo and the Great Bear Lake region of Canada. Radium is present in all uranium minerals, and could be extracted, if desired, from the extensive wastes of uranium processing. Large uranium deposits are located in Ontario, New Mexico, Utah, Australia, and elsewhere. Radium is obtained commercially as the bromide or chloride; it is doubtful if any appreciable stock of the isolated element now exists. The pure metal is brilliant white when freshly prepared, but blackens on exposure to air, probably due to formation of the nitride. It exhibits luminescence, as do its salts; it decomposes in water and is somewhat more volatile than barium. It is a member of the alkaline-earth group of metals. Radium imparts a carmine red color to a flame. Radium emits alpha, beta, and gamma rays and when mixed with beryllium produces neutrons. One g of Ra^{226} undergoes 3.7×10^{10} disintegrating per sec. The *curie* is defined as that amount of radioactivity which has the same disintegration rate as 1 g of Ra^{226}. Sixteen isotopes are now known; radium—226, the common isotope, has a half-life of 1620 years. One g of radium produces about 0.0001 ml (stp) of emanation, or radon gas, per day. This is pumped from the radium and sealed in minute tubes, which are used in the treatment of cancer and other diseases. One g of radium yields about 1000 cal/year. Radium is used in producing self-luminous paints, neutron sources, and in medicine for the treatment of disease. Some of the more recently discovered radioisotopes, such as Co^{60}, are now being used in place of radium. Some of these sources are much more powerful, and others are safer to use. Radium loses about 1% of its activity in 25 years, being transformed into elements of lower atomic weight. Lead is a final product of disintegration. The study of radium has greatly altered our ideas of the structure of the atom. Radium is a radiological hazard. (Stored radium should be ventilated to prevent build-up of radon.) Inhalation, injection, or body exposure to radium can cause cancer and other body disorders. The recommended maximum allowable concentration for total body content is 0.1 μg and exposure to 2 R/month.

Radon (from *radium*; called *niton* at first, L. *nitens*, shining) — Rn; at. wt (~222); at. no. 86; mp −71°C; bp −61.8°C; density of gas 9.73 g/l; sp gr liquid 4.4 at −62°C, solid 4; valence usually 0. The element was discovered in 1900 by Dorn, who called it *radium emanation*. In 1908 Ramsay and Gray, who named in *niton*, isolated the element and determined its density, finding it to be the heaviest known gas. It is essentially inert and occupies the last place in the zero group of gases in the Periodic Table. Since 1923, it has been called radon. Twenty isotopes are known. Radon—222, coming from radium, has a half-life of 3.823 days and is an alpha emitter; radon—220, emanating naturally from thorium and called *thoron*, has a half-life of 54.5 sec and is also an alpha emitter. Radon—219 emanates from actinium and is called *actinon*. It has a half-life of 3.92 sec and is also an alpha emitter. It is estimated that every square mile of soil to a depth of 6 in. contains about 1 g of radium, which releases radon in tiny amounts to the atmosphere. Radon is present in some spring waters, such as those at Hot Springs, Arkansas. On the average, one part of radon is present to 1 sextillion parts of air. At ordinary temperatures radon is a colorless gas; when cooled below the freezing point, radon exhibits a brilliant phosphorescence which becomes yellow as the temperature is lowered and orange-red at the temperature of liquid air. It has been reported that fluorine reacts with radon, forming radon fluoride. Radon clathrates have also been reported. Radon is still produced for therapeutic use by a few hospitals by pumping it from a radium source and sealing it in minute tubes, called seeds or needles, for application to patients. This practice is now largely discontinued as hospitals can order the seeds directly from suppliers, who make up the seeds with the desired activity for the day of use. Radon is available at a cost of about $4/mCi. Care must be taken in handling radon, as with other radioactive materials. The main hazard is from inhalation of the element and its solid daughters, which are collected on dust in the air. The maximum permissible level in air has been given at 10^{-8} mCi/ml. Good ventilation should be provided where radium, thorium, or actinium is stored to prevent buildup of this element. Radon buildup is also a health consideration in uranium mines.

Rhenium (L. *Rhenus,* Rhine) — Re; at. wt 186.2; at. no. 75; mp 3180°C; bp 5627°C (est.); sp gr 21.02 (20°C); valence − 1, 2, 3, 4, 5, 6, 7. Discovery of rhenium is generally attributed to Noddack, Tacke, and Berg, who announced in 1925 they had detected the element in platinum ores and *columbite.* They also found the element in *gadolinite* and *molybdenite.* By working up 660 kg of molybdenite they were able in 1928 to extract 1 g of rhenium. The price in 1928 was $10,000/g. Rhenium does not occur free in nature or as a compound in a distinct mineral species. It is, however, widely spread throughout the earth's crust to the extent of about 0.001 ppm. Commercial rhenium in the U.S. today is obtained from molybdenite roaster-flue dusts obtained from copper-sulfide ores mined in the vicinity of Miami, Arizona, and elsewhere in Arizona and Utah. Some molybdenites contain from 0.002 to 0.2% rhenium. More than 120,000 troy oz. of rhenium are now being produced yearly in the United States. The total estimated free-world reserve of rhenium metal is 100 tons. Natural rhenium is a mixture of two stable isotopes. Sixteen other unstable isotopes are recognized. Rhenium metal is prepared by reducing ammonium perrhenate with hydrogen at elevated temperatures. The element is silvery white with a metallic luster; its density is exceeded only by that of platinum, iridium, and osmium, and its melting point is exceeded only by that of tungsten and carbon. It has other useful properties. The usual commercial form of the element is a powder, but it can be consolidated by pressing and resistance-sintering in a vacuum or hydrogen atmosphere. This produces a compact shape in excess of 90% of the density of the metal. Annealed rhenium is very ductile, and can be bent, coiled, or rolled. Rhenium is used as an additive to tungsten and molybdenum-based alloys to impart useful properties. It is widely used for filaments for mass spectrographs and ion gages. Rhenium-molybdenum alloys are superconductive at 10°K. Rhenium is also used as an electrical contact material as it has good wear resistance and withstands arc corrosion. Thermocouples made of Re-W are used for measuring temperatures up to 2200°C, and rhenium wire is used in photoflash lamps for photography. Rhenium catalysts are exceptionally resistant to poisoning from nitrogen, sulfur, and phosphorus, and are used for hydrogenation of fine chemicals, hydrocracking, reforming, and the disproportionation of olefins. Rhenium costs about $40/troy oz. Little is known of its toxicity; therefore it should be handled with care until more data are available.

Rhodium (Gr. *rhodon,* rose) — Rh; at. wt 102.9055; at. no. 45; mp 1966 ± 3°C; bp 3727 ± 100°C; sp gr 12.41 (20°C); valence 2, 3, 4, 5, and 6. Wollaston discovered rhodium in 1803-4 in crude platinum ore he presumably obtained from South America. Rhodium occurs native with other platinum metals in river sands of the Urals and in North and South America. It is also found with other platinum metals in the copper-nickel sulfide ores of the Sudbury, Ontario, region. Although the quantity occurring here is very small, the large tonnages of nickel processed make the recovery commercially feasible. The annual world production of rhodium is only two or three tons. The metal is silvery white and at red heat slowly changes in air to the sesquioxide. At higher temperatures it converts back to the element. Rhodium has a higher melting point and lower density than platinum. Its major use is as an alloying agent to harden platinum and palladium. Such alloys are used for furnace windings, thermocouple elements, bushings for glass fiber production, electrodes for aircraft spark plugs, and laboratory crucibles. It is useful as an electrical contact material as it has a low electrical resistance, a low and stable contact resistance, and is highly resistant to corrosion. Plated rhodium, produced by electroplating or evaporation, is exceptionally hard and is used for optical instruments. It has a high reflectance and is hard and durable. Rhodium is also used for jewelry, for decoration, and as a catalyst. Rhodium in small quantities costs about $9/g or $225/troy oz.

Rubidium (L. *rubidius,* deepest red) — Rb; at.

wt 85.4678; at. no. 37; mp 38.89°C; bp 688°C; sp gr (solid) 1.532 (20°C), (liquid) 1.475 (39°C); valence 1, 2, 3, 4. Discovered in 1861 by Bunsen and Kirchoff in the mineral *lepidolite* by use of the spectroscope. The element is much more abundant that was thought several years ago. It is now considered to be the 16th most abundant element in the earth's crust. Rubidium occurs in *pollucite*, *carnallite*, *leucite*, and *zinnwaldite*, which contains traces up to 1%, in the form of the oxide. It is found in lepidolite to the extent of about 1.5%, and is recovered commercially from this source. Potassium minerals, such as those found at Searles Lake, California, and potassium chloride recovered from brines in Michigan also contain the element and are commercial sources. It is also found along with cesium in the extensive deposits of *pollucite* at Bernic Lake, Manitoba. Rubidium can be liquid at room temperature. It is a soft, silvery-white metallic element of the alkali group and is the second most electropositive and alkaline element. It ignites spontaneously in air and reacts violently in water, setting fire to the liberated hydrogen. As with other alkali metals, it forms amalgams with mercury and it alloys with gold, cesium, sodium, and potassium. It colors a flame yellowish violet. Rubidium metal can be prepared by reducing rubidium chloride with calcium, and by a number of other methods. It must be kept under a dry mineral oil or in a vacuum or inert atmosphere. Seventeen isotopes of rubidium are known. Natural-occurring rubidium is made of two isotopes Rb^{85} and Rb^{87}. Rubidium-87 is present to the extent of 27.85% in natural rubidium and is a beta emitter with a half-life of 5×10^{11} years. Ordinary rubidium is sufficiently radioactive to expose a photographic film in about 30 to 60 days. Rubidium forms four oxides: Rb_2O, Rb_2O_2, Rb_2O_3, Rb_2O_4. Because rubidium can be easily ionized, it is being considered for use in "ion engines" for space vehicles; however, cesium is somewhat more efficient for this purpose. It is also proposed for use as a working fluid for vapor turbines and for use in a thermoelectric generator using the magnetohydrodynamic principle where rubidium ions are formed by heat at high temperature and passed through a magnetic field. These conduct electricity and act like an armature of a generator and cause electricity to be generated. Rubidium is used as a getter in vacuum tubes and as a photocell component. It has been used in making special glasses.

$RbAg_4I_5$ is important as it has the highest room conductivity of any known ionic crystal. At 20°C its conductivity is about the same as dilute sulfuric acid. This suggests use in thin film batteries and other applications. The present cost in small quantities is about $10/g (99.9%), or $300/lb.

Ruthenium (L. *Ruthenia*, Russia) — Ru; at. wt 101.07; at. no. 44; mp 2310°C; bp 3900°C; sp gr 12.41 (20°C); valence 0, 1, 2, 3, 4, 5, 6, 7, 8. Berzelius and Osann in 1827 examined the residues left after dissolving crude platinum from the Ural Mts. in *aqua regia*. While Berzelius found no unusual metals, Osann thought he found three new metals, one of which he named ruthenium. In 1844 Klaus, generally recognized as the discoverer, showed that Osann's ruthenium oxide was very impure and that it contained a new metal. Klaus obtained 6 g of ruthenium from the portion of crude platinum that is insoluble in aqua regia. A member of the platinum group, ruthenium occurs native with other members of the group in ores found in the Ural Mts. and in North and South America. It is also found along with other platinum metals in small but commercial quantities in *pentlandite* of the Sudbury, Ontario, nickel-mining region, and in *pyroxinite* deposits of South Africa. The metal is isolated commercially by a complex chemical process, the final stage of which is the hydrogen reduction of ammonium ruthenium chloride, which yields a powder. The powder is consolidated by powder metallurgy techniques or by argon-arc welding. Ruthenium is a hard, white metal and has four crystal modifications. It does not tarnish at room temperatures, but oxidizes in air at about 800°C. The metal is not attacked by hot or cold acids or aqua regia, but when potassium chlorate is added to the solution, it oxidizes explosively. It is attacked by halogens, hydroxides, etc. Ruthenium can be plated by electrodeposition or by thermal decomposition methods. The metal is one of the most effective hardeners for platinum and palladium, and is alloyed with these metals to make electrical contacts for severe wear resistance. A ruthenium-molybdenum alloy is said to be superconductive at 10.6°K. The corrosion resistance of titanium is improved a hundredfold by addition of 0.1% ruthenium. It is a versatile catalyst. Compounds in at least eight oxidation states have been found, but of these, the +2, +3, and +4 states are most common. Ruthenium tetroxide, like osmium tetroxide, is highly toxic. Ruthenium compounds show a marked resemblance to those of osmium. The metal is priced at about $4/g or $60/troy oz.

Rutherfordium (see Element 104)

Samarium (*Samarskite, a mineral*) — Sm; at. wt 150.4; at. no. 62; mp 1077 ± 5°C; bp 1791°C; sp gr (α) 7.520, (β) 7.40; valence 2 or 3. Discovered spectroscopically by its sharp absorption lines in 1879 by Lecoq de Boisbaudran in the mineral *samarskite*, named in honor of a Russian mine official, Col. Samarski. Samarium is found along with other members of the rare-earth elements in many minerals, including *monazite* and *bastnasite*, which are commercial sources. It occurs in monazite to the extent of 2.8%. While *misch metal*, containing about 1% of samarium metal, has long been used, samarium has not been isolated in relatively pure form until recent years. Ion exchange and solvent extraction techniques have recently simplified separation of the rare earths from one another; more recently, electrochemical deposition, using an electrolytic solution of lithium citrate and a mercury electrode, is said to be a simple, fast, and highly specific way to separate the rare earths. Samarium metal can be produced by reducing the oxide with barium or lanthanum. Samarium has a bright silver luster and is reasonably stable in air. Two crystal modifications of the metal exist, with a transformation point at 917°C. The metal ignites in air at about 150°C. Sixteen isotopes of samarium exist. Natural samarium is a mixture of seven isotopes, three of which are unstable with long half-lives. Samarium, along with other rare earths, is used for carbon-arc lighting for the motion-picture industry. The sulfide has excellent high-temperature stability and good thermoelectric efficiencies up to 1100°C. $SmCo_5$ has been used in making a new permanent magnet material with the highest resistance to demagnetization of any known material. It is said to have an intrinsic coercive force as high as 28,000 Oe. Samarium oxide has been used in optical glass to absorb the infrared. Samarium is used to dope calcium fluoride crystals for use in optical masers or lasers. Compounds of the metal act as sensitizers for phosphors excited in the infrared; the oxide exhibits catalytic properties in the dehydration and dehydrogenation of ethyl alcohol. It is used in infrared absorbing glass and as a neutron absorber in nuclear reactors. The metal is priced at about $1/g or $250/lb. Little is known of the toxicity of

samarium; therefore it should be handled carefully.

Scandium (L. *Scandia, Scandinavia*) — Sc; at. wt 44.9559; at. no. 21; mp 1541°C; bp 2831°C; sp gr 2989 (25°C); valence 3. On the basis of the Periodic System, Mendeleev predicted the existence of *ekaboron*, which would have an atomic weight between 40 of calcium and 48 of titanium. The element was discovered by Nilson in 1876 in the minerals *euxenite* and *gadolinite*, which had not yet been found anywhere except in Scandinavia. By processing 10 kg of euxenite and other residues of rare-earth minerals, Nilson was able to prepare about 2 g of scandium oxide of high purity. Cleve later pointed out that Nilson's scandium was identical to Mendeleev's ekaboron. Scandium is apparently a much more abundant element in the sun and certain stars than here on earth. It is about the 23rd most abundant element in the sun compared to the 50th most abundant on earth. It is widely distributed on earth, occurring in very minute quantities in over 800 mineral species. The blue color of beryl (aquamarine variety) is said to be due to scandium. It occurs as a principal component in the rare mineral *thortveitite*, found in Scandinavia and Malagasy. It is also found in the residues remaining after the extraction of tungsten from Zinnwald *wolframite*, and in *wiikite* and *bazzite*. Most scandium is presently being recovered from *thortveitite* or as a by-product of the extraction of uranium from *davidite*, which contains about 0.02% Sc_2O_3. Metallic scandium was first prepared in 1937 by Fischer, Brunger, and Grieneisen, who electrolyzed a eutectic melt of potassium, lithium, and scandium chlorides at 700 to 800°C. Tungsten wire and a pool of molten zinc served as the electrodes in a graphite crucible. Methods of producing the metal are now somewhat more complicated. The production of the first pound of 99% pure scandium metal was announced in 1960 as having been made under a U.S. Air Force contract. Scandium is a silvery white metal which develops a slightly yellowish or pinkish cast upon exposure to air. It is relatively soft, and is reported to resemble yttrium and the rare-earth metals more than it resembles aluminum or titanium. It is a very light metal and has a higher melting point than aluminum, making it of interest to designers of space missiles. Scandium is not attacked by a 1:1 mixture of conc. HNO_3 and 48% HF. This mixture can be used to dissolve tantalum from scandium. Scandium reacts rapidly with many acids. Eleven isotopes of scandium are recognized. The metal is still expensive, costing about $20 to $60 or more/g, or about $7500/lb with a purity of about 99.9%. It has been reported that several hundred pounds of scandium have now been produced (1970). Scandium oxide costs about $8/g. Little is yet known about the toxicity of scandium; therefore it should be handled with care.

Selenium (Gr. *Selene,* moon) – Se; at. wt 78.96; at. no 34; mp (gray) 217°C; bp (gray) 684.9 ± 1.0°C; sp gr (gray) 4.79, (vitreous) 4.28; valence –2, +4, or +6. Discovered by Berzelius in 1817, who found it associated with tellurium, named for the earth. Selenium is found in a few rare minerals, such as *crooksite* and *clausthalite*. In years past it has been obtained from flue dusts remaining from processing copper sulfide ores, but the anode muds from electrolytic copper refineries now provide the source of most of the world's selenium. Selenium is recovered by roasting the muds with soda or sulfuric acid, or by smelting them with soda and niter. Selenium exists in several allotropic forms. Three are generally recognized, but as many as six have been claimed. Selenium can be prepared with either an amorphous or crystalline structure. The color of amorphous selenium is either red, in powder form, or black, in vitreous form. Crystalline monoclinic selenium is a deep red; crystalline hexagonal selenium, the most stable variety, is a metallic gray. Natural selenium contains six stable isotopes. Fourteen other nuclides and isomers have been characterized. The element is a member of the sulfur family and resembles sulfur both in its various forms and in its compounds. Selenium exhibits both photovoltaic action, where light is converted directly into electricity, and photoconductive action, where the electrical resistance decreases with increased illumination. These properties make selenium useful in the production of photo-cells and exposure meters for photographic use, as well as solar cells. Selenium is also able to convert a.c. electricity to d.c., and is extensively used in rectifiers. Below its melting point selenium is a p-type semiconductor, and is finding many uses in electronic and solid state applications. It is used in Xerography for reproducing and copying documents, letters, etc. It is used by the glass industry to decolorize glass and to make ruby-colored glasses and enamels. It is

also used as a photographic toner, and as an additive to stainless steel. Elemental selenium has been said to be practically nontoxic; however, hydrogen selenide and other selenium compounds are extremely toxic, and resemble arsenic in its physiological reactions. Hydrogen selenide in a concentration of 1.5 ppm is intolerable to man. Selenium occurs in some soils in amounts sufficient to produce serious effects on animals feeding on plants, such as locoweed, grown in such soils. The maximum allowable concentration of selenium compounds in air has been recommended to be 0.1 mg/m². Selenium is priced at about $5/lb. It is also available in high-purity form at a somewhat higher cost.

Silicon (L. *silex, silicis,* flint) – Si; at. wt 28.086; at. no. 14; mp 1410°C; bp 2355°C; sp gr 2.33 (25°C); valence 4. Davy in 1800 thought silica to be a compound and not an element; later in 1811, Gay Lussac and Thenard probably prepared impure amorphous silicon by heating potassium with silicon tetrafluoride. Berzelius, generally credited with the discovery, in 1824 succeeded in preparing amorphous silicon by the same general method as used earlier, but he purified the product by removing the fluosilicates by repeated washings. Deville in 1854 first prepared crystalline silicon, the second allotropic form of the element. Silicon is present in the sun and stars and is a principal component of a class of meteorites known as *aerolites*. It is also a component of *tektites*, a natural glass of uncertain origin, but believed by some to be ejected lunar material from the crater *Tycho*. Silicon makes up 25.7% of the earth's crust, by weight, and is the second most abundant element, being exceeded only by oxygen. Silicon is not found free in nature, but occurs chiefly as the oxide, and as silicates. *Sand, quartz, rock crystal, amethyst, agate, flint, jasper,* and *opal* are some of the forms in which the oxide appears. *Granite, hornblende, asbestos, feldspar, clay, mica,* etc. are but a few of the numerous silicate minerals. Silicon is prepared commercially by heating silica and carbon in an electric furnace, using carbon electrodes. Several other methods can be used for preparing the element. Amorphous silicon can be prepared as a brown powder, which can be easily melted or vaporized. Crystalline silicon has a metallic luster and grayish color. The Czochralski process is commonly used to produce single crystals of silicon used for solid-state or semiconductor devices. Hyper-pure silicon can be prepared by the thermal decomposition of ultra-pure trichlorosilane in a hydrogen atmosphere, and by a vacuum float zone process. This product can be doped with boron, gallium, phosphorus, or arsenic, etc. to produce silicon for use in transistors, solar cells, rectifiers, and other solid-state devices which are used extensively in the electronics and space-age industries. Silicon is a relatively inert element, but it is attacked by halogens and dilute alkali. Most acids, except hydrofluoric, do not affect it. Silicones are important products of silicon. They may be prepared by hydrolyzing a silicon organic chloride, such as dimethyl silicon chloride. Hydrolysis and condensation of various substituted chlorosilanes can be used to produce a very great number of polymeric products, or silicones, ranging from liquids to hard, glass-like solids with many useful properties. Elemental silicon transmits more than 95% of all wavelengths of infrared, from 1.3 to 6.7 μm. Silicon is one of man's most useful elements. In the form of sand and clay it is used to make concrete and brick; it is a useful refractory material for high-temperature work, and in the form of silicates it is used in making enamels, pottery, etc. Silica, as sand, is a principal ingredient of glass, one of the most inexpensive of materials with excellent mechanical, optical, thermal, and electrical properties. Glass can be made in a very great variety of shapes, and is used as containers, window glass, insulators, and thousands of other uses. Silicon tetrachloride can be used to iridize glass. Silicon is important in plant and animal life. Diatoms in both fresh and salt water extract silica from the water to build up their cell walls. Silica is present in ashes of plants and in the human skeleton. Silicon is an important ingredient in steel; silicon carbide is one of the most important abrasives and has been used in lasers to produce coherent light of 4560 Å. Regular grade silicon (97%) costs about $0.20/lb. Silicon 99.7% pure costs about $7/lb; hyper-pure silicon may cost as much as $100/lb. Miners, stonecutters, and others engaged in work where siliceous dust is breathed in large quantities often develop a serious lung disease known as *silicosis*.

Silver (Anglo-Saxon, *Seolfor siolfur*) — Ag (L. argentum); at. wt 107.868; at. no. 47; mp 961.93°C; bp 2212°C; sp gr 10.50 (20°C); valence 1, 2. Silver has been known since ancient times. It is mentioned in *Genesis*. Slag dumps in Asia Minor and on islands in the Aegean Sea indicate that man learned to separate silver from lead as early as 3000 B.C. Silver occurs native and in ores, such as *argentite* (Ag_2S) and *horn silver* (AgCl); lead, lead-zinc, copper, gold, and copper-nickel ores are principal sources. Mexico, Canada, Peru, and the United States are the principal silver producers in the Western Hemisphere. Silver is also recovered during electrolytic refining of copper. Commercial fine silver contains at least 99.9% silver. Purities of 99.999 + % are available commercially. Pure silver has a brilliant white metallic luster. It is a little harder than gold and is very ductile and malleable, being exceeded only by gold, and perhaps palladium. Pure silver has the highest electrical and thermal conductivity of all metals, and possesses the lowest contact resistance. It is stable in pure air and water, but tarnishes when exposed to ozone, hydrogen sulfide, or air containing sulfur. The alloys of silver are important. Sterling silver is used for jewelry, silverware, etc. where appearance is paramount. This alloy contains 92.5% silver, the remainder being copper or some other metal. Silver is of utmost importance in photography — about 30% of the U.S. industrial consumption going into this application. It is used for dental alloys. Silver is used in making solder and brazing alloys, electrical contacts, and high capacity silver − zinc and silver − cadmium batteries. Silver paints are used for making printed circuits. It is used in mirror production and may be deposited on glass or metals by chemical deposition, electro-deposition, or by evaporation. When freshly deposited, it is the best reflector of visible light known, but it rapidly tarnishes and loses much of its reflectance. It is a poor reflector of ultraviolet. Silver fulminate ($Ag_2C_2N_2O_2$), a powerful explosive, is sometimes formed during the silvering process. Silver iodide is used in seeding clouds to produce rain. Silver chloride has interesting optical properties as it can be made transparent; it also is a cement for glass. Silver nitrate, or *lunar caustic,* the most important silver compound, is used extensively in photography. While silver itself is not considered to be toxic, most of its salts are poisonous due to the anions present. Silver compounds can be absorbed in the circulatory system and reduced silver deposited in the various tissues of the body. A condition, known as *argyria,* results with a greyish pigmentation of the skin and mucous membranes. Silver has germicidal effects and kills many lower organisms effectively without harm to higher animals. Silver for centuries has been used traditionally for coinage by many countries of the world. In recent times, however, consumption of silver has greatly exceeded the output. In 1939, the price of silver was fixed by the U.S. Treasury at 71¢/troy oz and at 90.5¢/troy oz in 1946. In November 1961 the U.S. Treasury suspended sales of nonmonetized silver, and the price stabilized for a time at about $1.29, the melt-down value of silver U.S. coins. The Coinage Act of 1965 authorized a change in the metallic composition of the three U.S. subsidiary denominations to clad or composite type coins. This is the first change in U.S. coinage since the monetary system was established in 1792. Clad dimes and quarters are made of an outer layer of 75% Cu and 25% Ni bonded to a central core of pure Cu. The clad half dollars, with an overall silver content of 40%, contain an outer layer of 80% Ag and 20% Cu bonded to an inner core of about 20% Ag and 80% Cu. The U.S. Treasury has minted, as of 1970, about 1¼ billion of these half dollars and placed them into circulation. The composition of the one-cent and five-cent pieces remain unchanged. One-cent coins are 95% Cu and 5% Zn. Five-cent coins are 75% Cu and 25% Ni. Old silver dollars are 90% Ag and 10% Cu. Earlier subsidiary coins of 90% Ag and 10% Cu officially are to circulate alongside the clad coins; however, in practice they have largely disappeared (*Gresham's Law*), as the value of the silver is now greater than their exchange value. Silver coins of other countries have largely been replaced with coins made of other metals. On June 24, 1968, the U.S. Government ceased to redeem U.S. Silver Certificates with silver. Since that time, the price of silver has fluctuated widely, reaching levels as high as $2/troy oz or more. The U.S. Government discontinued selling silver to domestic users and foreign buyers on November 10, 1970.

Sodium (English, *soda*; Medieval Latin, *sodanum*, headache remedy) — Na (La. *natrium*); at. wt 22.9898; at. no. 11; mp 97.81 ± 0.03°C; bp 82.9°C; sp gr 0.971 (20°C); valence 1. Long recognized in compounds, sodium was first isolated by Davy in 1807 by electrolysis of caustic soda. Sodium is present in fair abundance in the sun and stars. The D lines of sodium are among the most prominent in the solar spectrum. Sodium is the sixth most abundant element on earth, comprising about 2.6% of the earth's crust; it is the most abundant of the alkali group of metals of which it is a member. The most common compound is sodium chloride, but it occurs in many other minerals, such as *soda niter, cryolite, amphibole, zeolite, sodalite,* etc. It is a very reactive element and is never found free in nature. It is now obtained commercially by the electrolysis of absolutely dry fused sodium chloride. This method is much cheaper than that of electrolyzing sodium hydroxide, as was used several years ago. Sodium is a soft, bright, silvery metal which floats on water, decomposing it with the evolution of hydrogen and the formation of the hydroxide. It may or may not ignite spontaneously on water, depending on the amount of oxide and metal exposed to the water. It normally does not ignite in air at temperatures below 115°C. Sodium should be handled with respect as it can be dangerous when improperly handled. Metallic sodium is vital in the manufacture of sodamide and sodium cyanide, sodium peroxide, and sodium hydride. It is used in preparing tetraethyl lead, in the reduction of organic esters, and in the preparation of organic compounds. The metal may be used to improve the structure of certain alloys, to descale metal, to purify molten metals, and as a heat transfer agent. An alloy of sodium with potassium, NaK, is also an important heat transfer agent. Sodium compounds are important to the paper, glass, soap, textile, petroleum, chemical and metal industries. Soap is generally a sodium salt of certain fatty acids. The importance of common salt to animal nutrition has been recognized since prehistoric times. Among the many compounds that are of the greatest industrial importance are common salt ($NaCl$), soda ash (Na_2CO_3), baking soda ($NaHCO_3$), caustic soda ($NaOH$), Chile saltpeter ($NaNO_3$), di- and tri-sodium phosphates, sodium thiosulfate (hypo, $Na_2S_2O_3 \cdot 5H_2O$), and borax ($Na_2B_4O_7 \cdot 10H_2O$). Seven isotopes of sodium are recognized. Metallic sodium is priced at about 15 to 20¢/lb in quantity. On a per in.³ basis, it is the cheapest of all metals. Sodium metal should be handled with great care. It should be maintained in an inert atmosphere and contact with water and other substances with which sodium reacts should be avoided.

Strontium (*Strontian*, town in Scotland) — Sr; at. wt 87.62; at. no. 38; mp 769°C; bp 1384°C; sp gr 2.54; valence 2. Discovered by Davy by electrolysis in 1808. Strontium is found chiefly as *celestite* ($SrSO_4$) and *strontianite* ($SrCO_3$). The metal can be prepared by electrolysis of the fused chloride mixed with potassium chloride, or is made by reducing strontium oxide with aluminum in a vacuum at a temperature at which strontium distills off. Three allotropic forms of the metal exist, with transition points at 235°C and 540°C. Strontium is softer than calcium and decomposes water more vigorously. It does not absorb nitrogen below 380°C. It should be kept under kerosene to prevent oxidation. Freshly cut strontium has a silvery appearance, but rapidly turns a yellowish color with the formation of the oxide. The finely divided metal ignites spontaneously in air. Volatile strontium salts impart a beautiful crimson color to flames, and these salts are used in pyrotechnics. Natural strontium is a mixture of four stable isotopes. Twelve other unstable isotopes are known to exist. Of greatest importance is Sr^{90} with a half-life of 28 years. It is a product of nuclear fallout and presents a health problem. This isotope is one of the best long-lived high-energy beta emitters known, and is used in SNAP devices (Systems for Nuclear Auxiliary Power). These devices hold promise for use in space vehicles, remote weather stations, navigational buoys, etc. where a lightweight, long-lived, nuclear-electric power source is needed. Strontium hydroxide has been used in sugar refining; however, lime is replacing its use as it is cheaper. Strontium titanate is an interesting optical material as it has an extremely high refractive index and an optical dispersion greater than that of diamond. It has been used as a gemstone, but is very soft. It does not occur naturally. The applications of strontium are similar to those of barium and calcium, but there are few advantages and the cost is much higher. Strontium metal costs about $6 to $8/lb.

Sulfur (Sanskrit, *sulvere;* L. *sulphurium*) — S; at wt 32.064; at. no. 16; mp (rhombic) 112.8°C, (monoclinic) 119.0°C; bp 444.674°C; sp gr (rhombic) 2.07, (monoclinic) 1.957 (20°C); valence 2, 4, or 6. Known to the ancients; referred to in Genesis as *brimstone.* Sulfur is found in meteorites. A dark area near the crater Aristarchus on the moon has been studied by R. W. Wood with ultraviolet light. This study suggests strongly that it is a sulfur deposit. Sulfur occurs native in the vicinity of volcanoes and hot springs. It is widely distributed in nature as *iron pyrites, galena, sphalerite, cinnabar, stibnite, gypsum, epson salts, celestite, barite,* etc. Sulfur is commercially recovered from wells sunk into the salt domes along the Gulf Coast of the U.S. It is obtained from these wells by the Frasch process, which forces heated water into the wells to melt the sulfur, that is then brought to the surface. Sulfur also occurs in natural gas and petroleum crudes and must be removed from these products. Formerly this was done chemically, which wasted the sulfur. New processes now permit recovery, and these sources promise to be very important. Large amounts of sulfur are being recovered from Alberta gas fields. Sulfur is a pale yellow, odorless, brittle solid, which is insoluble in water, but soluble in carbon disulfide. In every state, whether gas, liquid, or solid, elemental sulfur occurs in more than one allotropic form or modification; these present a confusing multitude of forms whose relations are not yet fully understood. Amorphous or "plastic" sulfur is obtained by fast cooling of the crystalline form. X-ray studies indicate that amorphous sulfur may have a helical structure with eight atoms per spiral. Crystalline sulfur seems to be made of rings, each containing eight sulfur atoms, which fit together to give a normal x-ray pattern. Ten isotopes of sulfur exist. Four occur in natural sulfur, none of which is radioactive. A finely divided form of sulfur, known as *flowers of sulfur,* is obtained by sublimation. Sulfur readily forms sulfides with many elements. Sulfur is a component of black gunpowder, is used in the vulcanization of natural rubber, and is used as a fungicide. It is also used extensively in making phosphatic fertilizers. A tremendous tonnage is used to produce sulfuric acid, the most important manufactured chemical. It is used in making sulfite paper and other papers, is used as a fumigant, and in the bleaching of dried fruits. The element is a good electrical insulator. Organic compounds containing sulfur are very important. Calcium sulfate, ammonium sulfate, carbon disulfide, sulfur dioxide, and hydrogen sulfide are but a few of the other many important compounds of sulfur. Sulfur is essential to life. It is a minor constituent of fats, body fluids, and skeletal minerals. Carbon disulfide, hydrogen sulfide, and sulfur dioxide should be handled carefully. Hydrogen sulfide in small concentrations can be metabolized, but in higher concentrations it quickly can cause death by respiratory paralysis. It is insidious in that it quickly deadens the sense of smell. Sulfur dioxide is a dangerous component in atmospheric air pollution. High-purity sulfur is commercially available in purities of 99.999 + %.

Tantalum (Gr. *Tantalos,* mythological character

— father of *Niobe*) — Ta; at. wt 180.9479; at. no 73; mp 2996°C; bp 5425 ± 100°C; sp gr 16.654; valence 2?, 3, 4?, or 5. Discovered in 1802 by Ekeberg, but many chemists thought niobium and tantalum were identical elements until Rose, in 1844, and Marignac, in 1866, indicated and showed that niobic and tantalic acids were two different acids. The early investigators only isolated the impure metal. The first relatively pure ductile tantalum was produced by von Bolton in 1903. Tantalum occurs principally in the mineral *columbite-tantalite* (Fe, Mn) (Nb, Ta)$_2$O$_6$. Tantalum ores are found in the Republic of Congo, Brazil, Mozambique, Thailand, Portugal, Nigeria, and Canada. The new mine at Bernic Lake, Manitoba, has reserves of more than a million tons of ore averaging about ¼% tantalum oxide. Separation of tantalum from niobium requires several complicated steps. Several methods are commercially used to produce the element including: electrolysis of molten potassium fluotantalate, reduction of potassium fluotantalate with sodium, or reacting tantalum carbide with tantalum oxide. Sixteen isotopes of tantalum are known to exist. Natural tantalum contains two isotopes; one of these, Ta^{180}, is present in very small quantity (0.0123%) and is unstable with a very long half-life of >10^{13} years. Tantalum is a gray, heavy, and very hard metal. When pure, it is ductile and can be drawn into fine wire, which is used as a filament for evaporating metals, such as aluminum. Tantalum is almost completely immune to chemical attack at temperatures below 150°C, and is attacked only by hydrofluoric acid, acidic solutions containing the fluoride ion, and free sulfur trioxide. Alkalis attack it only slowly. At higher temperatures, tantalum becomes much more reactive. The element has a melting point exceeded only by tungsten and rhenium. Tantalum is used to make a variety of alloys with desirable properties, such as high-melting point, high strength, good ductility, etc. The metal has good "gettering" ability at high temperatures, and tantalum oxide films are stable, and have good rectifying and dielectric properties. Tantalum is used to make electrolytic capacitors and vacuum furnace parts, which account for about 70% of its use. The metal is also widely used to fabricate chemical process equipment, nuclear reactors, and aircraft and missile parts. Tantalum is completely immune to body liquids and is a nonirritating metal. It has, therefore, found wide use in making surgical appliances. Tantalum oxide is used to make special glass with a high index of refraction for camera lenses. The metal has many other uses. The metal in powdered form costs about $35/lb. Sheet tantalum and fabricated forms are more expensive.

Technetium (Gr. *technetos*, artificial) — Tc; at. wt 98.9062; at. no. 43; mp 2172°C; bp 4877°C; sp gr 11.50 (calc.); valence 0, +2, +4, +5, +6, and +7. Element 43 was predicted on the basis of the periodic table, and was erroneously reported as having been discovered in 1925, at which time it was named *masurium*. The element was actually discovered by Perrier and Segrè in Italy in 1937. It was found in a sample of molybdenum, which was bombarded by deuterons in the Berkeley cycloton, and which E. Lawrence sent to these investigators. Technetium was the first element to be produced artificially. Since its discovery, searches for the element in terrestrial materials have been made without success. If it does exist, the concentration must be very small. Surprisingly, it has been found in the spectrum of S, M, and N type stars, and its presence in stellar matter is leading to new theories of the production of heavy elements in the stars. Sixteen isotopes of technetium, with atomic masses ranging from 92 to 107, are known. Tc^{97} has a half-life of 2.6 X 10^6 years, Tc^{98} has a half-life of 1.5 X 10^6 years. The isomeric isotope Tc^{95m}, with a half-life of 61 days, is useful for tracer work, as it produces energetic gamma rays. Technetium metal has been produced in Kg quantities. The metal was first prepared by passing hydrogen gas at 1100°C over Tc$_2$S$_7$. It is now conveniently prepared by the reduction of ammonium pertechnetate with hydrogen. Technetium is a silvery-gray metal that tarnishes slowly in moist air. Until 1960, technetium was available only in small amounts and the price was as high as $2800/g. It is now offered commercially to holders of A.E.C. permits at a price of $90 to $100/g. The chemistry of technetium is said to be similar to that of rhenium. Technetium dissolves in nitric acid, aqua regia, and conc. sulfuric acid, but is not soluble in hydrochloric acid of any strength. The element is a remarkable corrosion inhibitor for steel. It is reported that mild carbon steels may be effectively protected by as little as 5 ppm of KTcO$_4$ in aerated distilled water at temperatures up to 250°C. This corrosion protection is limited to closed systems, since technetium is radioactive and must be confined. Tc^{99} has a specific activity of 6.2 X 10^8 disintegrations/sec/g. Activity of this level must not be allowed to spread. Tc^{99} is a contamination hazard and should be handled in a glove box. The metal is an excellent superconductor at 11°K and below.

Tellurium (L. *tellus,* earth) — Te; at. wt 127.60; at. no 52; mp 449.5 ± 0.3°C; bp 989.8 ± 3.8°C; sp gr 6.24 (20°C); valence 2, 4, or 6. Discovered by Müller von Reichenstein in 1782; named by Klaproth who isolated it in 1798. Tellurium is occasionally found native, but is more often found as the telluride of gold (*calaverite*), and combined with other metals. It is recovered commercially from the anode muds produced during the electrolytic refining of blister copper. The U.S., Canada, Peru, and Japan are the largest Free World producers of the element. Crystalline tellurium has a silvery white appearance, and when pure exhibits a metallic luster. It is brittle and easily pulverized. Amorphous tellurium is formed by precipitating tellurium from a solution of telluric or tellurous acid. Whether this form is truly amorphous, or made of minute crystals, is open to question. Tellurium is a p-type semiconductor, and shows greater conductivity in certain directions, depending on alignment of the atoms. Its conductivity increases slightly with exposure to light. It can be doped with silver, copper, gold, tin, or other elements. In air, tellurium burns with a greenish-blue flame, forming the dioxide. Molten tellurium corrodes iron, copper, and stainless steel. Tellurium and its compounds are probably toxic and should be handled with care. Workmen exposed to as little as 0.01 mg/m^3 of air, or less, develop "tellurium breath," which has a garlic-like odor. Twenty-one isotopes of tellurium are known, with atomic masses ranging from 115 to 135. Natural tellurium consists of eight isotopes, one of which, Te^{127}, is unstable. It is present to the extent of 0.87% and has a half-life of 1.2 X 10^{13} years. Tellurium improves the machinability of copper and stainless steel, and its addition to lead decreases the corrosive action of sulfuric acid to lead and improves its strength and hardness. Tellurium is used as a basic ingredient in blasting caps, and is added to cast iron for chill control. Tellurium is used in ceramics. Bismuth telluride has been used in thermoelectric devices. One such device, using two Bi-Te semiconductors, is reportedly capable of freezing or boiling water in seconds with the power from two flash-light batteries. The unit is said to be capable of bringing

the temperature down to −75°C, using only two amperes of current. Tellurium with a purity of 99.7% costs about $6/lb. It is also available with purities of 99.999 + % at a cost of $20 to $30/lb.

Terbium (*Ytterby*, village in Sweden) — Tb; at. wt 158.9254; at. no. 65; mp 1356°C; bp 3123°C; sp gr 8.229; valence 3, 4. Discovered by Mosander in 1843. Terbium is a member of the lanthanide or "rare earth" group of elements. It is found in *cerite, gadolinite,* and other minerals along with other rare earths. It is recovered commercially from *monazite* in which it is present to the extent of 0.03%, from *xenotime,* and from *euxenite,* a complex oxide containing 1% or more of terbia. Terbium has been isolated only in recent years with the development of ion-exchange techniques for separating the rare-earth elements. As with other rare earths, it can be produced by reducing the anhydrous chloride or fluoride with calcium metal in a tantalum crucible. Calcium and tantalum impurities can be removed by vacuum remelting. Other methods of isolation are possible. Terbium is reasonably stable in air. It is a silvery gray metal, and is malleable, ductile and soft enough to be cut with a knife. Two crystal modifications exist, with a transformation temperature of 1315°C. Nineteen isotopes with atomic masses ranging from 147 to 164 are recognized. The oxide is a chocolate or dark maroon color. Sodium terbium borate is used as a laser material and emits coherent light at 5460 Å. Terbium is used to dope calcium fluoride, calcium tungstate, and strontium molybdate, used in solid-state devices. The oxide has potential application as an activator for green phosphors used in color T.V. tubes. It can be used with ZrO_2 as a crystal stabilizer of fuel cells which operate at elevated temperature. Few other uses have been found. The element is priced at about $3/g or $750/lb. Little is known of the toxicity of terbium. It should be handled with care, as with other lanthanide elements.

Thallium (Gr. *thallos,* a green shoot or twig) — Tl; at. wt 204.37; at. no. 81; mp 303.5°C; bp 1457 ± 10°C; sp gr 11.85 (20°C); valence 1 or 3. Discovered spectroscopically in 1861 by Crookes. The element was named after the beautiful green spectral line, which identified the element. The metal was isolated both by Crookes and Lamy in 1862 about the same time. Thallium occurs in *crooksite, lorandite,* and *hutchinsonite.* It is also present in *pyrites* and is recovered from the roasting of this ore in connection with the production of sulfuric acid. It is also obtained from the smelting of lead and zinc ores. Extraction is somewhat complex and depends on the source of the thallium. When freshly exposed to air thallium exhibits a metallic luster, but soon develops a bluish gray tinge, resembling lead in appearance. A heavy oxide builds up on thallium if left in air, and in the presence of water the hydroxide is formed. The metal is very soft and malleable. It can be cut with a knife. Twenty isotopic forms of thallium, with atomic masses ranging from 191 to 210, are recognized. Natural thallium is a mixture of two isotopes. The element and its compounds are toxic and should be carefully handled. Contact of the metal with the skin is dangerous, and when melting the metal, adequate ventilation should be provided. The maximum allowable concentration of soluble thallium compounds in air is 0.1 mg/m^3. Thallium sulfate is widely employed as a rodenticide and ant killer. It is odorless and tasteless, giving no warning of its presence. The electrical conductivity of thallium sulfide changes with exposure to infrared light and this compound is used in photocells. Thallium bromide-iodide crystals have been used in infrared detectors. Thallium has been used, with sulfur or selenium and arsenic, to produce low melting glasses which become fluid between 125 and 150°C. These glasses have properties at room temperatures similar to ordinary glasses and are said to be durable and insoluble in water. Thallium oxide has been used to produce glasses with a high index of refraction. A mercury-thallium alloy, which forms a eutectic at 8.5% thallium, is reported to freeze at -60°C, some 20°C below the freezing point of mercury. Commercial thallium metal costs about $8/lb. It is available also in high-purity form at a cost of about $4/oz.

Thorium (*Thor,* Scandinavian god of war) — Th; at. wt 232.0381; at. no. 90; mp 1750°C; bp ~4790°C; sp gr 11.72; valence +2(?), +3(?), +4. Discovered by Berzelius in 1828. Thorium occurs in *thorite* (ThSiO$_4$) and in *thorianite* (ThO$_2$ + UO$_2$). Large deposits of thorium minerals have been reported in New England and elsewhere, but these have not yet been exploited. Thorium is now thought to be about three times as abundant as uranium and about as abundant as lead or molybdenum. The metal is a source of nuclear power. There is probably more energy available for use from thorium in the minerals of the earth's crust than from both uranium and fossil fuels. Any sizable demand for thorium as a nuclear fuel is still a decade or more in the future; however, work is progressing with several thorium cycle converte-reactor systems. Several prototypes, including the HTGR (High-temperature gas-cooled reactor) and MSRE (Molten salt converte-reactor experiment), have operated successfully. Thorium is recovered commercially from the mineral *monazite,* which contains from 3 to 9% ThO$_2$ along with most rare-earth minerals. Much of the internal heat of the earth has been attributed to thorium and uranium. Several methods are available for producing thorium metal; it can be obtained by reducing thorium oxide with calcium; by electrolysis of anhydrous thorium chloride in a fused mixture of sodium and potassium chlorides; by calcium reduction of thorium tetrachloride mixed with anhydrous zinc chloride; and by reduction of thorium tetrachloride with an alkali metal. Thorium was originally assigned a position in Group IV of the periodic table. Because of its atomic weight, valence, etc., it is now considered to be the second member of the *actinide* series of elements. When pure, thorium is a silvery-white metal which is air-stable and retains its luster for several months. When contaminated with the oxide, thorium slowly tarnishes in air, becoming gray and finally black. The physical properties of thorium are greatly influenced by the degree of contamination with the oxide. The purest specimens often contain several tenths of a percent of the oxide. High-purity thorium has been made. Pure thorium is soft, very ductile, and can be cold-rolled, swaged, and drawn. Thorium is dimorphic, changing at 1400°C from a cubic to a body-centered cubic structure. Thorium oxide has a melting point of 3300°C which is the highest of all oxides. Only a few elements, such as tungsten, and a few compounds, such as tantalum carbide, have higher melting points. Thorium is slowly attacked by water, but does not dissolve readily in most common acids, except hydrochloric. Powdered thorium metal is often pyrophoric and should be carefully handled. When heated in air, thorium turnings ignite and burn brilliantly with a white light. The principal use of thorium has been in the preparation of the Welsbach mantle, used for portable gas lights. These mantles, consisting of thorium oxide with about 1% cerium oxide and other ingredients, glow with a dazzling light when heated in a gas flame. Thorium is an important alloying element in magnesium, imparting high strength and creep resistance at elevated temperatures. Because thorium has a low work-function and high electron emission, it is used to coat tungsten wire used in electronic equipment. The oxide is also used to control the grain size of tungsten used for electric lamps; it is also used for high-temperature laboratory crucibles. Glasses containing thorium oxide have a high refractive index and low dispersion. Consequently, they find application in high quality lenses for cameras and scientific instruments. Thorium oxide has also found use as a catalyst in the conversion of ammonia to nitric acid, in petroleum cracking, and in producing sulfuric acid. Twelve isotopes of thorium are known with atomic masses ranging from 223 to 234. All are unstable. Th^{232} occurs naturally and has a half-live of 1.41 $\times$ 10^{10} years. It is an alpha emitter. Th^{232} goes through six alpha and four beta decay steps before becoming the stable isotope Pb^{208}. Th^{232} is sufficiently radioactive to expose a photographic plate in a few hours. Thorium disintegrates with the production of thoron (radon^{220}), which is an alpha emitter and presents a radiation hazard. Good ventilation of areas where thorium is stored or handled is therefore recommended. Thorium and its compounds are subject to licensing and control by the U.S. Atomic Energy Commission. Thorium metal in pellet or powder form costs about 25¢/g or $15/lb.

Thulium (*Thule*, **the earliest name for Scandinavia**) – Tm; at. wt 168.934; at. no. 69; mp 1545 ± 15°C; bp 1747°C; sp gr 9.321 (25°C); valence 2, 3. Discovered in 1897 by Cleve. Thulium occurs in small quantities along with other rare earths in a number of minerals. It is obtained commercially from *monazite*, which contains about 0.007% of the element. Thulium is the least abundant of the rare-earth elements, but with new sources recently discovered, it is now considered to be about as rare as silver, gold or cadmium. Ion-exchange and solvent extraction techniques have recently permitted much easier separation of the rare earths, with much lower costs. Thulium metal, only a few years ago, was not obtainable at any cost; in 1950 the oxide sold for $450/g. Thulium metal now costs from $3 to $20/g depending on the purity, quantity, and supplier. Thulium can be isolated by reduction of the oxide with lanthanum metal or by calcium reduction of the anhydrous fluoride. The pure

metal has a bright, silvery luster. It is reasonably stable in air, but the metal should be protected from moisture in a closed container. The element is hard, silver-grey, soft, malleable and ductile, and can be cut with a knife. Sixteen isotopes are known, with atomic masses ranging from 161 to 176. Natural thulium, Tm^{169}, is stable. Because of the relatively high price of the metal, thulium has not yet found many practical applications. Tm^{169} bombarded in a nuclear reactor can be used as a radiation source in portable x-ray equipment. Tm^{171} is potentially useful as an energy source. Natural thulium also has possible use in *ferrites* (ceramic magnetic materials) used in microwave equipment. As with other lanthanides, thulium has a low to moderate acute toxic rating. It should be handled with care.

Tin (Anglo-Saxon, *tin*) – Sn (L. *stannum*); at. wt 118.69; at. no. 50; mp 231.9681; bp 2270°C; sp gr (gray) 5.75, (white) 7.31; valence, 2, 4. Known to the ancients. Tin is found chiefly in *cassiterite* (SnO_2). Most of the world's supply comes from Malaya, Bolivia, Indonesia, the Republic of the Congo, Thailand, and Nigeria. The U.S. produces almost none, although occurrences have been found in Alaska and California. Tin is obtained by reducing the ore with coal in a reverberatory furnace. Ordinary tin is composed of nine stable isotopes. Thirteen unstable isotopes are also known. Ordinary tin is a silvery white metal, is malleable, somewhat ductile, and has a highly crystalline structure. Due to the breaking of these crystals, a "tin cry" is heard when a bar is bent. The element has two or perhaps three allotropic forms. On warming, gray or α tin, with a cubic structure, changes at 13.2°C into white or β tin, the ordinary form of the metal. White tin has a tetragonal structure. Some authorities believe a γ form exists between 161°C and the melting point; however, other authorities discount its existence. When tin is cooled below 13.2°C, it changes slowly from white to gray. This change is affected by impurities, such as aluminum and zinc, and can be prevented by small additions of antimony or bismuth. This change from the α to β form is called " the tin pest." There are few if any uses for gray tin. Tin takes a high polish and is used to coat other metals to prevent corrosion or other chemical action. Such tin plate over steel is used in the so-called tin can for preserving food. Alloys of tin are very important. Soft solder, type metal, fusible metal, pewter, bronze, bell metal, Babbitt metal, White metal, die casting alloy, and phosphor bronze are some of the important alloys using tin. Tin resists distilled, sea, and soft tap water, but is attacked by strong acids, alkalis, and acid salts. Oxygen in solution accelerates the attack. When heated in air, tin forms SnO_2, which is feebly acid, forming stannate salts with basic oxides. The most important salt is the chloride ($SnCl_2 \cdot H_2O$), which is used as a reducing agent and as a mordant in calico printing. Tin salts sprayed onto glass are used to produce electrically conductive coatings on the glass. These have been used for panel lighting and for frost-free windshields. Of recent interest is a crystalline tin-niobium alloy that is superconductive at very low temperatures. This promises to be important in the construction of superconductive magnets that generate enormous field strengths, but use practically no power. Such magnets, made of tin-niobium wire, weigh but a few pounds and produce magnetic fields, when started with a small battery, they are comparable to that of a 100-ton electromagnet operated continuously with a large power supply. The small amount of tin used in canned foods is quite harmless. The agreed limit of tin content in U.S. foods is 300 mg/kg. The trialkyl and triaryl tin compounds are used as biocides and must be handled carefully. Tin prices have varied from 50c to about $2.00/lb over the past 25 years. It presently costs about $1.75/lb.

Titanium (L. *Titans*, the first sons of the Earth, myth.) — Ti; at. wt 47.90; at. no. 22; mp 1660 ± 10°C; bp 3287°C; sp gr 4.54; valence 2, 3, or 4. Discovered by Gregor in 1791; named by Klaproth in 1795. Impure titanium was prepared by Nilson and Pettersson in 1887; however, the pure metal (99.9%) was not made until 1910 by Hunter by heating $TiCl_4$ with sodium in a steel bomb. Titanium is present in meteorites and in the sun. Rocks obtained during the Apollo II lunar mission showed presence of 7 to 12% TiO_2. Analysis of rocks obtained during later missions show larger percentages. Titanium oxide bands are prominent in the spectra of M Type stars. The element is the ninth most abundant in the crust of the earth. Titanium is almost always present in igneous rocks and in the sediments derived from them. It occurs in the minerals *rutile, ilmenite,* and *sphene,* and is present in titanates and in many iron ores. Titanium is present in the ash of coal, in plants, and in the human body. The metal was a laboratory curiosity until Kroll, in 1946, showed that titanium could be produced commercially by reducing titanium tetrachloride with magnesium. This method is largely used for producing the metal today. The metal can be purified by decomposing the iodide. Titanium, when pure, is a lustrous, white metal. It has a low density, good strength, is easily fabricated, and has excellent corrosion resistance. It is ductile only when it is free of oxygen. The metal burns in air and is the only element that burns in nitrogen. Titanium is resistant to dilute sulfuric and hydrochloric acid, most organic acids, moist chlorine gas, and chloride solutions. Natural titanium consists of five isotopes with atomic masses from 46 to 50. All are stable. Four other unstable isotopes are known. Natural titanium is reported to become very radioactive after bombardment with deuterons. The emitted radiations are mostly positrons and hard gamma rays. The metal is dimorphic. The hexagonal α form changes to the cubic β form very slowly at about 880°C. The metal combines with oxygen at red heat, and with chlorine at 550°C. Titanium is important as an alloying agent with aluminum, molybdenum, manganese, iron, and other metals. Alloys of titanium are principally used for aircraft and missiles where light-weight, strength, and ability to withstand extremes of temperature are important. Titanium is as strong as steel, but 45% lighter. It is 60% heavier than aluminum, but twice as strong. The A-11 jet plane, which flies at 2000 mph, is largely constructed of titanium. Each SST jet plane was expected to consume 600,000 lb. The cancellation of the SST and overall decline in the aerospace industry have introduced considerable uncertainty and a significantly weakened market for titanium. Titanium has potential use in desalination plants for converting sea water into fresh water. The metal has excellent resistance to sea water and is used for propeller shafts, rigging, and other parts of ships exposed to salt water. A titanium anode, coated with platinum, has been used to provide cathodic protection from corrosion by salt water. Titanium metal is considered to be physiologically inert. When pure, titanium dioxide is relatively clear and has an extremely high index of refraction with an optical dispersion higher than diamond. It is produced artificially for use as a gemstone, but is relatively soft. Star sapphires and rubies exhibit their asterism as a result of the presence of TiO_2. Titanium dioxide is extensively used for both house paint and artist's paint, as it is permanent and has good covering power. Titanium paint is an excellent reflector of infrared, and is extensively used in solar observatories where heat causes poor seeing conditions. Titanium tetrachloride is used to iridize glass. This compound fumes strongly in air and is used to produce smoke screens. The price of titanium mill products is about $6/lb. Its use is growing rapidly, and by 1978, it is expected that 10 million lb of the metal will be used.

Tungsten (Swedish, *tung sten,* **heavy stone); also known as WOLFRAM (from** *wolframite,* **said to be named from** *wolf rahm* **or** *spumi lupi,* **because the ore interfered with the smelting of tin and was supposed to devour the tin)** — W; at. wt 183.85; at. no 74; mp 3410 ± 20°C; bp 5660°C; sp gr 19.3 (20°C); valence 2, 3, 4, 5, or 6. In 1779 Peter Woulfe examined the mineral now known as *wolframite* and concluded it must contain a new substance. Scheele, in 1781, found that a new acid could be made from *tung sten* (a name first applied about 1758 to a mineral now known as *scheelite*). Scheele and Bergman suggested the possibility of obtaining a new metal by reducing this acid. The de Elhuyar brothers found an acid in *wolframite* in 1783 that was identical to the acid of *tung sten* (tungstic acid) of Scheele, and in that year they succeeded in obtaining the element by reduction of this acid with charcoal. Tungsten occurs in *wolframite,* (Fe, Mn)WO_4; *scheelite,* $CaWO_4$; *huebnerite,* $MnWO_4$; and *ferberite,* $FeWO_4$. Important deposits of tungsten occur in California, North Carolina, South Korea, Bolivia, U.S.S.R., and Portugal. China is reported to have about 75% of the world's tungsten resources. Natural tungsten contains five stable isotopes. Twelve other unstable isotopes are recognized. The metal is obtained commercially by reducing tungstic oxide with hydrogen or carbon. Pure tungsten is a steel gray to tin-white metal. Very pure tungsten can be cut with a hacksaw, and can be forged, spun, drawn, and extruded. The impure metal is brittle and can be worked only with difficulty. Tungsten has the highest melting point and lowest vapor pressure of all metals, and at temperatures over 1650°C has the highest tensile strength. The metal oxidizes in air and must be protected at elevated temperatures. It has excellent corrosion resistance and is attacked only slightly by most mineral acids. The thermal expansion is about the same as boro-silicate glass, which makes the metal useful for glass-to-metal seals.

Tungsten and its alloys are used extensively for filaments for electric lamps, electron and television tubes, and for metal evaporation work; for electrical contact points for automobile distributors; x-ray targets; windings and heating elements for electrical furnaces; and for numerous space missile and high-temperature applications. High-speed tool steels, Hastelloys, Stellite, and many other alloys contain tungsten. Tungsten carbide is of great importance to the metalworking, mining, and petroleum industries. Calcium and magnesium tungstates are widely used in fluorescent lighting; other salts of tungsten are used in the chemical, and tanning industries. Tungsten disulfide is a dry, high-temperature lubricant, stable to 500°C. Tungsten bronze and other tungsten compounds are used in paints. Hydrogen-reduced tungsten powder costs about $5/lb.

Uranium (Planet *Uranus*) — U; at. wt 238.029; at. no. 92; mp 1132.3 ± 0.8°C; bp 3818°C; sp gr ~18.95; valence 2, 3, 4, 5, or 6. Yellow-colored glass, containing more than 1% uranium oxide and dating back to 79 A.D., has been found near Naples, Italy. Klaproth recognized an unknown element in *pitchblende* and attempted to isolate the metal in 1789. The metal apparently was first isolated in 1841 by Peligot, who reduced the anhydrous chloride with potassium. Uranium is not as rare as it was once thought. It is now considered to be more plentiful than mercury, antimony, silver, or cadmium, and is about as abundant as molybdenum or arsenic. It occurs in numerous minerals, such as *pitchblende, uraninite, carnotite, autunite, uranophane, davidite,* and *tobernite.* It is also found in *phosphate rock, lignite,* and *monazite sands,* and can be recovered commercially from these sources. The major producers are the U.S., South Africa, and Canada, with over 80% of the output. An important new mining area has been opened in the Northern Territory of Austria. The A.E.C. purchases uranium in the form of acceptable U_3O_8 concentrates. This incentive program has greatly increased the known uranium reserves. Uranium can be prepared by reducing uranium halides with alkai or alkaline earth metals or by reducing uranium oxides by calcium, aluminum, or carbon at high temperatures. The metal can also be produced by electrolysis of KUF_5 or UF_4, dissolved in a molten mixture of $CaCl_2$-NaCl. High-purity uranium can be prepared by the thermal decomposition of uranium halides on a hot filament. Uranium exhibits three crystallographic modifications as follows:

$$\alpha \xrightarrow{667^\circ C} \beta \xrightarrow{772^\circ C} \gamma$$

Uranium is a heavy, silvery white metal, which is pyrophoric when finely divided. It is a little softer than steel, and is attacked by cold water in a finely divided state. It is malleable, ductile, and slightly paramagnetic. In air, the metal becomes coated with a layer of oxide. Acids dissolve the metal, but it is unaffected by alkalis. Uranium has fourteen isotopes, all of which are radioactive. Naturally occurring uranium nominally contains 99.2830% by weight U^{238}, 0.7110% U^{235}, and 0.0054% U^{234}. Studies show that the percentage weight of U^{235} in natural uranium varies by as much as 0.1%, depending on the source. The A.E.C. has adopted the value of 0.711 as being their "official" percentage of U^{235} in natural uranium. Natural uranium is sufficiently radioactive to expose a photographic plate in an hour or so. Much of the internal heat of the earth is thought to be attributable to the presence of uranium and thorium. U^{238}, with a half-life of 4.51×10^9 years, has been used to estimate the age of igneous rocks. The origin of uranium, the highest member of the natural-occurring elements — except perhaps for traces of neptunium or plutonium — is not clearly understood, although it may be presumed that uranium is a decay product of elements of higher atomic weight, which may have once been present on earth or elsewhere in the universe. These original elements may have been created as a result of a primordial "creation," known as "the big bang," in a supernova, or in some other stellar processes. Uranium is of great importance as a nuclear fuel. U^{238} can be converted into fissionable plutonium by the following reactions:

$$U^{238}(n, \gamma)U^{239} \xrightarrow{\beta-} Np^{239} \xrightarrow{\beta-} Pu^{239}$$

This nuclear conversion can be brought about in "breeder" reactors where it is possible to produce more new fissionable material than the fissionable material used in maintaining the chain reaction. U^{235} is of even greater importance, for it is the key to the utilization of uranium. U^{235}, while occurring in natural uranium to the extent of only 0.71%, is so fissionable with slow neutrons that a self-sustaining fission chain reaction can be made to occur in a reactor constructed from natural uranium and a suitable moderator, such as heavy water or graphite, alone. U^{235} can be concentrated by gaseous diffusion and other physical processes, if desired, and used directly as a nuclear fuel, instead of natural uranium, or used as an explosive. Natural uranium, slightly enriched with U^{235} by a small percentage is used to fuel nuclear power reactors for the generation of electricity. Natural thorium can be irradiated with neutrons as follows to produce the important isotope U^{233}:

$$Th^{232}(n, \gamma)Th^{233} \xrightarrow{\beta-} Pa^{233} \xrightarrow{\beta-} U^{233}.$$

While thorium itself is not fissionable, U^{233} is, and in this way may be used as a nuclear fuel. One pound of completely fissioned uranium has the fuel value of over 1500 tons of coal. The uses of nuclear fuels to generate electrical power, to make isotopes for peaceful purposes, and to make explosives are well known. The estimated worldwide capacity of nuclear power reactors in 1970 amounted to about 25 million kW and is expected to grow to between 200 and 250 million kW by 1980. Uranium in the U.S.A. is controlled by the Atomic Energy Commission. New uses are being found for "depleted" uranium (i.e., uranium with the percentage of U^{235} lowered to about 0.2%). It has found use in inertial guidance devices, gyro compasses, counterweights for aircraft control surfaces, as ballast for missile reentry vehicles, and as a shielding material. Uranium metal is used for x-ray targets for production of high-energy x-rays; the nitrate has been used as photographic toner; and the acetate is used in analytical chemistry. Crystals of uranium nitrate are triboluminescent. Uranium salts have also been used for producing yellow "vaseline" glass and glazes. Uranium and its compounds are highly toxic, both from a chemical and radiological standpoint. Finely divided uranium metal, being pyrophoric, presents a fire hazard. The maximum recommended allowable concentration of soluble uranium compounds in air (based on chemical toxicity)is 0.05 mg/m^3; for insoluble compounds the concentration is set at 0.25 mg/m^3 of air. The permissible body level of natural uranium (based on radiotoxicity) is 0.2 mCi for soluble compounds; for insoluble compounds the level is 0.009 mCi, or in air 1.7×10^{-11} mCi/ml.

Vanadium (Scandinavian goddess, *Vanadis*) — at wt 50.9414; at. no. 23; mp 1890° ± 10°C; bp 3380°C; sp gr 6.11 (18.7°C); valence 2,3,4, or 5.

Vanadium was first discovered by del Rio in 1801. Unfortunately a French chemist incorrectly declared del Rio's new element was only impure chromium; del Rio thought himself to be mistaken and accepted the French chemist's statement. The element was rediscovered in 1830 by Sefström who named the element in honor of the Scandinavian goddess *Vanadis* because of its beautiful multi-colored compounds. It was isolated in nearly pure form by Roscoe, in 1867, who reduced the chloride with hydrogen. Vanadium of 99.3 to 99.8% purity was not produced until 1927. Vanadium is found in about 65 different minerals among which are *carnotite, roscoelite, vanadinite,* and *patronite* – important sources of the metal. Vanadium is also found in phosphate rock, certain iron ores, and is present in some crude oils in the form of organic complexes. It is also found in small percentages in meteorites. Commercial production from petroleum ash holds promise as an important source of the element. High-purity ductile vanadium can be obtained by reduction of vanadium trichloride with magnesium or with magnesium-sodium mixtures. Much of the vanadium metal being produced is now made by calcium reduction of V_2O_5 in a pressure vessel, an adaption of a process developed by McKechnie and Seybolt. Natural vanadium is a mixture of two isotopes V^{50} (0.24%) and V^{51} (99.76%). V^{50} is slightly radioactive, having a half-life of 6×10^{15} yr. Seven other unstable isotopes are recognized. Pure vanadium is a bright white metal, and is soft and ductile. It has good corrosion resistance to alkalis, sulfuric and hydrochloric acid, and salt waters, but the metal oxidizes readily above 660°C. The metal has good structural strength and a low-fission neutron cross section, making it useful in nuclear applications. Vanadium is used in producing rust-resistant, spring, and high-speed tool steels. It is an important carbide stabilizer in making steels. About 80% of the vanadium now produced is used as ferrovanadium or as a steel additive. Vanadium foil is used as a bonding agent in cladding titanium to steel. Vanadium pentoxide is used in ceramics and as a catalyst. It is also used as a mordant in dyeing and printing fabrics and in the manufacture of aniline black. Vanadium and its compounds are toxic and should be handled with care. It has been reported that small doses of vanadium salts have reversed hardening of arteries in experimental animals, and that workers engaged in vanadium mining and milling operations have a lower serum cholestrol than unexposed workers in the same area. The threshold limit value of vanadium (V_2O_5 fume) as V has been suggested to be 0.05 mg/m³. Ductile vanadium is commercially available at a cost of about $40/lb. Commercial vanadium metal, of about 95% purity, costs about $5/lb.

Wolfram – (see Tungsten).

Xenon (Gr. *xenon,* stranger) – Xe; at. wt 131.30; at. no. 54; mp -111.9°C; bp -107.1 ± 3°C; density (gas) 5.887 ± 0.009 g/l, sp gr (liquid) 3.52 (-109°C); valence usually 0. Discovered by Ramsay and Travers in 1898 in the residue left after evaporating liquid air components. Xenon is a member of the so-called noble or "inert" gases. It is present in the atmosphere to the extent of about one part in twenty million. The element is found in the gases evolved from certain minerals springs, and is commercially obtained by extraction from liquid air. Natural xenon is composed of nine stable isotopes. In addition to these 22 unstable nuclides and isomers have been characterized. Until recently, xenon has been considered inert and unable to form compounds with other elements. Those compounds that were occasionally reported in the literature were considered not to be true compounds. Evidence has been mounting in the past few years that xenon, as well as other members of the zero valence elements, do form compounds. Among the "compounds" of xenon now reported are xenon hydrate, sodium perxenate, xenon deuterate, difluoride, tetrafluoride, hexafluoride, and $XePtF_6$ and $XeRhF_6$. More recently, xenon trioxide, which is highly explosive has been prepared. The structure of these substances is still open to question. Xenon in a vacuum tube produces a beautiful blue glow when excited by an electrical discharge. The gas is used in making electron tubes, stroboscopic lamps, bactericidal lamps, and lamps used to excite ruby lasers for generating coherent light. Xenon is used in the atomic energy field in bubble chambers, probes, and other applications where its high molecular weight is of value. It is also potentially useful as a gas for ion engines. The perxenates are used in analytical chemistry as oxidizing agents. Xe^{133} and Xe^{135} are produced by neutron irradiation in air-cooled nuclear reactors. Xe^{133} has useful applications as a radio isotope. The element is available in sealed glass containers for about $20/l of gas at standard pressure. Xenon is not toxic, but its compounds are highly toxic because of their strong oxidizing characteristics.

Ytterbium (*Ytterby*, village in Sweden) — Yb; at. wt 173.04; at. no. 70; mp 819°C; bp 1194°C; sp gr (α) 6.965, (β) 6.54; valence 2,3. Marignac in 1878 discovered a new component, which he called *ytterbia*, in the earth then known as *erbia*. In 1907, Urbain separated *ytterbia* into two components, which he called *neoytterbia* and *lutecia*. The elements in these earths are now known as *ytterbium* and *lutetium* respectively. These elements are identical with *aldebaranium* and *cassiopeium* discovered independently and at about the same time by von Welsbach. Ytterbium occurs along with other rare earths in a number of rare minerals. It is commercially recovered principally from *monazite* sand, which contains about 0.03%. Ion-exchange and solvent extraction techniques developed in recent years have greatly simplified the separation of the rare earths from one another. The element was first prepared by Klemm and Bonner in 1937 by reducing ytterbium trichloride with potassium. Their metal was mixed, however, with KCl. Daane, Dennison, and Spedding prepared a much purer form in 1953 from which the chemical and physical properties of the element could be determined. Ytterbium has a bright silvery luster, is soft, malleable, and quite ductile. While the element is fairly stable, it should be kept in closed containers to protect it from air and moisture. Ytterbium is readily attacked and dissolved by dilute and concentrated mineral acids and reacts slowly with water. Ytterbium normally has two allotropic forms with a transformation point at 798°C. The alpha form is a room-temperature, face-centered, cubic modification, while the high-temperature beta form is a body-centered cubic form. Another body-centered cubic phase has recently been found to be stable at high pressures at room temperatures. The alpha form ordinarily has metallic-type conductivity, but becomes a semiconductor when the pressure is increased above 16,000 atm. The electrical resistance increases tenfold as the pressure is increased to 39,000 atm and drops to about 80% of its standard temperature-pressure resistivity at a pressure of 40,000 atm. Natural ytterbium is a mixture of seven stable isotopes. Nine other unstable isotopes are known. Ytterbium metal has possible use in improving the grain refinement, strength, and other mechanical properties of stainless steel. One isotope is reported to have been used as a radiation source as a substitute for a portable x-ray machine where electricity is unavailable. Few other uses have been found. Ytterbium metal is commercially available with a purity of about 99 + % for about $1.50/g, or $300/lb. Ytterbium has a low acute toxic rating.

Yttrium (*Ytterby*, village in Sweden near Vauxholm) — Y; at. wt 88.9059; at. no. 39; mp 1522 ± 8°C; bp 3338°C; sp gr 4469 (25°C); valence 3. *Yttria* which is an earth containing yttrium, was discovered by Gadolin in 1794. *Ytterby* is the site of a quarry which yielded many unusual minerals containing rare earths and other elements. This small town, near Stockholm, bears the honor of giving names to *erbium, terbium*, and *ytterbium,* as well as *yttrium*. In 1843 Mosander showed that yttria could be resolved into the earths of three elements. The name yttria was reserved for the most basic one; the others were named *erbia* and *terbia*. Yttrium occurs in nearly all of the rare-earth minerals. Preliminary analysis of lunar rock samples obtained during the Apollo II mission showed a relatively high yttrium content. It is recovered commercially from *monazite sand,* which contains about 3%, and from *bastnasite*, which contains about 0.2%. Wöhler obtained the impure element in 1828 by reduction of the anhydrous chloride with potassium. The metal is now produced commercially by reduction of the fluoride with calcium metal. It can also be prepared by other techniques. Yttrium has a silvery metallic cluster and is relatively stable in air. Turnings of the metal, however, ignite in air if their temperature exceeds 400°C, and finely divided yttrium is very unstable in air. Yttrium oxide is one of the most important compounds of yttrium and accounts for the largest use. It is widely used in making YVO_4: Europium, and Y_2O_3: Europium phosphors to give the red color in color television tubes. Many hundreds of thousands of pounds are now used in this application. Yttrium oxide also is used to produce yttrium-iron-garnets, which are very effective microwave filters. Yttrium iron, aluminum and gadolinium garnets, with formulas such as $Y_3Fe_5O_{12}$ and $Y_3Al_5O_{12}$, have interesting magnetic properties. Yttrium iron garnet is also exceptionally efficient as both a transmitter and transducer of acoustic energy. Yttrium aluminum garnet, with a hardness of 8.5, is also finding use as a gemstone (simulated diamond). Small amounts of yttrium (0.1 to 0.2%) can be used to reduce the grain size in chromium, molybdenum, zirconium, and titanium, and to increase strength of aluminum and magnesium alloys. Alloys with other useful properties can be obtained by using yttrium as an additive. The metal can be used as a deoxidizer for vanadium and other nonferrous metals. The metal has a low cross section for nuclear capture. Y^{90}, one of the isotopes of yttrium, exists in equilibrium with its parent Sr^{90}, a product of atomic explosions. Yttrium has been considered for use as a nodulizer for producing nodular cast iron, in which the graphite forms compact nodules instead of the usual flakes. Such iron has increased ductility. Yttrium is also finding application in laser systems and as a catalyst for ethylene polymerization. It has also potential use in ceramic and glass formulas as the oxide has a high melting point and imparts shock resistance and low expansion characteristics to glass. Natural yttrium contains but one isotope, Y^{89}. Twenty other unstable nuclides and isomers have been characterized. Yttrium metal of 99 + % purity is commercially available at a cost of about $1/g or $200/lb.

Zinc (Ger. *Zink,* of obscure origin) — Zn; at. wt 65.38; at. no. 30; mp 419.58°C; bp, 907°C; sp gr 7.133 (25°C); valence 2. Centuries before zinc was recognized as a distinct element, zinc ores were used for making brass. Tubal-Cain, seven generations from Adam, is mentioned as being an "instructor in every artificer in brass and iron." An alloy containing 87% zinc has been found in prehistoric ruins in Transylvania. Metallic zinc was produced in the 13th Century A.D. in India by reducing calamine with organic substances, such as wool. The metal was rediscovered in Europe by Marggraf in 1746, who showed that the metal could be obtained by reducing *calamine* with charcoal. The principal ores of zinc are *sphalerite* or *blende* (sulfide), *smithsonite* (carbonate), *calamine* (silicate), and *franklinite* (zinc, manganese, iron oxide). Zinc can be obtained by roasting its ores to form the oxide and by reduction of the oxide with coal or carbon, with subsequent distillation of the metal. Other methods of extraction are possible. Naturally occurring zinc contains five stable isotopes. Ten other unstable nuclides and isomers are recognized. Zinc is a bluish-white, lustrous metal. It is brittle at ordinary temperatures but malleable at 100° to 150°C. It is a fair conductor of electricity and burns in air at high red heat with

evolution of white clouds of the oxide. The metal is employed to form numerous alloys with other metals. Brass, nickel silver, typewriter metal, commercial bronze, spring brass, German silver, soft solder and aluminum solder are some of the more important alloys. Large quantities of zinc are used to produce die castings, used extensively by the automotive, electrical, and hardware industries. An alloy called *Prestal,* consisting of 78% zinc and 22% aluminum, is reported to be almost as strong as steel, but as easy to mold as plastic. It is said to be so plastic that it can be molded into form by relatively inexpensive die casts made of ceramics and cement. It exhibits *superplasticity.* Zinc is also extensively used to galvanize other metals, such as iron, to prevent corrosion. Neither zinc nor zirconium is ferromagnetic, but $ZrZn_2$ exhibits ferromagnetism at temperatures below 35°K. Zinc oxide is a unique and very useful material to modern civilization. It is widely used in the manufacture of paints, rubber products, cosmetics, pharmaceuticals, floor coverings, plastics, printing inks, soap, storage batteries, textiles, electrical equipment, and other products. It has unusual electrical, thermal, optical, and solid-state properties that have not yet been fully investigated. Lithopone, a mixture of zinc sulfide and barium sulfate, is an important pigment. Zinc sulfide is used in making luminous dials, x-ray and TV screens, and fluorescent lights. The chloride and chromate are also important compounds. Zinc is an essential element in the growth of human beings and animals. Tests show that zinc-deficient animals require 50% more food to gain the same weight of an animal supplied with sufficient zinc. Zinc is not considered to be toxic, but when freshly formed ZnO is inhaled a disorder known as the *oxide shakes* or *zinc chills* sometimes occurs. It is recommended that where zinc oxide is encountered good ventilation be provided to avoid concentrations exceeding 5 mg of zinc oxide/m^3 over an 8-hr exposure. Zinc roughly costs 20¢/lb.

Zirconium (Arabic *zargun,* gold color) — Zr; at. wt 91.22; at. no. 40; mp 1852 ± 2°C; bp 4377°C; sp gr 6.506 (20°C); valence +2, +3, and †4. The name *zircon* probably originated from the Arabic word *zargun* which describes the color of the gemstone now known as *zircon, jargon, hyacinth, jacinth,* or *ligure.* This mineral, or its variations, is mentioned in biblical writings. The mineral was not known to contain a new element until Klaproth, in 1789, analyzed a jargon from Ceylon and found a new earth, which Werner named zircon (*silex circonius*), and Klaproth called *Zirkonerde* (*zirconia*). The impure metal was first isolated by Berzelius in 1824 by heating a mixture of potassium and potassium zirconium fluoride in a small iron tube. Pure zirconium was first prepared in 1914. Very pure zirconium was first produced in 1925 by van Arkel and de Boer by an iodide decomposition process they developed. Zirconium is found in abundance in S-type stars, and has been identified in the sun and meteorites. Analyses of lunar rock samples obtained during the various apollo missions, to the moon showed a surprisingly high zirconium oxide content, compared with terrestrial rocks. Naturally occurring zirconium contains five isotopes, one of which, Zr^{96} (abundant to the extent of 2.80%), is unstable with a very long half-life of $>3.6 \times 10^{17}$ years. Fifteen other unstable nuclides and isomers of zirconium have been characterized. Zircon, $ZrSiO_4$, the principal ore, is found in deposits in Florida, South Carolina, Australia, and Brazil. *Baddeleyite,* found in Brazil, is an important zirconium mineral. It is principally pure ZrO_2 in crystalline form having a hafnium content of about 1%. Zirconium also occurs in some 30 other recognized mineral species. Zirconium is produced commercially by reduction of the chloride with magnesium (the Kroll Process), and by other methods. It is a grayish-white lustrous metal. When finely divided, the metal may ignite spontaneously in air, especially at elevated temperatures. The solid metal is much more difficult to ignite. The inherent toxicity of zirconium compounds is low. Hafnium is invariably found in zirconium ores, and the separation is difficult. Commercial-grade zirconium contains from 1 to 3% hafnium. Zirconium has a low absorption cross section for neutrons, and is therefore used for nuclear energy applications, such as for cladding fuel elements. Zirconium has been found to be extremely resistant to the corrosive environment inside atomic reactors, and it allows neutrons to pass through the internal zirconium construction material without appreciable absorption of energy. Reactors of the size now being made may use as much as a half-million lineal feet of zirconium alloy tubing. Reactor-grade zirconium is essentially free of hafnium. *Zircaloy* is an important alloy developed specifically for nuclear applications. Zirconium is exceptionally resistant to corrosion by many common acids and alkalis, by sea water, and by other agents. It is used extensively by the chemical industry where corrosive agents are employed. Zirconium is used as a getter in vacuum tubes, as an alloying agent in steel, in making surgical appliances, photoflash bulbs, explosive primers, rayon spinnerets, lamp filaments, etc. It is used in poison ivy lotions in the form of the carbonate as it combines with *urushiol.* With niobium, zirconium is superconductive at low temperatures and is used to make superconductive magnets, which offer hope of direct large-scale generation of electric power. Alloyed with zinc, zirconium becomes magnetic at temperatures below 35°K. Zirconium oxide (zircon) has a high index of refraction and is used as a gem material. The impure oxide, zirconia, is used for laboratory crucibles that will withstand heat shock, for linings of metallurgical furnaces, and by the glass and ceramic industries as a refractory material. Its use as a refractory material accounts for a large share of all zirconium consumed. Commercial zirconium metal sponge is priced at about $5/lb. Fabricated zirconium parts are higher in cost.

Modified from Hammond, C. R., in *Handbook of Chemistry and Physics,* 55th ed., Weast, R. C., Ed., CRC Press, Cleveland, 1974, B-4.

Section 2

Elemental Properties

Table 2–1
PROPERTIES OF THE CHEMICAL ELEMENTS[a]

Atmospheric Pressure at Room Temperature

Name	Symbol	Atomic number	Inter-national at. wt.[b]	Specific gravity (or density)	Melting point, °C	Boiling point, °C	Specific heat at 25° C	Thermal con-ductivity, watt/cm °C
Actinium	Ac	89	(227)	(10.02)	1050.	3200.	–	–
Aluminum	Al	13	26.9815	2.70	660.	2441.	0.215	2.37
Americium	Am	95	(243)	11.7	994.	2607.	–	–
Antimony (Stibium)	Sb	51	121.75	6.69	630.	1750.	0.050	0.185
Argon	Ar	18	39.948	1.78 g/l	– 189.	– 186.	0.125	1.75×10^{-4}
Arsenic	As	33	74.9216	5.73 (gray)	815[c]	613. (subl.)	0.079	–
Astatine	At	85	(210)	–	729.	2125.	–	–
Barium	Ba	56	137.34	3.5	725.	1630.	0.046	–
Berkelium	Bk	97	(247)	–	–	–	–	–
Beryllium	Be	4	9.0122	1.85	1285.	2475.	0.436	2.18
Bismuth	Bi	83	208.980	9.75	271.4	1560.	0.030	0.084
Boron	B	5	10.811	2.35	2300.	2550.	0.245	–
Bromine	Br	35	79.904	3.12 (liq.)	– 7.2	58.8	0.11	0.45×10^{-4}
Cadmium	Cd	48	112.40	8.65	321.	767.	0.055	0.92
Calcium	Ca	20	40.08	1.55	840.	1485.	–	1.3
Californium	Cf	98	(251)	–	–	–	–	–
Carbon	C	6	12.01115					
Diamond				3.5	> 3800.	4827.	0.124	1.5 (0°)
Graphite				2.1	> 3500.	4200.	0.170	0.24
Cerium	Ce	58	140.12	6.77	798.	3257.	0.047	0.11
Cesium	Cs	55	132.905	1.87	28.6	678.	0.057	–
Chlorine	Cl	17	35.453	3.21 g/l	– 101.	– 34.6	0.114	0.86×10^{-4}
Chromium	Cr	24	51.996	7.2	1860.	2670.	0.110	0.91
Cobalt	Co	27	58.9332	8.9	1495.	2870.	0.10	0.69
Copper	Cu	29	63.546	8.96	1084.	2575.	0.092	3.98
Curium	Cm	96	(247)	–	–	–	–	–
Dysprosium	Dy	66	162.50	8.54	1409.	2335.	0.0414	0.10
Einsteinium	Es	99	(254)	–	–	–	–	–
Erbium	Er	68	167.26	9.05	1522.	2510.	0.04	0.096
Europium	Eu	63	151.96	5.25	822.	1597.	0.042	–
Fermium	Fm	100	(257)	–	–	–	–	–
Fluorine	F	9	18.9984	1.11 (liq.)	– 219.6	– 188.	0.197	2.63×10^{-4}
Francium	Fr	87	(223)	–	27.	677.	–	–
Gadolinium	Gd	64	157.25	7.90	1311.	3233.	0.055	0.088
Gallium	Ga	31	69.72	5.91	29.8	2300.	0.089	0.29 – 0.38
Germanium	Ge	32	72.59	5.32	937.	2830.	0.077	0.59
Gold (Aurum)	Au	79	196.967	19.32	1063.	2857.	0.031	3.15
Hafnium	Hf	72	178.49	13.29	2220.	4700.	0.035	0.220
Helium	He	2	4.0026	0.177 g/l	–	– 269.	1.24	14.8×10^{-4}
Holmium	Ho	67	164.930	8.78	1470.	2720.	0.039	–
Hydrogen	H	1	1.00797	0.0899 g/l	– 259.	– 253.	3.41	18.4×10^{-4}
Indium	In	49	114.82	7.31	156.	2050.	0.056	0.24
Iodine	I	53	126.9044	4.93	113.5	184.4	0.102	43.5×10^{-4}
Iridium	Ir	77	192.2	22.42	2450.	4390.	0.031	1.47
Iron (Ferrum)	Fe	26	55.847	7.87	1536.	2870.	0.108	0.803
Krypton	Kr	36	83.80	3.73 g/l	– 157.	– 152.	0.059	0.94×10^{-4}
Lanthanum	La	57	138.91	6.17	920.	3454.	0.047	0.14
Lawrencium	Lr	103	(257)	–	–	–	–	–
Lead (Plumbum)	Pb	82	207.19	11.35	327.5	1750.	0.031	0.352
Lithium	Li	3	6.939	0.53	180.	1342.	0.84	0.71
Lutetium	Lu	71	174.97	9.84	1656.	3315.	0.037	–
Magnesium	Mg	12	24.312	1.74	650.	1090.	0.243	1.56

Table 2–1 (continued)
PROPERTIES OF THE CHEMICAL ELEMENTS

Name	Symbol	Atomic number	Inter-national at. wt.[b]	Specific gravity (or density)	Melting point, °C	Boiling point, °C	Specific heat at 25° C	Thermal con-ductivity, watt/cm °C
Manganese	Mn	25	54.9380	7.21–7.44	1244.	2060.	0.114	—
Mendelevium	Md	101	(256)	—	—	—	—	—
Mercury (Hydrargyrum)	Hg	80	200.59	13.546	−38.86	356.55	0.033	0.0839
Molybdenum	Mo	42	95.94	10.22	2620.	4651.	0.060	1.38
Neodymium	Nd	60	144.24	7.00	1010.	3127.	0.049	0.13
Neon	Ne	10	20.183	0.90 g/l	−249.	−246.	0.246	4.77×10^{-4}
Neptunium	Np	93	(237)	18.0–20.45	640.	3902.	0.296	—
Nickel	Ni	28	58.71	8.90	1453.	2914.	0.106	0.905
Niobium (Columbium)	Nb	41	92.906	8.57	2467.	4740.	0.064	0.53
Nitrogen	N	7	14.0067	1.251 g/l	−210.	−196.	0.249	2.55×10^{-4}
Nobelium	No	102	(254)	—	—	—	—	—
Osmium	Os	76	190.2	22.57	3025.	4225.	0.031	0.61
Oxygen	O	8	15.9994	1.43 g/l	−218.4	−183.	0.220	2.61×10^{-4}
Palladium	Pd	46	106.4	12.02	1550.	2927.	0.058	0.71
Phosphorus, white	P	15	30.9738	1.82	44.1	280.	0.18	—
Platinum	Pt	78	195.09	21.45	1770.	3825.	0.032	0.73
Plutonium	Pu	94	(244)	19.84	640.	3230.	0.032	0.08
Polonium	Po	84	(209)	9.32	254.	962.	0.030	—
Potassium (Kalium)	K	19	39.102	0.86	63.3	760.	0.180	0.99
Praseodymium	Pr	59	140.907	6.77	931.	3212.	0.046	0.12
Promethium	Pm	61	(145)	—	1080.	2460.	0.044	—
Protactinium	Pa	91	(231)	(15.37)	—	—	0.029	—
Radium	Ra	88	(226)	—	700.	1700.	0.029	—
Radon	Rn	86	(222)	9.73 g/l	−71.	−62.	0.0224	—
Rhenium	Re	75	186.2	21.0	3180.	5650.	0.033	0.71
Rhodium	Rh	45	102.905	12.41	1965.	3700.	0.058	1.50
Rubidium	Rb	37	85.47	1.532	39.	700.	0.086	—
Ruthenium	Ru	44	101.07	12.4	2400.	4100.	0.057	—
Samarium	Sm	62	150.35	7.54	1072.	1778.	0.047	—
Scandium	Sc	21	44.956	2.99	1539.	2832.	0.135	—
Selenium	Se	34	78.96	4.8	217.	700.	0.077	0.005
Silicon	Si	14	28.086	2.33	1411.	3280.	0.17	0.835
Silver (Argentum)	Ag	47	107.868	10.50	961.	2212.	0.057	4.27
Sodium (Natrium)	Na	11	22.9898	0.97	97.83	884.	0.293	1.34
Strontium	Sr	38	87.62	2.55	770.	1375.	0.072	—
Sulfur	S	16	32.064	1.96–2.07	113.	445.	0.175	26.4×10^{-4}
Tantalum	Ta	73	180.948	16.6	2980.	5365.	0.034	0.575
Technetium	Tc	43	(97)	(11.50)	2172.	4877.	0.058	—
Tellurium	Te	52	127.60	6.24	450.	990.	0.05	0.059
Terbium	Tb	65	158.924	8.23	1360.	3041.	0.0435	—
Thallium	Tl	81	204.37	11.85	304.	1480.	0.031	0.39
Thorium	Th	90	232.038	11.7	1750.	4800.	0.03	0.41
Thulium	Tm	69	168.934	9.31	1545.	1727.	0.0385	—
Tin (Stannum)	Sn	50	118.69	7.31	232.	2600.	0.054	0.67
Titanium	Ti	22	47.90	4.54	1670.	3290.	0.125	0.22
Tungsten (Wolfram)	W	74	183.85	19.3	3400.	5550.	0.032	1.78
Uranium	U	92	238.03	18.8	1132.	4140.	0.028	0.25
Vanadium	V	23	50.942	6.1	1900.	3400.	0.116	0.60
Xenon	Xe	54	131.30	5.89 g/l	−112.	−107.	0.038	5.2×10^{-4}
Ytterbium	Yb	70	173.04	6.97	824.	1193.	0.071	—
Yttrium	Y	39	88.905	4.46	1523.	3337.	0.0925	0.15
Zinc	Zn	30	65.37	7.	419.5	910.	0.093	1.21
Zirconium	Zr	40	91.22	6.53	1852.	4400.	0.067	0.227

Table 2–1 (continued)
PROPERTIES OF THE CHEMICAL ELEMENTS

[a]Table 2–2 gives additional properties of the chemical elements. See also Weast, R. C., Ed., *Handbook of Chemistry and Physics*, 55th ed., CRC Press, Cleveland, 1974.
[b]A value in parentheses is the mass number of the most stable isotope of the element.
[c]At 28 atm.

From Bolz, R. E. and Tuve, G. L., Eds., *Handbook of Tables for Applied Engineering Science*, 2nd ed., CRC Press, Cleveland, 1973, 329.

Table 2–2
ADDITIONAL PROPERTIES OF THE CHEMICAL ELEMENTS[a]

Name	Atomic number	Latent heat of fusion, cal/g	Coef of linear thermal expansion × 10^6, K^{-1}			Elasticity modulus, psi × 10^{-6}	First ionization potential, eV	Thermal neutron absorption cross section, barns[b]
			100	300	500			
Actinium (227)	89	11	—	—	—	—	6.9	510.
Aluminum	13	95	12.5	24	27	10.0	5.984	0.24
Americium (243)	95	10	—	—	—	—	—	—
Antimony	51	38.5	9	9.5	10.5	11.3	8.639	5.7
Argon	18	6.7	—	—	—	—	15.755	0.66
Arsenic	33	88.5	—	4.7	—	—	9.81	4.3
Astatine	85	—	—	—	—	—	9.5	—
Barium	56	13.4	—	16	24	—	5.21	1.2
Berkelium	97	—	—	—	—	—	—	—
Beryllium	4	324	—	12	15	40–44	9.32	0.01
Bismuth	83	12.4	12	13	13.5	4.6	7.287	0.034
Boron	5	400	—	2	—	64	8.296	755
Bromine	35	16.2	—	—	—	—	11.84	6.7
Cadmium	48	13.2	26	30	38	8	8.991	2 450.
Calcium	20	52	17.5	23	26	3.2–3.8	6.111	0.44
Californium (251)	98	—	—	—	—	—	—	—
Carbon (Graphite)	6	—	—	—	—	0.7	11.256	0.004
Cerium	58	9	—	8	—	4.4	5.6	0.73
Cesium	55	3.8	—	97	—	—	3.893	30.0
Chlorine	17	2.16	—	—	—	—	13.01	34.
Chromium	24	79	3.5	6	9.5	36	6.764	3.1
Cobalt	27	66	—	12	13	30	7.86	38.
Columbium See Niobium								
Copper	29	49	10.5	16.5	18	17	7.724	3.8
Curium (247)	96	—	—	—	—	—	—	—
Dysprosium	66	26.4	—	9.0	—	9.2	6.8	950.
Einsteinium (254)*	99	—	—	—	—	—	—	—
Erbium	68	24.6	—	9.0	—	10.6	6.08	170.
Europium	63	16.9	—	26	—	2.1	5.67	4 300.
Fermium	—	—	—	—	—	—	—	—
Fluorine	9	10.1	—	—	—	—	17.418	0.01
Francium	87	—	—	—	—	—	4	—
Gadolinium	64	16.4	—	4	—	8.1	6.16	46 000
Gallium	31	19.2	—	18	—	—	6	2.8
Germanium	32	114	2.5	5.6	6.5	—	7.88	2.45
Gold	79	15	11.5	14	15	10.8	9.22	98.8
Hafnium	72	34	—	6	—	20	7	105.
Helium	2	1.2	—	—	—	—	24.481	0.007

Table 2–2 (continued)
ADDITIONAL PROPERTIES OF THE CHEMICAL ELEMENTS

Name	Atomic number	Latent heat of fusion, cal/g	Coef of linear thermal expansion × 10⁶, K⁻¹			Elasticity modulus, psi × 10⁻⁶	First ionization potential, eV	Thermal neutron absorption cross section, barns[b]
			100	300	500			
Holmium	67	—	—	—	—	9.7	—	65.
Hydrogen	1	15.0	—	—	—	—	13.595	0.33
Indium	49	6.8	25	33	—	—	5.785	191.
Iodine	53	15	—	93	—	—	10.454	7.0
Iridium	77	33	4	6.5	7.5	75	9	425.
Iron	26	65	6	12	14.5	28.5	7.87	2.6
Krypton	36	4.7	—	—	—	—	13.996	31.
Lanthanum	57	10	—	5	6.5	5.5	5.61	8.9
Lawrencium	103	—	—	—	—	—	—	—
Lead	82	5.5	25	29	32	2.0	7.415	0.18
Lithium	3	103	23	50	—	—	5.39	71.
Lutetium	71	26.4	—	—	—	12.2	—	112.
Magnesium	12	88.0	15	25	29	6.4	7.644	0.07
Manganese	25	64	11.5	23	28	23	7.432	13.3
Mendelevium	101	—	—	—	—	—	—	—
Mercury	80	2.7	—	—	—	—	10.43	375.
Molybdenum	42	69	3	5	5.5	40	7.10	2.7
Neodymium	60	13	—	7	7.5	5.5	5.51	46.
Neon	10	4.0	—	—	—	—	21.559	<2.8
Neptunium (237)	93	9.7	—	—	—	—	—	(170)
Nickel	28	71	6.5	13	15.5	31	7.633	4.6
Niobium	41	68	5	7	7.5	15	6.88	1.15
Nitrogen	7	6.2	—	—	—	—	14.53	1.9
Nobelium	102	—	—	—	—	—	—	—
Osmium	76	34	—	5	5.5	80	8.5	15.3
Oxygen	8	3.3	—	—	—	—	13.614	<0.000 2
Palladium	46	38	8.5	12	13	17	8.33	8.
Phosphorus	15	4.8	—	125	—	—	10.484	0.2
Platinum	78	24	6.8	8.9	9.5	21.3	9.0	8.8
Plutonium (244)	94	3	—	54	—	14	5.1	—
Polonium	84	11	—	—	—	—	8.43	—
Potassium	19	14.5	—	83	—	—	4.339	2.1
Praseodymium	59	17	—	5.	5.3	4.7	5.46	11.3
Promethium	61	—	—	—	—	6.1	—	—
Protactinium (231)	91	17	—	—	—	—	—	(200)
Radium (226)	88	10	—	—	—	—	5.277	(20)
Radon (222)	86	3.1	—	—	—	—	10.746	(0.7)
Rhenium	75	42	—	7	—	66.7	7.87	85.
Rhodium	45	50	5.0	8.3	9.3	42	7.46	150.
Rubidium	37	6.3	—	90	—	—	4.176	0.7
Ruthenium	44	60	—	9	—	60	7.364	2.6
Samarium	62	24.7	—	—	—	4.9	5.6	5 600.
Scandium	21	87	—	—	—	11.5	6.54	24.
Selenium	34	16	—	35	—	8.4	9.75	12.3
Silicon	14	430	—	2.5	3.5	16	8.149	0.160
Silver	47	26.5	14.3	19.0	20.6	10.5	7.574	63.
Sodium	11	27	45.7	70.0	—	—	5.138	.53
Strontium	38	25	—	—	—	—	5.692	1.21
Sulfur	16	9.2	42	63	—	—	10.357	0.52
Tantalum	73	41	5.2	6.6	6.9	27	7.88	21.
Technetium	43	56.7	—	—	—	—	7.28	22.
Tellurium	52	33	—	17	—	17	9.01	4.7

Table 2–2 (continued)
ADDITIONAL PROPERTIES OF THE CHEMICAL ELEMENTS

Name	Atomic number	Latent heat of fusion, cal/g	Coef of linear thermal expansion $\times 10^6$, K^{-1}			Elasticity modulus, psi $\times 10^{-6}$	First ionization potential, eV	Thermal neutron absorption cross section, barns[b]
			100	300	500			
Terbium	65	23.6	—	7.0	—	8.3	5.98	46.
Thallium	81	5.0	24	29	32	—	6.106	3.4
Thorium	90	17	8.7	11.4	12.5	8.5	6.95	7.5
Thulium	69	26.0	—	—	—	11.0	5.81	127.
Tin	50	14.1	15.5	21	27.5	6	7.342	0.63
Titanium	22	100	4.4	8.6	9.8	16	6.82	5.8
Tungsten	74	46	2.7	4.4	4.6	50	7.98	19.
Uranium	92	12	10.6	13.5	17	24	6.08	7.7
Vanadium	23	98	4	8	—	19	6.74	5.
Xenon	54	4.2	—	—	—	—	12.127	35.
Ytterbium	70	12.7	—	25	26.3	2.6	6.2	37.
Yttrium	39	45	—	—	—	9.4	6.38	1.3
Zinc	30	27	23	30	32	12	9.391	1.10
Zirconium	40	54	3.9	5.5	6.2	13.7	6.84	0.18

[a]Based largely on Weast, R. C., Ed., *Handbook of Chemistry and Physics*, 55th ed., CRC Press, Cleveland, 1974; see this source for other properties of the chemical elements; also refer to Table 2–1. For electron work functions, see Table 2–16. For electrical resistivities, see Table 2–39.

[b]Values in parentheses apply to that isotope for which the mass number is given following the name of the element. All other values of neutron cross section apply to the naturally occurring mixture of isotopes.

From Bolz, R. E. and Tuve, G. L., Eds., *Handbook of Tables for Applied Engineering Science,* 2nd ed., CRC Press, Cleveland, 1973, 331.

Table 2–3
PERIODIC TABLE OF THE ELEMENTS

KEY TO CHART

Atomic Number → 50 +2 ← Oxidation States
Symbol → Sn +4
Atomic Weight → 118.69
18 18 4 ← Electron Configuration

Transition Elements — Group 8

Z	Symbol	Atomic Weight	Oxidation States	Electron Configuration
1	H	1.0079	+1, −1	1
2	He	4.00260	0	2
3	Li	6.94	+1, −1	2-1
4	Be	9.01218	+2	2-2
5	B	10.81	+3	2-3
6	C	12.011	+2, +4, −4	2-4
7	N	14.0067	+1, +2, +3, +4, +5, −1, −3	2-5
8	O	15.9994	−2	2-6
9	F	18.99840	−1	2-7
10	Ne	20.17	0	2-8
11	Na	22.98977	+1	2-8-1
12	Mg	24.305	+2	2-8-2
13	Al	26.98154	+3	2-8-3
14	Si	28.086	+2, +4, −4	2-8-4
15	P	30.97376	+3, +5, −3	2-8-5
16	S	32.06	+4, +6, −2	2-8-6
17	Cl	35.453	+1, +5, +7, −1	2-8-7
18	Ar	39.948	0	2-8-8
19	K	39.09	+1	-8-8-1
20	Ca	40.08	+2	-8-8-2
21	Sc	44.9559	+3	-8-9-2
22	Ti	47.90	+2, +3, +4	-8-10-2
23	V	50.941	+2, +3, +4, +5	-8-11-2
24	Cr	51.996	+2, +3, +6	-8-13-1
25	Mn	54.9380	+2, +3, +4, +7	-8-13-2
26	Fe	55.847	+2, +3	-8-14-2
27	Co	58.9332	+2, +3	-8-15-2
28	Ni	58.71	+2	-8-16-2
29	Cu	63.546	+1, +2	-8-18-1
30	Zn	65.38	+2	-8-18-2
31	Ga	69.72	+3	-8-18-3
32	Ge	72.59	+2, +4	-8-18-4
33	As	74.9216	+3, +5, −3	-8-18-5
34	Se	78.96	+4, +6, −2	-8-18-6
35	Br	79.904	+1, +5, −1	-8-18-7
36	Kr	83.80	0	-8-18-8
37	Rb	85.467	+1	-18-8-1
38	Sr	87.62	+2	-18-8-2
39	Y	88.9059	+3	-18-9-2
40	Zr	91.22	+4	-18-10-2
41	Nb	92.9064	+3, +5	-18-12-1
42	Mo	95.94	+6	-18-13-1
43	Tc	98.9062	+4, +6, +7	-18-13-2
44	Ru	101.07	+3	-18-15-1
45	Rh	102.9055	+3	-18-16-1
46	Pd	106.4	+2, +4	-18-18-0
47	Ag	107.868	+1	-18-18-1
48	Cd	112.40	+2	-18-18-2
49	In	114.82	+3	-18-18-3
50	Sn	118.69	+2, +4	-18-18-4
51	Sb	121.75	+3, +5, −3	-18-18-5
52	Te	127.60	+4, +6, −2	-18-18-6
53	I	126.9045	+1, +5, +7, −1	-18-18-7
54	Xe	131.30	0	-18-18-8
55	Cs	132.9054	+1	-18-8-1
56	Ba	137.34	+2	-18-8-2
57	La	138.9055	+3	-18-9-2
58	Ce	140.12	+3, +4	-20-8-2
59	Pr	140.9077	+3	-21-8-2
60	Nd	144.24	+3	-22-8-2
61	Pm	(145)	+3	-23-8-2
62	Sm	150.4	+2, +3	-24-8-2
63	Eu	151.96	+2, +3	-25-8-2
64	Gd	157.25	+3	-25-9-2
65	Tb	158.9254	+3	-27-8-2
66	Dy	162.50	+3	-28-8-2
67	Ho	164.9304	+3	-29-8-2
68	Er	167.26	+3	-30-8-2
69	Tm	168.9342	+3	-31-8-2
70	Yb	173.04	+2, +3	-32-8-2
71	Lu	174.97	+3	-32-9-2
72	Hf	178.49	+4	-32-10-2
73	Ta	180.947	+5	-32-11-2
74	W	183.85	+6	-32-12-2
75	Re	186.2	+4, +6, +7	-32-13-2
76	Os	190.2	+3, +4	-32-14-2
77	Ir	192.22	+3, +4	-32-15-2
78	Pt	195.09	+2, +4	-32-16-2
79	Au	196.9665	+1, +3	-32-18-1
80	Hg	200.59	+1, +2	-32-18-2
81	Tl	204.37	+1, +3	-32-18-3
82	Pb	207.2	+2, +4	-32-18-4
83	Bi	208.9808	+3, +5	-32-18-5
84	Po	(209)	+2, +4	-32-18-6
85	At	(210)	−1	-32-18-7
86	Rn	(222)	0	-32-18-8
87	Fr	(223)	+1	-18-8-1
88	Ra	226.0254	+2	-18-8-2
89	Ac	(227)	+3	-18-9-2
90	Th	232.0381	+4	-18-10-2
91	Pa	231.0359	+4, +5	-20-9-2
92	U	238.029	+3, +4, +5, +6	-21-9-2
93	Np	237.0482	+3, +4, +5, +6	-22-9-2
94	Pu	(244)	+3, +4, +5, +6	-24-8-2
95	Am	(243)	+3, +4, +5, +6	-25-8-2
96	Cm	(247)	+3	-25-9-2
97	Bk	(247)	+3, +4	-27-8-2
98	Cf	(251)	+3	-28-8-2
99	Es	(254)		-29-8-2
100	Fm	(257)		-30-8-2
101	Md	(256)		-31-8-2
102	No	(254)		-32-8-2
103	Lr	(254)		-32-9-2
104		(257)		-32-10-2
105		—		

Orbit labels: K; K-L; K-L-M; L-M-N; M-N-O; M-N-O; N-O-P; O-P; O-P-Q; N-O-P-Q; O-P-Q

•Lanthanides
••Actinides

Each number in parentheses represents the mass number of the most stable isotope of the given element.

Modified from Weast, R. C., Ed., *Handbook of Chemistry and Physics*, 55th ed., CRC Press, Cleveland, 1974.

Table 2–4
AVAILABLE STABLE ISOTOPES OF THE ELEMENTS

Element and mass No.	Natural abundance, percent	Element and mass No.	Natural abundance, percent	Element and mass No.	Natural abundance, percent
Hydrogen		Phosphorus		Iron	
1	99.985	31	100.0	54	5.82
2	0.015	Sulfur		56	91.66
Helium		32	95.0	57	2.19
3	0.00013	33	0.76	58	0.33
4	~100.0	34	4.22	Cobalt	
Lithium		36	0.014	59	100.0
6	7.42	Chlorine		Nickel	
7	92.58	35	75.53	58	67.84
Beryllium		37	24.47	60	26.23
9	100.0	Argon		61	1.19
Boron		36	0.34	62	3.66
10	19.78	38	0.06	64	1.08
11	80.22	40	99.60	Copper	
Carbon		Potassium		63	69.09
12	98.89	39	93.1	65	30.91
13	1.11	40^a	0.01	Zinc	
Nitrogen		41	6.9	64	48.89
14	99.63	Calcium		66	27.81
15	0.37	40	96.97	67	4.11
Oxygen		42	0.64	68	18.57
16	99.76	43	0.14	70	0.62
17	0.04	44	2.06	Gallium	
18	0.20	46	0.003	69	60.4
Fluorine		48	0.18	71	39.6
19	100.0	Scandium		Germanium	
Neon		45	100.0	70	20.52
20	90.92	Titanium		72	27.43
21	0.26	46	7.93	73	7.76
22	8.82	47	7.28	74	36.54
Sodium		48	73.94	76	7.76
23	100.0	49	5.51	Arsenic	
Magnesium		50	5.34	75	100.0
24	78.70	Vanadium		Selenium	
25	10.13	50^b	0.24	74	0.87
26	11.17	51	99.76	76	9.02
Aluminum		Chromium		77	7.58
27	100.0	50	4.31	78	23.52
Silicon		52	83.76	80	49.82
28	92.21	53	9.55	82	9.19
29	4.70	54	2.38	Bromine	
30	3.09	Manganese		79	50.54
		55	100.0	81	49.46
				Krypton	
				78	0.35

Table 2–4 (continued)
AVAILABLE STABLE ISOTOPES OF THE ELEMENTS

Element and mass No.	Natural abundance, percent	Element and mass No.	Natural abundance, percent	Element and mass No.	Natural abundance, percent
Krypton (continued)		**Palladium (continued)**		**Xenon (continued)**	
80	2.27	106	27.33	129	26.44
82	11.56	108	26.71	130	4.08
83	11.55	110	11.81	131	21.18
84	56.90	**Silver**		132	26.89
86	17.37	107	51.82	134	10.44
Rubidium		109	48.18	136	8.87
85	72.15	**Cadmium**		**Cesium**	
87	27.85	106	1.22	133	100.0
		108	0.88		
Strontium		110	12.39	**Barium**	
84	0.56	111	12.75	130	0.101
86	9.86	112	24.07	132	0.097
87	7.02	113	12.26	134	2.42
88	82.56	114	28.86	135	6.59
		116	7.58	136	7.81
Yttrium				137	11.30
89	100.0	**Indium**		138	71.66
		113	4.28		
Zirconium		115^c	95.72	**Lanthanum**	
90	51.46			138	0.09
91	11.23	**Tin**		139	99.91
92	17.11	112	0.96		
94	17.40	114	0.66	**Cerium**	
96	2.80	115	0.35	136	0.193
		116	14.30	138	0.250
Niobium		117	7.61	140	88.48
93	100.0	118	24.03	142^d	11.07
		119	8.58		
Molybdenum		120	32.85	**Praseodymium**	
92	15.84	122	4.72	141	100.0
94	9.04	124	5.94		
95	15.72			**Neodymium**	
96	16.53	**Antimony**		142	27.11
97	9.46	121	57.25	143	12.17
98	23.78	123	42.75	144	23.85
100	9.63			145	8.30
		Tellurium		146	17.22
Ruthenium		120	0.09	148	5.73
96	5.51	122	2.46	150	5.62
98	1.87	123	0.87		
99	12.72	124	4.61	**Samarium**	
100	12.62	125	6.99	144	3.09
101	17.07	126	18.71	147^e	14.97
102	31.61	128	31.79	148^f	11.24
104	18.60	130	34.48	149^g	13.83
				150	7.44
Rhodium		**Iodine**		152	26.72
103	100.0	127	100.0	154	22.71
Palladium		**Xenon**		**Europium**	
102	0.96	124	0.096	151	47.82
104	10.97	126	0.090	153	52.18
105	22.23	128	1.92		

Table 2–4 (continued)
AVAILABLE STABLE ISOTOPES OF THE ELEMENTS

Element and mass No.	Natural abundance, percent	Element and mass No.	Natural abundance, percent	Element and mass No.	Natural abundance, percent
Gadolinium		Lutetium		Platinum (continued)	
152[h]	0.20	175	97.40	192	0.78
154	2.15	176[j]	2.60	194	32.9
155	14.73			195	33.8
156	20.47	Hafnium		196	25.3
157	15.68	174[k]	0.18	198	7.2
158	24.87	176	5.20		
160	21.90	177	18.50	Gold	
		178	27.14	197	100.0
Terbium		179	13.75		
159	100.0	180	35.24	Mercury	
				196	0.146
Dysprosium				198	10.02
156[i]	0.052			199	16.84
158	0.090	Tantalum		200	23.13
160	2.29	180	0.012	201	13.22
161	18.88	181	99.988	202	29.80
162	25.53			204	6.85
163	24.97	Tungsten			
164	28.18	180	0.14	Thallium	
		182	26.41	203	29.50
		183	14.40	205	70.50
Holmium		184	30.64		
165	100.0	186	28.41	Lead	
				204	1.48
Erbium		Rhenium		206	23.6
162	0.136	185	37.07	207	22.6
164	1.56	187[l]	62.93	208	52.3
166	33.41				
167	22.94	Osmium		Bismuth	
168	27.07	184	0.018	209	100.0
170	14.88	186	1.59		
		187	1.64		
Thulium		188	13.3	Thorium	
169	100.0	189	16.1	232[n]†	100.0
		190	26.4		
Ytterbium		192	41.0	Uranium	
168	0.135			234[o]†	0.0006
170	3.03	Iridium		235[p]†	0.72
171	14.31	191	37.3	238[q]†	99.27
172	21.82	193	62.7		
173	16.13				
174	31.84	Platinum			
176	12.73	190[m]	0.013		

[a]Half-life = 1.3×10^9 y.
[b]Half-life > 10^{15} y.
[c]Half-life = 5×10^{14} y.
[d]Half-life = 5×10^{14} y.
[e]Half-life = 1.06×10^{11} y.
[f]Half-life = 1.2×10^{13} y.

[g]Half-life = 4×10^{14} y.
[h]Half-life - 1.1×10^{14} y.
[i]Half-life = 2×10^{14} y.
[j]Half-life = 2.2×10^{10} y.
[k]Half-life = 4.3×10^{15} y.
[l]Half-life = 4×10^{10} y.

[m]Half-life = 6×10^{11} y.
[n]Half-life = 1.4×10^{10} y.
[o]Half-life = 2.5×10^5 y.
[p]Half-life = 7.1×10^8 y.
[q]Half-life = 4.5×10^9 y.
†Naturally occurring.

From Wang, Y., Ed., *Handbook of Radioactive Nuclides,* The Chemical Rubber Co., Cleveland, 1969, 25.

Table 2–5
VAPOR PRESSURE OF THE ELEMENTS

This table lists the temperature in degrees Celsius (Centigrade) at which an element has a vapor pressure indicated by the headings of the columns. To convert pressures to SI units, 1 mm Hg (torr) = 133.3 N/m^2 and 1 atm = 0.1013 MN/m^2.

| Element | | mm Hg | | | | | atm | | | | |
		1	10	100	400	760	2	5	10	20	40
Aluminum	Al	1540	1780	2080	2320	2467	2610	2850	3050	3270	3530
Antimony	Sb		960	1280	1570	1750	1960	2490			
Arsenic	As	380	440	510	580	610					
Barium	Ba	860	1050	1300	1520	1640	1790	2030	2230		
Beryllium	Be	1520	1860	2300	2770	2970	3240	3730	4110	4720	5610
Bismuth	Bi		1060	1280	1450	1560	1660	1850	2000	2180	
Boron	B	2660	3030	3460	3810	4000					
Bromine	Br	−60	−30	+9	39	59	78	110			
Cadmium	Cd	393	486	610	710	765	830	930	1030	1120	1240
Calcium	Ca	800	970	1200	1390	1490	1630	1850	2020	2290	
Cesium	Cs		373	513	624	690					
Chlorine	Cl	−123	−101	−71	−46	−34	−17	+9	30	55	97
Chromium	Cr	1610	1840	2140	2360	2480	2630	2850	3010	3180	
Cobalt	Co	1910	2170	2500	2760	2870	3040	3270			
Copper	Cu		1870	2190	2440	2600	2760	3010	3500	3460	3740
Fluorine	F			−203	−193	−188	−180.7	−169.1	−159.6		
Gallium	Ga	1350	1570	1850	2060	2180	2320	2560	2730		
Germanium	Ge		2080	2440	2710	2830	2970	3200	3430		
Gold	Au	1880	2160	2520	2800	2940	3120	3490	3630	3890	
Indium	In				1960	2080	2230	2440	2600		
Iodine	I	40	72	115	160	185	216	265			
Iridium	Ir	2830	3170	3630	3960	4130	4310	4650			
Iron	Fe	1780	2040	2370	2620	2750	2900	3150	3360	3570	
Lanthanum	La				3230	3420	3620	3960	4270		
Lead	Pb	970	1160	1420	1630	1740	1880	2140	2320	2620	
Lithium	Li	750	890	1080	1240	1310	1420	1518			
Magnesium	Mg	620	740	900	1040	1110	1190	1330	1430	1560	
Manganese	Mn		1510	1810	2050	2100	2360	2580	2850		
Mercury	Hg			260	330	356.9	398	465	517	581	657
Molybdenum	Mo	3300	3770	4200	4580	4830	5050	5340	5680	5980	
Neodymium	Nd				2870	3100	3300	3680	3990		
Nickel	Ni	1800	2090	2370	2620	2730	2880	3120	3300	3310	
Palladium	Pd	1470	2290	2670	2950	3140	3270	3560	3840		
Phosphorus	P		127	199	253	283	319				
Platinum	Pt	2600	2940	3360	3650	3830	4000	4310	4570	4860	
Polonium	Po	472	587	752	890	960	1060	1200	1340		
Potassium	K			590	710	770	850	950	1110	1240	1420
Rhodium	Rh	2530	2850	3260	3590	3760	3930	4230	4440		
Rubidium	Rb		390	527	640	700					
Selenium	Se		429	547	640	685	750	850	920	1010	1120
Silver	Ag	1310	1540	1850	2060	2210	2360	2600	2850	3050	3300
Sodium	Na	440	546	700	830	890	980	1120	1230	1370	
Strontium	Sr	740	900	1100	1280	1380	1480	1670	1850	2030	
Sulfur	S		246	333	407	445	493	574	640	720	
Tellurium	Te	520	633	792	900	962	1030	1160	1250		
Thallium	Tl		1000	1210	1370	1470	1560	1750	1900	2050	2260
Tin	Sn	1610	1890	2270	2580	2750	2950	3270	3540	3890	
Titanium	Ti	2180	2480	2860	3100	3260	3400	3650	3800		
Tungsten	W	3980	4490	5160	5470	5940	6260	6670	7250	7670	
Uranium	U	2450	2800	3270	3620	3800	4040	4420			
Vanadium	V	2290	2570	2950	3220	3380	3540	3800			
Zinc	Zn		590	730	840	907	970	1090	1180	1290	

From Loebel, R., in *Handbook of Chemistry and Physics,* 55th ed., Weast, R. C., Ed., CRC Press, Cleveland, 1974, D-190.

Table 2–6
PROPERTIES OF LIQUID METALS[a]

At Atmospheric or Pumping Pressures

Metal (Melting point, °F)	Temperature °F	Temperature °C	Specific gravity	Specific heat	Thermal conductivity Btu/hr ft °F	Thermal conductivity cal/sec cm °C[b]	Absolute viscosity lb$_m$/ft sec	Absolute viscosity centipoises
Aluminum (1220)	1250	677	2.38	.259				
	1300	704	2.37	.259	60.2	.249	1.88×10^{-3}	2.8
	1350	732	2.36	.259	63.4	.262	1.61×10^{-3}	2.4
	1400	760	2.35	.259	64.3	.266	1.34×10^{-3}	2.0
	1450	788	2.34	.259	69.9	.289	1.08×10^{-3}	1.6
Bismuth (520)	600	316	10.0	.0345	9.5	.039	1.09×10^{-3}	1.62
	800	427	9.87	.0357	9.0	.037	9.0×10^{-4}	1.34
	1000	538	9.74	.0369	9.0	.037	7.4×10^{-4}	1.10
	1200	649	9.61	.0381	9.0	.037	6.2×10^{-4}	.923
Cesium (83)	83	28	1.84	.060	10.6	.044		
	150	66					3.84×10^{-4}	.571
	250	121					2.95×10^{-4}	.439
	350	177					2.47×10^{-4}	.368
	400	204					2.30×10^{-4}	.343
Lead (621)	700	371	10.5	.038	9.3	.038	1.61×10^{-3}	2.39
	850	454	10.4	.037	9.0	.037	1.38×10^{-3}	2.05
	1000	538	10.4	.037	8.9	.036	1.17×10^{-3}	1.74
	1150	621	10.2	.037	8.7	.036	1.02×10^{-3}	1.52
	1300	704	10.1		8.6	.035	9.20×10^{-4}	1.37
Lithium (355)	400	204	.506	1.0	24.	.10	4.0×10^{-4}	.595
	600	316	.497	1.0	23.	.095	3.4×10^{-4}	.506
	800	427	.489	1.0	22.	.090	3.7×10^{-4}	.551
	1200	649	.471				2.9×10^{-4}	.432
	1800	942	.442				2.8×10^{-4}	.417
Magnesium (1203)	1250	677	1.55	.318				
	1301	705	1.53	.320				
	1350	732	1.49	.322				
Mercury (−38)	50	10	13.6	.033	4.8	.020	1.07×10^{-3}	1.59
	200	93	13.4	.033	6.0	.025	8.4×10^{-4}	1.25
	300	149	13.2	.033	6.7	.028	7.4×10^{-4}	1.10
	400	204	13.1	.032	7.2	.030	6.7×10^{-4}	.997
	600	316	12.8	.032	8.1	.033	5.8×10^{-4}	.863
Tin (449)	500	260	6.94	.058	19	.079	1.22×10^{-3}	1.82
	700	371	6.86	.060	19.4	.080	9.8×10^{-4}	1.46
	850	454	6.81	.062	19	.079	8.5×10^{-4}	1.26
	1000	538	6.74	.064	19	.079	7.6×10^{-4}	1.13
	1200	649	6.68	.066	19	.079	6.7×10^{-4}	.997
Zinc (787)	600	316	6.97	.123	35.4	.146		
	850	454	6.90	.119	33.7	.139	2.10×10^{-3}	3.12
	1000	538	6.86	.116	33.2	.137	1.72×10^{-3}	2.56
	1200	649	6.76	.113	32.8	.136	1.39×10^{-3}	2.07
	1500	816	6.74	.107	32.6	.135	9.83×10^{-4}	1.46

Table 2—6 (continued)

PROPERTIES OF LIQUID METALS

At Atmospheric or Pumping Pressures (continued)

[a]Based largely on *Liquid Metals Handbook,* 3rd ed., Office of Naval Research, U.S. Government Printing Office, Washington, D.C., 1955.
[b]For watt/cm °C multiply by 418.4.

From Bolz, R. E. and Tuve, G. L., Eds., *Handbook of Tables for Applied Engineering Science,* 2nd ed., CRC Press, Cleveland, 1973, 98.

Table 2—7

PHYSICAL PROPERTIES OF SODIUM POTASSIUM MIXTURES[a]

Symbols and Units:

M.P. = melting point, °C. For °K, add 273.15.

ρ = density, lbm/ft^3. For specific gravity, multiply by 0.016018. For kg/m^3, multiply by 16.018.

c_p = specific heat, Btu/lbm·°R or cal/g·°C. For J/kg·°K, multiply by 4184.0

k = thermal conductivity, Btu/hr·ft·°R. For

W/m·°K, multiply by 1.7296.

μ = absolute viscosity, lbm/hr·ft. For centipoises, multiply by 0.41338. For N·s/m^2, multiply value in lbm/hr·ft by 0.00041338 or value in lbm/sec·ft by 1.4882.

% potassium	M.P.	Property	°C 100 °F 212	200 392	300 572	400 752	500 932	600 1112	700 1292
0	98	ρ	57.9	56.4	55.1	53.6	52.1	50.5	48.9
		c_p	.331	.320	.312	.306	.302	.300	.300
		k	—	46.7	43.7	41.1	38.9	36.9	35.3
		μ	1.71	1.09	0.83	0.69	0.59	0.51	0.45
44	17	ρ	55.6	54.1	52.6	51.1	49.6	48.1	46.5
		c_p	.269	.261	.255	.251	.249	.248	.250
		k	—	—	—	—	—	—	—
		μ	1.31	0.92	0.72	0.59	0.50	0.43	0.62
56[b]	7	ρ	53.8	52.6	51.4	49.2	48.0	46.8	45.6
		c_p	.254	.248	.241	.237	.233	.233	.236
		k	13.5	14.5	15.0	15.6	15.9	—	—
		μ	—	—	—	—	—	—	—
78	−10	ρ	53.1	51.6	50.1	48.6	47.1	45.7	44.0
		c_p	.225	.217	.212	.210	.209	.209	.211
		k	13.8	14.3	15.0	15.1	15.1	15.0	14.7
		μ[c]	1.28	0.86	0.67	0.55	0.47	0.41	0.35
100[b]	64	ρ	51.1	49.5	48.2	46.8	45.5	43.6	42.1
		c_p	.193	.190	.188	.184	.182	.183	.186
		k	—	25.8	24.6	23.4	22.2	21.0	19.8
		μ	1.07	0.74	0.55	0.44	0.40	0.35	0.33

Temperature

Table 2–7 (continued)
PHYSICAL PROPERTIES OF SODIUM POTASSIUM MIXTURES

Note: The surface heat-transfer coefficient for liquid metals in long tubes is approximately as follows:

$$h = k/D \left[7 + 0.025 \, (RePr)^{0.8} \right].$$

For details see *Liquid Metals Handbook*.

[a]Adapted from *Liquid Metals Handbook,* 3rd ed., Office of Naval Research, U.S. Government Printing Office, Washington, D.C., 1955.
[b]Data approximated from graph.
[c]Viscosity at 66.9% potassium.

From Bolz, R. E. and Tuve, G. L., Eds., *Handbook of Tables for Applied Engineering Science,* 2nd ed., CRC Press, Cleveland, 1973, 136.

Table 2–8
CRYSTAL IONIC RADII OF THE ELEMENTS

Numerical values of the radii of the ions may vary, depending on how they were measured. They may have been calculated from wave functions or determined from the lattice spacings or crystal structure of various salts. Different values are obtained, depending on the kind of salt used or on the method of calculation. Data for many of the rare-earth ions were furnished by F. H. Spedding and K. Gschneidner.

Element	Charge	Atomic number	Radius in A	Element	Charge	Atomic number	Radius in A	Element	Charge	Atomic number	Radius in A
Ac	+3	89	1.18	Ca	+1	20	1.18	Fr	+1	87	1.80
Ag	+1	47	1.26		+2		0.99	Ga	+1	31	0.81
	+2		0.89	Cd	+1	48	1.14		+3		0.62
Al	+3	13	0.51		+2		0.97	Gd	+3	64	0.938
Am	+3	95	1.07	Ce	+1	58	1.27	Ge	−4	32	2.72
	+4		0.92		+3		1.034		+2		0.73
Ar	+1	18	1.54		+4		0.92		+4		0.53
As	−3	33	2.22	Cl	−1	17	1.81	H	−1	1	1.54
	+3		0.58		+5		0.34	Hf	+4	72	0.78
	+5		0.46		+7		0.27	Hg	+1	80	1.27
At	+7	85	0.62	Co	+2	27	0.72		+2		1.10
Au	+1	79	1.37		+3		0.63	Ho	+3	67	0.894
	+3		0.85	Cr	+1	24	0.81	I	−1	53	2.20
B	+1	5	0.35		+2		0.89		+5		0.62
	+3		0.23		+3		0.63		+7		0.50
Ba	+1	56	1.53		+6		0.52	In	+3	49	0.81
	+2		1.34	Cs	+1	55	1.67	Ir	+4	77	0.68
Be	+1	4	0.44	Cu	+1	29	0.96	K	+1	19	1.33
	+2		0.35		+2		0.72	La	+1	57	1.39
Bi	+1	83	0.98	Dy	+3	66	0.908		+3		1.016
	+3		0.96	Er	+3	68	0.881	Li	+1	3	0.68
	+5		0.74	Eu	+3	63	0.950	Lu	+3	71	0.85
Br	−1	35	1.96		+2		1.09	Mg	+1	12	0.82
	+5		0.47	F	−1	9	1.33		+2		0.66
	+7		0.39		+7		0.08	Mn	+2	25	0.80
C	−4	6	2.60	Fe	+2	26	0.74		+3		0.66
	+4		0.16		+3		0.64		+4		0.60
									+7		0.46

Table 2–8 (continued)
CRYSTAL IONIC RADII OF THE ELEMENTS

Element	Charge	Atomic number	Radius in A	Element	Charge	Atomic number	Radius in A	Element	Charge	Atomic number	Radius in A
Mo	+1	42	0.93	Po	+6	84	0.67	Sr	+2	38	1.12
	+4		0.70	Pr	+3	59	1.013	Ta	+5	73	0.68
	+6		0.62		+4		0.90	Tb	+3	65	0.923
N	−3	7	1.71	Pt	+2	78	0.80		+4		0.84
	+1		0.25		+4		0.65	Tc	+7	43	0.979
	+3		0.16	Pu	+3	94	1.08	Te	−2	52	2.11
	+5		0.13		+4		0.93		−1		2.50
NH₄	+1		1.43	Ra	+2	88	1.43		+1		0.82
Na	+1	11	0.97	Rb	+1	37	1.47		+4		0.70
Nb	+1	41	1.00	Re	+4	75	0.72		+6		0.56
	+4		0.74		+7		0.56	Th	+4	90	1.02
	+5		0.69	Rh	+3	45	0.68	Ti	+1	22	0.96
Nd	+3	60	0.995	Ru	+4	44	0.67		+2		0.94
Ne	+1	10	1.12	S	−2	16	1.84		+3		0.76
Ni	+2	28	0.69		+2		2.19		+4		0.68
Np	+3	93	1.10		+4		0.37	Tl	+1	81	1.47
	+4		0.95		+6		0.30		+3		0.95
	+7		0.71	Sb	−3	51	2.45	Tm	+3	69	0.87
O	−2	8	1.32		+3		0.76	U	+4	92	0.97
	−1		1.76		+5		0.62		+6		0.80
	+1		0.22	Sc	+3	21	0.732	V	+2	23	0.88
	+6		0.09	Se	−2	34	1.91		+3		0.74
Os	+4	76	0.88		−1		2.32		+4		0.63
	+6		0.69		+1		0.66		+5		0.59
P	−3	15	2.12		+4		0.50	W	+4	74	0.70
	+3		0.44		+6		0.42		+6		0.62
	+5		0.35	Si	−4	14	2.71	Y	+3	39	0.893
Pa	+3	91	1.13		−1		3.84	Yb	+2	70	0.93
	+4		0.98		+1		0.65		+3		0.858
	+5		0.89		+4		0.42	Zn	+1	30	0.88
Pb	+2	82	1.20	Sm	+3	62	0.964		+2		0.74
	+4		0.84	Sn	−4	50	2.94	Zr	+1	40	1.09
Pd	+2	46	0.80		−1		3.70		+4		0.79
	+4		0.65		+2		0.93				
Pm	+3	61	0.979		+4		0.71				

From Weast, R. C., Ed., *Handbook of Chemistry and Physics,* 55th ed., CRC Press, Cleveland, 1974, F-198.

Table 2–9
STRENGTHS OF CHEMICAL BONDS

The strength of a chemical bond, D(R – X), often known as the bond dissociation energy, is defined as the heat of the reaction: RX → R + X. It is given by: $D(R - X) = \Delta Hf°(R) + \Delta Hf°(X) - \Delta Hf°(RX)$. Some authors list bond strengths for $0°K$, but here the values for $298°K$ are given because more thermodynamic data are available for this temperature. Bond strengths, or bond dissociation energies, are not equal to, and may differ considerable from, mean bond energies derived solely from thermochemical data on molecules and atoms.

Bond Strengths in Diatomic Molecules

These have usually been measured spectroscopically or by mass spectrometric analysis of hot gases effusing from a Knudsen cell. Excellent accounts of these and other methods are given in Reference 58. The errors quoted are those given in the original paper or review article. The references have been chosen primarily as a key to the literature. It should not be assumed that the author referred to was responsible for the determination quoted, as the reference may be only to a review article.

Molecule	kcal mol^{-1}	Ref.	Molecule	kcal mol^{-1}	Ref.
H–H	104.207 ± 0.001	67	H–Sn	63 ± 1	89
H–D	105.030 ± 0.001	67	H–Te	64 ± 1	63
D–D	106.010 ± 0.001	67	H–I	71.4 ± 0.2	51
H–Li	56.91 ± 0.01	125	H–Cs	42.6 ± 0.9	107
H–Be	54	66	H–Ba	42 ± 4	51
H–B	79 ± 1	74	H–Yb	38 ± 1	64
H–C	80.9	66	H–Pt	84 ± 9	51
H–N	75 ± 4	51	H–Au	75 ± 3	106
H–O	102.34 ± 0.30	19	H–Hg	9.5	51
H–F	135.9 ± 0.3	51	H–Tl	45 ± 2	51
H–Na	48 ± 5	51	H–Pb	42 ± 5	51
H–Mg	47 ± 12	46, 51	H–Bi	59 ± 7	51
H–Al	68 ± 2	51	Li–Li	24.55 ± 0.14	126
H–Si	71.4 ± 1.2	51	Li–O	78 ± 6	19
H–P	82 ± 7	51	Li–F	137.5 ± 1	18
H–S	82.3 ± 2.9	75	Li–Cl	111.9 ± 2	18
H–Cl	103.1	51	Li–Br	100.2 ± 2	18
H–K	43.8 ± 3.5	51	Li–I	84.6 ± 2	18
H–Ca	40.1	51	Be–Be	17	45
H–Cr	67 ± 12	51	Be–O	98 ± 7	19
H–Mn	56 ± 7	51	Be–F	136 ± 2	70
H–Ni	61 ± 7	51	Be–S	89 ± 14	51
H–Cu	67 ± 2	23	Be–Cl	92.8 ± 2.2	73
H–Zn	20.5 ± 0.5	51	Be–Au	∿67	11
H–Ga	68 ± 5	51	B–B	∿67 ± 5	112
H–Ge	76.8 ± 0.2	10	B–N	93 ± 12	51
H–As	65 ± 3	41	B–O	192.7 ± 1.2	120
H–Se	73 ± 1	63	B–F	180 ± 3	100
H–Br	87.4 ± 0.5	51	B–S	138.8 ± 2.2	120
H–Rb	40 ± 5	51	B–Cl	119	51
H–Sr	39 ± 2	51	B–Se	110 ± 4	120
H–Ag	59 ± 1	108	B–Br	101 ± 5	51
H–Cd	16.5 ± 0.1	51	B–Ru	107 ± 5	124
H–In	59 ± 2	51	B–Rh	114 ± 5	124

Table 2–9 (continued)
STRENGTHS OF CHEMICAL BONDS

Bond Strengths in Diatomic Molecules (continued)

Molecule	kcal mol^{-1}	Ref.	Molecule	kcal mol^{-1}	Ref.
B–Pd	79 ± 5	124	O–Sc	155 ± 5	19
B–Te	85 ± 5	120	O–Ti	158 ± 8	19
B–Ce	∿100	118	O–V	154 ± 5	19
B–Ir	123 ± 4	124	O–Cr	110 ± 10	19
B–Pt	114 ± 4	95	O–Mn	96 ± 8	19
B–Au	82 ± 4	124	O–Fe	96 ± 5	19
B–Th	71	55	O–Co	88 ± 5	19
C–C	144 ± 5	45	O–Ni	89 ± 5	19
C–N	184 ± 1	35	O–Cu	82 ± 15	19
C–O	257.26 ± 0.77	19	O–Zn	≤66	19
C–F	128 ± 5	51	O–Ga	68 ± 15	19
C–Si	104 ± 5	43	O–Ge	158.2 ± 3	19
C–P	139 ± 23	51	O–As	115 ± 3	19
C–S	175 ± 7	92	O–Se	101	44
C–Cl	93	102	O–Br	56.2 ± 0.6	19
C–Ti	128?	51	O–Rb	(61) ± 20	19
C–V	133	50	O–Sr	93 ± 6	19
C–Ge	110 ± 5	51	O–Y	162 ± 5	19
C–Se	139 ± 23	51	O–Zr	181 ± 10	19
C–Br	67 ± 5	51	O–Nb	189 ± 10	19
C–Ru	152 ± 3	95	O–Mo	115 ± 12	19
C–Rh	139 ± 2	122	O–Ru	115 ± 15	19
C–I	50 ± 5	51	O–Rh	90 ± 15	19
C–Ce	109 ± 7	56	O–Pd	56 ± 7	19
C–Ir	149 ± 3	95	O–Ag	51 ± 20	19
C–Pt	146 ± 2	122	O–Cd	≤67	19
C–U	111 ± 7	56	O–In	≤77	19
N–N	226.8 ± 1.5	49	O–Sn	127 ± 2	19
N–O	150.8 ± 0.2	19	O–Sb	89 ± 20	19
N–F	62.6 ± 0.8	5	O–Te	93.4 ± 2	99
N–Al	71 ± 23	51	O–I	47 ± 7	19
N–Si	105 ± 9	51	O–Xe	9 ± 5	19
N–P	148 ± 5	57	O–Cs	67 ± 8	19
N–S	∿120 ± 6	101	O–Ba	131 ± 6	19
N–Cl	93 ± 12	51	O–La	188 ± 5	19
N–Ti	111	116	O–Ce	188 ± 6	19
N–As	116 ± 23	51	O–Pr	183.7	48
N–Se	105 ± 23	51	O–Nd	168 ± 8	19
N–Br	67 ± 5	97	O–Sm	134 ± 8	19
N–Sb	72 ± 12	51	O–Eu	130 ± 10	19
N–I	∿38	104	O–Gd	162 ± 6	19
N–Xe	55	65	O–Tb	165 ± 8	19
N–Th	138 ± 1	54	O–Dy	146 ± 10	19
N–U	127 ± 1	53	O–Ho	149 ± 10	19
O–O	118.86 ± 0.04	20	O–Er	147 ± 10	19
O–F	56 ± 9	51	O–Tm	122 ± 15	19
O–Na	61 ± 4	72	O–Yb	98 ± 15	19
O–Mg	79 ± 7	19	O–Lu	159 ± 8	19
O–Al	116 ± 5	19	O–Hf	185 ± 10	19
O–Si	184 ± 3	71	O–Ta	183 ± 15	19
O–P	119.6 ± 3	19	O–W	156 ± 6	19
O–S	124.69 ± 0.03	19	O–Os	<142	19
O–Cl	64.29 ± 0.03	19	O–Ir	≤94	19
O–K	57 ± 8	19	O–Pt	83 ± 8	19
O–Ca	84 ± 7	19	O–Pb	90.3 ± 1.0	119

Table 2–9 (continued)
STRENGTHS OF CHEMICAL BONDS

Bond Strengths in Diatomic Molecules (continued)

Molecule	kcal mol^{-1}	Ref.	Molecule	kcal mol^{-1}	Ref.
O–Bi	81.9 ± 1.5	119	Mg–I	~68	14
O–Th	192 ± 10	19	Mg–Au	59 ± 23	51
O–U	182 ± 8	19	Al–Al	44	62
O–Np	172 ± 7	4	Al–P	52 ± 3	38
O–Pu	163 ± 15	19	Al–S	79	91
O–Cm	≤134	113	Al–Cl	119.0 ± 1	73
F–F	37.5 ± 2.3	36	Al–Br	103.1	8
F–Na	114 ± 1	18	Al–I	88	7
F–Mg	110 ± 1	68	Al–Au	65	12
F–Al	159 ± 3	100	Al–U	78 ± 7	59
F–Si	116 ± 12	51	Si–Si	76 ± 5	43
F–P	105 ± 23	51	Si–S	148 ± 3	51
F–Cl	59.9 ± 0.1	51	Si–Cl	105 ± 12	51
F–K	118.9 ± 0.6	9	Si–Fe	71 ± 6	123
F–Ca	125 ± 5	17	Si–Co	66 ± 4	123
F–Sc	141 ± 3	136	Si–Ni	76 ± 4	123
F–Ti	136 ± 8	137	Si–Ge	72 ± 5	51
F–Cr	104.5 ± 4.7	88	Si–Se	127 ± 6	51
F–Mn	101.2 ± 3.5	86	Si–Br	82 ± 12	51
F–Ni	89 ± 4	51	Si–Ru	95 ± 5	124
F–Cu	88 ± 9	87	Si–Rh	95 ± 5	124
F–Ga	138 ± 4	100	Si–Pd	75 ± 4	124
F–Ge	116 ± 5	47	Si–Te	121 ± 9	51
F–Br	55.9	21	Si–Ir	110 ± 5	124
F–Rb	116.1 ± 1	18	Si–Pt	120 ± 5	124
F–Sr	129.5 ± 1.6	68	Si–Au	75 ± 3	124
F–Y	144 ± 5	136	P–P	117 ± 3	57
F–Ag	84.7 ± 3.9	51	P–S	70	98
F–Cd	73 ± 5	16	P–Ga	56	61
F–In	121 ± 4	100	P–W	73 ± 1	52
F–Sn	111.5 ± 3	133	P–Th	90	55
F–Sb	105 ± 23	51	S–S	101.9 ± 2.5	44
F–I	67?	51	S–Ca	75 ± 5	29
F–Xe	11	76	S–Sc	114 ± 3	30
F–Cs	119.6 ± 1	18	S–Mn	72 ± 4	130
F–Ba	140.3 ± 1.6	68	S–Fe	78	96
F–Nd	130 ± 3	134	S–Cu	72 ± 12	51
F–Sm	126.9 ± 4.4	135	S–Zn	49 ± 3	39
F–Eu	126.1 ± 4.4	135	S–Ge	131.7 ± 0.6	30, 44
F–Gd	141.1 ± 6.5	135	S–Se	91 ± 5	44
F–Hg	31 ± 9	51	S–Sr	75 ± 5	29
F–Tl	106.4 ± 4.6	15	S–Y	127 ± 3	30
F–Pb	85 ± 2	133	S–Cd	48	96
F–Bi	62 ± 7	51	S–In	69 ± 4	28
F–Pu	129 ± 7	85	S–Sn	111 ± 1	44
Na–Na	18.4	112	S–Te	81 ± 5	44
Na–Cl	97.5 ± 0.5	18	S–Ba	96 ± 5	29
Na–K	15.2 ± 0.7	51	S–La	137 ± 3	26, 30
Na–Br	86.7 ± 1	18	S–Ce	137 ± 3	30
Na–Rb	14 ± 1	51	S–Pr	122.7	48
Na–I	72.7 ± 1	18	S–Nd	113 ± 4	114
Mg–Mg	8?	51	S–Eu	87 ± 4	114
Mg–S	56?	51	S–Gd	126 ± 4	114
Mg–Cl	76 ± 3	69	S–Ho	102 ± 4	114
Mg–Br	75 ± 23	51	S–Lu	121 ± 4	114

Table 2–9 (continued)
STRENGTHS OF CHEMICAL BONDS

Bond Strengths in Diatomic Molecules (continued)

Molecule	kcal mol^{-1}	Ref.	Molecule	kcal mol^{-1}	Ref.
S–Au	100 ± 6	58	Mn–Au	44 ± 3	115
S–Hg	51	96	Fe–Fe	24 ± 5	94
S–Pb	82.7 ± 0.4	119	Fe–Ge	50 ± 7	83
S–Bi	75.4 ± 1.1	119	Fe–Br	59 ± 23	51
S–U	135 ± 2	25	Fe–Au	45 ± 4	79
Cl–Cl	58.066 ± 0.001	93	Co–Co	40 ± 6	82
Cl–K	101.3 ± 0.5	18	Co–Cu	39 ± 5	84
Cl–Ca	95 ± 3	69	Co–Ge	57 ± 6	83
Cl–Sc	79	111	Co–Au	51 ± 3	79
Cl–Ti	26 ± 2	51	Ni–Ni	55.5 ± 5	77
Cl–Cr	87.5 ± 5.8	51	Ni–Cu	48 ± 5	84
Cl–Mn	86.2 ± 2.3	51	Ni–Ge	67.3 ± 4	51
Cl–Fe	84?	51	Ni–Br	86 ± 3	51
Cl–Ni	89 ± 5	51	Ni–I	70 ± 5	51
Cl–Cu	84 ± 6	51	Ni–Au	59 ± 5	79
Cl–Zn	54.7 ± 4.7	31	Cu–Cu	46.6 ± 2.2	2
Cl–Ga	114.5	7	Cu–Ge	49 ± 5	83
Cl–Ge	82?	51	Cu–Se	70 ± 9	51
Cl–Br	52.3 ± 0.2	51	Cu–Br	79 ± 6	51
Cl–Rb	100.7 ± 1	18	Cu–Ag	41.6 ± 2.2	2
Cl–Sr	97 ± 3	69	Cu–Sn	42.3 ± 4	1
Cl–Y	82 ± 23	51	Cu–Te	42 ± 9	51
Cl–Ag	75 ± 9	51	Cu–I	47?	51
Cl–Cd	49.9	22	Cu–Au	55.4 ± 2.2	2
Cl–In	103.3	7	Zn–Zn	7	112
Cl–Sn	75?	51	Zn–Se	33 ± 3	39
Cl–Sb	86 ± 12	51	Zn–Te	49?	51
Cl–I	50.5 ± 0.1	51	Zn–I	33 ± 7	51
Cl–Cs	106.2 ± 1	18	Ga–Ga	∿33	112
Cl–Ba	106 ± 3	69	Ga–As	50.1 ± 0.3	40
Cl–Au	82 ± 2	51	Ga–Br	101 ± 4	51
Cl–Hg	24 ± 2	51	Ga–Ag	43	24
Cl–Tl	89.0 ± 0.5	15	Ga–Te	60 ± 6	121
Cl–Pb	72 ± 7	51	Ga–I	81 ± 2	51
Cl–Bi	72 ± 1	32	Ga–Au	51 ± 23	51
Cl–Ra	82 ± 18	51	Ge–Ge	65.8 ± 3	78
Ar–Ar	0.2	117	Ge–Se	114 ± 5	51
K–K	12.8	112	Ge–Br	61 ± 7	51
K–Br	90.9 ± 0.5	18	Ge–Te	93 ± 2	51
K–I	76.8 ± 0.5	18	Ge–Au	70 ± 23	51
Ca–I	70 ± 23	51	As–As	91.7	51
Ca–Au	18	109	As–Se	23	103
Sc–Sc	25.9 ± 5	129	Se–Se	79.5 ± 0.1	119
Ti–Ti	34 ± 5	80	Se–Cd	75?	51
V–V	58 ± 5	80	Se–In	59 ± 4	28
Cr–Cr	<37	127	Se–Sn	95.9 ± 1.4	27
Cr–Cu	37 ± 5	84	Se–Te	64 ± 2	44
Cr–Ge	41 ± 7	83	Se–La	114 + 4	13
Cr–Br	78.4 ± 5.8	51	Se–Nd	92 + 4	13
Cr–I	68.6 ± 5.8	51	Se–Eu	72 ± 4	13
Cr–Au	51.3 ± 3.5	3	Se–Gd	103 ± 4	13
Mn–Mn	4 ± 3	81	Se–Ho	80 ± 4	13
Mn–Se	48 ± 3	131	Se–Lu	100 + 4	13
Mn–Br	75.1 ± 23	51	Se–Pb	72.4 ± 1	119
Mn–I	67.6 ± 2.3	51	Se–Bi	67.0 ± 1.5	119

Table 2–9 (continued)
STRENGTHS OF CHEMICAL BONDS

Bond Strengths in Diatomic Molecules (continued)

Molecule	kcal mol^{-1}	Ref.	Molecule	kcal mol^{-1}	Ref.
Br–Br	46.336 ± 0.001	93	Te–Te	63.2 ± 0.2	119
Br–Rb	90.4 ± 1	18	Te–La	91 ± 4	13
Br–Ag	70 ± 7	51	Te–Nd	73 ± 4	13
Br–Cd	38?	51	Te–Eu	58 ± 4	13
Br–In	93	7	Te–Gd	82 ± 4	13
Br–Sn	47 ± 23	51	Te–Ho	62 ± 4	13
Br–Sb	75 ± 14	51	Te–Lu	78 ± 4	13
Br–I	42.8 ± 0.1	51	Te–Au	59 ± 16	51
Br–Cs	96.5 ± 1	18	Te–Pb	60 ± 3	119
Br–Hg	17.3	132	Te–Bi	56 ± 3	119
Br–Tl	79.8 ± 0.4	15	I–I	36.460 ± 0.002	93
Br–Pb	59 ± 9	51	I–Cs	82.4 ± 1	18
Br–Bi	63.9 ± 1	33	I–Hg	9	132
Rb–Rb	12.2	112	I–Tl	65 ± 2	14
Rb–I	76.7 ± 1	18	I–Pb	47 ± 9	51
Sr–Au	63 ± 23	51	I–Bi	52 ± 1	34
Y–Y	38.3	128	Xe–Xe	∿0.7	110
Y–La	48.3	128	Cs–Cs	11.3	112
Pd–Pd	33?	51	Ba–Au	38 ± 14	51
Pd–Au	34.2 ± 5	3	La–La	58.6	128
Ag–Ag	41 ± 2	51	La–Au	80 ± 5	60
Ag–Sn	32.5 ± 5	1	Ce–Ce	66 ± 1	6
Ag–Te	70 ± 23	51	Ce–Au	76 ± 4	60
Ag–I	56 ± 7	51	Pr–Au	74 ± 5	60
Ag–Au	48.5 ± 2.2	2	Nd–Au	70 ± 6	60
Cd–Cd	2.7 ± 0.2	51	Au–Au	52.4 ± 2.2	2
Cd–I	33 ± 5	51	Au–Pb	31 ± 23	51
In–In	23.3 ± 2.5	37	Au–U	76 ± 7	59
In–Sb	36.3 ± 2.5	37	Hg–Hg	4.1 ± 0.5	51
In–Te	52 ± 4	28	Hg–Tl	1	66
In–I	80	7	Tl–Tl	15?	45
Sn–Sn	46.7 ± 4	1	Pb–Pb	24 ± 5	45
Sn–Te	76 ± 1	51	Pb–Bi	32 ± 5	42
Sn–Au	58.4 ± 4	1	Bi–Bi	45 ± 2	42
Sb–Sb	71.5 ± 1.5	37	Po–Po	44.4 ± 2.3	51
Sb–Te	61 ± 4	105	At–At	19	45
Sb–Bi	60 ± 1	90	Th–Th	≤69	55

From Kerr, J. A., Parsonage, M. J., and Trotman-Dickenson, A. F., in *Handbook of Chemistry and Physics*, 55th ed., Weast, R. C., Ed., CRC Press, Cleveland, 1974, F-204.

REFERENCES

1. **Ackerman, M., Drowart, J., Stafford, F. E., and Verhaegen, G.,** *J. Chem. Phys.,* 36, 1557, 1962.
2. **Ackerman, M., Stafford, F. E., and Drowart, J.,** *J. Chem. Phys.,* 33, 1784, 1960.
3. **Ackerman, M., Stafford, F. E., and Verhaegen, G.,** *J. Chem. Phys.,* 36, 1560, 1962.
4. **Ackermann, R. J., Faircloth, R. F., Rauh, E. G., and Thorn, R. J.,** *J. Inorg. Nucl. Chem.,* 28, 111, 1966.
5. **Armstrong, D. R., Marantz, S., and Coyle, C. E.,** *J. Am. Chem. Soc.,* 81, 3798, 1959.
6. **Balducci, G., De Maria, G., and Guido, M.,** *J. Chem. Phys.,* 50, 5424, 1969.
7. **Balfour, W. J. and Whitlock, R. E.,** *J. Chem. Soc. Sect. D.,* p. 1231, 1971.
8. **Barrow, R. E.,** *Trans. Faraday Soc.,* 56, 952, 1960.

Table 2–9 (continued)
STRENGTHS OF CHEMICAL BONDS

Bond Strengths in Diatomic Molecules (continued)

9. Barrow, R. E., *Nature*, 189, 480, 1961.
10. Barrow, R. F. and Caunt, A. D., *Proc. R. Soc.*, *Ser. A*, 219, 120, 1953.
11. Barrow, R. F., and Deutsch, E. W., *Proc. Chem. Soc.*, p. 122, 1960.
12. Barrow, R. F., Gissane, W. J. M., and Travis, D. N., *Proc. R. Soc.*, *Ser. A*, 287, 240, 1965.
13. Barrow, R. F. and Travis, D. N., *Proc. R. Soc.*, *Ser. A*, 273, 133, 1963.
14. Bergman, C., Coppens, P., Drowart, J., and Smoes, S., *Trans. Faraday Soc.*, 66, 800, 1970.
15. Berkowitz, J. and Chupka, W. A., *J. Chem. Phys.*, 45, 1287, 1966.
16. Berkowitz, J. and Walter, T., *J. Chem. Phys.*, 49, 1184, 1968.
17. Besenbruch, G., Kana'an, A. S., and Margrave, J. L., *J. Phys. Chem.*, 69, 3174, 1965.
18. Blue, G. D., Green, J. W., Bautista, R. G., and Margrave, J. L., *J. Phys. Chem.*, 67, 877, 1963.
19. Brewer, L. and Brackett, E., *Chem. Rev.*, 61, 425, 1961.
20. Brewer, L. and Rosenblatt, G. M., *Adv. High Temp. Sci.*, 2, 1, 1969.
21. Brix, P. and Herzberg, G., *J. Chem. Phys.*, 21, 2240, 1953.
22. Broderson, P. H. and Sicre, J. E., *Z. Phys.*, 141, 515, 1955.
23. Bruner, B. L. and Corbett, J. D., *U.S.A.E.C.I.S.*, p. 739, 1963.
24. Bulewicz, E. M. and Sugden, T. M., *Trans. Faraday Soc.*, 52, 1475, 1956.
25. Carbonel, M. and Laffitte, M., *C.R. Acad. Sci., Paris, Ser. C*, 270, 2105, 1970.
26. Cater, E. D., Rauh, E. G., and Thorn, R. J., *J. Chem. Phys.*, 44, 3106, 1966.
27. Cater, E. D. and Steiger, R. P., *J. Phys. Chem.*, 72, 2231, 1968.
28. Cocke, D. L. and Gingerich, K. A., *J. Phys. Chem.*, 75, 3264, 1971.
29. Cocke, D. L. and Gingerich, K. A., *J. Phys. Chem.*, 76, 2332, 1972.
30. Colin, R. and Drowart, J., *Trans. Faraday Soc.*, 60, 673, 1964.
31. Colin, R. and Drowart, J., *Trans. Faraday Soc.*, 64, 2611, 1968.
32. Colin, R., Goldfinger, P., and Jeunehomme, M., *Trans. Faraday Soc.*, 60, 306, 1964.
33. Coppens, P., Smoes, S., and Drowart, J., *Trans. Faraday Soc.*, 63, 2140, 1967.
34. Corbett, J. D. and Lynde, R. A., *Inorg. Chem.*, 6, 2199, 1967.
35. Cubicciotti, D., *J. Phys. Chem.*, 71, 3066, 1967.
36. Cubicciotti, D., *Inorg. Chem.*, 7, 208, 1968.
37. Cubicciotti, D., *Inorg. Chem.*, 7, 211, 1968.
38. Davis, D. D. and Okabe, H., *J. Chem. Phys.*, 49, 5526, 1968.
39. De Corpo, J. J., Steiger, R. P., Franklin, J. L., and Margrave, J. L., *J. Chem. Phys.*, 53, 936, 1970.
40. De Maria, G., Drowart, J., and Inghram, M. G., *J. Chem. Phys.*, 31, 1076, 1959.
41. De Maria, G., Gingerich, K. A., Malaspina, L., and Piacente, V., *J. Chem. Phys.*, 44, 2531, 1966.
42. De Maria, G., Goldfinger, P., Malaspina, L., and Piacente, V., *Trans. Faraday Soc.*, 61, 2146, 1965.
43. De Maria, G., Malaspina, L., and Piacente, V., *J. Chem. Phys.*, 52, 1019, 1970.
44. De Maria, G., Malaspina, L., and Piacente, V., *J. Chem. Phys.*, 56, 1978, 1972.
45. Dixon, R. N. and Lambertson, H. M., *J. Mol. Spectrosc.*, 25, 12, 1968.
46. Drowart, J., *Phase Stabil. Metals Alloys*, p. 305, 1967.
47. Drowart, J., De Maria, G., and Inghram, M. G., *J. Chem. Phys.*, 29, 1015, 1958.
48. Drowart, J. and Goldfinger, P., *Q. Rev.* (London), 20, 545, 1966.
49. Drowart, J. and Honig, R. E., *J. Phys. Chem.*, 61, 980, 1957.
50. Drowart, J., Myers, C. E., Szwarc, R., Vander Auwera-Mahieu, A., and Uy, O. M., *J. Chem. Soc. Faraday Trans. II*, 68, 1749, 1972.
51. Edwards, J. G., Franklin, H. F., and Gilles, P. W., *J. Chem. Phys.*, 54, 545, 1971.
52. Ehlert, T. C., Hilmer, R. M., and Beauchamp, E. A., *J. Inorg. Nucl. Chem.*, 30, 3112, 1968.
53. Ehlert, T. C. and Margrave, J. L., *J. Chem. Phys.*, 41, 1066, 1964.
54. Faure, F. M., Mitchell, M. J., and Bartlett, R. W., *High Temp. Sci.*, 4, 181, 1972.
55. Fries, J. A. and Cater, E. D., Report COO-1182-27, supported under USAEC Contract AT(11-1)-1182.
56. Frost, D. C. and McDowell, C. A., *Proc. R. Soc.*, *Ser. A*, 236, 278, 1956.

Table 2–9 (continued)
STRENGTHS OF CHEMICAL BONDS

Bond Strengths in Diatomic Molecules (continued)

57. Fujishiro, S., *Trans. Jap. Inst. Metals*, 1, 125, 1960.
58. Gaydon, A. G., *Dissociation Energies and Spectra of Diatomic Molecules*, 3rd ed., Chapman and Hall, London, 1968.
59. Gingerich, K. A., *J. Phys. Chem.*, 68, 768, 1964.
60. Gingerich, K. A., *J. Chem. Phys.*, 47, 2192, 1967.
61. Gingerich, K. A., *J. Chem. Phys.*, 49, 19, 1968.
62. Gingerich, K. A., *High Temp. Sci.*, 1, 258, 1969.
63. Gingerich, K. A., *J. Chem. Phys.*, 50, 2255, 1969.
64. Gingerich, K. A., *J. Phys. Chem.*, 73, 2734, 1969.
65. Gingerich, K. A., *Chem. Commun.*, 10, 580, 1970.
66. Gingerich, K. A., *J. Chem. Phys.*, 54, 2646, 1971.
67. Gingerich, K. A., *J. Chem. Phys.*, 54, 3720, 1971.
68. Gingerich, K. A., *High Temp. Sci.*, 3, 415, 1971.
69. Gingerich, K. A., *J. Chem. Phys.*, 56, 4239, 1972.
70. Gingerich, K. A. and Blue, G. D., *J. Chem. Phys.*, 47, 5447, 1967.
71. Gingerich, K. A. and Cocke, D. L., *J. Chem. Soc. Chem. Commun.*, p. 536, 1972.
72. Gingerich, K. A. and Finkbeiner, H. C., *Proc. Rare Earth Res. Conf. 9th*, 2, 795, 1971.
73. Gingerich, K. A. and Finkbeiner, H. C., *J. Chem. Phys.*, 54, 2621, 1971.
74. Gingerich, K. A. and Piacente, V., *J. Chem. Phys.*, 54, 2498, 1971.
75. Ginter, D. S., Ginter, M. L., and Innes, K. K., *Astrophys. J.*, 139, 365, 1964.
76. Graeber, P. and Weil, K. G., *Ber. Bunsenges. Phys. Chem.*, 76, 417, 1972.
77. Guido, M., Gigli, G., and Balducci, G., *J. Chem. Phys.*, 57, 3731, 1972.
78. Gunn, S. R., *J. Phys. Chem.*, 68, 949, 1964.
79. Hagland, I., Kopp, I., and Auslund, N., *Ark. Fys.*, 32, 321, 1966.
80. Herman, R. and Herman, L., *J. Phys. Radium*, 24, 73, 1963.
81. Herzberg, G., *Molecular Spectra and Molecular Structure. I. Spectra of Diatomic Molecules*, 2nd ed., Van Nostrand, New York, 1950.
82. Herzberg, G., *J. Mol. Spectrosc.*, 33, 147, 1970.
83. Hildenbrand, D. L., *J. Chem. Phys.*, 48, 3657, 1968.
84. Hildenbrand, D. L., *J. Chem. Phys.*, 52, 5751, 1970.
85. Hildenbrand, D. L., *Chem. Phys. Lett.*, 15, 379, 1972.
86. Hildenbrand, D. L. and Murad, E., *J. Chem. Phys.*, 44, 1524, 1966.
87. Hildenbrand, D. L. and Murad, E., *J. Chem. Phys.*, 53, 3403, 1970.
88. Hildenbrand, D. L. and Theard, L. P., *J. Chem. Phys.*, 50, 5350, 1969.
89. Hurley, A. C., *Proc. R. Soc., Ser. A*, 261, 237, 1961.
90. Johns, J. W. C. and Ramsey, D. A., *Can. J. Phys.*, 39, 210, 1961.
91. Johnston, H. S. and Woolfolk, R., *J. Chem. Phys.*, 41, 269, 1964.
92. Kant, A., *J. Chem. Phys.*, 41, 1872, 1964.
93. Kant, A., *J. Chem. Phys.*, 49, 5144, 1968.
94. Kant, A. and Lin, S-S., *J. Chem. Phys.*, 51, 1644, 1969.
95. Kant, A., Lin, S-S., and Strauss, B., *J. Chem. Phys.*, 49, 1983, 1968.
96. Kant, A. and Strauss, B. H., *J. Chem. Phys.*, 41, 3806, 1964.
97. Kant, A. and Strauss, B., *J. Chem. Phys.*, 49, 3579, 1968.
98. Kant, A., Strauss, B., and Lin, S-S., *J. Chem. Phys.*, 52, 2384, 1970.
99. Kent, R. A., *J. Am. Chem. Soc.*, 90, 5657, 1968.
100. Kent, R. A., Ehlert, T. C., and Margrave, J. L., *J. Am. Chem. Soc.*, 86, 5090, 1964.
101. Kent, R. A., McDonald, J. D., and Margrave, J. L., *J. Phys. Chem.*, 70, 874, 1966.
102. Kent, R. A. and Margrave, J. L., *J. Am. Chem. Soc.*, 87, 5382, 1965.
103. Klynning, L., Lundgrew, B., and Auslund, N., *Ark. Fys.*, 30, 141, 1965.
104. Kohl, F. J. and Carlson, K. D., *J. Am. Chem. Soc.*, 90, 4814, 1968.
105. Kordis, J. and Gingerich, K. A., *J. Phys. Chem.*, 76, 2336, 1972.
106. Kronekvist, M. and Lagerqvist, A., *Ark. Fys.*, 39, 133, 1969.
107. Kronekvist, M., Lagerqvist, A., and Neuhaus, H., *J. Mol. Spectrosc.*, 39, 516, 1971.
108. LeRoy, R. J., *J. Chem. Phys.*, 57, 573, 1972.
109. LeRoy, R. J. and Bernstein, R. B., *Chem. Phys. Lett.*, 5, 42, 1970.
110. Lin, S-S. and Kant, A., *J. Phys. Chem.*, 73, 2450, 1969.
111. Liu, B., *Phys. Rev. Lett.*, 27, 1251, 1971.

Table 2–9 (continued)
STRENGTHS OF CHEMICAL BONDS

Bond Strengths in Diatomic Molecules (continued)

112. McIntyre, N. S., Vander Auwera-Mahieu, A., and Drowart, J., *Trans. Faraday Soc.*, 64, 3006, 1968.
113. Marquart, J. R. and Berkowitz, J., *J. Chem. Phys.*, 39, 283, 1963.
114. Milton, E. R. V., Dunford, H. B., and Douglas, A. E., *J. Chem. Phys.*, 35, 1202, 1961.
115. Mole, M. F. and McCoubrey, J. C., *Nature*, 202, 450, 1964.
116. Muenow, D. W., Hastie, J. W., Hauge, R., Bautista, R., and Margrave, J. L., *Trans. Faraday Soc.*, 65, 3210, 1969.
117. Murad, E., Hildenbrand, D. L., and Main, R. P., *J. Chem. Phys.*, 45, 263, 1966.
118. Neckel, A. and Sodeck, G., *Monatsh. Chem.*, 103, 367, 1972.
119. O'Hare, P. A. G., *J. Chem. Phys.*, 52, 2992, 1970.
120. Ovcharenko, I. E., Ya. Kuzyankov, Y., and Tatevaskii, V. M., *Opt. Spektrosk.*, 19, 528, 1965.
121. Peeters, R., Vander Auwera-Mahieu, A., and Drowart, J., *Z. Naturforsch., Teil A*, 26, 327, 1971.
122. Pelevin, O. V., Mil'vidskii, M. G., Belyaev, A. I., and Khotin, B. A., *Izv. Akad. Nauk SSSR Neorgan. Mat.*, 2, 942, 1966.
123. Phillips, L. F., *Can. J. Chem.*, 46, 1429, 1968.
124. Piacente, V. and Desideri, A., *J. Chem. Phys.*, 57, 2213, 1972.
125. Piacente, V. and Gingerich, K. A., *High Temp. Sci.*, 4, 312, 1972.
126. Piacente, V. and Malaspina, L., *J. Chem. Phys.*, 56, 1780, 1972.
127. Porter, R. F. and Spencer, C. W. J., *J. Chem. Phys.*, 32, 943, 1960.
128. Ringstrom, U., *Ark. Fys.*, 27, 227, 1964.
129. Ringstrom, U., *J. Mol. Spectrosc.*, 36, 232, 1970.
130. Ringstrom, U. and Auslund, N., *Ark. Fys.*, 32, 19, 1966.
131. Schultz, J., *Compt. Rend.*, 252, 1750, 1961.
132. Shardanand, *Phys. Rev.*, 160, 67, 1967.
133. Shenyavskaya, E. A., Mal'tsev, A. A., Kataev, D. I., and Gurvich, L. V., *Opt. Spektrosk.*, 26, 937, 1969.
134. Siegel, B., *Q. Rev.* (London), 19, 77, 1965.
135. Smith, P. K. and Peterson, D. E., U.S.A.E.C. 1968 DM-MS-67-110.
136. Smoes, S., Coppens, P., Bergman, C., and Drowart, J., *Trans. Faraday Soc.*, 65, 682, 1969.
137. Smoes, S. and Drowart, J., *Chem. Commun.*, 534, 1968.
138. Smoes, S., Huguet, R., and Drowart, J., *Z. Naturforsch., Teil A.*, 26, 1934, 1971.
139. Smoes, S., Myers, C. E., and Drowart, J., *Chem. Phys. Lett.*, 8, 10, 1971.
140. Stearns, C. A. and Kohl, F. J., *High Temp. Sci.*, 2, 146, 1970; NASA Tech. Note 1969 NASA-TN-D-5027.
141. Torshina, V. V., Smolina, G. N., and Dobychin, S. L., *Zh. Prikl. Khim.*, 39, 1468, 1966.
142. Uy, O. M. and Drowart, J., *Trans. Faraday Soc.*, 65, 3221, 1969.
143. Uy, O. M. and Drowart, J., *High Temp. Sci.*, 2, 293, 1970.
144. Uy, O. M., Muenow, D. W., Ficalora, P. J., and Margrave, J. L., *Trans. Faraday Soc.*, 64, 2998, 1968.
145. Vander Auwera-Mahieu, A. and Drowart, J., *Chem. Phys. Lett.*, 1, 311, 1967.
146. Vander Auwera-Mahieu, A., McIntyre, N. S., and Drowart, J., *Chem. Phys. Lett.*, 4, 198, 1969.
147. Vander Auwera-Mahieu, A., Peeters, R., McIntyre, N. S., and Drowart, J., *Trans. Faraday Soc.*, 66, 809, 1970.
148. Velasco, R., *Can. J. Phys.*, 35, 1204, 1957.
149. Velasco, R., Ottinger, C., and Zare, R. N., *J. Chem. Phys.*, 51, 5522, 1969.
150. Verhaegen, G., Ph.D. thesis, University of Brussels, 1965.
151. Verhaegen, G., Smoes, S., and Drowart, J., U.S.A.E.C. WADD TR60-782, Part 19, 1963.
152. Verhaegen, G., Smoes, S., and Drowart, J., *J. Chem. Phys.*, 40, 239, 1964.
153. Wiedemeier, H. and Gilles, P. W., *J. Chem. Phys.*, 42, 2765, 1965.
154. Wiedemeier, H. and Goyette, W. J., *J. Chem. Phys.*, 48, 2936, 1968.
155. von Wieland, K., *Z. Elektrochem.*, 64, 761, 1960.

Table 2–9 (continued)
STRENGTHS OF CHEMICAL BONDS

Bond Strengths in Diatomic Molecules (continued)

156. Wu, Y. H. and Wahlbeck, P. G., *J. Chem. Phys.*, 56, 4534, 1972.
157. Zmbov, K. F., Hastie, J. W., and Margrave, J. L., *Trans. Faraday Soc.*, 64, 861, 1968.
158. Zmbov, K. F. and Margrave, J. L., *J. Chem. Phys.*, 45, 3167, 1966.
159. Zmbov, K. F. and Margrave, J. L., *J. Inorg. Nucl. Chem.*, 29, 59, 1976.
160. Zmbov, K. F. and Margrave, J. L., *J. Phys. Chem.*, 47, 3122, 1967.
161. Zmbov, K. F. and Margrave, J. L., *J. Phys. Chem.*, 71, 2893, 1967.

Table 2–10
STRENGTHS OF CHEMICAL BONDS

Heats of Formation of Gaseous Atoms from Elements in Their Standard States

For elements that are diatomic gases (g) in their standard states, these values are readily obtained from the bond strength; for elements that are crystalline (c) in their standard states, they are derived from vapor pressure data. All values are given in kcal mol^{-1} at 298°K.

Element	kcal mol^{-1}	Ref.	Element	kcal mol^{-1}	Ref.
H_2 (g)	52.104 ± 0.001	1	Zr(c)	145.5 ± 1	2
Li(c)	38.6 ± 0.4	2	Nb(c)	172.4 ± 1	2
Be(c)	77.5 ± 1.5	2	Mo(c)	157.3 ± 0.5	2
B(c)	136.5 ± 3	2	Ru(c)	155.5 ± 1.5	2
C(c)	171.29 ± 0.11	1	Rh(c)	133 ± 1	2
N_2 (g)	112.97 ± 0.10	1	Pd(c)	90.0 ± 0.5	2
O_2 (g)	59.553 ± 0.024	1	Ag(c)	67.9 ± 0.2	2
F_2 (g)	18.9	2	Cd(c)	26.72 ± 0.15	2
Na(c)	25.85 ± 0.15	2	In(c)	58 ± 1	2
Mg(c)	35.0 ± 0.3	2	Sn(c)	72.2 ± 0.5	2
Al(c)	78.7 ± 0.5	2	Sb(c)	63.2 ± 0.6	2
Si(c)	108.9 ± 1	2	Te(c)	47.0 ± 0.5	2
P(c) yellow	79.4 ± 10	2	I_2 (c)	25.517 ± 0.010	1
S(c)	66.2 ± 2	2	Cs(c)	18.7 ± 0.1	2
Cl_2 (g)	28.989 ± 0.002	1	Ba(c)	42.5 ± 1	2
K(c)	21.42 ± 0.05	2	Ce(c)	101 ± 3	2
Ca(c)	42.6 ± 0.4	2	Sm(c)	49.4 ± 0.5	2
Sc(c)	90.3 ± 1	2	Er(c)	75.8 ± 1	2
Ti(c)	112.3 ± 0.5	2	Yb	36.35 ± 0.2	2
V(c)	122.9 ± 0.3	2	Hf(c)	148 ± 1	2
Cr(c)	95 ± 1	2	Ta(c)	186.9 ± 0.6	2
Mn(c)	67.7 ± 1	2	W(c)	203.1 ± 1	2
Fe(c)	99.3 ± 0.3	2	Re(c)	185 ± 1.5	2
Co(c)	102.4 ± 1	2	Os(c)	188 ± 1.5	2
Ni(c)	102.8 ± 0.5	2	Ir(c)	160 ± 1	2
Cu(c)	80.7 ± 0.3	2	Pt(c)	135.2 ± 0.3	2
Zn(c)	31.25 ± 0.1	2	Au(c)	88.0 ± 0.5	2
Ga(c)	65.4 ± 0.5	2	Hg(c)	14.69 ± 0.03	2
Ge(c)	89.5 ± 0.5	2	Tl(c)	43.55 ± 0.1	2
As(c)	72.3 ± 3	2	Pb(c)	46.62 ± 0.3	2
Se(c)	54.3 ± 1	2	Bi(c)	50.1 ± 0.5	2
Br_2 (e)	26.730 ± 0.029	1	U(c)	126 ± 3	2
Rb(c)	19.6 ± 0.1	2	Th(c)	137.5 ± 0.5	2
Sr(c)	39.1 ± 0.5	2	Pu(c)	87.1 ± 4	2
Y(c)	101.5 ± 0.5	2			

Table 2–10 (continued)
STRENGTHS OF CHEMICAL BONDS

Heats of Formation of Gaseous Atoms from Elements in Their Standard States (continued)

From Kerr, J. A., Parsonage, M. J., and Trotman-Dickenson, A. F., in *Handbook of Chemistry and Physics,* 55th ed., Weast, R. C., Ed., CRC Press, Cleveland, 1974, F-212.

REFERENCES

1. *Tentative Set of Key Values for Thermodynamics – Part 1*, International Council of Scientific Unions, Committee on Data for Science and Technology, CODATA Bulletin No. 2, November 1970.
2. **Brewer, L. and Rosenblatt, G. M.,** *Adv. High Temp. Chem.,* 2, 1, 1969.

Table 2–11
STRENGTHS OF CHEMICAL BONDS

Bond Strengths in Polyatomic Molecules

The values below refer to a temperature of 298°K and have mostly been determined by kinetic methods (see References 8 and 44 for a full discussion of the methods).

Some have been calculated from the heats of formation of the species involved according to the equations:

$$D(R-X) = \Delta H_f^{\circ}(\dot{R}) + \Delta H_f^{\circ}(\dot{X}) - \Delta H_f^{\circ}(RX)$$

or

$$D(R-R) = 2\Delta H_f^{\circ}(\dot{R}) - \Delta H_f^{\circ}(RR)$$

The sources of the data on the heats of formation are given in the references.

An attempt has been made to list all the important values obtained by methods that are considered to be valid. The references are intended to serve as a guide to the literature.

Bond	kcal mol^{-1}	Ref.
H–CH	102 ± 2	15, 71
H–CH$_2$	110 ± 2	15, 71
H–CH$_3$	104 ± 1	39
H-ethynyl	128 ± 5	79
H-vinyl	≥ 108 ± 2	39
H–C$_2$H$_5$	98 ± 1	39
H–propargyl	93.9 ± 1.2	76
H–allyl	89 ± 1	39
H–cyclopropyl	100.7 ± 1	27
H–*n*-C$_3$H$_7$	98 ± 1	39
H–*i*-C$_3$H$_7$	95 ± 1	39
H–cyclobutyl	96.5 ± 1	27, 56
H–cyclopropylcarbinyl	97.4 ± 1.6	55
H–methallyl	83 ± 1	39
H–*s*-C$_4$H$_9$	95 ± 1	39
H–*t*-C$_4$H$_9$	92 ± 1	39
H–cyclopentadien-1,3-yl–5	81.2 ± 1.2	37
H–pentadien–1,4-yl-3	80 ± 1	39

Table 2–11 (continued)
STRENGTHS OF CHEMICAL BONDS

Bond Strengths in Polyatomic Molecules (continued)

Bond	kcal mol^{-1}	Ref.
H–cyclopentenyl-3	82.3 ± 1	36
H–spiropentyl	98.8 ± 1	27
H–cyclopentyl	94.5 ± 1	27, 35
H–dimethylallyl	77.2 ± 1.5	72
H–neo-C_5H_{11}	100.3 ± 1	51
H–C_6H_5	110.2 ± 2.0	14
H–cyclohexadieny-1,3-yl-5	70 ± 5	42
H–cyclohexyl	95.5 ± 1	27
H–$CH_2C_6H_5$	85 ± 1	39
H–cyclohcptatrien-1,3-yl-7	73.4	73
H–norbornyl	96.7 ± 2.5	57
H–cycloheptyl	92.5 ± 1	27
H–CN	120 ± 1	23
H–CHO	87 ± 1	39
H–CH_2OH	94 ± 2	39
H–$COCH_3$	86.0 ± 0.8	24
H–CH_2OCH_3	93 ± 1	39
H–$CH(CH_3)OH$	93.0 ± 1.0	1
H–$COCH:CH_2$	87.1 ± 1.0	2
H–CH_2COCH_3	98.3 ± 1.8	46, 69
H–COC_2H_5	87.4 ± 1.0	78
H–$CH(OH)CH:CH_2$	81.6 ± 1.8	4
H–$C(CH_3)_2OH$	91 ± 1	39
H–tetrahydrofuran-2-yl	92 ± 1	39
H–$CH(CH_3)COCH_3$	92.3 ± 1.4	70
H–COC_6H_5	86.9 ± 1	68
H–$COOCH_3$	92.7 ± 1	66
H–$CH_2OCOC_6H_5$	100.2 ± 1.3	67
H–$COCF_3$	91.0 ± 2	5
H–CH_2F	101 ± 2	45
H–CHF_2	101 ± 2	45
H–CF_3	106 ± 1	39
H–CF_2Cl	104 ± 1	54
H–CH_2Cl	100.9	34
H–$CHCl_2$	99.0	34
H–CCl_3	95.7 ± 1	44
H–CH_2Br	102.0	34
H–$CHBr_2$	103.7	34
H–CBr_3	96.0 ± 1.6	47
H–CH_2I	103 ± 2	39
H–CHI_2	103 ± 2	39
H–C_2F_5	103.1 ± 1.5	7
H–CCl_2CHCl_2	94 ± 2	33, 71
H–C_2Cl_5	95 ± 2	32, 71
H–n-C_3F_7	104 ± 2	44
H–$CHClCH:CH_2$	88.6	3
H–NH_2	110 ± 2	40
H–$NHCH_3$	103 ± 2	40
H–$N(CH_3)_2$	95 ± 2	40
H–NHC_6H_5	80 ± 3	10, 44
H–$N(CH_3)C_6H_5$	74 ± 3	10, 44
H–NO	⩽ 49.5	16
H–NF_2	75.7 ± 2.5	59, 71
H–N_3	85	41

Table 2—11 (continued)
STRENGTHS OF CHEMICAL BONDS

Bond Strengths in Polyatomic Molecules (continued)

Bond	kcal mol^{-1}	Ref.
H–OH	119 ± 1	44
H–OCH$_3$	103.6 ± 1	6, 10, 19, 71
H–OC$_2$H$_5$	103.9 ± 1	6, 10, 19, 71
H–OC(CH$_3$)$_3$	104.7 ± 1	6, 10, 19, 71
H–OC$_6$H$_5$	88 ± 5	13
H–O$_2$H	90 ± 2	44
H–O$_2$CCH$_3$	112 ± 4	44
H–O$_2$CC$_2$H$_5$	110 ± 4	44
H–O$_2$Cn-C$_3$H$_7$	103 ± 4	44
H–ONO	78.3 ± 0.5	9, 71
H–ONO$_2$	101.2 ± 0.5	9, 71
H–SH	90 ± 2	44
H–SCH$_3$	$\geqslant 88$	44
H–SiH$_3$	94 ± 3	64
H–Si(CH$_3$)$_3$	90 ± 3	77
BH$_3$–BH$_3$	35	10
HC≡CH	230 ± 2	15, 71
H$_2$C=CH$_2$	172 ± 2	15, 71
H$_3$C–CH$_3$	88 ± 2	44
CH$_3$–C(CH$_3$)$_2$CH:CH$_2$	69.4	10
C$_6$H$_5$CH$_2$–C$_2$H$_5$	69 ± 2	44
C$_6$H$_5$CH(CH$_3$)–CH$_3$	71	10
C$_6$H$_5$CH$_2$–n-C$_3$H$_7$	67 ± 2	44
CH$_3$–CH$_2$CN	72.7 ± 2	10, 44
CH$_3$–C(CH$_3$)$_2$CN	70.2 ± 2	10, 44
C$_6$H$_5$C(CH$_3$)(CN)–CH$_3$	59.9	10
NC–CN	128 ± 1	23
C$_6$H$_5$CH$_2$CO–CH$_2$C$_6$H$_5$	65.4	10
C$_6$H$_5$CO–CF$_3$	73.8	10
CH$_3$CO–COCH$_3$	67.4 ± 2.3	48
C$_6$H$_5$CH$_2$–COOH	68.1	10
C$_6$H$_5$CH$_2$–O$_2$CCH$_3$	67	10
C$_6$H$_5$CO–COC$_6$H$_5$	66.4	10
C$_6$H$_5$CH$_2$–O$_2$CC$_6$H$_5$	69	10
(C$_6$H$_5$CH$_2$)$_2$CH–COOH	59.4	10
CH$_2$F–CH$_2$F	88 ± 2	45
CF$_2$=CF$_2$	76.3 ± 3	80
CF$_3$–CF$_3$	96.9 ± 2	17
C$_6$H$_5$CH$_2$–NH$_2$	71.9 ± 1	40
C$_6$H$_5$NH–CH$_3$	67.7	10
C$_6$H$_5$CH$_2$–NHCH$_3$	68.7 ± 1	40
C$_6$H$_5$N(CH$_3$)–CH$_3$	65.2	10
C$_6$H$_5$CH$_2$–N(CH$_3$)$_2$	60.9 ± 1	40
CF$_3$–NF$_2$	65 ± 2.5	71, 74
CH$_2$=N$_2$	$\leqslant 41.7 \pm 1$	52
CH$_3$N:N–CH$_3$	52.5	10, 58
C$_2$H$_5$N:N–C$_2$H$_5$	50.0	10, 58
i-C$_3$H$_7$N:N-i-C$_3$H$_7$	47.5	10, 58
n-C$_4$H$_9$N:N-n-C$_4$H$_9$	50.0	10, 58
i-C$_4$H$_9$N:N-i-C$_4$H$_9$	49.0	10, 58
s-C$_4$H$_9$N:N-s-C$_4$H$_9$	46.7	10, 58
t-C$_4$H$_9$N:N-t-C$_4$H$_9$	43.5	10, 58

Table 2–11 (continued)
STRENGTHS OF CHEMICAL BONDS

Bond Strengths in Polyatomic Molecules (continued)

Bond	kcal mol^{-1}	Ref.
$C_6H_5CH_2N{:}N{-}CH_2C_6H_5$	37.6	10, 58
$CF_3N{:}N{-}CF_3$	55.2	10, 58
$C_2H_5{-}NO_2$	62	65, 71
$O{=}CO$	127.2 ± 0.1	21
$CH_3{-}O_2SCH_3$	66.8	10, 44
Allyl$-O_2SCH_3$	49.6	10, 44
$C_6H_5CH_2{-}O_2SCH_3$	52.9	10, 44
$C_6H_5S{-}CH_3$	60	44
$C_6H_5CH_2{-}SCH_3$	53.8	10
$F{-}CH_3$	103 ± 3	25, 44
$Cl{-}CN$	97 ± 1	23
$Cl{-}COC_6H_5$	74 ± 3	44
$Cl{-}CF_3$	86.1 ± 0.8	18
$Cl{-}CCl_2F$	73 ± 2	31
$Cl{-}C_2F_5$	82.7 ± 1.7	18
$Br{-}CH_3$	70.0 ± 1.2	26
$Br{-}CN$	83 ± 1	23
$Br{-}COC_6H_5$	64.2	10
$Br{-}CF_3$	70.6 ± 1.0	28, 30
$Br{-}CBr_3$	56.2 ± 1.8	47
$Br{-}C_2F_5$	68.7 ± 1.5	18, 29, 30
$Br{-}n{-}C_3F$	66.5 ± 2.5	18
$I{-}CH_3$	56.3 ± 1	44
$I{-}norbornyl$	62.5 ± 2.5	57
$I{-}CN$	73 ± 1	23
$I{-}CF_3$	53.5 ± 2	12
$CH_3{-}Ga(CH_3)_2$	59.5	10, 58
$CH_3{-}CdCH_3$	54.4	10, 58
$CH_3{-}HgCH_3$	57.5	49
$C_2H_5{-}HgC_2H_5$	43.7 ± 1	50
$n{-}C_3H_7{-}Hgn{-}C_3H_7$	47.1	10, 58
$i{-}C_3H_7{-}Hgi{-}C_3H_7$	40.7	10, 58
$C_6H_5{-}HgC_6H_5$	68	10, 58
$CH_3{-}Tl(CH_3)_2$	36.4 ± 0.6	63
$CH_3{-}Pb(CH_3)_3$	49.4 ± 1	38
$NH_2{-}NH_2$	70.8 ± 2	10, 58
$NH_2{-}NHCH_3$	64.8	10, 58
$NH_2{-}N(CH_3)_2$	62.7	10, 58
$NH_2{-}NHC_6H_5$	51.1	10, 58
$NO{-}NO_2$	9.5 ± 0.5	44
$NO_2{-}NO_2$	12.9 ± 0.5	44
$NF_2{-}NF_2$	21 ± 1	21
$O{-}N_2$	40	9, 71
$O{-}NO$	73	8, 71
$HO{-}N{:}CHCH_3$	49.7	10, 58
$Cl{-}NF_2$	$\sim$ 32	61, 71
$HO{-}OH$	51 ± 1	44
$CH_3O{-}OCH_3$	36.9 ± 1	10, 58
$HO{-}OC(CH_3)_3$	42.5	10, 11
$C_2H_5O{-}OC_2H_5$	37.3 ± 1.2	53
$n{-}C_3H_7O{-}On{-}C_3H_7$	37.2 ± 1	10, 58
$i{-}C_3H_7O{-}Oi{-}C_3H_7$	37.0 ± 1	43

Table 2–11 (continued)
STRENGTHS OF CHEMICAL BONDS

Bond Strengths in Polyatomic Molecules (continued)

Bond	kcal mol^{-1}	Ref.
s-C_4H_9O–Os-C_4H_9	36.4 ± 1	75
t-C_4H_9O–Ot-C_4H_9	37.4 ± 1	10, 44
$(CH_3)_3CCH_2O$–$OCH_2C(CH_3)_3$	36.4 ± 1	60
O–O_2ClF	58.4	10, 58
CH_3CO_2–O_2CCH_3	30.4 ± 2	10, 44
$C_2H_5CO_2$–$O_2CC_2H_5$	30.4 ± 2	10, 44
n-$C_3H_7CO_2$–O_2Cn-C_3H_7	30.4 ± 2	10, 44
O–SO	132 ± 2	21
F–OCF_3	43.5 ± 0.5	20
Cl–OH	60 ± 3	44
O–ClO	59 ± 3	21
Br–OH	56 ± 3	44
I–OH	56 ± 3	44
ClO_3–ClO_4	58.4	10, 58
$O{=}PF_3$	130 ± 5	44
$O{=}PCl_3$	122 ± 5	44
$O{=}PBr_3$	119 ± 5	44
SiH_3–SiH_3	81 ± 4	62, 64
$(CH_3)_3Si$–$Si(CH_3)_3$	80.5	22

From Kerr, J. A., Parsonage, M. J., and Trotman-Dickenson, A. F., in *Handbook of Chemistry and Physics,* 55th ed., Weast, R. C., Ed., CRC Press, Cleveland, 1974, F-213.

REFERENCES

1. **Alfassi, Z. B. and Golden, D. M.,** *J. Phys. Chem.,* 76, 3314, 1972.
2. **Alfassi, Z. B. and Golden, D. M.,** *J. Am. Chem. Soc.,* 95, 319, 1973.
3. **Alfassi, Z. B., Golden, D. M., and Benson, S. W.,** *Int. J. Chem. Kinet.,* 5, 155, 1973; Trenwith, A. B., *Int. J. Chem. Kinet.,* 5, 67, 1973.
4. **Alfassi, Z. B. and Golden, D. M.,** *Int. J. Chem. Kinet.,* 5, 295, 1973; Trenwith, A. B., *J. Chem. Soc. Faraday Trans. I,* 69, 1737, 1973.
5. **Amphlett, J. C. and Whittle, E.,** *Trans. Faraday Soc.,* 66, 2016, 1970.
6. **Baker, G., Littlefair J. H., Shaw, R., and Thynne, J. C. J.,** *J. Chem. Soc.,* p. 6970, 1965.
7. **Bassett J. E. and Whittle, E.,** *J. Chem. Soc. Faraday Trans. I,* 68, 492, 1972.
8. **Benson, S. W.,** *J. Chem. Educ.,* 42, 502, 1965.
9. **Benson, S. W.,** *Thermochemical Kinetics,* John Wiley and Sons, New York, 1968.
10. **Benson, S. W. and O'Neal, H. E.,** Kinetic Data on Gas Phase Unimolecular Reactions, NSRDS-NBS 21, 1970.
11. **Benson, S. W. and Spokes, G. N.,** *J. Phys. Chem.,* 72, 1182, 1968.
12. **Boyd, R. K., Downs, G. W., Gow, J. S., and Horrex, C.,** *J. Phys. Chem.,* 67, 719, 1963.
13. **Carson, A. S., Fine, D. H., Gray, P., and Laye, P. G.,** *J. Chem. Soc., Sect. B,* p. 1611, 1971.
14. **Chamberlain, G. A. and Whittle, E.,** *Trans. Faraday Soc.,* 67, 2077, 1971.
15. **Chupka, W. A. and Lifshitz, C.,** *J. Chem. Phys.,* 48, 1109, 1968.
16. **Clement, M. J. Y. and Ramsay, D. A.,** *Can. J. Phys.,* 39, 205, 1961.
17. **Coomber, J. W. and Whittle, E.,** *Trans. Faraday Soc.,* 63, 1394, 1967.

Table 2–11 (continued)
STRENGTHS OF CHEMICAL BONDS

Bond Strengths in Polyatomic Molecules (continued)

18. Coomber, J. W. and Whittle, E., *Trans. Faraday Soc.*, 63, 2656, 1967.
19. Cox, J. D. and Pilcher, G., *Thermochemistry of Organic and Organometallic Compounds*, Academic Press, New York, 1970.
20. Czarnarski, J., Castellano, E., and Shumacher, H. J., *Chem. Commun.*, p. 1255, 1968.
21. Darwent, B. de B., Bond Dissociation Energies in Simple Molecules, NSRDS-NBS, 31, 1970.
22. Davidson, I. M. T. and Howard, A. B., *J. Chem. Soc. Chem. Commun.*, p. 323, 1973.
23. Davis, D. D. and Okabe, H., *J. Chem. Phys.*, 49, 5526, 1968.
24. Devore, J. A. and O'Neal, H. E., *J. Phys. Chem.*, 73, 2644, 1969.
25. Dibeler, V. H. and Reese, R. M., *J. Res. Nat. Bur. Stand.*, 54, 127, 1955.
26. Ferguson, K. C., Okafo, E. N., and Whittle, E., *J. Chem. Soc. Faraday Trans. I*, 69, 295, 1973.
27. Ferguson, K. C. and Whittle, E., *Trans. Faraday Soc.*, 67, 2618, 1971; Jones, S. H. and Whittle, E., *Int. J. Chem. Kinet.*, 2, 479, 1970.
28. Ferguson, K. C. and Whittle, E., *J. Chem. Soc. Faraday Trans. I*, 68, 295, 1972.
29. Ferguson, K. C. and Whittle, E., *J. Chem. Soc. Faraday Trans. I*, 68, 306, 1972.
30. Ferguson, K. C. and Whittle, E., *J. Chem. Soc. Faraday Trans. I*, 68, 641, 1972.
31. Foon, R. and Tait, K. B., *J. Chem. Soc. Faraday Trans. I*, 68, 104, 1972.
32. Franklin, J. A., Huybrechts, G. H., and Cillien, C., *Trans. Faraday Soc.*, 65, 2094, 1969.
33. Franklin, J. A. and Huybrechts, G. H., *Int. J. Chem. Kinet.*, 1, 1, 1969.
34. Furuyama, S., Golden, D. M., and Benson, S. W., *J. Am. Chem. Soc.*, 91, 7564, 1969.
35. Furuyama, S., Golden, D. M., and Benson, S. W., *Int. J. Chem. Kinet.*, 2, 83, 1970.
36. Furuyama, S., Golden, D. M., and Benson, S. W., *Int. J. Chem. Kinet.*, 2, 93, 1970.
37. Furuyama, S., Golden, D. M., and Benson, S. W., *Int. J. Chem. Kinet.*, 3, 237, 1971.
38. Gilroy, K. M., Price, S. J., and Webster, N. J., *Can. J. Chem.*, 50, 2639, 1972.
39. Golden, D. M. and Benson, S. W., *Chem. Rev.*, 69, 125, 1969.
40. Golden, D. M., Solly, R. K., Gac, N. A., and Benson, S. W., *J. Am. Chem. Soc.*, 94, 363, 1972.
41. Gray, P., *Q. Rev.* (London), 17, 441, 1963.
42. James, D. G. L. and Suart, R. D., *Trans. Faraday Soc.*, 64, 2752, 1968.
43. Kerr, J. A., *Annu. Rep. Chem. Soc., Sect. A*, 64, 73, 1967.
44. Kerr, J. A., *Chem. Rev.*, 66, 465, 1966.
45. Kerr, J. A. and Timlin, D. M., *Int. J. Chem. Kinet.*, 3, 427, 1971.
46. King, K. D., Golden, D. M., and Benson, S. W., *J. Am. Chem. Soc.*, 92, 5541, 1970.
47. King, K. D., Golden, D. M., and Benson, S. W., *J. Phys. Chem.*, 75, 987, 1971.
48. Knoll, H., Scherker, K., and Geiseler, G., *Int. J. Chem. Kinet.*, 5, 271, 1973.
49. Kominar, R. J. and Price, S. J., *Can. J. Chem.*, 47, 991, 1969.
50. LaLonde, A. C. and Price, S. J. W., *Can. J. Chem.*, 49, 3367, 1971.
51. Larson, C. W., Hardwidge, E. A., and Rabinovitch, B. S., *J. Chem. Phys.*, 50, 2769, 1969.
52. Laufer, A. H. and Okabe, H., *J. Am. Chem. Soc.*, 93, 4137, 1971.
53. Leggett, C. and Thynne, J. C. J., *Trans. Faraday Soc.*, 63, 2504, 1967.

Table 2–11 (continued)
STRENGTHS OF CHEMICAL BONDS

Bond Strengths in Polyatomic Molecules (continued)

54.   Leyland, L. M., Majer, J. R., and Robb, J. C., *Trans. Faraday Soc.*, 66, 898, 1970.

55.   McMillen, D. F., Golden, D. M., and Benson, S. W., *Int. J. Chem. Kinet.*, 3, 359, 1971.

56.   McMillen, D. F., Golden, D. M., and Benson, S. W., *Int. J. Chem. Kinet.*, 4, 487, 1972.

57.   O'Neal, H. E., Bagg, J. W., and Richardson, W. H., *Int. J. Chem. Kinet.*, 2, 493, 1970.

58.   O'Neal, H. E. and Benson, S. W., in *Free Radicals*, Kochi, J. K., Ed., John Wiley and Sons, New York, 1973, 275.

59.   Pankratov, A. V., Zercheninov, A. N., Chesnokov, V. I., and Zhdanova, N. N., *Zh. Fiz. Khim.*, 43, 394, 1969.

60.   Perona, M. J. and Golden, D. M., *Int. J. Chem. Kinet.*, 5, 55, 1973.

61.   Petry, R. C., *J. Am. Chem. Soc.*, 89, 4600, 1967.

62.   Potzinger, P. and Lampe, F. W., *J. Phys. Chem.*, 74, 719, 1970.

63.   Price, S. J., Richard, J. P., Rufeldt, R. C., and Jacko, M. G., *Can. J. Chem.*, 51, 1397, 1973.

64.   Ring, M. A., Puentes, M. J., and O'Neal, H. E., *J. Am. Chem. Soc.*, 92, 4845, 1970; Steele, W. C. and Stone, F. G. A., *J. Am. Chem. Soc.*, 84, 3599, 1962; Steele, W. C., Nichols, L. D., and Stone, F. G. A., *J. Am. Chem. Soc.*, 84, 441, 1962.

65.   Shaw, R., *Int. J. Chem. Kinet.*, 5, 261, 1973.

66.   Solly, R. K. and Benson, S. W., *Int. J. Chem. Kinet.*, 1, 427, 1969.

67.   Solly, R. K. and Benson, S. W., *Int. J. Chem. Kinet.*, 3, 509, 1971.

68.   Solly, R. K. and Benson, S. W., *J. Am. Chem. Soc.*, 93, 1592, 1971.

69.   Solly, R. K., Golden, D. M., and Benson, S. W., *Int. J. Chem. Kinet.*, 2, 11, 1970.

70.   Solly, R. K., Golden, D. M., and Benson, S. W., *Int. J. Chem. Kinet.*, 2, 381, 1970.

71.   Thermochemical calculation.

72.   Trenwith, A. B., *Trans. Faraday Soc.*, 66, 2805, 1970.

73.   Vincow, G., Dauben, H. J., Hunter, F. R., and Volland, W. V., *J. Am. Chem. Soc.*, 91, 2823, 1969.

74.   Walker, L. C., *J. Chem. Thermodynamics*, 4, 219, 1972.

75.   Walker, R. F. and Phillips, L., *J. Chem. Soc., Sect. A*, p. 2103, 1968.

76.   Walsh, R., *Trans. Faraday Soc.*, 67, 2085, 1971.

77.   Walsh, R. and Wells, J. M., *J. Chem. Soc. Chem. Commun.*, 513, 1973.

78.   Watkins, K. W. and Thompson, W. W., *Int. J. Chem. Kinet.*, 5, 791, 1973.

79.   Wyatt, J. R. and Stafford, F. E., *J. Phys. Chem.*, 76, 1913, 1972.

80.   Zmbov, K. F., Uy, O. M., and Margrave, J. L., *J. Am. Chem. Soc.*, 90, 5090, 1968.

Table 2–12

BOND LENGTHS BETWEEN CARBON AND OTHER ELEMENTS

The following tables are based on bond distance determinations by experimental methods, mainly X-ray and electron diffraction, and include values published up to January 1, 1956. Values are given in Angstrom units. In the present tables, for the sake of completeness, individual values of bond distances of lower accuracy are quoted, with limits of error indicated where possible. Values for tungsten and bismuth should be treated with particular caution.

According to the statistical theory of errors, if an average quantity $\bar{\mu}$ and a standard deviation σ can be evaluated, there is a 95% probability that the true value lies within the interval $\bar{\mu} \pm 2\sigma$. Too much reliance should, however, not be placed on σ values in bond distance determinations, since the derivation of these certain sources of error may have been neglected.

Group	Bond type	Element						
I	All types	H[a] 1.056–1.115						
II		Be 1.93	Hg 2.07 ± 0.01					
III		B 1.56 ± 0.01	Al 2.24 ± 0.04	In 2.16 ± 0.04				
IV	All types	C[a] 1.54 –1.20	Si 1.865 ± 0.008	Ge 1.98 ± 0.03	Sn 2.143 ± 0.008	Pb 2.29 ± 0.05		
	Alkyls (CH_3XH_3)		Si 1.84 ± 0.01		Sn 2.18 ± 0.02			
	Aryl $(C_6H_5XH_3)$		Si 1.88 ± 0.01					
	Negatively substituted (CH_3XCl_3)							
V	All types	N[a] 1.47 –1.1	P 1.87 ± 0.02	As 1.98 ± 0.02	Sb 2.202 ± 0.016	Bi 2.30[b]		
	Paraffinic $(CH_3)_3X$							
VI		O[a] 1.43 –1.15	S[a] 1.81 –1.55	Cr 1.92 ± 0.04	Se 1.98 –1.71	Te 2.05 ± 0.14	Mo 2.08 ± 0.04	W 2.06 ± 0.01[b]

Table 2–12 (continued)
BOND LENGTHS BETWEEN CARBON AND OTHER ELEMENTS

Group	Bond type	Element			
		F	Cl	Br	I
VII	Paraffinic (monosubstituted) (CH_3X)	1.381 ± 0.005	1.767 ± 0.002	1.937 ± 0.003	$2.13_5 \pm 0.01$
	Paraffinic (disubstituted) (CH_2X_2)	1.334 ± 0.004	1.767 ± 0.002	1.937 ± 0.003	$2.13_5 \pm 0.1$
	Olefinic (CH_2:CHX)	$1.32_5 \pm 0.1$	1.72 ± 0.01	1.89 ± 0.01	2.092 ± 0.005
	Aromatic (C_6H_5X)	1.30 ± 0.01	1.70 ± 0.01	1.85 ± 0.01	2.05 ± 0.01
	Acetylenic (HC:CX)		1.635 ± 0.004	$1.79_5 \pm 0.01$	1.99 ± 0.02
		Fe	Co	Ni	Pd
VIII		1.84 ± 0.02	1.83 ± 0.02	1.82 ± 0.03	2.27 ± 0.04

[a]See following individual tables.
[b]Error uncertain.

Reproduced from International Tables for X-ray Crystallography.

Table 2–12 (continued)
BOND LENGTHS BETWEEN CARBON AND OTHER ELEMENTS

Carbon–Carbon

Single Bond

1. Paraffinic	1.541	± 0.003
2. In diamond (at 18°C)	1.54452	± 0.00014

Partial Double Bond

1. Shortening of a single bond in the presence of one carbon–carbon double bond, e.g., $(CH_3)_2.C:CH_2$, or of an aromatic ring, e.g., $C_6H_5.CH_3$	1.53	± 0.01
2. Shortening in the presence of two carbon–oxygen double bond, e.g., CH_3CHO	1.516	± 0.005
3. Shortening in th presence of two carbon–oxygen double bonds, e.g., $(CO_2H)_2$	1.49	± 0.01
4. Shortening in the presence of one carbon–carbon triple bond, e.g., $CH_3.C:CH$	1.460	± 0.003
5. In compounds with a tendency to dipole formation, e.g., $C:C.C:N$	1.44	± 0.01
6. In graphite (at 15°C)	1.4210	± 0.0001
7. In aromatic compounds	1.395	± 0.003
8. In the presence of two carbon–carbon triple bonds, e.g., $HC:C.C:CH$	1.373	± 0.004

Double Bond

1. Simple	1.337	± 0.006
2. Partial triple bond, e.g., $CH_2:C:CH_2$	1.309	± 0.005

Triple Bond

1. Simple, e.g., C_2H_2	1.204	± 0.002
2. Conjugated, e.g., $CH_3.(C:C)_2.H$	1.206	± 0.004

Carbon–Hydrogen

1. Paraffinic		
(a) in methane	1.091	
(b) in monosubstituted carbon	1.101	± 0.003
(c) in disubstituted carbon	1.073	± 0.004
(d) in trisubstituted carbon	1.070	± 0.007
2. Olefinic, e.g., $CH_2:CH_2$	1.07	± 0.01
3. Aromatic in C_6H_6	1.084	± 0.006
4. Acetylenic, e.g., $CH:C.X$	1.056	± 0.003
5. Shortening in the presence of a carbon triple bond, e.g., CH_3CN	1.115	± 0.004
6. In small rings, e.g., $(CH_2)_2S$	1.081	± 0.007

Carbon–Nitrogen

Single Bond

1. Paraffinic		
(a) 4-covalent nitrogen	1.479	± 0.005
(b) 3-covalent nitrogen	1.472	± 0.005
2. In C–N=, e.g., CH_3NO_2	1.475	± 0.010
3. Aromatic in $C_6H_5NHCOCH_3$	1.426	± 0.012

Table 2–12 (continued)
BOND LENGTHS BETWEEN CARBON AND OTHER ELEMENTS

Carbon–Nitrogen (continued)

Single Bond (continued)

4. Shortened (partial double bond) in heterocyclic systems, e.g., C_5H_5N	1.352	± 0.005
5. Shortened (partial double bond) in N–C=O, e.g., $HCONH_2$	1.322	± 0.003

Triple Bond

1. In R.C⋮N	1.158	± 0.002

Carbon–Oxygen

Single Bond

1. Paraffinic	1.43	± 0.01
2. Strained, e.g., epoxides	1.47	± 0.01
3. Shortened (partial double bond) as in carboxylic acids or through the influence of an aromatic ring, e.g., salicylic acid	1.36	± 0.01

Double Bond

1. In aldehydes, ketones, carboxylic acids, esters	1.23	± 0.01
2. In zwitterion forms, e.g., DL-serine	1.26	± 0.01
3. Shortened (partial triple bond) as in conjugated systems	1.207	± 0.006
4. Partial triple bond as in acyl halides or isocyanates	1.17	± 0.01

Carbon–Sulfur

Single Bond

1. Paraffinic, e.g., CH_3SH	1.81(5)	± 0.01
2. Lengthened in the presence of fluorine, e.g., $(CF_3)_2S$	1.83(5)	± 0.01
3. Shortened (partial double bond) as in heterocyclic systems, e.g.,	1.73	± 0.01

Double Bond

1. In ethylene thiourea	1.71	± 0.02
2. Shortened (partial triple bond) in the presence of a second carbon double bond, e.g., COS	1.558	± 0.003

From Kennard, O., in *Handbook of Chemistry and Physics*, 55th ed., Weast R. C., Ed., CRC Press, Cleveland, 1974, F-200.

Table 2–13
BOND LENGTHS OF ELEMENTS

Element	Bond	Bond length (Å)
Ac	Ac–Ac	3.756
Ag (25°C)	Ag–Ag	2.8894
Al (25°C)	Al–Al	2.863
As	As–As	2.49
As$_4$	As–As	2.44 ± 0.03
Au (25°C)	Au–Au	2.8841
B$_2$	B–B	1.589
Ba (room temperature)	Ba–Ba	4.347
Be	Be–Be	
α-form (20°C)		2.2260
Bi (25°C)	Bi–Bi	3.09
Br$_2$	Br–Br	2.290
Ca	Ca–Ca	
α-form (18°C)		3.947 (f.c.c.)
β-form (500°C)		3.877 (b.c.c.)
Cd (21°C)	Cd–Cd	2.9788
Cl$_2$	Cl–Cl	1.988
Ce	Ce–Ce	3.650
Co	Co–Co	2.5061
Cr	Cr–Cr	
α-form (20°C)		2.4980
β-form (>1,850°C)		2.61
Cs (–10°C)	Cs–Cs	5.309
Cu (20°C)	Cu–Cu	2.5560
Dy	Dy–Dy	3.503
Er	Er–Er	3.468
Eu	Eu–Eu	3.989
F$_2$	F–F	1.417 ± 0.001
Fe	Fe–Fe	
α-form (20°C)		2.4823 (b.c.c.)
γ-form (916°C)		2.578 (f.c.c.)
δ-form (1,394°C)		2.539 (b.c.c.)
Ga (20°C)	Ga–Ga	2.442
Gd (20°C)	Gd–Gd	3.573
Ge (20°C)	Ge–Ge	2.4498
H$_2$	H–H in H$_2$	0.74611
	H–D in HD	0.74136
	D–D in D$_2$	0.74164
He	He–He in (He$_2$)$^+$	1.08$_0$
Hf	Hf–Hf	
α-form (24°C)		3.1273 (h.c.p.)
Hg (–46°C)	Hg–Hg	3.005
Ho	Ho–Ho	3.486
I$_2$	I–I	2.662
In (20°C)	In–In	3.2511
Ir (room temperature)	Ir–Ir	2.714
K (195°C)	K–K	4.544
La	La–La	
α-form		3.739 (h.c.p.)
β-form		3.745 (f.c.c.)
Li (20°C)	Li–Li	3.0390
Lu	Lu–Lu	3.435
Mg (25°C)	Mg–Mg	3.1971
Mn	Mn–Mn	
γ-form (1,095°C)		2.7311 (f.c.c.)
δ-form (1,134°C)		2.6679 (b.c.c.)

Table 2–13 (continued)
BOND LENGTHS OF ELEMENTS

Element	Bond	Bond length (Å)
Mo (20°C)	Mo–Mo	2.7251
N_2	N–N	$1.0975_8 \pm 0.0001$
Na (20°C)	Na–Na	3.7157
Nb (20°C)	Nb–Nb	2.8584
Nd	Nd–Nd	3.628
Ni (18°C)	Ni–Ni	2.4916
Np	Np–Np	
α-form (20°C)		2.60 (orthorh.)
β-form (313°C)		2.76 (tetra.)
γ-form 600°C		3.05 (b.c.c.)
O_2	O–O	1.208
O_3 (angle = 116.8 ± 0.5°)	O–O	1.278 ± 0.003
Os (20°C)	Os–Os	2.6754
P (black)	P–P	2.18
P_4	P–P	2.21 ± 0.02
Pa	Pa–Pa	3.212
Pb (25°C)	Pb–Pb	3.5003
Pd (25°C)	Pd–Pd	2.7511
Po	Po–Po	
α-form (10°C)		3.345 (cub.)
β-form (75°C)		3.359 (rhbdr.)
Pr	Pr–Pr	
α-form		3.640 (tetra.)
β-form		3.649 (f.c.c.)
Pt (20°C)	Pt–Pt	2.746
Pu	Pu–Pu	
γ-form (235°C)		3.026 (f.c.c.)
δ-form (313°C)		3.279 (f.c.c.)
ε-form		3.150 (b.c.c.)
Rb (20°C)	Rb–Rb	4.95
Re (room temperature)	Re–Re	2.741
Rh (20°C)	Rh–Rh	2.6901
Ru (25°C)	Ru–Ru	2.6502
S_2	S–S	1.887
S_8	S–S	2.07 ± 0.02
Sb (25°C)	Sb–Sb	2.90
Sc (room temperature)	Sc–Sc	3.212
Se (20°C)	Se–Se	2.321
Se_2	Se–Se	2.152 ± 0.003
Se_8	Se–Se	2.32 ± 0.003
Si (20°C)	Si–Si	2.3517
Sn	Sn–Sn, diamond-type lattice	
α-form (20°C)		2.8099
β-form (25°C)		3.022 (tetra.)
Sr	Sr–Sr	
α-form (25°C)		4.302 (f.c.c.)
β-form (248°C)		4.32 (h.c.p.)
γ-form (614°C)		4.20 (b.c.c.)
Ta (20°C)	Ta–Ta	2.86
Tb	Tb–Tb	3.525
Tc (room temperature)	Tc-Tc	2.703
Te (25°C)	Te–Te	2.846
Th	Th–Th	
α-form (25°C)		3.595 (f.c.c.)
β-form (1,450°C)		3.56 (b.c.c.)

Table 2–13 (continued)
BOND LENGTHS OF ELEMENTS

Element	Bond	Bond length (A)
Ti	Ti–Ti	
α-form (25°C)		2.8956 (h.c.p.)
β-form (900°C)		2.8636 (b.c.c.
Tl	Tl–Tl	
α-form (18°C)		3.4076 (h.c.p.)
β-form (262°C)		3.362 (b.c.c.)
Tm	Tm–Tm	3.447
U	U–U	
α-form		2.77
β-form (805°C)		3.058 (b.c.c.)
V (30°C)	V–V	2.6224
W (25°C)	W–W	2.7409
Y	Y–Y	3.551
Yb	Yb–Yb	3.880
Zn (25°C)	Zn–Zn	2.6694
Zr	Zr–Zr	3.179

From Kennard, O., in *Handbook of Chemistry and Physics*, 55th ed., Weast, R. C., Ed., CRC Press, Cleveland, 1974, F-201.

Table 2–14
BOND LENGTH AND ANGLE VALUES BETWEEN ELEMENTS

Elements	In	Bond length (A)	Bond angle (°)
		Boron	
B–B	B_2H_6	1.770 ± 0.013	H–B–H 121.5 ± 7.5
B–Br	B Br	1.88_7	–
	B Br$_3$	1.87 ± 0.02	Br–B–Br 120 ± 6
B–Cl	BCl	1.715_7	–
	BCl$_3$	1.72 ± 0.01	Cl–B–Cl 120 ± 3
B–F	BF	1.262	–
	BF$_3$	1.29_5 ± 0.01	F–B–F 120
B–H	Hydrides	1.21 ± 0.02	–
B–H bridge	Hydrides	1.39 ± 0.02	–
B–N	(BClNH)$_3$	1.42 ± 0.01	B–N–B 121
B–O	BO	1.2049	–
	B(OH)$_3$	1.362 ± 0.005 (av.)	O–B–O 119.7
		Nitrogen	
N–Cl	NO$_2$Cl	1.79 ± 0.02	–
N–F	NF$_3$	1.36 ± 0.02	F–N–F 102.5 ± 1.5
N–H	[NH$_4$]$^+$	1.034 ± 0.003	–
	NH	1.038	–
	ND	1.041	–
	HNCS	1.013 ± 0.005	H–N–C 130.25 ± 0.25
N–N	N$_3$H	1.02 ± 0.01	H–N–N' 112.65 ± 0.5
	N$_2$O	1.126 ± 0.002	–
	[N$_2$]$^+$	1.116_2	–
N–O	NO$_2$Cl	1.24 ± 0.01	O–N–O 126 ± 2
	NO$_2$	1.188 ± 0.005	O–N–O 134.1 ± 0.25

Table 2–14 (continued)

BOND LENGTH AND ANGLE VALUES BETWEEN ELEMENTS

Elements	In	Bond length (Å)	Bond angle (°)
		Nitrogen (continued)	
N=O	N_2O	1.186 ± 0.002	–
	$[NO]^+$	1.0619	–
N–Si	SiN	1.572	–
		Oxygen	
O–H	$[OH]^+$	1.0289	–
	OD	0.9699	–
	H_2O_2	0.960 ± 0.005	O–O–H 100 ± 2
O–O	H_2O_2	1.48 ± 0.01	–
	$[O_2]^+$	1.227	–
	$[O_2]^-$	1.26 ± 0.02	–
	$[O_2]^{--}$	1.49 ± 0.02	–
		Phosphorus	
P–D	PD	1.429	–
P–H	$[PH_4]^+$	1.42 ± 0.02	–
P–N	PN	1.4910	–
P–S	$PSBr_3 (Cl_3, F_3)$	1.86 ± 0.02	–
		Sulfur	
S–Br	$SOBr_2$	2.27 ± 0.02	Br–S–Br 96 ± 2
S–F	SOF_2	1.585 ± 0.005	F–S–F 92.8 ± 1
S–D	SD	1.3473	–
	SD_2	1.345	–
S–O	SO_2	1.4321	O–S–O 119.54
	$SOCl_2$	1.45 ± 0.02	–
S–S	S_2Cl_2	2.04 ± 0.01	–
		Silicon	
Si–Br	$SiBr_4$	2.17 ± 1.01	–
Si–Cl	$SiCl_4$	2.03 ± 1.01 (av.)	–
Si–F	SiF_4	1.561 ± 0.003 (av.)	–
SiH	SiH_4	1.480 ± 0.005	–
Si–O	$[SiO]^+$	1.504	–
Si–Si	Si_2Cl_2	2.30 ± 0.02	–

From Kennard, O., in *Handbook of Chemistry and Physics*, 55th ed., Weast, R. C., Ed., CRC Press, Cleveland, 1974, F-202.

Table 2–15
BOND LENGTHS AND ANGLES OF CHEMICAL COMPOUNDS

A. Inorganic Compounds

Compound	Formula	Bond lengths (Å)			Bond angles (°)		
Ammonia	NH$_3$	N–H	1.008	± 0.004	H–N–H	107.3	± 0.2
Antimony tribromide	SbBr$_3$	Sb–Br	2.51	± 0.02	Br–Sb–Br	97	± 2
Antimony trichloride	SbCl$_3$	Sb–Cl	2.352	± 0.005	Cl–Sb–Cl	99.5	± 1.5
Antimony triiodide	SbI$_3$	Sb–I	2.67	± 0.03	I–Sb–I	99.0	± 1
Arsenic tribromide	AsBr$_3$	As–Br	2.33	± 0.02	Br–As–Br	100.5	± 1.5
Arsenic trichloride	AsCl$_3$	As–Cl	2.161	± 0.004	Cl–As–Cl	98.4	± 0.5
Arsenic trifluoride	AsF$_3$	As–F	1.712	± 0.005	F–As–F	102.0	± 2
Arsenic triiodide	AsI$_3$	As–I	2.55	± 0.03	I–As–I	101.0	± 1.5
Arsenic trioxide	As$_4$O$_6$	As–O	1.78	± 0.02	O–As–O	99.0	± 2
					As–O–As	128.0	± 2
Arsine	AsH$_3$	As–H	1.5192	± 0.002	H–As–H	91.83	± 0.33
Bismuthum tribromide	BiBr$_3$	Bi–Br	2.63	± 0.02	Br–Bi–Br	100.0	± 4
Bismuthum trichloride	BiCl$_3$	Bi–Cl	2.48	± 0.02	Cl–Bi–Cl	100.0	± 6
Bromosilane	SiH$_3$Br	Si–H	1.57	± 0.03	H–Si–H	111.3	± 1
Chlorine dioxide	ClO$_2$	Cl–O	1.49		O–Cl–O	118.5	
Chlorogermane	GeH$_3$Cl	Ge–H	1.52	± 0.03	H–Ge–H	110.9	± 1.5
Chlorosilane	SiH$_3$Cl	Si–H	1.483	± 0.001	H–Si–H	110	± 0.03
		Si–Cl	2.0479	± 0.0007			
Chromium oxychloride	Cr(OCl)$_2$	Cr–O	1.57	± 0.03	O–Cr–O	105	± 4
		Cr–Cl	2.12	± 0.02	Cl–Cr–Cl	113	± 3
					Cl–Cr–O	109	± 3
Cyanuric triazide	C$_3$N$_{12}$	C–N	1.38		C–N=C	113.0	
Dichlorosilane	SiH$_2$Cl$_2$	Si–H	1.46				
		SiCl	2.02	± 0.03	Cl–Si–Cl	110	± 1
Difluorodiazine	N$_2$F$_2$	N–F	1.44	± 0.04			
		N–N	1.25	± 0.02	F–N=N	115	± 5
Difluoromethylsilane	Ch$_3$SiHf$_2$				F–Si–F	106	± 0.5
					H–Si–C	116.2	± 1
					C–Si–F	109.8	± 0.5
Disilicon hexachloride	Si$_2$Cl$_6$	Si–Cl	2.02	± 0.02	Cl–Si–Cl	109.5	± 1
(hexachlorosilane)		Si–Si	2.34	± 0.06			
Fluorosilane	SiH$_3$F	Si–H	1.460	± 0.01	H–Si–H	109.3	± 0.3
		Si–F	1.595	± 0.002			
Hydrogen phosphide	PH$_3$	P–H	1.415	± 0.003	H–P–H	93.3	± 0.2
Hydrogen selenide	SeH$_2$	Se–H	1.47		H–Se–H	91.0	
Hydrogen sulfide	SH$_2$	S–H	1.3455		H–S–H	93.3	
Hydrogen telluride	Te–H$_2$				H–Te–H	89.5	± 1
Iodo silane	Si–H$_3$I	Si–H	1.48	± 0.01	H–Si–H	109.9	± 0.4
		Si–I	2.45	± 0.09			
Methylgermane	Ch$_3$GeH$_3$				H–C–H	108.2	± 0.5
					H–Ge–H	108.6	± 0.5
Nitrosyl bromide	NOBr	O–N	1.15	± 0.04	Br–N=O	117	± 3
		N–Br	2.14	± 0.02			
Nitrosyl chloride	NOCl	N–O	1.14	± 0.02	Cl–N=O	113.0	± 2
		N–Cl	1.97	± 0.01			
Nitrosyl fluoride	NOF	N–O	1.13		F–N=O	110.0	
		N–F	1.52				

Table 2–15 (continued)
BOND LENGTHS AND ANGLES OF CHEMICAL COMPOUNDS

A. Inorganic Compounds (continued)

Compound	Formula	Bond lengths (Å)			Bond angles (°)		
cis-Nitrous acid	NO(OH)	H–O	0.98		O–N=O	114	± 2
		O–N	1.46				
		N–O'	1.20				
trans-Nitrous acid	NO(OH)	H–O	0.98		O–N=O	118	± 2
		O–N	1.46				
		NO'	1.20				
Oxygen chloride	OCl$_2$	O–Cl	1.70$_1$	± 0.02	Cl–O–Cl	110.8	± 1
Oxygen fluoride	OF$_2$	O–F	1.418		F–O–F	103.2	
Phosphorus oxychloride	POCl$_3$	P–Cl	1.99$_5$	± 0.02	Cl–P–Cl	103.5	± 1
Phosphorus oxysulfide	P$_4$O$_6$S$_4$	P–O	1.61	± 0.02	P–O–P	128.5	± 1.5
					O–P–O	101.5	± 1
		P–S	1.85	± 0.02	O–P–S	116.5	± 1
Phosphorus pentoxide	P$_4$O$_{10}$	P–O	1.62	± 0.02	O–P–O	101.5	± 1
		P–O'	1.38	± 0.02	O–P–O'	116.5	± 1
					P–O–P	123.5	± 1
Phosphorus tribromide	PBr$_3$	P–Br	2.18	± 0.03	Br–P–Br	101.5	± 1.5
Phosphorus trichloride	PCl$_3$	P–Cl	2.043	± 0.003	Cl–P–Cl	100.1	± 0.3
Phosphorus trifluoride	PF$_3$	P–F	1.535		F–P–F	100.0	
Phosphorus trioxide	P$_4$O$_6$	P–O	1.65	± 0.02	O–P–O	99.0	± 1
					P–O–P	127.5	± 1
Stibine	SbH$_3$	Sb–H	1.7073	± 0.0025	H–Sb–H	91.3	± 0.33
Sulfur dichloride	SCl$_2$	S–Cl	1.99	± 0.03	Cl–S–Cl	101.0	± 4
Sulfur dioxide	SO$_2$	S–O	1.4321		O–S–O	119.536	
Sulfur monochloride	S$_2$Cl$_2$	S–Cl	2.01	± 0.07	Cl–S–S	104.5	± 0.25
Sulfurylchloride	SO$_2$Cl$_2$	S–O	1.43	± 0.02	O–S–O	119.75	± 5
		S–Cl	1.99	± 0.02	Cl–S–Cl	111.20	± 2
					Cl–O–O	106.5	± 2
Tellurium bromide	TeBr$_2$	Te–Br	2.51	± 0.02	Br–Te–Br	98.0	± 3
Tribromo silane	SiHBr$_3$	Si–Br	2.16	± 0.03	Br–Si–Br	110.5	± 1.5
Trichloro germane	GeHCl$_3$	Ge–H	1.55	± 0.04	Cl–Ge–Cl	108.3	± 0.2
Trichloro silane	Si–HCl$_3$	Si–H	1.47		Cl–Si–Cl	109.4	± 0.3
Trifluorochlorosilane	SiClF$_3$	Si–F	1.560	± 0.005	F–Si–F	108.5	± 1
		Si–Cl	1.989	± 0.018			
Trifluorochlorogermane	GeClF$_3$	Ge–F	1.688	± 0.0017	F–Ge–F	107.7	± 1.5
		Gr–Cl	2.067	± 0.005			
Trifluorosilane	SiHF$_3$	Si–H	1.455	± 0.01			
		Si–F	1.565	± 0.005	F–Si–F	108.3	± 0.5
Vanadium oxytrichloride	VOCl$_3$	V–O	1.56	± 0.04	Cl–V–Cl	111.2	± 2
		V–Cl	2.12	± 0.03	Cl–V–O	108.2	± 2
Water	H$_2$O	O–H	0.958$_4$		H–O–H	104.45	

Table 2–15 (continued)
BOND LENGTHS AND ANGLES OF CHEMICAL COMPOUNDS

B. Organic Compounds

Compound	Formula	Bond lengths (Å)			Bond angles (°)		
Acetaldehyde	CH_3COH	C–H	1.09		C–C=O	121	± 2
		C–C	1.50	± 0.02			
		C–O	1.22	± 0.02			
Bromomethane	Ch_3Br	C–H	1.11	± 0.01	H–C–H	111.2	± 0.5
		C–Br	1.929				
Carbon tetrachloride	CCl_4	C–Cl	1.766	± 0.003	Cl–C–Cl	109.5	
		Cl–Cl	2.887	± 0.004			
Chloromethane	CH_3Cl	C–H	1.11	± 0.01	H–C–H	110	± 2
		C–Cl	1.784	± 0.003			
Dichloromethane	CH_2Cl_2	C–H	1.068	± 0.005	H–C–H	112	± 0.3
		C–Cl	1.7724	± 0.0005	Cl–C–Cl	111.8	
Difluorochloromethane	$CHCLF_2$	C–H	1.06		F–C–F	110.5	± 1
		C–Cl	1.73	± 0.03	Cl–C–Cl	110.5	± 1
		C–F	1.36	± 0.03			
1,1-Difluoroethylene	$C_2H_2F_2$	C–H	1.07	± 0.02	F–C=C	125.2	± 0.2
		C–F	1.321	± 0.015	H–C–H	117	± 7
		C–C	1.311	± 0.035			
Difluoromethane	CH_2F_2	C–F	1.360	± 0.005	F–C–F	108.2	± 0.8
		C–H	1.09	± 0.03	H–C–H	112.5	± 6
p-Dinitrobenzene	$C_6H_4(NO_2)_2$	C–C	1.38		O–N=O	124.0	
		C–N	1.48				
		N–O	1.21				
Dithio oxamide	$NH_2CSCSNH_2$	C–C	1.54		N–C=S	124.8	
		C–N	1.30		S–C–C	124.87	
		C–S	1.66		N–C–C	115.25	
Ethane	C_2H_6	C–H	1.107		H–C–H	109.3	
		C–C	1.536				
Ethylidene fluoride	CH_3CHF_2	C–F	1.345	± 0.001	F–C–F	109.15	± 0.001
		C–C	1.540		C–C–F	109.4	
		C–H	1.100		H–C–C	110.2	
Fluorochloromethane	CH_2ClF	C–H	1.078	± 0.005	Cl–C–F	100.0	± 0.1
		C–Cl	1.759	± 0.003			
		C–F	1.378	± 0.006			
Fluorotrichloromethane	$CFCl_3$	C–Cl	1.76	± 0.02	Cl–C–Cl	113	± 3
		C–F	1.44	± 0.04			
Formaldehyde	CH_2O	C–H	1.060	± 0.038	H–C–H	125.8	± 7
		C–O	1.230	± 0.017			
Formamide	$HCONH_2$	C–O	1.25_5		N–C=O	121.5	
		C–N	1.300				
Formic acid	$HCOOH$	C–O	1.245		O–C=O′	124.3	
		C–O	1.312		H–C–O′	117.8	
		C–H	1.085		C–O–H	107.8	
		O–H	0.95				
Glycine	NH_2CH_2COOH	C–C	1.52		C–C–O	119.0	
		C–O	1.27		O–C–O	122.0	
		C–N	1.39		C–C–N	112.0	

Table 2–15 (continued)
BOND LENGTHS AND ANGLES OF CHEMICAL COMPOUNDS

B. Organic Compounds (continued)

Compound	Formula	Bond lengths (Å)			Bond angles (°)		
Hexachloroethane	C_2Cl_6	C–Cl	1.74	± 0.01	Cl–C–Cl	109.3	± 0.01
		C–C	1.57	± 0.06			
Iodomethane	CH_3I	C–H	1.11	± 0.01	H–C–H	111.4	± 0.1
		C–I	2.139				
Methane	CH_4	C–H	1.091				
Methanethiol	CH_3SH	C–H	1.1039 ± 0.002			110.3	± 0.2
		C–S	1.8177 ± 0.0002		H–S–C	100.3	± 0.2
		S–H	1.329	± 0.004			
Methanol	CH_3OH	C–H	1.096	± 0.01	H–C–H	109.3	± 0.75
		C–O	1.427	± 0.007	C–O–H	108.9	± 2
		O–H	0.956	± 0.015			
Methylamine	CH_3NH_2	C–H	1.093		H–C–H	109.5	± 1
		C–N	1.474	± 0.005	H–N–H	105.8	± 1
		N–H	1.014				
Methylether	$(CH_3)_2O$	C–O	1.43	± 0.03	C–O–C	110.0	± 3
Methylnitrite	CH_3NO_2	C–H	1.00		O–N=O	127	± 4
		C–N	1.49	± 0.02			
		N–O	1.22	± 0.1			
Methylsulfide	$(CH_3)_2S$	C–S	1.82	± 0.01	C–S–C	105	± 3
		C–H	1.06		H–C–H	109.5	
Oxamide	$NH_2COCONH_2$	C–C	1.54		N–C=O	125.7	± 0.3
		C–O	1.24				
		C–N	1.32				
Phosgene	CCl_2O	C–Cl	1.746	± 0.004	Cl–C=O	124.3	± 0.1
		C–O	1.166	± 0.002	Cl–C–Cl	111.3	± 0.1
Propylene	C_3H_6				C–C=C	124.75	± 0.3
Propynal	CHC.COH	$C_1–C_2$	1.204		C–C=O	123.0	
		$C_2–C_3$	1.46		C–C–H	120.0	
		C–O	1.21				
		$C_1–H$	1.06				
		$C_3–H$	1.08				
Tribromomethane	$CHBr_3$	C–H	1.068	± 0.01	Br–C–Br	110.8	± 0.3
		C–Br	1.930	± 0.003			
Trichlorobromomethane	$CBrCl_3$	C–Br	1.936		Cl–C–Cl	111.2	± 1
		C–Cl	1.764				
Trifluorochloromethane	$CClF_3$	C–Cl	1.751	± 0.004	F–C–F	108.6	± 0.4
		C–F	1.328	± 0.02			
Trifluoromethane	CHF_3	C–H	1.098		F–C–F	108	± 0.75
		C–F	1.332	± 0.008			
Triiodomethane	CHI_3	C–I	2.12	± 0.04	I–C–I	113.0	± 1
Trimethylamine	$(CH_3)_3N$	C–N	1.47	± 0.01	C–N–C	108.0	± 4
		C–H	1.06		H–C–H	109.5	
Trimethylarsine	$(CH_3)_3As$	C–H	1.09		C–As–C	96	± 5
		C–As	1.98	± 0.02			
Trimethylphosphine	$(CH_3)_3P$	C–H	1.09		C–P–C	100.0	± 4
		C–P	1.87	± 0.02			

From Kennard, O., in *Handbook of Chemistry and Physics,* 55th ed., Weast, R. C., Ed., CRC Press, Cleveland, 1974, F-202.

Table 2–16
ELECTRICAL RESISTIVITY AND TEMPERATURE COEFFICIENTS
OF ELEMENTS

Element	Temperature (°C)	Microhm-cm	Temperature coefficient per °C
Aluminum, 99.996%	20	2.6548	0.00429[20]
Antimony	0	39.0	
Arsenic	20	33.3	
Beryllium[a]	20	4.0	0.025[20]
Bismuth	0	106.8	
Boron	0	1.8×10^{12}	
Cadmium	0	6.83	0.0042[0]
Calcium	0	3.91	0.00416[0]
Carbon[b]	0	1375.0	
Cerium	25	75.0	0.00087[0-25]
Cesium	20	20	
Chromium	0	12.9	0.003[0]
Cobalt	20	6.24	0.006604[0-100]
Copper	20	1.6730	0.0068[20]
Dysprosium[c]	25	57.0	0.00119[0-25]
Erbium	25	107.0	0.00201[0-25]
Europium	25	90.0	
Gadolinium	25	140.5	0.00176[0-25]
Gallium[d]	20	17.4	
Germanium[e]	22	46×10^6	
Gold	20	2.35	0.004[0-100]
Hafnium	25	35.1	0.0038[25]
Holmium	25	87.0	0.00171[0-25]
Indium	20	8.37	
Iodine	20	1.3×10^{15}	
Iridium	20	5.3	0.003925[0-100]
Iron, 99.99%	20	9.71	0.00651[20]
Lanthanum	25	5.70	0.00218[0-25]
Lead	20	20.648	0.00336[20-40]
Lithium	0	8.55	
Lutetium	25	79.0	0.00240[0-25]
Magnesium[f]	20	4.45	0.0165[20]
Manganese α	23–100	185.0	
Mercury	50	98.4	
Molybdenum	0	5.2	
Neodymium	25	64.0	0.00164[0-25]
Nickel	20	6.84	0.0069[0-100]
Niobium (Columbium)[g]	0	12.5	
Osmium	20	9.5	0.0042[0-100]
Palladium	20	10.8	0.00377[0-100]
Phosphorus, white	11	1×10^{17}	
Platinum, 99.85%	20	10.6	0.003927[0-100]
Plutonium	107	141.4	
Potassium	0	6.15	
Praseodymium	25	68	0.00171[0-25]
Rhenium	20	19.3	0.00395[0-100]
Rhodium	20	4.51	0.0042[0-100]
Rubidium	20	12.5	
Ruthenium	0	7.6	
Samarium	25	88.0	0.00184[0-25]
Scandium[h]	22	61.0	0.00282[0-25]
Selenium	0	12.0	
Silicon	0	10.0	
Silver	20	1.59	0.0041[0-100]

Table 2–16 (continued)
ELECTRICAL RESISTIVITY AND TEMPERATURE COEFFICIENTS
OF ELEMENTS

Element	Temperature (°C)	Microhm-cm	Temperature coefficient per °C
Sodium	0	4.2	
Strontium	20	23.0	
Sulfur, yellow	20	2×10^{23}	
Tantalum	25	12.45	0.00383^{0-100}
Tellurium	25	4.36×10^{5}	
Thallium	0	18.0	
Thorium	0	13.0	0.0038^{0-100}
Thulium	25	79.0	0.00195^{0-25}
Tin	0	11.0	0.0047^{0-100}
Titanium	20	42.0	
Tungsten	27	5.65	
Uranium		30.0	
Vanadium	20	24.8–26.0	
Ytterbium	25	29.0	0.0013^{0-25}
Yttrium	25	57.0	0.0027^{0-25}
Zinc	20	5.916	0.00419^{0-100}
Zirconium	20	40.0	0.0044^{20}

[a]Annealed, comm. pure.
[b]Graphite.
[c]Polycrystalline.
[d]Hard wire.
[e]Intrinsic Ge.

[f]Polycrystalline.
[g]High purity.
[h]Zone-refined bar.
[z]Data not available to indicate range over which coefficient is valid.

From Weast, R. C., Ed., *Handbook of Chemistry and Physics,* 55th ed., CRC Press, Cleveland, 1974, F-159.

Table 2-17
SURFACE TENSION OF LIQUID ELEMENTS

The following data were collected from many sources; as a result, their accuracy varies. Users of the data in this table are advised that:

1. As a rule, results obtained from the "sessile drop" and "maximum bubble pressure" as well as from the "pendant drop" methods are preferable to results obtained from other methods for metals with very high melting points.

2. Values of single measurements are usually not as well supported by experiments as those of serial measurements at various temperatures.

3. Values in parentheses can be considered improbable.

Element	Purity (wt. %)	σ_{mp} (dyn/cm)	Atm.	σ_{t1} t_1 °C	σ (dyn/cm)	σ_{t2} t_2 °C	σ (dyn/cm)	σ_{t3} t_3 °C	σ (dyn/cm)	Method	Ref.
Ag	99.99		H_2	1000	916					Bubble pressure	134
	–	(785)	vac.							Pendant drop	182
	99.96		H_2	1000	893	1150	862	1250	849	Bubble pressure	127
			vac.	1000	908					Sessile drop	51
	99.995		H_2	1000	907	1100	894	1200	876	Bubble pressure	128
	99.999	(828)	vac.							Pendant drop	65
	99.99		Ar, H_2	1000	890					Bubble pressure	169
	spect. pure	921		$\sigma = 1136-0.174\ T\ (°K)\ (1300-2200°K)$						Bubble pressure	26
		918	–	$\sigma = 918-0.149\ (t-t_{mp})\ (t°C)$						Bubble pressure	204
	99.999		Ar	980	905±10	1108	890±10			Sessile drop	Z3
Al	99.99	860±20	Ar	950	840					Bubble pressure	102
	99.72		vac.							Bubble pressure	46
	99.7	863±25	Ar	$\sigma_t = (863\pm25)-0.33\ (t-t_{mp})\ (t°C)$						Bubble pressure	113
	99.99	865	vac.	$\sigma_t = 865-0.14\ (t-t_{mp})\ (t°C)$						Sessile drop	137, 56
	99.99	(825)	Ar	$(\sigma_t = 825-0.05\ (T-993))\ (T°K)$						Bubble pressure	38
	99.99	866	He	$\sigma_t = 866-0.15\ (t-t_{mp})\ (t°C)$						Sessile drop	10
	99.999	873	He	1600	725	$\sigma_t = 873-0.15\ (t-t_{mp})\ (t°C)$				Sessile drop	199
	99.99	(760)	vac.	$(\sigma_t = 948-0.202T)\ (T°K.980-1090°K)$						Sessile drop	Z4
	99.998	(915)	Ar	$(\sigma_t = 915-0.51\ (t-t_{mp}))\ (t°C)$ $(660-800°C)$						Sessile drop	Z6
	99.996	855±6	Ar	$\sigma_t = 855-0.104\ (t-t_{mp})\ (t°C)$ $(660-911°C)$						Bubble pressure	Z13

Table 2–17 (continued)
SURFACE TENSION OF LIQUID ELEMENTS

Element	Purity (wt. %)	σmp (dyn/cm)	Atm.	σ_{t1}		σ_{t2}		σ_{t3}		Method	Ref.
				t_1 °C	σ (dyn/cm)	t_2 °C	σ (dyn/cm)	t_3 °C	σ (dyn/cm)		
Au	99.999	(754)	vac.							Pendant drop	182
	99.999	1130	He	1200	1070	1300	1020			Sessile drop	86
	99.999-	(731)	vac.							Pendant drop	65
			Ar	1108	1130±10					Sessile drop	Z3
B	99.8	1060±50	vac.							Sessile & pendant drop	187
Ba	-		Ar	720	224					Bubble pressure	3
	99.5	276		$\sigma = 351-0.075\ T\ (°K)\ (1410-1880°K)$						Bubble pressure	25
Be	99.98		vac.	1500	1100					Sessile drop	58
Bi	99.99	376	vac.	800	343					Drop pressure	155
		376	vac.	450	(382)					Drop pressure	156
	99.90		H₂			1000	328			Bubble pressure	134
	99.9	380±10	vac.							Electro-capillarity	118
			Ar	350	362					Bubble pressure	103
			–	700	350					Drop pressure	76
			vac.							Sessile drop	29
	99.98	380±10	Ar	450	380					Bubble pressure	105
			–	300	379					Drop pressure	120
			vac., Ar, H₂							Sessile drop	64
		378								Drop weight	4
	99.99995	375		$\sigma = 423-0.088\ T\ (T°K)\ (1352-1555°K)$						Bubble pressure	26
	99.999	380±3	Ar	$\sigma_t = 380-0.142\ (t-t_{mp})\ (MP-555°C)\ (t°C)$						Bubble pressure	Z13a
Ca	-	360	Ar	850	337					Bubble pressure	3
	p.a.			$\sigma = 472-0.100\ T\ (T°K)\ (1445-1655°K)$						Bubble pressure	25
Cd			–	450	600					Drop pressure	119

Table 2–17 (continued)
SURFACE TENSION OF LIQUID ELEMENTS

Element	Purity (wt. %)	σmp (dyn/cm)	Atm.	σt1 t₁ °C	σt1 σ (dyn/cm)	σt2 t₂ °C	σt2 σ (dyn/cm)	σt3 t₃ °C	σt3 σ (dyn/cm)	Method	Ref.
Cd (cont.)			–	400	600					Bubble pressure	13
	99.9	(550±10)	–	350	586					Drop pressure	69
			Ar	390	604					Bubble pressure	103
		(525±30)	H₂							Bubble pressure	1
	99.9999	590±5	–							Solid state curvature	Z8
							(non linear)			Sessile drop	Z12
Co	99.99		Ar	1550	1836					Sessile drop	108
	99.99		vac., Al₂O₃	1520	1800					Sessile drop	46
	99.99		He, Al₂O₃	1520	(1630)					Sessile drop	46
	99.99		He, BeO	1520	(1640)					Sessile drop	46
	99.99		He, MgO	1520	(1560)					Sessile drop	46
	99.99		H₂, Al₂O₃	1520	1780					Sessile drop	46
	99.99		He	1520	(1620)					Bubble pressure	46
	99.99		H₂	1520	(1590)					Bubble pressure	46
			vac.	1500	1870					Sessile drop	7
				1600	(1640)					Sessile drop	135
				1600	(1600)					Sessile drop	135
			vac.	1600	1815					Sessile drop	53
			vac., Al₂O₃	1600	1812					Sessile drop	54
	99.99	(1520)	H₂, He							Bubble pressure	61
			H₂, He	1550	1845					Bubble pressure	63
	99.9983	1880	vac.							Pendant drop	5
	99.99		vac.	1550	1780					Bubble pressure	59
Cr	–	1700±50	vac.	1950	1590±50					Sessile drop	47
	99.9997		Ar							Dynam. drop weight	6
Cs			Ar	62	68.4					Pendant drop	195
			Ar	62	67.5	146	62.9	642	34.6	Bubble pressure	193
	99.95		Ar	39	69.5	494	42.8			Bubble pressure	24

Table 2–17 (continued)
SURFACE TENSION OF LIQUID ELEMENTS

Element	Purity (wt. %)	σ_{mp} (dyn/cm)	Atm.	σ_{t1} t_1 °C	σ (dyn/cm)	σ_{t2} t_2 °C	σ (dyn/cm)	σ_{t3} t_3 °C	σ (dyn/cm)	Method	Ref.
Cs (cont.)											
	99.995	68.6	He	$\sigma = 68.6 - 0.047\,(t-t_{mp})$ (t°C) (52–1100°C)						Bubble pressure	121
Cu	–	(1150)	Ar	1120	1269±20					Sessile drop	11
			vac.							Pendant drop	35
			Ar	1120	1285±10					Sessile drop	108
	99.99	73.74	Ar	$\sigma = 73.74 - 1.791\cdot10^{-2}\,(t-t_{mp}) - 9.610\cdot10^{-5}\,(t-t_{mp})^2 + 6.629\cdot10^{-8}\,(t-t_{mp})^3$ (t°C) (71–1011°C)						Bubble pressure	26
	99.98	1270	H_2	1100	1301	1165	1295	1255	1287	Bubble pressure	134
			vac.							Sessile drop	7
			vac.	1120	1285					Sessile drop	198
				1440	1298					Bubble pressure	129
		(1085)	vac.							Pendant drop	182
	99.99		He	1250	1290					Bubble pressure	62
	99.99		H_2	1250	1300					Bubble pressure	62
	99.9	(1180±40)	Ar							Bubble pressure	103
	–			1100	1220					Sessile drop	136
	–			1183	(1130)					Sessile drop	200
	–			1150	1370					Sessile drop	52
	99.997	1355	vac. / He, H_2							Sessile drop + Bubble pressure	138
	99.997	1352	vac.	$\sigma_t = 1352 - 0.17\,(t-t_{mp})$ (t°C)						Sessile drop	137
	99.997	1358	Ar	$\sigma_t = 1358 - 0.20\,(t-t_{mp})$ (t°C)						Bubble pressure	137
	–		Ar, He	1120	1285±10					Sessile drop	110
	99.99		H_2, He	1550	1265					Bubble pressure	63
	99.99999	1300	vac.	1130	1268±60					Pendant drop	5
	99.98		Ar	1600	1230					Sessile drop	15
	99.999		N_2	1100	1341	1150	1338	1200	1335	Bubble pressure	205
			vac.							Bubble pressure	128
	99.9	(1127)								Pendant drop	65
	99.99		Ar, H_2	1100	1320					Bubble pressure	169

Table 2–17 (continued)

SURFACE TENSION OF LIQUID ELEMENTS

Element	Purity (wt. %)	σ_{mp} (dyn/cm)	Atm.	σ_{t1} t_1°C	σ_{t1} σ (dyn/cm)	σ_{t2} t_2°C	σ_{t2} σ (dyn/cm)	σ_{t3} t_3°C	σ_{t3} σ (dyn/cm)	Method	Ref.
Fe	–			1570	(1731)					Sessile drop	95
		1720		1550	1860					Sessile drop	74
										Sessile drop	74
			He	1580–1760	(880)					Bubble pressure	100
	99.99			1570	(1632)					Sessile drop	93
	99.99		He	1650	(1610)					Sessile drop	46
	99.99		He	1650	(1430)					Sessile drop	46
	99.99		H$_2$	1650	(1400)					Sessile drop	46
	–	(1384)	vac.							Pendant drop	182
		(1700)	vac.							Sessile drop	7
			vac., He	1550	1865					Sessile drop	209
	99.99		He	1650	(1430)					Bubble pressure	62
	99.99		H$_2$	1650	(1400)					Bubble pressure	62
			He	1650	(1640)					Sessile drop	136
		(1650)	He, H$_2$	1550	1788					Bubble pressure	61
	99.985		Ar							Sessile drop	110
	–	(1560)		1560	(1710)					Sessile drop	68
	99.94		vac., Al$_2$O$_3$							Sessile drop	207
	99.9998	1880	vac.							Pendant drop	5
	99.93	1860±40	He							Sessile drop	116
		(1510)	vac.							Drop weight	139
	99.97		vac., BeO	1550	1830±6					Sessile drop	45
	Armco		Ar, N$_2$	1550	1795					Bubble pressure	164
	–		vac.	1550	1754					Sessile drop	42
	99.987		vac.	1550	(1730)					Sessile drop	159
	99.85	(1619)	vac.	1550	(1727)					Pendant drop	65
	99.69		He, Al$_2$O$_3$	1550	(1727)					Sessile drop	158
	99.69		H$_2$, Al$_2$O$_3$	1550	(1734)					Sessile drop	158
	–	1760±20	He, H$_2$		$\sigma = 1760 - 0.35\,(t - t_{mp})\ (t\,°C)$					Sessile drop	157
	99.9992	1773	He, H$_2$		$\sigma = 773 + 0.65\ t\ (t\,°C)\ (1550{-}1780\,°C)$					Oscillating drop	Z2
	–		–	1550	1780					Sessile drop	Z11

Table 2–17 (continued)
SURFACE TENSION OF LIQUID ELEMENTS

Element	Purity (wt.%)	σ_{mp} (dyn/cm)	Atm.	t_1°C	σ_{t1} (dyn/cm)	t_2°C	σ_{t2} (dyn/cm)	t_3°C	σ_{t3} (dyn/cm)	Method	Ref.
Fr	–		–	100	58.4					calculated	145
Ga	–	704	Ar, He							Bubble pressure	197
	–	725±10	Ar							Bubble pressure	104
			vac.	350	718					Sessile drop	64
	–		He, Al_2O_3	1500	559					Sessile drop	57
Ge	99.9998	718	vac., Al_2O_3		$\sigma = 718 - 0.101\,(t - t_{mp})$ (t°C)					Sessile drop	85
		621.4	vac.	1200	530					Drop weight	126
		650	vac.	1000	650					Sessile drop	99
										Sessile drop	Z5
Hf	97.5+2.5 Zr	632±5	N_2, He							Solid state curvature	Z8
		(1460)	vac.							Pendant drop	151
		1630	vac.							Pendant drop	5
Hg			H_2	20	(542)					Pendant drop	165
			air	16	(410)					Sessile drop	181
			air	20	(435.5)					Oscillating jet	175
			vac.	20	472					Oscillating jet	73
			vac.	20	(402)					Drop pressure	146
				25	476					Drop weight	75
			vac.	20	(436)					Sessile drop	160
			vac.	20	(432)					Sessile drop	170
			H_2	25	476					Drop pressure	78
				25	472					Drop weight	81
				25	(464)					Sessile drop	82
			H_2	19	473					Bubble pressure	173
				20	(437)					Capillary depression	144
			vac.	20	480					Drop weight	21
				25	(516)					Sessile drop	37
				25	(435)					Sessile drop	91

Table 2–17 (continued)
SURFACE TENSION OF LIQUID ELEMENTS

Element	Purity (wt. %)	σ_{mp} (dyn/cm)	Atm.	t_1 °C	σ_{t1} (dyn/cm)	t_2 °C	σ_{t2} (dyn/cm)	t_3 °C	σ_{t3} (dyn/cm)	Method	Ref.
Hg (cont.)											
			vac.	25	473					Drop weight	31
				25	488					Sessile drop	32
				25	(498)					Sessile drop	30
			vac.	20	(420)					Sessile drop	174
				25	476					Sessile drop	90
			vac.	20	(410)					Drop pressure	178
			vac.	20	(455)					Drop pressure	39
				25	484±1.5					Sessile drop	89
			vac.	22	(468)					Drop pressure	162
			vac.	20	(465.2)	103	449.7	350	387.1	Drop pressure	162
				25	484.9±1.8					Sessile drop	27
			vac.	20	485.5±1.0	$\sigma_t = 489.5 - 0.20\,(t°C)$				Drop pressure	17
			Ar	20	(454.7)					Bubble pressure	188
				21	(350.5)					Bubble pressure	23
	99.99		He, H$_2$	20	475					Bubble pressure	62
	99.9		Ar	20	(500±15)					Bubble pressure	103
			vac.	−10	487					Drop pressure	69
			vac.	25	483.5±1.0					Sessile drop	140
				22	(465)					Bubble pressure	194
				25	485.1					Sessile drop	142
				16.5	487.3	25	485.4±1.2			Pendant drop	171
				23–25	482.8±9.7					Contact angle	18
				$\sigma_t = 468.7 - 1.61\cdot10^{-1}t - 1.815\cdot10^{-4}t^2\,(t°C)$						—	206
			vac.	20	484.6±1.3					Pendant drop	132
			Ar	25	480					Bubble pressure	172
			vac.	20	482.5±3.0					Bubble pressure	177
			Ar	$\sigma_t = 485.5 - 0.149t - 2.84\cdot10^{-4}t^2\,(t°C)$							Bubble pressure	Z13a
				21.5	484.9±0.3						
In	99.95	559	H$_2$	600	515					Capillary method	133
				623	540					Capillary method	77
	99.995	556.0	Ar, He							Bubble pressure	196

Table 2–17 (continued)
SURFACE TENSION OF LIQUID ELEMENTS

Element	Purity (wt.%)	σ_{mp} (dyn/cm)	Atm.	σ_{t1} t₁°C	σ (dyn/cm)	σ_{t2} t₂°C	σ (dyn/cm)	σ_{t3} t₃°C	σ (dyn/cm)	Method	Ref.
In (cont.)			vac.	185	592					Drop pressure	69
			H₂	600	514					–	16
			Ar	300	541					Sessile drop	83
	99.999		Ar	200	556	400	535	550	527.8	Bubble pressure	123
	99.9994		vac.	350	539					Drop pressure	106
	99.9999	560±5	–		$\sigma = 568.0-0.04t-7.08 \cdot 10^{-5}t^2$ (t°C)					Sessile drop	Z12
Ir	99.9980	2250	vac.							Pendant drop	5
K	99.895	101	Ar							Bubble pressure	189
		110.3±1	–							–	84
		117	vac.		$\sigma = 117-0.66\,(t-t_{mp})$					Drop weight	185
			Ar	87	112	457	80	677	64.8	Bubble pressure	24
	99.986	116.95	Ar		$\sigma = 116.95-6.742 \cdot 10^{-2}\,(t-t_{mp})-3.836 \cdot 10^{-5}\,(t-t_{mp})^2$ $+3.707 \cdot 10^{-8}(t-t_{mp})^3$ (t°C) (77–983°C)					Bubble pressure	172
	99.936	(79.2)	He		$(\sigma = 76.8-70.3 \cdot 10^{-4}(t-400))$ (t°C) (600–1126°C)					Bubble pressure	121
		95±9.5	–							Drop weight	29
	99.97	111.35 ±0.64	He		$\sigma \div 115.51-0.0653 \cdot$ (t°C) (70–713°C)					Bubble pressure	Z10
Li	99.95		Ar	180	397.5	300	380	500	351.5	Bubble pressure	189
	99.98		Ar	287	386	922	275	1077	253	Bubble pressure	24
Mg	99.8		Ar	681	563	789	532	894	502	Bubble pressure	208
	99.9		N₂	670	552	700	542	740	528	Bubble pressure	149
	99.91	(525±10)	Ar	700	550±15					Bubble pressure	114
	99.5	583	Ar		$\sigma = 721-0.149\ T$ (T°K) (1125–1326°K)					Bubble pressure	105
										Bubble pressure	25
Mn	99.9985	1100 ± 50	Ar	1550	1030					Dynam. drop weight	6
	99.94		vac.	1550	1010					Sessile drop	159
	–		–							Sessile drop	Z11

Table 2–17 (continued)
SURFACE TENSION OF LIQUID ELEMENTS

Element	Purity (wt. %)	σ_{mp} (dyn/cm)	Atm.	σ_{t1} t_1°C	σ (dyn/cm)	σ_{t2} t_2°C	σ (dyn/cm)	σ_{t3} t_3°C	σ (dyn/cm)	Method	Ref.
Mo	99.7	(1915)	vac.							Pendant drop	148
		2080	vac.							Drop weight	139
	99.9996	2250	vac.							Pendant drop	5
	99.98	2049	vac.							Pendant drop	65
	–	2130	vac.							Pendant drop	Z1
Na	99.995	191	Ar	110	205.7	263	198.2			Bubble pressure	208
			Ar	123	198	129	198.5			Bubble pressure	188
			vac.	140	190					Drop volume	2
		200.2±0.6	vac.							Sessile drop	28
		202	vac.							—	84
			vac.	$\sigma = 202-0.092\,(t-t_{mp})$: 100–1000C						Drop weight	185
	p.a.	210.12	Ar	617	144	764	130	855	120.4	Bubble pressure	24
	99.96		Ar	$\sigma = 144-0.108(t-500)$ (t°C) (400–1125°C)						Bubble pressure	172
	99.982	187.4	He							Bubble pressure	121
Nb, Cb	99.9986	1900	vac.							Pendant drop	5
	99.99	(1827)	vac.							Pendant drop	65
	–	2020	vac.							Pendant drop	Z1
Nd	–	688	Ar	1186	674					Bubble pressure	124
Ni	99.7		He	1470	(1615)					Sessile drop	94
	99.7		H_2	1470	(1570)					Sessile drop	94
	99.7		vac.	1470	1735					Sessile drop	94
		1725	vac.	1475	1725					Sessile drop	141
			vac.	1550	(1934)					Sessile drop	115
	–		Ar							Sessile drop	108
	99.99		vac., Al_2O_3	1520	1740					Sessile drop	46
	99.99		He, Ar, Al_2O_3	1520	1770					Sessile drop	46
	99.99		H_2, Al_2O_3	1520	(1600)					Sessile drop	46
	99.99		He, MgO	1470	(1530)					Sessile drop	46
	99.99		He, BeO	1470	(1500)					Sessile drop	46

Na (ref. 24): $\sigma = 210.12-9.105 \cdot 10^{-2}\,(t-t_{mp})-8.064 \cdot 10^{-5}\,(t-t_{mp})^2 +3.380 \cdot 10^{-8}\,(t-t_{mp})^3$ (t°C) (141–992°C)

Table 2–17 (continued)
SURFACE TENSION OF LIQUID ELEMENTS

Element	Purity (wt. %)	σ_{mp} (dyn/cm)	Atm.	t_1 °C	σ_{t1} (dyn/cm)	t_2 °C	σ_{t2} (dyn/cm)	t_3 °C	σ_{t3} (dyn/cm)	Method	Ref.
Ni (cont.)	99.99		H₂	1530	(1650)					Bubble pressure	46
	99.99		H₂	1470	(1530)					Bubble pressure	46
	99.99		He	1470	(1490)					Bubble pressure	46
		1725	vac.	1600	(1600)					Sessile drop	7
	99.99		vac.	1500	1720					Sessile drop	135
	99.99		vac., Al₂O₃	1550	1780					Sessile drop	50
	99.99		vac., Al₂O₃	1550	1735					Sessile drop	54
			H₂, He	1470	1700					Bubble pressure	60
	99.99		vac., Al₂O₃	1640	1705					Bubble pressure	61
			vac.	1560	1810					Sessile drop	56
			vac.							Sessile drop	207
	99.9991	1770±13	vac.							Drop weight	5
	99.9991	1728±10	vac.							Drop weight	5
	99.9991	1822±8	vac.							Pendant drop	5
		(1670)	vac.							Drop weight	139
		1760	vac.							Sessile drop	55
		(1687)	vac.							Pendant drop	65
			He	1500	1745					Sessile drop	57
	—	1809±20	H₂, He Al₂O₃		$\sigma = 1770 - 0.39\,(t-1550)\ (t°C)$					Sessile drop	157
Os	99.99975	(1977)	He		$\sigma = 1665 + 0.215t\ (t°C)\ (1475{-}1650°C)$					Oscillating drop	Z2
	99.9998	2500	vac.							Pendant drop	5
P(white)				50	69.7	68.7	64.95			Bubble pressure	80
Pb	99.98	451	H₂, N₂	340	448	390	442	440	439	Bubble pressure	71
			air	360	452					Ring removal	97
		450	vac.	425	440					Drop pressure	101
			He	350–450	450					Pendant drop	87
										Ring removal	13
	99.998	480	H₂	623	474					Capillary method	133
										Capillary method	77

Table 2–17 (continued)
SURFACE TENSION OF LIQUID ELEMENTS

Element	Purity (wt. %)	σ_{mp} (dyn/cm)	Atm.	t_1 °C	σ_{t1} (dyn/cm)	t_2 °C	σ_{t2} (dyn/cm)	t_3 °C	σ_{t3} (dyn/cm)	Method	Ref.
Pb (cont.)			vac.	362	455					Drop pressure	69
				700	428					Sessile drop	27
	99.9		H₂	1000	388					Bubble pressure	134
	99.9	(410±5)	Ar							Bubble pressure	104
	99.98		vac.	350	445					Drop pressure	153
			vac.	340	442	400	435			Drop pressure	70
	99.9995	470		$\sigma = 538 - 0.114\,T$ (T°K) (1440–1970°K)						Bubble pressure	26
	99.9994		vac.	450	438					Drop pressure	107
	99.999		He	1600	310					Sessile drop	199
				390	456					Bubble pressure	1
		424±10	Air							Solid state curvature	Z8
	99.999	470	Ar	$\sigma_t = 470 - 0.164\,(t - t_{mp})$ (MP–535°C) (t°C)						Bubble pressure	Z13a
Pd	99.998	1470	vac.							Sessile drop	50
	99.998	1500	vac.							Pendant drop	5
	99.998	1460	He							Sessile drop	199
Pt	99.84	1869	CO₂							Drop weight	167
		(1740±20)	vac.							Drop volume	48
	99.999	1865	Ar	1800	(1699±20)					Sessile drop	109
	99.9980		vac.							Pendant drop	5
Pu		550±55								—	186
Rb		(77±5)	vac.							Drop diffusion in quartz tube	201
	99.8		Ar	52	84	477	55	632	46.8	Bubble pressure	24
	99.92	91.17	Ar	$\sigma = 91.17 - 9.189 \cdot 10^{-2}(t - t_{mp}) + 7.228 \cdot 10^{-5}(t - t_{mp})^2 - 3.830 \cdot 10^{-8}(t - t_{mp})^3$ (t°C) (104–1006°C)						Bubble pressure	172
	99.997	85.7	He	$\sigma = 85.7 - 0.054\,(t - t_{mp})$ (t°C) (53–1115°C)						Bubble pressure	121

Table 2–17 (continued)
SURFACE TENSION OF LIQUID ELEMENTS

Element	Purity (wt. %)	σ_{mp} (dyn/cm)	Atm.	σ_{t1} t_1 °C	σ_{t1} σ (dyn/cm)	σ_{t2} t_2 °C	σ_{t2} σ (dyn/cm)	σ_{t3} t_3 °C	σ_{t3} σ (dyn/cm)	Method	Ref.
Re	99.4	2610	vac.							Pendant drop	148
	99.9999	2700	vac.							Pendant drop	5
Ru	99.9980	2250	vac.							Pendant drop	5
Rh	99.9975	1940	vac.							Sessile drop	50
		2000	vac.							Pendant drop	5
S	–	60.9	vac.	250	51.1					Pendant drop	143
Sb	99.5	383	H_2	640	349	700	349	974	342	Drop weight	20
	99.99	395±20	H_2	750	368	900	361	1100	348	Bubble pressure	40
	99.15	395±20	vac.	640	367.9	762	364.9			Drop weight	131
			H_2, N_2	675	384	800	380			Bubble pressure	71
			Ar							Bubble pressure	104
			Ar							Bubble pressure	105
	99.999		N_2	800	359	1000	351	1100	345	Bubble pressure	128
	99.995		Ar	650	350.2	700	347.6	800	345.0	Bubble pressure	123
	99.999		He	1600	320					Sessile drop	199
Se	–		Ar	230–250	88.0±5					Bubble pressure	105
Si			He	1450	725					Pendant drop	87
			vac.	1550	720					Sessile drop	68
	99.99		vac.	1550	750					Sessile drop	41
	99.9999		Ar	1500	825					Pendant drop	43
Sn	–		N_2	275	612	500	572	800	520	Bubble pressure	149
			air	280	523	340	520			Ring removal	97
	99.99	537	vac.	500	524	600	508			Drop pressure	154
		530	He							Pendant drop	87
			H_2	489	543	572	528	692	503	Conical capillaries	9

Table 2–17 (continued)
SURFACE TENSION OF LIQUID ELEMENTS

Element	Purity (wt. %)	σ_{mp} (dyn/cm)	Atm.	σ_{t1} t_1 °C	σ_{t1} σ (dyn/cm)	σ_{t2} t_2 °C	σ_{t2} σ (dyn/cm)	σ_{t3} t_3 °C	σ_{t3} σ (dyn/cm)	Method	Ref.
Sn (cont.)											
			—	250	536					Drop pressure	179
			—	450	530					Drop pressure	119
			—	250	545					Drop pressure	112
	99.93		vac.	250	549	400	539	600	526	Drop pressure	156
	99.998	566	H_2	623	559					Capillary method	133
		610	vac.							Capillary method	77
			—							Pendant drop	182
			—	800	500					Sessile drop	7
			—	300	538					Drop pressure	190
			—	300	(527)					Drop pressure	76
				290	546					Sessile drop	190
	99.99		H_2, He	600	530					Bubble pressure	62
	99.9	(526±10)	Ar	290	600					Bubble pressure	104
			vac.							Sessile drop	147
	99.965	543.7	H_2	740	508	950	489.5	1115	479.5	Bubble pressure	127
	99.89	562	vac.							Bubble pressure	33
			vac.	300	554					Sessile drop	55
			vac.							Sessile drop	64
			vac.							Sessile drop	86
	99.999	590	H_2	290	(520)	290	(524)	(vac.)		Sessile drop	67
	99.9999		H_2	246	552.7					Sessile drop	203
	99.9994		vac.	350	537					Drop pressure	106
Sr	99.999	555.8±1.9		\multicolumn{6}{l}{$\sigma = 566.84 - 4.76 \cdot 10^{-2}\, t$ (t°C)}	Bubble pressure	176
	99.96	552	vac.	1000	470					Sessile drop	Z5
	99.96	552	Ar	\multicolumn{6}{l}{$\sigma_t = 552 - 0.167\,(t - t_{mp}) \cdot (MP{-}500°C)$ (t°C)}	Bubble pressure	Z13a
			Ar	775	288	830	282	893	282	Bubble pressure	125
	99.5	303	Ar	\multicolumn{6}{l}{$\sigma = 392 - 0.085\,T$ (T°K) (1152–1602°K)}	Bubble pressure	25
Ta		2360	vac.							Pendant drop	88
		2030	vac.							Pendant drop	88
		1910	vac.							Drop weight	139

Table 2–17 (continued)
SURFACE TENSION OF LIQUID ELEMENTS

Element	Purity (wt. %)	σ_{mp} (dyn/cm)	Atm.	t_1 °C	σ_{t1} (dyn/cm)	t_2 °C	σ_{t2} (dyn/cm)	t_3 °C	σ_{t3} (dyn/cm)	Method	Ref.
Ta (cont.)											
	99.9983	2150	vac.							Pendant drop	5
	99.9	(1884)	vac.							Pendant drop	65
Te	99.4	186±2	Ar							Bubble pressure	105
	–		vac.	460	178±1.5					Capillary method	184
			vac.	475	(162)					Electro-capillarity	117
		178	vac.		$\sigma = 178 - 0.024\,(t - t_{mp})\ (t°C)$					Bubble pressure	204
Ti	98.7	1510	vac.							Capillary method	44
	99.92	1390	Ar							Pendant drop	151
		1460	vac.							Drop weight	139
	99.9991	1650	vac.							Pendant drop	5
	99.0		vac.	1680	1576					Drop weight	191
	99.99999		vac.	1680	1588					Drop weight	191
	99.85	(1880)	vac.							Pendant drop	65
	99.69	1402	vac.							Pendant drop	65
Tl		464.5	Ar	450	452					Bubble pressure	192
			–	450	450					Drop pressure	119
	–		vac.							Electro-capillarity	117
	99.999	467	vac.							Bubble pressure	26
	99.999		vac.	450	450		$\sigma = 536 - 0.119T\ (T°K)\ (1270–1695°K)$			Drop pressure	107
U		1500±75	Ar							Bubble pressure	186
	99.94	1550	vac.							Pendant drop	34
	–	(1294)	vac.							Pendant drop	65
V	99.9977	1950	vac.							Pendant drop	5
	–	(1760)	vac.							Pendant drop	Z1

Table 2–17 (continued)
SURFACE TENSION OF LIQUID ELEMENTS

Element	Purity (wt. %)	σ_{mp} (dyn/cm)	Atm.	σ_{t1}		σ_{t2}		σ_{t3}		Method	Ref.
				t_1 °C	σ (dyn/cm)	t_2 °C	σ (dyn/cm)	t_3 °C	σ (dyn/cm)		
W	–	2310	vac.							Pendant drop	35
	99.9999	2500	vac.							Pendant drop	5
	99.8	2220	vac.							Pendant drop	148
	99.9	(2000)	vac.							Pendant drop	65
Zn	99.9	750±20	Ar							Bubble pressure	104
	99.99	757.0±5	vac.							Sessile drop	202
	99.999	761.0	vac.							Sessile drop	202
	99.9999	767.5	vac.							Sessile drop	202
Zr	99.5	1400	Ar							Drop weight	151
		1411±70	vac.							Drop weight	180
	99.9998	1480	vac.							Pendant drop	5
	99.7	(1533)	vac.							Pendant drop	65

From Lang, G., in *Handbook of Chemistry and Physics*, 55th ed., Weast, R. C., Ed., CRC Press, Cleveland, 1974, F-23.

REFERENCES

1. **Abdel-Aziz Abol Hassan,** *Neue Hütte,* 15, 304, 1970.
2. **Addison, Addison, Kerridge, and Lewis,** *J. Chem. Soc.,* p. 2262, 1955.
3. **Addison, Coldrey, and Pulham,** *J. Chem. Soc.,* p. 1227, 1963.
4. **Addison and Raymor,** *J. Chem. Soc.,* p. 965, 1966.
5. **Allen,** *Trans. Met. Soc. AIME,* 227, 1175, 1963.
6. **Allen,** *Trans. Met. Soc. AIME,* 230, 1357, 1964.
7. **Allen and Kingery,** *Trans Met. Soc. AIME,* 215, 30, 1959.
8. **Astakhov, Penin, and Dobkina,** *Zh. Fiz. Khim.,* 20, 403, 1946.
9. **Atterton and Hoar,** *J. Inst. Metals,* 81, 541, 1953.
10. **Ayushina, Levin, and Geld,** *Zh. Fiz. Khim.,* 42, 2799, 1968.
11. **Baes and Kellogg,** *J. Metals,* 15, 643, 1953.
12. **Baker and Gilbert,** *J. Am. Chem. Soc.,* 62, 2479, 1940.
13. **Bakradse and Pines,** *Zh. Tekhn. Fix.,* 23, 1548, 1953.
14. **Becker, Harders, and Kornfeld,** *Arch. Eisenhüttenw.,* 20, 363, 1949.
15. **Belforti and Lepie,** *Trans Met. Soc. AIME,* 227, 80, 1963.

Table 2–17 (continued)

SURFACE TENSION OF LIQUID ELEMENTS

16. Bergh, *J. Electrochem. Soc.*, 109, 1199, 1962.
17. Bring and Ioileva, *Dokl. Akad. Nauk*, 93, 85, 1953.
18. Biery and Oblak, *Ind. Eng. Chem. Fundam.*, 5, 121, 1966.
19. Bircumshaw, *Philos. Mag.*, 2, 341, 1926.
20. Bircumshaw, *Philos. Mag.*, 3, 1286, 1927.
21. Bircumshaw, *Philos. Mag.*, 6, 510, 1928.
22. Bircumshaw, *Philos. Mag.*, 12, 596, 1931.
23. Bobyk, *Przem. Chem.*, 39, 423, 1960.
24. Bohdanski and Schins, *J. Inorg. Nucl. Chem.*, 29, 2173, 1967.
25. Bohdanski and Schins, *J. Inorg. Nucl. Chem.*, 30, 2331, 1968.
26. Bohdanski and Schins, *J. Inorg. Nucl. Chem.*, 30, 3362, 1968.
27. Bradhurst and Buchanan, *J. Phys. Chem.*, 63, 1486, 1959.
28. Bradhurst and Buchanan, *Aust. J. Chem.*, 14, 397, 1961.
29. Bradhurst and Buchanan, *Aust. J. Chem.*, 14, 409, 1961.
30. Bradley, *J. Phys. Chem.*, 38, 234, 1934.
31. Brown, *Philos. Mag.*, 13, 578, 1932.
32. Burdon, *Trans. Faraday Soc.*, 28, 866, 1932.
33. Cahill and Kirshenbaum, *J. Inorg. Nucl. Chem.*, 26, 206, 1964.
34. Cahill and Kirshenbaum, *J. Inorg. Nucl. Chem.*, 27, 73, 1965.
35. Calverley, *Proc. Phys. Soc.*, 70, 1040, 1957.
36. Coffman and Parr, *Ind. Eng. Chem.*, 19, 1308, 1927.
37. Cook, *Phys. Rev.*, 34, 513, 1929.
38. de L. Davies and West, *J. Inst. Metals*, 92, 208, 1964.
39. Didenko and Pokrovskii, *Dokl. Akad. Nauk*, 31, 233, 1941.
40. Drath and Sauerwald, *Z. Allg. Anorg. Chem.*, 162, 301, 1927.
41. Dshemilev, Popel, and Zarevskii, *Fiz. Metal. Metalloved.*, 18, 83, 1964.
42. Dyson, *Trans. Met. Soc. AIME*, 227, 1098, 1963.
43. Eljutin, Kostikov, and Levin, *Izv. Vyssh. Uchebn. Zaved. Tsvet. Met.*, (2), 131, 1970.
44. Eljutin and Maurakh, *Izv. Akad. Nauk OTN*, (4), 129, 1956.
45. Eremenko, Ivashchenko, and Bogatyrenko, in *The Role of Surface Phenomena in Metallurgy*, Eremenko, V. N., Ed., Consultants Bureau Enterprises, New York, 1963, 37.
46. Eremenko, Ivashchenko, Fessenko, and Nichenko, *Izv. Akad. Nauk OTN*, 7, 144, 1958.
47. Eremenko and Naidich, *Izv. Akad. Nauk OTN*, (2), 111, 1959.
48. Eremenko and Naidich, *Izv. Akad. Nauk OTN*, (6), 129, 1959.
49. Eremenko and Naidich, *Izv. Akad. Nauk OTN*, (6), 100, 1961.
50. Eremenko and Naidich, *Izv. Akad. Nauk OTN*, (2), 53, 1960.
51. Eremenko and Naidich, in *The Role of Surface Phenomena in Metallurgy*, Eremenko, V. N., Ed., Consultants Bureau Enterprises, New York, 1963, 65.
52. Eremenko, Naidich, and Nossonovich, *Zh. Fiz. Khim.*, 34, 1018, 1960.
53. Eremenko and Nishenko, *Urk. Khim. Zh.*, 26, 423, 1960.
54. Eremenko and Nishenko, *Zh. Fiz. Khim.*, 35, 1301, 1961.
55. Eremenko and Nishenko, *Ukr. Khim. Zh.*, 30, 125, 1964.

Table 2–17 (continued)

SURFACE TENSION OF LIQUID ELEMENTS

56. Eremenko, Nishenko, and Naidich, *Izv. Akad. Nauk OTN*, (3), 150, 1961.
57. Eremenko, Nishenko, and Skljarenko, *Izv. Akad. Nauk OTN*, (2), 188, 1966.
58. Eremenko, Nishenko, and Taj-Shou-Vej, *Izv. Akad. Nauk OTN*, (3), 116, 1960.
59. Eremenko and Vassiliu, *Ukr. Khim. Zh.*, 31, 557, 1965.
60. Eremenko, Vassiliu, and Fessenko, *Zh. Fiz. Khim.*, 35, 1750, 1961.
61. Fessenko, *Zh. Fiz. Khim.*, 35, 707, 1961.
62. Fessenko and Eremenko, *Ukr. Khim Zh.*, 26, 198, 1960.
63. Fessenko, Vassiliu, and Eremenko, *Zh. Fiz. Khim.*, 36, 518, 1962.
64. Flechsig, Thesis, Technical University of Berlin, Germany, 1964.
65. Fling, *J. Nucl. Mater.*, 16, 260, 1965.
66. Gans, Pawlek, and Roepenack, *Z. Metallk.*, 54, 147, 1963.
67. Gans and Parthey, *Z. Metallk.*, 57, 19, 1966.
68. Geld and Petrushevski, *Izv. Akad Nauk OTN*, (3), 160, 1961.
69. Gratzianski and Rjabov, *Zh. Fiz. Khim.*, 33, 487, 1253, 1959.
70. Gratzianski, Rjabov, and Tobolich, *Ukr. Khim. Zh.*, 29, 1219, 1963.
71. Greenaway, *J. Inst. Metals*, 74, 133, 1947.
72. Grunmach, *Ann. Phys.*, 3, 660, 1900.
73. Hagemann, Thesis, University of Freiburg, Germany, 1914.
74. Halden and Kingery, *J. Phys. Chem.*, 59, 557, 1955.
75. Harkins and Ewing, *J. Am. Chem. Soc.*, 42, 2539, 1920.
76. Herczynska, *Z. Phys. Chem.*, 214, 355, 1960.
77. Hoar and Melford, *Trans. Faraday Soc.*, 53, 315, 1957.
78. Hogness, *J. Am. Chem. Soc.*, 43, 1621, 1921.
79. Humenik and Kingery, *J. Am. Ceram. Soc.*, 37, 18, 1954.
80. Hutchinson, *Trans Faraday Soc.*, 39, 229, 1943.
81. Iredale, *Philos Mag*, 45, 1088, 1923.
82. Iredale, *Philos Mag*, 48, 177, 1924.
83. Jacobj, Thesis, Technical University of Braunschweig, Germany, 1962.
84. Jordan and Lane, *Aust. J. Chem.*, 18, 1711, 1965.
85. Karasaev, Sadumkin, and Kukhno, *Zh. Fiz. Khim.*, 41, 654, 1967.
86. Kaufman and Whalen, *Acta Met.*, 13, 797, 1965.
87. Keck and van Horn, *Phys Rev.*, 91, 512, 1953.
88. Kelly and Calverley, *SERL Rep.*, 80, 53, 1959.
89. Kemball, *Trans Faraday Soc.*, 42, 526, 1946.
90. Kernaghan, *Phys. Rev.*, 37, 990, 1931.
91. Kernaghan, *Phys. Rev.*, 49, 414, 1936.
92. Kingery, *J. Am. Ceram. Soc.*, 37, 42, 1954.
93. Kingery, *Kolloid-Z. Z. Polym.*, 161, 95, 1958.
94. Kingery and Humenik, *J. Phys. Chem.*, 57, 359, 1953.
95. Kingery and Norton, AEC Progress Report NYO-629, U.S. Atomic Energy Commission, Washington, D.C., 1954.

Table 2–17 (continued)
SURFACE TENSION OF LIQUID ELEMENTS

96. Klyachko, *Zavod. Lab.*, 6, 1376, 1937.
97. Klyachko and Kunin, *Zavod. Lab.*, 14, 66, 1948.
98. Klyachko and Kunin, *Dokl. Akad. Nauk*, 64, 64, 1949.
99. Kolesnikova, *Izv. Vyssh. Uchebn. Zaved. Tsvet. Met.*, (9), 1960.
100. Kolesnikova and Samarin, *Izv. Akad. Nauk OTN*, (5), 63, 1956.
101. Konstantinov, Thesis, State University of Moscow, U.S.S.R., 1950.
102. Korolkov, *Izv. Akad. Nauk OTN*, (2), 35, 1956.
103. Korolkov, *Litein'e Svojstva Metallov i Splavov*, Isdatelstvo, Akad. Nauk, S.S.S.R., 1960, 37.
104. Korolkov and Bychkova, *Issled. Splav. Tsvet. Met.*, (2), 122, 1960.
105. Korolkov and Igumnova, *Izv. Akad. Nauk OTN*, (6), 95, 1961.
106. Kovalchuk, Kusnezov, and Kotlovanova, *Zh. Fiz. Khim.*, 42, 1754, 1968.
107. Kovalchuk, Kusnezov, and Butuzova, *Zh. Fiz. Khim.*, 42, 2265, 1968.
108. Kozakevitch and Urbain, *J. Iron Steel Inst.*, 186, 167, 1957.
109. Kozakevitch and Urbain, *Compt. Rend.* (Paris), 253, 167, 1957.
110. Kozakevitch and Urbain, *Mem. Sci. Rev. Met.*, 58, 401, 1961.
111. Krause, Sauerwald, and Michalke, *Z. Anorg. Allg. Chem.*, 181, 353, 1929.
112. Kristian, Thesis, State University of Moscow, U.S.S.R., 1954.
113. Kubichek, *Izv. Akad. Nauk OTN*, (2), 96, 1959.
114. Kubichek and Malzev, *Izv. Akad. Nauk OTN*, (3), 144, 1959.
115. Kurkjian and Kingery, *J. Phys. Chem.*, 60, 961, 1956.
116. Kurochkin, Baum, and Borodulin, *Fiz. Metal Metalloved.*, 15, 461, 1963.
117. Kusnezov, in *The Role of Surface Phenomena in Metallurgy*, Eremenko, V. N., Ed., Consultants Bureau Enterprises, New York, 1963, 72.
118. Kusnezov, Djakova, and Malzeva, *Zh. Fiz. Khim.*, 33, 1551, 1959.
119. Kusnezov, Kochergin, Tishchenko, and Posdynsheva, *Dokl. Akad. Nauk*, 92, 1197, 1953.
120. Kusnezov, Popova, and Duplina, *Zh. Fiz. Khim.*, 36, 880, 1962.
121. Kyrianenko and Solovev, *Teplofiz. Vys. Temp.*, 8, 537, 1970.
122. Lasarev, *Zh. Fiz. Khim.*, 36, 405, 1962.
123. Lasarev, *Zh. Fiz. Khim.*, 38, 325, 1964.
124. Lasarev and Pershikov, *Dokl. Akad. Nauk*, 146, 143, 1962.
125. Lasarev and Pershikov, *Zh. Fiz Khim.*, 37, 907, 1963.
126. Lasarev and Pugachevich, *Dokl. Akad. Nauk*, 134, 132, 1960.
127. Lauermann, Metzger, and Sauerwald, *Z. Phys. Chem.*, 216, 42, 1961.
128. Lauermann and Sauerwald, *Z. Metallk.*, 55, 605, 1964.
129. Lucas, *Compt. Rend.* (Paris), 248, 2336, 1959.
130. Mack, Davis, and Bartell, *J. Phys. Chem.*, 45, 846, 1941.
131. Matuyama, *Sci. Rep. Res. Inst. Tohoku Univ.*, 16, 555, 1927.
132. Melik-Gajkazan, Woronchikhina, and Sakharova, *Elektrokhimiya*, 4, 1420, 1968.
133. Melford and Hoar, *J. Inst. Metals*, 85, 197, 1957.
134. Metzger, *Z. Phys. Chem.*, 211, 1, 1959.
135. Monma and Suto, *J. Jap. Inst. Metals*, 24, 167, 1960.

Table 2–17 (continued)

SURFACE TENSION OF LIQUID ELEMENTS

136. Monma and Suto, *Trans. Jap. Inst. Metals*, 1, 69, 1960.
137. Naidich and Eremenko, *Fiz. Metal. Metalloved.*, 11, 883, 1961.
138. Naidich, Eremenko, Fessenko, Vassiliu, and Kirichenko, *Zh. Fiz. Kihm.*, 35, 694, 1961.
139. Namba and Isobe, *Sci. Pap. Inst. Phys. Chem. Res* (Tokyo), 57, 5154, 1963.
140. Nicholas, Joyner, Tessem, and Olson, *J. Phys. Chem.*, 65, 1375, 1961.
141. Norton and Kingery, AEC Progress Report NYO 4632, U.S. Atomic Energy Commission, Washington, D.C., 1955.
142. Olson and Johnson, *J. Phys. Chem.*, 67, 2529, 1963.
143. Ono and Matsushima, *Sci. Rep. Res. Inst. Tohoku Univ.*, 9, 309, 1957.
144. Oppenheimer, *Z. Anorg. Allg. Chem.*, 171, 98, 1928.
145. Osminin, *Zh. Fiz. Khim.*, 43, 2610, 1969.
146. Palacios, *An. Soc. Espan. Fis*, 18, 294, 1920.
147. Parthey, Thesis, Technical University of Berlin, Germany, 1961.
148. Pekarev, *Izv. Vyssh. Uchebn. Zaved. Tsvet. Met.*, 6, 111, 1963.
149. Pelzel, *Berg Hüttenmännische Monatsh.*, 93, 248, 1948.
150. Pelzel, *Berg Hüttenmännische Montash.*, 94, 10, 1949.
151. Peterson, Kedesdy, Keck, and Schwarz, *J. Appl. Phys.*, 29, 213, 1958.
152. Poindexter, *Phys. Rev.*, 27, 820, 1926.
153. Pokrovski, *Ukr. Khim. Zh.*, 7, 845, 1962.
154. Pokrovski and Galanina, *Zh. Fiz. Khim.*, 23, 324, 1949.
155. Pokrovski and Kristian, *Zh. Fiz. Khim.*, 28, 1954.
156. Pokrovski and Saidov, *Fiz Metal. Metalloved.*, 2, 546, 1956.
157. Popel, Shergin, and Zarevski, *Zh. Fiz. Khim.*, 43, 2365, 1969.
158. Popel, Smirnov, Zarevski, Dshemilev, and Pastukhov, *Izv. Akad. Nauk.* (1), 62, 1965.
159. Popel, Zarevski, and Dshemilev, *Fiz. Metal. Metalloved.*, 18, 468, 1964.
160. Popesco, *Compt. Rend.* (Paris), 172, 1474, 1921.
161. Portevin and Bastien, *Compt. Rend.* (Paris), 202, 1072, 1936.
162. Pugachevich, *Zh. Fiz. Khim.*, 25, 1365, 1951.
163. Pugachevich and Altynov, *Dokl. Akad. Nauk.*, 86, 117, 1952.
164. Pugachevich and Yashkichev, in *The Role of Surface Phenomena in Metallurgy*, Ermenko, V. N., Ed., Consultants Bureau Enterprises, New York, 1963, 46.
165. Quincke, *Ann. Phys.*, 134, 356, 1868.
166. Quincke, *Ann. Phys.*, 135, 621, 1868.
167. Quincke, *Ann. Phys.*, 138, 141, 1869.
168. Quincke, *Ann. Phys.*, 61, 267, 1897.
169. Raue, Metzger, and Sauerwald, *Metall*, 20, 1040, 1966.
170. Richards and Boyer, *J. Am. Chem. Soc.*, 43, 290, 1921.
171. Roberts, *J. Chem. Soc.*, p. 1907, 1964.
172. Roehlich, Jr., Tepper, and Rankin, *J. Chem. Eng. Data*, 13, 518, 1968.
173. Sauerwald and Drath, *Z. Anorg. Allg. Chem.*, 154, 79, 1926.
174. Sauerwald, Schmidt, and Pelka, *Z. Anorg. Allg. Chem.*, 223, 84, 1935.
175. Schmidt, *Ann. Phys.*, 39, 1108, 1912.

Table 2–17 (continued)
SURFACE TENSION OF LIQUID ELEMENTS

176. Schwaneke and Falke, Investigation Report No. 7372, U.S. Bureau of Mines, Washington, D.C., 1970.
177. Schwaneke, Falke, and Miller, Investigation Report No. 7340, U.S. Bureau of Mines, Washington, D.C., 1970.
178. Semenchenko and Pokrovski, *Usp. Khim.*, 6, 945, 1937.
179. Semenchenko, Pokrovski, and Lasarev, *Dokl. Akad. Nauk*, 89, 1021, 1953.
180. Shunk and Burr, *Trans. Am. Soc. Metals*, 55, 786, 1962.
181. Siedentopf, *Ann. Phys.*, 61, 235, 1897.
182. Smirnova and Ormont, *Zh. Fiz. Khim.*, 33, 771, 1959.
183. Smith, *J. Inst. Metals*, 12, 20, 168, 1914.
184. Smith and Spitzer, *J. Phys. Chem.*, 66, 946, 1962.
185. Solovev and Makarova, *Teplofiz. Vys. Temp.*, 4, 189, 1966.
186. Spriet, *Mem. Sci. Rev. Met.*, 60, 531, 1963.
187. Tavadse, Bairamashvili, Khantadse, and Zagareishvili, *Dokl. Akad. Nauk*, 150, 544, 1963.
188. Taylor, *J. Inst. Metals*, 83, 143, 1954.
189. Taylor, *Philos. Mag.*, 46, 867, 1955.
190. Thyssen, Thesis, Humboldt University, Berlin, Germany, 1960.
191. Tille and Kelly, *Br. J. Appl. Phys.*, 146, 717, 1963.
192. Timofeyevicheva and Lasarev, *Dokl. Akad. Nauk*, 138, 412, 1961.
193. Timofeyevicheva and Lasarev, *Dokl. Akad. Nauk*, 142, 358, 1962.
194. Timofeyevicheva and Lasarev, *Kolloid. Zh.*, 24, 227, 1962.
195. Timofeyevicheva, Lasarov, and Pershikov, *Dokl. Akad. Nauk*, 143, 618, 1962.
196. Timofeyevicheva and Pugachevich, *Dokl. Akad. Nauk*, 124, 1093, 1959.
197. Timofeyevicheva and Pugachevich, *Dokl. Akad. Nauk*, 134, 840, 1960.
198. Urbain and Lucas, Paper 4E, Proceedings, National Physical Laboratory, 1959.
199. Watolin, Esin, Ukhov, and Dubinin, *Tr. Inst. Met. Sverdlovsk*, 18, 73, 1969.
200. Whalen and Humenik, *Trans. Am. Met. Soc. AIME*, 218, 952, 1960.
201. Wegener, *Z. Phys.*, 143, 548, 1956.
202. White, *Trans. Am. Met. Soc. AIME*, 236, 796, 1966.
203. White, *Met. Rev.*, 124, 73, 1968.
204. Wobst and Rentzch, *Z. Phys. Chem.*, 240, 36, 1969.
205. Yashkichevich and Lasarev, *Izv. Akad. Nauk OTN*, (1), 170, 1964.
206. Yung Lee, *Ind. Eng. Chem. Prod. Res. Dev.*, 7, 66, 1968.
207. Zarevski and Popel, *Fiz. Metal. Metalloved.*, 13, 451, 1962.
208. Zhivov, *Tr. Vses Assos. Met. Inzhenerov*, 14, 99, 1937.
209. Zsin-Tan Wan, Karassev, and Samarin, *Izv. Akad. Nauk OTN*, (1), 30, 49, 1960.

Table 2–17 (continued)
SURFACE TENSION OF LIQUID ELEMENTS

Addendum

Z1. Eljutin, Kostikow, and Penkow, *Proshk. Met.*, 9, 46, 1970.
Z2. Fraser, Lu, Hamielee, and Muraka, *Met. Trans.*, 2, 817, 1971.
Z3. Bernard and Lupis, *Met. Trans.*, 2, 555, 1971.
Z4. Rhee, *J. Am. Ceram. Soc.*, 53, 386, 1970.
Z5. Naidich, Pervertailo, and Shuravlev, *Zh. Fiz. Khim.*, 45, 991, 1971.
Z6. Korber and Lohberg, *Giessereiforschung*, 23, 173, 1971.
Z7. Ziesing, *Aust. J. Phys.*, 6, 86, 1953.
Z8. Sangster and Carman, *J. Chem. Phys.*, 23, 1142, 1955.
Z9. Primak and Quarterman, *J. Chem. Phys.*, 58, 1051, 1954.
Z10. Cooke, HTLMHTTM, Vol. 1, Oak Ridge National Laboratory, Oak Ridge, Tenn., 1964, 66.
Z11. Ofzerow, *Izv. Akad. Nauk Metal.*, 4, 91, 1971.
Z12. White, *Met. Trans.*, 3, 1933, 1972.
Z13. Lang, *Aluminium* (Germany), 49, 231, 1872.
Z13a. Lang, *J. Inst. Metals*, to be published.

Table 2—18
SURFACE TENSION OF VARIOUS LIQUIDS

Substance	Formula	In contact with	Temperature (°C)	Surface tension (dyn/cm)
Acetaldehyde	C_2H_4O	vapor	20	21.2
Acetaldoxime	C_2H_5NO	vapor	35	30.1
Acetamide	C_2H_5NO	vapor	85	39.3
Acetanilide	C_2H_5NO	vapor	120	35.6
Acetic acid	$C_2H_4O_2$	vapor	10	28.8
		vapor	20	27.8
		vapor	50	24.8
Acetic anhydride	$C_4H_6O_3$	vapor	20	32.7
Acetone	C_3H_6O	air or vapor	0	26.21
		air or vapor	20	23.70
		air or vapor	40	21.16
Acetonitrile	C_2H_3N	vapor	20	29.30
Acetophenone	C_8H_8O	vapor	20	39.8
Acetyl chloride	C_2H_3ClO	vapor	14.8	26.7
Acetylene	C_2H_2	vapor	−70.5	16.4
Acetylsalicylic acid (in aqueous solution)	$C_9H_8O_4$	vapor	25.9	60.06
Allyl alcohol	C_3H_6O	air or vapor	20	25.8
Allyl isothiocyanate	C_4H_5NS	air or vapor	20	34.5
Ammonia	NH_3	vapor	11.1	23.4
		vapor	34.1	18.1
Aniline	C_6H_7N	air	10	44.10
		vapor	20	42.9
		air	50	39.4
Argon	A	vapor	−188	13.2
Azoxybenzene	$C_{12}H_{10}N_2O$	vapor	51	43.34
Benzaldehyde	C_7H_6O	air	20	40.04
Benzene	C_6H_6	air	10	30.22
		air	20	28.85
		vapor (saturated)	20	28.89
		air	30	27.56
Benzonitrile	C_7H_5N	air	20	39.05
Benzophenone	$C_{13}H_{10}O$	air or vapor	20	45.1
Benzylamine	C_7H_9N	vapor	20	39.5
Benzyl alcohol	C_7H_8O	air or vapor	20	39.0
Bromine	Br_2	air or vapor	20	41.5
Bromobenzene	C_6H_5Br	air	20	36.5
Bromoform	$CHBr_3$	vapor	20	41.53
p-Bromophenol	C_6H_5BrO	vapor	74.4	42.36
d-sec-Butyl alcohol	$C_4H_{10}O$	vapor	10	23.5
n-Butyl alcohol	$C_4H_{10}O$	air or vapor	0	26.2
		air or vapor	20	24.6
		air or vapor	50	22.1
tert-Butyl alcohol	$C_4H_{10}O$	air or vapor	20	20.7
n-Butylamine	$C_4H_{11}N$	nitrogen	41	19.7
n-Butyric acid	$C_4H_8O_2$	air	20	26.8
Carbon bisulfide	CS_2	vapor	20	32.33
Carbon dioxide	CO_2	vapor	20	1.16
		vapor	−25	9.13
Carbon tetrachloride	CCl_4	vapor	20	26.95
		vapor	100	17.26
		vapor	200	6.53

Table 2–18 (continued)
SURFACE TENSION OF VARIOUS LIQUIDS

Substance	Formula	In contact with	Temperature (°C)	Surface tension (dyn/cm)
Carbon monoxide	CO	vapor	−193	9.8
		vapor	−203	12.1
Chloral	C_2HCl_3O	vapor	19.4	25.34
Chlorine	Cl_2	vapor	20	18.4
		vapor	−30	25.4
		vapor	−40	27.3
		vapor	−50	29.2
		vapor	−60	31.2
Chloroacetic acid	$C_2H_2Cl_2O_2$	nitrogen	25.7	35.4
Chlorobenzene	C_4H_5Cl	vapor	20	33.56
Chloroform	$CHCl_3$	air	20	27.14
o-Chlorophenol	C_6H_5ClO	vapor	12.7	42.25
Cyclohexane	C_6H_{12}	air	20	25.5
Dichloroacetic acid	$C_2H_2Cl_2O_2$	nitrogen	25.7	35.4
Dichloroethane	$C_2H_4Cl_2$	air	35.0	23.4
Diethylamine	$C_4H_{11}N$	air	56	16.4
Diethylaniline	$C_{10}H_{15}N$	vapor	20	34.2
Diethyl carbonate	$C_5H_{10}O$	air	20	26.31
Diethyl oxalate	$C_6H_{10}O_4$	vapor	20	32.0
Diethyl phthalate	$C_{12}H_{14}O_4$	vapor	20	37.5
Diethyl sulfate	$C_4H_{12}O_4S$	air	13	34.61
Dimethylamine	C_2H_7N	nitrogen	0	18.1
		nitrogen	5	17.7
Dimethylaniline	C_8H_{11}	air or vapor	20	36.6
1,5-Dimethyl-2-phenyl-3-pyrazolone	$C_{11}H_{12}N_2O$	vapor	25.9	63.63
Dimethyl sulfate	$C_2H_6O_4S$	air	18	40.12
Diphenylamine	$C_{12}H_{11}N$	air or vapor	80	37.7
Ethyl acetate	$C_4H_8O_2$	air	0	26.5
		air	20	23.9
		air	50	20.2
Ethyl acetoacetate	$C_6H_{10}O_3$	air or vapor	20	32.51
Ethyl alcohol	C_2H_6O	air	0	24.05
		vapor	10	23.61
		vapor	20	22.75
		vapor	30	21.89
Ethylamine	C_2H_7N	nitrogen	0	21.3
		nitrogen	9.9	20.4
Ethylaniline	$C_8H_{11}N$	air or vapor	20	36.6
Ethylbenzene	C_8H_{10}	vapor	20	29.20
Ethylbenzoate	$C_9H_{10}O_2$	vapor	20	35.5
Ethyl bromide	C_2H_5Br	vapor	20	24.15
Ethyl chloroformate	$C_3H_5ClO_2$	vapor	15.1	27.5
Ethyl cinnamate	$C_{11}H_{12}O_2$	air	20	38.37
Ethylene bromide	$C_2H_4Br_2$	vapor	20	38.37
Ethylene chloride	$C_2H_4Cl_2$	air	20	24.15
Ethylene oxide	C_2H_4O	vapor	−20	30.8
		vapor	0.0	27.6
		vapor	20	24.3
Ethyl ether	$C_4H_{10}O$	vapor	20	17.01
		vapor	50	13.47
Ethyl formate	$C_3H_6O_2$	air or vapor	20	23.6
Ethyl iodide	C_2H_5I	vapor	20	29.4
Ethyl nitrate	$C_2H_5NO_3$	air or vapor	20	28.7

Table 2–18 (continued)
SURFACE TENSION OF VARIOUS LIQUIDS

Substance	Formula	In contact with	Temperature (°C)	Surface tension (dyn/cm)
dl-Ethyl lactate	$C_5H_{10}O_3$	air	20	29.9
Ethyl mercaptan	C_2H_6S	air or vapor	20	22.5
Ethyl salicylate	$C_9H_{10}O_3$	vapor	20.5	38.33
Formamide	CH_3NO	vapor	20	58.2
Formic acid	CH_2O_2	air	20	37.6
Furfural	$C_5H_4O_2$	air or vapor	20	43.5
Gelatin solution (1%)		water	2.85	8.3
Glycerol	$C_3H_8O_3$	air	20	63.4
		air	90	58.6
		air	150	51.9
Glycol	$C_2H_6O_2$	air or vapor	20	47.7
Helium	He	vapor	−269	.12
		vapor	−270	.239
		vapor	−271.5	.353
n-Hexane	C_6H_{14}	air	20	18.43
Hydrazine	N_2H_4	vapor	25	91.5
Hydrogen	H_2	vapor	−255	2.31
Hydrogen cyanide	HCN	vapor	17	18.2
Hydrogen peroxide	H_2O_2	vapor	18.2	76.1
Isobutyl alcohol	$C_4H_{10}O$	vapor	20	23.0
Isobutylamine	$C_4H_{11}N$	air	68	17.6
Isobutyl chloride	C_4H_9Cl	air	20	21.94
Isobutyric acid	$C_4H_8O_2$	air or vapor	20	25.2
Isopentane	C_5H_{12}	air	20	13.72
Isopropyl alcohol	C_3H_8O	air or vapor	20	21.7
Methyl acetate	$C_3H_6O_2$	air or vapor	20	24.6
Methyl alcohol	CH_4O	air	0	24.49
		air	20	22.61
		vapor	50	20.14
Methylamine	CH_3NH_2	nitrogen	−12	22.2
		vapor	−20	23.0
		nitrogen	−70	29.2
N-Methylaniline	C_7H_9N	air or vapor	20	39.6
Methyl benzoate	$C_8H_8O_2$	air or vapor	20	37.6
Methyl chloride	CH_3Cl	air	20	16.2
Methyl ether	C_2H_6O	vapor	−10	16.4
		vapor	−40	21
Methylene chloride	CH_2Cl_2	air	20	26.52
Methylene iodide	CH_2I_2	air	20	50.76
Methyl ethyl ketone	C_4H_8O	air or vapor	20	24.6
Methyl formate	$C_2H_4O_2$	vapor	20	25.08
Methyl iodide	CH_3I	air	43.5	25.8
Methyl propionate	$C_4H_8O_2$	air or vapor	20	24.9
Methyl salicylate	$C_8H_8O_3$	nitrogen	94	31.9
Methyl sulfide	C_2H_9S	vapor	11.1	26.50
Naphthalene	$C_{10}H_8$	air or vapor	127	28.8
Neon	Ne	vapor	−248	5.50
Nitric acid (98.8%)	HNO_3	air	11.6	42.7
Nitrobenzene	$C_6H_5NO_2$	air or vapor	20	43.9

Table 2–18 (continued)
SURFACE TENSION OF VARIOUS LIQUIDS

Substance	Formula	In contact with	Temperature (°C)	Surface tension (dyn/cm)
Nitroethane	$C_2H_5NO_2$	air or vapor	20	32.2
Nitrogen	N_2	vapor	−183	6.6
		vapor	−193	8.27
		vapor	−203	10.53
Nitrogen tetroxide	N_2O_4	vapor	19.8	27.5
Nitromethane	CH_3NO_2	vapor	20	36.82
Nitrous oxide	N_2O	vapor	20	1.75
n-Octane	C_8H_{18}	vapor	20	21.80
n-Octyl alcohol	$C_8H_{18}O$	air	20	27.53
Oleic acid	$C_{18}H_{34}O_2$	air	20	32.50
Oxygen	O_2	vapor	−183	13.2
Oxygen (65%)	O_2	air	−190.5	12.2
		vapor	−193	15.7
		vapor	−203	18.3
Paraldehyde	$C_6H_{12}O_3$	air	20	25.9
Phenetole	$C_8H_{10}O$	vapor	20	32.74
Phenol	C_6H_6O	air or vapor	20	40.9
		air or vapor	30	39.88
Phenylhydrazine	$C_6H_8N_2$	vapor	20	46.1
Phosphorus tribromide	PBr_3	air	24	45.8
Phosphorus trichloride	PCl_3	vapor	20	29.1
Phosphorus triiodide	PI_3	vapor	75.3	56.5
Propionic acid	$C_3H_6O_2$	vapor	20	26.7
n-Propyl acetate	$C_5H_{10}O_2$	air or vapor	20	24.3
n-Propyl alcohol	C_3H_8O	vapor	20	23.78
n-Propylamine	C_3H_9N	air	20	22.4
n-Propyl bromide	C_3H_7Br	vapor	71	19.65
n-Propyl chloride	C_3H_7Cl	air	47	18.2
n-Propyl formate	$C_4H_8O_2$	vapor	20	24.5
Pyridine	C_5H_5N	air	20	38.0
Quinoline	C_9H_7N	air	20	45.0
Ricinoleic acid	$C_{18}H_{34}O_3$	air	16	35.81
Selenium	Se	air	217	92.4
Styrene	C_8H_8	air	19	32.14
Sulfuric acid (98.5%)	H_2SO_4	air or vapor	20	55.1
Tetrabromoethane, 1,1,2,2-	$C_2H_2Br_4$	air	20	49.67
Tetrachloroethane, 1,1,2,2-	$C_2H_2Cl_4$	air	22.5	36.03
Tetrachloroethylene	C_2Cl_4	vapor	20	31.74
Toluene	C_7H_8	vapor	10	27.7
		vapor	20	28.5
		vapor	30	27.4
m-Toluidine	C_7H_9N	vapor	20	36.9
o-Toluidine	C_7H_9N	air or vapor	20	40.0
p-Toluidine	C_7H_9N	air	50	34.6
Trichloroacetic acid	$C_2HCl_3O_2$	nitrogen	80.2	27.8
Trichloroethane, 1,1,2-	$C_2H_3Cl_3$	air	114	22.0
Triethyl phosphate	$C_6H_{15}O_4P$	air	15.5	30.61
Trimethylamine	C_3H_9N	nitrogen	−4	17.3
Triphenylcarbinol	$C_{19}H_{16}O$	vapor	165.8	30.38

Table 2–18 (continued)
SURFACE TENSION OF VARIOUS LIQUIDS

Substance	Formula	In contact with	Temperature (°C)	Surface tension (dyn/cm)
Vinyl acetate	$C_4H_6O_2$	vapor	20	23.95
		vapor	25	23.16
		vapor	30	22.54
Water	H_2O	air	18	73.05
m-Xylene	C_8H_{10}	vapor	20	28.9
o-Xylene	C_8H_{10}	air	20	30.10
p-Xylene	C_8H_{10}	vapor	20	28.37

From Weast, R. C., Ed., *Handbook of Chemistry and Physics,* 55th ed., CRC Press, Cleveland, 1974, F-44.

Table 2–19
RADIOACTIVE TRACER DIFFUSION DATA FOR PURE METALS

The data in these tables are the most reliable set of radioactive tracer diffusion data for pure metals published in the literature from 1938 through December 1970. For a complete listing of all published data on this subject up to December 1968 see Askill, J., *Tracer Diffusion Data for Metals, Alloys and Simple Oxides,* Plenum Press, New York, 1970.

The diffusion coefficient D_T at a temperature $T(K)$ is given by the following:

$$D_T = D_0 e^{-Q/RT}$$

Abbreviations:

A.R.G. =	autoradiography	P =	polycrystalline
R.A. =	residual activity	S =	single crystal
S.D. =	surface decrease	⊥ c =	perpendicular to c direction
S.S. =	serial sectioning	∥ c =	parallel to c direction

Solute (tracer)	Crystalline form, purity (%)		Temperature range (°C)	Form of analysis	Activation energy, Q (kcal/mol)	Frequency factor, D_0 (cm²/sec)	Reference
Aluminum							
Ag^{110}	S	99.999	371–655	S.S.	27.83	0.118	1
Al^{27}	S		450–650	S.S.	34.0	1.71	2
Au^{198}	S	99.999	423–609	S.S.	27.0	0.077	3
Cd^{115}	S	99.999	441–631	S.S.	29.7	1.04	3
Ce^{141}	P	99.995	450–630	R.A.	26.60	1.9×10^{-6}	5
Co^{60}	S	99.999	369–655	S.S.	27.79	0.131	1
Cr^{51}	S	99.999	422–654	S.S.	41.74	464	1
Cu^{64}	S	99.999	433–652	S.S.	32.27	0.647	1
Fe^{59}	S	99.99	550–636	S.S.	46.0	135	3
Ga^{72}	S	99.999	406–652	S.S.	29.24	0.49	1
Ge^{71}	S	99.999	401–653	S.S.	28.98	0.481	1
In^{114}	P	99.99	400–600	S.S., R.A.	27.6	0.123	4
La^{140}	P	99.995	500–630	R.A.	27.0	1.4×10^{-6}	5
Mn^{54}	P	99.99	450–650	S.S.	28.8	0.22	2
Mo^{99}	P	99.995	400–630	R.A.	13.1	1.04×10^{-9}	6
Nb^{95}	P	99.95	350–480	R.A.	19.65	1.66×10^{-7}	7

Table 2-19 (continued)
RADIOACTIVE TRACER DIFFUSION DATA FOR PURE METALS

Solute (tracer)	Crystalline form, purity (%)		Temperature range (°C)	Form of analysis	Activation energy, Q (kcal/mol)	Frequency factor, D_0 (cm²/sec)	Reference
Aluminum (cont.)							
Nd¹⁴⁷	P	99.995	450–630	R.A.	25.0	4.8×10^{-7}	5
Ni⁶³	P	99.99	360–630	R.A.	15.7	2.9×10^{-8}	8
Pd¹⁰³	P	99.995	400–630	R.A.	20.2	1.92×10^{-7}	9
Pr¹⁴²	P	99.995	520–630	R.A.	23.87	3.58×10^{-7}	5
Sb¹²⁴	P		448–620	R.A.	29.1	0.09	10
Sm¹⁵³	P	99.995	450–630	R.A.	22.88	3.45×10^{-7}	5
Sn¹¹³	P		400–600	S.S., R.A.	28.5	0.245	4
V⁴⁸	P	99.995	400–630	R.A.	19.6	6.05×10^{-8}	11
Zn⁶⁵	S	99.999	357–653	S.S.	28.86	0.259	1
Beryllium							
Ag¹¹⁰	S⊥c	99.75	650–900	R.A.	43.2	1.76	12
Ag¹¹⁰	S∥c	99.75	650–900	R.A.	39.3	0.43	12
Be⁷	S⊥c	99.75	565–1065	R.A.	37.6	0.52	13
Be⁷	S∥c	99.75	565–1065	R.A.	39.4	0.62	13
Fe⁵⁹	S	99.75	700–1076	R.A.	51.6	0.67	12
Ni⁶³	P		800–1250	R.A.	58.0	0.2	14
Cadmium							
Ag¹¹⁰	S	99.99	180–300	—	25.4	2.21	15
Cd¹¹⁵	S	99.95	110–283	R.A.	19.3	0.14	16
Zn⁶⁵	S	99.99	180–300	—	19.0	0.0016	15
Calcium							
C¹⁴		99.95	550–800	R.A.	29.8	3.2×10^{-5}	17
Ca⁴⁵		99.95	500–800	R.A.	38.5	8.3	17
Fe⁵⁹		99.95	500–800	R.A.	23.3	2.7×10^{-3}	17
Ni⁶³		99.95	550–800	—	28.9	1.0×10^{-5}	17
U²³⁵		99.95	500–700	R.A.	34.8	1.1×10^{-5}	17
Carbon							
Ag¹¹⁰	⊥c		750–1050	R.A.	64.3	9280	18
C¹⁴			2000–2200	—	163	5	19
Ni⁶³	⊥c		540–920	R.A.	47.2	102	18
Ni⁶³	∥c		750–1060	R.A.	53.3	2.2	18
Th²²⁸	⊥c		1400–2200	R.A.	145.4	1.33×10^{-5}	18
Th²²⁸	∥c		1800–2200	R.A.	114.7	2.48	18
U²³²	⊥c		1400–2200	R.A.	115.0	6760	18
U²³²	∥c		1400–1820	R.A.	129.5	385	18
Chromium							
C¹⁴	P		1200–1500	R.A.	26.5	9.0×10^{-3}	20
Cr⁵¹	P	99.98	1030–1545	S.S.	73.7	0.2	21
Fe⁵⁹	P	99.8	980–1420	R.A.	79.3	0.47	22
Mo⁹⁹	P		1100–1420	R.A.	58.0	2.7×10^{-3}	20
Cobalt							
C¹⁴	P	99.82	600–1400	R.A.	34.0	0.21	23
Co⁶⁰	P	99.9	1100–1405	S.S.	67.7	0.83	24
Fe⁵⁹	P	99.9	1104–1303	S.S.	62.7	0.21	24
Ni⁶³	P		1192–1297	R.A.	60.2	0.10	25
S³⁵	P	99.99	1150–1250	R.A.	5.4	1.3	26

Table 2–19 (continued)
RADIOACTIVE TRACER DIFFUSION DATA FOR PURE METALS

Solute (tracer)	Crystalline form, purity (%)		Temperature range (°C)	Form of analysis	Activation energy, Q (kcal/mol)	Frequency factor, D_o (cm²/sec)	Reference
Copper							
Ag[110]	S, P		580–980	R.A.	46.5	0.61	27
As[76]	P		810–1075	R.A.	42.13	0.20	28
Au[198]	S, P		400–1050	S.S.	42.6	0.03	29
Cd[115]	S	99.98	725–950	S.S.	45.7	0.935	30
Ce[141]	P	99.999	766–947	R.A.	27.6	2.17×10^{-8}	31
Cr[51]	S, P		800–1070	R.A.	53.5	1.02	32
Co[60]	S	99.998	701–1077	S.S.	54.1	1.93	33
Cu[67]	S	99.999	698–1061	S.S.	50.5	0.78	34
Eu[152]	P	99.999	750–970	S.S., R.A.	26.85	1.17×10^{-7}	31
Fe[59]	S, P		460–1070	R.A.	52.0	1.36	32
Ga[72]			—	—	45.90	0.55	35
Ge[68]	S	99.998	653–1015	S.S.	44.76	0.397	36
Hg[203]	P		—	—	44.0	0.35	35
Lu[177]	P	99.999	857–1010	R.A.	26.15	4.3×10^{-9}	31
Mn[54]	S	99.99	754–950	S.S.	91.4	10^7	37
Nb[95]	P	99.999	807–906	R.A.	60.06	2.04	38
Ni[63]	P		620–1080	R.A.	53.8	1.1	39
Pd[102]	S	99.999	807–1056	S.S.	54.37	1.71	40
Pm[147]	P	99.999	720–955	R.A.	27.5	3.62×10^{-8}	31
Pt[195]	P		843–997	S.S.	37.5	4.8×10^{-4}	41
S[35]	S	99.999	800–1000	R.A.	49.2	23	42
Sb[124]	S	99.999	600–1000	S.S.	42.0	0.34	43
Sn[113]	P		680–910	—	45.0	0.11	44
Tb[160]	P	99.999	770–980	R.A.	27.45	8.96×10^{-9}	31
Tl[204]	S	99.999	785–996	S.S.	43.3	0.71	45
Tm[170]	P	99.999	705–950	R.A.	24.15	7.28×10^{-9}	31
Zn[65]	P	99.999	890–1000	S.S.	47.50	0.73	46
Germanium							
Cd[115]	S		750–950	R.A.	102.0	1.75×10^9	47
Fe[59]	S		775–930	R.A.	24.8	0.13	48
Ge[71]	S		766–928	S.S.	68.5	7.8	49
In[114]	S		600–920	—	39.9	2.9×10^{-4}	50
Sb[124]	S		720–900	—	50.2	0.22	51
Te[125]	S		770–900	S.S.	56.0	2.0	52
Tl[204]	S		800–930	S.S.	78.4	1700	53
Gold							
Ag[110]	S	99.99	699–1007	S.S.	40.2	0.072	54
Au[198]	S	99.97	850–1050	S.S.	42.26	0.107	224
Co[60]	P	99.93	702–948	R.A.	41.6	0.068	55
Fe[59]	P	99.93	701–948	R.A.	41.6	0.082	55
Hg[203]	S	99.994	600–1027	—	37.38	0.116	56
Ni[63]	P	99.96	880–940	S.S.	46.0	0.30	57
Pt[195]	P, S	99.98	800–1060	S.S.	60.9	7.6	58
β-Hafnium							
Hf[181]	P	97.9	1795–1995	S.S.	38.7	1.2×10^{-3}	59
Indium							
Ag[110]	S⊥c	99.99	25–140	S.S.	12.8	0.52	60
Ag[110]	S∥c	99.99	25–140	S.S.	11.5	0.11	60
Au[198]	S	99.99	25–140	S.S.	6.7	9×10^{-3}	60
In[114]	S⊥c	99.99	44–144	S.S.	18.7	3.7	61
In[114]	S∥c	99.99	44–144	S.S.	18.7	2.7	61
Tl[204]	S	99.99	49–157	S.S.	15.5	0.049	62

Table 2–19 (continued)
RADIOACTIVE TRACER DIFFUSION DATA FOR PURE METALS

Solute (tracer)	Crystalline form, purity (%)		Temperature range (°C)	Form of analysis	Activation energy, Q (kcal/mol)	Frequency factor, D_o (cm²/sec)	Reference
α-Iron							
Ag¹¹⁰	P		748–888	S.S.	69.0	1950	63
Au¹⁹⁸	P	99.999	800–900	R.A.	62.4	31	64
C¹⁴	P	99.98	616–844	R.A.	29.3	2.2	65
Co⁶⁰	P	99.995	638–768	R.A.	62.2	7.19	62
Cr⁵¹	P	99.95	775–875	R.A.	57.5	2.53	66
Cu⁶⁴	P	99.9	800–1050	R.A.	57.0	0.57	67
Fe⁵⁵	P	99.92	809–889	—	60.3	5.4	68
K⁴²	P	99.92	500–800	R.A.	42.3	0.036	69
Mn⁵⁴	P	99.97	800–900	R.A.	52.5	0.35	70
Mo⁹⁹	P		750–875	R.A.	73.0	7800	71
Ni⁶³	P	99.97	680–800	R.A.	56.0	1.3	72
P³²	P		860–900	R.A.	55.0	2.9	73
Sb¹²⁴	P		800–900	R.A.	66.6	1100	74
V⁴⁸	P		755–875	R.A.	55.4	1.43	75
W¹⁸⁵	P		755–875	R.A.	55.1	0.29	75
γ-Iron							
Be⁷	P	99.9	1100–1350	R.A.	57.6	0.1	76
C¹⁴	P	99.34	800–1400	—	34.0	0.15	23
Co⁶⁰	P	99.98	1138–1340	S.S.	72.9	1.25	77
Cr⁵¹	P	99.99	950–1400	R.A.	69.7	10.8	78
Fe⁵⁹	P	99.98	1171–1361	S.S.	67.86	0.49	79
Hf¹⁸¹	P	99.99	1110–1360	R.A.	97.3	3600	78
Mn⁵⁴	P	99.97	920–1280	R.A.	62.5	0.16	70
Ni⁶³	P	99.97	930–2050	R.A.	67.0	0.77	72
P³²	P	99.99	950–1200	R.A.	43.7	0.01	80
S³⁵	P		900–1250	R.A.	53.0	1.7	81
V⁴⁸	P	99.99	1120–1380	R.A.	69.3	0.28	78
W¹⁸⁵	P	99.5	1050–1250	R.A.	90.0	1000	82
δ-Iron							
Co⁶⁰	P	99.995	1428–1521	R.A.	61.4	6.38	83
Fe⁵⁹	P	99.95	1428–1492	S.S.	57.5	2.01	83
P³²	P	99.99	1370–1460	R.A.	55.0	2.9	73
Lanthanum							
Au¹⁹⁸	P	99.97	600–800	S.S.	45.1	1.5	84
La¹⁴⁰	P	99.97	690–850	S.S.	18.1	2.2×10^{-2}	84
Lead							
Ag¹¹⁰	P	99.9	200–310	R.A.	14.4	0.064	85
Au¹⁹⁸	S	99.999	190–320	S.S.	10.0	8.7×10^{-3}	86
Cd¹¹⁵	S	99.999	150–320	S.S.	21.23	0.409	87
Cu⁶⁴	S		150–320	S.S.	14.44	0.046	88
Pb²⁰⁴	S	99.999	150–320	S.S.	25.52	0.887	87
Tl²⁰⁵	P	99.999	207–322	S.S.	24.33	0.511	89
Lithium							
Ag¹¹⁰	P	92.5	65–161	S.S.	12.83	0.37	90
Au¹⁹⁵	P	92.5	47–153	S.S.	10.49	0.21	90
Bi	P	99.95	141–177	S.S.	47.3	5.3×10^{13}	91
Cd¹¹⁵	P	92.5	80–174	S.S.	16.05	2.35	90
Cu⁶⁴	P	99.98	51–120	S.S.	9.22	0.47	93
Ga⁷²	P	99.98	58–173	S.S.	12.9	0.21	93

Table 2–19 (continued)
RADIOACTIVE TRACER DIFFUSION DATA FOR PURE METALS

Solute (tracer)	Crystalline form, purity (%)		Temperature range (°C)	Form of analysis	Activation energy, Q (kcal/mol)	Frequency factor, D_0 (cm²/sec)	Reference
Lithium (cont.)							
Hg^{203}	P	99.98	58–173	S.S.	14.18	1.04	93
In^{114}	P	92.5	80–175	S.S.	15.87	0.39	90
Li^6	P	99.98	35–178	S.S.	12.60	0.14	94
Na^{22}	P	92.5	52–176	S.S.	12.61	0.41	90
Pb^{204}	P	99.95	129–169	S.S.	25.2	160	91
Sb^{124}	P	99.95	141–176	S.S.	41.5	1.6×10^{10}	91
Sn^{113}	P	99.95	108–174	S.S.	15.0	0.62	91
Zn^{65}	P	92.5	60–175	S.S.	12.98	0.57	92
Magnesium							
Ag^{110}	P	99.9	476–621	S.S.	28.50	0.34	95
Fe^{59}	P	99.95	400–600	R.A.	21.2	4×10^{-6}	96
In^{114}	P	99.9	472–610	S.S.	28.4	5.2×10^{-2}	95
Mg^{28}	S⊥c		467–635	S.S.	32.5	1.5	97
Mg^{28}	S‖c		467–635	S.S.	32.2	1.0	97
Ni^{63}	P	99.95	400–600	R.A.	22.9	1.2×10^{-5}	96
U^{235}	P	99.95	500–620	R.A.	27.4	1.6×10^{-5}	96
Zn^{65}	P	99.9	467–620	S.S.	28.6	0.41	95
Molybdenum							
C^{14}	P	99.98	1200–1600	R.A.	41.0	2.04×10^{-2}	99
Co^{60}	P	99.98	1850–2350	S.S.	106.7	18	100
Cr^{51}	P		1000–1500	R.A.	54.0	2.5×10^{-4}	20
Cs^{134}	S	99.99	1000–1470	R.A., A.R.G.	28.0	8.7×10^{-11}	101
K^{42}	S		800–1100	R.A.	25.04	5.5×10^{-9}	102
Mo^{99}	P		1850–2350	S.S.	96.9	0.5	103
Na^{24}	S		800–1100	R.A.	21.25	2.95×10^{-9}	102
Nb^{95}	P	99.98	1850–2350	S.S.	108.1	14	100
P^{32}	P	99.97	2000–2200	S.S.	80.5	0.19	104
Re^{186}	P		1700–2100	A.R.G.	94.7	0.097	105
S^{35}	S	99.97	2220–2470	S.S.	101.0	320	106
Ta^{182}	P		1700–2150	R.A.	83.0	3.5×10^{-4}	20
U^{235}	P	99.98	1500–2000	R.A.	76.4	7.6×10^{-3}	107
W^{185}	P	99.98	1700–2260	S.S.	110	1.7	108
Nickel							
Au^{198}	S, P	99.999	700–1075	S.S.	55.0	0.02	109
Be^7	P	99.9	1020–1400	R.A.	46.2	0.019	76
C^{14}	P	99.86	600–1400	—	34.0	0.012	23
Co^{60}	P	99.97	1149–1390	R.A.	65.9	1.39	110
Cr^{51}	P	99.95	1100–1270	S.S.	65.1	1.1	111
Cu^{64}	P	99.95	1050–1360	S.S.	61.7	0.57	111
Fe^{59}	P		1020–1263	S.S.	58.6	0.074	112
Mo^{99}	P		900–1200	R.A.	51.0	1.6×10^{-3}	20
Ni^{63}	P	99.95	1042–1404	S.S.	68.0	1.9	111
Pu^{238}	P		1025–1125	A.R.G.	51.0	0.5	113
Sb^{124}	P	99.97	1020–1220	—	27.0	1.8×10^{-5}	114
Sn^{113}	P	99.8	700–1350	A.R.G.	58.0	0.83	115
V^{48}	P	99.99	800–1300	R.A.	66.5	0.87	11
W^{185}	P	99.95	1100–1300	S.S.	71.5	2.0	116

Table 2–19 (continued)
RADIOACTIVE TRACER DIFFUSION DATA FOR PURE METALS

Solute (tracer)	Crystalline form, purity (%)		Temperature range (°C)	Form of analysis	Activation energy, Q (kcal/mol)	Frequency factor, D_0 (cm²/sec)	Reference
Niobium							
C^{14}	P		800–1250	R.A.	32.0	1.09×10^{-5}	117
Co^{60}	P	99.85	1500–2100	A.R.G.	70.5	0.74	118
Cr^{51}	S		943–1435	S.S.	83.5	0.30	119
Fe^{55}	P	99.85	1400–2100	A.R.G.	77.7	1.5	118
K^{42}	S		900–1100	R.A.	22.10	2.38×10^{-7}	102
Nb^{95}	P, S	99.99	878–2395	S.S.	96.0	1.1	120
P^{32}	P	99.0	1300–1800	S.S.	51.5	5.1×10^{-2}	104
S^{35}	S	99.9	1100–1500	R.A.	73.1	2600	121
Sn^{113}	P	99.85	1850–2400	S.S.	78.9	0.14	122
Ta^{182}	P, S	99.997	878–2395	S.S.	99.3	1.0	120
Ti^{44}	S		994–1492	S.S.	86.9	0.099	123
U^{235}	P	99.55	1500–2000	R.A.	76.8	8.9×10^{-3}	107
V^{48}	S	99.99	1000–1400	R.A.	85.0	2.21	124
W^{185}	P	99.8	1800–2200	R.A.	91.7	5×10^{-4}	125
Palladium							
Pd^{103}	S	99.999	1060–1500	S.S.	63.6	0.205	126
Phosphorus							
P^{32}	P		0–44	S.S.	9.4	1.07×10^{-3}	127
Platinum							
Co^{60}	P	99.99	900–1050	—	74.2	19.6	129
Cu^{64}	P		1098–1375	S.S.	59.5	0.074	41
Pt^{195}	P	99.99	1325–1600	S.S.	68.2	0.33	130
Potassium							
Au^{198}	P	99.95	5.6–52.5	S.S.	3.23	1.29×10^{-3}	131
K^{42}	S	99.7	– 52–61	S.S.	9.36	0.16	132
Na^{22}	P	99.7	0–62	S.S.	7.45	0.058	133
Rb^{86}	P	99.95	0.1–59.9	S.S.	8.78	0.090	134
γ-Plutonium							
Pu^{238}	P		190–310	S.S.	16.7	2.1×10^{-5}	135
δ-Plutonium							
Pu^{238}	P		350–440	S.S.	23.8	4.5×10^{-3}	136
ε-Plutonium							
Pu^{238}	P		500–612	R.A.	18.5	2.0×10^{-2}	137
α-Praseodymium							
Ag^{110}	P	99.93	610–730	S.S.	25.4	0.14	138
Au^{195}	P	99.93	650–780	S.S.	19.7	4.3×10^{-2}	138
Co^{60}	P	99.93	660–780	S.S.	16.4	4.7×10^{-2}	138
Zn^{65}	P	99.96	766–603	S.S.	24.8	0.18	139
β-Praseodymium							
Ag^{110}	P	99.93	800–900	S.S.	21.5	3.2×10^{-2}	138
Au^{195}	P	99.93	800–910	S.S.	20.1	3.3×10^{-2}	138
Ho^{166}	P	99.96	800–930	S.S.	26.3	9.5	140
In^{114}	P	99.96	800–930	S.S.	28.9	9.6	140
La^{140}	P	99.96	800–930	S.S.	25.7	1.8	140

Table 2–19 (continued)
RADIOACTIVE TRACER DIFFUSION DATA FOR PURE METALS

Solute (tracer)	Crystalline form, purity (%)		Temperature range (°C)	Form of analysis	Activation energy, Q (kcal/mol)	Frequency factor, D_0 (cm²/sec)	Reference
β-Praseodymium (cont.)							
Pr^{142}	P	99.93	800–900	S.S.	29.4	8.7	140
Zn^{65}	P	99.96	822–921	S.S.	27.0	0.63	139
Selenium							
Fe^{59}	P		40–100	R.A.	8.88	—	141
Hg^{203}	P	99.996	25–100	R.A.	1.2	—	141
S^{35}	S⊥c		60–90	S.D.	29.9	1700	142
S^{35}	S∥c		60–90	S.D.	15.6	1100	142
Se^{75}	P		35–140	—	11.7	1.4×10^{-4}	143
Silicon							
Au^{198}	S		700–1300	S.S.	47.0	2.75×10^{-3}	145
C^{14}	P		1070–1400	R.A.	67.2	0.33	146
Cu^{64}	P		800–1100	R.A.	23.0	4×10^{-2}	147
Fe^{59}	S		1000–1200	R.A.	20.0	6.2×10^{-3}	148
Ni^{63}	P		450–800	—	97.5	1000	149
P^{32}	S		1100–1250	R.A.	41.5	—	150
Sb^{124}	S		1190–1398	R.A.	91.7	12.9	151
Si^{31}	S	99.99999	1225–1400	S.S.	110.0	1800	146
Silver							
Au^{198}	P	99.99	718–942	S.S.	48.28	0.85	54
Ag^{110}	S	99.999	640–955	S.S.	45.2	0.67	152
Cd^{115}	S	99.99	592–937	S.S.	41.69	0.44	153
Co^{60}	S	99.999	700–940	—	48.75	1.9	154
Cu^{64}	P	99.99	717–945	S.S.	46.1	1.23	155
Fe^{59}	S	99.99	720–930	S.S.	49.04	2.42	156
Ge^{77}	P		640–870	S.S.	36.5	0.084	157
Hg^{203}	P	99.99	653–948	S.S.	38.1	0.079	155
In^{114}	S	99.99	592–937	S.S.	40.80	0.41	153
Ni^{63}	S	99.99	749–950	S.S.	54.8	21.9	158
Pb^{210}	P		700–865	S.S.	38.1	0.22	159
Pd^{102}	S	99.999	736–939	S.S.	56.75	9.56	140
Ru^{103}	S	99.99	793–945	S.S.	65.8	180	160
S^{35}	S	99.999	600–900	R.A.	40.0	1.65	161
Sb^{124}	P	99.999	780–950	S.S., R.A.	39.07	0.234	162
Sn^{113}	S	99.99	592–937	S.S.	39.30	0.255	153
Te^{125}	P		770–940	R.A.	38.90	0.47	163
Tl^{204}	P		640–870	S.S.	37.9	0.15	157
Zn^{65}	S	99.99	640–925	S.S.	41.7	0.54	164
Sodium							
Au^{198}	P	99.99	1.0–77	S.S.	2.21	3.34×10^{-4}	165
K^{42}	P	99.99	0–91	S.S.	8.43	0.08	133
Na^{22}	P	99.99	0–98	S.S.	10.09	0.145	166
Rb^{86}	P	99.99	0–85	S.S.	8.49	0.15	133
Tantalum							
C^{14}	P		1450–2200	S.S.	40.3	1.2×10^{-2}	167
Fe^{59}	P		930–1240	—	71.4	0.505	168
Mo^{99}	P		1750–2220	R.A.	81.0	1.8×10^{-3}	20
Nb^{95}	P, S	99.996	921–2484	S.S.	98.7	0.23	169
S^{35}	P	99.0	1970–2110	R.A.	70.0	100	170
Ta^{182}	P, S	99.996	1250–2200	S.S.	98.7	1.24	226

Table 2–19 (continued)
RADIOACTIVE TRACER DIFFUSION DATA FOR PURE METALS

Solute (tracer)	Crystalline form, purity (%)		Temperature range (°C)	Form of analysis	Activation energy, Q (kcal/mol)	Frequency factor, D_0 (cm²/sec)	Reference
Tellurium							
Hg[203]	P		270–440	—	18.7	3.14×10^{-5}	171
Se[75]	P		320–440	—	28.6	2.6×10^{-2}	171
Tl[204]	P		360–430	—	41.0	320	172
Te[127]	S⊥c	99.9999	300–400	S.S.	46.7	3.91×10^4	173
Te[127]	S∥c	99.9999	300–400	S.S.	35.5	130	173
α-Thallium							
Ag[110]	P⊥c	99.999	80–250	S.S.	11.8	3.8×10^{-2}	174
Ag[110]	P∥c	99.999	80–250	S.S.	11.2	2.7×10^{-2}	174
Au[198]	P⊥c	99.999	110–260	S.S.	2.8	2.0×10^{-5}	174
Au[198]	P∥c	99.999	110–260	S.S.	5.2	5.3×10^{-4}	174
Tl[204]	S⊥c	99.9	135–230	S.S.	22.6	0.4	175
Tl[204]	S∥c	99.9	135–230	S.S.	22.9	0.4	175
β-Thallium							
Ag[110]	P	99.999	230–310	S.S.	11.9	4.2×10^{-2}	174
Au[198]	P	99.999	230–310	S.S.	6.0	5.2×10^{-4}	174
Tl[204]	S	99.9	230–280	S.S.	20.7	0.7	175
α-Thorium							
Pa[231]	P	99.85	770–910	—	74.7	126	176
Th[228]	P	99.85	720–880	—	71.6	395	176
U[233]	P	99.85	700–880	—	79.3	2210	176
Tin							
Ag[110]	S⊥c		135–225	S.S.	18.4	0.18	177
Ag[110]	S∥c		135–225	S.S	12.3	7.1×10^{-3}	177
Au[198]	S⊥c		135–225	S.S.	17.7	0.16	177
Au[198]	S∥c		135–225	S.S.	11.0	5.8×10^{-3}	177
Co[60]	S, P		140–217	R.A.	22.0	5.5	178
In[114]	S⊥c	99.998	181–221	S.S.	25.8	34.1	179
In[114]	S∥c	99.998	181–221	S.S.	25.6	12.2	179
Sn[113]	S⊥c	99.999	160–226	S.S.	25.1	10.7	180
Sn[113]	S∥c	99.999	160–226	S.S.	25.6	7.7	180
Tl[204]	P	99.999	137–216	S.S.	14.7	1.2×10^{-3}	181
α-Titanium							
Ti[44]	P	99.99	700–850	R.A.	35.9	8.6×10^{-6}	182
β-Titanium							
Ag[110]	P	99.95	940–1570	S.S.	43.2	3×10^{-3}	183
Be[7]	P	99.96	915–1300	R.A.	40.2	0.8	184
C[14]	P	99.62	1100–1600	R.A.	20.0	3.02×10^{-3}	185
Cr[51]	P	99.7	950–1600	A.R.G.	35.1 / 61.0	5×10^{-3} / 4.9	186
Co[60]	P	99.7	900–1600	S.S.	30.6 / 52.5	1.2×10^{-2} / 2.0	186
Fe[59]	P	99.7	900–1600	A.R.G.	31.6 / 55.0	7.8×10^{-3} / 2.7	186
Mo[99]	P	99.7	900–1600	S.S.	43.0 / 73.0	8.0×10^{-3} / 20	186
Mn[54]	P	99.7	900–1600	S.S.	33.7 / 58.0	6.1×10^{-3} / 20	186

Table 2–19 (continued)
RADIOACTIVE TRACER DIFFUSION DATA FOR PURE METALS

Solute (tracer)	Crystalline form, purity (%)		Temperature range (°C)	Form of analysis	Activation energy, Q (kcal/mol)	Frequency factor, D_0 (cm² /sec)	Reference
β-Titanium (cont.)							
Nb[95]	P	99.7	1000–1600	A.R.G.	39.3	5.0×10^{-3}	186
					73.0	20	
Ni[63]	P	99.7	925–1600	A.R.G.	29.6	9.2×10^{-3}	186
					52.5	2.0	
P[32]	P	99.7	950–1600	S.S.	24.1	3.62×10^{-3}	187
					56.5	5	
Sc[46]	P	99.95	940–1590	S.S.	32.4	4.0×10^{-3}	183
Sn[113]	P	99.7	950–1600	S.S.	31.6	3.8×10^{-4}	187
					69.2	10	
Ti[44]	P	99.95	900–1540	S.S.	31.2	3.58×10^{-4}	188
					60.0	1.09	
U[235]	P	99.9	900–1400	R.A.	29.3	5.1×10^{-4}	189
V[48]	P	99.95	900–1545	S.S.	32.2	3.1×10^{-4}	190
					57.2	1.4	
W[185]	P	99.94	900–1250	R.A.	43.9	3.6×10^{-3}	191
Zr[95]	P	98.94	920–1500	R.A.	35.4	4.7×10^{-3}	191
Tungsten							
C[14]	P	99.51	1200–1600	R.A.	53.5	8.91×10^{-3}	99
Fe[59]	P		940–1240	—	66.0	1.4×10^{-2}	168
Mo[99]	P		1700–2100	R.A.	101.0	0.3	20
Nb[95]	P	99.99	1305–2367	S.S.	137.6	3.01	192
Re[186]	S		2100–2400	R.A.	141.0	19.5	193
Ta[182]	P	99.99	1305–2375	S.S.	139.9	3.05	192
W[185]	P	99.99	1800–2403	S.S.	140.3	1.88	192
α-Uranium							
U[234]	P		580–650	—	40.0	2×10^{-3}	194
β-Uranium							
Co[60]	P	99.999	692–763	S.S.	27.45	1.5×10^{-2}	195
U[235]	P		690–750	R.A.	44.2	2.8×10^{-3}	196
γ-Uranium							
Au[195]	P	99.99	785–1007	S.S.	30.4	4.86×10^{-3}	197
Co[60]	P	99.99	783–989	S.S.	12.57	3.51×10^{-4}	198
Cr[51]	P	99.99	797–1037	S.S.	24.46	5.37×10^{-3}	198
Cu[64]	P	99.99	787–1039	S.S.	24.06	1.96×10^{-3}	198
Fe[55]	P	99.99	787–990	S.S.	r2.0	2.69×10^{-4}	198
Mn[54]	P	99.99	787–939	S.S.	13.88	1.81×10^{-4}	198
Nb[95]	P	99.99	791–1102	S.S.	39.65	4.87×10^{-2}	198
Ni[63]	P	99.99	787–1039	S.S.	15.66	5.36×10^{-4}	198
U[233]	P	99.99	800–1070	S.S.	28.5	2.33×10^{-3}	227
Zr[95]	P		800–1000	R.A.	16.5	3.9×10^{-4}	228
Vanadium							
C[14]	P	99.7	845–1130	S.S.	27.3	4.9×10^{-3}	199
Cr[51]	P	99.8	960–1200	R.A.	64.6	9.54×10^{-3}	200
Fe[59]	P		960–1350	S.S.	71.0	0.373	201
P[32]	P	99.8	1200–1450	R.A.	49.8	2.45×10^{-2}	202
S[35]	P	99.8	1320–1520	R.A.	34.0	3.1×10^{-2}	184
V[48]	S, P	99.99	880–1360	S.S.	73.65	0.36	203
V[48]	S, P	99.99	1360–1830	S.S.	94.14	214.0	203

Table 2–19 (continued)
RADIOACTIVE TRACER DIFFUSION DATA FOR PURE METALS

Solute (tracer)	Crystalline form, purity (%)		Temperature range (°C)	Form of analysis	Activation energy, Q (kcal/mol)	Frequency factor, D_0 (cm²/sec)	Reference
Yttrium							
Y[90]	S⊥c		900–1300	R.A.	67.1	5.2	204
Y[90]	S‖c		900–1300	R.A.	60.3	0.82	204
Zinc							
Ag[110]	S⊥c	99.999	271–413	S.S.	27.6	0.45	205
Ag[110]	S‖c	99.999	271–413	S.S.	26.0	0.32	205
Au[198]	S⊥c	99.999	315–415	S.S.	29.72	0.29	206
Au[198]	S‖c	99.999	315–415	S.S.	29.73	0.97	206
Cd[115]	S⊥c	99.999	225–416	S.S.	20.12	0.117	206
Cd[115]	S‖c	99.999	225–416	S.S.	20.54	0.114	206
Cu[64]	S⊥c	99.999	338–415	S.S.	29.92	2.0	206
Cu[64]	S‖c	99.999	338–415	S.S.	29.53	2.22	207
Ga[72]	S⊥c		240–403	S.S.	18.15	0.018	207
Ga[72]	S‖c		240–403	S.S.	18.4	0.016	207
Hg[203]	S⊥c		260–413	S.S.	20.18	0.073	208
Hg[203]	S‖c		260–413	S.S.	19.70	0 056	208
In[114]	S⊥c		271–413	S.S.	19.60	0.14	205
In[114]	S‖c		271–413	S.S.	19.10	0.062	205
Sn[113]	S⊥c		298–400	S.S.	18.4	0.13	209
Sn[113]	S‖c		298–400	S.S.	19.4	0.15	209
Zn[65]	S⊥c	99.999	240–418	S.S.	23.0	0.18	210
Zn[65]	S‖c	99.999	240–418	S.S.	21.9	0.13	210
α-Zirconium							
Cr[51]	P	99.9	700–850	R.A.	18.0	1.19×10^{-8}	211
Fe[55]	P		750–840	—	48.0	2.5×10^{-2}	212
Mo[99]	P		600–850	R.A.	24.76	6.22×10^{-8}	213
Nb[95]	P	99.99	740–857	R.A.	31.5	6.6×10^{-6}	182
Sn[113]	P		300–700	A.R.G.	22.0	1.0×10^{-8}	214
Ta[182]	P	99.6	700–800	R.A.	70.0	100	215
V[48]	P	99.99	600–850	R.A.	22.9	1.12×10^{-8}	124
Zr[95]	P	99.95	750–850	S.S.	45.5	5.6×10^{-4}	216
β-Zirconium							
Be[7]	P	99.7	915–1300	R.A.	31.1	8.33×10^{-2}	184
C[14]	P	96.6	1100–1600	R.A.	34.2	3.57×10^{-2}	217
Ce[141]	P		880–1600	R.A.	41.4	3.16	218
					74.1	42.17	
Co[60]	P	99.99	920–1600	S.S.	21.82	3.26×10^{-3}	219
Cr[51]	P	99.9	700–850	R.A.	18.0	1.19×10^{-8}	211
Fe[55]	P		750–840	—	48.0	2.5×10^{-2}	212
Mo[99]	P		900–1635	R.A.	35.2	1.99×10^{-6}	218
					68.55	2.63	
Nb[95]	P		1230–1635	R.A.	36.6	7.8×10^{-4}	220
P[32]	P	99.94	950–1200	R.A.	33.3	0.33	221
Sn[113]	P		300–700	A.R.G.	22.0	1×10^{-8}	214
Ta[182]	P	99.6	900–1200	R.A.	27.0	5.5×10^{-5}	215
U[235]	P		900–1065	S.S.	30.5	5.7×10^{-4}	222
V[48]	P	99.99	870–1200	R.A.	45.8	7.59×10^{-3}	223
V[48]	P	99.99	1200–1400	R.A.	57.7	0.32	223
W[185]	P	99.7	900–1250	R.A.	55.8	0.41	223
Zr[95]	P		1100–1500	S.S.	30.1	2.4×10^{-4}	225

From Askill, J., in *Handbook of Chemistry and Physics*, 55th ed., Weast, R. C., Ed., CRC Press, Cleveland, 1974, F-61.

Table 2–19 (continued)
RADIOACTIVE TRACER DIFFUSION DATA FOR PURE METALS

REFERENCES

1. Peterson, N. L. and Rothman, S. J., *Phys. Rev.,* B1(8), 3264, 1970.
2. Lundy, T. S. and Murdock, J. F., *J. Appl. Phys.,* 33(5), 1671, 1962.
3. Alexander, W. B. and Slifkin, L. M., *Phys. Rev.,* B-1(8), 3274, 1970.
4. Anand, M. S. and Agarwala, R. P., *Phys. Status Solidi,* A-1(1), K-41, 1970.
5. Murarka, S. P. and Agarwala, R. P., Report BARC-368, Indian Atomic Energy Commission, 1968.
6. Paul, A. R. and Agarwala, R. P., *J. Appl. Phys.,* 38(9), 3790, 1967.
7. Tiwari, G. P. and Sharma, B. D., *Trans. Indian Inst. Metals,* 20, 83, 1967.
8. Hirano, K., Agarwala, R. P., and Cohen, M., *Acta Met.,* 10(9), 857, 1962.
9. Anand, M. S. and Agarwala, R. P., *Trans. Met. Soc. AIME,* 239(11), 1848, 1967.
10. Badrinarayanan, S. and Mathers, H. B., *Int. J. Appl. Radiat. Isot.,* 19(4), 353, 1968.
11. Murarka, S. P., Anand, M. S., and Agarwala, R. P., *Acta Met.,* 16(1), 69, 1968.
12. Naik, M. C., Dupony, J. M., and Adda, Y., *Mem. Sci. Rev. Met.,* 63, 488, 1966.
13. Dupony, J. M., Mathie, J., and Adda, Y., *Mem. Sci. Rev. Met.,* 63, 481, 1966.
14. Ananyn, V. M., Gladkov, V. P., Zotov, V. S., and Skorov, D. M., *At. Energ.,* 29(3), 220, 1970.
15. Hirschwald, W. and Schroedter, W., *Z. Phys. Chem. N. F.,* 53, 392, 1967.
16. Chomka, W., *Zesz. Nauk. Politech. Gdansk Fiz.,* 1, 39, 1967.
17. Pavlinov, L. V., Gladyshev, A. M., and Bikov, V. N., *Fiz. Metal. Metalloved.,* 26(5), 823, 1968.
18. Wolfe, J. R., McKenzie, D. R., and Borg, R. J., *J. Appl. Phys.,* 36(6), 1906, 1965.
19. Kanter, M. A., *Phys. Rev.,* 107, 655, 1957.
20. Borisov, E. V., Gruzin, P. L., and Zemskii, S. V., *Zashch. Pokryt. Metal.,* 2, 104, 1968.
21. Askill, J. and Tomlin, D. H., *Philos. Mag.,* 11(111), 467, 1965.
22. Paxton, W. H. and Wolfe, R. A., *Trans. Met. Soc. AIME,* 230, 1426, 1964.
23. Kovenski, I. I., *Fiz. Metal. Metalloved.,* 16, 613, 1963.
24. Mead, H. W. and Birchenall, C. E., *Trans. Met. Soc. AIME,* 203(9) 994, 1955.
25. Hirano, K., Agarwala, R. P., Averbach, B. L., and Cohen, M., *J. Appl. Phys.,* 33(10), 3049, 1962.
26. Pavlyuchenko, M. M. and Konoyuk, I. F., *Dokl. Akad. Nauk Beloruss. S.S.R.,* 8, 157, 1964.
27. Barreau, G., Brunel, G., Azeron, G., and Lacombe, P., *C. R. Acad. Sci. Ser. C* (Paris), 270(6), 514, 1964.
28. Klotsman, S. M., Rabovskii, A. Ya., Talinskii, V. K., and Timofeev, A. N., *Fiz. Metal. Metalloved.,* 29(4), 803, 1970.
29. Chatterjee, A. and Fabian, D. J., *Acta Met.,* 17(9), 1141, 1969.
30. Hirone, T., Kunitomi, N., Sakamoto, M., and Yamaki, H., *J. Phys. Soc. Jap.,* 13(8), 838, 1958.
31. Badrinarayanan, S. and Mathur, H. B., *Indian J. Pure Appl. Phys.,* 8(6), 324, 1970.
32. Barreau, G., Brunel, G., and Cizeron, G., *C. R. Acad. Sci. Ser. C.* (Paris) 272(7), 618, 1971.
33. Machlet, C. A., *Phys. Rev.,* 109(6), 1964, 1958.
34. Rothman, S. J. and Peterson, N. L., *Phys. Status Solidi,* 35, 305, 1969.
35. Tomizuka, C. T., cited by Lazarus, D., *Solid State Physics,* Vol. 10, 1970.
36. Reinke, F. D. and Dahlstrom, C. E., *Philos. Mag.,* 22(175), 57, 1970.
37. Ikushima, A., *J. Phys. Soc. Jap.,* 14, 111, 1959.
38. Saxena, M. C. and Sharma, B. D., *Trans. Indian Inst. Metals,* 23(3), 16, 1970.
39. Brunel, G., Cizeron, G., and Lacombe, P., *C. R. Acad. Sci. Ser. C* (Paris), 270(4), 393, 1970.
40. Peterson, N. L., *Phys. Rev.,* 132(6), 2471, 1963.
41. Johnson, R. D. and Faulkenberry, B. H., ASD-TDR-63-625, 1963.
42. Moya, F., Moya-Gontier, G. E., and Cabane-Brouty, F., *Phys. Status Solidi,* 35(2), 893, 1969.
43. Inman, M. C. and Barr, L. W., *Acta Met.,* 8(2), 112, 1960.
44. Kuzmenko, P. P., Ostrovskii, L. F., and Kovalchuk, V. S., *Fiz. Tverd, Tela,* 4, 490, 1962.
45. Komura, S. and Kunitomi, N., *J. Phys. Soc. Jap.,* 18(Suppl. 2), 208, 1963.
46. Klotsman, S. M., Rabovskii, A. Ya., Talinskii, V. K., and Timofeev, A. N., *Fiz. Metal. Metalloved.,* 28(6), 1025, 1969.
47. Kosenko, V. E., *Fiz. Tverd. Tela,* 1, 1622, 1959; *Sov. Phys. Solid State* (English transl.), 1, 1481, 1959.
48. Bugai, A. A., Kosenko, V. E., and Miselynuk, E. G., *Zh. Tekh. Fiz.* 27(1), 207, 1957; *NP-tr-*448, p. 219, 1960.
49. Letaw, H., Portnoy, W. M., and Slifkin, L., *Phys. Rev.,* 102, 636, 1956.
50. Sandulova, A. V., Droniuk, M. I., and P'dak, V. M., *Fiz. Tverd. Tela,* 3, 2913, 1961.
51. Boltaks, B. I., Grabtchak, V. P., and Dzafarov, T. D., *Fiz. Tverd. Tela,* 6, 3181, 1964.
52. Ignatkov, V. D. and Kosenko, V. E., *Fiz. Tverd. Tela,* 4(6), 1627, 1927; *Sov. Phys. Solid State* (English transl.), 4(6), 1627, 1962.
53. Tagirov, V. I. and Kuliev, A. A., *Fiz. Tverd. Tela,* 4(1), 272, 1962; *Sov. Phys. Solid State* (English transl.), 4(1), 196, 1962.

Table 2–19 (continued)
RADIOACTIVE TRACER DIFFUSION DATA FOR PURE METALS

54. Mallard, W. C., Gardner, A. B., Bass, R. F., and Slifkin, L. M., *Phys. Rev.*, 129(2), 617, 1963.
55. Duhl, D., Hirano, K., and Cohen, M., *Acta Met.*, 11(1), 1, 1963.
56. Mortlock, A. J. and Rowe, A. H., *Philos. Mag.*, 11(114), 115, 1965.
57. Reynolds, J. E. Averbach, B. L., and Cohen, M., *Acta Met.*, 5, 29, 1957.
58. Mortlock, A. J., Rowe, A. H., and LeClaire, A. D., *Philos. Mag.*, 5, 803, 1960.
59. Winslow, F. R. and Lundy, T. S., *Trans. Met. Soc. AIME*, 233, 1790, 1965.
60. Anthony, R. R. and Turnbull, D., *Phys. Rev.*, 151, 495, 1966.
61. Amonenko, V. M., Blinkin, A. M., and Ivantsov. I. G., *Fiz. Metal. Metalloved.*, 17(1), 56, 1964; *Phys. Metals Metallogr.*, 17(1), 56, 1964.
62. James, D. W. and Leak, G. M., *Philos. Mag.*, 14, 701, 1966.
63. Bondy, A. and Levy, V., *C. R. Acad. Sci. Ser. C* (Paris), 272(1), 19, 1971.
64. Borg, R. J. and Lai, D. Y. F., *Acta Met.*, 11(8), 861, 1963.
65. Homan, C. G., *Acta Met.*, 12, 1071, 1964.
66. Huntz, A. M., Aucouturier, M., and Lacombe, P., *C. R. Sci. Ser. C* (Paris), 265(10), 554, 1967.
67. Anand, M. S. and Agarwala, R. P., *J. Appl. Phys.*, 37, 4248, 1966.
68. Angers, R. and Claisse, F., *Can. Met. Quart.*, 7(2), 73, 1968.
69. Tomilov, A. V. and Shcherbedinskii, G. V., *Fiz. Khim. Mekh. Mater.*, 3(3), 261, 1967.
70. Nohara, K. and Hirano, K., Proceedings, International Conference on Science and Technology for Iron and Steel, Tokyo, 1970.
71. Borisov, V. T., Golikov, V. M., and Shcherbedinskii, G. V., *Fiz. Metal. Metalloved.*, 22(1), 159, 1966.
72. Hirano, K., Cohen, M., and Averbach, B. L., *Acta Met.*, 9(5), 440, 1961.
73. Seibel, G., *Compt. Rend.*, 256(22), 4661, 1963.
74. Bruggeman, G. and Roberts, J., *J. Metals*, 20(8), 54, 1968.
75. Lyubimov, V. D., *Izv. Akad. Nauk S.S.S.R. Met.*, 3, 201, 1969; Lyubimov, V. D., Tskhai, V. A., and Bogomolov, G. B., *Tr. Inst. Akad. Nauk S.S.S.R. Ural Filial.* 17, 44, 48, 1970.
76. Grigorev, G. V. and Pavlinov, L. V., *Fiz. Metal. Metalloved.*, 25(5), 836, 1968.
77. Suzuoka, T., *Trans. Jap. Inst. Metals*, 2, 176, 1961.
78. Bowen, A. W. and Leak, G. M., *Met. Trans.*, 1(6), 1695, 1970.
79. Heumann, T. and Imm, R., *J. Phys. Chem. Solids*, 29(9), 1613, 1968.
80. Gruzin, P. L. and Mural, V. V., *Fiz. Metal. Metalloved.*, 16(4), 551, 1963; *Phys. Metals Metallogr.* (English transl.), 16(4), 50, 1963.
81. Hoshino. A. and Ataki, R., *Tetsu to Hagane*, 56(2), 252, 1970.
82. Sato, K., *Trans. Jap. Inst. Metals*, 5, 91, 1964.
83. James, D. W. and Leak, G. M., *Philos. Mag.*, 14, 701, 1966.
84. Dariel, M. P., Erez, G., and Schmidt, G. M. J., *Philos. Mag.*, 19(161), 1053, 1969.
85. Kuzmenko, P. P., Grinevich, G. P., and Danilchenko, B. A., *Fiz. Metal. Metalloved.*, 29(2), 318, 1970.
86. Kidson, G. V., *Philos. Mag.*, 13, 247, 1966.
87. Miller, J. W., *Phys. Rev.*, 181(3), 1095, 1969.
88. Dyson, B. F., Anthony, T., and Turnbull, D., *J. Appl. Phys.*, 37, 2370, 1966.
89. Resing, H. A. and Nachtrieb, N. H., *Phys. Chem. Solids*, 21(1/2), 40, 1969.
90. Ott, A. and Lodding, A., International Conference on Vacancies and Interstitials in Metals, Julich, Germany, Vol. 1, p. 43, 1968; *Z. Naturforsch.*, 23A, 1683, 2126, 1968.
91. Ott, A., Lodding, A., and Lazarus, D., *Phys. Rev.*, 188(3), 1088, 1969.
92. Ott, A., *J. Appl. Phys.*, 40(6), 2395, 1969; Mundy, J. N., Ott, A., Lowenberg, L., and Lodding, A., *Phys. Status Solidi*, 35(1), 359, 1969.
93. Ott, A., *Z. Naturforsch.*, 25A(10), 1477, 1970.
94. Lodding, A., Mundy, J. N., and Ott, A., *Phys. Status Solidi*, 38(2), 559, 1970.
95. Lai, K., Report CEA-R 3136, Commissariat à l'Energie Atomique, Paris, 1967.
96. Pavlinov, L. V., Gladyshev, A. M., and Bikov, V. N., *Fiz. Metal. Metalloved.*, 26(5), 823, 1968.
97. Shewmon, P. G., *Trans. Met. Soc. AIME*, 206, 918, 1956.
98. Borisov, Y. V., Gruzin, P. L., and Pavlinov, L. V., *Met. Metalloved. Chistykh. Metal.*, 1, 213, 1959, translated in JPRS-5195.
99. Nakonechnikov, A. Y., Pavlinov, L. V., and Bikov, V. N., *Fiz. Metal. Metalloved.*, 22, 234, 1966.
100. Askill, J., *Phys. Status Solidi*, 9(2), K-113, 1965.
101. Pavlinov, L. V. and Kordev, A. A., *Fiz. Metal. Metalloved.*, 29(6), 1326, 1970.
102. Dubinin, G. N., Benediktova, G. P., Karpman, M. G., and Shcherbedinskii, G. V., *Khim. Term. Obrab. Stali Splavov*, 6, 129, 1969.
103. Askill, J. and Tomlin, D. H., *Philos. Mag.*, 8(90), 997, 1963.

Table 2–19 (continued)
RADIOACTIVE TRACER DIFFUSION DATA FOR PURE METALS

104. Vandyshev, B. A. and Panov, A. S., *Fiz. Metal. Metalloved.*, 26(3), 517, 1968.
105. Benediktova, S. Z., Dubinin, G. N., Kapman, M. G., and Shcherbedinskii, G. V., *Metalloved. Term. Obrab. Metal.*, 5, 5, 1966.
106. Vandyshev, B. A. and Panov, A. S., *Fiz. Metal. Metalloved*, 25(2), 321, 1968.
107. Pavlinov, L. V., Nakonechnikov, A. Y., and Bikov, V. N., *At. Energ.*, (U.S.S.R.), 19, 521, 1965.
108. Askill, J., *Phys. Status Solidi*, 23, K-21, 1967.
109. Chatterjee, A. and Fabian, D. J., *J. Inst. Metals*, 96(6), 186, 1968.
110. Hassner, A. and Lange, W., *Phys. Status Solidi*, 8, 77, 1965.
111. Monma, K., Suto, H., and Oikawa, H., *J. Jap. Inst. Metals*, 28, 188, 1964.
112. Shinyaev, A. Ya., *Izv. Akad. Nauk S.S.S.R. Metal.*, 4, 182, 1969.
113. Blechet, J. J., VanGryeynest, A., and Calais, D., *J. Nucl. Mater.*, 28(2), 177, 1968.
114. Kuzmenko, P. P. and Grinevich, G. P., *Fiz. Tverd. Tela*, 4(11), 3266, 1962; *Sov. Phys. Solid State* (English transl.), 4(11), 2390, 1962.
115. Bokshtein, S. Z., Kishkin, S. T., and Moroz, L. M., *Investigation of the Structure of Metals by Radioactive Isotope Methods*, State Publishing House of the Ministry of Industry, Moscow, 1959; (AEC-tr-4505), 1961.
116. Anand, M. S., Murarka, S. P., and Agarwala, R. P., *J. Appl. Phys.*, 36(12), 3860, 1965.
117. Lyubimov, V. D., *Izv. Akad. Nauk S.S.S.R. Metal.*, 3, 209, 1969; Lyubimov, V. D., Tskhai, V. A., and Bogmolov, G. B., *Tr. Inst. Khim. Akad. Nauk S.S.S.R. Ural Filial*, 17, 44, 48, 1970.
118. Peart, R. F., Graham, D., and Tomlin, D. H., *Acta Met.*, 10, 519, 1962.
119. Pelleg, J., *J. Metals*, 20(8), 54, 1968; *J. Less Common Metals*, 17, 319, 1969; *Philos. Mag.*, 19(157), 25, 1969.
120. Lundy, T. S., Winslow, F. E., Pawel, R. E., and McHargue, C. J., *Trans. Met. Soc. AIME*, 223, 1533, 1965.
121. Vandyshev, B. A. and Panov, A. S., *Izv. Akad. Nauk S.S.S.R.*, 1, 206, 1968.
122. Askill, J. *Phys. Status Solidi*, 9(3), K-167, 1965.
123. Pelleg, J., *Philos. Mag.*, 21(172), 735, 1970.
124. Agarwala, R. P., Murarka, S. P., and Anand, M. S., *Acta. Met.*, 16(1), 61, 1968.
125. Fedorov, G B., Zhomov, F. I., and Smirnov, E. A., *Met. Metalloved. Chistykh. Metal.*, 8, 145, 1969.
126. Peterson, N. L., *Phys. Rev.*, 136(2A), A-568, 1964.
127. Nachtrieb, N. H. and Handler, G. S., *J. Chem. Phys.*, 23, 1187, 1955.
129. Kucera, J. and Zemcik, T. *Can. Met. Q.*, 7(2), 83, 1968.
130. Kidson, G. V. and Ross, R., Proceedings of the 1st International Conference on Isotopes in Scientific Research, United Nations Educational and Cultural Organization, New York, 1958, 185.
131. Smith, F. A. and Barr, L. W., *Philos. mag.*, 21(171), 633, 1970.
132. Mundy, J. N., Miller, T. E., and Porte, R. J., *Phys. Rev.*, B-3(8), 2445, 1971.
133. Barr, L. W., Mundy, J. N., and Smith F. A., *Philos. Mag.*, 16, 1139, 1967.
134. Smith, F. A. and Barr, L. W., *Philos. Mag.*, 20(163), 205, 1969.
135. Tate, R. E. and Edwards, G. R., *Symposium on Thermodynamics with Emphasis on Nuclear Materials and Atomic Solids*, Vol. 2, International Atomic Energy Agency, Vienna, Austria, 1966, 105.
136. Tate, R. E. and Cramer, E. M., *Trans. Met. Soc. AIME*, 230, 639, 1964.
137. Dupuy, M. and Calais, D., *Trans. Met. Soc. AIME*, 242, 1679, 1968.
138. Dariel, M. P., Erez, G., and Schmidt, G. M. J., *J. Appl. Phys.*, 40(7), 2746, 1969.
139. Dariel, M. P., *Philos. Mag.*, 22(177), 653, 1970.
140. Dariel, M. P., Erez, G., and Schmidt, G. M. J., *Philos. Mag.*, 18(161), 1045, 1069.
141. Kuliev, A. A. and Nasledov, D. N., *Zh. Tekh. Fiz.*, 28(2), 259, 1958; *Sov. Phys. Tech. Phys.* (English transl.), 3(2), 235, 1958.
142. Braetter, P. and Gobrect, H., *Phys. Status Solidi*, 41(2), 631, 1970.
143. Boltaks, B. I. and Plachenov, B. T., *Sov. Phys. Tech. Phys.* (English transl.), 27(10), 2071, 1957.
144. Peart, R. F., *Phys. Status Solidi*, 15, K-119, 1966.
145. Wilcox, W. R. and LaChapelle, T. J., *J. Appl. Phys.*, 35(1), 240, 1964.
146. Newman, R. C. and Wakefield, J., *Phys. Chem. Solids*, 19(3), 230, 1961.
147. Boltaks, B. I. and Sozinov, I. I., *Zh. Tekh. Fiz.*, 28(3), 679, 1958; *Sov. Phys. Tech. Phys.* (English transl.), 3(3), 636, 1958.
148. Struthers, J. D., *J. Appl. Phys.*, 27(12), 1560, 1956; Errata, 28(4), 516, 1957.
149. Bonzel, H. P., *Phys. Status Solidi*, 20, 493, 1967.
150. Mackawa, S., *J. Phys. Soc. Jap.*, 17(10), 1592, 1962.
151. Rahan, J. J., Pickering, M. E., and Kennedy, J., *J. Electrochem. Soc.*, 106(8), 705, 1959.
152. Rothman, S. J., Peterson, N. L., and Robinson, J. T., *Phys. Status Solidi*, 39(2), 635, 1970.
153. Tomizuka, C. T. and Slifkin, L. M., *Phys. Rev.*, 96, 610, 1954.
154. Bernardini, J., Combe-Brun, A., and Cabane, J., *Ser. Met.*, 4(12), 985, 1970.

Table 2–19 (continued)
RADIOACTIVE TRACER DIFFUSION DATA FOR PURE METALS

155. Sawatskii, A. and Jaumot, F. E., *Trans. Met. Soc. AIME*, 209, 1207, 1957.
156. Mullen, J. G., *Phys. Rev.*, 121, 1649, 1961.
157. Hoffman, R. E., *Acta Met.*, 6, 95, 1958.
158. Hirone, T., Miura, S., and Suzuoka, T., *J. Phys. Soc. Jap.*, 16(12), 2456, 1961.
159. Hoffman, R. E., Turnbull, D., and Hart, E. W., *Acta Met.*, 3, 417, 1955.
160. Pierce, C. B. and Lazarus, D., *Phys. Rev.*, 114, 686, 1959.
161. Barbouth, N., Ouder, J., and Cabane, J., *C. R. Acad. Sci. Ser. C* (Paris), 264(12), 1029, 1967.
162. Kaigorodov, V. N., Rabovskii, A. Ya., and Talinskii, V. K., *Fiz. Metal. Metalloved.*, 24(4), 661, 1967.
163. Kaigorodov, V. N., Klotsman, S. M., Timofeev, A. N., and Traktenberg, I. Sh., *Fiz. Metal. Metalloved.*, 28(1), 120, 1969.
164. Sawatskii, A. and Jaumot, F. E., *Phys. Rev.*, 100, 1627, 1955.
165. Barr, L. W., Mundy, J. N., and Smith, F. A., *Philos. Mag.*, 20(164) 389, 1969.
166. Mundy, J. N., Barr, L. W., and Smith, F. A., *Philos. Mag.*, 14, 785, 1966.
167. Son, P., Ihara, S., Miyake, M., and Sano, T., *J. Jap. Inst. Metals*, 33(1), 1, 1969.
168. Vasilev, V. P., Kamardin, I. F., Skatskill, V. I., Chernomorchenko, S. G., and Shuppe, G. N., *Tr. Sred. Gos. Univ. V. I. Lenina*, 65, 47, 1955; English translation, (AEC-trans.-4272),
169. Pawel, R. E. and Lundy, T. S., *J. Phys. Chem. Solids*, 26, 937, 1965.
170. Vandyshev, B. A. and Panov, A. S., *Izv. Akad. Nauk S.S.S.R. Metal.*, 1, 244, 1969.
171. Molanov, S. and Kuliev, A. A., *Fiz. Tverd. Tela*, 4(2), 542, 1962.
172. Ibraginov, N. I., Shachtachtinskii, M. G., and Kuliev, A. A., *Fiz. Tverd. Tela*, 4, 3321, 1962.
173. Ghoshtagore, R. N., *Phys. Rev.*, 155(3), 698, 1967.
174. Anthony, T. R., Dyson, B. F., and Turnbull, D., *J. Appl. Phys.*, 39(3), 1391, 1968.
175. Shirn, G. A., *Acta Met.*, 3, 87, 1955.
176. Schmitz, F. and Fock, M., *J. Nucl. Mater.*, 21, 317, 1967.
177. Dyson, B. F., *J. Appl. Phys.*, 37, 2375, 1966.
178. Chomka, W. and Andruszkiewicz, J., *Nukleonika*, 5(10), 611, 1960.
179. Sawatskii, A., *J. Appl. Phys.*, 29(9), 1303, 1958.
180. Yolokoff, D., May, S., and Adda, Y., *Compt. Rend.*, 251(3), 2341, 1960.
181. Bartha, L. and Szalay, T., *Int. J. Appl. Radiat. Isot.*, 20(2), 825, 1969.
182. Dyment, F. and Libanati, C. M., *J. Mater. Sci.*, 3(4), 349, 1968.
183. Askill, J., *Phys. Status Solidi*, B-43(1), K-1, 1971.
184. Pavlinov, L. V., Grigorev, G. V., and Gromyko, G. O., *Izv. Akad. Nauk S.S.S.R. Metal.*, 3, 207, 1969.
185. Nakonechnikov, A. Y., Pavlinov, L. V., and Bikov, V. N., *Fiz. Metal. Metalloved.*, 22, 234, 1966.
186. Gibbs, G. B., Graham, D., and Tomlin, D. H., *Philos. Mag.*, 8(92), 1269, 1963.
187. Askill, J. and Gibbs, G. B., *Phys. Status Solidi*, 11, 557, 1965.
188. Murdock, J. F., Lundy, T. S., and Stansbury, E. E., *Acta Met.*, 12(9), 1033, 1964.
189. Pavlinov, L. V., *Fiz. Metal. Metalloved.*, 30(4), 800, 1970.
190. Murdock, J. F., Lundy, T.S., and Stansbury, E. E., *Acta Met.*, 12(9), 1033, 1964.
191. Pavlinov, L. V., *Fiz. Metal. Metalloved.*, 24(2), 272, 1967.
192. Pawel, R. E. and Lundy, T. S., *Acta Met.*, 17(8), 979, 1969.
193. Larikov, L. N., Tyshkevich, V. M., and Chorna, L. F., *Ukr. Fiz. Zh.*, 12(6), 983, 1967.
194. Adda, Y., Kirianenko, A., and Mairy, C., *Compt. Rend.*, 253, 445, 1961; *J. Nucl. Mater.*, 6(1), 130, 1962.
195. Dariel, M. P., Blumenfeld, M., and Kimmel, G., *J. Appl. Phys.*, 41(4), 1480, 1970.
196. Federov, G. B., Smirnov, E. A., and Moiseenko, S. S., *Met. Metalloved. Chistykh. Metal.*, 7, 124, 1968.
197. Rothman, S. J., *J. Nucl. Mater.*, 3(1), 77, 1961.
198. Peterson, N. L. and Rothman, S. J., *Phys. Rev.*, 136(3A), A-842, 1964.
199. Son, P., Ihara, S., Miyake, M., and Sano, T., *J. Jap. Inst. Metals*, 33(1), 1, 1969.
200. Paxton, H. W. and Wolfe, R. A., *Trans. Met. Soc. AIME*, 230, 1426, 1964.
201. Coleman, M. G., Ph.D. thesis, University of Illinois, Chicago, 1967; University microfilm 68-1725, 1968.
202. Vandychev, B. A. and Panov, A. S., *Izv. Akad. Nauk S.S.S.R. Metal.*, 2, 231, 1970.
203. Peart, R. F., *J. Phys. Chem. Solids*, 26, 1853, 1965.
204. Gorney, D. S. and Altovskii, R. M., *Fiz. Metal. Metalloved.*, 30(1), 85, 1970.
205. Rosolowski, J. H., *Phys. Rev.*, 124(6), 1828, 1961.
206. Ghate, P. B., *Phys. Rev.*, 130(1), 174, 1963.
207. Batra, A. P. and Huntington, H. B., *Phys. Rev.*, 145, 542, 1966.
208. Batra, A. P. and Huntington, H. B., *Phys. Rev.*, 154(3), 569, 1967.
209. Warford, J. S. and Huntington, M. B., *Phys. Rev.*, B-1(4), 1867, 1970.
210. Peterson, N. L. and Rothman, S. J., *Phys. Rev.*, 163(3), 645, 1967.

Table 2−19 (continued)
RADIOACTIVE TRACER DIFFUSION DATA FOR PURE METALS

211. **Agarwala, R. P., Murarka, S. P., and Anand, M. S.,** *Trans. Met. Soc. AIME,* 233, 986, 1965.
212. **Blinkin, A. M. and Vorobiov, V. V.,** *Ukr. Fiz. Zh.,* 9(1), 91, 1964.
213. **Agarwala, R. P. and Paul, A. R.,** Proceedings, Nuclear Radiation Chemistry Symposium, Poona, India, 1967, p. 542.
214. **Gruzin, P. L., Emelyanov, V. S., Ryabova, G. G., and Federov, G. B.,** Proceedings, 2nd International Conference on Peaceful Uses of Atomic Energy, Vol. 19, 1958, 187.
215. **Borisov, Y. V., Godin, Y. G., Gruzin, P. L., Evstyukhin, A. I., and Yemelyanov, V. S.,** *Met. Metal. Izdatel. Akad. Nauk S.S.S.R. Moscow,* 1958; English translation NP-TR-448, p. 196, 1960.
216. **Flubacher, P.,** *E. I. R. Ber.,* 49, 1963.
217. **Andriyevskii, R. A., Zagraykin, V. N. and Meshcheryakov, G. Ya.,** *Fys. Metal. Metalloved.,* 19(3), 146, 1966.
218. **Agarwala, R. P. and Paul, A. R.,** Report CONF-670335, International Conference on Vacancies and Interstitials in Metals, Julich, Germany, Vol. 1, p. 105, 1968; Report BARC-377, 1968.
219. **Kidson, G. V. and Young, G. J.,** *Philos. Mag.,* 20(167), 1047, 1969.
220. **Federov, G. B., Smirnov, E. A., and Novikov, S. M.,** *Met. Metalloved. Chistykh. Metal.,* 8, 41, 1969.
221. **Vandychev, B. A. and Pavoc, A. S.,** *Izv. Akad. Nauk S.S.S.R. Metal.,* 2, 231, 1970.
222. **Federov, G. B., Smirnov. E. A., and Zhomov. F. I.,** *Met. Metalloved. Chistykh. Metal.,* 7, 116, 1968.
223. **Pavlinov. L. V.,** *Fiz. Metal. Metalloved.,* 24(2), 272, 1967.
224. **Gilder, H. M. and Lazarus, D.,** *J. Phys. Chem. Solids,* 26, 2081, 1965.
225. **Kidson, G. V. and McGurn, J.,** *Can. J. Phys.,* 39(8), 1146, 1961.
226. **Pawel, R. E. and Lundy, T. S.,** *J. Phys. Chem. Solids,* 26, 937, 1965.
227. **Rothman, S. J., Lloyd, L. T., and Harkness, A. L.,** *Trans. Met. Soc. AIME,* 218(4), 605, 1960.
228. **Federov, G. B., Smirnov, E. A., and Gusev, V. N.,** *At. Energ.,* 27(2), 149, 1969.

Table 2−20
DIFFUSION OF METALS INTO METALS

Diffusing metal	Metal diffused into	Temperature (°C)	D (cm^2/hr)
Ag	Al	466	$6.84-8.1 \times 10^{-7}$
		500	$7.2-3.96 \times 10^{-8}$
		573	1.26×10^{-5}
	Pb	220	5.40×10^{-5}
		250	1.08×10^{-4}
		285	3.29×10^{-4}
	Sn	500	1.73×10^{-1}
Al	Cu	500	6.12×10^{-9}
		850	7.92×10^{-6}
As	Si		$0.32^{-82,000/RT}$
Au	Ag	456	1.76×10^{-9}
		491	$0.92-2.38 \times 10^{-13}$
		585	3.6×10^{-8}
		601	3.96×10^{-8}
		624	$2.5-5 \times 10^{-11}$
		717	$1.04-2.25 \times 10^{-9}$
		729	1.76×10^{-9}
		767	1.15×10^{-6}
		847	2.30×10^{-6}
		858	3.63×10^{-8}
		861	3.92×10^{-8}
		874	3.92×10^{-8}
		916	5.40×10^{-6}
		1040	1.17×10^{-6}

Table 2–20 (continued)
DIFFUSION OF METALS INTO METALS

Diffusing metal	Metal diffused into	Temperature (°C)	D (cm²/hr)
Au (cont.)			
		1120	2.29×10^{-5}
		1189	5.42×10^{-6}
	Au	800	1.17×10^{-8}
		900	9×10^{-8}
		1020	5.4×10^{-7}
	Bi	500	1.88×10^{-1}
	Cu	970	5.04×10^{-6}
	Hg	11	3×10^{-2}
	Pb	100	8.28×10^{-6}
		150	1.80×10^{-4}
		200	3.10×10^{-4}
		240	1.58×10^{-3}
		300	5.40×10^{-3}
		500	1.33×10^{-1}
	Si		$0.0011^{-25,800/RT}$
	Sn	500	1.94×10^{-1}
B	Si		$10.5^{-85,000/RT}$
Ba	Hg	7.8	2.17×10^{-2}
Bi	Si		$1030^{-107,000/RT}$
	Pb	220	1.73×10^{-7}
		250	1.33×10^{-6}
		285	1.58×10^{-6}
C	W	1700	1.87×10^{-9}
	Fe	930	$7.51–9.18 \times 10^{-9}$
Ca	Hg	10.2	2.25×10^{-2}
Cd	Ag	650	9.36×10^{-7}
		800	4.68×10^{-6}
		900	2.23×10^{-5}
	Hg	8.7	6.05×10^{-2}
		15	6.51×10^{-2}
		20	5.47×10^{-2}
		99.1	1.23×10^{-1}
	Pb	200	4.59×10^{-7}
		252	3.10×10^{-6}
Cd, 1 atom %	Pb	167	1.66×10^{-7}
Ce	W	1727	3.42×10^{-6}
Cs	Hg	7.3	1.88×10^{-2}
	W	27	4.32×10^{-8}
		227	5.40×10^{-4}
		427	2.88×10^{-2}
		540	1.44×10^{-1}
Cu	Al	440	1.8×10^{-7}
		457	2.88×10^{-7}
		540	5.04×10^{-6}

Table 2−20 (continued)
DIFFUSION OF METALS INTO METALS

Diffusing metal	Metal diffused into	Temperature (°C)	D (cm²/hr)
Cu (cont.)			
		565	$4.68–5.00 \times 10^{-4}$
	Ag	650	1.04×10^{-6}
		760	1.30×10^{-6}
		895	3.38×10^{-6}
	Au	301	5.40×10^{-10}
		443	8.64×10^{-9}
		560	3.38×10^{-7}
		604	5.10×10^{-7}
		616	7.92×10^{-7}
		740	3.35×10^{-6}
	Cu	650	1.15×10^{-8}
		750	2.34×10^{-8}
		830	1.44×10^{-7}
		850	9.36×10^{-7}
		950	2.30×10^{-6}
		1030	1.01×10^{-5}
	Ge	700–900	$1.01 \pm 0.1 \times 10^{-1}$
	Pt	1041	$7.83–9 \times 10^{-8}$
		1213	5.04×10^{-7}
		1401	6.12×10^{-6}
Fe	Au	753	1.94×10^{-6}
		1003	2.70×10^{-5}
	Si		$0.0062^{-20,000}/RT$
Ga	Si		$3.6^{-81,000}/RT$
Ge	Al	630	3.31×10^{-1}
	Au	529	1.84×10^{-1}
		563	2.80×10^{-1}
	Ge	766–928	$7.8^{-68,509}/RT$
		1060–1200° K	$87^{-73,000}/RT$
Hg	Cd	156	9.36×10^{-7}
		176	2.55×10^{-6}
		202	9×10^{-6}
	Pb	177	8.34×10^{-8}
		197	2.09×10^{-5}
In	Ag	650	1.04×10^{-6}
		800	6.84×10^{-6}
		895	4.68×10^{-5}
	Si		$16.5^{-90,000}/RT$
K	Hg	10.5	2.21×10^{-2}
	W	207	2.05×10^{-2}
		317	3.6×10^{-1}
		507	$1.1 \times 10^{+1}$
Li	Hg	8.2	2.75×10^{-2}

Table 2—20 (continued)
DIFFUSION OF METALS INTO METALS

Diffusing metal	Metal diffused into	Temperature (°C)	D (cm²/hr)
Mg	Al	365	3.96×10^{-8}
		395	$1.98-2.41 \times 10^{-7}$
		420	$2.38-2.74 \times 10^{-7}$
		440	1.19×10^{-7}
		447	9.36×10^{-7}
		450	6.84×10^{-6}
		500	$3.96-7.56 \times 10^{-6}$
		577	1.58×10^{-5}
	Pb	220	4.32×10^{-7}
Mn	Cu	400	7.2×10^{-10}
		850	4.68×10^{-7}
Mo	W	1533	9.36×10^{-10}
		1770	4.32×10^{-9}
		2010	7.92×10^{-8}
		2260	2.81×10^{-7}
Na	Hg	9.6	2.67×10^{-2}
	W	20	2.88×10^{-2}
		227	1.80
		417	9.72
		527	1.19×10^{-1}
Ni	Au	800	2.77×10^{-6}
		1003	2.48×10^{-5}
	Cu	550	2.56×10^{-9}
		950	7.56×10^{-7}
		320	1.26×10^{-6}
	Pt	1043	1.81×10^{-8}
		1241	1.73×10^{-6}
		1401	5.40×10^{-6}
Ni, 1 atom %	Pb	285	8.34×10^{-7}
Ni, 3 atom %	Pb	252	1.25×10^{-7}
Pb	Cd	252	2.88×10^{-8}
	Pb	250	5.42×10^{-8}
		285	2.92×10^{-7}
	Sn	500	1.33×10^{-1}
Pb, 2 atom %	Hg	9.4	6.46×10^{-2}
		15.6	5.71×10^{-2}
		99.2	8×10^{-2}
Pd	Ag	444	4.68×10^{-9}
		571	1.33×10^{-7}
		642	4.32×10^{-7}
		917	4.32×10^{-6}
	Au	727	2.09×10^{-8}
		970	1.15×10^{-6}
	Cu	490	3.24×10^{-9}
		950	$9.0-10.44 \times 10^{-7}$

Table 2–20 (continued)
DIFFUSION OF METALS INTO METALS

Diffusing metal	Metal diffused into	Temperature (°C)	D (cm²/hr)
Po	Au	470	4.59×10^{-11}
	Al	20	1.08×10^{-9}
		500	1.80×10^{-7}
	Bi	150	1.80×10^{-7}
		200	1.80×10^{-6}
	Pb	150	4.59×10^{-11}
		200	4.59×10^{-9}
		310	5.41×10^{-7}
Pt	Au	740	1.69×10^{-8}
		986	$6.12–10.08 \times 10^{-7}$
	Cu	490	2.01×10^{-9}
		960	$3.96–8.28 \times 10^{-7}$
	Pb	490	7.04×10^{-2}
Ra	Au	470	1.42×10^{-8}
	Pt	470	3.42×10^{-8}
Ra ($B + C$)	Ag	470	1.57×10^{-8}
Rb	Hg	7.3	1.92×10^{-2}
Rh	Pb	500	1.27×10^{-1}
Sb	Ag	650	1.37×10^{-6}
		760	5.40×10^{-6}
		895	1.55×10^{-5}
	Si		$5.6^{-91,000}/RT$
Si	Al	465	1.22×10^{-6}
		510	7.2×10^{-6}
		600	3.35×10^{-5}
		667	1.44×10^{-1}
		697	3.13×10^{-1}
	Fe + C[a]	1400–1600	$3.24–5.4 \times 10^{-2}$
Sn	Ag	650	2.23×10^{-6}
		895	2.63×10^{-5}
	Cu	400	1.69×10^{-9}
		650	2.48×10^{-7}
		850	1.40×10^{-5}
	Hg	10.7	6.38×10^{-2}
	Pb	245	1.12×10^{-7}
		250	1.83×10^{-7}
		285	5.76×10^{-7}
Sr	Hg	9.4	1.96×10^{-2}
Th	Mo	1615	1.30×10^{-6}
		2000	3.60×10^{-3}
	Tl	285	8.76×10^{-7}
	W	1782	3.96×10^{-7}
		2027	4.03×10^{-6}
		2127	1.29×10^{-5}
		2227	2.45×10^{-5}

Table 2–20 (continued)
DIFFUSION OF METALS INTO METALS

Diffusing metal	Metal diffused into	Temperature (°C)	D (cm²/hr)
Th (B)	Pb	165	2.54×10^{-12}
		260	2.54×10^{-8}
		324	5.84×10^{-6}
Tl	Hg	11.5	3.63×10^{-2}
	Pb	220	1.01×10^{-7}
		250	7.92×10^{-7}
		270	3.96×10^{-7}
		285	1.12×10^{-6}
		315	2.09×10^{-6}
	Si		$16.5^{-90,000/RT}$
U	W	1727	4.68×10^{-8}
Y	W	1727	6.55×10^{-5}
Zn	Ag	750	1.66×10^{-5}
		850	4.37×10^{-5}
	Al	415	9×10^{-7}
		473	1.91×10^{-6}
		500	$7.2-13.68 \times 10^{-6}$
		555	1.8×10^{-5}
	Hg	11.5	9.09×10^{-2}
		15	8.72×10^{-2}
		99.2	1.20×10^{-1}
	Pb	285	5.84
Zr	W	1727	1.17×10^{-5}

[a]Saturated FeC alloy.

From Loebel, R., in *Handbook of Chemistry and Physics,* 51st ed., Weast, R. C., Ed., Chemical Rubber, Cleveland, 1970, F-55.

Table 2–21
INDICES OF REFRACTION FOR SOME ELEMENTS

Element	Formula	Index[a]
Bromine	Br_2	
liquid		1.661^{15}
Cadmium	Cd	
liquid		0.82 (579 nm)
solid		1.13
Chlorine	Cl_2	
liquid		1.385
gas		1.000768
Hydrogen	H_2	
liquid		$1.10974^{-252.83}$ (579 nm)
Iodine	I_2	
solid		3.34
gas		1.001920
Lead	Pb	
solid		2.6 (579 nm)
Mercury	Hg	
liquid		1.6–1.9
Nitrogen	N_2	
liquid		1.2053^{-190}
Oxygen	O_2	
liquid		1.221^{-181}
Phosphorus	P	
solid, yellow		2.1442^{25}
Selenium	Se_8	
solid, crystalline		3.00, 4.04
solid, amorphous		2.92
Sodium	Na	
liquid		0.0045
solid		4.22
Sulfur	S_8	
liquid		1.929^{110}
solid, amorphous		1.998
solid, rhombic, α		1.957, 2.0377, 2.2454
Tin	Sn	
liquid		2.1

[a]Unless otherwise indicated, the indices are for sodium light, λ =589.3 nm; other wavelengths are given in parentheses following the index value. Temperatures are understood to be room temperature for solids and 20°C for liquids; temperatures other than these are shown as superior numbers following the index value.

From Weast, R. C., Ed., *Handbook of Chemistry and Physics*, 55th ed., CRC Press, Cleveland, 1974, E-218.

Table 2–22
INDICES OF REFRACTION FOR INORGANIC COMPOUNDS

Compound	Formula	Index[a]
Aluminum		
carbide	AlC_3	2.7, 2.75 (700 nm)
chloride	$AlCl_3 \cdot 6H_2O$	1.560, 1.057
oxide	Al_2O_3	1.665–1.680, 1.63–1.65
Ammonium		
antimonyl tartrate	$2(NH_4 \cdot SbO \cdot C_4H_4O_6) \cdot H_2O$	β1.6229 (C)
o-arsenate, di-H-	$NH_4H_2AsO_4$	1.5766, 1.5217
bromide	NH_4Br	1.7108
perchlorate	NH_4ClO_4	1.4818, 1.4833, 1.4881
chloroplatinate	$(NH_4)_2PtCl_6$	1.8
fluoride	NH_4F	ω<1.328
fluoride, acid	NH_4HF_2	1.385, 1.390, 1.394
hydrogen malate (d)	$NH_4C_4H_5O_5$	β1.503
nitrate	NH_4NO_3	1.413, 1.611 (He), 1.63
sulfate, acid	NH_4HSO_4	1.463, 1.473, 1.510
tartrate (dl)	$(NH_4)_2C_4H_4O_6 \cdot 2H_2O$	β1.564
thiocyanate	NH_4CNS	1.546, 1.685, 1.692
uranyl acetate	$NH_4C_2H_3O_2 \cdot UO_2(C_2H_3O_2)_2$	1.4808, 1.4933
Antimony		
bromide, tri-	$SbBr_3$	>1.74+
iodide, tri-	SbI_3	2.78 (Li), 2.36
Barium		
cadmium bromide	$BaCdBr_4 \cdot 4H_2O$	β1.702
cadmium chloride	$BaCdCl_4 \cdot 4H_2O$	β1.651
calcium propionate	$BaCa_2(C_3H_5O_2)_6$	1.4442
fluochloride	$BaCl_2 \cdot BaF_2$	1.640, 1.633
fluoride	BaF_2	1475, 1471
oxide	BaO	1.980
o-phosphate, di-	$BaHPO_4$	1.617, 1.63±, 1.635
propionate	$Ba(C_2H_5CO_2)_2 \cdot H_2O$	β1.5175
sulfide, mono-	BaS	2.155
Cadmium		
ammonium chloride	$CdCl_2 \cdot 4NH_4Cl$	1.6038, 1.6042
cesium sulfate	$CdSO_4 \cdot Cs_2SO_4 \cdot 6H_2O$	1.498, 1.500, 1.506
fluoride	CdF_2	1.56
magnesium chloride	$(CdCl_2)_2 \cdot MgCl_2 \cdot 12H_2O$	1.49, 1.5331, 1.5769
oxide	CdO	2.49 (Li)
potassium chloride	$CdCl_2 \cdot 4KCl$	1.5906, 1.5907
potassium cyanide	$Cd(CN)_2 \cdot 2KCN$	1.4213
rubidium sulfate	$CdSO_4 \cdot Rb_2SO_4 \cdot 6H_2O$	1.4798, 1.4848, 1.4948
Calcium		
aluminate	$Ca_3Al_2O_6$	1.710
borate	$CaO \cdot B_2O_3$	1.540, 1.656, 1.682
carbide	CaC_2	>1.75
copper acetate	$CaCu(C_2H_3O_2)_4 \cdot 6H_2O$	1.436, 1.478
cyanamide	$CaCN_2$	1.60, >1.95
dithionate	$CaS_2O_6 \cdot 4H_2O$	1.5516, 1.5414
pyrophosphate	$Ca_2P_2O_7$	1.585, 1.60±, 1.605
platinocyanide	$CaPt(CN)_4 \cdot 5H_2O$	1.623, 1.644, 1.767
strontium propionate	$Ca_2Sr(C_3H_5O_2)_6$	1.4871, 1.4956
sulfide (oldhamite)	CaS	2.137
sulfite	$CaSO_3 \cdot 2H_2O$	1.590, 1.595, 1.628
thiosulfate	$CaS_2O_3 \cdot 6H_2O$	1.545, 1.560, 1.605
Carbon		
dioxide, liquid	CO_2	1.195[15]
Cerium		
dithionate	$Ce_2(S_2O_6)_3 \cdot 15H_2O$	β1.507

Table 2–22 (continued)
INDICES OF REFRACTION FOR INORGANIC COMPOUNDS

Compound	Formula	Index[a]
Cesium		
perchlorate	$CsClO_4$	1.4752, 1.4788, 1.4804
nitrate	$CsNO_3$	1.55, 1.56
selenate	Cs_2SeO_4	1.5989, 1.5999, 1.6003
thallium chloride	$Cs_3Tl_2Cl_9$	1.784, 1.774
Chromium		
cesium sulfate	$CrCs(SO_4)_2 \cdot 12H_2O$	1.4810
oxide (chromic)	Cr_2O_2	2.5
potassium cyanide (chromic)	$CrK_3(CN)_6$	1.5221, 1.5244, 1.5373
sulfate (chromic)	$Cr_2(SO_4)_3 \cdot 18H_2O$	1.564
thallium sulfate	$CrTl(SO_4)_2 \cdot 12H_2O$	1.5228
Cobalt		
acetate	$Co(C_2H_3O_2)_2 \cdot 4H_2O$	β1.542
aluminate (Thenard's blue)	$Co(AlO_2)_2$	>1.78 (red), 1.74 (blue)
ammonium selenate	$CoSeO_4 \cdot (NH_4)_2SeO_4 \cdot 6H_2O$	1.5246, 1.5311, 1.5396
cesium sulfate	$CoCs_2(SO_4)_2 \cdot 6H_2O$	1.5057, 1.5085, 1.5132
chloride (cobaltous)	$CoCl_2\ 2H_2O$	<1.625, <1.671, >1.67
potassium selenate	$CoSeO_4 \cdot K_2SeO_4 \cdot 6H_2O$	1.5135, 1.5195, 1.5358
rubidium sulfate	$CoSO_4 \cdot Rb_2SO_4 \cdot 6H_2O$	1.4859, 1.4916, 1.5014
selenate	$CoSeO_4 \cdot 6H_2O$	β1.5225, γ1.5227
Copper		
ammonium selenate	$CuSeO_4 \cdot (NH_4)_2SeO_4 \cdot 6H_2O$	1.5213, 1.5355, 1.5395
ammonium sulfate	$CuSO_4 \cdot (NH_4)_2SO_4 \cdot 6H_2O$	1.4910, 1.5007, 1.5054
cesium sulfate	$CuSO_4 \cdot Cs_2SO_4 \cdot 6H_2O$	1.5048, 1.5061, 1.5153
chloride (cupric)	$CuCl_2 \cdot 2H_2O$	1.644, 1.684, 1.742
formate	$Cu(CHO_2)_2 \cdot 4H_2O$	1.4133, 1.5423, 1.5571
oxide (cuprous)	Cu_2O	2.705
potassium chloride	$CuCl_2 \cdot 2KCl \cdot 2H_2O$	1.6365, 1.6148
potassium cyanide (cuprous)	$CuK_3(CN)_4$	1.5215
potassium selenate	$CuSeO_4 \cdot K_2SeO_4 \cdot 6H_2O$	1.5096, 1.5235, 1.5387
potassium sulfate	$CuSO_4 \cdot K_2SO_4 \cdot 6H_2O$	1.4836, 1.4864, 1.5020
strontium formate	$Cu(HCO_2)_2 \cdot 2[Sr(HCO_2)_2] \cdot 8H_2O$	1.4995, 1.5199, 1.5801
sulfate (cupric)	$CuSO_4$	1.724, 1.733, 1.739
Cyanogen, liquid	C_2N_2	1.327[18]
Germanium		
bromide, tetra-	$GeBr_4$	1.6269
Gold		
sodium chloride	$AuNaCl_4 \cdot 2H_2O$	α1.545, γ1.75+
Hafnium		
oxychloride	$HfOCl_2 \cdot 8H_2O$	1.557, 1.543
Ice	H_2O	1.3049, 1.3062 (A), 1.3001, 1.3104 (D), 1.3133, 1.3147 (F)
Iron		
ammonium chloride	$Fe(NH_4)_2Cl_4$	1.6439
ammonium selenate	$FeSeO_4 \cdot (NH_2)_2SeO_4 \cdot 6H_2O$	1.5201, 1.5260, 1.5356
cesium sulfate (ferric)	$FeCs(SO_4)_2 \cdot 12H_2O$	1.4839
cesium sulfate (ferrous)	$FeSO_4 \cdot Cs_2SO_4 \cdot 6H_2O$	1.5003, 1.5035, 1.5094
rubidium sulfate	$FeRb(SO_4)_2 \cdot 12H_2O$	1.48234
sulfate (ferric)	$Fe_2(SO_4)_3$	1.802, 1.814, 1.818
thallium sulfate	$FeTl(SO_4)_2 \cdot 12H_2O$	1.52365
Lanthanum		
sulfate	$La_2(SO_4)_3 \cdot 9H_2O$	1.564, 1.569
Lead		
o-arsenate, di-	$PbHAsO_4$	1.8903, 1.9097, 1.9765
nitrate	$Pb(NO_3)_2$	1.782

Table 2–22 (continued)
INDICES OF REFRACTION FOR INORGANIC COMPOUNDS

Compound	Formula	Index[a]
Lithium		
ammonium sulfate	$LiNH_4 SO_4$	β1.437 (Li)
ammonium tartrate (d)	$LiNH_4 (C_4 H_4 O_6) \cdot H_2 O$	β1.567, γ1.5673
ammonium tartrate (dl)	$LiNH_4 (C_4 H_4 O_6) \cdot H_2 O$	β1.5287
bromide	$LiBr$	1.784
chloride	$LiCl$	1.662
dithionate	$Li_2 S_2 O_6 \cdot 2H_2 O$	1.5487, 1.5602, 1.5788
oxide	$Li_2 O$	1.644
potassium sulfate	$LiKSO_4$	1.4723, 1.4717
potassium tartrate	$LiK(C_4 H_4 O_6) \cdot H_2 O$	β1.5226 (red)
rubidium tartrate (d)	$LiRb(C_4 H_4 O_6) \cdot H_2 O$	β1.552
sodium tartrate (dl)	$LiNa(C_4 H_4 O_6) \cdot 2H_2 O$	β1.4904
Magnesium		
ammonium selenate	$MgSeO_4 \cdot (NH_4)_2 SeO_4 \cdot 6H_2 O$	1.5070, 1.5093, 1.5169
ammonium sulfate	$Mg(NH_4)_2 \cdot (SO_4)_2 \cdot 6H_2 O$	1.4716, 1.4730, 1.4786
o-borate	$3MgO \cdot B_2 O_3$	1.6527, 1.6537, 1.6748
cesium sulfate	$MgCs_2 (SO_4)_2 \cdot 6H_2 O$	1.4857, 1.4858, 1.4916
chlorostannate	$MgSnCl_6 \cdot 6H_2 O$	1.5885, 1.5970
fluosilicate	$MgSiF_6 \cdot 6H_2 O$	1.3439, 1.3602
platinocyanide	$MgPt(CN)_4 \cdot 7H_2 O$	1.5608, 1.91
potassium selenate	$MgK_2 (SeO_4)_2 \cdot 6H_2 O$	1.4969, 1.4991, 1.5139
potassium sulfate	$MgK_2 (SO_4)_2 \cdot 6H_2 O$	1.4607, 1.4629, 1.4755
rubidium sulfate	$MgRb_2 (SO_4)_2 \cdot 6H_2 O$	1.4672, 1.4689, 1.4779
silicate	$MgSiO_2$	1.651, 1.654 (calc.), 1.660
sulfide	MgS	2.271, 2.268
Manganese		
borate	$Mn_3 B_4 O_9$	1.617, 1.738, 1.776
cesium sulfate	$MnCs_2 (SO_4)_2^2 \cdot 6H_2 O$	1.4946, 1.4966, 1.5025
chloride	$MnCl_2 \cdot 4H_2 O$	1.555, 1.575, 1.607
rubidium sulfate	$MnRb_2 (SO_4)_2 \cdot 6H_2 O$	1.4767, 1.4807, 1.4907
sulfate (manganous)	$MnSO_4 \cdot 4H_2 O$	1.508, 1.518, 1.522
sulfate (manganous)	$MnSO_4 \cdot 5H_2 O$	1.495, 1.508, 1.514
Mercury		
chloride (mercuric)	$HgCl_2$	1.725, 1.859, 1.965
cyanide (mercuric)	$Hg(CN)_2$	1.645, 1.492
iodide (mercuric), red	HgI_2	2.748, 2.455
Nickel		
ammonium selenate	$Ni(NH_4)_2 \cdot (SeO_4)_2 \cdot 6H_2 O$	1.5291, 1.5372, 1.5466
cesium sulfate	$NiCs_2 (SO_4)_2 \cdot 6H_2 O$	1.5087, 1.5129, 1.5162
chloride	$NiCl_2 \cdot 6H_2 O$	α1.535, γ1.61
fluoride, acid	$NiF_2 \cdot 5HF \cdot 6H_2 O$	1.392, 1.408
potassium selenate	$NiK_2 (SeO_4)_2 \cdot 6H_2 O$	1.5199, 1.5248, 1.5339
rubidium sulfate	$NiRb_2 (SO_4)_2 \cdot 6H_2 O$	1.4895, 1.4961, 1.5052
selenate	$NiSeO_4 \cdot 6H_2 O$	1.5393, 1.5125
Platinum		
potassium dibromonitrite	$PtK_2 (NO_2)_2 Br_2 \cdot H_2 O$	1.626, 1.6684, 1.757
Potassium		
carbonate	$K_2 CO_3$	1.426, 1.531, 1.541
carbonate, acid	$KHCO_3$	1.380, 1.482, 1.578
perchlorate	$KClO_4$	1.4731, 1.4737, 1.4769
chloroplatinate	$K_2 PtCl_6$	1.827 (577 nm)
chloroplatinite	$K_2 PtCl_4$	1.64, 1.67
dichromate	$K_2 Cr_2 O_7$	1.7202, 1.7380, 1.8197
cyanide	KCN	1.410
fluoborate	KBF_4	1.3239, 1.3245, 1.3247
fluoride	KF	1.352 (1.361)
fluoride	$KF \cdot 2H_2 O$	1.345, 1.352, 1.363

Table 2—22 (continued)
INDICES OF REFRACTION FOR INORGANIC COMPOUNDS

Compound	Formula	Index[a]
Potassium (cont.)		
fluosilicate	K_2SiF_6	1.3391
periodate	KIO_4	1.6205, 1.6479
lithium ferrocyanide	$K_2Li_2Fe(CN)_6 \cdot 3H_2O$	1.5883, 1.6007, 1.6316
hypophosphate	$K_2H_2P_2O_6 \cdot 2H_2O$	1.4893, 1.5314, 1.5363
hypophosphate	$K_2H_2P_2O_6 \cdot 3H_2O$	1.4768, 1.4843, 1.4870
ruthenium cyanide	$K_4Ru(CN)_6 \cdot 3H_2O$	$\beta 1.5837$
silicate	K_2SiO_3	1.520, 1.521, 1.528
thiocyanate	$KCNS$	1.532, 1.660, 1.730
thionate, tetra-	$K_2S_4O_6$	1.5896, 1.6075, 1.6435
thionate, penta-	$2K_2S_5O_6 \cdot 3H_2O$	1.565, 1.63, 1.655
Rhodium		
cesium sulfate	$RhCs(SO_4)_2 \cdot 12H_2O$	1.5077
Rubidium		
perchlorate	$RbClO_4$	1.4692, 1.4701, 1.4731
chromate	Rb_2CrO_4	$\beta 1.71, \gamma 1.72$
dithionate	$Rb_2S_2O_6$	1.4574, 1.5078
fluoride	RbF	1.396
selenate	Rb_2SeO_4	1.5515, 1.5537, 1.5582
Ruthenium		
sodium nitrate	$RuNa_2(NO_2)_5 \cdot 2H_2O$	1.5889, 1.5943, 1.7163
Selenium		
oxide	SeO_2	>1.76
Silver		
cyanide	$AgCN$	1.685, 1.94
nitrate	$AgNO_3$	1.729, 1.744, 1.788
phosphate	Ag_2HPO_4	1.8036, 1.7983
potassium cyanide	$AgK(CN)_2$	1.625, 1.63
Sodium		
ammonium tartrate (d)	$NaNH_4(C_4H_4O_6) \cdot 4H_2O$	1.495, 1.498, 1.499
ammonium tartrate (dl)	$NaNH_4(C_4H_4O_6) \cdot H_2O$	$\beta 1.473$ (red)
o-arsenate	$NaH_2AsO_4 \cdot H_2O$	1.5382, 1.5535, 1.5607
o-arsenate	$NaH_2AsO_4 \cdot 2H_2O$	1.4794, 1.5021, 1.5265
bromide	$NaBr$	1.6412
carbonate	Na_2CO_3	1.415, 1.535, 1.546
carbonate, acid	$NaHCO_3$	1.376, 1.500, 1.582
cyanide	$NaCN$	1.452
iodide	NaI	1.7745
molybdate	$3Na_2O \cdot 7MoO_3 \cdot 22H_2O$	$\beta 1.627$
nitrate	$NaNO_3$	1.5874, 1.3361
phosphate	$NaH_2PO_4 \cdot 2H_2O$	1.4401, 1.4629, 1.4815
phosphate	$NaH_2PO_4 \cdot 7H_2O$	1.4412, 1.4424, 1.4526
hypophosphate	$Na_3HP_2O_6 \cdot 9H_2O$	1.4653, 1.4738, 1.4804
silicate	Na_2SiO_3	1.513, 1.520, 1.528
sulfate, acid	$NaHSO_4 \cdot H_2O$	1.43, 1.46, 1.47
sulfite	Na_2SO_3	1.565, 1.515
sulfite, acid	Na_2HSO_3	1.474, 1.526, 1.685
tartrate, acid (d)	$NaH(C_4H_4O_6) \cdot H_2O$	$\beta 1.533$
thiocyanate	$NaCNS$	1.545, 1.625, 1.695
tungstate	$Na_2WO_4 \cdot 2H_2O$	1.5526, 1.5533, 1.5695
vanadate	$Na_2VO_4 \cdot 10H_2O$	1.5305, $\omega 1.5398$, $\epsilon 1.5475$
vanadate	$Na_2VO_4 \cdot 12H_2O$	1.5095, 1.5232
Strontium		
dichromate	$SrCr_2O_7 \cdot 3H_2O$	1.7146, 1.7174, 1.812
fluoride	SrF_2	1.442 (1.438)
oxide	SrO	1.870

Table 2−22 (continued)
INDICES OF REFRACTION FOR INORGANIC COMPOUNDS

Compound	Formula	Index[a]
Strontium (cont.)		
o-phosphate, acid	$SrHPO_4$	1.608, 1.62+, 1.625
sulfide, mono-	SrS	2.107
Sulfur		
nitride	S_4N_4	$\alpha 1.908, \beta 2.046$
Thallium		
chloride, mono-	$TlCl$	2.247
iodide, mono-	TlI	2.78
Tin		
iodide (stannic)	SnI_4	2.106
Uranyl		
potassium sulfate	$UO_2 \cdot SO_4 \cdot K_2SO_4 \cdot 2H_2O$	1.5144, 1.5266, 1.5705 (580 nm)
Vanadium		
ammonium sulfate	$VNH_4(SO_4)_2 \cdot 12H_2O$	1.475
Zinc		
ammonium selenate	$Zn(SeO_4) \cdot (NH_4)_2 SeO_4 \cdot 6H_2O$	1.5240, 1.5300, 1.5385
bromate	$Zn(BrO_3)_2 \cdot 6H_2O$	1.5452
cesium sulfate	$ZnCs_2(SO_4)_2 \cdot 6H_2O$	1.5022, 1.5048, 1.5093
chloride	$ZnCl_2$	1.687, 1.713
fluosilicate	$ZnSiF_6 \cdot 6H_2O$	1.3824, 1.3956
potassium cyanide	$ZnK_2(CN)_4$	1.4115
potassium selenate	$ZnK_2(SeO_4)_2 \cdot 6H_2O$	1.5121, 1.5181, 1.5335
potassium sulfate	$ZnK_2(SO_4)_2 \cdot 6H_2O$	1.4775, 1.4833, 1.4969
rubidium sulfate	$ZnRb_2(SO_4)_2 \cdot 6H_2O$	1.4833, 1.4884, 1.4975
silicate	$ZnSiO_2$	1.616, 1.62±, 1.623
Zirconium		
ammonium fluoride	$Zr(NH_4)_3F_7$	1.433

[a]The indices given are understood to be for the solid form of the compound unless otherwise indicated. Temperatures are understood to be room temperature for solids and 20°C for liquids; temperatures other than these are shown as superior numbers following the index value. Unless otherwise indicated, the indices are for sodium light, λ = 589.3 nm. Other wavelengths, expressed either in nanometers or as symbols, are given in parentheses following the index value. They are as follows: He, λ = 587.6 nm; Li, λ = 670.8 nm; Hg, λ = 589.1 nm; A, λ = 759.4 nm; C, λ = 656.3 nm; D, λ = 589.3 nm; F, λ = 486.1 nm.

From Weast, R. C., Ed., *Handbook of Chemistry and Physics,* 55th ed., CRC Press, Cleveland, 1974, E-218.

Table 2–23

THERMODYNAMIC PROPERTIES OF ELEMENTS AND OXIDES

Thermodynamic calculations over a wide range of temperatures are generally made with the aid of algebraic equations representing the characteristic properties of the substances being considered. The necessary integrations and differentiations, or other mathematical manipulations, are then most easily effected.

The most convenient starting point in making such calculations for a given substance is the heat capacity at constant pressure. From this quantity and a knowledge of the properties of any phase transitions, the other thermodynamic properties may be computed by the well-known equations given in standard texts on thermodynamics.

Users of the following equations and relevant tables are cautioned that the units for a, b, c, and d are cal/g mole, whereas those for A are kcal/g mole. The necessary adjustment must be made when the data are substituted into the equations.

Empirical heat capacity equations are generally in the form of a power series, with the absolute temperature T as the independent variable:

$$C_p = a' + (b' \times 10^{-3})T + (c' \times 10^{-6})T_2$$

or

$$C_p = a'' + (b'' \times 10^{-3})T + \frac{d \times 10^5}{T^2}.$$

Since both forms are used in the ensuing, let

$$C_p = a + (b \times 10^{-3})T + (c \times 10^{-6})T^2 + \frac{d \times 10^5}{T^2}. \qquad (1)$$

The constants a, b, c, and d are to be determined either experimentally or by some theoretical or semi-empirical approach.

The heat content, or enthalpy (H), is determined from the heat capacity by a simple integration of the range of temperatures for which (1) is applicable. Thus, if 298K is taken as a reference temperature,

$$H_T - H_{298} = \int_{298}^{T} C_p dT \qquad (2)$$

$$= a(T - 298) + \tfrac{1}{2}(b \times 10^{-3})(T_2 - 298^2) + \frac{1}{3}(c \times 10^{-6})(T^3 - 298^3) - (d \times 10^5)\left(\frac{1}{T} - \frac{1}{298}\right)$$

$$= aT + \tfrac{1}{2}(b \times 10^{-3})T^2 + \frac{1}{3}(c \times 10^{-6})T^3 - \frac{d \times 10^5}{T} - A,$$

where all the constants on the right-hand side of the equation have been incorporated in the term $-A$.

In general, the enthalpy is given by a sum of terms such as (2) for each phase of the substance involved in the temperature range considered plus terms that represent the heats of transitions:

$$H_T - H_{298} = \Sigma \int_{T_1}^{T_2} C_p dT + \Sigma \, \Delta H_{tr}.$$

In a similar manner, the entropy S is obtained from (1) by performing the integration

$$S_T - S_{298} = \int_{298}^{T} (C_p/T)dt \qquad (3)$$

$$= a\ln(T/298) + (b \times 10^{-3})(T - 298) + \tfrac{1}{2}(c \times 10^{-6})(T^2 - 298^2) - \tfrac{1}{2}(d \times 10^5)\left(\frac{1}{T^2} - \frac{1}{298^2}\right)$$

$$= a\ln T + (b \times 10^{-3})T + \tfrac{1}{2}(c \times 10^{-6})T^2 - \frac{\tfrac{1}{2}(d \times 10^5)}{T^2} - B'$$

or

$$S_T = 2.303\, a \log T + (b \times 10^{-3})T + \tfrac{1}{2}(c \times 10^{-6})T^2 - \frac{\tfrac{1}{2}(d \times 10^5)}{T^2} - B \qquad (4)$$

Table 2–23 (continued)
THERMODYNAMIC PROPERTIES OF ELEMENTS AND OXIDES

where

$$B = B' - S_{298}.$$ (5)

the quantity

$$F_T - H_{298} = (H_T - H_{298}) - TS_T$$

From the definition of free energy (F):

is obtained from (2) and (4):

$$F = H - TS$$

$$F_T - H_{298} = -2.303aT \log T - \tfrac{1}{2}(b \times 10^{-3})T^2 - \frac{1}{6}(c \times 10^{-6})T^3 - \frac{\tfrac{1}{2}(d \times 10^5)}{T} + (B + a)T - A$$ (6)

and also the free energy function

$$\frac{F_T - H_{298}}{T} = -2.303a \log T - \tfrac{1}{2}(b \times 10^{-3})T - \frac{1}{6}(c \times 10^{-6})T^2 - \frac{\tfrac{1}{2}(d \times 10^5)}{T^2} + (B + a) - \frac{A}{T}$$ (7)

Values of the constants for elements are given in Table 2–24; corresponding values for oxides are given in Table 3–7. The first column in each table lists the element or the oxide. The second column gives the phase to which they are applicable. The third, fourth, and fifth columns specify the thermodynamic properties for the transition to the succeeding phase. In column 6, the value of the entropy at 298.15K, the reference temperature, is given. The remaining columns, except for the last,

give the values of the constants a, b, c, d, A, and B required in the thermodynamic equations.

All values that represent estimates are enclosed in parentheses. The heat capacities at temperatures beyond the range of experimental determination were estimated by extrapolation. Where no experimental values were found, analogy with compounds of neighboring elements in the Periodic Table was employed.

From Weast, R. C., Ed., *Handbook of Chemistry and Physics*, 55th ed., CRC Press, Cleveland, 1974, D-55.

Table 2–24
THERMODYNAMIC PROPERTIES OF THE ELEMENTS

Element	Phase	Temperature of transition (K)	Heat of transition (kcal/g mole)	Entropy of transition (e.u.)	Entropy at 298K (e.u.)	a (cal/g mole)	b (cal/g mole)	c (cal/g mole)	d (cal/g mole)	A (kcal/g mole)	B (e.u.)
Ac	solid	(1090)	(2.5)	(2.3)	(13)	(5.4)	(3.0)	—	—	(1.743)	(18.7)
	liquid	(2750)	(70)	(25)	—	(8)	—	—	—	(0.295)	(31.3)
Ag	solid	1234	2.855	2.313	10.20	5.09	1.02	—	—	1.488	19.21
	liquid	2485	60.72	24.43	—	7.30	—	—	0.36	0.164	30.12
	gas	—	—	—	—	(4.97)	—	—	—	(–66.34)	(–12.52)
Al	solid	931.7	2.57	2.76	6.769	4.94	2.96	—	—	1.604	22.26
	liquid	2600	67.9	26	—	7.0	—	—	—	0.33	30.83
Am	solid	(1200)	(2.4)	(2.0)	(13)	(4.9)	(4.4)	—	—	(1.657)	(16.2)
	liquid	2733	51.7	18.9	—	(8.5)	—	—	—	(0.409)	(34.5)
As	solid	883	31/4	35.1/4	8.4	5.17	2.34	—	—	1.646	21.8
Au	solid	1336.16	3.03	2.27	11.32	6.14	–0.175	0.92	—	1.831	23.65
	liquid	2933	74.21	25.30	—	7.00	—	—	—	–0.631	26.99
B	solid	2313	(3.8)	(1.6)	1.42	1.54	4.40	—	—	0.655	8.67
	liquid	2800	75	27	—	(6.0)	—	—	—	(–4.599)	(31.4)
Ba	solid, α	648	0.14	0.22	16	5.55	4.50	—	—	1.722	16.1
	solid, β	977	1.83	1.87	—	5.55	1.50	—	—	1.582	15.9
	liquid	1911	35.665	18.63	—	(7.4)	—	—	—	(0.843)	(25.3)
	gas	—	—	—	—	(4.97)	—	—	—	(–39.65)	(–11.7)
Be	solid	1556	2.919	1.501	2.28	5.07	1.21	—	–1.15	1.951	27.62
	liquid	—	—	—	—	5.27	—	—	—	–1.611	25.68
Bi	solid	544.2	2.63	4.83	13.6	5.38	2.60	—	—	1.720	17.8
	liquid	1900	41.1	21.6	—	7.60	—	—	—	–0.087	25.6
	gas	—	—	—	—	(4.97)	—	—	—	(–46.19)	(–15.9)
C	solid	—	—	—	1.3609	4.10	1.02	—	–2.10	1.972	23.484

Table 2-24 (continued)

THERMODYNAMIC PROPERTIES OF THE ELEMENTS

Element	Phase	Temperature of transition (K)	Heat of transition (kcal/g mole)	Entropy of transition (e.u.)	Entropy at 298K (e.u.)	a (cal/g mole)	b (cal/g mole)	c (cal/g mole)	d (cal/g mole)	A (kcal/g mole)	B (e.u.)
Ca	solid, α	723	0.24	0.33	9.95	5.24	3.50	—	—	1.718	20.95
	solid, β	1123	2.2	1.96	—	6.29	1.40	—	—	1.689	26.01
	liquid	1755	38.6	22.0	—	7.4	—	—	—	-0.147	30.28
	gas	—	°	—	—	(4.97)	—	—	—	(-43.015;)	(-9.88)
Cd	solid	594.1	1.46	2.46	12.3	5.31	2.94	—	—	1.714	18.8
	liquid	1040	23.86	22.94	—	7.10	—	—	—	0.798	26.1
	gas	—	—	—	—	(4.97)	—	—	—	(-25.28)	(-11.7)
Ce	solid	1048	2.1	2.0	13.8	4.40	6.0	—	—	1.579	13.1
	liquid	2800	73	26	—	(7.9)	—	—	—	(-0.148)	(29.1)
Cl_2	gas	—	—	—	53.286	8.76	0.27	—	-0.65	2.845	-2.929
Co	solid, α	723	0.005	0.007	6.8	4.72	4.30	—	—	1.598	21.4
	solid, β	1398	0.095	0.068	—	3.30	5.86	—	—	0.974	3.1
	solid, γ	1766	3.7	2.1	—	9.60	—	—	—	3.961	50.5
	liquid	3370	93	28	—	8.30	—	—	—	-2.034	38.7
Cr	solid	2173	3.5	1.6	5.68	5.35	2.36	—	—	1.848	25.75
	liquid	2495	72.97	29.25	—	9.40	—	—	—	1.556	50.13
	gas	—	—	—	—	(4.97)	—	—	-0.44	(-82.47)	(-13.8)
Cs	solid	301.9	0.50	1.7	19.8	7.42	—	—	—	2.212	22.5
	liquid	963	16.32	17.0	—	8.00	—	—	—	1.887	24.1
	gas	—	—	—	—	(4.97)	—	—	—	(-17.35)	(-13.6)
Cu	solid	1356.2	3.11	2.29	7.97	5.41	1.50	—	—	1.680	23.30
	liquid	2868	72.8	25.4	—	7.50	—	—	—	0.024	34.05
F_2	gas	—	—	—	48.58	8.29	0.44	—	-0.80	2.760	-0.76
Fe	solid, α	1033	0.410	0.397	6.491	3.37	7.10	—	0.43	1.176	14.59
	solid, β	1180	0.217	0.184	—	10.40	—	—	—	4.281	55.66
	solid, γ	1673	0.15	0.084	—	4.85	3.00	—	—	0.396	19.76
	solid, δ	1808	3.86	2.14	—	10.30	—	—	—	4.382	55.11
	liquid	3008	84.62	28.1	—	10.00	—	—	—	-0.021	50.73

Table 2–24 (continued)
THERMODYNAMIC PROPERTIES OF THE ELEMENTS

Element	Phase	Temperature of transition (K)	Heat of transition (kcal/g mole)	Entropy of transition (e.u.)	Entropy at 298K (e.u.)	a (cal/g mole)	b (cal/g mole)	c (cal/g mole)	d (cal/g mole)	A (kcal/g mole)	B (e.u.)
Ga	solid	302.94	1.335	4.407	9.82	5.237	3.33	—	—	1.710	21.01
	liquid	2700	—	—	—	(6.645)	—	—	—	(0.648)	(23.64)
Ge	solid	1232	8.3	6.7	10.1	5.90	1.13	—	—	1.764	23.8
	liquid	2980	68	23	—	(7.3)	—	—	—	(-5.668)	(25.7)
H₂	gas	—	—	—	31.211	6.62	0.81	—	—	2.010	6.75
Hf	solid	(2600)	(6.0)	(2.3)	13.1	(6.00)	(0.52)	—	—	(1.812)	(21.2)
Hg	liquid	629.73	13.985	22.208	18.46	—	—	—	—	1.971	19.20
	gas	—	—	—	—	4.969	—	—	—	-13.048	-13.54
In	solid	430	0.775	1.80	13.88	5.81	2.50	—	—	1.844	19.97
	liquid	2440	53.8	22.0	—	7.50	—	—	—	1.564	27.34
	gas	—	—	—	—	(4.97)	—	—	—	(-58.42)	(-14.46)
Ir	solid	2727	6.6	2.4	8.7	5.56	1.42	—	—	1.721	23.4
K	solid	336.4	0.5575	1.657	15.2	1.3264	19.405	—	—	1.258	-1.86
	liquid	1052	18.88	17.95	—	8.8825	-4.565	2.9369	—	1.923	32.55
	gas	—	—	—	—	(4.97)	—	—	—	(-19.689)	(-9.46)
La	solid	1153	(2.3)	(2.0)	13.7	6.17	1.60	—	—	1.911	21.9
	liquid	3000	80	27	—	(7.3)	—	—	—	(-0.15)	(26.0)
Li	solid	459	0.69	1.5	6.70	3.05	8.60	—	—	1.292	12.92
	liquid	1640	32.48	19.81	—	7.0	—	—	—	1.509	32.00
	gas	—	—	—	—	(4.97)	—	—	—	(-34.30)	(-2.84)
Mg	solid	923	2.2	2.4	7.77	5.33	2.45	—	-0.103	1.733	23.39
	liquid	1393	31.5	22.6	—	(8.0)	—	—	—	0.942	36.967
	gas	—	—	—	—	(4.97)	—	—	—	(-34.78)	(-7.60)

Table 2–24 (continued)

THERMODYNAMIC PROPERTIES OF THE ELEMENTS

Element	Phase	Temperature of transition (K)	Heat of transition (kcal/g mole)	Entropy of transition (e.u.)	Entropy at 298K (e.u.)	a (cal/g mole)	b (cal/g mole)	c (cal/g mole)	d (cal/g mole)	A (kcal/g mole)	B (e.u.)
Mn	solid, α	1000	0.535	0.535	7.59	6.70	3.38	—	-0.37	1.974	26.11
	solid, β	1374	0.545	0.397	—	8.33	0.66	—	—	2.672	41.02
	solid, γ	1410	0.430	0.305	—	10.70	—	—	—	4.760	56.84
	solid, δ	1517	3.5	2.31	—	11.30	—	—	—	5.176	60.88
	liquid	2368	53.7	22.7	—	11.00	—	—	—	1.221	56.38
	gas	—	—	—	—	6.26	—	—	—	-63.704	-3.13
Mo	solid	2883	(5.8)	(2.0)	6.83	5.48	1.30	—	—	1.692	24.78
N$_2$	gas	—	—	—	45.767	6.76	0.606	0.13	—	2.044	-7.064
Na	solid	371	0.63	1.7	12.31	5.657	3.252	0.5785	—	1.836	20.92
	liquid	1187	23.4	20.1	—	8.954	-4.577	2.540	—	1.924	36.0
	gas	—	—	—	—	(4.97)	—	—	—	(-24.40)	(-8.7)
Nb	solid	2760	(5.8)	(2.1)	8.3	5.66	0.96	—	—	1.730	24.24
Nd	solid	1297	(2.55)	(1.97)	13.9	5.61	5.34	—	—	1.910	19.7
	liquid	(2750)	(61)	(22)	—	(9.1)	—	—	—	(-0.606)	35.8
Ni	solid, α	626	0.092	0.15	7.137	4.06	7.04	—	—	1.523	18.095
	solid, β	1728	4.21	2.44	—	6.00	1.80	—	—	1.619	27.16
	liquid	3110	90.48	29.0	—	9.20	—	—	—	0.251	45.47
Np	solid	913	(2.3)	(2.5)	(14)	(5.3)	(3.4)	—	—	(1.731)	(17.9)
	liquid	(2525)	(55)	(22)	—	(9.0)	—	—	—	(1.392)	(37.5)
O$_2$	gas	—	—	—	49.003	8.27	0.258	—	-1.877	3.007	-0.750
Os	solid	2970	(6.4)	(2.2)	7.8	5.69	0.88	—	—	1.736	24.9
P$_4$	solid, white	317.4	0.601	1.89	42.4	13.62	28.72	—	—	5.338	43.8
	liquid	553	11.9	21.5	—	19.23	0.51	—	-2.98	6.035	66.7
	gas	—	—	—	—	(19.5)	(-0.4)	(1.3)	—	(-6.32)	(46.1)
Pa	solid	(18.25)	(4.0)	(2.2)	(13.5)	(5.2)	(4.0)	—	—	(1.728)	(17.3)
	liquid	(4500)	(115)	(26)	—	(8.0)	—	—	—	(-3.823)	(28.8)

Table 2–24 (continued)

THERMODYNAMIC PROPERTIES OF THE ELEMENTS

Element	Phase	Temperature of transition (K)	Heat of transition (kcal/g mole)	Entropy of transition (e.u.)	Entropy at 298K (e.u.)	a (cal/g mole)	b (cal/g mole)	c (cal/g mole)	d (cal/g mole)	A (kcal/g mole)	B (e.u.)
Pb	solid	600.6	1.141	1.900	15.49	5.64	2.30	—	—	1.784	17.33
	liquid	2023	42.5	21.0	—	7.75	-0.73	—	—	1.362	27.11
	gas	—	—	—	—	(4.97)	—	—	—	(-45.25)	(-13.6)
Pd	solid	1828	4.12	2.25	8.9	5.80	1.38	—	—	1.791	24.6
	liquid	3440	89	26	—	(9.0)	—	—	—	(1.215)	(43.8)
Po	solid	525	(2.4)	(4.6)	13	(5.2)	(3.2)	—	—	(1.693)	(17.6)
	liquid	(1235)	(24.6)	(19.9)	—	(9.0)	—	—	—	(0.847)	(35.2)
	gas	—	—	—	—	(4.97)	—	—	—	(-28.73)	(-13.5)
Pr	solid	1205	(2.5)	(2.1)	(13.5)	(5.0)	(4.6)	—	—	(1.705)	(16.4)
	liquid	3563	—	—	—	(8.0)	—	—	—	(-0.519)	(30.0)
Pt	solid	2042.5	5.2	2.5	10.0	5.74	1.34	—	0.10	1.737	23.0
	liquid	4100	122	29.8	—	(9.0)	—	—	—	(0.406)	(42.6)
Pu	solid	913	(2.26)	(2.48)	(13.0)	(5.2)	(3.6)	—	—	(1.710)	(17.7)
	liquid	—	—	—	—	(8.0)	—	—	—	(0.506)	(31.0)
Ra	solid	1233	(2.3)	(1.9)	(17)	(5.8)	(1.2)	—	—	(1.783)	(16.4)
	liquid	(1700)	(35)	(21)	—	(8.0)	—	—	—	(1.284)	(28.6)
	gas	—	—	—	—	(4.97)	—	—	—	(-38.87)	(-14.5)
Rb	solid	312.0	0.525	1.68	16.6	3.27	13.1	—	—	1.557	5.9
	liquid	952	18.11	19.0	—	7.85	—	—	—	1.814	26.5
	gas	—	—	—	—	(4.97)	—	—	—	(-19.04)	(-12.3)
Re	solid	3440	(7.9)	(2.3)	(8.89)	(5.85)	(0.8)	—	—	(1.780)	(24.7)
Rh	solid	2240	(5.2)	(2.3)	7.6	5.40	2.19	—	—	1.707	23.8
	liquid	4150	127	30.7	—	(9.0)	—	—	—	(-0.923)	(44.4)

Table 2–24 (continued)
THERMODYNAMIC PROPERTIES OF THE ELEMENTS

Element	Phase	Temperature of transition (K)	Heat of transition (kcal/g mole)	Entropy of transition (e.u.)	Entropy at 298K (e.u.)	a (cal/g mole)	b (cal/g mole)	c (cal/g mole)	d (cal/g mole)	A (kcal/g mole)	B (e.u.)
Ru	solid, α	1308	0.034	0.026	6.9	5.25	1.50	—	—	1.632	23.5
	solid, β	1473	0	—	—	7.20	—	—	—	2.867	35.5
	solid, γ	1773	0.23	0.13	—	7.20	—	—	—	2.867	35.5
	solid, δ	2700	(6.1)	(2.3)	—	7.50	—	—	—	3.169	37.6
S	solid, α	368.6	0.088	0.24	7.62	3.58	6.24	—	—	1.345	14.64
	solid, β	392	0.293	0.747	—	3.56	6.95	—	—	1.298	14.54
	liquid	717.76	2.5	3.5	—	5.4	5.0	—	—	1.576	24.02
$\frac{1}{2}S_2$	gas	—	—	—	—	(4.25)	(0.15)	—	(−1.0)	(−2.859)	(9.57)
Sb	solid, α, β, γ	903.7	4.8	5.3	10.5	5.51	1.74	—	—	1.720	21.4
	liquid	1713	46.665	27.3	—	7.50	—	—	—	−1.992	28.1
$\frac{1}{2}Sb_2$	gas	—	—	—	—	4.47	—	—	−0.11	−53.876	−21.7
Sc	solid	1670	(4.0)	(2.4)	(9.0)	(5.13)	(3.0)	—	—	1.663	21.1
	liquid	3000	80	27	—	(7.50)	—	—	—	(−2.563)	31.3
Se	solid	490.6	1.25	2.55	10.144	3.30	8.80	—	—	1.375	11.28
	liquid	1000	14.27	14.27	—	7.0	—	—	—	0.881	27.34
Si	solid	1683	11.1	6.60	4.50	5.70	1.02	—	—	2.100	28.88
	liquid	2750	71	26	—	7.4	—	—	−1.06	−7.646	33.17
Sm	solid	1623	3.7	2.3	(15)	(6.7)	(3.4)	—	—	(2.149)	(24.2)
	liquid	(2800)	(70)	(25)	—	(9.0)	—	—	—	(−2.296)	(33.4)
Sn	solid, α, β	505.1	1.69	3.35	12.3	4.42	6.30	—	—	1.598	14.8
	liquid	2473	(55)	(22)	—	7.30	—	—	—	0.559	26.2
	gas	—	—	—	—	(4.97)	—	—	—	(−60.21)	(−14.3)
Sr	solid	1043	2.2	2.1	13.0	(5.60)	(1.37)	—	—	(1.731)	(19.3)
	liquid	1657	33.61	20.28	—	(7.7)	—	—	—	(0.976)	(30.4)
	gas	—	—	—	—	(4.97)	—	—	—	(−37.16)	(−10.2)

Table 2–24 (continued)
THERMODYNAMIC PROPERTIES OF THE ELEMENTS

Element	Phase	Temperature of transition (K)	Heat of transition (kcal/g mole)	Entropy of transition (e.u.)	Entropy at 298K (e.u.)	a (cal/g mole)	b (cal/g mole)	c (cal/g mole)	d (cal/g mole)	A (kcal/g mole)	B (e.u.)
Ta	solid	3250	7.5	2.3	9.9	5.82	0.78	—	—	1.770	23.4
Tc	solid	(2400)	(5.5)	(2.3)	(8.0)	(5.6)	(2.0)	—	—	(1.759)	(24.5)
	liquid	(3800)	(120)	(32)	—	(11)	—	—	—	(3.459)	(59.4)
Te	solid, α	621	0.13	0.21	11.88	4.58	5.25	—	—	1.599	15.78
	solid, β	723	4.28	5.92	—	4.58	5.25	—	—	1.469	15.57
	liquid	1360	11.9	8.75	—	9.0	—	—	—	−0.988	34.96
½Te₂	gas	—	—	—	—	4.47	—	—	—	−19.048	−6.47
Th	solid	2173	(4.6)	(2.1)	12.76	8.2	−0.77	2.04	—	2.591	33.64
	liquid	4500	(130)	(29)	—	(8.0)	—	—	—	(−7.602)	(26.84)
Ti	solid, α	1155	0.950	0.822	7.334	5.25	2.52	—	—	1.677	23.33
	solid, β	2000	(4.6)	(2.3)	—	7.50	—	—	—	1.645	35.46
	liquid	3550	(101)	(28)	—	(7.8)	—	—	—	(−2.355)	(35.45)
Tl	solid, α	508.3	0.082	0.16	15.4	5.26	3.46	—	—	1.722	15.6
	solid, β	576.8	1.03	1.79	—	7.30	—	—	—	2.230	26.4
	liquid	1730	38.81	22.4	—	7.50	—	—	—	1.315	25.9
	gas	—	—	—	—	(4.97)	—	—	—	(−41.88)	(−15.4)
U	solid, α	938	0.665	0.709	12.03	3.25	8.15	—	0.80	1.063	8.47
	solid, β	1049	1.165	1.111	—	10.28	—	—	—	3.493	48.27
	solid, γ	1405	(3.0)	(2.1)	—	9.12	—	—	—	1.110	39.09
	liquid	3800	—	—	—	(8.99)	—	—	—	(−2.073)	36.01
V	solid	2003	(4.0)	(2.0)	7.05	5.57	0.97	—	—	1.704	24.97
	liquid	3800	—	—	—	(8.6)	—	—	—	1.827	44.06
W	solid	3650	8.42	2.3	8.0	5.74	0.76	—	—	1.745	24.9

Table 2–24 (continued)
THERMODYNAMIC PROPERTIES OF THE ELEMENTS

Element	Phase	Temperature of transition (K)	Heat of transition (kcal/g mole)	Entropy of transition (e.u.)	Entropy at 298K (e.u.)	a (cal/g mole)	b (cal/g mole)	c (cal/g mole)	d (cal/g mole)	A (kcal/g mole)	B (e.u.)
Y	solid	1750	(4.0)	(2.3)	(11)	(5.6)	(2.2)	—	—	(1.767)	(21.6)
	liquid	3500	(90)	(26)	–	(7.5)	–	—	—	(−2.277)	(29.6)
Zn	solid	692.7	1.595	2.303	9.95	5.35	2.40	—	—	1.702	21.25
	liquid	1180	27.43	23.24	–	7.50	–	—	—	1.020	31.35
	gas	–	–	–	–	(4.97)	–	—	—	(−29.407)	(−9.81)
Zr	solid, α	1135	0.920	0.811	9.29	6.83	1.12	—	−0.87	2.378	30.45
	solid, β	2125	(4.9)	(2.3)	–	7.27	–	—	—	1.159	31.43
	liquid	(3900)	(100)	(26)	–	(8.0)	–	—	—	(−2.190)	(34.7)

From Weast, R. C., Ed., *Handbook of Chemistry and Physics*, 55th ed., CRC Press, Cleveland, 1974, D-56.

Table 2–25A

THERMODYNAMIC FUNCTIONS OF COPPER, SILVER, AND GOLD

1 cal = 4.1840 J; $H°_0$ is the enthalpy of the solid at an absolute temperature of 0 and a pressure of 1 atm.

T (K)	$C°_r$ (J/deg·mol) Cu	Ag	Au	$H°_T - H°_0$ (J/mol) Cu	Ag	Au	$(H°_T - H°_0)/T$ (J/deg·mol) Cu	Ag	Au
1.00	0.000743	0.000818	0.00118	0.000359	0.000367	0.000478	0.000359	0.000367	0.000478
2.00	0.00177	0.00265	0.00504	0.00158	0.00197	0.00326	0.000790	0.000987	0.00163
3.00	0.00337	0.00650	0.0141	0.00409	0.00633	0.0123	0.00136	0.00211	0.00410
4.00	0.00582	0.0134	0.0306	0.00860	0.0160	0.0340	0.00215	0.00399	0.00849
5.00	0.00943	0.0243	0.0570	0.0161	0.0344	0.0768	0.00322	0.00689	0.0154
6.00	0.0145	0.0403	0.0955	0.0279	0.0663	0.152	0.00466	0.0110	0.0253
7.00	0.0213	0.0626	0.149	0.0456	0.117	0.273	0.00652	0.0167	0.0390
8.00	0.0301	0.0927	0.220	0.0712	0.194	0.456	0.00889	0.0243	0.0570
9.00	0.0414	0.132	0.313	0.107	0.306	0.720	0.0119	0.0340	0.0800
10.00	0.0555	0.183	0.431	0.155	0.462	1.090	0.0155	0.0462	0.109
11.00	0.0727	0.247	0.577	0.219	0.676	1.592	0.0199	0.0614	0.145
12.00	0.0936	0.325	0.755	0.302	0.961	2.255	0.0251	0.0801	0.188
13.00	0.119	0.421	0.963	0.407	1.332	3.112	0.0313	0.102	0.239
14.00	0.149	0.535	1.203	0.541	1.809	4.193	0.0386	0.129	0.299
15.00	0.184	0.670	1.474	0.706	2.409	5.529	0.0471	0.161	0.369
16.00	0.225	0.826	1.772	0.910	3.155	7.149	0.0569	0.197	0.447
17.00	0.273	1.002	2.096	1.158	4.067	9.081	0.0681	0.239	0.534
18.00	0.328	1.199	2.442	1.458	5.166	11.35	0.0810	0.287	0.630
19.00	0.390	1.414	2.807	1.816	6.471	13.97	0.0956	0.341	0.735
20.00	0.462	1.647	3.187	2.242	8.001	16.97	0.112	0.400	0.848
25.00	0.963	3.066	5.245	5.703	19.62	37.97	0.228	0.785	1.519
30.00	1.693	4.774	7.375	12.25	39.14	69.53	0.408	1.305	2.318
35.00	2.638	6.612	9.395	22.99	67.58	111.5	0.657	1.931	3.186
40.00	3.740	8.419	11.22	38.89	105.2	163.2	0.972	2.630	4.079
45.00	4.928	10.11	12.86	60.54	151.6	223.4	1.345	3.368	4.965
50.00	6.154	11.66	14.29	88.23	206.1	291.4	1.765	4.121	5.828
55.00	7.385	13.04	15.52	122.1	267.9	366.0	2.220	4.871	6.654
60.00	8.595	14.27	16.59	162.0	336.2	446.3	2.701	5.604	7.438
65.00	9.759	15.35	17.51	208.0	410.4	531.6	3.199	6.313	8.179
70.00	10.86	16.30	18.31	259.5	489.5	621.2	3.708	6.993	8.874
75.00	11.89	17.14	19.01	316.4	573.2	714.6	4.219	7.642	9.528
80.00	12.85	17.87	19.63	378.4	660.7	811.2	4.729	8.259	10.14
85.00	13.74	18.53	20.17	444.9	751.8	910.7	5.234	8.844	10.71
90.00	14.56	19.11	20.64	515.7	845.9	1013.	5.730	9.399	11.25
95.00	15.31	19.63	21.06	590.4	942.8	1117.	6.215	9.924	11.76
100.00	16.01	20.10	21.44	668.7	1042.	1223.	6.687	10.42	12.23
105.00	16.64	20.52	21.77	750.3	1144.	1331.	7.146	10.89	12.68
110.00	17.22	20.89	22.06	835.0	1247.	1441.	7.591	11.34	13.10
115.00	17.76	21.23	22.33	922.5	1353.	1552.	8.021	11.76	13.49
120.00	18.25	21.54	22.56	1013.	1460.	1664.	8.438	12.16	13.87
125.00	18.70	21.82	22.78	1105.	1568.	1777.	8.839	12.54	14.22
130.00	19.12	22.07	22.97	1199.	1678.	1892.	9.227	12.91	14.55
135.00	19.51	22.31	23.15	1296.	1789.	2007.	9.601	13.25	14.87
140.00	19.87	22.52	23.31	1395.	1901.	2123.	9.961	13.58	15.17
145.00	20.20	22.72	23.45	1495.	2014.	2240.	10.31	13.89	15.45
150.00	20.51	22.90	23.59	1597.	2128.	2358.	10.64	14.19	15.72
155.00	20.79	23.07	23.70	1700.	2243.	2476.	10.97	14.47	15.97
160.00	21.05	23.22	23.81	1804.	2358.	2595.	11.28	14.74	16.22
165.00	21.30	23.37	23.91	1910.	2475.	2714.	11.58	15.00	16.45
170.00	21.53	23.50	24.00	2017.	2592.	2834.	11.87	15.25	16.67
175.00	21.74	23.63	24.08	2125.	2710.	2954.	12.15	15.49	16.88
180.00	21.94	23.75	24.15	2235.	2828.	3075.	12.42	15.71	17.08
185.00	22.13	23.86	24.22	2345.	2947.	3196.	12.68	15.93	17.27
190.00	22.31	23.96	24.29	2456.	3067.	3317.	12.93	16.14	17.46
195.00	22.47	24.06	24.35	2568.	3187.	3438.	13.17	16.34	17.63
200.00	22.63	24.16	24.41	2681.	3308.	3650.	13.40	16.54	17.80
205.00	22.77	24.24	24.48	2794.	3429.	3683.	13.63	16.72	17.96
210.00	22.91	24.33	24.54	2908.	3550.	3805.	13.85	16.90	18.12
215.00	23.04	24.41	24.60	3023.	3672.	3928.	14.06	17.08	18.27
220.00	23.17	24.49	24.65	3139.	3794.	4051.	14.27	17.25	18.41
225.00	23.28	24.56	24.71	3255.	3917.	4174.	14.47	17.41	18.55
230.00	23.39	24.63	24.76	3372.	4040.	4298.	14.66	17.56	18.69
235.00	23.50	24.69	24.82	3489.	4163.	4422.	14.85	17.71	18.82
240.00	23.60	24.76	24.87	3607.	4287.	4546.	15.03	17.86	18.94
245.00	23.69	24.82	24.92	3725.	4411.	4671.	15.20	18.00	19.06

Table 2–25A (continued)
THERMODYNAMIC FUNCTIONS OF COPPER, SILVER, AND GOLD

T (K)	C°_r (J/deg·mol)			$H^\circ_T - H^\circ_0$ (J/mol)			$(H^\circ_T - H^\circ_0)/T$ (J/deg·mol)		
	Cu	Ag	Au	Cu	Ag	Au	Cu	Ag	Au
250.00	23.78	24.88	24.97	3844.	4535.	4796.	15.37	18.14	19.18
255.00	23.86	24.93	25.02	3963.	4659.	4921.	15.54	18.27	19.30
260.00	23.94	24.99	25.07	4082.	4784.	5046.	15.70	18.40	19.41
265.00	24.02	25.04	25.12	4202.	4909.	5171.	15.86	18.53	19.51
270.00	24.09	25.09	25.17	4322.	5035.	5297.	16.01	18.65	19.62
273.15	24.13	25.12	25.20	4398.	5114.	5376.	16.10	18.72	19.68
275.00	24.15	25.14	25.21	4443.	5160.	5423.	16.16	18.76	19.72
280.00	24.22	25.19	25.26	4564.	5286.	5549.	16.30	18.88	19.82
285.00	24.28	25.24	25.31	4685.	5412.	5676.	16.44	18.99	19.91
290.00	24.34	25.28	25.35	4807.	5538.	5802.	16.57	19.10	20.01
295.00	24.40	25.32	25.39	4929.	5665.	5929.	16.71	19.20	20.10
298.15	24.44	25.35	25.42	5005.	5745.	6009.	16.79	19.27	20.15
300.00	24.46	25.37	25.43	5051.	5792.	6056.	16.84	19.31	20.19

From Weast, R. C., Ed., *Handbook of Chemistry and Physics*, 55th ed., CRC Press, Cleveland, 1974, D-74.

Table 2–25B
THERMODYNAMIC FUNCTIONS OF COPPER, SILVER, AND GOLD

T (K)	S°_T (J/deg·mol)			$-(G^\circ_T - H^\circ_0)$ (J/mol)			$-(G^\circ_T - H^\circ_0)/T$ (J/deg·mol)		
	Cu	Ag	Au	Cu	Ag	Au	Cu	Ag	Au
1.00	0.000711	0.000706	0.000880	0.000351	0.000339	0.000402	0.000351	0.000339	0.000402
2.00	0.00152	0.00175	0.00266	0.00145	0.00152	0.00206	0.000727	0.000762	0.00103
3.00	0.00251	0.00347	0.00620	0.00345	0.00406	0.00631	0.00115	0.00135	0.00210
4.00	0.00379	0.00619	0.0123	0.00657	0.00879	0.0153	0.00164	0.00220	0.00383
5.00	0.00546	0.0103	0.0218	0.0112	0.0169	0.0321	0.00223	0.00338	0.00641
6.00	0.00760	0.0160	0.0354	0.0176	0.0299	0.0603	0.00294	0.00498	0.0100
7.00	0.0103	0.0238	0.0539	0.0265	0.0496	0.104	0.00379	0.00709	0.0149
8.00	0.0137	0.0341	0.0782	0.0385	0.0783	0.170	0.00481	0.00979	0.0212
9.00	0.0179	0.0472	0.109	0.0542	0.119	0.263	0.00602	0.0132	0.0292
10.00	0.0229	0.0636	0.148	0.0746	0.174	0.391	0.00746	0.0174	0.0391
11.00	0.0290	0.0839	0.196	0.100	0.247	0.562	0.00913	0.0225	0.0511
12.00	0.0362	0.109	0.253	0.133	0.343	0.786	0.0111	0.0286	0.0655
13.00	0.0447	0.138	0.322	0.173	0.466	1.073	0.0133	0.0359	0.0825
14.00	0.0545	0.174	0.402	0.223	0.622	1.434	0.0159	0.0444	0.102
15.00	0.0660	0.215	0.494	0.283	0.815	1.880	0.0189	0.0544	0.125
16.00	0.0791	0.263	0.598	0.355	1.054	2.426	0.0222	0.0659	0.152
17.00	0.0941	0.318	0.715	0.442	1.344	3.081	0.0260	0.0790	0.181
18.00	0.111	0.381	0.845	0.544	1.693	3.861	0.0302	0.0940	0.214
19.00	0.131	0.452	0.987	0.665	2.109	4.775	0.0350	0.111	0.251
20.00	0.152	0.530	1.140	0.806	2.599	5.838	0.0403	0.130	0.292
25.00	0.305	1.043	2.069	1.917	6.446	13.76	0.0767	0.258	0.550
30.00	0.541	1.750	3.214	3.995	13.35	26.89	0.133	0.445	0.896
35.00	0.871	2.623	4.505	7.487	24.22	46.14	0.214	0.692	1.318
40.00	1.294	3.625	5.881	12.86	39.79	72.08	0.322	0.995	1.802
45.00	1.802	4.715	7.299	20.57	60.61	105.0	0.457	1.347	2.334
50.00	2.385	5.862	8.729	31.01	87.04	145.1	0.620	1.741	2.902
55.00	3.029	7.040	10.15	44.52	119.3	192.3	0.809	2.169	3.496
60.00	3.724	8.228	11.55	61.38	157.5	246.5	1.023	2.624	4.109
65.00	4.458	9.414	12.91	81.82	201.6	307.7	1.259	3.101	4.734
70.00	5.222	10.59	14.24	106.0	251.6	375.6	1.514	3.594	5.366
75.00	6.007	11.74	15.53	134.1	307.4	450.0	1.788	4.099	6.001
80.00	6.806	12.87	16.78	166.1	368.9	530.8	2.076	4.612	6.635
85.00	7.612	13.97	17.98	202.1	436.1	617.7	2.378	5.130	7.267
90.00	8.421	15.05	19.15	242.2	508.6	710.6	2.691	5.652	7.895
95.00	9.229	16.10	20.28	286.4	586.5	809.1	3.014	6.174	8.517
100.00	10.03	17.12	21.37	334.5	669.6	913.3	3.345	6.696	9.133
105.00	10.83	18.11	22.42	386.7	757.7	1023.	3.683	7.216	9.740
110.00	11.62	19.07	23.44	442.8	850.6	1137.	4.025	7.733	10.34
115.00	12.39	20.01	24.43	502.8	948.3	1257.	4.372	8.246	10.93
120.00	13.16	20.92	25.38	566.7	1051.	1382.	4.723	8.755	11.51

Table 2–25B (continued)

THERMODYNAMIC FUNCTIONS OF COPPER, SILVER, AND GOLD

T (K)	S°_T (J/deg·mol)			$-(G^{\circ}_T - H^{\circ}_0)$ (J/mol)			$-(G^{\circ}_T - H^{\circ}_0)/T$ (J/deg·mol)		
	Cu	Ag	Au	Cu	Ag	Au	Cu	Ag	Au
125.00	13.91	21.80	26.31	634.4	1157.	1511.	5.075	9.260	12.09
130.00	14.66	22.66	27.20	705.8	1269.	1645.	5.429	9.759	12.65
135.00	15.39	23.50	28.07	780.9	1384.	1783.	5.785	10.25	13.21
140.00	16.10	24.32	28.92	859.7	1504.	1925.	6.140	10.74	13.75
145.00	16.80	25.11	29.74	941.9	1627.	2072.	6.496	11.22	14.29
150.00	17.49	25.88	30.54	1028.	1755.	2223.	6.851	11.70	14.82
155.00	18.17	26.64	31.31	1117.	1886.	2377.	7.206	12.17	15.34
160.00	18.84	27.37	32.07	1209.	2021.	2536.	7.559	12.63	15.85
165.00	19.49	28.09	32.80	1305.	2160.	2698.	7.910	13.09	16.35
170.00	20.13	28.79	33.52	1404.	2302.	2864.	8.260	13.54	16.85
175.00	20.75	29.47	34.21	1506.	2448.	3033.	8.608	13.99	17.33
180.00	21.37	30.14	34.89	1612.	2597.	3206.	8.954	14.43	17.81
185.00	21.97	30.79	35.55	1720.	2749.	3382.	9.298	14.86	18.28
190.00	22.57	31.43	36.20	1831.	2904.	3561.	9.639	15.29	18.74
195.00	23.15	32.05	36.83	1946.	3063.	3744.	9.978	15.71	19.20
200.00	23.72	32.66	37.45	2063.	3225.	3930.	10.31	16.12	19.65
205.00	24.28	33.26	38.05	2183.	3390.	4118.	10.65	16.54	20.09
210.00	24.83	33.85	38.64	2306.	3558.	4310.	10.98	16.94	20.52
215.00	25.37	34.42	39.22	2431.	3728.	4505.	11.31	17.34	20.95
220.00	25.90	34.98	39.79	2559.	3902.	4702.	11.63	17.74	21.37
225.00	26.42	35.53	40.34	2690.	4078.	4903.	11.96	18.12	21.79
230.00	26.94	36.07	40.89	2824.	4257.	5106.	12.28	18.51	22.20
235.00	27.44	36.60	41.42	2960.	4439.	5312.	12.59	18.89	22.60
240.00	27.94	37.12	41.94	3098.	4623.	5520.	12.91	19.26	23.00
245.00	28.42	37.63	42.46	3239.	4810.	5731.	13.22	19.63	23.39
250.00	28.90	38.14	42.96	3382.	4999.	5945.	13.53	20.00	23.78
255.00	29.37	38.63	43.46	3528.	5191.	6161.	13.83	20.36	24.16
260.00	29.84	39.11	43.94	3676.	5386.	6379.	14.14	20.71	24.53
265.00	30.30	39.59	44.42	3826.	5582.	6600.	14.44	21.07	24.91
270.00	30.75	40.06	44.89	3979.	5782.	6823.	14.74	21.41	25.27
273.15	31.02	40.35	45.18	4076.	5908.	6965.	14.92	21.63	25.50
275.00	31.19	40.52	45.35	4134.	5983.	7049.	15.03	21.76	25.63
280.00	31.62	40.97	45.81	4291.	6187.	7277.	15.32	22.10	25.99
285.00	32.05	41.42	46.25	4450.	6393.	7507.	15.61	22.43	26.34
290.00	32.48	41.86	46.69	4611.	6601.	7739.	15.90	22.76	26.69
295.00	32.89	42.29	47.13	4775.	6811.	7974.	16.19	23.09	27.03
298.15	33.15	42.56	47.40	4879.	6945.	8123.	16.36	23.29	27.24
300.00	33.30	42.72	47.56	4940.	7024.	8211.	16.47	23.41	27.37

From Weast, R. C., Ed., *Handbook of Chemistry and Physics,* 55th ed., CRC Press, Cleveland, 1974, D-75.

Table 2–26A
ELECTROCHEMICAL SERIES

Alphabetical Listing

Reaction	Potential, volts[a]	Reaction	Potential, volts[a]
$Ag^+ + e^- \rightarrow Ag$	0.7996	$Cd(OH)_2 + 2e^- \rightarrow Cd(Hg) + 2OH^-$	−0.761
$Ag^{+2} + e^- \rightarrow Ag^{+1}(4f\ HClO_4)$	1.987		(−0.81)
$AgAc + e^- \rightarrow Ag + Ac^-$	0.64	$CdSO_4 \cdot 8/3H_2O + 2e^- \rightarrow Cd(Hg) + CdSO_4(sat'd\ aq)$	−0.4346
$AgBr + e^- \rightarrow Ag + Br^-$	0.0713	$Ce^{+3} + 3e^- \rightarrow Ce$	−2.335
$AgBrO_3 + e^- \rightarrow Ag + BrO_3^-$	0.680	$Ce^{+3} + 3e^- \rightarrow Ce(Hg)$	−1.4373
$AgC_2O_4 + 2e^- \rightarrow Ag + C_2O_4^{-2}$	0.4776	$Ce^{+4} + e^- \rightarrow Ce^{+3}$	1.4430
$AgCl + e^- \rightarrow Ag + Cl^-$	0.2223		(1.61)
$AgCN + e^- \rightarrow Ag + CN^-$	−0.02	$Ce^{+4} + e^- \rightarrow Ce^{+3}(0.5f\ H_2SO_4)$	1.4587
$Ag_2CO_3 + 2e^- \rightarrow 2Ag + CO_3^-$	0.4769	$CeOH^{+3} + H^+ + e^- \rightarrow Ce^{+3} + H_2O$	1.7134
$Ag_2CrO_4 + 2e^- \rightarrow 2Ag + CrO_4^{-2}$	0.4463	$Cl_2(g) + 2e^- \rightarrow 2Cl^-$	1.3583
$Ag_4Fe(CN)_6 + 4e^- \rightarrow 4Ag + Fe(CN)_6^{-4}$	0.1943	$HClO + H^+ + e^- \rightarrow 1/2Cl_2 + H_2O$	1.63
$AgI + e^- \rightarrow Ag + I^-$	−0.1519	$HClO + H^+ + 2e^- \rightarrow Cl^- + H_2O$	1.49
$AgIO_3 + e^- \rightarrow Ag + IO_3^-$	0.3551	$ClO^- + H_2O + 2e^- \rightarrow Cl^- + 2OH^-$	0.90
$Ag_2MoO_4 + 2e^- \rightarrow 2Ag + MoO_4^-$	0.49	$ClO_2 + e^- \rightarrow ClO_2^-$	1.15
$AgNO_2 + e^- \rightarrow Ag + NO_2^-$	0.59	$ClO_2 + H^+ + e^- \rightarrow HClO_2$	1.27
$Ag_2O + H_2O + 2e^- \rightarrow 2Ag + 2OH^-$	0.342	$HClO_2 + 2H^+ + 2e^- \rightarrow HClO + H_2O$	1.64
$Ag_2O_3 + H_2O + 2e^- \rightarrow 2AgO + 2OH^-$	0.74	$HClO_2 + 3H^+ + 3e^- \rightarrow 1/2Cl_2 + 2H_2O$	1.63
$2AgO + H_2O + 2e^- \rightarrow Ag_2O + 2OH^-$	0.599	$HClO_2 + 4H^+ + 4e^- \rightarrow Cl^- + 2H_2O$	1.56
$AgOCN + e^- \rightarrow Ag + OCN^-$	0.41	$ClO_2^- + H_2O + 2e^- \rightarrow ClO^- + 2OH^-$	0.59
$Ag_2S + 2e^- \rightarrow 2Ag + S^{-2}$	−0.7051	$ClO_2^- + 2H_2O + 4e^- \rightarrow Cl^- + 4OH^-$	0.76
$Ag_2S + 2H^+ + 2e^- \rightarrow 2Ag + H_2S$	−0.0366	$ClO_2(aq) + e^- \rightarrow ClO_2^-$	0.954
$AgSCN + e^- \rightarrow Ag + SCN^-$	0.0895	$ClO_3^- + 2H^+ + e^- \rightarrow ClO_2 + H_2O$	1.15
$Ag_2SeO_3 + 2e \rightarrow 2Ag + SeO_3^{-2}$	0.3629	$ClO_3^- + 3H^+ + 2e \rightarrow HClO_2 + H_2O$	1.21
$Ag_2SO_4 + 2e^- \rightarrow 2Ag + SO_4^{-2}$	0.653		(1.23)
$Ag_2(WO_4) + 2e^- \rightarrow 2Ag + WO_4^{-2}$	0.466	$ClO_3^- + 6H^+ + 5e^- \rightarrow \frac{1}{2}Cl_2 + 3H_2O$	1.47
$Al^{+3} + 3e^- \rightarrow Al\ (0.1f\ NaOH)$	−1.706	$ClO_3^- + 6H^+ + 6e^- \rightarrow Cl^- + 3H_2O$	1.45
$H_2AlO_3^- + H_2O + 3e^- \rightarrow Al + 4OH^-$	−2.35	$ClO_3^- + H_2O + 2e^- \rightarrow ClO_2^- + 2OH^-$	0.35
$As + 3H^+ + 3e^- \rightarrow AsH_3$	−0.54	$ClO_3^- + 3H_2O + 6e^- \rightarrow Cl^- + 6OH^-$	0.62
$As_2O_3 + 6H^+ + 6e^- \rightarrow 2As + 3H_2O$	0.234	$ClO_4^- + 2H^+ + 2e^- \rightarrow ClO_3^- + H_2O$	1.19
$HAsO_2 + 3H^+ + 3e^- \rightarrow As + 2H_2O$	0.2475	$ClO_4^- + 8H^+ + 7e^- \rightarrow \frac{1}{2}Cl_2 + 4H_2O$	1.34
$AsO_2^- + 2H_2O + 3e^- \rightarrow As + 4OH^-$	−0.68	$ClO_4^- + 8H^+ + 8e- \rightarrow Cl^- + 4H_2O$	1.37
$H_3AsO_4 + 2H^+ + 2e^- \rightarrow HAsO_2 + 2H_2O(1f\ HCl)$	0.58	$ClO_4^- + H_2O + 2e^- \rightarrow ClO_3^- + 2OH^-$	0.17
$AsO_4^{-3} + 2H_2O + 2e^- \rightarrow AsO_2^- + 4OH^-$	−0.71	$(CN)_2 + 2H^+ + 2e^- \rightarrow 2HCN$	0.37
$AsO_4^{-3} + 2H_2O + 2e^- \rightarrow AsO_2^- + 4OH^-\ (1f\ NaOH)$	−0.08	$2HCNO + 2H^+ + 2e \rightarrow (CN)_2 + 2H_2O$	0.33
$Au^+ + e^- \rightarrow Au$	1.68	$(CNS)_2 + 2e^- \rightarrow 2CNS^-$	0.77
$Au^{+3} + 2e^- \rightarrow Au^{+1}$	1.29	$Co^{+2} + 2e^- \rightarrow Co$	−0.28
$Au^{+3} + 3e^- \rightarrow Au$	1.42	$Co^{+3} + e^- \rightarrow Co^{+2}(3f\ HNO_3)$	1.842
$AuBr_2^- + e^- \rightarrow Au + 2Br^-$	0.963	$CO_2 + 2H^+ + 2e^- \rightarrow HCOOH$	−0.2
$AuBr_4^- + 3e^- \rightarrow Au + 4Br^-$	0.858	$2CO_2 + 2H^+ + 2e^- \rightarrow H_2C_2O_4$	−0.49
$AuCl_4^- + 3e^- \rightarrow Au + 4Cl^-$	0.994	$Co(NH_3)_6^{+3} + e^- \rightarrow Co(NH_3)_6^{+2}$	0.1
$Au(OH)_3 + 3H^+ + 3e^- \rightarrow Au + 3H_2O$	1.45	$Co(OH)_2 + 2e^- \rightarrow Co + 2OH^-$	−0.73
$H_2BO_3^- + 5H_2O + 8e^- \rightarrow BH_4^- + 8OH^-$	−1.24	$Co(OH)_3 + e^- \rightarrow Co(OH)_2 + OH^-$	0.2 (0.17)
$H_2BO_3^- + H_2O + 3e^- \rightarrow B + 4OH^-$	−2.5	$Cr^{+2} + 2e \rightarrow Cr$	−0.557
$H_3BO_3 + 3H^+ + 3e^- \rightarrow B + 3H_2O$	−0.73	$Cr^{+3} + e^- \rightarrow Cr^{+2}$	−0.41
$Ba^{+2} + 2e^- \rightarrow Ba$	−2.90	$Cr^{+3} + 3e^- \rightarrow Cr$	−0.74
$Ba^{+2} + 2e^- \rightarrow Ba(Hg)$	−1.570	$Cr^{+6} + 3e^- \rightarrow Cr^{+3}(2f\ H_2SO_4)$	1.10
$Ba(OH)_2 \cdot 8H_2O + 2e^- \rightarrow Ba + 2OH^- + 8H_2O$	−2.97	$Cr^{+6} + 3e^- \rightarrow Cr^{+3}(1f\ NaOH)$	−0.12
$Be^{+2} + 2e^- \rightarrow Be$	−1.70	$Cr_2O_7^{-2} + 14H^+ + 6e^- \rightarrow 2Cr^{+3} + 7H_2O$	1.33
	−1.85)	$CrO_2^- + 2H_2O + 3e^- \rightarrow Cr + 4OH^-$	−1.2
$Be_2O_3^{-2} + 3H_2O + 4e^- \rightarrow 2Be + 6OH^-$	−2.28	$HCrO_4^- + 7H^+ + 3e^- \rightarrow Cr^{+3} + 4H_2O$	1.195
$Bi(Cl)_4^- + 3e^- \rightarrow Bi + 4Cl^-$	0.168	$CrO_4^{-2} + 4H_2O + 3e^- \rightarrow Cr(OH)_3 + 5OH^-$	−0.12
$Bi_2O_3 + 3H_2O + 6e^- \rightarrow 2Bi + 6OH^-$	−0.46	$Cr(OH)_3 + 3e^- \rightarrow Cr + 3OH^-$	−1.3
$Bi_2O_4 + 4H^+ + 2e^- \rightarrow 2BiO^+ + 2H_2O$	1.59	$Cs^+ + e^- \rightarrow Cs$	−2.923
$BiO^+ + 2H^+ + 3e^- \rightarrow Bi + H_2O$	0.32	$Cu^+ + e^- \rightarrow Cu$	0.522
$BiOCl + 2H^+ + 3e^- \rightarrow Bi + Cl^- + H_2O$	0.1583	$Cu^{+2} + 2CN^- + e^- \rightarrow Cu(CN)_2^-$	1.12
$BiOOH + H_2O + 3e^- \rightarrow Bi + 3OH^-$	−0.46	$Cu^{+2} + e \rightarrow Cu^+$	0.158
$Br_2(aq) + 2e^- \rightarrow 2Br^-$	1.087		(0.167)
$Br_2(l) + 2e^- \rightarrow 2Br^-$	1.065	$Cu^{+2} + 2e^- \rightarrow Cu$	0.3402
$HBrO + H^+ + e^- \rightarrow 1/2Br_2 + H_2O$	1.59	$Cu^{+2} + 2e^- \rightarrow Cu(Hg)$	0.345
$HBrO + H^+ + 2e^- \rightarrow Br^- + H_2O$	1.33	$CuI_2^- + e^- \rightarrow Cu + 2I^-$	0.00
$2HBrO + 2H^+ + 2e^- \rightarrow Br_2(l) + 2H_2O$	1.6	$Cu_2O + H_2O + 2e^- \rightarrow 2Cu + 2OH^-$	−0.361
$BrO^- + H_2O + 2e^- \rightarrow Br^- + 2OH^-\ (1f\ NaOH)$	0.70	$Cu(OH)_2 + 2e^- \rightarrow Cu + 2OH^-$	−0.224
$BrO_3^- + 6H^+ + 5e^- \rightarrow 1/2Br_2 + 3H_2O$	1.52	$2Cu(OH)_2 + 2e^- \rightarrow Cu_2O + 2OH^- + H_2O$	−0.09
$BrO_3^- + 6H^+ + 6e^- \rightarrow Br^- + 3H_2O$	1.44	$D^+ + e^- \rightarrow 1/2D_2$	−0.0034
$BrO_3^- + 3H_2O + 6e^- \rightarrow Br^- + 6OH^-$	0.61	$2D^+ + 2e^- \rightarrow D_2$	−0.044
$C_6H_4O_2 + 2H^+ + 2e \rightarrow C_6H_4(OH)_2$	0.6992	$Eu^{+3} + e^- \rightarrow Eu^{+2}$	−0.43
$Ca^+ + e^- \rightarrow Ca$	−3.02	$1/2F_2 + e^- \rightarrow F^-$	2.85
$Ca^{+2} + 2e^- \rightarrow Ca$	−2.76	$1/2F_2 + H^+ + e^- \rightarrow HF$	3.03
Calomel Electrode, Molal KCl	0.2800	$F_2 + 2e^- \rightarrow 2F^-$	2.87
Calomel Electrode, N KCl	0.2807	$F_2O + 2H^+ + 4e^- \rightarrow H_2O + 2F^-$	2.1
Calomel Electrode 0.1 N KCl	0.3337	$Fe^{+2} + 2e^- \rightarrow Fe$	−0.409
Calomel Electrode, Sat'd. KCl	0.2415	$Fe^{+3} + 3e^- \rightarrow Fe$	−0.036
Calomel Electrode, Sat'd NaCl	0.2360	$Fe^{+3} + e^- \rightarrow Fe^{+2}$	0.770
$Ca(OH)_2 + 2e^- \rightarrow Ca + 2OH^-$	−3.02	$Fe^{+3} + e^- \rightarrow Fe^{+2}(1f\ HCl)$	0.770
$Cb_2O_5 + 10H^+ + 10e^- \rightarrow 2Cb + 5H_2O$	−0.62	$Fe^{+3} + e^- \rightarrow Fe^{+2}(1f\ HClO_4)$	0.747
$Cd^{+2} + 2e^- \rightarrow Cd$	−0.4026	$Fe^{+3} + e^- \rightarrow Fe^{+2}(1f\ H_3PO_4)$	0.438
$Cd^{+2} + 2e^- \rightarrow Cd(Hg)$	−0.3521	$Fe^{+3} + e^- \rightarrow Fe^{+2}(0.5f\ H_2SO_4)$	0.679

Table 2–26A (continued)
ELECTROCHEMICAL SERIES

Alphabetical Listing (continued)

Reaction	Potential, volts[a]	Reaction	Potential, volts[a]
$Fe(CN)_6^{-3} + e^- \rightarrow Fe(CN)_6^{-4}(0.01f\ NaOH)$	0.46	$N_2O_4 + 2H^+ + 2e^- \rightarrow 2HNO_2$	1.07
$Fe(CN)_6^{-3} + e^- \rightarrow Fe(CN)_6^{-4}(1f\ H_2SO_4)$	0.69	$N_2O_4 + 4H^+ + 4e^- \rightarrow 2NO + 2H_2O$	1.03
$FeO_4^{-2} + 8H^+ + 3e^- \rightarrow Fe^{+3} + 4H_2O$	1.9	$Na^+ + e^- \rightarrow Na$	−2.7109
$Fe(OH)_3 + e^- \rightarrow Fe(OH)_2 + OH^-$	−0.56		(−2.712)
$Fe\ (phenanthroline)_3^{+3} + e^- \rightarrow Fe(ph)_3^{+2}$	1.14	$Nb^{+5} + 2e^- \rightarrow Nb^{+3}(2fHCl)$	0.344
$Fe\ (phenanthroline)_3^{+3} + e^- \rightarrow Fe(ph)_3^{+2}(2f\ H_2SO_4)$	1.056	$Nd^{+3} + 3e^- \rightarrow Nd$	−2.246
$Ga^{+3} + 3e^- \rightarrow Ga$	−0.560	$2NH_3OH^+ + H^+ + 2e^- \rightarrow N_2H_5^+ + 2H_2O$	1.42
$H_2GaO_3^- + H_2O + 3e^- \rightarrow Ga + 4OH^-$	−1.22	$Ni^{+2} + 2e^- \rightarrow Ni$	−0.23
$GeO_2 + 2H^+ + 2e^- \rightarrow GeO + H_2O$	−0.12	$Ni(OH)_2 + 2e^- \rightarrow Ni + 2OH^-$	−0.66
$H_2GeO_3 + 4H^+ + 4e^- \rightarrow Ge + 3H_2O$	−0.13	$NiO_2 + 4H^+ + 2e^- \rightarrow Ni^{+2} + 2H_2O$	1.93
$2H^+ + 2e^- \rightarrow H_2$	0.0000	$NiO_2 + 2H_2O + 2e^- \rightarrow Ni(OH)_2 + 2OH^-$	0.49
$1/2H_2 + e^- \rightarrow H^-$	−2.23	$2NO + 2e^- \rightarrow N_2O_2^{-2}$	0.10
$2H_2O + 2e^- \rightarrow H_2 + 2OH^-$	−0.8277	$2NO + 2H^+ + 2e^- \rightarrow N_2O + H_2O$	1.59
$H_2O_2 + 2H^+ + 2e^- \rightarrow 2H_2O$	1.776	$2NO + H_2O + 2e^- \rightarrow N_2O + 2OH^-$	0.76
$HfO^{+2} + 2H^+ + 4e^- \rightarrow Hf + H_2O$	−1.68	$HNO_2 + H^+ + e^- \rightarrow NO + H_2O$	0.99
$HfO_2 + 4H^+ + 4e^- \rightarrow Hf + 2H_2O$	−1.57	$2HNO_2 + 4H^+ + 4e^- \rightarrow H_2N_2O_2 + 2H_2O$	0.80
$HfO(OH)_2 + H_2O + 4e^- \rightarrow Hf + 4OH^-$	−2.60	$2HNO_2 + 4H^+ + 4e^- \rightarrow N_2O + 3H_2O$	1.27
$Hg^{+2} + 2e^- \rightarrow Hg$	0.851	$NO_2^- + H_2O + e^- \rightarrow NO + 2OH^-$	(1.29)
$2Hg^{+2} + 2e^- \rightarrow Hg_2^{+2}$	0.905	$2NO_2^- + 2H_2O + 4e^- \rightarrow N_2O_2^{-2} + 4OH^-$	−0.18
$1/2Hg_2^{+2} + e^- \rightarrow Hg$	0.7986	$2NO_2^- + 3H_2O + 4e^- \rightarrow N_2O + 6OH^-$	0.15
$Hg_2^{+2} + 2e^- \rightarrow 2Hg$	0.7961	$NO_3^- + 3H^+ + 2e^- \rightarrow HNO_2 + H_2O$	0.94
$Hg_2(AcO)_2 + 2e^- \rightarrow 2Hg + 2AcO^-$	0.5113	$NO_3^- + 4H^+ + 3e^- \rightarrow NO + 2H_2O$	0.96
$Hg_2Br_2 + 2e^- \rightarrow 2Hg + 2Br^-$	0.1396	$2NO_3^- + 4H^+ + 2e^- \rightarrow N_2O_4 + 2H_2O$	0.81
$Hg_2Cl_2 + 2e^- \rightarrow 2Hg + 2Cl^-$	0.2682	$NO_3^- + 2H_2O + 2e^- \rightarrow NO_2^- + 2OH^-$	0.01
$Hg_2Cl_2 + 2e \rightarrow 2Hg + 2Cl^-(0.1f\ NaOH)$	0.3419	$2NO_3^- + 2H_2O + 2e^- \rightarrow N_2O_4 + 4OH^-$	−0.85
	(0.268)	$Np^{+3} + 3e^- \rightarrow Np$	−1.9
$Hg_2HPO_4 + H^+ + 2e^- \rightarrow 2Hg + H_2PO_4^-$	0.639	$Np^{+4} + e^- \rightarrow Np^{+3}(1fHClO_4)$	0.155
$Hg_2I_2 + 2e^- \rightarrow 2Hg + 2I^-$	−0.0405	$Np^{+5} + e^- \rightarrow Np^{+4}(1fHClO_4)$	0.739
$Hg_2O + H_2O + 2e^- \rightarrow 2Hg + 2OH^-$	0.123	$Np^{+6} + e^- \rightarrow Np^{+5}(1fHClO_4)$	1.137
$HgO + H_2O + 2e^- \rightarrow Hg + 2OH^-$	0.0984	$1/2O_2 + 2H^+(10^{-7}M) + 2e^- \rightarrow H_2O$	0.815
$Hg_2SO_4 + 2e^- \rightarrow 2Hg + SO_4^{-2}$	0.6158	$O_2 + 2H^+ + 2e^- \rightarrow H_2O_2$	0.682
$HO_2 + H^+ + e^- \rightarrow H_2O_2$	1.5	$O_2 + 4H^+ + 4e^- \rightarrow 2H_2O$	1.229
$I_2 + 2e^- \rightarrow 2I^-$	0.535	$O_2 + H_2O + 2e^- \rightarrow HO_2^- + OH^-$	−0.076
$I_3^- + 2e^- \rightarrow 3I^-$	0.5338	$O_2 + 2H_2O + 2e^- \rightarrow H_2O_2 + 2OH^-$	−0.146
$In^{+2} + e^- \rightarrow In^{+1}$	−0.40	$O_2 + 2H_2O + 4e^- \rightarrow 4OH^-$	0.401
$In^{+3} + e^- \rightarrow In^{+2}$	−0.49	$O_3 + 2H^+ + 2e^- \rightarrow O_2 + H_2O$	2.07
$In^{+3} + 2e^- \rightarrow In^{+1}$	−0.40	$O_3 + H_2O + 2e^- \rightarrow O_2 + 2OH^-$	1.24
$In^{+3} + 3e^- \rightarrow In$	−0.338	$O_{(g)} + 2H^+ + 2e^- \rightarrow H_2O$	2.42
$H_3IO_6^{-2} + 2e^- \rightarrow IO_3^- + 3OH^-$	−0.70	$OH + e^- \rightarrow OH^-$	1.4
$H_5IO_6 + H^+ + 2e^- \rightarrow IO_3^- + 3H_2O$	~1.7	$HO_2^- + H_2O + 2e^- \rightarrow 3OH^-$	0.87
$HIO + H^+ + e^- \rightarrow 1/2I_2 + H_2O$	1.45	$OsO_4 + 8H^+ + 8e^- \rightarrow Os + 4H_2O$	0.85
$HIO + H^+ + 2e^- \rightarrow I^- + H_2O$	0.99	$P + 3H^+ + 3e^- \rightarrow PH_3(g)$	−0.04
$IO^- + H_2O + 2e^- \rightarrow I^- + 2OH^-$	0.49	$P + 3H_2O + 3e^- \rightarrow PH_3(g) + 3OH^-$	−0.87
$IO_3^- + 6H^+ + 5e^- \rightarrow 1/2I_2 + 3H_2O$	1.195	$Pb^{+2} + 2e^- \rightarrow Pb$	−0.1263
$IO_3^- + 6H^+ + 6e^- \rightarrow I^- + 3H_2O$	1.085		(−0.126)
$2IO_3^- + 12H^+ + 10e^- \rightarrow I_2 + 6H_2O$	1.19	$Pb^{+2} + 2e^- \rightarrow Pb(Hg)$	−0.1205
$IO_3^- + 2H_2O + 4e^- \rightarrow IO^- + 4OH^-$	0.56	$PbBr_2 + 2e^- \rightarrow Pb(Hg) + 2Br^-$	−0.275
$IO_3^- + 3H_2O + 6e^- \rightarrow I^- + 6OH^-$	0.26	$PbCl_2 + 2e^- \rightarrow Pb(Hg) + 2Cl^-$	−0.262
$IrCl_6^{-2} + e^- \rightarrow IrCl_6^{-3}$	1.02	$PbF_2 + 2e^- \rightarrow Pb(Hg) + 2F^-$	−0.3444
$IrCl_6^{-3} + 3e^- \rightarrow Ir + 6Cl^-$	0.77	$PbHPO_4 + H^+ + 2e^- \rightarrow Pb(Hg) + HPO_4^-$	−0.2448
$Ir_2O_3 + 3H_2O + 6e^- \rightarrow 2Ir + 6OH^-$	0.1	$PbI_2 + 2e^- \rightarrow Pb(Hg) + 2I^-$	−0.358
$K^+ + e^- \rightarrow K$	−2.924	$PbO + H_2O + 2e^- \rightarrow Pb + 2OH^-$	−0.576
	(−2.923)	$PbO_2 + 4H^+ + 2e^- \rightarrow Pb^{+2} + 2H_2O$	1.46
$La^{+3} + 3e^- \rightarrow La$	−2.37	$HPbO_2^- + H_2O + 2e^- \rightarrow Pb + 3OH^-$	−0.54
$La(OH)_3 + 3e \rightarrow La + 3OH^-$	−2.76	$PbO_2 + H_2O + 2e^- \rightarrow PbO + 2OH^-$	0.28
$Li^+ + e^- \rightarrow Li$	−3.045	$PbO_2 + SO_4^{-2} + 4H^+ + 2e^- \rightarrow PbSO_4 + 2H_2O$	1.685
	(−3.02)	$PbSO_4 + 2e^- \rightarrow Pb + SO_4^{-2}$	−0.356
$Mg^{++} + 2e^- \rightarrow Mg$	−2.375	$PbSO_4 + 2e^- \rightarrow Pb(Hg) + SO_4^{-2}$	−0.3505
$Mg(OH)_2 + 2e^- \rightarrow Mg + 2OH^-$	−2.67	$Pd^{+2} + 2e^- \rightarrow Pd$	0.83
$Mn^{+2} + 2e^- \rightarrow Mn$	−1.029	$Pd^{+2} + 2e^- \rightarrow Pd(1fHCl)$	0.623
$Mn^{+3} + e^- \rightarrow Mn^{+2}$	1.51	$Pd^{+2} + 2e^- \rightarrow Pd(4fHClO_4)$	0.987
$MnO_2 + 4H^+ + 2e^- \rightarrow Mn^{+2} + 2H_2O$	1.208	$PdCl_4^{-2} + 2e^- \rightarrow Pd + 4Cl^-$	0.623
$MnO_4^- + e^- \rightarrow MnO_4^{-2}$	0.564	$PdCl_6^{-2} + 2e^- \rightarrow PdCl_4^{-2} + 2Cl^-$	1.29
$MnO_4^- + 4H^+ + 3e^- \rightarrow MnO_2 + 2H_2O$	+1.679	$Pd(OH)_2 + 2e^- \rightarrow Pd + 2OH^-$	0.1
$MnO_4^- + 8H^+ + 5e^- \rightarrow Mn^{+2} + 4H_2O$	1.491	$H_2PO_2^- + e^- \rightarrow P + 2OH^-$	−1.82
$MnO_4^- + 2H_2O + 3e^- \rightarrow MnO_2 + 4OH^-$	0.588	$H_3PO_2 + H^+ + e^- \rightarrow P + 2H_2O$	−0.51
$MnO_4^{-1} + 2H_2O + 3e^- \rightarrow MnO_2 + 4OH^-$	0.58	$H_3PO_3 + 2H^+ + 2e^- \rightarrow H_3PO_2 + H_2O$	−0.50
$Mn(OH)_2 \rightarrow Mn + 2OH^-$	−1.47		(−0.59)
$Mn(OH)_3 + e^- \rightarrow Mn(OH)_2 + OH^-$	−0.40	$H_3PO_3 + 3H^+ + 3e^- \rightarrow P + 3H_2O$	−0.49
$H_2MoO_4 + 6H^+ + 6e^- \rightarrow Mo + 4H_2O$	0.0	$HPO_3^{-2} + 2H_2O + 2e^- \rightarrow H_2PO_2^- + 3OH^-$	−1.65
$N_2 + 2H_2O + 4H^+ + 2e^- \rightarrow 2NH_3OH^+$	−1.87	$HPO_3^{-2} + 2H_2O + 3e^- \rightarrow P + 5OH^-$	−1.71
$3N_2 + 2H^+ + 2e^- \rightarrow 2HN_3$	−3.1	$H_3PO_4 + 2H^+ + 2e^- \rightarrow H_3PO_3 + H_2O$	−0.276
$N_2H_5^+ + 3H^+ + 2e^- \rightarrow 2NH_4^+$	1.27		(−0.2)
$N_2O + 2H^+ + 2e^- \rightarrow N_2 + H_2O$	1.77	$PO_4^{-3} + 2H_2O + 2e^- \rightarrow HPO_3^{-2} + 3OH^-$	−1.05
$H_2N_2O_2 + 2H^+ + 2e^- \rightarrow N_2 + 2H_2O$	2.65	$Pt^{+2} + 2e^- \rightarrow Pt$	~1.2
$N_2O_4 + 2e^- \rightarrow 2NO_2^-$	0.88	$PtCl_4^{-2} + 2e^- \rightarrow Pt + 4Cl^-$	0.73

Table 2–26A (continued)
ELECTROCHEMICAL SERIES

Alphabetical Listing (continued)

Reaction	Potential, volts[a]	Reaction	Potential, volts[a]
$PtCl_6^{-2} + 2e^- \rightarrow PtCl_4^{-2} + 2Cl^-$	0.74	$2SO_4^{-2} + 4H^+ + 2e^- \rightarrow S_2O_6^{-2} + H_2O$	-0.2
$Pt(OH)_2 + 2e^- \rightarrow Pt + 2OH^-$	0.16	$SO_4^{-2} + H_2O + 2e^- \rightarrow SO_3^{-2} + 2OH^-$	-0.92
$Pu^{+4} + e^- \rightarrow Pu^{+3}(1f\,HClO_4)$	0.982	$Sr^{+2} + 2e^- \rightarrow Sr$	-2.89
$Pu^{+5} + e^- \rightarrow Pu^{+4}(0.5f\,HCl)$	1.099	$Sr^{+2} + 2e^- \rightarrow Sr(Hg)$	-1.793
$Pu^{+6} + e^- \rightarrow Pu^{+5}(1f\,HClO_4)$	0.9184	$Sr(OH)_2 \cdot 8H_2O + 2e^- \rightarrow Sr + 2OH^- + 8H_2O$	-2.99
$Pu^{+6} + 2e^- \rightarrow Pu^{+4}(1f\,HCl)$	1.052	$Ta_2O_5 + 10H^+ + 10e^- \rightarrow 2Ta + 5H_2O$	-0.71
Quinhydrone Elec. $H^+, a = 1$	0.6995	$TcO_4^- + 4H^+ + 3e^- \rightarrow TcO_{2(c)} + 2H_2O$	0.738
$Rb^+ + e^- \rightarrow Rb$	-2.925 (-2.99)	$Te + 2e^- \rightarrow Te^{-2}$	-0.92
		$Te + 2H^+ + 2e^- \rightarrow H_2Te(Ag)$	-0.69 (-0.72)
$Re^{+3} + 3e^- \rightarrow Re$	0.3 ~		
$ReO_4^- + 4H^+ + 3e^- \rightarrow ReO_2 + 2H_2O$	0.51	$Te^{+4} + 4e^- \rightarrow Te(2.5f\,HCl)$	0.63
$ReO_2 + 4H^+ + 4e^- \rightarrow Re + 2H_2O$	0.26	$TeO_2 + 4H_2 + 4e^- \rightarrow Te + 2H_2O$	0.593
$ReO_4^- + 2H^+ + e^- \rightarrow ReO_{3(c)} + 2H_2O$	0.768	$TeO_3^{-2} + 3H_2O + 4e^- \rightarrow Te + 6OH^-$	-0.02
$ReO_4^- + 4H_2O + 7e^- \rightarrow Re + 8OH^-$	-0.81	$TeO_4^- + 8H^+ + 7e^- \rightarrow Te + 4H_2O$	0.472
$ReO_4^- + 8H^+ + 7e^- \rightarrow Re + 4H_2O$	0.367	$H_6TeO_{6(s)} + 2H^+ + 2e^- \rightarrow TeO_{2(s)} + 4H_2O$	1.02
$Rh^{+4} + e^- \rightarrow Rh^{+3}$	1.43	$Th^{+4} + 4e^- \rightarrow Th$	-1.90
$Rh(Cl)_6^{-3} + 3e^- \rightarrow Rh + 6Cl^-$	0.44	$ThO_2 + 4H^+ + 4e^- \rightarrow Th + 2H_2O$	-1.80
$Ru^{+3} + e^- \rightarrow Ru^{+2}(0.1f\,HClO_4)$	-0.11	$ThO_2 + 2H_2O + 4e^- \rightarrow Th + 4OH^-$	-2.64
$Ru^{+3} + e^- \rightarrow Ru^{+2}(1-6f\,HCl)$	-0.084	$Ti^{+2} + 2e^- \rightarrow Ti$	-1.63
$Ru^{+4} + e^- \rightarrow Ru^{+3}(0.1f\,HClO_4)$	0.49	$Ti^{+3} + e^- \rightarrow Ti^{+2}$	-2.0
$Ru^{+4} + e^- \rightarrow Ru^{+3}(2f\,HCl)$	0.858	$TiO_2 + 4H^+ + 4e^- \rightarrow Ti + 2H_2O$	-0.86
$RuO_2 + 4H^+ + 4e^- \rightarrow Ru + 2H_2O$	-0.8	$Ti(OH)^{+3} + H^+ + e^- \rightarrow Ti^{+3} + H_2O$	0.06
$RuO_4^- + e^- \rightarrow RuO_4^{-2}$	0.59	$Tl^+ + e^- \rightarrow Tl$	-0.3363
$RuO_{4(c)} + e^- \rightarrow RuO_4^-$	1.00	$Tl^+ + e^- \rightarrow Tl(Hg)$	-0.3338
$S + 2e^- \rightarrow S^{-2}$	-0.508	$Tl^{+3} + e^- \rightarrow Tl^{+2}$	-0.37
$S + 2H^+ + 2e^- \rightarrow H_2S_{(aq)}$	0.141	$Tl^{+3} + 2e^- \rightarrow Tl^{+1}$	1.247
$S + H_2O + 2e^- \rightarrow HS^- + OH^-$	-0.478	$Tl^{+3} + 2e^- \rightarrow Tl^{+1}(1f\,HCl)$	0.783
$S_2O_6^{-2} + 4H^+ + 2e^- \rightarrow 2H_2SO_3$	0.6	$TlBr + e^- \rightarrow Tl(Hg) + Br^-$	-0.606
$S_2O_8^{-2} + 2e^- \rightarrow 2SO_4^{-2}$	2.0 (2.05)	$TlCl + e^- \rightarrow Tl(Hg) + Cl^-$	-0.555
		$TlI + e^- \rightarrow Tl(Hg) + I^-$	-0.769
$S_4O_6^- + 2e^- \rightarrow 2S_2O_3^{-2}$	0.09 (0.10)	$Tl_2O_3 + 3H_2O + 4e^- \rightarrow 2Tl^+ + 6OH^-$	0.02
$Sb + 3H^+ + 3e^- \rightarrow H_3Sb$	-0.51	$TlOH + e^- \rightarrow Tl + OH^-$	-0.3445
$Sb^{+5} + 2e^- \rightarrow Sb^{+3}(3.5f\,HCl)$	0.75	$Tl(OH)_3 + 2e^- \rightarrow TlOH + 2OH^-$	-0.05
$Sb_2O_3 + 6H^+ + 6e^- \rightarrow 2Sb + 3H_2O$	0.1445 (0.152)	$Tl_2SO_4 + 2e^- \rightarrow Tl(Hg) + SO_4^{-2}$	-0.4360
$Sb_2O_5 + 4H^+ + 4e^- \rightarrow Sb_2O_3 + 2H_2O$	0.69	$U^{+3} + 3e^- \rightarrow U$	-1.8
$Sb_2O_5 + 6H^+ + 4e^- \rightarrow 2SbO^+ + 3H_2O$	0.64	$U^{+4} + e^- \rightarrow U^{+3}$	-0.61
$SbO^+ + 2H^+ + 3e^- \rightarrow Sb + H_2O$	0.212	$U^{+4} + e^- \rightarrow U^{+3}(1f\,HClO_4)$	-0.631
$SbO_2^- + 2H_2O + 3e^- \rightarrow Sb + 4OH^-$	-0.66	$U^{+5} + e^- \rightarrow U^{+4}(1f\,HCl)$	1.02
$SbO_3^- + H_2O + 2e^- \rightarrow SbO_2^- + 2OH^-$	-0.59	$U^{+6} + e^- \rightarrow U^{+5}(1f\,HClO_4)$	0.063
$Sc^{+3} + 3e^- \rightarrow Sc$	-2.08	$UO_2^+ + 4H^+ + e^- \rightarrow U^{+4} + 2H_2O$	0.62
$Se + 2e^- \rightarrow Se^{-2}$	-0.78	$UO_2^{+2} + e^- \rightarrow UO_2^+$	0.062
$Se + 2H^+ + 2e^- \rightarrow H_2Se(aq)$	-0.36	$UO_2^{+2} + 4H^+ + 2e^- \rightarrow U^{+4} + 2H_2O$	0.334
$H_2SeO_3 + 4H^+ + 4e^- \rightarrow Se + 3H_2O$	0.74	$UO_2^{+2} + 4H^+ + 6e^- \rightarrow U + 2H_2O$	-0.82
$SeO_3^{-2} + 3H_2O + 4e^- \rightarrow Se + 6OH^-$	-0.35	$V^{+2} + 2e^- \rightarrow V$	-1.2
$SeO_4^{-2} + 4H^+ + 2e^- \rightarrow H_2SeO_3 + H_2O$	1.15	$V^{+3} + e^- \rightarrow V^{+2}$	-0.255
$SeO_4^{-2} + H_2O + 2e^- \rightarrow SeO_3^{-2} + 2OH^-$	0.03	$V^{+5} + e^- \rightarrow V^{+4}(1f\,NaOH)$	-0.74
$SiF_6^{-2} + 4e^- \rightarrow Si + 6F^-$	-1.2	$VO^{+2} + 2H^+ + e^- \rightarrow V^{+3} + H_2O$	0.337
$SiO_2 + 4H^+ + 4e^- \rightarrow Si + 2H_2O$	-0.84	$VO_2^+ + 2H^+ + e^- \rightarrow VO^{+2} + H_2O$	1.00
$SiO_3^{-2} + 3H_2O + 4e^- \rightarrow Si + 6OH^-$	-1.73	$V(OH)_4^+ + 2H^+ + e^- \rightarrow VO^{+2} + 3H_2O$	1.00
$Sn^{+2} + 2e^- \rightarrow Sn$	-0.1364	$V(OH)_4^+ + 4H^+ + 5e^- \rightarrow V + 4H_2O$	-0.25
$Sn^{+4} + 2e^- \rightarrow Sn^{+2}$	0.15	$W_2O_5 + 2H^+ + 2e^- \rightarrow 2WO_2 + H_2O$	-0.04
$Sn^{+4} + 2e^- \rightarrow Sn^{+2}(0.1f\,HCl)$	0.070	$WO_2 + 4H^+ + 4e^- \rightarrow W + 2H_2O$	-0.12
$Sn^{+4} + 2e^- \rightarrow Sn^{+2}(1f\,HCl)$	0.139	$WO_3 + 6H^+ + 6e^- \rightarrow W + 3H_2O$	-0.09
$HSnO_2^- + H_2O + 2e^- \rightarrow Sn + 3OH^-$	-0.79	$2WO_3 + 2H^+ + 2e^- \rightarrow W_2O_5 + H_2O$	-0.03
$Sn(OH)_6^{-2} + 2e^- \rightarrow HSnO_2^- + 3OH^- + H_2O$	-0.96	$Y^{+3} + 3e^- \rightarrow Y$	-2.37
$2H_2SO_3 + H^+ + 2e^- \rightarrow HS_2O_4^- + 2H_2O$	-0.08	$Zn^{+2} + 2e^- \rightarrow Zn$	-0.7628
$H_2SO_3 + 4H^+ + 4e^- \rightarrow S + 3H_2O$	0.45	$Zn^{+2} + 2e^- \rightarrow Zn(Hg)$	-0.7628
$2SO_3^{-2} + 2H_2O + 2e^- \rightarrow S_2O_4^{-2} + 4OH^-$	-1.12	$ZnO_2^- + 2H_2O + 2e^- \rightarrow Zn + 4OH^-$	-1.216
$2SO_3^{-2} + 3H_2O + 4e^- \rightarrow S_2O_3^{-2} + 6OH^-$	-0.58	$ZnSO \cdot 7H_2O + 2e^- \rightarrow Zn(Hg) + SO_4^{-2}(Sat'd\,ZnSO_4)$	-0.7993
$SO_4^{-2} + 4H^+ + 2e^- \rightarrow H_2SO_3 + H_2O$	0.20	$ZrO_2 + 4H^+ + 4e^- \rightarrow Zr + 2H_2O$	-1.43
		$ZrO(OH)_2 + H_2O + 4e^- \rightarrow Zr + 4OH^-$	-2.32

[a]Values listed are standard reduction potentials.

From Hunsberger, J. F., in *Handbook of Chemistry and Physics*, 55th ed., Weast, R. C., Ed., CRC Press, Cleveland, 1974, D-120.

Table 2–26B
ELECTROCHEMICAL SERIES

Positive Reduction Reactions

This table lists reduction reactions that are positive with respect to the potential of the standard hydrogen electrode. The reduction reactions are listed in the order of increasing positive potential, beginning with 0.00 and ending with +3.03 volts.

Reaction	Potential, volts[a]	Reaction	Potential, volts[a]
$2H^+ + 2e^- \to H_2$	0.0000	$Pd(OH)_2 + 2e^- \to Pd + 2OH^-$	0.1
$CuI_2^- + e^- \to Cu + 2I^-$	0.00	$Co(NH_3)_6^{+3} + e^- \to Co(NH_3)_6^{+2}$	0.1
$H_2MoO_4 + 6H^+ + 6e^- \to Mo + 4H_2O$	0.0	$Hg_2O + H_2O + 2e^- \to 2Hg + 2OH^-$	0.123
$NO_3^- + H_2O + 2e^- \to NO_2^- + 2OH^-$	0.01	$Sn^{+4} + 2e^- \to Sn^{+2}(If\,HCl)$	0.139
$Tl_2O_3 + 3H_2O + 4e^- \to 2Tl^+ + 6OH^-$	0.02	$Hg_2Br_2 + 2e^- \to 2Hg + 2Br^-$	0.1396
$SeO_4^{-2} + H_2O + 2e^- \to SeO_3^{-2} + 2OH^-$	0.03	$S + 2H^+ + 2e^- \to H_2S_{(aq)}$	0.141
$Ti(OH)^{+3} + H^+ + e^- \to Ti^{+3} + H_2O$	0.06	$Sb_2O_3 + 6H^+ + 6e^- \to 2Sb + 3H_2O$	0.1445
$UO_2^{+2} + e^- \to UO_2^+$	0.062		(0.152)
$U^{+6} + e^- \to U^{+5}(If\,HClO_4)$	0.063	$2NO_2^- + 3H_2O + 4e^- \to N_2O + 6OH^-$	0.15
$Sn^{+4} + 2e^- \to Sn^{+2}(0.1f\,HCl)$	0.070	$ReO_4^- + 8H^+ + 7e^- \to Re + 4H_2O$	0.15
$AgBr + e^- \to Ag + Br^-$	0.0713	$Sn^{+4} + 2e^- \to Sn^{+2}$	0.15
$AgSCN + e^- \to Ag + SCN^-$	0.0895	$Np^{+4} + e^- \to Np^{+3}(If\,HClO_4)$	0.155
$S_4O_6^= + 2e^- \to 2S_2O_3^{-2}$	0.09	$Cu^{+2} + e \to Cu^+$	0.158
	(0.10)		(0.167)
$HgO + H_2O + 2e^- \to Hg + 2OH^-$	0.0984	$BiOCl + 2H^+ + 3e^- \to Bi + Cl^- + H_2O$	0.1583
$2NO + 2e^- \to N_2O_2^{-2}$	0.10	$Pt(OH)_2 + 2e^- \to Pt + 2OH^-$	0.16
$Ir_2O_3 + 3H_2O + 6e^- \to 2Ir + 6OH^-$	0.1	$Bi(Cl)_4^- + 3e^- \to Bi + 4Cl^-$	0.168
$ClO_4^- + H_2O + 2e^- \to ClO_3^- + 2OH^-$	0.17	$PtCl_4^{-2} + 2e^- \to Pt + 4Cl^-$	0.73
$Ag_4Fe(CN)_6 + 4e^- \to 4Ag + Fe(CN)_6^{-4}$	0.1943	$TcO_4^- + 4H^+ + 3e^- \to TcO_{2(c)} + 2H_2O$	0.738
$SO_4^{-2} + 4H^+ + 2e^- \to H_2SO_3 + H_2O$	0.20	$Np^{+5} + e^- \to Np^{+4}(If\,HClO_4)$	0.739
$Co(OH)_3 + e^- \to Co(OH)_2 + OH^-$	0.2	$Ag_2O_3 + H_2O + 2e^- \to 2AgO + 2OH^-$	0.74
	(0.17)	$H_2SeO_3 + 4H^+ + 4e^- \to Se + 3H_2O$	0.74
$SbO^+ + 2H^+ + 3e^- \to Sb + 2H_2O$	0.212	$PtCl_6^{-2} + 2e^- \to PtCl_4^{-2} + 2Cl^-$	0.74
$AgCl + e^- \to Ag + Cl^-$	0.2223	$Fe^{+3} + e^- \to Fe^{+2}(If\,HClO_4)$	0.747
Calomel Electrode, Sat'd NaCl	0.2360	$Sb^{+5} + 2e^- \to Sb^{+3}(3.5f\,HCl)$	0.75
$As_2O_3 + 6H^+ + 6e^- \to 2As + 3H_2O$	0.234	$ClO_3^- + 2H_2O + 4e^- \to Cl^- + 4OH^-$	0.76
Calomel Electrode, Sat'd. KCl	0.2415	$NiO_2 + 2H_2O + 2e^- \to Ni(OH)_2 + 2OH^-$	0.76
$HAsO_2 + 3H^+ + 3e^- \to As + 2H_2O$	0.2475	$2NO + H_2O + 2e^- \to N_2O + 2OH^-$	0.76
$IO_3^- + 3H_2O + 6e^- \to I^- + 6OH^-$	0.26	$ReO_4^- + 2H^+ + e^- \to ReO_{3(cc)} + 2H_2O$	0.768
$ReO_2 + 4H^+ + 4e^- \to Re + 2H_2O$	0.26	$(CNS)_2 + 2e^- \to 2CNS^-$	0.77
$Hg_2Cl_2 + 2e^- \to 2Hg + 2Cl^-$	0.2682	$Fe^{+3} + e^- \to Fe^{+2}$	0.770
Calomel Electrode, Molal KCl	0.2800	$Fe^{+3} + e^- \to Fe^{+2}(If\,HCl)$	0.770
$PbO, H_2O + 2e^- \to PbO + 2OH^-$	0.28	$IrCl_6^{-3} + 3e^- \to Ir + 6Cl^-$	0.77
Calomel Electrode, N KCl	0.2807	$Tl^{+3} + 2e^- \to Tl^{+1}(If\,HCl)$	0.783
$Re^{+3} + 3e^- \to Re$	0.3 ~	$Hg_2^{+2} + 2e^- \to 2Hg$	0.7961
$BiO^+ + 2H^+ + 3e^- \to Bi + H_2O$	0.32	$1/2Hg_2^{+2} + e^- \to Hg$	0.7986
$2HCNO + 2H^+ + 2e^- \to (CN)_2 + 2H_2O$	0.33	$Ag^+ + e^- \to Ag$	0.7996
Calomel Electrode 0.1 N KCl	0.3337	$2HNO_2 + 4H^+ + 4e^- \to H_2N_2O_2 + 2H_2O$	0.80
$UO_2^{+2} + 4H^+ + 2e^- \to U^{+4} + 2H_2O$	0.334	$2NO_3^- + 4H^+ + 2e^- \to N_2O_4 + 2H_2O$	0.81
$VO^{+2} + 2H^+ + e^- \to V^{+3} + H_2O$	0.337	$1/2O_2 + 2H^+(10^{-7}M) + 2e^- \to H_2O$	0.815
$Cu^{+2} + 2e^- \to Cu$	0.3402	$Pd^{+2} + 2e^- \to Pd$	0.83
$Hg_2Cl_2 + 2e^- \to 2Hg + 2Cl^-(0.1f\,NaOH)$	0.3419	$OsO_4 + 8H^+ + 8e^- \to Os + 4H_2O$	0.85
	(0.268)	$Hg^{+2} + 2e^- \to Hg$	0.851
$Ag_2O + H_2O + 2e^- \to 2Ag + 2OH^-$	0.342	$AuBr_4^- + 3e^- \to Au + 4Br^-$	0.858
$Nb^{+5} + 2e \to Nb^{+3}(2f\,HCl)$	0.344	$Ru^{+4} + e^- \to Ru^{+3}(2f\,HCl)$	0.858
$Cu^{+2} + 2e \to Cu(Hg)$	0.345	$TiO_2 + 4H^+ + 4e^- \to Ti + 2H_2O$	0.86
$ClO_3^- + H_2O + 2e^- \to ClO_2^- + 2OH^-$	0.35	$HO_2^- + H_2O + 2e^- \to 3OH^-$	0.87
$AgIO_3 + e^- \to Ag + IO_3$	0.3551	$N_2O_4 + 2e^- \to 2NO_2^-$	0.88
$Ag_2SeO_3 + 2e^- \to 2Ag + SeO_3^{-2}$	0.3629	$ClO^- + H_2O + 2e^- \to Cl^- + 2OH^-$	0.90
$ReO_4^- + 8H^+ + 7e^- \to Re + 4H_2O$	0.367	$2Hg^{+2} + 2e^- \to Hg_2^{+2}$	0.905
$(CN)_2 + 2H^+ + 2e^- \to 2HCN$	0.37	$Pu^{+6} + e^- \to Pu^{+5}(If\,HClO_4)$	0.9184
$O_2 + 2H_2O + 4e^- \to 4OH^-$	0.401	$NO_3^- + 3H^+ + 2e^- \to HNO_2 + H_2O$	0.94
$AgOCN + e^- \to Ag + OCN^-$	0.41	$ClO_2(aq) + e^- \to ClO_2^-$	0.954
$Fe^{+3} + e^- \to Fe^{+2}(If\,H_3PO_4)$	0.438	$NO_3^- + 4H^+ + 3e^- \to NO + 2H_2O$	0.96
$Rh(Cl)_6^{-3} + 3e^- \to Rh + 6Cl^-$	0.44	$AuBr_2^- + e^- \to Au + 2Br^-$	0.963
$Ag_2CrO_4 + 2e^- \to 2Ag + CrO_4^-$	0.4463	$Pu^{+4} + e^- \to Pu^{+3}(If\,HClO_4)$	0.982
$H_2SO_3 + 4H^+ + 4e^- \to S + 3H_2O$	0.45	$Pd^{+2} + 2e^- \to Pd(4f\,HClO_4)$	0.987
$Fe(CN)_6^{-3} + e^- \to Fe(CN)_6^{-4}(0.01f\,NaOH)$	0.46	$HIO + H^+ + 2e^- \to I^- + H_2O$	0.99
$Ag_2(WO_4) + 2e^- \to 2Ag + WO_4^{-2}$	0.466	$HNO_2 + H^+ + e^- \to NO + H_2O$	0.99
$TeO_2^- + 8H^+ + 7e^- \to Te + 4H_2O$	0.472	$AuCl_4^- + 3e^- \to Au + 4Cl^-$	0.994
$Ag_2CO_3 + 2e^- \to 2Ag + CO_3^{-2}$	0.4769	$RuO_{4(c)} + e^- \to RuO_4^-$	1.00
$AgC_2O_4 + 2e^- \to Ag + C_2O_4^{-2}$	0.4776	$VO_2^+ + 2H^+ + e^- \to VO^{+2} + H_2O$	1.00
$Ag_2MoO_4 + 2e^- \to 2Ag + MoO_4^{-2}$	0.49	$V(OH)_4^+ + 2H^+ + e^- \to VO^{+2} + 3H_2O$	1.00
$IO^- + H_2O + 2e^- \to I^- + 2OH^-$	0.49	$H_6TeO_{6(s)} + 2H^+ + 2e^- \to TeO_{2(s)} + 4H_2O$	1.02
$NiO_2 + 2H_2O + 2e^- \to Ni(OH)_2 + 2OH^-$	0.49	$IrCl_6^{-2} + e^- \to IrCl_6^{-3}$	1.02
$Ru^{+4} + e^- \to Ru^{+3}(0.1f\,HClO_4)$	0.49	$U^{+5} + e^- \to U^{+4}(If\,HCl)$	1.02
$ReO_4^- + 4H^+ + 3e^- \to ReO_2 + 2H_2O$	0.51	$N_2O_4 + 4H^+ + 4e^- \to 2NO + 2H_2O$	1.03
$Hg_2(AcO)_2 + 2e^- \to 2Hg + 2AcO^-$	0.5113	$Pu^{+6} + 2e^- \to Pu^{+4}(If\,HCl)$	1.052

Table 2–26 B (continued)
ELECTROCHEMICAL SERIES

Positive Reduction Reactions (continued)

Reaction	Potential, volts[a]	Reaction	Potential, volts[a]
$Cu^+ + e^- \rightleftharpoons Cu$	0.522	$Fe(phenanthroline)_3^{+3} + e^- \rightleftharpoons Fe(ph)_3^{+2}(2f\,H_2SO_4)$	1.056
$I_3^- + 2e^- \rightleftharpoons 3I^-$	0.5338	$Br_{2(l)} + 2e^- \rightleftharpoons 2Br^-$	1.065
$I_2 + 2e^- \rightleftharpoons 2I^-$	0.535	$N_2O_4 + 2H^+ + 2e^- \rightleftharpoons 2HNO_2$	1.07
$IO_3^- + 2H_2O + 4e^- \rightleftharpoons IO^- + 4OH^-$	0.56	$IO_3^- + 6H^+ + 6e^- \rightleftharpoons I^- + 3H_2O$	1.085
$MnO_4^- + e^- \rightleftharpoons MnO_4^{-2}$	0.564	$Br_{2(aq)} + 2e^- \rightleftharpoons 2Br^-$	1.087
$MnO_4^{-1} + 2H_2O + 3e^- \rightleftharpoons MnO_2 + 4OH^-$	0.58	$Pu^{+5} + e^- \rightleftharpoons Pu^{+4}(0.5f\,HCl)$	1.099
$H_3AsO_4 + 2H^+ + 2e^- \rightleftharpoons HAsO_2 + 2H_2O(1f\,HCl)$	0.58	$Cr^{+6} + 3e^- \rightleftharpoons Cr^{+3}(2f\,H_2SO_4)$	1.10
$MnO_4^- + 2H_2O + 3e^- \rightleftharpoons MnO_2 + 4OH^-$	0.588	$Cu^{+2} + 2CN^- + e^- \rightleftharpoons Cu(CN)_2^-$	1.12
$AgNO_2 + e^- \rightleftharpoons Ag + NO_2^-$	0.59	$Np^{+6} + e^- \rightleftharpoons Np^{+5}(1f\,HClO_4)$	1.137
$ClO_3^- + H_2O + 2e^- \rightleftharpoons ClO_2^- + 2OH^-$	0.59	$Fe(phenanthroline)_3^{+3} + e^- \rightleftharpoons Fe(ph)_3^{+2}$	1.14
$RuO_4^- + e^- \rightleftharpoons RuO_4^{-2}$	0.59	$SeO_4^{-2} + 4H^+ + 2e^- \rightleftharpoons H_2SeO_3 + H_2O$	1.15
$TeO_2 + 4H_2 + 4e^- \rightleftharpoons Te + 2H_2O$	0.593	$ClO_2 + e^- \rightleftharpoons ClO_2^-$	1.15
$2AgO + H_2O + 2e^- \rightleftharpoons Ag_2O + 2OH^-$	0.599	$ClO_3^- + 2H^+ + e^- \rightleftharpoons ClO_2 + H_2O$	1.15
$S_2O_6^{-2} + 4H^+ + 2e^- \rightleftharpoons 2H_2SO_3$	0.6	$ClO_4^- + 2H^+ + 2e^- \rightleftharpoons ClO_3^- + H_2O$	1.19
$BrO_3^- + 3H_2O + 6e^- \rightleftharpoons Br^- + 6OH^-$	0.61	$2IO_3^- + 12H^+ + 10e^- \rightleftharpoons I_2 + 6H_2O$	1.19
$Hg_2SO_4 + 2e^- \rightleftharpoons 2Hg + SO_4^{-2}$	0.6158	$HCrO_4^- + 7H^+ + 3e^- \rightleftharpoons Cr^{+3} + 4H_2O$	1.195
$ClO_3^- + 3H_2O + 6e^- \rightleftharpoons Cl^- + 6OH^-$	0.62	$IO_3^- + 6H^+ + 5e^- \rightleftharpoons 1/2I_2 + 3H_2O$	1.195
$UO_2^+ + 4H^+ + e^- \rightleftharpoons U^{+4} + 2H_2O$	0.62	$Pt^{+2} + 2e^- \rightleftharpoons Pt$	~1.2
$Pd^{+2} + 2e^- \rightleftharpoons Pd(1f\,HCl)$	0.623	$MnO_2 + 4H^+ + 2e^- \rightleftharpoons Mn^{+2} + 2H_2O$	1.208
$PdCl_4^{-2} + 2e^- \rightleftharpoons Pd + 4Cl^-$	0.623	$ClO_3^- + 3H^+ + 2e^- \rightleftharpoons HClO_2 + H_2O$	1.21
$Te^{+4} + 4e^- \rightleftharpoons Te(2.5f\,HCl)$	0.63		(1.23)
$Hg_2HPO_4 + H^+ + 2e^- \rightleftharpoons 2Hg + H_2PO_4^-$	0.639	$O_2 + 4H^+ + 4e^- \rightleftharpoons 2H_2O$	1.229
$AgAc + e^- \rightleftharpoons Ag + Ac^-$	0.64	$O_3 + H_2O + 2e^- \rightleftharpoons O_2 + 2OH^-$	1.24
$Sb_2O_{5(s)} + 6H^+ + 4e^- \rightleftharpoons 2SbO^+ + 3H_2O$	0.64	$Tl^{+3} + 2e^- \rightleftharpoons Tl^{+1}$	1.247
$Ag_2SO_4 + 2e^- \rightleftharpoons 2Ag + SO_4^{-2}$	0.653	$ClO_2 + H^+ + e^- \rightleftharpoons HClO_2$	1.27
$Fe^{+3} + e^- \rightleftharpoons Fe^{+2}(0.5f\,H_2SO_4)$	0.679	$2HNO_2 + 4H^+ + 4e^- \rightleftharpoons N_2O + 3H_2O$	1.27
$AgBrO_3 + e^- \rightleftharpoons Ag + BrO_3^-$	0.680		(1.29)
$O_2 + 2H^+ + 2e^- \rightleftharpoons H_2O_2$	0.682	$N_2H_5^+ + 3H^+ + 2e^- \rightleftharpoons 2NH_4^+$	1.27
$Fe(CN)_6^{-3} + e^- \rightleftharpoons Fe(CN)_6^{-4}(1f\,H_2SO_4)$	0.69	$Au^{+3} + 2e^- \rightleftharpoons Au^{+1}$	~1.29
$Sb_2O_5 + 4H^+ + 4e^- \rightleftharpoons Sb_2O_3 + 2H_2O$	0.69	$PdCl_6^{-2} + 2e^- \rightleftharpoons PdCl_4^{-2} + 2Cl^-$	1.29
$C_6H_4O_2 + 2H^+ + 2e^- \rightleftharpoons C_6H_4(OH)_2$	0.6992	$HBrO + H^+ + 2e^- \rightleftharpoons Br^- + H_2O$	1.33
Quinhydrone Elec. H^+, a = 1	0.6995	$Cr_2O_7^{-2} + 14H^+ + 6e^- \rightleftharpoons 2Cr^{+3} + 7H_2O$	1.33
$BrO^- + H_2O + 2e^- \rightleftharpoons Br^- + 2OH^-(1f\,NaOH)$	0.70	$ClO_4^- + 8H^+ + 7e^- \rightleftharpoons 1/2Cl_2 + 4H_2O$	1.34
$H_3IO_6^{-2} + 2e^- \rightleftharpoons IO_3^- + 3OH^-$	~0.70	$Cl_{2(g)} + 2e^- \rightleftharpoons 2Cl^-$	1.3583
		$ClO_4^- + 8H^+ + 8e^- \rightleftharpoons Cl^- + 4H_2O$	1.37
$OH + e^- \rightleftharpoons OH^-$	1.4	$HClO_2 + 3H^+ + 3e^- \rightleftharpoons 1/2Cl_2 + 2H_2O$	1.63
$Au^{+3} + 3e^- \rightleftharpoons Au$	1.42	$HClO + H^+ + e^- \rightleftharpoons 1/2Cl_2 + H_2O$	1.63
$2NH_3OH^+ + H^+ + 2e^- \rightleftharpoons N_2H_5^+ + 2H_2O$	1.42	$HClO_2 + 2H^+ + 2e^- \rightleftharpoons HClO + H_2O$	1.64
$Rh^{+4} + e^- \rightleftharpoons Rh^{+3}$	1.43	$MnO_4^- + 4H^+ + 3e^- \rightleftharpoons MnO_2 + 2H_2O$	1.679
$BrO_3^- + 6H^+ + 6e^- \rightleftharpoons Br^- + 3H_2O$	1.44	$Au^+ + e^- \rightleftharpoons Au$	1.68
$Ce^{+4} + e^- \rightleftharpoons Ce^{+3}$	1.4430	$PbO_2 + SO_4^{-2} + 4H^+ + 2e^- \rightleftharpoons PbSO_4 + 2H_2O$	1.685
	(1.61)	$H_5IO_6 + H^+ + 2e^- \rightleftharpoons IO_3^- + 3H_2O$	~1.7
$Au(OH)_3 + 3H^+ + 3e^- \rightleftharpoons Au + 3H_2O$	1.45	$CeOH^{+3} + H^+ + e^- \rightleftharpoons Ce^{+3} + H_2O$	1.7134
$ClO_3^- + 6H^+ + 6e^- \rightleftharpoons Cl^- + 3H_2O$	1.45	$N_2O + 2H^+ + 2e^- \rightleftharpoons N_2 + H_2O$	1.77
$HIO + H^+ + e^- \rightleftharpoons 1/2I_2 + H_2O$	1.45	$H_2O_2 + 2H^+ + 2e^- \rightleftharpoons 2H_2O$	1.776
$Ce^{+4} + e^- \rightleftharpoons Ce^{+3}(0.5f\,H_2SO_4)$	1.4587	$Co^{+3} + e^- \rightleftharpoons Co^{+2}(3f\,HNO_3)$	1.842
$PbO_2 + 4H^+ + 2e^- \rightleftharpoons Pb^{+2} + 2H_2O$	1.46	$FeO_4^{-2} + 8H^+ + 3e^- \rightleftharpoons Fe^{+3} + 4H_2O$	1.9
$ClO_3^- + 6H^+ + 5e^- \rightleftharpoons 1/2Cl_2 + 3H_2O$	1.47	$NiO_2 + 4H^+ + 2e^- \rightleftharpoons Ni^{+2} + 2H_2O$	1.93
$HClO + H^+ + 2e^- \rightleftharpoons Cl^- + H_2O$	1.49	$Ag^{+2} + e^- \rightleftharpoons Ag^{+1}(4f\,HClO_4)$	1.987
$MnO_4^- + 8H^+ + 5e^- \rightleftharpoons Mn^{+2} + 4H_2O$	1.491	$S_2O_8^{-2} + 2e^- \rightleftharpoons 2SO_4^{-2}$	2.0
$HO_2 + H^+ + e^- \rightleftharpoons H_2O_2$	1.5		(2.05)
$Mn^{+3} + e^- \rightleftharpoons Mn^{+2}$	1.51	$O_3 + 2H^+ + 2e^- \rightleftharpoons O_2 + H_2O$	2.07
$BrO_3^- + 6H^+ + 5e^- \rightleftharpoons 1/2Br_2 + 3H_2O$	1.52	$F_2O + 2H^+ + 4e^- \rightleftharpoons H_2O + 2F^-$	2.1
$HClO_2 + 3H^+ + 4e^- \rightleftharpoons Cl^- + 2H_2O$	1.56	$O_{(g)} + 2H^+ + 2e^- \rightleftharpoons H_2O$	2.42
$Bi_2O_4 + 4H^+ + 2e^- \rightleftharpoons 2BiO^+ + 2H_2O$	1.59	$H_2N_2O_2 + 2H^+ + 2e^- \rightleftharpoons N_2 + 2H_2O$	2.65
$HBrO + H^+ + e^- \rightleftharpoons 1/2Br_2 + H_2O$	1.59	$1/2F_2 + e^- \rightleftharpoons F^-$	2.85
$2NO + 2H^+ + 2e^- \rightleftharpoons N_2O + H_2O$	1.59	$F_2 + 2e^- \rightleftharpoons 2F^-$	2.87
$2HBrO + 2H^+ + 2e^- \rightleftharpoons Br_{2(l)} + 2H_2O$	1.6	$1/2F_2 + H^+ + e^- \rightleftharpoons HF$	3.03

[a] Values listed are standard reduction potentials.

From Hunsberger, J. F., in *Handbook of Chemistry and Physics*, 55th ed., Weast, R. C., Ed., CRC Press, Cleveland, 1974, D-122.

Table 2–26C
ELECTROCHEMICAL SERIES

Negative Reduction Reactions

This table lists reduction reactions that are negative with respect to the potential of the standard hydrogen electrode. The reduction reactions are listed in the order of increasing negative potential, beginning with 0.00 and ending with -3.1 volts.

Reaction	Potential, volts[a]
$2H^+ + 2e^- \rightarrow H_2$	0.0000
$D^+ + e^- \rightarrow 1/2D_2$	−0.0034
$AgCN + e^- \rightarrow Ag + CN^-$	−0.02
$TeO_3^{-2} + 3H_2O + 4e^- \rightarrow Te + 6OH^-$	−0.02
$2WO_3 + 2H^+ + 2e^- \rightarrow W_2O_5 + H_2O$	−0.03
$Fe^{+3} + 3e^- \rightarrow Fe$	−0.036
$Ag_2S + 2H^+ + 2e^- \rightarrow 2Ag + H_2S$	−0.0366
$P + 3H^+ + 3e^- \rightarrow PH_3(g)$	−0.04
$W_2O_5 + 2H^+ + 2e^- \rightarrow 2WO_2 + H_2O$	−0.04
$Hg_2I_2 + 2e^- \rightarrow 2Hg + 2I^-$	−0.0405
$2D^+ + 2e^- \rightarrow D_2$	−0.044
$Tl(OH)_3 + 2e^- \rightarrow TlOH + 2OH^-$	−0.05
$O_2 + H_2O + 2e^- \rightarrow HO_2^- + OH^-$	−0.076
$AsO_4^{-3} + 2H_2O + 2e^- \rightarrow AsO_2^- + 4OH^-$ (If NaOH)	−0.08
$2H_2SO_3 + H^+ + 2e^- \rightarrow HS_2O_4^- + 2H_2O$	−0.08
$Ru^{+3} + e^- \rightarrow Ru^{+2}$(1 -6f HCL)	−0.084
$2Cu(OH)_2 + 2e^- \rightarrow Cu_2O + 2OH^- + H_2O$	−0.09
$WO_3 + 6H^+ + 6e^- \rightarrow W + 3H_2O$	−0.09
$Ru^{+3} + e^- \rightarrow Ru^{+2}$(0.1f HClO_4)	−0.11
$Cr^{+6} + 3e^- \rightarrow Cr^{+3}$(If NaOH)	−0.12
$CrO_4^{-2} + 4H_2O + 3e^- \rightarrow Cr(OH)_3 + 5OH^-$	−0.12
$GeO_2 + 2H^+ + 2e^- \rightarrow GeO + H_2O$	−0.12
$WO_2 + 4H^+ + 4e^- \rightarrow W + 2H_2O$	−0.12
$Pb^{+2} + 2e^- \rightarrow Pb(Hg)$	−0.1205
$Pb^{+2} + 2e^- \rightarrow Pb$	−0.1263
	(0.126)
$H_2GeO_3 + 4H^+ + 4e^- \rightarrow Ge + 3H_2O$	−0.13
$Sn^{+2} + 2e^- \rightarrow Sn$	−0.1364
$O_2 + 2H_2O + 2e^- \rightarrow H_2O_2 + 2OH^-$	−0.146
$AgI + e^- \rightarrow Ag + I^-$	−0.1519
$2NO_2^- + 2H_2O + 4e^- \rightarrow N_2O_2^{-2} + 4OH^-$	−0.18
$CO_2 + 2H^+ + 2e^- \rightarrow HCOOH$	−0.2
$2SO_4^{-2} + 4H^+ + 2e^- \rightarrow S_2O_6^{-2} + 2H_2O$	−0.2
$Cu(OH)_2 + 2e^- \rightarrow Cu + 2OH^-$	−0.224
$Ni^{+2} + 2e^- \rightarrow Ni$	−0.23
$PbHPO_4 + H^+ + 2e^- \rightarrow Pb(Hg) + HPO_4^-$	−0.2448
$V(OH)_4^+ + 4H^+ + 5e^- \rightarrow V + 4H_2O$	−0.25
$V^{+3} + e^- \rightarrow V^{+2}$	−0.255
$PbCl_2 + 2e^- \rightarrow Pb(Hg) + 2Cl^-$	−0.262
$PbBr_2 + 2e^- \rightarrow Pb(Hg) + Br^-$	−0.275
$H_3PO_4 + 2H^+ + 2e^- \rightarrow H_3PO_3 + H_2O$	−0.276
	(−0.2)
$Co^{+2} + 2e^- \rightarrow Co$	−0.28
$Tl^+ + e^- \rightarrow Tl(Hg)$	−0.3338
$Tl^+ + e^- \rightarrow Tl$	−0.3363
$In^{+3} + 3e^- \rightarrow In$	−0.338
$PbF_2 + 2e^- \rightarrow Pb(Hg) + 2F^-$	−0.3444
$TlOH + e^- \rightarrow Tl + OH^-$	−0.3445
$SeO_3^- + 3H_2O + 4e^- \rightarrow Se + 6OH^-$	−0.35
$PbSO_4 + 2e^- \rightarrow Pb(Hg) + SO_4^{-2}$	−0.3505
$Cd^{+2} + 2e^- \rightarrow Cd(Hg)$	−0.3521
$PbSO_4 + 2e^- \rightarrow Pb + SO_4^-$	−0.356
$PbI_2 + 2e^- \rightarrow Pb(Hg) + 2I^-$	−0.358
$Se + 2H^+ + 2e^- \rightarrow H_2Se(aq)$	−0.36
$Cu_2O + H_2O + 2e^- \rightarrow 2Cu + 2OH^-$	−0.361
$Tl^{+3} + e^- \rightarrow Tl^{+2}$	−0.37
$UO_2^+ + 4H^+ + 6e^- \rightarrow U + 2H_2O$	−0.82
$2H_2O + 2e \rightarrow H_2 + 2OH$	−0.8277
$SiO_2 + 4H^+ + 4e^- \rightarrow Si + 2H_2O$	−0.84
$2NO_3^- + 2H_2O + 2e^- \rightarrow N_2O_4 + 4OH$	−0.85
$TiO_2 + 4H^+ + 4e^- \rightarrow Ti + 2H_2O$	−0.86
$P + 3H_2O + 3e^- \rightarrow PH_3(g) + 3OH^-$	−0.87
$SO_4^- + H_2O + 2e^- \rightarrow SO_3^{-2} + 2OH^-$	−0.92
$Te + 2e^- \rightarrow Te^{-2}$	−0.92
$Sn(OH)_6^{-2} + 2e^- \rightarrow HSnO_2^- + 3OH^- + H_2O$	−0.96
$Mn^{+2} + 2e^- \rightarrow Mn$	−1.029
$PO_4^{-3} + 2H_2O + 2e^- \rightarrow HPO_3^{-2} + 3OH^-$	−1.05
$2SO_3^- + 2H_2O + 2e^- \rightarrow S_2O_4^{-2} + 4OH^-$	−1.12
$CrO_2^- + 2H_2O + 3e^- \rightarrow Cr + 4OH^-$	−1.2
$SiF_6^{-2} + 4e^- \rightarrow Si + 6F^-$	−1.2

Reaction	Potential, volts[a]
$In^{+2} + e^- \rightarrow In^{+1}$	−0.40
$In^{+3} + 2e^- \rightarrow In^{+1}$	−0.40
$Mn(OH)_3 + e^- \rightarrow Mn(OH)_2 + OH^-$	−0.40
$Cd^{+2} + 2e^- \rightarrow Cd$	−0.4026
$Fe^{+2} + 2e^- \rightarrow Fe$	−0.409
$Cr^{+3} + e^- \rightarrow Cr^{+2}$	−0.41
$Eu^{+3} + e^- \rightarrow Eu^{+2}$	−0.43
$CdSO_4 \cdot 8/3H_2O + 2e^- \rightarrow Cd(Hg) + CdSO_4$ (sat'd aq)	−0.4346
$Tl_2SO_4 + 2e^- \rightarrow Tl(Hg) + SO_4^{-2}$	−0.4360
$Bi_2O_3 + 3H_2O + 6e^- \rightarrow 2Bi + 6OH^-$	−0.46
$BiOOH + H_2O + 3e^- \rightarrow Bi + 3OH^-$	−0.46
$NO_2^- + H_2O + e^- \rightarrow NO + 2OH^-$	−0.46
$S + H_2O + 2e^- \rightarrow HS^- + OH^-$	−0.478
$H_3PO_3 + 3H^+ + 3e^- \rightarrow P + 3H_2O$	−0.49
$In^{+3} + e^- \rightarrow In^{+2}$	−0.49
$2CO_2 + 2H^+ + 2e^- \rightarrow H_2C_2O_4$	−0.49
$H_3PO_3 + 2H^+ + 2e^- \rightarrow H_3PO_2 + H_2O$	−0.50
	(−0.59)
$S + 2e^- \rightarrow S^{-2}$	−0.508
$H_3PO_2 + H^+ + e^- \rightarrow P + 2H_2O$	−0.51
$Sb + 3H^+ + 3e^- \rightarrow H_3Sb$	−0.51
$As + 3H^+ + 3e^- \rightarrow AsH_3$	−0.54
$HPbO_2^- + H_2O + 2e^- \rightarrow Pb + 3OH^-$	−0.54
$TlCl + e^- \rightarrow Tl(Hg) + Cl^-$	−0.555
$Cr^{+2} + 2e^- \rightarrow Cr$	−0.557
$Ga^{+3} + 3e^- \rightarrow Ga$	−0.560
$Fe(OH)_3 + e^- \rightarrow Fe(OH)_2 + OH^-$	−0.56
$PbO + H_2O + 2e^- \rightarrow Pb + 2OH^-$	−0.576
$2SO_3^{-2} + 3H_2O + 4e^- \rightarrow S_2O_3^{-2} + 6OH^-$	−0.58
$SbO_2^- + H_2O + 2e^- \rightarrow SbO_2^- + 2OH^-$	−0.59
$TlBr + e^- \rightarrow Tl(Hg) + Br^-$	−0.606
$U^{+4} + e^- \rightarrow U^{+3}$	−0.61
$Cb_2O_5 + 10H^+ + 10e^- \rightarrow 2Cb + 5H_2O$	−0.62
$U^{+4} + e^- \rightarrow U^{+3}$(1 fHCLO_4)	−0.631
$Ni(OH)_2 + 2e^- \rightarrow Ni + 2OH^-$	−0.66
$SbO_2^- + 2H_2O + 3e^- \rightarrow Sb + 4OH^-$	−0.66
$AsO_2^- + 2H_2O + 3e^- \rightarrow As + 4OH^-$	−0.68
$*Te + 2H^+ + 2e^- \rightarrow H_2Te(Ag)$	−0.69
	(−0.72)
$Ag_2S + 2e^- \rightarrow 2Ag + S^{-2}$	−0.7051
$Ta_2O_5 + 10H^+ + 10e^- \rightarrow 2Ta + 5H_2O$	−0.71
$AsO_4^{-3} + 2H_2O + 2e^- \rightarrow AsO_2^- + 4OH^-$	−0.71
$Co(OH)_2 + 2e^- \rightarrow Co + 2OH^-$	−0.73
$H_3BO_3 + 3H^+ + 3e^- \rightarrow B + 3H_2O$	−0.73
$Cr^{+3} + 3e^- \rightarrow Cr$	−0.74
$V^{+5} + e^- \rightarrow V^{+4}$(If NaOH)	−0.74
$Cd(OH)_2 + 2e^- \rightarrow Cd(Hg) + 2OH^-$	−0.761
	(−0.81)
$Zn^{+2} + 2e^- \rightarrow Zn$	−0.7628
$Zn^{+2} + 2e^- \rightarrow Zn(Hg)$	−0.7628
$TlI + e^- \rightarrow Tl(Hg) + I^-$	−0.769
$Se + 2e^- \rightarrow Se^{-2}$	−0.78
$HSnO_2^- + H_2O + 2e^- \rightarrow Sn + 3OH^-$	−0.79
$ZnSO_4 \cdot 7H_2O + 2e^- \rightarrow Zn(Hg) + SO_4^{-2}$ (Sat'd ZnSO_4)	−0.7993
$RuO_2 + 4H^+ + 4e^- \rightarrow Ru + 2H_2O$	−0.8
$ReO_4^- + 4H_2O + 7e^- \rightarrow Re + 8OH^-$	−0.81
$N_2 + 2H_2O + 4H^+ + 2e^- \rightarrow 2NH_3OH^+$	−1.87
$Th^{+4} + 4e^- \rightarrow Th$	−1.90
$Np^{+3} + 3e^- \rightarrow Np$	−1.9
$Ti^{+3} + e^- \rightarrow Ti^{+2}$	−2.0
$Sc^{+3} + 3e^- \rightarrow Sc$	−2.08
$1·2H_2 + e^- \rightarrow H^-$	−2.23
$Nd^{+3} + 3e^- \rightarrow Nd$	−2.246
$Be_2O_3^{-2} + 3H_2O + 4e^- \rightarrow 2Be + 6OH^-$	−2.28
$ZrO(OH)_2 + H_2O + 4e^- \rightarrow Zr + 4OH^-$	−2.32
$Ce^{+3} + 3e^- \rightarrow Ce$	−2.335
$H_2AlO_3^- + H_2O + 3e^- \rightarrow Al + 4OH^-$	−2.35
$Y^{+3} + 3e^- \rightarrow Y$	−2.37
$La^{+3} + 3e^- \rightarrow La$	−2.37
$Mg^{+2} + 2e^- \rightarrow Mg$	−2.375

Table 2–26C (continued)
ELECTROCHEMICAL SERIES

Negative Reduction Reactions (continued)

Reaction	Potential, volts[a]	Reaction	Potential, volts[a]
$V^{+2} + 2e^- \to V$	-1.2	$H_2BO_3^- + H_2O + 3e \to B + 4OH^-$	-2.5
$ZnO_2^{-2} + 2H_2O + 2e^- \rightleftharpoons Zn + 4OH^-$	-1.216	$HfO(OH)_2 + H_2O + 4e \to Hf + 4OH^-$	-2.60
$H_2GaO_3^- + H_2O + 3e^- \to Ga + 4OH^-$	-1.22	$ThO_2 + 2H_2O + 4e \to Th + 4OH^-$	-2.64
$H_2BO_3^- + 5H_2O + 8e^- \to BH_4^- + 8OH^-$	-1.24	$Mg(OH)_2 + 2e^- \to Mg + 2OH^-$	-2.67
$Cr(OH)_3 + 3e^- \to Cr + 3OH^-$	-1.3	$Na^+ + e^- \to Na$	-2.7109
$ZrO_2 + 4H^+ + 4e^- \to Zr + 2H_2O$	-1.43		(-2.712)
$Ce^{+3} + 3e^- \to Ce(Hg)$	-1.4373	$Ca^{+2} + 2e^- \to Ca$	-2.76
$Mn(OH)_2 \to Mn + 2OH^-$	-1.47	$La(OH)_3 + 3e^- \to La + 3OH^-$	-2.76
$Ba^{+2} + 2e^- \to Ba(Hg)$	-1.570	$Sr^{+2} + 2e^- \to Sr$	-2.89
$HfO_2 + 4H^+ + 4e^- \to Hf + 2H_2O$	-1.57	$Ba^{+2} + 2e^- \to Ba$	-2.90
$Ti^{+2} + 2e^- \to Ti$	-1.63	$Cs^+ + e^- \to Cs$	-2.923
$HPO_3^{-2} + 2H_2O + 2e^- \to H_2PO_2^- + 3OH^-$	-1.65	$K^+ + e^- \to K$	-2.924
$HfO^{+2} + 2H^+ + 4e^- \to Hf + H_2O$	-1.68		(-2.923)
$Be^{+2} + 2e^- \to Be$	-1.70	$Rb^+ + e^- \to Rb$	-2.925
	(-1.85)		(-2.99)
$Al^{+3} + 3e^- \to Al(0.1f\ NaOH)$	-1.706	$Ba(OH)_2 \cdot 8H_2O + 2e^- \to Ba + 2OH^- + 8H_2O$	-2.97
$HPO_3^{-2} + 2H_2O + 3e^- \to P + 5OH^-$	-1.71	$Sr(OH)_2 \cdot 8H_2O + 2e^- \to Sr + 2OH^- + 8H_2O$	-2.99
$SiO_3^{-2} + 3H_2O + 4e^- \to Si + 6OH^-$	-1.73	$Ca(OH)_2 + 2e^- \to Ca + 2OH^-$	-3.02
$Sr^{+2} + 2e^- \to Sr(Hg)$	-1.793	$Ca^+ + e^- \to Ca$	-3.02
$ThO_2 + 4H^+ + 4e^- \to Th + 2H_2O$	-1.80	$Li^+ + e^- \to Li$	-3.045
$U^{+3} + 3e^- \to U$	-1.8		(-3.02)
$H_2PO_2^- + e^- \to P + 2OH^-$	-1.82	$3N_2 + 2H^+ + 2e^- \to 2HN_3$	-3.1

[a]Values are standard reduction potentials.

From Hunsberger, J. F., in *Handbook of Chemistry and Physics,* 55th ed., Weast, R. C., Ed., CRC Press, Cleveland, 1974, D-124.

Table 2-27
SPECIFIC HEAT OF THE ELEMENTS AT 25°C

$$C_p = cal\ g^{-1}\ °K^{-1}$$

Element	Reference 1	Reference 2	Reference 3
Aluminum	0.215	0.215	0.2154
Antimony	0.049	0.0495	0.0501
Argon	0.124		0.124
Arsenic	0.0785		0.0796
Barium	0.046	0.0362	0.0458
Beryllium	0.436	0.436	0.4733
Bismuth	0.0296	0.0238	0.0292
Boron	0.245		0.2463
Bromine (Br₂)	0.113	0.0537	
Cadmium	0.0555	0.0552	0.0554
Calcium	0.156	0.155	0.1566
Carbon, diamond	0.124		0.120
Carbon, graphite	0.170		0.172
Cerium	0.049	0.0459	0.0442
Cesium	0.057	0.0575	0.0558
Chlorine (Cl₂)	0.114		0.114
Chromium		0.107	0.1073
Cobalt	0.109	0.107	0.1037
Columbium (see Niobium)			
Copper	0.092	0.0924	0.0920
Dysprosium	0.0414	0.0414	
Erbium	0.0401	0.0401	
Europium	0.0421	0.0326	
Fluorine (F₂)	0.197	0.197	
Gadolinium	0.055	0.056	
Gallium	0.089	0.088	0.0911
Germanium	0.077		
Gold	0.0308	0.0308	0.0305
Hafnium	0.035	0.028	0.0344
Helium	1.24		1.242
Holmium	0.0393	0.0394	
Hydrogen (H₂)	3.41		3.42
Indium	0.056	0.0556	0.0570
Iodine (I₂)	0.102		0.034
Iridium	0.0317	0.0312	0.0305
Iron (α)	0.106	0.1075	0.1078
Krypton	0.059		0.059
Lanthanum	0.047	0.0479	0.0475
Lead	0.038	0.0305	0.0308
Lithium	0.85	0.834	0.814
Lutetium	0.037	0.0285	
Magnesium	0.243	0.245	0.235
Manganese, α	0.114	0.114	0.1147
Manganese, β	0.119		0.1120
Mercury	0.0331	0.0333	0.0331
Molybdenum	0.0599	0.0597	0.0584
Neodymium	0.049	0.0453	0.0499
Neon	0.246		0.246
Nickel	0.106	0.1061	0.1057
Niobium	0.064	0.0633	
Nitrogen (N₂)	0.249		0.249
Osmium	0.03127	0.0310	0.0310
Oxygen (O₂)	0.219	0.219	
Palladium	0.0584	0.0583	0.0590

Table 2–27 (continued)
SPECIFIC HEAT OF THE ELEMENTS AT 25°C

Element	Reference 1	Reference 2	Reference 3
Phosphorus, white	0.181		0.178
Phosphorus, red, triclinic	0.160		
Platinum	0.0317	0.0317	0.0325
Polonium	0.030		
Potassium	0.180	0.180	0.1787
Praseodymium	0.046	(0.0467)	0.0482
Promethium	0.0442		
Protactinium	0.029		
Radium	0.0288		
Radon	0.0224		0.0224
Rhenium	0.0329	0.0330	0.0327
Rhodium	0.0583	0.0580	0.0592
Rubidium	0.0861	0.0860	0.0850
Ruthenium	0.057	0.0569	
Samarium	0.043	0.0469	
Scandium	0.133	0.1173	
Selenium (Se$_2$)	0.0767		0.0535
Silicon	0.168		0.169
Silver	0.0566	0.0562	0.0564
Sodium	0.293	0.292	0.2952
Strontium	0.0719	0.0719	0.0684
Sulfur, yellow	0.175		0.177
Tantalum	0.0334	0.0334	0.0335
Technetium	0.058		
Tellurium	0.0481		0.482
Terbium	0.0437	0.0435	
Thallium	0.0307	0.0307	0.0310
Thorium	0.0271	0.0281	0.0331
Thulium	0.0382		
Tin (α)	0.0510	0.0519	0.0518
Tin (β)	0.0530	0.0543	0.0530
Titanium	0.125	0.1248	0.1231
Tungsten	0.0317	0.0322	0.0324
Uranium	0.0276	0.0278	0.0276
Vanadium	0.116	0.116	0.1147
Xenon	0.0378		0.0379
Ytterbium	0.0346	0.0287	
Yttrium	0.068	0.0713	
Zinc	0.0928	0.0922	0.0916
Zirconium	0.0671	0.0660	

From Weast, R. C., Ed., *Handbook of Chemistry and Physics,* 55th ed., CRC Press, Cleveland, 1974, D-144.

REFERENCES

1. **Kelly, K. K.,** Bulletin 592, Bureau of Mines, Washington, D. C., 1961.
2. **Hultgren, R., Orr, R. L., Anderson, P. D., and Kelly, K. K.,** *Selected Values of Thermodynamic Properties of Metals and Alloys,* John Wiley & Sons, New York, 1963.
3. NBS Circular 500, National Bureau of Standards, Washington, D.C., 1952.

Table 2−28

SPECIFIC HEAT AND ENTHALPY OF SOME METALS AT LOW TEMPERATURES

$J/g \times 453.6 = J/lb$; $Jg \times 0.239 = cal/g$; $J/g \times 0.4299 = Btu/lb$

T (K)	Aluminum		Beryllium		Bismuth		Cadmium		Chromium	
	C_p (1/Jg × 1/K)	$H - H_0$ (1/Jg)	C_p (1/Jg × 1/K)	$H - H_0$ (1/Jg)	C_p (1/Jg × 1/K)	$H - H_0$ (1/Jg)	C_p (1/Jg × 1/K)	$H - H_0$ (1/Jg)	C_p (1/Jg × 1/K)	$H - H_0$ (1/Jg)
1	0.000 10[a]	0.000 025								
1	0.000 051		0.000 025	0.000 013	0.000 005 98	0.000 001 58	0.000 008	0.000 003	0.000 028 5	0.000 014 2
2	0.000 108	0.000 105	0.000 051	0.000 051	0.000 046 1	0.000 023 3	0.000 033	0.000 022	0.000 058	0.000 057 3
2										
2										
3	0.000 176	0.000 246	0.000 079	0.000 116	0.000 170	0.000 123	0.000 090	0.000 082	0.000 089	0.000 131
3										
3.40[b]										
3.40										
3.72[c]										
3.72										
4	0.000 261	0.000 463	0.000 109	0.000 209	0.000 493	0.000 432	0.000 21	0.000 22	0.000 124	0.000 237
4										
4.39[d]										
4.39										
5										
5										
6	0.000 50	0.001 21	0.000 180	0.000 496	0.002 14	0.002 88	0.001 30	0.001 5	0.000 206	0.000 567
7										
7										
8	0.000 88	0.002 6	0.000 271	0.000 944	0.005 47	0.010 2	0.004 3	0.007 0	0.000 312	0.001 07
9										
10	0.001 4	0.004 9	0.000 389	0.001 60	0.015 9	0.025 9	0.008 0	0.019	0.000 451	0.001 82
12										
14										
15	0.004 0	0.018	0.000 842	0.004 57	0.023 8	0.111	0.025	0.102	0.001 02	0.005 28
16										
18										
20	0.008 9	0.048	0.001 61	0.010 5	0.036 3	0.262	0.046	0.28	0.002 20	0.012 8
25	0.017 5	0.112	0.002 79	0.021 2	0.047 7	0.472	0.066	0.56	0.003 92	0.027 4
30	0.031 5	0.232	0.004 50	0.039 2	0.057 2	0.734	0.086	0.94	0.006 83	0.053 2
35	0.051 5	0.436								
40	0.077 5	0.755	0.009 96	0.109	0.072 7	1.38	0.117	1.96	0.017 1	0.163
50	0.142 5	1.85	0.019 2	0.253	0.084 6	2.17	0.141	3.26	0.035 8	0.421
60	0.214	3.64	0.034 1	0.523	0.093 5	3.06	0.159	4.76	0.062 1	0.904
70	0.287	6.15	0.056 2	0.971	0.100	4.03	0.172	6.43	0.092	1.68
80	0.357	9.37	0.090 6	1.69	0.105	5.05	0.182	8.20	0.127	2.77
90	0.422	13.25	0.139	2.82	0.108	6.12	0.190	10.1	0.161	4.21
100	0.481	17.76	0.199	4.51	0.111	7.21	0.196	12.0	0.193	5.98

Table 2–28 (continued)
SPECIFIC HEAT AND ENTHALPY OF SOME METALS AT LOW TEMPERATURES

T (K)	Copper		Germanium[e]		Gold		Indium		α-Iron[f]	
	C_p (J/Jg × 1/K)	$H - H_o$ (J/Jg)	C_p (J/Jg × 1/K)	$H - H_o$ (J/Jg)	C_p (J/Jg × 1/K)	$H - H_o$ (J/Jg)	C_p (J/Jg × 1/K)	$H - H_o$ (J/Jg)	C_p (J/Jg × 1/K)	$H - H_o$ (J/Jg)
1	0.000 012	0.000 006	0.000 000 528	0.000 000 132	0.000 006	0.000 002	0.000 029	0.000 011	0.000 090	0.000 045
1							0.000 019[a]	0.000 006[a]		
2	0.000 028	0.000 025	0.000 004 23	0.000 002 11	0.000 025	0.000 016	0.000 138	0.000 085	0.000 183	0.000 181
2							0.000 141[a]	0.000 073[a]		
2										
3	0.000 053	0.000 064	0.000 014 4	0.000 010 7	0.000 070	0.000 061	0.000 410	0.000 341	0.000 279	0.000 412
3							0.000 464[a]	0.000 357[a]		
3.40							0.000 584	0.000 537		
3.72							0.000 669[a]	0.000 581[a]		
3.40[b]										
3.72										
4	0.000 091	0.000 13	0.000 034 4	0.000 034 3	0.000 16	0.000 17	0.000 95	0.000 99	0.000 382	0.000 742
4										
4.39[d]										
4.39										
5										
5										
6	0.000 23	0.000 44	0.000 125	0.000 179	0.000 ,50	0.000 78	0.003 59	0.005 20	0.000 615	0.001 73
6										
7										
7										
8	0.000 47	0.001 12	0.000 335	0.000 612	0.001 2	0.002 4	0.008 55	0.017 0	0.000 90	0.003 23
9										
10	0.000 86	0.002 4	0.000 813	0.001 69	0.002 2	0.002 2	0.015 5	0.040 8	0.001 24	0.005 37
12										
14										
15	0.002 7	0.010 7	0.004 45	0.013 6	0.007 4	0.028	0.036 7	0.170	0.002 49	0.014 5
16										
18										
20	0.007 7	0.034	0.012 5	0.054 0	0.015 9	0.086	0.060 8	0.413	0.004 5	0.031 6
25	0.016	0.090	0.024 0	0.145	0.026 3	0.191	0.085 7	0.778	0.007 5	0.061
30	0.027	0.195	0.036 6	0.296	0.037 1	0.349	0.108	1.265	0.012 4	0.110
35										
40	0.060	0.61	0.061 7	0.786	0.056 2	0.821	0.141	2.52	0.029	0.31
50	0.099	1.40	0.085 8	1.52	0.072 6	1.47	0.162	4.04	0.055	0.73
60	0.137	2.58	0.108	2.50	0.084 2	2.25	0.176	5.73	0.087	1.43
70	0.173	4.13	0.131	3.70	0.092 8	3.14	0.186	7.53	0.121	2.46
80	0.205	6.02	0.153	5.12	0.099 2	4.10	0.193	9.42	0.154	3.84
90	0.232	8.22	0.173	6.74	0.104 3	5.12	0.198	11.38	0.186	5.55
100	0.254	10.6	0.191	8.55	0.108 3	6.18	0.203	13.39	0.216	7.56

Table 2–28 (continued)
SPECIFIC HEAT AND ENTHALPY OF SOME METALS AT LOW TEMPERATURES

T (K)	γ-Iron[g]		Lead		Molybdenum		Nickel		Palladium	
	C_p (1/Jg × 1/K)	$H-H_0$ (1/Jg)	C_p (1/Jg × 1/K)	$H-H_0$ (1/Jg)	C_p (1/Jg × 1/K)	$H-H_0$ (1/Jg)	C_p (1/Jg × 1/K)	$H-H_0$ (1/Jg)	C_p (1/Jg × 1/K)	$H-H_0$ (1/Jg)
1			0.000 026	0.000 010	0.000 022 9	0.000 010 5	0.000 120	0.000 060	0.000 099	0.000 049 3
1			0.000 012[a]	0.000 003[a]						
2			0.000 12	0.000 07	0.000 047 2	0.000 044 5	0.000 242	0.000 241	0.000 203	0.000 200
2			0.000 09[a]	0.000 05[a]						
3			0.000 31[a]	0.000 28	0.000 074 5	0.000 105	0.000 369	0.000 546	0.000 318	0.000 459
3			0.000 33	0.000 23[a]						
3.40[b]										
3.40										
3.72[c]										
3.72										
4			0.000 7	0.000 8						
4			0.000 7[a]	0.000 7[a]						
4.39[d]					0.000 106	0.000 194	**0.000 503**	**0.000 98**	0.000 447	0.000 840
4.39										
5			0.001 5	0.001 8						
5			0.001 5[a]	0.001 8[a]						
6			0.002 9	0.003 9	0.000 191	0.000 484	0.000 82	0.002 28	0.000 891	0.002 31
6			0.003 0[a]	0.004 0[a]						
7			0.004 8	0.008						
7			0.005 0[a]	0.008[a]						
8			0.007 3	0.014	0.000 317	0.000 981	0.001 19	0.004 28	0.001 41	0.004 60
9										
10			0.013 7	0.034	0.000 498	0.001 78	0.001 62	0.007 1	0.002 10	0.008 07
12										
14										
15			0.33 5	0.150	0.001 31	0.006 10	0.003 1	0.018 5	0.004 71	0.024 5
16										
18										
20	0.007	0	0.053 1	0.358	0.002 87	0.016 1	0.005 8	0.041	0.009 22	0.058 6
25			0.068 1	0.672	0.005 77	0.037 5	0.010 1	0.079	0.016 0	0.120
30	0.016	0.11	0.079 6	1.042	0.009 60	0.072 9	0.016 7	0.145	0.025 8	0.223
35										
40	0.041	0.39	0.094 4	1.920	0.023 6	0.232	0.038 1	0.413	0.050 7	0.600
50	0.090	1.0_2	0.103	2.91	0.041 0	0.554	0.068 2	0.937	0.077 7	1.24
60	0.13_7	2.1_6	0.108	3.97	0.061 9	1.07	0.103	1.79	0.101	2.14
70	0.18_0	3.7_5	0.112	5.07	0.083 8	1.80	0.139	3.00	0.122	3.26
80	0.21_8	5.7_4	0.114	6.20	0.104	2.74	0.173	4.56	0.139	4.56
90	0.25_5	8.1_1	0.116	7.35	0.123	3.88	0.204	6.45	0.154	6.03
100	0.28_8	$10._3$	0.118	8.53	0.193	5.20	0.232	8.63	0.167	7.63

Table 2–28 (continued)
SPECIFIC HEAT AND ENTHALPY OF SOME METALS AT LOW TEMPERATURES

T (K)	Platinum		Rhodium		Silicon[h]		Silver		Sodium[i]	
	C_p (1/Jg × 1/K)	H − H$_0$ (1/Jg)	C_p (1/Jg × 1/K)	H − H$_0$ (1/Jg)	C_p (1/Jg × 1/K)	H − H$_0$ (1/Jg)	C_p (1/Jg × 1/K)	H − H$_0$ (1/Jg)	C_p (1/Jg × 1/K)	H − H$_0$ (1/Jg)
1	0.000 035	0.000 017 5	0.000 048	0.000 024	0.000 000 263	0.000 000 0658	0.000 007 2	0.000 003 2	0.000 081	0.000 035
2	0.000 074	0.000 071	0.000 097	0.000 096	0.000 002 10	0.000 001 05	0.000 023 9	0.000 017 6	0.000 289	0.000 204
3	0.000 122	0.000 168	0.000 147	0.000 218	0.000 007 09	0.000 005 32	0.000 059 5	0.000 057 4	0.000 76	0.000 70
3.40[h]										
3.72[c]										
4	0.000 186	0.000 320	0.000 201	0.000 392	0.000 016 8	0.000 016 8	0.000 124	0.000 146	0.001 60	0.001 84
4.39[d]										
5										
6	0.000 37	0.000 85	0.000 32	0.000 91	0.000 059 6	0.000 085 3	0.000 39	0.000 62	0.005 1	0.008 1
7										
8	0.000 67	0.001 88	0.000 47	0.000 70	0.000 140	0.000 279	0.000 91	0.001 87	0.012 2	0.024 7
9										
10	0.001 12	0.003 65	0.000 65	0.002 81	0.000 275	0.000 679	0.001 8	0.004 52	0.023 8	0.060 2
12										
14										
15	0.003 3	0.013 5	0.001 35	0.007 65	0.001 09	0.003 74	0.006 4	0.023 3	0.093	0.380
16										
18										
20	0.007 4	0.039 5	0.002 71	0.017 4	0.003 37	0.013 8	0.015 5	0.076	0.124	0.597
25	0.013 7	0.092	0.005 61	0.037 3	0.008 49	0.042 3	0.028 7	0.185	0.155	0.875
30	0.021 2	0.182	0.010 6	0.077 1	0.017 1	0.105	0.044 2	0.368	0.259	1.90
35									0.364	3.45
40	0.038	0.48	0.026 6	0.256	0.044 0	0.400	0.078	0.979	0.544	8.03
50	0.055	0.95	0.048 9	0.633	0.078 5	1.00	0.108	1.91	0.695	14.2
60	0.068	1.56	0.072 4	1.238	0.115	1.97	0.133	3.12	0.793	21.7
70	0.079	2.29	0.094	2.07	0.152	3.31	0.151	4.54	0.86	30.0
80	0.088	3.12	0.114	3.11	0.188	5.01	0.166	6.13	0.91	38.9
90	0.094	4.02	0.132	4.34	0.224	7.06	0.177	7.85	0.95	48.2
100	0.100	5.01	0.147	5.74	0.259	9.47	0.187	9.67	0.98	57.9

Table 2–28 (continued)
SPECIFIC HEAT AND ENTHALPY OF SOME METALS AT LOW TEMPERATURES

T (K)	Tantalum C_p (1/Jg × 1/K)	Tantalum $H - H_0$ (1/Jg)	Tin (white) C_p (1/Jg × 1/K)	Tin (white) $H - H_0$ (1/Jg)	Titanium C_p (1/Jg × 1/K)	Titanium $H - H_0$ (1/Jg)	Tungsten C_p (1/Jg × 1/K)	Tungsten $H - H_0$ (1/Jg)	Zinc C_p (1/Jg × 1/K)	Zinc $H - H_0$ (1/Jg)
1	0.000 032	0.000 016	0.000 017 0	0.000 007 9	0.000 071	0.000 035	0.000 007 4	0.000 003 7	0.000 011	0.000 005
1	0.000 006 3[a]	0.000 002 1[a]	0.000 004 1[a]	0.000 000 9[a]						
2	0.000 068	0.000 065	0.000 047	0.000 038 3	0.000 146	0.000 143	0.000 015 8	0.000 015 2	0.000 028	0.000 023
2	0.000 054[a]	0.000 026[a]	0.000 048[a]	0.000 022 8[a]						
2										
3	0.000 112	0.000 155	0.000 109	0.000 113	0.000 226	0.000 329	0.000 026 2	0.000 036 0	0.000 058	0.000 065
3	0.000 178[a]	0.000 138[a]	0.000 151[a]	0.000 116[a]						
3.40[b]										
3.40										
3.72[c]			0.000 198	0.000 221						
3.72			0.000 285[a]	0.000 270[a]						
4	0.000 171	0.000 295	0.000 245	0.000 283	0.000 317	0.000 599	0.000 039 3	0.000 068 5	0.000 11	0.000 14
4	0.000 352[a]	0.000 400[a]								
4.39[d]	0.000 201	0.000 368								
4.39	0.000 433[a]	0.000 553[a]								
5			0.000 54	0.000 65						
5										
6	0.000 333	0.000 776	0.001 27	0.001 51	0.000 54	0.001 45	0.000 078 3	0.000 182	0.000 29	0.000 53
6										
7										
7										
8	0.000 648	0.001 73	0.004 2	0.006 8	0.000 84	0.002 81	0.000 141	0.000 396	0.000 96	0.001 6
9										
10	0.001 17	0.003 52	0.008 1	0.019 0	0.001 26	0.004 89	0.000 234	0.000 765	0.002 5	0.005 0
12										
14										
15	0.003 60	0.014 5	0.022 6	0.093	0.003 3	0.015 6	0.000 725	0.002 97	0.011	0.034
16										
18										
20	0.008 23	0.032 2	0.040	0.251	0.007 0	0.040	0.001 89	0.009 27	0.026	0.125
25	0.015 3	0.102	0.058	0.498	0.013 4	0.090	0.004 21	0.023 7	0.049	0.31
30	0.024 0	0.202	0.076	0.834	0.024 5	0.182	0.007 83	0.053 4	0.076	0.62
35										
40	0.043 0	0.540	0.106	1.75	0.057 1	0.581	0.018 4	0.181	0.125	1.62
50	0.060 4	1.06	0.130	2.93	0.099 2	1.358	0.033 2	0.436	0.171	3.11
60	0.075 4	1.74	0.148	4.33	0.146 7	2.592	0.048 3	0.843	0.208	5.01
70	0.087 9	2.56	0.162	5.88	0.189	4.27	0.060 5	1.39	0.236	7.23
80	0.097 6	3.49	0.173	7.55	0.230	6.37	0.071 5	2.05	0.258	9.70
90	0.105	4.50	0.182	9.33	0.267	8.86	0.081 0	2.81	0.277	12.38
100	0.111	5.58	0.189	11.18	0.300	11.69	0.088 8	3.66	0.293	15.24

Table 2–28 (continued)
SPECIFIC HEAT AND ENTHALPY OF SOME METALS AT LOW TEMPERATURES

[a] Superconducting.

[b] Superconducting transition temperature of indium.

[c] Superconducting transition temperature of tin.

[d] Superconducting transition temperature of tantalum.

[e] In germanium the electronic specific heat, γT, is markedly dependent on impurities. The values given are for pure germanium (negligible electronic specific heat).

[f] α-Iron is the form that is thermodynamically stable at low temperatures. It has the body-centered cubic lattice that is the basis of the ferritic steels.

[g] γ-Iron is stable between 910 and 1,400°C. It has the face-centered cubic structure that is the basis of the austenitic steels. Since pure γ-iron is not stable at low temperatures, the above values are calculated by application of the Kopp-Neumann rule to experimental data on two austenitic Fe-Mn alloys and are of uncertain accuracy.

[h] In silicon the electronic specific heat, γT, is markedly dependent on impurities. Values of the coefficient, γ, from 0 to 2.4×10^{-1} 1/Jg $\times$ 1/K² have been reported. The values in the above table are for pure silicon ($\gamma = 0$).

[i] It has been shown (Barrett, 1956; Hull and Rosenberg, 1959) that sodium partially transforms at low temperatures from the normal body-centered cubic structure to close-packed hexagonal. The transformation is of the martensitic type and is promoted by cold-working at the low temperature. Inasmuch as none of the calorimetric measurements on sodium were accompanied by crystallographic analysis, the tabulated data below 100K are to some degree ambiguous.

From Corruccini, R. J. and Gniewek, J. J., in *Handbook of Chemistry and Physics*, 55th ed., CRC Press, Cleveland, 1974, D-146.

Table 2–29
CONSTANTS OF THE DEBYE-SOMMERFELD EQUATION

$C_v = \gamma T + \alpha T^3$; $\alpha = 12\,\pi^4 R/5\theta_0{}^3$; $0 \lessgtr T \lessgtr T_{max}$; T_{max} = maximum temperature to which the equation can be used with limiting value of θ.

Substance	$10^6\,\gamma$ $(1/Jg \times 1/K^2)$	γ $(1/mJg\text{-atom} \times 1/K^2)$	$10^6\,\alpha$ $(1/Jg \times 1/K^4)$	θ_0 (K)	T_{max} (K)
		Metals			
Aluminum	50.4	1.36	0.93	426	4
Beryllium	25	0.226	0.138	1,160	20
Bismuth	0.32	0.067	5.66	118	2
Cadmium	5.6	0.63	2.69	186	3
Chromium	28.3	1.47	0.165	610	4
Copper	10.81	0.687	0.746	344.5	10
Germanium	Superconducting	Superconducting	0.528	370	2
Gold	3.75	0.74	2.19	165	15
Indium	15.8	1.81	13.1	109	2
α-Iron	90	5.0	0.349	464	10
Lead	15.1	3.1	10.6	96	4
Magnesium	54	1.32	1.19	406	4
α-Manganese	251	13.8	0.328	476	12
Molybdenum	23	2.18	0.238	440	4
Nickel	120	7.0	0.39	440	4
Niobium	85	7.9	0.64	320	1
Palladium	98	10.5	0.89	274	4
Platinum	34.1	6.7	0.72	240	3
Rhodium	48	4.9	0.173	478	4
Silicon	Superconducting	Superconducting	0.263	640	4
Silver	5.65	0.610	1.58	225	4
Sodium[a]	60	1.37	21.4	158	4
Tantalum	31.5	5.7	0.69	250	4
Tin (white)	14.7	1.75	2.21	195	2
Titanium	71	3.4	0.54	420	10
Tungsten	7	1.3	0.16	405	4
Zinc	9.5	0.63	1.10	300	4
		Alloys			
Constantan[a]	113	6.9	0.56	384	15
Monel[a]	108	6.5	0.62	374	20
		Other Inorganic Substances			
Diamond			0.0152	2,200	50
Ice			15.2	192	10
Pyrex			3.14		5
		Organic Substances			
Glyptal			27		4
Lucite			35		4
Polystyrene			63		4

[a]Superconducting.

From Weast, R. C., Ed., *Handbook of Chemistry and Physics,* 55th ed., CRC Press, Cleveland, 1974, D-148.

Table 2–30
MELTING POINTS OF MIXTURES OF METALS

Melting points, °C

Proportion of second metal

Metals	0%[a]	10%	20%	30%	40%	50%	60%	70%	80%	90%	100%[a]
Pb + Sn	326	295	276	262	240	220	190	185	200	216	232
Pb + Bi	322	290			179	145	126	168	205		268
Pb + Te	322	710	790	880	917	760	600	480	410	425	446
Pb + Ag	328	460	545	590	620	650	705	775	840	905	959
Pb + Na		360	420	400	370	330	290	250	200	130	96
Pb + Cu	326	870	920	925	945	950	955	985	1005	1020	1084
Pb + Sb	326	250	275	330	395	440	490	525	560	600	632
Al + Sb	650	750	840	925	945	950	970	1000	1040	1010	632
Al + Cu	650	630	600	560	540	580	610	755	930	1055	1084
Al + Au	655	675	740	800	855	915	970	1025	1055	675	1062
Al + Ag	650	625	615	600	590	580	575	570	650	750	954
Al + Zn	654	640	620	600	580	560	530	510	475	425	419
Al + Fe	653	860	1015	1110	1145	1145	1220	1315	1425	1500	1515
Al + Sn	650	645	635	625	620	605	590	570	560	540	232
Sb + Bi	632	610	590	575	555	540	520	470	405	330	268
Sb + Ag	630	595	570	545	520	500	505	545	680	850	959
Sb + Sn	622	600	570	525	480	430	395	350	310	255	232
Sb + Zn	632	555	510	540	570	565	540	525	510	470	419
Ni + Sn	1455	1380	1290	1200	1235	1290	1305	1230	1060	800	232
Na + Bi	96	425	520	590	645	690	720	730	715	570	268
Na + Cd	96	125	185	245	285	325	330	340	360	390	322
Na + Hg	96.5	90	80	70	60	45	22	55	95	215	
Cd + Ag	322	420	520	610	700	760	805	850	895	940	954
Cd + Tl	321	300	285	270	262	258	245	230	210	235	302
Ca + Zn	322	280	270	295	313	327	340	355	370	390	419
Au + Cu	1063	910	890	895	905	925	975	1000	1025	1060	1084
Au + Ag	1064	1062	1061	1058	1054	1049	1039	1025	1006	982	963
Au + Pt	1075	1125	1190	1250	1320	1380	1455	1530	1610	1685	1775
K + Na	62	17.5	−10	−3.5	5	11	26	41	58	77	97.5
K + Hg						90	110	135	162	265	
K + Tl	62.5	133	165	188	205	215	220	240	280	305	301
Cu + Ni	1080	1180	1240	1290	1320	1355	1380	1410	1430	1440	1455
Cu + Ag	1082	1035	990	945	910	870	830	788	814	875	960
Cu + Sn	1084	1005	890	755	725	680	630	580	530	440	232
Cu + Zn	1084	1040	995	930	900	880	820	780	700	580	419
Ag + Zn	959	850	755	705	690	660	630	610	570	505	419
Ag + Sn	959	870	750	630	550	495	450	420	375	300	232

[a]The data in this table were compiled from various sources; this accounts for the variations in the melting points of the pure metals.

From Weast, R. C., Ed., *Handbook of Chemistry and Physics,* 55th ed., CRC Press, Cleveland, 1974, D-149.

Table 2–31
MELTING AND BOILING POINTS OF THE ELEMENTS

Element	Melting point (°C)	Boiling point (°C)
Actinium	1050	3200 ± 300
Aluminum	660.37	2467
Americium	994 ± 4	2607
Antimony	630.74	1750
Argon	−189.2	−185.7
Arsenic, gray	817 (28 atm)	613 (sublimes)
Astatine	302	337
Barium	725	1640
Berkelium	–	–
Beryllium	1278 ± 5	2970 (5 mm)
Bismuth	271.3	1560 ± 5
Boron	2300	2550 (sublimes)
Bromine	−7.2	58.78
Cadmium	320.9	765
Calcium	839 ± 2	1484
Californium	–	
Carbon	~3550	4827
Cerium	799 ± 3	3426
Cesium	28.40 ± 0.01	678.4
Chlorine	−100.98	−34.6
Chromium	1857 ± 20	2672
Cobalt	1495	2870
Columbium (see Niobium)		
Copper	1083.4 ± 0.2	2567
Curium	1340 ± 40	–
Deuterium (see Hydrogen)		
Dysprosium	1412	2562
Einsteinium	–	–
Element 104	–	–
Element 105	–	–
Erbium	1529	2863
Europium	822	1597
Fermium	–	–
Fluorine	−219.62	−188.14
Francium	(27)	(677)
Gadolinium	1313 ± 1	3266
Gallium	29.78	2403
Germanium	937.4	28.30
Gold	1064.43	2807
Hafnium	2227 ± 20	4602
Helium	−272.2	−268.934
Holmium	1474 (26 atm)	2695
Hydrogen	−259.14	−252.87
Indium	156.61	2080
Iodine	113.5	184.35
Iridium	2410	4130
Iron	1535	2750
Krypton	−156.6	−152.30 ± 0.10
Lanthanum	921 ± 5	3457
Lawrencium	–	–
Lead	327.502	1740
Lithium	180.54	1347
Lutetium	1663 ± 5	3395
Magnesium	648.8 ± 0.5	1090

Table 2–31 (continued)
MELTING AND BOILING POINTS OF THE ELEMENTS

Element	Melting point (°C)	Boiling point (°C)
Manganese	1244 ± 3	1962
Mendelevium	–	–
Mercury	–38.87	356.58
Molybdenum	2617	4612
Neodymium	1021	3068
Neon	–248.67	–246.048
Neptunium	640 ± 1	3902
Nickel	1453	2732
Niobium	2468 ± 10	4742
Nitrogen	–209.86	–195.8
Nobelium	–	–
Osmium	3045 ± 30	5027 ± 10ʋ
Oxygen	–218.4	–182.962
Ozone	–192.7 ± 2	–111.9
Palladium	1552	3140
Phosphorus, white	44.1	280
Platinum	1772	3827 ± 100
Plutonium	641	3232
Polonium	254	962
Potassium	63.65	774
Praseodymium	931	3512
Promethium	~1080	2460 (?)
Protactinium	<1600	–
Radium	700	1140
Radon	–71	–61.8
Rhenium	3180	5627 (estimated)
Rhodium	1966 ± 3	3727 ± 100
Rubidium	38.89	688
Ruthenium	2310	3900
Samarium	1077 ± 5	1791
Scandium	1541	2831
Selenium, gray	217	684.9 ± 1.0
Silicon	1410	2355
Silver	961.93	2212
Sodium	97.81 ± 0.03	882.9
Strontium	769	1384
Sulfur, rhombic	112.8	444.674
Sulfur, monoclinic	119.0	444.674
Tantalum	2996	5425 ± 100
Technetium	2172	4877
Tellurium	449.5 ± 0.3	989.8 ± 3.8
Terbium	1356 ± 4	3123
Thallium	303.5	1457 ± 10
Thorium	1750	~4790
Thulium	1545	1947
Tin	231.9681	2270
Titanium	1660 ± 10	3287
Tungsten	3410 ± 20	5660
Uranium	1132.3 ± 0.8	3818
Vanadium	1890 ± 10	3380
Wolfram (see Tungsten)		
Xenon	–111.9	–107.1 ± 3
Ytterbium	819 ± 5	1194
Yttrium	1522 ± 8	3338
Zinc	419.58	907
Zirconium	1852 ± 2	4377

From Weast, R. C., Ed., *Handbook of Chemistry and Physics,* 55th ed., CRC Press, Cleveland, 1974, D-153.

Table 2–32
VAPOR PRESSURE OF MERCURY

Temperature (°C)	0	2	4	6	8
−30	0.00000179	0.00000359	0.00000266	0.00000197	0.00000145
−20	0.0000181	0.0000140	0.0000108	0.00000828	0.00000630
−10	0.0000606	0.0000481	0.0000380	0.0000298	0.0000232
0−	0.000185	0.000149	0.000119	0.0000954	0.0000762
0+	0.000185	0.000228	0.000276	0.000335	0.000406
10	0.000490	0.000588	0.000706	0.000846	0.001009
20	0.001201	0.001426	0.001691	0.002000	0.002359
30	0.002777	0.003261	0.003823	0.004471	0.005219
40	0.006079	0.007067	0.008200	0.009497	0.01098
50	0.01267	0.01459	0.01677	0.01925	0.02206
60	0.02524	0.02883	0.03287	0.03740	0.04251
70	0.04825	0.05469	0.06189	0.06993	0.07889
80	0.08880	0.1000	0.1124	0.1261	0.1413
90	0.1582	0.1769	0.1976	0.2202	0.2453
100	0.2729	0.3032	0.3366	0.3731	0.4132
110	0.4572	0.5052	0.5576	0.6150	0.6776
120	0.7457	0.8198	0.9004	0.9882	1.084
130	1.186	1.298	1.419	1.551	1.692
140	1.845	2.010	2.188	2.379	2.585
150	2.807	3.046	3.303	3.578	3.873
160	4.189	4.528	4.890	5.277	5.689
170	6.128	6.596	7.095	7.626	8.193
180	8.796	9.436	10.116	10.839	11.607
190	12.423	13.287	14.203	15.173	16.200
200	17.287	18.437	19.652	20.936	22.292
210	23.723	25.233	26.826	28.504	30.271
220	32.133	34.092	36.153	38.318	40.595
230	42.989	45.503	48.141	50.909	53.812
240	56.855	60.044	63.384	66.882	70.543
250	74.375	78.381	82.568	86.944	91.518
260	96.296	101.28	106.48	111.91	117.57
270	123.47	129.62	136.02	142.69	149.64
280	156.87	164.39	172.21	180.34	188.79
290	197.57	206.70	216.17	226.00	236.21
300	246.80	257.78	269.17	280.98	293.21
310	305.89	319.02	332.62	346.70	361.26
320	376.33	391.92	408.04	424.71	441.94
330	459.74	478.14	497.12	516.74	537.00
340	557.90	579.45	601.69	624.64	648.30
350	672.69	697.83	723.73	750.43	777.92
360	806.23	835.38	865.36	896.23	928.02
370	960.66	994.34	1028.9	1064.4	1100.9
380	1138.4	1177.0	1216.6	1257.3	1299.1
390	1341.9	1386.1	1431.3	1477.7	1525.2
400	1574.1				

Note: Values are given in mm of Hg for temperatures ranging from −38°C to 400°C.

From Weast, R. C. Ed., *Handbook of Chemistry and Physics,* 55th ed., CRC Press, Cleveland, 1974, D-161.

Table 2–33
RELATIONSHIP BETWEEN VAPOR PRESSURE AND TEMPERATURE FOR SOME ELEMENTS[a]

Element	Melting point (°C)	Pressure (mm Hg)					
		10^{-5}	10^{-4}	10^{-3}	10^{-2}	10^{-1}	1
Ag	961	767	848	936	1047	1184	1353
Al	660	724	808	889	996	1123	1279
Au	1063	1083	1190	1316	1465	1646	1867
Ba	717	418	476	546	629	730	858
Be	1284	942	1029	1130	1246	1395	1582
Bi	271	474	536	609	698	802	934
C		2129	2288	2471	2681	2926	3214
Cd	321	148	180	220	264	321	
Co	1478	1249	1362	1494	1649	1833	2056
Cr	1900	907	992	1090	1205	1342	1504
Cu	1083	946	1035	1141	1273	1432	1628
Fe	1535	1094	1195	1310	1447	1602	1783
Hg	−38.9	−23.9	−5.5	18.0	48.0	82.0	126
In	157	667	746	840	952	1088	1260
Ir	2454	1993	2154	2340	2556	2811	3118
Mg	651	287	331	383	443	515	605
Mn	1244	717	791	878	980	1103	1251
Mo	2622	1923	2095	2295	2533		
Ni	1455	1157	1257	1371	1510	1679	1884
Os	2697	2101	2264	2451	2667	2920	3221
Pb	328	483	548	625	718	832	975
Pd	1555	1156	1271	1405	1566	1759	2000
Pt	1774	1606	1744	1904	2090	2313	2582
Sb	630	466	525	595	678	779	904
Si	1410	1024	1116	1223	1343	1485	1670
Sn	232	823	922	1042	1189	1373	1609
Ta	2996	2407	2599	2820			
W	3382	2554	2767	3016	3309		
Zn	419	211	248	292	343	405	
Zr	2127	1527	1660	1816	2001	2212	2459

[a]The values given in this table are from a variety of sources that are not always in agreement; for that reason, the table should be used only as a general guide.

Reprinted from Dushman, S., *Scientific Foundations of Vacuum Technique,* John Wiley & Sons, New York, 1949, 752. With permission.

Table 2–34
VAPOR PRESSURE OF ICE

Temperature (°C)	0	2	4	6	8
−90	0.000070	0.000048	0.000033	0.000022	0.000015
−80	0.00040	0.00029	0.00020	0.00014	0.00010
−70	0.00194	0.00143	0.00105	0.00077	0.00056
−60	0.00808	0.00614	0.00464	0.00349	0.00261
−50	0.02955	0.0230	0.0178	0.0138	0.0106
−40	0.0966	0.0768	0.0609	0.0481	0.0378
−30	0.2859	0.2318	0.1873	0.1507	0.1209

Temperature (°C)	0.0	0.2	0.4	0.6	0.8
−29	0.317	0.311	0.304	0.298	0.292
−28	0.351	0.344	0.337	0.330	0.324
−27	0.389	0.381	0.374	0.366	0.359
−26	0.430	0.422	0.414	0.405	0.397
−25	0.476	0.467	0.457	0.448	0.439
−24	0.526	0.515	0.505	0.495	0.486
−23	0.580	0.569	0.558	0.547	0.536
−22	0.640	0.627	0.615	0.603	0.592
−21	0.705	0.691	0.678	0.665	0.652
−20	0.776	0.761	0.747	0.733	0.719
−19	0.854	0.838	0.822	0.806	0.791
−18	0.939	0.921	0.904	0.887	0.870
−17	1.031	1.012	0.993	0.975	0.956
−16	1.132	1.111	1.091	1.070	1.051
−15	1.241	1.219	1.196	1.175	1.153
−14	1.361	1.336	1.312	1.288	1.264
−13	1.490	1.464	1.437	1.411	1.386
−12	1.632	1.602	1.574	1.546	1.518
−11	1.785	1.753	1.722	1.691	1.661
−10	1.950	1.916	1.883	1.849	1.817
−9	2.131	2.093	2.057	2.021	1.985
−8	2.326	2.285	2.246	2.207	2.168
−7	2.537	2.493	2.450	2.408	2.367
−6	2.765	2.718	2.672	2.626	2.581
−5	3.013	2.962	2.912	2.862	2.813
−4	3.280	3.225	3.171	3.117	3.065
−2	3.880	3.816	3.753	3.691	3.630
−1	4.217	4.147	4.079	4.012	3.946
0−	4.579	4.504	4.431	4.359	4.287

Note: Values are given in mm of Hg for temperatures ranging from −98°C to 0°C.

From Weast, R. C., Ed., *Handbook of Chemistry and Physics,* 55th ed., CRC Press, Cleveland, 1974, D-158.

Table 2–35A
VAPOR PRESSURE OF WATER BELOW 100°C

°C	0.0	0.2	0.4	0.6	0.8	°C	0.0	0.2	0.4	0.6	0.8
−15	1.436	1.414	1.390	1.368	1.345	42	61.50	62.14	62.80	63.46	64.12
−14	1.560	1.534	1.511	1.485	1.460	43	64.80	65.48	66.16	66.86	67.56
−13	1.691	1.665	1.637	1.611	1.585	44	68.26	68.97	69.69	70.41	71.14
−12	1.834	1.804	1.776	1.748	1.720						
−11	1.987	1.955	1.924	1.893	1.863	45	71.88	72.62	73.36	74.12	74.88
						46	75.65	76.43	77.21	78.00	78.80
−10	2.149	2.116	2.084	2.050	2.018	47	79.60	80.41	81.23	82.05	82.87
− 9	2.326	2.289	2.254	2.219	2.184	48	83.71	84.56	85.42	86.28	87.14
− 8	2.514	2.475	2.437	2.399	2.362	49	88.02	88.90	89.79	90.69	91.59
− 7	2.715	2.674	2.633	2.593	2.553						
− 6	2.931	2.887	2.843	2.800	2.757	50	92.51	93.5	94.4	95.3	96.3
						51	97.20	98.2	99.1	100.1	101.1
− 5	3.163	3.115	3.069	3.022	2.976	52	102.09	103.1	104.1	105.1	106.2
− 4	3.410	3.359	3.309	3.259	3.211	53	107.20	108.2	109.3	110.4	111.4
− 3	3.673	3.620	3.567	3.514	3.461	54	112.51	113.6	114.7	115.8	116.9
− 2	3.956	3.898	3.841	3.785	3.730						
− 1	4.258	4.196	4.135	4.075	4.016	55	118.04	119.1	120.3	121.5	122.6
						56	123.80	125.0	126.2	127.4	128.6
0 −	4.579	4.513	4.448	4.385	4.320	57	129.82	131.0	132.3	133.5	134.7
						58	136.08	137.3	138.5	139.9	141.2
0	4.579	4.647	4.715	4.785	4.855	59	142.60	143.9	145.2	146.6	148.0
1	4.926	4.998	5.070	5.144	5.219						
2	5.294	5.370	5.447	5.525	5.605	60	149.38	150.7	152.1	153.5	155.0
3	5.685	5.766	5.848	5.931	6.015	61	156.43	157.8	159.3	160.8	162.3
4	6.101	6.187	6.274	6.363	6.453	62	163.77	165.2	166.8	168.3	169.8
						63	171.38	172.9	174.5	176.1	177.7
5	6.543	6.635	6.728	6.822	6.917	64	179.31	180.9	182.5	184.2	185.8
6	7.013	7.111	7.209	7.309	7.411						
7	7.513	7.617	7.722	7.828	7.936	65	187.54	189.2	190.9	192.6	194.3
8	8.045	8.155	8.267	8.380	8.494	66	196.09	197.8	199.5	201.3	203.1
9	8.609	8.727	8.845	8.965	9.086	67	204.96	206.8	208.6	210.5	212.3
						68	214.17	216.0	218.0	219.9	221.8
10	9.209	9.333	9.458	9.585	9.714	69	223.73	225.7	227.7	229.7	231.7
11	9.844	9.976	10.109	10.244	10.380						
12	10.518	10.658	10.799	10.941	11.085	70	233.7	235.7	237.7	239.7	241.8
13	11.231	11.379	11.528	11.680	11.833	71	243.9	246.0	248.2	250.3	252.4
14	11.987	12.144	12.302	12.462	12.624	72	254.6	256.8	259.0	261.2	263.4
						73	265.7	268.0	270.2	272.6	274.8
15	12.788	12.953	13.121	13.290	13.461	74	277.2	279.4	281.8	284.2	286.6
16	13.634	13.809	13.987	14.166	14.347						
17	14.530	14.715	14.903	15.092	15.284	75	289.1	291.5	294.0	296.4	298.8
18	15.477	15.673	15.871	16.071	16.272	76	301.4	303.8	306.4	308.9	311.4
19	16.477	16.685	16.894	17.105	17.319	77	314.1	316.6	319.2	322.0	324.6
						78	327.3	330.0	332.8	335.6	338.2
20	17.535	17.753	17.974	18.197	18.422	79	341.0	343.8	346.6	349.4	352.2
21	18.650	18.880	19.113	19.349	19.587						
22	19.827	20.070	20.316	20.565	20.815	80	355.1	358.0	361.0	363.8	366.8
23	21.068	21.324	21.583	21.845	22.110	81	369.7	372.6	375.6	378.8	381.8
24	22.377	22.648	22.922	23.198	23.476	82	384.9	388.0	391.2	394.4	397.4
						83	400.6	403.8	407.0	410.2	413.6
25	23.756	24.039	24.326	24.617	24.912	84	416.8	420.2	423.6	426.8	430.2
26	25.209	25.509	25.812	26.117	26.426						
27	26.739	27.055	27.374	27.696	28.021	85	433.6	437.0	440.4	444.0	447.5
28	28.349	28.680	29.015	29.354	29.697	86	450.9	454.4	458.0	461.6	465.2
29	30.043	30.392	30.745	31.102	31.461	87	468.7	472.4	476.0	479.8	483.4
						88	487.1	491.0	494.7	498.5	502.2
30	31.824	32.191	32.561	32.934	33.312	89	506.1	510.0	513.9	517.8	521.8
31	33.695	34.082	34.471	34.864	35.261						
32	35.663	36.068	36.477	36.891	37.308	90	525.76	529.77	533.80	537.86	541.95
33	37.729	38.155	38.584	39.018	39.457	91	546.05	550.18	554.35	558.53	562.75
34	39.898	40.344	40.796	41.251	41.710	92	566.99	571.26	575.55	579.87	584.22
						93	588.60	593.00	597.43	601.89	606.38
35	42.175	42.644	43.117	43.595	44.078	94	610.90	615.44	620.01	624.61	629.24
36	44.563	45.054	45.549	46.050	46.556						
37	47.067	47.582	48.102	48.627	49.157	95	633.90	638.59	643.30	648.05	652.82
38	49.692	50.231	50.774	51.323	51.879	96	657.62	662.45	667.31	672.20	677.12
39	52.442	53.009	53.580	54.156	54.737	97	682.07	687.04	692.05	697.10	702.17
						98	707.27	712.40	717.56	722.75	727.98
40	55.324	55.91	56.51	57.11	57.72	99	733.24	738.53	743.85	749.20	754.58
41	58.34	58.96	59.58	60.22	60.86						
						100	760.00	765.45	770.93	776.44	782.00
						101	787.57	793.18	798.82	804.50	810.21

Note: Values are given in mm of Hg for temperatures ranging from −15.8°C to 101.8°C. The values for fractional degrees between 50 and 89 were obtained by interpolation.

From Weast, R. C., Ed., *Handbook of Chemistry and Physics*, 55th ed., CRC Press, Cleveland, 1974, D-159.

Table 2–35B
VAPOR PRESSURE OF WATER BELOW 100°C

Temp. °C	mm	Pounds per sq. in.	Temp. °F	Temp. °C	mm	Pounds per sq. in.	Temp. °F	Temp. °C	mm	Pounds per sq. in.	Temp. °F	Temp. °C	mm	Pounds per sq. in.	Temp. °F
100	760.	14.696	212.0	170	5940.92	114.879	338.0	240	25100.52	485.365	464.0	310	74024.00	1431.390	590.0
101	787.51	15.228	213.8	171	6085.32	117.671	339.8	241	25543.60	493.933	465.8	311	75042.40	1451.083	591.8
102	815.86	15.776	215.6	172	6233.52	120.537	341.6	242	25994.28	502.647	467.6	312	76076.00	1471.070	593.6
103	845.12	16.342	217.4	173	6383.24	123.432	343.4	243	26449.52	511.450	469.4	313	77117.20	1491.203	595.4
104	875.06	16.921	219.2	174	6535.28	126.430	345.2	244	26912.36	520.400	471.2	314	78166.00	1511.484	597.2
105	906.07	17.521	221.0	175	6694.08	129.442	347.0	245	27381.28	529.467	473.0	315	79230.00	1532.058	599.0
106	937.92	18.136	222.8	176	6852.92	132.514	348.8	246	27855.52	538.638	474.8	316	80294.00	1552.632	600.8
107	970.60	18.768	224.6	177	7015.56	135.659	350.6	247	28335.84	547.926	476.6	317	81373.20	1573.501	602.6
108	1004.42	19.422	226.4	178	7180.48	138.848	352.4	248	28823.76	557.360	478.4	318	82467.60	1594.663	604.4
109	1038.92	20.089	228.2	179	7349.20	142.110	354.2	249	29317.00	566.898	480.2	319	83569.60	1615.972	606.2
110	1074.56	20.779	230.0	180	7520.20	145.417	356.0	250	29817.84	576.583	482.0	320	84686.80	1637.575	608.0
111	1111.20	21.487	231.8	181	7694.24	148.782	357.8	251	30324.00	586.370	483.8	321	85819.20	1659.472	609.8
112	1148.74	22.213	233.6	182	7872.08	152.221	359.6	252	30837.76	596.305	485.6	322	86959.20	1681.516	611.6
113	1187.42	22.961	235.4	183	8052.96	155.719	361.4	253	31356.84	606.342	487.4	323	88114.40	1703.854	613.4
114	1227.25	23.731	237.2	184	8236.88	159.275	363.2	254	31885.04	616.556	489.2	324	89277.20	1726.339	615.2
115	1267.98	24.519	239.0	185	8423.84	162.890	365.0	255	32417.80	626.858	491.0	325	90447.60	1748.971	617.0
116	1309.94	25.330	240.8	186	8616.12	166.609	366.8	256	32957.40	637.292	492.8	326	91633.20	1771.897	618.8
117	1352.95	26.162	242.6	187	8809.92	170.356	368.6	257	33505.36	647.888	494.6	327	92826.40	1794.969	620.6
118	1397.18	27.017	244.4	188	9007.52	174.177	370.4	258	34059.40	658.601	496.4	328	94042.40	1818.483	622.4
119	1442.63	27.896	246.2	189	9208.16	178.057	372.2	259	34618.76	669.417	498.2	329	95273.60	1842.291	624.2
120	1489.14	28.795	248.0	190	9413.36	182.025	374.0	260	35188.00	680.425	500.0	330	96512.40	1866.245	626.0
121	1536.80	29.717	249.8	191	9620.08	186.022	375.8	261	35761.80	691.520	501.8	331	97758.80	1890.346	627.8
122	1586.04	30.669	251.6	192	9831.36	190.107	377.6	262	36343.20	702.763	503.6	332	99020.40	1914.742	629.6
123	1636.36	31.642	253.4	193	10047.20	194.281	379.4	263	36932.20	714.152	505.4	333	100297.20	1939.431	631.4
124	1687.81	32.637	255.2	194	10265.32	198.499	381.2	264	37529.56	725.703	507.2	334	101581.60	1964.267	633.2
125	1740.93	33.664	257.0	195	10488.76	202.819	383.0	265	38133.00	737.372	509.0	335	102881.20	1989.398	635.0
126	1795.12	34.712	258.8	196	10715.24	207.199	384.8	266	38742.52	749.158	510.8	336	104196.00	2014.822	636.8
127	1850.83	35.789	260.6	197	10944.76	211.637	386.6	267	39361.92	761.135	512.6	337	105526.00	2040.540	638.6
128	1907.83	36.891	262.4	198	11179.60	216.178	388.4	268	39986.44	773.215	514.4	338	106871.20	2066.552	640.4
129	1966.35	38.023	264.2	199	11417.48	220.778	390.2	269	40619.72	785.457	516.2	339	108224.00	2092.710	642.2
130	2026.16	39.180	266.0	200	11659.16	225.451	392.0	270	41261.16	797.861	518.0	340	109592.00	2119.163	644.0
131	2087.42	40.364	267.8	201	11905.40	230.213	393.8	271	41910.20	810.411	519.8	341	110967.60	2145.763	645.8
132	2150.42	41.582	269.6	202	12155.44	235.048	395.6	272	42566.08	823.094	521.6	342	112358.40	2172.657	647.6
133	2214.64	42.824	271.4	203	12408.52	239.942	397.4	273	43229.56	835.923	523.4	343	113749.20	2199.550	649.4
134	2280.76	44.103	273.2	204	12666.16	244.924	399.2	274	43902.16	848.929	525.2	344	115178.00	2227.179	651.2
135	2347.26	45.389	275.0	205	12929.12	250.008	401.0	275	44580.84	862.053	527.0	345	116614.40	2254.954	653.0
136	2416.34	46.724	276.8	206	13197.40	255.196	402.8	276	45269.40	875.367	528.8	346	118073.60	2283.171	654.8
137	2488.16	48.113	278.6	207	13467.96	260.428	404.6	277	45964.04	888.799	530.6	347	119532.80	2311.387	656.6
138	2560.67	49.515	280.4	208	13742.32	265.733	406.4	278	46669.32	902.437	532.4	348	121014.80	2340.044	658.4
139	2634.84	50.950	282.2	209	14022.76	271.156	408.2	279	47382.20	916.222	534.2	349	122504.40	2368.848	660.2
140	2710.92	52.421	284.0	210	14305.48	276.623	410.0	280	48104.20	930.183	536.0	350	124001.60	2397.799	662.0
141	2788.44	53.920	285.8	211	14595.04	282.222	411.8	281	48833.80	944.291	537.8	351	125521.60	2427.191	663.8
142	2867.48	55.448	287.6	212	14888.40	287.895	413.6	282	49570.24	958.532	539.6	352	127049.20	2456.730	665.6
143	2948.80	57.020	289.4	213	15184.80	293.626	415.4	283	50316.16	972.963	541.4	353	128599.60	2486.710	667.4
144	3031.64	58.622	291.2	214	15488.04	299.490	417.2	284	51072.76	987.586	543.2	354	130157.60	2516.837	669.2
145	3116.76	60.268	293.0	215	15792.80	305.383	419.0	285	51838.08	1002.385	545.0	355	131730.80	2547.258	671.0
146	3203.40	61.944	294.8	216	16104.40	311.408	420.8	286	52611.76	1017.345	546.8	356	133326.80	2578.119	672.8
147	3292.32	63.663	296.6	217	16420.56	317.522	422.6	287	53395.32	1032.497	548.6	357	134945.60	2609.422	674.6
148	3382.76	65.412	298.4	218	16741.84	323.735	424.4	288	54187.24	1047.810	550.4	358	136579.60	2641.018	676.4
149	3476.24	67.220	300.2	219	17067.32	330.028	426.2	289	54989.04	1063.314	552.2	359	138228.80	2672.908	678.2
150	3570.48	69.042	302.0	220	17395.64	336.377	428.0	290	55799.20	1078.980	554.0	360	139893.20	2705.093	680.0
151	3667.00	70.908	303.8	221	17761.36	342.872	429.8	291	56612.40	1094.705	555.8	361	141572.80	2737.571	681.8
152	3766.56	72.833	305.6	222	18072.80	349.471	431.6	292	57448.40	1110.871	557.6	362	143275.20	2770.490	683.6
153	3868.44	74.773	307.4	223	18396.24	356.143	433.4	293	58284.40	1127.036	559.4	363	144992.80	2803.703	685.4
154	3970.24	76.772	309.2	224	18766.68	362.888	435.2	294	59135.60	1143.496	561.2	364	146733.20	2837.357	687.2
155	4075.88	78.815	311.0	225	19123.12	369.781	437.0	295	59994.40	1160.102	563.0	365	148519.20	2871.892	689.0
156	4183.80	80.901	312.8	226	19482.60	376.732	438.8	296	60860.80	1176.836	564.8	366	150320.40	2906.722	690.8
157	4293.24	83.018	314.6	227	19848.92	383.815	440.6	297	61742.40	1193.903	566.6	367	152129.20	2941.698	692.6
158	4404.96	85.178	316.4	228	20219.80	390.987	442.4	298	62608.00	1210.950	568.4	368	153960.80	2977.116	694.4
159	4519.72	87.397	318.2	229	20596.76	398.276	444.2	299	63528.40	1228.439	570.2	369	155815.20	3012.974	696.2
160	4636.00	89.646	320.0	230	20978.28	405.654	446.0	300	64432.80	1245.927	572.0	370	157692.40	3049.273	698.0
161	4755.32	91.953	321.8	231	21365.12	413.134	447.8	301	65360.00	1263.709	573.8	371	159584.80	3085.866	699.8
162	4876.92	94.304	323.6	232	21757.28	420.717	449.6	302	66279.60	1281.638	575.6	372	161507.60	3123.047	701.6
163	5000.04	96.685	325.4	233	22154.00	428.388	451.4	303	67214.40	1299.714	577.4	373	163468.40	3160.963	703.4
164	5126.96	99.139	327.2	234	22558.32	436.207	453.2	304	68156.80	1317.937	579.2	374	165467.20	3199.613	705.2
165	5256.16	101.638	329.0	235	22967.96	444.128	455.0	305	69114.40	1336.454	581.0				
166	5386.88	104.165	330.8	236	23382.92	452.152	456.8	306	70072.00	1354.971	582.8				
167	5521.40	106.766	332.6	237	23802.44	460.264	458.6	307	71052.40	1373.929	584.6				
168	5658.20	109.412	334.4	238	24229.56	468.523	460.4	308	72048.00	1393.181	586.4				
169	5798.04	112.116	336.2	239	24661.24	476.871	462.2	309	73028.40	1412.139	588.2				

From Weast, R. C., Ed., *Handbook of Chemistry and Physics*, 55th ed., CRC Press, Cleveland, 1974, D-160.

Table 2–36
VAPOR PRESSURE OF CARBON DIOXIDE

Critical temperature of carbon dioxide: 31.0°C
Triple point of carbon dioxide: −56.602 ± 0.005°C, 3885.2 ± 0.4 mm Hg
Density of mercury column: 13.5951 g/cm^3, G = 980.665 dyn

A. Solid CO_2

μm Hg

°C	0	1	2	3	4	5	6	7	8	9
−180	0.013	0.008	0.006	0.004	0.003	0.0017	0.0011	0.0007	0.0005	0.0003
−170	0.37	0.27	0.20	0.14	0.10	0.074	0.052	0.037	0.026	0.018
−160	5.9	4.6	3.6	2.7	2.1	1.58	1.19	0.90	0.67	0.50
−150	60.5	48.8	39.2	31.4	25.1	19.9	15.8	12.4	9.8	7.6
−140	431	359	298	247	204	168	138	113	92	75

mm Hg

°C	0	1	2	3	4	5	6	7	8	9
−130	2.31	1.97	1.68	1.43	1.22	1.03	0.87	0.73	0.61	0.51
−120	9.81	8.57	7.46	6.49	5.63	4.88	4.22	3.64	3.13	2.69
−110	34.63	30.76	27.27	24.14	21.34	18.83	16.58	14.58	12.80	11.22
−100	104.81	94.40	84.91	76.27	68.43	61.30	54.84	48.99	43.71	38.94
−90	279.5	254.7	231.8	210.8	191.4	173.6	157.3	142.4	128.7	116.2
−80	672.2	618.3	568.2	521.7	478.5	438.6	401.6	367.4	335.7	306.5
−70	1,486.1	1,377.3	1,275.6	1,180.5	1,091.7	1,008.9	931.7	859.7	792.7	730.3
−60	3,073.1	2,865.1	2,669.7	2,486.3	2,314.2	2,152.8	2,001.5	859.7	1,726.9	3,294.6
−50								3,780.9	3,530.2	3,294.6

B. Liquid CO_2

mm Hg

°C	0	1	2	3	4	5	6	7	8	9
−50	5,127.8	4,922.7	4,723.9	4,531.1	4,344.3	4,163.2	3,987.9	3,818.2[a]	3,653.9[a]	3,495.0[a]
−40	7,545	7,271	7,005	6,746	6,494	6,250	6,012	5,781	5,557	5,339
−30	10,718	10,363	10,017	9,679	9,350	9,029	8,716	8,412	8,115	7,826
−20	14,781	14,331	13,891	13,461	13,040	12,630	12,229	11,838	11,455	11,082
−10	19,872	19,312	18,764	18,288	17,703	17,189	16,686	16,194	15,712	15,241

Table 2–36 (continued)
VAPOR PRESSURE OF CARBON DIOXIDE

B. Liquid CO_2 (cont.)

mm Hg

°C	0	1	2	3	4	5	6	7	8	9
0–	26,142	25,457	24,786	24,127	23,482	22,849	22,229	21,622	21,026	20,443
0+	26,142	26,840	27,552	28,277	29,017	29,771	30,539	31,323	32,121	32,934
10	33,763	34,607	35,467	36,343	37,236	38,146	39,073	40,017	40,980	41,960
20	43,959	43,977	45,014	46,072	47,150	48,250	49,370	50,514	51,680	52,871
30	54,086	55,327								

[a]Undercooled liquid.

From Weast, R. C., Ed., *Handbook of Chemistry and Physics*, 55th ed., CRC Press, Cleveland, 1974, D-161.

Table 2–37
IONIZATION POTENTIALS OF THE ELEMENTS

Different methods were employed to measure ionization potentials. Abbreviation of the methods used are as follows: S = vacuum ultraviolet spectroscopy; SI = surface ionization, mass spectrometric; EI = electron impact with mass analysis.

Element	At. No.	Ionization potential (volts)								Method
		I	II	III	IV	V	VI	VII	VIII	
Ar	18	15.755	27.62	40.9	59.79	75	91.3	124	143.46	S
Ac	89	6.9	12.1	20						S
Ag	47	7.574	21.48	34.82						S
Al	13	5.984	18.823	28.44	119.96	153.77	190.42	241.38	284.53	S
As	33	9.81	18.63	28.34	50.1	62.6	127.5			S
At	85	9.5								S
Au	79	9.22	20.5							S
B	5	8.296	25.149	37.92	259.298	340.127				S
Ba	56	5.21	10.001	35.5						S
Be	4	9.32	18.206	153.85	217.657					S
Bi	83	7.287	16.68	25.56	45.3	56	88.3			S
Br	35	11.84	21.6	35.9	47.3	59.7	88.6	103	193	S
C	6	11.256	24.376	47.871	64.476	391.986	489.84			S
Ca	20	6.111	11.868	51.21	67	84.39	109	128	143.3	S
Cd	48	8.991	16.904	37.47						S
Ce	58	5.6	12.3	20	33.3					SI
Cl	17	13.01	23.8	39.9	53.5	67.8	96.7	114.27	348.3	S
Co	27	7.86	17.05	33.49	83.1					S
Cr	24	6.764	16.49	30.96	50	73	91	161	185	S
Cs	55	3.893	25.1	35						S
Cu	29	7.724	20.29	36.83						S
Dy	66	6.8								S
Er	68	6.08								SI
Eu	63	5.67	11.24							S
F	9	17.418	34.98	62.646	87.14	114.214	157.117	185.139	953.6	S
Fe	26	7.87	16.18	30.643	56.8				151	S
Fr	87	4								S
Ga	31	6	20.57	30.7	64.2					S
Gd	64	6.16	12							S
Ge	32	7.88	15.93	34.21	44.7	93.4				S
H	2	13.595								S
He	2	24.481	54.403							S
Hf	72	7	14.9	23.2	33.3					S
Hg	80	10.43	18.751	34.2	49.5[a]		67[a]			S
I	53	10.454	19.13						170	S
In	49	5.785	18.86	28.03	54.4					S
Ir	77	9								S
K	19	4.339	31.81	46	60.9	82.6	99.7	118	155	S
Kr	36	13.996	24.56	36.9	43.5[a]	63[a]	94[a]			S
La	57	5.61	11.43	19.17						

Table 2–37 (continued)
IONIZATION POTENTIALS OF THE ELEMENTS

Element	At. No.	Ionization potential (volts)								Method
		I	II	III	IV	V	VI	VII	VIII	
Li	3	5.39	75.619	122.419						S
Lu	71		14.7							S
Mg	12	7.644	15.031	80.14	109.29	141.23	186.49	224.9	265.957	S
Mn	25	7.432	15.636	33.69	52	76		119	196	S
Mo	42	7.10	16.15	27.13	46.4	61.2	68	126	153	S
N	7	14.53	29.593	47.426	77.45	97.863	551.925	666.83		S
Na	11	5.138	47.29	71.715	98.88	138.37	172.09	208.444	264.155	S
Nb	41	6.88	14.32	25.04	38.3	50	103	125		S
Nd	60	5.51								SI
Ne	10	21.559	41.07	63.5	97.02	126.3	157.91			S
Ni	28	7.633	18.15	35.16						S
O	8	13.614	35.108	54.886	77.394	113.873	138.08	739.114	871.12	S
Os	76	8.5	17							S
P	15	10.484	19.72	30.156	51.354	65.007	220.414	263.31	309.26	S
Pb	82	7.415	15.028	31.93	42.31	68.8				S
Pd	46	8.33	19.42	32.92						S
Po	84	8.43								S
Pr	59	5.46								SI
Pt	78	9.0	18.56							S
Pu	94	5.1								S
Ra	88	5.277	10.144							S
Rb	37	4.176	27.5	40						S
Re	75	7.87	16.6							S
Rh	45	7.46	18.07	31.05						S
Rn	86	10.746								S
Ru	44	7.364	16.76	28.46						S
S	16	10.357	23.4	35	47.29	72.5	88.029	280.99	328.8	S
Sb	51	8.639	16.5	25.3	44.1	56	108			S
Sc	21	6.54	12.8	24.75	73.9	92	111	139	159	S
Se	34	9.75	21.5	32	43	68	82	155		S
Si	14	8.149	16.34	33.488	45.13	166.73	205.11	246.41	303.07	S
Sm	62	5.6	11.2							S
Sn	50	7.342	14.628	30.49	40.72	72.3				S
Sr	38	5.692	11.027		57					S
Ta	73	7.88	16.2							S
Tb	65	5.98								SI
Tc	43	7.28	15.26	29.54						S
Te	52	9.01	18.6	31	38	60	72	12.7		S
Th	90	6.95			29.38					SI
Ti	22	6.82	13.57	27.47	43.24	99.8	120	141	172	S
Tl	81	6.106	20.42	29.8	50.7					S
Tm	69	5.81								S
U	92	6.08								SI
V	23	6.74	14.65	29.31	48	65	129	151	170	S
W	74	7.98	17.7							S

Table 2–37 (continued)
IONIZATION POTENTIALS OF THE ELEMENTS

Element	At. No.	Ionization potential (volts)								Method
		I	II	III	IV	V	VI	VII	VIII	
Xe	54	12.127[b]	21.2	31.3	42	53	58	135		EI
Y	39	6.38	12.23	20.5		77				S
Yb	70	6.2	12.10							S
Zn	30	9.391	17.96	39.7						S
Zr	40	6.84	13.13	22.98	34.33		99			S

[a]Measured by the EI method.
[b]Measured by the S method.

From Weast, R. C., Ed., *Handbook of Chemistry and Physics,* 55th ed., CRC Press, Cleveland, 1974, E-68.

Table 2–38

NUCLEAR SPINS, MOMENTS, AND MAGNETIC RESONANCE FREQUENCIES

This table contains the published values for the nuclear spins, magnetic moments, and quadrupole moments, and the calculated values for the nuclear magnetic resonance (NMR) frequency and for the relative sensitivities. Only those isotopes with both published spin and magnetic moment values are tabulated. The magnetic and quadrupole moment values were selected from results published during the period from January, 1955 to June, 1967. Earlier references were obtained from H. E. Walchli, A Table of Nuclear Moment Data, U.S. Atomic Energy Commission Report ORNL—1469, Supplement I (1953) and Supplement II (1955), and D. Strominger, J. M. Hollander, and G. T. Seaborg, Table of Isotopes, Rev. Mod. Phys. 30, 585 (1958). A table containing the known (1963) spin and electromagnetic moment values of nuclear ground and excited states has been compiled by I. Lindgren, Perturbed Angular Correlations; E. Karlsson, E. Mathias, and K. Siegbahn, editors; North-Holland Publishing Co. (1964). A more complete list of spin and moment results for nuclei in excited states are included in Lindgren's table.

In general, the results chosen for this table were selected with an inclination to NMR measurements and to the precision of the measurement. Only six significant figures are used in this table; therefore, the number of figures may be less than those published. The experimental methods employed in determining the moments are indicated by the following symbols:

Ab = atomic beam magnetic resonance (hyperfine structure, double or triple resonance or other method)
E = electron spin resonance or electron-nuclear double resonance
M = microwave absorption in gases
Mb = molecular (or diamagnetic) beam magnetic resonance
Mc = miscellaneous
Mo = Mössbauer effect
N = nuclear magnetic resonance
No = nuclear orientation
O = optical spectroscopy (hyperfine structure, band structure, double resonance, or optical pumping)
Qr = quadrupole resonance

Other symbols used in the table are:

A = atomic weight (mass number)
El = element
I = nuclear spin in units of $h/2\pi$
μ = magnetic moment in units of the nuclear magneton $eh/4\pi Mc$
n = metastable excited state
Q = quadrupole moment in units of barns (10^{-24} cm^2)
Z = atomic number
* = radioactive isotope
● = magnetic moment observed by NMR
( ) = assumed or estimated values

Assuming a nuclear magneton value of 5.0505×10^{-24} erg/gauss, the NMR frequency was calculated for a total field of 10^4 gauss. The sensitivities, relative to the proton, are calculated from the following expressions:

$$\text{Sensitivity at constant field} = 7.652 \times 10^{-3}\,\mu^3\,(I+1)/I^2$$
$$\text{Sensitivity at constant frequency} = 0.2387\,\mu/I(I+1).$$

These expressions assume an equal number of nuclei, a constant temperature, and $T_1 = T_2$ (the longitudinal relaxation time equals the transverse relaxation time). These sensitivities represent the ideal induced voltage in the receiver coil at saturation and with a constant noise source. The calculated values are therefore determined under complete optimum conditions and should be regarded as such.

Isotopes Z = 0 to 13

Z	El	A	I	NMR Freq (MHz, 10 kG)	Nat. Abund. %	Sens. const. field	Sens. const. freq	μ (nucl. magneton)	μ Ref	μ Method	Q (barns)	Q Ref	Q Method
0	n	1*	1/2	29.167	—	0.322	0.685	-1.91315	1	Ab			
1	H	1*	1/2	42.5759	99.985	1.00	1.00	2.79268	2	N			
1	H	2*	1	6.53566	1.5×10^{-2}	9.65×10^{-3}	0.409	0.857387	3	N	2.73×10^{-3}	4	Ab
1	H	3*	1/2	45.4129	—	1.21	1.07	2.97877	5	N			
2	He	3	1/2	32.433	1.3×10^{-4}	0.442	0.762	-2.1274	6	N			
3	Li	6	1	6.2658	7.42	8.50×10^{-3}	0.392	0.82192	7	N	6.9×10^{-4}	9	N
3	Li	7	3/2	16.546	92.58	0.293	1.94	3.2560	8	N	-3×10^{-2}	10	O
3	Li	8*	2	6.300	—	2.59×10^{-2}	1.184	1.653	11	Ab			
4	Be	9	3/2	5.9834	100	1.39×10^{-2}	0.703	-1.1774	12	N	5.2×10^{-2}	14	Ab
5	B	10	3	4.5754	19.58	1.99×10^{-2}	1.72	1.8007	13	N	7.4×10^{-2}	16	Ab
5	B	11	3/2	13.660	80.42	0.165	1.60	2.6880	15	N	3.55×10^{-2}	16	Ab
6	C	13*	1/2	10.7054	1.108	1.59×10^{-2}	0.251	0.702199	17	N			
7	N	13*	1/2	4.91	—	1.53×10^{-3}	0.115	(-)0.322	18	N			
7	N	14	1	3.0756	99.63	1.01×10^{-3}	0.193	0.40347	21	N	1.6×10^{-2}	30	Mc
7	N	15	1/2	4.3142	0.37	1.04×10^{-3}	0.101	-0.28398	20	N			
8	O	15*	1/2	11.0	—	1.70×10^{-3}	0.257	0.719	23	N			
8	O	17*	5/2	5.772	3.7×10^{-2}	2.91×10^{-2}	1.58	-1.8930	24	N	-2.6×10^{-2}	25	M
9	F	17*	5/2	14.40	—	0.451	3.94	4.720	26	Ab			
9	F	19	1/2	40.0641	100	0.833	0.941	2.62727	17	N			
9	F	20*	2	7.977	—	5.25×10^{-3}	1.50	2.093	28	N			
10	Ne	19*	1/2	28.75	—	0.308	0.675	-1.886	29	N			
10	Ne	21*	3/2	3.3611	0.257	2.50×10^{-3}	0.395	-0.66140	30	N			
11	Na	21*	3/2	12.126	—	0.116	1.42	2.3861	31	Ab	0.14-0.15	33	O
11	Na	22*	3	4.436	—	1.81×10^{-2}	1.67	1.746	17	N			
11	Na	23	3/2	11.262	100	9.25×10^{-2}	1.32	2.2161	34	N	0.149	36	Ab
11	Na	24*	4	3.221	—	1.15×10^{-2}	2.02	1.690	34	Ab			
12	Mg	25	5/2	2.6054	10.13	2.67×10^{-3}	0.714	-0.85449	35	N			
13	Al	27	5/2	11.094	—	0.206	3.04	3.6385	35	N			

Isotopes Z = 14 to 25

Z	El	A	I	NMR Freq (MHz, 10 kG)	Nat. Abund. %	Sens. const. field	Sens. const. freq	μ (nucl. magneton)	μ Ref	μ Method	Q (barns)	Q Ref	Q Method
14	Si	29	1/2	8.4578	4.70	7.84×10^{-3}	0.199	-0.55477	37	N			
15	P	31	1/2	17.235	100	6.63×10^{-2}	0.405	1.1305	38	N			
15	P	32*	1	1.923	—	2.46×10^{-3}	0.120	-0.2523	39	E			
16	S	33	3/2	3.2654	0.76	2.26×10^{-3}	0.383	0.64257	37	N	-6.4×10^{-2}	40	M
16	S	35*	3/2	5.08	—	8.50×10^{-3}	0.597	1.00	41	M	4.5×10^{-2}	40	M
17	Cl	35	3/2	4.1717	75.53	4.70×10^{-3}	0.490	0.83991	8	N	-7.89×10^{-2}	42	Ab
17	Cl	36*	2	4.8931	—	1.21×10^{-2}	0.919	1.2838	43	N	-1.72×10^{-2}	44	Ab
17	Cl	37	3/2	3.472	24.47	2.71×10^{-3}	0.408	0.6833	45	N	-6.21×10^{-2}	42	Ab
18	Ar	37*	3/2	3.491	—		0.597	(1.0)		O			
19	K	39	3/2	1.9868	93.10	5.08×10^{-4}	0.233	0.39097	47	N	0.11	49	O
19	K	40*	4	2.470	1.18×10^{-2}	5.21×10^{-3}	1.55	-1.296	51	Mb			
19	K	41	3/2	1.0905	6.88	8.40×10^{-5}	0.128	0.21459	53	Ab	0.11	50	O
19	K	42*	2	4.345	—	8.50×10^{-3}	0.816	-1.140	53	Ab			
20	Ca	41*	7/2	3.4681	—	3.68×10^{-3}	9.73×10^{-2}	-1.5924	54	Ab			
20	Ca	43	7/2	2.8646	0.145	1.14×10^{-4}	1.71	-1.3183	55	Ab			
21	Sc	43*	7/2	10.04	—	0.275	4.95	4.61	56	Ab			
21	Sc	44*	2	9.76	—	9.63×10^{-2}	1.83	2.56	57	Ab			
21	Sc	44m*	6	5.03	—	9.24×10^{-2}	6.62	3.96	57	Ab			
21	Sc	45	7/2	10.343	100	0.301	5.10	4.7492	58	Ab	-0.22	59	Ab
21	Sc	46*	4	5.77	—	6.65×10^{-2}	3.62	3.03	56	Ab	0.14	56	Ab
21	Sc	47*	7/2	11.6	—	0.426	5.73	5.33	61	Ab	0.37	57	Ab
22	Ti	47	5/2	2.4000	7.28	2.40×10^{-3}	0.102	-0.78710	62	N	-0.22	57	Ab
22	Ti	49	7/2	2.4005	5.51	2.09×10^{-3}	0.658	-1.1022	63	N	0.12	59	Ab
23	V	49*	7/2	9.71	—	3.76×10^{-1}	1.18	4.46	63	E	-0.22	60	Ab
23	V	50*	6	4.2460	0.24	0.249	4.79	3.3413	64	N		61	Ab
23	V	51	7/2	11.19	99.76	5.55×10^{-1}	5.58	5.139	66	N	1.5×10^{-1}	61	Ab
24	Cr	53	3/2	2.4065	9.55	9.03×10^{-4}	0.382	-0.47354	67	N			
25	Mn	53*	7/2	3.007	—	5.14	0.283	3.075	68	Ab			
25	Mn	55	5/2	0.030	100	4.33×10^{-10}	5.73×10^{-2}	0.008	69	E			
25	Mn	55	7/2	11.0	—	0.362	5.42	5.05		N	-6×10^{-2}	65	O

Table 2–38 (continued)
NUCLEAR SPINS, MOMENTS, AND MAGNETIC RESONANCE FREQUENCIES

Z	El	A	Spin I	NMR Frequency in MHz for a 10 kilogauss field	Natural Abundance %	Relative Sensitivity (Equal Number of Nuclei) At constant field	Relative Sensitivity At constant frequency	Magnetic Moment μ (in multiples of nuclear magneton eh/4πMc)	Ref	Method	Electric Quadrupole Moment Q (10^{-24} cm²) in multiples of barns	Ref	Method
25	Mn	54*	(2)	8.4	—	6.11×10^{-3}	1.58	(2.2)	70	No			
25	*Mn	55	(3)	6.0	100	5.68×10^{-2}	2.48	(2.0)	71	E			
25	Mn	56*	5/2	10.501	—	0.116	2.88	3.444	72	E	0.55	72	M
26	*Fe	57	1/2	1.3758	2.19	3.37×10^{-5}	3.23×10^{-2}	0.09024	73	Ab			
27	Co	55*	7/2	10.0	—	0.274	4.94	4.5	74	No			
27	Co	56*	4	7.34	—	0.136	4.60	3.85	75	E			
27	Co	57*	7/2	10.1	—	0.283	4.99	4.65	76	E			
27	Co	58*	2	15.4	—	0.381	2.90	4.05	77	E			
27	*Co	59	7/2	10.054	100	0.277	4.96	4.6183	241	N	0.40	78	Ab
27	Co	60*	5	5.793	—	0.101	5.44	3.800	79	E			
28	*Ni	61	3/2	3.8047	1.19	3.57×10^{-2}	0.447	-0.74668	80	N			
29	Cu	61*	3/2	10.8	—	8.23×10^{-2}	1.27	2.13	81	E			
29	*Cu	63	3/2	11.285	69.09	9.31×10^{-2}	1.33	2.2206	8	N	-0.16	82	E
29	Cu	64*	1	3.1	—	9.79×10^{-4}	0.191	0.40	83	E			
29	*Cu	65	3/2	12.089	30.91	0.114	1.42	2.3789	8	N	-0.15	82	E
29	Cu	66*	1	1.65	—	1.54×10^{-4}	0.103	-0.216	81	E			
30	Zn	65*	5/2	2.345	—	1.05×10^{-2}	0.643	0.7692	83	O	-2.4×10^{-1}	83	O
30	*Zn	67	5/2	2.663	4.11	2.85×10^{-3}	0.730	0.8733	37	N	0.15	83	O
30	Zn	69*	1/2	0.0992	—	5.59×10^{-5}	5.59×10^{-2}	0.0117	91	M			
31	Ga	68*	1	10.22	—	6.91×10^{-2}	1.20	-1.02	92	M	0.9	92	M
31	*Ga	69	3/2	10.22	60.4	9.01×10^{-2}	1.25	2.011	8	N	0.27	93	Ab
31	*Ga	71	3/2	12.984	39.6	0.142	1.52	2.5549	8	N	0.178	94	Ab
31	Ga	72*	3	0.33691	—	7.89×10^{-3}	0.126	-0.13220	94	N	0.112	94	Ab
32	Ge	71*	1/2	8.4	—	7.64×10^{-3}	0.197	0.55	87	M	0.72	86	N
32	*Ge	73	9/2	3.1	7.76	1.40×10^{-3}	1.15	-0.87679	62	Ab	-0.2	88	M
33	As	75	3/2	7.2919	100	2.51×10^{-1}	0.856	1.4349	15	N	0.3	89	M
34	Se	75*	5/2	3.45	—	4.27×10^{-2}	0.649	-0.006	90	No			
34	*Se	77	1/2	4.76	7.58	6.93×10^{-3}	0.191	0.5325	90	M	0.9	40	M
35	*Br	79	3/2	8.118	50.54	7.85×10^{-2}	1.10	2.0990	8	N	0.27	92	Ab
35	Br	80*	1	2.22	—	2.08×10^{-2}	1.25	0.514	94	N	0.33	93	N
35	Br	80m*	5	4.18	—	6.20×10^{-2}	2.15	1.317	94	N	0.20	94	N
35	*Br	81	3/2	10.667	49.46	9.85×10^{-2}	1.19	2.2626	8	N	0.28	93	Ab
36	Kr	79*	1/2	3.93	—	1.33×10^{-2}	0.945	(+)1.626	95	Ab			
36	Kr	83	9/2	2.008	11.55	1.05×10^{-2}	1.13	-0.9671	96	Ab	0.15	96	Ab
37	*Rb	85	5/2	4.1108	72.15	1.05×10^{-2}	1.21	1.3483	100	N	0.27	101	Ab
37	*Rb	87	3/2	13.931	27.85	1.77×10^{-1}	1.64	2.7414	102	N	0.13	101	E
37	Rb	86*	2	1.8453	—	2.09×10^{-2}	1.43	-1.0893	103	N	0.2	104	Ab
38	*Sr	87	9/2	2.0896	7.02	1.18×10^{-2}	1.16	-1.0893	47	N			
39	*Y	89	1/2	6.17	100	2.44×10^{-4}	0.163	-0.13682	105	N	-0.16	105	Ab
39	Y	90*	2	2.49	—	1.90×10^{-4}	1.90	0.163	60	N			
40	*Zr	91	5/2	3.97240	11.23	9.48×10^{-3}	0.482	-1.30284	107	N			
41	*Nb	93	9/2	10.407	100	3.23×10^{-3}	8.07	6.1435	107	O	-0.2	108	O
42	*Mo	95	5/2	2.774	15.72	3.43×10^{-3}	0.760	0.9097	45	N	0.12	109	N
42	*Mo	97	5/2	2.333	9.46	3.43×10^{-3}	0.776	-0.0289	112	N	1.1	109	N
43	*Tc	99*	9/2	9.5830	100	1.95×10^{-1}	7.43	5.6572	113	Mo	0.3	111	O
44	*Ru	99	5/2	1.44	12.72	1.95×10^{-2}	0.169	-0.294					
44	*Ru	101	5/2	2.1	17.07	1.41×10^{-1}	0.576	-0.69					
45	*Rh	103	1/2	1.3401	100	3.11×10^{-5}	3.16×10^{-3}	-0.08790	43	N	—		
46	*Pd	105	5/2	1.05	22.23	1.12×10^{-3}	0.534	-0.639	114	N	—		
47	Ag	104*	5	6.1	—	0.118	5.73	4.0	116	Ab	—		
47	Ag	105*	1/2	14.0	—	0.291	2.65	3.7	116	Ab	—		
47	Ag	106*	1/2	1.54	—	4.73×10^{-3}	3.62×10^{-3}	0.101	117	Ab			
47	*Ag	107	1/2	1.7229	51.82	6.62×10^{-3}	4.05×10^{-3}	-0.11301	47	N			
47	Ag	108*	1	32.0	—	1.13	2.01	4.2	119	Ab			
47	*Ag	109	1/2	1.9807	48.18	1.01×10^{-2}	4.65×10^{-3}	-0.12992	47	N			
47	Ag	110*	6	4.557	—	6.87×10^{-2}	5.99	3.587	120	N			
47	Ag	111*	1/2	2.21	—	1.40×10^{-2}	5.19×10^{-3}	-0.145	121	Ab			
47	Ag	112*	2	0.2077	—	9.00×10^{-4}	3.90×10^{-3}	0.0545	122	Ab			
47	Ag	113*	1/2	2.41	—	1.81×10^{-3}	5.66×10^{-3}	0.158	122	Ab			
48	Cd	109*	5/2	1.870	—	1.00×10^{-3}	0.515	-0.8162	123	O	0.8	123	O
48	Cd	111*	5/2	2.539	—	2.44×10^{-3}	0.693	-0.5293	124	O	0.8	124	O
48	*Cd	111	1/2	9.028	12.75	9.54×10^{-3}	0.212	-0.5922	125	N			
48	*Cd	113	1/2	9.445	12.26	1.09×10^{-2}	0.222	-0.6195	126	N			
48	Cd	115*	11/2	1.51	—	2.13×10^{-2}	1.69	-1.09	239	N	-0.79	239	O
48	Cd	115m*	1/2	9.862	—	1.24×10^{-2}	0.232	-0.6469	127	N	-0.61	127	Ab
49	In	113	9/2	9.3099	4.28	0.345	7.22	5.4960	45	O	1.14	97	Ab
49	In	114*	1	3.209	—	0.191	6.73	-0.2105	128	Ab			
49	*In	115	9/2	9.3301	95.72	0.347	7.23	5.5079	130	N	1.16	97	Ab
49	In	115m*	1/2	7.2	—			4.7	129	Ab			
49	In	116*	5	3.715	—	6.64×10^{-2}	6.03	-0.2437	131	N			
49	In	116*	5	6.43	—	0.137	4.21	4.21	132	N			
50	Sn	115	1/2	13.922	0.35	3.50×10^{-2}	0.327	-0.91320	45	N			
50	Sn	117	1/2	15.168	7.61	4.52×10^{-2}	0.356	-0.94490	125	N			
50	Sn	119	1/2	15.869	8.58	5.18×10^{-2}	0.373	-1.0409	125	N			
51	Sb	122*	2	10.189	57.25	0.160	2.79	3.3415	133	No	-0.5	134	O
51	Sb	123	7/2	7.34	42.75	3.94×10^{-3}	1.36	-1.90	136	M	-0.7	134	O
52	Te	119*	7/2	5.5176	—	4.57×10^{-3}	2.72	2.5334	45	M			
52	*Te	123	1/2	4.12	0.87	9.04×10^{-3}	9.67×10^{-3}	0.27	137	Ab			
52	*Te	125	1/2	11.16	6.99	1.8×10^{-2}	0.262	-0.7319	37	N			
53	I	125*	5/2	13.45	—	3.15×10^{-2}	0.316	-0.8584	37	M	-0.66	138	M
53	*I	127	5/2	9.0	100	0.116	2.51	3	138	N	-0.69	134	N
53	*I	129*	7/2	8.5183	—	9.34×10^{-3}	2.33	2.7937	139	N	-0.48	140	Ab
53	I	131*	7/2	5.6694	—	4.95×10^{-3}	2.80	2.6031	139	N	-0.41	141	Ab
54	*Xe	129	1/2	5.963	26.44	5.77×10^{-3}	2.94	2.738	141	N			
54	*Xe	131	3/2	11.777	21.18	2.12×10^{-2}	0.277	-0.77247	142	Ab	-0.12	143	O
55	Cs	127*	1/2	3.4911	—	2.76×10^{-2}	0.410	0.68697	142	Ab			
55	Cs	129*	5/2	21.8	—	0.134	0.512	1.43	143	N			
55	Cs	130*	1	22.4	—	0.146	0.526	1.47	144	N			
55	Cs	131*	5/2	10.7	—	4.20×10^{-3}	0.668	1.4	144	N			
55	Cs	132*	2	10.73	—	0.186	2.94	3.517	145	Ab			
55	*Cs	133	7/2	8.48	100	6.26×10^{-2}	1.59	2.22	144	N			
55	Cs	134*	4	5.58469		4.74×10^{-2}	2.75	2.56422	8	N	-3×10^{-3}	146	Ab
55	Cs	134m*	8	5.666	—	6.26×10^{-2}	3.55	2.973	148	N	0.43	147	O
55	Cs			1.0447	—	1.43×10^{-2}	2.36	1.0964	150	Ab		149	

Table 2–38 (continued)
NUCLEAR SPINS, MOMENTS, AND MAGNETIC RESONANCE FREQUENCIES

Isotope			Spin I	NMR Frequency in MHz for a 10 kilogauss field	Natural Abundance %	Relative Sensitivity for Equal Number of Nuclei		Magnetic Moment μ			Electric Quadrupole Moment Q		
Z	El	A				At constant field	At constant frequency	In multiples of the nuclear magneton (eh/4πMc)	Reference	Method	In multiples of barns ($10^{-24}cm^2$)	Reference	Method
55	Cs	135*	7/2	5.9096	—	5.62×10^{-2}	2.91	2.7134	148	Ab			
55	Cs	137*	7/2	6.1459	—	6.32×10^{-2}	3.03	2.8219	148	Ab			
56	•Ba	135	3/2	4.2296	6.59	4.90×10^{-3}	0.497	0.83229	151 152	N	0.25	153	O
56	•Ba	137	3/2	4.7315	11.32	6.86×10^{-3}	0.556	0.93107	151 152	N	0.2	153	O
57	•La	138	5*	5.6171	0.089	9.19×10^{-2}	5.28	3.6844	154	N	2.7	154	N
57	•La	139	7/2	6.0144	99.911	5.92×10^{-2}	2.97	2.7615	107	N	0.21	155	Ab
58	Ce	137*	3/2	4.6	—	6.20×10^{-3}	0.537	0.9	156	No			
58	Ce	137$_m^*$	11/2	0.96	—	5.40×10^{-4}	1.07	0.69	157	No			
58	Ce	139*	3/2	5.1	—	8.50×10^{-3}	0.597	1.0	156	No			
58	Ce	141*	7/2	2.1	—	2.57×10^{-3}	1.04	0.97	158	E			
58	Ce	143*	7/2	2.2	—	2.81×10^{-3}	1.07	1.0	156	No			
59	•Pr	141	5/2	12.5	100	0.293	3.42	4.09	159 160	O	-5.9×10^{-2}	158	Ab
59	Pr	142*	2	1.1	—	1.55×10^{-4}	0.215	0.30	161	Ab	4×10^{-2}	161	Ab
60	Nd	143	7/2	2:315	12.17	3.38×10^{-3}	1.14	−1.063	162	Ab	−0.48	162	Ab
60	Nd	145	7/2	1.42	8.30	7.86×10^{-4}	0.703	−0.654	162	Ab	−0.25	162	Ab
60	Nd	147*	5/2	1.77	—	8.32×10^{-4}	0.484	0.579	158	E			
61	Pm	143*	(5/2) (7/2)	11.6 8.5	—	0.235 0.167	3.17 4.19	(3.8) (3.9)	163	No			
61	Pm	144*	(5) (6)	2.6 2.3	—	9.02×10^{-3} 8.68×10^{-3}	2.43 3.01	(1.7) (1.8)	164	No			
61	Pm	147*	7/2	5.62	—	4.83×10^{-2}	2.77	2.58	165	O	0.7	165	O
61	Pm	148*	1	16	—	0.142	1.00	2.1	166	Ab	0.2	166	Ab
61	Pm	148$_m^*$	6	2.3	—	8.68×10^{-3}	3.01	1.8	163	No			
61	Pm	149*	7/2	7.2	—	0.101	3.54	3.3	163	No			
61	Pm	151*	5/2	5.5	—	2.50×10^{-2}	1.50	1.8	167	Ab	1.9	167	Ab
62	Sm	147	7/2	1.76	14.97	1.48×10^{-3}	0.867	−0.807	168	Ab	−0.208	158	Ab
62	Sm	149	7/2	1.40	13.83	7.47×10^{-4}	0.691	−0.643	158	Ab	6.0×10^{-2}	158	Ab
63	•Eu	151	5/2	10.559	47.82	0.178	2.89	3.4630	169	Ab	1.16	170	O
63	Eu	152*	3	4.858	—	2.38×10^{-2}	1.83	1.912	171	Ab			
63	•Eu	153	5/2	4.6627	52.18	1.53×10^{-2}	1.28	1.5292	169 172	Ab	2.9	170	O
63	Eu	154*	3	5.084	—	2.72×10^{-2}	1.91	2.001	158	E			
64	•Gd	155	3/2	1.6	14.73	2.79×10^{-4}	0.191	−0.32	173 174	O	1.6	173	O
64	•Gd	157	3/2	2.0	15.68	5.44×10^{-4}	0.239	−0.40	173 174	O	2	173	O
65	Tb	156*	3	3.8	—	1.15×10^{-2}	1.43	1.5	175	No	1.4	175	No
65	Tb	159	3/2	9.66	100	5.83×10^{-2}	1.13	1.90	158	E	1.3	176	O
65	Tb	160*	3	4.1	—	1.39×10^{-2}	1.53	1.6	177	No	1.9	177	No
66	Dy	155*	(3/2)	1.1	—	7.87×10^{-5}	0.125	0.21	178	No			
66	Dy	157*	(3/2)	1.6	—	2.79×10^{-4}	0.191	0.32	178	No			
66	Dy	161	5/2	1.4	18.88	4.17×10^{-4}	0.384	−0.46	158	E	1.4	158	E
66	Dy	163	5/2	2.0	24.97	1.12×10^{-3}	0.535	0.64	158	E	1.6	158	E
67	Ho	165	7/2	8.73	100	0.181	4.31	4.01	158	Ab	2.82	158	Ab
68	Er	165*	5/2	2.0	—	1.18×10^{-3}	0.543	0.65	166	Ab	2.2	166	Ab
68	Er	167	7/2	1.23	22.94	5.07×10^{-4}	0.607	−0.565	162	Ab	2.83	162	Ab
68	Er	169*	1/2	7.8	—	6.09×10^{-3}	0.183	0.51	179	Ab	—		
68	Er	171*	5/2	2.1	—	1.47×10^{-3}	0.585	0.70	180	Ab			
69	Tm	166*	2	0.19	—	7.17×10^{-7}	3.58×10^{-2}	0.05	181	Ab	4.6	181	Ab
69	•Tm	169	1/2	3.52	100	5.66×10^{-4}	8.27×10^{-2}	−0.231	182 160	Ab	—		
69	Tm	170*	1	2.0	—	2.69×10^{-4}	0.124	0.26	183	Ab	0.61	183	Ab
69	Tm	171*	1/2	3.46	—	5.37×10^{-4}	8.13×10^{-2}	0.227	180	Ab	—		
70	•Yb	171	1/2	7.4990	14.31	5.46×10^{-3}	0.176	0.49188	184 185	O	—		
70	Yb	173	5/2	2.0659	16.13	1.33×10^{-3}	0.566	−0.67755	184 185	O	2.8	186	O
70	Yb	175*	(7/2)	0.33	—	9.40×10^{-6}	0.161	−0.15	187	No			
71	•Lu	175	7/2	4.86	97.41	3.12×10^{-2}	2.40	2.23	188	N	5.68	189	Ab
71	Lu	176*	7	3.4	2.59	3.72×10^{-2}	5.92	3.1	190	Ab	8.0	190	Ab
71	Lu	177*	7/2	4.84	—	3.08×10^{-2}	2.38	2.22	191	Ab	5.51	191	Ab
72	Hf	177	7/2	1.3	18.50	6.38×10^{-4}	0.655	0.61	192 193	O	3	192 193	O
72	Hf	179	9/2	0.80	13.75	2.16×10^{-4}	0.617	−0.47	193	O	3	193	O
73	•Ta	181	7/2	5.096	99.988	3.60×10^{-2}	2.51	2.340	194	N	3	195	O
74	•W	183	1/2	1.7716	14.40	7.20×10^{-5}	4.16×10^{-2}	0.116205	196	N	—		
75	•Re	185	5/2	9.5855	37.07	0.133	2.63	3.1437	13	N	2.8	97 241	O
75	Re	186*	1	13.17	—	7.90×10^{-2}	0.825	1.728	197	Ab			
75	•Re	187*	5/2	9.6837	62.93	0.137	2.65	3.1759	13	N	2.6	97 241	O
75	Re	188*	1	13.55	—	8.59×10^{-2}	0.848	1.777	197	Ab			
76	•Os	187	1/2	0.98059	1.64	1.22×10^{-5}	2.30×10^{-2}	0.06432	198	N	—		
76	•Os	189	3/2	3.3034	16.1	2.34×10^{-3}	0.388	0.65004	199	N	0.8	195	O
77	•Ir	191	3/2	0.7318	37.3	2.53×10^{-5}	8.59×10^{-2}	0.1440	200 201	N	1.5	202	O
77	•Ir	193	3/2	0.7968	62.7	3.27×10^{-5}	9.36×10^{-2}	0.1568	200 201	N	1.5	202	O
78	•Pt	195	1/2	9.153	33.8	9.94×10^{-3}	0.215	0.6004	45	N	—		
79	Au	190*	1	0.496	—	4.20×10^{-6}	3.10×10^{-2}	0.065	203	Ab			
79	Au	194*	1	0.56	—	5.95×10^{-6}	3.49×10^{-2}	0.073	204	Ab			
79	Au	195*	3/2	0.742	—	2.65×10^{-5}	8.71×10^{-2}	0.146	204	Ab			
79	Au	196*	2	2.3	—	1.24×10^{-3}	0.430	0.6	204	Ab			
79	•Au	197	3/2	0.729188	100	2.51×10^{-5}	8.56×10^{-2}	0.143489	205 201	Ab	0.59	206	Ab
79	Au	198*	2	2.227	—	1.14×10^{-3}	0.418	0.5842	207	Ab			
79	Au	199*	3/2	1.358	—	1.62×10^{-4}	0.160	0.2673	207	Ab			
80	Hg	193*	3/2	3.1	—	1.93×10^{-3}	0.364	−0.61	208	O			
80	Hg	193$_m^*$	13/2	1.23	—	1.57×10^{-3}	1.88	−1.05	209	O	1.37	209	O
80	Hg	195*	1/2	8.1	—	6.84×10^{-3}	0.190	0.53	210 211	O	—		
80	Hg	195$_m^*$	13/2	1.22	—	1.53×10^{-3}	1.86	−1.04	209	O	1.41	209	O
80	Hg	197*	1/2	7.9	—	6.46×10^{-3}	0.186	0.52	212	O	—		
80	•Hg	199	1/2	7.59012	16.84	5.67×10^{-3}	0.178	0.497859	213 45	No			
80	•Hg	201	3/2	2.8099	13.22	1.44×10^{-3}	0.330	−0.55293	214 215	O	0.50	216	O
80	Hg	203*	5/2	2.5	—	2.45×10^{-3}	0.693	0.83	217	O	0.5	217	O
81	Tl	197*	1/2	23.6	—	0.171	0.555	1.55	218	O	—		
81	Tl	199*	1/2	23.9	—	0.178	0.562	1.57	219	O	—		
81	Tl	200*	2	0.57	—	1.94×10^{-5}	0.107	(0.15)	219	O			
81	Tl	201*	1/2	24.1	—	0.181	0.566	1.58	219	O	—		
81	Tl	202*	2	0.57	—	1.94×10^{-5}	0.107	(0.15)	219	O			
81	•Tl	203	1/2	24.332	29.50	0.187	0.571	1.5960	220	N	—		
81	Tl	204*	2	0.34	—	4.05×10^{-6}	6.37×10^{-2}	0.089	221	Ab			
81	•Tl	205	1/2	24.570	70.50	0.192	0.577	1.6116	8	N	—		
82	•Pb	207	1/2	8.90771	22.6	9.16×10^{-3}	0.209	0.584284	222	N	—		
83	Bi	203*	9/2	7.78	—	0.201	6.03	4.59	223	Ab	−0.64	223	Ab
83	Bi	204*	6	5.40	—	0.114	7.10	4.25	223	Ab	−0.41	223	Ab
83	Bi	205*	9/2	9.3	—	0.346	7.22	(5.5)	223	Ab			
83	Bi	206*	6	5.79	—	0.141	7.62	4.56	223	Ab	−0.19	223	Ab

Table 2–38 (continued)
NUCLEAR SPINS, MOMENTS, AND MAGNETIC RESONANCE FREQUENCIES

Isotope			Spin I	NMR Frequency in MHz for a 10 kilogauss field	Natural Abundance %	Relative Sensitivity for Equal Number of Nuclei At constant field	At constant frequency	Magnetic Moment μ In multiples of the nuclear magneton ($eh/4\pi Mc$)	Reference	Method	Electric Quadrupole Moment Q In multiples of barns ($10^{-24}\,cm^2$)	Reference	Method
Z	El	A											
83	*Bi	209*	9/2	6.84178	100	0.137	5.30	4.03896	224	N	-0.4	97	O
83	Bi	210*	1	0.337	—	1.32×10^{-4}	2.11×10^{-1}	0.0442	225	Ab	0.13	225	Ab
84	Po	205*	5/2	0.79	—	7.55×10^{-3}	0.217	0.26	226	Ab	0.17	226	Ab
84	Po	207*	5/2	0.82	—	8.43×10^{-3}	0.226	0.27	226	Ab	0.28	226	Ab
89	Ac	227*	3/2	5.6	—	1.13×10^{-2}	0.656	1.1	227	O	-1.7	227	O
90	Th	229*	5/2	1.2	—	2.74×10^{-4}	0.334	0.4	228	O	4.6	228	O
91	Pa	231*	3/2	9.96	—	6.40×10^{-2}	1.17	1.96	229	E			
91	Pa	233*	3/2	2.03	—	0.334		3.4	230	Ab	-3.0	230	Ab
92	U	233*	5/2	1.6	—	6.75×10^{-4}	0.451	0.54	231	E	3.5	231	E
92	U	235*	7/2	0.76	0.72	1.21×10^{-4}	0.376	0.35	231	E	4.1	231	E
93	Np	237*	5/2	18	—	0.926	5.01	(6)	232	E	—		
94	Pu	239*	1/2	3.05	—	3.67×10^{-4}	7.16×10^{-2}	0.200	233	Ab	—		
94	Pu	241*	5/2	2.09	—	1.38×10^{-3}	0.573	-0.686	234	O	4.9	236	O
95	Am	241*	5/2	4.83	—	1.69×10^{-2}	1.32	1.58	235	Ab	-2.8	237	Ab
95	Am	242*	1	2.90	—	8.46×10^{-4}	0.182	0.381	235	Ab	4.9	236	Ab
95	Am	243*	5/2	4.79	—	1.65×10^{-4}	1.31	1.57	238	O			
Free electron with g=2.00232			1/2	2.80246×10^{4}	—	2.84×10^{-8}	657	-1836.09	—		—		

From Lee, K. and Anderson, W. A., in *Handbook of Chemistry and Physics*, 55th ed., Weast, R. C., Ed., CRC Press, Inc., Cleveland, 1974, E-69.

Table 2–38 (continued)
NUCLEAR SPINS, MOMENTS, AND MAGNETIC RESONANCE FREQUENCIES

REFERENCES

1. Cohen, V. W. et al., *Phys. Rev.,* 104, 283, 1956.
2. Sommer, H. et al., *Phys. Rev.,* 82, 697, 1951.
3. Wimett, T. F., *Phys. Rev.,* 91, A499, 1953.
4. Kopfermann, H. et al., *Z. Phys.,* 144, 9, 1956.
5. Bloch, F. et al., *Phys. Rev.,* 71, 551, 1947.
6. Anderson, H. L., *Phys. Rev.,* 76, 1460, 1949.
7. Klein, M. P. et al., *Phys. Rev.,* 106, 837, 1957.
8. Walchli, H. E., M.S., thesis, University of Tennessee, Knoxville, 1954; AEC Report ORNL-1775.
9. Schuster, N. A. et al., *Phys. Rev.,* 81, 157, 1951.
10. Brog, K. C. et al., *Phys. Rev.,* 153, 91, 1967.
11. Connor, D., *Phys. Rev. Lett.,* 3, 429, 1959.
12. Brown, L. C. et al., *J. Chem. Phys.,* 24, 751, 1956.
13. Alder, F. et al., *Phys. Rev.,* 82, 105, 1951.
14. Blachman, A. G. et al., *Bull. Am. Phys. Soc.,* 11, 343, 1966.
15. Ting, Y. et al., *Phys. Rev.,* 89, 595, 1953.
16. Wessel, G., *Phys. Rev.,* 92, 1581, 1953.
17. Lindström, G., *Ark. Fys.,* 4, 1, 1951.
18. Royden, V., *Phys. Rev.,* 96, 543, 1954.
19. Bernstein, A. M. et al., *Phys. Rev.,* 136, B27, 1964.
20. Anderson, L. W. et al., *Phys. Rev.,* 116, 87, 1959.
21. Baker, M. R. et al., *Phys. Rev.,* 133, A1533, 1964.
22. Bassompiere, A., *Compt. Rend.,* 240, 285, 1955.
23. Commins, E. D. et al., *Phys. Rev.,* 131, 700, 1963.
24. Alder, F. et al., *Phys. Rev.,* 81, 1067, 1951.
25. Stevenson, M. J. et al., *Phys. Rev.,* 107, 635, 1957.
26. Sugimoto, K. et al., *J. Phys. Soc. Jap.,* 21, 213, 1966.
27. Tsang, T. et al., *Phys. Rev.,* 132, 1141, 1963.
28. Commins, E. D. et al., *Phys. Rev. Lett.,* 10, 347, 1963.
29. LaTourette, J. T. et al., *Phys. Rev.,* 107, 1202, 1957.
30. Ames, O. et al., *Phys. Rev.,* 137, B1157, 1965.
31. Davis, L. et al., *Phys. Rev.,* 76, 1068, 1949.
32. Ackermann, H. et al., *Z. Phys.,* 194, 253, 1966.
33. Baumann, M. et al., *Z. Phys.,* 194, 270, 1966.
34. Chan, Y. W. et al., *Phys. Rev.,* 150, 933, 1966.
35. Brown, L. C. et al., *J. Chem. Phys.,* 24, 751, 1956.
36. Lew, H. et al., *Phys. Rev.,* 90, 1, 1953.
37. Weaver, H. E., *Phys. Rev.,* 89, 923, 1953.
38. Kanda, T. et al., *Phys. Rev.,* 85, 938, 1952.
39. Feher, G. et al., *Phys. Rev.,* 107, 1462, 1957.
40. Bird, G. R. et al., *Phys. Rev.,* 94, 1203, 1954.
41. Burke, B. F. et al., *Phys. Rev.,* 93, 193, 1954.
42. Jaccarino, V. et al., *Phys. Rev.,* 83, 471, 1951.
43. Sogo, P. B. et al., *Phys. Rev.,* 98, 1316, 1955.
44. Townes, C. H. et al., *Phys. Rev.,* 76, 691, 1949.
45. Proctor, W. G. et al., *Phys. Rev.,* 81, 20, 1951.
46. Phillips, E. A. et al., *Phys. Rev.,* 138, B773, 1965.
47. Brun, E. et al., *Phys. Rev.,* 93, 172, 1954.
48. Lutz, O. et al., *Phys. Lett.,* 24A, 122, 1967.
49. Series, G. W., *Phys. Rev.,* 105, 1128, 1957.
50. Ritter, G. J. et al., *Proc. R. Soc. London,* 238, 473, 1957.
51. Eisinger, J. T. et al., *Phys. Rev.,* 86, 73, 1952.
52. Kahn, J. M. et al., *Phys. Rev.,* 134, A45, 1964.
53. Peterson, F. R. et al., *Phys. Rev.,* 116, 734, 1959.
54. Brun, E. et al., *Phys. Rev. Lett.,* 9, 166, 1962.
55. Jeffries, C. D., *Phys. Rev.,* 90, 1130, 1953.
56. Cornwall, R. G. et al., *Phys. Rev.,* 141, 1106, 1966.
57. Harris, D. L. et al., *Phys. Rev.,* 132, 310, 1963.

Table 2–38 (continued)
NUCLEAR SPINS, MOMENTS, AND MAGNETIC RESONANCE FREQUENCIES

58. Hunten, D. M., *Can. J. Phys.,* 29, 463, 1951.
59. Fricke, G. et al., *Z. Phys.,* 156, 416, 1959.
60. Petersen, F. R. et al., *Phys. Rev.,* 128, 1740, 1962.
61. Cornwall, R. G. et al., *Phys. Rev.,* 148, 1157, 1966.
62. Jeffries, C. D., *Phys. Rev.,* 92, 1262, 1953.
63. Weiss, M. M. et al., *Bull. Am. Phys. Soc.,* 2, 31, 1957.
64. Walchli, H. E. et al., *Phys. Rev.,* 87, 541, 1952.
65. Murakawa, K. et al., *J. Phys. Soc. Jap.,* 21, 1466, 1966.
66. Jeffries, C. D. et al., *Phys. Rev.,* 91, 1286, 1953.
67. Adelroth, K. E. et al., *Ark. Fys.,* 31, 549, 1966.
68. Phillips, E. A. et al., *Phys. Rev.,* 140, B555, 1965.
69. Dobrowolski, W. et al., *Phys. Rev.,* 104, 1378, 1956.
70. Bauer, R. W. et al., *Phys. Rev.,* 120, 946, 1960.
71. Mims, W. B. et al., *Phys. Lett.,* 24A, 481, 1967.
72. Javan, A. et al., *Phys. Rev.,* 96, 649, 1954.
73. Childs, W. J. et al., *Phys. Rev.,* 122, 891, 1961.
74. Locher, P. R. et al., *Phys. Rev.,* 139, A991, 1965.
75. Bauer, R. W. et al., *Nucl. Phys.,* 16, 264, 1960.
76. Baker, J. M. et al., *Proc. Phys. Soc.* (London), 69A, 354, 1956.
77. Dobrov, W. et al., *Phys. Rev.,* 108, 60, 1957.
78. Ehrenstein, D. V. et al., *Z. Phys.,* 159, 230, 1960.
79. Dobrowolski, W. et al., *Phys. Rev.,* 101, 1001, 1956.
80. Drain, L. E., *Phys. Lett.,* 11, 114, 1964.
81. Dodsworth, B. M. et al., *Phys. Rev.,* 142, 638, 1966.
82. Bleaney, B. et al., *Proc. R. Soc. London,* 228A, 166, 1955.
83. Lemonick, A. et al., *Phys. Rev.,* 95, 1356, 1954.
84. Ehlers, V. J. et al., *Phys. Rev.,* 127, 529, 1962.
85. Daly, R. T. et al., *Phys. Rev.,* 96, 539, 1954.
86. Childs, W. J. et al., *Phys. Rev.,* 120, 2138, 1960.
87. Childs, W. J. et al., *Phys. Rev.,* 141, 15, 1966.
88. Mays, J. M. et al., *Phys. Rev.,* 81, 940, 1951.
89. Dailey, B. P. et al., *Phys. Rev.,* 74, 1245, 1948.
90. Pipkin, F. M. et al., *Phys. Rev.,* 106, 1102, 1957.
91. Hardy, W. A. et al., *Phys. Rev.,* 92, 1532, 1953.
92. Lipworth, E. et al., *Phys. Rev.,* 119, 1053, 1960.
93. King, J. G. et al., *Phys. Rev.,* 94, 1610, 1954.
94. White, M. B. et al., *Phys. Rev.,* 136, B584, 1964.
95. Garvin, H. L. et al., *Phys. Rev.,* 116, 393, 1959.
96. Brun, E. et al., *Helv. Phys. Acta,* 27, A173, 1954.
97. Mack, J. E., *Rev. Mod. Phys.,* 22, 64, 1950.
98. Rasmussen, E. et al., *Z. Phys.,* 141, 160, 1955.
99. Hubbs, J. C. et al., *Phys. Rev.,* 107, 723, 1957.
100. Walchli, H. et al., *Phys. Rev.,* 85, 922; 1952.
101. Senitzky, B. et al., *Phys. Rev.,* 103, 315, 1956.
102. Bellamy, E. H. et al., *Philos. Mag.,* 44, 33, 1953.
103. Yasaitis, E. et al., *Phys. Rev.,* 82, 750, 1951.
104. Culvahouse, J. W. et al., *Phys. Rev.,* 140, A1181, 1965.
105. Peterson, R. F. et al., *Phys. Rev.,* 125, 284, 1962.
106. Brun, E. et al., *Phys. Rev.,* 105, 1929, 1957.
107. Sheriff, R. E. et al., *Phys. Rev.,* 82, 651, 1951.
108. Murakawa, K., *Phys. Rev.,* 98, 1285, 1955.
109. Narath, A. et al., *Phys. Rev.,* 143, 328, 1966.
110. Walchli, H. et al., *Phys. Rev.,* 85, 479, 1952.
111. Kessler, K. G. et al., *Phys. Rev.,* 92, 303, 1953.
112. Matthias, E. et al., *Phys. Rev.,* 139, B532, 1965.
113. Murakawa, K., *J. Phys. Soc. Jap.,* 10, 919, 1955.
114. Seitchik, J. A. et al., *Phys. Rev.,* 138, A148, 1965.
115. Seitchik, J. A. et al., *Phys. Rev.,* 136, A1119, 1964.
116. Ames, O. et al., *Phys. Rev.,* 123, 1793, 1960.

Table 2–38 (continued)
NUCLEAR SPINS, MOMENTS, AND MAGNETIC RESONANCE FREQUENCIES

117. Ewbank, W. B. et al., *Phys. Rev.*, 129, 1617, 1963.
118. Sogo, P. B. et al., *Phys. Rev.*, 93, 174, 1954.
119. Rochester, G. K. et al., *Phys. Lett.*, 8, 266, 1964.
120. Schmelling, S. G. et al., *Phys. Rev.*, 154, 1142, 1967.
121. Woodgate, G. K. et al., *Proc. Phys. Soc.* (London), 69, 581, 1956.
122. Chan, Y. W. et al., *Phys. Rev.*, 133, B1138, 1964.
123. Byron, F. W. et al., *Phys. Rev.*, 132, 1181, 1963.
124. Thaddeus, P. et al., *Phys. Rev.*, 132, 1186, 1963.
125. Proctor, W. G., *Phys. Rev.*, 79, 35, 1950.
126. Leduc, M. et al., *Compt. Rend.*, 262B, 736, 1966.
127. McDermott, M. N. et al., *Phys. Rev.*, 134, B25, 1964.
128. Childs, W. J. et al., *Phys. Rev.*, 118, 1578, 1960.
129. Goodman, L. S. et al., *Phys. Rev.*, 108, 1524, 1957.
130. Rice, M. et al., *Phys. Rev.*, 106, 953, 1957.
131. Cameron, J. A. et al., *Can. J. Phys.*, 40, 931, 1962.
132. Nutter, P. B., *Philos. Mag.*, 1, 587, 1956.
133. Cohen, V. W. et al., *Phys. Rev.*, 79, 191, 1950.
134. Murakawa, K., *Phys. Rev.*, 100, 1369, 1955.
135. Pipkin, F. M., *Bull. Am. Phys. Soc.*, 3, 8, 1958.
136. Fernando, P. C. B. et al., *Philos. Mag.*, 5, 1309, 1960.
137. Adelroth, K. E. et al., *Ark. Fys.*, 30, 111, 1965.
138. Fletcher, P. C. et al., *Phys. Rev.*, 110, 536, 1958.
139. Walchli, H. et al., *Phys. Rev.*, 82, 97, 1951.
140. Livingston, R. et al., *Phys. Rev.*, 90, 609, 1953.
141. Lipworth, E. et al., *Phys. Rev.*, 119, 2022, 1960.
142. Brinkmann, D., *Helv. Phys. Acta*, 36, 413, 1958.
143. Bohr, A. et al., *Ark. Fys.*, 4, 455, 1952.
144. Nierenberg, W. A. et al., *Phys. Rev.*, 112, 186, 1958.
145. Worley, R. D. et al., *Phys. Rev.*, 140, B1483, 1965.
146. Buck, P. et al., *Phys. Rev.*, 104, 553, 1956.
147. Althoff, K. H. et al., *Naturwissenschaften*, 41, 368, 1954.
148. Stroke, H. H. et al., *Phys. Rev.*, 105, 590, 1957.
149. Heinzelmann, G. et al., *Phys. Lett.*, 21, 162, 1966.
150. Cohen, V. W. et al., *Phys. Rev.*, 127, 517, 1962.
151. Walchli, H. E. et al., *Phys. Rev.*, 102, 1334, 1956.
152. Olschewski, L. et al., *Z. Phys.*, 196, 77, 1966.
153. Kaliteevskii, N. I. et al., *Sov. Phys. J.E.T.P.*, 12, 661, 1961.
154. Sogo, P. B. et al., *Phys. Rev.*, 99, 613, 1955.
155. Murakawa, K. et al., *J. Phys. Soc. Jap.*, 16, 2533, 1961.
156. Haag, J. N. et al., *Phys. Rev.*, 129, 1601, 1963.
157. Blok, J. et al., *Phys. Rev.*, 143, 78, 1966.
158. Bleaney, B., in *Quantum Electronics, Proceedings of the Third International Conference, Paris, 1963*, Grivet, P. and Bloembergen, N. Eds., Columbia University Press, New York, 1964.
159. Reader, J. et al., *Phys. Rev.*, 137, B784, 1965.
160. Jones, E. D., *Phys. Rev. Lett.*, 19, 432, 1967.
161. Cabezas, A. Y. et al., *Phys. Rev.*, 126, 1004, 1962.
162. Smith, K. F. et al., *Proc. Phys. Soc.* (London), 86, 1249, 1965.
163. Grant, R. W. et al., *Phys. Rev.*, 130, 1100, 1963.
164. Shirley, D. A. et al., *Phys. Rev.*, 121, 558, 1961. o165.
165. Reader, J., *Phys. Rev.*, 141, 1123, 1966.
166. Ali, D. et al., *Phys. Rev.*, 138, B1356, 1965.
167. Budick, B. et al., *Phys. Rev.*, 132, 723, 1963.
168. Woodgate, G. D., *Proc. R. Soc. London*, 293A, 117, 1966.
169. Evans, L. et al., *Proc. R. Soc. London*, 289A, 114, 1965.
170. Müller, W. et al., *Z. Phys.*, 183, 303, 1965.
171. Alpert, S. S., *Phys. Rev.*, 129, 1344, 1963.
172. Boyd, E. L., *Phys. Rev.*, 145, 174, 1966.
173. Kaliteevskii, N. I. et al., *Sov. Phys. J.E.T.P.*, 37, 629, 1960.

174. Boyd, E. L. et al., *Phys. Rev. Lett.*, 12, 20, 1964.
175. Lovejoy, C. A. et al., *Nucl. Phys.*, 30, 452, 1962.
176. Arnoult, C. et al., *J. Opt. Soc. Am.*, 56, 177, 1966.
177. Johnson, C. E. et al., *Phys. Rev.*, 120, 2108, 1960.
178. Navarro, Q. O. et al., *Phys. Rev.*, 123, 186, 1961.
179. Doyle, W. M. et al., *Phys. Rev.*, 131, 1586, 1963.
180. Budick, B. et al., *Phys. Rev.*, 135, B1281, 1964.
181. Walker, J. C., *Phys. Rev.*, 127, 1739, 1962.
182. Giglberger, D. et al., *Z. Phys.*, 199, 244, 1967.
183. Cabezas, A. Y. et al., *Phys. Rev.*, 120, 920, 1960.
184. Olschewski, L. et al., *Z. Phys.*, 200, 224, 1967.
185. Gossard, A. C. et al., *Phys. Rev.*, 133, A881, 1964.
186. Ross, S. J. et al., *Phys. Rev.*, 128, 1159, 1962.
187. Grace, M. A. et al., *Philos. Mag.*, 2, 1079, 1957.
188. Reddoch, A. H. et al., *Phys. Rev.*, 126, 1493, 1962.
189. Ritter, G. J., *Phys. Rev.*, 126, 240, 1962.
190. Spalding, I. J. et al., *Proc. Phys. Soc.* (London), 79, 787, 1962.
191. Peterson, F. R. et al., *Phys. Rev.*, 126, 252, 1962.
192. Speck, D. R., *Bull. Am. Phys. Soc.*, 1, 282, 1956.
193. Speck, D. R et al., *Phys. Rev.*, 101, 1831, 1956.
194. Bennett, L. H. et al., *Phys. Rev.*, 120, 1812, 1960.
195. Murakawa, K. et al., *Phys. Rev.*, 105, 671, 1957.
196. Klein, M. P. et al., *Bull. Am. Phys. Soc.*, 6, 104, 1961.
197. Armstrong, L. et al., *Phys. Rev.*, 138, B310, 1965.
198. Kaufmann, J. et al., *Phys. Lett.*, 24A, 115, 1967.
199. Loeliger, H. R. et al., *Phys. Rev.*, 95, 291, 1954.
200. Narath, A., *Phys. Rev.*, 165, 506, 1968.
201. Narath, A. et al., *Bull. Am. Phys. Soc.*, 12, 314, 1967.
202. von Siemens, W., *Ann. Phys.*, 13, 136, 1953.
203. Chan, Y. W. et al., *Phys. Rev.*, 144, 1020, 1966.
204. Chan, Y. W. et al., *Phys. Rev.*, 137, B1129, 1965.
205. Dahmen, H. et al., *Z. Phys.*, 200, 456, 1967.
206. Childs, W. J. et al., *Phys. Rev.*, 141, 176, 1966.
207. Vanden Bout, P. A. et al., *Phys. Rev.*, 158, 1078, 1967.
208. Kleiman, H. et al., *Phys. Lett.*, 13, 212, 1964.
209. Tomlinson, W. J. et al., *Nucl. Phys.*, 60, 614, 1964.
210. Kleiman, H. et al., *J. Opt. Soc. Am.*, 53, 822, 1963.
211. Tomlinson, W. J. et al., *J. Opt. Soc. Am.*, 53, 828, 1963.
212. Bitter, F. et al., *Phys. Rev.*, 96, 1531, 1954.
213. Cagnac, B. et al., *Compt. Rend.*, 249, 77, 1959.
214. Cagnac, B., *Ann. Phys.*, 6, 467, 1961.
215. Lehmann, J. C. et al., *Compt. Rend.*, 257, 3152, 1963.
216. Blaise, J. et al., *J. Phys. Radium*, 18, 193, 1957.
217. Redi, O. et al., *Phys. Lett.*, 8, 257, 1964.
218. Davis, S. P. et al., *J. Opt. Soc. Am.*, 56, 1604, 1966.
219. Hull, R. J. et al., *Phys. Rev.*, 122, 1574, 1961.
220. Poss, H. L., *Phys. Rev.*, 75, 600, 1949.
221. Brink, G. O. et al., *Phys. Rev.*, 107, 189, 1957.
222. Baker, E. B., *J. Chem. Phys.*, 26, 960, 1957.
223. Lindgren, I. et al., *Ark. Fys.*, 15, 445, 1959.
224. Ting, Y. et al., *Phys. Rev.*, 89, 595, 1953.
225. Alpert, S. S. et al., *Phys. Rev.*, 125, 256, 1962.
226. Olsmats, C. M. et al., *Ark. Fys.*, 19, 469, 1961.
227. Fred, M. et al., *Phys. Rev.*, 98, 1514, 1955.
228. Egorov, V. N., *Opt. Spektrosk.* (U.S.S.R.), 16, 301, 1964.
229. Axe, J. D. et al., *Phys. Rev.*, 121, 1630, 1961.
230. Marrus, R. et al., *Nucl. Phys.*, 23, 90, 1961.
231. Dorain, P. B. et al., *Phys. Rev.*, 105, 1307, 1957.

232. Bleaney, B. et al., *Philos. Mag.,* 45, 992, 1954.
233. Faust, J. et al., *Phys. Lett.,* 16, 71, 1965.
234. Champeau, R. J. et al., *Compt. Rend.,* 257, 1238, 1963.
235. Armstrong, L. et al., *Phys. Rev.,* 144, 994, 1966.
236. Manning, T. E. et al., *Phys. Rev.,* 102, 1108, 1956.
237. Marrus, R. et al., *Phys. Rev.,* 124, 1904, 1961.
238. Fred, M. et al., *J. Opt. Soc. Am.,* 47, 1076, 1957.
239. Perry, B. et al., *Bull. Am. Phys. Soc.,* 8, 345, 1963.
240. Robertson, M. M. et al., *Phys. Rev.,* 140, B820, 1965.
241. Walstedt, R. E. et al., *Phys. Rev.,* 162, 301, 1967.
242. Freeman, R. et al., *Proc. R. Soc. London,* 242A, 455, 1957.

Table 2–39
ELECTRON WORK FUNCTIONS OF THE ELEMENTS

Preferred values for well-outgassed poly-crystalline materials are indicated by an asterisk (*). Because of the anisotropy and allotropy of work function, however, the designation "pre-ferred value" has a rather limited meaning.

Values are expressed in electron-volts. Each value is followed by a reference to the literature.

Element	Thermionic work function	Ref.	Photoelectric work function	Ref.	Work function by contact potential method	Ref.
Ag	3.09	1	3.67	4	4.21	116
	3.56	2	4.1–4.75	5	4.33	8
	4.08	3	4.50–4.52	117	4.35	115
	4.31	114	4.56 (600°)*	6	4.44	9
			4.73 (20°)*	6	4.47	10
			4.75 (111) face	7		
			4.81 (100) face	7		
Al			2.98	4	3.38	9
			3.43	11	4.25	115
			4.08*	12		
			4.20	113		
			4.36	13		
As			4.72	118		
			5.11	14		
Au	4.0–4.58	2	4.73 (740°C)*	15	4.46	9
	4.25	114	4.82 (20°C)*	15		
	4.32	3	4.86–4.92	5		
B			4.4–4.6	16		
Ba	2.11	17	2.48*	19	1.73	9
			2.49	20	2.39	18
			2.52	21		
Be			3.17	14	3.10	9
			3.30	22		
			3.92*	23		

Table 2–39 (continued)
ELECTRON WORK FUNCTIONS OF THE ELEMENTS

Element	Thermionic work function	Ref.	Photoelectric work function	Ref.	Work function by contact potential method	Ref.
Bi			4.14	11	4.17	9
			4.22–4.25*	24		
			4.31	25		
			4.34	118		
			4.44	22		
			4.46	47		
C	4.00	26	4.81	29		
	4.34	27				
	4.35	119				
	4.39	28				
	4.60*	120				
	4.82	121				
	4.84	122				
Ca	2.24	30	2.42	14	3.33	9
			2.706*	21		
			2.76	31		
			3.20	32		
			3.21	33		
Cd			3.68	14	4.00	9
			3.73	4	4.49	123
			3.94	11		
			4.07*	36		
			4.099	113		
Ce	2.6	37	2.84	32		
Co	4.40*	38	3.90	31	4.21	9
	4.41	124	4.12–4.25	39		
Cr	4.58	124	4.37	25	4.38	9
	4.60*	40				
	4.7	41				
Cs	1.81*	42	1.9	43		
			1.96	44		
	3.85	1	4.07	4	4.46*	9, 126
	4.26	2	4.18	31	4.61	115
	4.38	3	4.70	125	4.86	123
	4.50	114	4.86 (111) face	46		
	4.55ᵃ	45	5.61 (100) face	46		
Fe	4.04	48	3.91	31	4.40	9
	4.21 (γ)	38	3.92	53		
			4.62 (β)	128		
	4.31 (γ)	124	4.68 (γ)	128		
	4.47	49	4.70 (α)	128		
	4.48 (β)	38	4.72	50		
	4.76	127	4.77	51		
Ga	4.12	14			3.80	9

Table 2–39 (continued)
ELECTRON WORK FUNCTIONS OF THE ELEMENTS

Element	Thermionic work function	Ref.	Photoelectric work function	Ref.	Work function by contact potential method	Ref.
Ge			4.29	31	4.50	9
			4.5	52		
			4.73	14		
			4.80	16		
Hf	3.53	54				
Hg			4.50	55	4.50	9
			4.52	56		
			4.63*	57, 58		
I			2.8 (rhombic)	129		
			5.4 (monoclinic)	129		
			6.8 (amorphous)	129		
Ir	5.3	130			4.57	9
K			2.0	59	1.60	9
			2.12	22		
			2.24*	44, 43		
			2.26	60		
La	3.3	37				
Li			2.28	43	1.40	9
			2.42	14	2.49*	61
Mg			2.74	14	3.58	9
			<3.0	62	3.78	66
			3.59	63		
			3.62	64		
			3.68*	23		
			3.79	65		
Mn	3.83	124	3.76	14	4.14	9
Mo	4.15	67	4.15	67	4.08	75
	4.17	68	4.34	25	4.48	9
	4.19	69				
	4.20*	70				
	4.23	121				
	4.32	71				
	4.33	72				
	4.38	73				
	4.44	74				
Na			2.06	14	1.60	9
			2.25	76	2.26	77
			2.28*	19		
Nb	3.96	34	2.29	23		
	4.01	35	2.47	43		
Nd	3.3	37				

Table 2-39 (continued)
ELECTRON WORK FUNCTIONS OF THE ELEMENTS

Element	Thermionic work function	Ref.	Photoelectric work function	Ref.	Work function by contact potential method	Ref.
Ni	4.50	124	3.67	4	4.32	9
	4.61	38	4.06	31	4.96 (300K)	79
	4.63	48	4.87	25		
	5.03*	78	5.01*	51		
	5.1[a]	45	5.05 (623K)	131		
	5.24 (1150K)	131	5.20 (1108K)	131		
Os					4.55	9
Pb			3.97	4	3.94	9
			4.14	11		
Pd	4.99*	80	4.97*	80	4.49	9
Pr	2.7	37				
Pt	4.72	132	4.09	4	4.52	9
	5.08	26	6.35	84	5.36*	85
	5.29	81				
	5.32*	82				
	6.27	83				
Rb			2.09*	44		
			2.16	43		
Re	4.74	133				
	5.1*	86	~5.0	87		
Rh	4.58	88	4.57	88	4.52	9
	4.80*	89				
	4.9	130				
Ru					4.52	9
Sb			4.01	90	4.14	9
			4.60	118		
Se			4.62	11	4.42	9
			5.11	14		
Si	3.59	28			4.2	91
	4.02	134	4.37-4.67	117		
Sm	3.2	37				
Sn			3.62	4	4.09	9
			3.87	11	4.64	123
			4.21 (liquid)	92		
			4.38* (γ)	92		
			4.50* (β)	92		
Sr			2.06	93		
			2.24	14		
			2.74*	64		

Table 2–39 (continued)
ELECTRON WORK FUNCTIONS OF THE ELEMENTS

Element	Thermionic work function	Ref.	Photoelectric work function	Ref.	Work function by contact potential method	Ref.
Ta	4.03	135	4.05	63		
	4.07	74	4.12	32	3.96	9
	4.10	94	4.16	96	4.25	97
	4.19*	95				
Te			4.04	117	4.70	9
			4.76	16		
Th	3.35*	73	3.38	32	3.46	9
			3.47	98		
			3.57	11		
Ti	3.95	124	3.95	14	4.14	9
			4.17	98		
			4.45	136		
Tl			3.68	107	3.84	9
U	3.27	99	3.63	32	4.32	9
V	4.12	124	3.77	14	4.44	9
W	4.25	100	4.35 (310) face	104	4.38	9
	4.29 (116) face	142	4.49	143		
	4.38 (111) face	142	4.50 (211) face	104		
	4.39 (111) face	140	4.54	105		
	4.39 (116) face	140	4.60	32		
	4.45 (doped)	138				
	4.46	101				
	4.52* (001)face	26, 102, 142				
	4.53	103, 138				
	4.56 (001) face	140				
	4.58 (110) face	71, 73, 142				
	4.59 (001) face	139				
	4.65 (112) face	142				
	4.68 (110) face	140				
	4.69 (112) face	140				
	5.01	137				
Zn			3.08	4		111
			3.28	106		9
			3.32, 3.57	108		112
			3.60	11		123
			3.98	31		
			4.24	109		
			4.26 (0001) face	110		
			4.307	113		
Zr	4.12	54	3.73	32	3.60	9
	4.21*	141	4.33	136		

[a]By field current method.

From Michaelson, H. B., in *Handbook of Chemistry and Physics,* 55th ed., Weast, R. C., Ed., CRC Press, Cleveland, 1974, E-81.

Table 2–39 (continued)
ELECTRON WORK FUNCTIONS OF THE ELEMENTS

REFERENCES

1. Wehnelt, A. and Seliger, S., *Z. Phys.,* 38, 443, 1926.
2. Ameiser, I., *Z. Phys.,* 69, 111, 1931.
3. Goetz, A., *Z. Phys.,* 43, 531, 1927.
4. Lukirsky, P. and Prilesaev, S., *Z. Phys.,* 49, 236, 1928.
5. Fowler, R. H. *Phys. Rev.,* 38, 45, 1931.
6. Winch, R. P., *Phys. Rev.,* 37, 1269, 1931.
7. Farnsworth, H. E. and Winch, R. P., *Phys. Rev.,* 58, 812, 1940.
8. Anderson, P. A., *Phys. Rev.,* 50, 320, 1936.
9. Klein, O. and Lange, E., *Z. Elektrochem.,* 44, 542, 1938.
10. Anderson, P. A., *Phys. Rev.,* 59, 1034, 1941.
11. Hamer, R., *J. Opt. Soc. Am.,* 9, 251, 1924.
12. Brady, J. and Jakobsmeyer, P., *Phys. Rev.,* 49, 670, 1936.
13. Gaviola, E. and Strong, J., *Phys. Rev.,* 49, 441, 1936.
14. Schulze, R., *Z. Phys.,* 92, 212, 1934.
15. Morris, L. W., *Phys. Rev.,* 37, 1263, 1931.
16. Apker, L., Taft, E., and Dickey, J., *Phys. Rev.,* 74, 1462, 1948.
17. Reimann, A. L., *Thermionic Emission,* Chapman and Hall, London, 1934, 37.
18. Anderson, P. A., *Phys. Rev.,* 47, 958, 1935.
19. Maurer, R. J., *Phys. Rev.,* 57, 653, 1940.
20. Cashman, R. J. and Bassoe, E., *Phys. Rev.,* 55, 63, 1939.
21. Jamison, N. C. and Cashman, R. J., *Phys. Rev.,* 50, 624, 1936.
22. Suhrmann, R. and Schallamach, A., *Z. Phys.,* 91, 775, 1934.
23. Mann, M. M., Jr. and DuBridge, L. A., *Phys. Rev.,* 51, 120, 1937.
24. Jupnik, H., *Phys. Rev.,* 60, 884, 1941.
25. Rentschler, H. C. and Henry, D. E., *J. Opt. Soc. Am.,* 26, 30, 1936.
26. Dushman, S., Thermal emission of electrons, in *International Critical Tables,* Vol. 6, McGraw-Hill, New York, 1959, 53.
27. Reiman, A. L., *Proc. Phys. Soc.,* 50, 496, 1938.
28. Braun, A. and Busch, G., *Helv. Phys. Acta,* 20(1), 33, 1947.
29. Roy, S. C., *Proc. R. Soc. London,* A112, 599, 1926.
30. Dushman, S., *Phys. Rev.,* 21, 623, 1923.
31. Welch, G. B., *Phys. Rev.,* 32, 657, 1928.
32. Rentschler, H. C., Henry, D. E., and Smith, K. O., *Rev. Sci. Instrum.,* 3, 794, 1932.
33. Liben, I., *Phys. Rev.,* 51, 642, 1937.
34. Wahlin, H. B. and Sordahl, L. O., *Phys. Rev.,* 45, 886, 1934.
35. Reimann, A. L. and Grant, C. K., *Philos. Mag.,* 22, 34, 1936.
36. Bomke, H., *Ann. Phys.,* 10, 579, 1931.
37. Schumaker, E. E. and Harris, J. E., *J. Am. Chem. Soc.,* 48, 3108, 1926.
38. Wahlin, H. B., *Phys. Rev.,* 61, 509, 1942.
39. Cardwell, A. B., *Phys. Rev.,* 38, 2033, 1931.
40. Wahlin, H. B., *Phys. Rev.,* 73, 1458, 1948.
41. Koesters, H., *Z. Phys.,* 66, 807, 1930.
42. Kingdon, K. H., *Phys. Rev.,* 25, 892, 1925.
43. Olpin, A. R., quoted by Hughes, A. L. and DuBridge L. A., Eds., in *Photoelectric Phenomena,* McGraw-Hill, New York, 1932.
44. Brady, J. J., *Phys. Rev.,* 41, 613, 1932.
45. Dyke, W. P., Thesis, University of Washington, Seattle, 1946.
46. Underwood, N., *Phys. Rev.,* 47, 502, 1935.
47. Weber, A. H. and Eisele, C. J., *Phys. Rev.,* 59, A473, 1941.
48. Distler, W. and Monch, G., *Z. Phys.,* 84, 271, 1933.
49. Siljeholm, G., *Ann. Phys.,* 10, 178, 1931.
50. Cardwell, A. B., *Proc. Natl. Acad. Sci. U.S.A.,* 14, 439, 1928.
51. Glascoe, G. N., *Phys. Rev.,* 38, 1490, 1931.
52. Smith, A. H., *Phys. Rev.,* 75, 953, 1949.
53. Welch, G. B., *Phys. Rev.,* 31, A709, 1928.
54. Zwikker, C., *Phys. Z.,* 30, 578, 1929.

Table 2–39 (continued)
ELECTRON WORK FUNCTIONS OF THE ELEMENTS

55. Cassel, H. and Schneider, A., *Naturwissenschaften,* 22, 464, 1934.
56. Roller, D., Jordan, W. H., and Woodward, C. S., *Phys. Rev.,* 38, 396, 1931.
57. Kazda, C. B., *Phys. Rev.,* 26, 643, 1925.
58. Hales, W. B., *Phys. Rev.,* 32, 950, 1928.
59. Ives, H. E., *J. Opt. Soc. Am.,* 8, 551, 1924.
60. Mayer, H., *Ann. Phys.,* 29, 129, 1937.
61. Anderson, P. A., *Phys. Rev.,* 75, 1205, 1949.
62. Kenty, C., *Phys. Rev.,* 43, A776, 1933.
63. Cashman, J. and Huxford, S., *Phys. Rev.,* 48, 734, 1935.
64. Cashman, R. J. and Bassoe, E., *Phys. Rev.,* 53, A919, 1938.
65. Cashman, R. J., *Phys. Rev.,* 54, 971, 1938.
66. Anderson, P. A., *Phys. Rev.,* 54, 753, 1938.
67. DuBridge, L. A. and Roehr, W. W., *Phys. Rev.,* 42, 52, 1932.
68. Wahlin, H. B. and Reynolds, J. A., *Phys. Rev.,* 48, 751, 1935.
69. Grover, H., *Phys. Rev.,* 52, 982, 1937.
70. Wright, R. W., *Phys. Rev.,* 60, 465, 1941.
71. Ahearn, A. J., *Phys. Rev.,* 44, 277, 1933.
72. Freitag, H. and Kruger, F., *Ann. Phys.,* 21, 697, 1934.
73. Zwikker, C., *Proc. R. Acad. Amsterdam,* 29, 792, 1926.
74. Dushman, S., Rowe, H. N., Ewald, J., and Kidner, C. A., *Phys. Rev.,* 25, 338, 1925.
75. Oatley, C. W., *Proc. R. Soc. London,* A155, 218, 1936.
76. Berkes, Z., *Math. Phys. Lapok,* 41, 131, 1934.
77. Patai, E., *Z. Phys.,* 59, 697, 1930.
78. Fox, G. W. and Bowie, R. M., *Phys. Rev.,* 44, 345, 1933.
79. Bosworth, R. C. L., *Trans. Faraday Soc.,* 35, 397, 1939.
80. DuBridge, L. A. and Roehr, W. W., *Phys. Rev.,* 39, 99, 1932.
81. Van Velzer, H. L., *Phys. Rev.,* 44, 831, 1933.
82. Whitney, L. V., *Phys. Rev.,* 50, 1154, 1936.
83. DuBridge, L. A., *Phys. Rev.,* 32, 961, 1928.
84. DuBridge, L. A., *Phys. Rev.,* 31, 236, 1928.
85. Oatley, C. W., *Proc. Phys. Soc.,* 51, 318, 1939.
86. Agte, C., Alterthum, H., Becker, K., Heyne, G., and Moers, K., *Naturwissenschaften,* 19, 108, 1931.
87. Engelmann, A., *Ann. Phys.,* 17, 185, 1933.
88. Dixon, E. H., *Phys. Rev.,* 37, 60, 1931.
89. Wahlin, H. B. and Whitney, L. V., *J. Chem. Phys.,* 6, 594, 1938.
90. Middel, V., *Z. Phys.,* 105, 358, 1957.
91. Meyerhof, W. E., *Phys. Rev.,* 71, 727, 1947.
92. Goetz, A., *Phys. Rev.,* 33, 373, 1929.
93. Doepel, R., *Z. Phys.,* 33, 237, 1925.
94. Cardwell, A. B., *Phys. Rev.,* 47, 628, 1935.
95. Fiske, M. D., *Phys. Rev.,* 61, 513, 1942.
96. Cardwell, A. B., *Phys. Rev.,* 38, 2041, 1931.
97. Heinze, W., *Z. Phys.,* 109, 459, 1938.
98. Rentschler, H. C. and Henry, D. E., *Trans. Electrochem. Soc.,* 87, 289, 1945.
99. Hole, W. L. and Wright, R. W., *Phys. Rev.,* 56, 785, 1939.
100. Warner, A. H., *Proc. Natl. Acad. Sci. U.S.A.,* 13, 56, 1927.
101. Fleming, G. M. and Henderson, J. E., *Phys. Rev.,* 58, 887, 1940.
102. Nottingham, W. B., *Phys. Rev.,* 47, A806, 1935.
103. Freitag, H. and Krueger, F., *Ann. Phys.,* 21, 697, 1934.
104. Mendenhall, C. E. and DeVoe, C. F., *Phys. Rev.,* 51, 346, 1937.
105. Warner, A. H., *Phys. Rev.,* 38, 1871, 1931.
106. Nitsche, A., *Ann. Phys.,* 14, 463, 1932.
107. Suhrmann, R. and Csesch, H., *Z. Chem.,* B28, 215, 1935.
108. Dillon, J. H., *Phys. Rev.,* 38, 408, 1931.
109. DeVoe, C. F., *Phys. Rev.,* 50, 481, 1936.
110. Klug, W. and Steyskal, H., *Z. Phys.,* 116, 415, 1940.

213

Table 2–39 (continued)
ELECTRON WORK FUNCTIONS OF THE ELEMENTS

111. Oatley, C. W., *Proc. R. Soc. London*, A155, 218, 1936.
112. Anderson, P. A., *Phys. Rev.*, 57, 122, 1940.
113. Suhrmann, R. and Pietrzyk, J., *Z. Phys.*, 122, 600, 1944.
114. Jain, S. C. and Krishman, K. S., *Proc. R. Soc. London*, A217, 451, 1953.
115. Mitchell, E. W. J. and Mitchell, J. W., *Proc. R. Soc. London*, A210, 70, 1952.
116. Weissler, G. L. and Wilson, T. N., *J. Appl. Phys.*, 24, 472, 1953.
117. Fainshtein, S. M., *Zavod. Lab.*, 14, 64, 1948.
118. Apker, L., Taft, E., and Dickey, J., *Phys. Rev.*, 76, 270, 1949.
119. Glockler, G. and Sausville, J. W., *J. Electrochem. Soc.*, 95, 292, 1949.
120. Ivey, H. F., *Phys. Rev.*, 76, 567, 1949.
121. Mathur, S. B. L., *Proc. Natl. Inst. Sci. India*, 19, 153, 1953.
122. Bhatnager, A. S., *Proc. Natl. Acad. Sci. India*, A14, 5, 1944.
123. Hirschberg, R. and Lange, E., *Naturwissenschaften*, 39, 131, 1952.
124. Krishman, K. S. and Jain, S. C., *Nature*, 170, 759, 1952.
125. Ito, R., *J. Phys. Soc. Jap.*, 6, 188, 1951.
126. Anderson, P. A., *Phys. Rev.*, 76, 388, 1949.
127. Mathur, S. B. L., *Proc. Natl. Inst. Sci. India*, 19, 165, 1953.
128. Cardwell, A. B., *Phys. Rev.*, 92, 554, 1953.
129. West, D. C., *Can. J. Phys.*, 31, 691, 1953.
130. Weinreich, O. A., *Phys. Rev.*, 82, 573, 1951.
131. Cardwell, A. B., *Phys. Rev.*, 76, 125, 1949.
132. Kondo, K., *Rep. Inst. Sci. Technol. Tokyo Univ.*, 1, 24, 1947.
133. Levi, R. and Espersen, G. A., *Phys. Rev.*, 78, 231, 1950.
134. Esaki, L., *J. Phys. Soc. Jap.*, 8, 347, 1953.
135. Munick, R. J., LaBerge, W. B., and Coomes, E. A., *Phys. Rev.*, 80, 887, 1950.
136. Malamud, H. and Krumbein, A. D., *J. Appl. Phys.*, 25, 591, 1954.
137. Fleming, G. M. and Henderson, J. E., *Phys. Rev.*, 58, 887, 1940.
138. Nichols, M. H., *Phys. Rev.*, 78, 158, 1950.
139. Brown, A. A., Neelands, L. J., and Farnsworth, H. E., *J. Appl. Phys.*, 21, 1, 1950.
140. Herring, C. and Nichols, M. H., *Rev. Mod. Phys.*, 21, 185, 1949.
141. Wahl, A., *Phys. Rev.*, 82, 574, 1951.
142. Smith, G. F., *Phys. Rev.*, 94, 295, 1954.
143. Apker, L., Taft, E., and Dickey, J., *Phys. Rev.*, 73, 46, 1948.

Table 2–40
MAGNETIC SUSCEPTIBILITIES OF THE ELEMENTS AND OF SOME INORGANIC COMPOUNDS

The following table lists the magnetic susceptibilities of one gram formula weight of a number of
paramagnetic and diamagnetic inorganic compounds as well as the magnetic susceptibilities of
the elements. In each instance the magnetic
moment is expressed in cgs units.

A more extensive listing of the magnetic suscep-

tibilities of inorganic compounds, as well as those
for organic compounds, can be found in *Con-
stantes Sélectioneés – Diamagnétisme et Para-
magnétisme – Relaxation Paramagnétique*
(Volume 7). The data presented here are taken
from the above publication.

Substance	Formula	Temperature (K)	Susceptibility (10^{-6} cgs)
Aluminum, liquid	Al		+12.0
Aluminum, solid	Al	ord.	+16.5
fluoride	AlF_2	302	−13.4
oxide	Al_2O_3	ord.	−37.0
sulfate	$Al_2(SO_4)_3$	ord.	−93.0
sulfate	$Al_2(SO_4)_3 \cdot 18H_2O$	ord.	−323.0
Ammonia, gaseous	NH_3	ord.	−17.0
Ammonia, aqueous solution	NH_3	ord.	−18.0
Ammonium			
acetate	$NH_4C_2H_3O_2$	ord.	−41.1
bromide	NH_4Br	ord.	−47.0
carbonate	$(NH_4)_2CO_3$	ord.	−42.50
chlorate	NH_4ClO_3	ord.	−42.1
chloride	NH_4Cl	ord.	−36.7
fluoride	NH_4F	ord.	−23.0
hydroxide, aqueous solution	NH_4OH	ord.	−31.5
iodate	NH_4IO_3	ord.	−62.3
iodide	NH_4I	ord.	−66.0
nitrate	NH_4NO_3	ord.	−33.6
sulfate	$(NH_4)_2SO_4$	ord.	−67.0
thiocyanate	NH_4SCN	ord.	−48.1
Americium, solid	Am	300	+1,000.0
Antimony, liquid	Sb		−2.5
Antimony, solid	Sb	293	−99.0
bromide	$SbBr_3$	ord.	−115.0
chloride, tri-	$SbCl_3$	ord.	−86.7
chloride, penta-	$SbCl_5$	ord.	−120.0
fluoride	SbF_3	ord.	−46.0
iodide	SbI_3	ord.	−147.0
oxide	Sb_2O_3	ord.	−69.4
sulfide	Sb_2S_3	ord.	−86.0
Argon, gaseous	A	ord.	−19.6
Arsenic (α)	As	293	−5.5
Arsenic (β)	As	293	−23.7
Arsenic (γ)	As	293	−23.0
bromide	$AsBr_3$	ord.	−106.0
chloride	$AsCl_3$	ord.	−79.9
iodide	AsI_3	ord.	−142.0
sulfide	As_2S_3	ord.	−70.0
Arsenious acid	H_3AsO_3	ord.	−51.2
Barium	Ba	ord.	+20.6
acetate	$Ba(C_2H_3O_2)_2 \cdot H_2O$	ord.	−100.1
bromate	$Ba(BrO_3)_2$	ord.	−105.8
bromide	$BaBr_2$	ord.	−92.0
bromide	$BaBr_2 \cdot 2H_2O$	ord.	−119.0
carbonate	$Ba(CO_3)$	ord.	−58.9

Table 2–40 (continued)

MAGNETIC SUSCEPTIBILITIES OF THE ELEMENTS AND OF SOME INORGANIC COMPOUNDS

Substance	Formula	Temperature (K)	Susceptibility (10^{-6} cgs)
Barium (cont.)			
chlorate	$Ba(ClO_3)_2$	ord.	−87.5
chloride	$BaCl_2$	ord.	−72.6
chloride	$BaCl_2 \cdot 2H_2O$	ord.	−100.0
fluoride	BaF_2	ord.	−51.0
hydroxide	$Ba(OH)_2$	ord.	−53.2
hydroxide	$Ba(OH)_2 \cdot 8H_2O$	ord.	−157.0
iodate	$Ba(IO_3)_2$	ord.	−122.5
iodide	BaI_2	ord.	−124.0
iodide	$BaI_2 \cdot 2H_2O$	ord.	−163.0
nitrate	$Ba(NO_3)_2$	ord.	−66.5
oxide	BaO	ord.	−29.1
oxide	BaO_2	ord.	−40.6
sulfate	$BaSO_4$	ord.	−71.3
Beryllium, solid	Be	ord.	−9.0
chloride	$BeCl_2$	ord.	−26.5
hydroxide	$Be(OH)_2$	ord.	−23.1
nitrate, aqueous solution	$Be(NO_3)_2$	298	−41.0
oxide	BeO	ord.	−11.9
sulfate	$BeSO_4$	ord.	−37.0
Bismuth, liquid	Bi		−10.5
Bismuth, solid	Bi	ord.	−280.1
bromide	$BiBr_3$	ord.	−147.0
chloride	$BiCl_3$	ord.	−26.5
chromate	$Bi_2(CrO_4)_3$	ord.	+154.0
fluoride	BiF_3	303	−61.0
hydroxide	$Bi(OH)_3$	ord.	−65.8
iodide	BiI_3	ord.	−200.5
nitrate	$Bi(NO_3)_3$	ord.	−91.0
nitrate	$Bi(NO_3)_3 \cdot 5H_2O$	ord.	−159.0
oxide	BiO	ord.	−110.0
oxide	Bi_2O_3	ord.	−83.0
phosphate	$BiPO_4$	ord.	−77.0
sulfate	$Bi_2(SO_4)_3$	ord.	−199.0
sulfide	Bi_2S_3	ord.	−123.0
Boric acid	H_3BO_3	ord.	−34.1
Boron, solid	B	ord.	−6.7
chloride	BCl_3	ord.	−59.9
oxide	B_2O_3	ord.	−39.0
Bromine, gaseous	Br_2		−73.5
Bromine, liquid	Br_2		−56.4
fluoride	BrF_3	ord.	−33.9
fluoride	BrF_5	ord.	−45.1
Cadmium, liquid	Cd		−18.0
Cadmium, solid	Cd	ord.	−19.8
acetate	$Cd(C_2H_3O_2)_2$	ord.	−83.7
bromide	$CdBr_2$	ord.	−87.3
bromide	$CdBr_2 \cdot 4H_2O$	ord.	−140.0
carbonate	$CdCO_3$	ord.	−46.7
chloride	$CdCl_2$	ord.	−68.7
chloride	$CdCl_2 \cdot 2H_2O$	ord.	−99.0
chromate	$CdCrO_4$	ord.	−16.8
cyanide	$Cd(CN)_2$	ord.	−54.0
fluoride	CdF_2	ord.	−40.6
hydroxide	$Cd(OH)_2$	ord.	−41.0
iodate	$Cd(IO_3)_2$	ord.	−108.4

Table 2—40 (continued)
MAGNETIC SUSCEPTIBILITIES OF THE ELEMENTS AND OF SOME INORGANIC COMPOUNDS

Substance	Formula	Temperature (K)	Susceptibility (10^{-6} cgs)
Cadmium, solid (cont.)			
iodide	CdI_2	ord.	−117.2
nitrate	$Cd(NO_3)_2$	ord.	−55.1
nitrate	$Cd(NO_3)_2 \cdot 4H_2O$	ord.	−140.0
oxide	CdO	ord.	−30.0
phosphate	$Cd_3(PO_4)_2$	ord.	−159.0
sulfate	$CdSO_4$	ord.	−59.2
sulfide	CdS	ord.	−50.0
Calcium, solid	Ca		+40.0
acetate	$Ca(C_2H_3O_2)_2$	ord.	−70.5
bromate	$Ca(BrO_3)_2$	ord.	−84.9
bromide	$CaBr_2$	ord.	−73.8
bromide	$CaBr_2 \cdot 3H_2O$	ord.	−115.0
carbonate	$CaCO_3$	ord.	−38.2
chloride	$CaCl_2$	ord.	−54.7
fluoride	CaF_2	ord.	−28.0
hydroxide	$Ca(OH)_2$	ord.	−22.0
iodate	$Ca(IO_3)_2$	ord.	−101.4
iodide	CaI_2	ord.	−109.0
nitrate, aqueous solution	$Ca(NO_3)_2$	ord.	−45.9
oxide	CaO	ord.	−15.0
oxide	CaO_2	ord.	−23.8
sulfate	$CaSO_4$	ord.	−49.7
sulfate	$CaSO_4 \cdot 2H_2O$	ord.	−74.0
Carbon, diamond	C	ord.	−5.9
Carbon, graphite	C	ord.	−6.0
dioxide	CO_2	ord.	−21.0
monoxide	CO	ord.	−9.8
Cerium (α)	Ce	80.5	+5,160.0
Cerium (β)	Ce	293	+2,450.0
Cerium (β)	Ce	80.5	+6,230.0
Cerium (γ)	Ce	287.9	+2,420.0
Cerium (γ)	Ce	125.6	+4,640.0
Cerium (γ)	Ce	80.5	+5,200.0
chloride	$CeCl_3$	287	+2,490.0
fluoride	CeF_3	293	+2,190.0
nitrate	$Ce(NO_3)_3 \cdot 5H_2O$	292	+2,310.0
oxide	CeO_2	293	+26.0
sulfate	$CeSO_4$	ord.	+37.0
sulfate	$Ce(SO_4)_2 \cdot 4H_2O$	293	−97.0
sulfate	$Ce_2(SO_4)_3 \cdot 5H_2O$	293	+4,540.0
sulfide	CeS	ord.	+2,110.0
sulfide	Ce_2S_3	292	+5,080.0
Cesium, liquid	Cs		+26.5
Cesium, solid	Cs	ord.	+29.0
bromate	$CsBrO_3$	ord.	−75.1
bromide	CsBr	ord.	−67.2
carbonate	Cs_2CO_3	ord.	−103.6
chlorate	$CsClO_3$	ord.	−65.0
chloride	CsCl	ord.	−86.7
fluoride	CsF	ord.	−44.5
iodate	$CsIO_3$	ord.	−83.1
iodide	CsI	ord.	−82.6
oxide	CsO_2	293	+1,534.0

Table 2–40 (continued)
MAGNETIC SUSCEPTIBILITIES OF THE ELEMENTS AND OF SOME INORGANIC COMPOUNDS

Substance	Formula	Temperature (K)	Susceptibility (10^{-6} cgs)
Cesium, solid (cont.)			
oxide	CsO_2	90	+4,504.0
sulfate	Cs_2SO_4	ord.	−116.0
sulfide	Cs_2S	ord.	−104.0
Chlorine, liquid	Cl_2	ord.	−40.5
fluoride, tri-	ClF_3	ord.	−26.5
Chromium	Cr	273	+180.0
Chromium	Cr	1,713	+224.0
acetate	$Cr(C_2H_3O_2)_3$	293	+5,104.0
chloride	$CrCl_2$	293	+7,230.0
chloride	$CrCl_3$	293	+6,890.0
fluoride	CrF_3	293	+4,370.0
oxide	CrO_3	ord.	+40.0
oxide	Cr_2O_3	300	+1,960.0
sulfate	$CrSO_4 \cdot 6H_2O$	293	+9,690.0
sulfate	$Cr_2(SO_4)_3$	293	+11,800.0
sulfate	$Cr_2(SO_4)_3 \cdot 8H_2O$	290	+12,700.0
sulfate	$Cr_2(SO_4)_3 \cdot 10H_2O$	290	+12,600.0
sulfate	$Cr_2(SO_4)_3 \cdot 14H_2O$	290	+12,160.0
sulfide	CrS	ord.	+2,390.0
Cobalt	Co		ferro
acetate	$Co(C_2H_3O_2)_2$	293	+11,000.0
bromide	$CoBr_2$	293	+13,000.0
chloride	$CoCl_2$	293	+12,660.0
chloride	$CoCl_2 \cdot 6H_2O$	293	+9,710.0
cyanide	$Co(CN)_2$	303	+3,825.0
fluoride	CoF_2	293	+9,490.0
fluoride	CoF_3	293	+1,900.0
iodide	CoI_2	293	+10,760.0
oxide	CoO	260	+4,900.0
oxide	Co_2O_3	ord.	+4,560.0
oxide	Co_3O_4	ord.	+7,380.0
phosphate	$Co_3(PO_4)_2$	291	+28,110.0
sulfate	$CoSO_4$	293	+10,000.0
sulfate	$Co_2(SO_4)_3$	297	+1,000.0
sulfide	CoS	688	+251.0
sulfide	CoS	293	+225.0
thiocyanate	$Co(SCN)_2$	303	+11,090.0
Copper, liquid	Cu		−6.16
Copper, solid	Cu	296	−5.46
bromide	CuBr	ord.	−49.0
bromide	$CuBr_2$	341.6	+653.3
bromide	$CuBr_2$	292.7	+685.5
bromide	$CuBr_2$	189	+736.9
bromide	$CuBr_2$	90	+658.7
chloride	CuCl	ord.	−40.0
chloride	$CuCl_2$	373.3	+1,030.0
chloride	$CuCl_2$	289	+1,080.0
chloride	$CuCl_2$	170	+1,815.0
chloride	$CuCl_2$	69.25	+2,370.0
chloride	$CuCl_2 \cdot 2H_2O$	293	+1,420.0
cyanide	CuCN	ord.	−24.0
fluoride	CuF_2	293	+1,050.0
fluoride	CuF_2	90	+1,420.0

Table 2–40 (continued)
MAGNETIC SUSCEPTIBILITIES OF THE ELEMENTS AND OF SOME INORGANIC COMPOUNDS

Substance	Formula	Temperature (K)	Susceptibility (10^{-6} cgs)
Copper, solid (cont.)			
fluoride	$CuF_2 \cdot 2H_2O$	293	+1,600.0
hydroxide	$Cu(OH)_2$	292	+1,170.0
iodide	CuI	ord.	−63.0
nitrate	$Cu(NO_3)_2 \cdot 3H_2O$	293	+1,570.0
nitrate	$Cu(NO_3)_2 \cdot 6H_2O$	293	+1,625.0
oxide	Cu_2O	293	−20.0
oxide	CuO	780	+259.6
oxide	CuO	561	+267.3
oxide	CuO	397	+256.9
oxide	CuO	289.6	+238.9
oxide	CuO	120	+156.2
phosphide	Cu_3P	ord.	−33.0
phosphide	CuP_2	ord.	−35.0
sulfate	$CuSO_4$	293	+1,330.0
sulfate	$CuSO_4 \cdot H_2O$	293	+1,520.0
sulfate	$CuSO_4 \cdot 3H_2O$	ord.	+1,480.0
sulfate	$CuSO_4 \cdot 5H_2O$	293	+1,460.0
sulfide	CuS	293	−2.0
thiocyanate	$CuSCN$	ord.	−48.0
Dysprosium	Dy	293.2	+103,500.0
oxide	Dy_2O_3	287.2	+89,600.0
sulfate	$Dy_2(SO_4)_3$	293	+91,400.0
sulfate	$Dy_2(SO_4)_3 \cdot 8H_2O$	291.2	+92,760.0
sulfide	Dy_2S_3	292	+95,200.0
Erbium	Er	291	+44,300.0
oxide	Er_2O_3	286	+73,920.0
sulfate	$Er_2(SO_4)_3 \cdot 8H_2O$	293	+74,600.0
sulfide	Er_2S_3	292	+77,200.0
Europium	Eu	293	+34,000.0
bromide	$EuBr_2$	292	+26,800.0
chloride	$EuCl_2$	292	+26,500.0
fluoride	EuF_2	292	+23,750.0
iodide	EuI_2	292	+26,000.0
oxide	Eu_2O_3	298	+10,100.0
sulfate	$EuSO_4$	293	+25,730.0
sulfate	$Eu_2(SO_4)_3$	293	+10,400.0
sulfate	$Eu_2(SO_4)_3 \cdot 8H_2O$	293	+9,540.0
sulfide	EuS	293	+23,800.0
sulfide	EuS	195	+35,400.0
Gadolinium	Gd	300.6	+755,000.0
chloride	$GdCl_2$	293	+27,930.0
oxide	Gd_2O_3	293	+53,200.0
sulfate	$Gd_2(SO_4)_3$	285.5	+54,200.0
sulfate	$Gd_2(SO_4)_3 \cdot 8H_2O$	293	+53,280.0
sulfide	Gd_2S_3	292	+55,500.0
Gallium, liquid	Ga	313	+2.5
Gallium, solid	Ga	80	−24.4
Gallium, solid	Ga	290	−21.6
chloride	$GaCl_3$	ord.	−63.0
iodide	GaI_3	ord.	−149.0
oxide	Ga_2O	ord.	−34.0
sulfide	GaS	ord.	−23.0
sulfide	Ga_2S	ord.	−36.0

Table 2-40 (continued)
MAGNETIC SUSCEPTIBILITIES OF THE ELEMENTS AND OF SOME INORGANIC COMPOUNDS

Substance	Formula	Temperature (K)	Susceptibility (10^{-6} cgs)
Gallium, solid (cont.)			
sulfide	Ga_2S_3	ord.	−80.0
Germanium	Ge	293	−76.84
chloride	$GeCl_4$	ord.	−72.0
fluoride	GeF_4	ord.	−50.0
iodide	GeI_4	ord.	−174.0
oxide	GeO	ord.	−28.8
oxide	GeO_2	ord.	−34.3
sulfide	GeS	ord.	−40.9
sulfide	GeS_2	ord.	−53.3
Gold, liquid	Au		−34.0
Gold, solid	Au	296	−28.0
bromide	AuBr	ord.	−61.0
chloride	AuCl	ord.	−67.0
chloride	$AuCl_3$	ord.	−112.0
fluoride	AuF_3	ord.	+74.0
iodide	AuI	ord.	−91.0
phosphide	AuP_3	ord.	−107.0
Hafnium, solid	Hf	1,673	+104.0
Hafnium, solid	Hf	289	+75.0
oxide	HfO_2	ord.	−23.0
Helium, gaseous	He	ord.	−1.88
Holmium	Ho		
oxide	Ho_2O_3	293	+88,100.0
sulfate	$Ho_2(SO_4)_3$	293	+91,700.0
sulfate	$Ho_2(SO_4)_3 \cdot 8H_2O$	293	+91,600.0
Hydrogen, gaseous	H_2		−3.98
chloride, aqueous solution	HCl	300	−22.0
chloride, liquid	HCl	273	−22.6
fluoride, aqueous solution	HF	ord.	−9.3
fluoride, liquid	HF	287	−8.6
iodide, aqueous solution	HI	ord.	−50.2
iodide, liquid	HI	281	−47.7
iodide, liquid	HI	233	−48.3
iodide, solid	HI	195	−47.3
oxide − see Water			
peroxide	H_2O_2	ord.	−17.7
sulfide	H_2S	ord.	−25.5
Indium	In	ord.	−64.0
bromide	$InBr_3$	ord.	−107.0
chloride	InCl	ord.	−30.0
chloride	$InCl_2$	ord.	−56.0
chloride	$InCl_3$	ord.	−86.0
fluoride	InF_2	ord.	−61.0
oxide	In_2O	ord.	−47.0
oxide	In_2O_3	ord.	−56.0
sulfide	InS	ord.	−28.0
sulfide	In_2S	ord.	−50.0
sulfide	In_2S_3	ord.	−98.0
Iodic acid	HIO_3	ord.	−48.0
m-per-	HIO_4	ord.	−56.5
o-per-	H_5IO_6	ord.	−71.4
Iodine, atomic	I	1,400	+1,120.0
Iodine, atomic	I	1,303	+869.0

Table 2—40 (continued)
MAGNETIC SUSCEPTIBILITIES OF THE ELEMENTS AND OF SOME INORGANIC COMPOUNDS

Substance	Formula	Temperature (K)	Susceptibility (10^{-6} cgs)
Iodine, solid	I_2	ord.	−88.7
chloride	ICl	ord.	−54.6
chloride	ICl_3	ord.	−90.2
fluoride	IF_5	ord.	−58.1
oxide	I_2O_5	ord.	−79.4
Iridium	Ir	698	+32.1
Iridium	Ir	298	+25.6
chloride	$IrCl_3$	ord.	−14.4
oxide	IrO_2	298	+224.0
Iron	Fe		ferro
bromide	$FeBr_2$	ord.	+13,600.0
carbonate	$FeCO_3$	293	+11,300.0
chloride	$FeCl_2$	293	+14,750.0
chloride	$FeCl_2 \cdot 4H_2O$	293	+12,900.0
chloride	$FeCl_3$	398	+9,980.0
chloride	$FeCl_3$	293	+13,450.0
chloride	$FeCl_3 \cdot 6H_2O$	290	+15,250.0
fluoride	FeF_2	293	+9,500.0
fluoride	FeF_3	305	+13,760.0
fluoride	$FeF_3 \cdot 3H_2O$	293	+7,870.0
iodide	FeI_2	ord.	+13,600.0
nitrate	$Fe(NO_3)_3 \cdot 9H_2O$	293	+15,200.0
oxide	FeO	293	+7,200.0
oxide	Fe_2O_3	1,033	+3,586.0
phosphate	$FePO_4$	ord.	+11,500.0
sulfate	$FeSO_4$	293	+10,200.0
sulfate	$FeSO_4 \cdot H_2O$	290	+10,500.0
sulfate	$FeSO_4 \cdot 7H_2O$	293	+11,200.0
sulfide	FeS	293	+1,074.0
Krypton	Kr		−28.8
Lanthanum	La	ord.	+118.0
oxide	La_2O_3	ord.	−78.0
sulfate	$La_2(SO_4)_3 \cdot 9H_2O$	293	−262.0
sulfide	La_2S_3	292	−37.0
sulfide	La_2S_4	293	−100.0
Lead, liquid	Pb	330	−15.5
Lead, solid	Pb	289	−23.0
acetate	$Pb(C_2H_3O_2)_2$	ord.	−89.1
bromide	$PbBr_2$	ord.	−90.6
carbonate	$PbCO_3$	ord.	−61.2
chloride	$PbCl_2$	ord.	−73.8
chromate	$PbCrO_4$	ord.	−18.0
fluoride	PbF_2	ord.	−58.1
iodate	$Pb(IO_3)_2$	ord.	−131.0
iodide	PbI_2	ord.	−126.5
nitrate	$Pb(NO_3)_2$	ord.	−74.0
oxide	PbO	ord.	−42.0
phosphate	$Pb_3(PO_4)_2$	ord.	−182.0
sulfate	$PbSO_4$	ord.	−69.7
sulfide	PbS	ord.	−84.0
thiocyanate	$Pb(CNS)_2$	ord.	−82.0
Lithium	Li	ord.	+14.2
acetate	$LiC_2H_3O_2$	ord.	−34.0
bromate	$LiBrO_3$	ord.	−39.0
bromide	LiBr	ord.	−34.7

Table 2–40 (continued)

MAGNETIC SUSCEPTIBILITIES OF THE ELEMENTS AND OF SOME INORGANIC COMPOUNDS

Substance	Formula	Temperature (K)	Susceptibility (10^{-6} cgs)
Lithium (cont.)			
carbonate	Li_2CO_3	ord.	−27.0
chlorate, aqueous solution	$LiClO_3$	ord.	−28.8
fluoride	LiF	ord.	−10.1
hydride	LiH	ord.	−4.6
hydroxide	$LiOH$	ord.	−12.3
iodate	$LiIO_3$	ord.	−47.0
iodide	LiI	ord.	−50.0
nitrate	$LiNO_3 \cdot 3H_2O$	ord.	−62.0
sulfate	Li_2SO_4	ord.	−40.0
Lutetium	Lu	ord.	>0.0
Magnesium	Mg	ord.	+13.1
acetate	$Mg(C_2H_3O_2)_2 \cdot 4H_2O$	ord.	−116.0
bromide	$MgBr_2$	ord.	−72.0
carbonate	$MgCO_3$	ord.	−32.4
carbonate	$MgCO_3 \cdot 3H_2O$	ord.	−72.7
chloride	$MgCl_2$	ord.	−47.4
fluoride	MgF_2	ord.	−22.7
hydroxide	$Mg(OH)_2$	288	−22.1
iodide	MgI_2	ord.	−111.0
oxide	MgO	ord.	−10.2
phosphate	$Mg_3(PO_4)_2 \cdot 4H_2O$	ord.	−167.0
sulfate	$MgSO_4$	294	−50.0
sulfate	$MgSO_4 \cdot H_2O$	ord.	−61.0
sulfate	$MgSO_4 \cdot 5H_2O$	ord.	−109.0
sulfate	$MgSO_4 \cdot 7H_2O$	ord.	−135.7
Manganese (α)	Mn	293	+529.0
Manganese (β)	Mn	293	+483.0
acetate	$Mn(C_2H_3O_2)_2$	293	+13,650.0
bromide	$MnBr_2$	294	+13,900.0
carbonate	$MnCO_3$	293	+11,400.0
chloride	$MnCl_2$	293	+14,350.0
chloride	$MnCl_2 \cdot 4H_2O$	293	+14,600.0
fluoride	MnF_2	290	+10,700.0
fluoride	MnF_3	293	+10,500.0
hydroxide	$Mn(OH)_2$	293	+13,500.0
iodide	MnI_2	293	+14,500.0
oxide	MnO	293	+4,850.0
oxide	MnO_2	293	+2,280.0
oxide	Mn_2O_3	293	+14,100.0
oxide	Mn_3O_4	298	+12,400.0
sulfate	$MnSO_4$	293	+13,660.0
sulfate	$MnSO_4 \cdot H_2O$	293	+14,200.0
sulfate	$MnSO_4 \cdot 4H_2O$	293	+14,600.0
sulfate	$MnSO_4 \cdot 5H_2O$	293	+14,700.0
sulfide (α)	MnS	293	+5,630.0
sulfide (β)	MnS	293	+3,850.0
Mercury, gaseous	Hg		−78.3
Mercury, liquid	Hg	293	−33.44
Mercury, solid	Hg		−24.1
acetate	$HgC_2H_3O_2$	ord.	−70.5
acetate	$Hg(C_2H_3O_2)_2$	ord.	−100.0
bromate	$HgBrO_3$	ord.	−57.7
bromide	$HgBr$	ord.	−57.2
bromide	$HgBr_2$	ord.	−94.2

Table 2–40 (continued)
MAGNETIC SUSCEPTIBILITIES OF THE ELEMENTS AND OF SOME INORGANIC COMPOUNDS

Substance	Formula	Temperature (K)	Susceptibility (10^{-6} cgs)
Mercury, solid (cont.)			
chloride	$HgCl$	ord.	−52.0
chloride	$HgCl_2$	ord.	−82.0
chromate	$HgCrO_4$	ord.	−12.5
chromate	Hg_2CrO_4	ord.	−63.0
cyanide	$Hg(CN)_2$	ord.	−67.0
fluoride	HgF	ord.	−53.0
fluoride	HgF_2	302	−62.0
hydroxide	$Hg_2(OH)_2$	ord.	−100.0
iodate	$HgIO_3$	ord.	−92.0
iodide	HgI	ord.	−83.0
iodide	HgI_2	ord.	−128.6
nitrate	$HgNO_3$	ord.	−55.9
nitrate	$Hg(NO_3)_2$	ord.	−74.0
oxide	HgO	ord.	−44.0
oxide	Hg_2O	ord.	−76.3
sulfate	$HgSO_4$	ord.	−78.1
sulfate	Hg_2SO_4	ord.	−123.0
sulfide	HgS	ord.	−55.4
thiocyanate	$Hg(SCN)_2$	ord.	−96.5
Molybdenum	Mo	298	+89.0
Molybdenum	Mo	63.8	+108.0
Molybdenum	Mo	20.4	+149.2
bromide	$MoBr_3$	293	+525.0
bromide	$MoBr_4$	293	+520.0
bromide	Mo_3Br_6	290.5	−46.0
chloride	$MoCl_3$	290	+43.0
chloride	$MoCl_4$	291	+1,750.0
chloride	$MoCl_5$	289	+990.0
fluoride	MoF_6	ord.	−26.0
oxide	MoO_2	289	+41.0
oxide	MoO_3	292.5	+3.0
oxide	Mo_2O_3	ord.	−42.0
oxide	Mo_3O_8	ord.	+42.0
sulfide	MoS_3	289	−63.0
Neodymium	Nd	287.7	+5,628.0
fluoride	NdF_3	293	+4,980.0
nitrate	$Nd(NO_3)_3$	293	+5,020.0
oxide	Nd_2O_3	292.0	+10,200.0
sulfate	$Nd_2(SO_4)_3$	293	+9,990.0
sulfide	Nd_2S_3	292	+5,550.0
Neon	Ne	ord.	−6.74
Nickel	Ni		ferro
acetate	$Ni(C_2H_3O_2)_2$	293	+4,690.0
bromide	$NiBr_2$	293	+5,600.0
chloride	$NiCl_2$	293	+6,145.0
chloride	$NiCl_2 \cdot 6H_2O$	293	+4,240.0
fluoride	NiF_2	293	+2,410.0
hydroxide	$Ni(OH)_2$	ord.	+4,500.0
iodide	NiI_2	293	+3,875.0
nitrate	$Ni(NO_3)_2 \cdot 6H_2O$	293.5	+4,300.0
oxide	NiO	293	+660.0
sulfate	$NiSO_4$	293	+4,005.0
sulfide	NiS	293	+190.0
sulfide	Ni_3S_2	ord.	+1,030.0

Table 2–40 (continued)
MAGNETIC SUSCEPTIBILITIES OF THE ELEMENTS AND OF SOME INORGANIC COMPOUNDS

Substance	Formula	Temperature (K)	Susceptibility (10^{-6} cgs)
Niobium	Nb	298	+195.0
oxide	Nb_2O_5	ord.	−10.0
Nitric acid	HNO_3	ord.	−19.9
Nitrogen	N_2	ord.	−12.0
oxide, gaseous	NO	293	+1,461.0
oxide, gaseous	NO	203.8	+1,895.0
oxide, gaseous	NO	146.9	+2,324.0
oxide, liquid	NO	117.64	+114.2
oxide, solid	NO	90	+19.8
oxide	NO_2	408	+150.0
oxide	N_2O	285	−18.9
oxide	N_2O_3	291	−16.0
oxide	N_2O_4	303.6	−22.1
oxide	N_2O_4	295.1	−23.0
oxide	N_2O_4	257	−25.4
oxide, aqueous solution	N_2O_5	289	−35.6
Osmium	Os	298	+9.9
chloride	$OsCl_2$	ord.	+41.3
Oxygen, gaseous	O_2	293	+3,449.0
Oxygen, liquid	O_2	90.1	+7,699.0
Oxygen, liquid	O_2	70.8	+8,685.0
Oxygen, solid (α)	O_2	23.7	+1,760.0
Oxygen, solid (γ)	O_2	54.3	+10,200.0
Ozone, liquid	O_3		+6.7
Palladium	Pd	288	+567.4
chloride	$PdCl_2$	291.3	−38.0
fluoride	PdF_3	293	+1,760.0
hydride	PdH	ord.	+1,077.0
hydride	Pd_4H	ord.	+2,353.0
Phosphoric acid, aqueous solution	H_3PO_4	ord.	−43.8
Phosphorous acid, aqueous solution	H_3PO_3	ord.	−42.5
Phosphorus, black	P		−26.6
Phosphorus, red	P		−20.8
chloride	PCl_3	ord.	−63.4
Platinum	Pt	290.3	+201.9
chloride	$PtCl_2$	298	−54.0
chloride	$PtCl_3$	ord.	−66.7
chloride	$PtCl_4$	ord.	−93.0
fluoride	PtF_4	293	+455.0
oxide	Pt_2O_3	ord.	−37.70
Plutonium	Pu	293	+610.0
fluoride	PuF_4	301	+1,760.0
fluoride	PuF_6	295	+173.0
oxide	PuO_2	300	+730.0
Potassium	K	ord.	+20.8
acetate, aqueous solution	$KC_2H_3O_2$	28	−45.0
bromate	$KBrO_3$	ord.	−52.6
bromide	KBr	ord.	−49.1
carbonate	K_2CO_3	ord.	−59.0
chlorate	$KClO_3$	ord.	−42.8
chloride	KCl	ord.	−39.0
chromate	K_2CrO_4	ord.	−3.9
chromate	$K_2Cr_2O_7$	293	+29.4
cyanide	KCN	ord.	−37.0

Table 2-40 (continued)
MAGNETIC SUSCEPTIBILITIES OF THE ELEMENTS AND OF SOME INORGANIC COMPOUNDS

Substance	Formula	Temperature (K)	Susceptibility (10^{-6} cgs)
Potassium (cont.)			
ferricyanide	$K_3 Fe(CN)_6$	297	+2,290.0
ferrocyanide	$K_4 Fe(CN)_6$	ord.	−130.0
ferrocyanide	$K_4 Fe(CN)_6 \cdot 3H_2 O$	ord.	−172.3
fluoride	KF	ord.	−23.6
hydroxide, aqueous solution	KOH	ord.	−22.0
iodate	KIO_3	ord.	−63.1
iodide	KI	ord.	−63.8
nitrate	KNO_3	ord.	−33.7
nitrite	KNO_2	ord.	−23.3
oxide	KO_2	293	+3,230.0
oxide	KO_3	ord.	+1,185.0
permanganate	$KMnO_4$	ord.	+20.0
sulfate	$KHSO_4$	ord.	−49.8
sulfate	$K_2 SO_4$	ord.	−67.0
sulfide	$K_2 S$	ord.	−60.0
sulfide	$K_2 S_2$	ord.	−71.0
sulfide	$K_2 S_3$	ord.	−80.0
sulfide	$K_2 S_4$	ord.	−89.0
sulfide	$K_2 S_5$	ord.	−98.0
sulfite	$K_2 SO_3$	ord.	−64.0
thiocyanate	$KSCN$	ord.	−48.0
Praseodymium	Pr	293	+5,010.0
chloride	$PrCl_3$	307.1	+44.5
oxide	PrO_2	293	+1,930.0
oxide	$Pr_2 O_3$	827	+4,000.0
oxide	$Pr_2 O_3$	294.5	+8,994.0
sulfate	$Pr_2 (SO_4)_3$	291	+9,660.0
sulfate	$Pr_2 (SO_4)_3 \cdot 8H_2 O$	289	+9,880.0
sulfide	$Pr_2 S_3$	292	+10,770.0
Rhenium	Re	293	+67.6
chloride	$ReCl_5$	293	+1,225.0
oxide	ReO_2	ord.	+44.0
oxide	$ReO_2 \cdot 2H_2 O$	295	+74.0
oxide	ReO_3	ord.	+16.0
oxide	$Re_2 O_7$	ord.	−16.0
sulfide	ReS_2	ord.	+38.0
Rhodium	Rh	723	+123.0
Rhodium	Rh	298	+111.0
chloride	$RhCl_3$	298	−7.5
fluoride	RhF_4	293	+500.0
oxide	$Rh_2 O_3$	298	+104.0
sulfate	$Rh_2 (SO_4)_3 \cdot 6H_2 O$	298	+104.0
sulfate	$Rh_2 (SO_4)_3 \cdot 14H_2 O$	298	+149.0
Rubidium	Rb	303	+17.0
bromide	$RbBr$	ord.	−56.4
carbonate	$Rb_2 CO_3$	ord.	−75.4
chloride	$RbCl$	ord.	−46.0
fluoride	RbF	ord.	−31.9
iodide	RbI	ord.	−72.2
nitrate	$RbNO_3$	ord.	−41.0
oxide	RbO_2	293	+1,527.0
sulfate	$Rb_2 SO_4$	ord.	−88.4
sulfide	$Rb_2 S$	ord.	−80.0
sulfide	$Rb_2 S_2$	ord.	−90.0

Table 2–40 (continued)
MAGNETIC SUSCEPTIBILITIES OF THE ELEMENTS AND OF SOME INORGANIC COMPOUNDS

Substance	Formula	Temperature (K)	Susceptibility (10^{-6} cgs)
Ruthenium	Ru	723	+50.2
Ruthenium	Ru	298	+43.2
chloride	$RuCl_2$	290.9	+1,998.0
oxide	RuO_2	298	+162.0
Samarium	Sm	291	+1,860.0
Samarium	Sm	195	+2,230.0
bromide	$SmBr_2$	293	+5,337.0
bromide	$SmBr_3$	293	+972.0
oxide	Sm_2O_3	292	+1,988.0
oxide	Sm_2O_3	170	+1,960.0
oxide	Sm_2O_3	85	+2,282.0
sulfate	$Sm_2(SO_4)_3 \cdot 8H_2O$	293	+1,710.0
sulfide	Sm_2S_3	292	+3,300.0
Scandium	Sc	292	+315.0
Selenic acid	H_2SeO_4	ord.	−51.2
Selenous acid	H_2SeO_3	ord.	−45.4
Selenium, liquid	Se	900	−24.0
Selenium, solid	Se	ord.	−25.0
bromide	Se_2Br_2	ord.	−113.0
chloride	Se_2Cl_2	ord.	−94.8
fluoride	SeF_6	ord.	−51.0
oxide	SeO_2	ord.	−27.2
Silicon	Si	ord.	−3.9
bromide	$SiBr_4$	ord.	−128.6
carbide	SiC	ord.	−12.8
chloride	$SiCl_4$	ord.	−88.3
hydroxide	$Si(OH)_4$	ord.	−42.6
oxide	SiO_2	ord.	−29.6
Silver, liquid	Ag		−24.0
Silver, solid	Ag	296	−19.5
acetate	$AgC_2H_3O_2$	ord.	−60.4
bromide	AgBr	283	−59.7
carbonate	Ag_2CO_3	ord.	−80.90
chloride	AgCl	ord.	−49.0
chromate	Ag_2CrO_4	ord.	−40.0
cyanide	AgCN	ord.	−43.2
fluoride	AgF	ord.	−36.5
iodide	AgI	ord.	−80.0
nitrate	$AgNO_3$	ord.	−45.7
nitrite	$AgNO_2$	ord.	−42.0
oxide	AgO	287	−19.6
oxide	Ag_2O	ord.	−134.0
permanganate	$AgMnO_4$	300	−63.0
phosphate	Ag_3PO_4	ord.	−120.0
sulfate	Ag_2SO_4	ord.	−92.90
thiocyanate	AgSCN	ord.	−61.8
Sodium	Na	ord.	+16.0
acetate	$NaC_2H_3O_2$	ord.	−37.6
borate, tetra-	$Na_2B_4O_7$	ord.	−85.0
bromate	$NaBrO_3$	ord.	−44.2
bromide	NaBr	ord.	−41.0
carbonate	Na_2CO_3	ord.	−41.0
chlorate	$NaClO_3$	ord.	−34.7
chloride	NaCl	ord.	−30.3

Table 2–40 (continued)

MAGNETIC SUSCEPTIBILITIES OF THE ELEMENTS AND OF SOME INORGANIC COMPOUNDS

Substance	Formula	Temperature (K)	Susceptibility (10^{-6} cgs)
Sodium (cont.)			
chromate, di-	$Na_2Cr_2O_7$	ord.	+55.0
fluoride	NaF	ord.	−16.4
hydroxide, aqueous solution	$NaOH$	300	−16.0
iodate	$NaIO_3$	ord.	−53.0
iodide	NaI	ord.	−57.0
nitrate	$NaNO_3$	ord.	−25.6
nitrite	$NaNO_2$	ord.	−14.5
oxide	Na_2O	ord.	−19.8
oxide	Na_2O_2	ord.	−28.10
phosphate, m-	$NaPO_3$	ord.	−42.5
phosphate	Na_2HPO_4	ord.	−56.6
sulfate	Na_2SO_4	ord.	−52.0
sulfate	$Na_2SO_4 \cdot 10H_2O$	ord.	−184.0
sulfide	Na_2S	ord.	−39.0
sulfide	Na_2S_2	ord.	−53.0
sulfide	Na_2S_3	ord.	−68.0
sulfide	Na_2S_4	ord.	−84.0
sulfide	Na_2S_5	ord.	−99.0
Strontium	Sr	ord.	+92.0
acetate	$Sr(C_2H_3O_2)_2$	ord.	−79.0
bromate	$Sr(BrO_3)_2$	ord.	−93.5
bromide	$SrBr_2$	ord.	−86.6
bromide	$SrBr_2 \cdot 6H_2O$	ord.	−160.0
carbonate	$SrCO_3$	ord.	−47.0
chlorate	$Sr(ClO_3)_2$	ord.	−73.0
chloride	$SrCl_2$	ord.	−63.0
chloride	$SrCl_2 \cdot 6H_2O$	ord.	−145.0
chromate	$SrCrO_4$	ord.	−5.1
fluoride	SrF_2	ord.	−37.2
hydroxide	$Sr(OH)_2$	ord.	−40.0
hydroxide	$Sr(OH)_2 \cdot 8H_2O$	ord.	−136.0
iodate	$Sr(IO_3)_2$	ord.	−108.0
iodide	SrI_2	ord.	−112.0
nitrate	$Sr(NO_3)_2$	ord.	−57.2
nitrate	$Sr(NO_3)_2 \cdot 4H_2O$	ord.	−106.0
oxide	SrO	ord.	−35.0
oxide	SrO_2	ord.	−32.3
sulfate	$SrSO_4$	ord.	−57.9
Sulfur, gaseous	S	1,023	+464.0
Sulfur, gaseous	S	828	+700.0
Sulfur, liquid	S		−15.4
Sulfur, solid (α)	S	ord.	−15.5
Sulfur, solid (β)	S	ord.	−14.9
chloride	SCl_2	ord.	−49.4
chloride	SCl_3	ord.	−49.4
chloride	S_2Cl_2	ord.	−62.2
fluoride	SF_6	ord.	−44.0
iodide	SI	ord.	−52.7
oxide, liquid	SO_2	ord.	−18.2
Sulfuric acid	H_2SO_4	ord.	−39.8
Tantalum	Ta	2,143	+124.0
Tantalum	Ta	293	+154.0
chloride	$TaCl_5$	304	+140.0

Table 2–40 (continued)
MAGNETIC SUSCEPTIBILITIES OF THE ELEMENTS AND OF SOME INORGANIC COMPOUNDS

Substance	Formula	Temperature (K)	Susceptibility (10^{-6} cgs)
Tantalum (cont.)			
fluoride	TaF_3	293	+795.0
oxide	Ta_2O_5	ord.	−32.0
Technetium	Tc	402	+250.0
Technetium	Tc	298	+270.0
Technetium	Tc	78	+290.0
oxide	$TcO_2 \cdot 2H_2O$	300	+244.0
oxide	Tc_2O_7	298	−40.0
Tellurium, liquid	Te		−6.4
Tellurium, solid	Te	ord.	−39.5
bromide	$TeBr_2$	ord.	−106.0
chloride	$TeCl_2$	ord.	−94.0
fluoride	TeF_6	ord.	−66.0
Terbium	Tb	273	+146,000.0
oxide	Tb_2O_3	288.1	+78,340.0
sulfate	$Tb_2(SO_4)_3$	293	+78,200.0
sulfate	$Tb_2(SO_4)_3 \cdot 8H_2O$	293	+76,500.0
Thallium, liquid	Tl	573	−26.8
Thallium, solid (α)	Tl	ord.	−50.9
Thallium, solid (β)	Tl	>508	−32.3
acetate	$TlC_2H_3O_2$	ord.	−69.0
bromate	$TlBrO_3$	ord.	−75.9
bromide	TlBr	ord.	−63.9
carbonate	Tl_2CO_3	ord.	−101.6
chlorate	$TlClO_3$	ord.	−65.5
chloride	TlCl	ord.	−57.8
chromate	Tl_2CrO_4	ord.	−39.3
cyanide	TlCN	ord.	−49.0
fluoride	TlF	ord.	−44.4
iodate	$TlIO_3$	ord.	−86.8
iodide	TlI	ord.	−82.2
nitrate	$TlNO_3$	ord.	−56.5
nitrite	$TlNO_2$	ord.	−50.8
oxide	Tl_2O_3	ord.	+76.0
phosphate	Tl_3PO_4	ord.	−145.2
sulfate	Tl_2SO_4	ord.	−112.6
sulfide	Tl_2S	ord.	−88.8
thiocyanate	TlCNS	ord.	−66.7
Thorium	Th	293	+132.0
Thorium	Th	90	+153.0
chloride	$ThCl_4 \cdot 8H_2O$	305.2	−180.0
nitrate	$Th(NO_3)_4$	ord.	−108.0
oxide	ThO_2	ord.	−16.0
Thulium	Tm	291	+25,500.0
oxide	Tm_2O_3	296.5	+51,444.0
Tin, liquid	Sn		−4.5
Tin, solid, gray	Sn	280	−37.0
Tin, solid, gray	Sn	100	−31.7
Tin, solid, white	Sn	ord.	+3.1
bromide	$SnBr_4$	ord.	−149.0
chloride	$SnCl_2$	ord.	−69.0
chloride	$SnCl_2 \cdot 2H_2O$	ord.	−91.4
chloride, liquid	$SnCl_4$	ord.	−115.0
hydroxide	$Sn(OH)_4$	ord.	−60.0

Table 2–40 (continued)

MAGNETIC SUSCEPTIBILITIES OF THE ELEMENTS AND OF SOME INORGANIC COMPOUNDS

Substance	Formula	Temperature (K)	Susceptibility (10^{-6} cgs)
Tin, solid, white (cont.)			
oxide	SnO	ord.	−19.0
oxide	SnO_2	ord.	−41.0
Titanium	Ti	293	+153.0
Titanium	Ti	90	+150.0
bromide	$TiBr_2$	288	+640.0
bromide	$TiBr_3$	441	+520.0
bromide	$TiBr_3$	291	+660.0
bromide	$TiBr_3$	195	+680.0
bromide	$TiBr_3$	90	+220.0
carbide	TiC	ord.	+8.0
chloride	$TiCl_2$	288	+570.0
chloride	$TiCl_3$	685	+705.0
chloride	$TiCl_3$	373	+1,030.0
chloride	$TiCl_3$	292	+1,110.0
chloride	$TiCl_3$	212	+690.0
chloride	$TiCl_3$	90	+220.0
chloride	$TiCl_4$	ord.	−54.0
fluoride	TiF_3	293	+1,300.0
iodide	TiI_2	288	+1,790.0
iodide	TiI_3	434	+221.0
iodide	TiI_3	292	+160.0
iodide	TiI_3	195	+159.0
iodide	TiI_3	90	+167.0
oxide	TiO_2	ord.	+5.9
oxide	TiO_3	248	+132.4
oxide	Ti_2O_3	382	+152.0
oxide	Ti_2O_3	298	+125.6
sulfide	TiS	ord.	+432.0
Tungsten	W	298	+59.0
bromide	WBr_5	293	+250.0
carbide	WC	ord.	+10.0
chloride	WCl_2	293	−25.0
chloride	WCl_5	293	+387.0
chloride	WCl_6	ord.	−71.0
fluoride	WF_6	ord.	−40.0
oxide	WO_2	ord.	+57.0
oxide	WO_3	ord.	−15.8
sulfide	WS_2	303	+5,850.0
Tungstic acid, *o-*	H_2WO_4	ord.	−28.0
Uranium (α)	U	623	+440.0
Uranium (α)	U	298	+409.0
Uranium (α)	U	78	+396.0
Uranium (γ)	U	1,393	+514.0
bromide	UBr_3	294	+4,740.0
bromide	UBr_4	293	+3,530.0
chloride	UCl_3	300	+3,460.0
chloride	UCl_4	294	+3,680.0
fluoride	UF_4	300	+3,530.0
fluoride	UF_6	ord.	+43.0
hydride	UH_3	462	+2,821.0
hydride	UH_3	391	+3,568.0
hydride	UH_3	295	+6,244.0
hydride	UH_3	255	+9,306.0
iodide	UI_3	293	+4,460.0

Table 2–40 (continued)
MAGNETIC SUSCEPTIBILITIES OF THE ELEMENTS AND OF SOME INORGANIC COMPOUNDS

Substance	Formula	Temperature (K)	Susceptibility (10^{-6} cgs)
Uranium (γ) (cont.)			
oxide	UO	293	+1,600.0
oxide	UO_2	293	+2,360.0
oxide	UO_3	ord.	+128.0
sulfate	$U(SO_4)_2$	ord.	+31.0
sulfide (α)	US_2	290.5	+3,137.0
sulfide (β)	US_2	290.5	+3,470.0
sulfide	U_2S_3	ord.	+5,206.0
sulfide	U_3S_5	ord.	+11,220.0
Vanadium	V	298	+255.0
bromide	VBr_2	293	+3,230.0
bromide	VBr_2	195	+3,760.0
bromide	VBr_2	90	+4,470.0
bromide	VBr_3	293	+2,890.0
bromide	VBr_3	195	+4,110.0
bromide	VBr_3	90	+8,540.0
chloride	VCl_2	293	+2,410.0
chloride	VCl_3	293	+3,030.0
chloride	VCl_4	293	+1,130.0
chloride	VCl_4	195	+1,700.0
chloride	VCl_4	90	+4,360.0
fluoride	VF_3	293	+2,730.0
oxide	VO_2	290	+270.0
oxide	V_2O_3	293	+1,976.0
oxide	V_2O_5	ord.	+128.0
sulfide	VS	ord.	+600.0
sulfide	V_2S_3	293	+1,560.0
Water, gaseous	H_2O	>373	−13.1
Water, liquid	H_2O	373	−13.09
Water, liquid	H_2O	293	−12.97
Water, liquid	H_2O	273	−12.93
Water, liquid	DHO	302	−12.97
Water, liquid	D_2O	293	−12.76
Water, liquid	D_2O	276.8	−12.66
Water, solid	H_2O	273	−12.65
Water, solid	H_2O	223	−12.31
Water, solid	D_2O	276.8	−12.54
Water, solid	D_2O	213	−12.41
Xenon	Xe	ord.	−43.9
Ytterbium	Yb	292	+249.0
Ytterbium	Yb	90	+639.0
sulfide	Yb_2S_3	292	+18,300.0
Yttrium	Y	292	+2.15
Yttrium	Y	90	+2.43
oxide	Y_2O_3	293	+44.4
sulfide	Y_2S_3	ord.	+100.0
Zinc, liquid	Zn		−7.8
Zinc, solid	Zn	ord.	−11.4
acetate	$Zn(C_2H_3O_2)_2 \cdot 2H_2O$	ord.	−101.0
carbonate	$ZnCO_3$	ord.	−34.0
chloride	$ZnCl_2$	296	−65.0
cyanide	$Zn(CN)_2$	ord.	−46.0
fluoride	ZnF_2	299.6	−38.2
hydroxide	$Zn(OH)_2$	ord.	−67.0

Table 2–40 (continued)
MAGNETIC SUSCEPTIBILITIES OF THE ELEMENTS AND OF SOME INORGANIC COMPOUNDS

Substance	Formula	Temperature (K)	Susceptibility (10^{-6} cgs)
Zinc, solid (cont.)			
iodide	ZnI_2	ord.	–98.0
nitrate, aqueous solution	$Zn(NO_3)_2$	ord.	–63.0
oxide	ZnO	ord.	–46.0
phosphate	$Zn_3(PO_4)_2$	ord.	–141.0
sulfate	$ZnSO_4$	ord.	–45.0
sulfate	$ZnSO_4 \cdot H_2O$	ord.	–63.0
sulfate	$ZnSO_4 \cdot 7H_2O$	ord.	–143.0
sulfide	ZnS	ord.	–25.0
Zirconium	Zr	293	+122.0
Zirconium	Zr	90	+119.0
carbide	ZrC	ord.	–26.0
nitrate	$Zr(NO_3)_4 \cdot 5H_2O$	ord.	–77.0
oxide	ZrO_2	ord.	–13.8

From Weast, R. C., Ed., *Handbook of Chemistry and Physics*, 55th ed., CRC Press, Cleveland, 1974, E-121.

Section 3

*Miscellaneous Tables of
Physical Properties*

Table 3–1
PHYSICAL CONSTANTS OF MINERALS

The following table presents data for many of the more common minerals.

In order to avoid duplication and save space, very few cross references are given in the body of the table. If the name sought is not found in the table, consult the synonym index given below.

Specific gravities are given at normal atmospheric temperatures, a more precise statement being valueless considering the large variations in natural minerals.

Hardness is given in terms of Mohs' scale (see under Hardness).

Indices of refraction for the sodium line: $\lambda = 5893$ Å, unless otherwise indicated. Li: $\lambda = 6708$ Å. Indices will invariably be given in the order ω, ϵ, or α, β, γ. Uniaxial crystals are considered positive if $\epsilon > \omega$ and negative if $\omega > \epsilon$. Biaxial crystals are considered positive if β is nearer α in value than it is γ, and negative if β is nearer γ than α.

Abbreviations

Abbreviation	Meaning of abbreviation	Abbreviation	Meaning of abbreviation	Abbreviation	Meaning of abbreviation
bl	blue	grn	green	rhbdr	rhombohedral
blk	black	grnsh	greenish	rhomb	rhombic
blksh	blackish	hex	hexagonal	somet	sometimes
blsh	bluish	iridesc	iridescent	tarn	tarnishes
br	brown	monocl	monoclinic	tetr	tetragonal
brnsh	brownish	oft	often	tricl	triclinic
ccl	colorless	pa	pale	vlt	violet
cub	cubic	purp	purple	wh	white
dk	dark	(R)	radioactive	yel	yellow
Fe	Fe, ferrous iron	redsh	redish	yelsh	yellowish
Fe^{+3}	Fe, ferric iron				

Synonym Index

Compound sought	Listed	Compound sought	Listed
Acmite	Aegirine	Garnierite	Serpentine (Ni-bearing)
Agate	Quartz (impure)	Glauber salt	Mirabilite
Aluminum hydroxide	Boehmite, Diaspore, Gibbsite	Hyacinth	Zircon
Amphibole	Actinolite, Anthophyllite, Cummingtonite, Glaucophane, Hornblende, Riebeckite, Tremolite	Iceland spar	Calcite
		Idocrase	Vesuvianite
Antimony oxide	Senarmontite, Valentinite	Iron carbonate	Siderite
Antimony sulfide	Stibnite	Iron hydroxide	Goethite, Lepidocrocite
Arsenic oxide	Arsenolite, Claudetite	Iron oxide	Hematite, Magnetite
Arsenic sulfide	Orpiment, Realgar	Iron spinel	Hercynite
Barium carbonate	Witherite	Iron sulfide	Marcasite, Pyrite, Pyrrhotite
Barium sulfate	Barite	Lapis lazuli	Lazurite
Barytes	Barite	Lead carbonate	Cerussite
Bauxite	Gibbsite, Boehmite, Diaspore	Lead chloride	Cotunnite
Brimstone	Sulfur	Lead chromate	Crocoite
Bronzite	Orthopyroxene	Lead oxide	Litharge, Minium
Cadmium sulfide	Greenockite	Lead sulfate	Anglesite
Calamine	Hemimorphite	Lead sulfide	Galena
Calcium carbonate	Aragonite, Calcite, Vaterite	Limonite	Goethite (impure)
Calcium sulfate	Anhydrite, Gypsum	Lithiophyllite	Triphylite
Calcium sulfide	Oldhamite	Lithium mica	Lepidolite
Carborundum	Moissanite	Lodestone	Magnetite
Chalcedony	Quartz (impure, fibrous)	Magnesium carbonate	Magnesite
Chinaclay	Kaolinite	Magnesium hydroxide	Brucite
Chloanthite	Skutterodite	Magnesium oxide	Periclase
Chromespinel	Chromite	Magnesium sulfate	Kieserite
Chrysolite	Serpentine	Manganese carbonate	Rhodochrosite
Clinoptolite	Heulandite	Manganese hydroxide	Pyrochroite
Clayminerals	Illite, Kaolinite, Montmorillonite	Manganese oxide	Hausmannite, Manganosite, Pyrolusite
Clinochlore	Chlorite	Manganese sulfide	Alabandite
Cobaltbloom	Erythrite	Meerschaum	Serpentine
Copper chloride	Nantokite	Mica	Muscovite, Paragonite, Phlogopite, Biotite, Lepidolite
Copper oxide	Cuprite		
Copper sulfide	Chalcocite, Covellite, Digenite	Native copper	Copper
Emerald	Beryl	Native gold	Gold
Emery	Mixture of Corundum, Magnetite and other minerals	Nickel oxide	Bunsenite
		Nickel sulfide	Millerite
Epsom salt	Epsomite	Orthite	Allanite
Feldspar	Orthoclase, Microcline, Anorthoclase, Albite, Oligoclase, Andesine, Anorthite	Penninite	Chlorite
		Peridote	Olivine
		Pistacite	Epidote
Fibrolite	Sillimanite	Pitchblende	Uraninite
Flint	Quartz (impure)	Plagioclase	Albite, Oligoclase, Andesine, Anorthite
Fluorapatite	Apatite		
Fluorspar	Fluorite	Potassium chloride	Sylvite
Garnet	Almandine, Pyrope, Spessartite, Andradite, Grossularite, Uvarovite, Hydrogrossularite	Potassium sulfate	Arcanite
		Pyroxene	Diopside, Augite, Aegirine, Jadeite, Pigeonite, Eustatite, Orthopyroxene

Table 3–1 (continued)
PHYSICAL CONSTANTS OF MINERALS

Synonym Index

Compound sought	Listed	Compound sought	Listed
Rocksalt	Halite	Tin oxide	Cassiterite
Ruby	Corundum	Titanite	Sphene
Sapphire	Corundum	Titanium oxide	Anatase, Brookite, Rutile
Silica	Christobalite, Quartz, Tridymite	Uranium oxide	Uraninite
Silver chloride	Cerargyrite	Zeolite	Natrolite, Mesolite, Scolecite, Thomasonite, Harmatome, Eddingtonite, Heulandite, Stilbite, Phillipsite, Chabazite, Gmelinite, Levyn, Laumontite, Mordenite
Silver iodide	Jodyrite, Miersite		
Silver sulfide	Acanthite, Argentite		
Smalltite	Skutterotite		
Soapstone	Mixture of Talc and other minerals		
Sodium chloride	Halite	Zincblende	Sphalerite
Sodium sulfate	Thenardite	Zinc carbonate	Smithsonite
Strontium carbonate	Strontianite	Zinc oxide	Zincite
Strontium sulfate	Celestite	Zinc spinel	Gahnite
Thorium oxide	Thorianite	Zinc sulfide	Sphalerite, Wurtzite
		Zirconium oxide	Baddeleyite

Table 3–1 (continued)
PHYSICAL CONSTANTS OF MINERALS

Name	Formula	Sp. gr.	Hardness	Crystalline form and color	Index of refraction (Na) η; ω ε / α β γ
Acanthite	Ag₂S	7.2–7.3	2–2.5	rhomb.(?), iron-blk.	
Actinolite	Ca₂(Mg,Fe)₅Si₈O₂₂(OH,F)₂	3.02–3.44	5–6	monocl., pa. to dk. grn.	1.599–1.688, 1.612–1.697, 1.622–1.705
Aegirine	NaFe⁺³Si₂O₆	3.55–3.60	6	monocl., dk. grn. to grnsh. blk.	1.750–1.776, 1.780–1.820, 1.800–1.836
Akermanite	Ca₂MgSi₂O₇	2.944	5–6	tetr., col., gray-grn., br.	1.632, 1.640
Alabandite	MnS	4.050	3.5–4	cub., iron-blk., tarn., br.	
Albite	NaAlSi₃O₈	2.63	6–6.5	tricl., col., wh., somet. yel., pink, grn.	1.527, 1.531, 1.538
Allanite	(Ca,Mn,Ce,La,Y,Th)₂(Fe,Fe⁺²,Ti)(Al,Fe⁺³)₂Si₃O₁₂(OH)	3.4–4.2	5–6.5	monocl., pa. br. to blk.	1.690–1.791, 1.700–1.815, 1.706–1.828
Allemontite	AsSb	5.8–6.2	3–4	hex., tin-wh. to redsh. gray, tarn. gray-brnsh. blk.	
Almandine	Fe₃Al₂Si₃O₁₂	4.318	6–7.5	cub., red, dk. red, blk.	1.830
Altaite	PbTe	8.15	3	cub., tin-wh., yelsh., tarn. bronze-yel.	
Aluminite	Al₂(SO₄)(OH)₄·7H₂O	1.66–1.82	1–2	monocl.(?), wh.	
Alunite	(K,Na)Al₃(SO₄)₂(OH)₆	2.6–2.9	3.5–4	rhbdr., wh., gray, yel., redsh., br.	1.459, 1.464, 1.470
Alunogen	Al₂(SO₄)₃·18H₂O	1.77	1.5–2	tricl., col., wh., yelsh. wh., redsh. wh.	1.459–1.475, 1.461–1.478, 1.470–1.485
Amblygonite	(Li,Na)Al(PO₄)(F,OH)	3.0–3.1	5.5–6	tricl., wh., yelsh. wh., grnsh. wh., blsh. wh., gray	1.591, 1.604, 1.613
Analcite	NaAlSi₂O₆·H₂O	2.24–2.29	5.5	cub., wh., pink, gray	1.479–1.493
Anatase	TiO₂	3.90	5.5–6	tetr., br., yelsh. br., redsh. br., bl., blk., grn., gray	2.5612, 2.4880
Andalusite	Al₂OSiO₄	3.13–3.16	6.5–7.5	rhomb., pink, wh., red	1.629–1.640, 1.633–1.644, 1.638–1.650
Andesine	((NaSi)₀.₇–₀.₅(CaAl)₀.₃–₀.₅)AlSi₂O₈	2.65–2.68	6–6.5	tricl., wh., gray, red	1.544–1.555, 1.546–1.558, 1.551–1.563
Andorite	PbAgSb₃S₆	5.33–5.37	3–3.5	rhomb., dk. steel gray, somet. tarn. yel. or iridesc.	
Andradite	Ca₃Fe⁺²₂Si₃O₁₂	3.859	6.5–7.5	cub., brnsh. red, blk., somet. yel., grn.	1.887
Anglesite	PbSO₄	6.37–6.39	2.5–3	rho.nb., col., wh., somet. gray, yelsh., grn. tinge	1.8771, 1.8826, 1.8937
Anhydrite	CaSO₄	2.98	3.5	rhomb., col., blsh. wh., vlt.	1.5698, 1.5754, 1.6136
Ankerite	Ca(Fe,Mg,Mn)(CO₃)₂	2.8–3.1	3.5–4	rhbdr., br., yelsh. br., grnsh. br., pink	1.690–1.750, 1.510–1.548
Anorthite	CaAl₂Si₂O₈	2.76	6–6.5	tricl., wh., yel., grn., blk.	1.577, 1.585, 1.590
Anorthoclase	(Na,K)AlSi₃O₈	2.56–2.60	6	tricl., col., wh.	1.523, 1.528, 1.529
Anthophyllite	(Mg,Fe)₇Si₈O₂₂(OH,F)₂	2.85–3.57	5.5–6	rhomb., wh., gray, grn., br., yelsh. br., dk. br.	1.596–1.694, 1.605–1.710, 1.615–1.722
Antimony	Sb	6.61–6.72	3–3.5	hex., tin-wh.	
Apatite	Ca₅(PO₄)₃(OH,F,Cl)	3.1–3.35	5	hex., grn., wh., yel., br., red, bl.	1.629–1.667, 1.624–1.666
Apophyllite	KFCa₄Si₈O₂₀·8H₂O	2.33–2.37	4.5–5	tetr., col., wh., pink, pa. yel., pa. grn.	1.534–1.535, 1.535–1.537
Aragonite	CaCO₃	2.94–2.95	3.5–4	rhomb., col., wh.	1.530–1.531, 1.680–1.681, 1.685–1.686
Arcanite	K₂SO₄	2.663	2–2.5	rhom., col., wh.	1.4935, 1.4947, 1.4973
Argentite	Ag₂S	7.2–7.4	2–2.5	cub., blkish, lead gray	
Arsenic	As	5.63–5.78	3.5	hex., tin-wh., tarn. dk. gray	1.755
Arsenolite	As₂O₃	3.86–3.88	1.5	cub., wh., somet. blsh., yelsh., redsh. tinge	
Arsenopyrite	FeAsS	5.9–6.2	5.5–6	monocl., silver-wh., to steel gray	
Atacamite	Cu₂(OH)₃Cl	3.74–3.78	3–3.5	rhomb., grn., dk. grn., blkish. grn.	1.831, 1.861, 1.880
Augelite	Al₂(PO₄)(OH)₃	2.696	4.5–5	monocl., col., wh., yelsh. wh., rose	1.5736, 1.5759, 1.5877
Augite	(Ca,Mg,Fe,Fe⁺²,Ti,Al)₂(Si,Al)₂O₆	3.23–3.52	5–6	monocl., pa. br., br., purp. br., grn., blk.	1.671–1.735, 1.672–1.741, 1.703–1.761
Autunite	Ca(UO₂)₂(PO₄)₂·10–12H₂O	3.1–3.2	2–2.5	tetr., yel., somet. grnsh. yel. to pa. grn.	1.577, 1.553
Axinite	(Ca,Mn,Fe)₃Al₂BO₃Si₄O₁₂(OH)	3.26–3.36	6.5–7	tricl., br., yelsh.	1.674–1.693, 1.681–1.701, 1.684–1.704
Azurite	Cu₃(OH)₂(CO₃)₂	3.77	3.5–4	monocl., azure bl., dk. bl., pa. bl.	1.730, 1.758, 1.838
Baddeleyite	ZrO₂	5.4–6.02	6.5	monocl., col., yel., gr., redsh. br., br., blk.	2.13, 2.19, 2.20
Barite	BaSO₄	4.50	3–3.5	rhomb., col., wh., somet. br., dk. br., gray	1.6362, 1.6373, 1.6482
Benitoite	BaTi(SiO₃)₃	3.65	6–6.5	rhbdr., bl., purp., col.	1.757, 1.804
Bertrandite	Be₄Si₂O₇(OH)₂	2.6	6	rhomb., col.	1.589, 1.602, 1.613
Beryl	Be₃Al₂Si₆O₁₈	2.66–2.83	7.5–8	hex., col., wh., blsh. grn., grnsh. gray, yel., yel., bl.	1.565–1.590, 1.567–1.598
Beryllonite	NaBe(PO₄)	2.81	5.5–6	monocl., col., wh., pa. yel.	1.5520, 1.5579, 1.561
Biotite	K(Mg,Fe)₃AlSi₃O₁₀(OH,F)₂	2.7–3.3	2.5–3	monocl., blk., dk. br., redsh. br.	1.565–1.625, 1.605–1.696, 1.605–1.696
Bismuth	Bi	9.70–9.83	2–2.5	rhomb., silver-wh. to redsh. wh.	
Bismuthinite	Bi₂S₃	6.75–6.81	2	rhomb., lead gray to tin-wh., tarn. yel. or irilesc.	

Table 3–1 (continued)
PHYSICAL CONSTANTS OF MINERALS

Name	Formula	Sp. gr.	Hardness	Crystalline form and color	Index of refraction (Na) γ; ω, ϵ / α, β, γ
Bixbyite	$(Mn,Fe)_2O_3$	4.945	6–6.5	cub., blk.	
Bloedite	$Na_2Mg(SO_4)_2 \cdot 4H_2O$	2.22–2.28	2.5–3	monocl., col., somet. blsh.-grn. or redsh.	1.483, 1.486, 1.487
Boehmite	$AlO(OH)$	3.01–3.06	3.5–4	rhomb., wh.	1.64–1.65, 1.65–1.66, 1.65–1.67
Boracite	$Mg_3B_7O_{13}Cl$	2.91–2.97	7–7.5	rhomb., col., wh., gray, yel., blsh.-grn., grn.	1.66, 1.66, 1.67
Borax	$Na_2B_4O_7 \cdot 10H_2O$	1.715	2–2.5	monocl., col., wh., gray, blsh. or grnsh-wh.	1.4466, 1.4687, 1.4717
Bornite	Cu_5FeS_4	5.06–5.08	3	cub., copper red to pinchbeck br., tarn. purp., iridesc.	
Boulangerite	$Pb_5Sb_4S_{11}$	6.0–6.2	2.5–3	monocl., blsh. lead gray, oft. with yel. spots	
Bournonite	$PbCuSbS_3$	5.80–5.86	2.5–3	rhomb., steel gray to blk.	
Braggite	PtS	10.0		tetr., steel gray	
Braunite	$(Mn,Si)_2O_3$	4.72–4.83	6–6.5	tetr., brnsh. blk. to steel gray	
Bravoite	$(Ni,Fe)S_2$	4.62	5.5–6	cub., steel gray	
Breithauptite	$NiSb$	8.23	5.5	hex., pa. copper red to vit., tarn.	
Brochantite	$Cu_4(SO_4)(OH)_6$	3.79	3.5–4	monocl., emerald-grn. to blksh. grn., pa. grn.	1.728, 1.771, 1.800
Bromyrite	$AgBr$	6.47	2.5	cub., col., gray, yelsh., grnsh.-br.	2.253
Brookite	TiO_2	4.08–4.20	5.5–6	rhomb., br., yelsh. br., redsh. br., blk.	2.5831, 2.5843, 2.7004
Brucite	$Mg(OH)_2$	2.38–3.40	2.5	hex., wh., pa. grn., gray, bl., yel., br.	1.560–1.590, 1.580–1.600 (Li) 2.37
Bunsenite	NiO	6.898	5.5	cub., dk. pistachio-grn.	
Cacoxenite	$Fe(PO_4)_3(OH)_3 \cdot 12H_2O$	2–2.4	3–4	hex., yel. to brnsh.-yel., redsh. yel., somet. grnsh.	1.575–1.585, 1.635–1.656
Calcite	$CaCO_3$	2.715–2.94	3	rhbdr., col., wh., somet. gray, yel., pink, bl.	1.659–1.740, 1.486–1.550
Caledonite	$Cu_2Pb_5(SO_4)_3(CO_3)(OH)_6$	5.75–5.77	2.5–3	rhomb., dk. blsh. grn.	1.815–1.821, 1.863–1.869, 1.906–1.912
Calomel	$HgCl$	7.15	1.5	tetr., col., wh., gray, yelsh. wh., br.	1.973, 2.656
Cancrinite	$(Na,Ca)_7Al_6Si_6O_{24}(CO_3SO_4Cl)_{1.5-2.1} \cdot 5H_2O$	2.51–2.42	5–6	hex., col., wh., pa. bl., pa. grn., yel., redsh.	1.528–1.507, 1.503–1.495
Carnallite	$KMgCl_3 \cdot 6H_2O$	1.602	2.5	rhomb., col., wh., oft. redsh., somet. yel., bl.	1.466, 1.475, 1.494
Carnotite	$K_2(UO_2)_2(VO_4)_2 \cdot 3H_2O$		1–2	rhomb. or monocl., bright yel., yel., grnsh. yel.	1.75, 1.92, 1.95
Cassiterite	SnO_2	6.99	6–7	tetr., yelsh. or redsh. br., brnsh.-blk.	2.006, 2.0972
Celestite	$SrSO_4$	3.96	3–3.5	rhomb., col., wh., pa. bl., redsh., grnsh., brnsh.	1.621–1.622, 1.623–1.624, 1.630–1.631
Celsian	$BaAl_2Si_2O_8$	3.10–3.39	6–6.5	monocl., col., wh., yel.	1.579–1.587, 1.583–1.593, 1.588–1.600
Cervantite	$Sb_2O_4(?)$	6.64	4–5	rhomb.(?), yel., wh., somet. redsh.-wh.	
Cerargyrite	$AgCl$	5.55	2.5	cub., col., gray, grnsh.-br., tarn. purp., yelsh.	2.071
Cerussite	$PbCO_3$	6.53–6.57	3–3.5	rhomb., col., wh., gray, somet. bl., blk., grn.	1.8036, 2.0765, 2.0786
Chabazite	$(Ca,Na_2)Al_2Si_4O_{12} \cdot 6H_2O$	2.05–2.10	4.5	rhbdr., redsh.-wh., wh., yelsh., grnsh.	1.470–1.494
Chalcocite	Cu_2S	5.5–5.8	2.5–3	rhomb., blksh. lead gray	
Chalcanthite	$CuSO_4 \cdot 5H_2O$	2.28	2.5	tricl., dk. bl. to sky bl., somet. grnsh.	1.514, 1.537, 1.543
Chalcopyrite	$CuFeS_2$	4.1–4.3	3.5–4	tetr., brass-yel., tarn., iridisc.	
Chiolite	$Na_5Al_3F_{14}$	3.00	3.5–4	tetr., wh. to col.	1.349, 1.342
Chlorite	$(Mg,Al,Fe)_{12}(Si,Al)_8O_{20}(OH)_{16}$	2.6–3.3	2–3	monocl., wh., yel., yel., pink, br., red	1.57–1.66, 1.57–1.67, 1.57–1.67
Chloritoid	$(Fe,Mg,Mn)_2(AlFe^{+3})Al_3O_2SiO_4(OH)_4$	3.51–3.80	6.5	monocl., tricl., dk. grn.	1.713–1.730, 1.719–1.734, 1.723–1.740
Chondrodite	$Mg(OH,F)_2 \cdot 2MgSiO_4$	3.16–3.26	6.5	monocl., yel., br., red	1.592–1.615, 1.602–1.627, 1.621–1.646
Chromite	$FeCr_2O_4$	4.5–5.1	5.5	cub., blk.	2.16
Chrysoberyl	$BeAl_2O_4$	3.65–3.85	8.5	rhomb., grn., yel., br., blk.	1.746, 1.748, 1.756
Chrysocolla	$CuSiO_3 \cdot 2H_2O$	~2.4	2–4	rhomb. (?), grn., bl., br., gray	1.575–1.597, 1.597–1.598
Cinnabar	HgS	8.090	2–2.5	hex., red, brnsh. red, gray	(Li) 2.814, 3.143
Claudetite	As_2O_3	4.15	2.5	monocl., col. to wh.	1.87, 1.92, 2.01
Clinozoisite	$Ca_2Al_3Si_3O_{12}(OH)$	3.21–3.38	6.5	monocl., col., pa. yel., gray, grn.	1.670–1.715, 1.674–1.725, 1.690–1.734
Cobaltite	$CoAsS$	6.33	5.5	cub., silver wh., redsh., steel gray, blk.	
Colemanite	$Ca_2B_6O_{11} \cdot 5H_2O$	2.42–2.43	4.5	monocl., col., wh., yelsh. wh., gray	1.586, 1.592, 1.614
Columbite	$(Fe,Mn)(Cb,Ta)_2O_6$	5.15–5.25	6	rhomb., iron blk. to br. blk.	
Connellite	$Cu_{19}(SO_4)Cl_4(OH)_{32} \cdot 3H_2O(?)$	3.36	3	hex., azure bl.	1.724–1.738, 1.746–1.758
Copiapite	$(Fe,Mg)Fe_4^{+3}(SO_4)_6(OH)_2 \cdot 20H_2O$	2.08–2.17	2.5–3	tricl., yel., grnsh. yel.	1.51–1.53, 1.53–1.55, 1.58–1.60
Copper	Cu	8.95	2.5–3	cub., red	
Coquimbite	$Fe_2(SO_4)_3 \cdot 9H_2O$	2.10–2.12	2.5	hex., pa. vlt. to dk. amethystine, yelsh., grnsh.	1.53–1.55, 1.55–1.57
Cordierite	$Al_3(Mg,Fe)_2Si_5AlO_{18}$	2.53–2.78	7	rhomb., gray-bl., bl., dk. bl.	1.522–1.558, 1.524–1.574, 1.527–1.578
Corundum	Al_2O_3	4.022	9	hex., col., bl., yel., purp., grn., pink, red	1.767–1.772, 1.759–1.763
Cotunnite	$PbCl_2$	5.80	2.5	rhomb., col. to wh., somet. yelsh., grnsh.	2.199, 2.217, 2.260
Covellite	CuS	4.6–4.76	1.5–2	hex., indigo bl., dk. bl., iridesc. brass yel. to red	

Table 3–1 (continued)
PHYSICAL CONSTANTS OF MINERALS

Name	Formula	Sp. gr.	Hardness	Crystalline form and color	Index of refraction (Na) η; ε / α β γ
Cristobalite	SiO_2	2.33	6-7	tetr.(?), col., wh., yel.	1.487, 1.484
Crocoite	$PbCrO_4$	5.96-6.02	2.5-3	monocl., red, orange red, orange yel.	2.29, 2.36, 2.66
Cryolite	Na_3AlF_6	2.95-2.98	2.5	monocl., col. to wh., brnsh., redsh., blk.	1.338, 1.338, 1.339
Cryolithionite	$Na_3Li_3Al_2F_{12}$	2.77	2.5-3	cub., col. to wh.	1.3395
Cubanite	$CuFeS_2$	4.03-4.18	3.5	rhomb., brass to bronze yel.	
Cummingtonite	$(Mg,Fe)_7Si_8O_{22}(OH)_2$	3.2-3.5	5-6	monocl., dk. grn., br.	1.635-1.665, 1.644-1.675, 1.655-1.698
Cuprite	Cu_2O	6.14	3.5-4	cub., red, somet. blk.	
Danburite	$CaSi_2B_2O_8$	3.0	7	rhomb., pa. yel., col., dk. yel., yelsh. br.	1.63, 1.63-1.64, 1.63-1.64
Datolite	$CaBSiO_4(OH)$	2.96-3.00	5-5.5	monocl., col., wh., yelsh., grnsh., pinksh.	1.622-1.626, 1.649-1.654, 1.666-1.670
Daubreelite	Cr_2FeS_4	3.80-3.82	?	cub., blk.	
Derbylite	$Fe_4Ti_3Sb_2O_{13}(?)$	4.53	5	rhomb., pitch blk.	2.45, 2.45, 2.51
Diamond	C	3.50-3.53	10	cub., col., pa. yel. to dk. yel., pa. br. to dk. br., wh., blsh. wh.	2.4175
Diaspore	$AlO(OH)$	3-3.5	6.5-7	rhomb., wh., graysh. wh., col.	1.682-1.706, 1.705-1.725, 1.730-1.752
Digenite	Cu_9S_5	5.546	2.5-3	cub., bl. to blk.	
Diopside	$CaMgSi_2O_6$	3.22-3.38	5.5-6.5	monocl., wh., pa. grn., dk. grn.	1.664-1.695, 1.672-1.701, 1.695-1.721
Dioptase	$CuSiO_2 \cdot 6H_2O$	3.5	5	rhbdr., emerald grn.	1.64-1.66, 1.70-1.71
Dolomite	$CaMg(CO_3)_2$	2.86	3.5-4	rhbdr., wh., oft. yel. or br. tinge, col.	1.679, 1.500
Douglasite	$K_2FeCl_4 \cdot 2H_2O(?)$	2.16		pa. grn., tarn. brnsh. red	1.485-1.491, 1.497-1.503
Dyscrasite	Ag_3Sb	9.67-9.81	3.5-4	rhomb., silver wh., tarn. gray, yelsh. or blksh.	
Eddingtonite	$BaAl_2Si_3O_{10} \cdot 4H_2O$	2.7-2.8	2.5	rhomb. or monocl., col., pink, br. wh.	1.541, 1.553, 1.557
Eglestonite	Hg_4OCl_2	8.4	2-3	cub., yel., orange-yel. to dk. brnsh., tarn. bl.	2.47-2.51
Emplectite	$CuBiS_2$	6.38	2	rhomb., gray to tin wh.	
Empressite	$AgTe$	7.510	3-3.5	pa. bronze	
Enargite	Cu_3AsS_4	4-4.5	3	rhomb., gray-blk. to iron-blk.	
Enstatite	$MgSiO_3$	3.209	5-6	rhomb., col., gray, grn., yel., brn.	1.650-1.662, 1.653-1.671, 1.658-1.680
Epidote	$Ca_2Fe^{3+}Al_2Si_3O_{12}(OH)$	3.38-3.49	6	monocl., grn., yel., gray	1.715-1.751, 1.725-1.784, 1.734-1.797
Epsomite	$MgSO_4 \cdot 7H_2O$	1.675-1.679	2-2.5	rhomb., col., wh., pink, grn.	1.4325, 1.4354, 1.4609
Erythrite	$(Co,Ni)_3(AsO_4)_2 \cdot 8H_2O$	3.06	1.5-2.5	monocl., crimson-red, red, pa. pink	1.626, 1.661, 1.699
Eucairite	$CuAgSe$	7.6-7.8	2.5	silver wh. to lead gray	
Euclase	$BeAlSiO_4(OH)$	3.0-3.1	7.5	monocl., col., pa. grn., bl.	1.651, 1.655, 1.671
Eudialyte	$(Na,Ca,Fe)_6ZrSi_6O_{18}(OH,Cl)(?)$	2.8-3.1	5-6	hex., pa. pink, red, br.	1.59-1.61, 1.59-1.61
Eulytite	$Bi_4Si_3O_{12}$	6.6	4.5	cub., br., yel., gray	2.05
Euxenite	$(Y,Ca,Ce,U,Th)(Cb,Ta,Ti)_2O_6$	5.0-5.9	5.5-6.5	rhomb., blk., grnsh. or brnsh. tint.	~2.2
Fayalite	Fe_2SiO_4	4.392	6.5	rhomb., grnsh., yelsh.	1.827, 1.869, 1.879
Ferberite	$FeWO_4$	7.51	4-4.5	monocl., br. to blk.	(Li)2.37-2.43
Fergusonite	$(Y,Er,Ce,Fe)(Cb,Ta,Ti)O_4$	5.6-5.8	5.5-6.5	tetr., gray, yel., br., dk. br.	2.1
Fluorite	CaF_2	3.18	4	cub., bl., purp., wh., col., yel., grn.	1.433-1.435
Forsterite	Mg_2SiO_4	3.222	7	rhomb., wh., grnsh., yelsh.	1.635, 1.651, 1.670
Franklinite	$(Zn,Fe)Fe_2O_4$	5.07-5.34	5.5-6.5	cub., dk. bl.-grn., somet. yelsh. or brnsh.	(Li)~2.36
Gahnite	$ZnAl_2O_4$	4.62	7.5-8	Cub., blk. to br.-blk.	1.79-1.81
Galena	PbS	7.57-7.59	2.5-2.75	cub., lead gray	
Galenabismuthite	$PbBi_2S_4$	7.04	2.5-3.5	rhomb., pa. gray to tin-wh., lead gray, somet. tarn., yel. or irid.	
Ganomalite	$(Ca,Pb)_{10}(OH,Cl)_2(Si_2O_7)_3$	5.4-5.7	3-4	hex., col., gray	1.910, 1.945
Gaylussite	$Na_2Ca(CO_3)_2 \cdot 5H_2O$	1.991	2.5-3	monocl., col. to yelsh. wh., graysh. wh., wh.	1.4435, 1.5156, 1.5233
Gehlenite	$Ca_2Al_2SiO_7$	3.038	5-6	tetr., col., gray-grn., br.	1.669, 1.658
Geikielite	$MgTiO_3$	4.05	5-6	rhbdr., brnsh blk., blsh.	2.31, 1.95
Gibbsite	$Al(OH)_3$	2.38-2.42	2.5-3.5	monocl., wh., graysh., grnsh. or redsh.-wh.	1.56-1.58, 1.56-1.58, 1.58-1.60
Glauberite	$Na_2Ca(SO_4)_2$	2.75-2.85	2.5-3	monocl., gray, yelsh., somet. col., redsh.	1.515, 1.535, 1.536
Glauconite	$(K,Na,Ca)_{1.2-2}(Fe^{3+},Al,Fe,Mg)_4Si_{7-7.6}Al_{1-0.4}O_{22}(OH)_2 \cdot nH_2O$	2.4-2.95	2	monocl., col., yelsh. grn., blsh. gray	1.592-1.610, 1.614-1.641, 1.614-1.641
Glaucophane	$Na_2Mg_3Al_2Si_8O_{22}(OH)_2$	3.08-3.30	6	monocl., gray, lavender bl.	1.606-1.661, 1.622-1.667, 1.627-1.670
Gmelinite	$(Ca,Na_2)Al_2Si_4O_{12} \cdot 6H_2O$	~2.1	4.5	rhbdr., wh., redsh.-wh., yelsh., grnsh.	1.476-1.494, 1.474-1.480
Goethite	$FeO(OH)$	3.3-4.3	5-5.5	rhomb., blksh.-br., yelsh. or redsh.-br., yel.	2.260-2.275, 2.393-2.409, 2.398-2.515
Gold	Au	19.3	2.5-3	cub., yel.	
Goslarite	$ZnSO_4 \cdot 7H_2O$	1.978	2-2.5	rhomb., col., wh., somet. br., grn., bl.	1.4568, 1.4801, 1.4844
Graphite	C	2.09-2.23	1-2	hex., iron-blk. to steel gray	
Greenockite	CdS	4.9	3-3.5	hex., yel. to orange	2.506, 2.529

Table 3–1 (continued)
PHYSICAL CONSTANTS OF MINERALS

Name	Formula	Sp. gr.	Hardness	Crystalline form and color	Index of refraction (Na) η; ω; ε / ω β γ
Grossularite	$Ca_3Al_2Si_3O_{12}$	3.594	6–7.5	cub., wh., yel., grn., br., red	1.734
Gummite (R)	$UO_3 \cdot H_2O$	3.9–6.4	2.5–5	yel., orange, redsh.-yel., red, br. blk.	
Gypsum	$CaSO_4 \cdot 2H_2O$	2.30–2.37	2.5	monocl., wh., col., somet. gray, red, yel., br.	1.519–1.521, 1.523–1.526, 1.529–1.531
Halite	$NaCl$	2.16–2.17	2.5	cub., col., wh., orange, red	1.544
Hambergite	$Be_2(OH)(BO_3)$	2.36	7.5	rhomb., col. to gray, wh., yel.	1.56, 1.59, 1.63
Hanksite	$Na_{22}K(SO_4)_9(CO_3)_2Cl$	2.562	3–3.5	hex., col., somet. pa.-yelsh. or gray	1.481, 1.461
Harmotome	$BaAl_2Si_5O_{14} \cdot 6H_2O$	2.41–2.47	4.5	monocl., or rhomb., col., wh., pink, gray, yel.	1.503–1.508, 1.505–1.509, 1.508–1.514
Hausmannite	Mn_3O_4	4.83–4.85	5.5	tetr., brnsh.-blk.	(Li) 2.46, 2.15
Haüyne	$(Na,Ca)_{4-8}Al_6Si_6O_{24}(SO_4,S)_{1-2}$	2.44–2.50	5.5	cub., wh., gray, grn., bl.	1.496–1.505
Hedenbergite	$CaFeSi_2O_6$	3.50–3.56	6	monocl., brnsh.-grn., dk. grn., blk.	1.716–1.726, 1.723–1.730, 1.741–1.751
Helvite	$Mn_4Be_3Si_3O_{12}S$	3.20–3.44	6	cub., yel., br., redsh.-brn.	1.728–1.749
Hematite	Fe_2O_3	5.26	5–6	rhbdr., steel gray, dull red to bright red	3.22, 2.94
Hemimorphite	$Zn_4Si_2O_7(OH)_2 \cdot H_2O$	3.45	5	rhomb., col., wh., pa. bl., pa. grn., br.	1.614, 1.617, 1.636
Hercynite	$FeAl_2O_4$	4.40	7.5–8	cub., blk.	1.835
Herderite	$CaBe(PO_4)(F,OH)$	2.95–3.01	5–5.5	monocl., col. to pa. yel. or grnsh.-wh.	1.592, 1.612, 1.621
Hessite	Ag_2Te	8.24–8.45	2–3	monocl. (<149.5°), cub. (>149.5°), gray, br.	
Heulandite	$(Ca,Na_2)Al_2Si_7O_{18} \cdot 6H_2O$	2.1–2.2	3.5–4	pseudo-monocl., col., wh., yel., pink, red, gray, br.	1.491–1.505, 1.493–1.503, 1.500–1.512
Hopeite	$Zn_3(PO_4)_2 \cdot 4H_2O$	3.0–3.1	3.25	rhomb., col. to grayish-wh., pa. yel.	1.57–1.59, 1.58–1.60, 1.58–1.60
Hornblende	$(Ca,Na,K)_{2-3}(Mg,Fe,Fe^{3+}Al)_5Si_6(Si,Al)_2O_{22}(OH,F)_2$	3.02–3.45	5–6	monocl., grn., dk. grn., blk.	1.615–1.705, 1.618–1.714, 1.632–1.730
Huebnerite	$MnWO_4$	7.12	4–4.5	monocl., yel.-br. to red br., somet. br., blk.	
Humite	$Mg(OH,F)_2 \cdot 3MgSiO_4$	3.2–3.32	6	rhomb., yel., orange	2.17, 2.22, 2.32
Huntite	$Mg_3Ca(CO_3)_4$	2.696		rhomb.(?), wh.	1.607–1.643, 1.619–1.653, 1.639–1.675
Hydrogrossularite	$Ca_3Al_2Si_2O_8(SiO_4)_{1-m}(OH)_{4m}$	3.594–3.3	6–7.5	cub., wh., buff, pa. grn., gray, pink	1.734–1.675
Hydromagnesite	$Mg_5(OH)_2(CO_3)_4 \cdot 3H_2O$	2.236	3.5	monocl., col. to wh.	1.520–1.526, 1.524–1.530, 1.544–1.546
Illite	$K_{1-1.5}Al_4Si_{7-6.5}Al_{1-1.5}O_{20}(OH)_4$	2.6–2.9	1–2	monocl., wh.	1.54–1.57, 1.57–1.61, 1.57–1.61
Ilmenite	$FeTiO_3$	4.68–4.76	5–6	rhbdr., iron-blk.	
Iodyrite	AgI	5.69	1.5	hex., col. on exposure to light, yel., br.	2.21, 2.22
Jadeite	$NaAlSi_2O_6$	3.24–3.43	6	monocl., col., wh., grn., grnsh. bl.	1.640–1.658, 1.645–1.663, 1.652–1.673
Jamesonite	$Pb_4FeSb_6S_{14}$	5.63	2.5	monocl., gray-blk., somet. tarn. iridesc.	
Jarosite	$KFe_3(SO_4)_2(OH)_6$	2.91–3.26	2.5–3.5	rhbdr., ocherous, amber yel. to dk. br.	1.820, 1.715
Kainite	$KMg(SO_4)Cl \cdot 3H_2O$	2.15	2.5–3	monocl., col., gray bl., vlt., yelsh., redsh.	1.494, 1.505, 1.516
Kaliophilite	$KAlSiO_4$	2.61	6	hex., col.	1.532, 1.537
Kaolinite	$Al_2Si_2O_5(OH)_4$	2.61–2.68	2–2.5	tricl. or monocl., wh., redsh.-wh., grnsh.-wh.	1.553–1.565, 1.559–1.569, 1.560–1.570
Kernite	$Na_2B_4O_7 \cdot 4H_2O$	1.908	2.5	monocl., col., wh.	1.454, 1.472, 1.488
Kieserite	$MgSO_4 \cdot H_2O$	2.571	3.5	monocl., col., gray, wh., yelsh.	1.520, 1.533, 1.584
Kyanite	Al_2OSiO_4	3.53–3.65	4.5	tricl., bl., wh., gray, grn., yel., pink	1.712–1.718, 1.721–1.723, 1.727–1.734
Lanarkite	$Pb_2(SO_4)O$	6.92	2–2.5	monocl., gray to grnsh. wh., pa. yel.	1.925–1.931, 2.004–2.010, 2.033–2.039
Lanthanite	$(La,Ce)_2(CO_3)_3 \cdot 8H_2O$	2.69–2.74	2.5–3	rhomb., col. to wh., pink, yelsh.	1.51–1.53, 1.584–1.590, 1.610–1.616
Laumontite	$CaAl_2Si_4O_{12} \cdot 4\text{-}3.5H_2O$	2.2–2.3	3–3.5	monocl., col., wh., red, yel., brn.	1.502–1.514, 1.512–1.522, 1.514–1.525
Laurionite	$Pb(OH)Cl$	6.24	3.5	rhomb., col., wh.	2.08, 2.12, 2.16
Lawsonite	$CaAl_2(OH)_2Si_2O_7 \cdot H_2O$	3.05–3.10	6	monocl., bl., blsh. wh., dk. bl., blsh. grn.	1.655, 1.674–1.675, 1.684–1.686
Lazulite	$(Mg,Fe)Al_2(PO_4)_2(OH)_2$	3.08–3.38	5–6	monocl., bl., azure bl., grnsh. bl., vlt.	1.604–1.626, 1.626–1.654, 1.637–1.663
Lazurite	$Na_4Si_3Al_3O_{12}S$	2.38–2.45	5–5.5	cub., berlin bl., azure bl., grn., pa. bl.	1.500
Leadhillite	$Pb_4(SO_4)(CO_3)_2(OH)_2$	6.55	2.5–3	monocl., col. to wh., gray, grn., pa. bl., yelsh.	1.87, 2.00, 2.01
Lepidocrocite	$FeO(OH)$	4.05–4.31	5	rhomb., ruby-red to red-br.	1.94, 2.20, 2.51
Lepidolite	$K_2(Li,Al)_{5-6}Si_{6-7}Al_{2-1}O_{20}(OH,F)_4$	2.80–2.90	2.5–4	monocl., col., pa. pink, pa. purp.	1.525–1.548, 1.551–1.585, 1.554–1.587
Leucite	$KAlSi_2O_6$	2.47–2.50	5.5–6	tetr. (pseudo-cub.) wh., gray	1.508–1.511
Levyne	$(Ca,Na_2)Al_2Si_5O_{14} \cdot 6H_2O$	~2.1	4.5	rhbdr., wh., redsh. wh., yelsh., grnsh.	1.496–1.505, 1.491–1.500
Litharge	PbO	9.14	2	tetr., red	(Li) 2.665, 2.535
Loellingite	$FeAs_2$	7.39–7.41	5–5.5	rhomb., silver wh. to steel-gray	
Magnesite	$MgCO_3$	2.98–3.44	3.5–4.5	rhbdr., wh., col., somet. yel., br.	1.700–1.782, 1.509–1.563
Magnetite	Fe_3O_4	5.175	5.5–6.5	cub., blk. to br.-blk.	2.42
Malachite	$Cu_2(OH)_2(CO_3)$	4.03–4.07	3.5–4	monocl., bright grn. to dk. grn., blksh. grn.	1.652–1.658, 1.872–1.878, 1.906–1.912
Manganite	$MnO(OH)$	4.32–4.43	4	monocl., dk. steel-gray to iron-blk.	(Li) 2.25, 2.25, 2.53
Manganosite	MnO	5.364	5.5	cub., emerald grn., tarn. bl.	
Marcasite	FeS_2	4.887	6–6.5	rhomb., pa. bronze-yel., tin-wh.	
Marialite	$Na_4Al_3Si_9O_{24}Cl$	2.50–2.62	5–6	tetr., col., wh., pa. grnsh. yel., gray, br.	1.546–1.550, 1.540–1.541
Marshite	CuI	5.68	2.5	cub., col. to pa. yel., on exposure to light, red	2.346

Table 3–1 (continued)
PHYSICAL CONSTANTS OF MINERALS

Name	Formula	Sp. gr.	Hardness	Crystalline form and color	Index of refraction (Na) η; ω ϵ / α β γ
Mascagnite	$(NH_4)_2SO_4$	1.768	2–2.5	rhomb., col., gray, yelsh.	1.5202, 1.5230, 1.5330
Mallockite	PbFCl	7.12	2.5–3	tetr., col. or yel. to pa. amber, grnsh.	2.145, 2.006
Meionite	$Ca_4Al_6Si_6O_{24}CO_3$	2.78	5–6	tetr., col., wh., pa. grnsh. yel., gray, br.	1.590–1.600, 1.556–1.562
Melanterite	$FeSO_4.7H_2O$	1.898	2	monocl., grn., grnsh. bl., grnsh. wh.	1.47, 1.48, 1.49
Melilite	$(Ca,Na,K)_4(Mg,Fe,Fe^{3+},Al,Si)_{12}O_{30}$?	2.95–3.05	5–6	tetr., yelsh., br., grn.-br.	1.624–1.666, 1.616–1.661
Mellite	$Al_2C_6O_{12}.18H_2O$	1.64	2–2.5	tetr., yel., redsh., brnsh., somet. wh.	1.5393, 1.5110
Mendipite	$Pb_3O_2Cl_2$	7.24	2.5	rhomb., col. to wh., gray, oft. yel., red, bl. tinge	2.22–2.26, 2.25–2.29, 2.29–2.33
Mesolite	$Na_2Ca_2(Al_2Si_3O_{10})_3.8H_2O$	~2.26	5	monocl., col., wh., gray, yel., pink, red	$\beta = 1.504$–1.508
Metacinnabar	HgS	7.65	5	cub., graysh.-blk.	
Microcline	$KAlSi_3O_8$	2.56–2.63	6–6.5	tricl., col., wh., pink, red, yel., grn.	1.514–1.529, 1.518–1.533, 1.521–1.539
Microlite	$(Na,Ca)_2Ta_2O_6(O,OH,F)$	4.2–6.4	5–5.5	cub., pa. yel. to br., somet. red, grn.	~2.0
Miersite	AgI	5.64–5.68	2.5	cub., canary-yel.	2.18–2.22
Millerite	NiS	5.3–5.7	3–3.5	hex., pa. brass-yel. to bronze-yel., gray, tarn. iridesc.	
Mimetite	$Pb_5(AsO_4)_3Cl$	7.24	3.5–4	hex., pa. yel. to yelsh. br., orange-yel., wh.	2.147, 2.128
Minium	Pb_3O_4	8.9–9.2	2.5	scarlet red, bl. red, somet. yel. tint.	(Li) 2.40–2.44
Mirabilite	$Na_2SO_4.10H_2O$	1.490	1.5–2	monocl., col. to wh.	1.391–1.397, 1.393–1.399, 1.395–1.401
Moissanite	SiC	3.218	9.5	hex., grn. to blk., somet. blsh., red	2.647–2.649, 2.689–2.693
Molybdenite	MoS_2	4.62–4.73	1–1.5	hex., lead-gray	
Monazite	$(Ce,La,Th)PO_4$	5.0–5.3	5.5	monocl., yel., br., redsh. br.	1.774–1.800, 1.777–1.801, 1.828–1.851
Monetite	$CaH(PO_4)$	2.929	3.5	tricl., wh., pa. yelsh.-wh.	1.587, ~1.615, 1.640
Monticellite	$CaMgSiO_4$	3.08–3.27	5.5	rhomb., col.	1.639–1.654, 1.646–1.664, 1.653–1.674
Montmorillonite	$(0.5Ca,Na)_{0.7}(Al,Mg,Fe)_4(Si,Al)_8O_{20}(OH)_4.nH_2O$	2–3	1–2	monocl., wh., yel., grn.	1.48–1.61, 1.50–1.64, 1.50–1.64
Montroydite	HgO	11.23	2.5	rhomb., dk. red to brnsh. red, br.	(Li) 2.37, 2.5, 2.65
Mordenite	$(Na_2,K,Ca)Al_2Si_9O_{22}.7H_2O$	2.12–2.15	3–4	rhomb., col., wh., red, yel., br.	1.472–1.483, 1.475–1.485, 1.477–1.487
Muscovite	$KAl_2Si_3AlO_{10}(OH,F)_2$	2.77–2.88	2.5–3	monocl., col., pa. grn., pa. red, pa. br.	1.552–1.574, 1.582–1.610, 1.587–1.616
Nantokite	CuCl	4.136	2.5	cub., col. to wh., grayish, grn.	1.925–1.935
Natrolite	$Na_2Al_2Si_3O_{10}.2H_2O$	2.20–2.26	5–5.5	rhomb., col., wh., gray, yel., pink, red	1.473–1.483, 1.476–1.486, 1.485–1.496
Nepheline	$NaKAl_4Si_4O_{16}$	2.56–2.665	5.5–6	hex., col., wh., gray	1.529–1.546, 1.526–1.542
Newberyite	$MgH(PO_4).3H_2O$	2.10	3.0–3.5	rhomb., col.	1.511–1.517, 1.514–1.520, 1.530–1.536
Niccolite	NiAs	7.784	5.5	hex., pa. copper-red, tarn. gray to blk.	
Nosean	$Na_8Al_6Si_6O_{24}SO_4$	2.30–2.40	5.5	cub., gray, bl., br.	1.495
Oldhamite	CaS	2.58	4	cub., pa. chestnut-br.	2.137
Oligoclase	$(Na_{1.0-0.7}Ca_{0.0-0.3})AlSi_3O_8$	2.63–2.65	6–6.5	tricl., col., wh., gray, grnsh., pink	1.533–1.544, 1.537–1.548, 1.543–1.552
Olivenite	$Cu_2(AsO_4)(OH)$	3.9–4.5	6.5–7	rhomb., olive grn., grnsh.-br., br., gray	1.75–1.78, 1.79–1.82, 1.83–1.87
Olivine	$(Mg,Fe)_2SiO_4$	3.22–4.39	~6	rhomb., olive grn., grayish grn. to yelsh. br.	1.63–1.83, 1.65–1.87, 1.67–1.88
Opal	$SiO_2.nH_2O$	1.73–2.16	1.5–2	col., wh., yel., br., red, grn., bl., blk., amorp.	1.41–1.46
Orpiment	As_2S_3	3.49	6–6.5	monocl., yel., brnsh. red.	(Li) 2.4, 2.81, 3.02
Orthoclase	$KAlSi_3O_8$	2.55–2.63	5–6	monocl., col., wh., pink, red, yel., grn.	1.518–1.529, 1.522–1.533, 1.522–1.539
Orthopyroxene	$(Mg,Fe)SiO_3$	3.209–3.96	2.5	rhomb., col., wh., pink, red, yel., grn.	1.650–1.768, 1.653–1.770, 1.658–1.788
Paragonite	$NaAl_2Si_3AlO_{10}(OH)_2$	2.85	4.5–5	rhomb., col., gray, yel., yel., dk. brn.	1.564–1.580, 1.594–1.609, 1.600–1.609
Parisite	$(Ce,La)_2Ca(FCO_3)_3.CaCO_3$	4.42	3.5–4	monocl., col., pa. yel.	1.672, 1.771
Pectolite	$Ca_2NaH(SiO_3)_3$	2.86–2.90	2.5	hex., brnsh., yel.	1.595–1.610, 1.605–1.615, 1.632–1.645
Penfieldite	$Pb_2Cl_3(OH)_2$	6.6	5.5	tricl., col., wh.	2.13, 2.21
Pentlandite	$(Fe,Ni)_9S_8$	4.6–5.0	6.5	hex., wh.	
Percylite	$Pb_2CuCl_2(OH)_4(?)$	3.55–3.68	2.5	cub., pa. bronze-yel.	
Periclase	MgO	3.97–4.26	7.5	cub.(?), sky bl.	2.04–2.06
Perovskite	$CaTiO_3$	3.97–4.26	5.5	cub., col. to gray-wh., yel., brnsh. yel., grn., bl.	1.7350
Petalite	$LiAlSi_4O_{10}$	2.412–2.422	6.5	pseudo cub., blk., gray-blk., brnsh. bl., redsh. br., br., yel.	2.30–2.38
Pharmacosiderite	$Fe_3(AsO_4)_2(OH)_3.5H_2O$	2.797	2.5	monocl., wh., gray, somet. pink, grn.	1.504–1.507, 1.510–1.513, 1.516–1.523
Phenakite	Be_2SiO_4	2.98	7.5	cub., olive-grn. to yel., br., redsh.	1.676–1.704
Phillipsite	$(0.5Ca,Na,K)_3Al_3Si_5O_{16}.6H_2O$	2.2	4–4.5	rhoder., col., rose, yel., br.	1.654, 1.670
Phlogopite	$KMg_3AlSi_3O_{10}(OH,F)_2$	2.76–2.90	2.5	monocl. or rhomb., col., wh., pink, gray, yel.	1.483–1.504, 1.484–1.509, 1.496–1.514
Phosgenite	$Pb_2(CO_3)Cl_2$	6.133	2–3	monocl., col., yelsh.-br., grn., redsh.-br., br.	1.530–1.590, 1.557–1.637, 1.558–1.637
				tetr., yelsh. wh. to yelsh. br., br., somet. wh., rose, gray	2.1181, 2.1446

Table 3-1 (continued)
PHYSICAL CONSTANTS OF MINERALS

Name	Formula	Sp. gr.	Hardness	Crystalline form and color	Index of refraction (N_a) η; ω ε; α β γ
Piemontite	$Ca_2(Mn,Fe^{+3},Al)_3Al_2Si_3O_{12}(OH)$	3.45–3.52	6	monocl., redsh. brn., blk.	1.732–1.794, 1.750–1.807, 1.762–1.829
Pigeonite	$(Mg,Fe,Ca)(Mg,Fe)Si_2O_6$	3.30–3.46	6	monocl., br., grnsh. br., blk.	1.682–1.722, 1.684–1.722, 1.705–1.751
Platinum	Pt	14–19	4–4.5	cub., whitish, steel gray to dk. gray	
Pollucite	$CsAlSi_2O_6$	2.9	6.5	tetr., (pseudo-cub.) col.	1.507–1.527
Polybasite	$(Ag,Cu)_{16}Sb_2S_{11}$	6.0–6.2	2–3	monocl., iron-blk.	
Powellite	$Ca(Mo,W)O_4$	4.21–4.25	3.5–4	tetr., straw-yel., br., grnsh., somet. gray, bl. blk.	1.959–1.982, 1.967–1.993
Prehnite	$Ca_2Al_2Si_3O_{10}(OH)_2$	2.90–2.95	6–6.5	rhomb., pa. grn., yel., gray, wh.	1.611–1.632, 1.615–1.642, 1.632–1.665
Proustite	Ag_3AsS_3	5.57	2–2.5	rhbdr., scarlet-vermilion	3.0877, 2.7924
Pseudobrookite	Fe_2TiO_5	4.33–4.39	5–6	rhomb., dk. red-br. to brnsh. blk. and blk.	(Li) 2.38, 2.39, 2.42
Psilomelane	$BaMn^{+2}Mn_8^{+4}O_{16}(OH)_4$	4.71	5–6	rhomb., iron-blk. to steel-gray	
Pumpellyite	$Ca_4(Mg,Fe,Mn)(Al,Fe^{+3},Ti)_5(OH)_3Si_6O_{21}\cdot2H_2O$	3.18–3.23	6	monocl., grn., blsh. grn., br.	1.674–1.702, 1.675–1.715, 1.688–1.722
Pyrargyrite	Ag_3SbS_3	5.85	2.5	rhdr., deep red	(Li) 3.084, 2.881
Pyrite	FeS_2	5.018	6–6.5	cub., pa. brass-yel., tarn. iridesc.	
Pyrochlore	$NaCaCb_2O_6F$	4.2–6.4	5–5.5	cub., br. to blk., yelsh., redsh. or blksh. br.	
Pyrochroite	$Mn(OH)_2$	3.23–3.27	2.5	hex., col. to pa. grn. or bl., tarn. br. to blk.	1.72, 1.68
Pyrolusite	MnO_2	4.73–5.08	6–6.5	tetr., pa. steel-gray, iron-gray, blk., blsh.	
Pyromorphite	$Pb_5(PO_4)_3Cl$	7.00–7.08	3.5–4	hex., grn., yel., br., orange, brnsh. red., gray	2.058, 2.048
Pyrope	$Mg_3Al_2Si_3O_{12}$	3.582	7–7.5	cub., red, pink	1.714
Pyrophyllite	$Al_2Si_4O_{10}(OH)_2$	2.65–2.90	1–2	monocl., wh., yel., pa. bl., gray-grn., brnsh. grn.	1.534–1.556, 1.568–1.589, 1.596–1.601
Pyrrhotite	$Fe_{1-x}S$	4.58–4.65	3.5–4.5	hex., bronze-yel. to br., tarn., somet. iridesc.	
Quartz	SiO_2	2.65	7	rhdr., col., wh., blk., purp., grn., bl., rose	1.544, 1.553
Rammelsbergite	$NiAs_2$	7.0–7.2	5.5–6	tin wh., redsh. tinge	
Raspite	$PbWO_4$	8.46	2.5–3	monocl., yelsh. br., pa. yel., gray	1.25–1.29, 1.25–1.29, 1.28–1.32
Realgar	AsS	3.56	1.5–2	monocl., aurora-red to orange-yel.	2.538, 2.684, 2.704
Riebeckite	$Na_2Fe_3Fe_2^{+3}Si_8O_{22}(OH,F)_2$	3.02–3.42	5	monocl., dk. bl., bl.	1.654–1.701, 1.662–1.711, 1.668–1.717
Rhodochrosite	$MnCO_3$	3.70	3.5–4	rhdr., pink, red, br., brnsh.-yel.	1.816, 1.597
Rhodonite	$(Mn,Fe,Ca)SiO_3$	3.57–3.76	5.5–6.5	tricl., pink to brnsh. red	1.711–1.738, 1.716–1.741, 1.724–1.751
Rutile	TiO_2	4.23–5.5	6–6.5	tetr., redsh. brn. to red, somet. yelsh., blsh.	2.605–2.613, 2.899–2.901
Safflorite	$(Co,Fe)As_2$	7.0–7.5	4.5–5.5	rhomb., tin-wh., tarn. dk. gray	
Samarskite	$(Y,Er,Ce,U,Ca,Fe,Pb,Th)(Cb,Ta,Ti,Sn)_2O_6$	5.69	5–6	rhomb., velvet blk., somet. brnsh. tint	~2.20
Sapphirine	$(Mg,Fe)_2Al_4O_6SiO_4$	3.40–3.58	7.5	monocl., pa. bl., pa. grn.	1.701–1.717, 1.703–1.720, 1.705–1.724
Scapolite	$(Na,Ca)_4(Al,Si)_3Si_6O_{24}(Cl,F,OH,CO_3,SO_4)$	2.50–2.78	5–6	tetr., col., wh., pa. grnsh. yel., gray, bl.	1.546–1.600, 1.600, 1.540–1.562
Scheelite	$CaWO_4$	6.06–6.12	4.5–5	tetr., yelsh. wh., pa. yel., brnsh., col., wh., gray	1.920, 1.936
Scolecite	$CaAl_2Si_3O_{10}\cdot3H_2O$	2.25–2.29	5	monocl., col., wh., gray, yel., pink, red	1.507–1.513, 1.516–1.520, 1.517–1.521
Scorodite	$Fe^{+3}(AsO_4)\cdot2H_2O$	3.28	3.5–4	rhomb., pa. grn., gray grn., br. somet. col., blsh., yel.	1.784, 1.795, 1.814
Sellaite	MgF_2	3.15	5	tetr., col. to wh.	1.378, 1.390
Senarmontite	Sb_2O_3	5.50	2–2.5	pseudo-cub., col., gray-wh.	2.087
Serpentine	$Mg_3Si_2O_5(OH)_4$	~2.55	2.5–3.5	monocl., wh., yel., gray, grn., blsh. grn.	1.53–1.57, 1.56, 1.54–1.57
Siderite	$FeCO_3$	3.96	4–4.5	rhbdr., yelsh. br., br., dk. br.	1.875, 1.635
Sillimanite	Al_2OSiO_4	3.23–3.27	6.5–7.5	rhomb., col., wh., gray or blk.	1.654–1.661, 1.658–1.662, 1.637–1.683
Silver	Ag	10.1–11.1	2.5	cub., wh., tarn. gray or blk.	
Skutterudite	$(Co,Ni)As_3$	6.1–6.9	5.5–6	cub., between tin-wh. and silver-gray, tarn. gray or iridesc.	
Smithsonite	$ZnCO_3$	4.42–4.44	4–4.5	rhdr., grayish wh. to dk. gray, grnsh., brnsh. wh.	1.848, 1.621
Sodalite	$Na_4Al_3Si_3O_{12}Cl$	2.27–2.33	5–6	cub., bl., grn., yel., gray, pink	1.483–1.487
Sperrylite	$PtAs_2$	10.58	6–7	cub., tin-wh.	
Spessartite	$Mn_3Al_2Si_3O_{12}$	4.190	6–7.5	cub., blk., dk. red, brnsh. red., bl., yelsh. orange	1.800
Sphalerite	ZnS	3.9–4.1	3.5–4	cub., br., blk., yel., red, wh.	2.369
Sphene	$CaTiSiO_4(O,OH,F)$	3.45–3.55	5	monocl., col., wh., yel., grn., br., blk.	1.843–1.950, 1.870–2.034, 1.943–2.110
Spinel	$MgAl_2O_4$	3.55	7.5–8	cub., grn., red, bl., br. to col.	1.719

Table 3–1 (continued)
PHYSICAL CONSTANTS OF MINERALS

Name	Formula	Sp. gr.	Hardness	Crystalline form and color	Index of refraction (Na) η; ω / ω β γ
Spodumene	LiAlSi₂O₆	3.03–3.22	6.5–7	monocl., col., gray-wh., pa., bl., pa. grn., yelsh.	1.648–1.663, 1.655–1.669, 1.662–1.679
Stannite	Cu₂FeSn₄	4.3–4.5	4	tetr., steel gray to iron blk.	
Staurolite	(Fe,Mg)₂(AlFe⁺³)₉O₆SiO₄(O,OH)₂	3.74–3.83	7.5	monocl., brn., redsh., yelsh.	1.739–1.747, 1.745–1.753, 1.752–1.761
Stercorite	Na(NH₄)H(PO₄).4H₂O	1.615	2	tricl., wh., yelsh., brnsh.	1.439, 1.442, 1.469
Stibiotantalite	Sb(Ta,Cb)O₄	5.7–7.5	5.5	rhomb., dk. br. to pa. yel.-br., red-br., grnsh.-yel.	2.38, 2.41, 2.46
Stibnite	Sb₂S₃	4.61–4.65	2	rhomb., lead-gray to steel-gray	
Stilbite	(Ca,Na₂K₂)Al₂Si₇O₁₆.7H₂O	2.1–2.2	3.5–4	monocl., col., wh., yel., pink, red, gray, br.	1.494–1.500, 1.492–1.507, 1.494–1.513
Stilpnomelane	(K,Na,Ca)₀₋₁.₄(Fe⁺²Fe,Mg,Al)₅₋₆Si₈O₂₀ (OH)₄(O,OH,H₂O)₄₋₁₂	2.59–2.96	3–4	monocl., br., dk. br., redsh. br., blk., dk. grn.	1.543–1.634, 1.576–1.745, 1.576–1.745
Stolzite	PbWO₄	7.9–8.4	2.5–3	tetr., redsh. br., yelsh. gray, straw-yel., grneh.	2.26–2.28, 2.18–2.20
Strengite	Fe⁺³(PO₄).2H₂O	2.90	3.5	rhomb., red, carmine, vlt., near col.	2.707, 1.719, 1.741
Strontianite	SrCO₃	3.72	3.5	rhomb., col., wh., yel., grnsh., brnsh.	1.516–1.520, 1.664–1.667, 1.666–1.669
Struvite	Mg(NH₄)(PO₄).6H₂O	1.71	1.5–2.5	rhomb., col., somet. yelsh., brnsh.	1.495, 1.496, 1.504
Sulfur	S	2.07	1.5–2.5	rhomb., yel., brnsh., grnsh., redsh., gray	1.9579, 2.0377, 2.2452
Sylvanite	(Ag,Au)Te₂	8.161	1.5–2	monocl., steel-gray to silver-wh.	
Sylvite	KCl	1.99	2	cub., col., wh., somet. grayish, blsh., yelsh., red	1.49031
Talc	Mg₃Si₄O₁₀(OH)₂	2.58–2.83	1	monocl., col., wh., pa. grn., dk. grn., br.	1.539–1.550, 1.589–1.594, 1.589–1.600
Tantalite	(Fe,Mn)(Ta,Cb)₂O₆	7.90–8.00	6.5	rhomb., iron-bl. to br.-blk.	2.26, 2.32, 2.43
Tapiolite	FeTa₂O₆	7.9	6–6.5	tetr., blk.	(Li) 2.27, 2.42
Tellurobismuthite	Bi₂Te₃	7.800–7.830	1.5–2	rhbdr., pa. lead-gray	
Terlinguaite	Hg₂OCl	8.725	2.5	monocl., yel. to grnsh.-yel., somet. br.	(Li) 2.33–2.37, 2.62–2.66, 2.64–2.68
Tetrahedrite	(Cu,Fe)₁₂Sb₄S₁₃	4.6–5.1	3–4	cub., flint-gray to iron-blk. to dull-blk.	
Thenardite	Na₂SO₄	2.664	2.5–3	rhomb., col., grayish-wh., yelsh., yelsh. br., redsh.	1.464–1.471, 1.473–1.477, 1.481–1.485
Thermonatrite	Na₂CO₃.H₂O	2.255	1–1.5	rhomb., col. to wh., grayish, yelsh.	1.420, 1.506, 1.524
Thomsenolite	NaCaAlF₆.H₂O	2.981	2	monocl., col. to wh., somet. brnsh., redsh.	1.4072, 1.4136, 1.4150
Thomsonite	NaCa₂(Al,Si)₅O₁₀)₂.6H₂O	2.10–2.39	5–5.5	rhomb., col., wh., pink, br.	1.497–1.530, 1.513–1.533, 1.518–1.544
Thorianite (R)	ThO₂	9.7	6.5	cub., dk. gray to brnsh-blk., blk.	~2.20
Thorite (R)	ThSiO₄	5.2–5.4	4.5–5	tetr., orange-yel., brnsh. to blk.	~1.8
Topaz	Al₂SiO₄(OH,F)₂	3.49–3.57	8	rhomb., col., wh., yel., gray, grn., red, bl.	1.606–1.629, 1.609–1.631, 1.616–1.638
Torbernite (R)	Cu(UO₂)₂(PO₄)₂.8–12H₂O	3.22	2–2.5	tetr., various shades of grn.	1.592, 1.582
Tourmaline (R)	Na₃(Mg,Fe,Mn,Li,Al)₃Al₆Si₆O₁₈(BO₃)₃ (OH,F)₄	3.03–3.25	7	rhbdr., blk., bl., grn., yel., red, col., br.	1.635–1.675, 1.610–1.650
Tremolite	Ca₂Mg₅Si₈O₂₂(OH,F)₂	3.0	5–6	monocl., col., gray, wh.	1.599, 1.612, 1.622
Tridymite	SiO₂	2.27	7	rhomb., col., wh.	1.471–1.479, 1.472–1.480, 1.474–1.483
Triphyllite-Litho-philite	Li(Fe,Mn)PO₄	3.34–3.58	4–5	rhomb., blsh. or grnsh. gray to yelsh. br., br.	1.66–1.70, 1.67–1.70, 1.68–1.71
Troegerite (R)	(UO₂)₃(AsO₄)₂.12H₂O	3.65	2–3	tetr., lemon-yel.	1.58–1.59, 1.625–1.635
Trona	Na₃H(CO₃)₂.2H₂O	2.14	2.5–3	monocl., gray or yelsh. wh., col.	1.412, 1.492, 1.540
Turquois	Cu(Al,Fe⁺³)₆(PO₄)₄(OH)₈.4H₂O	2.6–3.2	4.5–6	tricl., bl., grn., grnsh.-gray	1.61–1.78, 1.62–1.84, 1.65–1.84
Ullmannite	NiSbS	6.61–6.69	5–5.5	cub., steel-gray to silver-wh.	
Uraninite (R)	UO₂	8.0–11	5–6	cub., steel-blk., brnsh.-blk., grayish, brn.	1.86
Uvarovite	Ca₃Cr₂Si₃O₁₂	3.90	6–7.5	cub., emerald-grn.	2.18, 2.35, 2.35
Valentinite	Sb₂O₃	5.76	2.5–3	rhomb., col. to wh., somet. yelsh., redsh., gray, br.	
Vanadinite	Pb₅(VO₄)₃Cl	6.5–7.1	2.75–3	hex., orange-red, red, brnsh.-red, br., brnsh.-yel., yel.	2.416, 2.350
Variscite-Strengite	(AlFe⁺³)(PO₄).2H₂O	2.57–2.87	3.5–4.5	rhomb., pa. grn., grn., blah.-grn., red, vlt., col.	1.563–1.707, 1.588–1.719, 1.594–1.741
Vaterite	CaCO₃	2.645	~1.5	hex., col.	1.550, 1.640–1.650
Vermiculite	(Mg,Ca)₀₋₇(Mg,Fe⁺³Al)₆(Al,Si)₈O₂₀(OH)₄. 8H₂O	~2.3		monocl., col., yel., grn., br.	1.525–1.564, 1.545–1.583, 1.545–1.583
Vesuvianite	Ca₁₀(Mg,Fe)₂Al₄(Si₂O₇)₂(SiO₄)₅(OH,F)₄	3.33–3.43	6–7	tetr., yel., grn., br.	1.700–1.746, 1.703–1.752
Villiaumite	NaF	2.79	2–2.5	cub., carmine, (nat.), col. (artif.)	1.327
Vivianite	Fe₃(PO₄)₂.nH₂O	2.67–2.69	1.5–2	monocl., col., tarn. pa. bl., grnsh. bl., dk. bl., blsh. blk.	1.579–1.616, 1.602–1.656, 1.629–1.675

Table 3–1 (continued)
PHYSICAL CONSTANTS OF MINERALS

Name	Formula	Sp. gr.	Hardness	Crystalline form and color	Index of refraction (Na) η; ω ϵ / α β γ
Wagnerite	$Mg_2(PO_4)F$	3.15	5–5.5	monocl., yel., gray, somet. red, grn.	1.568, 1.572, 1.582
Wavellite	$Al_3(OH)_3(PO_4)_2 \cdot 5H_2O$	2.36	3.25–4	rhomb., grnsh. wh., grn. to yel., somet. br., bl., wh.	1.520–1.535, 1.526–1.543, 1.545–1.561
Whewellite	$CaC_2O_4 \cdot H_2O$	2.23	2.5–3	monocl., col., somet. yelsh., brnsh.	1.491, 1.554, 1.650
Willemite	Zn_2SiO_4	3.9–4.1	5.5	rhbdr., wh., yel., grn., red, gray, br.	1.691, 1.719
Witherite	$BaCO_3$	4.29–4.30	3.5	rhomb., col., wh., gray, yelsh. br.	1.529, 1.676, 1.677
Wolframite	$(Fe,Mn)WO_4$	7.12–7.51	4–4.5	monocl., dk. gray, brnsh. blk. to iron blk.	(Li) ~2.26, 2.32, 2.42
Wollastonite	$CaSiO_3$	2.87–3.09	4.5–5	tricl., wh., col., gray, pa. grn.	1.616–1.640, 1.628–1.650, 1.631–1.653
Wulfenite	$PbMoO_4$	6.5–7.0	2.75–3	tetr., orange-yel. to yel., gray, grn., br., red	2.403, 2.283
Wurtsite	ZnS	3.98	3.5–4	hex., brnsh. blk.	2.356, 2.378
Xenotime	$Y(PO_4)$	4.4–5.1	4–5	tetr., yelsh. br. to redsh. br., somet. gray, wh.	1.721, 1.816
Zeunerite (R)	$Cu(UO_2)_2(AsO_4)_2 \cdot 10\text{–}16H_2O$			tetr.	1.602–1.610
Zincite	ZnO	5.64–5.68	4	hex., orange-yel. to dk. red, somet. yel.	2.013, 2.029
Zircon	$ZrSiO_4$	4.6–4.7	7.5	tetr., redsh. br., yel., gray, grn., col.	1.923–1.960, 1.968–2.015
Zoisite	$Ca_2Al_3Si_3O_{12}(OH)$	3.15–3.365	6	rhomb., gray, grnsh., brnsh.	1.685–1.705, 1.688–1.710, 1.697–1.725

From Kretz, R., in *Handbook of Chemistry and Physics*, 55th ed., Weast, R. C., Ed., CRC Press, Cleveland, 1974, B-192.

Table 3–2
SOLUBILITY CHART

Abbreviations:

W = soluble in water

A = insoluble in water but soluble in acids

w = sparingly soluble in water but soluble in acids

a = insoluble in water and only sparingly soluble in acids

I = insoluble in both water and acids

d = decomposes in water

No.		Al	NH₄	Sb	Bi	Cd	Ba	Ca	Cr	Co	Cu	Au'	Au'''	H	Fe''	Fe'''
1	Acetates –(C₂H₃O₂)	Al(–)₃ W	NH₄(–) W	....	Bi(–)₃ A	Cd(–)₂ W	Ba(–)₂ W	Ca(–)₂ W	Cr(–)₃ W	Co(–)₂ W	Cu(–)₂ W	....	....	C₂H₃O₂ H W	Fe(–)₂ W	Fe(–)₃ W
2	Arsenate –(AsO₄)	Al(–) a	(NH₄)₃(–) W	Sb(–) A	Bi(–) A	Cd₃(–)₂ A	Ba₃(–)₂ A	Ca₃(–)₂ A		Co₃(–)₂ A	Cu₃(–)₂ A			H₃AsO₄ W	Fe₃(–)₂ A	Fe(–) A
3	Arsenite –(AsO₃)		NH₄AsO₂ W	Sb(–) A				Ca(–)₂								
4	Benzoate –(C₇H₅O₂)		NH₄(–) W		Bi(–)₃ A	Cd(–)₂ W	Ba(–)₂ W	Ca(–)₂ W		CₒH₄(–)₄ W	CuH(–)			C₇H₅O₂ H W	Fe(–)₂ W	Fe(–)₃ A
5	Bromide –(Br)	AlBr₃ W	NH₄Br W	SbBr₃ d	BiBr₃ d	CdBr₂ W	BaBr₂ W	CaBr₂ W	Cr₂Br₆ W	CoBr₂ W	CuBr A	AuBr W	AuBr₃ W	HBr W	FeBr₂ W	FeBr₃ W
6	Carbonate		(NH₄)₂CO₃ W			Cd(–)₂ A	BaCO₃ A	CaCO₃ A	CrCO₃ W	CoCO₃ A	Cu(–)₂ W				FeCO₃ A	Fe₂CO₃

Table 3–2 (continued)
SOLUBILITY CHART

No.		Al	NH₄	Sb	Ba	Bi	Cd	Ca	Cr	Co	Cu	Au′	Au‴	H	Fe″	Fe‴
7	Chlorate –(ClO₃)	W $Al(-)_3$	W $NH_4(-)$		W $Ba(-)_2$	W $Bi(-)_3$	W $Cd(-)_2$	W $Ca(-)_2$		W $Co(-)_2$	W $Cu(-)_2$			W $HClO_3$	W $Fe(-)_2$	W $Fe(-)_3$
8	Chloride	W $AlCl_3$	W NH_4Cl	W $SbCl_3$	W $BaCl_2$	d $BiCl_3$	W $CdCl_2$	W $CaCl_2$	I $CrCl_3$	W $CoCl_2$	W $CuCl_2$	W $AuCl$	W $AuCl_3$	W HCl	W $FeCl_2$	W $FeCl_3$
9	Chromate –(CrO₄)		W $(NH_4)_2(-)$		A $BaCr_4$	A $Bi(-)$	A $Cd(-)$	W $Ca(-)$		A $Co(-)$	A $Cu(-)$		W $Au(CN)_3$	W	A $Fe(-)$	A $Fe(-)_2$
10	Citrate –(C₆H₅O₇)	W $Al(-)$	W $(NH_4)_3(-)$		W $Ba(-)_2$	A $Bi(-)$	W $Cd(-)_2$	W $Ca(-)_2$	A $Cr(-)$	W $Co(-)_2$	A $Cu(-)_2$			W $C_6H_8O_7$	A $Fe(-)$	A $Fe(-)_2$
11	Cyanide		W NH_4CN		W $Ba(CN)_2$	W $Bi(CN)_3$	W $Cd(CN)_2$	W $Ca(CN)_2$	A $Cr(CN)_3$	W $Co(CN)_2$	I $Cu(CN)_2$	W $AuCN$	W $Au(CN)_3$	W HCN	a $Fe(CN)_2$	a $Fe(-)$
12	Ferricy′de –(Fe(CN)₆)	I $Al(-)_2$	W $(NH_4)_3(-)$		W $Ba(-)_2$	A $Bi(-)$	W $Cd(-)_3$	W $Ca(-)_2$		I $Co(-)_2$	I $Cu(-)_2$			W $H_6(-)$	I $Fe(-)_2$	I $Fe(-)$
13	Ferrocy′de –(Fe(CN)₆)	W $Al(-)_3$	W $(NH_4)_4(-)$		W $Ba(-)_2$	A $Bi(-)$	W $Cd(-)_2$	W $Ca(-)_2$		I $Co(-)$	I $Cu(-)$			W $H_4(-)$	I $Fe(-)_2$	a $Fe(-)_3$
14	Fluoride	W AlF_3	W NH_4F	W SbF_3	W BaF_2	W BiF_3	W CdF_2	I CaF_2	I CrF_3	W CoF_2	W CuF_2		A $Au(OH)_3$	W HF	W FeF_2	W FeF_3
15	Formate –(CHO₂)	W $Al(-)_3$	W $NH_4(-)$		W $Ba(-)_2$	A $Bi(-)_3$	W $Cd(-)_2$	W $Ca(-)_2$	W $Cr(-)_3$	W $Co(-)_2$	W $Cu(-)_2$	W $AuOH$	a AuI_3	W CH_2O_2	W $Fe(-)_2$	W $Fe(-)_3$
16	Hydroxide	A $Al(OH)_3$	W NH_4OH	d	A $Ba(OH)_2$	A $Bi(OH)_3$	A $Cd(OH)_2$	W $Ca(OH)_2$	A $Cr(OH)_3$	A $Co(OH)_2$	A $Cu(OH)_2$	A $AuOH$	A $Au(OH)_3$	W	A $Fe(OH)_2$	A $Fe(OH)_3$
17	Iodide	W AlI_3	W NH_4I	d SbI_3	W BaI_2	d BiI_3	W CdI_2	W CaI_2	W CrI_2	A CoI_2	I CuI	a AuI	I AuI_3	W HI	W FeI_2	W FeI_3
18	Nitrate	W $Al(NO_3)_3$	W NH_4NO_3		W $Ba(NO_3)_2$	W $Bi(NO_3)_3$	W $Cd(NO_3)_2$	W $Ca(NO_3)_2$	W $Cr(NO_3)_3$	W $Co(NO_3)_2$	W $Cu(NO_3)_2$		a	W HNO_3	W $Fe(NO_3)_2$	W $Fe(NO_3)_3$
19	Oxalate –(C₂O₄)	A $Al_2(-)_3$	W $(NH_4)_2(-)$		W $Ba(-)$	A $Bi_2(-)_3$	A $Cd(-)$	A $Ca(-)$	A $Cr(-)$	A $Co(-)$	A $Cu(-)$		A Au_2O_3	W $C_2H_2O_4$	A $Fe(-)$	W $Fe_2(-)_3$
20	Oxide	A Al_2O_3			W BaO	A Bi_2O_3	A CdO	A CaO	A Cr_2O_3	A CoO	A CuO	A Au_2O	A Au_2O_3	A H_2O_2	A FeO	A Fe_2O_3
21	Phosphate	A $AlPO_4$	W $NH_4H_2PO_4$	A Sb_2O_3	A $Ba_3(PO_4)_2$	A $BiPO_4$	A $Cd_3(PO_4)_2$	A $Ca_3(PO_4)_2$	A $Cr(PO_4)_2$	A $Co_3(PO_4)_2$	A $Cu_3(PO_4)_2$			W H_3PO_4	A $Fe(-)$	A $FePO_4$
22	Silicate –(SiO₃)	I $Al_2(-)_3$			W $Ba(-)$	d Bi_2S_3	W $Cd(-)$	W $Ca(-)$	W(II)	A Co_2SiO_4	I $Cu(-)$			I H_2SiO_3	A $Fe(-)$	d
23	Sulfate	A $Al_2(SO_4)_3$	W $(NH_4)_2SO_4$	A $Sb_2(SO_4)_3$	A $BaSO_4$	A $Bi_2(SO_4)_3$	W $CdSO_4$	A $CaSO_4$	A $Cr_2(SO_4)_3$	W $CoSO_4$	W $CuSO_4$	a Au_2S	I	W H_2SO_4	W $FeSO_4$	W $Fe_2(SO_4)_3$
24	Sulfide	d Al_2S_3	W $(NH_4)_2S$	d Sb_2S_3	a BaS	d Bi_2S_3	A CdS	W CaS	d Cr_2S_3	A CoS	A CuS	I Au_2S		A H_2S	A FeS	d FeS_2
25	Tartrate –(C₄H₄O₆)	W $Al_2(-)_3$	W $(NH_4)_2(-)$	W $Sb_2(-)$	W $Ba(-)$	A $Bi_2(-)_3$	A $Cd(-)$	A $Ca(-)$	W Cr_2S_3	A $Co(-)$	d $Cu(-)$	I Au_2S		W $C_4H_6O_6$	A $Fe(-)$	W $Fe(-)_2$
26	Thiocy′te	W $Al(-)_3$	W NH_4CNS		W $Ba(CNS)_2$	A $Bi(-)_3$	A $Cd(-)$	W $Ca(CNS)_2$		W $Co(CNS)_2$	d $CuCNS$			W $CNSH$	W $Fe(CNS)_2$	W $Fe(CNS)_3$

No.		Pb	Mg	Mn	Hg′	Hg″	Ni	K	Ag	Na	Sn″	Sn⁗	Sr	Zn	Pt
1	Acetate –(C₂H₃O₂)	W $Pb(-)_2$	W $Mg(-)_2$	W $Mn(-)_2$	W $Hg(-)_2$	W $Hg(-)_2$	W $Ni(-)_2$	W $K(-)$	A $Ag(-)$	W $Na(-)$	d $Sn(-)_2$	W $Sn(-)_4$	W $Sr(-)_2$	W $Zn(-)_2$	
2	Arsenate –(AsO₄)	A $Pb_3(-)_2$	A $Mg_3(-)$	A $MnH(-)$	A $Hg_3(-)_2$	A $Hg_3(-)_2$	A $Ni_3(-)_2$	W $K_3(-)$	A $Ag_3(-)$	W $Na_3(-)$	A $Sn(-)$		W $Sr(-)$	A $Zn(-)$	
3	Arsenite –(AsO₃)	A $PbBi(-)$	W $Mg_3(-)$	W $MnH(-)$	A $Hg_3(-)_2$	A $Hg_3(-)_2$	W $Ni_3(-)_2$	W K_3AsO_3	A $Ag_3(-)$	W $Na_3(-)$	A $Sn(-)$		W $Sr(-)$	W $Zn(-)_2$	
4	Benzoate –(C₇H₅O₂)	W $Pb(-)_2$	W $Mg(-)_2$	W $Mn(-)_2$	W $Hg(-)_2$	W $Hg(-)_2$	W $Ni(-)_2$	W $K(-)$	A $Ag(-)$	W $Na(-)$			W	W $Zn(-)_2$	
5	Bromide	W $PbBr_2$	W $MgBr_2$	W $MnBr_2$	W $HgBr$	W $HgBr_2$	W $NiBr_2$	W KBr	A $AgBr$	W $NaBr$	W $SnBr_2$	W $SnBr_4$	W $SrBr_2$	W $ZnBr_2$	W $PtBr_4$

Table 3–2 (continued)
SOLUBILITY CHART

No.	Pb	Mg	Mn	Hg'	Hg''	Ni	K	Ag	Na	Sn''''	Sn''	Sr	Zn	Pt
6 Carbonate	A $PbCO_3$	W $MgCO_3$	W $MnCO_3$	A Hg_2CO_3	......	W $NiCO_3$	W K_2CO_3	A Ag_2CO_3	W Na_2CO_3	......	......	W $SrCO_3$	W $ZnCO_3$	......
7 Chlorate —(ClO_3)	W $Pb(—)_2$	W $Mg(—)_2$	W $Mn(—)_2$	W $Hg(—)$	W $Hg(—)_2$	W $Ni(—)_2$	W $K(—)$	W $Ag(—)$	W $Na(—)$		W $Sn(—)_2$	W $Sr(—)_2$	W $Zn(—)_2$	W
8 Chloride	W $PbCl_2$	W $MgCl_2$	W $MnCl_2$	a $HgCl$	W $HgCl_2$	W $NiCl_2$	W KCl	a $AgCl$	W $NaCl$	W $SnCl_4$	W $SnCl_2$	W $SrCl_2$	W $ZnCl_2$	W $PtCl_4$
9 Chromate —(CrO_4)	A PbO_4	W $Mg(—)$	A $Mn(—)$	A $Hg_2(—)$	W $Hg(—)$	A $Ni(—)$	W $K_2(—)$	W $Ag_2(—)$	W $Na_2(—)$	W $Sn(—)_2$	A $Sn(—)$	W $Sr(—)$	W $Zn(—)$	
10 Citrate —$(C_6H_5O_7)$	A $Pb_2(—)_3$	W $Mg_3(—)_2$	W $Mn_3(—)_2$	W $Hg_2(—)$		W $Na_3(—)_2$	W $K_3(—)$	W $Ag_3(—)$	W $Na_3(—)$		A $Sn(—)$	W $Sr(—)$	W $Zn_3(—)_2$	I $Pt(CN)_6$
11 Cyanide	W $Pb(CN)_2$	W $Mg(CN)_2$	W $Mn(—)_2$	A Hg_2CN	W $Hg(CN)_2$	a $Ni(CN)_2$	W KCN	a $AgCN$	W $NaCN$			W $Sr(CN)_2$	W $Zn(CN)_2$	$Pt(CN)_6$
12 Ferricy'de —$Fe(CN)_6$	W $Pb_3(—)_2$	W $Mg_3(—)_2$	A $Mn_3(—)_2$	A $Hg_3(—)_2$	A $Hg_3(—)_2$	I $Ni_3(—)_2$	W $K_3(—)$	I $Ag_3(—)$	W $Na_3(—)$		A $Sn_3(—)_3$	W $Sr_3(—)_2$	W $Zn_3(—)_2$	A $Pt(OH)_4$
13 Ferrocy'de —$Fe(CN)_6$	a $Pb_2(—)$	W $Mg_2(—)$	A $Mn_2(—)$	I $Hg_2(—)$	I $Hg_2(—)$	W $Ni_2(—)$	W $K_4(—)$	I $Ag_4(—)$	W $Na_4(—)$	W $Sn(—)$	A $Sn_2(—)$	W $Sr_2(—)_2$	W $Zn_2(—)$	I
14 Fluoride	W PbF_2	W MgF_2	W MnF_2	d Hg_2F	d HgF_2	W NiF_2	W KF	W AgF	W NaF	W SnF_4	W SnF_2	W SrF_2	W ZnF_2	W PtF_4
15 Formate —(CHO_2)	W $Pb(—)_2$	W $Mg(—)_2$	W $Mn(—)_2$	W $Hg(—)$	W $Hg(—)_2$	W $Ni(—)_2$	W $K(—)$	W $Ag(—)$	W $Na(—)$	W $Sn(OH)_4$	W $Sn(—)_2$	W $Sr(—)_2$	W $Zn(—)_2$	A $Pt(OH)_4$
16 Hydroxide	W $Pb(OH)_2$	A $Mg(OH)_2$	A $Mn(OH)_2$	A $Hg(OH)$	A $Hg(OH)_2$	W $Ni(OH)_2$	W KOH		W $NaOH$	d $Sn(OH)_4$	A $Sn(OH)_2$	W $Sr(OH)_2$	W $Zn(OH)_2$	I
17 Iodide	W PbI_2	W MgI_2	W MnI_2	A Hg_2I	W HgI_2	W NiI_2	W KI	I AgI	W NaI	d SnI_4	W SnI_2	W SrI_2	W ZnI_2	W PtI_2
18 Nitrate	W $Pb(NO_3)_2$	W $Mg(NO_3)_2$	W $Mn(NO_3)_2$	W $HgNO_3$	W $Hg(NO_3)_2$	W $Ni(NO_3)_2$	W KNO_3	W $AgNO_3$	W $NaNO_3$	W $Sn(NO_3)_4$	d $Sn(NO_3)_2$	W $Sr(NO_3)_2$	W $Zn(NO_3)_2$	W $Pt(NO_3)_4$
19 Oxalate —(C_2O_4)	W $Pb(—)$	A $Mg(—)$	A $Mn(—)$	A $Hg_2(—)$	A $Hg(—)$	A $Ni(—)$	W $K_4(—)$	A $Ag_4(—)$	W $Na_4(—)$	W $Sn(—)$	A $Sn(—)$	A $Sr(—)$	W $Zn(—)$	W
20 Oxide	W PbO	A MgO	A MnO	A Hg_2O	A HgO	A NiO	W K_2O	W Ag_2O	d Na_2O	A SnO_2	A SnO	W SrO	W ZnO	A PtO
21 Phosphate	A $Pb_3(PO_4)_2$	A $Mg_3(PO_4)_2$	W $Mn_3(PO_4)_2$	A Hg_3PO_4	A $Hg_3(PO_4)_2$	A Ni_3PO_4	W K_3PO_4	A Ag_3PO_4	W Na_3PO_4		A $Sn_3(PO_4)_2$	A $Sr_3(PO_4)_2$	A $Zn_3(PO_4)_2$	
22 Silicate —(SiO_3)	A $Pb(—)$	W $Mg(—)$	A $Mn(—)$	W $Hg_2(—)$	d $Hg(—)$	A $Ni(—)$	W $K_2(—)$	A $Ag_2(—)$	W $Na_2(—)$		d $Sn(—)$	A $Sr(—)$	A $Zn(—)$	
23 Sulfate	A $PbSO_4$	W $MgSO_4$	W $MnSO_4$	W Hg_2SO_4	d $HgSO_4$	W $NiSO_4$	W K_2SO_4	W Ag_2SO_4	W Na_2SO_4	W $Sn(SO_4)_2$	W $SnSO_4$	W $SrSO_4$	W $ZnSO_4$	W $Pt(SO_4)_2$
24 Sulfide	A PbS	d MgS	W MnS	I Hg_2S	I HgS	W NiS	W K_2S	A Ag_2S	W Na_2S	A SnS_2	A SnS	W SrS	W ZnS	I PtS
25 Tartrate —$(C_4H_4O_6)$	A $Pb(—)$	W $Mg(—)$	W $Mn(—)$	A $Hg_2(—)$	A $Hg(—)$	A $Ni(—)$	W $K_4(—)$	I $Ag_4(—)$	W $Na_4(—)$		W $Sn(—)$	W $Sr(—)$	A $Zn(—)$	I
26 Thiocy'le	A $Pb(CNS)_2$	W $Mg(CNS)_2$	W $Mn(CNS)_2$	A Hg_2CNS	W $Hg(CNS)_2$	......	W $KCNS$	I $AgCNS$	W $NaCNS$		W $Sn(—)$	W $Sr(CNS)_2$	W $Zn(CNS)_2$	......

^{a}Certain salts occur in two modifications.

From Weast, R. C., Ed., *Handbook of Chemistry and Physics*, 55th ed., CRC Press, Cleveland, 1974, D-110.

Table 3-3

THERMAL CONDUCTIVITY OF GASES AT VARIOUS TEMPERATURES

The values in this table are given as $\dfrac{\text{cal}}{\text{sec cm }^{\circ}\text{C}} \times 10^{6}$. For W/m · K multiply by 0.0004184. For Btu/hr · ft · deg F multiply by 0.0002419.

Gas	°F 200	120	100	80	60	40	20	0	−20	−40	−100	−200	−300	−400
°C	93.3	48.9	37.8	26.7	15.6	4.4	−6.7	−17.8	−28.9	−40	−73.3	−128.9	−184.4	−240
Acetylene	69.43	56.62	53.72	50.83	47.94	45.04	42.15	39.67	37.19	34.71	28.10			
Air		66.04	64.22	62.20	60.34	58.31	56.24	54.22	52.15	50.09				
Ammonia		64.47	61.58	58.68	55.79	53.31	50.83	48.35	45.87	43.39				
Argon		45.46	44.22	42.57	41.33	40.09	38.85	37.19	35.95	34.30				
Bromine			11.57					9.09						
n-Butane	54.14	43.39	40.91	38.02	35.54	33.06	30.99							
i-Butane	55.79	44.22	41.74	38.85	36.37	33.89	32.65							
Carbon dioxide		43.81	41.74	39.67	37.61	35.62	33.68	31.70	29.75	27.90				
Carbon disulfide			19.84	19.01	17.77	16.53	15.29	14.05						
Carbon monoxide		63.89	61.99	59.92	57.86	55.87	53.85	51.95	50.00	47.94				
Chlorine		23.14	21.90	21.08	20.25	19.01	18.18	17.36	16.53	15.29				
Deuterium		355.40	343.01	334.74	322.34	309.95	305.81	295.07	285.15	274.82				
Ethane	74.39	58.27	54.55	51.24	47.94	44.63	41.33	38.43	35.54	32.65	23.97			
Ethanol		42.15	36.78	34.71	32.65	30.99	29.34							
Ethylamine			39.67	37.61	35.54	33.47	31.41							
Ethylene	68.19	54.96	52.07	49.18	46.29	43.39	40.50	38.02	35.54	33.06	26.86			
Fluorine	76.04	68.19	66.12	64.06	61.99	59.92	57.86	55.38	52.90	50.83	43.39	30.58	18.18	
Helium		376.07	368.63	360.36	352.10	343.42	333.50	324.00	314.49	304.99	274.8	221.51	142.57	59.92
Hydrogen		471.11	458.72	446.32	433.92	417.39	405.00	388.46	371.93	357.47	308.7	227.29	163.24	84.31
Hydrogen bromide			21.49	20.66	19.84	18.60	17.77	16.49	16.11	15.29				
Hydrogen chloride			35.12	33.89	32.23	30.99	29.75	28.51	26.86	25.62				
Hydrogen cyanide			30.99	29.75	28.10	26.86	25.62	23.97						
Hydrogen sulfide			36.78		33.47	31.41	29.75	28.10						
Krypton		32.65	23.56					19.84						
Methane	106.62	89.26	85.54	81.83	78.11	74.39	71.08	67.86	64.55	61.37	52.07	36.86	22.32	
Neon		121.09	118.19	115.71	112.82	109.93	107.03	104.14	100.84	97.94				
Nitric oxide	74.39	66.12	64.06	61.99	59.76	57.65	55.54	53.39	51.24	49.01	42.40	30.91		
Nitrogen		65.71	64.06	62.40	60.34	58.27	56.20	54.55	52.48	50.42	44.22	33.06	20.25	
Nitrous oxide		46.08	43.81	41.45	39.30	37.15	35.04	32.90	30.91	28.93				
Oxygen	76.87	68.19	65.91	63.64	61.58	59.43	57.24	54.96	52.81	50.54	43.72	31.66	18.84	
n-Propane	60.75	48.35	45.46	42.47	39.67	37.19	34.71	32.23	29.75	27.69				
R-11(CCl₃F)			18.60	17.77	16.53	15.70	14.88	13.64	12.81					
R-12(CCl₂F₂)			23.56	22.73	21.49	20.66	19.42	18.60	17.36					
R-21(CHCl₂F)			23.97	23.56	23.14	22.73	22.32	21.90						
R-22(CHClF₂)			28.93	28.10	27.28	26.45	25.62	24.80						
Water	54.96	46.70	44.63	42.57	40.50	38.85	36.78	34.71						

From Weast, R. C., Ed., *Handbook of Chemistry and Physics*, 55th ed., CRC Press, Cleveland, 1974, E-2.

Table 3–4A
THERMAL CONDUCTIVITY OF DIELECTRIC CRYSTALS

Name	Remarks	Conductivity mW/cm × K	
		83 K	273 K
Ammonium bromide	Pressed at 8,000 atm	67	25
Ammonium chloride	Pressed at 8,000 atm	109	25
Barium nitrate	Pressed at 8,000 atm	33	13
Beryl	Mineral	88	84
Calcite	Main crystal axis perpendicular to rod axis	180	46
	Main crystal axis parallel to rod axis	293	54
Chrome alum		13	21
Copper sulfate		29	21
Magnesium sulfate		25	25
Marble	Small crystals, 99.9% $CaCO_3$	42	33
	99.99% $CaCO_3$	54	38
	Large crystals	50	33
Mercuric chloride	Pressed at 8,000 atm	17	13
Potassium alum		13	21
Potassium bichromate	Main crystal axis perpendicular to rod axis	17	21
	Main crystal axis parallel to rod axis	17	17
Potassium bromide	Pressed at 8,000 atm	92	38
90% Potassium bromide, 10% potassium chloride	Pressed at 8,000 atm	50	29
75% Potassium bromide, 25% potassium chloride	Pressed at 8,000 atm	29	21
50% Potassium bromide, 50% potassium chloride	Pressed at 8,000 atm	25	25
25% Potassium bromide, 75% potassium chloride	Pressed at 8,000 atm	46	33
10% Potassium bromide, 90% potassium chloride	Pressed at 8,000 atm	80	50
Potassium chloride	Pressed at 8,000 atm	314	88
	From a melt	402	92
	Pressed at 1,250 atm	243	75
	Pressed at 2,500 atm	368	92
	Pressed at 8,900 atm	402	96
50% Potassium chloride, 50% sodium chloride	Pressed at 8,000 atm	188	71
Potassium ferrocyanide		17	17
Potassium fluoride	Pressed at 8,000 atm	234	71
Potassium iodide	Pressed at 8,000 atm	121	29
Potassium nitrate	Pressed at 8,000 atm	17	21
Rock salt	Pressed at 8,000 atm	180	63
Rubidium chloride	Pressed at 8,000 atm	29	21
Rubidium iodide	Pressed at 8,000 atm	59	33
Sodium bromide	Pressed at 8,000 atm	50	25
Sodium chloride	From a melt	343	92
	Pressed at 8,000 atm	251	71
Sodium fluoride	Pressed at 8,000 atm	519	105
Sylvite	Natural cyrstal	159	75
	Pressed at 8,000 atm	343	84
Topaz	Mineral		234
Tourmaline	Mineral	38	46
Zinc blende	Mineral	63	264

From Weast, R. C., Ed., *Handbook of Chemistry and Physics,* 55th ed., CRC Press, Cleveland, 1974, E-4.

Table 3–4B
THERMAL CONDUCTIVITY OF ORGANIC COMPOUNDS

Substance	Cal[a] sec · cm · °C	Temp, °C	Temp, °F
Acetaldehyde	0.0004089	21	69.8
Acetic acid	0.0004109	20	68
Acetic anhydride	0.0005286	21	69.8
Acetone	0.0004750	−80	−112
	0.0004543	16	61
	0.0004031	75	167
Allyl alcohol	0.0004295	30	86
Amyl acetate, *n*-	0.0003085	20	68
Amyl acetate, *iso*-	0.000310	20	68
Amyl alcohol, *n*-	0.0003874	30–100	86–212
Amyl alcohol, *iso*-	0.0003531	30	86
Amyl bromide, *n*-	0.0002350	18	64.4
Aniline	0.0004237	16.5	61.5
Benzene	0.0003780	22.5	72.5
	0.0003275	50	122
	0.0003630	60	140
	0.0002870	140	284
Bromobenzene	0.0002664	20	68
Butyl acetate, *n*-	0.000327	20	68
Butyl alcohol, *n*-	0.0003663	20	68
Carbon tetrachloride	0.0002470	20	68
	0.0002333	50	122
Chlorobenzene	0.0003457	30–100	86–212
Chlorotoluene, *p*-	0.000310	20	68
Chloroform	0.0002891	16	61
	0.000246	20	68
Cresol, *m*-	0.0003581	20	68
Cresol, *p*-	0.000345	20.1	68.2
Cumene	0.000298	20	68
Cymene, *p*-	0.0003217	30	86
Decane	0.0003349	30	86
Diethyl ether	0.0003283	30	86
Dichloroethane, 1,2	0.000302	20	68
Diisopropyl ether	0.000262	20	68
Ethyl acetate	0.0003560	16	60.8
Ethyl alcohol	0.0003995	20	68
Ethyl benzene	0.0003160	20	68
Ethyl bromide	0.0002862	30	86
Ethyl ether	0.0003283	30	86
Ethyl iodide	0.0002651	30	86
Ethylene glycol	0.0006236	20	68
	0.0006323	15	122
	0.0006443	80	176
Freon-12® (CCL$_2$F$_2$)	0.0002310	0–75	32–167
Freon-21® (CHCl$_2$F)	0.0003180	0–75	32–167
Freon-22® (CHClF$_2$)	0.0002309	40	104
Freon-113® (CCl$_2$FCCl$_2$F)	0.0002379	0–80	32–176
Freon-114® (C$_2$H$_2$F$_4$)	0.0002127	0–75	32–167
Glycerol	0.000703	20	68

Table 3–4B (continued)
THERMAL CONDUCTIVITY OF ORGANIC COMPOUNDS

Substance	Cal[a] sec · cm · °C	Temp, °C	Temp, °F
Heptane, *n*-	0.0003354	30	86
Heptyl alcohol	0.0003882	70–100	86–212
Hexane, *n*-	0.0003287	30–100	86–212
Hexyl alcohol, *n*-	0.0003857	30–100	86–212
Iodobenzene	0.0002874	30–100	86–212
Mesitylene	0.0003246	20	68
Methyl alcohol	0.0004832	20	68
Methyl aniline	0.0004419	21.5	70.5
Methyl chloride	0.0004597	–15(–)+30	5–86
Methyl cyclohexane	0.0003052	30	86
Methylene chloride	0.0002908	0	32
Nitrobenzene	0.0003907	30–100	86–212
Nitromethane	0.0005142	30	86
Nonane, *n*-	0.0003374	30–100	86–212
Nonyl alcohol, *n*-	0.0004014	30–100	86–212
Octane, *n*-	0.0003469	30	86
Octyl alcohol, *n*-	0.0003973	30–100	86–212
Oleic acid	0.0005514	26.5	79.7
Palmitic acid	0.0004097	72.5	162.5
Pentachloroethane	0.0002994	20	68
Pentane, *n*-	0.0003221	30	86
Phenetole	0.0003577	–20	–4
Phenyl hydrazine	0.0004121	25	69.8
Propyl acetate, *iso*-	0.000321	20	68
Propyl alcohol, *iso*-	0.0003362	20	68
Propylene chloride	0.0002994	20–50	68–122
Propylene glycol, 1,2-	0.0004799	20–80	68–176
Stearic acid	0.0003824	72.5	162.5
Tetrachloroethane, *sym*-	0.000272	20	68
Tetrachloroethylene	0.0003866	20	68
Toluene	0.0003804	–80	–112
	0.0003221	20	68
	0.0002808	80	176
Trichloroethylene	0.0003246	–60	–76
	0.0002775	20	68
Triethylamine	0.0003498	–80	–112
	0.0002891	20	68
	0.0002664	44.4	112
Xylene, *o*-	0.0003411	–20(–)+80	(–4)–176
Xylene, *m*-	0.0003767	25	77

[a]To convert these values to Btu/hr · ft · °F multiply by 242.08.

From Weast, R. C., Ed., *Handbook of Chemistry and Physics*, 55th ed., CRC Press, Cleveland, 1974, E-4.

Table 3–4C
THERMAL CONDUCTIVITY OF INORGANIC COMPOUNDS

Substance	Cal[a] sec · cm · °C	Temp, °C	Temp, °F
Ammonia	0.0001198	–15(–)+30	5–86
Argon	0.0002895	–183	–297
	0.0001677	–133	–207
	0.0000553	–105	–157.5
	0.0000409	–75	–102.5
Carbon dioxide	0.0002040	–50	–58
	0.0002412	–40	–40
	0.0002664	–30	–22
	0.0002746	–20	–4
	0.0002495	0	32
	0.0001677	30	86
Nitrogen	0.0003400	–196	–321.5
	0.0002961	–189	–308
	0.0002028	–158	–253
	0.0000640	–105	–155
Oxygen	0.0000500	(–207)–(–191)	(–340)–(–312)
	0.0000504	(–178)–(–182)	(–288)–(–295)
Water	0.001348	0	72
	0.001429	20	68
	0.001499	40	104
	0.001557	60	140
	0.001598	100	212
	0.001631	149	300
	0.001553	216	420
	0.001404	271	520
	0.001136	327	620

[a]To convert these values to Btu/hr · ft · °F multiply by 242.08.

From Weast, R. C., Ed., *Handbook of Chemistry and Physics*, 55th ed., CRC Press, Cleveland, 1974, E-4.

Table 3–4D
THERMAL CONDUCTIVITY OF MISCELLANEOUS SUBSTANCES

Substance	Cal[a] sec·cm·°C	Temp, °C	Temp, °C
Chlorinated diphenyl 1242	0.0002936	30–100	86–212
Chlorinated diphenyl 1248	0.0002808	30–100	86–212
Kerosene	0.0003572	30	86
Light heat transfer oil	0.0003159	30–100	86–212
Petroleum ether	0.0003118	30	86
Red oil	0.0003366	30	86
Transformer oil	0.0004242	70–100	86–212

[a]To convert these values to Btu/hr ·ft·°F multiply by 242.08.

From Weast, R. C., Ed., *Handbook of Chemistry and Physics*, 55th ed., CRC Press, Cleveland, 1974, E-5

Table 3—5
THERMAL CONDUCTIVITY OF MATERIALS

D = density in pounds per cubic foot.

K = thermal conductivity in Btu per hour, square foot, and temperature gradient of 1°F per inch thickness. The lower the conductivity, the greater the insulating values.

Soft Flexible Materials in Sheet Form

		D	K
Dry zero	Kapok between burlap or paper	1.0	0.24
		2.0	0.25
Cabots quilt	Eel grass between kraft paper	3.4	0.25
		4.6	0.26
Hair felt	Felted cattle hair	11.0	0.26
		13.0	0.26
Balsam wool	Chemically treated wood fiber	2.2	0.27
Hairinsul®	75% hair, 25% jute	6.3	0.27
	50% hair, 50% jute	6.1	0.26
Linofelt®	Flax fibers between paper	4.9	0.28
Thermofelt	Jute and asbestos fibers, felted	10.0	0.37
	Hair and asbestos fibers, felted	7.8	0.28

Loose Materials

		D	K
Rock wool	Fibrous material made from rock,	6.0	0.26
	also made in sheet form, felted	10.0	0.27
	and confined with wire netting	14.0	0.28
		18.0	0.29
Glass wool	Pyrex glass, curled	4.0	0.29
		10.0	0.29
Sil-O-Cel®	Powdered diatomaceous earth	10.6	0.31
Regranulated cork	Fine particles	9.4	0.30
	About 3/16-in. particles	8.1	0.31
Thermofill®	Gypsum in powdered form	26	0.52
		34	0.60
Sawdust	Various	12.0	0.41
	Redwood	10.9	0.42
Shavings	Various, from planer	8.8	0.41
Charcoal	From maple, beech and birch, coarse	13.2	0.36
	6 mesh	15.2	0.37
	20 mesh	19.2	0.39

Semiflexible Materials in Sheet Form

		D	K
Flaxlinum	Flax fiber	13.0	0.31
Fibrofelt®	Flax and rye fiber	13.6	0.32

Semirigid Materials in Board Form

		D	K
Corkboard	No added binder; very low density	5.4	0.25
Corkboard	No added binder; low density	7.0	0.27
Corkboard	No added binder; medium density	10.6	0.30
Corkboard	No added binder; high density	14.0	0.34
Eureka	Corkboard with asphaltic binder	14.5	0.32
Rock cork	Rock wool block with binder (also called "Tucork")	14.5	0.326
Lith	Board containing rock wool, flax and straw pulp	14.3	0.40

Table 3–5 (continued)
THERMAL CONDUCTIVITY OF MATERIALS

Stiff Fibrous Materials in Sheet Form		D	K
Insulite®	Wood pulp	16.2	0.34
		16.9	0.34
Celotex®	Sugar cane fiber	13.2	0.34
		14.8	0.34
Masonite®		K = 0.33	
Inso-board®		0.33	
Maizewood		0.33–0.39	
Cornstalk pith board		0.24–0.30	
Maftex		0.34	

Cellular Gypsum

	D	K
Insulex or Pyrocell®	8	0.35
	12	0.44
	18	0.59
	24	0.77
	30	1.00
Balsa	7.3	0.33
	8.8	0.38
	20	0.58
Cypress	29	0.67
White pine	32	0.78
Mahogany	34	0.90
Virginia pine	34	0.98
Oak	38	1.02
Maple	44	1.10

Miscellaneous Building Materials

Cinder concrete	2–3 Limestone	4–9
Building gypsum	About 3 Concrete	6–9
Plaster	2–5 Sandstone	8–16
Building brick	3–6 Marble	14–20
Glass	5–6 Granite	13–28

From Bureau of Standards Letter Circular No. 227.

Table 3–6
THERMAL CONDUCTIVITY DATA ON CERAMIC MATERIALS

Description[a]	Classification[b]	Water absorption,%	Bulk density, g/cm^3	Thermal conductivity[c] 100°F	200°F	300°F
Single crystals						
Silicon carbide	5	–	–	52.0	50.2	49.0
Periclase	5	–	–	26.7	22.5	19.5
Sapphire, c-axis	5	–	–	20.2	16.0	14.0
Sapphire, a-axis	5	–	–	18.7	15.0	12.9
Topaz, a-axis	5	–	–	10.8	9.4	7.9
Kyanite, c-axis	5	–	–	10.00	8.6	7.4
Kyanite, b-axis	5	–	–	9.6	8.3	7.1
Spinel, $MgO \cdot Al_2O_3$	5	–	–	6.80	6.20	5.50
Quartz, c-axis	4	–	–	6.40	5.40	5.02
Quartz, a-axis	4	–	–	3.40	3.00	2.60
Rutile, c-axis	5	–	–	5.60	4.80	4.40
Rutile, a-axis	5	–	–	3.20	3.20	3.20
Fluorite	5	–	–	5.30	4.37	3.45
Beryl, aquamarine, c-axis	4	–	–	3.18	3.15	3.12
Beryl, aquamarine, a-axis	4	–	–	2.52	2.52	2.52
Zircon, a-axis	4	–	–	2.45	2.45	2.45
Zircon, c-axis	4	–	–	2.34	2.34	2.35
Polycrystalline Oxide Ceramics						
Pure beryllium oxide, hot pressed	2	0.03	2.97	125.0	104.0	92.0
Magnesium oxide, spec. pure	1	0.83	3.21	21.2	18.4	16.0
Stannic oxide, 98%	1	0.03	6.62	17.5	15.0	12.7
Zinc oxide, yellow	1	0.00	5.28	16.8	14.6	12.5
Zinc oxide, gray	1	0.03	5.20	13.6	11.8	10.2
Cupric oxide, 100%	1	0.04	6.76	10.2	9.00	7.80
Thorium dioxide, hot pressed	2	–	9.58	8.00	7.02	6.50
Ceric oxide	1	0.00	6.20	6.63	6.29	5.20
Manganic-manganous oxide	1	0.02	4.21	4.18	3.80	3.41
Lead oxide, mono-#, 100%	1	0.38	7.98	1.6	1.25	0.98

[a]Composition: 90% MgO, 10% Al_2O_3 designates
weight percent. Li_2O: 4 B_2O_3 designates mole

[a]Composition: 90% MgO, 10% Al_2O_3 designates weight percent. Li_2O:$4B_2O_3$ designates mole composition, does not indicate compound formation.
[b]Classification: 1, research body; 2, industrial research body; 3, commercial body; 4, natural mineral; 5, synthetic mineral.
[c]Thermal conductivity: units given in Btu/hr · ft · °F; to convert to cal/sec · cm · °C, multiply by 0.00413.

From *Engineering Research Bulletin No. 40,* Rutgers, the State University, New Brunswick, N.J., 1958. With permission.

Table 3–7

THERMODYNAMIC PROPERTIES OF THE OXIDES

For description of headings and symbols see Table 2–23.

Oxide	Phase	Temperature of transition (K)	Heat of transition (kcal/mole)	Entropy of transition (e.u.)	Entropy at 298 K (e.u.)	a (cal/g mole)	b (cal/g mole)	c (cal/g mole)	d (cal/g mole)	A (kcal/mole)	B (e.u.)
Ac_2O_3	Solid	(2250)	(20)	(8.9)	(36.5)	(20.0)	(20.4)	–	–	(6.870)	(80.9)
	Liquid	–	–	–	–	(40)	–	–	–	(–19.767)	(180.5)
Ag_2O	Solid	dec. 460	–	–	29.09	13.26	7.04	–	–	4.266	48.56
Ag_2O_2	Solid	dec.	–	–	(20.4)	(16.4)	(12.2)	–	–	(5.432)	(76.7)
Al_2O_3	Solid	2300	26	11	12.186	26.12	4.388	–	–7.269	10.422	142.03
	Liquid	dec.	–	–	–	(33)	–	–	–	(–11.655)	(174.1)
Am_2O_3	Solid	(2225)	(17)	(7.6)	(37)	(20.0)	(15.6)	–	–	(6.657)	(81.6)
	Liquid	(3400)	(85)	(25)	–	(38.5)	–	–	–	(–7.796)	(181.8)
AmO_2	Solid	dec.	–	–	(20)	(14.0)	(6.8)	–	–	(4.477)	(61.8)
As_2O_3	Solid, α	503	4.1	8.2	25.6	8.37	48.6	–	–	4.656	36.6
	Solid, β	586	4.4	7.5	–	8.37	48.6	–	–	0.556	28.4
	Liquid	730	7.15	9.79	–	(39)	–	–	–	(5.760)	(187.6)
	Gas	–	–	–	–	(21.5)	–	–	–	(–14.164)	(62.5)
AsO_2	Solid	(1200)	(9.0)	(7.5)	(13)	(8.5)	(9.4)	–	–	(2.952)	(38.2)
	Liquid	(dec.)	–	–	–	(21)	–	–	–	(2.184)	(108.0)
As_2O_5	Solid	dec. >1100	–	–	25.2	(31.1)	(16.4)	–	(–5.4)	(11.813)	(159.9)
Au_2O_3	Solid	dec.	–	–	30	(23.5)	(4.8)	–	–	(7.220)	(105.3)
B_2O_3	Solid	723	5.27	7.29	12.91	8.73	25.40	–	–1.31	4.171	45.04
	Liquid	2520	(55)	(22)	–	30.50	–	–	–	7.822	161.59
Ba_2O	Solid	(880)	(5.2)	(5.9)	(23.5)	(20.0)	(2.2)	–	–	(6.061)	(91.1)
	Liquid	(1040)	(20)	(19)	–	(22)	–	–	–	(1.769)	(96.8)
	Gas	–	–	–	–	(15)	–	–	–	(–25.51)	(29.0)
BaO	Solid	2196	13.8	6.28	16.8	12.74	1.040	–	–1.984	4.510	57.2
	Liquid	3000	(62)	(21)	–	(13.9)	–	–	–	(–9.341)	(57.5)
BaO_2	Solid	723	(5.7)	(7.9)	(18.5)	(13.6)	(2.0)	–	–	(4.144)	(59.6)
	Liquid	dec. 1110	–	–	–	(21)	–	–	–	(3.241)	(99.0)

Table 3–7 (continued)
THERMODYNAMIC PROPERTIES OF THE OXIDES

Oxide	Phase	Temperature of transition (°K)	Heat of transition (kcal/mole)	Entropy of transition (e.u.)	Entropy at 298°K (e.u.)	a (cal/g mole)	b (cal/g mole)	c (cal/g mole)	d (cal/g mole)	A (kcal/mole)	B (e.u.)
BeO	Solid	dec.	—	—	3.37	8.69	3.65	—	-3.13	3.803	48.99
BiO	Solid	(1175)	(3.7)	(3.1)	(15)	(9.7)	(3.0)	—	—	(3.025)	(41.2)
	Liquid	(1920)	(54)	(28)	—	(14)	—	—	—	(2.306)	(64.9)
	Gas	—	—	—	—	(8.9)	—	—	—	(-61.49)	(-1.8)
Bi_2O_3	Solid	1090	6.8	6.2	36.2	23.27	11.05	—	—	7.429	99.7
	Liquid	(dec.)	—	—	—	(35.7)	—	—	—	(7.614)	(168.3)
CO	Gas	—	—	—	47.30	6.60	1.2	—	—	2.021	-9.34
CO_2	Gas	—	—	—	51.06	7.70	5.3	-0.83	—	2.490	-5.64
CaO	Solid	2860	(18)	(6.3)	9.5	10.00	4.84	—	-1.08	3.559	49.5
CdO	Solid	dec.	—	—	13.1	9.65	2.08	—	—	2.970	42.5
Ce_2O_3	Solid	1960	(20)	(10)	(33.5)	(23.0)	(9.0)	—	—	(7.258)	(100.2)
	Liquid	(3500)	(80)	(23)	(11.5)	(37)	—	—	—	(-2.591)	(178.5)
CeO_2	Solid	3000	(19)	(6.3)	17.7	15.0	2.5	—	—	4.579	68.5
CoO	Solid	2078	(12)	(5.8)	10.5	(9.8)	(2.2)	—	—	(3.020)	(46.0)
	Liquid	(2900)	(61)	(21)	—	(15.5)	—	—	—	(-1.886)	(79.2)
Co_3O_4	Solid	dec. 1240	—	—	(35.5)	(29.5)	(17.0)	—	—	(9.551)	(137.6)
Cr_2O_3	Solid	2538	(25)	(10)	19.4	28.53	2.20	—	-3.736	9.857	145.9
CrO_2	Solid	dec. 700	—	—	(11.5)	(16.1)	(3.0)	—	(-3.0)	(5.946)	(82.8)
CrO_3	Solid	460	(6.1)	(13)	(17.5)	(18.1)	(4.0)	—	(-2.0)	(6.245)	(87.9)
	Liquid	(1000)	(25)	(25)	—	(27)	—	—	—	(3.381)	(127.0)
	Gas	—	—	—	—	(20)	—	—	—	(-28.62)	(53.6)
Cs_2O	Solid	763	(4.58)	(6.0)	(23)	(16.5)	(5.4)	—	—	(5.160)	(72.6)
	Liquid	dec.	—	—	—	(22)	—	—	—	(3.205)	(99.0)
Cs_2O_2	Solid	867	(5.5)	(6.3)	(40)	(21.4)	(11.4)	—	—	(6.887)	(85.3)
	Liquid	dec.	—	—	—	(29.5)	—	—	—	(4.125)	(123.8)

Table 3–7 (continued)
THERMODYNAMIC PROPERTIES OF THE OXIDES

Oxide	Phase	Temperature of transition (°K)	Heat of transition (kcal/mole)	Entropy of transition (e.u.)	Entropy at 298°K (e.u.)	a (cal/g mole)	b (cal/g mole)	c (cal/g mole)	d (cal/g mole)	A (kcal/mole)	B (e.u.)
Cs_2O_3	Solid	775	(7.75)	(10)	(47)	(24.0)	(22.6)	—	—	(8.160)	(96.5)
	Liquid	dec.	—	—	—	(35)	—	—	—	(2.148)	(142.2)
Cu_2O	Solid	1503	13.4	8.92	22.44	(13.4)	(8.6)	—	—	(4.378)	(96.0)
	Liquid	dec.	—	—	—	(21.5)	—	—	—	(3.721)	(54.9)
CuO	Solid	1609	(8.9)	(5.5)	10.4	14.34	6.2	—	—	4.551	61.11
	Liquid	dec.	—	—	—	(22)	—	—	—	(−4.339)	(98.91)
FeO	Solid	1641	7.5	4.6	12.9	9.27	4.80	—	—	(2.977)	(43.8)
	Liquid	(2700)	(55)	(20)	—	(14.5)	—	—	—	(−3.721)	(69.2)
Fe_3O_4	Solid, α	900	(0)	(0)	35.0	12.38	1.62	—	−0.38	3.826	58.3
	Solid, β	dec.	—	—	—	(14.5)	—	—	—	(−2.399)	(66.7)
Fe_2O_3	Solid, α	950	0.16	0.17	21.5	21.88	48.20	—	—	8.666	104.0
	Solid, β	1050	0	0	—	48.00	18.6	—	—	12.652	238.3
	Solid, γ	dec.	—	—	—						
Ga_2O	Solid	(925)	(8.5)	(9.2)	(22.5)	23.49	18.6	—	−3.55	9.021	119.9
	Liquid	(1000)	(20)	(20)	—	36.00	—	—	—	11.979	187.6
	Gas	—	—	—	—	31.71	1.8	—	—	8.467	159.7
Ga_2O_3	Solid	2013	(22)	(11)	20.23	(13.8)	—	—	—	(4.497)	(58.7)
GeO	Liquid	(2900)	(75)	(26)	—	(21.5)	—	—	—	(−0.559)	(94.1)
	Solid	983	(50)	(51)	(12.5)	(14)	—	—	—	(−28.06)	(22.3)
	Gas	—	—	—	—	11.77	25.2	—	—	(4.630)	(54.35)
GeO_2	Solid (α,β)	1389	10.5	7.56	(12.5)	(35.5)	—	—	(−0.5)	(−20.66)	(173.2)
	Liquid	(2625)	(61)	(23)	—	(10.4)	(2.6)	—	—	(3.384)	(47.8)
In_2O	Solid	(600)	(4.5)	(7.5)	(28)	(14.7)	(7.8)	—	—	(4.730)	(58.1)
	Liquid	(800)	(16)	(20)	—	(22)	—	—	—	(3.206)	(92.6)
	Gas	—	—	—	—	(15)	—	—	—	(−18.39)	(25.8)
InO	Solid	(1325)	(4.0)	(3.0)	(14.5)	(10.0)	(3.2)	—	—	(3.124)	(43.4)
	Liquid	(2000)	(60)	(30)	—	(14)	—	—	—	(1.615)	(64.9)

Table 3–7 (continued)

THERMODYNAMIC PROPERTIES OF THE OXIDES

Oxide	Phase	Temperature of transition (°K)	Heat of transition (kcal/mole)	Entropy of transition (e.u.)	Entropy at 298°K (e.u.)	a (cal/g mole)	b (cal/g mole)	c (cal/g mole)	d (cal/g mole)	A (kcal/mole)	B (e.u.)
In_2O_3	Gas	—	—	—	—	(9.0)	—	—	—	(−68.38)	(−3.1)
	Solid	(2000)	(20)	(10)	30.1	(22.6)	(6.0)	—	—	(7.005)	(100.5)
	Liquid	(3600)	(85)	(24)	—	(35)	—	—	—	(−0.195)	(172.8)
Ir_2O_3	Solid	(1450)	(10)	(6.8)	(26.5)	(21.8)	(14.4)	—	—	(7.140)	(102.0)
	Liquid	(2250)	(50)	(22)	—	(35)	—	—	—	(0.706)	(170.3)
	Gas	—	—	—	—	(20)	(10)	—	—	(−57.73)	(54.8)
IrO_2	Solid	dec. 1373	—	—	(15.9)	9.17	15.20	—	—	3.410	40.9
K_2O	Solid	(980)	(6.8)	(6.9)	(23)	(15.9)	(6.4)	—	—	(5.025)	(69.5)
	Liquid	dec.	—	—	—	(22)	—	—	—	(1.130)	(98.3)
K_2O_2	Solid	763	(7.0)	(9.2)	(27)	(20.8)	(5.4)	—	—	(6.442)	(93.1)
	Liquid	(1800)	(45)	(25)	—	(29)	—	—	—	(4.127)	(134.2)
	Gas	—	—	—	—	(20)	—	—	—	(−57.07)	(41.7)
K_2O_3	Solid	703	(6.1)	(8.7)	(33.5)	(19.1)	(23.2)	—	—	(6.750)	(82.2)
	Liquid	(975)	(25)	(26)	—	(35.5)	—	—	—	(6.447)	(164.7)
	Gas	—	—	—	—	(20)	(5.0)	—	—	(−31.29)	(37.3)
KO_2	Solid	653	(4.9)	(7.5)	27.9	(15.0)	(12.0)	—	—	(5.006)	(61.1)
	Liquid	dec.	—	—	—	(24)	—	—	—	(3.424)	(105.5)
La_2O_3	Solid	2590	(18)	(7)	(36.5)	28.86	3.076	—	−3.275	9.840	(130.7)
Li_2O	Solid	2000	(14)	(7)	9.06	(11.4)	(5.4)	—	—	(3.639)	(57.5)
	Liquid	2600	(56)	(22)	—	(21)	—	—	—	(−1.961)	(112.7)
Li_2O_2	Solid	dec. 470	—	—	(16.5)	(17.0)	(5.4)	—	—	(5.309)	(82.0)
MgO	Solid	3075	18.5	5.8	6.4	10.86	1.197	—	−2.087	3.991	57.0
MgO_2	Solid	dec. 361	—	—	(20.5)	(12.1)	(2.4)	—	—	(3.714)	(49.2)
MnO	Solid	2058	13.0	6.32	14.27	11.11	1.94	—	−0.88	3.689	50.10
	Liquid	dec.	—	—	—	(13.5)	—	—	—	(−8.543)	(58.02)

Table 3–7 (continued)
THERMODYNAMIC PROPERTIES OF THE OXIDES

Oxide	Phase	Temperature of transition (°K)	Heat of transition (kcal/mole)	Entropy of transition (e.u.)	Entropy at 298°K (e.u.)	a (cal/g mole)	b (cal/g mole)	c (cal/g mole)	d (cal/g mole)	A (kcal/mole)	B (e.u.)
Mn_3O_4	Solid, α	1445	4.97	3.44	35.5	34.64	10.82	—	-2.20	11.312	166.3
	Solid, β	1863	(33)	(18)	—	50.20	—	—	—	17.376	260.4
	Liquid	(2900)	(75)	(26)	—	(49)	—	—	—	(-17.86)	(233.4)
Mn_2O_3	Solid	dec. 1620	—	—	26.4	24.73	8.38	—	-3.23	8.829	118.8
MnO_2	Solid	dec. 1120	—	—	12.7	16.60	2.44	—	-3.88	6.359	84.8
MoO_2	Solid	(2200)	(16)	(7.3)	(14.5)	(16.2)	(3.0)	—	(-3.0)	(5.973)	(80.4)
	Liquid	dec. 2250	—	—	—	(23)	—	—	—	(-2.463)	(118.4)
MoO_3	Solid	1068	12.54	11.74	18.68	13.6	13.5	—	—	4.655	62.83
	Liquid	1530	33	22	—	(28.4)	—	—	—	(0.222)	(139.88)
	Gas	—	—	—	—	(18.1)	—	—	—	(-48.54)	(42.8)
N_2O	Gas	—	—	—	52.58	(10.92)	2.06	—	-2.04	4.032	11.40
Na_2O	Solid	1193	(7.1)	(6.0)	17.4	15.70	5.40	—	—	4.921	73.7
	Liquid	dec.	—	—	—	(22)	—	—	—	(1.494)	(105.9)
Na_2O_2	Solid	dec. 919	—	—	22.6	(20.2)	(3.8)	—	—	(6.192)	(93.6)
NaO_2	Solid	(825)	(6.2)	(7.5)	27.7	(16.2)	(3.6)	—	—	(4.990)	(65.7)
	Liquid	(1300)	(28)	(22)	—	(23)	—	—	—	(3.175)	(100.9)
	Gas	—	—	—	—	(15)	—	—	—	(-35.22)	(22.0)
NbO	Solid	(2650)	(16)	(6.0)	(12)	(9.6)	(4.4)	—	—	(3.058)	(44.0)
NbO_2	Solid	(2275)	(16)	(7.0)	(12.7)	(17.1)	(1.6)	—	(-2.8)	(6.109)	(84.6)
	Liquid	(3800)	(85)	(22)	—	(24)	—	—	—	(1.033)	(127.2)
Nb_2O_5	Solid	1733	(28)	(16)	32.8	21.88	28.2	—	—	7.776	100.3
	Liquid	(3200)	(80)	(25)	—	(44.2)	—	—	—	(-24.09)	(201.6)
Nd_2O_3	Solid	2545	(22)	(8.8)	(35.3)	28.99	5.760	—	(-4.159)	10.295	(133.9)
NiO	Solid	2230	(12.1)	(5.43)	9.22	13.69	0.83	—	-2.915	5.097	70.67
	Liquid	dec.	—	—	—	(14.3)	—	—	—	(-7.861)	(67.91)

Table 3–7 (continued)

THERMODYNAMIC PROPERTIES OF THE OXIDES

Oxide	Phase	Temperature of transition (°K)	Heat of transition (kcal/mole)	Entropy of transition (e.u.)	Entropy at 298°K (e.u.)	a (cal/g mole)	b (cal/g mole)	c (cal/g mole)	d (cal/g mole)	A (kcal/mole)	B (e.u.)
NpO_2	Solid	(2600)	(15)	(5.7)	19.19	(17.7)	(3.2)	—	(−2.6)	(6.292)	(84.08)
Np_2O_5	Solid	dec. 800–900 K	—	—	(43)	(32.4)	(12.6)	—	—	(10.22)	(145.4)
OsO_2	Solid	dec. 923	—	—	(14.5)	(11.5)	(6.0)	—	—	(3.696)	(52.8)
OsO_4	Solid	313.3	3.41	10.9	34.7	(16.4)	(23.1)	—	(−2.4)	(6.726)	(67.0)
	Liquid	403	9.45	23.4	—	(33)	8.60	—	—	(6.612)	(143.0)
	Gas	—	—	—	—	16.46	—	—	−4.6	(−7.644)	(25.3)
P_2O_3	Liquid	448.5	4.5	10	(34)	(34.5)	—	—	—	(10.287)	(162.6)
	Gas	—	—	—	—	(15)	(10)	—	—	(−1.953)	(38.0)
PO_2	Solid	(350)	(2.7)	(7.7)	(11.5)	(11.3)	(5.0)	—	—	(3.591)	(54.4)
	Liquid	(dec.)	—	—	—	(20)	—	—	—	(3.640)	(95.9)
P_2O_5	Solid	631	8.8	13.9	33.5	8.375	5.40	—	—	4.897	30.3
	Gas	—	—	—	—	36.80	—	—	—	3.284	165.6
PaO_2	Solid	(2560)	(20)	(7.8)	(17.8)	(14.4)	(2.6)	—	—	(4.409)	(65.0)
Pa_2O_5	Solid	(2050)	(26)	(13)	(37.5)	(28.4)	(11.4)	—	—	(8.975)	(127.7)
	Liquid	(3350)	(95)	(28)	—	(48)	—	—	—	(−0.800)	(241.1)
PbO	Solid, red	762	(0.4)	(0.5)	16.2	10.60	4.00	—	—	3.338	45.4
	Solid, yellow	1159	2.8	2.4	—	9.05	6.40	—	—	2.454	36.4
	Liquid	1745	51	29	—	(14.6)	—	—	—	1.788	65.7
	Gas	—	—	—	—	(8.1)	(0.4)	—	—	(−59.94)	(−11.0)
Pb_2O_4	Solid	dec.	—	—	50.5	(31.1)	17.6	—	—	(10.055)	(132.0)
PbO_2	Solid	dec.	—	—	18.3	12.7	7.80	—	—	4.133	56.4
PdO	Solid	dec. 1150	—	—	(9.1)	3.30	14.2	—	—	1.615	(13.9)
PoO_2	Solid	(825)	(5.5)	(6.7)	(17)	(14.3)	(5.6)	—	—	(4.513)	(66.1)
	Liquid	(dec.)	—	—	—	(22)	—	—	—	(3.460)	(106.5)

Table 3–7 (continued)
THERMODYNAMIC PROPERTIES OF THE OXIDES

Oxide	Phase	Temperature of transition (°K)	Heat of transition (kcal/mole)	Entropy of transition (e.u.)	Entropy at 298°K (e.u.)	a (cal/g mole)	b (cal/g mole)	c (cal/g mole)	d (cal/g mole)	A (kcal/mole)	B (e.u.)
Pr_2O_3	Solid	(2200)	(22)	(10)	(35.5)	(29.0)	(4.0)	—	(−4.0)	(10.166)	(133.2)
	Liquid	(4000)	(90)	(23)	—	(36)	—	—	—	(−6.298)	(168.3)
PrO_2	Solid	dec. 700	—	—	(17)	(17.6)	(3.4)	—	(−2.8)	(6.338)	(85.9)
PtO	Solid	dec. 780	—	—	(13.5)	(9.0)	(6.4)	—	—	(2.968)	(39.7)
Pt_3O_4	Solid	(dec.)	—	—	(41)	(30.8)	(17.4)	—	—	(9.957)	(139.7)
PtO_2	Solid	723	(4.6)	(6.4)	(16.5)	(11.1)	(9.6)	—	—	(3.736)	(49.6)
	Liquid	dec. 750	—	—	—	(21)	—	—	—	(3.785)	(101.5)
PuO	Solid	(1290)	(7.2)	(5.6)	(20)	(12.0)	(2.4)	—	—	(3.685)	(49.1)
	Liquid	(2325)	(47)	(20)	—	(14.5)	—	—	—	(−2.287)	(58.3)
	Gas	—	—	—	—	(8.9)	—	—	—	(−62.307)	(−5.3)
Pu_2O_3	Solid	(1880)	(16)	(8.5)	(38)	(21.2)	(18.2)	—	—	(7.130)	(88.2)
	Liquid	(3250)	(75)	(23)	—	(40)	—	—	—	(−5.691)	(187.2)
PuO_2	Solid	(2400)	(15)	(6.2)	(19.7)	(17.1)	(3.4)	—	(−2.6)	(6.122)	(80.2)
	Liquid	(3500)	(90)	(26)	—	(20.5)	—	—	—	(−10.62)	(92.2)
RaO	Solid	(>2500)	—	—	(17)	(10.5)	(2.0)	—	—	(3.220)	(43.4)
Rb_2O	Solid	(910)	(5.7)	(6.3)	(27)	(15.4)	(5.8)	—	—	(4.850)	(62.5)
	Liquid	dec.	—	—	—	(22)	—	—	—	(2.754)	(95.9)
Rb_2O_2	Solid	843	(7.3)	(8.7)	(27.5)	(20.9)	(8.0)	—	—	(6.587)	(94.0)
	Liquid	(dec.)	—	—	—	(29)	—	—	—	(3.273)	(133.2)
Rb_2O_3	Solid	762	(7.6)	(10)	(32.5)	(20.5)	(13.0)	—	—	(6.690)	(88.2)
	Liquid	dec.	—	—	—	(34)	—	—	—	(5.603)	(157.8)
RbO_2	Solid	685	(4.1)	(6.0)	(21.5)	(13.8)	(6.4)	—	—	(4.399)	(59.0)
	Liquid	dec.	—	—	—	(21)	—	—	—	(3.720)	(95.7)
ReO_2	Solid	(1475)	(12)	(8.1)	(15)	(10.8)	(9.8)	—	—	(3.656)	(49.5)
ReO_2	Liquid	(3250)	(80)	(25)	—	(24.5)	—	—	—	(1.204)	(127.0)
ReO_3	Solid	433	5.2	12	19.8	(18.0)	(5.8)	—	—	(5.625)	(84.5)
	Liquid	dec.	—	—	—	29	—	—	—	(4.644)	(136.8)

Table 3–7 (continued)
THERMODYNAMIC PROPERTIES OF THE OXIDES

Oxide	Phase	Temperature of transition (°K)	Heat of transition (kcal/mole)	Entropy of transition (e.u.)	Entropy at 298°K (e.u.)	a (cal/g mole)	b (cal/g mole)	c (cal/g mole)	d (cal/g mole)	A (kcal/mole)	B (e.u.)
Re_2O_7	Solid	569	15.8	27.8	44	(41.8)	(14.8)	—	(–3.0)	(14.127)	(200.3)
	Liquid	635.5	17.7	27.9	—	(65.7)	—	—	—	(9.203)	(314.7)
	Gas	—			—	(38.2)	—	—	—	(–25.97)	(109.3)
ReO_4	Solid	420	(4.2)	(10)	(34.5)	(21.4)	(10.8)	—	(–2.0)	(7.531)	(91.8)
	Liquid	(460)	(9.3)	(20)	—	(33)	—	—	—	(6.775)	(146.7)
	Gas	—			—	(16.5)	(8.6)	—	(–5.0)	(–8.118)	(30.6)
Rh_2O	Solid	dec. 1400	—	—	(25.5)	15.59	6.47	—	—	4.936	(65.3)
RhO	Solid	dec. 1394	—	—	(12)	9.84	(5.53)	—	—	(3.179)	(45.7)
Rh_2O_3	Solid	dec. 1388	—	—	(23)	20.73	13.80	—	—	6.794	(99.2)
RuO_2	Solid	dec. 1400	—	—	(12.5)	(11.4)	(6.0)	—	—	3.666	(54.2)
RuO_4	Solid	300	(3.2)	(11)	(32.5)	(20)	—	—	—	(5.963)	(81.5)
	Liquid	dec.	—	—	—	(33)	—	—	—	(6.663)	(144.9)
SO_2	Gas	—	—	—	59.40	11.4	1.414	—	–2.045	4.148	7.12
Sb_2O_3	Solid	928	14.74	15.88	29.4	19.10	17.1	—	—	6.455	84.5
	Liquid	1698	8.92	5.25	—	(36)	—	—	—	(0.035)	(168.2)
	Gas	—			—	(20.8)	—	—	—	(–34.70)	(49.9)
SbO_2	Solid	dec.	—	—	15.2	11.30	8.1	—	—	3.725	51.6
Sb_2O_5	Solid	dec.	—	—	29.9	22.4	(23.6)	—	—	(7.723)	(104.8)
Sc_2O_3	Solid	(2500)	(23)	(9.3)	24.8	23.17	5.64	—	—	7.159	108.9
SeO	Solid	(1375)	(7.6)	(5.5)	(11)	(9.1)	(3.8)	—	—	(2.882)	(42.0)
	Liquid	(2075)	(45)	(22)	—	(15.5)	—	—	—	(0.490)	(77.5)
	Gas	—			—	8.20	0.50	—	–0.80	(–58.54)	(0.7)
SeO_2	Solid	603	(24.5)	(40.6)	(15)	(12.8)	(6.1)	—	(–0.2)	(4.150)	(59.9)
	Gas	—			—	(14.5)	—	—	—	(–20.45)	(26.4)
SiO	Solid	(2550)	(12)	(4.7)	(6.5)	(7.3)	(2.4)	—	—	(2.283)	(35.8)

Table 3–7 (continued)
THERMODYNAMIC PROPERTIES OF THE OXIDES

Oxide	Phase	Temperature of transition (°K)	Heat of transition (kcal/mole)	Entropy of transition (e.u.)	Entropy at 298°K (e.u.)	a (cal/g mole)	b (cal/g mole)	c (cal/g mole)	d (cal/g mole)	A (kcal/mole)	B (e.u.)
SiO$_2$	Solid, β	856	0.15	0.18	10.06	11.22	8.20	—	-2.70	4.615	57.83
	Solid, α	1883	2.04	1.08	—	14.41	1.94	—	—	4.602	73.67
	Liquid	dec. 2250	—	—	—	(20)	—	—	—	(9.649)	(111.08)
Sm$_2$O$_3$	Solid	(2150)	(20)	(9.3)	(36.5)	(25.9)	(7.0)	—	—	(8.033)	(113.2)
	Liquid	(3800)	(80)	(21)	—	(36)	—	—	—	(-6.431)	(166.3)
SnO	Solid	(1315)	(6.4)	(4.9)	13.5	9.40	3.62	—	—	2.964	41.1
	Liquid	(1800)	(60)	(33)	—	(14.5)	—	—	—	(0.141)	(68.1)
	Gas	—	—	—	—	(9.0)	—	—	—	(-69.76)	(-6.4)
SnO$_2$	Solid	1898	(11.39)	(5.95)	12.5	17.66	2.40	—	-5.16	7.103	91.7
	Liquid	(3200)	(75)	(23)	—	(22.5)	—	—	—	(0.304)	(117.7)
SrO	Solid	2703	16.7	6.2	13.0	12.34	1.120	—	-1.806	4.335	58.7
SrO$_2$	Solid	dec. 488	—	—	(14.8)	(16.8)	(2.2)	—	(-3.0)	(6.113)	(83.3)
Ta$_2$O$_5$	Solid	2150	(16)	(7.4)	34.2	29.2	10.0	—	—	9.151	135.2
	Liquid	—	—	—	—	(46)	—	—	—	(6.158)	(235.1)
TcO$_2$	Solid	(2400)	(18)	(7.5)	(13.5)	(10.4)	(9.2)	—	—	(3.510)	(48.6)
	Liquid	(4000)	(105)	(26)	—	(25)	—	—	—	(-5.946)	(132.7)
TcO$_3$	Solid	(dec. <1200)	—	—	(19.5)	(19.4)	(5.2)	—	(-2.0)	(6.686)	(93.7)
Tc$_2$O$_7$	Solid	392.7	(11)	(28)	(42.5)	(39.1)	(18.6)	—	(-2.4)	(13.29)	(187.2)
	Liquid	583.8	(14)	(24)	—	(64)	—	—	—	(10.02)	(299.8)
	Gas	—	—	—	—	(25)	(28)	—	—	(-21.98)	(43.8)
TeO	Solid	(1020)	(7.1)	(7.0)	(13)	(8.6)	(6.2)	—	—	(2.840)	(37.8)
	Liquid	(1775)	(50)	(28)	—	(15.5)	—	—	—	(-0.448)	(72.3)
	Gas	—	—	—	—	(8.9)	—	—	—	(-62.16)	(-5.2)
TeO$_3$	Solid	1006	3.2	3.2	16.99	13.85	6.87	—	—	4.435	63.97
	Liquid	dec.	—	—	—	(20)	—	—	—	(3.940)	(96.4)
TeO$_2$	Solid	(2150)	(13)	(6.0)	(16)	(11.0)	(2.4)	—	—	(3.386)	(47.4)
	Liquid	(3250)	(65)	(20)	—	(15)	—	—	—	(-6.561)	(66.9)

Table 3–7 (continued)
THERMODYNAMIC PROPERTIES OF THE OXIDES

Oxide	Phase	Temperature of transition (°K)	Heat of transition (kcal/mole)	Entropy of transition (e.u.)	Entropy at 298°K (e.u.)	a (cal/g mole)	b (cal/g mole)	c (cal/g mole)	d (cal/g mole)	A (kcal/mole)	B (e.u.)
ThO_2	Solid	3225	(18)	(5.6)	15.59	16.45	2.346	—	-2.124	5.721	80.03
TiO	Solid, α	1264	0.82	0.65	8.31	10.57	3.60	—	-1.86	3.935	54.03
	Solid, β	dec. 2010	—	—	—	11.85	3.00	—	—	4.108	61.71
Ti_2O_3	Solid, α	473[a]	0.215	0.455	18.83	7.31	53.52	—	—	4.559	38.78
	Solid, β	2400	(24)	(10)	—	34.68	1.30	—	-10.20	13.605	184.48
	Liquid	3300	—	—	—	(37.5)	—	—	—	(-7.796)	(193.2)
Ti_3O_5	Solid, α	450	2.24	4.98	30.92	35.47	29.50	—	—	11.887	179.98
	Solid, β	(2450)	(50)	(20)	—	41.60	8.00	—	—	10.230	202.80
	Liquid	(3600)	(85)	(24)	—	(60)	—	—	—	(-18.701)	(306.4)
TiO_2	Solid	2128	(16)	(7.5)	12.01	17.97	0.28	—	-4.35	6.829	92.92
	Liquid	dec. 3200	—	—	—	(21.4)	—	—	—	(-2.610)	(111.08)
Tl_2O	Solid	573	(5.0)	(8.7)	23.8	(15.8)	(6.0)	—	(-0.3)	(5.078)	(68.2)
	Liquid	773	(17)	(22)	—	(22.1)	—	—	—	(2.651)	(96.0)
	Gas	—	—	—	—	(13.7)	—	—	—	(-20.94)	(18.0)
Tl_2O_3	Solid	990	(12.4)	(13)	(33.5)	(23.0)	(5.0)	—	—	(7.080)	(99.0)
	Liquid	(dec.)	—	—	—	(35.5)	—	—	—	(4.604)	(167.8)
UO	Solid	(2750)	(14)	(5.1)	(16)	(10.6)	(2.0)	—	-3.957	(3.249)	(45.0)
UO_2	Solid	3000	—	—	18.63	19.20	1.62	—	(-10.9)	7.124	93.37
U_3O_8	Solid	dec.	—	—	(66)	(65)	(7.5)	—	—	(23.37)	(312.7)
UO_3	Solid	dec. 925	—	—	23.57	22.09	2.54	—	-2.973	7.969	104.72
VO	Solid	(2350)	(15)	(6.4)	9.3	11.32	1.61	—	-1.26	3.869	56.4
	Liquid	(3400)	(70)	(21)	—	(14.5)	—	—	—	(-8.157)	(70.9)
V_2O_3	Solid	2240	(24)	(11)	23.58	29.35	4.76	—	-5.42	10.780	148.12
	Liquid	dec. 3300	—	—	—	(38)	—	—	—	(-6.028)	(193.4)
V_3O_4	Solid	(2100)	(42)	(20)	(32)	(36)	(30)	—	—	(12.07)	(182.1)
	Liquid	(dec.)	—	—	—	(55.6)	—	—	—	(-54.72)	(249.1)

Table 3–7 (continued)
THERMODYNAMIC PROPERTIES OF THE OXIDES

Oxide	Phase	Temperature of transition (°K)	Heat of transition (kcal/mole)	Entropy of transition (e.u.)	Entropy at 298°K (e.u.)	a (cal/g mole)	b (cal/g mole)	c (cal/g mole)	d (cal/g mole)	A (kcal/mole)	B (e.u.)
VO_2	Solid, α	345	1.02	2.96	12.32	14.96	–	–	–	4.460	72.92
	Solid, β	1818	13.60	7.48	–	17.85	1.70	–	-3.94	5.680	89.09
	Liquid	dec. 3300	–	–	–	25.50	–	–	–	2.962	135.87
V_2O_5	Solid	943	15.56	16.50	313	46.54	-3.90	–	-13.22	18.136	240.2
	Liquid	(2325)	(63)	(27)	–	45.60	–	–	–	2.122	220.1
	Gas	–	–	–	–	(40)	–	–	–	(-73.90)	(149.6)
WO_2	Solid	(1543)	(11.5)	(7.45)	(15)	(17.6)	(4.2)	–	(-4.0)	(6.772)	(88.8)
	Liquid	dec. 2125	–	–	–	(24)	–	–	–	(-0.112)	(121.8)
WO_1	Solid	1743	(17)	(9.8)	19.90	17.33	7.74	–	–	5.511	81.15
	Liquid	(2100)	(43)	(20)	–	(30)	–	–	–	(-1.162)	(152.5)
	Gas	–	–	–	–	(18)	–	–	–	(-69.36)	(40.2)
Y_2O_3	Solid	(2500)	(25)	(10)	(29.5)	(26.0)	(8.2)	–	(-2.2)	(8.846)	(122.3)
ZnO	Solid	dec.	–	–	10.4	11.71	1.22	–	-2.18	4.277	57.88
ZrO_2	Solid, α	1478	1.420	0.961	12.03	16.64	1.80	–	-3.36	6.168	85.21
	Solid, β	2950	20.8	7.0	–	17.80	–	–	–	4.270	89.96

From Weast, R. C., Ed., *Handbook of Chemistry and Physics*, 55th ed., CRC Press, Cleveland, 1974, D-58.

Table 3–8
HEAT OF FORMATION OF SELECTED INORGANIC OXIDES[a]

The ΔH_o values are given in gram calories per mole. The a, b, and I values listed here make it possible for one to calculate the ΔF and ΔS values by use of the following equations:

$$\Delta F_t = \Delta H_o + 2.303aT \log T + b \times 10^{-3} T^2 + c \times 10^5 T^{-1} + IT$$
$$\Delta S_t = -a - 2.303a \log T - 2b \times 10^{-3} T - 2b \times 10^5 T^{-2} - I$$

Coefficients in Free-energy Equations

Reaction	Temperature range of validity	ΔH_o	2.303a	b	c	I
$2\,Ac(c) + 3/2\,O_2(g) = Ac_2O_3(c)$	298.16–1,000 K	-446,090	-16.12	—	—	+109.89
$2\,Al(c) + 1/2\,O_2(g) = Al_2O(g)$	298.16–931.7 K	-31,660	+14.97	—	—	-72.74
$2\,Al(l) + 1/2\,O_2(g) = Al_2O(g)$	931.7–2,000 K	-38,670	+10.36	—	—	-51.53
$Al(c) + 1/2\,O_2(g) = AlO(g)$	298.16–931.7 K	+10,740	+5.76	—	—	-37.61
$Al(l) + 1/2\,O_2(g) = AlO(g)$	931.7–2,000 K	+8,170	+5.76	—	—	-34.85
$2\,Al(c) + 3/2\,O_2(g) = Al_2O_3$ (corundum)	298.16–931.7 K	-404,080	-15.68	+2.18	+3.935	+123.64
$2\,Al(l) + 3/2\,O_2(g) = Al_2O_3$ (corundum)	931.7–2,000 K	-407,950	-6.19	-0.78	+3.935	+102.37
$2\,Sb(c) + 3/2\,O_2(g) = Sb_2O_3$ (cubic)	298.16–842 K	-169,450	+6.12	-6.01	-0.30	+52.21
$2\,Sb(c) + 3/2\,O_2(g) = Sb_2O_3$ (orthorhombic)	298.16–903 K	-168,060	+6.12	-6.01	-0.30	+50.56
$2\,As(c) + 3/2\,O_2(g) = As_2O_3$ (orthorhombic)	298.16–542 K	-154,870	+29.54	-21.33	-0.30	-8.83
$2\,As(c) + 3/2\,O_2(g) = As_2O_3$ (monoclinic)	298.16–586 K	-150,760	+29.54	-21.33	-0.30	-16.95
$2\,As(c) + 5/2\,O_2(g) = As_2O_5(c)$	298.16–883 K	-217,080	+12.32	-4.65	-0.50	+80.50
$Ba(\alpha) + 1/2\,O_2(g) = BaO(c)$	298.16–648 K	-134,590	-7.60	+0.87	+0.42	+45.76
$Ba(\beta) + 1/2\,O_2(g) = BaO(c)$	648–977 K	-134,140	-3.34	-0.56	+0.42	+34.01
$Be(c) + 1/2\,O_2(g) = BeO(c)$	298.16–1,556 K	-144,220	-1.91	-0.46	+1.24	+30.64
$Bi(c) + 1/2\,O_2(g) = BiO(c)$	298.16–544 K	-50,450	-4.61	—	—	+35.51
$Bi(l) + 1/2\,O_2(g) = BiO(c)$	544–1,600 K	-52,920	-4.61	—	—	+40.05
$2\,Bi(c) + 3/2\,O_2(g) = Bi_2O_3(c)$	298.16–544 K	-139,000	-11.56	+2.15	-0.30	+96.52
$2\,Bi(l) + 3/2\,O_2(g) = Bi_2O_3(c)$	544–1,090 K	-142,270	+2.30	-3.25	-0.30	+67.55
$2\,B(c) + 3/2\,O_2(g) = B_2O_3(c)$	298.16–723 K	-304,690	+11.72	-7.55	+0.355	+34.25

Table 3–8 (continued)
HEAT OF FORMATION OF SELECTED INORGANIC OXIDES

Coefficients in Free-energy Equations

Reaction	Temperature range of validity	ΔH_0	2.303_a	b	c	I
2 B(c) + 3/2 O₂(g) = B₂O₃(gl)	298.16–723 K	-298,670	+26.57	-15.90	-0.30	-10.40
Cd(c) + 1/2 O₂(g) = CdO(c)	298.16–594 K	-62,330	-2.05	+0.71	-0.10	+29.17
Cd(l) + 1/2 O₂(g) = CdO(c)	594–1,038 K	-63,240	+2.07	-0.76	-0.10	+20.14
Ca(α) + 1/2 O₂(g) = CaO(c)	298.16–673 K	-151,850	-6.56	+1.46	+0.68	+43.93
Ca(β) + 1/2 O₂(g) = CaO(c)	673–1,124 K	-151,730	-4.14	+0.41	+0.68	+37.63
C(graphite) + 1/2 O₂ (g) = CO(g)	298.16–2,000 K	-25,400	+2.05	+0.27	-1.095	-28.79
C(graphite) + O₂(g) = CO₂(g)	298.16–2,000 K	-93,690	+1.63	-0.7	-0.23	-5.64
2 Ce(c) + 3/2 O₂(g) = Ce₂O₃(c)	298.16–1,048 K	-435,600	-4.60	–	–	+92.84
2 Ce(l) + 3/2 O₂(g) = Ce₂O₃(c)	1,048–1,900 K	-440,400	-4.60	–	–	+97.42
Ce(c) + O₂(g) = CeO₂(c)	298.16–1,048 K	-245,490	-6.42	+2.34	-0.20	+67.79
Ce(l) + O₂(g) = CeO₂(c)	1,048–2,000 K	-247,930	+0.71	-0.66	-0.20	+51.73
2 Cs(c) + 1/2 O₂(g) = Cs₂O(c)	298.16–301.5 K	-75,900	–	–	–	+36.60
2 Cs(l) + 1/2 O₂(g) = Cs₂O(c)	301.5–763 K	-76,900	–	–	–	+39.92
2 Cs(l) + 1/2 O₂(g) = Cs₂O(l)	763–963 K	-75,370	-9.21	–	–	+64.47
2 Cs(g) + 1/2 O₂(g) = Cs₂O(l)	963–1,500 K	-113,790	-23.03	–	–	+145.60
2 Cs(c) + 3/2 O₂(g) = Cs₂O₃(c)	298.16–301.5 K	-112,690	-11.51	–	–	+110.10
2 Cs(l) + 3/2 O₂(g) = Cs₂O₃(c)	301.5–775 K	-113,840	-12.66	–	–	+116.77
2 Cs(l) + 3/2 O₂(g) = Cs₂O₃(l)	775–963 K	-110,740	-26.48	–	–	+152.70
2 Cs(g) + 3/2 O₂(g) = Cs₂O₃(l)	963–1,500 K	-148,680	-39.14	–	–	+229.87
Cl₂(g) + 1/2 O₂(g) = Cl₂O(g)	298.16–2,000 K	+17,770	-0.71	-0.12	+0.49	+16.81
1/2 Cl₂(g) + 1/2 O₂(g) = ClO(g)	298.16–1,000 K	+33,000	–	–	–	-0.24
1/2 Cl₂(g) + 3/2 O₂(g) = ClO₃(g)	298.16–500 K	+37,740	+5.76	–	–	+21.42
2 Cr(c) + 3/2 O₂(g) = Cr₂O₃(β)	298.16–1,823 K	-274,670	-14.07	+2.01	+0.69	+105.65
2 Cr(l) + 3/2 O₂(g) = Cr₂O₃(β)	1,823–2,000 K	-278,030	+2.33	-0.35	+1.57	+58.29
Cr(c) + O₂(g) = CrO₂(c)	298.16–1,000 K	-142,500	–	–	–	+42.00
Cr(c) + 3/2 O₂(g) = CrO₃(c)	298.16–471 K	-141,590	-13.82	–	–	+103.90
Cr(c) + 3/2 O₂(g) = CrO₃(l)	471–600 K	-141,580	-32.24	–	–	+153.14
Co(α, β) + 1/2 O₂(g) = CoO(c)	298.16–1,400 K	-56,910	+0.69	–	–	+16.03
Co(γ) + 1/2 O₂(g) = CoO(c)	1,400–1,763 K	-58,160	-1.15	–	–	+22.71
2 Cu(c) + 1/2 O₂(g) = Cu₂O(c)	298.16–1,357 K	-40,550	-1.15	-1.10	-0.10	+21.92

Table 3–8 (continued)
HEAT OF FORMATION OF SELECTED INORGANIC OXIDES

Coefficients in Free-energy Equations

Reaction	Temperature range of validity	ΔH_0	$2.303a$	b	c	I
$2\ Cu(l) + 1/2\ O_2\ (g) = Cu_2O(c)$	1,357–1,502 K	−43,880	+8.47	−2.60	−0.10	−3.72
$2\ Cu(l) + 1/2\ O_2\ (g) = Cu_2O(l)$	1,502–2,000 K	−37,710	−12.48	+0.25	−0.10	+54.44
$Cu(c) + 1/2\ O_2\ (g) = CuO(c)$	298.16–1,357 K	−37,740	−0.64	−1.40	−0.10	+24.87
$Cu(l) + 1/2\ O_2\ (g) = CuO(c)$	1,357–1,720 K	−39,410	+4.17	−2.15	−0.10	+12.05
$Cu(l) + 1/2\ O_2\ (g) = CuO(l)$	1,720–2,000 K	−41,060	−11.35	+0.25	−0.10	+59.09
$2\ Au(c) + 3/2\ O_2\ (g) = Au_2O_3\ (c)$	298.16–500 K	−2,160	−10.36	–	–	+95.14
$Hf(c) + O_2\ (g) = HfO_2\ (monoclinic)$	298.16–2,000 K	−268,380	−9.74	−0.28	+1.54	+78.16
$H_2\ (g) + 1/2\ O_2\ (g) = H_2O(l)$	298.16–373.16 K	−70,600	−18.26	+0.64	−0.04	+91.67
$H_2\ (g) + 1/2\ O_2\ (g) = H_2O(g)$	298.16–2,000 K	−56,930	+6.75	−0.64	−0.08	−8.74
$D_2\ (g) + 1/2\ O_2\ (g) = D_2O(l)$	298.16–374.5 K	−72,760	−18.10	–	–	+93.59
$D_2\ (g) + 1/2\ O_2\ (g) = D_2O(g)$	298.16–2,000 K	−58,970	+5.50	−0.75	+0.085	−3.74
$0.947\ Fe(\alpha) + 1/2\ O_2\ (g) = Fe_{0.947}O(c)$	298.16–1,033 K	−65,320	−11.26	+2.61	+0.44	+48.60
$0.947\ Fe(\beta) + 1/2\ O_2\ (g) = Fe_{0.947}O(c)$	1,033–1,179 K	−62,380	+4.08	−0.75	+0.235	+3.00
$0.947\ Fe(\gamma) + 1/2\ O_2\ (g) = Fe_{0.947}O(c)$	1,179–1,650 K	−66,750	−8.04	+0.67	−0.10	+42.28
$0.947\ Fe(\gamma) + 1/2\ O_2\ (g) = Fe_{0.947}O(l)$	1,650–1,674 K	−64,200	−18.72	+1.67	−0.10	+73.45
$0.947\ Fe(\delta) + 1/2\ O_2\ (g) = Fe_{0.947}O(l)$	1,647–1,803 K	−59,650	−6.84	+0.25	−0.10	+34.81
$0.947\ Fe(l) + 1/2\ O_2\ (g) = Fe_{0.947}O(l)$	1,803–2,000 K	−63,660	−7.48	+0.25	−0.10	+39.12
$3\ Fe(\alpha) + 2\ O_2\ (g) = Fe_3O_4\ (magnetite)$	298.16–900 K	−268,310	+5.87	−12.45	+0.245	+73.11
$3\ Fe(\alpha) + 2\ O_2\ (g) = Fe_3O_4\ (\beta)$	900–1,033 K	−272,300	−54.27	+11.65	+0.245	+233.52
$3\ Fe(\beta) + 2\ O_2\ (g) = Fe_3O_4\ (\beta)$	1,033–1,179 K	−262,990	−5.71	+1.00	−0.40	+89.19
$3\ Fe(\gamma) + 2\ O_2\ (g) = Fe_3O_4\ (\beta)$	1,179–1,674 K	−276,990	−44.05	+5.50	−0.40	+213.52
$2\ Fe(\alpha) + 3/2\ O_2\ (g) = Fe_2O_3\ (hematite)$	298.16–950 K	−200,000	−13.84	−1.45	+1.905	+108.26
$2\ Fe(\alpha) + 3/2\ O_2\ (g) = Fe_2O_3\ (\beta)$	950–1,033 K	−202,960	−42.64	+7.85	+0.13	+188.48
$2\ Fe(\beta) + 3/2\ O_2\ (g) = Fe_2O_3\ (\beta)$	1,033–1,050 K	−196,740	−10.27	+0.75	−0.30	+92.26
$2\ Fe(\beta) + 3/2\ O_2\ (g) = Fe_2O_3\ (\gamma)$	1,050–1,179 K	−193,200	−0.39	−0.13	−0.30	+59.96
$2\ Fe(\gamma) + 3/2\ O_2\ (g) = Fe_2O_3\ (\gamma)$	1,179–1,674 K	−202,540	−25.95	+2.87	−0.30	+142.85
$2\ Fe(\delta) + 3/2\ O_2\ (g) = Fe_2O_3\ (\gamma)$	1,674–1,800 K	−192,920	−0.85	−0.13	−0.30	+61.21
$Pb(c) + 1/2\ O_2\ (g) = PbO\ (red)$	298.16–600.5 K	−52,800	−2.76	−0.80	−0.10	+32.49
$Pb(l) + 1/2\ O_2\ (g) = PbO\ (red)$	600.5–762 K	−53,780	−0.51	−1.75	−0.10	+28.44
$Pb(c) + 1/2\ O_2\ (g) = PbO\ (yellow)$	298.16–600.5 K	−52,040	+0.81	−2.00	−0.10	+22.13

Table 3–8 (continued)
HEAT OF FORMATION OF SELECTED INORGANIC OXIDES

Coefficients in Free-energy Equations

Reaction	Temperature range of validity	ΔH_0	2.303$_a$	b	c	I
Pb(l) + 1/2 O_2(g) = PbO (yellow)	600.5–1,159 K	−53,020	+3.06	−2.95	−0.10	+18.08
I_2(c) + 5/2 O_2(g) = I_2O_5(c)	298.16–386.8 K	−42,040	+2.30	—	—	+113.71
I_2(l) + 5/2 O_2(g) = I_2O_5(c)	386.8–456 K	−43,490	+16.12	—	—	+81.70
I_2(g) + 5/2 O_2(g) = I_2O_5(c)	456–500 K	−58,020	−6.91	—	—	+174.79
Ir(c) + O_2(g) = IrO_2(c)	298.16–1,300 K	−39,480	+8.17	−6.39	−0.20	+20.33
3 Pb(c) + 2 O_2(g) = Pb_3O_4(c)	298.16–600.5 K	−174,920	+8.82	−8.20	−0.40	+72.78
Pb(c) + O_2(g) = PbO_2(c)	298.16–600.5 K	−66,120	+0.64	−2.45	−0.20	+45.58
2 Li(c) + 1/2 O_2(g) = Li_2O(c)	298.16–452 K	−142,220	−3.06	+5.77	−0.10	+34.19
Mg(c) + 1/2 O_2(g) = MgO (periclase)	298.16–923 K	−144,090	−1.06	+0.13	+0.25	+29.16
Mg(l) + 1/2 O_2(g) = MgO (periclase)	923–1,393 K	−145,810	+1.84	−0.62	+0.64	+23.07
Mg(g) + 1/2 O_2(g) = MgO (periclase)	1,393–2,000 K	−180,700	−3.75	−0.62	+0.64	+65.69
Mn(α) + 1/2 O_2(g) = MnO(c)	298.16–1,000 K	−92,600	−4.21	+0.97	+0.155	+29.66
Mn(β) + 1/2 O_2(g) = MnO(c)	1,000–1,374 K	−91,900	+1.84	−0.39	+0.34	+12.15
Mn(γ) + 1/2 O_2(g) = MnO(c)	1,374–1,410 K	−89,810	+7.30	−0.72	+0.34	−6.05
Mn(δ) + 1/2 O_2(g) = MnO(c)	1,410–1,517 K	−89,390	+8.68	−0.72	+0.34	−10.70
Mn(l) + 1/2 O_2(g) = MnO(c)	1,517–2,000 K	−93,350	+7.99	−0.72	+0.34	−5.90
3 Mn(α) + 2 O_2(g) = Mn_3O_4(α)	298.16–1,000 K	−332,400	−7.41	+0.66	+0.145	+106.62
2 Mn(α) + 3/2 O_2(g) = Mn_2O_3(c)	298.16–1,000 K	−230,610	−5.96	−0.06	+0.945	+80.74
Mn(α) + O_2(g) = MnO_2(c)	298.16–1,000 K	−126,400	−8.61	+0.97	+1.555	+70.14
2 Hg(l) + 1/2 O_2(g) = Hg_2O(c)	298.16–629.88 K	−22,400	−4.61	—	—	+43.29
Hg(l) + 1/2 O_2(g) = HgO (red)	298.16–629.88 K	−21,760	+0.85	−2.47	−0.10	+24.81
Mo(c) + O_2(g) = MoO_2(c)	298.16–2,000 K	−132,910	−3.91	—	—	+47.42
Mo(c) + 3/2 O_2(g) = MoO_3(c)	298.16–1,068 K	−182,650	−8.86	−1.55	+1.54	+90.07
Ni(α) + 1/2 O_2(g) = NiO(c)	298.16–633 K	−57,640	−4.61	+2.16	−0.10	+34.41
Ni(β) + 1/2 O_2(g) = NiO(c)	633–1,725 K	−57,460	−0.14	−0.46	−0.10	+23.27
2 Nb(c) + 2 O_2(g) = Nb_2O_4(c)	298.16–2,000 K	−382,050	−9.67	—	—	+116.23
2 Nb(c) + 5/2 O_2(g) = Nb_2O_5(c)	298.16–1,785 K	−458,640	−16.14	−0.56	+1.94	+157.66
2 Nb(c) + 5/2 O_2(g) = Nb_2O_5(l)	1,785–2,000 K	−463,630	−66.04	+2.21	−0.50	+317.84
N_2(g) + 1/2 O_2(g) = N_2O(g)	298.16–2,000 K	+18,650	−1.57	−0.27	+0.92	+23.47
3/2 O_2(g) = O_3(g)	298.16–2,000 K	+33,980	+2.03	−0.48	+0.36	+11.45

Table 3–8 (continued)
HEAT OF FORMATION OF SELECTED INORGANIC OXIDES

Coefficients in Free-energy Equations

Reaction	Temperature range of validity	ΔH_0	$2.303a$	b	c	I
P (white) + 1/2 O_2 (g) = PO(g)	298.16–317.4 K	−9,370	+2.53	—	—	−25.40
P(l) + 1/2 O_2 (g) = PO(g)	317.4–553 K	−9,390	+3.45	—	—	−27.63
4 P (white) + 5 O_2 (g) = P_4O_{10} (hexagonal)	298.16–317.4 K	−711,520	+95.67	−51.50	−1.00	−28.24
2 K(c) + 1/2 O_2 (g) = K_2O(c)	298.16–336.4 K	−86,400	—	—	—	+33.90
2 K(l) + 1/2 O_2 (g) = K_2O(c)	336.4–1,049 K	−87,380	+1.15	—	—	+33.90
2 K(g) + 1/2 O_2 (g) = K_2O(c)	1,049–1,500 K	−133,090	−16.12	—	—	+129.64
Ra(c) + 1/2 O_2 (g) = RaO(c)	298.16–1,000 K	−130,000	—	—	—	+23.50
Re(c) + 3/2 O_2 (g) = ReO_3 (c)	298.16–433 K	−149,090	−16.12	—	—	+110.49
Re(c) + 3/2 O_2 (g) = ReO_3 (l)	433–1,000 K	−146,750	−31.32	—	—	+145.16
2Re(c) + 7/2 O_2 (g) = Re_2O_7 (c)	298.16–569 K	−301,470	−34.64	—	—	+250.57
2 Re(c) + 7/2 O_7 (g) = Re_2O_7 (l)	569–635.5 K	−295,810	−73.68	—	—	+348.45
2 Re(c) + 4 O_2 (g) = Re_2O_8 (l)	420–600 K	−318,470	−87.50	—	—	+425.32
2 Rb(c) + 1/2 O_2 (g) = Rb_2O(c)	298.16–312.2 K	−78,900	—	—	—	+32.20
2 Rb(l) + 1/2 O_2 (g) = Rb_2O(c)	312.2–750 K	−79,950	—	—	—	+35.56
Se(c) + 1/2 O_2 (g) = SeO(g)	298.16–490 K	+9,280	−3.04	+4.40	+0.30	−14.78
Se(l) + 1/2 O_2 (g) = SeO(g)	490–1,027 K	+9,420	+8.70	—	+0.30	−44.50
1/2 Se_2 (g) + 1/2 O_2 (g) = SeO(g)	1,027–2,000 K	−7,400	−0.37	—	+0.19	−0.80
Si(c) + 1/2 O_2 (g) = SiO(g)	298.16–1,683 K	−21,090	+3.84	−0.16	−0.295	−33.14
Si(l) + 1/2 O_2 (g) = SiO(g)	1,683–2,000 K	−30,170	−7.78	−0.12	+0.25	−40.01
Si(c) + O_2 (g) = SiO_2 (α-quartz)	298.16–848 K	−210,070	+3.98	−3.32	+0.605	+34.59
Si(c) + O_2 (g) = SiO_2 (β-quartz)	848–1,683 K	−209,920	−3.36	−0.19	−0.745	+53.44
Si(l) + O_2 (g) = SiO_2 (l)	1,883–2,000 K	−228,590	−15.66	—	—	+103.97
Si(c) + O_2 (g) = SiO_2 (α-cristobalite)	298.16–523 K	−207,330	+19.96	−9.75	−0.745	−9.78
Si(c) + O_2 (g) = SiO_2 (β-cristobalite)	523–1,683 K	−209,820	−3.34	−0.24	−0.745	+53.35
Si(c) + O_2 (g) = SiO_2 (α-tridymite)	298.16–390 K	−207,030	+22.29	−11.62	−0.745	−15.64
Si(c) + O_2 (g) = SiO_2 (β-tridymite)	390–1,683 K	−209,350	−1.59	−0.54	−0.745	+47.86
2 Ag(c) + 1/2 O_2 (g) = Ag_2O(c)	298.16–1,000 K	−7,740	−4.14	—	—	+27.84
2 Ag(c) + O_2(g) = Ag_2O_2 (c)	298.16–500 K	−6,620	−3.22	—	—	+52.17
2 Na(c) + 1/2 O_2 (g) = Na_2O(c)	298.16–371 K	−99,820	−7.51	+5.47	−0.10	+50.43
2 Na(l) + 1/2 O_2 (g) = Na_2O(c)	371–1,187 K	−100,150	+4.97	−2.45	−0.10	+22.19

Table 3–8 (continued)
HEAT OF FORMATION OF SELECTED INORGANIC OXIDES

Reaction	Temperature range of validity	ΔH_0	2.303*a*	*b*	*c*	*I*
2 Na(c) + O_2(g) = Na_2O_2(c)	298.16–371 K	−122,500	−2.30	–	–	+57.51
Sr(c) + 1/2 O_2(g) = SrO(c)	298.16–1,043 K	−142,410	−6.79	+0.305	+0.675	+44.33
S(rhombohedral) + 1/2 O_2(g) = SO(g)	298.16–368.6 K	+19,250	−1.24	+2.95	+0.225	−18.84
S(monoclinic) + 1/2 O_2(g) = SO(g)	368.6–392 K	+19,200	−1.29	+3.31	+0.225	−18.72
S($l\lambda,\mu$) + 1/2 O_2(g) = SO(g)	392–718 K	+20,320	+10.22	−0.17	+0.225	−50.05
1/2 S_2(g) + 1/2 O_2(g) = SO(g)	298.16–2,000 K	+3,890	+0.07	–	–	−1.50
S(rhombohedral) + O_2(g) = SO_2(g)	298.16–368.6 K	−70,980	+0.83	+2.35	+0.51	−5.85
S(monoclinic) + O_2(g) = SO_2(g)	368.6–392 K	−71,020	+0.78	+2.71	+0.51	−5.74
S($l\lambda,\mu$) + O_2(g) = SO_2(g)	392–718 K	−69,900	+12.30	−0.77	+0.51	−37.10
1/2 S_2(g) + O_2(g) = SO_2(g)	298.16–2,000 K	−86,330	+2.42	−0.70	+0.31	+10.71
S(rhombohedral) + 3/2 O_2(g) = SO_3(c–I)	298.16–335.4 K	−111,370	−6.45	–	–	+88.32
S(rhombohedral) + 3/2 O_2(g) = SO_3(c–II)	298.16–305.7 K	−108,680	−11.97	–	–	+94.95
S(rhombohedral) + 3/2 O_2(g) = SO_3(l)	298.16–335.4 K	−107,430	−21.18	–	–	+113.76
S(rhombohedral) + 3/2 O_3(g) = SO_3(g)	298.16–368.6 K	−95,070	+1.43	+0.66	+1.26	+16.81
S(monoclinic) + 3/2 O_2(g) = SO_3(g)	368.6–392 K	−95,120	+1.38	+1.02	+1.26	+16.93
S($l\lambda,\mu$) + 3/2 O_2(g) = SO_3(g)	392–718 K	−94,010	+12.89	−2.46	+1.26	−14.40
1/2 S_2(g) + 3/2 O_2(g) = SO_3(g)	298.16–1,500 K	−110,420	+3.02	−2.39	+1.06	+33.41
2 Ta(c) + 5/2 O_2(g) = Ta_2O_5(c)	298.16–2,000 K	−492,790	−17.18	−1.25	+2.46	+161.68
Te(c) + 1/2 O_2(g) = TeO(g)	298.16–723 K	+43,110	+1.91	+0.84	+0.315	−27.22
Te(l) + 1/2 O_2(g) = TeO(g)	723–1,360 K	+39,750	+6.08	+0.09	+0.315	−33.94
2 Tl(α) + O_2(g) = Tl_2O(c)	298.16–505.5 K	−44,110	−6.91	–	–	+42.30
2 Tl(β) + O_2(g) = Tl_2O(c)	505.5–573 K	−44,260	−6.91	–	–	+42.60
2 Tl(α) + 3/2 O_2(g) = Tl_2O_3(c)	298.16–505.5 K	−99,410	−16.12	–	–	+119.09
Th(c) + O_2(g) = ThO_2(c)	298.16–2,000 K	−294,350	−5.25	+0.59	+0.775	+62.81
Sn(c) + 1/2 O_2(g) = SnO(c)	298.16–505 K	−68,600	−3.57	+1.65	−0.10	+32.59
Sn(l) + 1/2 O_2(g) = SnO(c)	505–1,300 K	−69,670	+3.06	−1.50	−0.10	+18.39
Sn(c) + O_2(g) = SnO_2(c)	298.16–505 K	−142,010	−14.00	+2.45	+2.38	+90.74
Ti(α) + 1/2 O_2(g) = TiO(α)	298.16–1,150 K	−125,010	−4.01	−0.29	+0.83	+36.28
Ti(β) + 1/2 O_2(g) = TiO(α)	1,150–1,264 K	−125,040	+1.17	−1.55	+0.83	+21.90
2 Ti(α) + 3/2 O_2(g) = Ti_2O_3(α)	298.16–473 K	−360,660	+32.08	−23.49	−0.30	−10.66
2 Ti(α) + 3/2 O_2(g) = Ti_2O_3(β)	473–1,150 K	−369,710	−30.95	+2.62	+4.80	+162.79
Ti(α) + O_2(g) = TiO_2 (rutile)	298.16–1,150 K	−228,360	−12.80	+1.62	+1.975	+82.81
Ti(β) + O_2(g) = TiO_2 (rutile)	1,150–2,000 K	−228,380	−7.62	+0.36	+1.975	+68.43

Table 3–8 (continued)
HEAT OF FORMATION OF SELECTED INORGANIC OXIDES

Reaction	Temperature range of validity	ΔH_0	2.303a	b	c	I
$W(c) + O_2(g) = WO_2(c)$	298.16–1,500 K	-137,180	-1.38	—	—	+45.56
$4\,W(c) + 11/2\,O_2(g) = W_4O_{11}(c)$	298.16–1,700 K	-745,730	-32.70	—	—	+321.84
$W(c) + 3/2\,O_2(g) = WO_3(c)$	298.16–1,743 K	-201,180	-2.92	-1.81	-0.30	+70.89
$W(c) + 3/2\,O_2(g) = WO_3(l)$	1,743–2,000 K	-203,140	-35.74	+1.13	-0.30	+173.27
$U(\alpha) + O_2(g) = UO_2(c)$	298.16–935 K	-262,880	-19.92	+3.70	+2.13	+100.54
$U(\beta) + O_2(g) = UO_2(c)$	935–1,045 K	-260,660	-4.28	-0.31	+1.78	+55.50
$U(\gamma) + O_2(g) = UO_2(c)$	1,045–1,405 K	-262,830	-6.54	-0.31	+1.78	+64.41
$U(l) + O_2(g) = UO_2(c)$	1,405–1,500 K	-264,790	-5.92	—	—	+63.50
$3\,U(\alpha) + 4\,O_2(g) = U_3O_8(c)$	298.16–935 K	-863,370	-56.57	+10.68	+5.20	+330.19
$3\,U(\beta) + 4\,O_2(g) = U_3O_8(c)$	935–1,045 K	-856,720	-9.67	-1.35	+4.15	+195.12
$3\,U(\gamma) + 4\,O_2(g) = U_3O_8(c)$	1,045–1,405 K	-863,230	-16.44	-1.35	+4.15	+221.79
$3\,U(l) + 4\,O_2(g) = U_3O_8(c)$	1,405–1,500 K	-869,460	-10.91	-1.35	+4.15	+208.82
$U(\alpha) + 3/2\,O_2(g) = UO_3$ (hexagonal)	298.16–935 K	-294,090	-18.33	+3.49	+1.535	+114.94
$U(\beta) + 3/2\,O_2(g) = UO_3$ (hexagonal)	935–1,045 K	-291,870	-2.69	-0.52	+1.185	+69.90
$U(\gamma) + 3/2\,O_2(g) = UO_3$ (hexagonal)	1,045–1,400 K	-294,040	-4.95	-0.52	+1.185	+78.80
$V(c) + 1/2\,O_2(g) = VO(c)$	298.16–2,000 K	-101,090	-5.39	-0.36	+0.53	+38.69
$V(c) + 1/2\,O_2(g) = VO(g)$	298.16–2,000 K	+52,090	+1.80	+1.04	+0.35	-28.42
$2\,V(c) + 3/2\,O_2(g) = V_2O_3(c)$	298.16–2,000 K	-299,910	-17.98	+0.37	+2.41	+118.83
$2\,V(c) + 2\,O_2(g) = V_2O_4(\alpha)$	209.16–345 K	-342,890	-11.03	+3.00	-0.40	+117.38
$2\,V(c) + 2\,O_2(g) = V_2O_4(\beta)$	345–1,818 K	-345,330	-24.36	+1.30	+3.545	+155.55
$6\,V(c) + 13/2\,O_2(g) = V_6O_{13}(c)$	298.16–1,000 K	-1,076,340	-95.33	—	—	+557.61
$2\,V(c) + 5/2\,O_2(g) = V_2O_5(c)$	298.16–943 K	-381,960	-41.08	+5.20	+6.11	+228.50
$2\,Y(c) + 3/2\,O_2(g) = Y_2O_3(c)$	298.16–1,773 K	-419,600	+2.76	-1.73	-0.30	+66.36
$Zn(c) + 1/2\,O_2(g) = ZnO(c)$	298.16–692.7 K	-84,670	-6.40	+0.84	+0.99	+43.25
$Zr(\alpha) + O_2(g) = ZrO_2(\alpha)$	298.16–1,135 K	-262,980	-6.10	+0.16	+1.045	+65.00
$Zr(\beta) + O_2(g) = ZrO_2(\alpha)$	1,135–1,478 K	-264,190	-5.09	-0.40	+1.48	+63.58
$Zr(\beta) + O_2(g) = ZrO_2(\beta)$	1,478–2,000 K	-262,290	-7.76	+0.50	-0.20	+69.50

[a]This table is a condensed version of the table "Heat of Formation of Inorganic Oxides" that appears in Section 2 of *CRC Handbook of Materials Science*, Volume II. Refer to this table for a more extensive listing.

From *Contributions to the Data on Theoretical Metallurgy*, Bulletin 542, U.S. Bureau of Mines, 1954, 60.

Table 3–9
HEATS AND FREE ENERGIES OF FORMATION, ENTROPIES, AND
HEAT CAPACITIES OF ELEMENTS AND INORGANIC COMPOUNDS

The table contains values of the enthalpy and Gibbs (formerly free) energy of formation, entropy, and heat capacity at 298.15K (25°C). No values are given in the table for metal alloys or other solid solutions, fused salts, or for substances of undefined chemical composition.

For a more complete listing of compounds see the tables of "Selected Values of Chemical Thermodynamic Properties," *NBS Notes* 270-3, 270-4, 270-5 by. D. D. Wagman et al., Washington, D.C.

The physical state of each substance is indicated in the column headed "State" as crystalline solid (c), liquid (liq), gaseous (g), or amorphous (amorp). Solutions in water are listed as aqueous (aq).

The values of the thermodynamic properties of the pure substances given in the table are, for the substances in their standard states, defined as follows. For a pure solid or liquid, the standard state is the substance in the condensed phase under a pressure of 1 atm. For a gas the standard state is the hypothetical ideal gas at unit fugacity, in which state the enthalpy is that of the real gas at the same temperature and at zero pressure.

The values of ΔHf° and ΔGf° given in the table represent the change in the appropriate thermodynamic quantity when one gram-formula weight of the substance in its standard state is formed, isothermally at the indicated temperature, from the elements, each in its appropriate standard reference state. The standard reference state at 25°C for each element (except phosphorus) has been chosen to be the standard state that is thermodynamically stable at 25°C and 1 atm

pressure. For phosphorus the standard reference state is the crystalline white form. The standard reference states are indicated in the table by the fact that the values of ΔHf° and ΔGf° are exactly zero.

The values of S° represent the virtual or "thermal" entropy of the substance in the standard state at 298.15 K, omitting contributions from nuclear spins. Isotope mixing effects are also excluded except in the case of the $(^1H\text{-}^2H)$ system.

Solutions in water are designated as aqueous, and the concentration of the solution is expressed in terms of the number of moles of solvent associated with 1 mol of the solute. If no concentration is indicated, the solution is assumed to be dilute. The standard state for a solute in aqueous solution is taken as the hypothetical ideal solution of unit molality (indicated as std. state, $m = 1$). In this state the partial molal enthalpy and the heat capacity of the solute are the same as in the infinitely dilute real solution (as, ∞).

The value of ΔHf° given for a solute in its standard state is the apparent molal enthalpy of formation of the substance in the infinitely dilute real solution. The experimental value for a heat of dilution is obtained directly as the difference between the two values of ΔHf° at the corresponding concentrations.

Values of ΔHf° and ΔGf° (or ΔFf°) in the table are expressed in kilocalories per mole; values of S° and C_p° are expressed in calories per degree per mole.

Formula and Description	State	ΔHf°	ΔGf°	S°	C_p°
Actinium					
Ac$_2$O$_3$	c	− 444			
Aluminum					
Al	c	0	0	6.77	5.82
	g	78.0	68.3	39.30	5.11
Al^{3+} std. state, $m = 1$	aq	− 127	− 116	− 76.9	
Al(BH$_4$)$_3$	liq	− 3.9	34.6	69.1	46.5
	g	3	35	90.6	
AlBr	g	−1	− 10	57.22	8.50
AlBr$_3$	c	− 126.0	− 120.7	44	24.3
std. state, $m = 1$	aq	− 214	− 191	− 17.8	
Al$_4$C$_3$	c	− 49.9	− 46.9	21.26	27.91
Al(CH$_3$)$_3$	liq	− 32.6	− 2.4	50.05	37.19
	g	− 17.7			

Table 3–9 (continued)
HEATS AND FREE ENERGIES OF FORMATION, ENTROPIES, AND HEAT CAPACITIES OF ELEMENTS AND INORGANIC COMPOUNDS

Formula and Description	State	$\Delta H f°$	$\Delta G f°$	$S°$	$C_p°$
$Al_2(CH_3)_6$	g	−55.19	−2.34	125.4	
$Al(OAc)_3$	c	−452.3			
AlCl	g	−11.4	−17.7	54.50	8.36
$AlCl_3$	c	−168.3	−150.3	26.45	21.95
std. state, $m = 1$	aq	−247	−210	−36.4	
$AlCl_3 \cdot 6H_2O$	c	−643.3	−542.4	90	
Al_2Cl_6	g	−308.5	−291.7	117	
AlF	g	−61.7	−67.8	51.36	7.63
AlF_3	c	−359.5	−340.6	15.88	17.95
	g	−287.9	−284.0	66.2	14.97
$AlF_3 \cdot 3H_2O$	c	−549.1	−490.4	50	
AlH	g	61.96	55.25	44.88	7.02
AlH_3	c	−11			
AlI	g	15.66			8.60
AlI_3	c	−75.0	−71.9	38	23.6
std. state, $m = 1$	aq	−167	−153	2.9	
AlN	c	−76.0	−68.6	4.82	7.20
$Al(NO_3)_3$ std. state, $m = 1$	aq	−276	−196	28.1	
$Al(NO_3)_3 \cdot 6H_2O$	c	−681.28	−526.74	111.8	103.5
$Al(NO_3)_3 \cdot 9H_2O$	c	−897.96	(−700.2)	(136.)	
AlP	c	−39.8			
$AlPO_4$ berlinite	c	−404.4	−382.7	21.70	22.27
AlO	g	21.8	15.6	52.17	7.38
AlO_2 std. state, $m = 1$	aq	−219.6	−196.8	−5	
Al_2O_3 α, corundum	c	−400.5	−378.2	12.17	18.89
δ	c	−398			
ρ	c	−391			
κ	c	−397			
γ	c	−395			
$Al_2O_3 \cdot H_2O$ boehmite	c	−472.0	−436.3	23.15	31.37
diaspore	c	−478	−440	16.86	25.22
$Al_2O_3 \cdot 3H_2O$ gibbsite	c	−612.5	−546.7	33.51	44.49
bayerite	c	−610.1			
$Al(OH)^{2+}$ std. state, $m = 1$	aq		−165.9		
$Al(OH)_3$	amorp	−305			
$Al(OH)_4$ std. state, $m = 1$	aq	−356.2	−310.2	28	
AlS	g	48.02	35.88	55.09	7.98
Al_2S_3	c	−173		(23)	
$Al_2(SO_4)_3$	c	−822.38	−740.95	57.2	62.00
std. state, $m = 1$	aq	−906	−766	−139.4	
$Al_2(SO_4)_3 \cdot 6H_2O$	c	−1269.53	−1104.82	112.1	117.8
Al_2Se_3	c	−135			
Al_2SiO_5 andalusite	c	−655.9	−620.8	22.28	29.33
kyanite	c	−656.4	−620.5	20.03	29.09
sillimanite	c	−662.6	−627.6	22.99	29.30
$Al_2Si_2O_7 \cdot 2H_2O$ kaolinite	c	−979.6	−903.0	48.5	58.62
halloysite	c	−975.1	−898.5	48.6	58.86
Al_2Te_3	c	−78			
Ammonium					
NH_3	g	−11.02	−3.94	45.97	8.38
undissoc; std. state, $m = 1$	aq	−19.19	−6.35	26.6	
aq, 1		−18.011	−18.011		
aq, 10		−19.074	−19.074		
aq, 100		−19.167	−19.167		
NH_4^+ std. state, $m = 1$	aq	−31.67	−18.97	27.1	19.1
NH_4OH	liq	−86.33	−60.74	39.57	37.02
undissoc; std. state, $m = 1$	aq	−87.505	−63.04	43.3	
ionized; std. state, $m = 1$	aq	−86.64	−56.56	24.5	−16.4
aq, 1		−86.875			
aq, 2		−87.078			
aq, 10		−87.396			
aq, 100		−87.483			

Table 3-9 (continued)
HEATS AND FREE ENERGIES OF FORMATION, ENTROPIES, AND HEAT CAPACITIES OF ELEMENTS AND INORGANIC COMPOUNDS

Formula and Description	State	$\Delta Hf°$	$\Delta f°$	$S°$	$C_p°$
$NH_4Al(SO_4)_2$	c	−562.2	−487.2	51.7	54.12
std. state, $m = 1$	aq	−593	−491	−40.2	
NH_4AsO_2 std. state, $m = 1$	aq	−134.21	−102.63	37.0	
$NH_4H_2AsO_3$ std. state, $m = 1$	aq	−202.51	−159.32	53.5	
$NH_4H_2AsO_4$	c	−523.3	−199.1	41.12	36.13
std. state, $m = 1$	aq	−249.06	−199.01	55.1	
$(NH_4)_2HAsO_4$	c	−282.4			
std. state, $m = 1$	aq	−279.9	−208.7	53.8	
$(NH_4)_3AsO_4$	c	−307.4			
std. state, $m = 1$	aq	−307.28	−211.91	42.4	
NH_4BO_2 std. state, $m = 1$	aq	−216.27	−181.24	18.2	
NH_4Br	c	−64.73	−41.9	27	23
std. state, $m = 1$	aq	−60.72	−43.82	46.8	−14.8
	aq, 100	−60.614			
	aq, 1000	−60.650			
NH_4BrO	aq	−54.2	−27.0	37	
NH_4BrO_3	aq	−51.7	−18.6	66.1	
$(NH_4)_2CO_3$ std. state, $m = 1$	aq	−225.18	−164.11	40.6	
NH_4HCO_3	c	−203.0	−159.2	28.9	
std. state, $m = 1$	aq	−197.06	−159.23	48.9	
NH_4 carbamate	c	−154.17	−107.09	31.9	
NH_4CN	c	0.10			32
std. state, $m = 1$	aq	4.3	22.2	49.6	
NH_4CNO cyanate	c	−72.75			
std. state, $m = 1$	aq	−66.6	−42.3	52.6	
NH_4CNS thiocyanate	c	−18.8			
std. state, $m = 1$	aq	−13.40	3.18	61.6	9.5
NH_4 formate	c	−135.63			
std. state, $m = 1$	aq	−133.38	−102.9	49	−1.9
NH_4 acetate	c	−147.26			
std. state, $m = 1$	aq	−147.83	−107.26	47.8	17.6
NH_4 chloroacetate	c	−153.7			
NH_4 trichloroacetate	c	−156.7			
$(NH_4)_2C_2O_4$	c	−268.72			
	aq, 2100	−260.6	−196.2		
NH_4 dithiocarbamate	c	−30.3			
NH_4Cl	c	−75.15	−48.51	22.6	20.1
std. state, $m = 1$	aq	−71.62	−50.34	40.6	−13.5
	aq, 10	−71.567			
	aq, 100	−71.487			
NH_4ClO std. state, $m = 1$	aq	−57.3	−27.8	37	
NH_4ClO_2 std. state, $m = 1$	aq	−47.6	−14.9	51.3	
NH_4ClO_3 std. state, $m = 1$	aq	−55.4	−19.8	65.9	
NH_4ClO_4	c	−70.58	−21.25	44.5	
std. state, $m = 1$	aq	−62.58	−21.03	70.6	
NH_4HCrO_4 std. state, $m = 1$	aq	−241.6	−201.8	71.1	
$(NH_4)_2CrO_4$	c	−279.0			
std. state, $m = 1$	aq	−273.5	−211.90	66.2	
$(NH_4)_2Cr_2O_7$	c	−431.8			
std. state, $m = 1$	aq	−419.5	−348.9	116.8	
$NH_4Cr(SO_4)_2 \cdot 12H_2O$	c			170.9	168.5
NH_4F	c	−110.89	−83.36	17.20	15.60
·std. state, $m = 1$	aq	−111.17	−85.61	3.8	−6.4
NH_4HF_2	c	−191.9	−155.6	27.61	25.50
std. state, $m = 1$	aq	−187.01	−157.15	49.2	
NH_4I	c	−48.14	−26.9	28	
std. state, $m = 1$	aq	−44.86	−31.30	53.7	−14.9
	aq, 100	−44.784			
NH_4IO std. state, $m = 1$	aq	−57.4	−28.2	25.8	
NH_4IO_3	c	−92.2			
std. state, $m = 1$	aq	−84.6	−49.6	55.4	
NH_4IO_4	aq	−66.9			

Table 3–9 (continued)
HEATS AND FREE ENERGIES OF FORMATION, ENTROPIES, AND
HEAT CAPACITIES OF ELEMENTS AND INORGANIC COMPOUNDS

Formula and Description	State	$\Delta H_f°$	$\Delta G_f°$	$S°$	$C_p°$
NH₄N₃ azide	c	27.6	65.5	26.9	
	aq	34.1	64.3	52.7	
NH₄NO₂	c	−61.3			
std. state, $m = 1$	aq	−56.7	−27.9	60.6	−4.2
NH₄NO₃	c	−87.37	−43.98	36.11	33.3
std. state, $m = 1$	aq	−81.23	−45.58	62.1	−1.6
	aq, 10	−82.470			
	aq, 100	−81.340			
(NH₄)₂O	liq	−102.94	−63.84	63.94	59.08
NH₄PO₃	aq	−265.2			
NH₄H₂PO₂ hypophosphite	c	−180.0			
NH₄H₂PO₄	c	−345.38	−289.33	36.32	34.00
std. state, $m = 1$	aq	−341.49	−289.14	48.7	
(NH₄)₂HPO₄	c	−374.50			45
std. state, $m = 1$	aq	−372.17	−298.28	46.2	
(NH₄)₃PO₄	c	−399.6			
std. state, $m = 1$	aq	−400.3	−300.4	28	
(NH₄)₄P₂O₇ std. state, $m = 1$	aq	−669.5	−534.6	80	
(NH₄)₂PoCl₆ std. state, $m = 1$	aq		−176		
(NH₄)₂PtCl₆	c	−192.0			56.8
NH₄ReO₄	c	−226.0	−185.2	55.6	
NH₄HS	c	−37.5	−12.1	23.3	
std. state, $m = 1$	aq	−35.9	−16.09	42.1	
(NH₄)₂S std. state, $m = 1$	aq	−55.4	−17.4	50.7	
(NH₄)₂S₂ std. state, $m = 1$	aq	−56.1	−18.9	61.0	
NH₄HSO₃	c	−183.7			
std. state, $m = 1$	aq	−181.34	−145.12	60.5	
NH₄HSO₄	c	−245.45			
std. state, $m = 1$	aq	−243.75	−199.66	58.6	−0.9
	aq, 200	−245.65			
(NH₄)₂SO₃	c	−211.6			
std. state, $m = 1$	aq	−215.2	−154.2	47.2	
(NH₄)₂SO₄	c	−282.23	−215.56	52.6	44.81
std. state, $m = 1$	aq	−280.66	−215.77	58.6	−31.8
	aq, 100	−280.407			
(NH₄)₂S₂O₃	aq	−219.2			
(NH₄)₂S₂O₄ std. state, $m = 1$	aq	−243.4	−181.4	76	
(NH₄)₂S₂O₆	aq	−349.7			
(NH₄)₂S₂O₇	aq	−398.2			
(NH₄)₂S₂O₈	c	−392.5			
std. state, $m = 1$	aq	−383.3	−303.3	113.5	
(NH₄)₂S₄O₆	aq	−355.92			
(NH₄)₂Sb₂S₄ std. state, $m = 1$	aq	−115.7	−61.7	41.7	
NH₄HSe	c	−31.8	−5.6	23.1	
std. state, $m = 1$	aq	−27.9	−8.5	46	
(NH₄)₂Se std. state, $m = 1$	aq		−7.0		
NH₄HSeO₃ std. state, $m = 1$	aq	−154.65	−117.33	60.2	
NH₄HSeO₄ std. state, $m = 1$	aq	−170.7	−127.1	62.8	
(NH₄)₂SeO₃ std. state, $m = 1$	aq	−185.0	−126.3	57.3	
(NH₄)₂SeO₄	c	−209.0			
std. state, $m = 1$	aq	−206.5	−143.4	67.1	
(NH₄)₂SiF₆ hexagonal	c	−640.94	−565.38	66.98	54.52
(NH₄)₂SnCl₆	c	−295.6			
NH₄HTe	c	0.3			
NH₄H₅TeO₆	aq	−333.2			
(NH₄)₂TeO₃	aq	−205.9			
NH₄VO₃	c	−251.7	−212.3	33.6	30.91
Antimony					
Sb III	c	0	0	10.92	6.03
IV explosive	amorp	2.54			
	g	62.7	53.1	43.06	4.97
Sb₂	g	56.3	44.7	60.90	8.70

Table 3–9 (continued)
HEATS AND FREE ENERGIES OF FORMATION, ENTROPIES, AND
HEAT CAPACITIES OF ELEMENTS AND INORGANIC COMPOUNDS

Formula and Description	State	$\Delta H_f°$	$\Delta G_f°$	$S°$	C_p
Sb_4	g	49.0	33.8	84	
$SbBr_3$	c	−62.0	−57.2	49.5	
SbCl	g	−6.22			8.49
$SbCl_2$	g	−18.5			
$SbCl_3$	c	−91.34	−77.37	44.0	25.8
	g	−75.0	−72.0	80.71	18.33
$SbCl_5$	liq	−105.2	−83.7	72	
	g	−94.25	−79.91	96.04	28.95
SbF	g	−11.29			7.97
SbF_3	c	−218.8			
SbH_3	g	34.681	35.31	55.61	9.81
Sb_2H_4	g	57.2			
SbI_3	c	−24.0			
SbN	g	63.66			7.41
SbO	g	47.67			
SbO std. state, $m = 1$	aq		−42.33		
SbO_2 std. state, $m = 1$	aq		−81.32		
Sb_2O_3	c	−164.9			
Sb_2O_4	c	−216.9	−190.2	30.4	27.39
Sb_2O_5	c	−232.3	−198.2	29.9	
Sb_4O_6 II, cubic	c	−344.3	−303.1	52.8	
I, orthorhombic	c	−338.7	−299.5	58.8	48.46
$HSbO_2$ undissoc; std. state, $m = 1$	aq	−116.6	−97.4	11.1	
$Sb(OH)_3$	c		−163.8		
undissoc; std. state, $m = 1$	aq	−184.9	−154.1	27.8	
H_3SbO_4	aq	−216.8			
$HSb(OH)_6$	aq	−353.4			
SbOCl	c	−89.4			
SbOF undissoc; std. state, $m = 1$	aq		−116.5		
Sb_2S_3 black	c	−41.8	−41.5	43.5	28.65
orange	amorp	−35.2			
$Sb_2S_4^{2}$ std. state, $m = 1$	aq	−52.4	−23.8	−12.5	
$Sb_2(SO_4)_3$	c	−574.2			
Sb_2Te_3	c	−13.5	−13.2	56	
Argon					
Ar	g	0	0	36.982	4.968
std. state, $m = 1$	aq	−2.9	3.9	14.2	
Arsenic					
As α, gray	c	0	0	8.4	5.89
γ, yellow, cubic	c	3.5			
β	amorp	1.0			
As_2	g	53.1	41.1	57.2	8.366
As_4	g	34.4	22.1	75	
$AsBr_3$	c	−47.2		(53)	
	liq	−43.1			
	g	−31	−38	86.94	18.92
$AsCl_3$	liq	−72.9	−62.0	51.7	
	g	−62.5	−59.5	78.17	18.10
AsF_3	liq	(−226)	−198	(43)	
AsH_3	g	15.88	16.47	53.22	9.10
As_2H_4	g	35.2		...	
AsI_3	c	−13.9	−14.2	50.92	25.28
	g			92.79	19.27
AsN	g	46.91	40.15	53.9	7.27
AsO	g	16.72			
AsO_2 std. state, $m = 1$	aq	−102.54	−83.66	9.9	
AsO_4^{3} std. state, $m = 1$	aq	−212.27	−55.00	−38.9	
As_2O_4	c	−189.72		(36)	
As_2O_5	c	−221.05	−187.0	25.2	27.85
As_4O_6 octahedral	c	−314.04	−275.46	51.2	45.72
monoclinic	c	−313.0	−275.82	56	
	g	−289.0	−262.4	91	

Table 3–9 (continued)
HEATS AND FREE ENERGIES OF FORMATION, ENTROPIES, AND HEAT CAPACITIES OF ELEMENTS AND INORGANIC COMPOUNDS

Formula and Description	State	$\Delta Hf°$	$\Delta Gf°$	$S°$	$C_p°$
$HAsO_2$ undissoc; std. state, $m = 1$	aq	−109.1	−96.25	30.1	
H_2AsO_3 undissoc; std. state, $m = 1$	aq	−170.84	−140.35	26.4	
H_3AsO_3 undissoc; std. state, $m = 1$	aq	−177.4	−152.94	46.6	
$HAsO_4^{2-}$ undissoc; std. state, $m = 1$	aq	−216.62	−170.82	−0.4	
$H_2AsO_4^-$ undissoc; std. state, $m = 1$	aq	−217.39	−180.04	28	
H_3AsO_4	c	−216.6			
undissoc; std. state, $m = 1$	aq	−215.7	−183.1	44	
As_2S_2	c	−34.1			
As_2S_3	c	−40.4	−40.3	39.1	27.8
Astatine					
At	c	0	0	29.0	
Barium					
Ba	c	0	0	16	
	g	42	34.60	40.70	5.10
Ba^{2+}	aq	−128.67	−134.0	3	
$Ba_3(AsO_4)_2$	c	−817.8			
$BaHAsO_4 \cdot H_2O$	c	−411.5			
$Ba(H_2AsO_4)_2 \cdot 2H_2O$	c	−694.7			
$Ba(acetate)_2$	c	−355.1			
	aq, 400	−361.5			
$Ba(acetate)_2 \cdot 3H_2O$	c	−567.3			
$BaBr_2$	c	−169		(35)	
std. state, $m = 1$	aq	−186.47	−183.1	42	
$BaBr_2 \cdot 2H_2O$	c	−326.3			
$Ba(HCO_2)_2$ formate	c	−326.5			
	aq, 400	−324.5			
$Ba(CN)_2$	c	−47.9			
	aq	−50.7			
$Ba(CN)_2 \cdot 2H_2O$	c	−191.1			
$Ba(CNO)_2$ cyanate	c	−209.9			
$BaCO_3$	c	−291.3	−272.2	26.8	
std. state, $m = 1$	aq	−290.30	−260.2	−10	
$Ba(HCO_3)_2$ std. state, $m = 1$	aq	−459.0	−414.6	48	
$BaC_2O_4 \cdot 2H_2O$	c	−470.1			
$BaCl_2$	c	−205.5	−193.8	30	
	aq, 50	−208.09			
std. state, $m = 1$	aq	−208.72	−196.7	29	
$BaCl_2 \cdot 2H_2O$	c	−349.35	−309.7	48.5	
$Ba(OCl)_2$	aq	−176.1			
$Ba(ClO_2)_2$	c	−158.2			
$Ba(ClO_3)_2$	c	−181.7			
	aq	−175.6			
$Ba(ClO_4)_2$	c	−194.3			
$Ba(ClO_4)_2 \cdot 3H_2O$	c	−405.4			
$BaCrO_4$	c	−341.3			
BaF_2	c	−286.9	−274.5	23.03	
	aq	−286.0	−265.3		
BaH	g	52	46	−33.7	52.97
BaH_2	c	−40.9	−31.5		
BaI_2	c	−144.6	−143	39	
std. state, $m = 1$	aq	−155.41	−158.7	55	
$BaI_2 \cdot 2H_2O$	c	−290.9			
$Ba(IO_3)_2$	c	−264.5			
	aq	−238.4			
$Ba(IO_3)_2 \cdot H_2O$	c	−319.6			
$BaMoO_4$	c	−373.8			
$Ba(N_3)_2$	c	−8.0			
	aq	0.2			
Ba_3N_2	c	−90.6	(−73.4)	36.4	
$Ba(NH_2)_2$	c	−78.9			
$Ba(NO_2)_2$	c	−174.0			
	aq, 800	−178.6	−150.75		
$Ba(NO_2)_2 \cdot H_2O$	c	−254.5			

Table 3–9 (continued)
HEATS AND FREE ENERGIES OF FORMATION, ENTROPIES, AND HEAT CAPACITIES OF ELEMENTS AND INORGANIC COMPOUNDS

Formula and Description	State	$\Delta H f°$	$\Delta G f°$	$S°$	C_p
$Ba(NO_3)_2$	c	−237.06	−190.0	51.1	
std. state, $m = 1$	aq	−227.41	−186.8	73	
BaO	c	−133.5	−126.3	16.8	
BaO_2	c	−151.9	−139.5	22.62	
$BaO_2 \cdot 8H_2O$	c	−719.3			
$Ba(OH)_2$	c	−226.2			
std. state, $m = 1$	aq	−238.58	−209.2	−2	
$Ba(OH)_2 \cdot H_2O$	c	−299.0			
$Ba_3(PO_4)_2$	c	−998.0			
$BaHPO_4$	c	−465.8			
$Ba(H_2PO_2)_2$ hypophosphite	aq, 400	−423.6			
$Ba(H_2PO_2)_2 \cdot H_2O$	c	−429.0			
$Ba(H_2PO_4)_2$	c	−749.6			
$BaPtCl_6$	c	−286.8			
	aq	−296.1			
BaS	c	−106.0			
	g	41			
	aq	−118.4			
$Ba(HS)_2$	aq	−134.8			
$Ba(HSO_3)_2$	aq	−430.7			
$BaSO_3$	c	−282.6			
$BaSO_4$	c	−350.2	−323.4	31.6	
std. state, $m = 1$	aq	−345.57	−311.3	7	
BaS_2O_6	aq	−409.3			
$BaS_2O_6 \cdot 2H_2O$	c	−552.5			
BaS_2O_8	aq	−454.0			
$BaS_2O_8 \cdot 4H_2O$	c	−738.7			
BaS_4O_6	aq	−401.3			
$BaSeO_4$	c	−280.0			
$BaSiF_6$	c	−691.6			
$BaSiO_3$	c	−359.5			
Ba_2SiO_4	c	−496.8			
$BaWO_4$	c	−407.7			
Beryllium					
Be	c	0	0	2.28	
	g	76.6	67.6	32.55	
$BeBr_2$	c	−79.4	(−76.5)	29	
	aq	−142	−127.9		
$BeCl_2$	c	−118	(−102.9)	(23)	
$BeCl_2 \cdot 4H_2O$	c	−436.8			
BeF_2	aq	−251.4			
	c	−241.2	(−216)	(17)	
BeH	g	78.1	71.3	40.84	6.96
BeI_2	c	−50.6	(−39.4)	(31)	
	aq	−112	−103.4		
$BeMoO_4$	c	−330			
Be_3N_2	c	−133.5	−121.4	12.0	
$Be(NO_3)_2$	aq	−188.3			
BeO	c	−143.1	−136.1	3.37	
	g	11.8	5.7	47.18	
$Be(OH)_2$	c	−216.8			
BeS	c	−55.9			
$BeSO_4$	c	−286.0			
	aq, 400	−304.1	−254.8		
$BeSO_4 \cdot 4H_2O$	c	−576.3			
Bismuth					
Bi	c	0	0	13.56	6.10
	g	49.5	40.2	44.669	4.968
Bi_2	g	52.5			8.83
$BiAsO_4$	c		−148		
$BiBr_3$	c	63	(−56)	(54)	26
BiCl	c	−31.2	−25.9	22.6	
$BiCl_3$	c	−90.6	−75.3	42.3	25

Table 3–9 (continued)
HEATS AND FREE ENERGIES OF FORMATION, ENTROPIES, AND
HEAT CAPACITIES OF ELEMENTS AND INORGANIC COMPOUNDS

Formula and Description	State	$\Delta Hf°$	$\Delta Gf°$	$S°$	$C_p°$
	g	−63.5	−61.2	85.74	19.04
BiF₃	c	(−216)	(−200)	(34)	
BiH₃	g	66.4			
BiI₃	c	−24.0	−41.9		
BiO· std. state, $m = 1$	aq		−35.0		
Bi₂O₃	c	−137.16	−118.0	36.2	27.13
BiO(OH)	c		−88.0		
Bi(OH)₃	c	−170.0			
BiOBr	c		−71.0		
BiOCl	c	−87.7	−77.0	28.8	
BiONO₃	c		−67.0		
BiS	g	43	29	68	
Bi₂S₃	c	−34.2	−33.6	47.9	29.2
Bi₂(SO₄)₃	c	−608.1			
BiSe	g	42.0			
BiTe	g	42.8			
Bi₂Te₃	c	−18.5	−18.4	62.36	28.8
Boron					
B	c	0	0	1.40	2.65
	amorp	0.9		1.56	2.86
	g	134.5	124.0	36.65	4.971
B₂	g	198.5	185.0	48.23	7.30
BBr	g	56.9	46.7	53.75	7.87
BBr₃	liq	−57.3	−57.0	54.9	
	g	−49.15	−55.56	77.47	16.20
B₄C	c	−17	−17	6.48	12.62
B(CH₃)₃	liq	−34.2	−7.7	57.1	
	g	−29.7	−8.6	75.2	21.15
BCl	g	35.73	28.90	50.94	7.57
BCl₃	liq	−102.1	−92.6	49.3	25.5
	g	−96.50	−92.91	69.31	14.99
B₂Cl₄	liq	−125.0	−111.1	62.7	32.9
BOCl	g	−75			
(BOCl)₃	g	−390.4	−370.5	91	
BClF₂	g	−212.8	−209.4	65	
BCl₂F	g	−154.2	−150.9	68	
BF	g	−29.2	−35.8	47.89	7.07
BF₃	g	−271.75	−267.77	60.71	12.06
BF₄⁻ std. state, $m = 1$	aq	−376.4	−355.4	43	
B₂F₄	g	−145			
BOF	g	−145			
HBF₄	aq	−375.5			
BH	g	107.46	100.29	41.05	6.97
BH₃	g	24			
BH₄⁻ std. state, $m = 1$	aq	11.51	27.31	26.4	
B₂H₆	g	8.5	20.7	55.45	13.60
B₄H₁₀	g	15.8			
B₅H₉	liq	10.20	41.03	44.03	36.12
BI₃	g	17.00	4.96	83.43	16.92
BN	c	−60.8	−54.6	3.54	4.71
	g	154.75	146.87	50.71	7.04
B₃N₃H₆	liq	−129.3	−93.88	47.7	
BO	g	6	−1	48.62	6.98
BO₂	g	−71.8	−73.1	54.84	10.28
BO₂⁻ std. state, $m = 1$	aq	−184.60	−162.27	−8.9	
B₂O₂	g	−108.7	−110.5	57.93	13.69
B₂O₃	c	−304.20	−285.30	12.90	15.04
	amorp	−299.84	−282.6	18.6	14.6
	g	−201.67	−198.85	66.85	15.98
B₄O₇²⁻ std. state, $m = 1$	aq		−622.6		
HBO₂ cubic	c	−192.17			
monoclinic	c	−189.83	−172.9	9	
orthorhombic	c	−188.52	−172.5	12	
	g	−134.3	−131.7	57.35	10.09

Table 3–9 (continued)
HEATS AND FREE ENERGIES OF FORMATION, ENTROPIES, AND HEAT CAPACITIES OF ELEMENTS AND INORGANIC COMPOUNDS

Formula and Description	State	$\Delta Hf°$	$\Delta Gf°$	$S°$	$C_p°$
H_3BO_3	c	− 261.55	− 231.60	21.23	19.45
	g	− 237.6			
unionized, std. state, $m = 1$	aq	− 256.29	− 231.56	38.8	
$B(OH)_4^-$ std. state, $m = 1$	aq	− 321.23	− 275.65	24.5	
BP cubic	c	− 19			
BS	b	81.74	69.02	51.65	7.18
B_2S_3	c	− 57.5			
Bromine					
Br	g	26.741	19.701	41.805	4.968
Br⁻	g	− 55.9			
Br⁻ std. state, $m = 1$	aq	− 29.05	− 24.85	19.7	− 33.9
Br_2	liq	0	0	36.384	18.090
	g	7.387	0.751	58.641	8.61
std. state, $m = 1$	aq	− 0.62	0.94	31.2	
	CCl_4	0.71	0.36	37.6	
Br_3^- std. state, $m = 1$	aq	− 31.17	− 25.59	51.5	
BrCl	g	3.50	− 0.23	57.36	8.36
Br_2Cl^- std. state, $m = 1$	aq	− 40.7	− 30.7	45.1	
BrF	g	− 22.43	− 26.09	54.70	7.88
BrF_3	liq	− 71.9	− 57.5	42.6	29.78
	g	− 61.09	− 54.84	69.89	15.92
BrF_5	liq	− 109.6	− 84.1	53.8	
BrO	g	30.06	25.87	56.75	7.67
BrO⁻ std. state, $m = 1$	aq	− 22.5	− 8.0	10	
BrO_2	c	11.6			
BrO_3^- std. state, $m = 1$	aq	− 20.0	0.4	39.0	
HBrO undissoc; std. state, $m = 1$	aq	− 27.0	− 19.7	34	
$HBrO_3$ std. state, $m = 1$	aq	− 20.0	0.4	39.0	
HBr std. state, $m = 1$	aq	− 29.05	− 24.85	19.7	− 33.9
	aq, 1	− 17.38			
	aq, 2	− 22.40			
	aq, 5	− 26.706			
	aq, 10	− 27.953			
	aq, 100	− 28.815			
	g	− 8.70	− 12.77	47.463	6.965
Cadmium					
Cd γ	c	0	0	12.37	6.21
α	c	− 0.14	− 0.14	12.37	
	g	26.77	18.51	40.066	4.968
$CdAs_2$	c	− 4.2			
Cd_3As_2	c	− 10.0			
$Cd_3(AsO_4)_2$	c		− 410.2		
$Cd(BO_2)_2$	c		− 354.87		
$CdBr_2$	c	− 75.57	− 70.82	32.8	18.32
std. state, $m = 1$	aq	− 76.24	− 68.24	21.9	
$CdBr_2 \cdot 4H_2O$	c	− 356.73	− 298.287	75.6	
$CdCl_2$	c	− 93.57	− 82.21	27.55	17.85
std. state, $m = 1$	aq	− 98.04	− 81.286	9.5	
$CdCl_2 \cdot \frac{5}{2}H_2O$	c	− 270.54	− 225.644	54.3	
$CdCl_3^-$ std. state, $m = 1$	a	− 134.1	− 116.4	48.5	
$Cd(ClO_4)_2$ std. state, $m = 1$	aq	− 79.96	− 22.66	69.5	
$Cd(ClO_4)_2 \cdot 6H_2O$	c	− 490.6			
$Cd(CN)_2$	c	38.8			
std. state, $m = 1$	aq	53.9	63.9	27.5	
$Cd(CN)_4^{2-}$ std. state, $m = 1$	aq	102.3	121.3	77	
$Cd(CNS)_2$ thiocyanate	c	12.43			
std. state, $m = 1$	aq	18.40	25.76	51.4	
$CdCO_3$	c	− 179.4	− 160.0	22.1	
CdC_2O_4	c	− 218.1			
std. state, $m = 1$	aq	− 215.3	− 179.6	− 6.6	
$Cd(acetate)_2$ std. state, $m = 1$	aq	− 250.46	− 195.12	23.9	
$Cd(formate)_2$ std. state, $m = 1$	aq	− 221.56	− 186.23	26	

Table 3–9 (continued)
HEATS AND FREE ENERGIES OF FORMATION, ENTROPIES, AND HEAT CAPACITIES OF ELEMENTS AND INORGANIC COMPOUNDS

Formula and Description	State	$\Delta Hf°$	$\Delta Gf°$	$S°$	$C_p°$
Cd fulminate	c	90			
CdF_2	c	−167.4	−154.8	18.5	
std. state, $m = 1$	aq	−172.14	−151.82	−24.1	
CdI_2	c	−48.6	−48.13	38.5	19.11
std. state, $m = 1$	aq	−44.52	−43.20	35.7	
CdI_3 std. state, $m = 1$	aq		−62.0		
CdI_4^2 std. state, $m = 1$	aq	−81.7	−75.5	78	
$Cd(IO_3)_2$	c		−90.13		
std. state, $m = 1$	aq	−123.9	−79.7	39.1	
$Cd(N_3)_2$	c	108			
std. state, $m = 1$	aq	113.38	147.9	34.1	
Cd_3N_2	c	38.7			
$Cd(NH_3)_4^2$ std. state, $m = 1$	aq	−107.6	−54.1	80.4	
$Cd(NO_3)_2$	c	−109.06			
std. state, $m = 1$	aq	−117.26	−71.76	52.5	
$Cd(NO_3)_2 \cdot 4H_2O$	c	−394.11			
Cd_3P_2	c	−27.4			
$Cd_3(PO_4)_2$	c		−587.1		
CdO	c	−61.7	−54.6	13.1	10.38
CdO_2^2 std. state, $m = 1$	aq		−68.0		
$CdOH·$ std. state, $m = 1$	aq		−62.4		
$HCdO_2$ std. state, $m = 1$	aq		−86.9		
$Cd(OH)_2$ precipitated	c	−134.0	−113.2	23	
std. state, $m = 1$	aq	−128.08	−93.73	−22.6	
$Cd(OH)_3$ std. state, $m = 1$	aq		−143.6		
$Cd(OH)_4^2$ std. state, $m = 1$	aq		−181.3		
CdS	c	−38.7	−37.4	15.5	
$CdSO_4$	c	−223.06	−196.65	29.407	23.80
std. state, $m = 1$	aq	−235.46	−196.51	−12.7	
$CdSO_4 \cdot H_2O$	c	−296.26	−255.46	36.814	32.16
$CdSO_4 \cdot {}^8_3H_2O$	c	−413.33	−350.224	54.883	50.97
$CdSb$	c	−3.44	−3.11	22.2	
Cd_3Sb_2	c	−13.9			
$CdSeO_3$	c	−137.5	−119.0	34.0	
std. state, $m = 1$	aq	−139.8	−106.9	−14.4	
$CdSeO_4$	c	−151.3	−127.1	39.3	
std. state, $m = 1$	aq	−161.3	−124.0	−4.6	
$CdSiO_3$	c	−284.20	−264.20	23.3	21.17
$CdTe$	c	−22.1	−22.0	24	
Calcium					
Ca	c	0	0	9.95	6.30
	g	42.2	34.14	36.99	4.97
$CaHAsO_4 \cdot H_2O$	c	−410	−363	35	
$CaBr_2$	c	−161.3	−157.5	(31)	
std. state, $m = 1$	aq	−187.57	−181.33	25.4	
$CaBr_2 \cdot 6H_2O$	c	−597.2			
$CaO \cdot B_2O_3$	c	−483.3	−457.7	25.1	
$CaO \cdot 2B_2O_3$	c	−798.8	−752.4	32.2	
CaC_2	c	−15.0	−16.2	16.8	
$CaCO_3$ calcite	c	−288.45	−269.78	22.2	
aragonite	c	−288.49	−269.53	21.2	
$Ca(HCO_3)_2$ std. state, $m = 1$	aq	−460.13	−412.80	32.2	
$Ca(formate)_2$	c	−323.5			
CaC_2O_4	c	−332.2			
$CaC_2O_4 \cdot 2H_2O$	c	−469.1	−416.9	47·	
$Ca(acetate)_2$	c	−355.0			
$Ca(CN)_2$	c	−44.2			
	aq		−54		
$CaCN_2$ cyanamide	c	−84.0			

Table 3–9 (continued)
HEATS AND FREE ENERGIES OF FORMATION, ENTROPIES, AND HEAT CAPACITIES OF ELEMENTS AND INORGANIC COMPOUNDS

Formula and Description	State	$\Delta Hf°$	$\Delta Gf°$	$S°$	$C_p°$
$CaCl_2$	c	−190.4	−179.5	27.2	
std. state, $m = 1$	aq	−209.82	−194.88	13.1	
	aq, 25	−208.51			
$CaCl_2 \cdot H_2O$	c	−265.1			
$CaCl_2 \cdot 2H_2O$	c	−335.5			
$CaCl_2 \cdot 4H_2O$	c	−480.2			
$CaCl_2 \cdot 6H_2O$	c	−623.15			
$CaOCl_2$	c	−178.6			
	aq	−189.1			
$Ca(OCl)_2$	aq	−180.0			
$CaCrO_4$	c	−329.6	−305.3	32	
	aq	−336.0			
CaF_2	c	−290.3	−277.7	16.46	
std. state, $m = 1$	aq	−287.09	−264.34	−17.8	
CaH	g	58.7			
CaH_2	c	−45.1	−35.8	10	
CaI_2	c	−127.5	(−126.4)	(34)	
std. state, $m = 1$	aq	−156.51	−156.88	39.1	
$CaI_2 \cdot 8H_2O$	c	−700.7			
Ca_3N_2	c	−108.2	−93.2	25.4	
$Ca(NO_2)_2$	c	−178.3			
	aq	−180.5			
$Ca(NO_3)_2$	c	−224.0	−177.34	46.2	
std. state, $m = 1$	aq	−228.51	−185.00	56.8	
	aq, 10	−229.48			
$Ca(NO_3)_2 \cdot 2H_2O$	c	−368.00	−293.51	64.3	
$Ca(NO_3)_2 \cdot 4H_2O$	c	−509.37	−406.5	81	
CaO	c	−151.79	−144.4	9.5	
CaO_2	c	(−156.5)	(−143.5)	(15.4)	
$Ca(OH)_2$	c	−235.80	−214.33	18.2	
std. state, $m = 1$	aq	−239.68	−207.37	−13.2	
Ca_3P_2	c	−120.5			
$Ca_3(PO_4)_2$ α	c	−986.2	−929.7	57.6	
β	c	−988.2	−932.0	56.4	
$CaHPO_4$	c	−435.2	−401.5	21	
$CaHPO_4 \cdot 2H_2O$	c	−576.0	−514.6	40	
$Ca(H_2PO_4)_2$	ppt	−744.4			
CaS	c	−115.3	−114.1	13.5	
$CaSO_3 \cdot 2H_2O$	c	−421.2	−374.1	44	
$CaSO_4$ anhydrite	c	−342.42	−315.56	25.5	
α	c	−340.27	−313.52	25.9	
β	c	−339.21	−312.46	25.9	
$CaSO_4 \cdot \frac{1}{2}H_2O$ α	c	−376.47	−343.02	31.2	
β	c	−375.97	−342.78	32.1	
$CaSO_4 \cdot 2H_2O$	c	−483.06	−429.19	46.36	
CaS_2O_3	aq, 1000	−283.4			
$CaS_2O_3 \cdot 6H_2O$	c		−602.2		
$CaSe$	c	−50.8		16	
$CaSi_2$	c	−36			
Ca_2Si	c	−50			
$CaSiO_3$ α	c	−377.4	−357.4	20.9	
β, wollastonite	c	−378.6	−358.2	19.6	
Ca_2SiO_4 β	c	−538.0			
γ	c	−539.0			
Ca_3SiO_5	c	−688.4			
$CaWO_4$	c	−392.5			
Carbon					
C graphite	c	0	0	1.372	2.038

Table 3–9 (continued)
HEATS AND FREE ENERGIES OF FORMATION, ENTROPIES, AND HEAT CAPACITIES OF ELEMENTS AND INORGANIC COMPOUNDS

Formula and Description	State	$\Delta Hf°$	$\Delta Gf°$	$S°$	$C_p°$
diamond	c	0.4533	0.6930	0.568	1.4617
	g	171.291	160.442	37.7597	4.9805
CBr_4　monoclinic	c	4.5	11.4	50.8	34.5
	g	19	16	85.55	21.79
$CHBr_3$	liq	−6.8	−1.2	52.8	31
	g	4	2	79.07	17.02
CH_3Br	g	−8.4	−6.2	58.86	10.14
CCl_4	liq	−32.37	−15.60	51.72	31.49
	g	−24.6	−14.49	74.03	19.91
$CHCl_3$	liq	−32.14	−17.62	48.2	27.2
	g	−24.65	−16.82	70.65	15.70
CH_2Cl_2	liq	−29.03	−16.09	42.5	23.9
	g	−22.10	−15.75	64.56	12.18
CH_3Cl	g	−19.32	−13.72	56.04	9.74
CCl_3Br	g	−11.0	−5.1	79.55	20.38
CF_3	g	−114			
CF_4	g	−221	−210	62.50	14.60
CF_3Br	g	−153.6	−147.3	71.14	16.57
CF_3Cl	g	−166	−156	68.16	15.98
CF_2Cl_2	g	−114	−105	71.86	17.27
$CFCl_3$	liq	−72.02	−56.61	53.86	29.05
CI_4	g			93.65	22.91
CHI_3	c	33.7			
	g			85.1	17.92
CH_2I_2	liq	16.0	21.6	41.6	32
	g	27.0	22.9	74.0	13.79
CH_3I	liq	−3.7	3.2	39.0	30
	g	3.1	3.5	60.71	10.54
CN	g	109	102	48.4	6.97
CN^-　std. state, $m = 1$	aq	36.0	41.2	22.5	
$CN \cdot N_3$　cyanogen azide	c	92.6			
HCN	liq	26.02	29.86	26.97	16.88
	g	32.3	29.8	48.20	8.57
std. state, $m = 1$	aq	36.0	41.2	22.5	
nonionized, std. state, $m = 1$	aq	25.6	28.6	29.8	
CNBr	c	33.58			
	g	44.5	39.5	59.32	11.22
CNCl	g	32.97	31.32	56.42	10.75
CNI	c	39.71	44.22	23.0	
	g	53.9	47.0	61.35	11.54
CO	g	−26.416	−32.780	47.219	6.959
CO_2	g	−94.051	−94.254	51.06	8.87
undissoc; std. state, $m = 1$	aq	−98.90	−92.26	28.1	
CO_3^{2-}　std. state, $m = 1$	aq	−161.84	−126.17	−13.6	
C_3O_2	liq	−28.03	−25.10	43.28	25.8
	g	−22.28	−26.08	65.852	16.029
HCO	g	−4.12	−7.76	53.68	8.26
$HCOO^-$　std. state, $m = 1$	aq	−101.71	−83.9	22	−21.0
HCO_3^-　std. state, $m = 1$	aq	−165.39	−140.26	21.8	
$HCHO^+$	g	224.2			
H_2CO_3　std. state, $m = 1$	aq	−167.22	−148.94	44.8	
$COBr_2$	g	−23.0	−26.5	73.85	14.78
$COCl_2$	g	−52.3	−48.9	67.74	13.78
COF_2	g	−151.7	−148.0	61.78	11.19
HCNO　cyanic acid, ionized	aq	−34.90	−23.3	25.5	
nonionized	aq	−36.90	−28.0	34.6	
HNCO　isocyanic acid	g			56.85	10.72
CNO^-　std. state, $m = 1$	aq	−34.9	−23.3	25.5	
CS	g	56	44	50.30	7.12

Table 3–9 (continued)
HEATS AND FREE ENERGIES OF FORMATION, ENTROPIES, AND HEAT CAPACITIES OF ELEMENTS AND INORGANIC COMPOUNDS

Formula and Description	State	$\Delta Hf°$	$\Delta Gf°$	$S°$	$C_p°$
CS_2	liq	21.44	15.60	36.17	18.1
	g	28.05	16.05	56.82	19.85
COS	g	−33.96	−40.47	55.32	9.92
$CSCl_2$	g			71.51	15.18
HCNS undissoc; std. state, $m = 1$	aq		23.31		
CNS⁻	aq	18.27	22.15	34.5	−9.6
HNCS isothiocyanic acid	g	30.5	27.0	59.2	11.2
NOSCN std. state, $m = 1$	aq	55.2	63.5	51.2	
$SC(SH)_2$ trithiocarbonic acid	liq	6.0	7.0	52	35.8
Cerium					
Ce	c	0	0	13.64	
$CeBr_3$	c	−192	(−185)	(45)	
$CeCl_3$	c	−260.3			
std. state, $m = 1$	aq	−293.9	−264.5	−5	
CeF_3	c	−391.0	(−372.1)	(24)	
CeF_4	c	−442.0	(−420)	(37)	
Ce_3H_8	c	−170			
CeI_3	c	−163.0	(−161)	(50)	
std. state, $m = 1$	aq	−213.9	−207.5	34	
CeN	c	−78.3	−70.8		
CeO_2	c	(−260)	−245.9	14.88	
Ce_2O_3	c	(435)	(−411.5)	(21.8)	
$CeO_3 \cdot 2H_2O$	c	−389			
CeS_2	c	−153.9			
Ce_2S_3	c	−298.7			
$Ce(SO_4)_2$	c	−560			
$Ce_2(SO_4)_3$ std. state, $m = 1$	aq	−998.3	−873.0	−76	
$Ce_2(SO_4)_3 \cdot 5H_2O$	c	−1308			
$Ce_2(SO_4)_3 \cdot 8H_2O$	c		−1340.2		
Cesium					
Cs	c	0	0	19.8	
	g	18.83	12.24	41.94	
$CsAl(SO_4)_2 \cdot 12H_2O$	c	−1449.5	−1218.5	164	
CsBr	c	−94.3	−91.6	29	
std. state, $m = 1$	aq	−88.1	−91.98	51.1	
$CsBrO_3$	c			38.8	
CsCN	c	−27			
CsCNO cyanate	c	−96			
CsCNS thiocyanate	c	−50			
Cs_2CO_3	c	−267.4			
	aq, 200	−280.0			
$CsHCO_3$	c	−228			
	aq	−224.4	−210.6		
CsCl	c	−103.5			
std. state, $m = 1$	aq	−99.2	−88.76	45.0	
$CsClO_4$	c	−103.86	−73.28	41.89	
std. state, $m = 1$	aq	−90.6	−69.98	75.3	
CsF	c	−126.9			
std. state, $m = 1$	aq	−135.9	−133.49	29.5	
$CsHF_2$	c	−216.1			
	aq	−212.8			
CsH	g	29.0	24.3	51.25	
CsI	c	−80.5	−79.7	31	
std. state, $m = 1$	aq	−72.6	−79.76	57.9	
$CsNH_2$	c	−25.4			
$CsNO_2$	c	−85			
$CsNO_3$	c	−118.11			
std. state, $m = 1$	aq	−108.6	−93.82	86.8	
Cs_2O	c	−75.9			

Table 3–9 (continued)
HEATS AND FREE ENERGIES OF FORMATION, ENTROPIES, AND HEAT CAPACITIES OF ELEMENTS AND INORGANIC COMPOUNDS

Formula and Description	State	$\Delta Hf°$	$\Delta Gf°$	$S°$	$C_p°$
Cs_2O_2	c	−96.2			
CsOH	c	−97.2			
std. state, $m = 1$	aq	−114.2	−105.00	29.3	
	aq, 200	−114.0			
$CsReO_4$	c	−257.2			
	aq	−249.5			
Cs_2S	c	−81.1			
CsHS	c	−62.9			
Cs_2SO_4	c	−349.8			
std. state, $m = 1$	aq	−335.3	−312.16	67.7	
$CsHSO_4$	c	−274.0			
CsHSe	c	−36.7			
Cs_2SiF_6	c	−669.5			
Chlorine					
Cl	g	29.082	25.262	39.457	5.220
Cl^-	g	−58.8			
Cl^- std. state, $m = 1$	aq	−39.952	−31.372	13.5	−32.6
Cl_2	g	0	0	53.288	8.104
ClF	g	−13.02	−13.37	52.05	7.66
ClF_3	liq	−45.3			
	g	−39.0	−29.4	67.28	15.26
Cl_2F_6	g	−81.1	−56.7	117	
HCl	g	−22.062	−22.777	44.646	6.96
std. state, $m = 1$	aq	−39.952	−31.372	13.5	−32.6
ClO	g	24.34	23.45	54.14	7.52
ClO^- std. state, $m = 1$	aq	−25.6	−8.8	10	
ClO_2	g	24.5	28.8	61.36	10.03
ClO_2^- std. state, $m = 1$	aq	−15.9	4.1	24.2	
ClO_3	g	37			
ClO_3^- std. state, $m = 1$	aq	−23.7	−0.8	38.8	
ClO_4^- std. state, $m = 1$	aq	−30.91	−2.06	43.5	
Cl_2O	g	19.2	23.4	63.60	10.85
Cl_2O_7	liq	56.9			
ClO_3F	g	−5.7	11.5	66.65	15.52
HClO	g			56.54	8.88
undissoc; std. state, $m = 1$	aq	−28.9	−19.1	34	
$HClO_2$ undissoc; std. state, $m = 1$	aq	−12.4	1.4	45.0	
$HClO_3$ std. state, $m = 1$	aq	−23.7	−0.8	38.8	
$HClO_4$	liq	−9.70			
std. state, $m = 1$	aq	−30.91	−2.06	43.5	
$HClO_4 \cdot H_2O$	c	−91.35			
$HClO_4 \cdot 2H_2O$	liq	−162.04			
$ClF_3 \cdot HF$	g	−107.7	−91.8	86	
Chromium					
Cr	c	0	0	5.68	5.58
	g	94.8	84.1	41.68	4.97
Cr^{2+} std. state, $m = 1$	aq	−34.3			
$CrBr_2$	c	−72.2			
	g	−17			
$[Cr(H_2O)_6]Br_3$ purple	c	−550.4			
Cr_3C_2	c	−19.3	−19.5	20.42	23.53
Cr_7C_3	c	−38.7	−39.9	48.0	49.92
$Cr_{23}C_6$	c	−87.2	−89.3	145.8	149.2
$CrCl_2$	c	−94.5	−85.1	27.56	17.01
	g	−30.7			
	aq	−114.2			
$CrCl_2 \cdot 2H_2O$	c	−237.1			
$CrCl_2 \cdot 3H_2O$	c	−308.9			
$CrCl_2 \cdot 4H_2O$	c	−384.4			

Table 3–9 (continued)
HEATS AND FREE ENERGIES OF FORMATION, ENTROPIES, AND HEAT CAPACITIES OF ELEMENTS AND INORGANIC COMPOUNDS

Formula and Description	State	$\Delta Hf°$	$\Delta Gf°$	$S°$	$C_p°$
$CrCl_3$	c	−133.0	−116.2	29.4	21.94
$CrCl_4$	g	−102	(−91.6)		
CrO_2Cl_2	liq	−138.5	−122.1	53.0	
	g	−128.6	−119.9	78.8	20.2
$[Cr(H_2O)_6]Cl_3$ violet	c	−586.2			
CrF_2	c	−186	(−172)		
	g	−99			
CrF_3	c	−277	−260	22.44	18.82
CrF_4	c	−298			
Cr_7H_2	c	−3.8			
CrI_2	c	−37.5			
	g	24			
	aq	−60.1			
CrI_3	c	−49.0			
CrN	c	−29.8			11.0
Cr_2N	c	−30.5			17.3
$Cr(NO_3)_3 \cdot 9H_2O$	c				109.2
CrO	g	53⁰ ᴷ			
CrO_2	c	−143			
	g	−14⁰ ᴷ			
CrO_3	c	−140.9			
	g	−92.2			
Cr_2O_3	c	−272.4	−252.9	19.4	28.38
$Cr_2O_3 \cdot H_2O$	c	−360			
$Cr_2O_3 \cdot 2H_2O$	c	−441			
$Cr_2O_3 \cdot 3H_2O$	c	−519			
Cr_3O_4	c	−366			
CrO_4^{2-} std. state, $m = 1$	aq	−210.60	−173.96	12.00	
$Cr_2O_7^{2-}$ std. state, $m = 1$	aq	−356.2	−311.0	62.6	
$HCrO_4^-$ std. state, $m = 1$	aq	−209.9	−182.8	44.0	
$CrO_2(OH)_2$	g	−174			
$Cr(OH)_3$	c	−254.3			
$[Cr(H_2O)_6]^{3+}$	aq	−477.8			
$Cr_2(SO_4)_3$	c				67.5
$Cr_2(SO_4)_3 \cdot 18H_2O$	c				223
$CrSb$	c				12.7
$CrSb_2$	c				19.7
$CrSi$	c	−15			9.2
$CrSi_2$	c	−23			12.7
Cr_3Si	c	−30			19.3
Cr_5Si_3	c	−64			34.9
Cobalt					
Co hexagonal	c	0	0	7.18	5.93
face-centered cubic	c	0.11	0.06	7.34	
Co^{2+} std. state, $m = 1$	aq	−13.9	−13.0	−27	
Co^{3+} std. state, $m = 1$	aq	22	32	−73	
$Co_3(AsO_4)_2$	c		−387.4		
$Co(BO_2)_2$	c		−325.8		
$CoBr_2$	c	−52.8			
std. state, $m = 1$	aq	−72.0	−62.7	12	
$CoBr_2 \cdot 6H_2O$	c	−482.8			
$CoCO_3$	c	−170.4			
$CoCl_2$	c	−74.7	−64.5	26.09	18.76
std. state, $m = 1$	aq	−93.8	−75.7	0	
$CoCl_2 \cdot H_2O$	c	−147			
$CoCl_2 \cdot 2H_2O$	c	−220.6	−182.8	45	
$CoCl_2 \cdot 6H_2O$	c	−505.6	−412.4	82	
$Co(ClO_4)_2$ std. state, $m = 1$	aq	−75.7	−17.1	60	
$Co(ClO_4)_2 \cdot 6H_2O$	c	−487.2			

Table 3–9 (continued)
HEATS AND FREE ENERGIES OF FORMATION, ENTROPIES, AND
HEAT CAPACITIES OF ELEMENTS AND INORGANIC COMPOUNDS

Formula and Description	State	$\Delta Hf°$	$\Delta Gf°$	$S°$	$C_p°$
Co(CNO)$_2$ cyanate	c	−51.8			
Co(CNS)$_2$ thiocyanate	c	24.2			
Co(formate)$_2$	c	−208.7			
CoC$_2$O$_4$	c	−203.5			
std. state, $m = 1$	aq	−211.1	−174.1	−16	
CoF$_2$	c	−165.4	−154.7	19.59	16.44
CoF$_3$	c	−193.8			
CoI$_2$	c	−21.2			
std. state, $m = 1$	aq	−40.3	−37.7	26	
Co(IO$_3$)$_2$ std. state, $m = 1$	aq	−119.7	−74.2	30	
Co(IO$_3$)$_2$ · 2H$_2$O	c	−258.6	−190.2	64	
Co(NH$_3$)$_6^{3+}$ std. state, $m = 1$	aq	−139.8	−38.9	40	
Co(NH$_3$)$_6^{2+}$ std. state, $m = 1$	aq		−45.3		
[Co(NH$_3$)$_6$]Br$_2$	c	−216.4			
[Co(NH$_3$)$_6$]Br$_3$	c	−239.7	−119.8	77.7	78.1
std. state, $m = 1$	aq	−227.0	−113.4	99	
[Co(NH$_3$)$_6$]Cl$_2$	c	−238.0			
[Co(NH$_3$)$_6$]Cl$_3$	c	−268.7			76.6
std. state, $m = 1$	aq	−259.7	−133.0	80	
[Co(NH$_3$)$_5$Cl]Cl$_2$	c	−243.1	−139.3	87.5	57.2
[Co(NH$_3$)$_6$](ClO$_4$)$_3$	c	−247.3	−54.3	152	
std. state, $m = 1$	aq	−232.5	−45.1	170	
[Co(NH$_3$)$_6$]I$_2$	c	−189.9			69.1
[Co(NH$_3$)$_6$]I$_3$	c	−104.8			74.3
[Co(NH$_3$)$_5$NO$_2$]$^{2+}$ std. state, $m = 1$	aq	−146.6	−41.3	43	
[Co(NH$_3$)$_5$NO$_2$](NO$_3$)$_2$	c	−260.2	−100.0	83	
	aq	−245.7	−94.5	113	
[Co(NH$_3$)$_6$](NO$_3$)$_3$ std. state, $m = 1$	aq	−288.5	−118.7	145	
	c	−306.4	−126.8	112	
Co(NO$_3$)$_2$	c	−100.5			
std. state, $m = 1$	aq	−113.0	−66.2	43	
Co(NO$_3$)$_2$ · 2H$_2$O	c	−244.2			
Co(NO$_3$)$_2$ · 6H$_2$O	c	−528.49			
CoO	c	−56.87	−51.20	12.66	13.20
Co$_3$O$_4$	c	−213	−185	24.5	29.5
Co(OH)$_2$ blue	c		−107.6		
pink	c	−129.0	−108.6	19	
std. state, $m = 1$	aq	−123.8	−88.2	−32	
Co(OH)$_3$	c	−171.3			
Co$_2$P	c	−45			
Co$_3$(PO$_4$)$_2$	c		−573.3		
CoHPO$_4$	c		−282.5		
CoS	c	−19.8			
Co$_2$S$_3$ pptd	c	−35.2			
CoSO$_4$	c	−212.3	−187.0	28.2	
std. state, $m = 1$	aq	−231.2	−191.0	−22	
CoSO$_4$ · 6H$_2$O	c	−641.4	−534.35	87.86	84.46
CoSO$_4$ · 7H$_2$O	c	−712.22	−591.26	97.05	93.33
CoSe	c	−14.6			
CoSeO$_3$ · 2H$_2$O	c	−266.5			
CoSi	c	−24.0	−23.6	10.3	10.6
Co$_2$SiO$_4$	c	−353			
CoTe$_2$	c	−31			
Co$_3$Te$_4$	c	−77			
Copper					
Cu	c	0	0	7.923	5.840
	g	80.86	71.37	39.74	4.968
Cu$^+$ std. state, $m = 1$	aq	17.13	11.95	9.7	
Cu^{2+} std. state, $m = 1$	aq	15.48	15.66	−23.8	
Cu$_2$	g	115.72	103.24	57.71	8.75

Table 3–9 (continued)
HEATS AND FREE ENERGIES OF FORMATION, ENTROPIES, AND HEAT CAPACITIES OF ELEMENTS AND INORGANIC COMPOUNDS

Formula and Description	State	$\Delta Hf°$	$\Delta Gf°$	$S°$	$C_p°$
Cu_3As	c	−2.8			
$Cu_3(AsO_4)_2$	c		−310.9		
std. state, $m = 1$	aq	−378.10	−263.02	−149.2	
CuBr	c	−25.0	−24.1	22.97	13.08
$CuBr_2$	c	−33.9			
$CuBr_2 \cdot 4H_2O$	c	−317.0			
$CuCO_3 \cdot Cu(OH)_2$ malachite	c	−251.3	−213.6	44.5	
$2CuCO_3 \cdot Cu(OH)_2$ azurite	c	−390.1			
CuCl	c	−32.8	−28.65	20.6	11.6
$CuCl_2^-$ std. state, $m = 1$	aq		−57.4		
$CuCl_3^{2-}$ std. state, $m = 1$	aq		−90		
$2CuCl \cdot CO \cdot 2H_2O$	c	−225			
$2CuCl \cdot C_2H_2$	c	−23.3	−7.63	50.7	
$3CuCl \cdot C_2H_2$	c	−56.4	−36.52	71.0	
$CuCl_2$	c	−52.6	−42.0	25.83	13.82
undissoc, std. state, $m = 1$	aq		−47.3		
$CuCl_2 \cdot 2H_2O$	c	−196.3	−156.8	40	
$Cu(ClO_4)_2$ std. state, $m = 1$	aq	−46.34	11.54	63.2	
$Cu(ClO_4)_2 \cdot 6H_2O$	c	−460.9			
CuCN	c	23.0	26.6	20.2	
$Cu(CN)_2^-$ std. state, $m = 1$	aq		61.6		
$Cu(CN)_3^{2-}$ std. state, $m = 1$	aq		96.5		
$Cu(CN)_4^{3-}$ std. state, $m = 1$	aq		135.4		
CuCNS thiocyanate	c		16.7		
std. state, $m = 1$	aq	35.40	34.10	44.2	
$Cu(CNS)_2$ std. state, $m = 1$	aq	52.02	59.96	45.2	
$Cu(acetate)_2$	c	−213.5			
std. state, $m = 1$	aq	−216.84	−160.92	17.6	
$Cu(formate)_2$	c	−186.7			
std. state, $m = 1$	aq	−187.94	−152.1	20	
Cu(I) fulminate	c	26.3			
CuC_2O_4	c		−158.2		
std. state, $m = 1$	aq	−181.7	−145.5	−12.9	
$Cu(C_2O_4)_2^{2-}$ std. state, $m = 1$	aq	−380.5	−319.3	35	
CuF^+ std. state, $m = 1$	aq	−62.4	−52.7	−16	
CuF_2	c	−129.7			
$CuF_2 \cdot 2H_2O$	c		−234.6		
$CuFeO_2$	c	−127.3	−114.7	21.2	19.13
$CuFe_2O_4$	c	−230.69	−205.26	33.7	35.52
CuH	c	5.1			
	g	70			
CuI	c	−16.2	−16.6	23.1	12.92
$Cu(IO_3)_2$ std. state, $m = 1$	aq	−90.3	−45.5	32.8	
$Cu(IO_3)_2 \cdot H_2O$	c	−165.4	−112.0	59.1	
$CuMoO_4$	c	−225			
CuN_3	c	66.7	82.4	24	
$Cu(N_3)_2$	c	143.0			
Cu_3N	c	17.8			22
$Cu(NH_3)^{2+}$ std. state, $m = 1$	aq	−9.3	3.72	2.9	
$Cu(NH_3)_2^{2+}$ std. state, $m = 1$	aq	−34.0	−7.28	26.6	
$Cu(NH_3)_3^{2+}$ std. state, $m = 1$	aq	−58.7	−17.48	47.7	
$Cu(NH_3)_4^{2+}$ std. state, $m = 1$	aq	−83.3	−26.60	65.4	
$Cu(NO_3)_2$	c	−72.4			
std. state, $m = 1$	aq	−83.64	−37.56	46.2	
$Cu(NO_3)_2 \cdot 6H_2O$	c	−504.5			
CuO	c	−37.6	−31.0	10.19	10.11
Cu_2O	c	−40.3	−34.9	22.26	15.21
$Cu(OH)_2$	c	−107.5			
std. state, $m = 1$	aq	−94.46	−59.53	−28.9	
CuP_2	c	−29			

Table 3–9 (continued)

Table 3–9 (continued)
HEATS AND FREE ENERGIES OF FORMATION, ENTROPIES, AND
HEAT CAPACITIES OF ELEMENTS AND INORGANIC COMPOUNDS

Formula and Description	State	$\Delta Hf°$	$\Delta Gf°$	$S°$	$C_p°$
Cu_3P	c	−36.2			
$Cu_2P_2O_7$	c		−448.0		
std. state, $m = 1$	aq	−511.8	−427.4	−76	
$Cu_3(PO_4)_2$	c		−490.3		
CuS	c	−12.7	−12.8	15.9	11.43
Cu_2S α	c	−19.0	−20.6	28.9	18.24
$CuSO_4$	c	−184.36	−158.2	26	23.9
std. state, $m = 1$	aq	−201.84	−162.31	−19.0	
	aq, 100	−200.374			
$CuSO_4 \cdot H_2O$	c	−259.52	−219.46	34.9	32
$CuSO_4 \cdot 3H_2O$	c	−402.56	−334.65	52.9	49
$CuSO_4 \cdot 5H_2O$	c	−544.85	−449.344	71.8	67
Cu_2SO_3 std. state, $m = 1$	aq	−117.6	−92.4	12	
Cu_2SO_4	c	−179.6			
CuSe	c	−9.45			
$CuSe_2$	c	−10.3			
Cu_2Se	c	−14.2			
$CuSeO_3$	c		−83.2		
std. state, $m = 1$	aq	−106.2	−72.7	−20.7	
$CuSeO_4$	c	−114.36			
$CuSeO_4 \cdot 5H_2O$	c	−466.96			
$CuWO_4$	c	−250.0			
Dysprosium					
Dy	c	0	0		
$DyBr_3$	c	−209			
$DyCl_3$ β	c	−237.8			
γ	c	−234.8			
std. state, $m = 1$	aq	−286.1	−255.2		
DyI_3	c	−144.5			
std. state, $m = 1$	aq	−206.1	−198.2		
$Dy(OH)_3$	c		−305.8		
$Dy_2(SO_4)_3$ std. state, $m = 1$	aq	−982.7	−854.4		
$Dy_2(SO_4)_3 \cdot 8H_2O$	c	−1322.0			
Erbium					
Er	c	0	0		
$ErBr_3$	c	−205			
$Er(acetate)_2$	aq	−512.8			
$Er(acetate)_2 \cdot 4H_2O$	c	−785.6			
$ErCl_3$	c	−213.8			
std. state, $m = 1$	aq	−282.4	−251.5		
ErI_3	c	−140.0			
std. state, $m = 1$	aq	−202.4	−194.5		
$Er(OH)_3$	c	−326.8			
$Er_2(SO_4)_3$ std. state, $m = 1$	aq	−975.3	−847.0		
$Er_2(SO_4)_3 \cdot 8H_2O$	c		−1313.9		
Europium					
Eu	c	0	0		
	g	87		45.10	
$EuBr_3$	c	−202			
$EuCl_3$	c	−247.1			
std. state, $m = 1$	aq	−289.4	−259.2		
$Eu_2(SO_4)_3$ std. state, $m = 1$	aq	−989.3	−862.2		
$Eu_2(SO_4)_3 \cdot 8H_2O$	c		−1331.0		
Fluorine					
F	g	18.88	14.80	37.917	5.436
F^-	g	−64.7			
F_2	g	0	0	48.44	7.48
HF	liq	−71.65		18.02	12.35
	g	−64.8	−65.3	41.508	6.963

Table 3–9 (continued)
HEATS AND FREE ENERGIES OF FORMATION, ENTROPIES, AND HEAT CAPACITIES OF ELEMENTS AND INORGANIC COMPOUNDS

Formula and Description	State	$\Delta H f^{\circ}$	$\Delta G f^{\circ}$	S°	C_p°
HF undissoc; std. state, $m = 1$	aq	−76.50	−70.95	21.2	
ionized; std. state, $m = 1$	aq	−79.50			
	aq, 2	−75.79			
	aq, 10	−76.235			
	aq, 100	−76.340			
HF_2^- std. state, $m = 1$	aq	−155.34	−138.18	22.1	
FO	g	41			
Gadolinium					
Gd	c	0	0	14	
	g	87	77	46.41	
$GdBr_3$	c	−214			
$GdCl_3$	c	−245.5			
std. state, $m = 1$	aq	−288.9	−256.6	−7.6	
GdI_3	c	−147.6			
std. state, $m = 1$	aq	−208.9	−201.6	−31.3	
$Gd(OH)_3$	c		−308.1		
$Gd_2(SO_4)_3$ std. state, $m = 1$	aq	−988.3	−861.2	−81.9	
$Gd_2(SO_4)_3 \cdot 8H_2O$	c	−1518.9	−1329.8	155.8	
Gallium					
Ga	c	0	0	9.77	6.18
	liq	1.33			
	g	66.2	57.1	40.38	6.06
Ga_2	g	104.8			
GaAs	c	−17	−16.2	15.34	11.05
Ga_2C_2	g	134			
$Ga(CH_3)_3$	liq	−18.7			
	g	−10.8			
GaBr	g	−11.9	−21.5	60.2	8.70
$GaBr_3$	c	−92.4	−86.0	43	
	g	−70			
$GaBr_4$ std. state, $m = 1$	aq	−158.2	−131.5	8.6	
GaCl	g	−19.1	−25.4	57.4	8.50
$GaCl_3$	c	−125.4	−108.7	34	
	g	−107.0			
GaF	g	−60.2			7.95
GaF_3	c	−278	−259.4	20	
GaH	g	52.7	46.3	46.69	7.00
GaI	g	6.9			8.76
GaI_3	c	−57.1			
GaN	c	−26.4			
	g	42	36	54	
GaO	g	66.8	60.6	55.2	7.66
Ga_2O_3 rhombic	c	−260.3	−238.6	20.31	22.00
GaOH	g	−27.4			
$Ga(OH)^{2+}$ std. state, $m = 1$	aq		−90.9		
$Ga(OH)_2^+$ std. state, $m = 1$	aq		−142.8		
GaO_3^{3-} std. state, $m = 1$	aq		−148		
$H_2GaO_3^-$ std. state, $m = 1$	aq		−178		
$Ga(OH)_3$	c	−230.5	−198.7	24	
GaP	c	−21			
$GaPO_4$	c		−310.1		
$Ga_2(SO_4)_3$	c				62.4
GaSb	c	−10.0	−9.3	18.18	11.60
Ga_2Te_3	c				41.2
Germanium					
Ge	c	0	0	7.43	5.580
	g	90.0	80.3	40.103	7.345
Ge_2	g	113.08	99 5	60.4	8.5
GeBr	g	56.32			8.87

Table 3–9 (continued)
HEATS AND FREE ENERGIES OF FORMATION, ENTROPIES, AND
HEAT CAPACITIES OF ELEMENTS AND INORGANIC COMPOUNDS

Formula and Description	State	$\Delta Hf°$	$\Delta Gf°$	$S°$	$C_p°$
GeBr$_2$	g	−15.0	−25.5	79.1	
GeBr$_4$	liq	−83.1	−79.2	67.1	
	g	−71.7	−76.0	94.66	24.34
GeH$_3$Br	g			65.66	13.47
GeC	g	151			
Ge(C$_2$H$_5$)$_4$	liq	−49.5			
GeCl	g	37	30	58.8	8.81
GeCl$_4$	liq	−127.1	−110.6	58.7	
	g	−118.5	−109.3	83.08	22.97
GeH$_3$Cl	g			63.00	13.08
GeHCl$_3$	liq			53.6	
GeF	g	−7.97			8.30
GeF$_4$	g	−284.4		72.36	19.56
GeH$_4$	g	21.7	27.1	51.87	10.76
Ge$_2$H$_6$	liq	32.82			
	g	38.8			
Ge$_3$H$_8$	liq	46.3			
	g	54.2			
GeI$_2$	c	−21	−20	32	
	g	11.2	−1.0	76	
GeI$_4$	c	−33.9	−34.5	64.8	
	g	−13.6	−25.4	102.49	24.89
GeH$_3$I	g			67.65	13.75
Ge$_3$N$_4$	c	−15.1			
GeO brown	c	−50.7	−56.7	12	
yellow	c		−49.5		
	g	−11.04	−17.49	53.58	7.39
GeO$_2$ hexagonal	c	−131.7	−118.8	13.21	12.45
tetragonal	c	−138.7			
Ge$_2$O$_2$	g	−112			
Ge$_2$O$_3$	g	−212			
H$_2$GeO$_3$	aq	−195.73			
GeP	c	−5	−4	15	
GeS	c	−16.5	−17.1	17	
	g	22	10	56	8.05
GeS$_2$	c	−45.3			
GeSe	c	−22.0			
	g	22.84			8.42
GeSi	g	127			
GeTe	c	−6			
	g	42			8.59
GeTe$_2$	g	44			
Gold					
Au	c	0	0	11.33	6.075
	g	87.5	78.0	43.115	4.968
Au$_2$	g	123.1			8.808
AuBr	c	−3.34			
AuBr$_3$	c	−12.73			
AuBr$_4^-$ std. state, $m = 1$	aq	−45.8	−40.0	80.3	
Au(CN)$_2^-$ std. state, $m = 1$	aq	57.9	68.3	41	
AuCl	c	−8.6			
AuCl$_3$	c	−28.1			
AuCl$_4^-$ std. state, $m = 1$	aq	−77.0	−56.72	63.8	
HAuCl$_4$ std. state, $m = 1$	aq	−77.0	−56.22	63.8	
AuF$_3$	c	−86.9			
AuH	g	70.5	63.5	50.441	6.968
AuI	c	0			
AuO$_2^3$ std. state, $m = 1$	aq		−12.4		
Au(OH)$_3$ precipitated	c	−101.5	−75.77	45.3	
Au$_2$P$_3$	g	−23.8			

Table 3–9 (continued)
HEATS AND FREE ENERGIES OF FORMATION, ENTROPIES, AND
HEAT CAPACITIES OF ELEMENTS AND INORGANIC COMPOUNDS

Formula and Description	State	$\Delta Hf°$	$\Delta Gf°$	$S°$	$C_p°$
Hafnium					
Hf hexagonal	c	0	0	10.41	6.15
	g	148.0	137.8	44.642	4.972
HfB	c	−47			
HfB$_2$	c	−80.3	−79.4	10.2	11.89
HfC	c	−60.1			
HfCl$_4$	c	−236.70	−215.42	45.6	28.80
	g	−211.4			
	aq	−302.7			
HfF$_4$ monoclinic	c	−461.4	−437.5	27	
	g	−399.1			
HfN	c	−88.3			
HfO	g	12			
HfO$_2$	c	−273.6	−245.5	14.18	14.40
hydrous ppt		−269.0			
HfOOH$^+$	aq	−279.5			
Helium					
He	g	0	0	30.124	4.9679
Holmium					
Ho	c	0	0		
HoBr$_3$	c	−225			
HoCl$_3$	c	−232.8			
std. state, $m = 1$	aq	−283.8	−253.2		
HoI$_3$	c	−141.7			
std. state, $m = 1$	aq	−203.8	−196.2		
Ho$_2$(SO$_4$)$_3$ std. state, $m = 1$	aq	−978.1	−850.4		
Ho$_2$(SO$_4$)$_3$ · 8H$_2$O	c		−1318.0		
Hydrogen					
H	g	52.095	48.581	27.391	4.9679
^{1}H	g	52.095	48.580	27.391	4.9679
^{2}H	g	52.981	49.360	29.455	4.9679
H$^+$ std. state, $m = 1$	aq	0	0	0	0
^{1}H$_2$	g	0	0	31.208	6.889
^{2}H$_2$	g	0	0	34.620	6.978
^{1}H^{2}H	g	0.076	−0.350	34.343	6.978
H$_2$ std. state, $m = 1$	aq	−1.0	4.2	13.8	
OH	g	9.31	8.18	43.890	7.143
OH$^+$	g	317.5			
OH$^-$	g	−33.67			
std. state, $m = 1$	aq	−54.970	−37.594	−2.57	−35.5
HO$_2$	g	5			
HO$_2^-$ std. state, $m = 1$	aq	−38.32	−16.1	5.7	
H$_2$O	liq	−68.315	−56.687	16.71	17.995
^{2}H$_2$O	liq	−70.411	−58.195	18.15	20.16
^{1}H^{2}HO	liq	−69.285	−57.817	18.95	
H$_2$O	g	−57.796	−54.634	45.104	8.025
^{2}H$_2$O deuterium oxide	g	−59.560	−56.059	47.378	8.19
H$_2$O$^+$	g	234.3			
H$_2$O$_2$	liq	−44.88	−28.78	26.2	21.3
	g	−32.58	−25.24	55.6	10.3
undissoc; std. state, $m = 1$	aq	−45.69	−32.05	34.4	
	aq, 1	−45.365			
	aq, 10	−45.670			
H$_2$O$_2^+$	g	220.7			
Indium					
In	c	0	0	13.82	6.39
	g	58.15	49.89	41.51	4.98
In$_2$	g	91.04			
InAs	c	−14.0	−12.8	18.1	11.42

Table 3–9 (continued)

HEATS AND FREE ENERGIES OF FORMATION, ENTROPIES, AND HEAT CAPACITIES OF ELEMENTS AND INORGANIC COMPOUNDS

Formula and Description	State	$\Delta Hf°$	$\Delta Gf°$	$S°$	$C_p°$
InBr	c	−41.9	−40.4	27	
	g	−13.6	−22.54	61.99	8.76
InBr₃	c	−102.5			
	g	−67.4			
InCl ll	c	−44.5			
	g	−18			
InCl₃	c	−128.4			
	g	−89.4			
In₂Cl₃	g	−103.6			
InF	g	−48.61			
InH	g	51.5	45.49	49.60	7.07
InI	g	1.8	−9.0	63.87	8.80
	c	−27.8	−28.8	31	
InI₃	c	−57			
InN	c	−4.2			
InO	g	92.5	87.1	56.5	7.78
In₂O₃	c	−221.27	−198.55	24.9	22
InOH	g	−19			
In(OH)²⁺ std. state, $m = 1$	aq	−88.5	−74.8	−21	
In(OH)₂⁺ std. state, $m = 1$	aq	−148	−125.5	6	
InP	c	−21.2	−18.4	14.3	10.86
InS	c	−33.0	−31.5	16	
	g	90			
In₂S	g	15	3.1	76	
In₂S₃	c	−102	−98.6	39.1	28.20
In₂(SO₄)₃	c	−666	−583	65	67
InSe	c	−28			
In₂Se₃	c	−82			
InSb	c	−7.3	−6.1	20.6	11.82
	g	82.3			
InTe	c	−23			
In₂Te₃	c	−47			
Iodine					
I	g	25.535	16.798	43.184	4.968
I	g	−47.0			
std. state, $m = 1$	aq	−13.19	−12.33	26.6	−34.0
I₂	c	0	0	27.757	13.011
	g	14.923	4.627	62.28	8.82
std. state, $m = 1$	aq	5.4	3.92	32.8	
std. state, $m = 1$	CCl₄	6.0	2.66	39.0	
I₃ std. state, $m = 1$	aq	−12.3	−12.3	57.2	
IBr	g	9.76	0.89	61.822	8.71
IBr₂⁻ std. state, $m = 1$	aq		−29.4		
BrI₂⁻ std. state, $m = 1$	aq	−30.6	−26.3	47.2	
IBrCl⁻ std. state, $m = 1$	aq		−35.0		
ICl	liq	−5.71	−3.25	32.3	
	g	4.25	−1.30	59.140	8.50
ICl₂⁻ std. state, $m = 1$	aq		−38.5		
ICl₃	c	−21.4	−5.34	40.0	
I₂Cl⁻ std. state, $m = 1$	aq		−31.7		
IF	g	−22.86	−28.32	56.42	7.99
IF₅	liq	−206.7			
	g	−196.58	−179.68	78.3	23.7
IF₇	g	−225.6	−195.6	82.8	32.6
HI	g	6.33	0.41	49.351	6.969
std. state, $m = 1$	aq	−13.19	−12.33	26.6	−34.0
HIO undissoc; std. state, $m = 1$	aq	−33.0	−23.7	22.8	
HIO₃	c	−55.0			
	aq, ∞	−52.2			

Table 3–9 (continued)
HEATS AND FREE ENERGIES OF FORMATION, ENTROPIES, AND
HEAT CAPACITIES OF ELEMENTS AND INORGANIC COMPOUNDS

Formula and Description	State	$\Delta Hf°$	$\Delta Gf°$	$S°$	$C_p°$
H_5IO_6	aq	−180.4			
IO	g	41.84	35.80	58.65	7.86
IO^- std. state, $m = 1$	aq	−25.7	−9.2	−1.3	
IO_3^- std. state, $m = 1$	aq	−52.9	−30.6	28.3	
IO_4^-	aq	−35.2			
I_2O_5	c	−37.78			
Iridium					
Ir	c	0	0	8.48	6.00
	g	159.0	147.7	46.240	4.968
$IrCl_3$	c	−58.7	−43	27	
	g	25	24	90	
$IrCl_6^{2-}$	aq	−148.1	(−129)	(50)	
$IrCl_6^{3-}$	aq	−179.5	(−109)	(70)	
IrF_6	c	−138.54	−110.34	59.2	
	g	−130	−110	85.5	28.94
IrO_2	c	−65.5	(−42)	(15)	
IrO_3	g	1.9	(7)	(69)	
IrS_2	c	−33	(−32)	(15)	
Ir_2S_3	c	−56	(−53)	(23)	
Iron					
Fe α	c	0	0	6.52	6.00
	g	99.5	88.6	43.112	6.137
Fe^{2+} std. state, $m = 1$	aq	−21.3	−18.85	−32.9	
Fe^{3+} std. state, $m = 1$	aq	−11.6	−1.1	−75.5	
$FeAl_2O_4$	c	−470	−442	25.4	29.53
FeAsS	c	−10	−12	29	
$FeBr_2$	c	−59.7			
	g	−11			
std. state, $m = 1$	aq	−79.4	−68.55	6.5	
$FeBr_3$	c	−64.1			
	g	−29.6			
std. state, $m = 1$	aq	−98.8	−75.7	−16.4	
Fe_3C α-cementite	c	6.0	4.8	25.0	25.3
$Fe(CN)_6^{3-}$ std. state, $m = 1$	aq	134.3	174.3	64.6	
$Fe(CN)_6^{4-}$ std. state, $m = 1$	aq	108.9	166.09	22.7	
$H_2Fe(CN)_6^{2-}$ std. state, $m = 1$	aq	108.9	157.37	52	
$FeCO_3$ siderite	c	−177.00	−159.35	22.2	19.63
$Fe(CO)_5$	liq	−185.0	−168.6	80.8	57.5
	g	−175.4	−166.65	106.4	
$Fe_2(C_2O_4)_3$ std. state, $m = 1$	aq	−614.8	−485.5	−118.3	
$Fe(acetate)_3$ std. state, $m = 1$	aq	−360.1	−266.0	−13.4	
$FeCNS^{2+}$ std. state, $m = 1$	aq	5.6	17.0	−31	
$FeCl^{2+}$ std. state, $m = 1$	aq	−43.1	−34.4	−27	
$FeCl_2$	c	−81.69	−72.26	28.19	18.32
std. state, $m = 1$	aq	−101.2	−81.59	−5.9	
	aq, 100	−99.88			
$FeCl_2 \cdot 2H_2O$	c	−227.8			
$FeCl_2 \cdot 4H_2O$	c	−370.3			
$FeCl_3$	c	−95.48	−79.84	34.0	23.10
std. state, $m = 1$	aq	−131.5	−95.2	−35.0	
$FeCl_3 \cdot 6H_2O$	c	−531.5			
Fe_2Cl_6	g	−156.5			
$Fe(ClO_4)_2$ std. state, $m = 1$	aq	−83.1	−22.97	54.1	
$Fe(ClO_4)_2 \cdot 6H_2O$	c	−494.4			
$FeCr_2O_4$	c	−345.3	−321.2	34.9	31.94
FeF_2	c			20.79	16.28
std. state, $m = 1$	aq	−180.3	−152.13	−39.5	
FeF_3 std. state, $m = 1$	aq	−250.1	−201.0	−85.4	
FeI_2	c	−27.0			
std. state, $m = 1$	aq	−47.7	−43.51	20.3	

Table 3-9 (continued)
HEATS AND FREE ENERGIES OF FORMATION, ENTROPIES, AND
HEAT CAPACITIES OF ELEMENTS AND INORGANIC COMPOUNDS

Formula and Description	State	$\Delta Hf°$	$\Delta Gf°$	$S°$	$C_p°$
FeI$_3$	g	17			
std. state, $m = 1$	aq	−51.2	−38.1	4.3	
FeMoO$_4$	c	−257	−233	30.9	28.31
Fe$_2$(MoO$_4$)$_3$	c	−702			
Fe$_4$N	c	−2.5	0.9	37	
Fe(NO$_3$)$_3$ std. state, $m = 1$	aq	−160.3	−80.9	29.5	
Fe(NO$_3$)$_3$ · 9H$_2$O	c	−785.2			
Fe$_{0.947}$O wistite	c	−63.64	−58.59	13.74	11.50
FeO	c	−65.0			
Fe$_2$O$_3$ hematite	c	−197.0	−177.4	20.89	24.82
Fe$_3$O$_4$ magnetite	c	−267.3	−242.7	35.0	34.28
FeOH$^+$ std. state, $m = 1$	aq	−77.6	−66.3	−7	
Fe(OH)$^{2+}$ std. state, $m = 1$	aq	−69.5	−54.83	−34	
Fe(OH)$_2$ pptd	c	−136.0	−116.3	21	
Fe(OH)$_2^+$ std. state, $m = 1$	aq		−104.7		
Fe(OH)$_3$ pptd	c	−196.7	−166.5	25.5	
FeP	c	−30			
FeP$_2$	c	−46			
Fe$_2$P	c	−39			
Fe$_3$P	c	−39			
FePO$_4$	c	−310.1			
FePO$_4$ · 2H$_2$O strengite	c	−451.3	−396.2	40.93	43.15
FeS iron-rich pyrrhotite	c	−23.9	−24.0	14.41	12.08
FeS$_2$ pyrite	c	−42.6	−39.9	12.65	14.86
Fe$_7$S$_8$ sulfur-rich pyrrhotite	c	−176.0	−178.9	116.1	95.26
FeSO$_4$	c	−221.9	−196.2	25.7	24.04
std. state, $m = 1$	aq	−238.6	−196.82	−28.1	
FeSO$_4$ · 7H$_2$O	c	−720.50	−599.97	97.8	94.28
Fe$_2$(SO$_4$)$_3$	c	−617.0			
std. state, $m = 1$	aq	−675.2	−536.1	−136.4	
FeSe	c	−18.0			
FeSe$_2$	c			20.75	17.42
FeSi	c	−17.6	−17.6	11.0	11.4
FeSi$_2$ β-lebanite	c	−19.4	−18.7	13.3	15.79
Fe$_3$Si	c	−22.4	−22.6	24.8	23.50
Fe$_2$SiO$_4$ fayalite	c	−353.7	−329.6	34.7	31.76
FeTe	c	−15.0			
FeWO$_4$	c	−276	−252	31.5	27.39
Fe$_2$(WO$_4$)$_3$ · 8H$_2$O	c	−1355			
Krypton					
Kr	g	0	0	39.191	4.968
KrF$_2$	c	14.4			
Lanthanum					
La	c	0	0	13.7	
	g	88	79	43.67	
LaBr$_3$	c	−233			
La$_2$(CN$_2$)$_3$	c	−229			
LaCl$_3$	c	−263.6			
std. state, $m = 1$	aq	−296.3	−266.9	−5	
La$_3$H$_8$	c	−382			
LaI$_3$	c	−167.4			
std. state, $m = 1$	aq	−216.3	−209.9	34	
LaN	c	−72.1	−64.6		
La$_2$O$_3$	c	−428.5			
La(OH)$_3$	c		−312.8		
LaS$_2$	c	−156.7			
La$_2$S$_3$	c	−306.8			
La$_2$(SO$_4$)$_3$ std. state, $m = 1$	aq	−1003.1	−877.8	−76	
La$_2$(SO$_4$)$_3$ · 8H$_2$O	c		−14.031		

Table 3–9 (continued)
HEATS AND FREE ENERGIES OF FORMATION, ENTROPIES, AND HEAT CAPACITIES OF ELEMENTS AND INORGANIC COMPOUNDS

Formula and Description	State	$\Delta Hf°$	$\Delta Gf°$	$S°$	$C_p°$
Lead					
Pb	c	0	0	15.49	6.32
	g	46.6	38.7	41.889	4.968
PbBr$_2$	c	−66.6	−62.60	38.6	19.15
Pb(BrO$_3$)$_2$	c		−11.95		
Pb(CH$_3$)$_4$	liq	23.4			
	g	32.48			
Pb(C$_2$H$_5$)$_4$	liq	12.6			
	g	26.19			
Pb(acetate)$_2$	c	−230.5			
PbCl$_2$	c	−85.90	−75.08	32.5	
PbCl$_4$	liq	−78.7			
Pb(ClO$_3$)$_2$ undissoc; std. state	aq		−6.6		
PbClF	c	−127.8	−116.7	29.1	
PbCO$_3$	c	−167.1	−149.5	31.3	20.89
PbC$_2$O$_4$	c	−203.5	−179.3	34.9	25.2
PbO · PbCO$_3$	c	−219.5	−195.2	48.8	
PbCrO$_4$	c	−222.5			
PbF$_2$	c	−158.7	−147.5	26.4	
PbF$_4$	c	−225.1			
PbI$_2$	c	−41.94	−41.50	41.79	18.49
Pb(IO$_3$)$_2$	c	−118.4	−84.0	74.8	
PbMoO$_4$	c	−251.4	−227.4	39.7	28.61
Pb(N$_3$)$_2$ monoclinic	c	114.3	149.3	35.4	
orthorhombic	c	113.8	148.7	35.7	
Pb(NO$_3$)$_2$	c	−108.0			
PbO yellow	c	−51.94	−44.91	16.42	10.94
red	c	−52.34	−45.16	15.9	10.95
PbO$_2$	c	−66.3	−51.95	16.4	15.45
Pb$_2$O$_3$	c			36.3	25.74
Pb$_3$O$_4$	c	−171.7	−143.7	50.5	35.1
HPbO$_2^-$ std. state, $m = 1$	aq		−80.90		
Pb(OH)$_2$	c		−108.1	21	
precipitated	c	−123.3			
H$_2$PbO$_2$ undissoc; std. state, $m = 1$	aq		−95.8		
Pb(OH)$_3^-$ std. state, $m = 1$	aq		−137.6		
Pb$_3$(PO$_4$)$_2$	c	−620.3	−581.4	84.4	61.25
Pb(ReO$_4$)$_2$ · 2H$_2$O	c	−534	−455	74	
PbS	c	−24.0	−23.6	21.8	11.83
Pb(SCN)$_2$	c		32.1		
PbSO$_3$	c	−160.1			
PbSO$_4$	c	−219.87	−194.36	35.51	24.667
PbSe	c	−24.6	−24.3	24.5	12.0
PbSeO$_3$	c	−128.5			
PbSeO$_4$	c	−145.6	−120.7	40.1	
PbSiO$_3$	c	−273.83	−253.86	26.2	21.52
Pb$_2$SiO$_4$	c	−325.8	−299.4	44.6	32.78
PbTe	c	−16.9	−16.6	26.3	12.08
PbWO$_4$	c				28.63
Lithium					
Li	c	0	0	6.70	
	g	38	29.19	33.14	
LiAlH$_4$	c	−24.1			18.2
LiBr	c	−83.72			
	b	−41	−50	53.78	
std. state, $m = 1$	aq	−95.45	−94.79	22.7	
LiBr · 2H$_2$O	c	−229.94			
LiBrO$_3$	aq	−77.9	−65.70		
Li$_2$C$_2$	c	−14.2			

Table 3–9 (continued)
HEATS AND FREE ENERGIES OF FORMATION, ENTROPIES, AND HEAT CAPACITIES OF ELEMENTS AND INORGANIC COMPOUNDS

Formula and Description	State	$\Delta Hf°$	$\Delta Gf°$	$S°$	$C_p°$
LiCN	aq, 200	−30.3	−31.35		
LiCNO cyanate	aq	−101.2	−94.12		
Li_2CO_3	c	−290.54	−270.66	21.60	
	aq, ∞	−294.74	−266.66	−5.9	
$LiHCO_3$	aq, ∞	−231.73	−210.53	29.5	
$LiC_2H_3O_2$	aq	−183.9	−160.0		
LiCl	g	−53	−58	51.01	
	c	−97.70			
	aq, 3	−102.62			
	aq, ∞	−106.58	−101.57	16.6	
$LiCl \cdot H_2O$	c	−170.31	−151.2	24.8	
$LiCl \cdot 2H_2O$	c	−242.1			
$LiCl \cdot 3H_2O$	c	−313.5			
$LiClO_3$	aq	−87.5	−70.95		
$LiClO_4$	aq	−106.3	−81.4		
LiF	c	−144.7	−139.5	8.57	
	aq, ∞	−145.21	−136.30	1.1	
$LiHF_2$	aq, 400	−220.4			
Li_2SiF_6	c	−688.9			
LiH	g	30.7	25.2	40.77	
	c	−21.61	−16.72	5.9	8.3
LiI	g	−16	−26	55.68	
	c	−71.2			
	aq, ∞	−79.92	−82.57	29.5	
$LiI \cdot \frac{1}{2}H_2O$	c	−103.8			
$LiI \cdot H_2O$	c	−141.16			
$LiI \cdot 2H_2O$	c	−213.03			
$LiI \cdot 3H_2O$	c	−285.02			
$LiIO_3$	aq	−121.3	−102.95		
Li_3N	c	−47.2	−37.33		
$LiNH_2$	c	−43.50			
$LiNO_2$	c	−96.6			
$LiNO_3$	c	−115.28			
	aq, 3	−115.1			
	aq, ∞	−115.93	−96.63	38.4	
$LiNO_3 \cdot 3H_2O$	c	−328.6			
Li_2O	c	−142.4			
Li_2O_2	c	−151.7	−138.0		
	aq	−159.0			
LiOH	c	−116.45	−106.1	12	
	aq, 11	−120.51			
	aq, ∞	−121.51	−107.82	0.9	
$LiOH \cdot H_2O$	c	−188.77	−164.8	22	
$LiReO_4$	c	−253.4			
$LiReO_4 \cdot H_2O$	c	−324.9			
$LiReO_4 \cdot 2H_2O$	c	−395.1			
Li_2SO_4	c	−342.83	−314.66		
	aq, 18	−348.40			
	aq, ∞	−350.01	−317.78	10.9	
$Li_2SO_4 \cdot H_2O$	c	−414.20	−375.07		
Li_2Se	c	−91.1			
Li_2SiO_3	gls	−376.7			
Lutetium					
Lu	c	0	0		
	g	87		44.14	
$LuBr_3$	c	−200			
$LuCl_3$ γ	c	−227.9			
	aq, ∞	−280.2	−249.0		
LuI_3 β	c	−133.2			
	aq, ∞	−200.2	−192.0		

Table 3–9 (continued)
HEATS AND FREE ENERGIES OF FORMATION, ENTROPIES, AND
HEAT CAPACITIES OF ELEMENTS AND INORGANIC COMPOUNDS

Formula and Description	State	$\Delta Hf°$	$\Delta Gf°$	$S°$	$C_p°$
$Lu_2(SO_4)_3$	aq, ∞	−970.9	−842.0		
$Lu_2(SO_4)_3 \cdot 8H_2O$	c		−1308.1		
Magnesium					
Mg	c	0	0	7.77	
	g	35.9	27.6	35.50	
$Mg(AlH_4)_2$	c	−23.1			
$Mg_3(AsO_4)_2$	c	−731.3			
	aq	−749	−630.14		
$MgHAsO_4$	aq	−349.2			
$Mg(H_2AsO_4)_2$	aq	−541.0			
$Mg(NH_4)AsO_4 \cdot 6H_2O$	c	−800.7			
Mg_3Bi_2	c	−36.5			
$MgBr_2$	c	−123.7			
	aq, 25	−166.66			
	aq, ∞	−168.21	−158.14	10.4	
$MgBr_2 \cdot 6H_2O$	c	−575.4	−491.0	95	
$Mg(CN)_2$	aq	−31.9	−29.08		
$MgCN_2$	c	−60.3			
$MgCO_3$	c	−266	−246	15.7	
$Mg(C_2H_3O_2)_2$	aq	−344.6	−286.38		
$MgCl_2$	c	−153.40	−141.57	21.4	
	aq, 10	−185.65			
	aq, ∞	−190.46	−171.69	−1.9	
$MgCl_2 \cdot H_2O$	c	−231.15	−206.11	32.8	
$MgCl_2 \cdot 2H_2O$	c	−305.99	−267.32	43.0	
$MgCl_2 \cdot 4H_2O$	c	−454.00	−390.49	63.1	
$MgCl_2 \cdot 6H_2O$	c	−597.42	−305.65	87.5	
$MgCl_2 \cdot KCl \cdot 6H_2O$	c	−702.0			
$MgCl_2 \cdot MgO$	c	−312.6			
$MgCl_2 \cdot MgO \cdot 6H_2O$	c	−742.1			
$MgCl_2 \cdot MgO \cdot 16H_2O$	c	−1440.8			
$Mg(OH)Cl$	c	−191.3	−175.0	19.8	
$Mg(ClO_4)_2$	c	−140.6			
	aq	−172.5			
$Mg(ClO_4)_2 \cdot 2H_2O$	c	−290.7			
$Mg(ClO_4)_2 \cdot 4H_2O$	c	−438.6			
$Mg(ClO_4)_2 \cdot 6H_2O$	c	−583.2			
$MgCrO_4$	c	−318.3			
	aq	−321.2			
MgF_2	c	−263.5	−250.8	13.68	
MgH	g	41	34	47.61	
MgI_2	c	−86.0			
	aq, ∞	−137.15	−133.69	24.1	
$MgMoO_4$	c	−329.9			
Mg_3N_2	c	−110.24	−100.8		
$Mg(NO_3)_2$	c	−188.72	−140.63	39.2	
	aq, 15	−208.410			
	aq, ∞	−209.15	−161.81	41.8	
$Mg(NO_3)_2 \cdot 2H_2O$	c	−336.63			
$Mg(NO_3)_2 \cdot 6H_2O$	c	−624.36			
MgO	c	−143.84	−136.13	6.4	
MgO_2	c	−148.9			
$Mg(OH)_2$	c	−221.00	−199.27	15.09	
$Mg_3(PO_4)_2$	c	−961.5			
$MgHPO_4$	aq	−422.2			
$Mg(NH_4)PO_4 \cdot 6H_2O$	c	−881.0			
MgS	c	−83.0			
	aq	−108			
$MgSO_3$	c	−241.0			

Table 3–9 (continued)
HEATS AND FREE ENERGIES OF FORMATION, ENTROPIES, AND
HEAT CAPACITIES OF ELEMENTS AND INORGANIC COMPOUNDS

Formula and Description	State	ΔHf°	ΔGf°	S°	C_p°
MgSO$_3$ · 3H$_2$O	c	−461.6			
MgSO$_3$ · 6H$_2$O	c	−673.4			
MgSO$_4$	c	−305.5	−280.5	21.9	
	aq, 200	−326.23			
	aq, ∞	−327.31	−286.33	−24.0	
MgSO$_4$ · 2H$_2$O	c	−381.9			
MgSO$_4$ · 4H$_2$O	c	−595.5			
MgSO$_4$ · 6H$_2$O	c	−736.6			
MgSO$_4$ · 7H$_2$O	c	−808.7			
MgS$_2$O$_3$	aq	−256.1			
MgS$_2$O$_3$ · 3H$_2$O	c	−454.2			
MgS$_2$O$_3$ · 6H$_2$O	c	−669.4			
Mg$_3$Sb$_2$	c	−68.1			
Mg$_2$Si	c	−18.6			
MgSiO$_3$	c	−357.9	−337.2	16.2	
MgSiO$_4$	c	−488.2	−459.8	22.7	
Mg$_2$Sn	c	−17.0			
MgTe	c	−50			
MgWO$_4$	c	−345.2			
Manganese					
Mn α	c	0	0	7.65	6.29
β	c			8.22	6.34
γ	c	0.37	0.34	7.75	9.59
	g	67.1	57.0	41.49	4.97
MnAs	c	−14			
Mn$_3$(AsO$_4$)$_2$	c	−512.8			
MnBr$_2$	c	−92.0	−89	(33)	
std. state, $m = 1$	aq	−110.9	−97.8		
MnBr$_2$ · H$_2$O	c	−168.5			
MnBr$_2$ · 4H$_2$O	c	−380.1			
Mn$_3$C	c	1.1	1.3	23.6	22.33
Mn$_7$C$_3$	c	−10			
MnCO$_3$ natural	c	−213.7	−195.2	20.5	19.48
pptd	c	−211	−194	27.0	
MnC$_2$O$_4$	c	−246.2			
undissoc; std. state, $m = 1$	aq	−248.5	−221.0	16.1	
MnC$_2$O$_4$ · 2H$_2$O	c	−389.2	−338.2	48	
MnC$_2$O$_4$ · 3H$_2$O	c	−459.1			
Mn(acetate)$_2$	c	−274.4			
Mn(acetate)$_2$ · 4H$_2$O	c	−558.8			
Mn$_2$(CO)$_{10}$	c	−401.0			
	g	−386.0			
MnCl	g	10.1			8.05
MnCl$_2$	c	−115.03	−105.29	28.26	17.43
std. state, $m = 1$	aq	−132.66	−117.3	9.3	−53
MnCl$_2$ · H$_2$O	c	−188.8	−166.4	41.6	
MnCl$_2$ · 2H$_2$O	c	−261.0	−225.2	52.3	
MnCl$_2$ · 4H$_2$O	c	−403.3	−340.3	72.5	
Mn^{2+} std. state, $m = 1$	aq	−52.76	−54.5	−17.6	12
MnF$_2$	c	−189	−179	22.05	15.96
MnI$_2$	c	−59.3	(−65)	(37)	
2H$_2$O	c	−201.4			
4H$_2$O	c	−343.9			
Mn(IO$_3$)$_2$	c	−160	−124.4	63	
MnMoO$_4$	c	−284.8	−261	(32)	
MnN$_6$ azide	c	92.2			
Mn$_5$N$_2$	c	−48.8			
Mn(NO$_3$)$_2$	c	−137.73			
std. state, $m = 1$	aq	−151.9	−107.8	52	−29

Table 3–9 (continued)
HEATS AND FREE ENERGIES OF FORMATION, ENTROPIES, AND HEAT CAPACITIES OF ELEMENTS AND INORGANIC COMPOUNDS

Formula and Description	State	$\Delta H f^\circ$	$\Delta G f^\circ$	S°	C_p°
$Mn(NO_3)_2 \cdot 6H_2O$ glassy	amorp	−566.9			
	liq	−557.27			
MnO	c	−92.07	−86.74	14.27	10.86
	g	29.6			
MnO_2	c	−124.29	−111.18	12.68	12.94
pptd	amorp	−120.1			
Mn_2O_3	c	−229.2	−210.6	26.4	25.73
Mn_3O_4	c	−331.7	−306.7	37.2	33.38
MnO_4^- std. state, $m = 1$	aq	−129.4	−106.9	45.7	
MnO_4^{2-} std. state, $m = 1$	aq	−156	−119.7	14	
$Mn(OH)_2$ pptd	amorp	−166.2	−147.0	23.7	
MnP	c	−27			
MnP_3	c	−51			
$Mn_3(PO_4)_2$	c	−744.9			
$MnHPO_4$	c		−332.5		
MnS green	c	−51.2	−52.2	18.7	11.94
pink pptd	amorp	−51.1			
$MnSO_4$	c	−254.60	−228.83	26:8	24.02
std. state, $m = 1$	aq	−270.1	−232.5		
$MnSO_4 \cdot H_2O$ α	c	−329.0	−289.9	(37)	
β	c	−322.2			
$MnSO_4 \cdot 4H_2O$	c	−539.7			
$MnSO_4 \cdot 5H_2O$	c	−610.2			78
$MnSO_4 \cdot 7H_2O$	c	−750.3			
$Mn_2(SO_4)_3$	c	−666.9			
$MnS_2O_6 \cdot 6H_2O$	c	−751.0			
MnSb	c	−12			
MnSe	c	−25.5	−26.7	21.7	12.20
$MnSiO_3$	c	−315.7	−296.5	21.3	20.66
Mn_2SiO_4	c	−413.6	−390.1	39.0	31.04
MnTe	c			22.4	17.49
$MnWO_4$	c	−311.9			29.7
Mercury					
Hg	liq	0	0	18.17	6.688
	g	14.655	7.613	41.79	4.968
$HgBr_2$	c	−40.8	−36.6	41	
undissoc; std. state, $m = 1$	aq	−38.4	−34.2	41	
$HgBr_3^-$ std. state, $m = 1$	aq	−70.1	−62.0	62	
$HgBr_4^{2-}$ std. state, $m = 1$	aq	−103.0	−88.7	74	
Hg_2Br_2	c	−49.45	−43.278	52	
HgBrCl std. state, $m = 1$	aq	−45.5	−38.7	40	
HgBrI undissoc; std. state, $m = 1$	aq	−28.1	−26.7	46	
$Hg(CN)_2$	c	63.0			
	g	91			
undissoc; std. state, $m = 1$	aq	66.5	74.6	37.3	
$Hg(CN)_3^-$ std. state, $m = 1$	aq	94.9	110.7	51.4	
$Hg(CN)_4^{2-}$ std. state, $m = 1$	aq	125.8	147.8	71	
Hg(II) fulminate	c	64			
Hg(I) acetate	c	−199.6	−153.3	78	
Hg_2CO_3	c	−132.3	−111.9	43	
HgC_2O_4	c	−162.1			
$Hg_2C_2O_4$	c		−141.8		
$Hg(CNS)_2$ undissoc; std. state	aq	46.9	60.1	37.6	
$Hg(CNS)_4^{2-}$ std. state, $m = 1$	aq	78.0	98.3	109	
$Hg_2(CNS)_2$ thiocyanate	c		54.4		
HgCl	g	20.1	15.0	62.09	8.68
$HgCl_2$	c	−53.6	−42.7	34.9	
undissoc; std. state, $m = 1$	aq	−51.7	−41.4	37	
$HgCl_3^-$ std. state, $m = 1$	aq	−92.9	−73.9	50	

Table 3–9 (continued)
HEATS AND FREE ENERGIES OF FORMATION, ENTROPIES, AND
HEAT CAPACITIES OF ELEMENTS AND INORGANIC COMPOUNDS

Formula and Description	State	$\Delta Hf°$	$\Delta Gf°$	$S°$	$C_p°$
HgCl$_4^{2-}$ std. state, $m = 1$	aq	−132.4	−106.8	70	
Hg$_2$Cl$_2$	c	−63.39	−50.377	46.0	
HgF	g	1.0	−2.4	53.98	8.26
Hg$_2$F$_2$	c		−104.1		
HgH	g	57.36	51.63	52.46	7.16
HgI	g	31.64	21.14	67.26	8.99
HgI$_2$ red	c	−25.2	−24.3	43	
yellow	c	−24.6			
	g	−4.1	−14.3	80.31	14.60
undissoc; std. state, $m = 1$	aq	−19.0	−18.0	42	
HgI$_3^-$ std. state, $m = 1$	aq	−36.5	−35.5	72	
HgI$_4^{2-}$ std. state, $m = 1$	aq	−56.2	−50.6	86	
Hg$_2$I$_2$	c	−29.00	−26.53	55.8	
Hg$_2$(N$_3$)$_2$	c	142.0	178.4	49	
Hg(NO$_3$)$_2$ · ½H$_2$O	c	−93.8			
Hg$_2$(NO$_3$)$_2$ · 2H$_2$O	c	−207.5			
HgO red, orthorhombic	c	−21.71	−13.995	16.80	10.53
yellow	c	−21.62	−13.964	17.0	
hexagonal	c	−21.4	−13.92	17.6	
Hg(OH)$_2$ undissoc; std. state	aq	−84.9	−65.7	34	
HgS red	c	−13.9	−12.1	19.7	11.57
black	c	−12.8	−11.4	21.1	
HgSO$_4$	c	−169.1			
Hg$_2$SO$_4$	c	−177.61	−149.589	47.96	31.54
Hg(HS)$_2$ undissoc; std. state, $m = 1$	aq		−6.4		
HgSe	c	−11			
	g	18.1	7.5	63.82	8.8
HgSeO$_3$	c		−68.0		
Hg$_2$SeO$_3$	c		−71.1		
HgTe	c	−10			
	g			65.73	
Molybdenum					
Mo	c	0	0	6.85	5.75
	g	157.3	146.4	43.461	4.968
MoBr$_2$	c	−62.4	−53		18.3
MoBr$_3$	c	(−34)			
MoBr$_4$	c	−76.8			
MoO$_2$Br$_2$	c	−150.4			
MoB	c	−21			
MoC	c	−2.4			
Mo$_2$C	c	−10.9			
Mo(CO)$_6$	c	−234.9	−209.8	77.9	57.90
	g	−218.0	−204.6	117	49
MoCl$_2$	c	−67.4	(−35)		
MoCl$_3$	c	−92.5		(32.6)	
MoCl$_4$	c	−114.8	−58.5	44.7	
	g	−90			
MoCl$_5$	c	−126.0	(−68.5)	53	
	g	−103			
MoOCl$_2$	c	−126			
MoO$_2$Cl$_2$	c	−171.4			
	g	−151.6			
	aq	−190.4			
MoO$_2$Cl$_2$ · H$_2$O	c	−245.4			
MoOCl$_4$	c	−153.0			
MoF$_6$	liq	−378.95	−352.08	62.06	40.58
	g	−372.29	−351.88	83.75	28.82
MoI$_2$	g	32			

Table 3–9 (continued)
HEATS AND FREE ENERGIES OF FORMATION, ENTROPIES, AND HEAT CAPACITIES OF ELEMENTS AND INORGANIC COMPOUNDS

Formula and Description	State	$\Delta Hf°$	$\Delta Gf°$	$S°$	$C_p°$
MoI_3		(−15)	(−15)		
Mo_2N	c	−19.50			
MoO	g	101			
MoO_2	c	−140.76	−127.40	11.06	13.38
MoO_3	c	−178.08	−159.66	18.58	17.92
	g	−78			
	aq	−172.5			
MoO_4	aq	−158.0			
MoO_5	aq	−139.6			
MoO_4^{2-} std. state	aq	−238.5	−199.9	6.5	
H_2MoO_4 white	c	−250.0			
	aq	−240.8			
$H_2MoO_4 \cdot H_2O$ yellow	c	−325			
MoS_2	c	−56.2	−54.0	14.96	15.19
Mo_2S_3	c	−87			
$MoSi_2$	c	−28			
Mo_3Si	c	−23	−23	25.4	22.23
Mo_5Si_3	c	−68			
Neodymium					
Nd	g	87			
	c	0	0		
$NdCl_3$ α	c	−254.3			
	aq, ∞	−291.3	−261.6		
$NdCl_3 \cdot 6H_2O$	c	−692.3			
$Nd(OH)_3$	c		−309.6		
NdI_3 α	c	−158.9			
	aq, ∞	−211.3	−204.6		
Nd_2O_3	c	−442.0			
Nd_2S_3	c	−281.8			
$Nd_2(SO_4)_3$	c	−948.1			
	aq, ∞	−993.1	−867.2		
$Nd_2(SO_4)_3 \cdot 5H_2O$	c	−1318.4			
$Nd_2(SO_4)_3 \cdot 8H_2O$	c	−1524.7	−1334.5		
Neon					
Ne	g	0	0	34.95	4.968
Neptunium					
Np	c	0	0		
$NpBr_3$	c	−174			
$NpBr_4$	c	−183			
$NpCl_3$	c	−216			
$NpCl_4$	c	−237			
$NpCl_5$	c	−246			
NpF_3	c	−360			
NpF_4	c	−428			
NpI_3	c	−120			
NpO_2	c	−246			
Nickel					
Ni	c	0	0	7.14	6.23
	g	102.7	91.9	43.519	5.583
Ni^{2+} std. state, $m = 1$	aq	−12.9	−10.9	−30.8	
$Ni_3(AsO_4)_2$	c		−377.5		
$Ni(BO_2)_2$	c		−347.3		
$NiBr_2$	c	−50.7			
std. state, $m = 1$	aq	−71.0	−60.6	8.6	
Ni_3C	c	16.1			
$Ni(CN)_2$ pptd	c	30.5			
$Ni(CN)_4^{2-}$ std. state, $m = 1$	aq	87.9	112.8	52	
$Ni(CNO)_2$ cyanate	c	−54.4			
$Ni(CNS)_2$ thiocyanate	c	22.8			

Table 3–9 (continued)
HEATS AND FREE ENERGIES OF FORMATION, ENTROPIES, AND
HEAT CAPACITIES OF ELEMENTS AND INORGANIC COMPOUNDS

Formula and Description	State	$\Delta Hf°$	$\Delta Gf°$	$S°$	$C_p°$
Ni(CO)$_4$	liq	−151.3	−140.6	74.9	48.9
	g	−144.10	−140.36	98.1	34.70
Ni(acetate)$_2$ std. state, $m = 1$	aq	−245.2	−187.5	10.6	
NiCO$_3$	c		−146.4		
NiC$_2$O$_4$	c	−204.8			
std. state, $m = 1$	aq	−210.1	−172.0	−19.9	
NiCl$_2$	c	−72.976	−61.918	23.34	17.13
std. state, $m = 1$	aq	−92.8	−73.6	−3.6	
NiCl$_2$ · 2H$_2$O	c	−220.4	−181.7	42	
NiCl$_2$ · 4H$_2$O	c	−362.5	−295.2	58	
NiCl$_2$ · 6H$_2$O	c	−502.67	−409.54	82.3	
Ni(ClO$_4$)$_2$ std. state, $m = 1$	aq	−74.7	−15.0	56.2	
Ni(ClO$_4$)$_2$ · 6H$_2$O	c	−486.6			
NiF$_2$	c	−155.7	−144.4	17.59	15.31
NiI$_2$	c	−18.7			
std. state, $m = 1$	aq	−39.3	−35.6	22.4	
Ni(IO$_3$)$_2$	c	−116.9	−78.0	51	
Ni(NO$_3$)$_2$	c	−99.2			
std. state, $m = 1$	aq	−112.0	−64.2	39.2	
Ni(NO$_3$)$_2$ · 6H$_2$O	c	−528.6			111
NiO	c	−57.3	−50.6	9.08	10.59
Ni$_2$O$_3$	c	−117.0			
Ni(OH)$_2$	c	−126.6	−106.9	21	
Ni(OH)$_3$ pptd	c	−160			
Ni$_3$P	c	−50.2			
Ni$_5$P$_2$	c	−97.7			
Ni$_2$P$_2$O$_7$	c		−497.9		
Ni$_3$(PO$_4$)$_2$	c		−562.4		
NiS	c	−19.6	−19.0	12.66	11.26
pptd	c	−18.5			
Ni$_3$S$_2$	c	−48.5	−47.1	32.0	28.12
NiSO$_4$	c	−208.63	−181.6	22	33
std. state, $m = 1$	aq	−230.2	−188.9	−26.0	
NiSO$_4$ · 6H$_2$O tetrahedral green	c	−641.21	−531.78	79.391	78.36
NiSO$_4$ · 7H$_2$O	c	−711.36	−588.49	90.57	87.14
NiSe	c	−14.1			
NiSeO$_3$ · 2H$_2$O	c	−271.11			
NiSi	c	−20.6			10.9
Ni$_2$Si	c	−33.6			16.8
NiTe	c	−12.8			12.5
Ni$_4$W	c	−43			29.0
NiWO$_4$	c	−270.9			29.1
Niobium					
Nb	c	0	0	8.70	5.88
	g	173.5	162.8	44.490	7.208
NbBr$_5$	c	−132.9			
	g	−104.8			
NbC	c	−33.2	−32.7	8.46	8.81
Nb$_2$C	c	−45.4	−44.4	15.3	14.48
NbCl$_3$	g	−86			
NbCl$_4$	c	−166.0			
	g	−134			
NbCl$_5$	c	−190.6	−163.3	50.3	35.4
	g	−168.2	−154.4	95.71	28.88
NbCo$_2$	c	−13.7	−13.2	22	
NbCo$_3$	c	−14.1	−13.7	29	
NbCr$_2$	c	−5.0	−5.0	19.97	17.45
NbF$_5$	c	−433.5	−406.1	38.3	32.2
	g	−415.8	−401.1	76.9	23.2

Table 3–9 (continued)
HEATS AND FREE ENERGIES OF FORMATION, ENTROPIES, AND HEAT CAPACITIES OF ELEMENTS AND INORGANIC COMPOUNDS

Formula and Description	State	$\Delta Hf°$	$\Delta Gf°$	$S°$	$C_p°$
NbFe$_2$	c	−11.1	−11.8	24	13.0
NbGe$_2$	c	−20.8			
NbI$_5$	c	−64.2			
NbN	c	−56.2	−49.2	8.25	9.32
Nb$_2$N	c	−59.9			
NbO	c	−97.0	−90.5	11.5	9.86
	g	51	44	57.09	7.36
NbO$_2$	c	−190.3	−177.0	13.03	13.74
	g	−51.3	−52.3	61.0	
NbO$_3^-$ ionic strength = 1	aq		−222.8		
Nb$_2$O$_5$ high-temperature form	c	−454.0	−422.1	32.80	31.57
Nb(OH)$_5$ undissoc; ionic strength = 1	aq		−346.2		
NbOCl$_2$	c	−185.1			
NbOCl$_3$	c	−210.2	−187	34	
	g	−179.8	−171.6	85.6	22.0
Nitrogen					
N	g	112.979	108.886	36.613	4.968
N$_2$	g	0	0	45.77	6.961
N$_3^-$ std. state, $m = 1$	aq	65.76	83.2	25.8	
HN$_3$	g	70.3	78.4	57.09	10.44
undissoc; std. state, $m = 1$	aq	62.16	76.9	34.9	
NCl$_3$	liq	55			
NH$_3$	g	−11.02	−3.94	45.97	8.38
undissoc; std. state, $m = 1$	aq	−19.19	−6.35	26.6	
	aq, 1	−18.011			
	aq, 10	−19.074			
NH$_4^+$ std. state, $m = 1$	aq	−31.67	−18.97	27.1	19.1
N$_2$H$_4$	liq	12.10	35.67	28.97	23.63
	g	22.80	38.07	56.97	11.85
undissoc; std. state, $m \doteq 1$	aq	8.20	30.6	33	
N$_2$H$_5^+$ std. state, $m = 1$	aq	−1.8	19.7	36	16.8
N$_2$H$_5$Br	c	−37.2			
std. state, $m = 1$	aq	−30.8	−5.2	55.7	−17.1
N$_2$H$_5$Br · HBr	c	−64.8			
N$_2$H$_5$Cl	c	−47.0			
std. state, $m = 1$	aq	−41.8	−11.7	49.5	−15.8
N$_2$H$_5$Cl · HCl	c	−87.8			
N$_2$H$_5$ClO$_4$	c	−42.2			
std. state, $m = 1$	aq	−32.7	17.6	79.7	
N$_2$H$_5$OH	liq	−58.01			
	g	−49.0	−18.9	63	
undissoc; std. state, $m = 1$	aq	−60.11	−26.1	49.7	17.5
N$_2$H$_5$NO$_3$	c	−60.13			
std. state, $m = 1$	aq	−51.41	−6.91	71	
(N$_2$H$_5$)$_2$SO$_4$	c	−229.2			
std. state, $m = 1$	aq	−221.0	−138.6	77	−36
HNF$_2$	g			60.40	10.37
NO	g	21.57	20.69	50.347	7.133
NO$_2$	g	7.93	12.26	57.35	8.89
NO$_2^-$ std. state, $m = 1$	aq	−25.0	−8.9	33.5	−23.3
NO$_3$	g	16.95	27.36	60.36	11.22
NO$_3^-$ nitrate, std. state, $m = 1$	aq	−49.56	−26.61	35.0	−20.7
peroxynitrite	aq	−10.7			
N$_2$O	g	19.61	24.90	52.52	9.19
HN$_2$O$_2^-$	aq	−9.4	18.2	34	
N$_2$O$_2^{2-}$ hyponitrite	aq	−4.1	33.2	6.6	
N$_2$O$_3$	g	20.01	33.32	74.61	15.68
N$_2$O$_4$	g	2.19	23.38	72.70	18.47
N$_2$O$_5$	g	2.7	27.5	85.0	20.2

Table 3–9 (continued)
HEATS AND FREE ENERGIES OF FORMATION, ENTROPIES, AND HEAT CAPACITIES OF ELEMENTS AND INORGANIC COMPOUNDS

Formula and Description	State	ΔH_f°	ΔG_f°	S°	C_p°
NOBr	g	19.64	19.70	65.38	10.87
NOCl	g	12.36	15.79	62.52	10.68
NO$_2$Cl	g	3.0	13.0	65.02	12.71
NOClO$_4$	c	-36.9			
NO$_2$ClO$_4$	c	8.7			
NOF	g	-15.9	-12.2	59.27	9.88
NO$_2$F	g	26		62.2	11.9
N$_2$O$_3$(SO$_3$)$_2$	c	-253			
HNO$_2$ *cis*	g	-18.64	-10.27	59.43	10.70
trans	g	-19.15	-10.82	59.54	11.01
undissoc; std. state, $m = 1$	aq	-28.5	-13.3	36.5	
HNO$_3$	g	-32.28	-17.87	63.64	12.75
	liq	-41.40	-19.10	37.19	
std. state, $m = 1$	aq	-49.56	-26.61	35.0	-20.7
	aq, 1	-44.845			
	aq, 2	-46.500			
	aq, 10	-49.192			
	aq, 100	-49.440			
NH$_2$OH	c	-25.5			
	aq	-21.7	-5.60	40	
NH$_2$OH$_2^+$	aq	-32.8			
NH$_2$OH $\cdot$ HCl	c	-75.9			
(NH$_2$OH)$_2 \cdot$ H$_2$SO$_4$	aq	-281.3			
NH$_2$OH $\cdot$ H$_2$SO$_4$	aq	-246.7			
H$_2$N$_2$O$_2$ hyponitrous acid	aq	-13.7	8.6	52	
NH$_2$NO$_2$ nitramide	c	-21.4			
N$_4$S$_4$	c	128.0			
NSF	g			62.07	10.55
NSF$_3$	g			68.48	17.18
NSe	c	42.3			
Osmium					
Os	c	0	0	7.8	5.9
	g	189	178	46.000	4.968
OsCl$_3$	c	-45.5	-29	31	
OsCl$_4$	c	-60.9	-38	37	
	g	-19			
OsF$_6$ cubic	c			58.8	
	g			85.56	28.88
OsO$_2$	c		-46		
OsO$_3$	g	-67.8			
OsO$_4$ yellow	c	-94.2	-72.9	34.4	
white	c	-92.2	-72.6	40.1	
	g	-80.6	-70.0	70.2	17.7
Os(OH)$_4$	amorp		-161.0		
OsS$_2$	c	-34.9	-32	13	
Oxygen					
O	g	59.553	55.389	38.467	5.237
O$_2$	g	0	0	49.003	7.016
O$_3$	g	34.1	39.0	57.08	9.37
OF$_2$	g	-5.2	-1.1	59.11	10.35
O$_2$F$_2$	g	4.3			
O$_2$F$_3$	g	3.8			
Palladium					
Pd	c	0	0	8.98	6.21
	g	90.4	81.2	39.90	4.968
Pd^{2+}	aq, ∞	40.5	42.2	-28	
PdBr$_2$	c	-24.9			
PdBr$_4^{2-}$	aq, ∞	-88.8	-76.0	70	
Pd(CN)$_2$	c	56.9			

Table 3–9 (continued)
HEATS AND FREE ENERGIES OF FORMATION, ENTROPIES, AND HEAT CAPACITIES OF ELEMENTS AND INORGANIC COMPOUNDS

Formula and Description	State	$\Delta Hf°$	$\Delta Gf°$	$S°$	$C_p°$
$PdCl_2$	c	−41.0	−29.9	25	
$PdCl_4^{2-}$ 1M HCl	aq, ∞	−124.8	−99.6	62	
$PdCl_6^{2-}$ 1M HCl	aq, ∞	−143	−102.8	65	
$Pd(CNS)_2$	c		56.0		
Pd_2H	c	−4.7	−1.2	21.9	
PdI_2	c	−15.2	−15.0	36	
PdO	c	−20.4			7.5
$Pd(OH)_2$ pptd	c	−89.0	−72		
$Pd(OH)_4$ pptd	c	−156.0	−115	(35)	
PdS	c	−18	−16	11	
PdS_2	c	−19.4	−17.8	19	
PdTe	c			21.42	12.23
$PdTe_2$	c			30.25	18.31
Phosphorus					
P α, white	c	0	0	9.82	5.698
red, triclinic	c	−4.2	−2.9	5.45	5.07
black	c	−9.4			
red	amorp	−1.8			
P_2	g	75.20	66.51	38.978	4.968
P_4	g	14.08	5.85	66.89	16.05
PBr_3	liq	−44.1	−42.0	57.4	
	g	−33.3	−38.9	83.17	18.16
PBr_5	c	−64.5			
PCl_3	liq	−76.4	−65.1	51.9	
	g	−68.6	−64.0	74.49	17.17
PCl_5	c	−106.0			
	g	−89.6	−72.9	87.11	26.96
PF	g			53.74	7.56
PF_3	g	−219.6	−214.5	65.28	14.03
PF_5	g	−381.4			
PH_3	g	1.3	3.2	50.22	8.87
std. state, $m = 1$	aq	−2.16	0.35	48.2	
PH_4^+ std. state, $m = 1$	aq		16.2		
P_2H_4	liq	−1.2	16.0	40	
	g	5.0			
PH_4Br	c	−30.5	−11.4	26.3	
PH_4Cl	c	−34.7			
PH_4I	c	−16.7	0.2	29.4	26.2
PH_4OH std. state, $m = 1$	aq	−70.48	−56.34	65	
PI_3	c	−10.9			
	g			89.45	18.73
PN	g	32.76	27.47	50.45	7.10
P_3N_5	c	−71.4			
$P_3N_3Cl_6$	c	−194.1			
	g	−175.9			
$P_4N_4Cl_8$	c	−259.2			
	g	−236.1			
PO	g			53.221	
PO_3^-	aq	−233.5			
PO_4^{3-} std. state, $m = 1$	aq	−305.3	−243.5	−53	
$P_2O_7^{4-}$ std. state, $m = 1$	aq	−542.8	−458.7	−28	
P_4O_6	c	−392.0			
P_4O_{10} hexagonal	c	−713.2	−644.8	54.70	50.60
	amorp	−727			
$POBr_3$	c	−109.6			
	g			85.97	21.48
$POCl_3$	liq	−142.7	−124.5	53.17	33.17
	g	−133.48	−122.60	77.76	20.30
POF_3	g	−289.5	−277.9	68.11	16.41

Table 3–9 (continued)
HEATS AND FREE ENERGIES OF FORMATION, ENTROPIES, AND
HEAT CAPACITIES OF ELEMENTS AND INORGANIC COMPOUNDS

Formula and Description	State	$\Delta Hf°$	$\Delta Gf°$	$S°$	$C_p°$
HPO_3	c	−226.7			
	aq	−233.5			
HPO_3^{2-}	aq	−231.6			
HPO_4^{2-} std. state, $m = 1$	aq	−308.83	−260.34	−8.0	
$H_2PO_4^-$ std. state, $m = 1$	aq .	−309.82	−270.17	21.6	
H_3PO_4	c	−305.7	−267.5	26.41	25.35
ionized, std. state, $m = 1$	aq	−305.3	−243.5	−53	
undissoc; std. state, $m = 1$	aq	−307.92	−273.10	37.8	
	aq, 1	−304.69			
	aq, 2	−305.60			
	aq, 5	−306.87			
$H_2PO_3^-$	aq	−231.7			
H_3PO_3	c	−230.5			
	aq	−230.6			
$H_2PO_2^-$	aq	−146.7			
H_3PO_2	c	−144.5			
$H_4P_2O_7$	c	−535.6			
undissoc; std. state, $m = 1$	aq	−542.2	−485.7	64	
$HP_2O_7^{3-}$ std. state, $m = 1$	aq	−543.7	−471.6	11	
$H_2P_2O_7^{2-}$ std. state, $m = 1$	aq	−544.6	−480.5	39	
$H_3P_2O_7^-$ std. state, $m = 1$	aq	−544.1	−483.6	51	
$H_4P_2O_5$	aq	−393.6			
H_2PO_3F undissoc; std. state, $m = 1$	aq		−287.5		
P_2S_3	c	−19.2			
$PSBr_3$	c			55.2	
	g			89.07	22.69
$PSCl_3$	g			80.47	21.39
Platinum					
Pt	c	0	0	9.95	6.18
	g	135.1	124.4	45.960	6.102
PtBr	c	−11	(−7)	(28)	
$PtBr_2$	c	−23.1		(44)	
$PtBr_3$	c	−30.9		(56)	
$PtBr_4$	c	−38.0		(68)	
$PtBr_4^{2-}$	aq	−89	(−69)	(47)	
$PtBr_6^{2-}$	aq	−114	(−89)	(67)	
PtCl	c	−9		(27)	
$PtCl_2$	c	−26.5		(28)	
std. state, $m = 1$	aq		−18.3		
$PtCl_3$	c	−41.6	(−32)	(36)	
$PtCl_4$	c	−56.5	(−41)	(42)	
	aq	−76.2			
$PtCl_4^{2-}$ std. state, $m = 1$	aq	−120.3	−88.1	40	
$PtCl_6^{2-}$ std. state, $m = 1$	aq	−161	−117	52.6	
PtF_6 cubic	c			56.3	
	g			83.23	29.35
PtI_4	c	−17.4			
PtO_2	g	41.0	40.1	62	
Pt_3O_4	c	−39			
$Pt(OH)_2$	c	−84.1	(−66)	(30)	
PtS	c	−19.5	−18.2	13.16	10.37
PtS_2	c	−26.0	−23.8	17.85	15.75
PtTe	c			19.41	11.93
$PtTe_2$	c			30.25	18.31
Polonium					
Po	c	0	0		
Po^{2+} std. state, $m = 1$	aq		17		
Po^{4+} std. state, $m = 1$	aq		70		
$PoCl_6^{2-}$ std. state, $m = 1$	aq		−138		

Table 3–9 (continued)
HEATS AND FREE ENERGIES OF FORMATION, ENTROPIES, AND HEAT CAPACITIES OF ELEMENTS AND INORGANIC COMPOUNDS

Formula and Description	State	$\Delta Hf°$	$\Delta Gf°$	$S°$	$C_p°$
Po(OH)$_2^{2+}$ std. state, $m = 1$	aq		-113		
Po(OH)$_4$	c		-130		
PoS	c		52		
Potassium					
K	g	21.51	14.62	38.30	
	c	0	0	15.2	
K$_2$O · Al$_2$O$_3$ · 4SiO$_2$ leucite	c	-1406.4			
K$_2$O · Al$_2$O$_3$ · 6SiO$_2$ microcline	c	-1816			
(adularia)	c	-1842			
K$_2$SiF$_6$	c	-671			
KAl(SO$_4$)$_2$	c	-589.24	-534.29	48.9	
	aq	-616.9			
KAl(SO$_4$)$_2$ · H$_2$O	c	-667.5			
KAl(SO$_4$)$_2$ · 2H$_2$O	c	-742			
KAl(SO$_4$)$_2$ · 3H$_2$O	c	-814			
KAl(SO$_4$)$_2$ · 12H$_2$O	c	-1447.74	-1227.8	164.3	
K$_3$AsO$_3$	aq	-323.0			
K$_3$AsO$_4$	aq	-390.3	-355.7		
KH$_2$AsO$_4$	c	-271.5	-237.0	37.08	
	aq	-276.2			
KBr	c	-93.73	-90.63	23.05	
	aq, 10	-89.77			
	aq, 50	-89.07			
	aq, 500	-88.87	-89		
	aq, 1000	-88.88			
	aq, ∞	-88.94	-92.04	43.8	
KBrO	aq	-82.0			
KBrO$_3$	c	-79.4	-58.2	35.65	
	aq, ∞	-69.6	-56.6	63.4	
K$_2$CO$_3$	c	-273.93			
	aq, 50	-281.02			
	aq, 1000	-281.56	-264		
K$_2$CO$_3$ · ½H$_2$O	c	-210.43			
K$_2$CO$_3$ · 1½H$_2$O	c	-283.40			
KHCO$_3$	c	-229.3			
	aq, 1500	-224.5	-207.7		
K(HCO$_2$) formate	c	-158.0			
	aq	-157.7			
KC$_2$H$_3$O$_2$	c	-173.2			
	aq, 10	-175.6			
	aq, ∞	-176.88	-156.7		
K$_2$C$_2$O$_4$	c	-320.8			
	aq, 100	-316.88			
	aq, 5000	-316.90	-393.1		
	aq, ∞	-317.1			
K$_2$C$_2$O$_4$ · H$_2$O	c	-392.17			
KCN	c	-26.90			
	aq	-24.1	-28		
KCNO cyanate	c	-98.5			
	aq	-93.5	-90.85		
KCNS thiocyanate	c	-48.62			
	aq, 2	-45.46			
	aq, 8	-44.32			
	aq, 50	-43.11			
	aq, ∞	-42.8	-44		
KCl	g	-51.6	-56.2	57.24	
	c	-104.18	-97.59	19.76	
	aq, 50	-100.11			
	aq, ∞	-100.06	-98.82	37.7	
KClO	aq	-85.4			

Table 3–9 (continued)
HEATS AND FREE ENERGIES OF FORMATION, ENTROPIES, AND HEAT CAPACITIES OF ELEMENTS AND INORGANIC COMPOUNDS

Formula and Description	State	$\Delta Hf°$	$\Delta Gf°$	$S°$	$C_p°$
$KClO_3$	c	−93.50	−69.29	34.17	
	aq, 100	−84.09			
	aq, ∞	−83.54	−68.09	63.5	
$KClO_4$	c	−103.6	−72.7	36.1	
	aq, 500	−91.58			
	aq, ∞	−91.45	−70.04	68.0	
K_2CrO_4	c	−330.49			
	aq, 17.2	−328.2			
	aq, 54	−327.0			
	aq, ∞	−326.0	−306		
$K_2Cr_2O_7$	c	−485.90			
	aq, 135	−470.41			
	aq, 800	−469.50	−441		
	aq, 2000	−468.7			
KF	c	−134.46	−127.42	15.91	
	aq, 4	−137.35			
	aq, 50	−138.51			
	aq, ∞	−138.70	−133		
$KF \cdot 2H_2O$	c	−277.0	−242.7	36	
$KF \cdot 4H_2O$	c	−418.0			
KHF_2	c	−219.98	−203.73	24.92	
	aq, 25	−213.73			
	aq, 200	−213.4			
$KF \cdot 2HF$	c	−296.7			
$KF \cdot 3HF$	c	−373.0			
KI	c	−78.31	−77.03	24.94	
	aq, 6	−74.94			
	aq, ∞	−73.41	−79.82	50.6	
KIO_3	c	−121.5	−101.7	36.20	
	aq, 200	−115.16			
	aq, ∞	−115.0	−99.9	52.2	
KIO_4	aq, 1200	−97.6			
$KMnO_4$	c	−194.4	−170.6	41.04	
	aq, 140	−184.44			
	aq, 400	−183.9	−168.0		
K_2MoO_4	aq, 880	−374.1			
KNH_2	c	−28.3			
KNO_2	c	−88.5			
	aq, 400	−85.2	−76		
KNO_3	c	−117.76	−93.96	31.77	
	aq, 400	−109.48	−93.68		
	aq, ∞	−109.41	−93.88	69.5	
K_2O	c	−86.4			
K_2O_2	c	−118			
K_2O_4	c	−134			
KOH	c	−101.78			
	aq, 3	−111.77			
	aq, 5	−113.31			
	aq, 25	−114.70			
	aq, 500	−114.86	−105		
	aq, ∞	−115.00	−105.06	22.0	
$KOH \cdot \frac{1}{2}H_2O$	c	−161.7			
$KOH \cdot H_2O$	c	−179.6			
$KOH \cdot 2H_2O$	c	−251.2			
K_2OsCl_6	c	−280			
K_3PO_3	aq	−397.5			
K_3PO_4	aq	−478.7	−443.3		
KH_2PO_4	c	−374.9	−326		
	aq, 35	−370.70			
	aq, 755	−370.4			

Table 3–9 (continued)
HEATS AND FREE ENERGIES OF FORMATION, ENTROPIES, AND
HEAT CAPACITIES OF ELEMENTS AND INORGANIC COMPOUNDS

Formula and Description	State	$\Delta Hf°$	$\Delta Gf°$	$S°$	$C_p°$
$KReO_4$	c	−264.02			
K_3RhCl_6	c	−343			
K_2S	c	−100			
	aq, 7	−107.04			
	aq, 10	−108.95			
	aq, 20	−110.17			
	aq, 25	−110.27			
	aq, 50	−110.17			
	aq, 200	−109.94	−111.4		
$K_2S \cdot 2H_2O$	c	−243.0			
$K_2S \cdot 5H_2O$	c	−456.7			
KHS	c	−63.2			
	aq, 10	−64.35			
	aq, 200	−64.1			
$KHS \cdot \frac{1}{4}H_2O$	c	−80.1			
K_2S_4	c	−113.0			
	aq	−114.6			
$K_2S_4 \cdot \frac{1}{2}H_2O$	c	−150.5			
$K_2S_4 \cdot 2H_2O$	c	−258.7			
K_2SO_3	c	−266.9			
	aq	−269.1	−251.3		
$KHSO_3$	aq, 385	−209.7			
K_2SO_4	c	−342.66	−314.62	42.0	
	aq, 1000	−336.75	−310		
	aq, 5000	−336.81			
$KHSO_4$	c	−276.8			
	aq, 20	−273.38			
	aq, 100	−273.44			
	aq, 800	−274.3			
$K_2S_2O_5$	c	−362.6			
	aq	−351.9			
$K_2S_2O_5 \cdot 1\frac{1}{2}H_2O$	c	−397.0			
$K_2S_2O_6$	c	−413.6			
	aq, 400	−401.0			
$K_2S_2O_8$	c	−458.3			
	aq	−445.3			
$K_2S_4O_6$	c	−422			
	aq	−410			
$K_2S_5O_6 \cdot 1\frac{1}{2}H_2O$	c	−515.4			
K_2Se	c	−79.3			
	aq	−88.5	−99		
$K_2Se \cdot 9H_2O$	c	−722.0			
$K_2Se \cdot 14H_2O$	c	−1066			
$K_2Se \cdot 19H_2O$	c	−1417			
KHSe	c	−35.9			
	aq	−35.4			
K_2SeO_4	aq, 440	−265.3	−240		
$KHSeO_4$	aq, 220	−202.8			
K_2TeO_3	aq	−261.6			
K_2TeO_4	aq	−290.3			
KVO_3	aq	−284.0			
KVO_4	c	−273.9			
	aq	−270.4			
KVO_5	aq	−254.9			
Praseodymium					
Pr	g	87			
	c	0	0		
$PrBr_3$	c	−225			
$PrCl_3$ α	c	−257.8			
	aq, ∞	−292.8	−263.1		

Table 3–9 (continued)
HEATS AND FREE ENERGIES OF FORMATION, ENTROPIES, AND HEAT CAPACITIES OF ELEMENTS AND INORGANIC COMPOUNDS

Formula and Description	State	$\Delta Hf°$	$\Delta Gf°$	$S°$	$C_p°$
$PrCl_3 \cdot H_2O$	c	−330.8			
$PrCl_3 \cdot 7H_2O$	c	−764.5			
PrI_3 n	c	−162.0			
	aq, ∞	−212.8	−206.1		
$Pr(NO_3)_3$	aq	−318.8			
PrO_2	c	−234.0			
Pr_2O_3	c	−444.5			
Pr_6O_{11}	c	−1391			
$Pr(OH)_3$	c		−310.7		
$Pr_2(SO_4)_3$	aq, ∞	−996.1	−870.2		
$Pr_2(SO_4)_3 \cdot 8H_2O$	c		−1337.0		
Promethium					
Pm	c	0	0		
$PmCl_3$	c	−251.9			
	aq, ∞	−290.5			
Radium					
Ra	g	31	23	42.15	
	c	0	0	17	
$RaBr_2$	c	−195			
$RaCl_2 \cdot 2H_2O$	c	−351	−311.7	50	
$Ra(NO_3)_2$	c	−237	−190.3	52	
RaO	c	−125			
$RaSO_4$	c	−352	−326.0	34	
Radon					
Rn	g	0	0	42.09	4.968
Rhenium					
Re	c	0	0	8.81	6.09
	g	184.0	173.2	45.131	4.968
Re^- std. state, $m = 1$	aq	11	2.4	55	
$ReAs_2$	c	1.3			
$ReBr_3$	c	−40			
Re_3Br_9	g	−69			
$ReCl_3$	c	−63	−45	29.6	22.08
$ReCl_5$	c	−89			
Re_3Cl_9	g	−137			
$ReCl_6^{2-}$ std. state, $m = 1$	aq	−182	−141	60	
H_2ReCl_4	c	−152			
ReF_6	g	−273			
ReO_2	c	−101	−88	41	
$ReO_2 \cdot 2H_2O$ pptd	c	−236			
ReO_3	c	−144.6	−127	61.5	
Re_2O_7	c	−296.4	−254.8	49.5	39.7
	g	−263	−237.6	108	
$HReO_4$	c	−182.2	−156.9	37.8	
	g	−159			
std. state, $m = 1$	aq	−188.2	−166.0	48.1	−3.2
ReS_2	c	−43			
Re_2S_7	c	−107			
Re_3Si	c	0			
Rhodium					
Rh	c	0	0	7.53	5.97
	g	133.1	122.1	44.383	5.022
$RhCl_2$	g	30.3			
$RhCl_3$	c	−71.5	(−54)	(21)	
	g	16			
Rh_2O_3	c	−82	(−65)	(27)	24.8
$Rh(OH)_3$	c		(−112)		
Rubidium					
Rb	g	20.51	13.35	40.63	
	c	0	0	16.6	

Table 3–9 (continued)
HEATS AND FREE ENERGIES OF FORMATION, ENTROPIES, AND HEAT CAPACITIES OF ELEMENTS AND INORGANIC COMPOUNDS

Formula and Description	State	$\Delta Hf°$	$\Delta Gf°$	$S°$	$C_p°$
$RbAl(SO_4)_2$	c	−569			
$RbAl(SO_4)_2 \cdot H_2O$	c	−647			
$RbAl(SO_4)_2 \cdot 2H_2O$	c	−722			
$RbAl(SO_4)_2 \cdot 3H_2O$	c	−797			
$RbAl(SO_4)_2 \cdot 12H_2O$	c	−1448.0			
RbBr	c	−93.03	−90.38	25.88	
	aq, 200	−87.77			
	aq, ∞	−87.8	−92.02	49.0	
Rb_2CO_3	c	−269.6			
	aq, 200	−279.4	−263.8		
	aq, 2000	−278.0			
$Rb_2CO_3 \cdot H_2O$	c	−344.2			
$Rb_2CO_3 \cdot 1\frac{1}{2}H_2O$	c	−381.5			
$Rb_2CO_3 \cdot 3\frac{1}{2}H_2O$	c	−521.7			
$RbHCO_3$	c	−228.5			
	aq	−224.1	−209.1		
RbCN	aq	−25.9			
RbCNS thiocyanate	c	−54			
	aq	−41.7			
RbCl	c	−102.9	−98.48		
	aq, 100	−98.90			
	aq, 1000	−98.84			
	aq, ∞	−98.9	−98.80	42.9	
$RbClO_3$	c	−93.8	−69.8	36.3	
	aq, ∞	−82.4	−68.07	68.7	
$RbClO_4$	c	−103.87	−73.19	38.4	
	aq, ∞	−90.3	−70.02	73.2	
RbF	c	−131.28			
	aq, 100	−137.45			
	aq, ∞	−137.6	−133.53	27.4	
$RbF \cdot \frac{1}{3}H_2O$	c	−156.12			
$RbF \cdot 1\frac{1}{2}H_2O$	c	−240.33			
$RbHF_2$	c	−217.3			
	aq	−212.5			
RbI	c	−78.5	−77.8	28.21	
	aq, ∞	−72.3	−79.80	55.8	
Rb_2IrCl_6	c	−290.9			
$RbNO_3$	·c	−117.04			
	aq, 60	−109.00			
	aq, 200	−108.54			
	aq, ∞	−108.5	−93.86	64.7	
$RbNH_2$	c	−25.7			
Rb_2O	c	−78.9			
Rb_2O_2	c	−101.7			
RbOH	c	−98.9			
	aq, 3.2	−110.5			
	aq, 200	−113.7			
	aq, ∞	−113.9	−105.05	27.2	
$RbOH \cdot H_2O$	c	−177.8			
$RbOH \cdot 2H_2O$	c	−250.8			
$RbReO_4$	c	−256.9			
	aq	−249.2			
Rb_2S	c	−83.2			
	aq, 500	−107.8			
RbHS	c	−62.4			
	aq	−63.1			
$RbHSO_4$	c	−273.7			
	aq, 300	−270.4			
Rb_2SO_4	c	−340.50			
	aq, 400	−334.51			
	aq, ∞	−334.7	−312.24	33.8	

Table 3–9 (continued)
HEATS AND FREE ENERGIES OF FORMATION, ENTROPIES, AND HEAT CAPACITIES OF ELEMENTS AND INORGANIC COMPOUNDS

Formula and Description	State	$\Delta Hf°$	$\Delta Gf°$	$S°$	$C_p°$
RbHSe	c	−35.5			
	aq	−34.3			
Rb_2SiF_6	c	−678.4			
Ruthenium					
Ru	c	0	0	6.82	5.75
	g	153.6	142.4	44.550	5.144
$RuBr_3$	c	−33			
$RuCl_3$ black	c	−49			
	g	−0.3			
$RuCl_4$	g	−12.4			
$RuCl_5(OH)^{2-}$	aq, ∞		−168.7		
RuF_5	c	−213.4			
	g	−189			
RuI_3	c	−15.7			
RuO_2	c	−72.9			
hydrated	amorp		−51.3		
RuO_3	g	−18.7			
RuO_4	c	−57.2	−36.4	35.0	
	liq	−54.6	−36.4	43.8	
	g	−44.0	−33.4	69.3	18.14
RuO_4^-	aq, ∞		−58.7		
RuO_4^{2-}	aq, ∞		−72.6		
RuS_2	c	−47	−44		
Samarium					
Sm	g	87		43.74	
	c	0	0		
$SmBr_3$	c	−216			
$SmCl_3$ α	c	−249.8			
	aq, ∞	−289.9	−259.9		
SmI_3 β	c	−153.4			
	aq, ∞	−209.9	−202.9		
$Sm(OH)_3$	c		−308.8		
$Sm_2(SO_4)_3$	aq, ∞	−990.3	−863.8		
$Sm_2(SO_4)_3 \cdot 8H_2O$	c		−1332.6		
Scandium					
Sc	c	0	0	8.28	6.10
	g	90.3	80.32	41.75	5.28
Sc_2	g	154.9	141.6	61	8.7
Sc^{3+} std. state, $m = 1$	aq	−146.8	−140.2	−61	
$ScBr_2$	g			77.6	13.0
$ScBr_3$	c	−177.6			
ScCl	g	26.9	20.6	56.00	8.40
$ScCl_2$	g			72.5	12.6
$ScCl_3$	c	−221.1			
	aq, 5500	−268.8			
$ScCl_3 \cdot 6H_2O$	c	−671.6			
$Sc(CNS)^{2+}$ std. state, $m = 1$	aq		−119.6		
ScF	g	−33.2	−39.3	53.11	7.74
ScF_2	g	−153.5	−156.6	67.0	11.5
ScF_3	c	−389.4	−371.8	22	
	g	−298	−295	71.8	16.2
ScI_2	g			81.6	13.3
ScO	g	−13.68	−19.90	53.65	7.38
Sc_2O	g	−6.9			11.2
Sc_2O_3	c	−456.22	−434.85	18.4	22.52
$Sc(OH)^{2+}$ std. state, $m = 1$	aq	−205.9˙	−191.5	−32	
$Sc(OH)_3$	c	−325.9	−294.8	24	
$Sc(OH)_2Cl$	c	−303	−276.3	26	
ScS	g	41.8	29.7	56.3	8.0

Table 3–9 (continued)
HEATS AND FREE ENERGIES OF FORMATION, ENTROPIES, AND
HEAT CAPACITIES OF ELEMENTS AND INORGANIC COMPOUNDS

Formula and Description	State	$\Delta Hf°$	$\Delta Gf°$	$S°$	$C_p°$
Sc(SO$_4$)' std. state, $m = 1$	aq		−321.7		
Sc(SO$_4$)$_2$ std. state, $m = 1$	aq		−501.5		
Sc$_2$(SeO$_3$)$_3$ · 10H$_2$O	c	−1326.5			
Selenium					
Se hexagonal, black	c	0	0	10.144	6.062
monoclinic, red	c	1.6			
	g	54.27	44.71	42.21	4.978
glassy	amorp	1.2			
Se$_2$	g	34.9	23.0	60.2	8.46
Se^2 std. state, $m = 1$	aq		30.9		
Se$_6$	g	39.2			
SeBr$_2$	g	−5			
Se$_2$Br$_2$	g	7			
SeCl$_2$	g	−7.6			
SeCl$_4$	c	−43.8			
Se$_2$Cl$_2$	liq	−19.7			
	g	4			
SeF$_6$	g	−267	−243	74.99	26.4
HSe std. state, $m = 1$	aq	3.8	10.5	19	
H$_2$Se	g	7.1	3.8	52.32	8.30
std. state, $m = 1$	aq	4.6	5.3	39.1	
SeO	g	12.75	6.41	55.9	7.47
SeO$_2$	c	−53.86			
SeO$_3$	c	−39.9			
SeO$_3^2$ std. state, $m = 1$	aq	−121.7	−88.4	3	
SeO$_4^2$ std. state, $m = 1$	aq	−143.2	−105.5	12.9	
Se$_2$O$_5$	c	−97.6			
SeOCl$_2$	g	−6			
HSeO$_3$ std. state, $m = 1$	aq	−122.98	−98.36	32.3	
HSeO$_4$ std. state, $m = 1$	aq	−139.0	−108.1	35.7	
H$_2$SeO$_3$	c	−125.35			
undissoc; std. state, $m = 1$	aq	−121.29	−101.87	49.7	
H$_2$SeO$_4$	c	−126.7			
	aq, ∞	−140.3			
Silicon					
Si	c	0	0	4.50	4.78
	amorp	1.0			
	g	108.9	98.3	40.12	5.318
Si$_2$	g	142	128	54.92	8.22
SiBr	g	50			9.23
SiBr$_4$	liq	−109.3	−106.1	66.4	
	g	−99.3	−103.2	90.29	23.21
SiC β, cubic	c	−15.6	−15.0	3.97	6.42
α, hexagonal	c	−15.0	−14.4	3.94	6.38
	g	177			
SiCl	g	45.39			8.81
SiCl$_2$	g	−39.59	−42.35	67.0	12.16
SiCl$_4$	liq	−164.2	−148.16	57.3	34.73
	g	−157.03	−147.47	79.02	21.57
SiH$_3$Cl	g			59.88	12.20
SiH$_2$Cl$_2$	g			68.26	14.45
SiHCl$_3$	liq	−128.9	−115.34	54.4	
SiF	g	1.7	−5.8	53.94	7.80
SiF$_2$	g	−148	−150	60.38	10.49
SiF$_4$	g	−385.98	−375.88	67.49	17.60
SiF$_6^2$ std. state, $m = 1$	aq	−571.0	−525.7	29.2	
SiH$_3$F	g			56.95	11.33
SiHF$_3$	g			64.96	14.47
SiH	g			47.42	6.98

Table 3–9 (continued)
HEATS AND FREE ENERGIES OF FORMATION, ENTROPIES, AND
HEAT CAPACITIES OF ELEMENTS AND INORGANIC COMPOUNDS

Formula and Description	State	$\Delta Hf°$	$\Delta Gf°$	$S°$	$C_p°$
SiH₄	g	8.2	13.6	48.88	10.24
Si₂H₆	g	19.2	30.4	65.14	19.31
Si₃H₈	liq	22.1			
	g	28.9			
SiI₄	c	−45.3			
SiN	g	116.28	109.01	51.78	7.21
Si₃N₄	c	−177.7	−153.6	24.2	
SiO₂ quartz	c	−217.72	−204.75	10.00	10.62
cristobalite	c	−217.37	−204.56	10.20	10.56
tridymite	c	−217.27	−204.42	10.4	10.66
	amorp	−215.94	−203.33	11.2	10.6
H₂SiO₃	c	−284.1	−261.1	32	
undissoc; std. state, $m = 1$	aq	−282.7	−258.0	26	
H₄SiO₄	c	−354.0	−318.6	46	
undissoc; std. state, $m = 1$	aq	−351.0	−314.7	43	
SiS	g	26.88	14.56	53.43	7.71
SiS₂	c	−49.5			
SiSe	g	23.78			8.04
SiSe₂	c	−7			
SiTe	g	30.99			8.31
Silver					
Ag	c	0	0	10.17	6.059
	g	68.01	58.72	41.321	4.968
Ag₂	g	97.99	85.75	61.43	8.84
Ag⁺ std. state, $m = 1$	aq	25.234	18.433	17.37	5.2
Ag²⁺ in $4M$ HClO₄, std. state	aq	64.2	64.3	−21	
Ag₃AsO₄	c		−129.7		
AgBr	c	−23.99	−23.16	25.6	12.52
std. state, $m = 1$	aq	−3.82	−6.42	37.1	−28.7
AgBrO₃	c	−6.5	13.0	36.5	
AgCl	c	−30.370	−26.244	23.0	12.14
	g			58.75	8.57
std. state, $m = 1$	aq	−14.718	−12.939	30.9	−27.4
AgClO₂	c	2.10	18.1	32.16	20.87
std. state, $m = 1$	aq	9.3	22.5	41.6	
AgClO₃	c	−6.1			
std. state, $m = 1$	aq	1.5	17.6	56.2	
AgClO₄	c	−7.44			
std. state, $m = 1$	aq	−5.68	16.37	60.9	
Ag₂CrO₄	c	−174.89	−153.40	52.0	34.00
AgCN	c	34.9	37.5	25.62	15.95
Ag(CN)₂⁻ std. state, $m = 1$	aq	64.6	73.0	46	
AgCN₂ cyanamide	c	56.2			
AgONC fulminate	c	43			
AgOCN cyanate	c	−22.8	−13.9	29	
AgSCN thiocyanate	c	21.0	24.23	31.3	15
Ag₂CO₃	c	−120.9	−104.4	40.0	26.83
Ag₂C₂O₄	c	−160.9	−139.6	50	
Ag acetate	c	−95.3	−73.56	35.8	
AgF	c	−48.9			
std. state, $m = 1$	aq	−54.27	−48.21	14.1	−20.3
AgF · 2H₂O	c	−191.4	−160.4	41.8	31
AgF₂	c	−87.3			
AgI	c	−14.78	−15.82	27.6	13.58
std. state, $m = 1$	aq	12.04	6.10	44.0	−28.8
AgIO₃	c	−40.9	−22.4	35.7	24.60
std. state, $m = 1$	aq	−27.7	−12.2	45.7	
Ag₂MoO₄	c	−200.9	−178.8	51	
AgN₃	c	73.8	89.9	24.9	

Table 3–9 (continued)
HEATS AND FREE ENERGIES OF FORMATION, ENTROPIES, AND
HEAT CAPACITIES OF ELEMENTS AND INORGANIC COMPOUNDS

Formula and Description	State	$\Delta Hf°$	$\Delta Gf°$	$S°$	$C_p°$
Ag_3N	c	47.6			
$Ag(NH_3)_2^+$ std. state, $m = 1$	aq	−26.60	−4.12	58.6	
$AgNO_2$	c	−10.77	4.56	30.64	19.17
$AgNO_3$	c	−29.73	−8.00	33.68	22.24
Ag_2O	c	−7.42	−2.68	29.0	15.74
AgO	c	−2.73	3.40	13.810	10.76
Ag_2O_2	c	−5.8	6.6	28	21
Ag_2O_3	c	8.1	29.0	24	
$AgOH$ std. state, $m = 1$	aq	−29.736	−19.161	14.80	−30.3
AgP_2	c	−11.0			
AgP_3	c	−16.6			
Ag_3PO_4	c		−210		
$Ag_4P_2O_7$	c	−453			
$AgReO_4$	c	−176	−151.9	36.6	
Ag_2S α, orthorhombic	c	−7.79	−9.72	34.42	18.29
β	c	−7.03	−9.43	36.0	
Ag_2SO_3	c	−117.3	−98.3	37.8	
std. state, $m = 1$	aq	−101.4	−79.4	27.8	
Ag_2SO_4	c	−171.10	−147.82	47.9	31.40
std. state, $m = 1$	aq	−166.85	−141.10	39.6	−60
Ag_2Se	c	−9	−10.6	36.02	19.54
Ag_2SeO_3	c	−87.3	−72.7	55.0	
Ag_2SeO_4	c	−100.5	−79.9	59.4	
Ag_2Te	c	−8.9	10.3	37.0	20.9
Ag_2WO_4	c	−221.2			
Sodium					
Na	g	25.98	18.67	36.72	
	c	0	0	12.2	
$NaAlO_2$	c	−273			
NaH_2AsO_3	aq, 400	−227.7			
NaH_2AsO_4	aq, 300	−273.4			
Na_2HAsO_4	aq, 400	−329.5			
Na_3AsO_3	aq, 500	−314.6			
Na_3AsO_4	c	−365			
	aq, 500	−381.5	−341.17		
$Na_3AsO_4 \cdot 12H_2O$	c	−1213.9			
$NaBO_2$	c	−253			
	aq, 300	−241.1			
$NaBO_3$ peroxyborate	aq	−220.0			
$NaBO_3 \cdot 4H_2O$	c	−504.8			
$Na_2B_4O_7$	c	−777.7			
	aq, 900	−787.9			
$Na_2B_4O_7 \cdot 4H_2O$	c	−1072.9			
$Na_2B_4O_7 \cdot 5H_2O$	c	−1143.5			
$Na_2B_4O_7 \cdot 10H_2O$	c	−1497.2			
Na_3BiO_4	c	−288			
NaBr	g	−36.33			
	c	−86.03			
	aq, 8	−86.89			
	aq, ∞	−86.18	−87.16	33.7	
$NaBr \cdot 2H_2O$	c	−227.25			
NaBrO	aq	−79.1			
$NaBrO_3$	aq, 400	−68.89	−57.59		
NaCN	c	−21.46			
$NaCN \cdot \frac{1}{2}H_2O$	c	−56.19			
$NaCN \cdot 2H_2O$	c	−162.25			
NaCNO cyanate	c	−95.6			
	aq, 2000	−90.9			
NaNHCN	aq	−34.5			

Table 3–9 (continued)
HEATS AND FREE ENERGIES OF FORMATION, ENTROPIES, AND HEAT CAPACITIES OF ELEMENTS AND INORGANIC COMPOUNDS

Formula and Description	State	$\Delta Hf°$	$\Delta Gf°$	$S°$	$C_p°$
NaCNS thiocyanate	c	−41.73			
	aq, 3	−41.13			
	aq, 100	−40.16			
	aq, ∞	−40.1			
Na$_2$CO$_3$	c	−270.3	−250.4	32.5	
	aq, 15	−278.13			
	aq, 40	−277.30			
	aq, 200	−276.17			
	aq, 400	−275.9	−251.4		
Na$_2$CO$_3$ · H$_2$O	c	−341.8			
Na$_2$CO$_3$ · 7H$_2$O	c	−765.1			
Na$_2$CO$_3$ · 10H$_2$O	c	−975.6			
NaHCO$_3$	c	−226.5	−203.6	24.4	
	aq	−222.5	−203.9		
Na$_2$CO$_3$ · NaHCO$_3$ · 2H$_2$O	c	−641.2			
NaHCO$_2$ formate	c	−155.03			
	aq, 400	−154.92			
NaHCO$_2$ · 2H$_2$O	c	−296.6			
NaHCO$_2$ · 3H$_2$O	c	−364.2			
NaC$_2$H$_3$O$_2$	c	−169.8			
	aq, 3	−171.20			
	aq, 6	−172.57			
	aq, 25	−173.63			
	aq, 6400	−174.07	−152.3		
	aq, ∞	−174.12			
NaC$_2$H$_3$O$_2$ · 3H$_2$O	c	−383.50			
Na$_2$C$_2$O$_4$	c	−314.3			
	aq, 600	−310.5	−283.4		
NaHC$_2$O$_4$	c	−257.8			
	aq, 400	−252.7			
NaHC$_2$O$_4$ · H$_2$O	c	−330.2			
NaCl	g	−43.50			
	c	−98.23	−91.79	17.30$^-$	
	aq, 8	−97.78			
	aq, 100	−97.25			
	aq, 400	−97.21	−93.92		
	aq, 1000	−97.23			
	aq, ∞	−97.30	−93.94	27.6	
NaClO	aq	−82.7			
NaClO$_2$	c	−72.65			
	aq, 1000	−73.80			
NaClO$_3$	c	−85.73			
	aq, 400	−80.74	−62.8		
	aq, 1000	−80.73			
	aq, 5000	−80.74			
	aq, ∞	−80.78	−63.21	53.4	
NaClO$_4$	c	−92.18			
	aq, 400	−88.76			
	aq, 5000	−88.67			
	aq, ∞	−88.69	−65.16	57.9	
Na$_2$CrO$_4$	c	−317.6			
	aq, 600	−320.6	−296		
Na$_2$CrO$_4$ · 4H$_2$O	c	−601.3			
Na$_2$Cr$_2$O$_7$	aq, 600	−463.4	−431		
NaF	c	−136.0	−129.3	14.0	
	aq, ∞	−135.94	−128.67	12.1	
NaHF$_2$	c	−216.6			
	aq	−211.2			
NaH	g	29.88	24.78	44.93	
	c	−13.7	−9.3		

Table 3–9 (continued)
HEATS AND FREE ENERGIES OF FORMATION, ENTROPIES, AND HEAT CAPACITIES OF ELEMENTS AND INORGANIC COMPOUNDS

Formula and Description	State	$\Delta Hf°$	$\Delta Gf°$	$S°$	$C_p°$
NaI	g	−20.94			
	c	−68.84			
	aq, 8	−71.50			
	aq, 15	−71.19			
	aq, 30	−70.82			
	aq, 100	−70.60			
	aq, 1000	−70.58			
	aq, 5000	−70.61			
	aq, ∞	−70.65	−70.94	40.5	
NaI · 2H$_2$O	c	−211.05			
NaIO$_3$	aq, 100	−111.93			
	aq, 500	−112.13	−94.8		
	aq, ∞	−112.2			
Na$_2$IrCl$_6$	c	−233.4			
Na$_2$MoO$_4$	c	−368			
	aq, 800	−368.6	−333		
NaNH$_2$	c	−28.4			
NaNO$_2$	c	−85.9			
	aq	−82.6	−71.0		
NaNO$_3$	c	−111.54	−87.45	27.8	
	aq, 6	−108.47			
	aq, 15	−107.81			
	aq, 100	−106.83			
	aq, 1000	−106.60			
	aq, 5000	−106.61			
	aq, ∞	−106.65	−89.00	49.4	
Na$_2$O	c	−99.4	−90.0	17.4	
Na$_2$O$_2$	c	−120.6	−105.0		
NaOH	c	−101.99	−90.60		
	aq, 3	−108.89			
	aq, 6	−111.52			
	aq, 25	−112.22			
	aq, 100	−112.11			
	aq, 300	−112.11			
	aq, 1000	−112.14			
	aq, ∞	−112.24	−100.18	11.9	
NaOH · H$_2$O	c	−175.17	−149.00	20.2	
NaPO$_3$	c	−288.6			
	aq, 600	−292.8			
Na$_3$PO$_3$	aq, 1000	−389.1			
NaH$_2$PO$_3$	c	−289.4			
	aq, 600	−290.5			
NaH$_2$PO$_3$ · 2½H$_2$O	c	−454.8			
Na$_2$HPO$_3$	c	−338.0			
	aq, 800	−347.5			
Na$_2$HPO$_3$ · 5H$_2$O	c	−684.2			
NaH$_2$PO$_4$	aq, 300	−367.7			
Na$_2$HPO$_4$	c	−417.4			
	aq, 200	−423.89			
	aq, 1000	−423.44			
Na$_2$HPO$_4$ · 2H$_2$O	c	−560.2			
Na$_2$HPO$_4$ · 7H$_2$O	c	−913.3			
Na$_2$HPO$_4$ · 12H$_2$O	c	−1266.4			
Na$_3$PO$_4$	c	−460			
	aq, 300	−475.0	−428.7		
	aq, 1000	−473.9			
Na$_3$PO$_4$ · 12H$_2$O	c	−1309.0			
NaNH$_4$HPO$_4$	aq, 500	−398.8			
NaNH$_4$HPO$_4$ · 4H$_2$O	c	−682.7			

Table 3–9 (continued)
Table 3–9 (continued)
HEATS AND FREE ENERGIES OF FORMATION, ENTROPIES, AND HEAT CAPACITIES OF ELEMENTS AND INORGANIC COMPOUNDS

Formula and Description	State	$\Delta Hf°$	$\Delta Gf°$	$S°$	$C_p°$
$NaH_2P_2O_7$	c	−602.7			
	aq, 120	−603.9			
$NaH_2P_2O_7 \cdot H_2O$	c	−670.6			
$Na_2H_2P_2O_7$	c	−663.4			
	aq, 1500	−661.6			
$Na_2H_2P_2O_7 \cdot 6H_2O$	c	−1085.5			
$Na_3HP_2O_7$	c	−711.4			
	aq, 1500	−718.5			
$Na_3HP_2O_7 \cdot H_2O$	c	−788.2			
$Na_3HP_2O_7 \cdot 6H_2O$	c	−1135.7			
$Na_4P_2O_7$	c	−760.8			
	aq, 1500	−772.9			
$Na_4P_2O_7 \cdot 10H_2O$	c	−1468.2			
Na_2PbO_3	c	−205			
Na_2PtBr_6	c	−221.8			
	aq	−231.7			
$Na_2PtBr_6 \cdot 6H_2O$	c	−650.1			
Na_2PtCl_4	aq	−240.0	−216.8		
Na_2PtCl_6	c	−273.6			
	aq	−282.0			
$Na_2PtCl_6 \cdot 2H_2O$	c	−418.5			
$Na_2PtCl_6 \cdot 6H_2O$	c	−702.5			
Na_2PtI_6	aq	−170.1			
$NaReO_4$	c	−249.4			
	aq	−247.6			
Na_2S	c	−89.2			
	aq, 20	−105.42			
	aq, 100	−104.55			
	aq, 800	−104.36	−101.8		
$Na_2S \cdot 4\frac{1}{2}H_2O$	c	−416.9			
$Na_2S \cdot 5H_2O$	c	−452.7			
$Na_2S \cdot 9H_2O$	c	−736.7			
$NaHS$	c	−56.5			
	aq, 4	−61.60			
	aq, 8	−62.26			
	aq, 20	−61.77			
	aq, 100	−61.36			
	aq, 800	−61.25			
$NaHS \cdot 2H_2O$	c	−199.27			
Na_2S_2	aq	−104.6			
Na_2S_3	aq	−106.5			
Na_2S_4	c	−98.4			
Na_2SO_3	c	−260.6	−239.5	34.9	
	aq	−263.8			
$Na_2SO_3 \cdot 7H_2O$	c	−753.4			
$NaHSO_3$	aq	−206.6			
Na_2SO_4	c	−330.90	−302.78	35.73	
	aq, 100	−331.64			
	aq, 1000	−331.22			
	aq, 3000	−331.25			
	aq, ∞	−331.46	−302.52	32.9	
$Na_2SO_4 \cdot 10H_2O$	c	−1033.48	−870.93	141.7	
$NaHSO_4$	c	−269.2			
	aq, 200	−270.6			
$NaHSO_4 \cdot H_2O$	c	−339.2			
$Na_2SO_4 \cdot (NH_4)_2SO_4 \cdot H_2O$	c	−691.5			
$Na_2S_2O_3$	c	−267.0			
	aq	−269			
$Na_2S_2O_3 \cdot 5H_2O$ I	c	−621.89			
II	c	−620.60			

Table 3–9 (continued)
HEATS AND FREE ENERGIES OF FORMATION, ENTROPIES, AND HEAT CAPACITIES OF ELEMENTS AND INORGANIC COMPOUNDS

Formula and Description	State	$\Delta Hf°$	$\Delta Gf°$	$S°$	$C_p°$
$Na_2S_2O_5$	c	−349.1			
	aq	−345			
$Na_2S_2O_6$	c	−399.9			
	aq	−394.6			
$Na_2S_2O_6 \cdot 2H_2O$	c	−542.5			
$Na_2S_3O_6$	aq	−409			
$Na_2S_3O_6 \cdot 3H_2O$	c	−623.0			
$Na_2S_4O_6$	aq	−405			
$Na_2S_4O_6 \cdot 2H_2O$	c	−550.0			
Na_3SbO_4	c	−352			
Na_2Se	c	−63.0			
	aq	−82.9	−89.4		
$Na_2Se \cdot 4\frac{1}{2}H_2O$	c	−398.2			
$Na_2Se \cdot 9H_2O$	c	−709.1			
$Na_2Se \cdot 16H_2O$	c	−1199.4			
NaHSe	c	−27.8			
	aq	−32.7			
Na_2SeO_3	aq	−236.6			
$NaHSeO_3$	aq	−180.7			
Na_2SeO_4	c	−258			
	aq	−259.7	−230.3		
$NaHSeO_4$	aq	−201.1			
Na_2SiO_3	c	−363	341	27.2	
$Na_2SiO_3 \cdot 5H_2O$	c	−720.0			
$Na_2SiO_3 \cdot 9H_2O$	c	−1002.0			
Na_2SiF_6	c	−677			
	aq, 600	−671.2			
$NaHSiF_6$	aq, 400	−614.1			
Na_2SnO_3	c	−276			
Na_4SnO_4	aq, 1200	−455.5			
Na_2Te	c	−84.0			
Na_2Te_2	c	−101.5			
Na_2TeO_4	c	−313			
Na_2UO_4	c	−501			
$Na_2U_2O_7 \cdot 1\frac{1}{2}H_2O$	c	−880			
$(Na_2O_2)_2 \cdot UO_4$	aq	−596			
$(Na_2O_2)_2 \cdot UO_4 \cdot 9H_2O$	c	−1225			
Na_3VO_4	c	−420			
Na_2WO_4	c	−395			
	aq, 220	−380.9	−345		
Na_2ZnO_2	c	−188			
Strontium					
Sr	g	39.2	26.3	39.33	
	c	0	0	13.0	
$Sr_3(AsO_4)_2$	c	−800.7			
$SrHAsO_4$	aq	−344.8			
$Sr(H_2AsO_4)_2$	aq	−561.4			
$SrBr_2$	c	−171.7			
	aq, 100	−187.56			
	aq, ∞	−188.2	−182.1	29.2	
$SrBr_2 \cdot H_2O$	c	−246.2			
$SrBr_2 \cdot 6H_2O$	c	−604.4			
$SrCO_3$	c	−291.2	−271.9	23.2	
$Sr(HCO_3)_2$	aq, ∞	−460.7	−413.8	36.0	
$Sr(HCO_2)_2$ formate	c	−325.5			
	aq	−326.4			
$Sr(HCO_2)_2 \cdot 2H_2O$	c	−468.3			
$Sr(C_2H_3O_2)_2$	c	−356.7			
	aq	−362.7	−311.8		

Table 3–9 (continued)
HEATS AND FREE ENERGIES OF FORMATION, ENTROPIES, AND HEAT CAPACITIES OF ELEMENTS AND INORGANIC COMPOUNDS

Formula and Description	State	$\Delta Hf°$	$\Delta Gf°$	$S°$	$C_p°$
$Sr(C_2H_3O_2)_2 \cdot \frac{1}{2}H_2O$	c	−391.2			
SrC_2O_4	aq, ∞	−326.1	−293.1	3	
$SrC_2O_4 \cdot H_2O$	c		−359.6		
$SrC_2O_4 \cdot 2\frac{1}{2}H_2O$	c	−502.9			
$Sr(CN)_2$	aq	−57.8	−54.5		
$Sr(CN)_2 \cdot 4H_2O$	c	−335.2			
$SrCl_2$	c	−198.0	−186.7	28	
	aq, ∞	−210.43	−195.7	16.9	
$SrCl_2 \cdot H_2O$	c	−271.7			
$SrCl_2 \cdot 2H_2O$	c	−343.7			
$SrCl_2 \cdot 6H_2O$	c	−627.1			
SrF_2	c	−290.3			
SrH	g	52.4	45.8	49.43	
SrH_2	c	−42.3			
SrI_2	c	−135.5			
	aq, ∞	−157.1	−157.7	42.9	
$SrI_2 \cdot H_2O$	c	−212.2			
$SrI_2 \cdot 2H_2O$	c	−282.8			
$SrI_2 \cdot 6H_2O$	c	−571.2			
$Sr(N_3)_2$	c	48.9			
Sr_3N_2	c	−93.4	−76.5		
$Sr(NO_2)_2$	c	−179.3			
	áq	−180.4			
$Sr(NO_3)_2$	c	−233.25			
	aq, 20	−231.95			
	aq, 200	−229.18			
	aq, 1000	−228.84			
	aq, 2000	−228.84			
	aq, ∞	−229.02	−185.8	60.06	
$Sr(NO_3)_2 \cdot 4H_2O$	c	−514.5			
SrO	c	−141.1	−133.8	13.0	
SrO_2	c	−153.6	−139.0		
$SrO_2 \cdot 8H_2O$	c	−722.6			
$Sr(OH)_2$	c	−229.3			
	aq, ∞	−240.29	−208.2	−14.4	
$Sr(OH)_2 \cdot H_2O$	c	−302.3			
$Sr(OH)_2 \cdot 8H_2O$	c	−801.2			
$Sr_3(PO_4)_2$	c	−987.3			
$SrHPO_4$	c	−431.3			
$Sr(H_2PO_4)_2 \cdot H_2O$	c	−819.4			
SrS	g	19			
	c	−108.1			
$SrSO_4$	c	−345.3	−318.9	29.1	
	aq, ∞	−347.38	−310.3	−5.3	
SrS_2O_6	aq	−410.8			
$SrS_2O_6 \cdot 4H_2O$	c	−693.1			
$SrSe$	c	−78.7			
$SrSi_2$	c	−150			
$SrSiO_3$	c	−371.2			
Sr_2SiO_4	c	−520.6			
$SrWO_4$	c	−398.3			
Sulfur					
S rhombic	c	0	0	7.60	5.41
monoclin	c	0.08			
	g	66.636	56.951	40.084	5.658
S_2	g	30.68	18.96	54.51	7.76
S_8	g	24.45	11.87	102.98	37.39
S^{2-} std. state, $m = 1$	aq	7.9	20.5	−3.5	
S_2Br_2	liq	−3			

Table 3–9 (continued)
HEATS AND FREE ENERGIES OF FORMATION, ENTROPIES, AND
HEAT CAPACITIES OF ELEMENTS AND INORGANIC COMPOUNDS

Formula and Description	State	$\Delta Hf°$	$\Delta Gf°$	$S°$	$C_p°$
SCl_2	liq	-12			
	g	-4.7			
S_2Cl_2	g	-4.4	-7.6	79.2	17.6
S_3Cl_2	liq	-12.4			
S_4Cl_2	liq	-10.2			
S_5Cl_2	liq	-8.8			
SCl_4	liq	-13.7			
SF_4	g	-185.2	-174.8	69.77	17.45
SF_6	g	-289	-264.2	69.72	23.25
std. state, $m = 1$	aq	-293.0	-259.3	39.8	
S_2F_{10}	liq	(-485)			
SF_2Cl	g	-250.5	-226.9	76.26	24.9
HS	g	34.10	27.08	46.74	7.72
HS^- std. state, $m = 1$	aq	-4.2	2.88	15.0	
H_2S	g	-4.93	-8.02	49.16	8.18
std. state, $m = 1$	aq	-9.5	-6.66	29	
H_2S_2	liq	-4.33			20.1
	g	3.71			
H_2S_3	liq	-3.57			
	g	7.29			
H_2S_4	liq	-2.99			
	g	10.57			
H_2S_5	liq	-2.49			
	g.	13.84			
H_2S_6	liq	-1.99			
SO	g	1.496	-4.741	53.02	7.21
SO_2	liq	-76.6			
	g	-70.944	-71.748	59.30	9.53
undissoc; std. state, $m = 1$	aq	-77.194	-71.871	38.7	
	aq	-80.584			
SO_3 β	c	-108.63	-88.19	12.5	
	liq	-105.41	-88.04	22.85	
	g	-94.58	-88.69	61.34	12.11
SO_3^{2-} std. state, $m = 1$	aq	-151.9	-116.3	-7	
SO_4^{2-} std. state, $m = 1$	aq	-217.32	-177.97	4.8	-70
$S_2O_3^{2-}$	aq	-155.9			
$S_2O_4^{2-}$ std. state, $m = 1$	aq	-180.1	-143.5	22	
$S_2O_5^{2-}$	aq	-286.4			
$S_2O_7^{2-}$	aq	-334.9			
$S_2O_8^{2-}$ std. state, $m = 1$	aq	-320.0	-265.4	59.3	
$S_3O_6^{2-}$	aq	-286.7			
$S_4O_6^{2-}$	aq	-292.58			
$SOBr_2$	g	-21.8			
$SOCl_2$	liq	-58.7			
	g	-50.8	-47.4	74.01	15.9
SO_2Cl_2	liq	-94.2			
	g	-87.0	-76.5	74.53	18.4
$S_2O_5Cl_2$	liq	-168.7			56
SOF_2	g			66.58	13.58
SO_2F_2	g			67.86	15.78
SO_3F^-	aq	-193.0			
HSO_3^- std. state, $m = 1$	aq	-149.67	-126.15	33.4	
HSO_4^- std. state, $m = 1$	aq	-212.08	-180.69	31.5	-20
H_2SO_3 undissoc; std. state, $m = 1$	aq	-145.51	-128.56	55.5	
	aq	-148.899			
H_2SO_4	liq	-194.548	-164.938	37.501	33.20
std. state, $m = 1$	aq	-217.32	-177.97	4.8	-70
	aq, 1	-201.193			
	aq, 2	-204.455			

Table 3–9 (continued)
HEATS AND FREE ENERGIES OF FORMATION, ENTROPIES, AND
HEAT CAPACITIES OF ELEMENTS AND INORGANIC COMPOUNDS

Formula and Description	State	$\Delta Hf°$	$\Delta Gf°$	$S°$	$C_p°$
	aq, 5	− 208.288			
	aq, 10	− 210.451			
	aq, 100	− 212.150			
	aq, 1000	− 213.275			
$H_2SO_4 \cdot H_2O$	liq	− 269.508	− 227.182	50.56	51.35
$H_2SO_4 \cdot 2H_2O$	liq	− 341.085	− 286.770	66.06	62.34
HS_2O_4 std. state, $m = 1$	aq		− 146.9		
$H_2S_2O_4$ std. state, $m = 1$	aq		− 147.4		
$H_2S_2O_6$	aq	− 286.4			
$H_2S_2O_7$	c	− 304.4			27
$H_2S_2O_8$ std. state, $m = 1$	aq	− 320.0	− 265.4	59.3	
HSO_3Cl	liq	− 143.7			
HSO_3F	liq	− 186			28.00
$SO_2(NH_2)_2$ sulfamide	c	− 129.3			
Tantalum					
Ta	c	0	0	9.92	6.06
	g	186.9	176.7	44.241	4.985
TaB_2	c	− 46			
TaC	c	− 35.0	− 34.6	10.11	8.79
Ta_2C	c	− 51.0	− 50.8	20.7	
$TaBr_5$	c	− 143.0			
	g	− 115.6			
$TaCl_3$	c	− 132.2			
$TaCl_4$	c	− 167.7			
	g	− 134.0			
$TaCl_5$	c	− 205.3			
	g	− 181.3			
TaF_5	c	− 454.97			
std. state, undissoc	aq		− 137.6		
TaF_6^- std. state	aq		− 209.2		
TaF_7^{2-} std. state	aq		− 280.2		
Ta_2H	c	− 7.8	− 16.5	18.9	21.7
TaN	c	− 60.1			9.7
Ta_2N	c	− 65			
TaO	g	60	53	57.6	7.31
TaO_2	g	− 41	− 43	64	12.5
Ta_2O_5 β	c	− 489.0	− 456.8	34.2	32.30
	aq	− 496.7			
$TaOCl_3$	g	− 186.6			
TaS_2	c	− 111			
$TaSi_2$	c	− 28			
Ta_5Si_3	c	− 76			
Technetium					
Tc	c	0	0		
	g	162		43.25	4.97
Tc_2O_7	c	− 266			
$HTcO_4$	c	− 167			
std. state, $m = 1$	aq	− 173			
Tellurium					
Te	c	0	0	11.88	6.15
	g	47.02	37.55	43.65	4.968
	amorp	2.7			
Te_2	g	40.2	28.2	64.06	8.78
$TeBr_4$	c	− 45.5			
$TeCl_4$	c	− 78.0			33.1
TeF_6	g	− 315			
H_2Te	g	23.8			
TeO	g	15.6	9.2	57.7	7.19
TeO_2	c	− 77.1	− 64.6	19.0	

Table 3–9 (continued)
HEATS AND FREE ENERGIES OF FORMATION, ENTROPIES, AND
HEAT CAPACITIES OF ELEMENTS AND INORGANIC COMPOUNDS

Formula and Description	State	$\Delta Hf°$	$\Delta Gf°$	$S°$	$C_p°$
TeO_3^{2-}	aq	−142.6			
H_2TeO_3 std. state, $m = 1$	aq		−76.2		
$Te(OH)_3^+$ std. state, $m = 1$	aq	−145.4	−118.6	26.7	
H_6TeO_6	c	−310.4			
TeSe	g	38.0	26.0	63.5	
Terbium					
Tb	g	87			
	c	0	0		
$TbCl_3$ β	c	−241.6			
	aq, ∞	−288.5	−257.9		
$Tb_2(SO_4)_3$	aq, ∞	−987.5	−859.8		
$Tb_2(SO_4)_3 \cdot 8H_2O$	c		−1328.2		
Thallium					
Tl	c	0	0	15.34	6.29
	g	43.55	35.24	43.225	4.968
TlBr	c	−41.4	−40.00	28.8	
	g	−9.0			
std. state, $m = 1$	aq	−27.77	−32.59	49.7	
$TlBr^{2+}$ std. state, $m = 1$	aq	9.0	13.5	−13	
$TlBr_2^-$ std. state, $m = 1$	aq	−26.1	−21.4	20	
$TlBr_2^-$ std. state, $m = 1$	aq	−53.8	−58.8	84	
$TlBr_3$ undissoc; std. state	aq	−59.7	−53.7	49	
std. state, $m = 1$	aq	−40.2	−23.2	13	
$TlBr_4^-$ std. state, $m = 1$	aq	−90.9	−84.2	80	
$TlBrO_3$	c	−32.6	−12.70	40.3	
std. state, $m = 1$	aq	−18.7	−7.3	69.0	
TlCl	c	−48.79	−44.20	26.59	12.17
	g	−16.2			
std. state, $m = 1$	aq	−36.67	−39.11	43.5	
undissoc; std. state, $m = 1$	aq	−41.10	−39.91	41.3	
$TlCl^{2+}$ std. state, $m = 1$	aq	1.0	9.7	−19	
$TlCl_2^-$ std. state, $m = 1$	aq	−43.0	−29.6	7	
$TlCl_2^-$ std. state, $m = 1$	aq		−70.7		
$TlCl_3$	c	−75.3			
std. state, $m = 1$	aq	−72.9	−42.8	−5.5	
undissoc; std. state, $m = 1$	aq	−84.0	−65.6	32	
$TlCl_4^-$ std. state, $m = 1$	aq	−124.1	−100.8	58	
$TlClO_3$	aq	−22.4	−8.5	68.8	
Tl_2CO_3	c	−167.3	−146.9	37.1	
Tl(I) acetate	c	−126.1			
Tl(I) fulminate	c	27.6			
TlCNS thiocyanate	c	6.8	9.21	39	
std. state, $m = 1$	aq	19.55	14.41	64.5	
Tl_2CrO_4	c	−225.8	−205.9	67.5	
TlF	c	−77.6			
	g	−43.6			
std. state, $m = 1$	aq	−78.22	−74.38	26.7	
$TlHF_2$	c			34.92	21.35
TlI	c	−29.6	−29.97	30.5	
	g	1.7			
std. state, $m = 1$	aq	−11.91	−20.07	56.6	
TlI_2^- std. state, $m = 1$	aq		−35.1		
TlI_4^- std. state, $m = 1$	aq		−39.3		
$TlIO_3$	c	−63.9	−45.86	42.2	
std. state, $m = 1$	aq	−51.6	−38.3	58.3	
TlN_3	c	55.8	70.38	35.1	
$TlNO_3$	c	−58.30	−36.44	38.4	23.78
std. state, $m = 1$	aq	−48.28	−34.35	65.0	
Tl_2O	c	−42.7	−35.2	30	

Formula and Description	State	$\Delta Hf°$	$\Delta Gf°$	$S°$	$C_p°$
Tl_2O_3	c	−74.5			
Tl_2O_4	c	−83.0			
TlOH	c	−57.1	−46.8	21	
std. state, $m = 1$	aq	−53.69	−45.33	27.4	
$Tl(OH)_3$	c		−121.2		
Tl_2S	c	−23.2	−22.4	36	
Tl_2SO_4	c	−222.7	−198.49	55.1	
Tl_2Se	c	−14	−14.1	41	
Tl_2SeO_4	c	−151	−126.4	56	
Tl_2Te	c	−22			
Thorium					
Th	c	0	0	13.6	
$ThBr_4$	c	−227.1			
	aq	−298.6			
$ThBr_4 \cdot 7H_2O$	c	−753.7			
$ThBr_4 \cdot 10H_2O$	c	−971.6			
$ThBr_4 \cdot 12H_2O$	c	−1116.0			
$ThOBr_2$	c	−252.8			
ThC_2	c	−45			
$ThCl_4$	c	−285			
	aq	−343.0			
$ThCl_4 \cdot 2H_2O$	c	−437.4			
$ThCl_4 \cdot 4H_2O$	c	−589.2			
$ThCl_4 \cdot 7H_2O$	c	−805.9			
$ThCl_4 \cdot 8H_2O$	c	−877.6			
$ThCl_4 \cdot NH_4Cl$	c	−373.6			
$ThCl_4 \cdot 2NH_4Cl \cdot 10H_2O$	c	−1172.6			
$ThOCl_2$	c	−274.8			
$Th(OH)Cl_3 \cdot H_2O$	c	−377.4			
ThF_4	c	−477			
ThH_4	c	−43			
ThI_4	c	−131			
$ThOI_2$	c	−228.2			
$ThOI_2 \cdot 3\frac{1}{2}H_2O$	c	−479.3			
$Th(OH)I_3 \cdot 10H_2O$	c	−952.4			
Th_3N_4	c	−308	−282		
$Th(NO_3)_4$	aq, 20	−379.10			
	aq, 50	−380.36			
	aq, 100	−380.48			
	aq, 500	−380.5			
ThO_2	c	−292	−280.1		
$Th(OH)_4$	c	−421.5			
Th_2S_3	c	−262.0			
$Th(SO_4)_2$	c	−602			
	aq	−616.8			
$Th(SO_4)_2 \cdot 4H_2O$	c	−882.7			
$Th(SO_4)_2 \cdot 8H_2O$	c	−1168.6			
$ThOSO_4$	c	−487			
Thulium					
Tm	c	0	0		
$TmCl_3$ γ	c	−229.5			
	aq, ∞	−281.4	−250.5		
TmI_3 β	c	−137.8			
	aq, ∞	−201.4	−193.5		
Tin					
Sn I, white	c	0	0	12.32	6.45
II, gray	c	−0.50	0.03	10.55	6.16
	g	72.2	63.9	40.243	5.081
$SnBr_2$	c	−58.2			

Table 3–9 (continued)
HEATS AND FREE ENERGIES OF FORMATION, ENTROPIES, AND
HEAT CAPACITIES OF ELEMENTS AND INORGANIC COMPOUNDS

Formula and Description	State	$\Delta Hf°$	$\Delta Gf°$	$S°$	$C_p°$
SnBr₄	c	−90.2	−83.7	63.2	
	g	−75.2	−79.2	98.43	24.71
SnCl₂	c	−79.3			
std. state, $m = 1$	aq	−78.8	−71.6	41	
SnCl₂ · 2H₂O	c	−220.2			
SnCl₃ in aq HCl, std. state	aq	−116.4	−102.8	62	
SnCl₄	liq	−122.2	−105.2	61.8	39.5
SnH₄	g	38.9	45.0	54.39	11.70
Sn₂H₆	g	65.6			
SnI₂	c	−34.3			
SnI₄	c				20.3
SnO	c	−68.3	−61.4	13.5	10.59
	g			55.45	7.55
SnO₂	c	−138.8	−124.2	12.5	12.57
Sn(OH)₂ pptd	c	−134.1	−117.5	37	
Sn(OH)₄ pptd	c	−265.3			
	g			106.6	25.2
SnS	c	−24	−23.5	18.4	11.77
	g	28.5			
SnS₂	c			20.9	16.76
Sn(SO₄)₂	c	−389.4			
std. state, $m = 1$	aq		−354.2		
SnSe	c	−21.7			
	g	30.8			
SnTe	c	−14.6			
	g	38.4			
Titanium					
Ti	c	0	0	7.32	5.98
TiAs	c	−35.8			
TiBr₂	c	−96			
TiBr₃	c	−131.1	−125.2	42.2	24.31
TiBr₄	c	−147.4	−140.9	58.2	31.43
	g	−131.3	−135.8	95.2	24.1
TiB₂	c	−77.4	−76.4	6.81	10.58
TiC	c	−44.1	−43.2	5.79	8.04
TiCl₂	c	−122.8	−111.0	20.9	16.69
TiCl₃	c	−172.3	−156.2	33.4	23.22
TiCl₄	liq	−192.2	−176.2	60.31	34.70
	g	−182.4	−173.7	84.8	22.8
	aq, 1600	−250.3			
TiF₄	amorp	−394.2	−372.7	32.02	27.31
	g	−371.0			
H₂TiF₆	aq	−573.7			
TiH₂	c	−28.6	−19.2	7.1	7.2
TiI₂	c	−63			
	g	−13			
TiI₄	c	−89.8	−88.8	59.6	30.03
	g	−66.4			
TiN	c	−80.8	−74.0	7.23	8.86
TiO α	c	−124.2	−118.3	8.31	9.55
	g	4	−3	56.0	7.81
TiO²⁺ in HClO₄ medium	aq	−164.9			
TiO₂ anatase	c	−224.6	−211.4	11.93	13.26
Brookite	c	−225.1			
rutile	c	−225.8	−212.6	12.03	13.15
	amorp	−210			
hydrated ppt		−219.8			
Ti₂O₃	c	−363.5	−342.8	18.83	23.27
Ti₃O₅	c	−587.8	−553.9	30.9	37.00

Table 3–9 (continued)
HEATS AND FREE ENERGIES OF FORMATION, ENTROPIES, AND HEAT CAPACITIES OF ELEMENTS AND INORGANIC COMPOUNDS

Formula and Description	State	$\Delta Hf°$	$\Delta Gf°$	$S°$	$C_p°$
TiOCl	c	− 180			
TiP	c	− 67.6			
TiS	c	− 57			
TiS$_2$	c			18.73	16.23
TiSi	c	− 31			
TiSi$_2$	c	− 32			
Ti$_5$Si$_3$	c	− 138			
Tungsten					
W	c	0	0	7.80	5.80
	g	203.0	192.9	41.549	5.093
WBr$_5$	c	− 75.6			
WBr$_6$	c	− 83.3			
WO$_2$Br$_2$	c	− 170.3			
WOBr$_4$	c	− 130.1			
WC	c	− 9.69			
W$_2$C	c	− 6.3			
W(CO)$_6$	c	− 227.9			
WCl$_2$	c	− 61			
WCl$_4$	c	− 112			
	g	− 73			
WCl$_5$	c	− 118.6			
	g	− 100.8			
WCl$_6$	c	− 144.0			
	g	− 122.8			
W$_2$Cl$_{10}$	g	− 210			
WO$_2$Cl$_2$	c	− 187.2			
	g	− 164			
WOCl$_4$	c	− 161.7			
	g	− 140			
WF$_6$	c	− 418.2			
	liq	− 417.7	− 389.93	60.1	
	g	− 411.5	− 390.1	81.49	28.45
WO$_2$	c	− 140.94	− 127.61	12.08	13.41
WO$_3$	c	− 201.45	− 182.62	18.14	17.63
WO$_4^{2-}$ std. state	aq	− 257.1			
H$_2$WO$_4$	c	− 270.5			
	g	− 229			
WS$_2$	c	− 50			
WSi$_2$	c	− 22			
Uranium					
U	g	125			
	c	0	0	12.03	
UBr$_3$	c	− 170.1	− 164.7	49	
UBr$_4$	c	− 196.6	− 188.5	58	
UC$_2$	c	− 42	− 42	14	
UCl$_3$	c	− 213.0	− 196.9	37.99	
UCl$_4$	c	− 251.2	− 230.0	47.4	
UCl$_5$	c	− 262.1	− 237.4	62	
UCl$_6$	c	− 272.4	− 241.5	68.3	
UF$_3$	c	− 357	− 339	26	
UF$_4$	c	− 443	− 421	36.1	
UF$_5$	c	− 488	− 461	43	
UF$_6$	g	− 505	− 485	90.76	
UH$_3$	c	− 30.4			
UI$_3$	c	− 114.7	− 115.3	56	
UI$_4$	c	− 127.0	− 126.1	65	
UIBr$_3$	c	− 177.1			
UICl$_3$	c	− 219.9	− 204.4	54	
UN	c	− 80	− 75	18	

Table 3–9 (continued)
HEATS AND FREE ENERGIES OF FORMATION, ENTROPIES, AND HEAT CAPACITIES OF ELEMENTS AND INORGANIC COMPOUNDS

Formula and Description	State	$\Delta Hf°$	$\Delta Gf°$	$S°$	$C_p°$
U_2N_3	c	−213	−194	29	
UO_2	c	−270	−257	18.6	
UO_3	c	−302	−283	23.57	
$UO_3 \cdot H_2O$	c	−375.4			
$UO_3 \cdot 2H_2O$	c	−446.2			
$UO_4 \cdot 2H_2O$	c	−436			
U_3O_8	c	−898			
$U(SO_4)_2$	c	−563			
UO_2Br_2	aq	−308.2			
$UO_2(C_2H_3O_2)_2$	aq	−484.0			
$UO_2(C_2H_3O_2)_2 \cdot 2H_2O$	c	−624.9			
$UO_2(C_2H_3O_2)_2 \cdot$					
$NH_4C_2H_3O_2 \cdot 6H_2O$	c	−1045.8			
UO_2Cl_2	aq	−331			
UO_2CrO_4	aq	−456.7			
$UO_2CrO_4 \cdot 5\tfrac{1}{2}H_2O$	c	−838.8			
$UO_2(NO_3)_2$	c	−329.2	−273.1	66	
	aq, ∞	−349.1	−289.2	53	
$UO_2(NO_3)_2 \cdot H_2O$	c	−404.8	−335.3	76	
$UO_2(NO_3)_2 \cdot 2H_2O$	c	−480.0	−396.6	85	
$UO_2(NO_3)_2 \cdot 3H_2O$	c	−552.2	−454.7	94	
$UO_2(NO_3)_2 \cdot 6H_2O$	c	−764.3	−625.0	120.85	
UO_2SO_4	aq, ∞	−467.3	−413.7	−13	
$UO_2SO_4 \cdot 3H_2O$	c	−666.8	−586.0	63	
Vanadium					
V	c	0	0	6.91	5.95
	g	122.90	108.32	43.544	6.217
VBr_2	c	−87.3			
	g	−37.1			
VBr_3	c	−103.6			
	g	−56.8			
VBr_4	g	−80.5			
VCl_2	c	−108	−97	23.2	17.26
	g	−61.3			
VCl_3	c	−138.8	−122.2	31.3	22.27
VCl_4	liq	−136.1	−120.4	61	
	g	−125.6	−117.6	86.6	23.0
$V(CO)_6$	g	−236			
VF_3	c			23.18	21.62
VF_4	c	−335.4			
VF_5	liq	−353.8	−328.2	42.0	
	g	−342.7	−327.4	76.67	23.56
VI_2	c	−60.1			
	g	−1			
VI_3	c	−64.7			
VI_4	g	−29.3			
VN	c	−51.9	−45.7	8.91	9.08
VO	c	−103.2	−96.6	9.3	10.86
	g	25	18	55.8	7.3
VO^{2+} std. state	aq	−116.3	−106.7	−32.0	
VO_2	g	−57			
VO_2^+ std. state	aq	−155.3	−140.3	−10.1	
VO_3^- std. state	aq	−212.3	−187.3	12	
V_2O_3	c	−293.5	−272.3	23.5	24.67
V_2O_4 α	c	−341.1	−315.1	24.5	27.96
V_2O_5	c	−370.6	−339.3	31.3	30.51
V_3O_5	c	−465	−434	39	
V_4O_7	c	−635	−591	52	
V_6O_{13}	c	−1062			

Table 3–9 (continued)
**HEATS AND FREE ENERGIES OF FORMATION, ENTROPIES, AND
HEAT CAPACITIES OF ELEMENTS AND INORGANIC COMPOUNDS**

Formula and Description	State	$\Delta Hf°$	$\Delta Gf°$	S	C_p
HVO_4^{2-} std. state	aq	−277.0	−233.0	4	
H_2VO_4 std. state	aq	−280.6	−244.0	29	
$[VO_2H_2O_2]^-$ std. state	aq		−178.4		
VOCl	c	−138.4			
VO_2Cl	c	−185.6			
$VOCl_2$	c	−165.0			
$VOCl_3$	liq	−175.6	−159.8	58.4	
	g	−166.25	−157.58	82.26	21.49
$VOSO_4$	c	−312.9	−279.6	26.0	
std. state, undissoc	aq		−282.6		
$VOSCN^-$ std. state	aq	−98	−86	8	
V_2S_3	c	−227			
VSi_2	c	−73			
V_2Si	c	−37			
V_3Si	c	−26			
V_5Si_3	c	−94			
Xenon					
Xe	g	0	0	40.529	4.968
XeF_4	c	−62.5	−29.4		
	g	−51.5	−50		28.334
XeF_2	c	−39.2			
	g	−25.9	−37		
XeF_6	s	−86			
	g	−71			
XeO_3	c	96			
$XeOF_4$	liq	35			
XeO_2F_2	c	35			
Ytterbium					
Yb	c	0	0		
	g	87		41.30	
$YbBr_3$	c	−185			
$YbCl_3$	c	−228.7			
	aq	−280.7	−249.5		
$Yb_2(SO_4)_3$	aq	−971.9	−843.0		
$Yb_2(SO_4)_3 \cdot 8H_2O$	c		−1308.8		
Yttrium					
Y	c	0	0	10.62	6.34
	g	100.7	91.1	42.87	6.18
Y_2	g	163.5	150.7	64	8.7
Y^{3+} std. state, $m = 1$	aq	−172.9	−156.8	−60	
YBr^{2+} std. state, $m = 1$	aq	−203.8	−191.6	−43	
YC_2	c	−26	−26	13	
	g	142.6	128.4	61	10.7
$Y_2(CO_3)_3$			−752.4		
$Y_2(C_2O_4)_3 \cdot 9H_2O$	c		−1363.8		
Y(acetate)$_3$	aq	−516.1			
$Y(SCN)^{2+}$ std. state, $m = 1$	aq		−144.7		
YCl	g	47.8	41.5	58.33	8.56
YCl^{2+} std. state, $m = 1$	aq	−214.0	−198.7	−46	
YCl_3	c	−239.0			
	g	−179.3			18
	aq, 4000	−291.96			
$YCl_3 \cdot 6H_2O$	c	−691.3	−592.1	92	
YF	g	−33	−39	55.38	7.92
YF_3	c	−410.8	−393.1	24	
	g	−308.0	−305.4	74.5	16.8
YH_2	c	−37.5	−27.8	9.17	8.24
YH_3	c	−47.3	−33.2	10.02	10.36
YI_3	c	−147.4			
$Y(IO_3)_3$	c		−271.2		

Table 3–9 (continued)
HEATS AND FREE ENERGIES OF FORMATION, ENTROPIES, AND HEAT CAPACITIES OF ELEMENTS AND INORGANIC COMPOUNDS

Formula and Description	State	$\Delta Hf°$	$\Delta Gf°$	$S°$	$C_p°$
YO	g	−9.3	−15.5	55.88	7.53
Y_2O_2	g	−127.4			15.8
Y_2O_3	c	−455.38	−434.19	23.68	24.50
$Y(OH)^{2+}$ std. state, $m = 1$	aq		−210.1		
$Y(OH)_3$	c		−308.6		
$Y(OH)_2Cl$	c		−297.9		
$Y(ReO_4)_3$	c	−701.9	−629.4	88	
YS	g	41.7	29.7	58	8.2
Zinc					
Zn	c	0	0	9.95	6.07
	g	31.245	22.748	38.450	4.968
$ZnAs_2$	c	−7.6			
Zn_3As_2	c	−0.8			
$Zn(BO_2)_2$	c		−373.58		
$ZnBr_2$	c	−78.55	−74.60	33.1	
std. state, $m = 1$	aq	−94.88	−84.84	12.6	−57
$ZnBr_2 \cdot 2H_2O$	c	−224.0	−191.1	47.5	
$ZnCO_3$	c	−94.26	−174.85	19.7	19.05
ZnC_2O_4 std. state, $m = 1$	aq	−234.0	−196.2	−15.9	
$ZnC_2O_4 \cdot 2H_2O$	c	−374.0	−321.7	46.7	
$Zn(C_2O_4)_2^{2-}$ std. state, $m = 1$	aq	−430.7	−367.5	31	
$Zn(formate)_2$	c	−235.8			
std. state, $m = 1$	aq	−240.20	−202.9	17	
$Zn(acetate)_2$	c	−257.8			
std. state, $m = 1$	aq	−269.10	−211.72	14.6	
$Zn(CN)_2$	c	22.9			
$Zn(CN)_4^{2-}$ std. state, $m = 1$	aq	81.8	106.8	54	
$ZnCl_2$	c	−99.20	−88.296	26.64	17.05
	g	−63.6			
std. state, $m = 1$	, aq	−116.68	−97.88	0.2	−54
$ZnCl_3^-$ std. state, $m = 1$	aq		−129.2		
$ZnCl_4^{2-}$ std. state, $m = 1$	aq		−159.2		
$Zn(ClO_4)_2$ std. state, $m = 1$	aq	−98.60	−39.26	60.2	
$Zn(ClO_4)_2 \cdot 6H_2O$	c	−509.89	−371.8	130.4	
ZnF_2	c	−182.7	−170.5	17.61	15.69
std. state, $m = 1$	aq	−195.78	−168.42	−33.4	−40
ZnI_2	c	−49.72	−49.94	38.5	
std. state, $m = 1$	aq	−63.16	−59.80	25.2	−57
$Zn(IO_3)_2$	c		−103.68		
std. state, $m = 1$	aq	−142.6	−96.3	29.8	
$Zn(N_3)_2$	c	52			
Zn_3N_2	c	−5.4			26
$Zn(NO_3)_2$	c	−115.6			
ionized, std. state	aq	−135.90	−88.36	43.2	−30
$Zn(NO_3)_2 \cdot 6H_2O$	c	−551.30	−423.79	109.2	77.2
$Zn(NH_3)_4^{2+}$ std. state, $m = 1$	aq	−127.5	−72.2	72	
ZnO	c	−83.24	−76.08	10.43	9.62
ZnO_2^{2-} std. state, $m = 1$	aq		−91.85		
$ZnO_2 \cdot 2H_2O$	c	−207.6			
$Zn(OH)^+$ std. state, $m = 1$	aq		−78.9		
$HZnO_2^-$ std. state, $m = 1$	aq		−109.26		
$Zn(OH)_2$ β	c	−153.42	−132.31	19.4	
ϵ	c	−152.74	−132.68	19.5	17.3
std. state, $m = 1$	aq	−146.72	−110.33	−31.9	−60
$Zn(OH)_3^-$ std. state, $m = 1$	aq		−165.95		
$Zn(OH)_4^{2-}$ std. state, $m = 1$	aq		−205.23		
Zn_3P_2	c	−113			
$Zn(PO_3)_2$	c	−497.9			
$Zn_2P_2O_7$	c	−600.0			
$Zn_3(PO_4)_2$	c	−691.3			

Table 3–9 (continued)
HEATS AND FREE ENERGIES OF FORMATION, ENTROPIES, AND HEAT CAPACITIES OF ELEMENTS AND INORGANIC COMPOUNDS

Formula and Description	State	$\Delta H f^\circ$	$\Delta G f^\circ$	S°	C_p°
ZnS wurtzite	c	−46.04			
sphalerite	c	−49.23	−48.11	13.8	11.0
ZnSO₄	c	−234.9	−209.0	28.6	
std. state, $m = 1$	aq	−254.10	−213.11	−22.0	−59
ZnSO₄·H₂O	c	−311.78	−270.58	33.1	
ZnSO₄·6H₂O	c	−663.83	−555.64	86.9	85.49
ZnSO₄·7H₂O	c	−735.60	−612.59	92.9	91.64
ZnS₂O₆	aq	−323.2			
ZnS₂O₆·6H₂O	c	−735.2			
ZnSb	c	−3.5			
ZnSe	c	39	39	20	
ZnSeO₃ std. state, $m = 1$	aq	−158.5	−123.5	−23.7	
ZnSeO₃·H₂O	c	−222.5	−189.5	39	
ZnSeO₄	c	−158.8			
std. state, $m = 1$	aq	−180.0	−140.6	−13.9	
ZnSeO₄·6H₂O	c	−587.5			
ZnSiO₃	c	−301.2			
Zn₂SiO₄	c	−391.19	−364.06	31.4	29.48
ZnTe	c	−28.1			
ZnWO₄	c	−293			30.0
Zirconium					
Zr α, hexagonal	c	0	0	9.32	6.06
	g	145.5	135.4	43.32	6.37
ZrB₂	c	−78.0	−77.0	8.59	11.53
ZrBr₄	c	−181.8			
	g	−153.8			
ZrC	c	−48.5	−47.7	7.96	9.06
ZrCl	c	−63			
	g	60			
ZrCl₂	c	−120			
ZrCl₃	c	−179			
ZrCl₄	c	−234.35	−212.7	43.4	28.63
	g	−208.0	−199.7	88.0	23.49
	aq	−293.3			
ZrF₂	g	−135			
ZrF₃	g	−265			
ZrF₄ β, monoclinic	c	−456.8	−432.6	25.00	24.79
	g	−400.0	−391.1	76.3	20.9
ZrH₂	c	−40.4	−30.8	8.37	7.40
ZrI₄	c	−115.1			
	g	−84.1			
ZrN	c	−87.2	−80.4	9.29	9.66
	g	134			
ZrO	g	15			
ZrO₂ α, monoclinic	c	−263.04	−249.24	12.04	13.43
hydrated precipitate			−260.4		
ZrO₃ precipitate		−241			
ZrO(OH)⁺	aq	−270.1			
ZrOBr₂	aq	−259.9			
ZrOBr₂·8H₂O	c	−808.4			
ZrOCl₂	aq	−280.3			
ZrOCl₂·8H₂O	c	−829.8			
ZrS₂	c	−135.3			
Zr(SO₄)₂	c	−529.9			
Zr(SO₄)₂·4H₂O	c	−825.6			
ZrO(SO₄)₂²⁻	aq	−630.1			
ZrSi	c	−37			
ZrSi₂	c	−38			
ZrSiO₄	c	−486.0	−458.7	20.1	23.58

From Dean, J. A., Ed., *Lange's Handbook of Chemistry,* 11th ed., Copyright © 1973 by McGraw-Hill, Inc. New York, 9-2. Used with permission of McGraw-Hill Book Company.

Table 3–10

DENSITY OF VARIOUS SOLIDS

Approximate Density of Various Solids at Ordinary Atmospheric Temperature

In the case of substances with voids such as paper or leather the bulk density rather than the density of the solid portion is indicated.

Substance	Grams per cu. cm	Pounds per cu. ft.	Substance	Grams per cu. cm	Pound per cu. ft.	Substance	Grams per cu. cm	Pounds per cu. ft.
Agate	2.5–2.7	156–168	Cork	0.22–0.26	14–16	Mica	2.6–3.2	165–200
Alabaster, carbonate	2.69–2.78	168–173	Cork linoleum	0.54	34	Muscovite	2.76–3.00	172–187
sulfate	2.26–2.32	141–145	Corundum	3.9–4.0	245–250	Ochre	3.5	218
Albite	2.62–2.65	163–165	Diamond	3.01–3.52	188–220	Opal	2.2	137
Amber	1.06–1.11	66–69	Dolomite	2.84	177	Paper	0.7–1.15	44–72
Amphiboles	2.9–3.2	180–200	Ebonite	1.15	72	Paraffin	0.87–0.91	54–57
Anorthite	2.74–2.76	171–172	Emery	4.0	250	Peat blocks	0.84	52
Asbestos	2.0–2.8	125–175	Epidote	3.25–3.50	203–218	Pitch	1.07	67
Asbestos slate	1.8	112	Feldspar	2.55–2.75	159–172	Porcelain	2.3–2.5	143–156
Asphalt	1.1–1.5	69–94	Flint	2.63	164	Porphyry	2.6–2.9	162–181
Basalt	2.4–3.1	150–190	Fluorite	3.18	198	Pressed wood pulp board	0.19	12
Beeswax	0.96–0.97	60–61	Galena	7.3–7.6	460–470	Pyrite	4.95–5.1	309–318
Beryl	2.69–2.7	168–169	Gamboge	1.2	75	Quartz	2.65	165
Biotite	2.7–3.1	170–190	Garnet	3.15–4.3	197–268	Resin	1.07	67
Bone	1.7–2.0	106–125	Gas carbon	1.88	117	Rock salt	2.18	136
Brick	1.4–2.2	87–137	Gelatin	1.27	79	Rubber, hard	1.19	74
Butter	0.86–0.87	53–54	Glass, common	2.4–2.8	150–175	Rubber, soft commercial	1.1	69
Calamine	4.1–4.5	255–280	flint	2.9–5.9	180–370	pure gum	0.91–0.93	57–58
Calcspar	2.6–2.8	162–175	Glue	1.27	79	Sandstone	2.14–2.36	134–147
Camphor	0.99	62	Granite	2.64–2.76	165–172	Serpentine	2.50–2.65	156–165
Caoutchouc	0.92–0.99	57–62	Graphite*	2.30–2.72	144–170	Silica, fused transparent	2.21	138
Cardboard	0.69	43	Gum arabic	1.3–1.4	81–87	translucent	2.07	129
Celluloid	1.4	87	Gypsum	2.31–2.33	144–145	Slag	2.0–3.9	125–240
Cement, set	2.7–3.0	170,190	Hematite	4.9–5.3	306–330	Slate	2.6–3.3	162–205
Chalk	1.9–2.8	118–175	Hornblende	3.0	187	Soapstone	2.6–2.8	162–175
Charcoal, oak	0.57	35	Ice	0.917	57.2	Spermaceti	0.95	59
pine	0.28–0.44	18–28	Ivory	1.83–1.92	114–120	Starch	1.53	95
Cinnabar	8.12	507	Leather, dry	0.86	54	Sugar	1.59	99
Clay	1.8–2.6	112–162	Lime, slaked	1.3–1.4	81–87	Talc	2.7–2.8	168–174
Coal, anthracite	1.4–1.8	87–112	Limestone	2.68–2.76	167–171	Tallow, beef	0.94	59
bituminous	1.2–1.5	75–94	Linoleum	1.18	74	mutton	0.94	59
Cocoa butter	0.89–0.91	56–57	Magnetite	4.9–5.2	306–324	Tar	1.02	66
Coke	1.0–1.7	62–105	Malachite	3.7–4.1	231–256			
Copal	1.04–1.14	65–71	Marble	2.6–2.84	160–177			
			Meerschaum	0.99–1.28	62–80			

Table 3–10 (continued)
DENSITY OF VARIOUS SOLIDS

Substance	Grams per cu. cm	Pounds per cu. ft.	Substance	Grams per cu. cm	Pound per cu. ft.	Substance	Grams per cu. cm	Pounds per cu. ft.
Topaz	3.5–3.6	219–223	cedar	0.49–0.57	30–35	maple	0.62–0.75	39–47
Tourmaline	3.0–3.2	190–200	cherry	0.70–0.90	43–56	oak	0.60–0.90	37–56
Wax, sealing	1.8	112	dogwood	0.76	47	pear	0.61–0.73	38–45
Wood (seasoned)			ebony	1.11–1.33	69–83	pine, pitch	0.83–0.85	52–53
alder	0.42–0.68	26–42	elm	0.54–0.60	34–37	white	0.35–0.50	22–31
apple	0.66–0.84	41–52	hickory	0.60–0.93	37–58	yellow	0.37–0.60	23–37
ash	0.65–0.85	40–53	holly	0.76	47	plum	0.66–0.78	41–49
balsa	0.11–0.14	7–9	juniper	0.56	35	poplar	0.35–0.5	22–31
bamboo	0.31–0.40	19–25	larch	0.50–0.56	31–35	satinwood	0.95	59
basswood	0.32–0.59	20–37	lignum vitae	1.17–1.33	73–83	spruce	0.48–0.70	30–44
beech	0.70–0.90	43–56	locust	0.67–0.71	42–44	sycamore	0.40–0.60	24–37
birch	0.51–0.77	32–48	logwood	0.91	57	teak, Indian	0.66–0.88	41–55
blue gum	1.00	62	mahogany			African	0.98	61
box	0.95–1.16	59–72	Honduras	0.66	41	walnut	0.64–0.70	40–43
butternut	0.38	24	Spanish	0.85	53	water gum	1.00	62
						willow	0.40–0.60	24–37

aSome values reported as low as 1.6.

From Weast, R. C., Ed., *Handbook of Chemistry and Physics*, 55th ed., CRC Press, Cleveland, 1974, F-1.

Table 3-11

COEFFICIENT OF FRICTION

The coefficient of friction between two surfaces is the ratio of the force required to move one over the other to the total force pressing the two together. If F is the force required to move one surface over another and W is the force pressing the surfaces together, the coefficient of friction, μ, equals $\frac{F}{W}$.

Static Friction

Materials	Condition	Temp, °C	μ (static)
Nonmetals			
Glass on glass	Clean	—	0.9–1.0
Glass on glass	Lubricated with paraffin oil	—	0.5–0.6
Glass on glass	Lubricated with liquid fatty acids	—	0.3–0.6
Glass on glass	Lubricated with solid hydrocarbons, alcohols, or fatty acids	—	0.1
Glass on metal	Clean	—	0.5–0.7
Glass on metal	Lubricated	—	0.2–0.3
Diamond on diamond	Clean	—	0.1
Diamond on diamond	Lubricated	—	0.05–0.1
Diamond on metal	Clean	—	0.1–0.15
Diamond on metal	Lubricated	—	0.1
Sapphire on sapphire	Clean or lubricated	—	0.2
Sapphire on steel	Clean or lubricated	—	0.15
Hard carbon on carbon	Clean	—	0.16
Hard carbon on carbon	Lubricated	—	0.12–0.14
Graphite on graphite	Clean or lubricated	—	0.1
Graphite on graphite	Outgassed	—	0.5–0.8
Graphite on steel	Clean or lubricated	—	0.1
Mica on mica	Freshly cleaved	—	1.0
Mica on mica	Contaminated	—	0.2–0.4
Crystals of $NaNO_3$, KNO_3, NH_4Cl on self	Clean	—	0.5
Crystals of $NaNO_3$, KNO_3, NH_4Cl on self	Lubricated with long chain polar compounds	—	0.12
Tungsten carbide on tungsten carbide	Clean	Room	0.17
Tungsten carbide on tungsten carbide	Outgassed	Room	0.58
Tungsten carbide on tungsten carbide	Clean	820	0.35
Tungsten carbide on tungsten carbide	Clean	970	0.40

Table 3–11 (continued)
COEFFICIENT OF FRICTION

Static Friction (continued)

Materials	Condition	Temp, °C	μ (static)
Tungsten carbide on tungsten carbide	Clean	1,010	0.45
Tungsten carbide on tungsten carbide	Clean	1,160	0.5
Tungsten carbide on tungsten carbide	Clean	1,220	0.7
Tungsten carbide on tungsten carbide	Clean	1,440	1.2
Tungsten carbide on tungsten carbide	Clean	1,600	1.8
Tungsten carbide on graphite	Outgassed	Room	0.62
Tungsten carbide on graphite	Clean	Room	0.15
Tungsten carbide on graphite	Clean	800	0.32
Tungsten carbide on graphite	Clean	910	0.30
Tungsten carbide on graphite	Clean	1,000	0.25
Tungsten carbide on graphite	Clean	1,120	0.29
Tungsten carbide on graphite	Clean	1,220	0.26
Tungsten carbide on graphite	Clean	1,300	0.25
Tungsten carbide on graphite	Clean	1,410	0.25
Tungsten carbide on graphite	Clean	1,800	0.24
Tungsten carbide on graphite	Clean	2,030	0.25
Tungsten carbide on steel	Clean	—	0.4–0.6
Tungsten carbide on steel	Lubricated	—	0.1–0.2
Polymethyl methacrylate on self	Clean	—	0.8
Polymethyl methacrylate on steel	Clean	—	0.4–0.5
Polystyrene on self	Clean	—	0.5
Polystyrene on steel	Clean	—	0.3–0.35
Polyethylene on self	Clean	—	0.2
Polyethylene on steel	Clean	—	0.2
Polytetrafluoroethylene on self	Clean	—	0.04
Polytetrafluoroethylene on steel	Clean	—	0.04
Nylon on nylon	Clean	—	0.15–0.25
Silk on silk	Commercially clean	—	0.2–0.3
Cotton on cotton (thread)	Commercially clean	—	0.3
Cotton on cotton (from cotton wool)	Commercially clean	—	0.6

Table 3–11 (continued)
COEFFICIENT OF FRICTION

Static Friction (continued)

Materials	Condition	Temp. °C	μ (static)
Rubber on solids	Commercially clean	—	1–4
Wood on wood	Commercially clean and dry	—	0.25–0.5
Wood on wood	Commercially clean and wet	—	0.2
Wood on metals	Commercially clean and dry	—	0.2–0.6
Wood on metals	Commercially clean and wet	—	0.2
Wood on brick	Commercially clean	—	0.6
Wood on leather	Commercially clean	—	0.3–0.4
Leather on metal	Commercially clean	—	0.6
Leather on metal	Commercially clean and wet	—	0.4
Leather on metal	Greasy	—	0.2
Brake material on cast iron	Commercially clean	—	0.4
Brake material on cast iron	Commercially clean and wet	—	0.2
Brake material on cast iron	Lubricated with mineral oil	—	0.1
Wool fiber on horn	Clean (against scales)	—	0.8–1.0
Wool fiber on horn	Clean (with scales)	—	0.4–0.6
Wool fiber on horn	Greasy (against scales)	—	0.5–0.8
Wool fiber on horn	Greasy (with scales)	—	0.3–0.4
Metals			
Steel on steel	Clean	20	0.58
Steel on steel	Vegetable oil lubricant		
	Castor oil	20	0.095
		100	0.105
	Rape	20	0.105
		100	0.105
	Olive	20	0.105
		100	0.105
	Coconut	20	0.08
		100	0.08
Steel on steel	Animal oil lubricant		
	Sperm	100	0.10

Table 3–11 (continued)
COEFFICIENT OF FRICTION

Static Friction (continued)

Materials	Condition	Temp. °C	μ (static)
Steel on steel (cont.)	Pale whale	20	0.095
		100	0.095
	Neat's-foot	20	0.095
		100	0.095
	Lard	20	0.085
		100	0.085
Steel on steel	Mineral oil lubricant		
	Light machine	20	0.16
		100	0.19
	Thick gear	20	0.125
		100	0.15
	Solvent refined	20	0.15
		100	0.20
	Heavy motor	20	0.195
		100	0.205
	Extreme pressure	20	0.09–0.1
		100	0.09–0.1
	Graphited oil	20	0.13
		100	0.15
	B.P. paraffin	20	0.18
		100	0.22
Steel on steel	Lubricated with trichloroethylene	20	0.33
Steel on steel	Lubricated with benzene	20	0.48
Steel on steel	Lubricated with glycerol	20	0.2
Steel on steel	Lubricated with ethyl alcohol	20	0.43
Steel on steel	Lubricated with butyl alcohol	Room	0.3
Steel on steel	Lubricated with octyl	Room	0.23
Steel on steel	Lubricated with decyl	Room	0.16
Steel on steel	Lubricated with cetyl	Room	0.10
Steel on steel	Lubricated with nonane	Room	0.26
Steel on steel	Lubricated with decane	Room	0.23
Steel on steel	Lubricated with acetic acid	Room	0.5
Steel on steel	Lubricated with propionic acid	Room	0.4

Table 3–11 (continued)
COEFFICIENT OF FRICTION

Static Friction (continued)

Materials	Condition	Temp. °C	μ (static)
Steel on steel	Lubricated with valeric acid	Room	0.17
Steel on steel	Lubricated with caproic acid	Room	0.12
Steel on steel	Lubricated with pelargonic acid	Room	0.11
Steel on steel	Lubricated with capric acid	Room	0.11
Steel on steel	Lubricated with lauric acid	Room	0.11
Steel on steel	Lubricated with myristic acid	Room	0.11
Steel on steel	Lubricated with oleic acid	20–100	0.08
Steel on steel	Lubricated with palmitic acid	Room	0.11
Steel on hard steel	Lubricated with stearic acid	Room	0.10
Steel on hard steel	Lubricated with rape oil	—	0.14
Steel on hard steel	Lubricated with castor oil	—	0.12
Steel on hard steel	Lubricated with mineral oil	—	0.16
Steel on hard steel	Lubricated with long chain fatty acid	—	0.09
Steel on cast iron	Lubricated with rape oil	—	0.11
Steel on cast iron	Lubricated with castor oil	—	0.15
Steel on cast iron	Lubricated with mineral oil	—	0.21
Steel on cast iron	Clean	—	0.4
Steel on gun metal	Lubricated with rape oil	—	0.15
Steel on gun metal	Lubricated with castor oil	—	0.16
Steel on gun metal	Lubricated with mineral oil	—	0.21
Steel on bronze	Lubricated with rape oil	—	0.12
Steel on bronze	Lubricated with castor oil	—	0.12
Steel on bronze	Lubricated with mineral oil	—	0.16
Steel on lead	Lubricated with mineral oil	—	0.5
Steel on lead	Lubricated with long chain fatty acid	—	0.22
Steel on base white metal	Lubricated with mineral oil	—	0.1
Steel on base white metal	Lubricated with long chain fatty acid	—	0.08
Steel on base white metal	Clean	—	0.55

Table 3–11 (continued)
COEFFICIENT OF FRICTION

Static Friction (continued)

Materials	Condition	Temp. °C	μ (static)
Steel on tin	Lubricated with mineral oil	—	0.6
Steel on tin	Lubricated with long chain fatty acid	—	0.21
Steel on white metal, tin base	Lubricated with mineral oil	—	0.1
Steel on white metal, tin base	Lubricated with long chain fatty acid	—	0.07
Steel on white metal, tin base	Clean	—	0.8
Steel on sintered bronze	Lubricated with mineral oil	—	0.13
Steel on brass	Lubricated with mineral oil	—	0.19
Steel on brass	Lubricated with castor oil	—	0.11
Steel on brass	Lubricated with long chain fatty acid	—	0.13
Steel on brass	Clean	—	0.35
Steel on copper-lead alloy	Clean	—	0.22
Steel on Wood's alloy	Clean	—	0.7
Steel on phosphor bronze	Clean	—	0.35
Steel on aluminum bronze	Clean	—	0.45
Steel on constantan	Clean	—	0.4
Steel on indium film deposited on steel[a]	4-kg load, clean	—	0.08
Steel on indium film deposited on steel[a]	8-kg load, clean	—	0.04
Steel on indium film deposited on silver[a]	4-kg load, clean	—	0.1
Steel on indium film deposited on silver[a]	8-kg load, clean	—	0.07
Steel on lead film deposited on copper[a]	4-kg load, clean	—	0.18
Steel on lead film deposited on copper[a]	8-kg load, clean	—	0.12
Steel on copper film deposited on steel[a]	4-kg load, clean	—	0.3
Steel on copper film deposited on steel[a]	8-kg load, clean	—	0.2
Aluminum on aluminum	In air or oxygen (O_2)	—	1.9
Aluminum on aluminum	In water vapor	—	1.1
Copper on copper[b]	In hydrogen (H_2) or nitrogen (N_2)	—	4.0
Copper on copper	In air or oxygen (O_2)	—	1.6
Gold on gold[b]	In hydrogen (H_2) or nitrogen (N_2)	—	4.0
Gold on gold	In air or oxygen (O_2)	—	2.8
Gold on gold	In water vapor	—	2.5

Table 3–11 (Continued)
COEFFICIENT OF FRICTION

Static Friction (continued)

Materials	Condition	Temp. °C	μ (static)
Iron on iron[b]	In air or oxygen (O_2)	—	1.2
Iron on iron	In water vapor	—	1.2
Molybdenum on molybdenum[b]	In air or oxygen (O_2)	—	0.8
Molybdenum on molybdenum	In water vapor	—	0.8
Nickel on nickel[b]	In hydrogen (H_2) or nitrogen (N_2)	—	5.0
Nickel on nickel	In air or oxygen (O_2)	—	3.0
Nickel on nickel	In water vapor	—	1.6
Platinum on platinum[b]	In air or oxygen (O_2)	—	3.0
Platinum on platinum	In water vapor	—	3.0
Silver on silver[b]	In air or oxygen (O_2)	—	1.5
Silver on silver	In water vapor	—	1.5
Various materials on snow and ice			
Ice on ice	Clean	0	0.05–0.15
Ice on ice	Clean	-12	0.3
Ice on ice	Clean	-71	0.5
Ice on ice	Clean	-82	0.5
Ice on ice	Clean	-110	0.5
Polymethylmethacrylate	On wet snow	0	0.5
Polymethylmethacrylate	On dry snow	0	0.3
Polymethylmethacrylate	On dry snow	-10	0.34
Polymethylmethacrylate	On dry snow	-32	0.4
Polyester of teraphthalic acid and ethylene glycol	On wet snow	0	0.5
Polyester of teraphthalic acid and ethylene glycol	On dry snow	0	0.35
Polyester of teraphthalic acid and ethylene glycol	On dry snow	-10	0.38
Nylon	On wet snow	0	0.4
Nylon	On dry snow	0	0.3
Nylon	On dry snow	-10	0.3
Polytetrafluoroethylene	On wet snow	0	0.05
Polytetrafluoroethylene	On dry snow	0	0.02
Polytetrafluoroethylene	On dry snow	-10	0.08

Table 3–11
COEFFICIENT OF FRICTION

Static Friction (continued)

Materials	Condition	Temp. °C	μ (static)
Polytetrafluoroethylene	On dry snow	-32	0.1
Paraffin wax	On wet snow	0	0.06
Paraffin wax	On dry snow	0	0.06
Paraffin wax	On dry snow	-10	0.35
Paraffin wax	On dry snow	-32	0.4
Swiss wax	On wet snow	0	0.05
Swiss wax	On dry snow	0	0.03
Swiss wax	On dry snow	-10	0.2
Swiss wax	On dry snow	-32	0.2
Ski wax	On wet snow	0	0.1
Ski wax	On dry snow	0	0.04
Ski wax	On dry snow	-10	0.2
Ski wax	On dry snow	-32	0.2
Ski lacquer	On wet snow	0	0.2
Ski lacquer	On dry snow	0	0.1
Ski lacquer	On dry snow	-10	0.4
Ski lacquer	On dry snow	-32	0.4
Aluminum	On wet snow	0	0.4
Aluminum	On dry snow	0	0.35
Aluminum	On dry snow	-10	0.38

Kinetic Friction

Materials	Condition	Temp, °C	μ (kinetic)
Various materials			
Unwaxed hickory	4 m/sec on dry snow	-3	0.08
Waxed hickory	0.1 m/sec on wet snow	0	0.14
Waxed hickory	0.1 m/sec on dry snow	0	0.04

Table 3–11
COEFFICIENT OF FRICTION

Kinetic Friction (continued)

Materials	Condition	Temp. °C	μ (kinetic)
Waxed hickory	0.1 m/sec on dry snow	−3	0.09
Waxed hickory	4 m/sec on dry snow	−3	0.03
Waxed hickory	0.1 m/sec on dry snow	−10	0.18
Waxed hickory	0.1 m/sec on dry snow	−40	0.4
Ice on ice	4 m/sec, clean	0	0.02
Ice on ice	4 m/sec, clean	−10	0.035
Ice on ice	4 m/sec, clean	−20	0.050
Ice on ice	4 m/sec, clean	−40	0.075
Ice on ice	4 m/sec, clean	−60	0.085
Ice on ice	4 m/sec, clean	−80	0.09
Ebonite	4 m/sec on ice	0	0.02
Ebonite	4 m/sec on ice	−10	0.05
Ebonite	4 m/sec on ice	−20	0.065
Ebonite	4 m/sec on ice	−40	0.085
Ebonite	4 m/sec on ice	−60	0.10
Ebonite	4 m/sec on ice	−80	0.11
Brass	4 m/sec on ice	0	0.02
Brass	4 m/sec on ice	−10	0.075
Brass	4 m/sec on ice	−20	0.085
Brass	4 m/sec on ice	−40	0.115
Brass	4 m/sec on ice	−60	0.14
Brass	4 m/sec on ice	−80	0.15
Natural rubber, vulcanized	10 m/min on ground glass, clean	—	1.07
Natural rubber, vulcanized	100 m/min on ground glass, wetted with water	—	0.94
Natural rubber, vulcanized	100 m/min on concrete, clean	—	1.02
Natural rubber, vulcanized	100 m/min on concrete, wetted with water	—	0.97
Natural rubber, vulcanized	100 m/min on bitumen, clean	—	1.07
Natural rubber, vulcanized	100 m/min on bitumen, wetted with water	—	0.95
Natural rubber, vulcanized	100 m/min on rubber flooring or rubber tread vulcanizate, clean	—	1.16

Table 3–11
COEFFICIENT OF FRICTION

Kinetic Friction (continued)

Materials	Condition	Temp. °C	μ (kinetic)
Natural rubber, vulcanized	100 m/min on bitumen containing rubber powder, clean	—	1.15 (varies with quantity of powder)
Natural rubber, vulcanized	100 m/min on bitumen containing rubber powder, wetted with water	—	1.03

[a]Hemispherical steel slider having 0.6-cm diameter. The 10^{-3} to 10^{-4} cm, thin metallic films were deposited on various substrates as indicated. Amonton's Law is not obeyed in this case.

[b]The metals that were spectroscopically pure were outgassed in a vacuum prior to other gases being admitted. When clean and in vacuum there is gross seizure.

From Minshall, H., in *Handbook of Chemistry and Physics*, 55th ed., Weast, R. C., Ed., CRC Press, Cleveland, 1974, F-19.

Table 3-12A
HARDNESS OF LOW MELTING POINT ALLOYS

Melting point, °C	Name	Composition, weight percent
-48	Binary eutectic	Cs, 77.0; K, 23.0
-40	Binary eutectic	Cs, 87.0; Rb, 13.0
-30	Binary eutectic	Cs, 95.0; Na, 5.0
-11	Binary eutectic	K, 78.0; Na, 22.0
-8	Binary eutectic	Rb, 92.0; Na, 8.0
10.7	Ternary eutectic	Ga, 62.5; In, 21.5; Sn, 16.0
10.8	Ternary eutectic	Ga, 69.8; In, 17.6; Sn, 12.5
17	Ternary eutectic	Ga, 82.0; Sn, 12.0; Zn, 6.0
33	Binary eutectic	Rb, 68.0; K, 32.0
46.5	Quinternary eutectic	Sn, 10.65; Bi, 40.63; Pb, 22.11; In, 18.1; Cd, 8.2
47	Quinternary eutectic	Bi, 44.7; Pb, 22.6; Sn, 8.3; Cd, 5.3; In, 19.1
58.2	Quaternary eutectic	Bi, 49.5; Pb, 17.6; Sn, 11.6; In, 21.3
60.5	Ternary eutectic	In, 51.0; Bi, 32.5; Sn, 16.5
70	Wood's metal	Bi, 50.0; Pb, 25.0; Sn, 12.5; Cd, 12.5
70	Lipowitz's metal	Bi, 50.0; Pb, 26.7; Sn, 13.3; Cd, 10.0
70	Binary eutectic	In, 67.0; Bi, 33.0
91.5	Ternary eutectic	Bi, 51.6; Pb, 40.2; Cd, 8.2
95	Ternary eutectic	Bi, 52.5; Pb, 32.0; Sn, 15.5
97	Newton's metal	Bi, 50.0; Sn, 18.8; Pb, 31.2
98	D'Arcet's metal	Bi, 50.0; Sn, 25.0; Pb, 25.0
100	Onion's or Lichtenberg's metal	Bi, 50.0; Sn, 20.0; Pb, 30.0
102.5	Ternary eutectic	Bi, 54.0; Sn, 26.0; Cd, 20.0
109	Rose's metal	Bi, 50.0; Pb, 28.0; Sn, 22.0
117	Binary eutectic	In, 52.0; Sn, 48.0
120	Binary eutectic	In, 75.0; Cd, 25.0
123	Malotte's metal	Bi, 46.1; Sn, 34.2; Pb, 19.7
124	Binary eutectic	Bi, 55.5; Pb, 44.5
130	Ternary eutectic	Bi, 56.0; Sn, 40.0; Zn, 4.0
140	Binary eutectic	Bi, 58.0; Sn, 42.0
140	Binary eutectic	Bi, 60.0; Cd, 40.0
183	Eutectic solder	Sn, 63.0; Pb, 37.0
185	Binary eutectic	Tl, 52.0; Bi, 48.0
192	Soft solder	Sn, 70.0; Pb, 30.0
198	Binary eutectic	Sn, 91.0; Zn, 9.0
199	Tinfoil	Sn, 92.0; Zn, 8.0
199	White metal	Sn, 92.0; Sb, 8.0
221	Binary eutectic	Sn, 96.5; Ag, 3.5
226	Matrix	Bi, 48.0; Pb, 28.5; Sn, 14.5; Sb, 9.0
227	Binary eutectic	Sn, 99.25; Cu, 0.75
240	Antimonial tin solder	Sn, 95.0; Sb, 5.0
245	Tin-silver solder	Sn, 95.0; Ag, 5.0

From Weast, R. C., Ed., *Handbook of Chemistry and Physics*, 55th ed., CRC Press, Cleveland, 1974, F-22.

Table 3−12B
HARDNESS OF MATERIALS

Agate	6−7	Indium	1.2
Alabaster	1.7	Iridium	6−6.5
Alum	2−2.5	Iridosmine	7
Aluminum	2−2.9	Iron	4−5
Alundum	9+	Kaolinite	2.0−2.5
Amber	2−2.5	Lead	1.5
Andalusite	7.5	Lithium	0.6
Anthracite	2.2	Loess (0°)	0.3
Antimony	3.0−3.3	Magnesium	2.0
Apatite	5	Magnetite	6
Argonite	3.5	Manganese	5.0
Arsenic	3.5	Marble	3−4
Asbestos	5	Meerschaum	2−3
Asphalt	1−2	Mica	2.8
Augite	6	Opal	4−6
Barite	3.3	Orthoclase	6
Bell metal	4	Osmium	7.0
Beryl	7.8	Palladium	4.8
Bismuth	2.5	Phosphorus	0.5
Boric acid	3	Phosphorbronze	4
Boron	9.5	Platinum	4.3
Brass	3−4	Platinum-iridium	6.5
Cadmium	2.0	Potassium	0.5
Calamine	5	Pumice	6
Calcite	3	Pyrite	6.3
Calcium	1.5	Quartz	7
Carbon	10.0	Rock salt (halite)	2
Carborundum	9−10	Ross' metal	2.5−3.0
Cesium	0.2	Rubidium	0.3
Chromium	9.0	Ruthenium	6.5
Copper	2.5−3	Selenium	2.0
Corundum	9	Serpentine	3−4
Diamond	10	Silicon	7.0
Diatomaceous earth	1−1.5	Silver	2.5−4
Dolomite	3.5−4	Silver chloride	1.3
Emery	7−9	Sodium	0.4
Feldspar	6	Steel	5−8.5
Flint	7	Stibnite	2
Fluorite	4	Strontium	1.8
Galena	2.5	Sulfur	1.5−2.5
Gallium	1.5	Talc	1
Garnet	6.5−7	Tellurium	2.3
Glass	4.5−6.5	Tin	1.5−1.8
Gold	2.5−3	Topaz	8
Graphite	0.5−1	Tourmaline	7.3
Gypsum	1.6−2	Wax (0°)	0.2
Hematite	6	Wood's metal	3
Hornblende	5.5	Zinc	2.5

From Weast, R. C., Ed., *Handbook of Chemistry and Physics*, 55th ed., CRC Press, Cleveland, 1974, F-22.

Table 3–12C
MOHS HARDNESS SCALE

Hardness number	Original scale	Modified scale
1	Talc	Talc
2	Gypsum	Gypsum
3	Calcite	Calcite
4	Fluorite	Fluorite
5	Apatite	Apatite
6	Orthoclase	Orthoclase
7	Quartz	Vitreous silica
8	Topaz	Quartz or stellite
9	Corundum	Topaz
10	Diamond	Garnet
11	–	Fused zirconia
12	–	Fused alumina
13	–	Silicon carbide
14	–	Boron carbide
15	–	Diamond

From Weast, R. C., Ed., *Handbook of Chemistry and Physics*, 55th ed., CRC Press, Cleveland, 1974, F-22.

Table 3–12D
COMPARISON OF MOHS AND KNOOP HARDNESS SCALES

Substance	Formula	Mohs value	Knoop value
Talc	$3MgO \cdot 4SiO_2 \cdot H_2O$	1	
Gypsum	$CaSO_4 \cdot 2H_2O$	2	32
Cadmium	Cd	–	37
Silver	Ag	–	60
Zinc	Zn	–	119
Calcite	$CaCO_3$	3	135
Fluorite	CaF_2	4	163
Copper	Cu	–	163
Magnesia	MgO	–	370
Apatite	$CaF_2 \cdot 3Ca_3(PO_4)_2$	5	430
Nickel	Ni	–	557
Glass (soda lime)	–	–	530
Feldspar (orthoclase)	$K_2O \cdot Al_2O_3 \cdot 6SiO_2$	6	560
Quartz	SiO_2	7	820
Chromium	Cr	–	935
Zirconia	ZrO_2	–	1160
Beryllia	BeO	–	1250
Topaz	$(AlF)_2 SiO_4$	8	1340
Garnet	$Al_2O_3 \cdot 3FeO \cdot 3SiO_2$	–	1360
Tungsten carbide alloy	WC, Co	–	1400–1800
Zirconium boride	ZrB_2	–	1550
Titanium nitride	TiN	9	1800
Tungsten carbide	WC	–	1880
Tantalum carbide	TaC	–	2000
Zirconium carbide	ZrC	–	2100

Table 3–12D (continued)
COMPARISON OF MOHS AND KNOOP HARDNESS SCALES

Substance	Formula	Mohs value	Knoop value
Alumina	Al_2O_3	–	2100
Beryllium carbide	Be_2C	–	2410
Titanium carbide	TiC	–	2470
Silicon carbide	SiC	–	2480
Aluminum boride	AlB	–	2500
Boron carbide	B_4C	–	2750
Diamond	C	10	7000

From Foster, L. S., in *Handbook of Chemistry and Physics,* 55th ed., Weast, R. C., Ed., CRC Press, Cleveland, 1974, F-22.

Table 3–13
ELECTRICAL RESISTIVITY OF THE ALKALI METALS

Metal	Atomic number	Atomic weight	Specific gravity	Melting point, °C	Resistivity Microhm-cm	Temp,°C
Li	3	6.939	0.534^{20}	179	12.17	86.6
					13.36	120.5
					14.73	153.3
					15.54	178.2
Na	11	22.9898	0.971^{20}	97.81	5.23	29.4
					5.72	50.6
					6.53	83.3
					6.71	91.6
					6.83	97.1
K	19	39.102	0.86^{20}	63.65	7.01	22.8
					7.32	35.5
					7.54	41.1
					8.05	56.1
Rb	37	85.47	$1.475^{38.89}$	38.89	11.28	0
					12.52	53
Cs	55	132.905	1.8785^{15}	28.5	20.29	18.2
					37.39	28.1

From Weast, R. C., Ed., *Handbook of Chemistry and Physics,* 55th ed., CRC Press, Cleveland, 1974, F-141.

Table 3–14
ELECTRICAL RESISTIVITY OF ALLOYS

Composition	Name	Resistivity				Temp coef of expansion/ deg C	Melting point, deg C
		Microhm-cm	Ohm/mil-ft	Relative, pure copper = 1	Temp coef of resist/ deg C		
ALUMINUM ALLOYS							
97 Al, 2 Mg, .5 Cr	Alloy 5052	5	30	2.9	.004	.000 025	625
COPPER ALLOYS							
	Soft copper wire	1.72	10.3	1.01	.004	.000 017	1 080
Hard copper wire	6101-B317-64	1.8	10.8	1.06	.004		1 085
Copper-clad steel	30 HS, B227-65	35	210	20.6	.005		
98 Cu, 2 Ni	Alloy 30	5	30	2.9	.001 4	.000 017	1 090
94 Cu, 6 Ni	Cuprothal 60[a]	10	60	5.9	.001 4	.000 018	1 100
91 Cu, 8 Sn, .25 P	Phosphor bronze	13	65	7.65		.000 018	1 050
89 Cu, 11 Ni	Alloy 90	15	90	8.8	.000 5	.000 02	1 110
87 Cu, 13 Mn	Manganin	48	290	28.4	.000 01	.000 02	1 020
78 Cu, 22 Ni	Midohm[a] Cuprothal 180[a]	30	180	17.6	.000 2	.000 02	1 130
65 Cu, 35 Zn	Brass	7	42	4.1	.002	.000 018	940
64 Cu, 18 Zn, 18 Ni	Nickel, silver	28	168	16.5	.000 3	.000 018	1 110
57 Cu, 43 Ni	Constantan	49	294	28.8	.000 01	.000 015	1 220
IRON ALLOYS							
	Soft steel 1010	12	72	7.1	.006	.000 011	1 450
99 Fe, 1 C	Carbon steel	20	120	11.8	.005	.000 011	1 430
96 Fe, 4 Si	High silicon iron	59	354	34.7	.002	.000 013	1 410
81 Fe, 15 Cr, 4 Al	Alloy 750	125	750	73.5	.000 15	.000 015	1 520
74 Fe, 18 Cr, 8 Ni	18-8 Stainless steel	73	440	43.0	.000 94	.000 018	1 400
72 Fe, 22.5 Cr, 5.5 Al	Alloy 875	146	875	85.8	.000 02	.000 017	1 520
65 Fe, 35 Ni	Invar	81	485	47.6	.001 35	.000 001	1 425
62 Fe, 21 Ni, 12 Al, 5 Co	Alnico I[a]	75	450	44.1	.002	.000 015	
58 Fe, 42 Ni	Alloy 142	67	400	39.3	.001 2	.000 005	1 425
55 Fe, 37.5 Cr, 7.5 Al	High-resistance alloy	166	1 000	97.6	.001	.000 015	1 500
45 Fe, 35 Ni, 20 Cr	Chromax,[a] Chromel D[a]	100	600	58.8	.000 36	.000 016	1 380
NICKEL ALLOYS (> 50%)							
80 Ni, 20 Cr	Nichrome V[a] Nikrothal 8[a] Chromel A[a]	108	650	63.6	.000 1	.000 016	1 375
72 Ni, 38 Fe	Hytemco[a]	20	120	11.8	.004 2	.000 015	1 425
68 Ni, 20 Cr, 8 Fe	Chromel AA[a]	117	700	68.7	.000 11	.000 014	1 390
60 Ni, 16 Cr, 24 Fe	Nichrome[a] Chromel C[a] Nicrothal 6[a]	112	675	66.0	.000 15	.000 015	1 350
95 Ni, 4 Mn, 1 Si	Alloy R63	22	135	13.0	.003	.000 015	1 400
67 Ni, 30 Cu, 1.4 Fe, 1 Mn	Monel	42	252	24.7	.002	.000 014	1 330
SILVER ALLOYS							
92.5 Ag, 7.5 Cu	Sterling silver	2	12	1.18	.004	.000 018	905
85 Ag, 15 Cd	Contact alloy	5	30	2.9	.004	.000 019	875

[a]Proprietary name.

From Bolz, R. E. and Tuve, G. L., Eds., *Handbook of Tables for Applied Engineering Science,* 2nd ed., CRC Press, Cleveland, 1973, 224.

Table 3–15
MELTING POINTS OF METALLIC COMPOUNDS

Refractory and Ceramic Materials and Salts

Metal	Boride		Bromide		Carbide		Chloride		Fluoride		Iodide	
	K	deg F	K	deg F	K	deg F	K	deg F	K	deg F	K	deg F
Ag			AgBr 703	806			AgCl 728	851	AgF 708	815	AgI 831	1036
Al			AlBr$_3$ 371	207	Al$_4$C$_3$ 2000^a	3600^a	AlCl$_3$ 465	377	AlF$_3$ 1564^a	2356^a	AlI 464	376
B			BBr$_3$ 227	−51	B$_4$C 2720	4440	BCl$_3$ 166	−161	BF$_3$ 146	−196		
Ba	BaB$_4$ 2543	4118	BaBr$_2$ 1123	1562			BaCl$_2$ 1235	1764	BaF$_2$ 1627	2470	BaI$_2$ 1013	1364
Be	BeB$_2$ >2243	>3357	BeBr$_2$ 793	968	Be$_2$C >2375^a	>3815^a	BeCl$_2$ 713	824	BeF$_2$ 813	1004	BeI$_2$ 783	950
Bi			BiBr$_3$ 491	424			BiCl$_3$ 507	452	BiF$_3$ 1000	1341	BiI$_3$ 681	784
Ca			CaBr$_2$ 1003^a	1346^a			CaCl$_2$ 1055	1440	CaF$_2$ 1675	2555	CaI$_2$ 848	1067
Cd			CdBr$_2$ 841	1054			CdCl$_2$ 841	1054	CdF$_2$ 1373	2012	CdI$_2$ 423	302
Ce	CeB$_6$ 2463	3975					CeCl$_3$ 1095	1512	CeF$_3$ 1710	2618	CeI$_3$ 1025	1386
Cr	CrB$_2$ 2123	3362			Cr$_3$C$_2$ 2168	3440						
Cu			CuBr 777	939			CuCl 695	792	CuF$_2$ 1129	1573	CuI 878	1121
Fe			FeBr$_2$ 955	1754^a	Fe$_3$C 2110	3339	FeCl$_2$ 945	1242	FeF$_3$ >1275	>1835		
In			InBr$_3$ 709	817			InCl 498	437	InF$_3$ 1443	2138	InI$_3$ 483	410
K			KBr 1008	1355			KCl 1043	1418	KF 1131	1576	KI 958	1265
Li			LiBr 823	1022			LiCl 883	1130	LiF 1119	1554	LiI 722	840
Mg			MgBr$_2$ 984	1312			MgCl$_2$ 987	1317	MgF$_2$ 1536	2305	MgI$_2$ <910	<1078
Mn							MnCl$_2$ 923	1202	MnF$_2$ 1129	1573		
Mo	MoB 2625	4250			Mo$_2$C 2963	4875			MoF$_6$ 290	63	MoI$_4$ 373^a	212
Na			NaBr 1023	1382	NaC$_2$ 973	1292	NaCl 1073	1472	NaF 1267	1821	NaI 935	1224

Table 3–15 (continued)
MELTING POINTS OF METALLIC COMPOUNDS

Metal	Compound										
	Nitrate		Nitride		Oxide		Silicide		Sulfate		Sulfide	
	K	deg F	K	deg F	K	deg F	K	deg F	K	deg F	K	deg F
Ag	$AgNO_3$ 483	410			Ag_2O 573[a]	572			Ag_2SO_4 933	1220	Ag_2S 1098	1517
Al			AlN >2475	>4000	Al_2O_3 2322	3720			$Al_2(SO_4)_3$ 1043[a]	1418[a]	Al_2S_3 1373	2012
B			BN 3000	4945	B_2O_3 723	841					B_2S_4 663	734
Ba	$Ba(NO_3)_2$ 865	1098			BaO 2283	3649			$BaSO_4$ 1853	2876	BaS 1473	2192
Be			Be_3N_2 2513[a]	4064[a]	BeO 2725	4445			$BeSO_4$ 848[a]	1067[a]		
Bi					Bi_2O_3 1098	1516			$Bi(SO_4)_3$ 678[a]	761[a]	Bi_2S_3 1020	1377
Ca	$Ca(NO_3)_2$ 834	1042	Ca_3N_2 1468	2183	CaO 3183	5269			$CaSO_4$ 1723	3542		
Cd	$Cd(NO_3)_2$ 623	662			CdO 1773	2731			$CdSO_4$ 1273[b]	1832[b]	CdS 2023[b]	3182[b]
Ce					CeO_2 >2873	>4711			$Ce(SO_4)_2$ 468[a]	383[a]	CeS 2400	3860
Cr			CrN 1770[a]	2730[a]	Cr_2O_3 2603	4225	$CrSi_2$ 1843	2858				
Cu			Cu_3N 573[a]	572	Cu_2O 1508	2254	Cu_4Si 1123	1562	Cu_2S 1400	2060		
Fe					Fe_2O_3 1864	2895			$Fe_2(SO_4)_3$ 753[a]	896[a]	FeS 1468	2183
In					In_2O_3 2183	3469					In_2S_3 1323	1922
K	KNO_3 610	639			K_2O_3 703	806			K_2SO_4 1342	1956	K_2S 1113	1544
Li	$LiNO_3$ 527	489	Li_3N 1118[a]	1553	Li_2O >1975	>3095			Li_2SO_4 1132	1578	Li_2S 1198	1697
Mg					MgO 3098	5116	Mg_2Si 1375	2016	MgS >2275[a]	>3635[a]	$MgSO_4$ 1397[a]	2055[a]
Mn					MnO 1840	2852						
Mo					MoO_3 1068	1462	$MoSi_2$ 2553	3595			MoS_2 1458	2165
Na	$NaNO_3$ 583	590	Na_2N 573[a]	572[a]					Na_2SO_4 1157	1632	Na_2S 1453	2156

Table 3–15 (continued)
MELTING POINTS OF METALLIC COMPOUNDS

Metal	Compound										
	Boride		Bromide		Carbide		Chloride		Fluoride		Iodide	
	K	deg F	K	deg F	K	deg F	K	deg F	K	deg F	K	deg F
Nb	NbB >2270	>3630			NbC 3770	6330						
Ni			NiBr$_2$ 1236	1765			NiCl$_3$ 1274	1834	NiF$_2$ 1273^a	1832	NiI$_2$ 1070	1467
Pb			PbBr$_2$ 643	698			PbCl$_2$ 771	928	PbF$_2$ 1095	1512	PbI$_2$ 675	756
Pt			PtBr$_2$ 523^a	482^a			PtCl$_2$ 854^a	1078^a			PtI$_2$ 633^a	680^a
Sb			SbBr$_3$ 370	207			SbCl$_3$ 346	164	SbF$_3$ 565	558	SbI$_3$ 443	338
Si					SiC 2970	4890			SiF$_4$ 183	−130		
Sn			SnBr$_2$ 488	420			SnCl$_2$ 518	473	SnF$_4$ 978^a	1300^a	SnI$_2$ 593	608
Sr	SrB$_6$ 2508	4055	SrBr$_2$ 916	1189	SrC$_2$ >1970	>3100	SrCl$_2$ 1148	1607	SrF$_2$ 1736	2665	SrI$_2$ 788	959
Ta	TaB >2270^a	>3630^a	TaBr$_5$ 538	509	TaC 3813	6403	TaCl$_5$ 489	421	TaF$_5$ 370	206		
Te			TeBr$_2$ 612	642			TeCl$_2$ 448	347				
Th	ThB$_4$ >2770	>4530	ThBr$_4$ 883	1130^a	ThC 2898	5250	ThCl$_4$ 1043	1418	ThF$_4$ 1375	2015		
Ti	TiB$_2$ 3253	5396	TiBr$_4$ 312	102	TiC 3433	5720	TiCl$_4$ 250	−9	TiF$_3$ 1475	2195	TiI$_2$ 873	1112
U	UB$_2$ >1770	>2730	UBr$_4$ 789	961	UC 2863	4693	UCl$_4$ 843	1058	UF$_4$ 1233	1760	UI$_4$ 779	943
V	VB$_2$ 2373	3812			VC 3600	5120	VCl$_4$ 245	−18	VF$_3$ >1075	>1475	VI$_2$ 1048^a	1427
W	WB 3133	5180	WBr$_5$ 549	529	WC 2900^a	4760^a	WCl$_6$ 548	527				
Zn			ZnBr$_2$ 667	741			ZnCl$_2$ 548	527	ZnF$_2$ 1145	1602	ZnI$_2$ 719	835
Zr	ZrB$_2$ 3313	5505	ZrBr$_2$ >625^a	>660^a	ZrC 3533	5900	ZrCl$_2$ 623	662	ZrF$_4$ 873^a	1112^a	ZrI$_4$ 772	930

Table 3–15 (continued)
MELTING POINTS OF METALLIC COMPOUNDS

Metal	Nitrate		Nitride		Oxide		Silicide		Sulfate		Sulfide	
	K	deg F	K	deg F	K	deg F	K	deg F	K	deg F	K	deg F
Nb			NbN		Nb$_2$O$_5$		NbSi$_2$					
			2323	3722	1764	2715	2203	3505				
Ni					NiO				NiSO$_4$		NiS	
					2257	3603			1121^a	1558^a	1070	1466
Pb	Pb(NO$_3$)$_2$				PbO				PbSO$_4$		PbS	
	743^a	878^a			1159	1626			1443	2138	1387	2037
Pt											PtS$_2$	
											508^a	455^a
Sb					Sb$_2$O$_3$						SbS$_3$	
					928	1211					820	1016
Si			Si$_3$N$_4$		SiO$_2$							
			2175^b	3450	1978	3100						
Sn					SnO				SnSO$_4$		SnS	
					1353^a	1973^a			>635	>680	1153	1616
Sr	Sr(NO$_3$)$_2$				SrO				SrSO$_4$		SrS	
	643	1058			2933	4819			1878	2921	>2275	>3635
Ta			Ta$_2$N		Ta$_2$O$_5$		TaSi$_2$				TaS$_4$	
			3360	5595	2100	3325	2670	4350			>1575	>2375
Te					TeO$_2$							
					1006	1351						
Th			ThN		ThO$_2$						ThS$_2$	
			2903	4765	3493	5827					2198	3497
Ti			TiN		TiO$_2$		TiSi$_2$					
			3200^a	5790^a	2113	3344	1813	2804				
U			UN		UO$_2$		USi$_2$				US$_2$	
			3123	5161	3151	5212	1970	3090			>1375	>2015
V			VN		V$_2$O$_5$		VSi$_2$				V$_2$S$_3$	
			2593	4208	947	1245	2023	3182			>875^a	>1115^a
W					WO$_3$		WSi$_2$				WS$_2$	
					1744	2679	2320	3720			1523^a	2282^a
Zn					ZnO				ZnSO$_4$			
					2248	3586			873^a	1112^a		
Zr			ZrN		ZrO$_2$				Zr(SO$_4$)$_2$		ZrS$_2$	
			3250	5400	3123	5161			683	770	1823	2822

^{a}Decomposes or sublimes.
^{b}Pressure above 1 atm.

Table 3–15 (continued)
MELTING POINTS OF METALLIC COMPOUNDS

Other Compounds Melting Point Temperatures in K (with °F in Parentheses)

Ag_2CO_3—491 (424); $BaCO_3$—1653 (2516); $CaCO_3$—1613[b] (2444)[b]; $CdCO_3$—<775[a] (<930[a]); K_2CO_3—1170 (1647); Li_2CO_3—950 (1250); Na_2CO_3—1127 (1569); $PbCO_3$—588[a] (600[a]); $SrCO_3$—1770 (2726)

$AgNO_2$—413[a] (284[a]); $Ba(NO_2)_2$—540 (513); KNO_2—692 (786); $LiNO_2$—493 (428); $NaNO_2$—558 (545)

Ag_2Te—1228 (1750); Bi_2Te_3—861 (1091); $CdTe$—1314 (1906); $InTe$—965 (1277); $PtTe_2$—1523 (2282); Sb_2Te_3—891 (1145); $SnTe$—1053 (1436)

$BaTiO_3$—1891 (2944); $FeTiO_3$—1640 (2492)

$Ce_2(WO_4)_3$—1362 (1992); K_2WO_4—1203 (1706); $LiWO_4$—1015 (1368); $NaWO_4$—971 (1288)

REFERENCES

Weast, R. C., Ed., *Handbook of Chemistry and Physics,* 55th ed., CRC Press, Cleveland, 1974.

Charlesworth, J. H., *Melting Points of Metallic Elements and Selected Compounds,* Air Force Materials Laboratory Technical Report AFML-TR-70-137, 1970.

Janz, G. J., Dampier, F. W., Lakshminarayanan, G. R., Lorenz, P. K., and Tompkins, R. P. T., *Molten Salts: Electrical Conductance, Density, and Viscosity Data,* Vol. 1, NSRDS-NBS 15, National Bureau of Standards, October 1968.

From Bolz, R. E. and Tuve, G. L., Eds., *Handbook of Tables for Applied Engineering Science,* 2nd ed., CRC Press, Cleveland, 1973, 338.

Table 3–16
THERMAL CONDUCTIVITY OF METALS AT CRYOGENIC TEMPERATURES

As Recommended by National Standard Reference Data System-NBS

Values in this table are in watts/cm K. To convert to Btu/hr ft °R, multiply the tabular values by 57.818. These data apply only to metals of purity of at least 99.9%. In the table the third significant figure is for smoothness and is not indicative of the degree of accuracy.

Temperature		Aluminum	Cadmium	Chro-mium	Copper	Gold	Iron	Lead	Magne-sium	Molyb-denum
°K	°R									
1	1.8	7.8	48.7	0.401	28.7	4.4	0.75	27.7	1.30	0.146
2	3.6	15.5	89.3	0.802	57.3	8.9	1.49	42.4	2.59	0.292
3	5.4	23.2	104	1.20	85.5	13.1	2.24	34.0	3.88	0.438
4	7.2	30.8	92.0	1.60	113	17.1	2.97	22.4	5.15	0.584
5	9	38.1	69.0	1.99	138	20.7	3.71	13.8	6.39	0.730
6	10.8	45.1	44.2	2.38	159	23.7	4.42	8.2	7.60	0.876
7	12.6	51.5	28.0	2.77	177	26.0	5.13	4.9	8.75	1.02
8	14.4	57.3	18.0	3.14	189	27.5	5.80	3.2	9.83	1.17
9	16.2	62.2	12.2	3.50	195	28.2	6.45	2.3	10.8	1.31
10	18	66.1	8.87	3.85	196	28.2	7.05	1.78	11.7	1.45
11	19.8	69.0	6.91	4.18	193	27.7	7.62	1.46	12.5	1.60
12	21.6	70.8	5.56	4.49	185	26.7	8.13	1.23	13.1	1.74
13	23.4	71.5	4.67	4.78	176	25.5	8.58	1.07	13.6	1.88
14	25.2	71.3	4.01	5.04	166	24.1	8.97	0.94	14.0	2.01
15	27	70.2	3.55	5.27	156	22.6	9.30	0.84	14.3	2.15
16	28.8	68.4	3.16	5.48	145	20.9	9.56	0.77	14.4	2.28
18	32.4	63.5	2.62	5.81	124	17.7	9.88	0.66	14.3	2.53
20	36	56.5	2.26	6.01	105	15.0	9.97	0.59	13.9	2.77
25	45	40.0	1.79	6.07	68	10.2	9.36	0.507	12.0	3.25
30	54	28.5	1.56	5.58	43	7.6	8.14	0.477	9.5	3.55
35	63	21.0	1.41	5.03	29	6.1	6.81	0.462	7.4	3.62
40	72	16.0	1.32	4.30	20.5	5.2	5.55	0.451	5.7	3.51
45	81	12.5	1.25	3.67	15.3	4.6	4.50	0.442	4.57	3.26
50	90	10.0	1.20	3.17	12.2	4.2	3.72	0.435	3.75	3.00
60	108	6.7	1.13	2.48	8.5	3.8	2.65	0.424	2.74	2.60
70	126	5.0	1.08	2.08	6.7	3.58	2.04	0.415	2.23	2.30
80	144	4.0	1.06	1.82	5.7	3.52	1.68	0.407	1.95	2.09
90	162	3.4	1.04	1.68	5.14	3.48	1.46	0.401	1.78	1.92
100	180	3.0	1.03	1.58	4.83	3.45	1.32	0.396	1.69	1.79

Temperature		Nickel	Niobium	Plat-inum	Silver	Tan-talum	Tin	Tita-nium	Tung-sten	Zinc	Zirco-nium
°K	°R										
1	1.8	0.64	0.251	2.31	39.4	0.115		0.0144	14.4	19.0	0.111
2	3.6	1.27	0.501	4.60	78.3	0.230		0.0288	28.7	37.9	0.223
3	5.4	1.91	0.749	6.79	115	0.345	297	0.0432	42.6	55.5	0.333
4	7.2	2.54	0.993	8.8	147	0.459	181	0.0576	55.6	69.7	0.442
5	9	3.16	1.23	10.5	172	0.571	117	0.0719	67.1	77.8	0.549
6	10.8	3.77	1.46	11.8	187	0.681	76	0.0863	76.2	78.0	0.652
7	12.6	4.36	1.67	12.6	193	0.788	52	0.101	82.4	71.7	0.748
8	14.4	4.94	1.86	12.9	190	0.891	36	0.115	85.3	61.8	0.837
9	16.2	5.49	2.04	12.8	181	0.989	26	0.129	85.1	51.9	0.916
10	18	6.00	2.18	12.3	168	1.08	19.3	0.144	82.4	43.2	0.984
11	19.8	6.48	2.30	11.7	154	1.16	14.8	0.158	77.9	36.4	1.04
12	21.6	6.91	2.39	10.9	139	1.24	11.6	0.172	72.4	30.8	1.08
13	23.4	7.30	2.46	10.1	124	1.30	9.3	0.186	66.4	26.1	1.11
14	25.2	7.64	2.49	9.3	109	1.36	7.6	0.200	60.4	22.4	1.13
15	27	7.92	2.50	8.4	96	1.40	6.3	0.214	54.8	19.4	1.13
16	28.8	8.15	2.49	7.6	85	1.44	5.3	0.227	49.3	16.9	1.12
18	32.4	8.45	2.42	6.1	66	1.47	4.0	0.254	40.0	13.3	1.08
20	36	8.56	2.29	4.9	51	1.47	3.2	0.279	32.6	10.7	1.01
25	45	8.15	1.87	3.15	29.5	1.36	2.22	0.337	20.4	6.9	0.85
30	54	6.95	1.45	2.28	19.3	1.16	1.76	0.382	13.1	4.9	0.74
35	63	5.62	1.16	1.80	13.7	0.99	1.50	0.411	8.9	3.72	0.65
40	72	4.63	0.97	1.51	10.5	0.87	1.35	0.422	6.5	2.97	0.58
45	81	3.91	0.84	1.32	8.4	0.78	1.23	0.416	5.07	2.48	0.535
50	90	3.36	0.76	1.18	7.0	0.72	1.15	0.401	4.17	2.13	0.497
60	108	2.63	0.66	1.01	5.5	0.651	1.04	0.377	3.18	1.71	0.442

Table 3—16 (continued)

THERMAL CONDUCTIVITY OF METALS AT CRYOGENIC TEMPERATURES

As Recommended by National Standard Reference Data System-NBS

Temperature		Nickel	Niobium	Plat-inum	Silver	Tan-talum	Tin	Tita-nium	Tung-sten	Zinc	Zirco-nium
K	R										
70	126	2.21	0.61	0.90	4.97	0.616	0.96	0.356	2.76	1.48	0.403
80	144	1.93	0.58	0.84	4.71	0.603	0.91	0.339	2.56	1.38	0.373
90	162	1.72	0.563	0.81	4.60	0.596	0.88	0.324	2.44	1.34	0.350
100	180	1.58	0.552	0.79	4.50	0.592	0.85	0.312	2.35	1.32	0.332

From Ho, C. Y., Powell, R. W., and Liley, P. E., *Thermal Conductivity of Selected Materials,* NSRDS-NBS-8 and NSRDS-NBS-16, National Standard Reference Data System-National Bureau of Standards, Part 1, 1966; Part 2, 1968.

Table 3–17
THERMAL CONDUCTIVITY OF METALS

100 to 3000 K

Values in this table are in watts/cm K. To convert to Btu/hr ft °R, multiply the tabular values by 57.818. These data apply only to metals of purity of at least 99.9%. In the table the third significant figure is for smoothness and is not indicative of the degree of accuracy.

Metal	Temperature, °K									
	100°	200°	273°	300°	400°	500°	600°	700°	800°	900°
Aluminum	3.0	2.37	2.36	2.37	2.40	2.37	2.32	2.26	2.20	2.13
Cadmium	1.03	0.993	0.975	0.968	0.947	0.920	(0.420)	(0.490)	(0.559)	
Chromium	1.58	1.11	0.948	0.903	0.873	0.848	0.805	0.757	0.713	0.678
Copper	4.83	4.13	4.01	3.98	3.92	3.88	3.83	3.77	3.71	3.64
Gold	3.45	3.27	3.18	3.15	3.12	3.09	3.04	2.98	2.92	2.85
Iron	1.32	0.94	0.835	0.803	0.694	0.613	0.547	0.487	0.433	0.380
Lead	0.396	0.366	0.355	0.352	0.338	0.325	0.312	(0.174)	(0.190)	(0.203)
Magnesium	1.69	1.59	1.57	1.56	1.53	1.51	1.49	1.47	1.46	1.45
Mercury			(0.078)	(0.084)	(0.098)	(0.109)	(0.120)	(0.127)	(0.130)	
Molybdenum	1.79	1.43	1.39	1.38	1.34	1.30	1.26	1.22	1.18	1.15
Nickel	1.58	1.06	0.94	0.905	0.801	0.721	0.655	0.653	0.674	0.696
Niobium (Columbium)	0.552	0.526	0.533	0.537	0.552	0.567	0.582	0.598	0.613	0.629
Platinum	0.79	0.748	0.734	0.730	0.722	0.719	0.720	0.723	0.729	0.737
Silver	4.50	4.30	4.28	4.27	4.20	4.13	4.05	3.97	3.89	3.82
Tantalum	0.592	0.575	0.574	0.575	0.578	0.582	0.586	0.590	0.594	0.598
Tin	0.85	0.733	0.682	0.666	0.622	0.596	(0.323)	(0.343)	(0.364)	(0.384)
Titanium	0.312	0.245	0.224	0.219	0.204	0.197	0.194	0.194	0.197	0.202
Tungsten	2.35	1.97	1.82	1.78	1.62	1.49	1.39	1.33	1.28	1.24
Zinc	1.32	1.26	1.22	1.21	1.16	1.11	1.05	(0.499)	(0.557)	(0.615)
Zirconium	0.332	0.252	0.232	0.227	0.216	0.210	0.207	0.209	0.216	0.226

Metal	Temperature, °K									
	1000°	1100°	1200°	1400°	1600°	1800°	2000°	2200°	2600°	3000°
Aluminum	(0.93)	(0.96)	(0.99)							
Chromium	0.653	0.636	0.624	0.611						
Copper	3.57	3.50	3.42							
Gold	2.78	2.71	2.62							
Iron	0.326	0.297	0.282	0.309	0.327					
Lead	(0.215)									
Magnesium	(0.84)	(0.91)	(0.98)							
Molybdenum	1.12	1.08	1.05	0.996	0.946	0.907	0.880	0.858	0.825	
Nickel	0.718	0.739	0.761	0.804						
Niobium (Columbium)	0.644	0.659	0.675	0.705	0.735	0.764	0.791	0.815		
Platinum	0.748	0.760	0.775	0.807	0.842	0.877	0.913			
Silver	3.74	3.66	3.58							
Tantalum	0.602	0.606	0.610	0.618	0.626	0.634	0.640	0.647	0.658	0.665
Tin	(0.405)	(0.425)	(0.446)	(0.487)						
Titanium	0.207	0.213	0.220	0.236	0.253	0.271				
Tungsten	1.21	1.18	1.15	1.11	1.07	1.03	1.00	0.98	0.94	0.915
Zinc	(0.673)	(0.730)								
Zirconium	0.237	0.248	0.257	0.275	0.290	0.302	0.313			

Note: Values in parentheses are for liquid state.

From Ho, C. Y., Powell, R. W., and Liley, P. E., *Thermal Conductivity of Selected Materials*, NSRD-NBS-16, Part 2, National Standard Reference Data System-National Bureau of Standards, February 1968.

Table 3–18
HEAT OF FUSION OF SOME INORGANIC COMPOUNDS

For heat of fusion in J/kg, multiply values in cal/g by 4.184. For heat of fusion in J/mol, multiply values in cal/g·mol (=cal/mol) by 4.184. For melting point in K, add 273.15 to values in °C. Values in parentheses are of uncertain reliability.

Compound	Formula	Melting point, °C	Heat of fusion Btu/lb	Heat of fusion cal/g	Heat of fusion cal/g mole
Actinium[227]	Ac	1050 ± 50	(20.)	(11.0)	(3400)
Aluminum	Al	658.5	170.	94.5	2550
Aluminum bromide	Al_2Br_6	87.4	18.2	10.1	5420
Aluminum chloride	Al_2Cl_6	192.4	114.	63.6	19600
Aluminum iodide	Al_2I_6	190.9	17.6	9.8	7960
Aluminum oxide	Al_2O_3	2045.0	(461.)	(256.0)	(26000)
Antimony	Sb	630	70.4	39.1	4770
Antimony pentachloride	$SbCl_5$	4.0	14.4	8.0	2400
Antimony tribromide	$SbBr_3$	96.8	17.5	9.7	3510
Antimony trichloride	$SbCl_3$	73.3	23.9	13.3	3030
Antimony trioxide	Sb_4O_6	655.0	(83.3)	(46.3)	(26990)
Antimony trisulfide	Sb_4S_6	546.0	59.4	33.0	11200
Argon	Ar	-190.2	13.1	7.25	290
Arsenic	As	816.8	(39.6)	(22.0)	(6620)
Arsenic pentafluoride	AsF_5	-80.8	29.7	16.5	2800
Arsenic tribromide	$AsBr_3$	30.0	16.0	8.9	2810
Arsenic trichloride	$AsCl_3$	-16.0	23.9	13.3	2420
Arsenic trifluoride	AsF_3	-6.0	34.0	18.9	2486
Arsenic trioxide	As_4O_6	312.8	40.0	22.2	8000
Barium	Ba	725	23.9	13.3	1830
Barium bromide	$BaBr_2$	846.8	39.4	21.9	6000
Barium chloride	$BaCl_2$	959.8	46.6	25.9	5370
Barium fluoride	BaF_2	1286.8	30.8	17.1	3000
Barium iodide	BaI_2	710.8	(31.1)	(17.3)	(6800)
Barium nitrate	$Ba(NO_3)_2$	594.8	(40.7)	(22.6)	(5900)
Barium oxide	BaO	1922.8	168.	93.2	13800
Barium phosphate	$Ba_3(PO_4)_2$	1727	55.6	30.9	18600
Barium sulfate	$BaSO_4$	1350	74.9	41.6	9700
Beryllium	Be	1278	468.	260.0	—
Beryllium bromide	$BeBr_2$	487.8	(47.9)	(26.6)	(4500)
Beryllium chloride	$BeCl_2$	404.8	(54)	(30)	(3000)
Beryllium oxide	BeO	2550.0	1223.	679.7	17000
Bismuth	Bi	271	21.6	12.0	2505
Bismuth trichloride	$BiCl_3$	223.8	14.8	8.2	2600
Bismuth trifluoride	BiF_3	726.0	(41.9)	(23.3)	(6200)
Bismuth trioxide	Bi_2O_3	815.8	26.3	14.6	6800
Boron	B	2300	(882)	(490)	(5300)
Boron tribromide	BBr_3	-48.8	(5.2)	(2.9)	(700)
Boron trichloride	BCl_3	-107.8	(7.7)	(4.3)	(500)
Boron trifluoride	BF_3	-128.0	12.6	7.0	480
Boron trioxide	B_2O_3	448.8	142.	78.9	5500
Bromine	Br_2	-7.2	29.0	16.1	2580
Bromine pentafluoride	BrF_5	-61.4	12.7	7.07	1355
Cadmium	Cd	320.8	23.2	12.9	1460
Cadmium bromide	$CdBr_2$	567.8	(33.1)	(18.4)	(5000)
Cadmium chloride	$CdCl_2$	567.8	51.8	28.8	5300
Cadmium fluoride	CdF_2	1110	(64.6)	(35.9)	(5400)
Cadmium iodide	CdI_2	386.8	18.0	10.0	3660
Cadmium sulfate	$CdSO_4$	1000	41.2	22.9	4790
Calcium	Ca	851	100.	55.7	2230
Calcium bromide	$CaBr_2$	729.8	37.6	20.9	4180
Calcium carbonate	$CaCO_3$	1282	(227)	(126)	(12700)
Calcium chloride	$CaCl_2$	782	99	55	6100
Calcium fluoride	CaF_2	1382	94.5	52.5	4100
Calcium metasilicate	$CaSiO_3$	1512	208.	115.4	13400
Calcium nitrate	$Ca(NO_3)_2$	560.8	56.2	31.2	5120
Calcium oxide	CaO	2707	(393.)	(218.1)	(12240)
Calcium sulfate	$CaSO_4$	1297	88.6	49.2	6700
Carbon dioxide	CO_2	-57.6	77.8	43.2	1900
Carbon monoxide	CO	-205	12.8	7.13	199.7

Table 3–18 (continued)
HEAT OF FUSION OF SOME INORGANIC COMPOUNDS

Compound	Formula	Melting point, °C	Heat of fusion			Compound	Formula	Melting point, °C	Heat of fusion		
			Btu/lb	cal/g	cal/g mole				Btu/lb	cal/g	cal/g mole
Cyanogen	C_2N_2	-27.2	71.3	39.6	2060	Hydrogen fluoride	HF	-83.11	98.5	54.7	1094
Cyanogen chloride	CNCl	-5.2	65.5	36.4	2240	Hydrogen iodide	HI	-50.91	9.7	5.4	686.3
Cerium	Ce	775	27.2	15.1	2120	Hydrogen nitrate	HNO_3	-47.2	17.1	9.5	601
Cesium	Cs	28.3	6.7	3.7	500	Hydrogen oxide (water)	H_2O	0	138.	79.72	1436
Cesium chloride	CsCl	641.8	38.5	21.4	3600	Deuterium oxide	D_2O	3.78	136.	75.8	1516
Cesium nitrate	$CsNO_3$	406.8	29.9	16.6	3250	Hydrogen peroxide	H_2O_2	-0.7	15.4	8.58	2920
Chlorine	Cl_2	-103 ± 5	41.0	22.8	1531	Hydrogen selenate	H_2SeO_4	57.8	42.8	23.8	3450
Chromium	Cr	1890	112.	62.1	3660	Hydrogen sulfate	H_2SO_4	10.4	43.2	24.0	2360
Chromium (II) chloride	$CrCl_2$	814	119.	65.9	7700	Hydrogen sulfide	H_2S	-85.6	30.2	16.8	5683
Chromium (III) sequioxide	Cr_2O_3	2279	49.7	27.6	4200	Hydrogen sulfide, di-	H_2S_2	-89.7	49.1	27.3	1805
Chromium trioxide	CrO_3	197	67.9	37.7	3770	Hydrogen telluride	H_2Te	-49.0	23.2	12.9	1670
Cobalt	Co	1490	112.	62.1	3640	Indium	In	156.3	12.2	6.8	781
Cobalt (II) chloride	$CoCl_2$	727	102.	56.9	7390	Iodine	I_2	112.9	25.7	14.3	3650
Copper	Cu	1083	88.2	49.0	3110	Iodine chloride (α)	ICl	17.1	29.5	16.4	2660
Copper (II) chloride	$CuCl_2$	430	44.5	24.7	4890	Iodine chloride (β)	ICl	13.8	23.9	13.3	2270
Copper (I) chloride	CuCl	429	47.5	26.4	2620	Iron	Fe	1530.0	115.	63.7	3560
Copper (I) cyanide	$Cu_2(CN)_2$	473	(54.2)	(30.1)	(5400)	Iron carbide	Fe_3C	1226.8	123.	68.6	12330
Copper (I) iodide	CuI	587	(24.5)	(13.6)	(2600)	Iron (III) chloride	Fe_2Cl_6	303.8	114.	63.2	20500
Copper (II) oxide	CuO	1446	63.7	35.4	2820	Iron (II) chloride	$FeCl_2$	677	111.	61.5	7800
Copper (I) oxide	Cu_2O	1230	(168.)	(93.6)	(13400)	Iron (II) oxide	FeO	1380	(193.)	(107.2)	(7700)
Copper (I) sulfide	Cu_2S	1129	62.3	34.6	5500	Iron oxide	Fe_3O_4	1596	257.	142.5	33000
Dysprosium	Dy	1407	45.4	25.2	4100	Iron pentacarbonyl	$Fe(CO)_5$	-21.2	29.7	16.5	3250
Erbium	Er	1496	44.1	24.5	4100	Iron (II) sulfide	FeS	1195	102.	56.9	5000
Europium	Eu	826	29.5	16.4	2500	Lanthanum	La	920	31.3	17.4	2400
Europium trichloride	$EuCl_3$	622	(37.6)	(20.9)	(8000)	Lead	Pb	327.3	10.6	5.9	1224
Fluorine	F_2	-219.6	11.5	6.4	244.0	Lead bromide	$PbBr_2$	487.8	21.1	11.7	4290
Gadolinium	Gd	1312	42.8	23.8	3700	Lead chloride	$PbCl_2$	497.8	36.5	20.3	5650
Gallium	Ga	29	(34.4)	19.1	1336	Lead fluoride	PbF_2	823	13.7	7.6	1860
Germanium	Ge	959	(206.)	(114.3)	(8300)	Lead iodide	PbI_2	412	32.2	17.9	5970
Gold	Au	1063	(27.5)	15.3	3030	Lead molybdate	$PbMoO_4$	1065	(127.)	70.8	(25800)
Hafnium	Hf	2214	(61.4)	(34.1)	(6000)	Lead oxide	PbO	890	22.7	12.6	2820
Holmium	Ho	1461	44.6	24.8	4100	Lead sulfate	$PbSO_4$	1087	56.9	31.6	9600
Hydrogen	H_2	-259.25	24.8	13.8	28	Lead sulfide	PbS	1114	31.1	17.3	4150
Hydrogen bromide	HBr	-86.96	12.8	7.1	575.1	Lithium	Li	178.8	285.	158.5	1100
Hydrogen chloride	HCl	-114.3	23.4	13.0	476.0	Lithium bromide	LiBr	552	60.1	33.4	2900

Table 3–18 (continued)
HEAT OF FUSION OF SOME INORGANIC COMPOUNDS

Compound	Formula	Melting point, °C	Heat of fusion Btu/lb	cal/g	cal/g mole
Lithium chloride	LiCl	614	136.	75.5	3200
Lithium fluoride	LiF	896	(164.)	(91.1)	(2360)
Lithium hydroxide	LiOH	462	186.	103.3	2480
Lithium iodide	LiI	440	(19.1)	(10.6)	(1420)
Lithium metasilicate	Li$_2$SiO$_3$	1177	144.	80.2	7210
Lithium molybdate	Li$_2$MoO$_4$	705	43.4	24.1	4200
Lithium nitrate	LiNO$_3$	250	158.	87.8	6060
Lithium orthosilicate	Li$_4$SiO$_4$	1249	109.	60.5	7430
Lithium sulfate	Li$_2$SO$_4$	857	49.7	27.6	3040
Lithium tungstate	Li$_2$WO$_4$	742	(46.1)	(25.6)	(6700)
Lutetium	Lu	1651	47.3	26.3	4600
Magnesium	Mg	650	160.	88.9	2160
Magnesium bromide	MgBr$_2$	711	81.0	45.0	8300
Magnesium chloride	MgCl$_2$	712	149.	82.9	8100
Magnesium fluoride	MgF$_2$	1221	170.	94.7	5900
Magnesium oxide	MgO	2642	826.	459.0	18500
Magnesium silicate	MgSiO$_3$	1524	264.	146.4	14700
Magnesium sulfate	MgSO$_4$	1327	52.0	28.9	3500
Manganese	Mn	1220	113.	62.7	3450
Manganese dichloride	MnCl$_2$	650	105.	58.4	7340
Manganese metasilicate	MnSiO$_3$	1274	(113.)	(62.6)	(8200)
Manganese (II) oxide	MnO	1784	330.	183.3	13000
Manganese oxide	Mn$_3$O$_4$	1590	(307.)	(170.4)	(39000)
Mercury	Hg	−39	4.9	2.7	557.2
Mercury bromide	HgBr$_2$	241	19.6	10.9	3960
Mercury chloride	HgCl$_2$	276.8	27.5	15.3	4150
Mercury iodide	HgI$_2$	250	17.8	9.9	4500
Mercury sulfate	HgSO$_4$	850	8.6	(4.8)	(1440)
Molybdenum	Mo	2622	(123.)	(68.4)	(6600)
Molybdenum dichloride	MoCl$_2$	726.8	64.4	3.58	6000
Molybdenum hexafluoride	MoF$_6$	17	21.4	11.9	2500
Molybdenum trioxide	MoO$_3$	795	(31.1)	(17.3)	(2500)
Neodymium	Nd	1020	21.2	11.8	1700
Neon	Ne	−248.6	6.89	3.83	77.4
Nickel	Ni	1452	129.	71.5	4200

Compound	Formula	Melting point, °C	Heat of fusion Btu/lb	cal/g	cal/g mole
Nickel chloride	NiCl$_2$	1030	257.	142.5	18470
Nickel subsulfide	Ni$_3$S$_2$	790	46.4	25.8	5800
Niobium	Nb	2496	(124.)	(68.9)	(6500)
Niobium pentachloride	NbCl$_5$	211	55.4	30.8	8400
Niobium pentoxide	Nb$_2$O$_5$	1511	164.	91.0	24200
Nitric oxide	NO	−163.7	32.9	18.3	549.5
Nitrogen	N$_2$	−210	11.1	6.15	172.3
Nitrogen tetroxide	N$_2$O$_4$	−13.2	108.	60.2	5540
Nitrous oxide	N$_2$O	−90.9	63.9	35.5	1563
Osmium	Os	2700	(66.1)	(36.7)	(7000)
Osmium tetroxide (white)	OsO$_4$	41.8	16.6	9.2	2340
Osmium tetroxide (yellow)	OsO$_4$	55.8	27.9	15.5	4060
Oxygen	O$_2$	−218.8	5.9	3.3	106.3
Palladium	Pd	1555	69.5	38.6	4120
Phosphoric acid	H$_3$PO$_4$	42.3	46.4	25.8	2520
Phosphoric acid, hypo-	H$_4$P$_2$O$_6$	54.8	92.2	51.2	8300
Phosphorus acid, hypo-	H$_3$PO$_2$	17.3	63.0	35.0	2310
Phosphorus acid, ortho-	H$_3$PO$_3$	73.8	67.3	37.4	3070
Phosphorus oxychloride	POCl$_3$	1.0	36.5	20.3	3110
Phosphorus pentoxide	P$_4$O$_{10}$	569.0	108.	60.1	17080
Phosphorus trioxide	P$_4$O$_6$	23.7	27.5	15.3	3360
Phosphorus, yellow	P$_4$	44.1	8.6	4.8	600
Platinum	Pt	1770	43.4	24.1	4700
Potassium	K	63.4	26.3	14.6	574
Potassium borate, meta-	KBO$_2$	947	(124.)	(69.1)	(5660)
Potassium bromide	KBr	742	75.6	42.0	5000
Potassium carbonate	K$_2$CO$_3$	897	102.	56.4	7800
Potassium chloride	KCl	770	155.	85.9	6410
Potassium chromate	K$_2$CrO$_4$	984	64.1	35.6	6920
Potassium cyanide	KCN	623	(96.7)	(53.7)	(3500)
Potassium dichromate	K$_2$Cr$_2$O$_7$	398	53.6	29.8	8770
Potassium fluoride	KF	875	201.	111.9	6500
Potassium hydroxide	KOH	360	(63.5)	(35.3)	(1980)
Potassium iodide	KI	682	44.5	24.7	4100
Potassium nitrate	KNO$_3$	338	50.6	28.1	2840

Table 3–18 (continued)
HEAT OF FUSION OF SOME INORGANIC COMPOUNDS

Compound	Formula	Melting point, °C	Heat of fusion Btu/lb	Heat of fusion cal/g	Heat of fusion cal/g mole
Potassium peroxide	K_2O_2	490	99.5	55.3	6100
Potassium phosphate	K_3PO_4	1340	75.4	41.9	8900
Potassium pyrophosphate	$K_4P_2O_7$	1092	76.3	42.4	14000
Potassium sulfate	K_2SO_4	1074	83.5	46.4	8100
Potassium thiocyanate	KSCN	179	41.6	23.1	2250
Praseodymium	Pr	931	34.2	19.0	2700
Rhenium	Re	3167 ± 60	(76.3)	(42.4)	(7900)
Rhenium heptoxide	Re_2O_7	296	54.2	30.1	15340
Rhenium hexafluoride	ReF_6	19.0	29.9	16.6	5000
Rubidium	Rb	38.9	11.0	6.1	525
Rubidium bromide	RbBr	677	40.3	22.4	3700
Rubidium chloride	RbCl	717	65.5	36.4	4400
Rubidium fluoride	RbF	833	71.1	39.5	4130
Rubidium iodide	RbI	638	25.2	14.0	2990
Rubidium nitrate	$RbNO_3$	305	16.4	9.1	1340
Samarium	Sm	1072	31.1	17.3	2600
Scandium	Sc	1538	152.	84.4	3800
Selenium	Se	217	27.7	15.4	1220
Selenium oxychloride	$SeOCl_3$	9.8	11.0	6.1	1010
Silane, hexafluoro-	Si_2F_6	−28.6	41.2	22.9	3900
Silicon	Si	1427	607.	337.0	9470
Silicon dioxide (Cristobalite)	SiO_2	2100	63.0	35.0	2100
Silicon dioxide (Quartz)	SiO_2	1470	102.	56.7	3400
Silicon tetrachloride	$SiCl_4$	−67.7	19.4	10.8	1845
Silver	Ag	961	45.0	25.0	2700
Silver bromide	AgBr	430	20.9	11.6	2180
Silver chloride	AgCl	455	39.6	22.0	3155
Silver cyanide	AgCN	350	36.9	20.5	2750
Silver iodide	AgI	557	17.1	9.5	2250
Silver nitrate	$AgNO_3$	209	29.2	16.2	2755
Silver sulfate	Ag_2SO_4	657	(24.7)	(13.7)	(4280)
Silver sulfide	Ag_2S	841	24.3	13.5	3360
Sodium	Na	97.8	49.3	27.4	630
Sodium borate, meta-	$NaBO_2$	966	242.	134.6	8660
Sodium bromide	NaBr	747	107.	59.7	6140
Sodium carbonate	Na_2CO_3	854	119.	66.0	7000
Sodium chlorate	$NaClO_3$	255	89.5	49.7	5290
Sodium chloride	NaCl	800	222.	123.5	7220
Sodium cyanide	NaCN	562	(160.)	(88.9)	(4360)
Sodium fluoride	NaF	992	300.	166.7	7000
Sodium hydroxide	NaOH	322	90.0	50.0	2000
Sodium iodide	NaI	662	63.2	35.1	5340
Sodium molybdate	Na_2MoO_4	687	31.5	17.5	3600
Sodium nitrate	$NaNO_3$	310	79.6	44.2	3760
Sodium peroxide	Na_2O_2	460	135.	75.1	5860
Sodium phosphate, meta-	$NaPO_3$	988	(87.5)	(48.6)	(4960)
Sodium pyrophosphate	$Na_4P_2O_7$	970	(92.7)	(51.5)	(13700)
Sodium silicate, aluminum-	$NaAlSi_3O_8$	1107	90.2	50.1	13150
Sodium silicate, di-	$Na_2Si_2O_5$	884	83.5	46.4	8460
Sodium silicate, meta-	Na_2SiO_3	1087	152.	84.4	10300
Sodium sulfate	Na_2SO_4	884	73.8	41.0	5830
Sodium sulfide	Na_2S	920	(27.7)	15.4	(1200)
Sodium thiocyanate	NaSCN	323	98.6	54.8	4450
Sodium tungstate	Na_2WO_4	702	35.3	19.6	5800
Strontium	Sr	757	45.0	25.0	2190
Strontium bromide	$SrBr_2$	643	34.7	19.3	4780
Strontium chloride	$SrCl_2$	872	47.7	26.5	4100
Strontium fluoride	SrF_2	1400	61.2	34.0	4260
Strontium oxide	SrO	2430	290.	161.2	16700
Sulfur (monatomic)	S	119.	16.6	9.2	295
Sulfur dioxide	SO_2	−73.2	58.0	32.2	2060
Sulfur trioxide (α)	SO_3	16.8	46.4	25.8	2060
Sulfur trioxide (β)	SO_3	32.3	65.0	36.1	2890
Sulfur trioxide (γ)	SO_3	62.1	142.	79.0	6310
Tantalum	Ta	2996 ± 50	(62.3)	34.6 to 41.5	(7500)

Table 3—18 (continued)
HEAT OF FUSION OF SOME INORGANIC COMPOUNDS

Compound	Formula	Melting point, °C	Heat of fusion Btu/lb	Heat of fusion cal/g	Heat of fusion cal/g mole
Tantalum pentachloride	TaCl$_5$	206.8	45.2	25.1	9000
Tantalum pentoxide	Ta$_2$O$_5$	1877	195.	108.6	48000
Tellurium	Te	453	45.5	25.3	3230
Terbium	Tb	1356	44.3	24.6	3900
Thallium	Tl	302.4	9.0	5.0	1030
Thallium bromide, mono-	TlBr	460	37.8	21.0	5990
Thallium carbonate	Tl$_2$CO$_3$	273	17.1	9.5	4400
Thallium chloride, mono-	TlCl	427	31.9	17.7	4260
Thallium iodide, mono-	TlI	440	16.9	9.4	3125
Thallium nitrate	TlNO$_3$	207	15.5	8.6	2290
Thallium sulfate	Tl$_2$SO$_4$	632	19.6	10.9	5500
Thallium sulfide	Tl$_2$S	449	12.2	6.8	3000
Thorium	Th	1845	(<35.6)	(<19.8)	(<4600)
Thorium chloride	ThCl$_4$	765	111.	61.6	22500
Thorium dioxide	ThO$_2$	2952	1984.	1102.0	291100
Thulium	Tm	1545	46.8	26.0	4400
Tin	Sn	231.7	25.9	14.4	1720
Tin bromide, di-	SnBr$_2$	231.8	(11.0)	(6.1)	(1720)
Tin bromide, tetra-	SnBr$_4$	29.8	12.2	6.8	3000
Tin chloride, di-	SnCl$_2$	247	28.8	16.0	3050
Tin chloride, tetra-	SnCl$_4$	-33.3	15.]	8.4	2190
Tin iodide, tetra-	SnI$_4$	143.4	(12.4)	(6.9)	(4330)
Tin oxide	SnO	1042	(84.2)	(46.8)	(6400)
Titanium	Ti	1800	(188.)	(104.4)	(5000)
Titanium bromide, tetra-	TiBr$_4$	38	(10.1)	(5.6)	(2060)
Titanium chloride, tetra-	TiCl$_4$	-23.2	21.4	11.9	2240
Titanium dioxide	TiO$_2$	1825	(257.)	(142.7)	(11400)
Titanium oxide	TiO	991	394	219	14000
Tungsten	W	3387	(82.4)	(45.8)	(8420)
Tungsten dioxide	WO$_2$	1270	108.	60.1	13940
Tungsten hexafluoride	WF$_6$	-0.5	10.8	6.0	1800
Tungsten tetrachloride	WCl$_4$	327	33.1	18.4	6000
Tungsten trioxide	WO$_3$	1470	108.	60.1	13940
Uranium^{235}	U	~1133	36	20	3700
Uranium tetrachloride	UCl$_4$	590	48.8	27.1	10300
Vanadium	V	1917	(126)	(70)	(4200)
Vanadium dichloride	VCl$_2$	1027	118.	65.6	8000
Vanadium oxide	VO	2077	403.	224.0	15000
Vanadium pentoxide	V$_2$O$_5$	670	154.	85.5	15560
Xenon	Xe	-111.6	10.1	5.6	740
Ytterbium	Yb	823	22.9	12.7	2200
Yttrium	Y	1504	83.0	46.1	4100
Yttrium oxide	Y$_2$O$_3$	2227	199.	110.7	25000
Zinc	Zn	419.4	43.9	24.4	1595
Zinc chloride	ZnCl$_2$	283	(73.1)	(40.6)	(5540)
Zinc oxide	ZnO	1975	98.8	54.9	4470
Zinc sulfide	ZnS	1745	168.	93.3	(9100)
Zirconium	Zr	1857	168.	(60)	(5500)
Zirconium dichloride	ZrCl$_2$	727	81.0	45.0	7300
Zirconium oxide	ZrO$_2$	2715	304.	168.8	20800

REFERENCE

Weast, R. C., Ed., *Handbook of Chemistry and Physics*, 55th ed., CRC Press, Cleveland, 1974.

From Bolz, R. E. and Tuve, G. L., Eds., *Handbook of Tables for Applied Engineering Science*, 2nd ed., CRC Press, Cleveland, 1973, 479.

Table 3–19
SOLUBILITY PRODUCT

The solubility product (or ion product constant) is the product of the concentrations of the ions in the saturated solution of a difficultly soluble salt. The concentrations are expressed as moles per liter of solution. The number of cations (or anions) resulting from the dissociation of one molecule of the salt appears in the formula for calculations of the solubility product as the exponent of the concentration of the cation (or anion).

If two solutions, each containing one of the ions of a difficultly soluble salt, are mixed, no precipitation takes place unless the product of the ion concentrations in the mixture is greater than the solubility product.

In a solution containing two salts that yield a common ion, the ratio of solubilities of the two salts is the ratio of the solubility products.

Substance	Solubility product at temperature noted, °C	Substance	Solubility product at temperature noted, °C
Aluminum hydroxide	4×10^{-13} (15°)	Lead iodide	7.47×10^{-9} (15°)
Aluminum hydroxide	1.1×10^{-15} (18°)	Lead iodide	1.39×10^{-8} (25°)
Aluminum hydroxide	3.7×10^{-15} (25°)	Lead oxalate	2.74×10^{-11} (18°)
Barium carbonate	7×10^{-9} (16°)	Lead sulfate	1.06×10^{-8} (18°)
Barium carbonate	8.1×10^{-9} (25°)	Lead sulfide	3.4×10^{-28} (18°)
Barium chromate	1.6×10^{-10} (18°)	Lithium carbonate	1.7×10^{-3} (25°)
Barium chromate	2.4×10^{-10} (28°)	Magnesium ammonium phosphate	2.5×10^{-13} (25°)
Barium fluoride	1.6×10^{-6} (9.5°)	Magnesium carbonate	2.6×10^{-5} (12°)
Barium fluoride	1.7×10^{-6} (18°)	Magnesium fluoride	7.1×10^{-9} (18°)
Barium fluoride	1.73×10^{-6} (25.8°)	Magnesium fluoride	6.4×10^{-9} (27°)
Barium iodate, $Ba(IO_3) \cdot 2H_2O$	8.4×10^{-11} (10°)	Magnesium hydroxide	1.2×10^{-11} (18°)
Barium iodate, $Ba(IO_3) \cdot 2H_2O$	6.5×10^{-10} (25°)	Magnesium oxalate	8.57×10^{-5} (18°)
Barium oxalate, $BaC_2O_4 \cdot 3\frac{1}{2}H_2O$	1.62×10^{-7} (18°)	Manganese hydroxide	4×10^{-14} (18°)
Barium oxalate, $BaC_2O_4 \cdot 2H_2O$	1.2×10^{-7} (18°)	Manganese sulfide	1.4×10^{-15} (18°)
Barium oxalate, $BaC_2O_4 \cdot \frac{1}{2}H_2O$	2.18×10^{-7} (18°)	Mercuric sulfide	4×10^{-53} to 2×10^{-49} (18°)
Barium sulfate	0.87×10^{-10} (18°)		
Barium sulfate	1.08×10^{-10} (25°)	Mercurous bromide	1.3×10^{-21} (25°)
Barium sulfate	1.98×10^{-10} (50°)	Mercurous chloride	2×10^{-18} (25°)
Cadmium oxalate $CdC_2O_4 \cdot 3H_2O$	1.53×10^{-8} (18°)	Mercurous iodide	1.2×10^{-28} (25°)
Cadmium sulfide	3.6×10^{-29} (18°)	Nickel sulfide	1.4×10^{-24} (18°)
Calcium carbonate (calcite)	0.99×10^{-8} (15°)	Potassium acid tartrate $[K^+]$ $[HC_4H_4O_6{}^-]$	3.8×10^{-4} (18°)
Calcium carbonate (calcite)	0.87×10^{-8} (25°)		
Calcium fluoride	3.4×10^{-11} (18°)	Silver bromate	3.97×10^{-5} (20°)
Calcium fluoride	3.95×10^{-11} (26°)	Silver bromate	5.77×10^{-5} (25°)
Calcium iodate, $Ca(IO_3)_2 \cdot 6H_2O$	22.2×10^{-8} (10°)	Silver bromide	4.1×10^{-13} (18°)
		Silver bromide	7.7×10^{-13} (25°)
Calcium iodate, $Ca(IO_3)_2 \cdot 6H_2O$	64.4×10^{-8} (18°)	Silver carbonate	6.15×10^{-12} (25°)
Calcium oxalate, $CaC_2O_4 \cdot H_2O$	1.78×10^{-9} (18°)		
Calcium oxalate, $CaC_2O_4 \cdot H_2O$	2.57×10^{-9} (25°)	Silver chloride	0.21×10^{-10} (4.7°)
Calcium sulfate	1.95×10^{-4} (10°)	Silver chloride	0.37×10^{-10} (9.7°)
Calcium tartrate, $CaC_4H_4O_6 \cdot 2H_2O$	0.77×10^{-6} (18°)	Silver chloride	1.56×10^{-10} (25°)
		Silver chloride	13.2×10^{-10} (50°)
Cobalt sulfide	3×10^{-26} (18°)	Silver chloride	215×10^{-10} (100°)
Cupric iodate	1.4×10^{-7} (25°)		
Cupric oxalate	2.87×10^{-8} (25°)	Silver chromate	1.2×10^{-12} (14.8°)
Cupric sulfide	8.5×10^{-45} (18°)	Silver chromate	9×10^{-12} (25°)
Cuprous bromide	4.15×10^{-8} (18–20°)	Silver cyanide $[Ag^+][Ag(CN)_2{}^-]$	2.2×10^{-12} (20°)
Cuprous chloride	1.02×10^{-6} (18–20°)	Silver dichromate	2×10^{-7} (25°)
Cuprous iodide	5.06×10^{-12} (18–20°)	Silver hydroxide	1.52×10^{-8} (20°)
Cuprous sulfide	2×10^{-47} (16–18°)		
Cuprous thiocyanate	1.6×10^{-11} (18°)	Silver iodate	0.92×10^{-8} (9.4°)
Ferric hydroxide	1.1×10^{-36} (18°)	Silver iodide	0.32×10^{-16} (13°)
		Silver iodide	1.5×10^{-16} (25°)
Ferrous hydroxide	1.64×10^{-14} (18°)	Silver sulfide	1.6×10^{-49} (18°)
Ferrous oxalate	2.1×10^{-7} (25°)	Silver thiocyanate	0.49×10^{-12} (18°)
Ferrous sulfide	3.7×10^{-19} (18°)		
Lead carbonate	3.3×10^{-14} (18°)	Silver thiocyanate	1.16×10^{-12} (25°)
Lead chromate	1.77×10^{-14} (18°)	Strontium carbonate	1.6×10^{-9} (25°)
		Strontium fluoride	2.8×10^{-9} (18°)
Lead fluoride	2.7×10^{-8} (9°)	Strontium oxalate	5.61×10^{-8} (18°)
Lead fluoride	3.2×10^{-8} (18°)	Strontium sulfate	2.77×10^{-7} (2.9°)
Lead fluoride	3.7×10^{-8} (26.6°)		
Lead iodate	5.3×10^{-14} (9.2°)	Strontium sulfate	3.81×10^{-7} (17.4°)
Lead iodate	1.2×10^{-13} (18°)	Zinc hydroxide	1.8×10^{-14} (18–20°)
		Zinc oxalate, $ZnC_2O_4 \cdot 2H_2O$	1.35×10^{-9} (18°)
Lead iodate	2.6×10^{-13} (25.8°)	Zinc sulfide	1.2×10^{-23} (18°)

From Weast, R. C., Ed., *Handbook of Chemistry and Physics*, 55th ed., CRC Press, Cleveland, 1974, B-232.

Table 3–20
CONCENTRATION OF ACIDS AND BASES

Common Commercial Strengths

	Molecular weight	Moles per liter	Grams per liter	Percent by weight	Specific gravity [a]
ACIDS					
Acetic acid, glacial	60.05	17.4	1,045	99.5	1.05
Acetic acid	60.05	6.27	376	36	1.045
Butyric acid	88.1	10.3	912	95	0.96
Formic acid	46.02	23.4	1,080	90	1.20
	—	5.75	264	25	1.06
Hydriodic acid	127.9	7.57	969	57	1.70
	—	5.51	705	47	1.50
	—	0.86	110	10	1.1
Hydrobromic acid	80.92	8.89	720	48	1.50
	—	6.82	552	40	1.38
Hydrochloric acid	36.5	11.6	424	36	1.18
	—	2.9	105	10	1.05
Hydrocyanic acid	27.03	25	676	97	0.697
	—	0.74	19.9	2	0.996
Hydrofluoric acid	20.01	32.1	642	55	1.167
	—	28.8	578	50	1.155
Hydrofluosilicic acid	144.1	2.65	382	30	1.27
Hypophosphorous acid	66.0	9.47	625	50	1.25
	—	5.14	339	30	1.13
	—	1.57	104	10	1.04
Lactic acid	90.1	11.3	1,020	85	1.2
Nitric acid	63.02	15.99	1,008	71	1.42
	—	14.9	938	67	1.40
	—	13.3	837	61	1.37
Perchloric acid	100.5	11.65	1,172	70	1.67
	—	9.2	923	60	1.54
Phosphoric acid	98	14.7	1,445	85	1.70
Sulfuric acid	98.1	18.0	1,766	96	1.84
Sulfurous acid	82.1	0.74	61.2	6	1.02
BASES					
Ammonia water	17.0	14.8	252	28	0.898
Potassium hydroxide	56.1	13.5	757	50	1.52
	—	1.94	109	10	1.09
Sodium carbonate	106.0	1.04	110	10	1.10
Sodium hydroxide	40.0	19.1	763	50	1.53
	—	2.75	111	10	1.11

[a]For density in kg/m^3, multiply by 1000.

REFERENCE

For vapor-pressure curves for dilute solutions of hydrochloric, nitric, and sulfuric acids up to about 1000 psia, consult Staples, B. G., Procopio, J. M., Jr., and Su, G. J., Vapor-pressure data for common acids at high temperatures, *Chem. Eng.,* 77(25), 113, 1970. The acid concentrations represented by these curves are for hydrochloric acid, 10–35% by weight; for nitric acid, 30–70% by weight; and for sulfuric acid, 0–70% by weight.

Credit *The Merck Index,* copyright 1968, Merck & Co., Inc., Rahway, N. J., U.S.A. With permission.

Table 3–21
PROPERTIES OF COMMON LIQUIDS[a]

At 1.0 atm Pressure, 77°F (25°C), Except as Noted

For viscosity in N·s/m² (= kg/m·s), multiply values in centipoises by 0.001. For surface tension in N/m, multiply values in dyn/cm by 0.001.

Common name	Chemical formula	Molecular weight	Density, $\frac{lb}{ft^3}$	Specific gravity	Viscosity $lb_m/ft\,sec \times 10^4$	Viscosity cp	Sound velocity, $\frac{meters}{sec}$	Surface tension, $\frac{dynes}{cm}$	Dielectric constant	Refractive index
Acetic acid	$C_2H_4O_2$	60.0537	65.493	1.049	7.76	1.155	1584[50]	27.3	6.15	1.37
Acetone	C_3H_6O	58.081	48.98	.787	2.12	0.316	1174	23.1	20.7	1.36
Alcohol, ethyl	C_2H_5OH	46.070	49.01	.787	7.36	1.095	1144	22.33	24.3	1.36
Alcohol, methyl	CH_3OH	32.043	49.10	.789	3.76	0.56	1103	22.2	32.6	1.33
Alcohol, propyl	C_3H_8O	60.098	49.94	.802	12.9	1.92	1205	23.5	20.1	1.38
Ammonia (aqua)	—	17.698	51.411	.826	—	—	—	—	16.9	—
Benzene	C_6H_6	78.117	54.55	.876	4.04	0.601	1298	28.18	2.2	1.50
Bromine	Br_2	159.818	—	—	6.38	0.95	—	41.5	3.20	—
Carbon disulfide	CS_2	76.140	78.72	1.265	2.42	0.36	1149	32.33	2.64	1.63
Carbon tetrachloride	CCl_4	153.824	98.91	1.59	6.11	0.91	924	26.3	2.23	1.46
Castor oil	—	—	59.69	0.960	—	650	1474	—	4.7	—
Chloroform	$CHCl_3$	119.378	91.44	1.47	3.56	0.53	995	27.14	4.8	1.44
Decane	$C_{10}H_{22}$	142.290	45.34	.728	5.77	0.859	—	23.43	2.0	1.41
Dodecane	$C_{12}H_{26}$	170.345	47.11	—	9.23	1.374	—	—	—	1.41
Ether	$C_4H_{10}O$	74.125	44.54	0.715	1.50	0.223	985	16.42	4.3	1.35
Ethylene glycol	$C_2H_6O_2$	62.070	68.47	1.100	109	16.2	1644	48.2	37.7	1.43
Fluorine refrigerant R–11	CCl_3F	137.369	92.14	1.480	2.82	0.42	—	18.3	2.0	1.37
Fluorine refrigerant R–12	CCl_2F_2	120.914	81.84	1.315	—	—	—	8.87	2.0	1.29
Fluorine refrigerant R–22	CHF_2Cl	86.469	74.53	1.197	—	—	—	8.35	2.0	1.26
Glycerine	$C_3H_8O_3$	92.096	78.62	1.263	6380	950	1909	63.0	40	1.47
Heptane	C_7H_{16}	100.208	42.42	.681	2.53	0.376	1138	19.9	1.92	1.38
Hexane	C_6H_4	86.181	40.88	.657	2.00	0.297	1203	18.0	—	1.37
Iodine	I_2	253.809	—	—	—	—	—	—	11	—
Kerosene	—	—	51.2	0.823	11.0	1.64	1320	—	—	—
Linseed oil	—	—	58.0	0.93	222	33.1	—	—	3.3	—
Mercury	Hg	200.59	—	13.633	10.3	1.53	1450	484	—	—
Octane	C_8H_{18}	114.235	43.61	.701	3.43	0.51	1171	21.14	—	1.40
Phenol	C_6H_6O	94.116	66.94	1.071	54	8.0	1274[100]	40.4	9.8	—
Propane	C_3H_8	44.098	30.81	.495	0.74	0.11	—	6.6	1.27	1.34
Propylene	C_3H_6	42.082	32.11	.516	0.60	0.09	—	7.0	—	1.36
Propylene glycol	$C_3H_8O_2$	76.097	60.26	.968	—	42	—	36.3	—	1.43
Sea water	—	—	18.52	64.0	1.03	—	—	1535	—	—
Toluene	C_7H_8	92.144	53.83	0.865	3.70	0.550	1275[30]	27.3	2.4	1.49

Table 3–21 (continued)
PROPERTIES OF COMMON LIQUIDS

At 1.0 atm Pressure, 77°F (25°C), Except as Noted

Common name	Chemical formula	Molecular weight	Density, $\frac{lb}{ft^3}$	Specific gravity	Viscosity $lb_m/ft\ sec \times 10^4$	cp	Sound velocity, $\frac{meters}{sec}$	Surface tension, $\frac{dynes}{cm}$	Dielectric constant	Refractive index
Turpentine	$C_{10}H_{16}$	136.242	54.2	0.87	9.24	1.375	1240	—	—	1.47
Water	H_2O	18.0153	62.247	1.00	6.0	0.89	1498	71.97	78.54[a]	1.33

[a]For thermal properties see Table 3–22. For properties of liquids in SI units see Table 3–23.
[b]The dielectric constant of water near the freezing point is 87.8; it decreases with increase in temperature to about 55.6 near the boiling point.

From Bolz, R. E. and Tuve, G. L., Eds., *Handbook of Tables for Applied Engineering Science*, 2nd ed., CRC Press, Cleveland, 1973, 90.

Table 3–22
THERMAL PROPERTIES OF COMMON LIQUIDS[a]

At 1.0 atm Pressure, 77°F (25°C), Except as Noted

Common name	Specific heat, Btu/lbm °F	Thermal conductivity, Btu/ft hr °F	Freezing point, °F	Latent heat of fusion, Btu/lb	Boiling point, °F	Latent heat of evaporation, Btu/lb	Coefficient of cubical expansion per °F
Acetic acid	0.522	0.099	62	77.7	245	173	0.0006
Acetone	0.514	0.093	− 137.4	42.3	133	223	0.00082
Alcohol, ethyl	0.584	0.099	− 174.2	46.4	172.96	364	0.0006
Alcohol, methyl	0.606	0.117	− 143.7	42.5	148.4	474	0.00075
Alcohol, propyl	0.567	0.093	− 197	37.2	208	335	—
Ammonia (aqua)	1.047	0.204	—	—	—	—	—
Benzene	0.414	0.083	41.96	54.4	176.2	168	0.0007
Bromine	0.113	—	− 17.15	28.7	137.3	83	0.00065
Carbon disulfide	0.237	0.093	− 169.5	24.80	115.26	151.2	0.0007
Carbon tetrachloride	0.207	0.060	− 9.04	74.8	169.7	83.5	0.0007
Castor oil	0.47	0.104	14.1	—	—	—	—
Chloroform	0.25	0.068	− 82.3	33.14	142.2	106.4	0.00073
Decane	0.528	0.085	− 21.4	86.6	345.3	113	—
Dodecane	0.528	0.081	− 14.74	93.0	421.3	110	—
Ether	0.529	0.075	− 177	41.4	94.2	160	0.0009
Ethylene glycol	0.565	0.149	8.6	77.9	387	344	—
Fluorine refrigerant R–11	0.208[b]	0.054[b]	− 168	—	74.9	77.58[c]	—
Fluorine refrigerant R–12	0.232[b]	0.041[b]	− 252	14.8	− 21.6	71.04[c]	—
Fluorine refrigerant R–22	0.300[b]	0.050[b]	− 256	78.7	− 41.4	100.05[c]	—
Glycerine	0.627	0.166	17.0	86	554.5	419	0.0003

Table 3-22 (continued)
THERMAL PROPERTIES OF COMMON LIQUIDS[a]

At 1.0 atm Pressure, 77°F (25°C), Except as Noted

Common name	Specific heat, Btu/lbm °F	Thermal conductivity, Btu/ft hr °F	Freezing point, °F	Latent heat of fusion, Btu/lb	Boiling point, °F	Latent heat of evaporation, Btu/lb	Coefficient of cubical expansion per °F
Heptane	0.536	0.074	−131.1	60.2	209.1	137	—
Hexane	0.541	0.072	−139.3	65.3	155.65	157	—
Iodine	0.513	—	236.3	26.74	363.8	70.71	—
Kerosene	0.5	0.084	—	—	—	108	—
Linseed oil	0.44	—	−4	—	549	—	—
Mercury	0.0333	—	−38.0	5.0	674	126.9	0.0001
Octane	0.514	0.076	−70.2	78.0	257	128	0.0004
Phenol	0.342	0.11	109.4	52.2	360	—	0.0005
Propane, R–290	0.576†	—	−305.8	34.38	−43.73	184[c]	—
Propylene	0.682	—	−301.5	30.70	−53.86	147	—
Propylene glycol	0.598	—	−76	—	369	393	—
Sea water	0.90–.98	—	27.5	—	—	—	—
Toluene	0.4†	0.077	−139	30.90	230.8	156	—
Turpentine	0.425	0.070	−75	—	320	126	0.00055
Water	0.998	0.352	32	143.3	212	970.3	0.00011

[a]For other properties of liquids see Table 3−21. For properties of liquids in SI units see Table 3−23.
[b]At 75°F, liquid.
[c]At 14.7 psia, saturation temperature.

From Bolz, R. E. and Tuve, G. L., Eds., *Handbook of Tables for Applied Engineering Science,* 2nd ed., CRC Press, Cleveland, 1973, 91.

Table 3–23
PROPERTIES OF COMMON LIQUIDS – SI UNITS

At 1.0 atm Pressure (0.101 325 MN/m^2), 300 K, Except as Noted

Common name	Density, kg/m^3	Specific heat, kJ/kg·K	Viscosity, N·s/m^2	Thermal conductivity, W/m·K	Freezing point, K	Latent heat of fusion, kJ/kg	Boiling point, K	Latent heat of evaporation, kJ/kg	Coefficient of cubical expansion per K
Acetic acid	1 049	2.18	.001 155	0.171	290	181	391	402	0.001 1
Acetone	784.6	2.15	.000 316	0.161	179.0	98.3	329	518	0.001 5
Alcohol, ethyl	785.1	2.44	.001 095	0.171	158.6	108	351.46	846	0.001 1
Alcohol, methyl	786.5	2.54	.000 56	0.202	175.5	98.8	337.8	1 100	0.001 4
Alcohol, propyl	800.0	2.37	.001 92	0.161	146	86.5	371	779	
Ammonia (aqua)	823.5	4.38		0.353					
Benzene	873.8	1.73	.000 601	0.144	278.68	126	353.3	390	0.001 3
Bromine		.473	.000 95		245.84	66.7	331.6	193	0.001 2
Carbon disulfide	1 261	.992	.000 36	0.161	161.2	57.6	319.40	351	0.001 3
Carbon tetrachloride	1 584	.866	.000 91	0.104	250.35	174	349.6	194	0.001 3
Castor oil	956.1	1.97	.650	0.180	263.2				
Chloroform	1 465	1.05	.000 53	0.118	209.6	77.0	334.4	247	0.001 3
Decane	726.3	2.21	.000 859	0.147	243.5	201	447.2	263	
Dodecane	754.6	2.21	.001 374	0.140	247.18	216	489.4	256	
Ether	713.5	2.21	.000 223	0.130	157	96.2	307.7	372	0.001 6
Ethylene glycol	1 097	2.36	.016 2	0.258	260.2	181	470	800	
Fluorine refrigerant R-11	1 476	.870[a]	.000 42	0.093[a]	162		297.0	180[b]	
Fluorine refrigerant R-12	1 311	.971[a]		0.071[a]	115	34.4	243.4	165[b]	
Fluorine refrigerant R-22	1 194	1.26[a]		0.086[a]	113	183	232.4	232[b]	
Glycerine	1 259	2.62	.950	0.287	264.8	200	563.4	974	0.000 54
Heptane	679.5	2.24	.000 376	0.128	182.54	140	371.5	318	
Hexane	654.8	2.26	.000 297	0.124	178.0	152	341.84	365	
Iodine		2.15			386.6	62.2	457.5	164	
Kerosene	820.1	2.09	.001 64	0.145				251	
Linseed oil	929.1	1.84	.033 1		253		560		
Mercury		.139	.001 53		234.3	11.6	630	295	0.000 18
Octane	698.6	2.15	.000 51	0.131	216.4	181	398	298	0.000 72
Phenol	1 072	1.43	.008 0	0.190	316.2	121	455		0.000 90
Propane	493.5	2.41[a]	.000 11		85.5	79.9	231.08	428[b]	
Propylene	514.4	2.85	.000 09		87.9	71.4	225.45	342	
Propylene glycol	965.3	2.50	.042		213		460	914	
Sea water	1 025	3.76–4.10			270.6				
Toluene	862.3	1.72	.000 550	0.133	178	71.8	383.6	363	
Turpentine	868.2	1.78	.001 375	0.121	214		433	293	0.000 99
Water	997.1	4.18	.000 89	0.609	273	333	373	2 260	0.000 20

[a]For other data on these liquids see Tables 3–21 and 3–22.

Table 3–23 (continued)
PROPERTIES OF COMMON LIQUIDS – SI UNITS

At 1.0 atm Pressure ($0.101\ 325\ MN/m^2$), 300 K, Except as Noted

[b]At 297 K, liquid.
[c]At .101 325 MN, saturation temperature.

From Bolz, R. E. and Tuve, G. L., Eds., *Handbook of Tables for Applied Engineering Science,* 2nd ed., CRC Press, Cleveland, 1973, 92.

Table 3–24
ANTIFREEZE SOLUTIONS

Specific gravity is given at 60°F. Specific heat is in Btu/lbm ·°R = cal/g·K; for kJ/kg·K multiply by 4.184. Viscosity is in centipoises; for N·s/m² (=kg/m·s) divides by 1000; for lbm/s·ft multiply by 0.000 672. Heat conductivity is in Btu/hr·ft· °F; for W/m·K multiply by 1.729 6.

Solution and property	Percentage by weight—pure antifreeze agent†								
	5	10	15	20	25	30	35	40	50
ETHANOL									
Specific gravity	.991	.984	.977	.970	.963	.956	.947	.937	.915
Freezing point, °F	28.2	23.9	18.8	12.3	4.2	−4.9			
Specific heat, 68°F	1.00	1.00	1.00	0.99	0.98	0.96	0.92	0.90	
Specific heat, 32°F									
Viscosity, cp, 68°F									
Viscosity, cp, 32°F									
Heat conductivity, 68°F									
Heat conductivity, 32°F									
METHANOL									
Specific gravity	.991	.983	.976	.968	.960	.953	.945	.937	.916
Freezing point, °F	26.6	20.2	13.0	4.9	−4.2	−13.5	−25.3		
Specific heat, 68°F	1.00	1.00	0.99	0.98	0.97	0.95	0.93	0.91	
Specific heat, 32°F									
Viscosity, cp, 68°F									
Viscosity, cp, 32°F									
Heat conductivity, 68°F									
Heat conductivity, 32°F									
ETHYLENE GLYCOL									
Specific gravity		1.012		1.025		1.04		1.055	1.065
Freezing point, °F		24		15		4		−12	−32
Specific heat, 68°F		.97		.94		.89		.84	.79
Specific heat, 32°F		.96		.93		.87		.81	.76
Viscosity, cp, 68°F		1.4		1.9		2.4		3.1	4.1
Viscosity, cp, 32°F		2.5		3.0		4.0		5.3	8.
Heat conductivity, 68°F		.33		.31		.28		.26	.24
Heat conductivity, 32°F		.32		.30		.28		.26	.24
PROPYLENE GLYCOL									
Specific gravity		1.006		1.016		1.026		1.035	1.042
Freezing point, °F		27		18		8		−7	−25
Specific heat, 68°F		.99		.96		.94		.89	.86
Specific heat, 32°F		.99		.96		.94		.89	.85
Viscosity, cp, 68°F		1.5		2.0		3.0		6.1	9.
Viscosity, cp, 32°F		2.8		4.2		6.5		11.	30.
Heat conductivity, 68°F		.32		.29		.27		.24	.23
Heat conductivity, 32°F		.31		.28		.26		.24	.23
GLYCEROL									
Specific gravity	1.012	1.023	1.036	1.048	1.060	1.074	1.87	1.10	1.13
Freezing point, °F	31.	29.5	26.5	23.5	19.5	15.	10.	4.5	−9.5
Specific heat, 68°F	0.97	0.96	0.94	0.93	0.91	0.90	0.88	0.86	0.83
Specific heat, 32°F	.96	0.94	0.92	0.90	0.88	0.87	0.86	0.84	0.80
Viscosity, cp, 68°F		1.31		1.76		2.50		3.72	6.0
Viscosity, cp, 32°F		2.44		3.44		5.14		8.25	14.6
Heat conductivity, 68°F									
Heat conductivity, 32°F									

Table 3–24 (continued)
ANTIFREEZE SOLUTIONS

Solution and property	Percentage by weight—pure antifreeze agent†								
	5	10	15	20	25	30	35	40	50
SODIUM CHLORIDE									
Specific gravity	1.035	1.073	1.110	1.151	1.190				
Freezing point, °F	26.7	20.2	12.3	2.4	+16				
Specific heat, 68°F	.94	.89	.85	.81	.79				
Specific heat, 32°F	.93	.88	.84	.80	.78				
Viscosity, cp, 68°F		1.4	1.6	1.9	2.3				
Viscosity, cp, 32°F		2.0	2.3	2.7	3.4				
Heat conductivity, 68°F	.32	.31	.29	.28	.27				
Heat conductivity, 32°F	.31	.29	.28	.26	.25				
CALCIUM CHLORIDE									
Specific gravity	1.043	1.088	1.138	1.183	1.230	1.287			
Freezing point, °F	27.7	22.	13.5	−1.0	−21.	−48.			
Specific heat, 68°F	.91	.86	.80	.74	.69	.66			
Specific heat, 32°F	.90	.85	.79	.73	.68	.65			
Viscosity, cp, 68°F		1.3	1.6	2.0	3.0	3.8			
Viscosity, cp, 32°F		2.4	2.7	3.1	4.1	5.9			
Heat conductivity, 68°F	.34	.34	.33	.33	.33	.32			
Heat conductivity, 32°F	.32	.32	.31	.31	.31	.30			

[a]For commercial grades divide by decimal purity.

From Bolz, R. E. and Tuve, G. L., Eds., *Handbook of Tables for Applied Engineering Science*, 2nd ed., CRC Press, Cleveland, 1973, 96.

Table 3–25
CRYOGENIC AND REFRIGERATING LIQUIDS

Fluid	Boiling point				Density (lb/ft^3)	Volume ratio (to room temp.), gas/liq.	Latent heat of vaporization (Btu/lb_m)
	K	°R	°C	°F			
Air	79	142	−194	−318	54.6	740:1	88.2
Argon	87	157	−186	−303	87.6	840:1	69.5
Carbon dioxide[a]	195	350	−79	−110	97.6	730:1	246
Ethane	185	334	−88	−126	34.2	420:1	210
Fluorine	85	154	−188	−306	94.0	880:1	71.6
Helium 3	3.20	5.76	−270	−454	3.68	600:1	3.65
Helium 4	4.215	7.59	−269	−452	7.80	600:1	8.92
Hydrogen	20	36.7	−253	−423	4.43	800:1	192
Methane	111	201	−162	−259	26.5	550:1	219
Neon	27	48.8	−246	−411	75.2	1400:1	37.3
Nitrogen	77.4	139.2	−196	−320	50.6	700:1	85.3
Propane	231	416	−42	−44	36.3	310:1	183
Propylene	225	406	−48.	−54	38.3	330:1	188
Refrigerant 12	243	438	−30	−22	92.9	280:1	71.1
Refrigerant 13	192	346	−81	−114	95.0	350:1	64
Refrigerant 13B1	215	388	−58	−72	124	310:1	51.1
Refrigerant 22	232	419	−41	−41	88.2	380:1	100.5

Fluid	Specific heat, c_p		Viscosity		Thermal conductivity		Dielectric constant
	liquid $(Btu/lb_m\,°R)$	gas $(Btu/lb_m\,°R)$	liquid $(lb_m/ft\,hr)$	gas $(lb_m/ft\,hr)$	liquid $\dfrac{Btu}{hr\,ft\,°F}$	gas $\dfrac{Btu}{hr\,ft\,°F}$	
Air	.470	.245	.195	.016		.0043	1.52
Argon	.272	.127	.610	.0200	.0712	.00350	1.59[c]
Carbon Dioxide[a]	.318	.190				.0085[c]	
Ethane	.600	.250	.592		.078	.0145[d]	1.43
Fluorine	.37	.194		.0178		.00416	
Helium 3	1.10		.00392		.0099		
Helium 4	1.09	1.63	.00864	.00309	.0156	.00560	1.0492
Hydrogen	2.34	2.85	.0316	.00254	.0683	.0090	1.226
Methane	.825	.398	.287		.0642		1.68
Neon	0.44	.280	.30	.0110	.075	.0057	
Nitrogen	.487	.259	.382	.0134	.0804	.00415	1.434

Table 3–25 (continued)
CRYOGENIC AND REFRIGERATING LIQUIDS

Fluid	Specific heat, c_p		Viscosity		Thermal conductivity		Dielectric constant
	liquid (Btu/lb$_m$ °R)	gas (Btu/lb$_m$ °R)	liquid (lb$_m$/ft hr)	gas (lb$_m$/ft hr)	liquid $\frac{Btu}{hr\,ft\,°F}$	gas $\frac{Btu}{hr\,ft\,°F}$	
Propane	.526	.290					1.27
Propylene	.619	.363					
Refrigerant 12	.213	.109	.897	.0264			2.13
Refrigerant 13	.214	.0987			.035		
Refrigerant 13B1	.405	.203					
Refrigerant 22	.252	.111	.849	.0254	.0635[b]	.00450	2.44

[a]Sublimes.
[b]At −4°F.
[c]At 32° F.
[d]At 125°F

From Bolz, R. E. and Tuve, G. L., Eds., *Handbook of Tables for Applied Engineering Science*, 2nd ed., CRC Press, Cleveland, 1973, 97.

Table 3-26
BINARY AZEOTROPES

The following compounds form a binary azeotrope, a liquid of constant boiling point, when mixed in the given proportions.

No.	Component A	Per-cent of A	Component B	Per-cent of B	Boiling points, °C		
					A	B	Azeotrope
1	Acetic acid	3.0	Water	97.0	118.1	100.0	76.6
2	Acetic acid	2.0	Benzene	98.0	118.1	80.1	80.1
3	Acetic acid	58.5	Chlorobenzene	41.5	118.1	132.0	114.7
4	Acetone	88.5	Water	11.5	56.2	100.0	56.1
5	Acetone	33.0	Carbon disulfide	67.0	56.2	46.3	39.3
6	Acetone	59.0	Hexane	41.0	56.2	69.0	49.8
7	Acetone	88.0	Methanol	12.0	56.2	64.7	55.7
8	Allyl alcohol	72.9	Water	27.1	97.1	100.0	88.2
9	Allyl alcohol	17.4	Benzene	82.6	97.1	80.1	76.8
10	Allyl alcohol	4.5	Hexane	95.5	97.1	69.0	65.5
11	Benzene	91.1	Water	8.9	80.1	100.0	69.4
12	Benzene	67.6	Ethanol	32.4	80.1	78.5	67.8
13	Benzene	60.5	Methanol	39.5	80.1	64.7	58.3
14	l-Butanol	55.5	Water	44.5	117.7	100.0	93.0
15	Butyl acetate	72.9	Water	27.1	126.5	100.0	90.7
16	Butyl ether	66.6	Water	33.4	142.0	100.0	94.1
17	Butyl ether	93.6	Ethylene glycol	6.4	142.0	197.2	139.5
18	Carbon disulfide	98.0	Water	2.0	46.3	100.0	43.6
19	Carbon disulfide	1.0	Ethyl ether	99.0	46.3	34.6	34.4
20	Carbon tetra-chloride	95.9	Water	4.1	76.8	100.0	66.8
21	Carbon tetra-chloride	57.0	Ethyl acetate	43.0	76.8	77.1	74.8
22	Carbon tetra-chloride	79.4	Methanol	20.6	76.8	64.7	55.7
23	Chloroform	97.0	Water	3.0	61.2	100.0	56.3
24	Chloroform	93.0	Ethanol	7.0	61.2	78.5	59.4
25	Chloroform	87.0	Methanol	13.0	61.2	64.7	53.5
26	Ethanol	95.6	Water	4.4	78.5	100.0	78.2
27	Ethanol	31.0	Ethyl acetate	69.0	78.5	77.1	71.8
28	Ethanol	48.0	Heptane	52.0	78.5	98.4	72.0
29	Ethanol	21.0	Hexane	79.0	78.5	69.0	58.7
30	Ethanol	68.0	Toluene	32.0	78.5	110.6	76.7
31	Ethyl acetate	91.9	Water	8.1	77.1	100.0	70.4
32	Ethyl ether	98.8	Water	1.2	34.6	100.0	34.2
33	Formic acid	77.5	Water	22.5	100.7	100.0	107.1
34	Heptane	87.1	Water	12.9	98.4	100.0	79.2
35	Heptane	48.5	Methanol	51.5	98.4	64.7	59.1
36	Hexane	94.4	Water	5.6	69.0	100.0	61.6
37	Hexane	73.1	Methanol	26.9	69.0	64.7	50.0
38	Hexane	96.0	Propanol	4.0	69.0	97.2	65.7
39	Hydrogen chloride	20.2	Water	79.8	−83.7	100.0	108.6
40	Isobutyl alcohol	70.0	Water	30.0	108.4	100.0	89.7
41	Isopropyl ether	95.4	Water	4.6	67.5	100.0	62.2
42	Methanol	72.0	Octane	28.0	64.7	125.8	63.0
43	Methanol	72.4	Toluene	27.6	64.7	110.6	63.7
44	Nonane	60.2	Water	39.8	150.8	100.0	95.0
45	Pentane	98.6	Water	1.4	36.1	100.0	34.6
46	Phenol	9.2	Water	90.8	182.0	100.0	99.5
47	Propanol	71.8	Water	28.2	97.2	100.0	88.1
48	Propanol	49.0	Toluene	51.0	97.2	110.6	92.6
49	Styrene	59.1	Water	40.9	145.2	100.0	93.9
50	Toluene	79.8	Water	20.2	110.6	100.0	85.0
51	m-Xylene	60.0	Water	40.0	139.1	100.0	94.5

From Bolz, R. E. and Tuve, G. L., Eds., *Handbook of Tables for Applied Engineering Science*, 2nd ed., CRC Press, Cleveland, 1973, 99.

Table 3—27
DIELECTRIC CONSTANTS OF WATER

°C	ϵ^1	ϵ^2
0	87.74	87.90
5	85.76	85.90
10	83.83	83.95
15	81.95	82.04
18	80.84	80.93
20	80.10	80.18
25	78.30	78.36
30	76.55	76.58
35	74.83	74.85
38	73.82	73.83
40	73.15	73.15
45	71.51	71.50
50	69.91	69.88
55	68.34	68.30
60	66.81	66.76
65	65.32	65.25
70	63.86	63.78
75	62.43	62.34
80	61.03	60.93
85	59.66	59.55
90	58.32	58.20
95	57.01	56.88
100	55.72	55.58

REFERENCES

1. Malmberg, C. G. and Maryott, A. A., *J. Res. Natl. Bur. Stand.*, 56, 1, 1956.
2. Owen, B. B., Miller, R. C., Milner, C. E., and Cogan, H. L., *J. Phys. Chem.*, 65, 2065, 1961.

From Hamer, W. J. and DeWane, H. J., *NSRDS-NBS*, 33, 20, 1970.

Table 3–28
DIPOLE MOMENTS

A. Inorganic Compounds

Compound	Dipole moment $\times 10^{-18}$ esu	Method[a]
Aluminum		
bromide	5.14	B
iodide	2.48	B
Ammonia	1.3	
Carbon		
oxide, mono-	0.10	
oxide, di-	0.0	
Cesium		
chloride	10.42	St
fluoride	7.875	St
Hydrazine	1.84	
Hydrogen		
bromide	0.78	
chloride	1.084 ± 0.003–0.007	
chloride, deuterium	1.084 ± 0.003–0.007	
fluoride	1.92 ± 0.02	
iodide	0.38	
oxide (water)	1.87	
peroxide	2.13 ± 0.05	
selenide, deuterium	0.62	St
sulfide	1.10	
Mercury		
bromide (mercuric)	0.95	B
chloride (mercuric)	1.23	B
Nitric acid	2116	St
Nitrogen		
oxide	0.16	
oxide, di-	0.29	
oxide, tetra-	0.37	
oxybromide	1.87	
oxychloride	1.83	
Phosphorus		
chloride, tri-	0.90–1.16	
chloride, penta-	0.0	
Silicon		
fluoride, deuterium	1.53	St
fluoride, hydrogen	1.54	St
Sulfur		
oxide, di-	1.60	
oxide, tri-	0.00	
Sulfuryl		
fluoride	1.110	St
Tin		
chloride (stannic)	0.95	B
iodide (stannic)	0	B
Titanium		
chloride, tetra-	0	C

[a]Methods of measurement: B = benzene solution; C = carbon tetrachloride solution; St = measurement of Stark effect in microwave spectrum of gas.

Table 3–28 (continued)
DIPOLE MOMENTS

B. Organic Compounds

Compound	Dipole moments $\times 10^{-18}$ esu	Compound	Dipole moments $\times 10^{-18}$ esu
		Amino Acid Esters[a]	
α-Alanine ethyl ester	2.11	α-Aminocaproic acid ethyl ester	2.13
β-Alanine ethyl ester	2.14	α-Aminovaleric acid ethyl ester	2.13
α-Aminobutyric acid ethyl ester	2.13	Glycine ethyl ester	2.11
β-Aminobutyric acid ethyl ester	2.11	Valine ethyl ester	2.11
		Amides	
Acetamide	3.6	Sulfamide	3.9
Benzamide	3.6	Tetraethylurea	3.3
Caproamide	3.9	Thiourea	4.89
sym-Dimethylurea	4.8	Urea	4.56
Propylurea	4.1	Valeramide	3.7
		Hormones and Related Compounds in Doxane[b]	
Δ^4-Androstene-3:17-dione	3.32	Cholestane-3(β):7(α)-diol	2.31
Δ^5-Androstene-3(β):17(α)-diol	2.89	Cholestane-3(β):7(β)-diol	2.55
Δ^5-Androstene-3(β):17(β)-diol	2.69	Δ^5-Cholestane-3(β)-ol-7-one	3.79
Δ^5-Androstene-3(β)-ol-17-one	2.46	Isophorone	3.96
Androsterone	3.70	Testosterone	4.32
β-Androsterone	2.95	cis-Testosterone	5.17
Cholestane	2.98		

[a]Accurate to ±0.01 $\times 10^{-18}$ esu. (Wyman, J., *Chem. Rev.*, 19, 213, 1936.)

[b]Ethylenic >C=C< in a six-membered ring and conjugated with >C=O increases the dipole moment by approximately 1 Debye. Nonconjugated >C=C< in sterols decreases the dipole moments by approximately 0.49. Biologic activity is not correlated with dipole moment. (Kumler, W. D. and Fohlen, G. M., *J. Am. Chem. Soc.*, 67, 437, 1945.)

From Weast, R. C., Ed., *Handbook of Chemistry and Physics*, 55th ed., CRC Press, Cleveland, 1974, E-66.

Table 3–29
BUFFER SOLUTIONS

A. Operational Definition of pH

The operational definition of pH is

$$pH = pH(s) + E/k,$$

where E is the e.m.f. of the cell

$$H_2 \mid Solution, pH \mid Saturated\ KCl \mid Solution, pH(s) \mid H_2,$$

the half-cell on the left containing the solution whose pH is being measured and that on the right a standard buffer mixture of known pH; $k = 2.303\ RT/F$, where R is the gas constant, T the temperature (K), and F the value of the faraday.

Alternatively, the cell

$$Glass\ electrode \mid Solution, pH \mid Saturated\ calomel\ electrode$$

can be used, the glass electrode being calibrated by a standard buffer mixture or, if possible, two standard buffer mixtures whose pH values lie on either side of that of the solution being measured. Suitable standard buffer mixtures are

$0.05M$ potassium hydrogen phthalate (pH = 4.008 at 25°C)
$0.025M$ potassium dihydrogen phosphate
$0.025M$ disodium hydrogen phosphate (pH = 6.865 at 25°C)
$0.01M$ borax (pH = 9.180 at 25°C)

For most purposes pH can be equated to $-\log_{10}\gamma_H^+ m_H^+$, i.e., to the negative logarithm of the hydrogen ion activity. There is a small difference between those two quantities if pH > 9.2 or pH < 4.0, given by

$$-\log\gamma_H^+ m_H^+ = pH + 0.014(pH - 9.2)\ for\ pH > 9.2$$

$$-\log\gamma_H^+ m_H^+ = pH + 0.009(4.0 - pH)\ for\ pH < 4.0$$

It should be noted that in the table listing the round values of pH at 25°C the value for $-\log\gamma_H^+ m_H^+$ is quoted rather than pH, when there is a difference between these two values.

REFERENCES

Bates, R. G., *Electrometric pH Determinations: Theory and Practice*, Wiley & Sons, New York, 1954.
Robinson, R. A. and Stokes, R. H., *Electrolyte Solutions*, 2nd ed. Butterworths, London, 1959.
Bates, R. G., *J. Res. Natl. Bur. Stand.*, 66A, 179, 1962.

National Bureau of Standards
Bates, R. G. and Acree, S. F., *Research*, 34, 373, 1945.
Hamer, W. J., Pinching, C. D., and Acree, S. F., *Research*, 36, 47, 1946.
Manor, G. G., DeLollis, N. J., Lindwall, P. W., and Acree, S. F., *Research*, 36, 543, 1946.
Bates, R. G., *Research*, 39, 411, 1947.
Bates, R. G., Bower, V. E., Miller, R. G., and Smith, E. R., *Research*, 47, 433, 1951.
Bower, V. E., Bates, R. G., and Smith, E. R., *Research*, 51, 189, 1953.
Bower, V. E. and Bates, R. G., *Research*, 55, 197, 1955.
Bower, V. E. and Smith, R. E., *Research*, 56, 305, 1956.
Bower, V. E. and Bates, R. G., *Research*, 59, 261, 1956.
Bates, R. G. and Bower, V. E., *Anal. Chem.*, 28, 1322, 1956.

Table 3–29 (continued)
BUFFER SOLUTIONS

B. Properties of Standard Aqueous Buffer Solutions at 25°C

Buffer substance	Molality (m)	Weight of salt in air per liter solution	Density (g/ml)	Molarity (M)	Dilution value ($\triangle pH_{1/2}$)	$\triangle pH_s{}^a$	Buffer value, equiv. per pH	Temperature coefficient, dpH_s/dt (units per °C)
			Tetroxalate Solution					
$KH_3(C_2O_4)_2 \cdot 2H_2O$	0.05	12.61	1.0032	0.04962	+0.186	−0.0028	0.070	+0.001
			Tartrate Solution					
$KHC_4H_4O_6$, sat.	0.0341	–	1.0036	0.034	+0.049	−0.003	0.027	−0.0014
			Phthalate Solution					
$KHC_8O_4H_4$	0.05	10.12	1.0017	0.04958	+0.052	−0.0009	0.016	+0.0012
			Phosphate Solution					
KH_2PO_4 +	0.025	3.39	1.0028^b	0.0249	$+0.080^b$	$−0.0006^b$	0.029^b	$−0.0028^b$
Na_2HPO_4	0.025	3.53		0.0249				
KH_2PO_4 +	0.008695	1.179	1.0020^b	0.008665	$+0.07^c$	$−0.0005^b$	0.016^b	$−0.0028^b$
Na_2HPO_4	0.03043	4.30		0.03032				
			Borax Solution					
$Na_2B_4O_7 \cdot 10H_2O$	0.01	3.80	0.9996	0.009971	+0.01	−0.0001	0.020	−0.0082
			Calcium Hydroxide Solution					
$Ca(OH)_2$, saturated	0.0203	–	0.9991	0.02025	−0.28	+0.0014	0.09	−0.033

$^a \triangle pH_s = pH_s (M$ molar solution$) - pH_s (m$ molal solution$)$.
^{b}For mixture at given concentration.
^{c}Calculated value for mixture at given concentration.

C. Solutions Giving Round Values of pH at 25°C

Code for Solutions

A: 25 ml of 0.2M potassium chloride + x ml of 0.2M hydrochloric acid
B: 50 ml of 0.1M potassium hydrogen phthalate + x ml of 0.1M hydrochloric acid
C: 50 ml of 0.1M potassium hydrogen phthalate + x ml of 0.1M sodium hydroxide
D: 50 ml of 0.1M potassium dihydrogen phosphate + x ml of 0.1M sodium hydroxide
E: 50 ml of 0.1M tris(hydroxymethyl)aminomethane + x ml of 0.1M hydrochloric acid
F: 50 ml of 0.025M borax + x ml of 0.1M hydrochloric acid
G: 50 ml of 0.025M borax + x ml of 0.1M sodium hydroxide
H: 50 ml of 0.05M sodium bicarbonate + x ml of 0.1M sodium hydroxide
I: 50 ml of 0.05M disodium hydrogen phosphate + x ml of 0.1M sodium hydroxide
J: 25 ml of 0.2M potassium chloride + x ml of 0.2M sodium hydroxide

Table 3–29 (continued)
BUFFER SOLUTIONS

C. Solutions Giving Round Values of pH at 25°C (continued)

Solution A		Solution B		Solution C		Solution D		Solution E	
pH	x	pH	x	pH	x	pH	x	pH	x
1.00	67.0	2.20	49.5	4.10	1.3	5.80	3.6	7.00	46.6
1.10	52.8	2.30	45.8	4.20	3.0	5.90	4.6	7.10	45.7
1.20	42.5	2.40	42.2	4.30	4.7	6.00	5.6	7.20	44.7
1.30	33.6	2.50	38.8	4.40	6.6	6.10	6.8	7.30	43.3
1.40	26.6	2.60	35.4	4.50	8.7	6.20	8.1	7.40	42.0
1.50	20.7	2.70	32.1	4.60	11.1	6.30	9.7	7.50	40.3
1.60	16.2	2.80	28.9	4.70	13.6	6.40	11.6	7.60	38.5
1.70	13.0	2.90	25.7	4.80	16.5	6.50	13.9	7.70	36.6
1.80	10.2	3.00	22.3	4.90	19.4	6.60	16.4	7.80	34.5
1.90	8.1	3.10	18.8	5.00	22.6	6.70	19.3	7.90	32.0
2.00	6.5	3.20	15.7	5.10	25.5	6.80	22.4	8.00	29.2
2.10	5.1	3.30	12.9	5.20	28.8	6.90	25.9	8.10	26.2
2.20	3.9	3.40	10.4	5.30	31.6	7.00	29.1	8.20	22.9
		3.50	8.2	5.40	34.1	7.10	32.1	8.30	19.9
		3.60	6.3	5.50	36.6	7.20	34.7	8.40	17.2
		3.70	4.5	5.60	38.8	7.30	37.0	8.50	14.7
		3.80	2.9	5.70	40.6	7.40	39.1	8.60	12.2
		3.90	1.4	5.80	42.3	7.50	41.1	8.70	10.3
		4.00	0.1	5.90	43.7	7.60	42.8	8.80	8.5
						7.70	44.2	8.90	7.0
						7.80	45.3	9.00	5.7
						7.90	46.1		
						8.00	46.7		

Solution F		Solution G		Solution H		Solution I		Solution J	
pH	x	pH	x	pH	x	pH	x	pH	x
8.00	20.5	9.20	0.9	9.60	5.0	10.90	3.3	12.00	6.0
8.10	19.7	9.30	3.6	9.70	6.2	11.00	4.1	12.10	8.0
8.20	18.8	9.40	6.2	9.80	7.6	11.10	5.1	12.20	10.2
8.30	17.7	9.50	8.8	9.90	9.1	11.20	6.3	12.30	12.8
8.40	16.6	9.60	11.1	10.00	10.7	11.30	7.6	12.40	16.2
8.50	15.2	9.70	13.1	10.10	12.2	11.40	9.1	12.50	20.4
8.60	13.5	9.80	15.0	10.20	13.8	11.50	11.1	12.60	25.6
8.70	11.6	9.90	16.7	10.30	15.2	11.60	13.5	12.70	32.2
8.80	9.6	10.00	18.3	10.40	16.5	11.70	16.2	12.80	41.2
8.90	7.1	10.10	19.5	10.50	17.8	11.80	19.4	12.90	53.0
9.00	4.6	10.20	20.5	10.60	19.1	11.90	23.0	13.00	66.0
9.10	2.0	10.30	21.3	10.70	20.2	12.00	26.9		
		10.40	22.1	10.80	21.2				
		10.50	22.7	10.90	22.0				
		10.60	23.3	11.00	22.7				
		10.70	23.8						
		10.80	24.25						

From Robinson, R. A. and Stokes, R. H., *Electrolyte Solutions,* 2nd ed., Butterworths, London. 1959. With permission.

Table 3–29 (continued)
BUFFER SOLUTIONS

D. pH Values of Standard Solutions at Temperatures from 0 to 95°C

Temperature, °C	Tetroxalate, 0.05m	Tartrate, 0.0341m, saturated at 25°C	Phthalate, 0.05m	Phosphate[a]	Phosphate[b]	Borax, 0.01m	Calcium Hydroxide, saturated at 25°C
0	1.666	–	4.003	6.984	7.534	9.464	13.423
5	1.668	–	3.999	6.951	7.500	9.395	13.207
10	1.670	–	3.998	6.923	7.472	9.332	13.003
15	1.672	–	3.999	6.900	7.448	9.276	12.810
20	1.675	–	4.002	6.881	7.429	9.225	12.627
25	1.679	3.557	4.008	6.865	7.413	9.180	12.454
30	1.683	3.552	4.015	6.853	7.400	9.139	12.289
35	1.688	3.549	4.024	6.844	7.389	9.102	12.133
38	1.691	3.548	4.030	6.840	7.384	9.081	12.043
40	1.694	3.547	4.035	6.838	7.380	9.068	11.984
45	1.700	3.547	4.047	6.834	7.373	9.038	11.841
50	1.707	3.549	4.060	6.833	7.367	9.011	11.705
55	1.715	3.554	4.075	6.834	–	8.985	11.574
60	1.723	3.560	4.091	6.836	–	8.962	11.449
70	1.743	3.580	4.126	6.845	–	8.921	–
80	1.766	3.609	4.164	6.859	–	8.885	–
90	1.792	3.650	4.205	6.877	–	8.850	–
95	1.806	3.674	4.227	6.886	–	8.833	–

[a]Solution of 0.025m KH_2PO_4 and 0.025m Na_2HPO_4.
[b]Solution of 0.008695m KH_2PO_4 and 0.03043m Na_2HPO_4.

From Robinson, R. A., in *Handbook of Chemistry and Physics,* 55th ed., Weast, R. C., Ed., CRC Press, Cleveland, 1974, D-112.

Table 3–30
APPROXIMATE pH VALUES OF SOME ACIDS, BASES, FOODS, AND BIOLOGIC MATERIALS[a]

Substance	pH	Substance	pH
Acids			
Acetic acid, N	2.4	Hydrochloric acid, 0.01N	2.0
Acetic acid, 0.1N	2.9	Hydrocyanic acid, 0.1N	5.1
Acetic acid, 0.01N	3.4	Hydrogen sulfide, 0.1N	4.1
Alum, 0.1N	3.2	Lactic acid, 0.1N	2.4
Arsenious acid, saturated	5.0	Malic acid, 0.1N	2.2
Benzoic acid, 0.01N	3.1	Orthophosphoric acid, 0.1N	1.5
Boric acid, 0.1N	5.2	Oxalic acid, 0.1N	1.6
Carbonic acid, saturated	3.8	Sulfuric acid, N	0.3
Citric acid, 0.1N	2.2	Sulfuric acid, 0.1N	1.2
Formic acid, 0.1N	2.3	Sulfuric acid, 0.01N	2.1
Hydrochloric acid, N	0.1	Sulfurous acid, 0.1N	1.5
Hydrochloric acid, 0.1N	1.1	Tartaric acid, 0.1N	2.2
Bases			
Ammonia, N	11.6	Potassium hydroxide, 0.1N	13.0
Ammonia, 0.1N	11.1	Potassium hydroxide, 0.01N	12.0
Ammonia, 0.01N	10.6	Sodium bicarbonate, 0.1N	8.4
Borax, 0.1N	9.2	Sodium carbonate, 0.1N	11.6
Calcium carbonate, saturated	9.4	Sodium hydroxide, N	14.0
Ferrous hydroxide, saturated	9.5	Sodium hydroxide, 0.1N	13.0
Lime, saturated	12.4	Sodium hydroxide, 0.01N	12.0
Magnesia, saturated	10.5	Sodium metasilicate, 0.1N	12.6
Potassium cyanide, 0.1N	11.0	Sodium sesquicarbonate, 0.1M	10.1
Potassium hydroxide, N	14.0	Trisodium phosphate, 0.1N	12.0
Biologic Materials			
Bile, human	6.8–7.0	Gastric contents, human	1.0–3.0
Blood, plasma, human	7.3–7.5	Milk, human	6.6–7.6
Blood, whole, dog	6.9–7.2	Saliva, human	6.5–7.5
Duodenal contents, human	4.8–8.2	Spinal fluid, human	7.3–7.5
Feces, human	4.6–8.4	Urine, human	4.8–8.4
Foods			
Apples	2.9–3.3	Milk, cows	6.3–6.6
Apricots	3.6–4.0	Olives	3.6–3.8
Asparagus	5.4–5.8	Oranges	3.0–4.0
Bananas	4.5–4.7	Oysters	6.1–6.6
Beans	5.0–6.0	Peaches	3.4–3.6
Beers	4.0–5.0	Pears	3.6–4.0
Beets	4.9–5.5	Peas	5.8–6.4
Blackberries	3.2–3.6	Pickles, dill	3.2–3.6
Bread, white	5.0–6.0	Pickles, sour	3.0–3.4
Butter	6.1–6.4	Pimento	4.6–5.2
Cabbage	5.2–5.4	Plums	2.8–3.0
Carrots	4.9–5.3	Potatoes	5.6–6.0
Cheeses	4.8–6.4	Pumpkin	4.8–5.2
Cherries	3.2–4.0	Raspberries	3.2–3.6
Cider	2.9–3.3	Rhubarb	3.1–3.2
Corn	6.0–6.5	Salmon	6.1–6.3
Crackers	6.5–8.5	Sauerkraut	3.4–3.6
Dates	6.2–6.4	Shrimp	6.8–7.0
Eggs, fresh white	7.6–8.0	Soft drinks	2.0–4.0

Table 3–30 (continued)
APPROXIMATE pH VALUES OF SOME ACIDS, BASES, FOODS, AND BIOLOGIC MATERIALS

Substance	pH	Substance	pH
Foods (cont.)			
Flour, wheat	5.5–6.5	Spinach	5.1–5.7
Gooseberries	2.8–3.0	Squash	5.0–5.4
Grapefruit	3.0–3.3	Strawberries	3.0–3.5
Grapes	3.5–4.5	Sweet potatoes	5.3–5.6
Hominy (lye)	6.8–8.0	Tomatoes	4.0–4.4
Jams, fruit	3.5–4.0	Tuna	5.9–6.1
Jellies, fruit	2.8–3.4	Turnips	5.2–5.6
Lemons	2.2–2.4	Vinegar	2.4–3.4
Limes	1.8–2.0	Water, drinking	6.5–8.0
Maple syrup	6.5–7.0	Wines	2.8–3.8

[a]All values are based on measurements made at 25°C and are rounded off to the nearest tenth.

From *Modern pH and Chlorine Control,* Taylor Chemicals, Baltimore. With permission.

Table 3–31
IONIZATION CONSTANTS FOR WATER

°C	$-\log_{10} K_w$	°C	$-\log_{10} K_w$
0	14.9435	30	13.8330
5	14.7338	35	13.6801
10	14.5346	40	13.5348
15	14.3463	45	13.3960
20	14.1669	50	13.2617
24	14.0000	55	13.1369
25	13.9965	60	13.0171

From Weast, R. C., Ed., *Handbook of Chemistry and Physics,* 55th ed., CRC Press, Cleveland, 1974, D-131.

Table 3–32
COMPOSITION AND PROPERTIES OF SOME INORGANIC ACIDS AND BASES SUPPLIED AS CONCENTRATED AQUEOUS SOLUTIONS

Compound	Composition, weight %		Formula weight	Molarity	Specific gravity
Acetic acid	99–100	CH_3COOH	60.05	17.5	1.05
Ammonium hydroxide	28–30	NH_4OH	35.05	7.4	0.90
Hydriodic acid	47–47.5	HI	127.91	5.5	1.5
Hydrobromic acid	47–49	HBr	80.93	9.0	1.5
Hydrochloric acid	36.5–38	HCl	36.46	12.0	1.18
Hydrofluoric acid	48–51	HF	20.01	28.9	1.17
Phosphoric acid	85	H_3PO_4	98.00	14.7	1.7
Sulfuric acid	95–98	H_2SO_4	98.08	18.0	1.84

From Weast, R. C., Ed., *Handbook of Chemistry and Physics,* 55th ed., CRC Press, Cleveland, 1974, D-131.

Table 3–33
EQUIVALENT CONDUCTANCES OF SOME ELECTROLYTES IN AQUEOUS SOLUTIONS AT 25°C

Compound	Infinite dilution	Concentration in gram equivalents per 1000 cm³						
		0.0005	0.001	0.005	0.01	0.02	0.05	0.1
$AgNo_3$	133.36	131.36	130.51	127.20	124.76	121.41	115.24	109.14
$BaCl_2$	139.98	135.96	134.34	128.02	123.94	119.09	111.48	105.19
$CaCl_2$	135.84	131.93	130.36	124.25	120.36	115.65	108.47	102.4
$Ca(OH)_2$	257.9			232.9	225.9	213.9		
$CuSO_4$	133.6	121.6	115.26	94.07	83.12	72.20	59.05	50.58
HCl	426.16	422.74	421.36	415.80	412.00	407.24	399.09	391.32
KBr	151.9		146.09	143.43	140.48	135.68	131.39
KCl	149.86	147.81	146.95	143.35	141.27	138.34	133.37	128.96
$KClO_4$	140.04	138.76	137.87	134.16	131.46	127.92	121.62	115.20
$KaFe(CN)_6$	174.5	166.4	163.1	150.7				
$K_4Fe(CN)_6$	184.5		167.24	146.09	134.83	122.82	107.70	97.87
$KHCO_3$	118.0	116.10	115.34	112.24	110.08	107.22		
KI	150.38			144.37	142.18	139.45	134.97	131.11
KIO_4	127.92	125.80	124.94	121.24	118.51	114.14	106.72	98.12
KNO_3	144.96	142.77	141.84	138.48	132.82	132.41	126.31	120.40
$KReO_4$	128.20	126.03	125.12	121.31	118.49	114.49	106.40	97.40
$LaCl_3$	145.8	139.6	137.0	127.5	121.8	115.3	106.2	99.1
LiCl	115.03	113.15	112.40	109.40	107.40	104.65	100.11	95.86
$LiClO_4$	105.98	104.18	103.44	100.57	98.61	96.18	92.20	88.56
$MgCl_2$	129.40	125.61	124.11	118.31	114.55	110.04	103.08	97.10
NH_4Cl	149.7		146.8	143.5	141.28	138.33	133.29	128.75
NaCl	126.45	124.50	123.74	120.65	118.51	115.51	111.06	106.74
$NaClO_4$	117.48	115.64	114.87	111.75	109.59	106.96	102.40	98.43
NaI	126.94	125.36	124.25	121.25	119.24	116.70	112.79	108.78
$NaOOCCH_3$	91.0	89.2	88.5	85.72	83.76	81.24	76.92	72.80
$NaOOCC_2H_5$	85.9		83.5	80.9	79.1	76.6		
$NaOOCC_3H_7$	82.70	81.04	80.31	77.58	75.76	73.39	69.32	65.27

Table 3–33 (continued)
EQUIVALENT CONDUCTANCES OF SOME ELECTROLYTES IN AQUEOUS SOLUTIONS AT 25°C

Compound	Infinite dilution	Concentration in gram equivalents per 1000 cm³						
		0.0005	0.001	0.005	0.01	0.02	0.05	0.1
NaOH	247.8	245.6	244.7	240.8	238.0			
Na₂SO₄	129.9	125.74	124.15	117.15	112.44	106.78	97.75	89.98
SrCl₂	135.80	131.90	130.33	124.24	120.24	115.54	108.25	102.19
ZnSO₄	132.8	121.4	114.53	95.49	84.91	74.24	61.20	52.64

From Weast, R. C., Ed., *Handbook of Chemistry and Physics,* 55th ed., CRC Press, Cleveland, 1974, D-132.

Table 3–34
THE EQUIVALENT CONDUCTIVITY OF THE SEPARATE IONS

Ion	0°C	18°C	25°C	50°C	75°C	100°C	128°C	156°C
Ag	32.9	54.3	63.5	101	143	188	245	299
1/2Ba	33	55	65	104	149	200	262	322
C₂H₃O₂	20.3	34.6	40.8	67	96	130	171	211
1/2C₂O₄	39	63	73	115	163	213	275	336
1/3C₆H₅O₇	36	60	70	113	161	214		
1/2Ca	30	51	60	98	142	191	252	312
Cl	41.1	65.6	75.5	116	160	207	264	318
1/4Fe(CN)₆	58	95	111	173	244	321		
H	240	314	350	465	565	644	722	777
K	40.4	64.4	74.5	115	159	206	263	317
1/3La	35	61	72	119	173	235	312	388
NH₄	40.2	64.5	74.5	115	159	207	264	319
NO₃	40.4	61.7	70.6	104	140	178	222	263
Na	26	43.5	50.9	82	116	155	203	249
OH	105	172	192	284	360	439	525	592
1/2SO₄	41	68	79	125	177	234	303	370

From Forsythe, W. E., *Smithsonian Physical Tables,* 9th ed., Smithsonian Institution, Washington, D.C., 1956, 399.

Table 3–35
ACTIVITY COEFFICIENTS OF ACIDS, BASES, AND SALTS[a]

Compound	0.1	0.2	0.3	0.4	0.5	0.6	0.7	0.8	0.9	1.0
$AgNO_3$	0.734	0.657	0.606	0.567	0.536	0.509	0.485	0.464	0.446	0.429
$AlCl_3$	(0.337)	0.305	0.302	0.313	0.331	0.356	0.388	0.429	0.479	0.539
$Al_2(SO_4)_3$	(0.0350)	0.0225	0.0176	0.0153	0.0143	0.0140	0.0142	0.0149	0.0159	0.0175
$CdSO_4$	(0.150)	0.102	0.082	0.069	0.061	0.055	0.050	0.046	0.043	0.041
$CrCl_3$	(0.331)	0.298	0.294	0.300	0.314	0.335	0.362	0.397	0.436	0.481
$Cr(NO_3)_3$	(0.319)	0.285	0.279	0.281	0.291	0.304	0.322	0.344	0.371	0.401
$Cr_2(SO_4)_3$	(0.0458)	0.0300	0.0238	0.0207	0.0190	0.0182	0.0181	0.0185	0.0194	0.0208
CsBr	0.754	0.694	0.654	0.626	0.603	0.586	0.571	0.558	0.547	0.538
CsCl	0.756	0.694	0.656	0.628	0.606	0.589	0.575	0.563	0.553	0.544
CsI	0.754	0.692	0.651	0.621	0.599	0.581	0.567	0.554	0.543	0.533
$CsNO_3$	0.733	0.655	0.602	0.561	0.528	0.501	0.478	0.458	0.439	0.422
CsOAc	0.799	0.771	0.761	0.759	0.762	0.768	0.776	0.783	0.792	0.802
$CuSO_4$	(0.150)	0.104	0.083	0.071	0.062	0.056	0.052	0.048	0.045	0.043
HBr	0.805	0.782	0.777	0.781	0.789	0.801	0.815	0.832	0.850	0.871
HCl	0.796	0.767	0.756	0.755	0.757	0.763	0.772	0.783	0.795	0.809
$HClO_4$	0.803	0.778	0.768	0.766	0.769	0.776	0.785	0.795	0.808	0.823
HI	0.818	0.807	0.811	0.823	0.839	0.860	0.883	0.908	0.935	0.963
HNO_3	0.791	0.754	0.735	0.725	0.720	0.717	0.717	0.718	0.721	0.724
KBr	0.772	0.722	0.693	0.673	0.657	0.646	0.636	0.629	0.622	0.617
KCl	0.770	0.718	0.688	0.666	0.649	0.637	0.626	0.618	0.610	0.604
KCNS	0.769	0.716	0.685	0.663	0.646	0.633	0.623	0.614	0.606	0.599
KF	0.755	0.727	0.700	0.682	0.670	0.661	0.654	0.650	0.646	0.645
KI	0.778	0.733	0.707	0.689	0.676	0.667	0.660	0.654	0.649	0.645
KNO_3	0.739	0.663	0.614	0.576	0.545	0.519	0.496	0.476	0.459	0.443
KOAc	0.796	0.766	0.754	0.750	0.751	0.754	0.759	0.766	0.774	0.783
KOH	0.798	0.760	0.742	0.734	0.732	0.733	0.736	0.742	0.749	0.756
LiBr	0.796	0.766	0.756	0.752	0.753	0.758	0.767	0.777	0.789	0.803
LiCl	0.790	0.757	0.744	0.740	0.739	0.743	0.748	0.755	0.764	0.774
$LiClO_4$	0.812	0.794	0.792	0.798	0.808	0.820	0.834	0.852	0.869	0.887
LiI	0.815	0.802	0.804	0.813	0.824	0.838	0.852	0.870	0.888	0.910
$LiNO_3$	0.788	0.752	0.736	0.728	0.726	0.727	0.729	0.733	0.737	0.743
LiOAc	0.784	0.742	0.721	0.709	0.700	0.691	0.689	0.688	0.688	0.689
$MgSO_4$	(0.150)	0.108	0.088	0.076	0.068	0.062	0.057	0.054	0.051	0.049
$MnSO_4$	(0.150)	0.106	0.085	0.073	0.064	0.058	0.053	0.049	0.046	0.044
NaBr	0.782	0.741	0.719	0.704	0.697	0.692	0.689	0.687	0.687	0.687
NaCl	0.788	0.735	0.710	0.693	0.681	0.673	0.667	0.662	0.659	0.657
$NaClO_4$	0.775	0.729	0.701	0.683	0.668	0.656	0.648	0.641	0.635	0.629
NaCNS	0.787	0.750	0.731	0.720	0.715	0.712	0.710	0.710	0.711	0.712
NaF	0.765	0.710	0.676	0.651	0.632	0.616	0.603	0.592	0.582	0.573
NaH_2PO_4	0.744	0.675	0.629	0.593	0.563	0.539	0.517	0.499	0.483	0.468
NaI	0.787	0.751	0.735	0.727	0.723	0.723	0.724	0.727	0.731	0.736
$NaNO_3$	0.762	0.703	0.666	0.638	0.617	0.599	0.583	0.570	0.558	0.548
NaOAc	0.791	0.757	0.744	0.737	0.735	0.736	0.740	0.745	0.752	0.757
NaOH	0.766	0.727	0.708	0.697	0.690	0.685	0.681	0.679	0.678	0.678
$NiSO_4$	(0.150)	0.105	0.084	0.071	0.063	0.056	0.052	0.047	0.044	0.042
RbBr	0.763	0.706	0.673	0.650	0.632	0.617	0.605	0.595	0.586	0.578
RbCl	0.764	0.709	0.675	0.652	0.634	0.620	0.608	0.599	0.590	0.583
RbI	0.762	0.705	0.671	0.647	0.629	0.614	0.602	0.591	0.583	0.575

Table 3–35 (continued)
ACTIVITY COEFFICIENTS OF ACIDS, BASES, AND SALTS

Compound	0.1	0.2	0.3	0.4	0.5	0.6	0.7	0.8	0.9	1.0
$RbNO_3$	0.734	0.658	0.606	0.565	0.534	0.508	0.485	0.465	0.446	0.430
RbOAc	0.796	0.767	0.756	0.753	0.755	0.759	0.766	0.773	0.782	0.792
$TiNO_3$	0.702	0.606	0.545	0.500						
$ZnSO_4$	(0.150)	0.104	0.083	0.071	0.063	0.057	0.052	0.048	0.046	0.043

[a]The coefficients given are valid at 25°C. Concentrations are expressed as molalities.

From Weast, R. C., Ed., *Handbook of Chemistry and Physics,* 55th ed., CRC Press, Cleveland, 1974, D-132.

Table 3–36
DISSOCIATION CONSTANTS OF ORGANIC BASES IN AQUEOUS SOLUTIONS

Compound	°C	Step	pK_a	K_a
Acetamide	25		0.63	2.34×10^{-1}
Acridine	20		5.58	2.63×10^{-6}
α-Alanine	25		2.345	4.52×10^{-3}
Alanine				
glycyl	25		3.153	7.03×10^{-4}
methoxy-(DL)	25		2.037	9.18×10^{-3}
phenyl	25	2	9.19	6.61×10^{-10}
Allothreonine	25	1	2.108	7.80×10^{-3}
	25	2	9.096	8.02×10^{-10}
n-Amylamine	25		10.63	2.34×10^{-11}
Aniline	25		4.63	2.34×10^{-5}
n-allyl	25		4.17	6.76×10^{-5}
4-(p-aminobenzoyl)	25	1	2.932	1.17×10^{-3}
4-benzyl	25		2.17	6.76×10^{-3}
2-bromo	25		2.53	2.95×10^{-3}
3-bromo	25		3.58	2.63×10^{-4}
4-bromo	25		3.86	1.38×10^{-4}
4-bromo-N,N-dimethyl	25		4.232	5.86×10^{-5}
o-chloro	25		2.65	2.24×10^{-3}
m-chloro	25		3.46	3.47×10^{-4}
p-chloro	25		4.15	7.08×10^{-5}
3-chloro-N,N-dimethyl	20		3.837	1.46×10^{-4}
4-chloro-N,N-dimethyl	20		4.395	4.03×10^{-5}
3,5-dibromo	25		2.34	4.57×10^{-3}
2,4-dichloro	22		2.05	8.91×10^{-3}
N,N-diethyl	22		6.61	2.46×10^{-7}
N,N-dimethyl	25		5.15	7.08×10^{-6}
N,N-dimethyl-3-nitro	25		2.626	2.37×10^{-3}
N-ethyl	24		5.12	7.59×10^{-6}
2-fluoro	25		3.20	6.31×10^{-4}
3-fluoro	25		3.50	3.16×10^{-4}
4-fluoro	25		4.65	2.24×10^{-5}
2-iodo	25		2.60	2.51×10^{-3}
N-methyl	25		4.848	1.41×10^{-5}
4-methylthio	25		4.35	4.46×10^{-5}
3-nitro	25		2.466	3.42×10^{-3}
4-nitro	25		1.00	1.00×10^{-1}
2-sulfonic acid	25	2	2.459	3.47×10^{-3}

Table 3–36 (continued)
DISSOCIATION CONSTANTS OF ORGANIC BASES IN AQUEOUS SOLUTIONS

Compound	°C	Step	pK_a	K_a
3-sulfonic acid	25	2	3.738	1.82×10^{-4}
4-sulfonic acid	25	2	3.227	5.92×10^{-4}
o-Anisidine	25		4.52	3.02×10^{-5}
m-Anisidine	25		4.23	5.89×10^{-5}
p-Anisidine	25		5.34	4.57×10^{-6}
Arginine	25	1	1.8217	1.5×10^{-2}
	25	2	8.9936	1.01×10^{-9}
Asparagine	20	1	2.213	6.12×10^{-3}
	20	2	8.85	1.41×10^{-9}
glycyl	25	1	2.942	1.14×10^{-3}
	18	2	8.44	3.63×10^{-9}
DL-Aspartic acid	1	1	2.122	7.55×10^{-3}
	1	2	4.006	1.00×10^{-4}
Acetidine (trimethylimidine)	25		11.29	5.12×10^{-12}
Aziridine	25		8.01	9.77×10^{-9}
Benzene				
4-aminoazo	25		2.82	1.51×10^{-3}
2-aminoethyl (β-phenylamine)	25		9.84	1.45×10^{-10}
4-dimethylaminoazo	25		3.226	5.94×10^{-4}
Benzidine	30	1	4.66	2.19×10^{-5}
	30	2	3.57	2.69×10^{-4}
Benzimidazole	25		5.532	2.94×10^{-6}
2-ethyl	25		6.18	6.61×10^{-7}
2-methyl	25		6.19	6.46×10^{-7}
2-phenyl	25	1	5.23	5.89×10^{-6}
Benzoic acid				
2-amino (anthranilic acid)	25	1	2.108	7.80×10^{-3}
	25	2	4.946	1.13×10^{-5}
4-amino	25	1	2.501	3.15×10^{-3}
	25	2	4.874	1.33×10^{-5}
Benzylamine	25		9.33	4.67×10^{-10}
Betaine	0		1.83	1.48×10^{-2}
Biphenyl				
2-amino	22		3.82	1.51×10^{-4}
trans-Bornylamine	25		10.17	6.76×10^{-11}
Brucine	25	1	8.28	5.24×10^{-9}
Butane				
1-amino-3-methyl	25		10.60	2.51×10^{-11}
2-amino-2-methyl	19		10.85	1.41×10^{-11}
1,4-diamino (putrescine)	10	1	11.15	7.08×10^{-12}
n-Butylamine	20		10.77	1.69×10^{-11}
t-Butylamine	18		10.83	1.48×10^{-11}
Butyric acid				
4-amino	25	1	4.0312	9.31×10^{-5}
	25	2	10.5557	2.78×10^{-11}
n-Butyric acid				
glycyl-2-amino	25	1	3.1546	7.01×10^{-4}
Cacodylic acid	25	1	1.57	2.69×10^{-2}
		2	6.27	5.37×10^{-7}
β-Chlortriethylammonium	25		8.80	1.59×10^{-9}
Cinnoline	20		2.37	4.27×10^{-3}
Codeine	25		8.21	6.15×10^{-9}
Cyclohexaneamine				
n-butyl	25		11.23	5.89×10^{-12}

Table 3–36 (continued)
DISSOCIATION CONSTANTS OF ORGANIC BASES IN AQUEOUS SOLUTIONS

Compound	°C	Step	pK_a	K_a
Cyclohexylamine	24		10.66	2.19×10^{-11}
Cystine	30	1	1.90	1.25×10^{-2}
	30	2	8.24	5.76×10^{-9}
n-Decylamine	25		10.64	2.29×10^{-11}
Diethylamine	40		10.489	3.24×10^{-11}
Diisobutylamine	21		10.489	1.23×10^{-11}
Diisopropylamine	28.5		10.96	1.09×10^{-11}
Dimethylamine	25		10.732	1.85×10^{-11}
n-Diphenylamine	25		0.79	1.62×10^{-1}
n-Dodecaneamine (laurylamine)	25		10.63	2.35×10^{-11}
d-Ephedrine	10		10.139	7.26×10^{-11}
l-Ephedrine	10		9.958	1.10×10^{-10}
Ethane				
1-amino-3-methoxy	10		9.89	1.29×10^{-10}
1,2-bismethylamino	25	1	10.40	3.98×10^{-11}
	25	2	8.26	5.50×10^{-9}
Ethanol				
2-amino	25		9.50	3.16×10^{-10}
Ethylamine	20		10.807	1.56×10^{-11}
Ethylenediamine	0	1	10.712	1.94×10^{-11}
	0	2	7.564	2.73×10^{-8}
l-Glutamic acid	25	1	2.13	7.41×10^{-3}
	25	2	4.31	4.90×10^{-5}
Glutamic acid				
a-monoethyl	25	1	3.846	1.42×10^{-4}
	25	2	7.838	1.45×10^{-8}
l-Glutamine			9.28	5.25×10^{-10}
l-Glutathione	25	2	3.59	2.57×10^{-4}
Glycine	25	1	2.3503	4.46×10^{-3}
	25	2	9.7796	1.68×10^{-10}
n-acetyl	25		3.6698	2.14×10^{-4}
dimethyl	5		10.3371	4.60×10^{-11}
glycyl	25		3.1397	7.25×10^{-4}
glycylglycyl	25	1	3.225	5.96×10^{-4}
	25	2	8.090	8.13×10^{-9}
leucyl	25	1	3.25	5.62×10^{-4}
	25	2	8.28	5.25×10^{-9}
methyl (sarcosine)	25	1	2.21	6.16×10^{-3}
	25	2	10.12	7.58×10^{-11}
phenyl	25	1	1.83	1.48×10^{-2}
	25	2	4.39	4.07×10^{-5}
N,*n*-propyl	25	1	2.35	4.46×10^{-3}
	25	2	10.19	6.46×10^{-11}
tetraglycyl	20	1	3.10	7.94×10^{-4}
	20	2	8.02	9.55×10^{-9}
Glycylserine	25	1	2.9808	1.04×10^{-3}
	25	2	8.38	4.17×10^{-9}
Heptadecaneamine	25		10.63	2.35×10^{-11}
Heptane				
1-amino	25		10.66	2.19×10^{-11}
2-amino	19		10.88	1.58×10^{-11}
2-methylamino	17		10.99	1.02×10^{-11}
Hexadecaneamine	25		10.61	2.46×10^{-11}

Table 3–36 (continued)
DISSOCIATION CONSTANTS OF ORGANIC BASES IN AQUEOUS SOLUTIONS

Compound	°C	Step	pK_a	K_a
Hexamethylenediamine	0	1	111.857	1.39×10^{-12}
	0	2	0.762	1.73×10^{-11}
Hexanoic acid				
6-amino	25	1	4.373	4.23×10^{-5}
	25	2	10.804	1.57×10^{-11}
n-Hexylamine	25		10.56	2.75×10^{-11}
dl-Histidine	25	1	1.80	1.58×10^{-2}
	25	2	6.04	9.12×10^{-7}
	25	3	9.33	4.67×10^{-10}
Histidine				
β-alanyl (carnosine)	20	1	2.73	1.86×10^{-3}
	20	2	6.87	1.35×10^{-7}
	20	3	9.73	1.48×10^{-10}
Imidazol	25		6.953	1.11×10^{-7}
2,4-dimethyl	25		8.359	5.50×10^{-9}
1-methyl (oxalmethyline)	25		6.95	1.12×10^{-7}
Indane				
1-amino (*d*-1-hydrindamine)	22.5		9.21	6.17×10^{-10}
Isobutyric acid				
2-amino	25	1	2.357	4.30×10^{-3}
	25	2	10.205	6.23×10^{-11}
Isoleucine	25	1	2.318	4.81×10^{-3}
	25	2	9.758	1.74×10^{-10}
Isoquinoline (leucoline)	20		5.42	3.80×10^{-6}
1-amino	20		7.59	2.57×10^{-8}
7-hydroxy	20	1	5.68	2.09×10^{-6}
	20	2	8.90	1.26×10^{-9}
L-Leucine	25	1	2.328	4.70×10^{-3}
	25	2	9.744	1.80×10^{-10}
Leucine				
glycyl	25		3.18	6.61×10^{-4}
Methionine	25	1	2.22	6.02×10^{-3}
	25	2	9.27	5.37×10^{-10}
Methylamine	25		10.657	2.70×10^{-11}
Morphine	25		8.21	6.16×10^{-9}
Morpholine	25		8.33	4.67×10^{-9}
Naphthalene				
1-amino-6-hydroxy	25		3.97	1.07×10^{-4}
dimethylamino	25		4.566	2.72×10^{-5}
α-Naphthylamine	25		3.92	1.20×10^{-4}
n-methyl	27		3.67	2.13×10^{-4}
β-Naphthylamine	25		4.16	6.92×10^{-5}
cis-Neobornylamine	25		10.01	9.77×10^{-11}
Nicotine	25	1	8.02	9.55×10^{-9}
	25	2	3.12	7.59×10^{-4}
n-Nonylamine	25		10.64	2.29×10^{-11}
Norleucine	25		2.335	4.62×10^{-3}
Octadecaneamine	25		10.60	2.51×10^{-11}
Octylamine	25		10.65	2.24×10^{-11}
Ornithine	25	1	1.705	1.97×10^{-2}
	25	2	8.690	2.04×10^{-9}

Table 3—36 (continued)
DISSOCIATION CONSTANTS OF ORGANIC BASES IN AQUEOUS SOLUTIONS

Compound	°C	Step	pK_a	K_a
Papaverine	25		6.40	3.98×10^{-7}
Pentane				
3-amino	17		10.59	2.57×10^{-11}
3-amino-3-methyl	16		11.01	9.77×10^{-12}
n-Pentadecylamine	25		10.61	2.46×10^{-11}
Pentanoic acid				
5-amino (valeric acid)	25	1	4.270	5.37×10^{-5}
	25	2	10.766	1.71×10^{-11}
Perimidine	20		6.35	4.47×10^{-7}
Phenanthridine	20		5.58	2.63×10^{-6}
1,10-Phenanthroline	25		4.84	1.44×10^{-5}
o-Phenetidine (2-ethoxyaniline)	28		4.43	3.72×10^{-5}
m-Phenetidine (3-ethoxyaniline)	25		4.18	6.60×10^{-5}
p-Phenetidine (4-ethoxyaniline)	28		5.20	6.31×10^{-6}
a-Picoline	20		5.97	1.07×10^{-6}
β-Picoline	20		5.68	2.09×10^{-6}
γ-Picoline	20		6.02	9.55×10^{-7}
Pilocarpine	30		6.87	1.35×10^{-7}
Piperazine	23.5	1	9.83	1.48×10^{-10}
	23.5	2	5.56	2.76×10^{-6}
2,5-dimethyl (*trans*-)	25	1	9.66	2.19×10^{-10}
	25	2	5.20	6.31×10^{-6}
Piperidine	25		11.123	7.53×10^{-12}
3-acetyl	25		3.18	6.61×10^{-4}
1-*n*-butyl	23		10.47	3.39×10^{-11}
1,2-dimethyl	25		10.22	6.03×10^{-11}
1-ethyl	23		10.45	3.55×10^{-11}
1-methyl	25		10.08	8.32×10^{-11}
2,2,6,6-tetramethyl	25		11.07	8.51×10^{-12}
2,2,4-trimethyl	30		11.04	9.12×10^{-12}
Proline	25	1	1.952	1.11×10^{-10}
	25	2	10.640	2.29×10^{-11}
hydroxy	25	1	1.818	1.52×10^{-2}
	25	2	9.662	2.18×10^{-10}
Propane				
1-amino-2,2-dimethyl	25		10.15	7.08×10^{-11}
1,2-diamino	25	1	9.82	1.52×10^{-10}
	25	2	6.61	2.46×10^{-7}
1,3-diamino	10	1	10.94	1.15×10^{-11}
	10	2	9.03	9.33×10^{-10}
1,2,3-triamino	20	1	9.59	2.57×10^{-10}
	20	2	7.95	1.12×10^{-8}
Propanoic acid				
3-amino (*β*-alanine)	25	1	3.551	2.81×10^{-4}
	25	2	10.238	5.78×10^{-11}
Propylamine	20		10.708	1.96×10^{-11}
Pteridine	20		4.05	8.91×10^{-5}
2-amino-4,6-dihydroxy	20	2	6.59	2.57×10^{-7}
	20	3	9.31	4.90×10^{-10}
2-amino-4-hydroxy	20	1	2.27	5.37×10^{-3}
	20	2	7.96	1.10×10^{-8}
6-chloro	20		3.68	2.09×10^{-4}
6-hydroxy-4-methyl	20	1	4.08	8.32×10^{-5}
	20	2	6.41	3.89×10^{-7}
Purine	20	1	2.30	5.01×10^{-3}
	20	2	8.96	1.10×10^{-9}
6-amino (adenine)	25	1	4.12	7.59×10^{-5}

Table 3–36 (continued)
DISSOCIATION CONSTANTS OF ORGANIC BASES IN AQUEOUS SOLUTIONS

Compound	°C	Step	pK_a	K_a
6-amino (adenine) (cont.)				
	25	2	9.83	1.48×10^{-10}
2-dimethylamino	20	1	4.00	1.00×10^{-4}
	20	2	10.24	5.75×10^{-11}
8-hydroxy	20	1	2.56	2.75×10^{-3}
	20	2	8.26	9.49×10^{-9}
Pyrazine	27		0.65	2.24×10^{-1}
2-methyl	27		1.45	3.54×10^{-2}
methylamino	25		3.39	4.07×10^{-4}
Pyridazine	20		2.24	5.76×10^{-3}
Pyridine	25		5.25	5.62×10^{-6}
2-aldoxime	20	1	3.59	2.57×10^{-4}
	20	2	10.18	6.61×10^{-11}
2-amino	20		6.82	1.51×10^{-7}
4-amino	25		9.1141	7.69×10^{-10}
2-benzyl	25		5.13	7.41×10^{-6}
3-bromo	25		2.84	1.45×10^{-3}
3-chloro	25		2.84	1.45×10^{-3}
2,5-diamino	20		6.48	3.31×10^{-7}
2,3-dimethyl (2,3-lutidine)	25		6.57	2.69×10^{-7}
2,4-dimethyl (2,4-lutidine)	25		6.99	1.02×10^{-7}
3,5-dimethyl (3,5-lutidine)	25		6.15	7.08×10^{-7}
2-ethyl	25		5.89	1.28×10^{-6}
2-formyl	20		3.80	1.59×10^{-4}
2-hydroxy (2-pyridol)	20	1	0.75	9.82×10^{-1}
	20	2	11.65	2.24×10^{-12}
4-hydroxy	20	1	3.20	6.31×10^{-4}
	20	2	11.12	7.59×10^{-12}
methoxy	25		6.47	3.30×10^{-7}
4-methylamino	20		9.65	2.24×10^{-10}
2,4,6-trimethyl	25		7.43	3.72×10^{-8}
Pyrimidine				
2-amino	20		3.45	3.54×10^{-4}
2-amino-4,6-dimethyl	20		4.82	1.51×10^{-5}
2-amino-5-nitro	20		0.35	4.46×10^{-1}
Pyrrolidine	25		11.27	5.37×10^{-12}
1,2-dimethyl	26		10.20	6.31×10^{-11}
n-methyl	25		10.32	4.79×10^{-11}
Quinazoline	20		3.43	3.72×10^{-4}
5-hydroxy	20	1	3.62	2.40×10^{-4}
	20	2	7.41	3.89×10^{-8}
Quinine	25	1	8.52	3.02×10^{-9}
	25	2	4.13	7.41×10^{-5}
Quinoline	20		4.90	1.25×10^{-5}
3-amino	20		4.91	1.23×10^{-5}
3-bromo	25		2.69	2.04×10^{-3}
8-carboxy	25		1.82	1.51×10^{-2}
3-hydroxy (3-quinolinol)	20	1	4.28	5.25×10^{-5}
	20	2	8.08	8.32×10^{-9}
8-hydroxy (8-quinolinol)	20	1	5.017	1.21×10^{-6}
	25	2	9.812	1.54×10^{-10}
8-hydroxy-5-sulfo	25	1	4.112	7.73×10^{-5}
	25	2	8.757	1.75×10^{-9}
6-methoxy	20		5.03	9.33×10^{-6}
2-methyl (quinaldine)	20		5.83	1.48×10^{-6}
4-methyl (lepidine)	20		5.67	2.14×10^{-6}
5-methyl	20		5.20	6.31×10^{-6}

Table 3–36 (continued)
DISSOCIATION CONSTANTS OF ORGANIC BASES IN AQUEOUS SOLUTIONS

Compound	°C	Step	pK_a	pK_a
Quinoxaline (quinazine)	20		0.56	3.63×10^{-1}
Serine (2-amino-3-hydroxypropanoic acid)	25	1	2.186	5.49×10^{-3}
	25	2	9.208	6.19×10^{-10}
Strychnine	25		8.26	5.49×10^{-9}
Taurine (2-aminoethane sulfonic acid)	25	2	9.0614	8.69×10^{-10}
Tetradecaneamine (myristilamine)	25		10.62	2.40×10^{-11}
Thiazole	20		2.44	3.63×10^{-3}
2-amino	20		5.36	4.36×10^{-5}
Threonine	25	1	2.088	8.16×10^{-3}
	25	2	9.10	7.94×10^{-10}
o-Toluidine	25		4.44	3.63×10^{-5}
m-Toluidine	25		4.73	1.86×10^{-5}
p-Toluidine	25		5.08	8.32×10^{-6}
1,3,5-Triazine				
2,4,6-triamino	25		5.00	1.00×10^{-5}
Tridecaneamine	25		10.63	2.35×10^{-11}
Triethylamine	18		11.01	9.77×10^{-12}
Trimethylamine	25		9.81	1.55×10^{-10}
Tryptophan	25	1	2.43	3.72×10^{-3}
	25	2	9.44	3.63×10^{-10}
Tyrosine	25	2	9.11	7.76×10^{-10}
	25	3	10.13	7.41×10^{-11}
amide	25		7.33	4.68×10^{-8}
Urea	21		0.10	7.94×10^{-1}
Valine	25	1	2.286	5.17×10^{-3}
	25	2	9.719	1.91×10^{-10}

From Weast, R. C., Ed., *Handbook of Chemistry and Physics,* 55th ed., CRC Press, Cleveland, 1974, D-126.

Table 3–37
DISSOCIATION CONSTANTS OF INORGANIC BASES IN AQUEOUS SOLUTIONS[a]

Compound	°C	Step	pK_b	K_b
Ammonium hydroxide	25		4.75	1.79×10^{-5}
Arsenous oxide	25		3.96	1.1×10^{-4}
Beryllium hydroxide	25	2	10.30	5×10^{-11}
Calcium hydroxide	25	1	2.43	3.74×10^{-3}
	30	2	1.40	4.0×10^{-2}
Deuterammonium hydroxide	25		4.96	1.1×10^{-5}
Hydrazine	20		5.77	1.7×10^{-6}
Hydroxylamine	20		7.97	1.07×10^{-8}
Lead hydroxide	25		3.02	9.6×10^{-4}
Silver hydroxide	25		3.96	1.1×10^{-4}
Zinc hydroxide	25		3.02	9.6×10^{-4}

[a]Approximately 0.1–0.01N.

From Weast, R. C., Ed., *Handbook of Chemistry and Physics,* 55th ed., CRC Press, Cleveland, 1974, D-128.

Table 3–38
DISSOCIATION CONSTANTS OF ORGANIC ACIDS IN AQUEOUS SOLUTIONS

Compound	°C	Step	pK	K
Acetic acid	25		4.75	1.76×10^{-5}
Acetoacetic acid	18		3.58	2.62×10^{-4}
Acrylic acid	25		4.25	5.6×10^{-5}
Adipamic acid	25		4.63	2.35×10^{-5}
Adipic acid	25	1	4.43	3.71×10^{-5}
	25	2	4.41	3.87×10^{-5}
d-Alanine	25		9.87	1.35×10^{-10}
Allantoin	25		8.96	1.10×10^{-9}
Alloxanic acid	25		6.64	2.3×10^{-7}
a-Aminoacetic acid (Glycine)	25		9.78	1.67×10^{-10}
o-Aminobenzoic acid	25		6.97	1.07×10^{-7}
m-Aminobenzoic acid	25		4.78	1.67×10^{-5}
p-Aminobenzoic acid	25		4.92	1.2×10^{-5}
o-Aminobenzosulfonic acid	25		2.48	3.3×10^{-3}
m-Aminobenzosulfonic acid	25		3.73	1.85×10^{-4}
p-Aminobenzosulfonic acid	25		3.24	5.81×10^{-4}
Anisic acid	25		4.47	3.38×10^{-5}
o-β-Anisylpropionic acid	25		4.80	1.59×10^{-5}
m-β-Anisylpropionic acid	25		4.65	2.24×10^{-5}
p-β-Anisylpropionic acid	25		4.69	2.04×10^{-5}
Ascorbic acid	24	1	4.10	7.94×10^{-5}
	16	2	11.79	1.62×10^{-12}
DL-Aspartic acid	25	1	3.86	1.38×10^{-4}
	25	2	9.82	1.51×10^{-10}

Table 3–38 (continued)
DISSOCIATION CONSTANTS OF ORGANIC ACIDS IN AQUEOUS SOLUTIONS

Compound	°C	Step	pK	K
Barbituric acid	25		4.01	9.8×10^{-5}
Benzoic acid	25		4.19	6.46×10^{-5}
Benzosulfonic acid	25		0.70	2×10^{-1}
Bromoacetic acid	25		2.69	2.05×10^{-3}
o-Bromobenzoic acid	25		2.84	1.45×10^{-3}
m-Bromobenzoic acid	25		3.86	1.37×10^{-4}
n-Butyric acid	20		4.81	1.54×10^{-5}
iso-Butyric acid	18		4.84	1.44×10^{-5}
Cacodylic acid	25		6.19	6.4×10^{-7}
n-Caproic acid	18		4.83	1.43×10^{-5}
iso-Caproic acid	18		4.84	1.46×10^{-5}
Chloroacetic acid	25		2.85	1.40×10^{-3}
o-Chlorobenzoic acid	25		2.92	1.20×10^{-3}
m-Chlorobenzoic acid	25		3.82	1.5×10^{-4}
p-Chlorobenzoic acid	25		3.98	1.04×10^{-4}
a-Chlorobutyric acid	RT*a*		2.86	1.39×10^{-3}
β-Chlorobutyric acid	RT		4.05	8.9×10^{-5}
γ-Chlorobutyric acid	RT		4.52	3.0×10^{-5}
o-Chlorocinnamic acid	25		4.23	5.89×10^{-5}
m-Chlorocinnamic acid	25		4.29	5.13×10^{-5}
p-Chlorocinnamic acid	25		4.41	3.89×10^{-5}
o-Chlorophenoxyacetic acid	25		3.05	8.91×10^{-4}
m-Chlorophenoxyacetic acid	25		3.07	8.51×10^{-4}
p-Chlorophenoxyacetic acid	25		3.10	7.94×10^{-4}
o-Chlorophenylacetic acid	25		4.07	1.18×10^{-5}
m-Chlorophenylacetic acid	25		4.14	7.25×10^{-5}
p-Chlorophenylacetic acid	25		4.19	6.46×10^{-5}
β-(*o*-Chlorophenyl) propionic acid	25		4.58	2.63×10^{-4}
β-(*m*-Chlorophenyl) propionic acid	25		4.59	2.57×10^{-5}
β-(*p*-Chlorophenyl) propionic acid	25		4.61	2.46×10^{-5}
a-Chloropropionic acid	25		2.83	1.47×10^{-3}
β-Chloropropionic acid	25		3.98	1.04×10^{-4}
cis-Cinnamic acid	25		3.89	1.3×10^{-4}
trans-Cinnamic acid	25		4.44	3.65×10^{-5}
Citric acid	18	1	3.08	8.4×10^{-4}
	18	2	4.74	1.8×10^{-5}
	18	3	5.40	4.0×10^{-6}
o-Cresol	25		10.20	6.3×10^{-11}
m-Cresol	25		10.01	9.8×10^{-11}
p-Cresol	25		10.17	6.7×10^{-11}
trans-Crotonic acid	25		4.69	2.03×10^{-5}
Cyanoacetic acid	25		2.45	3.65×10^{-3}
γ-Cyanobutyric acid	25		2.42	3.80×10^{-3}
o-Cyanophenoxyacetic acid	25		2.98	1.05×10^{-3}
m-Cyanophenoxyacetic acid	25		3.03	9.33×10^{-4}
p-Cyanophenoxyacetic acid	25		2.93	1.18×10^{-3}
Cyanopropionic acid	25		2.44	3.6×10^{-3}
Cyclohexane-1:	25	1	3.45	3.55×10^{-4}
1-dicarboxylic acid	25	2	6.11	7.76×10^{-7}
Cyclopropane-1:	25	1	1.82	1.51×10^{-2}
1-dicarboxylic acid	25	2	7.43	3.72×10^{-8}
DL-Cysteine	30	1	8.14	7.25×10^{-9}
	30	2	10.34	4.6×10^{-11}
L-Cystine	25	1	7.85	1.4×10^{-8}
	25	2	9.85	1.4×10^{-10}

Table 3-38 (continued)
DISSOCIATION CONSTANTS OF ORGANIC ACIDS IN AQUEOUS SOLUTIONS

Compound	°C	Step	pK	K
Deuteroacetic acid (in D_2O)	25		5.25	5.5×10^{-6}
Dichloroacetic acid	25		1.48	3.32×10^{-2}
Dichloroacetylacetic acid	?		2.11	7.8×10^{-3}
2,3-Dichlorphenol	25		7.44	3.6×10^{-8}
2,2-Dihydroxybenzoic acid	25		2.94	1.14×10^{-3}
2,5-Dihydroxybenzoic acid	25		2.97	1.08×10^{-3}
3,4-Dihydroxybenzoic acid	25		4.48	3.3×10^{-5}
3,5-Dihydroxybenzoic acid	25		4.04	9.1×10^{-5}
Dihydroxymalic acid	25		1.92	1.2×10^{-2}
Dihydroxytartaric acid	25		1.92	1.2×10^{-2}
Dimethylglycine	25		9.89	1.3×10^{-10}
Dimethylmalic acid	25	1	3.17	6.83×10^{-4}
	25	2	6.06	8.72×10^{-7}
Dimethylmalonic acid	25		3.15	7.08×10^{-4}
Dinicotinic acid	25		2.80	1.6×10^{-3}
2,4-Dinitrophenol	15		3.96	1.1×10^{-4}
3,6-Dinitrophenol	15		5.15	7.1×10^{-6}
Diphenylacetic acid	25		3.94	1.15×10^{-4}
Ethylbenzoic acid	25		4.35	4.47×10^{-5}
Ethylphenylacetic acid	25		4.37	4.27×10^{-5}
Fluorobenzoic acid	17		2.90	1.25×10^{-3}
Formic acid	20		3.75	1.77×10^{-4}
trans-Fumaric acid	18	1	3.03	9.30×10^{-4}
	18	2	4.44	3.62×10^{-5}
Furancarboxylic acid	25		3.15	7.1×10^{-4}
Furoic acid	25		3.17·	6.76×10^{-4}
Gallic acid	25		4.41	3.9×10^{-5}
Glutaramic acid	25		4.60	3.98×10^{-5}
Glutaric acid	25	1	4.34	4.58×10^{-5}
	25	2	5.41	3.89×10^{-6}
Glycerol	25		14.15	7×10^{-15}
Glycine	25		9.87	1.67×10^{-10}
Glycol	25		14.22	6×10^{-15}
Glycollic acid	25		3.83	1.48×10^{-4}
Heptanoic acid	25		4.89	1.28×10^{-5}
Hexahydrobenzoic acid	25		4.90	1.26×10^{-5}
Hexanoic acid	25		4.88	1.31×10^{-5}
Hippuric acid	25		3.80	1.57×10^{-4}
Histidine	25		9.17	6.7×10^{-10}
Hydroquinone	20		10.35	4.5×10^{-11}
o-Hydroxybenzoic acid	19	1	2.97	1.07×10^{-3}
	18	2	13.40	4×10^{-14}
m-Hydroxybenzoic acid	19	1	4.06	8.7×10^{-5}
	19	2	9.92	1.2×10^{-10}
p-Hydroxybenzoic acid	19	1	4.48	3.3×10^{-5}
	19	2	9.32	4.8×10^{-10}
β-Hydroxybutyric acid	25		4.70	2×10^{-5}
γ-Hydroxybutyric acid	25		4.72	1.9×10^{-5}
β-Hydroxypropionic acid	25		4.51	3.1×10^{-5}
γ-Hydroxyquinoline	20		9.51	3.1×10^{-10}
Iodoacetic acid	25		3.12	7.5×10^{-4}

Table 3-38 (continued)
DISSOCIATION CONSTANTS OF ORGANIC ACIDS IN AQUEOUS SOLUTIONS

Compound	°C	Step	pK	K
o-Iodobenzoic acid	25		2.85	1.4×10^{-3}
m-Iodobenzoic acid	25		3.80	1.6×10^{-4}
Itaconic acid	25	1	3.85	1.40×10^{-4}
	25	2	5.45	3.56×10^{-6}
Lactic acid	100		3.08	8.4×10^{-4}
Lutidinic acid	25		2.15	7.0×10^{-3}
Lysine	25		10.53	2.95×10^{-11}
Maleic acid	25	1	1.83	1.42×10^{-2}
	25	2	6.07	8.57×10^{-7}
Malic acid	25	1	3.40	3.9×10^{-4}
	25	2	5.11	7.8×10^{-6}
Malonic acid	25	1	2.83	1.49×10^{-3}
	25	2	5.69	2.03×10^{-6}
DL-Mandelic acid	25		3.85	1.4×10^{-4}
Mesaconic acid	25	1	3.09	8.22×10^{-4}
	25	2	4.75	1.78×10^{-5}
Mesitylenic acid	25		4.32	4.8×10^{-5}
Methyl-o-aminobenzoic acid	25		5.34	4.6×10^{-6}
Methyl-m-aminobenzoic acid	25		5.10	8×10^{-6}
Methyl-p-aminobenzoic acid	25		5.04	9.2×10^{-6}
o-Methylcinnamic acid	25		4.50	3.16×10^{-5}
m-Methylcinnamic acid	25		4.44	3.63×10^{-5}
p-Methylcinnamic acid	25		4.56	2.76×10^{-5}
β-Methylglutaric acid	25		4.24	5.75×10^{-5}
n-Methylglycine	18		9.92	1.2×10^{-10}
Methylmalonic acid	25		3.07	1.17×10^{-4}
Methylsuccinic acid	25	1	4.13	7.4×10^{-5}
	25	2	5.64	2.3×10^{-6}
o-Monochlorophenol	25		8.49	3.2×10^{-9}
m-Monochlorophenol	25		8.85	1.4×10^{-9}
p-Monochlorophenol	25		9.18	6.6×10^{-10}
Naphthalenesulfonic acid	25		0.57	2.7×10^{-1}
a-Naphthoic acid	25		3.70	2×10^{-4}
β-Naphthoic acid	25		4.17	6.8×10^{-5}
a-Naphthol	25		9.34	4.6×10^{-10}
β-Naphthol	25		9.51	3.1×10^{-10}
Nitrobenzene	0		3.98	1.05×10^{-4}
o-Nitrobenzoic acid	18		2.16	6.95×10^{-3}
m-Nitrobenzoic acid	25		3.47	3.4×10^{-4}
p-Nitrobenzoic acid	25		3.41	3.93×10^{-4}
o-Nitrophenol	25		7.17	6.8×10^{-8}
m-Nitrophenol	25		8.28	5.3×10^{-9}
p-Nitrophenol	25		7.15	7×10^{-8}
o-Nitrophenylacetic acid	25		4.00	1.00×10^{-5}
m-Nitrophenylacetic acid	25		3.97	1.07×10^{-4}
p-Nitrophenylacetic acetic	25		3.85	1.41×10^{-4}
o-β-Nitrophenylpropionic acid	25		4.50	3.16×10^{-5}
p-β-Nitrophenylpropionic acid	25		4.47	3.39×10^{-5}
Nonanic acid	25		4.96	1.09×10^{-5}
Octanoic acid	25		4.89	1.28×10^{-5}
Oxalic acid	25	1	1.23	5.90×10^{-2}
	25	2	4.19	6.40×10^{-5}

Table 3—38 (continued)

DISSOCIATION CONSTANTS OF ORGANIC ACIDS IN AQUEOUS SOLUTIONS

Compound	°C	Step	pK	K
Phenol	20		9.89	1.28×10^{-10}
Phenylacetic acid	18		4.28	5.2×10^{-5}
o-Phenylbenzoic acid	25		3.46	3.47×10^{-4}
γ-Phenylbutyric acid	25		4.76	1.74×10^{-5}
a-Phenylpropionic acid	25		4.64	2.27×10^{-5}
β-Phenylpropionic acid	25		4.37	4.25×10^{-5}
o-Phthalic acid	25	1	2.89	1.3×10^{-3}
	25	2	5.51	3.9×10^{-6}
m-Phthalic acid	25	1	3.54	2.9×10^{-4}
	18	2	4.60	2.5×10^{-5}
p-Phthalic acid	25	1	3.51	3.1×10^{-4}
	16	2	4.82	1.5×10^{-5}
Picric acid	25		0.38	4.2×10^{-1}
Pimelic acid	25		4.71	3.09×10^{-5}
Propionic acid	25		4.87	1.34×10^{-5}
iso-Propylbenzoic acid	25		4.40	3.98×10^{-5}
2-Pyridinecarboxylic acid	25		5.52	3×10^{-6}
3-Pyridinecarboxylic acid	25		4.85	1.4×10^{-5}
4-Pyridinecarboxylic acid	25		4.96	1.1×10^{-5}
Pyrocatechol	20		9.85	1.4×10^{-10}
Quinolinic acid	25		2.52	3×10^{-3}
Resorcinol	25		9.81	1.55×10^{-10}
Saccharin	18		11.68	2.1×10^{-12}
Suberic acid	25		4.52	2.99×10^{-5}
Succinic acid	25	1	4.16	6.89×10^{-5}
	25	2	5.61	2.47×10^{-6}
Sulfanilic acid	25		3.23	5.9×10^{-4}
a-Tartaric acid	25	1	2.98	1.04×10^{-3}
	25	2	4.34	4.55×10^{-5}
meso-Tartaric acid	25	1	3.22	6×10^{-4}
	25	2	4.82	1.53×10^{-5}
Theobromine	18		7.89	1.3×10^{-8}
Terephthalic acid	25		3.51	3.1×10^{-4}
Thioacetic acid	25		3.33	4.7×10^{-4}
Thiophenecarboxylic acid	25		3.48	3.3×10^{-4}
o-Toluic acid	25		3.91	1.22×10^{-4}
m-Toluic acid	25		4.27	5.32×10^{-5}
p-Toluic acid	25		4.36	4.33×10^{-5}
Trichloroacetic acid	25		0.70	2×10^{-1}
Trichlorophenol	25		6.00	1×10^{-6}
2,4,6-Trihydroxybenzoic acid	25		1.68	2.1×10^{-2}
Trimethylacetic acid	18		5.03	9.4×10^{-6}
2,4,6-Trinitrophenol	25		0.38	4.2×10^{-1}
Tryptophan	25		9.38	4.2×10^{-10}
Tyrosine	17		8.40	3.98×10^{-9}
Uric acid	12		3.89	1.3×10^{-4}
n-Valeric acid	18		4.82	1.51×10^{-5}
iso-Valeric acid	25		4.77	1.7×10^{-5}
Veronal	25		7.43	3.7×10^{-8}

Table 3–38 (continued)
DISSOCIATION CONSTANTS OF ORGANIC ACIDS IN AQUEOUS SOLUTIONS

Compound	°C	Step	pK	K
Vinylacetic acid	25		4.34	4.57×10^{-5}
Xanthine	40		9.91	1.24×10^{-10}

[a]RT = room temperature.

From Weast, R. C., Ed., *Handbook of Chemistry and Physics,* 55th ed., CRC Press, Cleveland, 1974, D-129.

Table 3–39
DISSOCIATION CONSTANTS OF INORGANIC ACIDS IN AQUEOUS SOLUTIONS[a]

Compound	°C	Step	pK	K
Arsenic acid	18	1	2.25	5.62×10^{-3}
	18	2	6.77	1.70×10^{-3}
	18	3	11.60	3.95×10^{-12}
Arsenious acid	25		9.23	6×10^{-10}
o-Boric acid	20	1	9.14	7.3×10^{-10}
	20	2	12.74	1.8×10^{-13}
	20	3	13.80	1.6×10^{-14}
Carbonic acid	25	1	6.37	4.30×10^{-7}
	25	2	10.25	5.61×10^{-11}
Chromic acid	25	1	0.74	1.8×10^{-1}
	25	2	6.49	3.20×10^{-7}
Germanic acid	25	1	8.59	2.6×10^{-9}
	25	2	12.72	1.9×10^{-13}
Hydrocyanic acid	25		9.31	4.93×10^{-10}
Hydrofluoric acid	25		3.45	3.53×10^{-4}
Hydrogen sulfide	18	1	7.04	9.1×10^{-3}
	18	2	11.96	1.1×10^{-12}
Hydrogen peroxide	25		11.62	2.4×10^{-12}
Hypobromous acid	25		8.69	2.06×10^{-9}
Hypochlorous acid	18		7.53	2.95×10^{-8}
Hypoiodous acid	25		10.64	2.3×10^{-11}
Iodic acid	25		0.77	1.69×10^{-1}
Nitrous acid	12.5		3.37	4.6×10^{-4}
Periodic acid	25		1.64	2.3×10^{-2}
o-Phosphoric acid	25	1	2.12	7.52×10^{-3}
	25	2	7.21	6.23×10^{-8}
	18	3	12.67	2.2×10^{-13}
Phosphorous acid	18	1	2.00	1.0×10^{-2}
	18	2	6.59	2.6×10^{-7}
Pyrophosphoric acid	18	1	0.85	1.4×10^{-1}
	18	2	1.49	3.2×10^{-2}
	18	3	5.77	1.7×10^{-6}
	18	4	8.22	6×10^{-9}

Table 3–39 (continued)
DISSOCIATION CONSTANTS OF ORGANIC ACIDS IN AQUEOUS SOLUTIONS

Compound	°C	Step	pK	K
Selenic acid	25	2	1.92	1.2×10^{-2}
Selenous acid	25	1	2.46	3.5×10^{-3}
	25	2	7.31	5×10^{-8}
o-Silicic acid	30	1	9.66	2.2×10^{-10}
	30	2	11.70	2×10^{-12}
	30	3	12.00	1×10^{-12}
	30	4	12.00	1×10^{-12}
m-Silicic acid	RT[b]	1	9.70	2×10^{-10}
	RT	2	12.00	1×10^{-12}
Sulfuric acid	25	2	1.92	1.20×10^{-2}
Sulfurous acid	18	1	1.81	1.54×10^{-2}
	18	2	6.91	1.02×10^{-7}
Telluric acid	18	1	7.68	2.09×10^{-8}
	18	2	11.29	6.46×10^{-12}
Tellurous acid	25	1	2.48	3×10^{-3}
	25	2	7.70	2×10^{-8}
Tetraboric acid	25	1	4.00	$\sim 10^{-4}$
	25	1	9.00	$\sim 10^{-9}$

[a]Approximately 0.1–0.01N.
[b]RT = room temperature.

From Weast, R. C., Ed., *Handbook of Chemistry and Physics*, 55th ed., CRC Press, Cleveland, 1974, D-130.

Table 3–40
DISSOCIATION CONSTANTS (K_b) OF AQUEOUS AMMONIA FROM 0 TO 50°C[a]

°C	pK_b	K_b	°C	pK_b	K_b	°C	pK_b	K_b
0	4.862	1.374×10^{-5}	20	4.767	1.710×10^{-5}	35	4.733	1.849×10^{-5}
5	4.830	1.479×10^{-5}	25	4.751	1.774×10^{-5}	40	4.730	1.862×10^{-5}
10	4.804	1.570×10^{-5}	30	4.740	1.820×10^{-5}	45	4.726	1.879×10^{-5}
15	4.782	1.652×10^{-5}				50	4.723	1.892×10^{-5}

[a]Values of K_b were determined by the emf method and are accurate to ±0.005.

Reprinted with permission from Bates, R. G. and Pinching, G. D., *J. Am. Chem. Soc.*, 72, 1393, 1950. Copyright by the American Chemical Society.

Table 3–41
SPECIFIC HEAT AND ENTHALPY OF AIR-FREE WATER FROM
0 TO 100°C AT A PRESSURE OF 1 ATM[a]

°C	Heat capacity $cal_{IT}/g/°C$	$J/g/°C$	Enthalpy cal_{IT}/g	J/g	°C	Heat capacity $cal_{IT}/g/°C$	$J/g/°C$	Enthalpy cal_{IT}/g	J/g
0	1.00738	4.2177	0.0245	0.1026	45	0.99826	4.1795	45.0159	188.4726
1	1.00652	4.2141	1.0314	4.3184	46	0.99830	4.1797	46.0142	192.6522
2	1.00571	4.2107	2.0376	8.5308	47	0.99835	4.1799	47.0125	196.8320
3	1.00499	4.2077	3.0429	12.7400	48	0.99842	4.1802	48.0109	201.0120
4	1.00430	4.2048	4.0475	16.9462	49	0.99847	4.1804	49.0094	205.1923
5	1.00368	4.2022	5.0515	21.1498	50	0.99854	4.1807	50.0079	209.3729
6	1.00313	4.1999	6.0549	25.3508	51	0.99862	4.1810	51.0065	213.5538
7	1.00260	4.1977	7.0578	29.5496	52	0.99871	4.1814	52.0051	217.7350
8	1.00213	4.1957	8.0602	33.7463	53	0.99878	4.1817	53.0039	221.9166
9	1.00170	4.1939	9.0621	37.9410	54	0.99885	4.1820	54.0027	226.0984
10	1.00129	4.1922	10.0636	42.1341	55	0.99895	4.1824	55.0016	230.2806
11	1.00093	4.1907	11.0647	46.3255	56	0.99905	4.1828	56.0006	234.4632
12	1.00060	4.1893	12.0654	50.5155	57	0.99914	4.1832	56.9997	238.6462
13	1.00029	4.1880	13.0659	54.7041	58	0.99924	4.1836	57.9989	242.8296
14	1.00002	4.1869	14.0660	58.8916	59	0.99933	4.1840	58.9982	247.0134
15	0.99976	4.1858	15.0659	63.0779	60	0.99943	4.1844	59.9975	251.1976
16	0.99955	4.1849	16.0655	67.2632	61	0.99955	4.1849	60.9970	255.3822
17	0.99933	4.1840	17.0650	71.4476	62	0.99964	4.1853	61.9966	259.5673
18	0.99914	4.1832	18.0642	75.6312	63	0.99976	4.1858	62.9963	263.7529
19	0.99897	4.1825	19.0633	79.8141	64	0.99988	4.1863	63.9962	267.9390
20	0.99883	4.1819	20.0622	83.9963	65	1.00000	4.1868	64.9961	272.1256
21	0.99869	4.1813	21.0609	88.1778	66	1.00014	4.1874	65.9962	276.3127
22	0.99857	4.1808	22.0596	92.3589	67	1.00026	4.1879	66.9964	280.5003
23	0.99847	4.1804	23.0581	96.5395	68	1.00041	4.1885	67.9967	284.6885
24	0.99838	4.1800	24.0565	100.7196	69	1.00053	4.1890	68.9972	288.8772
25	0.99828	4.1796	25.0548	104.8994	70	1.00067	4.1896	69.9977	293.0665
26	0.99821	4.1793	26.0530	109.0788	71	1.00081	4.1902	70.9985	297.2564
27	0.99814	4.1790	27.0512	113.2580	72	1.00096	4.1908	71.9994	301.4469
28	0.99809	4.1788	28.0493	117.4369	73	1.00112	4.1915	73.0004	305.6381
29	0.99804	4.1786	29.0474	121.6157	74	1.00127	4.1921	74.0016	309.8299
30	0.99802	4.1785	30.0455	125.7943	75	1.00143	4.1928	75.0030	314.0224
31	0.99799	4.1784	31.0435	129.9727	76	1.00160	4.1935	76.0045	318.2155
32	0.99797	4.1783	32.0414	134.1510	77	1.00177	4.1942	77.0062	322.4094
33	0.99797	4.1783	33.0394	138.3293	78	1.00194	4.1949	78.0080	326.6039
34	0.99795	4.1782	34.0374	142.5076	79	1.00213	4.1957	79.0101	330.7992
35	0.99795	4.1782	35.0353	146.6858	80	1.00229	4.1964	80.0123	334.9952
36	0.99797	4.1783	36.0333	150.8641	81	1.00248	4.1972	81.0147	339.1920
37	0.99797	4.1783	37.0312	155.0423	82	1.00268	4.1980	82.0172	343.3897
38	0.99799	4.1784	38.0292	159.2207	83	1.00287	4.1988	83.0200	347.5881
39	0.99802	4.1785	39.0272	163.3991	84	1.00308	4.1997	84.0230	351.7873
40	0.99804	4.1786	40.0253	167.5777	85	1.00327	4.2005	85.0262	355.9874
41	0.99807	4.1787	41.0233	171.7563	86	1.00349	4.2014	86.0295	360.1883
42	0.99811	4.1789	42.0214	175.9351	87	1.00370	4.2023	87.0331	364.3902
43	0.99816	4.1791	43.0195	180.1141	88	1.00392	4.2032	88.0369	368.5929
44	0.99819	4.1792	44.0177	184.2933	89	1.00416	4.2042	89.0410	372.7966

Table 3–41 (continued)
SPECIFIC HEAT AND ENTHALPY OF AIR-FREE WATER FROM
0 TO 100°C AT A PRESSURE OF 1 ATM

°C	Heat capacity		Enthalpy		°C	Heat capacity		Enthalpy	
	$cal_{IT}/g/°C$	$J/g/°C$	cal_{IT}/g	J/g		$cal_{IT}/g/°C$	$J/g/°C$	cal_{IT}/g	J/g
90	1.00437	4.2051	90.0452	377.0012	95	1.00561	4.2103	95.0701	398.0395
91	1.00461	4.2061	91.0497	381.2068	96	1.00588	4.2114	96.0759	402.2503
92	1.00485	4.2071	92.0545	385.4135	97	1.00614	4.2125	97.0819	406.4622
93	1.00509	4.2081	93.0594	389.6211	98	1.00640	4.2136	98.0882	410.6753
94	1.00535	4.2092	94.0647	393.8297	99	1.00669	4.2148	99.0947	414.8895
					100	1.00697	4.2160	100.1015	419.1049

[a]Heat capacity and heat content are both given in International Table calories (cal_{IT}) per gram as well as in absolute joule (J) per gram. J = 0.238846 cal_{IT}.

From Osborne, Stimson and Ginnings, *J. Res. NBS*, 23, 238, 1939.

Table 3–42
MOLECULAR ELEVATION OF THE BOILING POINT[a]

Solvent	K_B	Barometric correction per millimeter	Solvent	K_B	Barometric correction per millimeter
Acetic acid	3.07	0.0008	Ethyl acetate	2.77	0.0007
Acetone	1.71	0.0004	Ethyl ether	2.02	0.0005
Aniline	3.52	0.0009	n-Hexane	2.75	0.0007
Benzene	2.53	0.0007	Methanol	0.83	0.0005
Bromobenzene	6.26	0.0016	Methyl acetate	2.15	0.0002
Carbon bisulfide	2.34	0.0006	Nitrobenzene	5.24	0.0013
Carbon tetrachloride	5.03	0.0013	n-Octane	4.02	0.0010
Chloroform	3.63	0.0009	Phenol	3.56	0.0009
Cyclohexane	2.79	0.0007	Toluene	3.33	0.0008
Ethanol	1.22	0.0003	Water	0.512	0.0001

[a]Elevation of the boiling point (K_B) due to the addition of 1 g molecular weight of solute to 1,000 g of solvent is given in °C. Barometric correction indicates the number of °C to be subtracted for each millimeter of difference between the barometric reading and 760 mm.

From Weast, R. C., Ed., *Handbook of Chemistry and Physics*, 55th ed., CRC Press, Cleveland, 1974, D-154.

Table 3–43
MOLECULAR DEPRESSION OF THE FREEZING POINT[a]

Solvent	Depression	Solvent	Depression
Acetic acid	39.0	Naphthalene	68–69
Benzene	49.0	Nitrobenzene	70.0
Benzophenone	98.0	Phenol	74.0
Diphenyl	80.0	Stearic acid	45.0
Diphenylamine	86.0	Triphenyl methane	124.5
Ethylene dibromide	118.0	Urethane	51.4
Formic acid	27.7	Water	18.5–18.7

[a]Depression of the freezing point due to the addition of 1 g molecular weight of solute to 100 g of solvent is given in °C.

From Weast, R. C., Ed., *Handbook of Chemistry and Physics,* 55th ed., CRC Press, Cleveland, 1974, D-154.

Table 3–44
LOWERING OF VAPOR PRESSURE BY SALTS IN AQUEOUS SOLUTION[a]

Substance	Gram molecules of salt								
	0.5	1.0	2.0	3.0	4.0	5.0	6.0	8.0	10.0
$Al_2(SO_4)_3$	12.8	36.5							
$AlCl_3$	22.5	61.0	179.0	318.0					
BaS_2O_6	6.6	15.4	34.4						
$Ba(OH)_2$	12.3	22.5	39.0						
$Ba(NO_3)_2$	13.5	27.0							
$Ba(ClO_3)_2$	15.8	33.3	70.5	108.2					
$BaCl_2$	16.4	36.7	77.6						
$BaBr_2$	16.8	38.8	91.4	150.0	204.7				
CaS_2O_3	9.9	23.0	56.0	106.0					
$Ca(NO_3)_2$	16.4	34.8	74.6	139.3	161.7	205.4			
$CaCl_2$	17.0	39.8	95.3	166.6	241.5	319.5			
$CaBr_2$	17.7	44.2	105.8	191.0	283.3	368.5			
$CdSO_4$	4.1	8.9	18.1						
CdI_2	7.6	14.8	33.5	52.7					
$CdBr_2$	8.6	17.8	36.7	55.7	80.0				
$CdCl_2$	9.6	18.8	36.7	57.0	77.3	99.0			
$Cd(NO_3)_2$	15.9	36.1	78.0	122.2					
$Cd(ClO_3)_2$	17.5								
$CoSO_4$	5.5	10.7	22.9	45.5					
$CoCl_2$	15.0	34.8	83.0	136.0	186.4				
$Co(NO_3)_2$	17.3	39.2	89.0	152.0	218.7	282.0	332.0		
$FeSO_4$	5.8	10.7	24.0	42.4					
H_3BO_3	6.0	12.3	25.1	38.0	51.0				
H_3PO_4	6.6	14.0	28.6	45.2	62.0	81.5	103.0	146.9	189.5
H_3AsO_4	7.3	15.0	30.2	46.4	64.9				
H_2SO_4	12.9	26.5	62.8	104.0	148.0	198.4	247.0	343.2	
KH_2PO_4	10.2	19.5	33.3	47.8	60.5	73.1	85.2		
NKO_3	10.3	21.1	40.1	57.6	74.5	88.2	102.1	126.3	148.0

Table 3–44 (continued)
LOWERING OF VAPOR PRESSURE BY SALTS IN AQUEOUS SOLUTION

	Gram molecules of salt								
Substance	0.5	1.0	2.0	3.0	4.0	5.0	6.0	8.0	10.0
$KClO_3$	10.6	21.6	42.8	62.1	80.0				
$KBrO_3$	10.9	22.4	45.0						
$KHSO_4$	10.9	21.9	43.3	65.3	85.5	107.8	129.2	170.0	
KNO_2	11.1	22.8	44.8	67.0	90.0	110.5	130.7	167.0	198.8
$KClO_4$	11.5	22.3							
KCl	12.2	24.4	48.8	74.1	100.9	128.5	152.2		
$KHCO_3$	11.6	23.6	59.0	77.6	104.2	132.0	160.0	210.0	255.0
KI	12.5	25.3	52.2	82.6	112.2	141.5	171.8	225.5	278.5
$K_2C_2O_4$	13.9	28.3	59.8	94.2	131.0				
K_2WO_4	13.9	33.0	75.0	123.8	175.4	226.4			
K_2CO_3	14.4	31.0	68.3	105.5	152.0	209.0	258.8	350.0	
KOH	15.0	29.5	64.0	99.2	140.0	181.8	223.0	309.5	387.8
K_2CrO_4	16.2	29.5	60.0						
$LiNO_3$	12.2	25.9	55.7	88.9	122.2	155.1	188.0	253.4	309.2
$LiCl$	12.1	25.5	57.1	95.0	132.5	175.5	219.5	311.5	393.5
$LiBr$	12.2	26.2	60.0	97.0	140.0	186.3	241.5	341.5	438.0
Li_2SO_4	13.3	28.1	56.8	89.0					
$LiHSO_4$	12.8	27.0	57.0	93.0	130.0	168.0			
LiI	13.6	28.6	64.7	105.2	154.5	206.0	264.0	357.0	445.0
Li_2SiF_6	15.4	34.0	70.0	106.0					
$LiOH$	15.9	37.4	78.1						
Li_2CrO_4	16.4	32.6	74.0	120.0	171.0				
$MgSO_4$	6.5	12.0	24.5	47.5					
$MgCl_2$	16.8	39.0	100.5	183.3	277.0	377.0			
$Mg(NO_3)_2$	17.6	42.0	101.0	174.8					
$MgBr_2$	17.9	44.0	115.8	205.3	298.5				
$MgH_2(SO_4)_2$	18.3	46.0	116.0						
$MnSO_4$	6.0	10.5	21.0						
$MnCl_2$	15.0	34.0	76.0	122.3	167.0	209.0			
NaH_2PO_4	10.5	20.0	36.5	51.7	66.8	82.0	96.5	126.7	157.1
$NaHSO_4$	10.9	22.1	47.3	75.0	100.2	126.1	148.5	189.7	231.4
$NaNO_3$	10.6	22.5	46.2	68.1	90.3	111.5	131.7	167.8	198.8
$NaClO_3$	10.5	23.0	48.4	73.5	98.5	123.3	147.5	196.5	223.5
$(NaPO_3)_6$	11.6								
$NaOH$	11.8	22.8	48.2	77.3	107.5	139.1	172.5	243.3	314.0
$NaNO_2$	11.6	24.4	50.0	75.0	98.2	122.5	146.5	189.0	226.2
Na_2HPO_4	12.1	23.5	43.0	60.0	78.7	99.8	122.1		
$NaHCO_3$	12.9	24.1	48.2	77.6	102.2	127.8	152.0	198.0	239.4
Na_2SO_4	12.6	25.0	48.9	74.2					
$NaCl$	12.3	25.2	52.1	80.0	111.0	143.0	176.5		
$NaBrO_3$	12.1	25.0	54.1	81.3	108.8	136.0			
$NaBr$	12.6	25.9	57.0	89.2	124.2	159.5	197.5	268.0	
NaI	12.1	25.6	60.2	99.5	136.7	177.5	221.0	301.5	370.0
$Na_4P_2O_7$	13.2	22.0							
Na_2CO_3	14.3	27.3	53.5	80.2	111.0				
$Na_2C_2O_4$	14.5	30.0	65.8	105.8	146.0				

Table 3—44 (continued)
LOWERING OF VAPOR PRESSURE BY SALTS IN AQUEOUS SOLUTION

Substance	Gram molecules by salt								
	0.5	1.0	2.0	3.0	4.0	5.0	6.0	8.0	10.0
Na_2WO_4	14.8	33.6	71.6	115.7	162.6				
Na_3PO_4	16.5	30.0	52.5						
$(NaPO_3)_3$	17.1	36.5							
NH_4NO_3	12.8	22.0	42.1	62.7	82.9	103.8	121.0	152.2	180.0
$(NH_4)_2SiF_6$	11.5	25.0	44.5						
NH_4Cl	12.0	23.7	45.1	69.3	94.2	118.5	138.2	179.0	213.8
NH_4HSO_4	11.5	22.0	46.8	71.0	94.5	118	139.0	181.2	218.0
$(NH_4)_2SO_4$	11.0	24.0	46.5	69.5	93.0	117.0	141.8		
NH_4Br	11.9	23.9	48.8	74.1	99.4	121.5	145.5	190.2	228.5
NH_4I	12.9	25.1	49.8	78.5	104.5	132.3	156.0	200.0	243.5
$NiSO_4$	5.0	10.2	21.5						
$NiCl_2$	16.1	37.0	86.7	147.0	212.8				
$Ni(NO_3)_2$	16.1	37.3	91.3	156.2	235.0				
$Pb(NO_3)_2$	12.3	23.5	45.0	63.0					
$Sr(SO_3)_2$	7.2	20.3	47.0						
$Sr(NO_3)_2$	15.8	31.0	64.0	97.4	131.4				
$SrCl_2$	16.8	38.8	91.4	156.3	223.3	281.5			
$SrBr_2$	17.8	42.0	101.1	179.0	267.0				
$ZnSO_4$	4.9	10.4	21.5	42.1	66.2				
$ZnCl_2$	9.2	18.7	46.2	75.0	107.0	153.0	195.0		
$Zn(NO_3)_2$	16.6	39.0	93.5	157.5	223.8				

[a]The reduction of vapor pressure in relation to the salt content of water is expressed in millimeters; salt content is expressed by the number of gram molecules per liter of pure water at a temperature of 100°C and a pressure of 760 mm.

From Forsythe, W. E., Ed., *Smithsonian Physical Tables*, 9th ed., Smithsonian Institution, Washington, D.C., 1956, 373.

Table 3—45
EFFICIENCY OF DRYING AGENTS

A. Drying Agents Depending on Chemical Action (Absorption)[a]

Substance	Formula	Residual water, milligrams per litter of dry air[b]	Reference
Phosphorus pentoxide	P_2O_5	<1 mg in 40,000 l	1
Magnesium perchlorate, anhydrous	$Mg(ClO_4)_2$	Unweighable in 210 l	2
Barium oxide	BaO	0.00065	3
Potassium hydroxide, fused	KOH	0.002	4
Calcium oxide	CaO	0.003	3
Sulfuric acid	H_2SO_4	0.003	4
Calcium sulfate, anhydrous	$CaSO_4$	0.005	3
Aluminum oxide	Al_2O_3	0.005	3
Potassium hydroxide, sticks	KOH	0.014	3
Sodium hydroxide, fused	NaOH	0.16	4
Calcium bromide	$CaBr_2$	0.18	5
Calcium chloride, fused	$CaCl_2$	0.34	4
Sodium hydroxide, sticks	NaOH	0.80	3
Barium perchlorate	$Ba(ClO_4)_2$	0.82	3
Zinc chloride	$ZnCl_2$	0.85	5
Zinc bromide	$ZnBr_2$	1.16	5
Calcium chloride, granular	$CaCl_2$	1.5	3
Cupric sulfate, anhydrous	$CuSO_4$	2.8	3

B. Drying Agents Depending on Physical Action (Adsorption)[a]

Alumina, low-temperature-fired	Clay and porcelain, low-temperature-fired
Asbestos	Kieselguhr
Charcoal	Silica gel
Glass wool	Refrigeration

[a]It should be noted that the efficiency of some drying agents depends on both absorption and adsorption, e.g., aluminum oxide and calcium chloride, and probably also barium oxide, magnesium perchlorate, barium perchlorate, and calcium sulfate.

[b]At 25°C or room temperature, except for the values determined in Reference 3, which were taken at 30°C.

From Yoe, J. H., in *Handbook of Chemistry and Physics,* 55th ed., Weast, R. C., Ed., CRC Press, Cleveland, 174, E-41.

REFERENCES

1. **Morley,** *Am. J. Sci.,* 34, 199, 1887; *J. Am. Chem. Soc.,* 26, 1171, 1904.
2. **Willard and Smith,** *J. Am. Chem. Soc.,* 44, 2255, 1922.
3. **Bower,** *Bur. Stand. J. Res.,* 12, 241, 1934.
4. **Baxter and Starkweather,** *J. Am. Chem. Soc.,* 38, 2038, 1916.
5. **Baxter and Warren,** *J. Am. Chem. Soc.,* 33, 340, 1911.

Table 3—46
CONSTANT HUMIDITY

The following table shows the percent humidity and the aqueous tension at the given temperature within a closed space when an excess of the substance indicated is in contact with a saturated aqueous solution of the given solid phase.

Solid Phase	°C	Percent Humidity	Aqueous Tension, mm Hg	Solid Phase	°C	Percent Humidity	Aqueous Tension, mm Hg
$H_3PO_4 \cdot \frac{1}{2}H_2O$	24	9	1.99	NH_4Cl and KNO_3	20	72.6	12.6
$KC_2H_3O_2$	168	13	738	$NaClO_3$	20	75	13.0
$LiCl \cdot H_2O$	20	15	2.60	$(NH_4)_2SO_4$	108	75	754
$KC_2H_3O_2$	20	20	3.47	$NaC_2H_3O_2 \cdot 3H_2O$	20	76	13.2
KF	100	22.9	174	$H_2C_2O_4 \cdot 2H_2O$	20	76	13.2
$NaBr$	100	22.9	174	NH_4Cl	30	77.5	24.4
$NaCl$, KNO_3 and $NaNO_3$	16.39	30.49	4.23	$Na_2S_2O_3 \cdot 5H_2O$	20	78	13.5
$CaCl_2 \cdot 6H_2O$	24.5	31	7.08	NH_4Cl	25	79.3	18.6
$CaCl_2 \cdot 6H_2O$	20	32.3	5.61	NH_4Cl	20	79.5	13.8
$CaCl_2 \cdot 6H_2O$	18.5	35	5.54	$(NH_4)_2SO_4$	20	81	14.1
CrO_3	20	35	6.08	$(NH_4)_2SO_4$	25	81.1	19.1
$CaCl_2 \cdot 6H_2O$	10	38	3.47	$(NH_4)_2SO_4$	30	81.1	25.6
$CaCl_2 \cdot 6H_2O$	5	39.8	2.59	KBr	20	84	14.6
$Zn(NO_3)_2 \cdot 6H_2O$	20	42	7.29	Tl_2SO_4	104.7	84.8	768
$K_2CO_3 \cdot 2H_2O$	24.5	43	9.82	$KHSO_2$	20	86	14.9
$K_2CO_3 \cdot 2H_2O$	18.5	44	6.96	$Na_2CO_3 \cdot 10H_2O$	24.5	87	20.9
KNO_2	20	45	7.81	$BaCl_2 \cdot 2H_2O$	24.5	88	20.1
$KCNS$	20	47	8.16	K_2CrO_4	20	88	15.3
NaI	100	50.4	383	$Pb(NO_3)_2$	103.5	88.4	760
$Ca(NO_3)_2 \cdot 4H_2O$	24.5	51	11.6	$ZnSO_4 \cdot 7H_2O$	20	90	15.6
$NaHSO_4 \cdot H_2O$	20	52	9.03	$Na_2CO_3 \cdot 10H_2O$	18.5	92	14.6
$Na_2Cr_2O_7 \cdot 2H_2O$	20	52	9.03	$NaBrO_3$	20	92	16.0
$Mg(NO_3)_2 \cdot 6H_2O$	24.5	52	11.9	K_2HPO_4	20	92	16.0
$NaClO_3$	100	54	410	$NH_4H_2PO_4$	30	92.9	29.3
$Ca(NO_3)_2 \cdot 4H_2O$	18.5	56	8.86	$NH_4H_2PO_4$	25	93	21.9
$Mg(NO_3)_2 \cdot 6H_2O$	18.5	56	8.86	$Na_2SO_4 \cdot 10H_2O$	20	93	16.1
KI	100	56.2	427	$NH_4H_2PO_4$	20	93.1	16.2
$NaBr \cdot 2H_2O$	20	58	10.1	$ZnSO_4 \cdot 7H_2O$	5	94.7	6.10
$Mg(C_2H_3O_2)_2 \cdot 4H_2O$	20	65	11.3	$Na_2SO_3 \cdot 7H_2O$	20	95	16.5
$NaNO_2$	20	66	11.5	$Na_2HPO_4 \cdot 12H_2O$	20	95	16.5
NH_4Cl and KNO_3	30	68.6	21.6	NaF	100	96.6	734
KBr	100	69.2	526	$Pb(NO_3)_2$	20	98	17.0
NH_4Cl and KNO_3	25	71.2	16.7	$CuSO_4 \cdot 5H_2O$	20	98	17.0
$TlCl$	100.1	99.7	761	$TlNO_3$	100.3	98.7	759

From Weast, R. C., Ed., *Handbook of Chemistry and Physics*, 55th ed., CRC Press, Cleveland, 1974, E-46.

Table 3–47
VELOCITY OF SOUND

A. Solids

Substance	Density (g/cm^3)	$V_l{}^a$ (m/sec)	$V_s{}^b$ (m/sec)	$V_{ext}{}^c$ (m/sec)
Metals				
Aluminum, rolled	2.7	6,420	3,040	5,000
Beryllium	1.87	12,890	8,880	12,870
Brass (70% copper, 30% zinc)	8.6	4,700	2,110	3,480
Copper, annealed	8.93	4,760	2,325	3,810
Copper, rolled	8.93	5,010	2,270	3,750
Duraluminum 17S	2.79	6,320	3,130	5,150
Gold, hard-drawn	19.7	3,240	1,200	2,030
Iron, Armco	7.85	5,960	3,240	5,200
Iron, electrolytic	7.9	5,950	3,240	5,120
Lead, annealed	11.4	2,160	700	1,190
Lead, rolled	11.4	1,960	690	1,210
Magnesium, drawn, annealed	1.74	5,770	3,050	4,940
Molybdenum	10.1	6,250	3,350	5,400
Monel metal	8.90	5,350	2,720	4,400
Nickel	8.9	6,040	3,000	4,900
Nickel, unmagnetized	8.85	5,480	2,990	4,800
Platinum	21.4	3,260	1,730	2,800
Silver	10.4	3,650	1,610	2,680
Steel (1% carbon)	7.84	5,940	3,220	5,180
Steel (1% carbon), hardened	7.84	5,854	3,150	5,070
Steel, mild	7.85	5,960	3,235	5,200
Steel, stainless 347	7.9	5,790	3,100	5,000
Tin, rolled	7.3	3,320	1,670	2,730
Titanium	4.5	6,070	3,125	5,080
Tungsten, annealed	19.3	5,220	2,890	4,620
Tungsten, drawn	19.3	5,410	2,640	4,320
Tungsten carbide	13.8	6,655	3,980	6,220
Zinc, rolled	7.1	4,210	2,440	3,850
Woods				
Ash, across the rings				1,390
Ash, along the fiber				4,670
Ash, along the rings				1,260
Beech, along the fiber				3,340
Elm, along the fiber				4,120
Maple, along the fiber				4,110
Oak, along the fiber				3,850
Various Materials				
Brick	1.8			3,650
Clay rock	2.2			3,480
Cork	0.25			500
Fused silica	2.2	5,869	3,764	5,760

Table 3–47 (continued)
VELOCITY OF SOUND

A. Solids
Various Materials (continued)

Substance	Density (g/cm^3)	$V_l{}^a$ (m/sec)	$V_s{}^b$ (m/sec)	$V_{ext}{}^c$ (m/sec)
Glass, heavy silicate flint	3.88	3,980	2,380	3,720
Glass, light borate crown	2.24	5,100	2,840	4,540
Glass, pyrex	2.32	5,640	3,280	5,170
Lucite	1.18	2,680	1,100	1,840
Marble	2.6			3,810
Nylon 6-6	1.11	2,620	1,070	1,800
Paraffin	0.9			1,300
Polyethylene	0.90	1,950	540	920
Polystyrene	1.06·	2,350	1,120	2,240
Rubber, butyl	1.07	1,830		
Rubber, gum	0.95	1,550		
Rubber, neoprene	1.33	1,600		
Tallow	0.9			1,300

$^a V_l$ = velocity of plane longitudinal wave in bulk material.
$^b V_s$ = velocity of plane transverse (shear) wave.
$^c V_{ext}$ = velocity of longitudinal (extensional) wave in thin rods.

B. Liquids

Substance	Formula	Density (g/cm^3)	Velocity at 25°C (m/sec)	$-\Delta v/\Delta t$ (m/sec °C)
Acetone	C_3H_6O	0.79	1,174	4.5
Benzene	C_6H_6	0.870	1,295	4.65
Carbon disulphide	CS_2	1.26	1,149	
Carbon tetrachloride	CCl_4	1.595	926	2.7
Castor oil	$C_{11}H_{10}O_{10}$	0.969	1,477	3.6
Chloroform	$CHCl_3$	1.49	987	3.4
Ethanol	C_2H_6O	0.79	1,207	4.0
Ethanol amide	C_2H_7NO	1.018	1,724	3.4
Ethyl ether	$C_4H_{10}O$	0.713	985	4.87
Ethylene glycol	$C_2H_6O_2$	1.113	1,658	2.1
Glycerol	$C_3H_6O_3$	1.26	1,904	2.2
Kerosene		0.81	1,324	3.6
Mercury	Hg	13.5	1,450	
Methanol	CH_4O	0.791	1,103	3.2
Nitrobenzene	$C_6H_5NO_2$	1.20	1,463	3.6
Turpentine		0.88	1,255	
Water, distilled	H_2O	0.998	1,496.7 ± 0.2	−2.4
Water, sea		1.025	1,531	−2.4
Xylene hexafluoride	$C_8H_4F_6$	1.37	879	

Table 3–47 (continued)
VELOCITY OF SOUND

C. Gases and Vapors

Substance	Formula	Density (g/liter)	Velocity (m/sec)	$\Delta v/\Delta t$ (m/sec °C)
Gases (0° C)				
Air, dry		1.293	331.45	0.59
Ammonia	NH_3	0.771	415	
Argon	Ar	1.783	319	0.56
Carbon dioxide	CO_2	1.977	259	0.4
Carbon monoxide	CO	1.25	338	0.6
Chlorine	Cl_2	3.214	206	
Coal gas, illuminating			453	
Deuterium	D_2		890	1.6
Ethane (10°C)	C_2H_6	1.356	308	
Ethylene	C_2H_4	1.260	317	
Helium	He	0.178	965	0.8
Hydrogen	H_2	0.0899	1,284	2.2
Hydrogen bromide	HBr	3.50	200	
Hydrogen chloride	HCl	1.639	296	
Hydrogen iodide	HI	5.66	157	
Hydrogen sulfide	H_2S	1.539	289	
Methane	CH_4	0.7168	430	
Neon	Ne	0.900	435	0.8
Nitric oxide (10°C)	NO	1.34	324	
Nitrogen	N_2	1.251	334	0.6
Nitrous oxide	N_2O	1.977	263	0.5
Oxygen	O_2	1.429	316	0.56
Sulfur dioxide	SO_2	2.927	213	0.47
Vapors (97.1°C)				
Acetone	C_3H_6O		239	0.32
Benzene	C_6H_6		202	0.3
Carbon tetrachloride	CCl_4		145	
Chloroform	$CHCl_3$		171	0.24
Ethanol	C_2H_6O		269	0.4
Ethyl ether	$C_4H_{10}O$		206	0.3
Methanol	CH_4O		335	0.46
Water (134°C)	H_2O		494	

From Becker, G. E., in *Handbook of Chemistry and Physics*, 55th ed., Weast, R. C., Ed., CRC Press, Cleveland, 1974, E-47.

Table 3-48
SOUND VELOCITY IN WATER ABOVE 212°F

Temperature, °F	Velocity, m/sec	Velocity, ft/sec
186.8[a]	1552	5092
200	1548	5079
210	1544	5066
220	1538	5046
230	1532	5026
240	1524	5000
250	1516	4974
260	1507	4944
270	1497	4911
280	1487	4879
290	1476	4843
300	1465	4806
310	1453	4767
320	1440	4724
330	1426	4678
340	1412	4633
350	1398	4587
360	1383	4537
370	1368	4488
380	1353	4439
390	1337	4386
400	1320	4331
410	1302	4272
420	1283	4209
430	1264	4147
440	1244	4081
450	1220	4010
460	1200	3940
470	1180	3880
480	1160	3800
490	1140	3730
500	1110	3650
510	1090	3570
520	1070	3500
530	1040	3410
540	1010	3320
550	980	3230

[a]Data at 186.8°F were taken from Randall, C. R., *NBS J. Res.*, Vol. 8, January 1932.

From McDade, J. C., Pardue, D. R., Gedrich, A. L., and Vrataric, F., *J. Acoust. Soc. Am.*, 31(10), 1381, 1959. With permission.

Section 4
Conversion Tables, Miscellaneous Materials Properties, and Binary Phase Information

Table 4–1

UNITS AND THEIR CONVERSION

Policy

In each table in this handbook, the numerical values are preferably expressed in those units most commonly used by U.S. engineers and scientists working in the specific field, but sometimes SI metric units have also been added. In some cases two tables are given, one in English units, one in metric. In other tables parallel columns showing figures in both units are used, or the conversion factors are listed.

In a general materials handbook complete consistency in units, abbreviations, and symbols is hardly possible, or even desirable. Such consistency would quickly defeat the objective of providing quick access to numbers of maximum immediate usefulness. Within each special field of engineering and science, the technical societies and industry associations have developed certain uniform practices and standards; if tables and data are given only in units that are foreign to these prevailing standards, convenience is sacrificed.

The present edition of this handbook reflects the changes in abbreviations, symbols, and forms that are resulting from the efforts to reduce the diversity of practices from one specialty to another and from one nation to another. Recommendations of the International Organization for Standardization (ISO-R 1000) and of the "Metric Practice Guide," adopted by ASTM, NBS, APL, and others, have focused attention on the diversity of so-called standards.

Since the United States is the only major industrial nation that has not yet converted to metric units, some legal requirements in that direction are to be expected. It is now a contradiction to speak of the "English" system of units, and for. some time to come U.S. engineers and scientists must accommodate to a wide use of conversions from one set of units to another. The extensive conversion tables that follow are offered with this expectation.

In spite of major efforts to unify engineering practices, there are many good reasons for retaining several means of expressing a physical quantity. For ease of learning and communication a descriptive name is better than one arbitrarily assigned, such as Hz for cps, celsius for centigrade,

and torr for mm Hg; an opposite trend is prevalent at this time. Numerical scales directly related to the physical phenomena and to the method of their measurement have an advantage in the laboratory or field and will not soon be abandoned. Examples are barometric pressure in millimeters or inches of mercury, viscosity in seconds Saybolt, the calorie or the Btu, and even the "coefficients" of expansion, friction, diffusion, attenuation, and reflection. Symbols, abbreviations, and even the units themselves are not infrequently subject to change; note, for example, the new preferred *dB* in place of the well-established *db*; elimination of widely used abbreviations, such as kwh, cps, gpm, cc, and psi; and revised values for the second, the calorie, or the atomic weights. Users of this handbook are invited to call attention to places where consistency could be improved without sacrificing the objectives.

For units that might be assigned more than one value, we prefer the thermochemical gram-calorie (4.184 J), the thermochemical Btu (1054.35 J), the avoirdupois pound and ounce, the statute mile (5,280 ft), the short ton (2,000 lb), the U.S. liquid gallon (231 in.^{3}), and the electrical horsepower (746 W).

Both a special condensed version of conversion factors for quick reference and a more extended table from the *Handbook of Chemistry and Physics* have been included. Certain specialized conversion factors and tables have also been included. This results in some redundancy, but the aim is to provide ease and utility in making unit conversions.

The Metric International System (SI)

Moves toward an international system of metric units are now following each other in quick succession, so a table of conversion factors for the most common units is given herewith (see also Table 4–3). Perhaps the most definite are the moves toward the SI standards already initiated by the National Bureau of Standards, the various military services, the National Aeronautics and Space Administration, and other U.S. Government research groups. The American Society for Testing and Materials has declared in favor of SI units and

Table 4—1 (continued)

UNITS AND THEIR CONVERSION

will give other units only a secondary place in all newly issued ASTM Standards.[a] Other major engineering societies have committees to explore the adoption of SI units and are holding many meetings for discussion among members.

Whatever the decisions about converting to the metric system, the actual process will require many years, as can readily be seen from the experiences of other countries; in Great Britain, for example, even the single conversion to decimal monetary units and coinage moves very slowly. The practices and standards among the metric system countries are far from uniform; no real international system exists among them.

Mere conversion of present U.S. specifications, drawings, tools, machines, and stock sizes to equivalent metric units (so-called "soft" conversion) will not in any sense result in an "international" system. Instead, a "hard" conversion representing the abandonment of the 1/2-fractional system in favor of a 1/10-fractional system is necessary to attain the real advantages of the metric system. This means resizing of all round and sheet stock, lumber, bolts, screws, nails, wires, gears, containers, modules, and subassemblies, plus all the tools and machines related thereto. A long period of double stocking must follow. The entire change is made the more difficult by the great penetration of U.S. products and materials into the markets of the world, e.g., airplanes and military equipment, production, and construction machinery. This is not to mention the problem of the individual engineer, technician, and user, who visualizes all his size relationships in inches and feet and his weights in pounds. Realistically, more than one generation will be required for the educational conversion alone.

In presenting data in international standard metric units throughout this edition, the practices and forms used in the "Metric Practice Guide" have been carefully followed.[a] Certain conventions used in the "Metric Practice Guide" are not consistent with those originally adopted for this handbook, nor with ANSI standards. Special attention is directed to the following conventions:

1. For degrees Kelvin the degree symbol is omitted; for example, 50 K, not 50°K.

2. For multiplication a center point is used; for example, the unit of dynamic viscosity is abbreviated as $N \cdot s/m^2$, not $N\ s/m^2$ or $N \times s/m^2$.

3. Symbols for SI units are not capitalized unless the unit is derived from a proper name, as N for Sir Isaac Newton; however, *unabbreviated* units are not capitalized, such as newton, kelvin, hertz.

Table 4–1 (continued)
UNITS AND THEIR CONVERSION

Conversion Factors to SI Standard Units[b]

To convert	To	Multiply by	To convert	To	Multiply by
Acceleration			Power		
feet/second^2	meters/second^2	0.3048	Btu/second	watt	1,054.350
Area			foot-pounds/second	watt	1.355818
square feet	square meters	0.09290304	horsepower	watt	746.
Energy			Pressure		
Btu (mean)	joule	1,055.87	atmosphere	newtons/meter^2	101,325.0
calorie (mean)	joule	4.19002	bar	newtons/meter^2	100,000.
electron volt	joule	1.60210×10^{-19}	kilograms/centimeter^2	newtons/meter^2	98,066.50
foot-pound	joule	1.355818	pounds/inch^2	newtons/meter^2	6,894.757
watthour	joule	3,600.	torr (mm Hg, 0°C)	newtons/meter^2	133.322
Force			Viscosity		
dyne	newton	0.00001	centipoise	newton-second/meter^2	0.001
kilogram	newton	9.80665	pounds/foot second	newton-second/meter^2	1.488164
pound	newton	4.448222	Volume		
Length			cubic foot	cubic meter	0.02831685
foot	meter	0.3048000	gallon (U.S. liquid)	cubic meter	0.003785412
mil	meter	0.0000254			
mile (U.S. statute)	meter	1,609.344			
Mass					
pound	kilogram	0.4535924			
slug	kilogram	14.59390			
ton (2,000 lb)	kilogram	907.1847			

[a]See *Metric Practice Guide*, ASTM Standard E 380-70, American Society for Testing and Materials, 1970.

[b]For more complete conversions to SI units, see Tables 4–3 and 4–4.

From Bolz, R. E. and Tuve, G. L., Eds., *Handbook of Tables for Applied Engineering Science*, 2nd ed., CRC Press, Cleveland, 1973, 803.

Table 4–2A

ABBREVIATED COMPARISON OF INTERNATIONAL SYSTEM OF UNITS (SI) WITH PRE-SI SYSTEM OF METRIC UNITS

Physical quantity	MKS nomenclature (SI)			cgs nomenclature (pre SI)		
	Symbol	Equivalents in basic MKS units	Name	Symbol	Equivalents in basic cgs units	Name
BASIC UNITS						
Length	m		meter	cm		centimeter
Mass	kg		kilogram	g		gram
Time	s		second	sec		second
Temperature	K		kelvin	°F		degrees fahrenheit
				°C		degrees centigrade (Celsius)
Angle						
-plane (2D)	rad		radian	° ′ ″		degrees, minutes, seconds
-solid (3D)	sr		steradian			
Luminous intensity	cd		candela			
Electric current	A		ampere			cf. Table 4–2B
DERIVED UNITS						
Length	Å	0.1 nm	angstrom	Å	10^{-8} cm	angstrom
Frequency	Hz	1/s	hertz	cps	1/sec	cycle per second
Velocity		m/s	meter per second		cm/sec	centimeter per second
Acceleration		m/s^2	meter per sq second		cm/sec^2	centimeter per sq second
Density		kg/m^3	kilogram per cubic meter		g/cm^3	gram per cubic centimeter
Momentum		kg·m/s	kilogram meter per second		g·cm/sec	gram centimeter per second
Force	N	$kg \cdot m/s^2$	newton	dyn	$g \cdot cm/sec^2$	dyne
Pressure	N/m^2	$kg/m \cdot s^2$	newton per sq meter	atm	$\sim 10^6$ dyn/cm²	atmosphere
Dynamic viscosity	N·s/m²	kg/m·s	newton second per sq meter	poise	g/cm·sec	poise
Energy (work, heat)	J, (N·m)	$kg \cdot m^2/s^2$	joule or newton meter	erg	$g \cdot cm^2/sec^2$	erg or stat voltcoulomb
Power	W, (J/s)	$kg \cdot m^2/s^3$	watt or joule per second	W (VA)	10^7 erg/sec	watt or 10^7 stat voltampere
Potential	V, (W/A)	$kg \cdot m^2/A \cdot s^3$	volt or watt per ampere			
Charge	C, (J/V)	A·s	coulomb or joule per volt			
Resistance	Ω, (V/A)	$kg \cdot m^2/A^2 \cdot s^3$	ohm or volt per ampere			
Capacitance	F, (C/V)	$A^2 \cdot s^4/kg \cdot m^2$	farad or coulomb per volt			
Magnetic flux	Wb, (V·s)	$kg \cdot m^2/A \cdot s^2$	weber or volt second			
Inductance	H, (Wb/A)	$kg \cdot m^2/A^2 \cdot s^2$	henry or weber per ampere			
Luminous flux	lm	cd·sr	lumen			
Luminous flux density	lx	cd·sr/m²	lux or illumination			

Table compiled by A. Pigeaud, University of Cincinnati.

Table 4—2B
CONVERSION FROM MKS TO cgs

MKS	cgs	
	esu (electrostatic units)	emu (electromagnetic units)
1 coulomb	$= 3 \times 10^9$ statcoulomb	$= 10^{-1}$ abcoulomb
1 ampere	$= 3 \times 10^9$ statampere	$= 10^{-1}$ abampere
1 volt	$= .333 \times 10^{-2}$ statvolt	$= 10^8$ abvolt
1 joule (V·C)	$= 10^7$ erg	$= 10^7$ erg

Table compiled by A. Pigeaud, University of Cincinnati.

Table 4-3
INTERNATIONAL SYSTEM (SI) METRIC UNITS

Basic Units — MKS

Property		Abbreviation
Length	meter	m
Mass	kilogram	kg
Time	second	s
Electric current	ampere	A
Thermodynamic temperature	kelvin	K
Luminous intensity	candela	cd

Derived Units

Property	Units[a]	Abbreviations and dimensions	
Acceleration	Meter per second squared	m/s^2	
Activity (of radioactive source)	1 per second	s^{-1}	
Angular acceleration	radian per second squared	rad/s^{-1}	
Angular velocity	radian per second	rad/s	
Area	square meter	m^2	
Density	kilogram per cubic meter	kg/m^3	
Dynamic viscosity	newton-second per square meter	$N \cdot s/m^2$	
Electric capacitance	farad	F	$(A \cdot s/V)$
Electric charge	coulomb	C	$(A \cdot s)$
Electric field strength	volt per meter	V/m	
Electric resistance	ohm		(V/A)
Entropy	joule per kelvin	J/K	
Force	newton	N	$(kg \cdot m/s^2)$
Frequency	hertz	hz	(s^{-1})
Illumination	lux	lx	(lm/m^2)
Inductance	henry	H	$(V \cdot s/A)$
Kinematic viscosity	square meter per second	m^2/s	
Luminance	candela per square meter	cd/m^2	
Luminous flux	lumen	lm	$(cd \cdot sr)$
Magnetomotive force	ampere	A	
Magnetic field strength	ampere per meter	A/m	
Magnetic flux	weber	Wb	$(V \cdot s)$
Magnetic flux density	tesla	T	(Wb/m^2)
Power	watt	W	(J/s)
Pressure	newton per square meter	N/m^2	

Table 4–3
INTERNATIONAL SYSTEM (SI) METRIC UNITS (continued)

Derived Units (continued)

Property	Units[a]	Abbreviations and dimensions	
Radiant intensity	watt per steradian	W/sr	
Specific heat	joule per kilogram kelvin	J/kg K	
Thermal conductivity	watt per meter kelvin	W/m K	
Velocity	meter per second	m/s	
Volume	cubic meter	m^3	
Voltage, potential difference, electromotive force	volt	V	(W/A)
Wave number	1 per meter	m^{-1}	
Work, energy, quantity of heat	joule	J	(N·m)

Prefix Names of Multiples and Submultiples of Units

Decimal equivalent	Prefix	Pronunciation	Symbol	Exponential expression
1,000,000,000,000	tera	ter'ȧ	T	10^{+12}
1,000,000,000	giga	ji'gȧ	G	10^{+9}
1,000,000	mega	meg'ȧ	M	10^{+6}
1,000	kilo	kil'ō	k	10^{+3}
100	hecto	hek'tō	h	10^{+2}
10	deka	dek'ȧ	da	10
0.1	deci	des'i	d	10^{-1}
0.01	centi	sen'ti	c	10^{-2}
0.001	milli	mil'i	m	10^{-3}
0.000001	micro	mī'krō	μ	10^{-6}
0.000000001	nano	nan'ō	n	10^{-9}
0.000000000001	pico	pē'kō	p	10^{-12}
0.000000000000001	femto	fem'tō	f	10^{-15}
0.000000000000000001	atto	at'tō	a	10^{-18}

Table 4–3 (continued)

INTERNATIONAL SYSTEM (SI) METRIC UNITS

Definitions of the Most Important International System (SI) Units

The *ampere* (unit of electric current) is the constant current that, if maintained in two straight parallel conductors of infinite length, of negligible circular sections, and placed 1 m apart in a vacuum, will produce between these conductors a force equal to 2×10^{-7} N per meter of length.

The *candela* is the luminous intensity, in the direction of the normal, of a black body surface $1/600,000$ m^2 in area, at the temperature of solidification of platinum under a pressure of 101,325 N per square meter.

The *coulomb* (unit of quantity of electricity) is the quantity of electricity transported in 1 s by a current of 1 A.

The *ephemeris second* (unit of time) is exactly $1/31,556,925.9747$ of the tropical year of 1900, January, 0 days, and 12 hr ephemeris time.

The *farad* (unit of electric capacitance) is the capacitance of a capacitor between the plates of which there appears a difference of potential of 1 V when it is charged by a quantity of electricity equal to 1 C.

The *henry* (unit of electric inductance) is the inductance of a closed circuit in which an electromotive force of 1 V is produced when the electric current in the circuit varies uniformly at a rate of 1 A per second.

The *International Practical Kelvin Temperature Scale* of 1960 and the *International Practical Celsius Temperature Scale* of 1960 are defined by a set of interpolation equations based on the following reference temperatures:

	K	deg C
Oxygen, liquid-gas equilibrium	90.18	−182.97
Water, solid-liquid equilibrium	273.15	0.00
Water, solid-liquid-gas equilibrium	273.16	0.01
Water, liquid-gas equilibrium	373.15	100.00
Zinc, solid-liquid equilibrium	692.655	419.505
Sulfur, liquid-gas equilibrium	717.75	444.6
Silver, solid-liquid equilibrium	1,233.95	960.8
Gold, solid-liquid equilibrium	1,336.15	1,063.0

The *joule* (unit of energy) is the work done when the point of application of 1 N is displaced a distance of 1 m in the direction of the force.

The *kelvin* (unit of thermodynamic temperature) is the fraction $1/273.16$ of the thermodynamic temperature of the triple point of water. The decision was made at the 13th General Conference on Weights and Measures on October 13, 1967, that the name of the unit of thermodynamic temperature would be changed from *degree Kelvin* (symbol: °K) to *kelvin* (symbol: K). The name (*kelvin*) and symbol (K) are to be used for expressing temperature intervals. The former convention that expressed a temperature interval in *degrees Kelvin* or, abbreviated, *deg K*, is dropped. However, the old designations are acceptable temporarily as alternatives to the new ones. One may also express temperature intervals in *degrees Celsius.*

The *kilogram* (unit of mass) is the mass of a particular cylinder of platinum iridium alloy, called the International Prototype Kilogram, which is preserved in a vault at Sevres, France, by the International Bureau of Weights and Measures.

Length: The name *micron*, for a unit of length equal to 10^{-6} m, and the symbol μ that has been used for it were dropped by action of the 13th General Conference on Weights and Measures on October 13, 1967. The symbol μ is to be used solely as an abbreviation for the prefix *micro-*, standing for the multiplication by 10^{-6}. Thus, the length previously designated as 1 μ should be designated 1 μm.

The *lumin* (unit of luminous flux) is the luminous flux emitted in a solid angle of 1 sr by a uniform point source having an intensity of 1 cd.

The *newton* (unit of force) is that force that gives to a mass of 1 kg an acceleration of 1 m per second.

The *ohm* (unit of electric resistance) is the

Table 4–3 (continued)
INTERNATIONAL SYSTEM (SI) METRIC UNITS

electric resistance between two points of a conductor when a constant difference of potential of 1 V, applied between these two points, produces in this conductor a current of 1 A, this conductor not being the source of any electromotive force.

The *meter* (unit of length) is the length of exactly 1,650,763.73 wavelengths of the radiation in vacuum corresponding to the unperturbed transition between the levels $2p_{10}$ and $5d_5$ of the atom of krypton 86, the orange-red line.

The *second* is the unit of time of the International System of Units. The definition adopted at the October 13, 1967, meeting of the 13th General Conference on Weights and Measures is "The second is the duration of 9,192,631,770 periods of the radiation corresponding to the transition between the two hyperfine levels of the fundamental state of the atom of cesium 133." The frequency (9,192,631,770 hz), which the definition assigns to the cesium radiation, was carefully chosen to make it impossible, by any existing experimental evidence, to distinguish the new second from the *ephemeris second* based on the earth's motion. Therefore, no changes need to be made in data stated in terms of the old standard in order to convert them to the new one. The atomic definition has two important advantages over the previous definition: (1) it can be realized (i.e., generated by a suitable clock) with sufficient precision, ± 1 part per hundred billion (10^{11}) or

better, to meet the most exacting demands of modern metrology, and (2) it is available to anyone who has access to or who can build an atomic clock controlled by the specified cesium radiation.[b] In addition, one can compare other high-precision clocks directly with such a standard in a relatively short time — an hour or so compared against years with the astronomical standard. Laboratory-type atomic clocks are complex and expensive, so that most clocks and frequency generators will continue to be calibrated against a standard such as the NBS Frequency Standard, controlled by a cesium atomic beam, at the Radio Standards Laboratory in Boulder, Colorado. In most cases the comparison will be by way of the standard-frequency and time-interval signals broadcast by NBS radio stations WWV, WWVH, WWVB, and WWVL.

The *volt* (unit of electric potential difference and electromotive force) is the difference of electric potential between two points of a conducting wire carrying a constant current of 1 A, when the power dissipated between these points is equal to 1 W.

The *watt* (unit of power) is the power that gives rise to the production of energy at the rate of 1 J per second.

The *weber* (unit of magnetic flux) is the magnetic flux that, linking a circuit of one turn, produces in it an electromotive force of 1 V as it is reduced to zero at a uniform rate in 1 s.

[a]According to SI terminology, the following should be treated as obsolete:

angstrom (now 100 picometers or 0.1 nanometer)	kiloton (now gigagram)
	liter (now cubic decimeter)
bar (now 100 kilonewtons/meter²)	metric ton (now megagram)
kiloliter (now cubic meter)	micron (now micrometer)

[b]A description of such clocks is given in "Atomic Frequency Standards," *NBS Tech. News Bull.*, 45, 8, January 1961.

Note: For more recent developments and technical details, see Beehler, R. E., Mockler, R. C., and Richardson, J. M., Cesium beam atomic time and frequency standards, *Metrologia*, 1, 114, July 1965.

From Bolz, R. E. and Tuve, G. L., Eds., *Handbook of Tables for Applied Engineering Science*, 2nd ed., CRC Press, Cleveland, 1973, 805.

Table 4–4
CONVERSIONS TO SI UNITS

International Metric System

This table can be used for conversion of any quantity in English units to corresponding SI units to five significant figures (without the use of a calculator). Exact values are shown in boldface. Unless otherwise stated, values are in thermochemical calorie, thermochemical Btu, and avoirdupois mass units.[a]

Instruction: Shift decimal as required for each digit in the original quantity and add the converted results.

Example: Convert an acceleration of 15.30 ft/s² to m/s².

Solution: From first line of table, 3.048 0 + 1.524 0 + 0.091 44 = 4.663 4 m/s².

	1	2	3	4	5	6	7	8	9
ACCELERATION									
foot/second² to meter/second², m/s²	**0.304 8**	0.609 6	0.914 4	1.219 2	1.524 0	1.828 8	2.133 6	2.438 4	2.743 2
g's (free fall, standard) to meter/second², m/s²	**9.806 65**	19.613	29.420	39.227	49.033	58.840	68.647	78.453	88.260
inch/second² to meter/second², m/s²	**0.025 4**	0.050 8	0.076 2	0.101 6	0.127 0	0.152 4	0.177 8	0.203 2	0.228 6
AREA									
acre to meter², m²	4 046.856	8 093.7	12 141	16 187	20 234	24 281	28 328	32 375	36 422
circular mil to meter², m²	$5.067\ 075 \times 10^{-10}$	10.134×10^{-10}	15.201×10^{-10}	20.268×10^{-10}	25.335×10^{-10}	30.402×10^{-10}	35.470×10^{-10}	40.537×10^{-10}	45.604×10^{-10}
foot² to meter², m²	**0.092 903 04**	0.185 81	0.278 71	0.371 61	0.464 52	0.557 42	0.650 32	0.743 22	0.836 13
inch² to meter², m²	**0.000 645 16**	**0.001 290 32**	**0.001 935 48**	**0.002 580 64**	**0.003 225 80**	**0.003 870 96**	**0.004 516 12**	**0.005 161 28**	**0.005 806 44**
mile² (U.S. statute) to meter², m²	2 589 988	5 180 000	7 770 000	10 360 000	12 950 000	15 540 000	18 130 000	20 720 000	23 310 000
yard² to meter², m²	**0.836 127 36**	1.672 3	2.508 4	3.344 5	4.180 6	5.016 8	5.852 9	6.689 0	7.525 1
BENDING MOMENT OR TORQUE									
ounce-force-inch to newton-meter, N·m	0.007 061 552	0.014 123	0.021 185	0.028 246	0.035 308	0.042 369	0.049 431	0.056 492	0.063 554
pound-force-inch to newton-meter, N·m	0.112 984 8	0.225 97	0.338 95	0.451 94	0.564 92	0.677 91	0.790 89	0.903 88	1.016 9
pound-force-foot to newton-meter, N·m	1.355 818	2.711 6	4.067 5	5.423 3	6.779 1	8.134 9	9.490 7	10.847	12.202

Table 4–4 (continued)
CONVERSIONS TO SI UNITS

	1	2	3	4	5	6	7	8	9
DENSITY (MASS/VOLUME)									
grain/gallon to kilogram/meter³, kg/m³	0.017 118 06	0.034 236	0.051 354	0.068 472	0.085 590	0.102 71	0.119 83	0.136 94	0.154 06
ounce/gallon to kilogram/meter³, kg/m³	7.489 152	14.978	22.467	29.957	37.446	44.935	52.424	59.913	67.402
ounce/inch³ to kilogram/meter³, kg/m³	1 729.994	3 460.0	5 190.0	6 920.0	8 650.0	10 380	12 110	13 840	15 570
pound-mass/foot³ to kilogram/meter³, kg/m³	16.018 46	32.037	48.055	64.074	80.092	96.111	112.13	128.15	144.17
pound-mass/inch³ to kilogram/meter³, kg/m³	27 679.90	55 360	83 040	110 720	138 400	166 080	193 760	221 440	249 120
pound-mass/gallon to kilogram/meter³, kg/m³	119.826 4	239.65	359.48	479.31	599.13	718.96	838.78	958.61	1 078.4
slug/foot³ to kilogram/meter³, kg/m³	515.378 8	1 030.8	1 546.1	2 061.5	2 576.9	3 092.3	3 607.7	4 123.0	4 638.4
ELECTRICITY AND MAGNETISM									
ampere-hour to coulomb, C	3 600	7 200	10 800	14 400	18 000	21 600	25 200	28 800	32 400
faraday (based on C-12) to coulomb, C	96 487.00	192 970	289 460	385 950	482 440	578 920	675 410	771 900	868 380
gauss to tesla, T	0.000 1	0.000 2	0.000 3	0.000 4	0.000 5	0.000 6	0.000 7	0.000 8	0.000 9
gilbert to ampere-turn	0.795 774 7	1.591 5	2.387 3	3.183 1	3.978 9	4.774 6	5.570 4	6.366 2	7.162 0
oersted to ampere-meter, A/m	79.577 47	159.15	238.73	318.31	397.89	477.46	557.04	636.62	716.20
unit pole to weber, Wb	$1.256\ 637 \times 10^{-7}$	$2.513\ 3 \times 10^{-7}$	$3.769\ 9 \times 10^{-7}$	$5.026\ 5 \times 10^{-7}$	$6.283\ 2 \times 10^{-7}$	$7.539\ 8 \times 10^{-7}$	$8.796\ 5 \times 10^{-7}$	10.053×10^{-7}	11.310×10^{-7}
ENERGY AND WORK									
British thermal unit to joule, J	1 054.350	2 108.7	3 163.1	4 217.4	5 271.8	6 326.1	7 380.5	8 434.8	9 489.2
British thermal unit (IT) to joule, Jᵃ	1 055.056	2 110.1	3 165.2	4 220.2	5 275.3	6 330.3	7 385.4	8 440.4	9 495.5
calorie to joule, J	4.184	8.368	12.552	16.736	20.920	25.104	29.288	33.472	37.656
calorie (IT) to joule, Jᵃ	4.186 8	8.373 6	12.560 4	16.747 2	20.934 0	25.120 8	29.307 6	33.494 4	37.681 2
electron volt to joule, J	$1.602\ 10 \times 10^{-19}$	$3.204\ 2 \times 10^{-19}$	$4.806\ 3 \times 10^{-19}$	$6.408\ 4 \times 10^{-19}$	$8.010\ 5 \times 10^{-19}$	$9.612\ 6 \times 10^{-19}$	11.215×10^{-19}	12.817×10^{-19}	14.419×10^{-19}
foot-pound-force to joule, J	1.355 818	2.711 6	4.067 5	5.423 3	6.779 1	8.134 9	9.490 7	10.847	12.202
kilowatt-hour to joule, J	3 600 000	7 200 000	10 800 000	14 400 000	18 000 000	21 600 000	25 200 000	28 800 000	32 400 000
horsepower-hour to joule, J	2 684 520	5 369 039	8 053 559	10 738 078	13 422 598	16 107 117	18 791 637	21 476 156	24 160 676

Table 4–4 (continued)
CONVERSIONS TO SI UNITS

	1	2	3	4	5	6	7	8	9
FLOW RATE									
foot³/minute to meter³/second, m³/s	0.000 471 947 4	0.000 943 89	0.001 415 8	0.001 887 8	0.002 359 7	0.002 831 7	0.003 303 6	0.003 775 6	0.004 247 5
foot³/second to meter³/second, m³/s	0.028 316 85	0.056 634	0.084 951	0.113 27	0.141 58	0.169 90	0.198 22	0.226 53	0.254 85
gallon (U.S. liquid)/day to meter³/second, m³/s	$4.381\ 264 \times 10^{-8}$	$8.762\ 5 \times 10^{-8}$	13.144×10^{-8}	17.525×10^{-8}	21.906×10^{-8}	26.288×10^{-8}	30.669×10^{-8}	35.050×10^{-8}	39.431×10^{-8}
gallon (U.S. liquid)/minute to meter³/second, m³/s	0.000 063 090 20	0.000 126 18	0.000 189 27	0.000 252 36	0.000 315 45	0.000 378 54	0.000 441 63	0.000 504 72	0.000 567 81
pound-mass/hour to kilogram/second, kg/s	0.000 125 997 9	0.000 252 00	0.000 377 99	0.000 503 99	0.000 629 99	0.000 755 99	0.000 881 99	0.001 007 98	0.001 133 98
pound-mass/minute to kilogram/second, kg/s	0.007 559 873	0.015 120	0.022 680	0.030 239	0.037 799	0.045 359	0.052 919	0.060 479	0.068 039
FORCE									
kilogram-force to newton, N	9.806 65	19.613	29.420	39.227	49.033	58.840	68.647	78.453	88.260
ounce-force to newton, N	0.278 014 0	0.556 03	0.834 04	1.112 1	1.390 1	1.668 1	1.946 1	2.224 1	2.502 1
pound-force to newton, N	4.448 222	8.896 4	13.345	17.793	22.241	26.689	31.138	35.586	40.034
HEAT									
SPECIFIC HEAT CAPACITY									
British thermal unit/pound-mass·deg F to joule/kilogram-kelvin, J/kg·K	4 184	8 368	12 552	16 736	20 920	25 104	29 288	33 472	37 656
British thermal unit (IT)/pound-mass·deg F to joule/kilogram-kelvin, J/kg·K[a]	4 186.8	8 373.6	12 560.4	16 747.2	20 934.0	25 120.8	29 307.6	33 494.4	37 681.2
calorie/gram·deg C to joule/kilogram-kelvin, J/kg·K	4 184	8 368	12 552	16 736	20 920	25 104	29 288	33 472	37 656
ENERGY/MASS (ENTHALPY, ETC.)									
British thermal unit/pound-mass to joule/kilogram, J/kg	2 324.444	4 648.9	6 973.3	9 297.8	11 622	13 947	16 271	18 596	20 920
British thermal unit (IT)/pound-mass to joule/kilogram, J/kg[a]	2 326	4 652	6 978	9 304	11 630	13 956	16 282	18 608	20 934
calorie/gram to joule/kilogram, J/kg	4 184	8 368	12 552	16 736	20 920	25 104	288	33 472	37 656

Table 4–4 (continued)
CONVERSIONS TO SI UNITS

	1	2	3	4	5	6	7	8	9
THERMAL CONDUCTIVITY									
British thermal unit/hour-foot-deg F to watt/meter-kelvin,[a] W/m·K	1.729 577	3.459 2	5.188 7	6.918 3	8.647 9	10.377	12.107	13.837	15.566
British thermal unit (IT)/hour-foot-deg F to watt/meter-kelvin, W/m·K	1.730 735	3.461 5	5.192 2	6.922 9	8.653 7	10.384	12.115	13.846	15.577
British thermal unit-inch/hour-foot²-deg F to watt/meter-kelvin, W/m·K	0.144 131 4	0.288 26	0.432 39	0.576 53	0.720 66	0.864 79	1.008 9	1.153 1	1.297 2
British thermal unit (IT)-inch/hour-foot²-deg F to watt/meter-kelvin, W/m·K	0.144 227 9	0.288 46	0.432 68	0.576 91	0.721 14	0.865 37	1.009 60	1.153 82	1.298 05
calorie/second-centimeter-deg C to watt/meter-kelvin, W/m·K[a]	418.4	836.8	1 255.2	1 673.6	2 092.0	2 510.4	2 928.8	3 347.2	3 765.6
ENERGY PER UNIT AREA									
British thermal unit/foot² to joule/meter², J/m²	11 348.93	22 698	34 047	45 396	56 745	68 094	79 443	90 791	102 140
calorie/centimeter² to joule/meter², J/m²	41 840	83 680	125 520	167 360	209 200	251 040	292 880	334 720	376 560
THERMAL DIFFUSIVITY foot²/hour to meter²/second, m²/s	0.000 025 806 4	0.000 051 612 8	0.000 077 419 2	0.000 103 256	0.000 129 032	0.000 154 838 4	0.000 180 644 8	0.000 206 451 2	0.000 232 257 6
THERMAL RESISTANCE deg F-hour-foot²/British thermal unit to kelvin-meter²/watt, K·m²/W	0.176 228 0	0.352 46	0.528 68	0.704 91	0.881 14	1.057 4	1.233 6	1.409 8	1.586 1
THERMAL CONDUCTANCE British thermal unit/hour-foot²-deg F to watt/meter²-kelvin, W/m²·K	5.674 466	11.349	17.023	22.698	28.372	34.047	39.721	45.396	51.070

Table 4–4 (continued)
CONVERSIONS TO SI UNITS

	1	2	3	4	5	6	7	8	9
THERMAL CONDUCTANCE (*continued*)									
British thermal unit/second-foot²-deg F to watt/meter²-kelvin, W/m²·K	20 428.08	40 856	61 284	81 712	102 140	122 570	143 000	163 420	183 850
calorie/second-centimeter²-deg C to watt/meter²-kelvin, W/m²·K	41 840	83 680	125 520	167 360	209 200	251 040	292 880	334 720	376 560
LENGTH									
caliber to meter, m	0.000 254	0.000 508	0.000 762	0.001 016	0.001 270	0.001 524	0.001 778	0.002 032	0.002 286
fathom to meter, m	1.828 8	3.657 6	5.486 4	7.315 2	9.144 0	10.972 8	12.801 6	14.630 4	16.459 2
foot to meter, m	0.304 8	0.609 6	0.914 4	1.219 2	1.524 0	1.828 8	2.133 6	2.438 4	2.743 2
inch to meter, m	0.025 4	0.050 8	0.076 2	0.101 6	0.127 0	0.152 4	0.177 8	0.203 2	0.228 6
light year to meter, m	$9.460\ 550 \times 10^{15}$	18.921×10^{15}	28.382×10^{15}	37.842×10^{15}	47.303×10^{15}	56.763×10^{15}	66.224×10^{15}	75.684×10^{15}	85.145×10^{15}
mil to meter, m	0.000 025 4	0.000 050 8	0.000 076 2	0.000 101 6	0.000 127 0	0.000 152 4	0.000 177 8	0.000 203 2	0.000 228 6
mile (U.S. nautical) to meter, m	1 852	3 704	5 556	7 408	9 260	11 112	12 964	14 816	16 668
mile (U.S. statute) to meter, m	1 609.344	3 218.7	4 828.0	6 437.4	8 046.7	9 656.1	11 265	12 875	14 484
rod to meter, m	5.029 2	10.058 4	15.087 6	20.116 8	25.146 0	30.175 2	35.204 4	40.233 6	45.262 8
yard to meter, m	0.914 4	1.828 8	2.743 2	3.657 6	4.572 0	5.486 4	6.400 8	7.315 2	8.229 6
MASS									
grain to kilogram, kg	0.000 064 798 91	0.000 129 60	0.000 194 40	0.000 259 20	0.000 324 00	0.000 388 80	0.000 453 60	0.000 518 40	0.000 583 20
ounce-mass to kilogram, kg	0.028 349 52	0.056 699	0.085 049	0.113 40	0.141 75	0.170 10	0.198 45	0.226 80	0.255 15
ounce-mass (troy or apothecary) to kilogram, kg	0.031 103 48	0.062 207	0.093 310	0.124 41	0.155 52	0.186 62	0.217 72	0.248 83	0.279 93
pound-mass to kilogram, kg	0.453 592 37	0.907 18	1.360 8	1.814 4	2.268 0	2.721 6	3.175 1	3.628 7	4.082 3
pound-mass (troy or apothecary) to kilogram, kg	0.373 241 7	0.746 48	1.119 7	1.493 0	1.866 2	2.239 5	2.612 7	2.985 9	3.359 2
slug to kilogram, kg	14.593 90	29.188	43.782	58.376	72.970	87.563	102.16	116.75	131.35
ton (long, 2 240 lb_m) to kilogram, kg	1 016.047	2 032.1	3 048.1	4 064.2	5 080.2	6 096.3	7 112.3	8 128.4	9 144.4
ton (short, 2 000 lb_m) to kilogram, kg	907.184 7	1 814.4	2 721.6	3 628.7	4 535.9	5 443.1	6 350.3	7 257.5	8 164.7

Table 4–4 (continued)
CONVERSIONS TO SI UNITS

	1	2	3	4	5	6	7	8	9
POWER									
British thermal unit/second to watt, W	1 054.350	2 108.7	3 163.1	4 217.4	5 271.8	6 326.1	7 380.5	8 434.8	9 489.2
British thermal unit/minute to watt, W	17.572 50	35.145	52.718	70.290	87.863	105.44	123.01	140.58	158.15
British thermal unit/hour to watt, W	0.292 875 1	0.585 75	0.878 63	1.171 5	1.464 4	1.757 3	2.050 1	2.343 0	2.635 9
British thermal unit (IT)/hour to watt, W[a]	0.293 071 1	0.586 14	0.879 21	1.172 3	1.465 4	1.758 4	2.051 5	2.344 6	2.637 6
calorie/second to watt, W	**4.184**	**8.368**	**12.552**	**16.736**	**20.920**	**25.104**	**29.288**	**33.472**	**37.656**
calorie/minute to watt, W	0.069 733 33	0.139 47	0.209 20	0.278 93	0.348 67	0.418 40	0.488 13	0.557 87	0.627 60
foot-pound-force/second to watt, W	1.355 818	2.711 6	4.067 5	5.423 3	6.779 1	8.134 9	9.490 7	10.847	12.202
foot-pound-force/minute to watt, W	0.022 596 97	0.045 194	0.067 791	0.090 388	0.112 98	0.135 58	0.158 18	0.180 78	0.203 37
foot-pound-force/hour to watt, W	0.000 376 616 1	0.000 753 23	0.001 129 8	0.001 506 5	0.001 883 1	0.002 259 7	0.002 636 3	0.003 012 9	0.003 389 5
horsepower (550 ft·lb$_f$/s) to watt, W	745.699 9	1 491.4	2 237.1	2 982.8	3 728.5	4 474.2	5 219.9	5 965.6	6 711.3
horsepower (electric) to watt, W	**746.**	**1 492.**	**2 238.**	**2 984.**	**3 730.**	**4 476.**	**5 222.**	**5 968.**	**6 714.**
tons of refrigeration to watt, W	3 516.853	7 033.7	10 551	14 067	17 584	21 101	24 618	28 135	31 652
POWER/AREA									
British thermal unit/foot²-second to watt/meter², W/m²	11 348.93	22 698	34 047	45 396	56 745	68 094	79 443	90 791	102 140
British thermal unit/foot²-minute to watt/meter², W/m²	189.148 9	378.30	567.45	756.60	945.74	1 134.9	1 324.0	1 513.2	1 702.3
British thermal unit/foot²-hour to watt/meter², W/m²	3.152 481	6.305 0	9.457 4	12.610	15.762	18.915	22.067	25.220	28.372
British thermal unit/inch²-second to watt/meter², W/m²	1 634 246	3 268 500	4 902 700	6 537 000	8 171 200	9 805 500	11 440 000	13 074 000	14 708 000
calorie/centimeter²-minute to watt/meter², W/m²	697.333 3	1 394.7	2 092.0	2 789.3	3 486.7	4 184.0	4 881.3	5 578.7	6 276.0

Table 4–4 (continued)
CONVERSIONS TO SI UNITS

	1	2	3	4	5	6	7	8	9
PRESSURE OR STRESS (FORCE/AREA)									
atmosphere (normal = 760 torr) to newton/meter², N/m²	101 325	202 650	303 975	405 300	506 625	607 950	709 275	810 600	911 925
bar to newton/meter², N/m²	100 000	200 000	300 000	400 000	500 000	600 000	700 000	800 000	900 000
foot of water (39.2 F) to newton/meter², N/m²	2 988.980	5 978.0	8 966.9	11 956	14 945	17 934	20 923	23 912	26 901
inch of mercury (32 F) to newton/meter², N/m²	3 386.389	6 772.8	10 159	13 546	16 932	20 318	23 705	27 091	30 478
inch of water (39.2 F) to newton/meter², N/m²	249.082 0	498.16	747.25	996.33	1 245.4	1 494.5	1 743.6	1 992.7	2 241.7
inch of water (60 F) to newton/meter², N/m²	248.840 0	497.68	746.52	995.36	1 244.2	1 493.0	1 741.9	1 990.7	2 239.6
kilogram-force/centimeter² to newton/meter², N/m²	98 066.5	196 133	294 199.5	392 266	490 332.5	588 399	686 465.5	784 532	882 598.5
millimeter of mercury (0 C), torr, to newton/meter², N/m²	133.322 4	266.64	399.97	533.29	666.61	799.93	933.26	1 066.6	1 199.9
pound-force/foot² to newton/meter², N/m²	47.880 26	95.761	143.64	191.52	239.40	287.28	335.16	383.04	430.92
pound-force/inch² (psi) to newton/meter², N/m²	6 894.757	13 790	20 684	27 579	34 474	41 369	48 263	55 158	62 053
TEMPERATURE (*see* Tables 4–12 – 4–15)									
VELOCITY									
foot/hour to meter/second, m/s	0.000 084 666 67	0.000 169 33	0.000 254 00	0.000 338 67	0.000 423 33	0.000 508 00	0.000 592 67	0.000 677 33	0.000 762 00
foot/minute to meter/second, m/s	0.005 08	0.010 16	0.015 24	0.020 32	0.025 40	0.030 48	0.035 56	0.040 64	0.045 72
foot/second to meter/second, m/s	0.304 8	0.609 6	0.914 4	1.219 2	1.524 0	1.828 8	2.133 6	2.438 4	2.743 2
inch/second to meter/second, m/s	0.025 4	0.050 8	0.076 2	0.101 6	0.127 0	0.152 4	0.177 8	0.203 2	0.228 6
kilometer/hour to meter/second, m/s	0.277 777 8	0.555 56	0.833 33	1.111 1	1.388 9	1.666 7	1.944 4	2.222 2	2.500 0
knot (international) to meter/second, m/s	0.514 444 4	1.028 9	1.543 3	2.057 8	2.572 2	3.086 7	3.601 1	4.115 6	4.630 0

CONVERSIONS TO SI UNITS

	1	2	3	4	5	6	7	8	9
VELOCITY (*continued*)									
mile/hour (U.S. statute) to meter/second, m/s	0.447 04	0.894 08	1.341 12	1.788 16	2.235 20	2.682 24	3.129 28	3.576 32	4.023 36
mile/minute (U.S. statute) to meter/second, m/s	26.822 4	53.644 8	80.467 2	107.289 6	134.112 0	160.934 4	187.756 8	214.579 2	241.401 6
mile/second (U.S. statute) to meter/second, m/s	1 609.344	3 218.7	4 828.0	6 437.4	8 046.7	9 656.1	11 265	12 875	14 484
VISCOSITY									
DYNAMIC OR ABSOLUTE, μ									
centipoise to newton-second/meter², N·s/m²	0.001	0.002	0.003	0.004	0.005	0.006	0.007	0.008	0.009
pound-mass/foot-second to newton-second/meter², N·s/m²	1.488 164	2.976 3	4.464 5	5.952 7	7.440 8	8.929 0	10.417	11.905	13.393
pound-force-second/foot² to newton-second/meter², N·s/m²	47.880 26	95.761	143.64	191.52	239.40	287.28	335.16	383.04	430.92
slug/foot-second to newton-second/meter², N·s/m²	47.880 26	95.761	143.64	191.52	239.40	287.28	335.16	383.04	430.92
KINEMATIC, ν									
centistoke to meter²/second, m²/s	$1. \times 10^{-6}$	$2. \times 10^{-6}$	$3. \times 10^{-6}$	$4. \times 10^{-6}$	$5. \times 10^{-6}$	$6. \times 10^{-6}$	$7. \times 10^{-6}$	$8. \times 10^{-6}$	$9. \times 10^{-6}$
foot²/second to meter²/second, m²/s	0.092 903 04	0.185 81	0.278 71	0.371 61	0.464 52	0.557 42	0.650 32	0.743 22	0.836 12
VOLUME									
acre-foot to meter³, m³	1 233.482	2 467.0	3 700.4	4 933.9	6 167.4	7 400.9	8 634.4	9 867.9	11 101
barrel (oil, 42 gal) to meter³, m³	0.158 987 3	0.317 97	0.476 96	0.635 95	0.794 94	0.953 92	1.112 9	1.271 9	1.430 9
board foot to meter³, m³	0.002 359 737	0.004 719 5	0.007 079 2	0.009 438 9	0.011 799	0.014 158	0.016 518	0.018 878	0.021 238
bushel (U.S.) to meter³, m³	0.035 239 07	0.070 478	0.105 72	0.140 96	0.176 20	0.211 43	0.246 67	0.281 91	0.317 15
foot³ to meter³, m³	0.028 316 85	0.056 634	0.084 951	0.113 27	0.141 58	0.169 90	0.198 22	0.226 53	0.254 85
gallon (U.S. liquid) to meter³, m³	0.003 785 412	0.007 570 8	0.011 356	0.015 142	0.018 927	0.022 712	0.026 498	0.030 283	0.034 069

Table 4–4 (continued)
CONVERSIONS TO SI UNITS

	1	2	3	4	5	6	7	8	9
VOLUME (*Continued*)									
inch³ to meter³, m³	0.000 016 387 06	0.000 032 774	0.000 049 161	0.000 065 548	0.000 081 935	0.000 098 322	0.000 114 71	0.000 131 10	0.000 147 48
ounce (U.S. fluid) to meter³, m³	0.000 029 573 53	0.000 059 147	0.000 088 721	0.000 118 29	0.000 147 87	0.000 177 44	0.000 207 01	0.000 236 59	0.000 266 16
peck (U.S.) to meter³, m³	0.008 809 768	0.017 620	0.026 429	0.035 239	0.044 049	0.052 859	0.061 668	0.070 478	0.079 288
quart (U.S. liquid) to meter³, m³	0.000 946 352 9	0.001 892 7	0.002 839 1	0.003 785 4	0.004 731 8	0.005 678 1	0.006 624 5	0.007 570 8	0.008 517 2
yard³ to meter³, m³	0.764 554 9	1.529 1	2.293 7	3.058 2	3.822 8	4.587 3	5.351 9	6.116 4	6.881 0
VOLUME/MASS (SPECIFIC VOLUME)									
foot³/pound to meter³/kilogram, m³/kg	0.062 427 96	0.124 86	0.187 28	0.249 71	0.312 14	0.374 57	0.437 00	0.499 42	0.561 85

[a]The thermochemical calorie is exactly 4.184 by definition. The international steam table (IT) calorie is exactly 4.1868 by definition. The thermochemical Btu is 1,054.350. Each Btu is defined in terms of the corresponding calorie by 1 Btu/lbm · R ≡ 1 cal/g · K.

From Bolz, R. E. and Tuve, G. L., Eds., *Handbook of Tables for Applied Engineering Science*, 2nd ed., CRC Press, Cleveland, 1973, 808.

Table 4–5
SELECTED MATHEMATICAL AND PHYSICAL CONSTANTS

Rounded to 3 to 5 Significant Figures

Name	Symbol (or equivalent)	Value	Units
Naperian base (natural)	e	2.718	
Natural logarithm of 10	$\ln_e 10$	2.303	
Common logarithm of 2	$\log_{10} 2$	0.301	
Semicircle	π	3.1416	radians
Cos 60°	½	0.500	
Cos 45°	$\sqrt{2}/2$	0.707	
Cos 30°	$\sqrt{3}/2$	0.866	
Ice point	T_0	273.15	K
Avogadro's number	N	6.022×10^{23}	atoms (particles)/mole
Gas constant	R	8.317×10^{7}	erg/ K-mole
		1.988	cal/ K-mole
		8.208	l-atm/ K-mole
Boltzmann constant	k (R/N)	8.617×10^{-5}	ev/ K-atom (particles)
		1.380×10^{-16}	erg/ K-atom
		0.33×10^{-23}	cal/ K-atom
Acceleration due to gravity	g_0	9.806	m/s²
Velocity of light (vacuum)	c	2.998×10^{8}	m/s
Planck constant	h	6.626×10^{-27}	erg-sec/atom (particles)
		1.58×10^{-34}	cal-sec/atom
Faraday constant	$\mathscr{F}$	9.649×10^{4}	coulomb/mole
		27 (approx.)	amp-hr/mole
Charge of one electron	e^- ($\mathscr{F}$/N)	4.803×10^{-10}	(erg-cm)½, or esu
		1.602×10^{-12}	erg/volt
		1.602×10^{-19}	joule/volt (J/V)
		1.602×10^{-20}	emu
Charge-to-mass ratio of electron	e^-/m_e	1.759×10^{8}	coulomb/gram
Rest mass of one electron	m_e $(\dfrac{\mathscr{F}/e^-}{N/m_e})$	9.109×10^{-28}	gram
Compton wavelength of free electron	λ_{ce} (h/m_ec)	2.426×10^{-10}	cm

Table compiled by A. Pigeaud, University of Cincinnati.

Table 4–6A
ABBREVIATED TABLE OF CONVERSION FACTORS

Length (meter)

1 light year	=	9,460	Tm (10^{12} m)	1 Tm	=	0.1058×10^{-3}	light years
1 astron. unit	=	149.5	Gm (10^{9} m)	1 Tm	=	6.69	astron. units
1 naut. mile	=	1.852	km (10^{3} m)	1 km	=	0.54	naut. miles
1 mile (U.S.)	=	1.609	km	1 km	=	0.622	mile (U.S.)
1 fathom	=	1.829	m	1 m	=	0.547	fathom
1 yard	=	0.9144	m	1 m	=	1.0936	yard
1 foot	=	0.3048	m	1 m	=	3.281	foot
1 inch	=	2.54	cm (10^{-2} m)	1 cm	=	0.3937	inch
1 mil	=	25.4	μm (10^{-6} m)	1 μm	$\cong$	0.04	mil
1 Kr^{86} wavelength	=	606	nm (10^{-9} m)	1 μm	=	1.65	Kr^{86} wavelength
1 angstrom	=	0.1	nm	1 nm	=	10	angstrom

Table 4—6A (continued)
ABBREVIATED TABLE OF CONVERSION FACTORS

Area (meter)²

1 sq mile (U.S.)	=	2.59	km² (10³ m)²	1 km²	=	0.3861	sq mile (U.S.)
1 hectare	=	1	hm² (10² m)²	1 km²	=	100	hectare
1 acre	=	4,047	m²	1 hm²	=	2.471	acre
1 are	=	100	m²	1 hm²	=	100	are
1 sq yard	=	0.8361	m²	1 m²	=	1.196	sq yard
1 sq foot	=	929	cm² (10⁻² m)²	1 m²	=	10.764	sq foot
1 sq inch	=	6.452	cm²	1 cm²	=	0.155	sq inch
1 circ. mil	≅	0.0005	mm² (10⁻³ m)²	1 mm²	=	1,973.5	circ. mil
1 sq angstrom	=	10,000	pm² (10⁻¹² m)²	1 nm²	=	100	sq angstrom
1 barn	=	0.0001	pm²	1 pm²	=	10,000	barns

Density (kilogram/m³)

1 lb mass/cu inch	=	27.68	Mg/m³	1 kg/m³	=	36.13	lb mass/cu inch
1 gram/cc	=	1,000	kg/m³	1 kg/m³	=	0.001	gram/cc
1 slug/cu ft	=	515.4	kg/m³	1 g/cm³	=	1.94	slug/cu ft
1 lb mass/USG	=	119.8	kg/m³	1 g/cm³	=	8.345	lb mass/USG
1 lb mass/cu ft	=	16.02	kg/m³	1 g/cm³	=	62.43	lb mass/cu ft

Volume (meter)³

1 acre foot	=	1,233.5	m³ (kl)	1 dam³	=	0.8107	acre foot
1 cord	=	3.625	m³	1 m³	=	0.2759	cord
1 register ton	=	2.832	m³	1 m³	=	0.3531	register ton
1 cu yard	=	764.5	dm³ (10⁻¹ m)³	1 m³	=	1.308	cu yard
1 barrel (42 USG)	=	159	l (10⁻¹ m)³	1 kl	=	6.3	barrel (USG)
1 bushel (U.S.)	=	35.24	l	1 kl	=	28.4	bushel (U.S.)
1 cu ft	=	28.32	dm³	1 m³	=	35.31	cu ft
1 board foot	=	2.36	dm³	1 dm³	=	0.424	board foot
1 gallon (Imp.)	=	4.546	l	1 l	=	0.220	(Imp.) gallon
1 gallon (U.S.)	=	3.785	l	1 l	=	0.2642	(U.S.) gallon
1 quart (U.S.)	=	0.9463	l	1 l	=	1.0567	(U.S.) quart
1 pint (U.S.)	=	473.2	ml (10⁻² m)³	1 l	=	2.113	(U.S.) pint
1 fl ounce (U.S.)	=	29.57	ml	1 l	=	33.81	(U.S.) fl ounce
1 cu inch	=	16.39	cc (10⁻² m)³	1 cc	=	0.061	cu inch
1 dram (U.S.)	=	3.697	ml	1 ml	=	0.27	dram (U.S.)
1 minim (U.S.)	=	61.6	μl (10⁻³ m)³	1 ml	=	16.23	minim (U.S.)

Mass (kilogram)

1 long ton	=	1.016	Mg (10⁶ g)	1 Mg	=	0.985	long ton
1 metric ton	=	1,000	kg (10³ g)	1 Mg	=	1.0	metric ton
1 short ton (2,000 lb)	=	907	kg	1 Mg	=	1.102	short ton (2,000 lb)
1 pound (avoir)	=	453.6	g	1 kg	=	2.204	pounds (avoir)
1 ounce (avoir)	=	28.35	g	1 kg	=	35.27	ounce (avoir)
1 dram (avoir)	=	1.77	g	1 g	=	0.564	dram (avoir)
1 carat (metric)	=	0.2	g	1 g	=	5	carat (metric)
1 grain	=	64.8	mg (10⁻³ g)	1 g	=	15.4	grain

Time (mean solar seconds)

1 calendar year	≈	30	Ms (10⁶ s)	1 Ms	≈	0.032	calendar year
1 calendar month	≈	2.5	Ms	1 Ms	≈	0.4	calendar month
1 day	=	86.40	ks	1 Ms	=	11.574	days
1 hour	=	3.60	ks (10³ s)	1 Ms	=	277.8	hours
1 minute	=	60.0	s	1 ks	=	16.67	minutes

Table 4–6A (continued)
ABBREVIATED TABLE OF CONVERSION FACTORS

Planar Angle (radian)

360° (2π)	=	6.283	rad	1 rad	=	57.296°	(degrees)
180° (π)	=	3.1416	rad	1 mrad	=	3.438'	(minutes)
90° ($\frac{\pi}{2}$)	=	1.5708	rad	1 μrad	=	0.206"	(seconds)
1° ($\frac{\pi}{180}$)	=	0.01745	rad				

Solid Angle (steradian)

1 sphere	=	12.566	sr	1 sr	=	0.0796	sphere
1 hemisphere	=	6.283	sr	1 sr	=	0.159	hemisphere
1 octant	=	1.5708	sr	1 sr	=	0.636	octant
1 square degree	=	0.3046	msr	1 sr	=	3,282.8	square degree

Velocity (meter/second)

1 (U.S.) mile/sec	=	1.609	km/s	1 km/s	=	0.622	(U.S.) mile/sec
1 (U.S.) mile/min	=	26.82	m/s	1 km/s	=	37.28	(U.S.) mile/min
1 (Int.) knot	=	0.514	m/s	1 m/s	=	1.94	(Int.) knot
1 (U.S.) mile/hr	=	0.447	m/s	1 m/s	=	2.237	(U.S.) mile/hr
1 foot/sec	=	0.305	m/s	1 m/s	=	3.28	ft/sec
1 km/hr	=	0.278	m/s	1 m/s	=	3.6	km/hr
1 inch/sec	=	2.54	cm/s	1 m/s	=	39.37	in./sec
1 foot/min	=	5.08	mm/s	1 m/s	=	196.85	ft/min
1 foot/hr	=	84.67	μm/s	1 cm/s	=	118.1	ft/hr

Acceleration (meter/second^2)

gravity (mean)	=	9.806	m/s^2	1 m/s^2	=	0.102	gravity (mean)
1 mile/hr-sec	=	0.447	m/s^2	1 m/s^2	=	2.237	mile/hr-sec
1 foot/sq sec	=	0.305	m/s^2	1 m/s^2	=	3.281	foot/sq sec
1 km/hr-sec	=	0.278	m/s^2	1 m/s^2	=	3.6	km/hr-sec
1 inch/sq sec	=	0.0254	m/s^2	1 m/s^2	=	39.37	inch/sq sec
1 galileo	=	0.01	m/s^2	1 m/s^2	=	100	galileo

Force (newton)

1 kg force (kgf)	=	9.806	N	1 N	=	0.102	kg force
1 pound force (lbf)	=	4.448	N	1 N	=	0.225	pound force (avoir)
1 ounce force	=	0.278	N	1 N	=	3.6	ounce force
1 poundal	=	0.138	N	1 N	=	7.24	poundal
1 dyne	=	10.0	μN	1 N	=	10^5	dyne

Pressure (newton/m^2 or pascal)

1 atmosphere (14.7 psi)	=	101.325	kN/m^2	1 MN/m^2	=	9.87	atmosphere
1 bar	=	100.0	kN/m^2	1 MN/m^2	=	10.0	bar
1 kgf/sq cm	=	98.06	kN/m^2	1 MN/m^2	=	10.197	kgf/sq cm
1 lbf/sq inch (psi)	=	6.895	kN/m^2	1 MN/m^2	=	145.	lbf/sq inch
1 inch Hg (60°F)	=	3.377	kN/m^2	1 kN/m^2	=	0.296	inch Hg (60°F)
1 cm Hg (0°F)	=	1.333	kN/m^2	1 kN/m^2	=	0.75	cm Hg (0°F)
1 inch H_2O (60°F)	=	248.84	N/m^2	1 kN/m^2	=	4.02	inch H_2O (60°F)
1 mm Hg (0°C) or torr	=	133.32	N/m^2	1 kN/m^2	=	7.57	mm Hg (0°C) or torr
1 millibar	=	100.0	N/m^2	1 kN/m^2	=	10.0	millibar
1 cm H_2O (4°C)	=	98.064	N/m^2	1 kN/m^2	=	10.2	cm H_2O (4°C)
1 lbf/sq f	=	47.88	N/m^2	1 kN/m^2	=	20.88	lbf/sq foot
1 kgf/sq meter	=	9.806	N/m^2	1 kN/m^2	=	101.97	kgf/sq meter
1 pascal (P)	=	1.0	N/m^2	1 N/m^2	=	1.0	P (pascal)

Table 4–6A (continued)
ABBREVIATED TABLE OF CONVERSION FACTORS

Pressure (newton/m² or pascal) (continued)

1 dyne/sq cm or barye	=	0.1	N/m^2	1 N/m^2	=	10.0	dyne/sq cm or barye
10^{-6} torr	=	133.32	$\mu N/m^2$	1 mN/m^2	=	7.57	10^{-6} torr

Dynamic Viscosity (newton-seconds/meter²)

1 lbf-sec/sq foot	=	47.88	$N \cdot s/m^2$	1 $N \cdot s/m^2$	=	0.0209	lbf-sec/sq foot
1 poundal-sec/sq foot	=	1.488	$N \cdot s/m^2$	1 $N \cdot s/m^2$	=	0.672	poundal-sec/sq foot
1 lbm/foot-sec	=	1.488	$N \cdot s/m^2$	1 $N \cdot s/m^2$	=	0.672	lbm/foot-sec
1 poise	=	0.1	$N \cdot s/m^2$	1 $N \cdot s/m^2$	=	10	poise
1 centipoise	=	0.001	$N \cdot s/m^2$	1 $N \cdot s/m^2$	=	1,000	centipoise

Energy (joule)

1 ton (TNT-equiv)	=	4.2	GJ	1 TJ	=	238	ton (TNT-equiv)
1 kWh	=	3.6	MJ	1 GJ	=	277.8	kWh
1 HP-hr	=	2.68	MJ	1 GJ	=	372.5	HP-hr
1 watt-hr	=	3.6	kJ	1 MJ	=	277.8	watt-hr
1 Btu (mean)	=	1.0559	kJ	1 MJ	=	947.0	Btu (mean)
1 l-atm	=	101.33	J	1 kJ	=	9.869	liter-atmosphere
1 kg-m	=	9.807	J	1 kJ	=	101.97	kilogram-meter
1 cal (mean)	=	4.190	J	1 kJ	=	238.89	calorie (mean)
1 cal (15°C)	=	4.1858	J	1 kJ	=	238.95	calorie (15°C)
1 cal (thermochem)	=	4.1840	J	1 kJ	=	239.00	calorie (thermochem)
1 cal (20°C)	=	4.1819	J	1 J	=	239.18	calorie (20°C)
1 foot-lbf	=	1.356	J	1 J	=	0.737	foot-lbf
1 watt-sec	=	1.0	J	1 J	=	1.0	watt-sec
1 foot-poundal	=	42.14	mJ	1 J	=	23.73	foot-poundal
1 erg (dyne-cm)	=	0.10	μJ	1 μJ	=	10.0	erg
1 eV	=	0.1602	aJ	1 aJ	=	6.242	eV

Power (watt)

1 boiler horsepower	=	9.809	kW	1 MW	=	102	boiler horsepower
1 kcal/sec	=	4.184	kW	1 MW	=	239	kcal/sec (thermochem)
1 Btu/sec	=	1.054	kW	1 kW	=	0.9485	Btu/sec (thermochem)
1 HP (electric/water)	=	746	W	1 kW	=	1.3405	HP (electric/water)
1 HP (U.K.)	=	745.7	W	1 kW	=	1.341	HP (550 ft-lbf/sec)
1 HP (metric)	=	735.5	W	1 kW	=	1.3596	HP (metric)
1 kcal/min	=	69.73	W	1 kW	=	14.33	kcal/min (thermochem)
1 Btu/min	=	17.57	W	1 kW	=	56.88	Btu/min (thermochem)
1 cal/sec	=	4.184	W	1 kW	=	239	cal/sec (thermochem)
1 ft-lbf/sec	=	1.356	W	1 kW	=	737.56	ft-lbf/sec
1 cal/min	=	69.73	mW	1 W	=	14.33	cal/min (thermochem)
1 ft-lbf/min	=	22.597	mW	1 W	=	44.25	ft-lbf/min
1 ft-lbf/hr	=	.3766	mW	1 mW	=	2.655	ft-lbf/hr
1 erg/sec	=	0.10	μW	1 μW	=	10	erg/sec

Charge (coulomb)

1 faraday	=	96,487	C	1 MC	=	10.363	faraday
1 amp-hr	=	3,600	C	1 MC	=	277.8	amp-hr
1 amp-sec	=	1.0	C	1 C	=	1.0	amp-sec
1 faraday	=	26.8	$A \cdot hr$	1 $A \cdot hr$	=	0.0336	faraday

Table compiled by A. Pigeaud, University of Cincinnati.

Table 4–6B
TEMPERATURE CONVERSIONS – SHORT TABLE[a]

Temperature (Kelvin)

$t_C = \frac{5}{9} t_F - 17.8$	$t_K = \frac{5}{9} t_F + 255.4$	$t_K = t_C + 273.15$
−524.8 °F	0 K	−273.2 °C
−328	73.2	−200
−279.8	100	−173.2
−148	173.2	−100
− 99.8	200	− 73.2
0	255.4	− 17.8
32 °F	273.2 K	0 °C
80.2	300	26.8
100	311	37.8
200	366.5	93.3
212	373.2	100
260.2	400	126.8
300	422.1	148.9
392 °F	473.2 K	200 °C
400	477.6	204.4
440.2	500	226.8
500	533.2	260
572	573.2	300
600	588.7	315.6
620.2	600	326.8
700	644.3	371.1
752 °F	673.2 K	400 °C
800.2	700	426.7
900	756.5	483.3
932	773.2	500
980.2	800	526.8
1,000	811	537.8
1,100	866.5	593.3
1,112 °F	873.2 K	600 °C
1,160.2	900	626.8
1,200	922.1	648.9
1,292	973.2	700
1,300	977.6	704.4
1,340.2	1,000	726.8
1,400	1,033.2	760
1,472 °F	1,073.2 K	800 °C
1,500	1,088.7	815.6
1,520.4	1,100	826.9
1,600	1,144.3	871
1,652	1,173.2	900
1,700.2	1,200	926.8
1,800	1,255.4	982.2
1,832 °F	1,273.2 K	1,000 °C
1,878	1,300	1,026.8
2,000	1,366.5	1,093.3
2,012	1,373.2	1,100
2,058	1,400	1,126.8
2,192	1,473.2	1,200
2,238	1,500	1,226.8

Table 4–6B (continued)
TEMPERATURE CONVERSIONS – SHORT TABLE[a]

Temperature (Kelvin)

$t_C = \frac{5}{9} t_F - 17.8$		$t_K = \frac{5}{9} t_F + 255.4$		$t_K = t_C + 273.15$	
2,372		1,573.2		1,300	
2,552	°F	1,673.2	K	1,400	°C
2,598		1,700		1,426.8	
2,732		1,773.2		1,500	
3,000		1,922		1,650	
3,092		1,973.2		1,700	
3,140.2		2,600		1,726.8	
3,500		2,199.8		1,926.8	
3,632	°F	2,273.2	K	2,000	°C
4,038		2,500		2,226.8	
4,532		2,773.2		2,500	
4,940		3,000		2,726.8	
5,000		3,033.2		2,760	
5,432		3,273.2		3,000	

[a]See also Tables 4–12 through 4–15.

Table compiled by A. Pigeaud, University of Cincinnati.

Table 4–7
CONVERSION OF ELECTRICAL TO
MECHANICAL UNITS (SI)

Electrical units	Mechanical equivalents[a]
Charge–coulomb (C)	$(J \cdot m)^{1/2}$
Potential–volt (V)	$(J/m)^{1/2}$
Current–ampere (A = C/s)	$\dfrac{(J \cdot m)^{1/2}}{s}$
Energy–volt-coulomb (VC)	$(J/m)^{1/2} \cdot (J \cdot m)^{1/2} = J$
Power–watt (W = VA)	$(J/m)^{1/2} \cdot \dfrac{(J \cdot m)^{1/2}}{s} = J/s$
Force–upon a unit charge in a unit electric field (V/m·C)	$(J/m^3)^{1/2} \cdot (J \cdot m)^{1/2} = J/m = N$
Force–between two unit charges according to law of electrostatic attraction-repulsion in a vacuum $\dfrac{C_1\, C_2}{m^2}$	$\dfrac{(J \cdot m)^{1/2} \cdot (J \cdot m)^{1/2}}{m^2} = J/m = N$

[a] $\dfrac{\text{joule}}{\text{meter}}$ (J/m) = newton (N) = $\dfrac{kg \cdot m}{s^2}$

Table compiled by A. Pigeaud, University of Cincinnati.

Table 4–8

Structural and material features	Log dimensions	Analytical and instrument resolutions

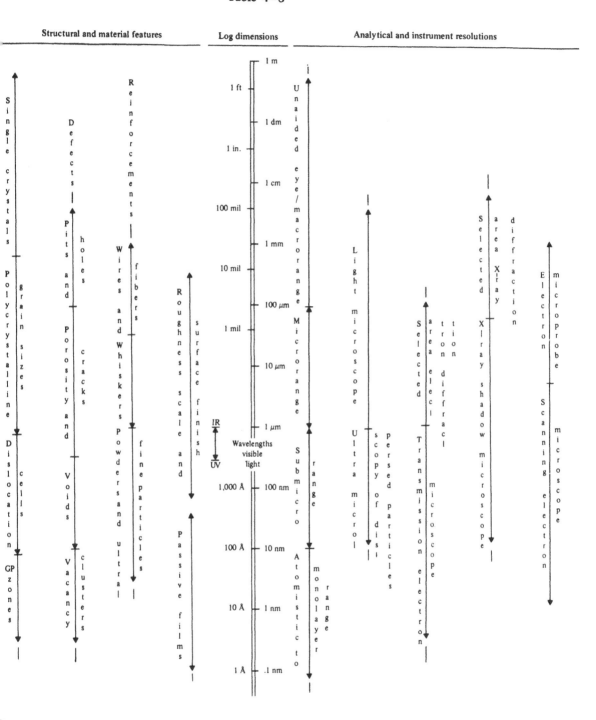

Table compiled by A. Pigeaud, University of Cincinnati.

Table 4—9

CONVERSION FACTORS

To convert from	To	Multiply by
Abamperes	Amperes	10
"	E.M. cgs. units of current	1
"	E.S. cgs. units	2.997930×10^{10}
"	Faradays (chem.)/sec	1.036377×10^{-4}
"	Faradays (phys.)/sec	1.036086×10^{-4}
"	Statamperes	2.997930×10^{10}
Abamperes/cm	E.M. cgs. units of surface charge density	1
"	E.S. cgs. units	2.997930×10^{10}
Abamperes/sq. cm	Amperes/circ. mil	5.0670748×10^{-6}
"	Amperes/sq. cm	10
"	Amperes/sq. inch	64.516
Abampere-turns	Ampere-turns	10
Abampere-turns/cm	Ampere-turns/cm	10
Abcoulombs	Ampere-hours	0.0027777
"	Coulombs	10
"	Electronic charges	6.24196×10^{19}
"	E.M. cgs. units of charge	1
"	E.S. cgs. units	2.997930×10^{10}
"	Faradays (chem.)	1.036377×10^{-4}
"	Faradays (phys.)	1.036086×10^{-4}
"	Statcoulombs	2.997930×10^{10}
Abfarads	E.M. cgs. units of capacitance	1
"	E.S. cgs. units	8.987584×10^{20}
"	Farads	1×10^{9}
"	Microfarads	1×10^{15}
"	Statfarads	8.987584×10^{20}
Abhenries	E.M. cgs. units of induction	1
"	E.S. cgs. units	1.112646×10^{-21}
"	Henries	1×10^{-9}
Abmhos	E.M. cgs. units of conductance	1
"	E.S. cgs. units	8.987584×10^{20}
"	Megamhos	1000
"	Mhos	1×10^{9}
"	Statmhos	8.987584×10^{20}
Abohms	E.M. cgs. units of resistance	1
"	Megohms	1×10^{-15}
"	Microhms	0.001
"	Ohms	1×10^{-9}
"	Statohms	1.112646×10^{-21}
Abohm-cm	Circ. mil-ohms/ft	0.0060153049
"	E.M. cgs. units of resistivity	1
"	Microhm-inches	0.00039370079
"	Ohm-cm	1×10^{-9}
Abvolts	Microvolts	0.01
"	Millivolts	1×10^{-5}
"	Volts	1×10^{-8}
"	Volts (Int.)	9.99670×10^{-9}
Abvolts/cm	E.M. cgs. units of electric field intensity	1
"	E.S. cgs. units	3.335635×10^{-11}
"	Volts/cm	1×10^{-8}
"	Volts/inch	2.54×10^{-8}
"	Volts/meter	1×10^{-6}
Acres	Sq. cm	40468564
"	Sq. ft	43560
"	Sq. ft. (U.S. Survey)	43559.826
"	Sq. inches	6272640
"	Sq. kilometers	0.0040468564
"	Sq. links (Gunter's)	1×10^{5}
"	Sq. meters	4046.8564
"	Sq. miles (statute)	0.0015625
"	Sq. perches	160
"	Sq. rods	160
"	Sq. yards	4840
Acre-feet	Cu. feet	43560
"	Cu. meters	1233.4818
"	Cu. yards	1613.333
Acre-inches	Cu. feet	3630
"	Cu. meters	102.79033

To convert from	To	Multiply by
Acre-inches	Gallons (U.S.)	27154.286
Amperes	Abamperes	0.1
"	Amperes (Int.)	1.000165
"	Cgs. units of current	1
"	Mks. units of current	1
"	Coulombs/sec	1
"	Coulombs (Int.)/sec	1.000165
"	Faradays (chem.)/sec	1.036377×10^{-5}
"	Faradays (phys.)/sec	1.036086×10^{-5}
"	Statamperes	2.997930×10^{9}
Amperes (Int.)	Amperes	0.999835
"	Coulombs/sec	0.999835
"	Coulombs (Int.)/sec	1
"	Faradays (chem.)/sec*	1.03623×10^{-5}
"	Faradays (phys.)/sec*	1.03592×10^{-5}
Amperes/meter	Cgs. units of surface current density	0.01
"	E.M. cgs. units	0.001
"	E.S. cgs. units	2.997930×10^{7}
"	Mks. units	1
Amperes/sq. meter	Cgs. units of volume current density	0.0001
"	E.M. cgs. units	1×10^{-5}
"	E.S. cgs. units	299793.0
"	Mks. units	1
Amperes/sq. mil	Abamperes/sq. cm	15500.031
"	Amperes/sq. cm	1.5500031×10^{5}
Ampere-hours	Abcoulombs	360
"	Coulombs	3600
"	Faradays (chem.)*	0.0373096
"	Faradays (phys.)*	0.0372991
Ampere-turns	Cgs. units of magnetomotive force	1.2566371
"	E.M. cgs. units	1.2566371
"	E.S. cgs. units	3.767310×10^{10}
"	Gilberts	1.2566371
Ampere-turns/weber	Cgs. units of reluctance	1.256637×10^{-8}
"	E.M. cgs. units	1.256637×10^{-8}
"	E.S. cgs. units	1.129413×10^{13}
"	Gilberts/maxwell	1.256637×10^{-8}
Ångström units	Centimeters	1×10^{-8}
"	Inches	3.9370079×10^{-9}
"	Microns	0.0001
"	Millimicrons	0.1
"	Wave length of orange-red line of krypton 86	0.000165076373
"	Wave length of red line of cadmium	0.000155316413
Ares	Acres	0.024710538
"	Sq. dekameters	1
"	Sq. feet	1076.3910
"	Sq. ft. (U.S. Survey)	1076.3867
"	Sq. meters	100
"	Sq. miles	3.8610216×10^{-5}
Atmospheres	Bars	1.01325
"	Cm. of Hg (0°C.)	76
"	Cm. of H_2O (4°C.)	1033.26
"	Dynes/sq. cm	1.01325×10^{6}
"	Ft. of H_2O (39.2°F.)	33.8995
"	Grams/sq. cm	1033.23
"	In. of Hg (32°F.)	29.9213
"	Kg./sq. cm	1.03323
"	Mm. of Hg (0°C.)	760
"	Pounds/sq. inch	14.6960
"	Tons (short)/sq. ft	1.05811
"	Torrs	760
Atomic mass units (chem.)*	Electron volts	9.31395×10^{8}
"	Grams*	1.66024×10^{-24}
Atomic mass units (phys.)*	Electron volts	9.31141×10^{8}
"	Grams*	1.65979×10^{-24}

Table 4–9

CONVERSION FACTORS

To convert from	To	Multiply by	To convert from	To	Multiply by
Bags (Brit.)	Bushels (Brit.)	3	B.t.u	Kw.-hours (Int.)	0.000292827
Barns	Sq. cm	1×10^{-24}	"	Liter-atm	10.4053
Barrels (Brit.)*	Bags (Brit.)	1.5	"	Tons of refrig. (U.S. std.)	3.46995×10^{-8}
"	Barrels (U.S., dry)	1.415404	"	Watt-seconds	1054.35
"	Barrels (U.S., liq.)	1.372513	"	Watt-seconds (Int.)	1054.18
"	Bushels (Brit.)	4.5	B.t.u. (IST.)	B.t.u.	1.00065
"	Bushels (U.S.)	4.644253	B.t.u. (mean)	B.t.u.	1.00144
"	Cu. feet	5.779568	"	B.t.u. (IST.)	1.00078
"	Cu. meters	0.1636591	"	B.t.u. (39°F.)	0.996415
"	Gallons (Brit.)	36	"	B.t.u. (60°F.)	1.00113
"	Liters	163.6546	"	Hp.-hours	0.000393317
Barrels (petroleum, U.S.)	Cu. feet	5.614583	"	Joules	1055.87
"	Gallons (U.S.)	42	"	Kg.-meters	107.669
"	Liters	158.98284	"	Kw.-hours	0.000293297
Barrels (U.S., dry)	Barrels (U.S. liq.)	0.969696	"	Kw.-hours (Int.)	0.000293248
"	Bushels (U.S.)	3.2812195	"	Liter-atm	10.4203
"	Cu. feet	4.083333	"	Watt-hours	0.293297
"	Cu. inches	7056	"	Watt-hours (Int.)	0.293248
"	Cu. meters	0.11562712	B.t.u. (39°F.)	B.t.u.	1.00504
"	Quarts (U.S., dry)	105	"	B.t.u. (IST.)	1.00439
Barrels (U.S., liq.)	Barrels (U.S., dry)	1.03125	"	B.t.u. (mean)	1.00360
"	Barrels (wine)	1	"	B.t.u. (60°F)	1.00473
"	Cu. feet	4.2109375	"	Joules	1059.67
"	Cu. inches	7276.5	B.t.u. (60°F.)	B.t.u.	1.00031
"	Cu. meters	0.11924047	"	B.t.u. (IST.)	0.999657
"	Gallons (Brit.)	26.22925	"	B.t.u. (mean)	0.998873
"	Gallons (U.S., liq.)	31.5	"	B.t.u. (39°F.)	0.995291
"	Liters	119.23713	B.t.u./hr	Cal., kg./hr.	0.251996
Bars	Atmospheres	0.986923	"	Ergs/sec	2.928751×10^{6}
"	Baryes	1×10^{6}	"	Foot-pounds/hr	777.649
"	Cm. of Hg (0°C.)	75.0062	"	Horsepower	0.000392752
"	Dynes/sq. cm	1×10^{6}	"	Horsepower (boiler)	2.98563×10^{-5}
"	Ft. of H₂O (60°F.)	33.4883	"	Horsepower (electric)	0.000392594
"	Grams/sq. cm	1019.716	"	Horsepower (metric)	0.000398199
"	In. of Hg (32°F.)	29.5300	"	Kilowatts	0.000292875
"	Kg./sq. cm	1.019716	"	Lb. ice melted/hr	0.0069714
"	Millibars	1000	"	Tons of refrig. (U.S. comm.)	8.32789×10^{-5}
"	Pounds/sq. inch	14.5038	"	Watts	0.292875
Baryes	Atmospheres	9.86923×10^{-7}	B.t.u./min	Cal., kg./min.	0.251996
"	Bars	1×10^{-6}	"	Ergs/sec	1.75725×10^{8}
"	Dynes/sq. cm	1	"	Foot-pounds/min	777.649
"	Grams/sq. cm	0.001019716	"	Horsepower	0.0235651
"	Millibars	0.001	"	Horsepower (boiler)	0.00179138
Bels	Decibels	10	"	Horsepower (electric)	0.0235556
Board feet	Cu. cm	2359.7372	"	Horsepower (metric)	0.0238920
"	Cu. feet	0.083333	"	Joules/sec	17.5725
"	Cu. inches	144	"	Kg.-meters/min	107.514
Bolts of cloth	Linear feet	120	"	Kilowatts	0.0175725
"	Meters	36.576	"	Lb. ice melted/hr	0.41828
Bougie decimales	Candles (Int.)	1.00	"	Tons of refrig. (U.S. comm.)	0.00499673
B.t.u.	B.t.u. (IST.)**	0.999346	"	Watts	17.5725
"	B.t.u. (mean)	0.998563	B.t.u. (mean)/min	B.t.u. (mean)/hr.	60
"	B.t.u. (39°F.)	0.994982	"	Cal., kg. (mean)/hr.	15.1197
"	B.t.u. (60°F.)	0.999689	"	Cal., kg. (mean)/min.	0.251996
"	Cal. gm.	251.99576	"	Ergs/sec	1.75978×10^{8}
"	Cal., gm. (IST.)	251.831	"	Foot-pounds/min	778.768
"	Cal., gm. (mean)	251.634	"	Horsepower	0.0235990
"	Cal., gm. (20°C.)	252.122	"	Horsepower (boiler)	0.00179396
"	Cu. cm.-atm.	10405.6	"	Horsepower (electric)	0.0235895
"	Ergs	1.05435×10^{10}	"	Horsepower (metric)	0.0239264
"	Foot-poundals	25020.1	"	Joules/sec	17.5978
"	Foot-pounds	777.649	"	Kg.-meters/min	107.669
"	Gram-cm	1.07514×10^{7}	"	Kilowatts	0.0175978
"	Hp.-hours	0.000392752	"	Lb. ice-melted/hr	0.41888
"	Hp.-years	4.48347×10^{-8}	B.t.u./lb	Cal., gm./gram	0.555555
"	Joules	1054.35	"	Cu. cm.-atm./gram	22.9405
"	Joules (Int.)	1054.18	"	Cu. ft.-atm./lb	0.367471
"	Kg.-meters	107.514	"	Cu. ft.-(lb./sq. in.)/lb	5.40034
"	Kw.-hours	0.000292875	"	Foot-pounds/lb	777.649
			"	Hp.-hr./lb	0.000392752

* Barrel (Brit., liq.) = Barrel (Brit., dry)
** International Steam Table.

Table 4–9 (continued)

CONVERSION FACTORS

To convert from	To	Multiply by	To convert from	To	Multiply by
B.t.u./lb	Joules/gram	2.32444	Calories, gm.**	B.t.u.	0.0039683207
B.t.u. (mean)/lb	Cal., gm. (mean)/gram	0.555555	"	B.t.u. (IST.)	0.00396573
"	Cu. cm.-atm./gram	22.9735	"	B.t.u. (mean)	0.00396262
"	Foot-pounds/lb	778.768	"	B.t.u. (39°F.)	0.00394841
"	Hp.-hr./lb	0.000393317	"	B.t.u. (60°F.)	0.00396709
"	Joules/gram	2.32779	"	Cal., gm. (IST.)	0.999346
B.t.u./sec	B.t.u./hr	3600	"	Cal., gm. (mean)	0.998563
"	B.t.u./min	60	"	Cal., gm. (15°C.)	0.999570
"	Cal., kg./hr	907.185	"	Cal., gm. (20°C.)	1.00050
"	Cal., kg./min	15.1197	"	Cal., kg	0.001
"	Cheval-vapeur	1.43352	"	Cal., kg. (IST.)	0.000999346
"	Ergs/sec	1.05435×10^{10}	"	Cal., kg. (mean)	0.000998563
"	Foot-pounds/sec	777.649	"	Cal., kg. (15°C.)	0.000999570
"	Horsepower	1.41391	"	Cal., kg. (20°C.)	0.00100050
"	Horsepower (boiler)	0.107483	"	Cu. cm.-atm	41.2929
"	Horsepower (electric)	1.41334	"	Cu. ft.-atm	0.00145824
"	Horsepower (metric)	1.43352	"	Ergs	4.184×10^7
"	Kg.-meters/sec	107.514	"	Foot-poundals	99.2878
"	Kilowatts	1.05435	"	Foot-pounds	3.08596
"	Kilowatts (Int.)	1.05418	"	Gram-cm	42664.9
"	Watts	1054.35	"	Hp.-hours	1.55857×10^{-6}
"	Watts (Int.)	1054.18	"	Joules	4.184
B.t.u. (mean)/sec	Ergs/sec	1.05587×10^{10}	"	Joules (Int.)	4.18331
"	Foot-pounds/sec	778.768	"	Kg.-meters	0.426649
"	Horsepower	1.41594	"	Kw.-hours	1.162222×10^{-6}
"	Horsepower (boiler)	0.107637	"	Liter-atm	0.0412917
"	Horsepower (electric)	1.41537	"	Watt-hours	0.001162222
"	Horsepower (metric)	1.43558	"	Watt-hours (Int.)	0.00116203
"	Watts	1055.87	"	Watt-seconds	4.184
B.t.u./sq. ft	Cal., gm./sq. cm	0.271246	Calories, gm. (mean)	B.t.u.	0.00397403
B.t.u./sq.ft. × min.)	Hp./sq. ft	0.0235651	"	Cal., gm	1.00144
"	Kw./sq. ft	0.0175725	"	Cal., gm. (IST.)	1.00078
"	Watts/sq. in	0.122031	"	Cal., gm. (20°C.)	1.00194
Buckets (Brit.)	Cu. cm	18184.35	"	Cal., kg. (mean)	0.001
"	Gallons (Brit.)	4	"	Cu. cm.-atm	41.3523
Bushels (Brit.)	Bags (Brit.)	0.333333	"	Cu. ft.-atm	0.00146034
"	Bushels (U.S.)	1.032056	"	Ergs	4.19002×10^7
"	Cu. cm	36368.70	"	Foot-poundals	99.4308
"	Cu. feet	1.284348	"	Foot-pounds	3.09040
"	Cu. inches	2219.354	"	Hp.-hours	1.56081×10^{-6}
"	Dekaliters	3.636768	"	Joules	4.19002
"	Gallons (Brit.)	8	"	Joules (Int.)	4.18933
"	Hectoliters	0.3636768	"	Kg.-meters	0.427263
"	Liters	36.36768	"	Kw.-hours	1.16390×10^{-6}
Bushels (U.S.)*	Barrels (U.S.), dry	0.3047647	"	Liter-atm	0.0413511
"	Bushels (Brit.)	0.9689395	"	Watt-seconds	4.19002
"	Cu. cm	35239.07	Calories, gm. (15°C.)	B.t.u.	0.00397003
"	Cu. feet	1.244456	"	Cal., gm	1.00043
"	Cu. inches	2150.42	"	Cal., gm. (IST.)	0.999776
"	Cu. meters	0.03523907	"	Cal., gm. (mean)	0.998992
"	Cu. yards	0.04609096	"	Cal., gm. (20°C.)	1.00093
"	Gallons (U.S., dry)	8	"	Joules	4.18580
"	Gallons (U.S., liq.)	9.309177	"	Joules (Int.)	4.18511
"	Liters	35.23808	Calories, gm. (20°C.)	B.t.u.	0.00396633
"	Ounces (U.S., fluid)	1191.575	"	Cal., gm	0.999498
"	Pecks (U.S.)	4	"	Cal., gm. (IST.)	0.998845
"	Pints (U.S., dry)	64	"	Cal., gm. (mean)	0.998061
"	Quarts (U.S., dry)	32	"	Cal., gm. (15°C.)	0.999068
"	Quarts (U.S., liq.)	37.23671	"	Joules	4.18190
Butts (Brit.)	Bushels (U.S.)	13.53503	"	Joules (Int.)	4.18121
"	Cu. feet	16.84375	Calories, kg	B.t.u.	3.9683207
"	Cu. meters	0.4769619	"	B.t.u. (IST.)	3.96573
"	Gallons (U.S.)	126	"	B.t.u. (mean)	3.96262
			"	B.t.u. (60°F.)	3.96709
			"	Cal., gm	1000
			"	Cal., kg. (mean)	0.998563
Cable lengths	Fathoms	120	"	Cal., kg. (15°C.)	0.999570
"	Feet	720	"	Cal., kg. (20°C.)	1.00050
"	Meters	219.456	"	Cu. cm.-atm	41292.86

* Stricken or struck bushel. A heaped bushel for apples of 2747.715 cu. inches was established by the U.S. Court of Customs Appeals on Feb. 15, 1912. A heaped bushel equal to 1¼ stricken bushels is also known.

** This is the calorie as defined by the U.S. National Bureau of Standards and is equal to 4.18400 joules.

Table 4–9 (continued)

CONVERSION FACTORS

To convert from	To	Multiply by
Calories, *kg.*	Ergs	4.184×10^{10}
"	Foot-poundals	99287.8
"	Foot-pounds	3085.96
"	Gram-cm	4.26649×10^{7}
"	Hp.-hours	0.00155857
"	Joules	4184
"	Kw.-hours	0.001162222
"	Liter-atm	41.2917
"	Watt-hours	1.162222
Calories, *kg.* (mean)	B.t.u.	3.97403
"	B.t.u. (IST.)	3.97144
"	B.t.u. (mean)	3.9683207
"	B.t.u. (60°F.)	3.97280
"	Cal., *gm*	1001.44
"	Cal., *gm*. (IST.)	1000.78
"	Cal., *gm*. (mean)	1000
"	Cal., *gm*. (15°C.)	1000.10
"	Cal., *gm*. (20°C.)	1001.94
"	Ergs	4.19002×10^{10}
"	Foot-poundals	99430.8
"	Foot-pounds	3090.40
"	Gram-cm	4.27263×10^{7}
"	Hp.-hours	0.00156081
"	Joules	4190.02
"	Kg.-meters	427.263
"	Kw.-hours (Int.)	0.00116370
"	Liter-atm	41.3511
"	Watt-hours	1.16390
Cal., *gm*./°C	B.t.u./°F	0.00220462
"	Joules/°F	2.324444
"	Joules (Int.)/°F	2.32406
Cal., *gm*./gram	B.t.u./lb	1.8
"	Foot-pounds/lb	1399.77
"	Joules/gram	4.184
"	Watt-hours/gram	0.001162222
Cal., *gm*./(gram × °C)	B.t.u./(lb. × °C.)	1.8
"	B.t.u./(lb. × °F.)	1
"	Cal., *kg*./(kg. × °C.)	1
"	Joules/(gram × °C.)	4.184
"	Joules/(lb. × °F.)	1054.35
Cal., *gm*./hr.	B.t.u./hr.	0.0039683207
"	Ergs/sec.	11622.222
"	Watts	0.001162222
Cal., *gm*. (mean)/hr.	B.t.u. (mean)/hr.	0.0039683207
"	Ergs/sec.	11639.0
"	Watts	0.00116390
Cal., *kg*./hr.	Watts	1.162222
Cal., *gm*./min.	B.t.u./min.	0.0039683207
"	Ergs/sec.	697333.3
"	Watts	0.069733
Cal., *gm*. (mean)/min.	B.t.u. (mean)/min.	0.0039683207
"	Ergs/sec.	698337
"	Joules/sec.	0.0698337
"	Watts	0.0698337
Cal., *kg*./min.	Kg. ice melted/min.	0.012548
"	Lb. ice melted/min.	0.027065
"	Watts	69.7333
Cal., *gm*./sec.	B.t.u./sec.	0.0039683207
"	Ergs/sec.	4.184×10^{7}
"	Foot-pounds/sec.	3.08596
"	Horsepower	0.00561084
"	Watts	4.184
Cal., *gm*. (mean)/sec.	Ergs/sec.	4.19002×10^{7}
"	Watts	4.19002
Cal., *gm*./(sec. × sq. cm.)	B.t.u./(hr. × sq. ft.)	13272.1
"	Cal., *gm*./(hr. × sq. cm.)	3600
"	Watts/sq. cm	4.184
Cal., *gm*./(sec. × sq. cm. × °C.)	B.t.u./(hr. × sq. ft. × °F.)	7373.38
Cal., *gm*./sq. cm	B.t.u./sq. ft.	3.68669

To convert from	To	Multiply by
$\dfrac{\text{Cal., } gm.\text{-cm.}}{(\text{hr.} \times \text{sq. cm.} \times °C.)}$	$\dfrac{\text{B.t.u.-ft.}}{(\text{hr.} \times \text{sq. ft.} \times °F_i)}$	0.0671969
"	$\dfrac{\text{B.t.u.-inch}}{(\text{hr.} \times \text{sq. ft.} \times °F.)}$	0.806363
Cal., *gm*.-cm./sq. cm	B.t.u.-inch/sq. ft.	1.4514530
Cal., *gm*.-sec.	Planck's constant	6.31531×10^{33}
Cal., *gm*.-sec./Avog. No. (chem.)*	Planck's constant	1.04849×10^{10}
Cal., *gm*.-sec./Avog. No. (phys.)*	Planck's constant	1.04821×10^{10}
Candles (English)	Candles (Int.)	1.04
"	Hefner units	1.16
Candles (German)	Candles (English)	1.01
"	Candles (Int.)	1.05
"	Hefner units	1.17
Candles (Int.)	Candles (English)	0.96
"	Candles (German)	0.95
"	Candles (pentane)	1.00
"	Hefner units	1.11
"	Lumens (Int.)/steradian	1
Candles (pentane)	Candles (Int.)	1.00
Candles/sq. cm	Candles/sq. inch	6.4516
"	Candles/sq. meter	10000
"	Foot-lamberts	2918.6351
"	Lamberts	3.1415927
Candles/sq. ft	Candles/sq. inch	0.0069444
"	Candles/sq. meter	10.763910
"	Foot-lamberts	3.1415927
"	Lamberts	0.0033815822
Candles/sq. inch	Candles/sq. cm.	0.15500031
"	Candles/sq. foot	144
"	Foot-lamberts	452.38934
"	Lamberts	0.48694784
Candle power (spher.)	Lumens	12.566370
Carats (parts of gold per 24 of mixture)	Milligrams/gram	41.6666
Carats (1877)	Grains	3.168
"	Milligrams	205.3
Carats (metric)	Grains	3.08647
"	Grams	0.2
"	Milligrams	200
Carcel units	Candles (Int.)	9.61
Centals	Kilograms	45.359237
"	Pounds	100
Centares	Ares	0.01
"	Sq. feet	10.763910
"	Sq. inches	1550.0031
"	Sq. meters	1
"	Sq. yards	1.1959900
Centigrams	Grains	0.15432358
"	Grams	0.01
Centiliters	Cu. cm	10
"	Cu. inches	0.6102545
"	Liters	0.01
"	Ounces (U.S., fluid)	0.3381497
Centimeters	Ångström units	1×10^{8}
"	Feet	0.032808399
"	Feet (U.S. Survey)	0.032808333
"	Hands	0.098425197
"	Inches	0.39370079
"	Links (Gunter's)	0.049709695
"	Links (Ramden's)	0.032808399
"	Meters	0.01
"	Microns	10000
"	Miles (naut., Int.)	5.3995680×10^{-6}
"	Miles (statute)	6.2137119×10^{-6}
"	Millimeters	10
"	Millimicrons	1×10^{7}
"	Mils	393.70079
"	Picas (printer's)	2.3710630
"	Points (printer's)	28.452756

Table 4–9 (continued)

CONVERSION FACTORS

To convert from	To	Multiply by	To convert from	To	Multiply by
Centimeters	Rods	0.0019883878	Circumferences	Minutes	21600
"	Wave length of orange-red line of krypton 86	16507.6373	"	Radians	6.2831853
"	Wave length of red line of cadmium	15531.6413	"	Seconds	1296000
"	Yards	0.010936133	Cords	Cord-feet	8
Cm. of Hg (0°C.)	Atmospheres	0.013157895	"	Cu. feet	128
"	Bars	0.0133322	"	Cu. meters	3.6245734
"	Dynes/sq. cm	13332.2	Cord-feet	Cords	0.125
"	Ft. of H_2O (4°C.)	0.446050	"	Cu. feet	16
"	Ft. of H_2O (60°F.)	0.446474	Coulombs	Abcoulombs	0.1
"	In. of Hg (0°C.)	0.39370079	"	Ampere-hours	0.0002777
"	Kg./sq. meter	135.951	"	Ampere-seconds	1
"	Pounds/sq. ft	27.8450	"	Coulombs (Int.)	1.000165
"	Pounds/sq. inch	0.193368	"	Electronic charge	6.24196×10^{18}
"	Torrs	10	"	E.M. cgs. units of electric charge	0.1
Cm. of H_2O (4°C.)	Atmospheres	0.000967814	"	E.S. cgs. units of electric charge	2.997930×10^9
"	Dynes/sq. cm	980.638	"	Faradays (chem.)	1.036377×10^{-5}
"	Pounds/sq. inch	0.0142229	"	Faradays (phys.)	1.036086×10^{-5}
Centimeters/sec	Feet/min	1.9685039	"	Mks. units of electric charge	1
"	Feet/sec	0.032808399	"	Statcoulombs	2.997930×10^9
"	Kilometers/hr	0.036	Coulombs/cu. meter	E.M. cgs. units of volume charge density	1×10^{-7}
"	Kilometers/min	0.0006	"	E.S. cgs. units	2997.930
"	Knots (Int.)	0.019438445	Coulombs/sq. cm	Abcoulombs/sq. cm	0.1
"	Meters/min	0.6	"	Cgs. units of polarization, *and* surface charge density	1
"	Miles/hr	0.022369363	Cubic centimeters	Board feet	0.00042377600
"	Miles/min	0.00037282272	"	Bushels (Brit.)	2.749617×10^{-5}
Cm./(sec. × sec.)	Kilometers/(hr. × sec.)	0.036	"	Bushels (U.S.)	2.837759×10^{-5}
"	Miles/(hr. × sec.)	0.022369363	"	Cu. feet	3.5314667×10^{-5}
Centimeters/year	Inches/year	0.39370079	"	Cu. inches	0.061023744
Centipoises*	Grams/(cm. × sec.)	0.01	"	Cu. meters	1×10^{-6}
"	Poises	0.01	"	Cu. yards	1.3079506×10^{-6}
"	Pound/(ft. × hr.)	2.4190883	"	Drachms (Brit., fluid)	0.28156080
"	Pounds/(ft. × sec.)	0.00067196898	"	Drams (U.S., fluid)	0.27051218
Centistokes*	Stokes	0.01	"	Gallons (Brit.)	0.0002199694
Chains (Gunter's)	Centimeters	2011.68	"	Gallons (U.S., dry)	0.00022702075
"	Chains (Ramden's)	0.66	"	Gallons (U.S., liq.)	0.00026417205
"	Feet	66	"	Gills (Brit.)	0.007039020
"	Feet (U.S. Survey)	65.999868	"	Gills (U.S.)	0.0084535058
"	Furlongs	0.1	"	Liters	0.001
"	Inches	792	"	Ounces (Brit., fluid)	0.03519510
"	Links (Gunter's)	100	"	Ounces (U.S., fluid)	0.033814023
"	Links (Ramden's)	66	"	Pints (U.S., dry)	0.0018161660
"	Meters	20.1168	"	Pints (U.S., liq.)	0.0021133764
"	Miles (statute)	0.0125	"	Quarts (Brit.)	0.0008798775
"	Rods	4	"	Quarts (U.S., dry)	0.00090808298
"	Yards	22	"	Quarts (U.S., liq.)	0.0010566882
Chains (Ramden's)	Centimeters	3048	Cu. cm./gram	Cu. ft./lb	0.016018463
"	Chains (Gunter's)	1.515151	Cu. cm./sec	Cu. ft./min	0.0021188800
"	Feet	100	"	Gal. (U.S.)/min	0.015850323
"	Feet (U.S. Survey)	99.999800	"	Gal. (U.S.)/sec	0.00026417205
Cheval-vapeur	Horsepower (metric)	1	Cu. cm.-atm	B.t.u.	9.61019×10^{-5}
Cheval-vapeur-heures	Joules	2647795	"	B.t.u. (mean)	9.59637×10^{-5}
Circles	Degrees	360	"	Cal., *gm.*	0.0242173
"	Grades	400	"	Cal., *gm.* (mean)	0.0241824
"	Minutes	21600	"	Cu. ft.-atm	3.5314667×10^{-5}
"	Radians	6.2831853	"	Joules	0.101325
"	Signs	12	"	Watt-hours	2.81458×10^{-5}
Circular inches	Circular mm	645.16	Cu. cm.-atm./gram	B.t.u./lb	0.0435911
"	Sq. cm	5.0670748	"	Cal., *gm.*/gram	0.0242173
"	Sq. inches	0.78539816	"	Cu. ft.-(lb./sq. in.)/lb	0.235406
Circular mm	Sq. cm	0.0078539816	"	Ft.-lb./lb	33.8985
"	Sq. inches	0.0012173696	"	Joules/gram	0.101325
"	Sq. mm	0.78539816	"	Kg.-meters/gram	0.0103323
Circular mils	Circular inches	1×10^{-6}	"	Kw.-hr./gram	2.81458×10^{-5}
"	Sq. cm	5.0670748×10^{-6}	Cubic decimeters	Cu. cm	1000
"	Sq. inches	7.8539816×10^{-7}	"	Cu. feet	0.035316667
"	Sq. mm	0.00050670748	"	Cu. inches	61.023744
"	Sq. mils	0.78539816	"	Cu. meters	0.001
Circumferences	Degrees	360			
"	Grades	400			

Table 4–9 (continued)

CONVERSION FACTORS

To convert from	To	Multiply by	To convert from	To	Multiply by
Cubic decimeters.....	Cu. yards................	0.0013079506	Cubic inches.........	Ounces (U.S., fluid).......	0.55411255
''	Liters....................	1	''	Pecks (U.S.)...............	0.0018601017
Cubic dekameters....	Cu. decimeters...........	1 × 10⁶	''	Pints (U.S., dry).........	0.029761628
''	Cu. feet.................	35314.667	''	Pints (U.S., liq.).........	0.034632035
''	Cu. inches...............	6.1023744 × 10⁷	''	Quarts (U.S., dry).........	0.014880814
''	Cu. meters...............	1000	''	Quarts (U.S., liq.).........	0.017316017
''	Liters..................	999972	Cu. in. of H₂O (4°C.).	Pounds of H₂O...........	0.0361263
Cubic feet..........	Acre-feet................	2.2956841 × 10⁻⁵	Cu. in. of H₂O (60°F.).	Pounds of H₂O...........	0.0360916
''	Board feet...............	12	Cubic meters.......	Acre-feet................	0.00081071319
''	Bushels (Brit.)...........	0.7786049	''	Barrels (Brit.)...........	6.110261
''	Bushels (U.S.)...........	0.80356395	''	Barrels (U.S., dry)........	8.648490
''	Cords (wood).............	0.0078125	''	Barrels (U.S., liq.)........	8.3864145
''	Cord-feet................	0.0625	''	Bushels (Brit.)...........	27.49617
''	Cu. centimeters..........	28316.847	''	Bushels (U.S.)...........	28.377593
''	Cu. meters...............	0.028316847	''	Cu. cm..................	1 × 10⁶
''	Gallons (U.S., dry).......	6.4285116	''	Cu. feet................	35.314667
''	Gallons (U.S., liq.)......	7.4805195	''	Cu. inches..............	61023.74
''	Liters...................	28.316847	''	Cu. yards...............	1.3079506
''	Ounces (Brit., fluid)......	996.6143	''	Gallons (Brit.)..........	219.9694
''	Ounces (U.S., fluid)......	957.50649	''	Gallons (U.S., liq.)......	264.17205
''	Pints (U.S., liq.).......	59.844156	''	Hogshead...............	4.1932072
''	Quarts (U.S., dry)........	25.714047	''	Liters..................	1000
''	Quarts (U.S., liq.).......	29.922078	''	Pints (U.S., liq.).......	2113.3764
Cu. ft. of H₂O			''	Quarts (U.S., liq.).......	1056.6882
(39.2°F.)........	Pounds of H₂O...........	62.4262	''	Steres.................	1
Cu. ft. of H₂O (60°F.)	Pounds of H₂O...........	62.3663	Cu. meters/min.....	Gal. (Brit.)/min.........	219.9694
Cu. ft./hr........	Acre-feet/hr.............	2.2956841 × 10⁻⁵	''	Gal. (U.S.)/min.........	264.1721
''	Cu. cm./sec.............	7.8657907	''	Liters/min..............	999.972
''	Cu. ft./day.............	24	Cu. millimeters	Cu. cm.................	0.001
''	Gal. (U.S.)/hr...........	7.4805195	''	Cu. inches..............	6.1023744 × 10⁻⁵
''	Liters/hr................	28.31605	''	Cu. meters..............	1 × 10⁻⁹
Cu. ft./min.......	Acre-feet/hr.............	0.0013774105	''	Minims (Brit.)..........	0.01689365
''	Acre-feet/min............	2.2956841 × 10⁻⁵	''	Minims (U.S.)..........	0.016230731
''	Cu. cm./sec.............	471.94744	Cu. yards.........	Bushels (Brit.)..........	21.02233
''	Cu. ft./hr..............	60	''	Bushels (U.S.)..........	21.696227
''	Gal. (U.S.)/min.........	7.4805195	''	Cu. cm.................	764554.86
''	Liters/sec..............	0.4719342	''	Cu. feet................	27
Cu. ft./lb........	Cu. cm./gram............	62.427961	''	Cu. inches..............	46656
''	Millimeters/gram.........	62.42621	''	Cu. meters..............	0.76455486
Cu. ft./sec........	Acre-inches/hr...........	0.99173553	''	Gallons (Brit.)..........	168.1787
''	Cu. cm./sec.............	28316.847	''	Gallons (U.S., dry).......	173.56981
''	Cu. yards/min...........	2.222222	''	Gallons (U.S., liq.)......	201.97403
''	Gal. (U.S.)/min.........	448.83117	''	Liters..................	764.55486
''	Liters/min..............	1698.963	''	Quarts (Brit.)..........	672.7146
''	Liters/sec..............	28.31605	''	Quarts (U.S., dry)........	694.27926
Cu. ft. of H₂O			''	Quarts (U.S., liq.).......	807.89610
(60°F.)/sec......	Lb. of H₂O/min.........	3741.98	Cu. yd./min.........	Cu. ft./sec.............	0.45
Cu. ft.-atm......	B.t.u...................	2.72130	''	Gal. (U.S.)/sec.........	3.3662338
''	Cal., gm...............	685.756	''	Liters/sec..............	12.74222
''	Cu. cm.-atm............	28316.847	Cubits............	Centimeters.............	45.72
''	Cu. ft.-(lb/sq. in.).......	14.6960	''	Feet...................	1.5
''	Foot-pounds.............	2116.22	''	Inches.................	18
''	Hp.-hours..............	0.00106880			
''	Joules.................	2869.20	Daltons (chem.).....	Grams.................	1.66024 × 10⁻²⁴
''	Kg.-meters.............	292.577	Daltons (phys.).....	Grams.................	1.65979 × 10⁻²⁴
''	Kw.-hours.............	0.000797001	Days (mean solar).....	Days (sidereal)..........	1.00273791
Cubic inches......	Barrels (Brit.)..........	0.0001001292	''	Hours (mean solar).....	24
''	Barrels (U.S., dry)........	0.00014172336	''	Hours (sidereal).........	24.065710
''	Board feet..............	0.0069444	''	Years (calendar)........	0.0027397260
''	Bushels (Brit.)..........	0.0004505815	''	Years (sidereal).........	0.0027378031
''	Bushels (U.S.)..........	0.00046502544	''	Years (tropical).........	0.0027379093
''	Cu. cm.................	16.387064	Days (sidereal)......	Days (mean solar)........	0.99726957
''	Cu. feet................	0.00057870370	''	Hours (mean solar).....	23.934470
''	Cu. meters..............	1.6387064 × 10⁻⁵	''	Hours (sidereal).........	24
''	Cu. yards...............	2.1433470 × 10⁻⁵	''	Minutes (mean solar).....	1436.0682
''	Drams (U.S., fluid).......	4.4329004	''	Minute (sidereal)........	1440
''	Gallons (Brit.)..........	0.003604652	''	Second (sidereal)........	86400
''	Gallons (U.S., dry).......	0.0037202035	''	Years (calendar)........	0.0027322454
''	Gallons (U.S., liq.)......	0.0043290043	''	Years (sidereal).........	0.0027303277
''	Liters..................	0.016387064	''	Years (tropical).........	0.0027304336
''	Milliliters..............	16.387064	Decibels...........	Bels...................	0.1
''	Ounces (Brit., fluid)......	0.5767444	Decimeters.........	Centimeters.............	10

Table 4–9 (continued)

CONVERSION FACTORS

To convert from	To	Multiply by	To convert from	To	Multiply by
Decimeters	Feet	0.32808399	Dynes/sq. cm	Cm. of H_2O (4°C.)	0.001019745
"	Feet (U.S. Survey)	0.328083333	"	Grams/sq. cm	0.001019716
"	Inches	3.9370079	"	In. of Hg (32°F.)	2.95300×10^{-5}
"	Meters	0.1	"	In. of H_2O (4°C.)	0.000401474
Decisteres	Cu. meters	0.1	"	Kg./sq. meter	0.01019716
Degrees	Circles	0.0027777	"	Poundals/sq. in	0.00046664510
"	Minutes	60	"	Pounds/sq. in	1.450377×10^{-5}
"	Quadrants	0.0111111	Dyne-centimeters	Ergs	1
"	Radians	0.017453293	"	Foot-poundals	2.3730360×10^{-6}
"	Seconds	3600	"	Foot-pounds	7.37562×10^{-8}
Degrees/cm	Radians/cm	0.017453293	"	Gram-cm	0.001019716
Degrees/foot	Radians/cm	0.00057261458	"	Inch-pounds	8.85075×10^{-7}
Degrees/inch	Radian/cm	0.0068713750	"	Kg.-meters	1.019716×10^{-8}
Degrees/min	Degrees/sec	0.0166666	"	Newton-meters	1×10^{-7}
"	Radians/sec	0.00029088821			
"	Revolutions/sec	4.629629×10^{-5}	Electron volts	Ergs	1.60219×10^{-12}
Degrees/sec	Radians/sec	0.017453293	"	Grams	1.78253×10^{-33}
"	Revolutions/min	0.166666	Electronic charges	Abcoulombs	1.60209×10^{-20}
"	Revolutions/sec	0.0027777	"	Coulombs	1.60209×10^{-19}
Dekaliters	Pecks (U.S.)	1.135136	"	Statcoulombs	4.80296×10^{-10}
"	Pints (U.S., dry)	18.16217	Electronic charges/kg.	Statcoulombs/dyne	4.89766×10^{-14}
Dekameters	Centimeters	1000	E.S. cgs. units of induction flux	E.M. cgs. units	2.997930×10^{10}
"	Feet	32.808399	E.S. cgs. units of magnetic charge	E.M. cgs. units	2.997930×10^{10}
"	Feet (U.S. Survey)	32.808333	E.S. cgs. units of magnetic field intensity	E.M. cgs. units	3.335635×10^{-11}
"	Inches	393.70079	Ells	Centimeters	114.3
"	Kilometers	0.01	"	Inches	45
"	Meters	10	Ergs	B.t.u.	9.48451×10^{-11}
"	Yards	10.93613	"	Cal., *gm.*	2.39006×10^{-8}
Demals	Gram-equiv./cu. decimeter	1	"	Cal., *kg.*	2.39006×10^{-11}
Drachms (Brit., fluid)	Cu. cm	3.551631	"	Cal., *kg.* (20°C.)	2.39126×10^{-11}
"	Cu. inches	0.2167338	"	Cu. cm.-atm.	9.86923×10^{-7}
"	Drams (U.S., fluid)	0.9607594	"	Cu. ft.-atm.	3.48529×10^{-11}
"	Milliliters	3.551531	"	Cu. ft.-(lb./sq. in.)	5.12196×10^{-10}
Drams (apoth. *or* troy)	Drams (avdp.)	2.1942857	"	Dyne-cm	1
"	Grains	60	"	Electron volts	6.24145×10^{11}
"	Grams	3.8879346	"	Foot-poundals	2.3730360×10^{-6}
"	Ounces (apoth. *or* troy)	0.125	"	Foot-pounds	7.37562×10^{-8}
"	Ounces (avdp.)	0.13714286	"	Gram-cm	0.001019716
"	Scruples (apoth.)	3	"	Joules	1×10^{-7}
Drams (avdp.)	Drams (apoth. *or* troy)	0.455729166	"	Joules (Int.)	9.99835×10^{-8}
"	Grains	27.34375	"	Kw.-hours	2.777777×10^{-14}
"	Grams	1.7718452	"	Kg.-meters	1.019716×10^{-8}
"	Ounces (apoth. *or* troy)	0.056966146	"	Liter-atm	9.86895×10^{-10}
"	Ounces (avdp.)	0.0625	"	Watt-sec	1×10^{-7}
"	Pennyweights	1.1393229	Ergs/(gram-mol. × °C.)	Foot-pounds/(lb.-mol. × °F.)	1.85863×10^{-5}
"	Pounds (apoth. *or* troy)	0.0047471788	Ergs/sec	B.t.u./min	5.69071×10^{-9}
"	Pounds (avdp.)	0.00390625	"	Cal., *gm.*/min	1.43403×10^{-6}
"	Scruples (apoth.)	1.3671875	"	Dyne-cm./sec	1
Drams (U.S., fluid)	Cu. cm	3.6967162	"	Foot-pounds/min	4.42537×10^{-6}
"	Cu. inches	0.22558594	"	Gram-cm./sec	0.001019716
"	Drachms (Brit., fluid)	1.040843	"	Horsepower	1.34102×10^{-10}
"	Gills (U.S.)	0.03125	"	Joules/sec	1×10^{-7}
"	Milliliters	3.696588	"	Kilowatts	1×10^{-10}
"	Minims (U.S.)	60	"	Watts	1×10^{-7}
"	Ounces (U.S., fluid)	0.125	Ergs/sq. cm	Dynes/cm	1
"	Pints (U.S., liq.)	0.0078125	"	Ergs/sq. mm	0.01
Dynes	Grains	0.01573663	Ergs/sq. mm	Dynes/cm	100
"	Grams	0.001019716	"	Ergs/sq. cm	100
"	Newtons	0.00001	Erg-sec	Planck's constant	1.50932×10^{26}
"	Poundals	7.2330138×10^{-5}	Farads	Abfarads	1×10^{-9}
"	Pounds	2.248089×10^{-6}	"	E.M. cgs. units	1×10^{-9}
Dynes/cm	Ergs/sq. cm	1	"	E.S. cgs. units	8.987584×10^{11}
"	Ergs/sq. mm	0.01	"	Farads (Int.)	1.000495
"	Grams/cm	0.001019716	"	Microfarads	1×10^{6}
"	Poundals/inch	0.00018371855	"	Statfarads	8.98758×10^{11}
Dynes/cu. cm	Grams/cu. cm	0.001019716	Farads (Int.)	Farads	0.999505
"	Poundals/cu. inch	0.0011852786	Fathoms	Centimeters	182.88
Dynes/sq. cm	Atmospheres	9.86923×10^{-7}			
"	Bars	1×10^{-6}			
"	Baryes	1			
"	Cm. of Hg (0°C.)	7.50062×10^{-5}			

Table 4–9 (continued) .

CONVERSION FACTORS

To convert from	To	Multiply by	To convert from	To	Multiply by
Fathoms	Feet	6	Feet/(sec. × sec.)	Meters/(sec. × sec.)	0.3048
"	Inches	72	"	Miles/(hr. × sec.)	0.68181818
"	Meters	1.8288	Firkins (Brit.)	Bushels (Brit.)	1.125
"	Miles (naut., Int.)	0.00098747300	"	Cu. cm	40914.79
"	Miles (statute)	0.001136363	"	Cu. feet	1.444892
"	Yards	2	"	Firkins (U.S.)	1.200949
Feet	Centimeters	30.48	"	Gallons (Brit.)	9
"	Chains (Gunter's)	0.01515151	"	Liters	40.91364
"	Fathoms	0.166666	"	Pints (Brit.)	72
"	Feet (U.S. Survey)	0.99999800	Firkins (U.S.)	Barrels (U.S., dry)	0.29464286
"	Furlongs	0.00151515	"	Barrels (U.S., liq.)	0.28571429
"	Inches	12	"	Bushels (U.S.)	0.96678788
"	Meters	0.3048	"	Cu. feet	1.203125
"	Microns	304800	"	Firkins (Brit.)	0.8326747
"	Miles (naut., Int.)	0.00016457883	"	Liters	34.06775
"	Miles (statute)	0.000189393	"	Pints (U.S., liq.)	72
"	Rods	0.060606	Foot-candles	Lumens/sq. ft	1
"	Ropes (Brit.)	0.05	"	Lumens/sq. meter	10.763910
"	Yards	0.333333	"	Lux	10.763910
Feet (U.S. Survey)	Centimeters	30.480061	"	Milliphots	1.0763910
"	Chains (Gunter's)	0.015151545	Foot-lamberts	Candles/sq. cm.	0.00034262591
"	Chains (Ramden's)	0.010000020	"	Candles/sq. ft.	0.31830989
"	Feet	1.0000020	"	Millilamberts	1.0763910
"	Inches	12.000024	"	Lamberts	0.0010763910
"	Links (Gunter's)	1.5151545	"	Lumens/sq. ft	1
"	Links (Ramden's)	1.0000020	Foot-poundals	B.t.u.	3.99678 × 10⁻⁵
"	Meters	0.30480061	"	B.t.u. (IST.)	3.99417 × 10⁻⁵
"	Miles (statute)	0.00018939432	"	B.t.u. (mean)	3.99104 × 10⁻⁵
"	Rods	0.060606182	"	Cal., gm.	0.0100717
"	Yards	0.33333400	"	Cal., gm. (IST.)	0.0100651
Feet of air (1 atm., 60°F.)			"	Cal., gm. (mean)	0.0100573
"	Atmospheres	3.6083 × 10⁻⁵	"	Cu. cm.-atm	0.415890
"	Ft. of Hg (32°F.)	0.00089970	"	Cu. ft.-atm	1.46870 × 10⁻⁵
"	Ft. of H₂O (60°F.)	0.0012244	"	Dyne-cm	4.2140110 × 10⁵
"	In. of Hg (32°F.)	0.0010796	"	Ergs	4.2140110 × 10⁵
"	Pounds/sq. inch	0.00053027	"	Foot-pounds	0.0310810
Feet of Hg (32°F.)	Cm. of Hg (0°C.)	30.48	"	Hp.-hours	1.56974 × 10⁻⁸
"	Ft. of H₂O (60°F.)	13.6085	"	Joules	0.042140110
"	In. of H₂O (60°F.)	163.302	"	Joules (Int.)	0.0421332
"	Ounces/sq. inch	94.3016	"	Kg.-meters	0.00429710
"	Pounds/sq. inch	7.89385	"	Kw.-hours	1.17056 × 10⁻⁸
Feet of H₂O (4°C.)	Atmospheres	0.0294990	"	Liter-atm	0.000415879
"	Cm. of Hg (0°C.)	2.24192	Foot-pounds	B.t.u.	0.00128593
"	Dynes/sq. cm	29889.8	"	B.t.u. (IST.)	0.00128509
"	Grams/sq. cm	30.4791	"	B.t.u. (mean)	0.00128408
"	In. of Hg (32°F.)	0.882646	"	Cal., gm.	0.324048
"	Kg./sq. meter	304.791	"	Cal., gm. (IST.)	0.323836
"	Pounds/sq. inch	0.433515	"	Cal., gm. (mean)	0.323582
Feet/hour	Cm./hr.	30.48	"	Cal., gm. (20°C.)	0.324211
"	Cm./min.	0.508	"	Cal., kg.	0.000324048
"	Cm./sec.	0.0084666	"	Cal., kg. (IST.)	0.000323836
"	Feet/min	0.0166666	"	Cal., kg. (mean)	0.000323582
"	Inches/hr.	12	"	Cu. ft.-atm	0.000472541
"	Kilometers/hr.	0.0003048	"	Dyne-cm	1.35582 × 10⁷
"	Kilometers/min	5.08 × 10⁻⁶	"	Ergs	1.35582 × 10⁷
"	Knots (Int.)	0.0001645788	"	Foot-poundals	32.1740
"	Miles/hr.	0.000189393	"	Gram-cm	13825.5
"	Miles/min	3.156565 × 10⁻⁶	"	Hp.-hours	5.05050 × 10⁻⁷
"	Miles/sec	5.2609428 × 10⁻⁸	"	Joules	1.35582
Feet/minute	Cm./sec.	0.508	"	Kg.-meters	0.138255
"	Feet/sec	0.0166666	"	Kw.-hours	3.76616 × 10⁻⁷
"	Kilometers/hr	0.018288	"	Kw.-hours (Int.)	3.76554 × 10⁻⁷
"	Meters/min	0.3048	"	Liter-atm	0.0133805
"	Meters/sec	0.00508	"	Newton-meters	1.3558180
"	Miles/hr	0.01136363	"	Lb. H₂O evap. from and at 212°F	1.3245 × 10⁻⁸
Feet/second	Cm./sec	30.48	"	Watt-hours	0.000376616
"	Kilometers/hr	1.09728	Foot-pounds/hr.	B.t.u./min	2.14321 × 10⁻⁵
"	Kilometers/min	0.018288	"	B.t.u. (mean)/min	2.14013 × 10⁻⁵
"	Meters/min	18.288	"	Cal., gm./min	0.00540080
"	Miles/hr	0.68181818	"	Cal., gm. (mean)/min	0.00539304
"	Miles/min	0.01136363	"	Ergs/min	2.25970 × 10⁵
Feet/(sec. × sec.)	Kilometers/(hr. × sec.)	1.09728			

Table 4–9 (continued)

CONVERSION FACTORS

To convert from	To	Multiply by	To convert from	To	Multiply by
Foot-pounds/hr	Foot-pounds/min	0.0166666	Gallons (U.S., dry)	Cu. inches	268.8025
"	Horsepower	5.050505×10^{-7}	"	Gallons (U.S., liq.)	1.16364719
"	Horsepower (metric)	5.12055×10^{-7}	"	Liters	4.404760
"	Kilowatts	3.76616×10^{-7}	Gallons (U.S., liq.)	Acre-feet	3.0688833×10^{-6}
"	Watts	0.000376616	"	Barrels (U.S., liq.)	0.031746032
"	Watts (Int.)	0.000376554	"	Barrels (petroleum, U.S.)	0.023809524
Foot-pounds/min	B.t.u./sec	2.14321×10^{-5}	"	Bushels (U.S.)	0.10742082
"	B.t.u. (mean)/sec	2.14013×10^{-5}	"	Cu. centimeters	3785.4118
"	Cal., gm./sec	0.00540080	"	Cu. feet	0.133680555
"	Cal., gm. (mean)/sec	0.00539304	"	Cu. inches	231
"	Ergs/sec	2.25970×10^{5}	"	Cu. meters	0.0037854118
"	Foot-pounds/sec	0.0166666	"	Cu. yards	0.0049511317
"	Horsepower	3.030303×10^{-5}	"	Gallons (Brit.)	0.8326747
"	Horsepower (metric)	3.07233×10^{-5}	"	Gallons (U.S., dry)	0.85936701
"	Joules/sec	0.0225970	"	Gallons (wine)	1
"	Joules (Int.)/sec	0.0225932	"	Gills (U.S.)	32
"	Kilowatts	2.25970×10^{-5}	"	Liters	3.7854118
"	Watts	0.0225970	"	Minims (U.S.)	61440
Foot-pounds/lb	B.t.u./lb	0.00128593	"	Ounces (U.S., fluid)	128
"	B.t.u. (IST.)/lb	0.00128509	"	Pints (U.S., liq.)	8
"	B.t.u. (mean)/lb	0.00128408	"	Quarts (U.S., liq.)	4
"	Cal., gm./gm	0.000714404	Gallons (U.S.) of H₂O (4°C.) in air	Lb. of H₂O	8.33585
"	Cal., gm. (IST.)/gram	0.000713937	Gallons (U.S.) of H₂O (60°F.) in air	Lb. of H₂O	8.32823
"	Cal., gm. (mean)/gram	0.000713377	Gallons (U.S.)/day	Cu. ft./hr	0.0055700231
"	Hp.-hr./lb	5.05050×10^{-7}	Gallons (Brit.)/hr	Cu. meters/min	7.576812×10^{-5}
"	Joules/gram	0.00298907	Gallons (U.S.)/hr	Acre-feet/hr	3.0688833×10^{-6}
"	Kg.-meters/gram	0.000304800	"	Cu. ft./hr	0.1336805
"	Kw.-hr./gram	8.30296×10^{-10}	"	Cu. meters/min	6.3090197×10^{-5}
Foot-pounds/sec	B.t.u./min	0.0771556	"	Cu. yd./min	8.2518861×10^{-5}
"	B.t.u. (mean)/min	0.0770447	"	Liters/hr	3.7854118
"	B.t.u./sec	0.00128593	Gal. (Brit.)/sec	Cu. cm./sec	4546.087
"	B.t.u. (mean)/sec	0.00128408	Gal. (U.S.)/sec	Cu. cm./sec	3785.4118
"	Cal., gm./sec	0.324048	"	Cu. ft./min	8.020833
"	Cal., gm. (mean)/sec	0.323582	"	Cu. yd./min	0.29706790
"	Ergs/sec	1.35582×10^{7}	"	Liters/min	227.1183
"	Gram-cm./sec	13825.5	Gammas	Grams	1×10^{-6}
"	Horsepower	0.00181818	"	Micrograms	1
"	Joules/sec	1.35582	Gausses	E.M. cgs. units of magnetic flux density	1
"	Kilowatts	0.00135582	"	E.S. cgs. units	3.335635×10^{-11}
"	Watts	1.35582	"	Gausses (Int.)	0.999670
"	Watts (Int.)	1.35559	"	Maxwells/sq. cm.	1
Furlongs	Centimeters	20116.8	"	Lines/sq. cm	1
"	Chains (Gunter's)	10	"	Lines/sq. inch	6.4516
"	Chains (Ramden's)	6.6	Gausses (Int.)	Gausses	1.000330
"	Feet	660	Gausses/oersted	E.M. cgs. units of permeability	1
"	Inches	7920	"	E.S. cgs. units	1.112646×10^{-21}
"	Meters	201.168	Geepounds	Slugs	1
"	Miles (naut., Int.)	0.10862203	"	Kilograms	14.5939
"	Miles (statute)	0.125	Gigameters	Meters	1×10^{9}
"	Rods	40	Gilberts	Abampere-turns	0.079577472
"	Yards	220	"	Ampere-turns	0.79577472
Gallons (Brit.)	Barrels (Brit.)	0.027777	"	E.M. cgs. units of mmf., *or* magnetic potential	1
"	Bushels (Brit.)	0.125	"	E.S. cgs. units	2.997930×10^{10}
"	Cu. centimeters	4546.087	"	Gilberts (Int.)	1.000165
"	Cu. feet	0.1605436	Gilberts (Int.)	Gilberts	0.999835
"	Cu. inches	277.4193	Gilberts/cm	Ampere-turns/cm	0.79577472
"	Drachms (Brit. fluid)	1280	"	Ampere-turns/in	2.0212678
"	Firkins (Brit.)	0.111111	"	Oersteds	1
"	Gallons (U.S., liq.)	1.200949	Gilberts/maxwell	Ampere-turns/weber	7.957747×10^{7}
"	Gills (Brit.)	32	"	E.M. cgs. units of reluctance	1
"	Liters	4.545960	"	E.S. cgs. units	8.987584×10^{20}
"	Minims (Brit.)	76800	Gills (Brit.)	Cu. cm	142.0652
"	Ounces (Brit., fluid)	160	"	Gallons (Brit.)	0.03125
"	Ounces (U.S., fluid)	153.7215	"	Gills (U.S.)	1.200949
"	Pecks (Brit.)	0.5	"	Liters	0.1420613
"	Lb. of H₂O (62°F.)	10	"	Ounces (Brit., fluid)	5
Gallons (U.S., dry)	Barrels (U.S., dry)	0.038095592			
"	Barrels (U.S., liq.)	0.036941181			
"	Bushels (U.S.)	0.125			
"	Cu. centimeters	4404.8828			
"	Cu. feet	0.15555700			

Table 4–9 (continued)

CONVERSION FACTORS

To convert from	To	Multiply by
Gills (Brit.)	Ounces (U.S., fluid)	4.803764
"	Pints (Brit.)	0.25
Gills (U.S.)	Cu. cm	118.29412
"	Cu. inches	7.21875
"	Drams (U.S., fluid)	32
"	Gallons (U.S., liq.)	0.03125
"	Gills (Brit.)	0.8326747
"	Liters	0.1182908
"	Minims (U.S.)	1920
"	Ounces (U.S., fluid)	4
"	Pints (U.S., liq.)	0.25
"	Quarts (U.S., liq.)	0.125
Grades	Circles	0.0025
"	Circumferences	0.0025
"	Degrees	0.9
"	Minutes	54
"	Radians	0.015707963
"	Revolutions	0.0025
"	Seconds	3240
Grains	Carats (metric)	0.32399455
"	Drams (apoth. or troy)	0.016666
"	Drams (avdp.)	0.036571429
"	Dynes	63.5460
"	Grams	0.06479891
"	Milligrams	64.79891
"	Ounces (apoth. or troy)	0.0020833
"	Ounces (avdp.)	0.0022857143
"	Pennyweights	0.041666
"	Pounds (apoth. or troy)	0.000173611
"	Pounds (avdp.)	0.00014285714
"	Scruples (apoth.)	0.05
"	Tons (metric)	6.479891×10^{-8}
Grains/cu. ft.	Grams/cu. meter	2.2883519
Grains/gal. (U.S.)	Parts/million*	17.11854
"	Pounds/million gal	142.8571
Grams	Carats (metric)	5
"	Decigrams	10
"	Dekagrams	0.1
"	Drams (apoth. or troy)	0.25720597
"	Drams (avdp.)	0.56438339
"	Dynes	980.665
"	Grains	15.432358
"	Kilograms	0.001
"	Micrograms	1×10^{6}
"	Myriagrams	0.0001
"	Ounces (apoth. or troy)	0.032150737
"	Ounces (avdp.)	0.035273962
"	Pennyweights	0.64301493
"	Poundals	0.0709316
"	Pounds (apoth. or troy)	0.0026792289
"	Pounds (avdp.)	0.0022046226
"	Scruples (apoth.)	0.77161792
"	Tons (metric)	1×10^{-6}
Grams/cm	Dynes/cm	980.665
"	Grams/inch	2.54
"	Kg./km	100
"	Kg./meter	0.1
"	Poundals/inch	0.180166
"	Pounds/ft	0.067196898
"	Pounds/inch	0.0055997415
"	Tons (metric)/km	0.1
Grams/(cm. × sec.)	Poises	1
"	Lb./(ft. × sec.)	0.06719690
Grams/cu. cm	Dynes/cu. cm	980.665
"	Grains/milliliter	15.43279
"	Grams/milliliter	1
"	Poundals/cu. inch	1.16236
"	Pounds/circ. mil-ft.	3.4049170×10^{-7}
"	Pounds/cu. ft	62.427961
"	Pounds/cu. inch	0.036127292
"	Pounds/gal. (Brit.)	10.02241
Grams/cu. cm	Pounds/gal. (U.S., dry)	9.7111064
"	Pounds/gal. (U.S., liq.)	8.3454044
Grams/cu. meter	Grains/cu. ft	0.43699572
Grams/liter	Parts/million*	1000
"	Lb./cu. ft	0.06242621
"	Lb./gal. (U.S.)	8.345171×10^{-3}
Grams/milliliter	Grams/cu. cm	1
"	Pounds/cu. ft	62.42621
"	Pounds/gallon (U.S.)	8.345171
Grams/sq. cm	Atmospheres	0.000967841
"	Bars	0.000980665
"	Cm. of Hg. (0°C.)	0.0735559
"	Dynes/sq. cm	980.665
"	In. of Hg (32°F.)	0.0289590
"	Kg./sq. meter	10
"	Mm. of Hg (0°C.)	0.735559
"	Poundals/sq. inch	0.457623
"	Pounds/sq. inch	0.014223343
Grams/ton (long)	Milligrams/kg	0.98420653
Grams/ton (short)	Milligrams/kg	1.1023113
Grams-cm	B.t.u.	9.30113×10^{-8}
"	B.t.u. (IST.)	9.29505×10^{-8}
"	B.t.u. (mean)	9.28776×10^{-8}
"	Cal., gm.	2.34385×10^{-5}
"	Cal., gm. (IST.)	2.34231×10^{-5}
"	Cal., gm. (mean)	2.34048×10^{-5}
"	Cal., gm. (15°C.)	2.34284×10^{-5}
"	Cal., gm, (20°C.)	2.34502×10^{-5}
"	Cal., kg.	2.34385×10^{-8}
"	Cal., kg. (IST.)	2.34231×10^{-8}
"	Cal., kg. (mean)	2.34048×10^{-8}
"	Dyne-cm	980.665
"	Ergs	980.665
"	Foot-poundals	0.00232715
"	Foot-pounds	7.2330138×10^{-5}
"	Hp.-hours	3.65303×10^{-11}
"	Joules	9.80665×10^{-5}
"	Kw.-hours	2.72407×10^{-11}
"	Kw.-hours (Int.)	2.72362×10^{-11}
"	Newton-meters	9.80665×10^{-5}
"	Watt-hours	2.72407×10^{-8}
Gram-cm./sec.	B.t.u./sec.	9.30113×10^{-8}
"	Cal., gm./sec.	2.34385×10^{-5}
"	Ergs-sec	980.665
"	Foot-pounds/sec.	7.2330138×10^{-5}
"	Horsepower	1.31509×10^{-7}
"	Joules/sec.	9.80665×10^{-5}
"	Kilowatts	9.80665×10^{-5}
"	Kilowatts (Int.)	9.80503×10^{-5}
"	Watts	9.80665×10^{-5}
Gram/sq. cm	Pounds/sq. inch	0.000341717
Gram wt.-sec./sq. cm	Poises	980.665
Gravitational constants	Cm./(sec. × sec.)	980.621
"	Ft./(sec. × sec.)	32.1725
Hands	Centimeters	10.16
"	Inches	4
Hectares	Acres	2.4710538
"	Ares	100
"	Sq. cm	1×10^{8}
"	Sq. feet	107639.10
"	Sq. meters	10000
"	Sq. miles	0.0038610216
"	Sq. rods	395.36861
Hectograms	Grams	100
"	Poundals	7.09316
"	Pounds (apoth or troy)	0.26792289
"	Pounds (avdp.)	0.22046226
Hectoliters	Bushels (Brit.)	2.749694
"	Bushels (U.S.)	2.837839

* Based on density of 1 gram/ml.

Table 4–9 (continued)

CONVERSION FACTORS

To convert from	To	Multiply by	To convert from	To	Multiply by
Hectoliters	Cu. cm	1.00028×10^5	Horsepower (boiler)	Horsepower (metric)	13.3372
"	Cu. feet	3.531566	"	Horsepower (water)	13.1487
"	Gallons (U.S., liq.)	26.41794	"	Joules/sec	9809.50
"	Liters	100	"	Kilowatts	9.80950
"	Ounces (U.S.) fluid	3381.497	"	Lb. H_2O evap. per hr. from	
"	Pecks (U.S.)	11.35136		and at 212°F	34.5
Hectometers	Centimeters	10000	Horsepower (electric)	B.t.u./hr	2547.16
"	Decimeters	1000	"	B.t.u. (IST.)/hr	2545.50
"	Dekameters	10	"	B.t.u. (mean)/hr	2543.50
"	Feet	328.08399	"	Cal., gm./sec	178.298
"	Meters	100	"	Cal., kg./hr	641.874
"	Rods	19.883878	"	Ergs/sec	7.46×10^9
"	Yards	109.3613	"	Foot-pounds/min	33013.3
Hectowatts	Watts	100	"	Foot-pounds/sec	550.221
Hefner units	Candles (English)	0.86	"	Horsepower	1.00040
"	Candles (German)	0.85	"	Horsepower (boiler)	0.0760487
"	Candles (Int.)	0.90	"	Horsepower (metric)	1.0142777
"	10-cp. pentane candles	0.090	"	Horsepower (water)	0.999942
Henries	Abhenries	1×10^9	"	Joules/sec	746
"	E.M. cgs. units	1×10^9	"	Kilowatts	0.746
"	E.S. cgs. units	1.112646×10^{-12}	"	Watts	746
"	Henries (Int.)	0.999505	Horsepower (metric)	B.t.u./hr	2511.31
"	Millihenries	1000	"	B.t.u. (IST.)/hr	2509.66
"	Mks. (r or nr) units	1	"	B.t.u. (mean)/hr	2507.70
"	Stathenries	1.112646×10^{-12}	"	Cal., gm./hr	6.32838×10^5
Henries (Int.)	Henries	1.000495	"	Cal., gm. (IST.)/hr	6.32425×10^5
Henries/meter	Cgs. units of permeability	795774.72	"	Cal., gm. (mean)/hr	6.31929×10^5
"	E.M. cgs. units	795774.72	"	Ergs/sec	7.35499×10^9
"	E.S. cgs. units	8.854156×10^{-16}	"	Foot-pounds/min	32548.6
"	Gausses/oersted	795774.72	"	Foot-pounds/sec	542.476
"	Mks. (nr) units	0.079577472	"	Horsepower	0.986320
"	Mks. (r) units	1	"	Horsepower (boiler)	0.0749782
Hogsheads	Butts (Brit.)	0.5	"	Horsepower (electric)	0.985923
"	Cu. feet	8.421875	"	Horsepower (water)	0.985866
"	Cu. inches	14553	"	Kg.-meters/sec	75
"	Cu. meters	0.23848094	"	Kilowatts	0.735499
"	Gallons (Brit.)	52.458505	"	Watts	735.499
"	Gallons (U.S.)	63	Horsepower (water)	Foot-pounds/min	33015.2
"	Gallons (wine)	63	"	Horsepower	1.00046
"	Liters	238.47427	"	Horsepower (boiler)	0.0760531
Horsepower[*]	B.t.u. (mean)/hr	2542.48	"	Horsepower (electric)	1.00006
"	B.t.u./min	42.4356	"	Horsepower (metric)	1.01434
"	B.t.u. (mean)/sec	0.706243	"	Kilowatts	0.746043
"	Cal., gm./hr	6.41616×10^5	Horsepower-hours	B.t.u	2546.14
"	Cal., gm. (IST.)/hr	6.41196×10^5	"	B.t.u. (IST.)	2544.47
"	Cal., gm. (mean)/hr	6.40693×10^5	"	B.t.u. (mean)	2542.48
"	Cal., gm./min	10693.6	"	Cal., gm	641616
"	Cal., gm. (IST.)/min	10686.6	"	Cal., gm. (IST.)	641196
"	Cal., gm. (mean)/min	10678.2	"	Cal., gm. (mean)	640693
"	Ergs/sec	7.45700×10^9	"	Foot-pounds	1.98×10^6
"	Foot-pounds/hr	1980000	"	Joules	2.68452×10^6
"	Foot-pounds/min	33000	"	Kg.-meters	273745
"	Foot-pounds/sec	550	"	Kw.-hours	0.745700
"	Horsepower (boiler)	0.0760181	"	Watt-hours	745.700
"	Horsepower (electric)	0.999598	Hp.-hr./lb	B.t.u./lb	2546.14
"	Horsepower (metric)	1.01387	"	Cal., gm./gram	1414.52
"	Joules/sec	745.700	"	Cu. ft.-(lb./sq. in.)/lb	13750
"	Kilowatts	0.745700	"	Foot-pounds/lb	1980000
"	Kilowatts (Int.)	0.745577	"	Joules/gram	5918.35
"	Tons of refrig. (U.S., comm.)	0.21204	Hours (mean solar)	Days (mean solar)	0.0416666
"	Watts	745.700	"	Days (sidereal)	0.041780746
Horsepower (boiler)	B.t.u. (mean)/hr	33445.7	"	Hours (sidereal)	1.00273791
"	Cal., gm./min	140671.6	"	Minutes (mean solar)	60
"	Cal., gm. (mean)/min	140469.4	"	Minutes (sidereal)	60.164275
"	Cal., gm. (15°C.)/min	140611.1	"	Seconds (mean solar)	3600
"	Cal., gm, (20°C.)/min	140742.2	"	Seconds (sidereal)	3609.8565
"	Ergs/sec	9.80950×10^{10}	"	Weeks (mean calendar)	0.0059523809
"	Foot-pounds/min	434107	Hours (sidereal)	Days (mean solar)	0.41552899
"	Horsepower	13.1548	"	Days (sidereal)	0.0416666
"	Horsepower (electric)	13.1495	"	Hours (mean solar)	0.99726957
			"	Minutes (mean solar)	59.836174

[*] Mechanical horsepower, equal to 550 ft.-lb./sec.

Table 4–9 (continued)

CONVERSION FACTORS

To convert from	To	Multiply by	To convert from	To	Multiply by
Hours (sidereal)	Minutes (sidereal)	60	Joules (abs)	Cal., *kg.* (mean)	0.000238662
Hundredweights (long)	Kilograms	50.802345	" "	Cu. ft.-atm	0.000348529
"	Pounds	112	" "	Ergs	1×10^7
"	Quarters (Brit., long)	4	" "	Foot-poundals	23.730360
"	Quarters (U.S., long)	0.2	" "	Foot-pounds	0.737562
"	Tons (long)	0.05	" "	Gram-cm	10197.16
Hundredweights (short)	Kilograms	45.359237	" "	Hp.-hours	3.72506×10^{-7}
"	Pounds (advp.)	100	" "	Joules (Int.)	0.999835
"	Quarters (Brit., short)	4	" "	Kg.-meters	0.1019716
"	Quarters (U.S., short)	0.2	" "	Kw.-hours	2.7777×10^{-7}
"	Tons (long)	0.044642857	" "	Liter-atm	0.00986895
"	Tons (metric)	0.045359237	" "	Volt-coulombs (Int.)	0.999835
"	Tons (short)	0.05	" "	Watt-hours (abs.)	0.0002777777
Inches	Ångström units	2.54×10^8	" "	Watt-hours (Int.)	0.000277732
"	Centimeters	2.54	" "	Watt-sec.	1
"	Chains (Gunter's)	0.00126262	" "	Watt-sec. (Int.)	0.999835
"	Cubits	0.055555	Joules (Int.)	B.t.u.	0.000948608
"	Fathoms	0.013888	"	B.t.u. (IST.)	0.000947988
"	Feet	0.083333	"	B.t.u. (mean)	0.000947244
"	Feet (U.S. Survey)	0.083333167	"	Cal. *gm*	0.239045
"	Links (Gunter's)	0.126262	"	Cal., *gm.* (IST.)	0.238888
"	Links (Ramden's)	0.083333	"	Cal., *gm.* (mean)	0.238702
"	Meters	0.0254	"	C.h.u.	0.000527004
"	Mils	1000	"	C.h.u. (IST.)	0.000526660
"	Picas (printer's)	6.0225	"	C.h.u. (mean)	0.000526247
"	Points (printer's)	72.27000	"	Cu. cm.-atm	9.87086
"	Wave length of orange-red line of krypton 86	41929.399	"	Cu. ft.-atm	0.000348586
"	Wave length of the red line of cadmium	39450.369	"	Dyne-cm	1.000165×10^7
"	Yards	0.027777	"	Ergs	1.000165×10^7
Inches of Hg (32°F.)	Atmospheres	0.0334211	"	Foot-poundals	23.73428
"	Bars	0.0338639	"	Foot-pounds	0.737684
"	Dynes/sq. cm.	33863.9	"	Gram-cm	10198.8
"	Ft. of air (1 atm., 60°F.)	926.24	"	Joules (abs.)	1.000165
"	Ft. of H₂O (39.2°F.)	1.132957	"	Kw.-hours	2.77824×10^{-7}
"	Grams/sq. cm.	34.5316	"	Liter-atm	0.00987058
"	Kg./sq. meter	345.316	"	Volt-coulombs	1.000165
"	Mm. of Hg (60°C.)	25.4	"	Volt-coulombs (Int.)	1
"	Ounces/sq. inch	7.85847	"	Watt-sec	1.000165
Inches of Hg (32°F.)	Pounds/sq. ft.	70.7262	"	Watt-sec. (Int.)	1
Inches of Hg (60°F.)	Atmospheres	0.0333269	Joules/(abcoulomb × °F.)	Joules/(coulomb × °C.)	0.18
"	Dynes/sq. cm.	39768.5	Joules/amp.-hr	Joules/abcoulomb	0.002777
"	Grams/sq. cm.	34.4343	"	Joules/statcoulomb	9.265653×10^{-14}
"	Mm. of Hg (60°C.)	25.4	Joules/coulomb	Joules/abcoulomb	10
"	Ounces/sq. inch	7.83633	"	Volts	1
"	Pounds/sq. ft.	70.5269	Joules/(coulomb × °F.)	Joules/(coulomb × °C.)	1.8
Inches of H₂O(4°C.)	Atmospheres	0.0024582	Joules/°C	B.t.u./°F.	0.000526917
"	Dynes/sq. cm.	2490.82	"	Cal., *gm.*/°C	0.239006
"	In. of Hg (32°F.)	0.0735539	"	Cal., *gm.* (mean)/°C	0.238662
"	Kg./sq. meter	25.3993	Joules/electronic charge	Joules/abcoulomb	6.24196×10^{19}
"	Ounces/sq. ft.	83.2350	Joules/(electronic charge × °C.)	Joules/(coulomb × °C.)	6.24196×10^{18}
"	Ounces/sq. inch	0.578020	Joules/(gram × °C.)	B.t.u./(lb. × °F.)	0.239006
"	Pounus/sq. ft.	5.20218	"	Cal., *gm.*/(gram × °C.)	0.239006
"	Pounds/sq. inch	0.03612628	Joules (Int.)/(gram °C.)	B.t.u./(lb. × °F.)	0.239045
Inches/hr.	Cm./hr.	2.54	"	Cal., *gm.* (mean)/(gram × °C.)	0.238702
"	Feet/hr.	0.0833333	Joules/sec. (abs.)	B.t.u./min	0.0569071
"	Miles/hr.	1.578282×10^{-5}	"	Cal., *gm.*/min	14.3403
Inches/min.	Cm./hr.	152.4	"	Cal., *kg.*/min	0.0143403
"	Feet/hr.	5	"	Cal., *kg.* (mean)/min	0.0143197
"	Miles/hr.	0.000946969	"	Dyne-cm./sec	1×10^7
Joules (abs.)	B.t.u.	0.000948451	"	Ergs/sec	1×10^7
" "	B.t.u. (IST.)	0.000947831	"	Foot-pounds/sec	0.737562
" "	B.t.u. (mean)	0.000947088	"	Gram-cm./sec	10197.16
" "	Cal., *gm*	0.239006	"	Horsepower	0.00134102
" "	Cal., *gm.* (IST.)	0.238849	"	Watts	1
" "	Cal., *gm.* (mean)	0.238662	"	Watts (Int.)	0.999835
" "	Cal., *gm.* (15°C.)	0.238903	Joules (Int.)/sec	B.t.u./min	0.0569165
" "	Cal., *gm.* (20°C.)	0.239126			

Table 4–9 (continued)

CONVERSION FACTORS

To convert from	To	Multiply by
Joules (Int.)/sec	B.t.u. (mean)/min	0.0568347
"	Cal., *gm.*/min	14.3427
"	Cal., *kg.*/min	0.0143427
"	Dyne-cm./sec	1.000165×10^7
"	Ergs/sec	1.000165×10^7
"	Foot-pounds/min	44.2610
"	Foot-pounds/sec	0.737684
"	Gram-cm./sec	10198.8
"	Horsepower	0.00134124
"	Watts	1.000165
"	Watts (Int.)	1
Kilderkins (Brit.)	Cu. cm	81829.57
"	Cu. feet	2.889784
"	Cu. inches	4993.55
"	Cu. meters	0.08182957
"	Gallons (Brit.)	18
Kilograms	Drams (apoth. *or* troy)	257.20597
"	Drams (avdp.)	564.38339
"	Dynes	980665
"	Grains	15432.358
"	Hundredweights (long)	0.019684131
"	Hundredweights (short)	0.022046226
"	Ounces (apoth. *or* troy)	32.150737
"	Ounces (avdp.)	35.273962
"	Pennyweights	643.01493
"	Poundals	70.931635
"	Pounds (apoth. *or* troy)	2.6792289
"	Pounds (avdp.)	2.2046226
"	Quarters (Brit., long)	0.078736522
"	Quarters (U.S. long)	0.0039368261
"	Scruples (apoth.)	771.61792
"	Slugs	0.06852177
"	Tons (long)	0.00098420653
"	Tons (metric)	0.001
"	Tons (short)	0.0011023113
Kilograms/cu. meter	Grams/cu. cm	0.001
"	Lb./cu. ft	0.062427961
"	Lb./cu. inch	3.6127292×10^{-5}
Kg. of ice melted/hr	Tons of refrig. (U.S., comm.)	0.026336
Kilograms/sq. cm	Atmospheres	0.967841
"	Bars	0.980665
"	Cm. of Hg (0°C.)	73.5559
"	Dynes/sq. cm	980665
"	Ft. of H_2O (39.2°F.)	32.8093
"	In. of Hg (32°F.)	28.9590
"	Pounds/sq. inch	14.223343
Kilograms/sq. meter	Atmospheres	9.67841×10^{-5}
"	Bars	9.80665×10^{-5}
"	Dynes/sq. cm	98.0665
"	Ft. of H_2O (39.2°F.)	0.00328093
"	Grams/sq. cm	0.1
"	In. of Hg. (32°F.)	0.00289590
"	Mm. of Hg (0°C.)	0.0735559
"	Pounds/sq. ft	0.20481614
"	Pounds/sq. in	0.0014223343
Kilograms/sq. mm	Pounds/sq. ft	204816.14
"	Pounds/sq. in	1422.3343
"	Tons (short)/sq. in	0.71116716
Kilogram sq. cm	Pounds sq. ft	0.0023730360
"	Pounds sq. in	0.34171719
Kilogram-meters	B.t.u. (mean)	0.00928776
"	Cal., *gm.* (mean)	2.34048
"	Cal., *kg.* (mean)	0.00234048
"	Cu. ft.-atm	0.00341790
"	Dynes-cm	9.80665×10^7
"	Ergs	9.80665×10^7
"	Foot-poundals	232.715
"	Foot-pounds	7.23301
"	Gram-cm	100000

To convert from	To	Multiply by
Kilogram-meters	Hp.-hours	3.65304×10^{-6}
"	Joules	9.80665
"	Joules (Int.)	9.80503
"	Kw.-hours	2.72407×10^{-6}
"	Liter-atm	0.0967814
"	Newton-meters	9.80665
"	Watt-hours	0.00272407
"	Watt-hours (Int.)	0.00272362
Kilogram-meters/sec.	Watts	9.80665
Kilolines	Maxwells	1000
"	Webers	1×10^{-5}
Kiloliters	Cu. centimeters	1×10^6
"	Cu. feet	35.31566
"	Cu. inches	61025.45
"	Cu. meters	1.000028
"	Cu. yards (Brit.)	1.307987
"	Gallons (Brit.)	219.9755
"	Gallons (U.S., dry)	227.0271
"	Gallons (U.S., liq.)	264.1794
"	Liters	1000
Kilometers	Astronomical units	6.68878×10^{-9}
"	Centimeters	100000
"	Feet	3280.8399
"	Feet (U.S. Survey)	3280.833
"	Light years	1.05702×10^{-13}
"	Meters	1000
"	Miles (naut., Int.)	0.53995680
"	Miles (statute)	0.62137119
"	Myriameters	0.1
"	Rods	198.83878
"	Yards	1093.6133
Kilometers/hr.	Cm./sec	27.7777
"	Feet/hr	3280.8399
"	Feet/min	54.680665
"	Knots (Int.)	0.53995680
"	Meters/sec	0.277777
"	Miles (statute)/hr	0.62137119
Kilometers/(hr. × sec.)	Cm./(sec. × sec.)	27.7777
"	Ft./(sec. × sec.)	0.91134442
"	Meters/(sec. × sec.)	0.277777
Kilometers/min	Cm./sec	1666.666
"	Feet/min	3280.8399
"	Kilometers/hr	60
"	Knots (Int.)	32.397408
"	Miles/hr	37.282272
"	Miles/min	0.62137119
Kilovolts/cm	Abvolts/cm	1×10^{11}
"	Microvolts/meter	1×10^{11}
"	Millivolts/meter	1×10^8
"	Statvolts/cm	3.335635
"	Volts/inch	2540
Kilowatts	B.t.u./hr	3414.43
"	B.t.u. (IST.)/hr	3412.19
"	B.t.u. (mean)/hr	3409.52
"	B.t.u. (mean)/min	56.8253
"	B.t.u. (mean)/sec	0.947088
"	Cal., *gm.* (mean)/hr	859184
"	Cal., *gm.* (mean)/min	14319.7
"	Cal., *gm.* (mean)/sec	238.662
"	Cal., *kg.* (mean)/hr	859.184
"	Cal., *kg.* (mean)/min	14.3197
"	Cal., *kg.* (mean)/sec	0.238662
"	Cu. ft.-atm./hr	1254.70
"	Ergs/sec	1×10^{10}
"	Foot-poundals/min	1.42382×10^6
"	Foot-pounds/hr	2.65522×10^6
"	Foot-pounds/min	44253.7
"	Foot-pounds/sec	737.562
"	Gram-cm./sec	1.019716×10^7
"	Horsepower	1.34102

Table 4–9 (continued)

CONVERSION FACTORS

To convert from	To	Multiply by	To convert from	To	Multiply by
Kilowatts	Horsepower (boiler)	0.101942	Lamberts	Candles/sq. cm	0.31830989
"	Horsepower (electric)	1.34048	"	Candles/sq. ft.	295.71956
"	Horsepower (metric)	1.35962	"	Candles/sq. inch	2.0536081
"	Joules/hr.	3.6×10^6	"	Foot-lamberts	929.0304
"	Joules (IST.)/hr.	3.59941×10^6	"	Lumens/sq. cm	1
"	Joules/sec.	1000	Lasts (Brit.)	Liters	2909.414
"	Kg.-meters/hr.	3.67098×10^5	Leagues (naut., Brit.)	Feet	18240
"	Kilowatts (Int.)	0.999835	"	Kilometers	5.559552
"	Watts (Int.)	999.835	"	Leagues (naut., Int.)	1.0006393
Kilowatts (Int.)	B.t.u./hr.	3414.99	"	Leagues (statute)	1.151515
"	B.t.u. (IST.)/hr.	3412.76	"	Miles (statute)	3.454545
"	B.t.u. (mean)/hr.	3410.08	Leagues (naut., Int.)	Fathoms	3038.0577
"	B.t.u. (mean)/min.	56.8347	"	Feet	18228.346
"	B.t.u. (mean)/sec.	0.947244	"	Kilometers	5.556
"	Cal., gm. (mean)/hr.	859326	"	Leagues (statute)	1.1507794
"	Cal., gm. (mean)/min.	14322.1	"	Miles (statute)	3.4523383
"	Cal., kg./hr.	860.563	Leagues (statute)	Fathoms	2640
"	Cal., kg. (IST.)/hr.	860	"	Feet	15840
"	Cal., kg. (mean)/hr.	859.326	"	Kilometers	4.828032
"	Cu. cm.-atm./hr.	3.55351×10^7	"	Leagues (naut., Int.)	0.86897625
"	Cu. ft.-atm./hr.	1254.91	"	Miles (naut., Int.)	2.6069287
"	Ergs/sec.	1.000165×10^{10}	"	Miles (statute)	3
"	Foot-poundals/min.	1.42406×10^6	Light years	Astronomical units	63279.5
"	Foot-pounds/min.	44261.0	"	Kilometers	9.46055×10^{12}
"	Foot-pounds/sec.	737.684	"	Miles (statute)	5.87851×10^{12}
"	Gram-cm./sec.	1.01988×10^7	Lines	Maxwells	1
"	Horsepower	1.34124	Lines (Brit.)	Centimeters	0.211666
"	Horsepower (boiler)	0.101959	"	Inches	0.083333
"	Horsepower (electric)	1.34070	Lines/sq. cm	Gausses	1
"	Horsepower (metric)	1.35985	Lines/sq. inch	Gausses	0.15500031
"	Joules/hr.	3.60059×10^6	"	Webers/sq. inch	1×10^{-8}
"	Joules (Int.)/hr.	3.6×10^6	Links (Gunter's)	Chains (Gunter's)	0.01
"	Kg.-meters/hr.	367158	"	Feet	0.66
"	Kilowatts	1.000165	"	Feet (U.S. Survey)	0.65999868
Kilowatt-hours	B.t.u. (mean)	3409.52	"	Inches	7.92
"	Cal., gm. (mean)	859184	"	Meters	0.201168
"	Foot-pounds	2.65522×10^6	"	Miles (statute)	0.000125
"	Hp.-hours	1.34102	"	Rods	0.04
"	Joules	3.6×10^6	Links (Ramden's)	Centimeters	30.48
"	Kg.-meters	367098	"	Chains (Ramdens)	0.01
"	Lb. H$_2$O evap. from and at 212°F	3.5168	"	Feet	1
"	Watt-hours	1000	"	Inches	12
"	Watt-hours (Int.)	999.835	Liters	Bushels (Brit.)	0.02749694
Kilowatt-hours (Int.)	B.t.u. (mean)	3410.08	"	Bushels (U.S.)	0.02837839
"	Cal., gm. (IST)	860000	"	Cu. centimeters	1000
"	Cal., gm. (mean)	859326	"	Cu. feet	0.03531566
"	Cu. cm.-atm	3.55351×10^7	"	Cu. inches	61.02545
"	Cu. ft.-atm	1254.91	"	Cu. meters	0.001
"	Foot-pounds	2.65566×10^6	"	Cu. yards	0.001307987
"	Hp.-hours	1.34124	"	Drams (U.S., fluid)	270.5198
"	Joules	3.60059×10^6	"	Gallons (Brit.)	0.2199755
"	Joules (Int.)	3.6×10^6	"	Gallons (U.S., dry)	0.2270271
"	Kg.-meters	367158	"	Gallons (U.S., liq.)	0.2641794
Kw.-hr./gram	B.t.u./(lb.	1.54876×10^6	"	Gills (Brit.)	7.039217
"	B.t.u. (IST.)/lb.	1.54774×10^6	"	Gills (U.S.)	8.453742
"	B.t.u. (mean)/lb.	1.54653×10^6	"	Hogsheads	0.004193325
"	Cal., gm./gram	860421	"	Minims (U.S.)	16231.19
"	Cal., gm. (mean)/gram	859184	"	Ounces (Brit., fluid)	35.19609
"	Cu. cm.-atm./gram	3.55292×10^7	"	Ounces (U.S., fluid)	33.81497
"	Cu. ft.-atm./lb.	569124	"	Pecks (Brit.)	0.1099878
"	Hp.-hr./lb.	608.277	"	Pecks (U.S.)	0.1135136
"	Joules/gram	3.6×10^6	"	Pints (Brit.)	1.759804
Knots (Int.)	Cm./sec.	51.4444	"	Pints (U.S., dry)	1.816217
"	Feet/hr.	6076.1155	"	Pints (U.S., liq.)	2.113436
"	Feet/min.	101.26859	"	Quarts (Brit.)	0.8799021
"	Feet/sec.	1.6878099	"	Quarts (U.S., dry)	0.9081084
"	Kilometers/hr.	1.852	"	Quarts (U.S., liq.)	1.056718
"	Meters/min.	30.8666	Liters/min.	Cu. ft./min.	0.03531566
"	Meters/sec.	0.514444	"	Cu. ft./sec.	0.0005885943
"	Miles (naut., Int.)/hr.	1	"	Gal. (U.S., liq.)/min.	0.2641794
"	Miles (statute)/hr.	1.1507794	Liters/sec.	Cu. ft./min.	2.118939
			"	Cu. ft./sec	0.03531566

Table 4–9 (continued)

CONVERSION FACTORS

To convert from	To	Multiply by	To convert from	To	Multiply by
Liters/sec	Cu. yards/min	0.07847923	Meters	Links (Ramden's)	3.2808399
"	Gal. (U.S., liq.)/min	15.85077	"	Megameters	1×10^{-6}
"	Gal. (U.S., liq.)/sec	0.2641794	"	Miles (naut., Brit.)	0.00053961182
Liter-atm	B.t.u.	0.0961045	"	Miles (naut., Int.)	0.00053995680
"	B.t.u. (IST.)	0.0960417	"	Miles (statute)	0.00062137119
"	B.t.u. (mean)	0.0959664	"	Millimeters	1000
"	Cal., gm.	24.2179	"	Millimicrons	1×10^{9}
"	Cal., gm. (IST.)	24.2021	"	Mils	39370.079
"	Cal., gm. (mean)	24.1831	"	Rods	0.19883878
"	Cu. ft.-atm	0.0353157	"	Yards	1.0936133
"	Foot-poundals	2404.55	Meters of Hg (0°C.)	Atmospheres	1.3157895
"	Foot-pounds	74.7356	"	Ft. of H_2O (60°F.)	44.6474
"	Hp.-hours	3.77452×10^{-5}	"	In. of Hg (32°F.)	39.376079
"	Joules	101.328	"	Kg./sq. cm	1.35951
"	Joules (Int.)	101.311	"	Pounds/sq. inch	19.3368
"	Kg.-meters	10.3326	Meters/hr	Feet/hr	3.2808399
"	Kw.-hours	2.81466×10^{-5}	"	Feet/min	0.054680665
Liter-atm. (lat. 45°)	Joules	101.323	"	Knots (Int.)	0.00053995680
Lumens	Candle power (spher.)	0.079577472	"	Miles (statute)/hr	0.00062137119
Lumens (at 5550 Å)	Watts	0.0014705882	Meters/min	Cm./sec	1.666666
Lumens/sq. cm	Lamberts	1	"	Feet/min	3.2808399
"	Phots	1	"	Feet/sec	0.054680665
Lumens/(sq. cm. × steradian)	Lamberts	3.1415927	"	Kilometers/hr	0.06
Lumens/sq. ft	Foot-candles	1	"	Knots (Int.)	0.032397408
"	Foot-lamberts	1	"	Miles (statute)/hr	0.037282272
"	Lumens/sq. meter	10.763910	Meters/sec	Feet/min	196.85039
Lumens/(sq. ft. × steradian)	Millilamberts	3.3815822	"	Feet/sec	3.2808399
Lumens/sq. meter	Foot-candles	0.09290304	"	Kilometers/hr	3.6
"	Lumens/sq. ft	0.09290304	"	Kilometers/min	0.06
"	Phots	0.0001	"	Miles (statute)/hr	2.2369363
Lux	Foot-candles	0.09290304	Meters/(sec. × sec.)	Kilometers/(hr. × sec.)	3.6
"	Lumens/sq. meter	1	"	Miles/(hr. × sec.)	2.2369363
"	Phots	0.0001	Meter-candles	Lumens/sq. meter	1
Maxwells	E.M. cgs. units of induction flux	1	Mhos	Abmhos	1×10^{-9}
"	E.S. cgs. units	3.335635×10^{-11}	"	Cgs. units of conductance	1
"	Gauss-sq. cm	1	"	E.M. cgs. units	1×10^{-9}
"	Lines	1	"	E.S. cgs. units	8.987584×10^{11}
"	Maxwells (Int.)	0.999670	"	Mhos (Int.)	1.000495
"	Volt-seconds	1×10^{-8}	"	Mks. (r or nr) units	1
"	Webers	1×10^{-8}	"	Ohms^{-1}	1
Maxwells (Int.)	Maxwells	1.000330	"	Siemen's units	1
Maxwells/sq. cm	Maxwells/sq. in	6.4516	"	Statmhos	8.987584×10^{11}
"	Maxwells (Int.)/sq. cm	0.999670	Mhos (Int.)	Abmhos	9.99505×10^{-10}
Maxwells (Int.)/sq. cm	Maxwells/sq. cm	1.000330	"	Mhos	0.999505
Maxwells/sq. inch	Maxwells/sq. cm	0.15500031	Mhos/meter	Abmhos/cm	1×10^{-11}
Megalines	Maxwells	1×10^{6}	"	Mhos (Int.)/meter	1.000495
Megmhos/cm	Abmhos/cm	0.001	Mho-ft./circ. mil	Mhos/cm	6.0153049×10^{6}
"	Megmhos/inch cube	2.54	Microfarads	Abfarads	1×10^{-15}
"	(Microhm-cm.)$^{-1}$	1	"	Farads	1×10^{-6}
Megmhos/inch	Megmhos/cm	0.39370079	"	Statfarads	8.987584×10^{5}
"	(Microhm-inches)$^{-1}$	1	Micrograms	Grams	1×10^{-6}
Megohms	Microhms	1×10^{12}	"	Milligrams	0.001
"	Ohms	1×10^{6}	Microhenries	Henries	1×10^{-6}
"	Statohms	1.112646×10^{-6}	"	Stathenries	1.112646×10^{-18}
Megohms^{-1}	Micromhos	1	Microhms	Abohms	1000
Meters	Ångström units	1×10^{10}	"	Megohms	1×10^{-12}
"	Centimeters	100	"	Ohms	1×10^{-6}
"	Chains (Gunter's)	0.049709695	"	Statohms	1.112646×10^{-18}
"	Chains (Ramden's)	0.032808399	Microhm-cm	Abohm-cm	1000
"	Fathoms	0.54680665	"	Circ. mil-óhms/ft	6.0153049
"	Feet	3.2808399	"	Microhm-inches	0.39370079
"	Feet (U.S. Survey)	3.280833	"	Ohm-cm	1×10^{-6}
"	Furlongs	0.0049709695	Microhm-inches	Circ. mil-ohms/ft	15.278875
"	Inches	39.370079	"	Michrom-cm	2.54
"	Kilometers	0.001	Micromicrofarads	Farads	1×10^{-12}
"	Links (Gunter's)	4.9709695	Micromicrons	Ångström units	0.01
			"	Centimeters	1×10^{-10}
			"	Inches	$3.9370079 \times 10^{-11}$
			"	Meters	1×10^{-12}
			"	Microns	1×10^{-6}
			Microns	Ångström units	10000

Table 4-9 (continued)

CONVERSION FACTORS

To convert from	To	Multiply by
Microns	Centimeters	0.0001
"	Feet	3.2808399×10^{-6}
"	Inches	3.9370079×10^{-5}
"	Meters	1×10^{-6}
"	Millimeters	0.001
"	Millimicrons	1000
Miles (naut., Brit.)	Cable lengths (Brit.)	8.4444
"	Fathoms	1013.333
"	Feet	6080
"	Meters	1853.184
"	Miles (Adm., Brit.)	1
"	Miles (naut., Int.)	1.0006393
"	Miles (statute)	1.151515
Miles (naut., Int.)	Cable lengths	8.4390493
"	Fathoms	1012.6859
"	Feet	6076.1155
"	Feet (U.S. Survey)	6076.1033
"	Kilometers	1.852
"	Leagues (naut., Int.)	0.333333
"	Meters	1852
"	Miles (geographical)	1
"	Miles (naut. Brit.)	0.99936110
"	Miles (statute)	1.1507794
Miles (statute)	Centimeters	160934.4
"	Chains (Gunter's)	80
"	Chains (Ramden's)	52.8
"	Feet	5280
"	Feet (U.S. Survey)	5279.9894
"	Furlongs	8
"	Inches	63360
"	Kilometers	1.609344
"	Light years	1.70111×10^{-13}
"	Links (Gunter's)	8000
"	Meters	1609.344
"	Miles (naut., Brit.)	0.86842105
"	Miles (naut., Int.)	0.86897624
"	Myriameters	0.1609344
"	Rods	320
"	Yards	1760
Miles/hr	Cm./sec	44.704
"	Feet/hr	5280
"	Feet/min	88
"	Feet/sec	1.466666
"	Kilometers/hr	1.609344
"	Knots (Int.)	0.86897624
"	Meters/min	26.8224
"	Miles/min	0.0166666
Miles/(hr. × min.)	Cm./(sec. × sec.)	0.7450666
Miles/(hr. × sec.)	Cm./(sec. × sec.)	44.704
"	Ft./(sec. × sec.)	1.466666
"	Kilometers/(hr. × sec.)	1.609344
"	Meters/(sec. × sec.)	0.44704
Miles/min	Cm./sec	2682.24
"	Feet/hr	316800
"	Feet/sec	88
"	Kilometers/min	1.609344
"	Knots (Int.)	52.138574
"	Meters/min	1609.344
"	Miles/hr	60
Millibars	Atmospheres	0.000986923
"	Bars	0.001
"	Baryes	1000
"	Dynes/sq. cm	1000
"	Grams/sq. cm	1.019716
"	In. of Hg (32°F.)	0.0295300
"	Pounds/sq. ft	2.088543
"	Pounds/sq. inch	0.0145038
Milligrams	Carats (1877)	0.004871
"	Carats (metric)	0.005
"	Drams (apoth. or troy)	0.00025720597
"	Drams (advp.)	0.00056438339

To convert from	To	Multiply by
Milligrams	Grains	0.015432358
"	Grams	0.001
"	Ounces (apoth. or troy)	3.2150737×10^{-5}
"	Ounces (avdp.)	3.5273962×10^{-5}
"	Pennyweights	0.00064301493
"	Pounds (apoth. or troy)	2.6792289×10^{-6}
"	Pounds (avdp.)	2.2046226×10^{-6}
"	Scruples (apoth.)	0.00077161792
Milligrams/assay ton	Milligrams/kg	34.285714
"	Ounces (troy)/ton (avdp.)	1
Milligrams/gm	Dynes/cm	0.980665
"	Pounds/inch	5.5997415×10^{-5}
Milligrams/gram	Carats (parts gold per 24 of mixture)	0.024
"	Grams/ton (short)	907.18474
"	Milligrams/assay ton	29.166666
"	Ounces (avdp.)/ton (long)	35.84
"	Ounces (avdp.)/ton (short)	32
"	Ounces (troy)/ton (long)	32.6666
"	Ounces (troy)/ton (short)	29.1666
Milligrams/inch	Dynes/cm	0.386089
"	Dynes/inch	0.980665
"	Grams/cm	0.00039370079
"	Grams/inch	0.0001
Milligrams/kg	Pounds (avdp.)/ton (short)	0.002
Milligrams/liter	Grains/gal. (U.S.)	0.05841620
"	Grams/liter	0.001
"	Parts/million*	1
"	Lb./cu. ft.	6.242621×10^{-5}
Milligrams/mm	Dynes/cm	9.80665
Millihenries	Abhenries	1×10^{6}
"	Henries	0.001
"	Stathenries	1.112646×10^{-15}
Millilamberts	Candles/sq. cm	0.00031830989
"	Candles/sq. inch	0.0020536081
"	Foot-lamberts	0.9290304
"	Lamberts	0.001
"	Lumens/sq. cm	0.001
"	Lumens/sq. ft	0.9290304
Milliliters	Cu. cm	1
"	Cu. inches	0.06102545
"	Drams (U.S., fluid)	0.2705198
"	Gills (U.S.)	0.008453742
"	Liters	0.001
"	Minims (U.S.)	16.23119
"	Ounces (Brit., fluid)	0.03519609
"	Ounces (U.S., fluid)	0.03381497
"	Pints (Brit.)	0.001759804
"	Pints (U.S., liq.)	0.002113436
Millimeters	Ångström units	1×10^{7}
"	Centimeters	0.1
"	Decimeters	0.01
"	Dekameters	0.0001
"	Feet	0.0032808399
"	Inches	0.039370079
"	Meters	0.001
"	Microns	1000
"	Mils	39.370079
"	Wave length of orange-red line of krypton 86	1650.76373
"	Wave length of red line of cadmium	1553.16413
Millimeters of Hg (0°C.)	Atmospheres	0.0013157895
"	Bars	0.00133322
"	Dynes/sq. cm	1333.224
"	Grams/sq. cm	1.35951
"	Kg./sq. meter	13.5951
"	Pounds/sq. ft	2.78450
"	Pounds/sq. inch	0.0193368
"	Torrs	1

* Density of 1 gram per milliliter of solvent.

Table 4–9 (continued)

CONVERSION FACTORS

To convert from	To	Multiply by
Millimicrons	Ångström units	10
"	Centimeters	1×10^{-7}
"	Inches	3.9370079×10^{-8}
"	Microns	0.001
"	Millimeters	1×10^{-6}
Milliphots	Foot-candles	0.9290304
"	Lumens/sq. ft	0.9290304
"	Lumens/sq. meter	10
"	Lux	10
"	Phots	0.001
Millivolts	Statvolts	3.335635×10^{-6}
"	Volts	0.001
Minims (Brit.)	Cu. cm	0.05919385
"	Cu. inches	0.003612230
"	Milliliters	0.05919219
"	Ounces (Brit., fluid)	0.0020833333
"	Scruples (Brit., fluid)	0.05
Minims (U.S.)	Cu. cm	0.061611520
"	Cu. inches	0.0037597656
"	Drams (U.S., fluid)	0.0166666
"	Gallons (U.S., liq.)	1.6276042×10^{-5}
"	Gills (U.S.)	0.0005208333
"	Liters	6.160979×10^{-5}
"	Milliliters	0.06160979
"	Ounces (U.S., fluid)	0.002083333
"	Pints (U.S., liq.)	0.0001302083
Minutes (angular)	Degrees	0.0166666
"	Quadrants	0.000185185
"	Radians	0.00029088821
"	Seconds (angular)	60
Minutes (mean solar)	Days (mean solar)	0.0006944444
"	Days (sidereal)	0.00069634577
"	Hours (mean solar)	0.0166666
"	Hours (sidereal)	0.016712298
"	Minutes (sidereal)	1.00273791
Minutes (sidereal)	Days (mean solar)	0.00069254831
"	Minutes (mean solar)	0.99726957
"	Months (mean calendar)	2.2768712×10^{-5}
"	Seconds (sidereal)	60
Minutes/cm	Radians/cm	0.00029088821
Months (lunar)	Days (mean solar)	29.530588
"	Hours (mean solar)	708.73411
"	Minutes (mean solar)	42524.047
"	Second (mean solar)	2.5514428×10^{6}
"	Weeks (mean calendar)	4.2186554
Months (mean calendar)	Days (mean solar)	30.416666
"	Hours (mean solar)	730
"	Months (lunar)	1.0300055
"	Weeks (mean calendar)	4.3452381
"	Years (calendar)	0.08333333
"	Years (sidereal)	0.083274845
"	Years (tropical)	0.083278075
Myriagrams	Grams	10000
"	Kilograms	10
"	Pounds (avdp.)	22.046226
Newtons	Dynes	100000
"	Pounds	0.22480894
Newton-meters	Dyne-cm	1×10^{7}
"	Gram-cm	10197.162
"	Kg.-meters	0.10197162
"	Pound-feet	0.73756215
Noggins (Brit.)	Cu. cm	142.0652
"	Gallons (Brit.)	0.03125
"	Gills (Brit.)	1
Oersteds	Ampere-turns/inch	2.0212678
"	Ampere-turns/meter	79.577472
"	E.M. cgs. units of magnetic field intensity	1
"	E.S. cgs. units	2.997930×10^{10}

To convert from	To	Multiply by
Oersteds	Gilberts/cm	1
"	Oersteds (Int.)	1.000165
Oersteds (Int.)	Oersteds	0.999835
Ohms	Abohms	1×10^{9}
"	Cgs. units of resistance	1
"	Megohms	1×10^{-6}
"	Microhms	1×10^{6}
"	Ohms (Int.)	0.999505
"	Statohms	1.112646×10^{-12}
Ohms (Int.)	Ohms	1.000495
Ohms (mil, foot)	Circ. mil-ohms/ft	1
"	Ohm-cm	1.6624261×10^{-7}
Ohm-cm	Circ. mil-ohms/ft	6.0153049×10^{6}
"	Microhm-cm	1×10^{6}
"	Ohm-inches	0.39370079
Ohm-inches	Ohm-cm	2.54
Ohm-meters	Abohms	1×10^{11}
"	E.M. cgs. units	1×10^{11}
"	E.S. cgs. units	1.112646×10^{-10}
"	Mks. units	1
"	Statohm-cm	1.112646×10^{-10}
Ounces (apoth. or troy)	Dekagrams	1.7554286
"	Drams (apoth. or troy)	8
"	Drams (avdp.)	17.554286
"	Grains	480
"	Grams	31.103486
"	Milligrams	31103.486
"	Ounces (avdp.)	1.0971429
"	Pennyweights	20
"	Pounds (apoth. or troy)	0.0833333
"	Pounds (avdp.)	0.068571429
"	Scruples (apoth.)	24
"	Tons (short)	3.4285714×10^{-5}
Ounces (avdp.)	Drams (apoth. or troy)	7.291666
"	Drams (avdp.)	16
"	Grains	437.5
"	Grams	28.349523
"	Hundredweights (long)	0.00055803571
"	Hundredweights (short)	0.000625
"	Ounces (apoth. or troy)	0.9114583
"	Pennyweights	18.229166
"	Pounds (apoth. or troy)	0.075954861
"	Pounds (avdp.)	0.0625
"	Scruples (apoth.)	21.875
"	Tons (long)	2.7901786×10^{-5}
"	Tons (metric)	2.8349527×10^{-5}
"	Tons (short)	3.125×10^{-5}
Ounces (Brit., fluid)	Cu. cm	28.41305
"	Cu. inches	1.733870
"	Drachms (Brit., fluid)	8
"	Drams (U.S., fluid)	7.686075
"	Gallons (Brit.)	0.00625
"	Milliliters	28.41225
"	Minims (Brit.)	480
"	Ounces (U.S., fluid)	0.9607594
Ounces (U.S., fluid)	Cu. cm	29.573730
"	Cu. inches	1.8046875
"	Cu. meters	2.9573730×10^{-5}
"	Drams (U.S., fluid)	8
"	Gallons (U.S., dry)	0.0067138047
"	Gallons (U.S., liq.)	0.0078125
"	Gills (U.S.)	0.25
"	Liters	0.029572702
"	Minims (U.S.)	480
"	Ounces (Brit., fluid)	1.040843
"	Pints (U.S., liq.)	0.0625
"	Quarts (U.S., liq.)	0.03125
Ounces/sq. inch	Dynes/sq. cm	4309.22
"	Grams/sq. cm	4.3941849
"	In. of H_2O (39.2°F.)	1.73004
"	In. of H_2O (60°F.)	1.73166

Table 4–9 (continued)

CONVERSION FACTORS

To convert from	To	Multiply by	To convert from	To	Multiply by
Ounces/sq. inch	Pounds/sq. ft	9	Pints (Brit.)	Minims (Brit.)	9600
"	Pounds/sq. inch	0.0625	"	Ounces (Brit., fluid)	20
Ounces (avdp.)/ton (long)	Milligrams/kg	27.901786	"	Pints (U.S., dry)	1.032056
Ounces (avdp.)/ton (short)	Milligrams/kg	31.25	"	Pints (U.S., liq.)	1.200949
			"	Quarts (Brit.)	0.5
			"	Scruples (Brit., fluid)	480
Paces	Centimeters	76.2	Pints (U.S., dry)	Bushels (U.S.)	0.015625
"	Chains (Gunter's)	0.0378788	"	Cu. cm	550.61047
"	Chains (Ramden's)	0.025	"	Cu. inches	33.6003125
"	Feet	2.5	"	Gallons, (U.S., dry)	0.125
"	Hands	7.5	"	Gallons (U.S., liq.)	0.14545590
"	Inches	30	"	Liters	0.5505951
"	Ropes (Brit.)	0.125	"	Pecks (U.S.)	0.0625
Palms	Centimeters	7.62	"	Quarts (U.S., dry)	0.5
"	Chains (Ramden's)	0.0025	Pints (U.S., liq.)	Cu. cm	473.17647
"	Cubits	0.1666666	"	Cu. feet	0.016710069
"	Feet	0.25	"	Cu. inches	28.875
"	Hands	0.75	"	Cu. yards	0.00061889146
"	Inches	3	"	Drams (U.S., fluid)	128
Parsecs	Kilometers	3.08572×10^{13}	"	Gallons (U.S., liq.)	0.125
"	Miles (statute)	1.91738×10^{13}	"	Gills (U.S.)	4
Parts/million*	Grains/gal. (Brit.)	0.07015488	"	Liters	0.4731632
"	Grains/gal. (U.S.)	0.05841620	"	Milliliters	473.1632
"	Grams/liter	0.001	"	Minims (U.S.)	7680
"	Milligrams/liter	1	"	Ounces (U.S., fluid)	16
Pecks (Brit.)	Bushels (Brit.)	0.25	"	Pints (Brit.)	0.8326747
"	Coombs (Brit.)	0.0625	"	Quarts (U.S., liq.)	0.5
"	Cu. cm	9092.175	Planck's constant	Erg-seconds	6.6255×10^{-27}
"	Cu. inches	554.8385	"	Joule-seconds	6.6255×10^{-34}
"	Gallons (Brit.)	2	"	Joule-sec./Avog. No. (chem.)	3.9905×10^{-10}
"	Gills (Brit.)	64	Points (printer's)	Centimeters	0.03514598
"	Hogsheads	0.03812537	"	Inches	0.013837
"	Kilderkins (Brit.)	0.111111	"	Picas	0.0833333
"	Liters	9.091920	Poises	Cgs. units of absolute viscosity	1
"	Pints (Brit.)	16	"	Grams/(cm. $\times$ sec.)	1
"	Quarterns (Brit., dry)	4	Poise-cu. cm./gram	Sq. cm./sec	1
"	Quarters (Brit., dry)	0.03125	Poise-cu. ft./lb	Sq. cm./sec	62.427960
"	Quarts (Brit.)	8	Poise-cu. in./gram	Sq. cm./sec	16.387064
"	Quarts (U.S., dry)	8.256449	Poles/sq. cm	E.M. cgs. units of magnetization	1
Pecks (U.S.)	Barrels (U.S., dry)	0.076191185	Pottles (Brit.)	Gallons (Brit.)	0.5
"	Bushels (U.S.)	0.25	"	Liters	2.272980
"	Cu. cm	8809.7675	Poundals	Dynes	13825.50
"	Cu. feet	0.311114005	"	Grams	14.09808
"	Cu. inches	537.605	"	Pounds (avdp.)	0.0310810
"	Gallons (U.S., dry)	2	Pounds (apoth. or troy)	Drams (apoth. or troy)	96
"	Gallons (U.S., liq.)	2.3272944	"	Drams (avdp.)	210.65143
"	Liters	8.809521	"	Grains	5760
"	Pints (U.S., dry)	16	"	Grams	373.24172
"	Quarts (U.S., dry)	8	"	Kilograms	0.37324172
Pennyweights	Drams (apoth. or troy)	0.4	"	Ounces (apoth. or troy)	12
"	Drams (avdp.)	0.87771429	"	Ounces (avdp.)	13.165714
"	Grains	24	"	Pennyweights	240
"	Grams	1.55517384	"	Pounds (avdp.)	0.8228571
"	Ounces (apoth. or troy)	0.05	"	Scruples (apoth.)	288
"	Ounces (avdp.)	0.054857143	"	Tons (long)	0.00036734694
"	Pounds (apoth. or troy)	0.0041666	"	Tons (metric)	0.00037324172
"	Pounds (avdp.)	0.0034285714	"	Tons (short)	0.00041142857
Perches (masonry)	Cu. feet	24.75	Pounds (avdp.)	Drams (apoth. or troy)	116.6666
Phots	Foot-candles	929.0304	"	Drams (avdp.)	256
"	Lumens/sq. cm	1	"	Grains	7000
"	Lumens/sq. meter	10000	"	Grams	453.59237
"	Lux	10000	"	Hundredweights (long)	0.00892857
Picas (printer's)	Centimeters	0.42175176	"	Hundredweights (short)	0.01
"	Inches	0.166044	"	Kilograms	0.45359237
Pints (Brit.)	Cu. cm	568.26092	"	Ounces (apoth. or troy)	14.583333
"	Gallons (Brit.)	0.125	"	Ounces (avdp.)	16
"	Gills (Brit.)	4			
"	Gills (U.S.)	4.803797			
"	Liters	0.5682450			

* Based on density of 1 gram/ml. for the solvent.

Table 4–9 (continued)

CONVERSION FACTORS

To convert from	To	Multiply by	To convert from	To	Multiply by
Pounds (avdp.)	Pennyweights	291.6666	Quarterns (Brit., liq.)	Gallons (Brit.)	0.03125
"	Poundals	32.1740	"	Liters	0.1420613
"	Pounds (apoth. or troy)	1.215277	Quarters (U.S., long)	Kilograms	254.0117272
"	Scruples (apoth.)	350	"	Pounds (avdp.)	560
"	Slugs	0.0310810	Quarters (U.S., short)	Kilograms	226.796185
"	Tons (long)	0.00044642857	"	Pounds	500
"	Tons (metric)	0.00045359237	Quarts (Brit.)	Cu. cm	1136.522
"	Tons (short)	0.0005	"	Cu. inches	69.35482
Pounds of H$_2$O evap. from and at 212°F.	B.t.u.	970.9	"	Gallons (Brit.)	0.25
"	B.t.u. (IST.)	970.2	"	Gallons (U.S., liq.)	0.3002373
"	B.t.u. (mean)	969.4	"	Liters	1.136490
"	Joules	1.0237 × 10⁶	"	Quarts (U.S., dry)	1.032056
"	Joules (Int.)	1.0234 × 10⁶	"	Quarts (U.S., liq.)	1.200949
Pounds/cu. ft.	Grams/cu. cm.	0.016018463	Quarts (U.S., dry)	Bushels (U.S.)	0.03125
"	Kg./cu. meter	16.018463	"	Cu. cm	1101.2209
Pounds/cu. inch	Grams/cu. cm.	27.679905	"	Cu. feet	0.038889251
"	Grams/liter	27.68068	"	Cu. inches	67.200625
"	Kg./cu. meter	27679.905	"	Gallons (U.S., dry)	0.25
Pounds/gal. (Brit.)	Pounds/cu. ft.	6.228839	"	Gallons (U.S., liq.)	0.29091180
Pounds/gal. (U.S., liq.)	Grams/cu. cm.	0.11982643	"	Liters	1.1011901
"	Pounds/cu. ft.	7.4805195	"	Pecks (U.S.)	0.125
Pounds/inch	Grams/cm	178.57967	"	Pints (U.S., dry)	2
"	Grams/ft.	5443.1084	Quarts (U.S., liq.)	Cu. cm	946.35295
"	Grams/inch	453.59237	"	Cu. feet	0.033420136
"	Ounces/cm.	6.2992	"	Cu. inches	57.75
"	Ounces/inch	16	"	Drams (U.S., fluid)	256
"	Pounds/meter	39.370079	"	Gallons (U.S., dry)	0.21484175
Pounds/minute	Kilograms/hr	27.2155422	"	Gallons (U.S., liq.)	0.25
"	Kilograms/min	0.45359237	"	Gills (U.S.)	8
Pounds of H$_2$O (39.2°F.)/min	Cu. ft./min	0.01601891	"	Liters	0.9463264
"	Gal. (U.S.)/min	0.1198298	"	Ounces (U.S., fluid)	32
"	Liters/min	0.45359237	"	Pints (U.S., liq.)	2
Pounds/sq. ft.	Atmospheres	0.000472541	"	Quarts (Brit.)	0.8326747
"	Bars	0.000478803	"	Quarts (U.S., dry)	0.8593670
"	Cm. of Hg (0°C.)	0.0359131	Quintals (metric)	Grams	100000
"	Dynes/sq. cm.	478.803	"	Hundredweights (long)	1.9684131
"	Ft. of air (1 atm., 60°F.)	13.096	"	Kilograms	100
"	Grams/sq. cm.	0.48824276	"	Pounds (avdp.)	220.46226
"	In. of Hg (32°F.)	0.0141390			
"	In. of H$_2$O (39.2°F.)	0.192227	Radians	Circumferences	0.15915494
"	Kg./sq. meter	4.8824276	"	Degrees	57.295779
"	Mm. of Hg (0°C.)	0.359131	"	Minutes	3437.7468
Pounds/sq. inch	Atmospheres	0.0680460	"	Quadrants	0.63661977
"	Bars	0.0689476	"	Revolutions	0.15915494
"	Cm. of Hg (0°C.)	5.17149	"	Seconds	206264.81
"	Cm. of H$_2$O (4°C.)	70.3089	Radians/cm	Degrees/cm	57.295779
"	Dynes/sq. cm.	68947.6	"	Degrees/ft	1746.3754
"	Grams/sq. cm.	70.306958	"	Degrees/inch	145.53128
"	In. of Hg (32°F.)	2.03602	"	Minutes/cm	3437.7468
"	In. of H$_2$O (39.2°F.)	27.6807	Radians/sec	Degrees/sec	57.295779
"	Kg./sq. cm.	0.070306958	"	Revolutions/min	9.5492966
"	Mm. of Hg (0°C.)	51.7149	"	Revolutions/sec	0.15915494
Pound wt.-sec./sq. ft.	Poises	478.803	Radians/(sec. × sec.)	Revolutions/(min. × min.)	572.95779
Pound wt.-sec./sq. in.	Poises	68947.6	"	Revolutions/(min. × sec.)	9.5492966
Puncheons (Brit.)	Cu. meters	0.31797510	"	Revolutions/(sec. × sec.)	0.15915494
"	Gallons (Brit.)	69.94467	Register tons	Cu. feet	100
"	Gallons (U.S.)	84	"	Cu. meters	2.8316847
			Revolutions	Degrees	360
Quadrants	Minutes	5400	"	Grades	400
"	Radians	1.5707963	"	Quadrants	4
Quarterns (Brit., dry)	Buckets (Brit.)	0.125	"	Radians	6.2831853
"	Bushels (Brit.)	0.0625	Reyns*	Centipoises	6.89476 × 10⁶
"	Cu. cm	2273.044	Rhes	Poises⁻¹	1
"	Gallons (Brit.)	0.5	Rods	Centimeters	502.92
"	Liters	2.272980	"	Chains (Gunter's)	0.25
"	Pecks (Brit.)	0.25	"	Chains (Ramden's)	0.165
Quarterns (Brit., liq.)	Cu. cm	142.0652	"	Feet	16.5
			"	Feet (U.S. Survey)	16.499967
			"	Furlongs	0.025
			"	Inches	198
			"	Links (Gunter's)	25

Table 4–9 (continued)

CONVERSION FACTORS

To convert from	To	Multiply by	To convert from	To	Multiply by
Rods	Links (Ramden's)	16.5	Sq. centimeters	Sq. mm	100
"	Meters	5.0292	"	Sq. mils	155000.31
"	Miles (statute)	0.003125	"	Sq. rods	3.9536861×10^{-6}
"	Perches	1	"	Sq. yards	0.00011959900
"	Yards	5.5	Sq. chains (Gunter's)	Acres	0.1
Rods (Brit., volume)	Cu. feet	1000	"	Sq. feet	4356
"	Cu. meters	28.316847	"	Sq. ft. (U.S. Survey)	4355.9826
Roods (Brit.)	Acres	0.25	"	Sq. inches	627264
"	Ares	10.117141	"	Sq. links (Gunter's)	10000
"	Sq. perches	40	"	Sq. meters	404.68564
"	Sq. yards	1210	"	Sq. miles	0.00015625
Ropes (Brit.)	Feet	20	"	Sq. rods	16
"	Meters	6.096	"	Sq. yards	484
"	Yards	6.6666666	Sq. chains (Ramden's)	Acres	0.22956841
			"	Sq. feet	10000
Scruples (apoth.)	Drams (apoth. or troy)	0.333333	"	Sq. ft. (U.S. Survey)	9999.9600
"	Drams (avdp.)	0.73142857	"	Sq. inches	1.44×10^{6}
"	Grains	20	"	Sq. links (Ramden's)	10000
"	Grams	1.2959782	"	Sq. meters	929.0304
"	Ounces (apoth. or troy)	0.041666	"	Sq. miles	0.00035870064
"	Ounces (avdp.)	0.045714286	"	Sq. rods	36.730946
"	Pennyweights	0.833333	"	Sq. yards	1111.111
"	Pounds (apoth. or troy)	0.003472222	Sq. decimeters	Sq. cm	100
"	Pounds (avdp.)	0.0028571429	"	Sq. inches	15.500031
Scruples (Brit., fluid)	Minims (Brit.)	20	Square degrees	Steradians	0.00030461742
Seams (Brit.)	Bushels (Brit.)	8	Sq. dekameters	Acres	0.024710538
"	Cu. feet	10.27479	"	Ares	1
"	Liters	290.9414	"	Sq. meters	100
Seconds (angular)	Degrees	0.000277777	"	Sq. yards	119.59900
"	Minutes	0.0166666	Sq. feet	Acres	2.295684×10^{-5}
"	Radians	4.8481368×10^{-6}	"	Ares	0.0009290304
Seconds (mean solar)	Days (mean solar)	1.1574074×10^{-5}	"	Sq. cm	929.0304
"	Days (sidereal)	1.1605763×10^{-5}	"	Sq. chains (Gunter's)	0.00022956841
"	Hours (mean solar)	0.0002777777	"	Sq. ft. (U.S. Survey)	0.99999600
"	Hours (sidereal)	0.00027853831	"	Sq. inches	144
"	Minutes (mean solar)	0.0166666	"	Sq. links (Gunter's)	2.2956841
"	Minutes (sidereal)	0.016712298	"	Sq. meters	0.09290304
"	Seconds (sidereal)	1.00273791	"	Sq. miles	3.5870064×10^{-8}
Seconds (sidereal)	Days (mean solar)	1.1542472×10^{-5}	"	Sq. rods	0.0036730946
"	Days (sidereal)	1.1574074×10^{-5}	"	Sq. yards	0.111111
"	Hours (mean solar)	0.00027701932	Sq. feet (U.S. Survey)	Acres	$2.29569330 \times 10^{-5}$
"	Hours (sidereal)	0.000277777	"	Sq. centimeters	929.03412
"	Minutes (mean solar)	0.016621159	"	Sq. chains (Ramden's)	0.00010000040
"	Minutes (sidereal)	0.0166666	"	Sq. feet	1.0000040
"	Seconds (mean solar)	0.99726957	Sq. hectometers	Sq. meters	10000
Siemen's units	Same as Mhos		Sq. inches	Circ. mils	1273239.5
Skeins	Feet	360	"	Sq. cm	6.4516
"	Meters	109.728	"	Sq. chains (Gunter's)	1.5942251×10^{-6}
Slugs	Geepounds	1	"	Sq. decimeters	0.064516
"	Kilograms	14.5939	"	Sq. feet	0.0069444
"	Pounds (avdp.)	32.1740	"	Sq. ft. (U.S. Survey)	0.0069444167
Slugs/cu. ft.	Grams/cu. cm.	0.515379	"	Sq. links (Gunter's)	0.01594225
Space (entire)	Hemispheres	2	"	Sq. meters	0.00064516
"	Steradians	12.566371	"	Sq. miles	$2.4909767 \times 10^{-10}$
Spans	Centimeters	22.86	"	Sq. mm	645.16
"	Fathoms	0.125	"	Sq. mils	1×10^{6}
"	Feet	0.75	Sq. inches/sec	Sq. cm./hr	23225.76
"	Inches	9	"	Sq. cm./sec	6.4516
"	Quarters (Brit. linear)	1	"	Sq. ft./min	0.416666
Spherical right angles	Hemispheres	0.25	Sq. kilometers	Acres	247.10538
"	Spheres	0.125	"	Sq. feet	1.0763910×10^{7}
"	Steradians	1.5707963	"	Sq. ft. (U.S. Survey)	1.0763867×10^{7}
Sq. centimeters	Ares	1×10^{-6}	"	Sq. inches	1.5500031×10^{9}
"	Circ. mm	127.32395	"	Sq. meters	1×10^{6}
"	Circ. mils	197352.52	"	Sq. miles	0.38610216
"	Sq. chains (Gunter's)	2.4710538×10^{-7}	"	Sq. yards	1.1959900×10^{6}
"	Sq. chains (Ramden's)	1.0763910×10^{-7}	Sq. links (Gunter's)	Acres	1×10^{-5}
"	Sq. decimeters	0.01	"	Sq. cm	404.68564
"	Sq. feet	0.0010763910	"	Sq. chains (Gunter's)	0.0001
"	Sq. ft. (U.S. Survey)	0.0010763867	"	Sq. feet	0.4356
"	Sq. inches	0.15500031	"	Sq. ft. (U.S. Survey)	0.43559826
"	Sq. meters	0.0001	"	Sq. Inches	62.7264

Table 4–9 (continued)

CONVERSION FACTORS

To convert from	To	Multiply by	To convert from	To	Multiply by
Sq. links (Ramden's)..	Acres.................	2.2956841×10^{-5}	Statfarads.........	Farads............	1.112646×10^{-12}
"	Sq. feet.............	1	"	Microfarads........	1.112646×10^{-6}
Sq. meters............	Acres.................	0.00024710538	Stathenries........	Abhenries..........	8.987584×10^{20}
"	Ares..................	0.01	"	E.M. cgs. units of induct-	
"	Hectares..............	0.0001		ance..............	8.987584×10^{20}
"	Sq. cm...............	10000	"	E.S. cgs. units......	1
"	Sq. feet.............	10.763910	"	Henries............	8.987584×10^{11}
"	Sq. inches............	1550.0031	"	Millihenries........	8.987584×10^{14}
"	Sq. kilometers........	1×10^{-6}	Statohms.........	Abohms............	8.987584×10^{20}
"	Sq. links (Gunter's)...	24.710538	"	E.S. cgs. units......	1
"	Sq. links (Ramden's)..	10.763910	"	Ohms..............	8.987584×10^{11}
"	Sq. miles.............	3.8610216×10^{-7}	Statvolts.........	Abvolts............	2.997930×10^{10}
"	Sq. mm...............	1×10^{6}	"	Volts..............	299.7930
"	Sq. rods.............	0.039536861	Statvolts/cm.......	Volts/cm...........	299.7930
"	Sq. yards............	1.1959900	"	Volts/inch..........	761.4742
Sq. miles.............	Acres.................	640	Statvolts/inch......	Volts/cm...........	118.0287
"	Hectares..............	258.99881	Steradians.........	Hemispheres........	0.15915494
"	Sq. chains (Gunter's)...	6400	"	Solid angles........	0.079577472
"	Sq. feet.............	2.7878288×10^{7}	"	Spheres............	0.079577472
"	Sq. ft. (U.S. Survey)...	2.78288×10^{7}	"	Spher. right angles...	0.63661977
"	Sq. kilometers........	2.5899881	"	Square degrees......	3282.8063
"	Sq. meters............	2589988.1	Steres............	Cubic meters.......	1
"	Sq. rods.............	102400	"	Decisteres..........	10
"	Sq. yards............	3.0976×10^{6}	"	Dekasteres.........	0.1
Sq. millimeters......	Circ. mm..............	1.2732395	"	Liters.............	999.972
"	Circ. mils............	1973.5252	Stilbs............	Candles/sq. cm......	1
"	Sq. cm...............	0.01	"	Candles/sq. inch.....	6.4516
"	Sq. inches............	0.0015500031	"	Lamberts...........	3.1415927
"	Sq. meters............	1×10^{-6}	Stokes*...........	Cgs. units of kinematic	
Sq. mils..............	Circ. mils............	1.2732395		viscosity..........	1
"	Sq. cm...............	6.4516×10^{-6}	"	Sq. cm./sec.........	1
"	Sq. inches............	1×10^{-6}	"	Sq. inches/sec......	0.15500031
"	Sq. mm...............	0.00064516	"	Poise cu. cm./gram...	1
Sq. rods.............	Acres.................	0.00625	Stones (Brit., legal)...	Centals (Brit.)......	0.14
"	Ares..................	0.2529285264			
"	Hectares..............	0.002529285264	Tons (long).........	Dynes.............	9.96402×10^{8}
"	Sq. cm...............	252928.5264	"	Hundredweights (long)....	20
"	Sq. feet.............	272.25	"	Hundredweights (short)..	22.4
"	Sq. ft. (U.S. Survey)...	272.24891	"	Kilograms..........	1016.0469
"	Sq. inches............	39204	"	Ounces (avdp.)......	35840
"	Sq. links (Gunter's)...	625	"	Pounds (apoth. or troy)..	2722.22
"	Sq. links (Ramden's)..	272.25	"	Pounds (avdp.)......	2240
"	Sq. meters............	25.29285264	"	Tons (metric).......	1.0160469
"	Sq. miles.............	9.765625×10^{-6}	"	Tons (short)........	1.12
"	Sq. yards............	30.25	Tons (metric).......	Dynes.............	9.80665×10^{8}
Sq. yards............	Acres.................	0.00020661157	"	Grams.............	1×10^{6}
"	Ares..................	0.0083612736	"	Hundredweights (short)...	22.046226
"	Hectares..............	8.3612736×10^{-5}	"	Kilograms..........	1000
"	Sq. cm...............	8361.2736	"	Ounces (avdp.)......	35273.962
"	Sq. chains (Gunter's)...	0.0020661157	"	Pounds (apoth. or troy)..	2679.2289
"	Sq. chains (Ramden's)...	0.0009	"	Pounds (avdp.)......	2204.6226
"	Sq. feet.............	9	"	Tons (long).........	0.98420653
"	Sq. ft. (U.S. Survey)...	8.9999640	"	Tons (short)........	1.1023113
"	Sq. inches............	1296	Tons (short).........	Dynes.............	8.89644×10^{8}
"	Sq. links (Gunter's)...	20.661157	"	Hundredweights (short)...	20
"	Sq. links (Ramden's)..	9	"	Kilograms..........	907.18474
"	Sq. meters............	0.83612736	"	Ounces (avdp.)......	32000
"	Sq. miles.............	$3.228305785 \times 10^{-7}$	"	Pounds (apoth. or troy)..	2430.555
"	Sq. perches (Brit.)....	0.033057851	"	Pounds (avdp.)......	2000
"	Sq. rods.............	0.033057851	"	Tons (long).........	0.89285714
Statamperes.........	Abamperes............	3.335635×10^{-11}	"	Tons (metric).......	0.90718474
"	Amperes..............	3.335635×10^{-10}	Tons of refrig. (U.S.,		
"	E.M. cgs. units of current.	3.335635×10^{-11}	comm.)...........	B.t.u. (IST.)/hr......	12000
"	E.S. cgs. units.........	1	"	B.t.u. (IST.)/min.....	200
Statcoulombs........	Ampere-hours.........	9.265653×10^{-14}	"	Cal., *kg*. (IST.)/hr...	3023.949
"	Coulombs............	3.335635×10^{-10}	"	Horsepower.........	4.71611
"	Electronic charges......	2.082093×10^{9}	"	Kg. of ice melted/hr...	37.971
"	E.M. cgs. units of electric		"	Lb. of ice melted/hr...	83.711
	charge............	3.335635×10^{-11}	Tons of refrig. (U.S.,		
Statfarads..........	E.M. cgs. units of capaci-		std.)...........	B.t.u. (IST.).......	288000
	tance..............	1.112646×10^{-21}	"	B.t.u. (mean).......	287774
"	E.S. cgs. units.........	1	"	Cal., *kg*. (IST.)....	72574.8

Table 4–9 (continued)

CONVERSION FACTORS

To convert from	To	Multiply by
Tons of refrig. (U.S., std.)	Cal., *kg.* (mean)	72517.9
"	Lb. of ice melted	2009.1
Tons (long)/sq. ft	Atmospheres	1.05849
"	Dynes/sq. cm	1.07252×10^6
"	Grams/sq. cm	1093.6638
"	Pounds/sq. ft	2240
Tons (short)/sq. ft	Atmospheres	0.945082
"	Dynes/sq. cm	957.605
"	Grams/sq. cm	976.486
"	Pounds/sq. inch	13.8888
Tons (long)/sq. in	Atmospheres	152.423
"	Dynes/sq. cm	1.54443×10^8
"	Grams/sq. cm	157487.59
Tons (short)/sq. in	Dynes/sq. cm	1.37895×10^8
"	Kg./sq. mm	1406.139
"	Pounds/sq. inch	2000
Torrs (*or* Tors)	Millimeters of Hg (0°C.)	1
Townships (U.S.)	Acres	23040
"	Sections	36
"	Sq. miles	36
Tuns	Gallons (U.S.)	252
"	Hogsheads	4
Volts	Abvolts	1×10^8
"	Mks. (r *or* nr) units	1
"	Statvolts	0.003335635
"	Volts (Int.)	0.999670
Volts (Int.)	Volts	1.000330
Volts/°C.	Joules/(coulomb × °C.)	1
Volt-coulombs	Joules (Int.)	0.999835
Volt-coulombs (Int.)	Joules	1.000165
Volt-electronic charge-seconds	Planck's constant*	2.41814×10^{14}
Volt-faraday (chem.)-seconds	Planck's constant*	1.45650×10^{38}
Volt-faraday (phys.)-seconds	Planck's constant	1.45690×10^{38}
Volt-seconds	Maxwells	1×10^8
Watts	B.t.u./hr	3.41443
"	B.t.u. (mean)/hr	3.40952
"	B.t.u. (mean)/min	0.0568253
"	B.t.u./sec	0.000948451
"	B.t.u. (mean)/sec	0.000947088
"	Cal., *gm.*/hr	860.421
"	Cal., *gm.* (mean)/hr	859.184
"	Cal., *gm.* (20°C.)/hr	860.853
"	Cal., *gm.*/min	14.3403
"	Cal., *gm.* (IST.)/min	14.3310
"	Cal., *gm.* (mean)/min	14.3197
"	Cal., *kh.*/min	0.0143403
"	Cal., *kg.* (IST.)/min	0.0143310
"	Cal., *kg.* (mean)/min	0.0143197
"	Ergs/sec	1×10^7
"	Foot-pounds/min	44.2537
"	Horsepower	0.00134102
"	Horsepower (boiler)	0.000101942
"	Horsepower (elec.)	0.00134048
"	Horsepower (metric)	0.00135962
"	Joules/sec	1
"	Kilowatts	0.001
"	Liter-atm./hr	35.5282
Watts (Int.)	B.t.u./hr	3.41499
"	B.t.u. (mean)/hr	3.41008
"	B.t.u./min	0.569165
"	B.t.u. (mean)/min	0.0568347
"	Cal., *gm.*/hr	860.563
"	Cal., *gm.* (mean)/hr	859.326
"	Cal., *kg.*/min	0.0143427
"	Cal., *kg.* (IST.)/min	0.0143333
"	Cal., *kg.* (mean)/min	0.0143221
Watts (Int.)	Ergs/sec	1.000165×10^7
"	Joules (Int.)/sec	1
"	Watts	1.000165
Watts/sq. cm	B.t.u./(hr. × sq. ft.)	3172.10
"	Cal., *gm.*/(hr. × sq. cm.)	860.421
"	Ft.-lb./(min. × sq. ft.)	41113.1
Watts/sq. in	B.t.u./(hr. × sq. ft.)	491.677
"	Cal., *gm.*/(hr. × sq. cm.)	133.365
"	Ft.-lb./(min. × sq. ft.)	6372.54
Watt-hours	B.t.u.	3.41443
"	B.t.u. (mean)	3.40952
"	Cal., *gm.*	860.421
Watt-hours	Cal., *kg.* (mean)	0.859184
"	Cal., *gm.* (mean)	859.184
"	Foot-pounds	2655.22
"	Hp.-hours	0.00134102
"	Joules	3600
"	Joules (Int.)	3599.41
"	Kg.-meters	367.098
"	Kw.-hours	0.001
"	Watt-hours (Int.)	0.999835
Watt-sec	Foot-pounds	0.737562
"	Gram-cm	10197.16
"	Joules	1
"	Liter-atm	0.00986895
"	Volt-coulombs	1
Wave length of orange-red line of krypton 86	Ångström units	6057.80211
"	Millimeters	0.000605780211
Wave length of red line of cadmium	Ångström units	6438.4696
"	Millimeters	0.00064384696
Webers	Cgs. units of induction flux	1×10^8
"	E.M. cgs. units of induction flux	1×10^8
"	Lines	1×10^8
"	Maxwells	1×10^8
"	Mks. units of induction flux	1
"	Mks. nr units of magnetic charge	0.079577472
"	Mks. r units of magnetic charge	1
"	Volt-seconds	1
Webers/sq. cm	Gausses	1×10^8
"	Lines/sq. cm	1×10^8
"	Lines/sq. inch	6.4516×10^8
Webers/sq. in	Gausses	1.5500031×10^7
Weeks (mean calendar)	Days (mean solar)	7
"	Days (sidereal)	7.0191654
"	Hours (mean solar)	168
"	Hours (sidereal)	168.45997
"	Minutes (mean solar)	10080
"	Minutes (sidereal)	10107.598
"	Months (lunar)	0.23704235
"	Months (mean calendar)	0.23013699
"	Years (calendar)	0.019178082
"	Years (sidereal)	0.019164622
"	Years (tropical)	0.019165365
Weys (Brit., mass.)	Pounds (avdp.)	252
Yards	Centimeters	91.44
"	Chains (Gunter's)	0.4545454
"	Chains (Ramden's)	0.03
"	Cubits	2
"	Fathoms	0.5
"	Feet	3
"	Feet (U.S. Survey)	2.9999940
"	Furlongs	0.00454545
"	Inches	36
"	Meters	0.9144
"	Poles (Brit.)	0.181818

Table 4–9 (continued)

CONVERSION FACTORS

To convert from	To	Multiply by	To convert from	To	Multiply by
Yards.............	Quarters (Brit., linear)....	4	Years (sidereal)......	Days (sidereal)..........	366.25640
"	Rods...................	0.181818	"	Years (calendar).........	1.0007024
"	Spans..................	4	"	Years (tropical).........	1.0000388
Years (calendar).....	Days (mean solar).......	365	Years (tropical)......	Days (mean solar).......	365.24219
"	Hours (mean solar).......	8760	"	Days (sidereal)..........	366.24219
"	Minutes (mean solar)......	525600	"	Hours (mean solar).......	8765.8126
"	Months (lunar)...........	12.360065	"	Hours (sidereal).........	8789.8126
"	Months (mean calendar)...	12	"	Months (mean calendar)...	12.007963
"	Seconds (mean solar).....	3.1536×10^7	"	Seconds (mean solar).....	3.1556926×10^7
"	Weeks (mean calendar)....	52.142857	"	Seconds (sidereal).........	3.1643326×10^7
"	Years (sidereal)..........	0.99929814	"	Weeks (mean calendar)....	52.177456
"	Years (tropical)..........	0.99933690	"	Years (calendar)..........	1.0006635
Years (leap)........	Days (mean solar).......	366	"	Years (sidereal)..........	0.99996121
Years (sidereal)......	Days (mean solar).......	365.25636			

Defined Values and Equivalents

Meter.................................	**(m)**	1 650 763.73 wave lengths in vacuo of the unperturbed transition $2p_{10} - 5d_5$ in ^{86}Kr
Kilogram..............................	**(kg)**	mass of the international kilogram at Sèvres, France
Second................................	**(s)**	1/31 556 925.974 7 of the tropical year at 12^h ET, 0 January 1900
Degree Kelvin.........................	**(°K)**	defined in the thermodynamic scale by assigning 273.16 °K to the triple point of water (freezing point, 273.15 °K = 0 °C)
Unified atomic mass unit...............	**(u)**	1/12 the mass of an atom of the ^{12}C nuclide
Mole.................................	**(mol)**	amount of substance containing the same number of atoms as 12 g of pure ^{12}C
Standard acceleration of free fall.......	**(gₙ)**	9.806 65 m s^{-2}, 980.665 cm s^{-2}
Normal atmospheric pressure............	**(atm)**	101 325 N m^{-2}, 1 013 250 dyn cm^{-2}
Thermochemical calorie.................	**(cal$_{th}$)**	4.1840 J, 4.1840×10^7 erg
International Steam Table calorie........	**(cal$_{IT}$)**	4.1868 J, 4.1868×10^7 erg
Liter.................................	**(l)**	0.001 m^3, 1000 cm^3 (recommended by GCWM, 1964)
Inch.................................	**(in)**	0.0254 m. 2.54 cm
Pound (avdp)..........................	**(lb)**	0.453 592 37 kg, 453.592 37 g

From various U.S. Government and IUPAC publications and from calculations based on values given in these publications, in Weast, R. C., Ed., *Handbook of Chemistry and Physics*, 55th ed., CRC Press, Cleveland, 1974, F-282.

Table 4–10
INCH–MILLIMETER CONVERSIONS–EXACT

$$\frac{1}{64} \text{ to } 10 \text{ in.}$$

Even Inches

Inches	1	2	3	4	5	6	7	8	9	10
Millimeters	25.4	50.8	76.2	101.6	127.0	152.4	177.8	203.2	228.6	254.0

Note: All of the above values are exact, based on the accepted conversion 1 in. = 25.4000 mm; hence the exact millimeter equivalent of any decimal-multiple or decimal-fractional value may be obtained by shifting the decimal point.

Fractional Inches

16ths	32nds	64ths	Decimal equivalents, inches (exact)	Millimeters (exact)	16ths	32nds	64ths	Decimal equivalents, inches (exact)	Millimeters (exact)
		1	0.015 625	0.396 875			33	0.515 625	13.096 875
	1	2	0.031 25	0.793 750		17	34	0.531 25	13.493 750
		3	0.046 875	1.190 625			35	0.546 875	13.890 625
1	2	4	0.062 5	1.587 500	9	18	36	0.562 5	14.287 500
		5	0.078 125	1.984 375			37	0.578 125	14.684 375
	3	6	0.093 75	2.381 250		19	38	0.593 75	15.081 250
		7	0.109 375	2.778 125			39	0.609 375	15.478 125
2	4	8	0.125 0	3.175 000	10	20	40	0.625 0	15.875 000
		9	0.140 625	3.571 875			41	0.640 625	16.271 875
	5	10	0.156 25	3.968 750		21	42	0.656 25	16.668 750
		11	0.171 875	4.365 625			43	0.671 875	17.065 625
3	6	12	0.187 5	4.762 500	11	22	44	0.687 5	17.462 500
		13	0.203 125	5.159 375			45	0.703 125	17.859 375
	7	14	0.218 75	5.556 250		23	46	0.718 75	18.256 250
		15	0.234 375	5.953 125			47	0.734 375	18.653 125
4	8	16	0.250 0	6.350 000	12	24	48	0.750 0	19.050 000
		17	0.265 625	6.746 875			49	0.765 625	19.446 875
	9	18	0.281 25	7.143 750		25	50	0.781 25	19.843 750
		19	0.296 875	7.540 625			51	0.796 875	20.240 625
5	10	20	0.312 5	7.937 500	13	26	52	0.812 5	20.637 500
		21	0.328 125	8.334 375			53	0.828 125	21.034 375
	11	22	0.343 75	8.731 250		27	54	0.843 75	21.431 250
		23	0.359 375	9.128 125			55	0.859 375	21.828 125
6	12	24	0.375 0	9.525 000	14	28	56	0.875 0	22.225 000
		25	0.390 625	9.921 875			57	0.890 625	22.621 875
	13	26	0.406 25	10.318 750		29	58	0.906 25	23.018 750
		27	0.421 875	10.715 625			59	0.921 875	23.415 625
7	14	28	0.437 5	11.112 500	15	30	60	0.937 5	23.812 500
		29	0.453 125	11.509 375			61	0.953 125	24.209 375
	15	30	0.468 75	11.906 250		31	62	0.968 75	24.606 250
		31	0.484 375	12.303 125			63	0.984 375	25.003 125
8	16	32	0.500 0	12.700 000	16	32	64	1.000 0	25.400 000

From Bolz, R. E. and Tuve, G.L., Eds., *Handbook of Tables for Applied Engineering Science*, 2nd ed., CRC Press, Cleveland, 1973, 842.

Table 4–11
DEGREES–RADIANS CONVERSIONS

Degrees–Radians

The table gives in radians the angle that is expressed in degrees and minutes at the side and top. Angles expressed to the nearest minute and second can readily be converted to radians by adding to the equivalent of the whole number of degrees the equivalents of the minutes and seconds found on the third and fourth pages of this table.

°	00′	10	20	30	40	50
0	0.00000	0.00291	0.00582	0.00873	0.01164	0.01454
1	0.01745	0.02036	0.02327	0.02618	0.02909	0.03200
2	0.03491	0.03782	0.04072	0.04363	0.04654	0.04945
3	0.05236	0.05527	0.05818	0.06109	0.06400	0.06690
4	0.06981	0.07272	0.07563	0.07854	0.08145	0.08436
5	0.08727	0.09018	0.09308	0.09599	0.09890	0.10181
6	0.10472	0.10763	0.11054	0.11345	0.11636	0.11926
7	0.12217	0.12508	0.12799	0.13090	0.13381	0.13672
8	0.13963	0.14254	0.14544	0.14835	0.15126	0.15417
9	0.15708	0.15999	0.16290	0.16581	0.16872	0.17162
10	0.17453	0.17744	0.18035	0.18326	0.18617	0.18908
11	0.19199	0.19490	0.19780	0.20071	0.20362	0.20653
12	0.20944	0.21235	0.21526	0.21817	0.22108	0.22398
13	0.22689	0.22980	0.23271	0.23562	0.23853	0.24144
14	0.24435	0.24725	0.25016	0.25307	0.25598	0.25889
15	0.26180	0.26471	0.26762	0.27053	0.27343	0.27634
16	0.27925	0.28216	0.28507	0.28798	0.29089	0.29380
17	0.29671	0.29961	0.30252	0.30543	0.30834	0.31125
18	0.31416	0.31707	0.31998	0.32289	0.32579	0.32870
19	0.33161	0.33452	0.33743	0.34034	0.34325	0.34616
20	0.34907	0.35197	0.35488	0.35779	0.36070	0.36361
21	0.36652	0.36943	0.37234	0.37525	0.37815	0.38106
22	0.38397	0.38688	0.38979	0.39270	0.39561	0.39852
23	0.40143	0.40433	0.40724	0.41015	0.41306	0.41597
24	0.41888	0.42179	0.42470	0.42761	0.43051	0.43342
25	0.43633	0.43924	0.44215	0.44506	0.44797	0.45088
26	0.45379	0.45669	0.45960	0.46251	0.46542	0.46833
27	0.47124	0.47415	0.47706	0.47997	0.48287	0.48578
28	0.48869	0.49160	0.49451	0.49742	0.50033	0.50324
29	0.50615	0.50905	0.51196	0.51487	0.51778	0.52069
30	0.52360	0.52651	0.52942	0.53233	0.53523	0.53814
31	0.54105	0.54396	0.54687	0.54978	0.55269	0.55560
32	0.55851	0.56141	0.56432	0.56723	0.57014	0.57305
33	0.57596	0.57887	0.58178	0.58469	0.58759	0.59050
34	0.59341	0.59632	0.59923	0.60214	0.60505	0.60796
35	0.61087	0.61377	0.61668	0.61959	0.62250	0.62541
36	0.62832	0.63123	0.63414	0.63705	0.63995	0.64286
37	0.64577	0.64868	0.65159	0.65450	0.65741	0.66032
38	0.66323	0.66613	0.66904	0.67195	0.67486	0.67777
39	0.68068	0.68359	0.68650	0.68941	0.69231	0.69522
40	0.69813	0.70104	0.70395	0.70686	0.70977	0.71268
41	0.71558	0.71849	0.72140	0.72431	0.72722	0.73013
42	0.73304	0.73595	0.73886	0.74176	0.74467	0.74758
43	0.75049	0.75340	0.75631	0.75922	0.76213	0.76504
44	0.76794	0.77085	0.77376	0.77667	0.77958	0.78249

Table 4-11 (continued)
DEGREES-RADIANS CONVERSIONS

Degrees-Radians

°	00'	10	20	30	40	50
45	0.78540	0.78831	0.79122	0.79412	0.79703	0.79994
46	0.80285	0.80576	0.80867	0.81158	0.81449	0.81740
47	0.82030	0.82321	0.82612	0.82903	0.83194	0.83485
48	0.83776	0.84067	0.84358	0.84648	0.84939	0.85230
49	0.85521	0.85812	0.86103	0.86394	0.86685	0.86976
50	0.87266	0.87557	0.87848	0.88139	0.88430	0.88721
51	0.89012	0.89303	0.89594	0.89884	0.90175	0.90466
52	0.90757	0.91048	0.91339	0.91630	0.91921	0.92212
53	0.92502	0.92793	0.93084	0.93375	0.93666	0.93957
54	0.94248	0.94539	0.94830	0.95120	0.95411	0.95702
55	0.95993	0.96284	0.96575	0.96866	0.97157	0.97448
56	0.97738	0.98029	0.98320	0.98611	0.98902	0.99193
57	0.99484	0.99775	1.00066	1.00356	1.00647	1.00938
58	1.01229	1.01520	1.01811	1.02102	1.02393	1.02684
59	1.02974	1.03265	1.03556	1.03847	1.04138	1.04429
60	1.04720	1.05011	1.05302	1.05592	1.05883	1.06174
61	1.06465	1.06756	1.07047	1.07338	1.07629	1.07920
62	1.08210	1.08501	1.08792	1.09083	1.09374	1.09665
63	1.09956	1.10247	1.10538	1.10828	1.11119	1.11410
64	1.11701	1.11992	1.12283	1.12574	1.12865	1.13156
65	1.13446	1.13737	1.14028	1.14319	1.14610	1.14901
66	1.15192	1.15483	1.15774	1.16064	1.16355	1.16646
67	1.16937	1.17228	1.17519	1.17810	1.18101	1.18392
68	1.18682	1.18973	1.19264	1.19555	1.19846	1.20137
69	1.20428	1.20719	1.21009	1.21300	1.21591	1.21882
70	1.22173	1.22464	1.22755	1.23046	1.23337	1.23627
71	1.23918	1.24209	1.24500	1.24791	1.25082	1.25373
72	1.25664	1.25955	1.26245	1.26536	1.26827	1.27118
73	1.27409	1.27700	1.27991	1.28282	1.28573	1.28863
74	1.29154	1.29445	1.29736	1.30027	1.30318	1.30609
75	1.30900	1.31191	1.31481	1.31772	1.32063	1.32354
76	1.32645	1.32936	1.33227	1.33518	1.33809	1.34099
77	1.34390	1.34681	1.34972	1.35263	1.35554	1.35845
78	1.36136	1.36427	1.36717	1.37008	1.37299	1.37590
79	1.37881	1.38172	1.38463	1.38754	1.39045	1.39335
80	1.39626	1.39917	1.40208	1.40499	1.40790	1.41081
81	1.41372	1.41663	1.41953	1.42244	1.42535	1.42826
82	1.43117	1.43408	1.43699	1.43990	1.44281	1.44571
83	1.44862	1.45153	1.45444	1.45735	1.46026	1.46317
84	1.46608	1.46899	1.47189	1.47480	1.47771	1.48062
85	1.48353	1.48644	1.48935	1.49226	1.49517	1.49807
86	1.50098	1.50389	1.50680	1.50971	1.51262	1.51553
87	1.51844	1.52135	1.52425	1.52716	1.53007	1.53298
88	1.53589	1.53880	1.54171	1.54462	1.54753	1.55043
89	1.55334	1.55625	1.55916	1.56207	1.56498	1.56789
90	1.57080	1.57371	1.57661	1.57952	1.58243	1.58534
91	1.58825	1.59116	1.59407	1.59698	1.59989	1.60279
92	1.60570	1.60861	1.61152	1.61443	1.61734	1.62025
93	1.62316	1.62607	1.62897	1.63188	1.63479	1.63770
94	1.64061	1.64352	1.64643	1.64934	1.65225	1.65515
95	1.65806	1.66097	1.66388	1.66679	1.66970	1.67261
96	1.67552	1.67842	1.68133	1.68424	1.68715	1.69006
97	1.69297	1.69588	1.69879	1.70170	1.70460	1.70751
98	1.71042	1.71333	1.71624	1.71915	1.72206	1.72497
99	1.72788	1.73078	1.73369	1.73660	1.73951	1.74242

Table 4–11 (continued)
DEGREES–RADIANS CONVERSIONS

Degrees–Radians

°	00'	10	20	30	40	50
100	1.74533	1.74824	1.75115	1.75406	1.75696	1.75987
101	1.76278	1.76569	1.76860	1.77151	1.77442	1.77733
102	1.78024	1.78314	1.78605	1.78896	1.79187	1.79478
103	1.79769	1.80060	1.80351	1.80642	1.80932	1.81223
104	1.81514	1.81805	1.82096	1.82387	1.82678	1.82969
105	1.83260	1.83550	1.83841	1.84132	1.84423	1.84714
106	1.85004	1.85296	1.85587	1.85878	1.86168	1.86459
107	1.86750	1.87041	1.87332	1.87623	1.87914	1.88205
108	1.88496	1.88786	1.89077	1.89368	1.89659	1.89950
109	1.90241	1.90532	1.90823	1.91114	1.91404	1.91695
110	1.91986	1.92277	1.92568	1.92859	1.93150	1.93441

Degrees	Radians	Degrees	Radians	Minutes	Radians	Seconds	Radians
90	1.57080	150	2.61799	0	0.00000	0	0.00000
91	1.58825	151	2.63545	1	0.00029	1	0.00000
92	1.60570	152	2.65290	2	0.00058	2	0.00001
93	1.62316	153	2.67035	3	0.00087	3	0.00001
94	1.64061	154	2.68781	4	0.00116	4	0.00002
95	1.65806	155	2.70526	5	0.00145	5	0.00002
96	1.67552	156	2.72271	6	0.00175	6	0.00003
97	1.69297	157	2.74017	7	0.00204	7	0.00003
98	1.71042	158	2.75762	8	0.00233	8	0.00004
99	1.72788	159	2.77507	9	0.00262	9	0.00004
100	1.74533	160	2.79253	10	0.00291	10	0.00005
101	1.76278	161	2.80998	11	0.00320	11	0.00005
102	1.78024	162	2.82743	12	0.00349	12	0.00006
103	1.79769	163	2.84489	13	0.00378	13	0.00006
104	1.81514	164	2.86234	14	0.00407	14	0.00007
105	1.83260	165	2.87979	15	0.00436	15	0.00007
106	1.85005	166	2.89725	16	0.00465	16	0.00008
107	1.86750	167	2.91470	17	0.00495	17	0.00008
108	1.88496	168	2.93215	18	0.00524	18	0.00009
109	1.90241	169	2.94961	19	0.00553	19	0.00009
110	1.91986	170	2.96706	20	0.00582	20	0.00010
111	1.93732	171	2.98451	21	0.00611	21	0.00010
112	1.95477	172	3.00197	22	0.00640	22	0.00011
113	1.97222	173	3.01942	23	0.00669	23	0.00011
114	1.98968	174	3.03687	24	0.00698	24	0.00012
115	2.00713	175	3.05433	25	0.00727	25	0.00012
116	2.02458	176	3.07178	26	0.00756	26	0.00013
117	2.04204	177	3.08923	27	0.00785	27	0.00013
118	2.05949	178	3.10669	28	0.00814	28	0.00014
119	2.07694	179	3.12414	29	0.00844	29	0.00014
120	2.09440	180	3.14159	30	0.00873	30	0.00015
121	2.11185	190	3.31613	31	0.00902	31	0.00015
122	2.12930	200	3.49066	32	0.00931	32	0.00016
123	2.14676	210	3.66519	33	0.00960	33	0.00016
124	2.16421	220	3.83972	34	0.00989	34	0.00016
125	2.18166	230	4.01426	35	0.01018	35	0.00017
126	2.19911	240	4.18879	36	0.01047	36	0.00017
127	2.21657	250	4.36332	37	0.01076	37	0.00018
128	2.23402	260	4.53786	38	0.01105	38	0.00018
129	2.25147	270	4.71239	39	0.01134	39	0.00019

Table 4–11 (continued)
DEGREES–RADIANS CONVERSIONS

Degrees–Radians

Degrees	Radians	Degrees	Radians	Minutes	Radians	Seconds	Radians
130	2.26893	**280**	4.88692	**40**	0.01164	**40**	0.00019
131	2.28638	290	5.06145	41	0.01193	41	0.00020
132	2.30383	300	5.23599	42	0.01222	42	0.00020
133	2.32129	310	5.41052	43	0.01251	43	0.00021
134	2.33874	320	5.58505	44	0.01280	44	0.00021
135	2.35619	**330**	5.75959	**45**	0.01309	**45**	0.00022
136	2.37365	340	5.93412	46	0.01338	46	0.00022
137	2.39110	350	6.10865	47	0.01367	47	0.00023
138	2.40855	360	6.28319	48	0.01396	48	0.00023
139	2.42601	370	6.45772	49	0.01425	49	0.00024
140	2.44346	**380**	6.63225	**50**	0.01454	**50**	0.00024
141	2.46091	390	6.80678	51	0.01484	51	0.00025
142	2.47837	400	6.98132	52	0.01513	52	0.00025
143	2.49582	410	7.15585	53	0.01542	53	0.00026
144	2.51327	420	7.33038	54	0.01571	54	0.00026
145	2.53073	**430**	7.50492	**55**	0.01600	**55**	0.00027
146	2.54818	440	7.67945	56	0.01629	56	0.00027
147	2.56563	450	7.85398	57	0.01658	57	0.00028
148	2.58309	460	8.02851	58	0.01687	58	0.00028
149	2.60054	470	8.20305	59	0.01716	59	0.00029
150	2.61799	**480**	8.37758	**60**	0.01745	**60**	0.00029

Degrees, Minutes, and Seconds to Radians

Units in degrees, minutes or seconds	Degrees to radians	Minutes to radians	Seconds to radians
10	0.174 5329	0.002 9089	0.000 0485
20	0.349 0659	0.005 8178	0.000 0970
30	0.523 5988	0.008 7266	0.000 1454
40	0.698 1317	0.011 6355	0.000 1939
50	0.872 6646	0.014 5444	0.000 2424
60	1.047 1976	0.017 4533	0.000 2909
70	1.221 7305	(0.020 3622)	(0.000 3394)
80	1.396 2634	(0.023 2711)	(0.000 3879)
90	1.570 7963	(0.026 1800)	(0.000 4364)
100	1.745 3293	. . . .	. . . .
200	3.490 6585	. . . .	. . . .
300	5.235 9878	. . . .	. . . .

where n = 1, 2, 3, 4, etc. n (100°) = n (1.745 3293)

Table 4–11 (continued)
DEGREES–RADIANS CONVERSIONS

Radians to Degrees and Decimals

Radians	Degrees	Radians	Degrees	Radians	Degrees	Radians	Degrees
1	57.2958	0.1	5.7296	0.01	0.5730	0.001	0.0573
2	114.5916	.2	11.4592	.02	1.1459	.002	.1146
3	171.8873	.3	17.1887	.03	1.7189	.003	.1719
4	229.1831	.4	22.9183	.04	2.2918	.004	.2292
5	286.4789	.5	28.6479	.05	2.8648	.005	.2865
6	343.7747	.6	34.3775	.06	3.4377	.006	.3438
7	401.0705	.7	40.1070	.07	4.0107	.007	.4011
8	458.3662	.8	45.8366	.08	4.5837	.008	.4584
9	515.6620	.9	51.5662	.09	5.1566	.009	.5157
10	572.9578	1.0	57.2958	.10	5.7296	.010	.5730

Radians–Degrees
Multiples and Fractions of π Radians in Degrees

Radians	Radians	Degrees	Radians	Radians	Degrees	Radians	Radians	Degrees
π	3.1416	180	$\pi/2$	1.5708	90	$2\pi/3$	2.0944	120
2π	6.2832	360	$\pi/3$	1.0472	60	$3\pi/4$	2.3562	135
3π	9.4248	540	$\pi/4$	0.7854	45	$5\pi/6$	2.6180	150
4π	12.5664	720	$\pi/5$	0.6283	36	$7\pi/6$	3.6652	210
5π	15.7080	900	$\pi/6$	0.5236	30	$5\pi/4$	3.9270	225
6π	18.8496	1080	$\pi/7$	0.4488	25.714	$4\pi/3$	4.1888	240
7π	21.9911	1260	$\pi/8$	0.3927	22.5	$3\pi/2$	4.7124	270
8π	25.1327	1440	$\pi/9$	0.3491	20	$5\pi/3$	5.2360	300
9π	28.2743	1620	$\pi/10$	0.3142	18	$7\pi/4$	5.4978	315
10π	31.4159	1800	$\pi/12$	0.2618	15	$11\pi/6$	5.7596	330

From Selby, S. M., Ed., *Handbook of Tables for Mathematics,* 4th ed., Chemical Rubber, Cleveland, 1970, 337.

Table 4–12
TEMPERATURE CONVERSION

This table permits one to convert from degrees Celsius to degrees Fahrenheit or from degrees Fahrenheit to degrees Celsius. The conversion is accomplished by first locating in a column printed in boldface type the number that is to be converted. If the number to be converted is in degrees Fahrenheit, one may find its equivalent in degrees Celsius by reading to the left. If the number to be converted is in degrees Celsius, one may find its equivalent in degrees Fahrenheit by

reading to the right. Degrees Celsius are identical to degrees Centigrade; however, the word Celsius is preferred for international use.

The approved international symbolic abbreviation for degrees Celsius is $^\circ$C; for degrees Fahrenheit it is $^\circ$F. Absolute zero on the Celsius scale is -273.15°C; on the Fahrenheit scale it is -459.67°F. The relation between degrees Fahrenheit and degrees Celsius may be expressed by

$$^\circ C = 5/9 \; (^\circ F - 32) \text{ or }$$
$$^\circ F = 9/5 \; (^\circ C) + 32.$$

To °C	←°F or °C→	To °F	To °C	←°F or °C→	To °F	To °C	←°F or °C→	To °F
−273.15	−459.67	—	−106.67	−160	−256	−33.89	−29	−20.2
−267.78	−450	—	−103.89	−155	−247	−33.33	−28	−18.4
−262.22	−440	—	−101.11	−150	−238	−32.78	−27	−16.6
−256.67	−430	—	−98.33	−145	−229	−32.22	−26	−14.8
−251.11	−420	—	−95.56	−140	−220	−31.67	−25	−13
−245.56	−410	—	−92.78	−135	−211	−31.11	−24	−11.2
−240	−400	—	−90	−130	−202	−30.56	−23	−9.4
−234.44	−390	—	−87.22	−125	−193	−30	−22	−7.6
−228.89	−380	—	−84.44	−120	−184	−29.44	−21	−5.8
−223.33	−370	—	−81.67	−115	−175	−28.89	−20	−4
−217.78	−360	—	−78.89	−110	−166	−28.33	−19	−2.2
−212.22	−350	—	−76.11	−105	−157	−27.78	−18	−0.4
−206.67	−340	—	−73.33	−100	−148	−27.22	−17	1.4
−201.11	−330	—	−70.56	−95	−139	−26.67	−16	3.2
−195.56	−320	—	−67.78	−90	−130	−26.11	−15	5
−190	−310	—	−65	−85	−121	−25.56	−14	6.8
−184.44	−300	—	−62.22	−80	−112	−25	−13	8.6
−178.89	−290	—	−59.44	−75	−103	−24.44	−12	10.4
−173.33	−280	—	−56.67	−70	−94	−23.89	−11	12.2
−167.78	−270	−454	−53.89	−65	−85	−23.33	−10	14
−162.22	−260	−436	−51.11	−60	−76	−22.78	−9	15.8
−156.67	−250	−418	−48.33	−55	−67	−22.22	−8	17.6
−151.11	−240	−400	−45.56	−50	−58	−21.67	−7	19.4
−145.56	−230	−382	−42.78	−45	−49	−21.11	−6	21.2
−140	−220	−364	−40	−40	−40	−20.56	−5	23
−134.44	−210	−346	−39.44	−39	−38.2	−20	−4	24.8
−131.67	−205	−337	−38.89	−38	−36.4	−19.44	−3	26.6
−128.89	−200	−328	−38.33	−37	−34.6	−18.89	−2	28.4
−126.11	−195	−319	−37.78	−36	−32.8	−18.33	−1	30.2
−123.33	−190	−310	−37.22	−35	−31	−17.78	0	32
−120.56	−185	−301	−36.67	−34	−29.2	−17.22	1	33.8
−117.78	−180	−292	−36.11	−33	−27.4	−16.67	2	35.6
−115	−175	−283	−35.56	−32	−25.6	−16.11	3	37.4
−112.22	−170	−274	−35	−31	−23.8	−15.56	4	39.2
−109.44	−165	−265	−34.44	−30	−22	−15	5	41

Table 4–12 (continued)
TEMPERATURE CONVERSION

To convert			To convert			To convert		
To °C	←°F or °C→	To °F	To °C	←°F or °C→	To °F	To °C	←°F or °C→	To °F
−14.44	6	42.8	12.78	55	131	40.56	105	221
−13.89	7	44.6	13.33	56	132.8	41.11	106	222.8
−13.33	8	46.4	13.89	57	134.6	41.67	107	224.6
−12.78	9	48.2	14.44	58	136.4	42.22	108	226.4
−12.22	10	50	15	59	138.2	42.78	109	228.2
−11.67	11	51.8	15.56	60	140	43.33	110	230
−11.11	12	53.6	16.11	61	141.8	43.89	111	231.8
−10.56	13	55.4	16.67	62	143.6	44.44	112	233.6
−10	14	57.2	17.22	63	145.4	45	113	235.4
−9.44	15	59	17.78	64	147.2	45.56	114	237.2
−8.89	16	60.8	18.33	65	149	46.11	115	239
−8.33	17	62.6	18.89	66	150.8	46.67	116	240.8
−7.78	18	64.4	19.44	67	152.6	47.22	117	242.6
−7.22	19	66.2	20	68	154.4	47.78	118	244.4
−6.67	20	68	20.56	69	156.2	48.33	119	246.2
−6.11	21	69.8	21.11	70	158	48.89	120	248
−5.56	22	71.6	21.67	71	159.8	49.44	121	249.8
−5	23	73.4	22.22	72	161.6	50	122	251.6
−4.44	24	75.2	22.78	73	163.4	50.56	123	253.4
−3.89	25	77	23.33	74	165.2	51.11	124	255.2
−3.33	26	78.8	23.89	75	167	51.67	125	257
−2.78	27	80.6	24.44	76	168.8	52.22	126	258.8
−2.22	28	82.4	25	77	170.6	52.78	127	260.6
−1.67	29	84.2	25.56	78	172.4	53.33	128	262.4
−1.11	30	86	26.11	79	174.2	53.89	129	264.2
−0.56	31	87.8	26.67	80	176	54.44	130	266
0	32	89.6	27.22	81	177.8	55	131	267.8
.56	33	91.4	27.78	82	179.6	55.56	132	269.6
1.11	34	93.2	28.33	83	181.4	56.11	133	271.4
1.67	35	95	28.89	84	183.2	56.67	134	273.2
2.22	36	96.8	29.44	85	185	57.22	135	275
2.78	37	98.6	30	86	186.8	57.78	136	276.8
3.33	38	100.4	30.56	87	188.6	58.33	137	278.6
3.89	39	102.2	31.11	88	190.4	58.89	138	280.4
4.44	40	104	31.67	89	192.2	59.44	139	282.2
5	41	105.8	32.22	90	194	60	140	284
5.56	42	107.6	32.78	91	195.8	60.56	141	285.8
6.11	43	109.4	33.33	92	197.6	61.11	142	287.6
6.67	44	111.2	33.89	93	199.4	61.67	143	289.4
7.22	45	113	34.44	94	201.2	62.22	144	291.2
7.78	46	114.8	35	95	203	62.78	145	293
8.33	47	116.6	35.56	96	204.8	63.33	146	294.8
8.89	48	118.4	36.11	97	206.6	63.89	147	296.6
9.44	49	120.2	36.67	98	208.4	64.44	148	298.4
			37.22	99	210.2	65	149	300.2
10	50	122	37.78	100	212	65.56	150	302
10.56	51	123.8	38.33	101	213.8	66.11	151	303.8
11.11	52	125.6	38.89	102	215.6	66.67	152	305.6
11.67	53	127.4	39.44	103	217.4	67.22	153	307.4
12.22	54	129.2	40	104	219.2	67.78	154	309.2

Table 4–12 (continued)
TEMPERATURE CONVERSION

To convert			To convert			To convert		
To °C	←°F or °C→	To °F	To °C	←°F or °C→	To °F	To °C	←°F or °C→	To °F
68.33	155	311	96.11	205	401	123.89	255	491
68.89	156	312.8	96.67	206	402.8	124.44	256	492.8
69.44	157	314.6	97.22	207	404.6	125	257	494.6
70	158	316.4	97.78	208	406.4	125.56	258	496.4
70.56	159	318.2	98.33	209	408.2	126.11	259	498.2
71.11	160	320	98.89	210	410	126.67	260	500
71.67	161	321.8	99.44	211	411.8	127.22	261	501.8
72.22	162	323.6	100	212	413.6	127.78	262	503.6
72.78	163	325.4	100.56	213	415.4	128.33	263	505.4
73.33	164	327.2	101.11	214	417.2	128.89	264	507.2
73.89	165	329	101.67	215	419	129.44	265	509
74.44	166	330.8	102.22	216	420.8	130	266	510.8
75	167	332.6	102.78	217	422.6	130.56	267	512.6
75.56	168	334.4	103.33	218	424.4	131.11	268	514.4
76.11	169	336.2	103.89	219	426.2	131.67	269	516.2
76.67	170	338	104.44	220	428	132.22	270	518
77.22	171	339.8	105	221	429.8	132.78	271	519.8
77.78	172	341.6	105.56	222	431.6	133.33	272	521.6
78.33	173	343.4	106.11	223	433.4	133.89	273	523.4
78.89	174	345.2	106.67	224	435.2	134.44	274	525.2
79.44	175	347	107.22	225	437	135	275	527
80	176	348.8	107.78	226	438.8	135.56	276	528.8
80.56	177	350.6	108.33	227	440.6	136.11	277	530.6
81.11	178	352.4	108.89	228	442.4	136.67	278	532.4
81.67	179	354.2	109.44	229	444.2	137.22	279	534.2
82.22	180	356	110	230	446	137.78	280	536
82.78	181	357.8	110.56	231	447.8	138.33	281	537.8
83.33	182	359.6	111.11	232	449.6	138.89	282	539.6
83.89	183	361.4	111.67	233	451.4	139.44	283	541.4
84.44	184	363.2	112.22	234	453.2	140	284	543.2
85	185	365	112.78	235	455	140.56	285	545
85.56	186	366.8	113.33	236	456.8	141.11	286	546.8
86.11	187	368.6	113.89	237	458.6	141:67	287	548.6
86.67	188	370.4	114.44	238	460.4	142.22	288	550.4
87.22	189	372.2	115	239	462.2	142.78	289	552.2
87.78	190	374	115.56	240	464	143.33	290	554
88.33	191	375.8	116.11	241	465.8	143.89	291	555.8
88.89	192	377.6	116.67	242	467.6	144.44	292	557.6
89.44	193	379.4	117.22	243	469.4	145	293	559.4
90	194	381.2	117.78	244	471.2	145.56	294	561.2
90.56	195	383	118.33	245	473	146.11	295	563
91.11	196	384.8	118.89	246	474.8	146.67	296	564.8
91.67	197	386.6	119.44	247	476.6	147.22	297	566.6
92.22	198	388.4	120	248	478.4	147.78	298	568.4
92.78	199	390.2	120.56	249	480.2	148.33	299	570.2
93.33	200	392	121.11	250	482	148.89	300	572
93.89	201	393.8	121.67	251	483.8	149.44	301	573.8
94.44	202	395.6	122.22	252	485.6	150	302	575.6
95	203	397.4	122.78	253	487.4	150.56	303	577.4
95.56	204	399.2	123.33	254	489.2	151.11	304	579.2

Table 4-12 (continued)
TEMPERATURE CONVERSION

To convert			To convert			To convert		
To °C	←°F or °C→	To °F	To °C	←°F or °C→	To °F	To °C	←°F or °C→	To °F
151.67	305	581	179.44	355	671	210	410	770
152.22	306	582.8	180	356	672.8	211.11	412	773.6
152.78	307	584.6	180.56	357	674.6	212.22	414	777.2
153.33	308	586.4	181.11	358	676.4	213.33	416	780.8
153.89	309	588.2	181.67	359	678.2	214.44	418	784.4
154.44	310	590	182.22	360	680	215.56	420	788
155	311	591.8	182.78	361	681.8	216.67	422	791.6
155.56	312	593.6	183.33	362	683.6	217.78	424	795.2
156.11	313	595.4	183.89	363	685.4	218.89	426	798.8
156.67	314	597.2	184.44	364	687.2	220	428	802.4
157.22	315	599	185	365	689	221.11	430	806
157.78	316	600.8	185.56	366	690.8	222.22	432	809.6
158.33	317	602.6	186.11	367	692.6	223.33	434	813.2
158.89	318	604.4	186.67	368	694.4	224.44	436	816.8
159.44	319	606.2	187.22	369	696.2	225.56	438	820.4
160	320	608	187.78	370	698	226.67	440	824
160.56	321	609.8	188.33	371	699.8	227.78	442	827.6
161.11	322	611.6	188.89	372	701.6	228.89	444	831.2
161.67	323	613.4	189.44	373	703.4	230	446	834.8
162.22	324	615.2	190	374	705.2	231.11	448	838.4
162.78	325	617	190.56	375	707	232.22	450	842
163.33	326	618.8	191.11	376	708.8	233.33	452	845.6
163.89	327	620.6	191.67	377	710.6	234.44	454	849.2
164.44	328	622.4	192.22	378	712.4	235.56	456	852.8
165	329	624.2	192.78	379	714.2	236.67	458	856.4
165.56	330	626	193.33	380	716	237.78	460	860
166.11	331	627.8	193.89	381	717.8	238.89	462	863.6
166.67	332	629.6	194.44	382	719.6	240	464	867.2
167.22	333	631.4	195	383	721.4	241.11	466	870.8
167.78	334	633.2	195.56	384	723.2	242.22	468	874.4
168.33	335	635	196.11	385	725	243.33	470	878
168.89	336	636.8	196.67	386	726.8	244.44	472	881.6
169.44	337	638.6	197.22	387	728.6	245.56	474	885.2
170	338	640.4	197.78	388	730.4	246.67	476	888.8
170.56	339	642.2	198.33	389	732.2	247.78	478	892.4
171.11	340	644	198.89	390	734	248.89	480	896
171.67	341	645.8	199.44	391	735.8	250	482	899.6
172.22	342	647.6	200	392	737.6	251.11	484	903.2
172.78	343	649.4	200.56	393	739.4	252.22	486	906.8
173.33	344	651.2	201.11	394	741.2	253.33	488	910.4
173.89	345	653	201.67	395	743	254.44	490	914
174.44	346	654.8	202.22	396	744.8	255.56	492	917.6
175	347	656.6	202.78	397	746.6	256.67	494	921.2
175.56	348	658.4	203.33	398	748.4	257.78	496	924.8
176.11	349	660.2	203.89	399	750.2	258.89	498	928.4
176.67	350	662	204.44	400	752	260	500	932
177.22	351	663.8	205.56	402	755.6	261.11	502	935.6
177.78	352	665.6	206.67	404	759.2	262.22	504	939.2
178.33	353	667.4	207.78	406	762.8	263.33	506	942.8
178.89	354	669.2	208.89	408	766.4	264.44	508	946.4

Table 4–12 (continued)
TEMPERATURE CONVERSION

To °C	←°F or °C→	To °F	To °C	←°F or °C→	To °F	To °C	←°F or °C→	To °F
	To convert			To convert			To convert	
265.56	510	950	321.11	610	1130	376.67	710	1310
266.67	512	953.6	322.22	612	1133.6	377.78	712	1313.6
267.78	514	957.2	323.33	614	1137.2	378.89	714	1317.2
268.89	516	960.8	324.44	616	1140.8	380	716	1320.8
270	518	964.4	325.56	618	1144.4	381.11	718	1324.4
271.11	520	968	326.67	620	1148	382.22	720	1328
272.22	522	971.6	327.78	622	1151.6	383.33	722	1331.6
273.33	524	975.2	328.89	624	1155.2	384.44	724	1335.2
274.44	526	978.8	330	626	1158.8	385.56	726	1338.8
275.56	528	982.4	331.11	628	1162.4	386.67	728	1342.4
276.67	530	986	332.22	630	1166	387.78	730	1346
277.78	532	989.6	333.33	632	1169.6	388.89	732	1349.6
278.89	534	993.2	334.44	634	1173.2	390	734	1353.2
280	536	996.8	335.56	636	1176.8	391.11	736	1356.8
281.11	538	1000.4	336.67	638	1180.4	392.22	738	1360.4
282.22	540	1004	337.78	640	1184	393.33	740	1364
283.33	542	1007.6	338.89	642	1187.6	394.44	742	1367.6
284.44	544	1011.2	340	644	1191.2	395.56	744	1371.2
285.56	546	1014.8	341.11	646	1194.8	396.67	746	1374.8
286.67	548	1018.4	342.22	648	1198.4	397.78	748	1378.4
287.78	550	1022	343.33	650	1202	398.89	750	1382
288.89	552	1025.6	344.44	652	1205.6	400	752	1385.6
290	554	1029.2	345.56	654	1209.2	401.11	754	1389.2
291.11	556	1032.8	346.67	656	1212.8	402.22	756	1392.8
292.22	558	1036.4	347.78	658	1216.4	403.33	758	1396.4
293.33	560	1040	348.89	660	1220	404.44	760	1400
294.44	562	1043.6	350	662	1223.6	405.56	762	1403.6
295.56	564	1047.2	351.11	664	1227.2	406.67	764	1407.2
296.67	566	1050.8	352.22	666	1230.8	407.78	766	1410.8
297.78	568	1054.4	353.33	668	1234.4	408.89	768	1414.4
298.89	570	1058	354.44	670	1238	410	770	1418
300	572	1061.6	355.56	672	1241.6	411.11	772	1421.6
301.11	574	1065.2	356.67	674	1245.2	412.22	774	1425.2
302.22	576	1068.8	357.78	676	1248.8	413.33	776	1428.8
303.33	578	1072.4	358.89	678	1252.4	414.44	778	1432.4
304.44	580	1076	360	680	1256	415.56	780	1436
305.56	582	1079.6	361.11	682	1259.6	416.67	782	1439.6
306.67	584	1083.2	362.22	684	1263.2	417.78	784	1443.2
307.78	586	1086.8	363.33	686	1266.8	418.89	786	1446.8
308.89	588	1090.4	364.44	688	1270.4	420	788	1450.4
310	590	1094	365.56	690	1274	421.11	790	1454
311.11	592	1097.6	366.67	692	1277.6	422.22	792	1457.6
312.22	594	1101.2	367.78	694	1281.2	423.33	794	1461.2
313.33	596	1104.8	368.89	696	1284.8	424.44	796	1464.8
314.44	598	1108.4	370	698	1288.4	425.56	798	1468.4
315.56	600	1112	371.11	700	1292	426.67	800	1472
316.67	602	1115.6	372.22	702	1295.6	427.78	802	1475.6
317.78	604	1119.2	373.33	704	1299.2	428.89	804	1479.2
318.89	606	1122.8	374.44	706	1302.8	430	806	1482.8
320	608	1126.4	375.56	708	1306.4	431.11	808	1486.4

Table 4–12 (continued)
TEMPERATURE CONVERSION

To °C	←°F or °C→	To °F	To °C	←°F or °C→	To °F	To °C	←°F or °C→	To °F
	To convert			To convert			To convert	
432.22	810	1490	487.78	910	1670	543.33	1010	1850
433.33	812	1493.6	488.89	912	1673.6	544.44	1012	1853.6
434.44	814	1497.2	490	914	1677.2	545.56	1014	1857.2
435.56	816	1500.8	491.11	916	1680.8	546.67	1016	1860.8
436.67	818	1504.4	492.22	918	1684.4	547.78	1018	1864.4
437.78	820	1508	493.33	920	1688	548.89	1020	1868
438.89	822	1511.6	494.44	922	1691.6	550	1022	1871.6
440	824	1515.2	495.56	924	1695.2	551.11	1024	1875.2
441.11	826	1518.8	496.67	926	1698.8	552.22	1026	1878.8
442.22	828	1522.4	497.78	928	1702.4	553.33	1028	1882.4
443.33	830	1526	498.89	930	1706	554.44	1030	1886
444.44	832	1529.6	500	932	1709.6	555.56	1032	1889.6
445.56	834	1533.2	501.11	934	1713.2	556.67	1034	1893.2
446.67	836	1536.8	502.22	936	1716.8	557.78	1036	1896.8
447.78	838	1540.4	503.33	938	1720.4	558.89	1038	1900.4
448.89	840	1544	504.44	940	1724	560	1040	1904
450	842	1547.6	505.56	942	1727.6	561.11	1042	1907.6
451.11	844	1551.2	506.67	944	1731.2	562.22	1044	1911.2
452.22	846	1554.8	507.78	946	1734.8	563.33	1046	1914.8
453.33	848	1558.4	508.89	948	1738.4	564.44	1048	1918.4
454.44	850	1562	510	950	1742	565.56	1050	1922
455.56	852	1565.6	511.11	952	1745.6	566.67	1052	1925.6
456.67	854	1569.2	512.22	954	1749.2	567.78	1054	1929.2
457.78	856	1572.8	513.33	956	1752.8	568.89	1056	1932.8
458.89	858	1576.4	514.44	958	1756.4	570	1058	1936.4
460	860	1580	515.56	960	1760	571.11	1060	1940
461.11	862	1583.6	516.67	962	1763.6	572.22	1062	1943.6
462.22	864	1587.2	517.78	964	1767.2	573.33	1064	1947.2
463.33	866	1590.8	518.89	966	1770.8	574.44	1066	1950.8
464.44	868	1594.4	520	968	1774.4	575.56	1068	1954.4
465.56	870	1598	521.11	970	1778	576.67	1070	1958
466.67	872	1601.6	522.22	972	1781.6	577.78	1072	1961.6
467.78	874	1605.2	523.33	974	1785.2	578.89	1074	1965.2
468.89	876	1608.8	524.44	976	1788.8	580	1076	1968.8
470	878	1612.4	525.56	978	1792.4	581.11	1078	1972.4
471.11	880	1616	526.67	980	1796	582.22	1080	1976
472.22	882	1619.6	527.78	982	1799.6	583.33	1082	1979.6
473.33	884	1623.2	528.89	984	1803.2	584.44	1084	1983.2
474.44	886	1626.8	530	986	1806.8	585.56	1086	1986.8
475.56	888	1630.4	531.11	988	1810.4	586.67	1088	1990.4
476.67	890	1634	532.22	990	1814	587.78	1090	1994
477.78	892	1637.6	533.33	992	1817.6	588.89	1092	1997.6
478.89	894	1641.2	534.44	994	1821.2	590	1094	2001.2
480	896	1644.8	535.56	996	1824.8	591.11	1096	2004.8
481.11	898	1648.4	536.67	998	1828.4	592.22	1098	2008.4
482.22	900	1652	537.78	1000	1832	593.33	1100	2012
483.33	902	1655.6	538.89	1002	1835.6	594.44	1102	2015.6
484.44	904	1659.2	540	1004	1839.2	595.56	1104	2019.2
485.56	906	1662.8	541.11	1006	1842.8	596.67	1106	2022.8
486.67	908	1666.4	542.22	1008	1846.4	597.78	1108	2026.4

Table 4–12 (continued)
TEMPERATURE CONVERSION

To °C	←°F or °C→	To °F	To °C	←°F or °C→	To °F	To °C	←°F or °C→	To °F
598.89	1110	2030	654.44	1210	2210	710	1310	2390
600	1112	2033.6	655.56	1212	2213.6	711.11	1312	2393.6
601.11	1114	2037.2	656.67	1214	2217.2	712.22	1314	2397.2
602.22	1116	2040.8	657.78	1216	2220.8	713.33	1316	2400.8
603.33	1118	2044.4	658.89	1218	2224.4	714.44	1318	2404.4
604.44	1120	2048	660	1220	2228	715.56	1320	2408
605.56	1122	2051.6	661.11	1222	2231.6	716.67	1322	2411.6
606.67	1124	2055.2	662.22	1224	2235.2	717.78	1324	2415.2
607.78	1126	2058.8	663.33	1226	2238.8	718.89	1326	2418.8
608.89	1128	2062.4	664.44	1228	2242.4	720	1328	2422.4
610	1130	2066	665.56	1230	2246	721.11	1330	2426
611.11	1132	2069.6	666.67	1232	2249.6	722.22	1332	2429.6
612.22	1134	2073.2	667.78	1234	2253.2	723.33	1334	2433.2
613.33	1136	2076.8	668.89	1236	2256.8	724.44	1336	2436.8
614.44	1138	2080.4	670	1238	2260.4	725.56	1338	2440.4
615.56	1140	2084	671.11	1240	2264	726.67	1340	2444
616.67	1142	2087.6	672.22	1242	2267.6	727.78	1342	2447.6
617.78	1144	2091.2	673.33	1244	2271.2	728.89	1344	2451.2
618.89	1146	2094.8	674.44	1246	2274.8	730	1346	2454.8
620	1148	2098.4	675.56	1248	2278.4	731.11	1348	2458.4
621.11	1150	2102	676.67	1250	2282	732.22	1350	2462
622.22	1152	2105.6	677.78	1252	2285.6	733.33	1352	2465.6
623.33	1154	2109.2	678.89	1254	2289.2	734.44	1354	2469.2
624.44	1156	2112.8	680	1256	2292.8	735.56	1356	2472.8
625.56	1158	2116.4	681.11	1258	2296.4	736.67	1358	2476.4
626.67	1160	2120	682.22	1260	2300	737.78	1360	2480
627.78	1162	2123.6	683.33	1262	2303.6	738.89	1362	2483.6
628.89	1164	2127.2	684.44	1264	2307.2	740	1364	2487.2
630	1166	2130.8	685.56	1266	2310.8	741.11	1366	2490.8
631.11	1168	2134.4	686.67	1268	2314.4	742.22	1368	2494.4
632.22	1170	2138	687.78	1270	2318	743.33	1370	2498
633.33	1172	2141.6	688.89	1272	2321.6	744.44	1372	2501.6
634.44	1174	2145.2	690	1274	2325.2	745.56	1374	2505.2
635.56	1176	2148.8	691.11	1276	2328.8	746.67	1376	2508.8
636.67	1178	2152.4	692.22	1278	2332.4	747.78	1378	2512.4
637.78	1180	2156	693.33	1280	2336	748.89	1380	2516
638.89	1182	2159.6	694.44	1282	2339.6	750	1382	2519.6
640	1184	2163.2	695.56	1284	2343.2	751.11	1384	2523.2
641.11	1186	2166.8	696.67	1286	2346.8	752.22	1386	2526.8
642.22	1188	2170.4	697.78	1288	2350.4	753.33	1388	2530.4
643.33	1190	2174	698.89	1290	2354	754.44	1390	2534
644.44	1192	2177.6	700	1292	2357.6	755.56	1392	2537.6
645.56	1194	2181.2	701.11	1294	2361.2	756.67	1394	2541.2
646.67	1196	2184.8	702.22	1296	2364.8	757.78	1396	2544.8
647.78	1198	2188.4	703.33	1298	2368.4	758.89	1398	2548.4
648.89	1200	2192	704.44	1300	2372	760	1400	2552
650	1202	2195.6	705.56	1302	2375.6	761.11	1402	2555.6
651.11	1204	2199.2	706.67	1304	2379.2	762.22	1404	2559.2
652.22	1206	2202.8	707.78	1306	2382.8	763.33	1406	2562.8
653.33	1208	2206.4	708.89	1308	2386.4	764.44	1408	2566.4

Table 4—12 (continued)
TEMPERATURE CONVERSION

To convert			To convert			To convert		
To °C	←°F or °C→	To °F	To °C	←°F or °C→	To °F	To °C	←°F or °C→	To °F
765.56	1410	2570	843.33	1550	2922	1121.11	2050	3722
766.67	1412	2573.6	848.89	1560	2840	1126.67	2060	3740
767.78	1414	2577.2	854.44	1570	2858	1132.22	2070	3758
768.89	1416	2580.8	860	1580	2876	1137.78	2080	3776
770	1418	2584.4	865.56	1590	2894	1143.33	2090	3794
771.11	1420	2588	871.11	1600	2912	1148.89	2100	3812
772.22	1422	2591.6	876.67	1610	2930	1154.44	2110	3830
773.33	1424	2595.2	882.22	1620	2948	1160	2120	3848
774.44	1426	2598.8	887.78	1630	2966	1165.56	2130	3866
775.56	1428	2602.4	893.33	1640	2984	1171.11	2140	3884
776.67	1430	2606	898.89	1650	3002	1176.67	2150	3902
777.78	1432	2609.6	904.44	1660	3020	1182.22	2160	3920
778.89	1434	2613.2	910	1670	3038	1187.78	2170	3938
780	1436	2616.8	915.56	1680	3056	1193.33	2180	3956
781.11	1438	2620.4	921.11	1690	3074	1198.89	2190	3974
782.22	1440	2624	926.67	1700	3092	1204.44	2200	3992
783.33	1442	2627.6	932.22	1710	3110	1210	2210	4010
784.44	1444	2631.2	937.78	1720	3128	1215.56	2220	4028
785.56	1446	2634.8	943.33	1730	3146	1221.11	2230	4046
786.67	1448	2638.4	948.89	1740	3164	1226.67	2240	4064
787.78	1450	2642	954.44	1750	3182	1232.22	2250	4082
788.89	1452	2645.6	960	1760	3200	1237.78	2260	4100
790	1454	2649.2	965.56	1770	3218	1243.33	2270	4118
791.11	1456	2652.8	971.11	1780	3236	1248.89	2280	4136
792.22	1458	2656.4	976.67	1790	3254	1254.44	2290	4154
793.33	1460	2660	982.22	1800	3272	1260	2300	4172
794.44	1462	2663.6	987.78	1810	3290	1265.56	2310	4190
795.56	1464	2667.2	993.33	1820	3308	1271.11	2320	4208
796.67	1466	2670.8	998.89	1830	3326	1276.67	2330	4226
797.78	1468	2674.4	1004.44	1840	3344	1282.22	2340	4244
798.89	1470	2678	1010	1850	3362	1287.78	2350	4262
800	1472	2681.6	1015.56	1860	3380	1293.33	2360	4280
801.11	1474	2685.2	1021.11	1870	3398	1298.89	2370	4298
802.22	1476	2688.8	1026.67	1880	3416	1304.44	2380	4316
803.33	1478	2692.4	1032.22	1890	3434	1310	2390	4334
804.44	1480	2696	1037.78	1900	3452	1315.56	2400	4352
805.56	1482	2699.6	1043.33	1910	3470	1321.11	2410	4370
806.67	1484	2703.2	1048.89	1920	3488	1326.67	2420	4388
807.78	1486	2706.8	1054.44	1930	3506	1332.22	2430	4406
808.89	1488	2710.4	1060	1940	3524	1337.78	2440	4424
810	1490	2714	1065.56	1950	3542	1343.33	2450	4442
811.11	1492	2717.6	1071.11	1960	3560	1348.89	2460	4460
812.22	1494	2721.2	1076.67	1970	3578	1354.44	2470	4478
813.33	1496	2724.8	1082.22	1980	3596	1360	2480	4496
814.44	1498	2728.4	1087.78	1990	3614	1365.56	2490	4514
815.56	1500	2732	1093.33	2000	3632	1371.11	2500	4532
821.11	1510	2750	1098.89	2010	3650	1385	2525	4577
826.67	1520	2768	1104.44	2020	3668	1398.89	2550	4622
832.22	1530	2786	1110	2030	3686	1412.78	2575	4667
837.78	1540	2804	1115.56	2040	3704	1426.67	2600	4712

Table 4–12 (continued)
TEMPERATURE CONVERSION

To °C	←°F or °C→	To °F	To °C	←°F or °C→	To °F	To °C	←°F or °C→	To °F
1440.56	2625	4757	1996.11	3625	6557	2551.67	4625	8357
1454.44	2650	4802	2010.00	3650	6602	2565.56	4650	8402
1468.33	2675	4847	2023.89	3675	6647	2579.44	4675	8447
1482.22	2700	4892	2037.78	3700	6692	2593.33	4700	8492
1496.11	2725	4937	2051.67	3725	6737	2607.22	4725	8537
1510	2750	4982	2065.56	3750	6782	2621.11	4750	8582
1523.89	2775	5027	2079.44	3775	6827	2635.00	4775	8627
1537.78	2800	5072	2093.33	3800	6872	2648.89	4800	8672
1551.67	2825	5117	2107.22	3825	6917	2662.78	4825	8717
1565.56	2850	5162	2121.11	3850	6962	2676.67	4850	8762
1579.44	2875	5207	2135.00	3875	7007	2690.55	4875	8807
1593.33	2900	5252	2148.89	3900	7052	2704.44	4900	8852
1607.22	2925	5297	2162.78	3925	7097	2718.33	4925	8897
1621.11	2950	5342	2176.67	3950	7142	2732.22	4950	8942
1635	2975	5387	2190.56	3975	7187	2746.11	4975	8987
1648.89	3000	5432	2204.44	4000	7232	2760.00	5000	9032
1662.78	3025	5477	2218.33	4025	7277	2787.78	5050	9122
1676.67	3050	5522	2232.22	4050	7322	2815.56	5100	9212
1690.56	3075	5567	2246.11	4075	7367	2843.33	5150	9302
1704.44	3100	5612	2260.00	4100	7412	2871:11	5200	9392
1718.33	3125	5657	2273.89	4125	7457	2898.89	5250	9482
1732.22	3150	5702	2287.78	4150	7502	2926.67	5300	9572
1746.11	3175	5747	2301.67	4175	7547	2954.44	5350	9662
1760	3200	5792	2315.56	4200	7592	2982.22	5400	9752
1773.89	3225	5837	2329.44	4225	7637	3010.00	5450	9842
1787.78	3250	5882	2343.33	4250	7682	3037.78	5500	9932
1801.67	3275	5927	2357.22	4275	7727	3065.56	5550	10022
1815.56	3300	5972	2371.11	4300	7772	3093.33	5600	10112
1829.44	3325	6017	2385.00	4325	7817	3121.11	5650	10202
1843.33	3350	6062	2398.89	4350	7862	3148.89	5700	10292
1857.22	3375	6107	2412.78	4375	7907	3176.67	5750	10382
1871.11	3400	6152	2426.67	4400	7952	3204.44	5800	10472
1885.00	3425	6197	2440.56	4425	7997	3232.22	5850	10562
1898.89	3450	6242	2454.44	4450	8042	3260.00	5900	10652
1912.78	3475	6287	2468.33	4475	8087	3287.78	5950	10742
1926.67	3500	6332	2482.22	4500	8132	3315.56	6000	10832
1940.56	3525	6377	2496.11	4525	8177	3593.33	6500	11732
1954.44	3550	6422	2510.00	4550	8222	3871.11	7000	12632
1968.33	3575	6467	2523.89	4575	8267	4148.89	7500	13532
1982.22	3600	6512	2537.78	4600	8312	4426.67	8000	14432

Condensed from Weast, R.C., Ed., *Handbook of Chemistry and Physics,* 53rd ed., Chemical Rubber, Cleveland, 1972.

Table 4–13

TEMPERATURE CONVERSION – DEGREES FAHRENHEIT TO KELVIN

$$K = (5/9)(deg\ F + 459.67)$$

Deg F	K	Deg F	K	Deg F	K	Deg F	K	Deg F	K	Deg F	K
−459.67	0	−412	26.48	−220	133.14	−28	239.82	20	266.48	68	293.15
−459	.37	−411	27.04	−215	135.91	−27	240.37	21	267.04	69	293.71
−458	.93	−410	27.59	−210	138.69	−26	240.93	22	267.59	70	294.26
−457	1.48	−409	28.15	−205	141.47	−25	241.48	23	268.15	71	294.82
−456	2.04	−408	28.70	−200	144.25	−24	242.04	24	268.71	72	295.37
−455	2.59	−407	29.26	−195	147.02	−23	242.59	25	269.26	73	295.93
−454	3.15	−406	29.81	−190	149.80	−22	243.15	26	269.82	74	296.48
−453	3.70	−405	30.37	−185	152.58	−21	243.71	27	270.37	75	297.04
−452	4.26	−404	30.92	−180	155.36	−20	244.26	28	270.93	76	297.59
−451	4.82	−403	31.48	−175	158.13	−19	244.82	29	271.48	77	298.15
−450	5.37	−402	32.04	−170	160.91	−18	245.37	30	272.04	78	298.71
−449	5.93	−401	32.59	−165	163.69	−17	245.93	31	272.59	79	299.26
−448	6.48	−400	33.15	−160	166.47	−16	246.48	32	273.15	80	299.82
−447	7.04	−395	35.92	−155	169.24	−15	247.04	33	273.71	81	300.37
−446	7.59	−390	38.70	−150	172.02	−14	247.59	34	274.26	82	300.93
−445	8.15	−385	41.48	−145	174.80	−13	248.15	35	274.82	83	301.48
−444	8.70	−380	44.26	−140	177.58	−12	248.71	36	275.37	84	302.04
−443	9.26	−375	47.03	−135	180.35	−11	249.26	37	275.93	85	302.59
−442	9.82	−370	49.81	−130	183.13	−10	249.82	38	276.48	86	303.15
−441	10.37	−365	52.59	−125	185.91	−9	250.37	39	277.04	87	303.71
−440	10.93	−360	55.37	−120	188.69	−8	250.93	40	277.59	88	304.26
−439	11.48	−355	58.14	−115	191.46	−7	251.48	41	278.15	89	304.82
−438	12.04	−350	60.92	−110	194.24	−6	252.04	42	278.71	90	305.37
−437	12.59	−345	63.70	−105	197.02	−5	252.59	43	279.26	91	305.93
−436	13.15	−340	66.48	−100	199.80	−4	253.15	44	279.82	92	306.48
−435	13.70	−335	69.25	−95	202.57	−3	253.71	45	280.37	93	307.04
−434	14.26	−330	72.03	−90	205.35	−2	254.26	46	280.93	94	307.59
−433	14.82	−325	74.81	−85	208.13	−1	254.82	47	281.48	95	308.15
−432	15.37	−320	77.59	−80	210.91	0	255.37	48	282.04	96	308.71
−431	15.93	−315	80.36	−75	213.68	1	255.93	49	282.59	97	309.26
−430	16.48	−310	83.14	−70	216.46	2	256.48	50	283.15	98	309.82
−429	17.04	−305	85.92	−65	219.24	3	257.04	51	283.71	99	310.37
−428	17.59	−300	88.70	−60	222.02	4	257.59	52	284.26	100	310.93
−427	18.15	−295	91.47	−55	224.79	5	258.15	53	284.82	101	311.48
−426	18.70	−290	94.25	−50	227.57	6	258.71	54	285.37	102	312.04
−425	19.26	−285	97.03	−45	230.35	7	259.26	55	285.93	103	312.59
−424	19.81	−280	99.81	−40	233.13	8	259.82	56	286.48	104	313.15
−423	20.37	−275	102.58	−39	233.71	9	260.37	57	287.04	105	313.71
−422	20.92	−270	105.36	−38	234.26	10	260.93	58	287.59	106	314.26
−421	21.48	−265	108.14	−37	234.82	11	261.48	59	288.15	107	314.82
−420	22.04	−260	110.92	−36	235.37	12	262.04	60	288.71	108	315.37
−419	22.59	−255	113.69	−35	235.93	13	262.59	61	289.26	109	315.93
−418	23.15	−250	116.47	−34	236.48	14	263.15	62	289.82	110	316.48
−417	23.70	−245	119.25	−33	237.04	15	263.71	63	290.37	111	317.04
−416	24.26	−240	122.03	−32	237.59	16	264.26	64	290.93	112	317.59
−415	24.81	−235	124.80	−31	238.15	17	264.82	65	291.48	113	318.15
−414	25.37	−230	127.58	−30	238.71	18	265.37	66	292.04	114	318.71
−413	25.92	−225	130.36	−29	239.26	19	265.93	67	292.59	115	319.26

Table 4–13 (continued)

TEMPERATURE CONVERSION – DEGREES FAHRENHEIT TO KELVIN

$$K = (5/9)(\deg F + 459.67)$$

Deg F	K	Deg F	K	Deg F	K	Deg F	K	Deg F	K	Deg F	K
116	319.82	166	347.59	216	375.37	266	403.15	460	510.93	960	788.71
117	320.37	167	348.15	217	375.93	267	403.71	470	516.48	970	794.26
118	320.93	168	348.71	218	376.48	268	404.26	480	522.04	980	799.82
119	321.48	169	349.26	219	377.04	269	404.82	490	527.59	990	805.37
120	322.04	170	349.82	220	377.59	270	405.37	500	533.15	1 000	810.93
121	322.59	171	350.37	221	378.15	271	405.93	510	538.71	1 010	816.48
122	323.15	172	350.93	222	378.71	272	406.48	520	544.26	1 020	822.04
123	323.71	173	351.48	223	379.26	273	407.04	530	549.82	1 030	827.59
124	324.26	174	352.04	224	379.82	274	407.59	540	555.37	1 040	833.15
125	324.82	175	352.59	225	380.37	275	408.15	550	560.93	1 050	838.71
126	325.37	176	353.15	226	380.93	276	408.71	560	566.48	1 060	844.26
127	325.93	177	353.71	227	381.48	277	409.26	570	572.04	1 070	849.82
128	326.48	178	354.26	228	382.04	278	409.82	580	577.59	1 080	855.37
129	327.04	179	354.82	229	382.59	279	410.37	590	583.15	1 090	860.93
130	327.59	180	355.37	230	383.15	280	410.93	600	588.71	1 100	866.48
131	328.15	181	355.93	231	383.71	281	411.48	610	594.26	1 110	872.04
132	328.71	182	356.48	232	384.26	282	412.04	620	599.82	1 120	877.59
133	329.26	183	357.04	233	384.82	283	412.59	630	605.37	1 130	883.15
134	329.82	184	357.59	234	385.37	284	413.15	640	610.93	1 140	888.71
135	330.37	185	358.15	235	385.93	285	413.71	650	616.48	1 150	894.26
136	330.93	186	358.71	236	386.48	286	414.26	660	622.04	1 160	899.82
137	331.48	187	359.26	237	387.04	287	414.82	670	627.59	1 170	905.37
138	332.04	188	359.82	238	387.59	288	415.37	680	633.15	1 180	910.93
139	332.59	189	360.37	239	388.15	289	415.93	690	638.71	1 190	916.48
140	333.15	190	360.93	240	388.71	290	416.48	700	644.26	1 200	922.04
141	333.71	191	361.48	241	389.26	291	417.04	710	649.82	1 210	927.59
142	334.26	192	362.04	242	389.82	292	417.59	720	655.37	1 220	933.15
143	334.82	193	362.59	243	390.37	293	418.15	730	660.93	1 230	938.71
144	335.37	194	363.15	244	390.93	294	418.71	740	666.48	1 240	944.26
145	335.93	195	363.71	245	391.48	295	419.26	750	672.04	1 250	949.82
146	336.48	196	364.26	246	392.04	296	419.82	760	677.59	1 260	955.37
147	337.04	197	364.82	247	392.59	297	420.37	770	683.15	1 270	960.93
148	337.59	198	365.37	248	393.15	298	420.93	780	688.71	1 280	966.48
149	338.15	199	365.93	249	393.71	299	421.48	790	694.26	1 290	972.04
150	338.71	200	366.48	250	394.26	300	422.04	800	699.82	1 300	977.59
151	339.26	201	367.04	251	394.82	310	427.59	810	705.37	1 310	983.15
152	339.82	202	367.59	252	395.37	320	433.15	820	710.93	1 320	988.71
153	340.37	203	368.15	253	395.93	330	438.71	830	716.48	1 330	994.26
154	340.93	204	368.71	254	396.48	340	444.26	840	722.04	1 340	999.82
155	341.48	205	369.26	255	397.04	350	449.82	850	727.59	1 350	1 005.37
156	342.04	206	369.82	256	397.59	360	455.37	860	733.15	1 360	1 010.93
157	342.59	207	370.37	257	398.15	370	460.93	870	738.71	1 370	1 016.48
158	343.15	208	370.93	258	398.71	380	466.48	880	744.26	1 380	1 022.04
159	343.71	209	371.48	259	399.26	390	472.04	890	749.82	1 390	1 027.59
160	344.26	210	372.04	260	399.82	400	477.59	900	755.37	1 400	1 033.15
161	344.82	211	372.59	261	400.37	410	483.15	910	760.93	1 410	1 038.71
162	345.37	212	373.15	262	400.93	420	488.71	920	766.48	1 420	1 044.26
163	345.93	213	373.71	263	401.48	430	494.26	930	772.04	1 430	1 049.82
164	346.48	214	374.26	264	402.04	440	499.82	940	777.59	1 440	1 055.37
165	347.04	215	374.82	265	402.59	450	505.37	950	783.15	1 450	1 060.93

Table 4–13 (continued)

TEMPERATURE CONVERSION – DEGREES FAHRENHEIT TO KELVIN

$$K = (5/9) (\deg F + 459.67)$$

Deg F	K	Deg F	K	Deg F	K	Deg F	K	Deg F	K	Deg F	K
1 460	1 066.48	1 960	1 344.26	3 150	2 005.37	4 400	2 699.82	6 300	3 755.37	8 800	5 144.26
1 470	1 072.04	1 970	1 349.82	3 175	2 019.26	4 425	2 713.70	6 350	3 783.15	8 850	5 172.04
1 480	1 077.59	1 980	1 355.37	3 200	2 033.15	4 450	2 727.59	6 400	3 810.93	8 900	5 199.82
1 490	1 083.15	1 990	1 360.93	3 225	2 047.04	4 475	2 741.48	6 450	3 838.71	8 950	5 227.59
1 500	1 088.71	2 000	1 366.48	3 250	2 060.93	4 500	2 755.37	6 500	3 866.48	9 000	5 255.37
1 510	1 094.26	2 025	1 380.37	3 275	2 074.82	4 525	2 769.26	6 550	3 894.26	9 050	5 283.15
1 520	1 099.82	2 050	1 394.26	3 300	2 088.70	4 550	2 783.15	6 600	3 922.04	9 100	5 310.93
1 530	1 105.37	2 075	1 408.15	3 325	2 102.59	4 575	2 797.04	6 650	3 949.82	9 150	5 338.71
1 540	1 110.93	2 100	1 422.04	3 350	2 116.48	4 600	2 810.93	6 700	3 977.59	9 200	5 366.48
1 550	1 116.48	2 125	1 435.93	3 375	2 130.37	4 625	2 824.82	6 750	4 005.37	9 250	5 394.26
1 560	1 122.04	2 150	1 449.82	3 400	2 144.26	4 650	2 838.70	6 800	4 033.15	9 300	5 422.04
1 570	1 127.59	2 175	1 463.70	3 425	2 158.15	4 675	2 852.59	6 850	4 060.93	9 350	5 449.82
1 580	1 133.15	2 200	1 477.59	3 450	2 172.04	4 700	2 866.48	6 900	4 088.71	9 400	5 477.59
1 590	1 138.71	2 225	1 491.48	3 475	2 185.93	4 725	2 880.37	6 950	4 116.48	9 450	5 505.37
1 600	1 144.26	2 250	1 505.37	3 500	2 199.82	4 750	2 894.26	7 000	4 144.26	9 500	5 533.15
1 610	1 149.82	2 275	1 519.26	3 525	2 213.70	4 775	2 908.15	7 050	4 172.04	9 550	5 560.93
1 620	1 155.37	2 300	1 533.15	3 550	2 227.59	4 800	2 922.04	7 100	4 199.82	9 600	5 588.71
1 630	1 160.93	2 325	1 547.04	3 575	2 241.48	4 825	2 935.93	7 150	4 227.59	9 650	5 616.48
1 640	1 166.48	2 350	1 560.93	3 600	2 255.37	4 850	2 949.82	7 200	4 255.37	9 700	5 644.26
1 650	1 172.04	2 375	1 574.82	3 625	2 269.26	4 875	2 963.70	7 250	4 283.15	9 750	5 672.04
1 660	1 177.59	2 400	1 588.70	3 650	2 283.15	4 900	2 977.59	7 300	4 310.93	9 800	5 699.82
1 670	1 183.15	2 425	1 602.59	3 675	2 297.04	4 925	2 991.48	7 350	4 338.71	9 850	5 727.59
1 680	1 188.71	2 450	1 616.48	3 700	2 310.93	4 950	3 005.37	7 400	4 366.48	9 900	5 755.37
1 690	1 194.26	2 475	1 630.37	3 725	2 324.82	4 975	3 019.26	7 450	4 394.26	9 950	5 783.15
1 700	1 199.82	2 500	1 644.26	3 750	2 338.70	5 000	3 033.15	7 500	4 422.04	10 000	5 810.93
1 710	1 205.37	2 525	1 658.15	3 775	2 352.59	5 050	3 060.93	7 550	4 449.82		
1 720	1 210.93	2 550	1 672.04	3 800	2 366.48	5 100	3 088.71	7 600	4 477.59		
1 730	1 216.48	2 575	1 685.93	3 825	2 380.37	5 150	3 116.48	7 650	4 505.37		
1 740	1 222.04	2 600	1 699.82	3 850	2 394.26	5 200	3 144.26	7 700	4 533.15		
1 750	1 227.59	2 625	1 713.70	3 875	2 408.15	5 250	3 172.04	7 750	4 560.93		
1 760	1 233.15	2 650	1 727.59	3 900	2 422.04	5 300	3 199.82	7 800	4 588.71		
1 770	1 238.71	2 675	1 741.48	3 925	2 435.93	5 350	3 227.59	7 850	4 616.48		
1 780	1 244.26	2 700	1 755.37	3 950	2 449.82	5 400	3 255.37	7 900	4 644.26		
1 790	1 249.82	2 725	1 769.26	3 975	2 463.70	5 450	3 283.15	7 950	4 672.04		
1 800	1 255.37	2 750	1 783.15	4 000	2 477.59	5 500	3 310.93	8 000	4 699.82		
1 810	1 260.93	2 775	1 797.04	4 025	2 491.48	5 550	3 338.71	8 050	4 727.59		
1 820	1 266.48	2 800	1 810.93	4 050	2 505.37	5 600	3 366.48	8 100	4 755.37		
1 830	1 272.04	2 825	1 824.82	4 075	2 519.26	5 650	3 394.26	8 150	4 783.15		
1 840	1 277.59	2 850	1 838.70	4 100	2 533.15	5 700	3 422.04	8 200	4 810.93		
1 850	1 283.15	2 875	1 852.59	4 125	2 547.04	5 750	3 449.82	8 250	4 838.71		
1 860	1 288.71	2 900	1 866.48	4 150	2 560.93	5 800	3 477.59	8 300	4 866.48		
1 870	1 294.26	2 925	1 880.37	4 175	2 574.82	5 850	3 505.37	8 350	4 894.26		
1 880	1 299.82	2 950	1 894.26	4 200	2 588.70	5 900	3 533.15	8 400	4 922.04		
1 890	1 305.37	2 975	1 908.15	4 225	2 602.59	5 950	3 560.93	8 450	4 949.82		
1 900	1 310.93	3 000	1 922.04	4 250	2 616.48	6 000	3 588.71	8 500	4 977.59		
1 910	1 316.48	3 025	1 935.93	4 275	2 630.37	6 050	3 616.48	8 550	5 005.37		
1 920	1 322.04	3 050	1 949.82	4 300	2 644.26	6 100	3 644.26	8 600	5 033.15		
1 930	1 327.59	3 075	1 963.70	4 325	2 658.15	6 150	3 672.04	8 650	5 060.93		
1 940	1 333.15	3 100	1 977.59	4 350	2 672.04	6 200	3 699.82	8 700	5 088.71		
1 950	1 338.71	3 125	1 991.48	4 375	2 685.93	6 250	3 727.59	8 750	5 116.48		

From Bolz, R. E. and Tuve, G. L., Eds., *Handbook of Tables for Applied Engineering Science,* 2nd ed., CRC Press, Cleveland, 1973, 855.

Table 4–14

WIDE-RANGE TEMPERATURE CONVERSIONS FROM FAHRENHEIT TO KELVIN OR CELSIUS

Table A (for kelvin) and Table B (for Celsius) enable the user to estimate quickly the kelvin or Celsius equivalent of any Fahrenheit temperature to 90,000 deg F. Conversion of any temperature to 100,000 deg F to any desired accuracy can be obtained by simple addition, using Table A or B for the first digit and Table C for additional digits (see examples).

Examples:

$$106 \text{ deg F} = (100 + 6) \text{ deg F} = ? \text{ K}$$

100,	Table A:	310.9277 . . .
6,	Table C:	3.3333 . . .

$$106 \text{ deg F} = \overline{314.2611 \ldots \text{K}}$$

$$3.96 \text{ deg F} = (3 + 0.9 + 0.06) \text{ deg F} = ? \text{ deg C}$$

3.00,	Table B:	−16.1111 . . .
0.90,	Table C:	0.5000 . . .
0.06,	Table C:	+ 0.0333 . . .

$$3.96 \text{ deg F} = \overline{-15.5777 \ldots \text{ deg C}}$$

$$-103 \text{ deg F} = (-100-3) \text{ deg F} = ? \text{ K}$$

−100,	Table A:	+199.8166 . . .
−3,	Table C:	−1.6666 . . .

$$-103 \text{ deg F} = \overline{198.1500 \ldots \text{K}}$$

Table A. Temperature Conversion from Fahrenheit to Kelvin[a]

T, K = (T, deg F + 459.67)(5/9)   0 deg F = 255.372222 . . . K

		Fahrenheit temperature								
Range, deg F		1	2	3	4	5	6	7	8	9
−100 to	−400	199.816	144.261	88.705	33.150					
−10 to	−90	249.816	244.261	238.705	233.150	227.594	222.038	216.483	210.927	205.372
−1 to	−9	254.816	254.261	253.705	253.150	252.594	252.038	251.483	250.927	250.372
1 to	9	255.927	256.483	257.038	257.594	258.150	258.705	259.261	259.816	260.372
10 to	90	260.927	266.483	272.038	277.594	283.150	288.705	294.261	299.816	305.372
100 to	900	310.927	366.483	422.038	477.594	533.150	588.705	644.261	699.816	755.372
1 000 to	9 000	810.927	1 366.483	1 922.038	2 477.594	3 033.150	3 588.705	4 144.261	4 699.816	5 255.372
10 000 to 90 000		5 810.927	11 366.483	16 922.038	22 477.594	28 033.150	33 588.705	39 144.261	44 699.816	50 255.372

Table B. Temperature Conversion from Fahrenheit to Celsius[a]

T, deg C = (T, deg F−32)(5/9)   0 deg F = −17.777 . . . deg C

		Fahrenheit temperature								
Range, deg F		1	2	3	4	5	6	7	8	9
−100 to	−400	−73.333	−128.888	−184.444	−240.000					
−10 to	−90	−23.333	−28.888	−34.444	−40.000	−45.555	−51.1H	−56.666	−62.222	−67.777
−1 to	−9	−18.333	−18.888	−19.444	−20.000	−20.555	−21.111	−21.666	−22.222	−22.777
1 to	9	−17.222	−16.666	−16.111	−15.555	−15.000	−14.444	−13.888	−13.333	−12.777
10 to	90	−12.222	−6.666	−1.111	+4.444	+10.000	+15.555	+21.111	+26.666	+32.222
100 to	900	+37.777	+93.333	+148.888	204.444	260.000	315.555	371.111	426.666	482.222
1 000 to	9 000	537.777	1 093.333	1 648.888	2 204.444	2 760.000	3 315.555	3 871.111	4 426.666	4 982.222
10 000 to 90 000		5 537.777	11 093.333	16 648.888	22 204.444	27 760.000	33 315.555	38 871.111	44 426.666	49 982.222

Table 4—14 (continued)
WIDE-RANGE TEMPERATURE CONVERSIONS FROM FAHRENHEIT TO KELVIN OR CELSIUS

Table C.   Temperature Increment Conversion from Fahrenheit to Kelvin or Celsius[a]
$$\Delta T, K = \Delta T, \deg C = (\Delta T, \deg F)(5/9)$$

	Fahrenheit temperature increment								
Range, deg F	**1**	**2**	**3**	**4**	**5**	**6**	**7**	**8**	**9**
0.01 to 0.09	0.005	0.011	0.016	0.022	0.027	0.033	0.038	0.044	0.050
0.1 to 0.9	0.055	0.111	0.166	0.222	0.277	0.333	0.388	0.444	0.500
1 to 9	0.555	1.111	1.666	2.222	2.777	3.333	3.888	4.444	5.000
10 to 90	5.555	11.111	16.666	22.222	27.777	33.333	38.888	44.444	50.000
100 to 900	55.555	111.111	166.666	222.222	277.777	333.333	388.888	444.444	500.000
1 000 to 9 000	555.555	1 111.111	1 666.666	2 222.222	2 777.777	3 333.333	3 888.888	4 444.444	5 000.000

[a]The final digit in every kelvin value, Celsius value, and kelvin/Celsius increment repeats infinitum. The user should carry them out and/or round them off to suit his needs.

From Domholdt, L. C., copyright © 1972, in *Handbook of Tables for Applied Engineering Science*, 2nd ed., Bolz, R. E. and Tuve, G. L., Eds., CRC Press, Cleveland, 1973, 858, With permission.

Table 4—15
WIDE-RANGE TEMPERATURE CONVERSION FROM CELSIUS TO RANKINE OR FAHRENHEIT

Table A (for Rankine) and Table B (for Fahrenheit) enable the user to estimate quickly the Rankine or Fahrenheit equivalent of any Celsius temperature to 90,000 deg C. Conversion of any temperature to 100,000 deg C to any desired accuracy can be obtained by simple addition, using Table A or B for the first digit and Table C for additional digits (see examples).

Examples:

$$106 \ C = (100 + 6) \ C = ? \ R$$

100,	Table A:	671.67
6,	Table C:	10.8
	106 C =	682.47 R

$$3.96 \ C = (3 + 0.9 + 0.6) \ C = ? \ F$$

3.00,	Table B;	37.4
0.90,	Table C:	1.62
0.06,	Table C:	0.108
	3.96 C =	39.128 F

$$-103 \ C = (-100-3) \ C = ? \ R = ? \ F$$

−100,	Table A:	311.67
−3,	Table C:	−5.4
	−103C =	+306.27 R
−100,	Table B:	−148.0
3,	Table C:	−5.4
	−103 C =	−153.4 F

Table 4–15 (continued)
WIDE-RANGE TEMPERATURE CONVERSION FROM CELSIUS TO RANKINE OR FAHRENHEIT

Table A. Temperature Conversion from Celsius to Rankine[a]
T, deg R = (T, deg C + 273.15(9/5)   0 deg C = 491.67 deg R (exactly)

Celsius temperature

Range, deg C	1	2	3	4	5	6	7	8	9
−100 to −200	311.67	131.67							
−10 to −90	473.67	455.67	437.67	419.67	401.67	383.67	365.67	347.67	329.67
−1 to −9	489.87	488.07	486.27	484.47	482.67	480.87	479.07	477.27	475.47
1 to 9	493.47	495.27	497.07	498.87	500.67	502.47	504.27	506.07	507.87
10 to 90	509.67	527.67	545.67	563.67	581.67	599.67	617.67	635.67	653.67
100 to 900	671.67	851.67	1 031.67	1 211.67	1 391.67	1 571.67	1 751.67	1 931.67	2 111.67
1 000 to 9 000	2 291.67	4 091.67	5 891.67	7 691.67	9 491.67	11 291.67	13 091.67	14 891.67	16 691.67
10 000 to 90 000	18 491.67	36 491.67	54 491.67	72 491.67	90 491.67	108 491.67	126 491.67	144 491.67	162 491.67

Table B. Temperature Conversion from Celsius to Fahrenheit[a]
T, deg F = (T, deg C)(9/5) + 32   0 deg C = 32 deg F

Celsius temperature

Range, deg C	1	2	3	4	5	6	7	8	9
−100 to −200	−148.0	−328.0							
−10 to −90	+14.0	−4.0	−22.0	−40.0	−58.0	−76.0	−94.0	−112.0	−130.0
−1 to −9	+30.2	+28.4	+26.6	+24.8	+23.0	+21.2	+19.4	+17.6	+15.8
1 to 9	33.8	35.6	37.4	39.2	41.0	42.8	44.6	46.4	48.2
10 to 90	50.0	68.0	86.0	104.0	122.0	140.0	158.0	176.0	194.0
100 to 900	212.0	392.0	572.0	752.0	932.0	1 112.0	1 292.0	1 472.0	1 652.0
1 000 to 9 000	1 832.0	3 632.0	5 432.0	7 232.0	9 032.0	10 832.0	12 632.0	14 432.0	16 232.0
10 000 to 90 000	18 032.0	36 032.0	54 032.0	72 032.0	90 032.0	108 032.0	126 032.0	144 032.0	162 032.0

Table C. Temperature Increment Conversion from Celsius to Rankine or Fahrenheit[a]
ΔT, deg R = ΔT, deg F = (ΔT, deg C)(9/5)

Celsius temperature increment

Range, deg C	1	2	3	4	5	6	7	8	9
0.01 to 0.09	0.018	0.036	0.054	0.072	0.090	0.108	0.126	0.144	0.162
0.1 to 0.9	0.18	0.36	0.54	0.72	0.90	1.08	1.26	1.44	1.62
1 to 9	1.8	3.6	5.4	7.2	9.0	10.8	12.6	14.4	16.2
10 to 90	18	36	54	72	90	108	126	144	162
100 to 900	180	360	540	720	900	1 080	1 260	1 440	1 620
1 000 to 9 000	1 800	3 600	5 400	7 200	9 000	10 800	12 600	14 400	16 200

[a]All values are exact.

From Domholdt, L. C., copyright © 1972, in *Handbook of Tables for Applied Engineering Science*, 2nd ed., Bolz, R. E. and Tuve, G. L., Eds., CRC Press, Cleveland, 1973, 859. With permission.

Table 4–16
STRESS CONVERSION

Metric (KG/mm²)	English (KSI)	SI (MN/m²)	Metric (KG/mm²)	English (KSI)	SI (MN/m²)	Metric (KG/mm²)	English (KSI)	SI (MN/m²)	Metric (KG/mm²)	English (KSI)	SI (MN/m²)
0.00	0.0	0.00	4.92	7.0	48.26	9.84	14.0	96.53	14.76	21.0	144.79
0.07	0.1	0.69	4.99	7.1	48.95	9.91	14.1	97.22	14.83	21.1	145.48
0.14	0.2	1.38	5.06	7.2	49.64	9.98	14.2	97.91	14.91	21.2	146.17
0.21	0.3	2.07	5.13	7.3	50.33	10.05	14.3	98.60	14.98	21.3	146.86
0.28	0.4	2.76	5.20	7.4	51.02	10.12	14.4	99.28	15.05	21.4	147.55
0.35	0.5	3.45	5.27	7.5	51.71	10.19	14.5	99.97	15.12	21.5	148.24
0.42	0.6	4.14	5.34	7.6	52.40	10.26	14.6	100.66	15.19	21.6	148.93
0.49	0.7	4.83	5.41	7.7	53.09	10.34	14.7	101.35	15.26	21.7	149.62
0.56	0.8	5.52	5.48	7.8	53.78	10.41	14.8	102.04	15.33	21.8	150.31
0.63	0.9	6.21	5.55	7.9	54.47	10.48	14.9	102.73	15.40	21.9	151.00
0.70	1.0	6.89	5.62	8.0	55.16	10.55	15.0	103.42	15.47	22.0	151.68
0.77	1.1	7.58	5.69	8.1	55.85	10.62	15.1	104.11	15.54	22.1	152.37
0.84	1.2	8.27	5.77	8.2	56.54	10.69	15.2	104.80	15.61	22.2	153.06
0.91	1.3	8.96	5.84	8.3	57.23	10.76	15.3	105.49	15.68	22.3	153.75
0.98	1.4	9.65	5.91	8.4	57.92	10.83	15.4	106.18	15.75	22.4	154.44
1.05	1.5	10.34	5.98	8.5	58.61	10.90	15.5	106.87	15.82	22.5	155.13
1.12	1.6	11.03	6.05	8.6	59.29	10.97	15.6	107.56	15.89	22.6	155.82
1.20	1.7	11.72	6.12	8.7	59.98	11.04	15.7	108.25	15.96	22.7	156.51
1.27	1.8	12.41	6.19	8.8	60.67	11.11	15.8	108.94	16.03	22.8	157.20
1.34	1.9	13.10	6.26	8.9	61.36	11.18	15.9	109.63	16.10	22.9	157.89
1.41	2.0	13.79	6.33	9.0	62.05	11.25	16.0	110.32	16.17	23.0	158.58
1.48	2.1	14.48	6.40	9.1	62.74	11.32	16.1	111.01	16.24	23.1	159.27
1.55	2.2	15.17	6.47	9.2	63.43	11.39	16.2	111.70	16.31	23.2	159.96
1.62	2.3	15.86	6.54	9.3	64.12	11.46	16.3	112.38	16.38	23.3	160.65
1.69	2.4	16.55	6.61	9.4	64.81	11.53	16.4	113.07	16.45	23.4	161.34
1.76	2.5	17.24	6.68	9.5	65.50	11.60	16.5	113.76	16.52	23.5	162.03
1.83	2.6	17.93	6.75	9.6	66.19	11.67	16.6	114.45	16.59	23.6	162.72
1.90	2.7	18.62	6.82	9.7	66.88	11.74	16.7	115.14	16.66	23.7	163.41
1.97	2.8	19.31	6.89	9.8	67.57	11.81	16.8	115.83	16.73	23.8	164.10
2.04	2.9	19.99	6.96	9.9	68.26	11.88	16.9	116.52	16.80	23.9	164.78
2.11	3.0	20.68	7.03	10.0	68.95	11.95	17.0	117.21	16.87	24.0	165.47
2.18	3.1	21.37	7.10	10.1	69.64	12.02	17.1	117.90	16.94	24.1	166.16
2.25	3.2	22.06	7.17	10.2	70.33	12.09	17.2	118.59	17.01	24.2	166.85
2.32	3.3	22.75	7.24	10.3	71.02	12.16	17.3	119.28	17.08	24.3	167.54
2.39	3.4	23.44	7.31	10.4	71.71	12.23	17.4	119.97	17.15	24.4	168.23
2.46	3.5	24.13	7.38	10.5	72.39	12.30	17.5	120.66	17.23	24.5	168.92
2.53	3.6	24.82	7.45	10.6	73.08	12.37	17.6	121.35	17.30	24.6	169.61
2.60	3.7	25.51	7.52	10.7	73.77	12.44	17.7	122.04	17.37	24.7	170.30
2.67	3.8	26.20	7.59	10.8	74.46	12.51	17.8	122.73	17.44	24.8	170.99
2.74	3.9	26.89	7.66	10.9	75.15	12.58	17.9	123.42	17.51	24.9	171.68
2.81	4.0	27.58	7.73	11.0	75.84	12.66	18.0	124.11	17.58	25.0	172.37
2.88	4.1	28.27	7.80	11.1	76.53	12.73	18.1	124.80	17.65	25.1	173.06
2.95	4.2	28.96	7.87	11.2	77.22	12.80	18.2	125.48	17.72	25.2	173.75
3.02	4.3	29.65	7.94	11.3	77.91	12.87	18.3	126.17	17.79	25.3	174.44
3.09	4.4	30.34	8.01	11.4	78.60	12.94	18.4	126.86	17.86	25.4	175.13
3.16	4.5	31.03	8.09	11.5	79.29	13.01	18.5	127.55	17.93	25.5	175.82
3.23	4.6	31.72	8.16	11.6	79.98	13.08	18.6	128.24	18.00	25.6	176.51
3.30	4.7	32.41	8.23	11.7	80.67	13.15	18.7	128.93	18.07	25.7	177.20
3.37	4.8	33.09	8.30	11.8	81.36	13.22	18.8	129.62	18.14	25.8	177.88
3.45	4.9	33.78	8.37	11.9	82.05	13.29	18.9	130.31	18.21	25.9	178.57
3.52	5.0	34.47	8.44	12.0	82.74	13.36	19.0	131.00	18.28	26.0	179.26
3.59	5.1	35.16	8.51	12.1	83.43	13.43	19.1	131.69	18.35	26.1	179.95
3.66	5.2	35.85	8.58	12.2	84.12	13.50	19.2	132.38	18.42	26.2	180.64
3.73	5.3	36.54	8.65	12.3	84.81	13.57	19.3	133.07	18.49	26.3	181.33
3.80	5.4	37.23	8.72	12.4	85.49	13.64	19.4	133.76	18.56	26.4	182.02
3.87	5.5	37.92	8.79	12.5	86.18	13.71	19.5	134.45	18.63	26.5	182.71
3.94	5.6	38.61	8.86	12.6	86.87	13.78	19.6	135.14	18.70	26.6	183.40
4.01	5.7	39.30	8.93	12.7	87.56	13.85	19.7	135.83	18.77	26.7	184.09
4.08	5.8	39.99	9.00	12.8	88.25	13.92	19.8	136.52	18.84	26.8	184.78
4.15	5.9	40.68	9.07	12.9	88.94	13.99	19.9	137.21	18.91	26.9	185.47
4.22	6.0	41.37	9.14	13.0	89.63	14.06	20.0	137.90	18.98	27.0	186.16
4.29	6.1	42.06	9.21	13.1	90.32	14.13	20.1	138.59	19.05	27.1	186.85
4.36	6.2	42.75	9.28	13.2	91.01	14.20	20.2	139.27	19.12	27.2	187.54
4.43	6.3	43.44	9.35	13.3	91.70	14.27	20.3	139.96	19.19	27.3	188.23
4.50	6.4	44.13	9.42	13.4	92.39	14.34	20.4	140.65	19.26	27.4	188.92
4.57	6.5	44.82	9.49	13.5	93.08	14.41	20.5	141.34	19.33	27.5	189.61
4.64	6.6	45.51	9.56	13.6	93.77	14.48	20.6	142.03	19.40	27.6	190.30
4.71	6.7	46.19	9.63	13.7	94.46	14.55	20.7	142.72	19.48	27.7	190.98
4.78	6.8	46.88	9.70	13.8	95.15	14.62	20.8	143.41	19.55	27.8	191.67
4.85	6.9	47.57	9.77	13.9	95.84	14.69	20.9	144.10	19.62	27.9	192.36

Table 4–16 (continued)
STRESS CONVERSION

Metric (KG/mm²)	English (KSI)	SI (MN/m²)	Metric (KG/mm²)	English (KSI)	SI (MN/m²)	Metric (KG/mm²)	English (KSI)	SI (MN/m²)	Metric (KG/mm²)	English (KSI)	SI (MN/m²)
19.69	28.0	193.05	24.61	35.0	241.32	29.53	42.0	289.58	34.45	49.0	337.84
19.76	28.1	193.74	24.68	35.1	242.01	29.60	42.1	290.27	34.52	49.1	338.53
19.83	28.2	194.43	24.75	35.2	242.70	29.67	42.2	290.96	34.59	49.2	339.22
19.90	28.3	195.12	24.82	35.3	243.38	29.74	42.3	291.65	34.66	49.3	339.91
19.97	28.4	195.81	24.89	35.4	244.07	29.81	42.4	292.34	34.73	49.4	340.60
20.04	28.5	196.50	24.96	35.5	244.76	29.88	42.5	293.03	34.80	49.5	341.29
20.11	28.6	197.19	25.03	35.6	245.45	29.95	42.6	293.72	34.87	49.6	341.98
20.18	28.7	197.88	25.10	35.7	246.14	30.02	42.7	294.41	34.94	49.7	342.67
20.25	28.8	198.57	25.17	35.8	246.83	30.09	42.8	295.10	35.01	49.8	343.36
20.32	28.9	199.26	25.24	35.9	247.52	30.16	42.9	295.79	35.08	49.9	344.05
20.39	29.0	199.95	25.31	36.0	248.21	30.23	43.0	296.47	35.15	50.0	344.74
20.46	29.1	200.64	25.38	36.1	248.90	30.30	43.1	297.16	35.22	50.1	345.43
20.53	29.2	201.33	25.45	36.2	249.59	30.37	43.2	297.85	35.29	50.2	346.12
20.60	29.3	202.02	25.52	36.3	250.28	30.44	43.3	298.54	35.36	50.3	346.81
20.67	29.4	202.71	25.59	36.4	250.97	30.51	43.4	299.23	35.43	50.4	347.50
20.74	29.5	203.40	25.66	36.5	251.66	30.58	43.5	299.92	35.51	50.5	348.19
20.81	29.6	204.08	25.73	36.6	252.35	30.65	43.6	300.61	35.58	50.6	348.87
20.88	29.7	204.77	25.80	36.7	253.04	30.72	43.7	301.30	35.65	50.7	349.56
20.95	29.8	205.46	25.87	36.8	253.73	30.79	43.8	301.99	35.72	50.8	350.25
21.02	29.9	206.15	25.94	36.9	254.42	30.86	43.9	302.68	35.79	50.9	350.94
21.09	30.0	206.84	26.01	37.0	255.11	30.94	44.0	303.37	35.86	51.0	351.63
21.16	30.1	207.53	26.08	37.1	255.80	31.01	44.1	304.06	35.93	51.1	352.32
21.23	30.2	208.22	26.15	37.2	256.48	31.08	44.2	304.75	36.00	51.2	353.01
21.30	30.3	208.91	26.22	37.3	257.17	31.15	44.3	305.44	36.07	51.3	353.70
21.37	30.4	209.60	26.29	37.4	257.86	31.22	44.4	306.13	36.14	51.4	354.39
21.44	30.5	210.29	26.37	37.5	258.55	31.29	44.5	306.82	36.21	51.5	355.08
21.51	30.6	210.98	26.44	37.6	259.24	31.36	44.6	307.51	36.28	51.6	355.77
21.58	30.7	211.67	26.51	37.7	259.93	31.43	44.7	308.20	36.35	51.7	356.46
21.65	30.8	212.36	26.58	37.8	260.62	31.50	44.8	308.89	36.42	51.8	357.15
21.72	30.9	213.05	26.65	37.9	261.31	31.57	44.9	309.57	36.49	51.9	357.84
21.80	31.0	213.74	26.72	38.0	262.00	31.64	45.0	310.26	36.56	52.0	358.53
21.87	31.1	214.43	26.79	38.1	262.69	31.71	45.1	310.95	36.63	52.1	359.22
21.94	31.2	215.12	26.86	38.2	263.38	31.78	45.2	311.64	36.70	52.2	359.91
22.01	31.3	215.81	26.93	38.3	264.07	31.85	45.3	312.33	36.77	52.3	360.60
22.08	31.4	216.50	27.00	38.4	264.76	31.92	45.4	313.02	36.84	52.4	361.29
22.15	31.5	217.18	27.07	38.5	265.45	31.99	45.5	313.71	36.91	52.5	361.97
22.22	31.6	217.87	27.14	38.6	266.14	32.06	45.6	314.40	36.98	52.6	362.66
22.29	31.7	218.56	27.21	38.7	266.83	32.13	45.7	315.09	37.05	52.7	363.35
22.36	31.8	219.25	27.28	38.8	267.52	32.20	45.8	315.78	37.12	52.8	364.04
22.43	31.9	219.94	27.35	38.9	268.21	32.27	45.9	316.47	37.19	52.9	364.73
22.50	32.0	220.63	27.42	39.0	268.90	32.34	46.0	317.16	37.26	53.0	365.42
22.57	32.1	221.32	27.49	39.1	269.58	32.41	46.1	317.85	37.33	53.1	366.11
22.64	32.2	222.01	27.56	39.2	270.27	32.48	46.2	318.54	37.40	53.2	366.80
22.71	32.3	222.70	27.63	39.3	270.96	32.55	46.3	319.23	37.47	53.3	367.49
22.78	32.4	223.39	27.70	39.4	271.65	32.62	46.4	319.92	37.54	53.4	368.18
22.85	32.5	224.08	27.77	39.5	272.34	32.69	46.5	320.61	37.61	53.5	368.87
22.92	32.6	224.77	27.84	39.6	273.03	32.76	46.6	321.30	37.68	53.6	369.56
22.99	32.7	225.46	27.91	39.7	273.72	32.83	46.7	321.99	37.75	53.7	370.25
23.06	32.8	226.15	27.98	39.8	274.41	32.90	46.8	322.67	37.83	53.8	370.94
23.13	32.9	226.84	28.05	39.9	275.10	32.97	46.9	323.36	37.90	53.9	371.63
23.20	33.0	227.53	28.12	40.0	275.79	33.04	47.0	324.05	37.97	54.0	372.32
23.27	33.1	228.22	28.19	40.1	276.48	33.11	47.1	324.74	38.04	54.1	373.01
23.34	33.2	228.91	28.26	40.2	277.17	33.18	47.2	325.43	38.11	54.2	373.70
23.41	33.3	229.60	28.33	40.3	277.86	33.26	47.3	326.12	38.18	54.3	374.39
23.48	33.4	230.28	28.40	40.4	278.55	33.33	47.4	326.81	38.25	54.4	375.07
23.55	33.5	230.97	28.47	40.5	279.24	33.40	47.5	327.50	38.32	54.5	375.76
23.62	33.6	231.66	28.54	40.6	279.93	33.47	47.6	328.19	38.39	54.6	376.45
23.69	33.7	232.35	28.61	40.7	280.62	33.54	47.7	328.88	38.46	54.7	377.14
23.76	33.8	233.04	28.69	40.8	281.31	33.61	47.8	329.57	38.53	54.8	377.83
23.83	33.9	233.73	28.76	40.9	282.00	33.68	47.9	330.26	38.60	54.9	378.52
23.90	34.0	234.42	28.83	41.0	282.69	33.75	48.0	330.95	38.67	55.0	379.21
23.97	34.1	235.11	28.90	41.1	283.37	33.82	48.1	331.64	38.74	55.1	379.90
24.04	34.2	235.80	28.97	41.2	284.06	33.89	48.2	332.33	38.81	55.2	380.59
24.12	34.3	236.49	29.04	41.3	284.75	33.96	48.3	333.02	38.88	55.3	381.28
24.19	34.4	237.18	29.11	41.4	285.44	34.03	48.4	333.71	38.95	55.4	381.97
24.26	34.5	237.87	29.18	41.5	286.13	34.10	48.5	334.40	39.02	55.5	382.66
24.33	34.6	238.56	29.25	41.6	286.82	34.17	48.6	335.09	39.09	55.6	383.35
24.40	34.7	239.25	29.32	41.7	287.51	34.24	48.7	335.77	39.16	55.7	384.04
24.47	34.8	239.94	29.39	41.8	288.20	34.31	48.8	336.46	39.23	55.8	384.73
24.54	34.9	240.63	29.46	41.9	288.89	34.38	48.9	337.15	39.30	55.9	385.42

Table 4–16 (continued)
STRESS CONVERSION

Metric (KG/mm²)	English (KSI)	SI (MN/m²)	Metric (KG/mm²)	English (KSI)	SI (MN/m²)	Metric (KG/mm²)	English (KSI)	SI (MN/m²)	Metric (KG/mm²)	English (KSI)	SI (MN/m²)
39.37	56.0	386.11	44.29	63.0	434.37	49.21	70.0	482.63	54.14	77.0	530.40
39.44	56.1	396.80	44.36	63.1	435.06	49.29	70.1	483.32	54.21	77.1	531.59
39.51	56.2	387.49	44.43	63.2	435.75	49.36	70.2	484.01	54.28	77.2	532.28
39.58	56.3	388.17	44.50	63.3	436.44	49.43	70.3	484.70	54.35	77.3	532.96
39.65	56.4	388.86	44.57	63.4	437.13	49.50	70.4	485.39	54.42	77.4	533.65
39.72	56.5	389.55	44.64	63.5	437.82	49.57	70.5	486.08	54.49	77.5	534.34
39.79	56.6	390.24	44.72	63.6	438.51	49.64	70.6	486.77	54.56	77.6	535.03
39.86	56.7	390.93	44.79	63.7	439.20	49.71	70.7	487.46	54.63	77.7	535.72
39.93	56.8	391.62	44.86	63.8	439.89	49.78	70.8	488.15	54.70	77.8	536.41
40.00	56.9	392.31	44.93	63.9	440.57	49.85	70.9	488.84	54.77	77.9	537.10
40.07	57.0	393.00	45.00	64.0	441.26	49.92	71.0	489.53	54.84	78.0	537.79
40.15	57.1	393.69	45.07	64.1	441.95	49.99	71.1	490.22	54.91	78.1	538.48
40.22	57.2	394.38	45.14	64.2	442.64	50.06	71.2	490.91	54.98	78.2	539.17
40.29	57.3	395.07	45.21	64.3	443.33	50.13	71.3	491.60	55.05	78.3	539.86
40.36	57.4	395.76	45.28	64.4	444.02	50.20	71.4	492.29	55.12	78.4	540.55
40.43	57.5	396.45	45.35	64.5	444.71	50.27	71.5	492.98	55.19	78.5	541.24
40.50	57.6	397.14	45.42	64.6	445.40	50.34	71.6	493.66	55.26	78.6	541.93
40.57	57.7	397.83	45.49	64.7	446.09	50.41	71.7	494.35	55.33	78.7	542.62
40.64	57.8	398.52	45.56	64.8	446.78	50.48	71.8	495.04	55.40	78.8	543.31
40.71	57.9	399.21	45.63	64.9	447.47	50.55	71.9	495.73	55.47	78.9	544.00
40.78	58.0	399.90	45.70	65.0	448.16	50.62	72.0	496.42	55.54	79.0	544.69
40.85	58.1	400.59	45.77	65.1	448.85	50.69	72.1	497.11	55.61	79.1	545.38
40.92	58.2	401.27	45.84	65.2	449.54	50.76	72.2	497.80	55.68	79.2	546.06
40.99	58.3	401.96	45.91	65.3	450.23	50.83	72.3	498.49	55.75	79.3	546.75
41.06	58.4	402.65	45.98	65.4	450.92	50.90	72.4	499.18	55.82	79.4	547.44
41.13	58.5	403.34	46.05	65.5	451.61	50.97	72.5	499.87	55.89	79.5	548.13
41.20	58.6	404.03	46.12	65.6	452.30	51.04	72.6	500.56	55.96	79.6	548.82
41.27	58.7	404.72	46.19	65.7	452.99	51.11	72.7	501.25	56.03	79.7	549.51
41.34	58.8	405.41	46.26	65.8	453.68	51.18	72.8	501.94	56.10	79.8	550.20
41.41	58.9	406.10	46.33	65.9	454.36	51.25	72.9	502.63	56.18	79.9	550.89
41.48	59.0	406.79	46.40	66.0	455.05	51.32	73.0	503.32	56.25	80.0	551.58
41.55	59.1	407.48	46.47	66.1	455.74	51.39	73.1	504.01	56.32	80.1	552.27
41.62	59.2	408.17	46.54	66.2	456.43	51.46	73.2	504.70	56.39	80.2	552.96
41.69	59.3	408.86	46.61	66.3	457.12	51.54	73.3	505.39	56.46	80.3	553.65
41.76	59.4	409.55	46.68	66.4	457.81	51.61	73.4	506.08	56.53	80.4	554.34
41.83	59.5	410.24	46.75	66.5	458.50	51.68	73.5	506.76	56.60	80.5	555.03
41.90	59.6	410.93	46.82	66.6	459.19	51.75	73.6	507.45	56.67	80.6	555.72
41.97	59.7	411.62	46.89	66.7	459.88	51.82	73.7	508.14	56.74	80.7	556.41
42.04	59.8	412.31	46.97	66.8	460.57	51.89	73.8	508.83	56.81	80.8	557.10
42.11	59.9	413.00	47.04	66.9	461.26	51.96	73.9	509.52	56.88	80.9	557.79
42.18	60.0	413.69	47.11	67.0	461.95	52.03	74.0	510.21	56.95	81.0	558.48
42.25	60.1	414.37	47.18	67.1	462.64	52.10	74.1	510.90	57.02	81.1	559.16
42.32	60.2	415.06	47.25	67.2	463.33	52.17	74.2	511.59	57.09	81.2	559.85
42.40	60.3	415.75	47.32	67.3	464.02	52.24	74.3	512.28	57.16	81.3	560.54
42.47	60.4	416.44	47.39	67.4	464.71	52.31	74.4	512.97	57.23	81.4	561.23
42.54	60.5	417.13	47.46	67.5	465.40	52.38	74.5	513.66	57.30	81.5	561.92
42.61	60.6	417.82	47.53	67.6	466.09	52.45	74.6	514.35	57.37	81.6	562.61
42.68	60.7	418.51	47.60	67.7	466.78	52.52	74.7	515.04	57.44	81.7	563.30
42.75	60.8	419.20	47.67	67.8	467.46	52.59	74.8	515.73	57.51	81.8	563.99
42.82	60.9	419.89	47.74	67.9	468.15	52.66	74.9	516.42	57.58	81.9	564.68
42.89	61.0	420.58	47.81	68.0	468.84	52.73	75.0	517.11	57.65	82.0	565.37
42.96	61.1	421.27	47.88	69.1	469.53	52.80	75.1	517.80	57.72	82.1	566.06
43.03	61.2	421.96	47.95	68.2	470.22	52.87	75.2	518.49	57.79	82.2	566.75
43.10	61.3	422.65	48.02	68.3	470.91	52.94	75.3	519.18	57.86	82.3	567.44
43.17	61.4	423.34	48.09	68.4	471.60	53.01	75.4	519.86	57.93	82.4	568.13
43.24	61.5	424.03	48.16	68.5	472.29	53.08	75.5	520.55	58.00	82.5	568.82
43.31	61.6	424.72	48.23	68.6	472.98	53.15	75.6	521.24	58.07	82.6	569.51
43.38	61.7	425.41	48.30	68.7	473.67	53.22	75.7	521.93	58.14	82.7	570.20
43.45	61.8	426.10	48.37	68.8	474.36	53.29	75.8	522.62	58.21	82.8	570.89
43.52	61.9	426.79	48.44	68.9	475.05	53.36	75.9	523.31	58.28	82.9	571.58
43.59	62.0	427.47	48.51	69.0	475.74	53.43	76.0	524.00	58.35	83.0	572.26
43.66	62.1	428.16	48.58	69.1	476.43	53.50	76.1	524.69	58.43	83.1	572.95
43.73	62.2	428.85	48.65	69.2	477.12	53.57	76.2	525.38	58.50	83.2	573.64
43.80	62.3	429.54	48.72	69.3	477.81	53.64	76.3	526.07	58.57	83.3	574.33
43.87	62.4	430.23	48.79	69.4	478.50	53.71	76.4	526.76	58.64	83.4	575.02
43.94	62.5	430.92	48.86	69.5	479.19	53.78	76.5	527.45	58.71	83.5	575.71
44.01	62.6	431.61	48.93	69.6	479.88	53.86	76.6	528.14	58.78	83.6	576.40
44.08	62.7	432.30	49.00	69.7	480.56	53.93	76.7	528.83	58.85	83.7	577.09
44.15	62.8	432.99	49.07	69.8	481.25	54.00	76.8	529.52	58.92	83.8	577.78
44.22	62.9	433.68	49.14	69.9	481.94	54.07	76.9	530.21	58.99	83.9	578.47

Table 4-16 (continued)
STRESS CONVERSION

Metric (KG/mm²)	English (KSI)	SI (MN/m²)	Metric (KG/mm²)	English (KSI)	SI (MN/m²)	Metric (KG/mm²)	English (KSI)	SI (MN/m²)	Metric (KG/mm²)	English (KSI)	SI (MN/m²)
59.06	84.0	579.16	63.99	91.0	627.42	68.90	98.0	675.69	73.82	105.0	723.95
59.13	84.1	579.85	64.05	91.1	628.11	68.97	98.1	676.38	73.89	105.1	724.64
59.20	84.2	580.54	64.12	91.2	628.80	69.04	98.2	677.07	73.96	105.2	725.33
59.27	84.3	581.23	64.19	91.3	629.49	69.11	98.3	677.75	74.03	105.3	726.02
59.34	84.4	581.92	64.26	91.4	630.18	69.18	98.4	678.44	74.10	105.4	726.71
59.41	84.5	582.61	64.33	91.5	630.87	69.25	98.5	679.13	74.17	105.5	727.40
59.48	84.6	583.30	64.40	91.6	631.56	69.32	98.6	679.82	74.24	105.6	728.09
59.55	84.7	583.99	64.47	91.7	632.25	69.39	98.7	680.51	74.31	105.7	728.78
59.62	84.8	584.68	64.54	91.8	632.94	69.46	98.8	681.20	74.38	105.8	729.47
59.69	84.9	585.36	64.61	91.9	633.63	69.53	98.9	681.89	74.46	105.9	730.15
59.76	85.0	586.05	64.68	92.0	634.32	69.60	99.0	682.58	74.53	106.0	730.84
59.83	85.1	586.74	64.75	92.1	635.01	69.67	99.1	683.27	74.60	106.1	731.53
59.90	85.2	587.43	64.82	92.2	635.70	69.74	99.2	683.96	74.67	106.2	732.22
59.97	85.3	588.12	64.89	92.3	636.39	69.81	99.3	684.65	74.74	106.3	732.91
60.04	85.4	588.81	64.96	92.4	637.08	69.89	99.4	685.34	74.81	106.4	733.60
60.11	85.5	589.50	65.03	92.5	637.77	69.96	99.5	686.03	74.88	106.5	734.29
60.18	85.6	590.19	65.10	92.6	638.45	70.03	99.6	686.72	74.95	106.6	734.99
60.25	85.7	590.88	65.17	92.7	639.14	70.10	99.7	687.41	75.02	106.7	735.67
60.32	85.8	591.57	65.24	92.8	639.83	70.17	99.8	688.10	75.09	106.8	736.36
60.39	85.9	592.26	65.32	92.9	640.52	70.24	99.9	688.79	75.16	106.9	737.05
60.46	86.0	592.95	65.39	93.0	641.21	70.31	100.0	689.48	75.23	107.0	737.74
60.53	86.1	593.64	65.46	93.1	641.90	70.38	100.1	690.17	75.30	107.1	738.43
60.60	86.2	594.33	65.53	93.2	642.59	70.45	100.2	690.85	75.37	107.2	739.12
60.67	86.3	595.02	65.60	93.3	643.28	70.52	100.3	691.54	75.44	107.3	739.81
60.75	86.4	595.71	65.67	93.4	643.97	70.59	100.4	692.23	75.51	107.4	740.50
60.82	86.5	596.40	65.74	93.5	644.66	70.66	100.5	692.92	75.58	107.5	741.19
60.89	86.6	597.09	65.81	93.6	645.35	70.73	100.6	693.61	75.65	107.6	741.88
60.96	86.7	597.78	65.88	93.7	646.04	70.80	100.7	694.30	75.72	107.7	742.57
61.03	86.8	598.46	65.95	93.8	646.73	70.87	100.8	694.99	75.79	107.8	743.25
61.10	86.9	599.15	66.02	93.9	647.42	70.94	100.9	695.68	75.86	107.9	743.94
61.17	87.0	599.84	66.09	94.0	648.11	71.01	101.0	696.37	75.93	108.0	744.63
61.24	87.1	600.53	66.16	94.1	648.80	71.08	101.1	697.06	76.00	108.1	745.32
61.31	87.2	601.22	66.23	94.2	649.49	71.15	101.2	697.75	76.07	108.2	746.01
61.38	87.3	601.91	66.30	94.3	650.18	71.22	101.3	698.44	76.14	108.3	746.70
61.45	87.4	602.60	66.37	94.4	650.87	71.29	101.4	699.13	76.21	108.4	747.39
61.52	87.5	603.29	66.44	94.5	651.55	71.36	101.5	699.82	76.28	108.5	748.08
61.59	87.6	603.98	66.51	94.6	652.24	71.43	101.6	700.51	76.35	108.6	748.77
61.66	87.7	604.67	66.58	94.7	652.93	71.50	101.7	701.20	76.42	108.7	749.46
61.73	87.8	605.36	66.65	94.8	653.62	71.57	101.8	701.89	76.49	108.8	750.15
61.80	87.9	606.05	66.72	94.9	654.31	71.64	101.9	702.58	76.56	108.9	750.84
61.87	88.0	606.74	66.79	95.0	655.00	71.71	102.0	703.27	76.63	109.0	751.53
61.94	88.1	607.43	66.86	95.1	655.69	71.78	102.1	703.95	76.70	109.1	752.22
62.01	88.2	608.12	66.93	95.2	656.38	71.85	102.2	704.64	76.78	109.2	752.91
62.08	88.3	608.81	67.00	95.3	657.07	71.92	102.3	705.33	76.85	109.3	753.60
62.15	88.4	609.50	67.07	95.4	657.76	71.99	102.4	706.02	76.92	109.4	754.29
62.22	88.5	610.19	67.14	95.5	658.45	72.06	102.5	706.71	76.99	109.5	754.98
62.29	88.6	610.88	67.21	95.6	659.14	72.13	102.6	707.40	77.06	109.6	755.67
62.36	88.7	611.56	67.28	95.7	659.83	72.21	102.7	708.09	77.13	109.7	756.35
62.43	88.8	612.25	67.35	95.8	660.52	72.28	102.8	708.78	77.20	109.8	757.04
62.50	88.9	612.94	67.42	95.9	661.21	72.35	102.9	709.47	77.27	109.9	757.73
62.57	89.0	613.63	67.49	96.0	661.90	72.42	103.0	710.16	77.34	110.0	758.42
62.64	89.1	614.32	67.56	96.1	662.59	72.49	103.1	710.85	77.41	110.1	759.11
62.71	89.2	615.01	67.64	96.2	663.28	72.56	103.2	711.54	77.48	110.2	759.80
62.78	89.3	615.70	67.71	96.3	663.97	72.63	103.3	712.23	77.55	110.3	760.49
62.85	89.4	616.39	67.78	96.4	664.65	72.70	103.4	712.92	77.62	110.4	761.18
62.92	89.5	617.08	67.85	96.5	665.34	72.77	103.5	713.61	77.69	110.5	761.87
63.00	89.6	617.77	67.92	96.6	666.03	72.84	103.6	714.30	77.76	110.6	762.56
63.07	89.7	618.46	67.99	96.7	666.72	72.91	103.7	714.99	77.83	110.7	763.25
63.14	89.8	619.15	68.06	96.8	667.41	72.98	103.8	715.68	77.90	110.8	763.94
63.21	89.9	619.84	68.13	96.9	668.10	73.05	103.9	716.37	77.97	110.9	764.63
63.28	90.0	620.53	68.20	97.0	668.79	73.12	104.0	717.05	78.04	111.0	765.32
63.35	90.1	621.22	68.27	97.1	669.48	73.19	104.1	717.74	78.11	111.1	766.01
63.42	90.2	621.91	68.34	97.2	670.17	73.26	104.2	718.43	78.18	111.2	766.70
63.49	90.3	622.60	68.41	97.3	670.86	73.33	104.3	719.12	78.25	111.3	767.39
63.56	90.4	623.29	68.48	97.4	671.55	73.40	104.4	719.81	78.32	111.4	768.08
63.63	90.5	623.98	68.55	97.5	672.24	73.47	104.5	720.50	78.39	111.5	768.77
63.70	90.6	624.66	68.62	97.6	672.93	73.54	104.6	721.19	78.46	111.6	769.45
63.77	90.7	625.35	68.69	97.7	673.62	73.61	104.7	721.88	78.53	111.7	770.14
63.84	90.8	626.04	68.76	97.8	674.31	73.68	104.8	722.57	78.60	111.8	770.83
63.91	90.9	626.73	68.83	97.9	675.00	73.75	104.9	723.26	78.67	111.9	771.52

Table 4–16 (continued)
STRESS CONVERSION

Metric (KG/mm²)	English (KSI)	SI (MN/m²)	Metric (KG/mm²)	English (KSI)	SI (MN/m²)	Metric (KG/mm²)	English (KSI)	SI (MN/m²)	Metric (KG/mm²)	English (KSI)	SI (MN/m²)
78.74	112.0	772.21	83.67	119.0	820.48	88.59	126.0	868.74	93.51	133.0	917.00
78.81	112.1	772.90	83.74	119.1	821.17	88.66	126.1	869.43	93.58	133.1	917.69
78.88	112.2	773.59	83.81	119.2	821.86	88.73	126.2	870.12	93.65	133.2	918.38
78.95	112.3	774.28	83.88	119.3	822.54	88.80	126.3	870.81	93.72	133.3	919.07
79.03	112.4	774.97	83.95	119.4	823.23	88.87	126.4	871.50	93.79	133.4	919.76
79.10	112.5	775.66	84.02	119.5	823.92	88.94	126.5	872.19	93.86	133.5	920.45
79.17	112.6	776.35	84.09	119.6	824.61	89.01	126.6	872.88	93.93	133.6	921.14
79.24	112.7	777.04	84.16	119.7	825.30	89.08	126.7	873.57	94.00	133.7	921.83
79.31	112.8	777.73	84.23	119.8	825.99	89.15	126.8	874.26	94.07	133.8	922.52
79.38	112.9	778.42	84.30	119.9	826.68	89.22	126.9	874.94	94.14	133.9	923.21
79.45	113.0	779.11	84.37	120.0	827.37	89.29	127.0	875.63	94.21	134.0	923.90
79.52	113.1	779.80	84.44	120.1	828.06	89.36	127.1	876.32	94.28	134.1	924.59
79.59	113.2	780.49	84.51	120.2	828.75	89.43	127.2	877.01	94.35	134.2	925.28
79.66	113.3	781.18	84.58	120.3	829.44	89.50	127.3	877.70	94.42	134.3	925.97
79.73	113.4	781.87	84.65	120.4	830.13	89.57	127.4	878.39	94.49	134.4	926.66
79.80	113.5	782.55	84.72	120.5	830.82	89.64	127.5	879.08	94.56	134.5	927.34
79.87	113.6	783.24	84.79	120.6	831.51	89.71	127.6	879.77	94.63	134.6	928.03
79.94	113.7	783.93	84.86	120.7	832.20	89.78	127.7	880.46	94.70	134.7	928.72
80.01	113.8	784.62	84.93	120.8	832.89	89.85	127.8	881.15	94.77	134.8	929.41
80.08	113.9	785.31	85.00	120.9	833.58	89.92	127.9	881.84	94.84	134.9	930.10
80.15	114.0	786.00	85.07	121.0	834.27	89.99	128.0	882.53	94.91	135.0	930.79
80.22	114.1	786.69	85.14	121.1	834.96	90.06	128.1	883.22	94.98	135.1	931.48
80.29	114.2	787.38	85.21	121.2	835.64	90.13	128.2	883.91	95.06	135.2	932.17
80.36	114.3	788.07	85.28	121.3	836.33	90.20	128.3	884.60	95.13	135.3	932.86
80.43	114.4	788.76	85.35	121.4	837.02	90.27	128.4	885.29	95.20	135.4	933.55
80.50	114.5	789.45	85.42	121.5	837.71	90.34	128.5	885.98	95.27	135.5	934.24
80.57	114.6	790.14	85.49	121.6	838.40	90.41	128.6	886.67	95.34	135.6	934.93
80.64	114.7	790.83	85.56	121.7	839.09	90.49	128.7	887.36	95.41	135.7	935.62
80.71	114.8	791.52	85.63	121.8	839.78	90.56	128.8	888.04	95.48	135.8	936.31
80.78	114.9	792.21	85.70	121.9	840.47	90.63	128.9	888.73	95.55	135.9	937.00
80.85	115.0	792.90	85.77	122.0	841.16	90.70	129.0	889.42	95.62	136.0	937.69
80.92	115.1	793.59	85.84	122.1	841.85	90.77	129.1	890.11	95.69	136.1	938.38
80.99	115.2	794.28	85.92	122.2	842.54	90.84	129.2	890.80	95.76	136.2	939.07
81.06	115.3	794.97	85.99	122.3	843.23	90.91	129.3	891.49	95.83	136.3	939.76
81.13	115.4	795.65	86.06	122.4	843.92	90.98	129.4	892.18	95.90	136.4	940.45
81.20	115.5	796.34	86.13	122.5	844.61	91.05	129.5	892.87	95.97	136.5	941.13
81.27	115.6	797.03	86.20	122.6	845.30	91.12	129.6	893.56	96.04	136.6	941.82
81.35	115.7	797.72	86.27	122.7	845.99	91.19	129.7	894.25	96.11	136.7	942.51
81.42	115.8	798.41	86.34	122.8	846.68	91.26	129.8	894.94	96.18	136.8	943.20
81.49	115.9	799.10	86.41	122.9	847.37	91.33	129.9	895.63	96.25	136.9	943.89
81.56	116.0	799.79	86.48	123.0	848.06	91.40	130.0	896.32	96.32	137.0	944.58
81.63	116.1	800.48	86.55	123.1	848.74	91.47	130.1	897.01	96.39	137.1	945.27
81.70	116.2	801.17	86.62	123.2	849.43	91.54	130.2	897.70	96.46	137.2	945.96
81.77	116.3	801.86	86.69	123.3	850.12	91.61	130.3	898.39	96.53	137.3	946.65
81.84	116.4	802.55	86.76	123.4	850.81	91.68	130.4	899.08	96.60	137.4	947.34
81.91	116.5	803.24	86.83	123.5	851.50	91.75	130.5	899.77	96.67	137.5	948.03
81.98	116.6	803.93	86.90	123.6	852.19	91.82	130.6	900.46	96.74	137.6	948.72
82.05	116.7	804.62	86.97	123.7	852.88	91.89	130.7	901.14	96.81	137.7	949.41
82.12	116.8	805.31	87.04	123.8	853.57	91.96	130.8	901.83	96.88	137.8	950.10
82.19	116.9	806.00	87.11	123.9	854.26	92.03	130.9	902.52	96.95	137.9	950.79
82.26	117.0	806.69	87.18	124.0	854.95	92.10	131.0	903.21	97.02	138.0	951.48
82.33	117.1	807.38	87.25	124.1	855.64	92.17	131.1	903.90	97.09	138.1	952.17
82.40	117.2	808.07	87.32	124.2	856.33	92.24	131.2	904.59	97.16	138.2	952.86
82.47	117.3	808.75	87.39	124.3	857.02	92.31	131.3	905.28	97.23	138.3	953.54
82.54	117.4	809.44	87.46	124.4	857.71	92.38	131.4	905.97	97.30	138.4	954.23
82.61	117.5	810.13	87.53	124.5	858.40	92.45	131.5	906.66	97.38	138.5	954.92
82.68	117.6	810.82	87.60	124.6	859.09	92.52	131.6	907.35	97.45	138.6	955.61
82.75	117.7	811.51	87.67	124.7	859.78	92.59	131.7	908.04	97.52	138.7	956.30
82.82	117.8	812.20	87.74	124.8	860.47	92.66	131.8	908.73	97.59	138.8	956.99
82.89	117.9	812.89	87.81	124.9	861.16	92.73	131.9	909.42	97.66	138.9	957.68
82.96	118.0	813.58	87.88	125.0	861.84	92.81	132.0	910.11	97.73	139.0	958.37
83.03	118.1	814.27	87.95	125.1	862.53	92.88	132.1	910.80	97.80	139.1	959.06
83.10	118.2	814.96	88.02	125.2	863.22	92.95	132.2	911.49	97.87	139.2	959.75
83.17	118.3	815.65	88.09	125.3	863.91	93.02	132.3	912.18	97.94	139.3	960.44
83.24	118.4	816.34	88.16	125.4	864.60	93.09	132.4	912.87	98.01	139.4	961.13
83.31	118.5	817.03	88.24	125.5	865.29	93.16	132.5	913.56	98.08	139.5	961.82
83.38	118.6	817.72	88.31	125.6	865.98	93.23	132.6	914.24	98.15	139.6	962.51
83.45	118.7	818.41	88.38	125.7	866.67	93.30	132.7	914.93	98.22	139.7	963.20
83.52	118.8	819.10	88.45	125.8	867.36	93.37	132.8	915.62	98.29	139.8	963.89
83.59	118.9	819.79	88.52	125.9	868.05	93.44	132.9	916.31	98.36	139.9	964.58

Table 4–16 (continued)
STRESS CONVERSION

Metric (KG/mm²)	English (KSI)	SI (MN/m²)	Metric (KG/mm²)	English (KSI)	SI (MN/m²)	Metric (KG/mm²)	English (KSI)	SI (MN/m²)	Metric (KG/mm²)	English (KSI)	SI (MN/m²)
98.43	140.0	965.27	103.35	147.0	1013.53	108.27	154.0	1061.79	113.19	161.0	1110.06
98.50	140.1	965.96	103.42	147.1	1014.22	108.34	154.1	1062.48	113.26	161.1	1110.75
98.57	140.2	966.64	103.49	147.2	1014.91	108.41	154.2	1063.17	113.33	161.2	1111.43
98.64	140.3	967.33	103.56	147.3	1015.60	108.48	154.3	1063.86	113.41	161.3	1112.12
98.71	140.4	968.02	103.63	147.4	1016.29	108.55	154.4	1064.55	113.48	161.4	1112.81
98.78	140.5	968.71	103.70	147.5	1016.98	108.62	154.5	1065.24	113.55	161.5	1113.50
98.85	140.6	969.40	103.77	147.6	1017.67	108.69	154.6	1065.93	113.62	161.6	1114.19
98.92	140.7	970.09	103.84	147.7	1018.36	108.76	154.7	1066.62	113.69	161.7	1114.88
98.99	140.8	970.78	103.91	147.8	1019.05	108.84	154.8	1067.31	113.76	161.8	1115.57
99.06	140.9	971.47	103.98	147.9	1019.73	108.91	154.9	1068.00	113.83	161.9	1116.26
99.13	141.0	972.16	104.05	148.0	1020.42	108.98	155.0	1068.69	113.90	162.0	1116.95
99.20	141.1	972.85	104.12	148.1	1021.11	109.05	155.1	1069.38	113.97	162.1	1117.64
99.27	141.2	973.54	104.19	148.2	1021.80	109.12	155.2	1070.07	114.04	162.2	1118.33
99.34	141.3	974.23	104.27	148.3	1022.49	109.19	155.3	1070.76	114.11	162.3	1119.02
99.41	141.4	974.92	104.34	148.4	1023.18	109.26	155.4	1071.45	114.18	162.4	1119.71
99.48	141.5	975.61	104.41	148.5	1023.87	109.33	155.5	1072.13	114.25	162.5	1120.40
99.55	141.6	976.30	104.48	148.6	1024.56	109.40	155.6	1072.82	114.32	162.6	1121.09
99.62	141.7	976.99	104.55	148.7	1025.25	109.47	155.7	1073.51	114.39	162.7	1121.78
99.70	141.8	977.68	104.62	148.8	1025.94	109.54	155.8	1074.20	114.46	162.8	1122.47
99.77	141.9	978.37	104.69	148.9	1026.63	109.61	155.9	1074.89	114.53	162.9	1123.16
99.84	142.0	979.06	104.76	149.0	1027.32	109.68	156.0	1075.58	114.60	163.0	1123.85
99.91	142.1	979.74	104.83	149.1	1028.01	109.75	156.1	1076.27	114.67	163.1	1124.53
99.98	142.2	980.43	104.90	149.2	1028.70	109.82	156.2	1076.96	114.74	163.2	1125.22
100.05	142.3	981.12	104.97	149.3	1029.39	109.89	156.3	1077.65	114.81	163.3	1125.91
100.12	142.4	981.81	105.04	149.4	1030.08	109.96	156.4	1078.34	114.88	163.4	1126.60
100.19	142.5	982.50	105.11	149.5	1030.77	110.03	156.5	1079.03	114.95	163.5	1127.29
100.26	142.6	983.19	105.18	149.6	1031.46	110.10	156.6	1079.72	115.02	163.6	1127.98
100.33	142.7	983.88	105.25	149.7	1032.15	110.17	156.7	1080.41	115.09	163.7	1128.67
100.40	142.8	984.57	105.32	149.8	1032.83	110.24	156.8	1081.10	115.16	163.8	1129.36
100.47	142.9	985.26	105.39	149.9	1033.52	110.31	156.9	1081.79	115.23	163.9	1130.05
100.54	143.0	985.95	105.46	150.0	1034.21	110.38	157.0	1082.48	115.30	164.0	1130.74
100.61	143.1	986.64	105.53	150.1	1034.90	110.45	157.1	1083.17	115.37	164.1	1131.43
100.68	143.2	987.33	105.60	150.2	1035.59	110.52	157.2	1083.86	115.44	164.2	1132.12
100.75	143.3	988.02	105.67	150.3	1036.28	110.59	157.3	1084.55	115.51	164.3	1132.81
100.82	143.4	988.71	105.74	150.4	1036.97	110.66	157.4	1085.23	115.58	164.4	1133.50
100.89	143.5	989.40	105.81	150.5	1037.66	110.73	157.5	1085.92	115.65	164.5	1134.19
100.96	143.6	990.09	105.88	150.6	1038.35	110.80	157.6	1086.61	115.73	164.6	1134.88
101.03	143.7	990.78	105.95	150.7	1039.04	110.87	157.7	1087.30	115.80	164.7	1135.57
101.10	143.8	991.47	106.02	150.8	1039.73	110.94	157.8	1087.99	115.87	164.8	1136.26
101.17	143.9	992.16	106.09	150.9	1040.42	111.01	157.9	1088.68	115.94	164.9	1136.95
101.24	144.0	992.85	106.16	151.0	1041.11	111.08	158.0	1089.37	116.01	165.0	1137.63
101.31	144.1	993.53	106.23	151.1	1041.80	111.16	158.1	1090.06	116.08	165.1	1138.32
101.38	144.2	994.22	106.30	151.2	1042.49	111.23	158.2	1090.75	116.15	165.2	1139.01
101.45	144.3	994.91	106.37	151.3	1043.18	111.30	158.3	1091.44	116.22	165.3	1139.70
101.52	144.4	995.60	106.44	151.4	1043.87	111.37	158.4	1092.13	116.29	165.4	1140.39
101.59	144.5	996.29	106.52	151.5	1044.56	111.44	158.5	1092.82	116.36	165.5	1141.08
101.66	144.6	996.98	106.59	151.6	1045.25	111.51	158.6	1093.51	116.43	165.6	1141.77
101.73	144.7	997.67	106.66	151.7	1045.93	111.58	158.7	1094.20	116.50	165.7	1142.46
101.80	144.8	998.36	106.73	151.8	1046.62	111.65	158.8	1094.89	116.57	165.8	1143.15
101.87	144.9	999.05	106.80	151.9	1047.31	111.72	158.9	1095.58	116.64	165.9	1143.84
101.95	145.0	999.74	106.87	152.0	1048.00	111.79	159.0	1096.27	116.71	166.0	1144.53
102.02	145.1	1000.43	106.94	152.1	1048.69	111.86	159.1	1096.96	116.78	166.1	1145.22
102.09	145.2	1001.12	107.01	152.2	1049.38	111.93	159.2	1097.65	116.85	166.2	1145.91
102.16	145.3	1001.81	107.08	152.3	1050.07	112.00	159.3	1098.33	116.92	166.3	1146.60
102.23	145.4	1002.50	107.15	152.4	1050.76	112.07	159.4	1099.02	116.99	166.4	1147.29
102.30	145.5	1003.19	107.22	152.5	1051.45	112.14	159.5	1099.71	117.06	166.5	1147.98
102.37	145.6	1003.88	107.29	152.6	1052.14	112.21	159.6	1100.40	117.13	166.6	1148.67
102.44	145.7	1004.57	107.36	152.7	1052.83	112.28	159.7	1101.09	117.20	166.7	1149.36
102.51	145.8	1005.26	107.43	152.8	1053.52	112.35	159.8	1101.78	117.27	166.8	1150.05
102.58	145.9	1005.95	107.50	152.9	1054.21	112.42	159.9	1102.47	117.34	166.9	1150.73
102.65	146.0	1006.63	107.57	153.0	1054.90	112.49	160.0	1103.16	117.41	167.0	1151.42
102.72	146.1	1007.32	107.64	153.1	1055.59	112.56	160.1	1103.85	117.48	167.1	1152.11
102.79	146.2	1008.01	107.71	153.2	1056.28	112.63	160.2	1104.54	117.55	167.2	1152.80
102.86	146.3	1008.70	107.78	153.3	1056.97	112.70	160.3	1105.23	117.62	167.3	1153.49
102.93	146.4	1009.39	107.85	153.4	1057.66	112.77	160.4	1105.92	117.69	167.4	1154.18
103.00	146.5	1010.08	107.92	153.5	1058.35	112.84	160.5	1106.61	117.76	167.5	1154.87
103.07	146.6	1010.77	107.99	153.6	1059.03	112.91	160.6	1107.30	117.83	167.6	1155.56
103.14	146.7	1011.46	108.06	153.7	1059.72	112.99	160.7	1107.99	117.90	167.7	1156.25
103.21	146.8	1012.15	108.13	153.8	1060.41	113.05	160.8	1108.68	117.98	167.8	1156.94
103.28	146.9	1012.84	108.20	153.9	1061.10	113.12	160.9	1109.37	118.05	167.9	1157.63

Table 4–16 (continued)
STRESS CONVERSION

Metric (KG/mm²)	English (KSI)	SI (MN/m²)	Metric (KG/mm²)	English (KSI)	SI (MN/m²)	Metric (KG/mm²)	English (KSI)	SI (MN/m²)	Metric (KG/mm²)	English (KSI)	SI (MN/m²)	Metric (KG/mm²)	English (KSI)	SI (MN/m²)
118.12	168.0	1158.32	123.04	175.0	1206.58	127.96	182.0	1254.85	132.88	189.0	1303.11
118.19	168.1	1159.01	123.11	175.1	1207.27	128.03	182.1	1255.54	132.95	189.1	1303.80
118.26	169.2	1159.70	123.18	175.2	1207.96	128.10	182.2	1256.22	133.02	189.2	1304.49
118.33	168.3	1160.39	123.25	175.3	1208.65	128.17	182.3	1256.91	133.09	189.3	1305.18
118.40	168.4	1161.08	123.32	175.4	1209.34	128.24	182.4	1257.60	133.16	189.4	1305.87
118.47	168.5	1161.77	123.39	175.5	1210.03	128.31	182.5	1258.29	133.23	189.5	1306.56
118.54	168.6	1162.46	123.46	175.6	1210.72	128.38	182.6	1258.98	133.30	189.6	1307.25
118.61	168.7	1163.15	123.53	175.7	1211.41	128.45	182.7	1259.67	133.37	189.7	1307.94
118.68	168.8	1163.83	123.60	175.8	1212.10	128.52	182.8	1260.36	133.44	189.8	1308.62
118.75	168.9	1164.52	123.67	175.9	1212.79	128.59	182.9	1261.05	133.51	189.9	1309.31
118.82	169.0	1165.21	123.74	176.0	1213.48	128.66	183.0	1261.74	133.58	190.0	1310.00
118.89	169.1	1165.90	123.81	176.1	1214.17	128.73	183.1	1262.43	133.65	190.1	1310.69
118.96	169.2	1166.59	123.88	176.2	1214.86	128.80	183.2	1263.12	133.72	190.2	1311.38
119.03	169.3	1167.28	123.95	176.3	1215.55	128.87	183.3	1263.81	133.79	190.3	1312.07
119.10	169.4	1167.97	124.02	176.4	1216.24	128.94	183.4	1264.50	133.86	190.4	1312.76
119.17	169.5	1168.66	124.09	176.5	1216.92	129.01	183.5	1265.19	133.93	190.5	1313.45
119.24	169.6	1169.35	124.16	176.6	1217.61	129.08	183.6	1265.88	134.01	190.6	1314.14
119.31	169.7	1170.04	124.23	176.7	1218.30	129.15	183.7	1266.57	134.08	190.7	1314.83
119.38	169.8	1170.73	124.30	176.8	1218.99	129.22	183.8	1267.26	134.15	190.8	1315.52
119.45	169.9	1171.42	124.37	176.9	1219.68	129.29	183.9	1267.95	134.22	190.9	1316.21
119.52	170.0	1172.11	124.44	177.0	1220.37	129.36	184.0	1268.64	134.29	191.0	1316.90
119.59	170.1	1172.80	124.51	177.1	1221.06	129.44	184.1	1269.32	134.36	191.1	1317.59
119.66	170.2	1173.49	124.58	177.2	1221.75	129.51	184.2	1270.01	134.43	191.2	1318.28
119.73	170.3	1174.18	124.65	177.3	1222.44	129.58	184.3	1270.70	134.50	191.3	1318.97
119.80	170.4	1174.87	124.72	177.4	1223.13	129.65	184.4	1271.39	134.57	191.4	1319.66
119.87	170.5	1175.56	124.79	177.5	1223.82	129.72	184.5	1272.08	134.64	191.5	1320.35
119.94	170.6	1176.25	124.87	177.6	1224.51	129.79	184.6	1272.77	134.71	191.6	1321.04
120.01	170.7	1176.94	124.94	177.7	1225.20	129.86	184.7	1273.46	134.78	191.7	1321.72
120.08	170.8	1177.62	125.01	177.8	1225.89	129.93	184.8	1274.15	134.85	191.8	1322.41
120.15	170.9	1178.31	125.08	177.9	1226.58	130.00	184.9	1274.84	134.92	191.9	1323.10
120.22	171.0	1179.00	125.15	178.0	1227.27	130.07	185.0	1275.53	134.99	192.0	1323.79
120.30	171.1	1179.69	125.22	178.1	1227.96	130.14	185.1	1276.22	135.06	192.1	1324.48
120.37	171.2	1180.38	125.29	178.2	1228.65	130.21	185.2	1276.91	135.13	192.2	1325.17
120.44	171.3	1181.07	125.36	178.3	1229.34	130.28	185.3	1277.60	135.20	192.3	1325.86
120.51	171.4	1181.76	125.43	178.4	1230.02	130.35	185.4	1278.29	135.27	192.4	1326.55
120.58	171.5	1182.45	125.50	178.5	1230.71	130.42	185.5	1278.98	135.34	192.5	1327.24
120.65	171.6	1183.14	125.57	178.6	1231.40	130.49	185.6	1279.67	135.41	192.6	1327.93
120.72	171.7	1183.83	125.64	178.7	1232.09	130.56	185.7	1280.36	135.48	192.7	1328.62
120.79	171.8	1184.52	125.71	178.8	1232.78	130.63	185.8	1281.05	135.55	192.8	1329.31
120.86	171.9	1185.21	125.78	178.9	1233.47	130.70	185.9	1281.74	135.62	192.9	1330.00
120.93	172.0	1185.90	125.85	179.0	1234.16	130.77	186.0	1282.42	135.69	193.0	1330.69
121.00	172.1	1186.59	125.92	179.1	1234.85	130.84	186.1	1283.11	135.76	193.1	1331.38
121.07	172.2	1187.28	125.99	179.2	1235.54	130.91	186.2	1283.80	135.83	193.2	1332.07
121.14	172.3	1187.97	126.06	179.3	1236.23	130.98	186.3	1284.49	135.90	193.3	1332.76
121.21	172.4	1188.66	126.13	179.4	1236.92	131.05	186.4	1285.18	135.97	193.4	1333.45
121.29	172.5	1189.35	126.20	179.5	1237.61	131.12	186.5	1285.87	136.04	193.5	1334.14
121.35	172.6	1190.04	126.27	179.6	1238.30	131.19	186.6	1286.56	136.11	193.6	1334.82
121.42	172.7	1190.72	126.34	179.7	1238.99	131.26	186.7	1287.25	136.18	193.7	1335.51
121.49	172.8	1191.41	126.41	179.8	1239.68	131.33	186.8	1287.94	136.25	193.8	1336.20
121.56	172.9	1192.10	126.48	179.9	1240.37	131.40	186.9	1288.63	136.33	193.9	1336.89
121.63	173.0	1192.79	126.55	180.0	1241.06	131.47	187.0	1289.32	136.40	194.0	1337.58
121.70	173.1	1193.48	126.62	180.1	1241.75	131.54	187.1	1290.01	136.47	194.1	1338.27
121.77	173.2	1194.17	126.69	180.2	1242.44	131.61	187.2	1290.70	136.54	194.2	1338.96
121.84	173.3	1194.86	126.76	180.3	1243.12	131.68	187.3	1291.39	136.61	194.3	1339.65
121.91	173.4	1195.55	126.83	180.4	1243.81	131.76	187.4	1292.08	136.68	194.4	1340.34
121.98	173.5	1196.24	126.90	180.5	1244.50	131.83	187.5	1292.77	136.75	194.5	1341.03
122.05	173.6	1196.93	126.97	180.6	1245.19	131.90	187.6	1293.46	136.82	194.6	1341.72
122.12	173.7	1197.62	127.04	180.7	1245.88	131.97	187.7	1294.15	136.89	194.7	1342.41
122.19	173.8	1198.31	127.11	180.8	1246.57	132.04	187.8	1294.84	136.96	194.8	1343.10
122.26	173.9	1199.00	127.19	180.9	1247.26	132.11	187.9	1295.52	137.03	194.9	1343.79
122.33	174.0	1199.69	127.26	181.0	1247.95	132.18	188.0	1296.21	137.10	195.0	1344.48
122.40	174.1	1200.38	127.33	181.1	1248.64	132.25	188.1	1296.90	137.17	195.1	1345.17
122.47	174.2	1201.07	127.40	181.2	1249.33	132.32	188.2	1297.59	137.24	195.2	1345.86
122.55	174.3	1201.76	127.47	181.3	1250.02	132.39	188.3	1298.28	137.31	195.3	1346.55
122.62	174.4	1202.45	127.54	181.4	1250.71	132.46	188.4	1298.97	137.38	195.4	1347.24
122.69	174.5	1203.14	127.61	181.5	1251.40	132.53	188.5	1299.66	137.45	195.5	1347.92
122.76	174.6	1203.82	127.68	181.6	1252.09	132.60	188.6	1300.35	137.52	195.6	1348.61
122.83	174.7	1204.51	127.75	181.7	1252.78	132.67	188.7	1301.04	137.59	195.7	1349.30
122.90	174.8	1205.20	127.82	181.8	1253.47	132.74	188.8	1301.73	137.66	195.8	1349.99
122.97	174.9	1205.89	127.89	181.9	1254.16	132.81	188.9	1302.42	137.73	195.9	1350.68

Table 4–16 (continued)
STRESS CONVERSION

Metric (KG/mm²)	English (KSI)	SI (MN/m²)	Metric (KG/mm²)	English (KSI)	SI (MN/m²)	Metric (KG/mm²)	English (KSI)	SI (MN/m²)	Metric (KG/mm²)	English (KSI)	SI (MN/m²)
137.80	196.0	1351.37	142.72	203.0	1399.64	147.64	210.0	1447.90	152.57	217.0	1496.16
137.87	196.1	1352.06	142.79	203.1	1400.33	147.71	210.1	1448.59	152.64	217.1	1496.85
137.94	196.2	1352.75	142.86	203.2	1401.01	147.79	210.2	1449.28	152.71	217.2	1497.54
138.01	196.3	1353.44	142.93	203.3	1401.70	147.86	210.3	1449.97	152.78	217.3	1498.23
138.08	196.4	1354.13	143.00	203.4	1402.39	147.93	210.4	1450.66	152.85	217.4	1498.92
138.15	196.5	1354.82	143.07	203.5	1403.08	148.00	210.5	1451.35	152.92	217.5	1499.61
138.22	196.6	1355.51	143.14	203.6	1403.77	148.07	210.6	1452.04	152.99	217.6	1500.30
138.29	196.7	1356.20	143.22	203.7	1404.46	148.14	210.7	1452.73	153.06	217.7	1500.99
138.36	196.8	1356.89	143.29	203.8	1405.15	148.21	210.8	1453.41	153.13	217.8	1501.68
138.43	196.9	1357.58	143.36	203.9	1405.84	148.28	210.9	1454.10	153.20	217.9	1502.37
138.50	197.0	1358.27	143.43	204.0	1406.53	148.35	211.0	1454.79	153.27	218.0	1503.06
138.58	197.1	1358.96	143.50	204.1	1407.22	148.42	211.1	1455.48	153.34	218.1	1503.75
138.65	197.2	1359.65	143.57	204.2	1407.91	148.49	211.2	1456.17	153.41	218.2	1504.44
138.72	197.3	1360.34	143.64	204.3	1408.60	148.56	211.3	1456.86	153.48	218.3	1505.13
138.79	197.4	1361.03	143.71	204.4	1409.29	148.63	211.4	1457.55	153.55	218.4	1505.81
138.86	197.5	1361.71	143.78	204.5	1409.98	148.70	211.5	1458.24	153.62	218.5	1506.50
138.93	197.6	1362.40	143.85	204.6	1410.67	148.77	211.6	1458.93	153.69	218.6	1507.19
139.00	197.7	1363.09	143.92	204.7	1411.36	148.84	211.7	1459.62	153.76	218.7	1507.88
139.07	197.8	1363.78	143.99	204.8	1412.05	148.91	211.8	1460.31	153.83	218.8	1508.57
139.14	197.9	1364.47	144.06	204.9	1412.74	148.98	211.9	1461.00	153.90	218.9	1509.26
139.21	198.0	1365.16	144.13	205.0	1413.43	149.05	212.0	1461.69	153.97	219.0	1509.95
139.28	198.1	1365.85	144.20	205.1	1414.11	149.12	212.1	1462.38	154.04	219.1	1510.64
139.35	198.2	1366.54	144.27	205.2	1414.80	149.19	212.2	1463.07	154.11	219.2	1511.33
139.42	198.3	1367.23	144.34	205.3	1415.49	149.26	212.3	1463.76	154.18	219.3	1512.02
139.49	198.4	1367.92	144.41	205.4	1416.18	149.33	212.4	1464.45	154.25	219.4	1512.71
139.56	198.5	1368.61	144.48	205.5	1416.87	149.40	212.5	1465.14	154.32	219.5	1513.40
139.63	198.6	1369.30	144.55	205.6	1417.56	149.47	212.6	1465.83	154.39	219.6	1514.09
139.70	198.7	1369.99	144.62	205.7	1418.25	149.54	212.7	1466.51	154.46	219.7	1514.78
139.77	198.8	1370.68	144.69	205.8	1418.94	149.61	212.8	1467.20	154.53	219.8	1515.47
139.84	198.9	1371.37	144.76	205.9	1419.63	149.68	212.9	1467.89	154.61	219.9	1516.16
139.91	199.0	1372.06	144.83	206.0	1420.32	149.75	213.0	1468.58	154.68	220.0	1516.85
139.98	199.1	1372.75	144.90	206.1	1421.01	149.82	213.1	1469.27	154.75	220.1	1517.54
140.05	199.2	1373.44	144.97	206.2	1421.70	149.89	213.2	1469.96	154.82	220.2	1518.23
140.12	199.3	1374.13	145.04	206.3	1422.39	149.96	213.3	1470.65	154.89	220.3	1518.91
140.19	199.4	1374.81	145.11	206.4	1423.08	150.04	213.4	1471.34	154.96	220.4	1519.60
140.26	199.5	1375.50	145.18	206.5	1423.77	150.11	213.5	1472.03	155.03	220.5	1520.29
140.33	199.6	1376.19	145.25	206.6	1424.46	150.18	213.6	1472.72	155.10	220.6	1520.98
140.40	199.7	1376.88	145.32	206.7	1425.15	150.25	213.7	1473.41	155.17	220.7	1521.67
140.47	199.8	1377.57	145.39	206.8	1425.84	150.32	213.8	1474.10	155.24	220.8	1522.36
140.54	199.9	1378.26	145.47	206.9	1426.53	150.39	213.9	1474.79	155.31	220.9	1523.05
140.61	200.0	1378.95	145.54	207.0	1427.21	150.46	214.0	1475.48	155.38	221.0	1523.74
140.68	200.1	1379.64	145.61	207.1	1427.90	150.53	214.1	1476.17	155.45	221.1	1524.43
140.75	200.2	1380.33	145.68	207.2	1428.59	150.60	214.2	1476.86	155.52	221.2	1525.12
140.82	200.3	1381.02	145.75	207.3	1429.28	150.67	214.3	1477.55	155.59	221.3	1525.81
140.90	200.4	1381.71	145.82	207.4	1429.97	150.74	214.4	1478.24	155.66	221.4	1526.50
140.97	200.5	1382.40	145.89	207.5	1430.66	150.81	214.5	1478.93	155.73	221.5	1527.19
141.04	200.6	1383.09	145.96	207.6	1431.35	150.88	214.6	1479.61	155.80	221.6	1527.88
141.11	200.7	1383.78	146.03	207.7	1432.04	150.95	214.7	1480.30	155.87	221.7	1528.57
141.18	200.8	1384.47	146.10	207.8	1432.73	151.02	214.8	1480.99	155.94	221.8	1529.26
141.25	200.9	1385.16	146.17	207.9	1433.42	151.09	214.9	1481.68	156.01	221.9	1529.95
141.32	201.0	1385.85	146.24	208.0	1434.11	151.16	215.0	1482.37	156.08	222.0	1530.64
141.39	201.1	1386.54	146.31	208.1	1434.80	151.23	215.1	1483.06	156.15	222.1	1531.33
141.46	201.2	1387.23	146.38	208.2	1435.49	151.30	215.2	1483.75	156.22	222.2	1532.02
141.53	201.3	1387.91	146.45	208.3	1436.18	151.37	215.3	1484.44	156.29	222.3	1532.70
141.60	201.4	1388.60	146.52	208.4	1436.87	151.44	215.4	1485.13	156.36	222.4	1533.39
141.67	201.5	1389.29	146.59	208.5	1437.56	151.51	215.5	1485.82	156.43	222.5	1534.08
141.74	201.6	1389.98	146.66	208.6	1438.25	151.58	215.6	1486.51	156.50	222.6	1534.77
141.81	201.7	1390.67	146.73	208.7	1438.94	151.65	215.7	1487.20	156.57	222.7	1535.46
141.88	201.8	1391.36	146.80	208.8	1439.63	151.72	215.8	1487.89	156.64	222.8	1536.15
141.95	201.9	1392.05	146.87	208.9	1440.31	151.79	215.9	1488.58	156.71	222.9	1536.84
142.02	202.0	1392.74	146.94	209.0	1441.00	151.86	216.0	1489.27	156.78	223.0	1537.53
142.09	202.1	1393.43	147.01	209.1	1441.69	151.93	216.1	1489.96	156.85	223.1	1538.22
142.16	202.2	1394.12	147.08	209.2	1442.38	152.00	216.2	1490.65	156.93	223.2	1538.91
142.23	202.3	1394.81	147.15	209.3	1443.07	152.07	216.3	1491.34	157.00	223.3	1539.60
142.30	202.4	1395.50	147.22	209.4	1443.76	152.14	216.4	1492.03	157.07	223.4	1540.29
142.37	202.5	1396.19	147.29	209.5	1444.45	152.21	216.5	1492.71	157.14	223.5	1540.98
142.44	202.6	1396.88	147.36	209.6	1445.14	152.28	216.6	1493.40	157.21	223.6	1541.67
142.51	202.7	1397.57	147.43	209.7	1445.83	152.36	216.7	1494.09	157.28	223.7	1542.36
142.58	202.8	1398.26	147.50	209.8	1446.52	152.43	216.8	1494.78	157.35	223.8	1543.05
142.65	202.9	1398.95	147.57	209.9	1447.21	152.50	216.9	1495.47	157.42	223.9	1543.74

Table 4–16 (continued)
STRESS CONVERSION

Metric (KG/mm²)	English (KSI)	SI (MN/m²)	Metric (KG/mm²)	English (KSI)	SI (MN/m²)	Metric (KG/mm²)	English (KSI)	SI (MN/m²)	Metric (KG/mm²)	English (KSI)	SI (MN/m²)
157.49	224.0	1544.43	162.41	231.0	1592.69	167.33	238.0	1640.95	172.25	245.0	1689.22
157.56	224.1	1545.12	162.48	231.1	1593.39	167.40	238.1	1641.64	172.32	245.1	1689.90
157.63	224.2	1545.80	162.55	231.2	1594.07	167.47	238.2	1642.33	172.39	245.2	1690.59
157.70	224.3	1546.49	162.62	231.3	1594.76	167.54	238.3	1643.02	172.46	245.3	1691.28
157.77	224.4	1547.18	162.69	231.4	1595.45	167.61	238.4	1643.71	172.53	245.4	1691.97
157.84	224.5	1547.87	162.76	231.5	1596.14	167.68	238.5	1644.40	172.60	245.5	1692.66
157.91	224.6	1548.56	162.83	231.6	1596.83	167.75	238.6	1645.09	172.67	245.6	1693.35
157.98	224.7	1549.25	162.90	231.7	1597.52	167.82	238.7	1645.78	172.74	245.7	1694.04
158.05	224.8	1549.94	162.97	231.8	1598.20	167.89	238.8	1646.47	172.81	245.8	1694.73
158.12	224.9	1550.63	163.04	231.9	1598.89	167.96	238.9	1647.16	172.88	245.9	1695.42
158.19	225.0	1551.32	163.11	232.0	1599.58	168.03	239.0	1647.85	172.96	246.0	1696.11
158.26	225.1	1552.01	163.18	232.1	1600.27	168.10	239.1	1648.54	173.03	246.1	1696.80
158.33	225.2	1552.70	163.25	232.2	1600.96	168.17	239.2	1649.23	173.10	246.2	1697.49
158.40	225.3	1553.39	163.32	232.3	1601.65	168.24	239.3	1649.92	173.17	246.3	1698.18
158.47	225.4	1554.08	163.39	232.4	1602.34	168.31	239.4	1650.60	173.24	246.4	1698.87
158.54	225.5	1554.77	163.46	232.5	1603.03	168.39	239.5	1651.29	173.31	246.5	1699.56
158.61	225.6	1555.46	163.53	232.6	1603.72	168.46	239.6	1651.98	173.38	246.6	1700.25
158.68	225.7	1556.15	163.60	232.7	1604.41	168.53	239.7	1652.67	173.45	246.7	1700.94
158.75	225.8	1556.84	163.67	232.8	1605.10	168.60	239.8	1653.36	173.52	246.8	1701.63
158.82	225.9	1557.53	163.74	232.9	1605.79	168.67	239.9	1654.05	173.59	246.9	1702.32
158.89	226.0	1558.22	163.82	233.0	1606.48	168.74	240.0	1654.74	173.66	247.0	1703.00
158.96	226.1	1558.90	163.89	233.1	1607.17	168.81	240.1	1655.43	173.73	247.1	1703.69
159.03	226.2	1559.59	163.96	233.2	1607.86	168.88	240.2	1656.12	173.80	247.2	1704.38
159.10	226.3	1560.28	164.03	233.3	1608.55	168.95	240.3	1656.81	173.87	247.3	1705.07
159.17	226.4	1560.97	164.10	233.4	1609.24	169.02	240.4	1657.50	173.94	247.4	1705.76
159.25	226.5	1561.66	164.17	233.5	1609.93	169.09	240.5	1658.19	174.01	247.5	1706.45
159.32	226.6	1562.35	164.24	233.6	1610.62	169.16	240.6	1658.88	174.08	247.6	1707.14
159.39	226.7	1563.04	164.31	233.7	1611.30	169.23	240.7	1659.57	174.15	247.7	1707.83
159.46	226.8	1563.73	164.38	233.8	1611.99	169.30	240.8	1660.26	174.22	247.8	1708.52
159.53	226.9	1564.42	164.45	233.9	1612.68	169.37	240.9	1660.95	174.29	247.9	1709.21
159.60	227.0	1565.11	164.52	234.0	1613.37	169.44	241.0	1661.64	174.36	248.0	1709.90
159.67	227.1	1565.80	164.59	234.1	1614.06	169.51	241.1	1662.33	174.43	248.1	1710.59
159.74	227.2	1566.49	164.66	234.2	1614.75	169.58	241.2	1663.02	174.50	248.2	1711.28
159.81	227.3	1567.18	164.73	234.3	1615.44	169.65	241.3	1663.70	174.57	248.3	1711.97
159.88	227.4	1567.87	164.80	234.4	1616.13	169.72	241.4	1664.39	174.64	248.4	1712.66
159.95	227.5	1568.56	164.87	234.5	1616.82	169.79	241.5	1665.08	174.71	248.5	1713.35
160.02	227.6	1569.25	164.94	234.6	1617.51	169.86	241.6	1665.77	174.78	248.6	1714.04
160.09	227.7	1569.94	165.01	234.7	1618.20	169.93	241.7	1666.46	174.85	248.7	1714.73
160.16	227.8	1570.63	165.08	234.8	1618.89	170.00	241.8	1667.15	174.92	248.8	1715.42
160.23	227.9	1571.32	165.15	234.9	1619.58	170.07	241.9	1667.84	174.99	248.9	1716.11
160.30	228.0	1572.00	165.22	235.0	1620.27	170.14	242.0	1668.53	175.06	249.0	1716.79
160.37	228.1	1572.69	165.29	235.1	1620.96	170.21	242.1	1669.22	175.13	249.1	1717.48
160.44	228.2	1573.38	165.36	235.2	1621.65	170.28	242.2	1669.91	175.20	249.2	1718.17
160.51	228.3	1574.07	165.43	235.3	1622.34	170.35	242.3	1670.60	175.28	249.3	1718.86
160.58	228.4	1574.76	165.50	235.4	1623.03	170.42	242.4	1671.29	175.35	249.4	1719.55
160.65	228.5	1575.45	165.57	235.5	1623.72	170.49	242.5	1671.98	175.42	249.5	1720.24
160.72	228.6	1576.14	165.64	235.6	1624.40	170.56	242.6	1672.67	175.49	249.6	1720.93
160.79	228.7	1576.83	165.71	235.7	1625.09	170.63	242.7	1673.36	175.56	249.7	1721.62
160.86	228.8	1577.52	165.78	235.8	1625.78	170.71	242.8	1674.05	175.63	249.8	1722.31
160.93	228.9	1578.21	165.85	235.9	1626.47	170.78	242.9	1674.74	175.70	249.9	1723.00
161.00	229.0	1578.90	165.92	236.0	1627.16	170.85	243.0	1675.43	175.77	250.0	1723.69
161.07	229.1	1579.59	165.99	236.1	1627.85	170.92	243.1	1676.12	175.84	250.1	1724.38
161.14	229.2	1580.28	166.07	236.2	1628.54	170.99	243.2	1676.80	175.91	250.2	1725.07
161.21	229.3	1580.97	166.14	236.3	1629.23	171.06	243.3	1677.49	175.98	250.3	1725.76
161.28	229.4	1581.66	166.21	236.4	1629.92	171.13	243.4	1678.18	176.05	250.4	1726.45
161.35	229.5	1582.35	166.28	236.5	1630.61	171.20	243.5	1678.87	176.12	250.5	1727.14
161.42	229.6	1583.04	166.35	236.6	1631.30	171.27	243.6	1679.56	176.19	250.6	1727.83
161.50	229.7	1583.73	166.42	236.7	1631.99	171.34	243.7	1680.25	176.26	250.7	1728.52
161.57	229.8	1584.42	166.49	236.8	1632.68	171.41	243.8	1680.94	176.33	250.8	1729.21
161.64	229.9	1585.10	166.56	236.9	1633.37	171.48	243.9	1681.63	176.40	250.9	1729.89
161.71	230.0	1585.79	166.63	237.0	1634.06	171.55	244.0	1682.32	176.47	251.0	1730.58
161.78	230.1	1586.48	166.70	237.1	1634.75	171.62	244.1	1683.01	176.54	251.1	1731.27
161.85	230.2	1587.17	166.77	237.2	1635.44	171.69	244.2	1683.70	176.61	251.2	1731.96
161.92	230.3	1587.86	166.84	237.3	1636.13	171.76	244.3	1684.39	176.68	251.3	1732.65
161.99	230.4	1588.55	166.91	237.4	1636.82	171.83	244.4	1685.08	176.75	251.4	1733.34
162.06	230.5	1589.24	166.98	237.5	1637.50	171.90	244.5	1685.77	176.82	251.5	1734.03
162.13	230.6	1589.93	167.05	237.6	1638.19	171.97	244.6	1686.46	176.89	251.6	1734.72
162.20	230.7	1590.62	167.12	237.7	1638.88	172.04	244.7	1687.15	176.96	251.7	1735.41
162.27	230.8	1591.31	167.19	237.8	1639.57	172.11	244.8	1687.84	177.03	251.8	1736.10
162.34	230.9	1592.00	167.26	237.9	1640.26	172.18	244.9	1688.53	177.10	251.9	1736.79

Table 4–16 (continued)
STRESS CONVERSION

Metric (KG/mm²)	English (KSI)	SI (MN/m²)	Metric (KG/mm²)	English (KSI)	SI (MN/m²)	Metric (KG/mm²)	English (KSI)	SI (MN/m²)	Metric (KG/mm²)	English (KSI)	SI (MN/m²)
177.17	252.0	1737.48	182.10	259.0	1785.74	187.02	266.0	1834.01	191.94	273.0	1882.27
177.24	252.1	1738.17	182.17	259.1	1786.43	187.09	266.1	1834.69	192.01	273.1	1882.96
177.31	252.2	1738.86	182.24	259.2	1787.12	187.16	266.2	1835.38	192.08	273.2	1883.65
177.38	252.3	1739.55	182.31	259.3	1787.81	187.23	266.3	1836.07	192.15	273.3	1884.34
177.45	252.4	1740.24	182.38	259.4	1788.50	187.30	266.4	1836.76	192.22	273.4	1885.03
177.53	252.5	1740.93	182.45	259.5	1789.19	187.37	266.5	1837.45	192.29	273.5	1885.72
177.60	252.6	1741.62	182.52	259.6	1789.88	187.44	266.6	1838.14	192.36	273.6	1886.41
177.67	252.7	1742.31	182.59	259.7	1790.57	187.51	266.7	1838.83	192.43	273.7	1887.09
177.74	252.8	1742.99	182.66	259.8	1791.26	187.58	266.8	1839.52	192.50	273.8	1887.78
177.81	252.9	1743.68	182.73	259.9	1791.95	187.65	266.9	1840.21	192.57	273.9	1888.47
177.88	253.0	1744.37	182.80	260.0	1792.64	187.72	267.0	1840.90	192.64	274.0	1889.16
177.95	253.1	1745.06	182.87	260.1	1793.33	187.79	267.1	1841.59	192.71	274.1	1889.85
178.02	253.2	1745.75	182.94	260.2	1794.02	187.86	267.2	1842.28	192.78	274.2	1890.54
178.09	253.3	1746.44	183.01	260.3	1794.71	187.93	267.3	1842.97	192.85	274.3	1891.23
178.16	253.4	1747.13	183.08	260.4	1795.39	188.00	267.4	1843.66	192.92	274.4	1891.92
178.23	253.5	1747.82	183.15	260.5	1796.08	188.07	267.5	1844.35	192.99	274.5	1892.61
178.30	253.6	1748.51	183.22	260.6	1796.77	188.14	267.6	1845.04	193.06	274.6	1893.30
178.37	253.7	1749.20	183.29	260.7	1797.46	188.21	267.7	1845.73	193.13	274.7	1893.99
178.44	253.8	1749.89	183.36	260.8	1798.15	188.28	267.8	1846.42	193.20	274.8	1894.68
178.51	253.9	1750.58	183.43	260.9	1798.84	188.35	267.9	1847.11	193.27	274.9	1895.37
178.58	254.0	1751.27	183.50	261.0	1799.53	188.42	268.0	1847.79	193.34	275.0	1896.06
178.65	254.1	1751.96	183.57	261.1	1800.22	188.49	268.1	1848.48	193.41	275.1	1896.75
178.72	254.2	1752.65	183.64	261.2	1800.91	188.56	268.2	1849.17	193.48	275.2	1897.44
178.79	254.3	1753.34	183.71	261.3	1801.60	188.63	268.3	1849.86	193.56	275.3	1898.13
178.86	254.4	1754.03	183.78	261.4	1802.29	188.70	268.4	1850.55	193.63	275.4	1898.82
178.93	254.5	1754.72	183.85	261.5	1802.98	188.77	268.5	1851.24	193.70	275.5	1899.51
179.00	254.6	1755.41	183.92	261.6	1803.67	188.84	268.6	1851.93	193.77	275.6	1900.20
179.07	254.7	1756.09	183.99	261.7	1804.36	188.91	268.7	1852.62	193.84	275.7	1900.88
179.14	254.8	1756.78	184.06	261.8	1805.05	188.98	268.8	1853.31	193.91	275.8	1901.57
179.21	254.9	1757.47	184.13	261.9	1805.74	189.06	268.9	1854.00	193.98	275.9	1902.26
179.28	255.0	1758.16	184.20	262.0	1806.43	189.13	269.0	1854.69	194.05	276.0	1902.95
179.35	255.1	1758.85	184.27	262.1	1807.12	189.20	269.1	1855.38	194.12	276.1	1903.64
179.42	255.2	1759.54	184.34	262.2	1807.81	189.27	269.2	1856.07	194.19	276.2	1904.33
179.49	255.3	1760.23	184.42	262.3	1808.49	189.34	269.3	1856.76	194.26	276.3	1905.02
179.56	255.4	1760.92	184.49	262.4	1809.18	189.41	269.4	1857.45	194.33	276.4	1905.71
179.63	255.5	1761.61	184.56	262.5	1809.87	189.48	269.5	1858.14	194.40	276.5	1906.40
179.70	255.6	1762.30	184.63	262.6	1810.56	189.55	269.6	1858.83	194.47	276.6	1907.09
179.77	255.7	1762.99	184.70	262.7	1811.25	189.62	269.7	1859.52	194.54	276.7	1907.78
179.85	255.8	1763.68	184.77	262.8	1811.94	189.69	269.8	1860.21	194.61	276.8	1908.47
179.92	255.9	1764.37	184.84	262.9	1812.63	189.76	269.9	1860.89	194.68	276.9	1909.16
179.99	256.0	1765.06	184.91	263.0	1813.32	189.83	270.0	1861.58	194.75	277.0	1909.85
180.06	256.1	1765.75	184.98	263.1	1814.01	189.90	270.1	1862.27	194.82	277.1	1910.54
180.13	256.2	1766.44	185.05	263.2	1814.70	189.97	270.2	1862.96	194.89	277.2	1911.23
180.20	256.3	1767.13	185.12	263.3	1815.39	190.04	270.3	1863.65	194.96	277.3	1911.92
180.27	256.4	1767.82	185.19	263.4	1816.08	190.11	270.4	1864.34	195.03	277.4	1912.61
180.34	256.5	1768.51	185.26	263.5	1816.77	190.18	270.5	1865.03	195.10	277.5	1913.30
180.41	256.6	1769.19	185.33	263.6	1817.46	190.25	270.6	1865.72	195.17	277.6	1913.98
180.48	256.7	1769.88	185.40	263.7	1818.15	190.32	270.7	1866.41	195.24	277.7	1914.67
180.55	256.8	1770.57	185.47	263.8	1818.84	190.39	270.8	1867.10	195.31	277.8	1915.36
180.62	256.9	1771.26	185.54	263.9	1819.53	190.46	270.9	1867.79	195.38	277.9	1916.05
180.69	257.0	1771.95	185.61	264.0	1820.22	190.53	271.0	1868.48	195.45	278.0	1916.74
180.76	257.1	1772.64	185.68	264.1	1820.91	190.60	271.1	1869.17	195.52	278.1	1917.43
180.83	257.2	1773.33	185.75	264.2	1821.59	190.67	271.2	1869.86	195.59	278.2	1918.12
180.90	257.3	1774.02	185.82	264.3	1822.28	190.74	271.3	1870.55	195.66	278.3	1918.81
180.97	257.4	1774.71	185.89	264.4	1822.97	190.81	271.4	1871.24	195.73	278.4	1919.50
181.04	257.5	1775.40	185.96	264.5	1823.66	190.88	271.5	1871.93	195.80	278.5	1920.19
181.11	257.6	1776.09	186.03	264.6	1824.35	190.95	271.6	1872.62	195.88	278.6	1920.88
181.18	257.7	1776.78	186.10	264.7	1825.04	191.02	271.7	1873.31	195.95	278.7	1921.57
181.25	257.8	1777.47	186.17	264.8	1825.73	191.09	271.8	1873.99	196.02	278.8	1922.26
181.32	257.9	1778.16	186.24	264.9	1826.42	191.16	271.9	1874.68	196.09	278.9	1922.95
181.39	258.0	1778.85	186.31	265.0	1827.11	191.23	272.0	1875.37	196.16	279.0	1923.64
181.46	258.1	1779.54	186.38	265.1	1827.80	191.31	272.1	1876.06	196.23	279.1	1924.33
181.53	258.2	1780.23	186.45	265.2	1828.49	191.38	272.2	1876.75	196.30	279.2	1925.02
181.60	258.3	1780.92	186.52	265.3	1829.18	191.45	272.3	1877.44	196.37	279.3	1925.71
181.67	258.4	1781.61	186.59	265.4	1829.87	191.52	272.4	1878.13	196.44	279.4	1926.40
181.74	258.5	1782.29	186.66	265.5	1830.56	191.59	272.5	1878.82	196.51	279.5	1927.08
181.81	258.6	1782.98	186.74	265.6	1831.25	191.66	272.6	1879.51	196.58	279.6	1927.77
181.88	258.7	1783.67	186.81	265.7	1831.94	191.73	272.7	1880.20	196.65	279.7	1928.46
181.95	258.8	1784.36	186.88	265.8	1832.63	191.80	272.8	1880.89	196.72	279.8	1929.15
182.02	258.9	1785.05	186.95	265.9	1833.32	191.87	272.9	1881.58	196.79	279.9	1929.84

Table 4–16 (continued)
STRESS CONVERSION

Metric (KG/mm²)	English (KSI)	SI (MN/m²)	Metric (KG/mm²)	English (KSI)	SI (MN/m²)	Metric (KG/mm²)	English (KSI)	SI (MN/m²)	Metric (KG/mm²)	English (KSI)	SI (MN/m²)
196.86	280.0	1930.53	201.78	287.0	1978.80	206.70	294.0	2027.06	211.42	301.0	2075.32
196.93	280.1	1931.22	201.85	287.1	1979.48	206.77	294.1	2027.75	211.69	301.1	2076.01
197.00	280.2	1931.91	201.92	287.2	1980.17	206.84	294.2	2028.44	211.76	301.2	2076.70
197.07	280.3	1932.60	201.99	287.3	1980.86	206.91	294.3	2029.13	211.83	301.3	2077.39
197.14	280.4	1933.29	202.06	287.4	1981.55	206.98	294.4	2029.82	211.91	301.4	2078.08
197.21	280.5	1933.98	202.13	287.5	1982.24	207.05	294.5	2030.51	211.98	301.5	2078.77
197.28	280.6	1934.67	202.20	287.6	1982.93	207.12	294.6	2031.20	212.05	301.6	2079.46
197.35	280.7	1935.36	202.27	287.7	1983.62	207.19	294.7	2031.88	212.12	301.7	2080.15
197.42	280.8	1936.05	202.34	287.8	1984.31	207.26	294.8	2032.57	212.19	301.8	2080.84
197.49	280.9	1936.74	202.41	287.9	1985.00	207.34	294.9	2033.26	212.26	301.9	2081.53
197.56	281.0	1937.43	202.48	288.0	1985.69	207.41	295.0	2033.95	212.33	302.0	2082.22
197.63	281.1	1938.12	202.55	288.1	1986.38	207.48	295.1	2034.64	212.40	302.1	2082.91
197.70	281.2	1938.81	202.62	288.2	1987.07	207.55	295.2	2035.33	212.47	302.2	2083.60
197.77	281.3	1939.50	202.69	288.3	1987.76	207.62	295.3	2036.02	212.54	302.3	2084.29
197.84	281.4	1940.18	202.77	288.4	1988.45	207.69	295.4	2036.71	212.61	302.4	2084.97
197.91	281.5	1940.87	202.84	288.5	1989.14	207.76	295.5	2037.40	212.68	302.5	2085.66
197.98	281.6	1941.56	202.91	288.6	1989.83	207.83	295.6	2038.09	212.75	302.6	2086.35
198.05	281.7	1942.25	202.98	288.7	1990.52	207.90	295.7	2038.78	212.82	302.7	2087.04
198.13	281.8	1942.94	203.05	288.8	1991.21	207.97	295.8	2039.47	212.89	302.8	2087.73
198.20	281.9	1943.63	203.12	288.9	1991.90	208.04	295.9	2040.16	212.96	302.9	2088.42
198.27	282.0	1944.32	203.19	289.0	1992.58	208.11	296.0	2040.85	213.03	303.0	2089.11
198.34	282.1	1945.01	203.26	289.1	1993.27	208.18	296.1	2041.54	213.10	303.1	2089.80
198.41	282.2	1945.70	203.33	289.2	1993.96	208.25	296.2	2042.23	213.17	303.2	2090.49
198.48	282.3	1946.39	203.40	289.3	1994.65	208.32	296.3	2042.92	213.24	303.3	2091.18
198.55	282.4	1947.08	203.47	289.4	1995.34	208.39	296.4	2043.61	213.31	303.4	2091.87
198.62	282.5	1947.77	203.54	289.5	1996.03	208.46	296.5	2044.30	213.38	303.5	2092.56
198.69	282.6	1948.46	203.61	289.6	1996.72	208.53	296.6	2044.98	213.45	303.6	2093.25
198.76	282.7	1949.15	203.68	289.7	1997.41	208.60	296.7	2045.67	213.52	303.7	2093.94
198.83	282.8	1949.84	203.75	289.8	1998.10	208.67	296.8	2046.36	213.59	303.8	2094.63
198.90	282.9	1950.53	203.82	289.9	1998.79	208.74	296.9	2047.05	213.66	303.9	2095.32
198.97	283.0	1951.22	203.89	290.0	1999.48	208.81	297.0	2047.74	213.73	304.0	2096.01
199.04	283.1	1951.91	203.96	290.1	2000.17	208.88	297.1	2048.43	213.80	304.1	2096.70
199.11	283.2	1952.60	204.03	290.2	2000.86	208.95	297.2	2049.12	213.87	304.2	2097.39
199.18	283.3	1953.28	204.10	290.3	2001.55	209.02	297.3	2049.81	213.94	304.3	2098.07
199.25	283.4	1953.97	204.17	290.4	2002.24	209.09	297.4	2050.50	214.01	304.4	2098.76
199.32	283.5	1954.66	204.24	290.5	2002.93	209.16	297.5	2051.19	214.08	304.5	2099.45
199.39	283.6	1955.35	204.31	290.6	2003.62	209.23	297.6	2051.88	214.15	304.6	2100.14
199.46	283.7	1956.04	204.38	290.7	2004.31	209.30	297.7	2052.57	214.23	304.7	2100.83
199.53	283.8	1956.73	204.45	290.8	2005.00	209.37	297.8	2053.26	214.30	304.8	2101.52
199.60	283.9	1957.42	204.52	290.9	2005.68	209.44	297.9	2053.95	214.37	304.9	2102.21
199.67	284.0	1958.11	204.59	291.0	2006.37	209.51	298.0	2054.64	214.44	305.0	2102.90
199.74	284.1	1958.80	204.66	291.1	2007.06	209.59	298.1	2055.33	214.51	305.1	2103.59
199.81	284.2	1959.49	204.73	291.2	2007.75	209.66	298.2	2056.02	214.58	305.2	2104.28
199.88	284.3	1960.18	204.80	291.3	2008.44	209.73	298.3	2056.71	214.65	305.3	2104.97
199.95	284.4	1960.87	204.87	291.4	2009.13	209.80	298.4	2057.40	214.72	305.4	2105.66
200.02	284.5	1961.56	204.94	291.5	2009.82	209.87	298.5	2058.08	214.79	305.5	2106.35
200.09	284.6	1962.25	205.02	291.6	2010.51	209.94	298.6	2058.77	214.86	305.6	2107.04
200.16	284.7	1962.94	205.09	291.7	2011.20	210.01	298.7	2059.46	214.93	305.7	2107.73
200.23	284.8	1963.63	205.16	291.8	2011.89	210.08	298.8	2060.15	215.00	305.8	2108.42
200.30	284.9	1964.32	205.23	291.9	2012.58	210.15	298.9	2060.84	215.07	305.9	2109.11
200.37	285.0	1965.01	205.30	292.0	2013.27	210.22	299.0	2061.53	215.14	306.0	2109.80
200.45	285.1	1965.70	205.37	292.1	2013.96	210.29	299.1	2062.22	215.21	306.1	2110.49
200.52	285.2	1966.38	205.44	292.2	2014.65	210.36	299.2	2062.91	215.28	306.2	2111.17
200.59	285.3	1967.07	205.51	292.3	2015.34	210.43	299.3	2063.60	215.35	306.3	2111.86
200.66	285.4	1967.76	205.58	292.4	2016.03	210.50	299.4	2064.29	215.42	306.4	2112.55
200.73	285.5	1968.45	205.65	292.5	2016.72	210.57	299.5	2064.98	215.49	306.5	2113.24
200.80	285.6	1969.14	205.72	292.6	2017.41	210.64	299.6	2065.67	215.56	306.6	2113.93
200.87	285.7	1969.83	205.79	292.7	2018.10	210.71	299.7	2066.36	215.63	306.7	2114.62
200.94	285.8	1970.52	205.86	292.8	2018.78	210.78	299.8	2067.05	215.70	306.8	2115.31
201.01	285.9	1971.21	205.93	292.9	2019.47	210.85	299.9	2067.74	215.77	306.9	2116.00
201.08	286.0	1971.90	206.00	293.0	2020.16	210.92	300.0	2068.43	215.84	307.0	2116.69
201.15	286.1	1972.59	206.07	293.1	2020.85	210.99	300.1	2069.12	215.91	307.1	2117.38
201.22	286.2	1973.28	206.14	293.2	2021.54	211.06	300.2	2069.81	215.98	307.2	2118.07
201.29	286.3	1973.97	206.21	293.3	2022.23	211.13	300.3	2070.50	216.05	307.3	2118.76
201.36	286.4	1974.66	206.28	293.4	2022.92	211.20	300.4	2071.19	216.12	307.4	2119.45
201.43	286.5	1975.35	206.35	293.5	2023.61	211.27	300.5	2071.87	216.19	307.5	2120.14
201.50	286.6	1976.04	206.42	293.6	2024.30	211.34	300.6	2072.56	216.26	307.6	2120.83
201.57	286.7	1976.73	206.49	293.7	2024.99	211.41	300.7	2073.25	216.33	307.7	2121.52
201.64	286.8	1977.42	206.56	293.8	2025.68	211.48	300.8	2073.94	216.40	307.8	2122.21
201.71	286.9	1978.11	206.63	293.9	2026.37	211.55	300.9	2074.63	216.48	307.9	2122.90

Table 4-16 (continued)
STRESS CONVERSION

Metric (KG/mm²)	English (KSI)	SI (MN/m²)	Metric (KG/mm²)	English (KSI)	SI (MN/m²)	Metric (KG/mm²)	English (KSI)	SI (MN/m²)	Metric (KG/mm²)	English (KSI)	SI (MN/m²)
216.55	308.0	2123.59	221.47	315.0	2171.85	226.39	322.0	2220.11	231.31	329.0	2268.38
216.62	308.1	2124.27	221.54	315.1	2172.54	226.46	322.1	2220.80	231.38	329.1	2269.06
216.69	308.2	2124.96	221.61	315.2	2173.23	226.53	322.2	2221.49	231.45	329.2	2269.75
216.76	308.3	2125.65	221.68	315.3	2173.92	226.60	322.3	2222.18	231.52	329.3	2270.44
216.83	308.4	2126.34	221.75	315.4	2174.61	226.67	322.4	2222.87	231.59	329.4	2271.13
216.90	308.5	2127.03	221.82	315.5	2175.30	226.74	322.5	2223.56	231.66	329.5	2271.82
216.97	308.6	2127.72	221.89	315.6	2175.99	226.81	322.6	2224.25	231.73	329.6	2272.51
217.04	308.7	2128.41	221.96	315.7	2176.67	226.88	322.7	2224.94	231.80	329.7	2273.20
217.11	308.8	2129.10	222.03	315.8	2177.36	226.95	322.8	2225.63	231.87	329.8	2273.89
217.18	308.9	2129.79	222.10	315.9	2178.05	227.02	322.9	2226.32	231.94	329.9	2274.58
217.25	309.0	2130.48	222.17	316.0	2178.74	227.09	323.0	2227.01	232.01	330.0	2275.27
217.32	309.1	2131.17	222.24	316.1	2179.43	227.16	323.1	2227.70	232.08	330.1	2275.96
217.39	309.2	2131.86	222.31	316.2	2180.12	227.23	323.2	2228.39	232.15	330.2	2276.65
217.46	309.3	2132.55	222.38	316.3	2180.81	227.30	323.3	2229.07	232.22	330.3	2277.34
217.53	309.4	2133.24	222.45	316.4	2181.50	227.37	323.4	2229.76	232.29	330.4	2278.03
217.60	309.5	2133.93	222.52	316.5	2182.19	227.44	323.5	2230.45	232.36	330.5	2278.72
217.67	309.6	2134.62	222.59	316.6	2182.88	227.51	323.6	2231.14	232.43	330.6	2279.41
217.74	309.7	2135.31	222.66	316.7	2183.57	227.58	323.7	2231.83	232.51	330.7	2280.10
217.81	309.8	2136.00	222.73	316.8	2184.26	227.65	323.8	2232.52	232.58	330.8	2280.79
217.88	309.9	2136.69	222.80	316.9	2184.95	227.72	323.9	2233.21	232.65	330.9	2281.48
217.95	310.0	2137.37	222.87	317.0	2185.64	227.79	324.0	2233.90	232.72	331.0	2282.16
218.02	310.1	2138.06	222.94	317.1	2186.33	227.86	324.1	2234.59	232.79	331.1	2282.85
218.09	310.2	2138.75	223.01	317.2	2187.02	227.94	324.2	2235.28	232.86	331.2	2283.54
218.16	310.3	2139.44	223.08	317.3	2187.71	228.01	324.3	2235.97	232.93	331.3	2284.23
218.23	310.4	2140.13	223.15	317.4	2188.40	228.08	324.4	2236.66	233.00	331.4	2284.92
218.30	310.5	2140.82	223.22	317.5	2189.09	228.15	324.5	2237.35	233.07	331.5	2285.61
218.37	310.6	2141.51	223.29	317.6	2189.77	228.22	324.6	2238.04	233.14	331.6	2286.30
218.44	310.7	2142.20	223.37	317.7	2190.46	228.29	324.7	2238.73	233.21	331.7	2286.99
218.51	310.8	2142.89	223.44	317.8	2191.15	228.36	324.8	2239.42	233.28	331.8	2287.68
218.58	310.9	2143.58	223.51	317.9	2191.84	228.43	324.9	2240.11	233.35	331.9	2288.37
218.65	311.0	2144.27	223.58	318.0	2192.53	228.50	325.0	2240.80	233.42	332.0	2289.06
218.72	311.1	2144.96	223.65	318.1	2193.22	228.57	325.1	2241.49	233.49	332.1	2289.75
218.80	311.2	2145.65	223.72	318.2	2193.91	228.64	325.2	2242.17	233.56	332.2	2290.44
218.87	311.3	2146.34	223.79	318.3	2194.60	228.71	325.3	2242.86	233.63	332.3	2291.13
218.94	311.4	2147.03	223.86	318.4	2195.29	228.78	325.4	2243.55	233.70	332.4	2291.82
219.01	311.5	2147.72	223.93	318.5	2195.98	228.85	325.5	2244.24	233.77	332.5	2292.51
219.08	311.6	2148.41	224.00	318.6	2196.67	228.92	325.6	2244.93	233.84	332.6	2293.20
219.15	311.7	2149.10	224.07	318.7	2197.36	228.99	325.7	2245.62	233.91	332.7	2293.89
219.22	311.8	2149.79	224.14	318.8	2198.05	229.06	325.8	2246.31	233.98	332.8	2294.58
219.29	311.9	2150.47	224.21	318.9	2198.74	229.13	325.9	2247.00	234.05	332.9	2295.26
219.36	312.0	2151.16	224.28	319.0	2199.43	229.20	326.0	2247.69	234.12	333.0	2295.95
219.43	312.1	2151.85	224.35	319.1	2200.12	229.27	326.1	2248.38	234.19	333.1	2296.64
219.50	312.2	2152.54	224.42	319.2	2200.81	229.34	326.2	2249.07	234.26	333.2	2297.33
219.57	312.3	2153.23	224.49	319.3	2201.50	229.41	326.3	2249.76	234.33	333.3	2298.02
219.64	312.4	2153.92	224.56	319.4	2202.19	229.48	326.4	2250.45	234.40	333.4	2298.71
219.71	312.5	2154.61	224.63	319.5	2202.87	229.55	326.5	2251.14	234.47	333.5	2299.40
219.78	312.6	2155.30	224.70	319.6	2203.56	229.62	326.6	2251.83	234.54	333.6	2300.09
219.85	312.7	2155.99	224.77	319.7	2204.25	229.69	326.7	2252.52	234.61	333.7	2300.78
219.92	312.8	2156.68	224.84	319.8	2204.94	229.76	326.8	2253.21	234.68	333.8	2301.47
219.99	312.9	2157.37	224.91	319.9	2205.63	229.83	326.9	2253.90	234.75	333.9	2302.16
220.06	313.0	2158.06	224.98	320.0	2206.32	229.90	327.0	2254.59	234.83	334.0	2302.85
220.13	313.1	2158.75	225.05	320.1	2207.01	229.97	327.1	2255.28	234.90	334.1	2303.54
220.20	313.2	2159.44	225.12	320.2	2207.70	230.04	327.2	2255.96	234.97	334.2	2304.23
220.27	313.3	2160.13	225.19	320.3	2208.39	230.11	327.3	2256.65	235.04	334.3	2304.92
220.34	313.4	2160.82	225.26	320.4	2209.08	230.18	327.4	2257.34	235.11	334.4	2305.61
220.41	313.5	2161.51	225.33	320.5	2209.77	230.26	327.5	2258.03	235.18	334.5	2306.30
220.48	313.6	2162.20	225.40	320.6	2210.46	230.33	327.6	2258.72	235.25	334.6	2306.99
220.55	313.7	2162.89	225.47	320.7	2211.15	230.40	327.7	2259.41	235.32	334.7	2307.68
220.62	313.8	2163.57	225.54	320.8	2211.84	230.47	327.8	2260.10	235.39	334.8	2308.36
220.69	313.9	2164.26	225.62	320.9	2212.53	230.54	327.9	2260.79	235.46	334.9	2309.05
220.76	314.0	2164.95	225.69	321.0	2213.22	230.61	328.0	2261.48	235.53	335.0	2309.74
220.83	314.1	2165.64	225.76	321.1	2213.91	230.68	328.1	2262.17	235.60	335.1	2310.43
220.90	314.2	2166.33	225.83	321.2	2214.60	230.75	328.2	2262.86	235.67	335.2	2311.12
220.97	314.3	2167.02	225.90	321.3	2215.29	230.82	328.3	2263.55	235.74	335.3	2311.81
221.05	314.4	2167.71	225.97	321.4	2215.97	230.89	328.4	2264.24	235.81	335.4	2312.50
221.12	314.5	2168.40	226.04	321.5	2216.66	230.96	328.5	2264.93	235.88	335.5	2313.19
221.19	314.6	2169.09	226.11	321.6	2217.35	231.03	328.6	2265.62	235.95	335.6	2313.88
221.26	314.7	2169.78	226.18	321.7	2218.04	231.10	328.7	2266.31	236.02	335.7	2314.57
221.33	314.8	2170.47	226.25	321.8	2218.73	231.17	328.8	2267.00	236.09	335.8	2315.26
221.40	314.9	2171.16	226.32	321.9	2219.42	231.24	328.9	2267.69	236.16	335.9	2315.95

Table 4–16 (continued)
STRESS CONVERSION

Metric (KG/mm²)	English (KSI)	SI (MN/m²)	Metric (KG/mm²)	English (KSI)	SI (MN/m²)	Metric (KG/mm²)	English (KSI)	SI (MN/m²)	Metric (KG/mm²)	English (KSI)	SI (MN/m²)
236.23	336.0	2316.64	241.15	343.0	2364.90	246.07	350.0	2413.16	251.00	357.0	2461.43
236.30	336.1	2317.33	241.22	343.1	2365.59	246.14	350.1	2413.85	251.07	357.1	2462.12
236.37	336.2	2318.02	241.29	343.2	2366.28	246.21	350.2	2414.54	251.14	357.2	2462.81
236.44	336.3	2318.71	241.36	343.3	2366.97	246.29	350.3	2415.23	251.21	357.3	2463.50
236.51	336.4	2319.40	241.43	343.4	2367.66	246.36	350.4	2415.92	251.28	357.4	2464.19
236.58	336.5	2320.09	241.50	343.5	2368.35	246.43	350.5	2416.61	251.35	357.5	2464.88
236.65	336.6	2320.78	241.57	343.6	2369.04	246.50	350.6	2417.30	251.42	357.6	2465.57
236.72	336.7	2321.46	241.65	343.7	2369.73	246.57	350.7	2417.99	251.49	357.7	2466.25
236.79	336.8	2322.15	241.72	343.8	2370.42	246.64	350.8	2418.68	251.56	357.8	2466.94
236.86	336.9	2322.84	241.79	343.9	2371.11	246.71	350.9	2419.37	251.63	357.9	2467.63
236.93	337.0	2323.53	241.86	344.0	2371.80	246.78	351.0	2420.06	251.70	358.0	2468.32
237.00	337.1	2324.22	241.93	344.1	2372.49	246.85	351.1	2420.75	251.77	358.1	2469.01
237.08	337.2	2324.91	242.00	344.2	2373.18	246.92	351.2	2421.44	251.84	358.2	2469.70
237.15	337.3	2325.60	242.07	344.3	2373.86	246.99	351.3	2422.13	251.91	358.3	2470.39
237.22	337.4	2326.29	242.14	344.4	2374.55	247.06	351.4	2422.82	251.98	358.4	2471.08
237.29	337.5	2326.98	242.21	344.5	2375.24	247.13	351.5	2423.51	252.05	358.5	2471.77
237.36	337.6	2327.67	242.28	344.6	2375.93	247.20	351.6	2424.20	252.12	358.6	2472.46
237.43	337.7	2328.36	242.35	344.7	2376.62	247.27	351.7	2424.89	252.19	358.7	2473.15
237.50	337.8	2329.05	242.42	344.8	2377.31	247.34	351.8	2425.58	252.26	358.8	2473.84
237.57	337.9	2329.74	242.49	344.9	2378.00	247.41	351.9	2426.26	252.33	358.9	2474.53
237.64	338.0	2330.43	242.56	345.0	2378.69	247.48	352.0	2426.95	252.40	359.0	2475.22
237.71	338.1	2331.12	242.63	345.1	2379.38	247.55	352.1	2427.64	252.47	359.1	2475.91
237.78	338.2	2331.81	242.70	345.2	2380.07	247.62	352.2	2428.33	252.54	359.2	2476.60
237.85	338.3	2332.50	242.77	345.3	2380.76	247.69	352.3	2429.02	252.61	359.3	2477.29
237.92	338.4	2333.19	242.84	345.4	2381.45	247.76	352.4	2429.71	252.68	359.4	2477.98
237.99	338.5	2333.88	242.91	345.5	2382.14	247.83	352.5	2430.40	252.75	359.5	2478.67
238.06	338.6	2334.56	242.98	345.6	2382.83	247.90	352.6	2431.09	252.82	359.6	2479.35
238.13	338.7	2335.25	243.05	345.7	2383.52	247.97	352.7	2431.78	252.89	359.7	2480.04
238.20	338.8	2335.94	243.12	345.8	2384.21	248.04	352.8	2432.47	252.96	359.8	2480.73
238.27	338.9	2336.63	243.19	345.9	2384.90	248.11	352.9	2433.16	253.03	359.9	2481.42
238.34	339.0	2337.32	243.26	346.0	2385.59	248.18	353.0	2433.85	253.11	360.0	2482.11
238.41	339.1	2338.01	243.33	346.1	2386.28	248.25	353.1	2434.54	253.18	360.1	2482.80
238.48	339.2	2338.70	243.40	346.2	2386.96	248.32	353.2	2435.23	253.25	360.2	2483.49
238.55	339.3	2339.39	243.47	346.3	2387.65	248.39	353.3	2435.92	253.32	360.3	2484.18
238.62	339.4	2340.08	243.54	346.4	2388.34	248.46	353.4	2436.61	253.39	360.4	2484.87
238.69	339.5	2340.77	243.61	346.5	2389.03	248.54	353.5	2437.30	253.46	360.5	2485.56
238.76	339.6	2341.46	243.68	346.6	2389.72	248.61	353.6	2437.99	253.53	360.6	2486.25
238.83	339.7	2342.15	243.75	346.7	2390.41	248.68	353.7	2438.68	253.60	360.7	2486.94
238.90	339.8	2342.84	243.82	346.8	2391.10	248.75	353.8	2439.37	253.67	360.8	2487.63
238.97	339.9	2343.53	243.89	346.9	2391.79	248.82	353.9	2440.05	253.74	360.9	2488.32
239.04	340.0	2344.22	243.97	347.0	2392.48	248.89	354.0	2440.74	253.81	361.0	2489.01
239.11	340.1	2344.91	244.04	347.1	2393.17	248.96	354.1	2441.43	253.88	361.1	2489.70
239.18	340.2	2345.60	244.11	347.2	2393.86	249.03	354.2	2442.12	253.95	361.2	2490.39
239.25	340.3	2346.29	244.18	347.3	2394.55	249.10	354.3	2442.81	254.02	361.3	2491.08
239.32	340.4	2346.98	244.25	347.4	2395.24	249.17	354.4	2443.50	254.09	361.4	2491.77
239.40	340.5	2347.66	244.32	347.5	2395.93	249.24	354.5	2444.19	254.16	361.5	2492.45
239.47	340.6	2348.35	244.39	347.6	2396.62	249.31	354.6	2444.88	254.23	361.6	2493.14
239.54	340.7	2349.04	244.46	347.7	2397.31	249.38	354.7	2445.57	254.30	361.7	2493.83
239.61	340.8	2349.73	244.53	347.8	2398.00	249.45	354.8	2446.26	254.37	361.8	2494.52
239.68	340.9	2350.42	244.60	347.9	2398.69	249.52	354.9	2446.95	254.44	361.9	2495.21
239.75	341.0	2351.11	244.67	348.0	2399.38	249.59	355.0	2447.64	254.51	362.0	2495.90
239.82	341.1	2351.80	244.74	348.1	2400.06	249.66	355.1	2448.33	254.58	362.1	2496.59
239.89	341.2	2352.49	244.81	348.2	2400.75	249.73	355.2	2449.02	254.65	362.2	2497.28
239.96	341.3	2353.18	244.88	348.3	2401.44	249.80	355.3	2449.71	254.72	362.3	2497.97
240.03	341.4	2353.87	244.95	348.4	2402.13	249.87	355.4	2450.40	254.79	362.4	2498.66
240.10	341.5	2354.56	245.02	348.5	2402.82	249.94	355.5	2451.09	254.86	362.5	2499.35
240.17	341.6	2355.25	245.09	348.6	2403.51	250.01	355.6	2451.78	254.93	362.6	2500.04
240.24	341.7	2355.94	245.16	348.7	2404.20	250.08	355.7	2452.47	255.00	362.7	2500.73
240.31	341.8	2356.63	245.23	348.8	2404.89	250.15	355.8	2453.15	255.07	362.8	2501.42
240.38	341.9	2357.32	245.30	348.9	2405.58	250.22	355.9	2453.84	255.14	362.9	2502.11
240.45	342.0	2358.01	245.37	349.0	2406.27	250.29	356.0	2454.53	255.21	363.0	2502.80
240.52	342.1	2358.70	245.44	349.1	2406.96	250.36	356.1	2455.22	255.28	363.1	2503.49
240.59	342.2	2359.39	245.51	349.2	2407.65	250.43	356.2	2455.91	255.35	363.2	2504.18
240.66	342.3	2360.08	245.58	349.3	2408.34	250.50	356.3	2456.60	255.43	363.3	2504.87
240.73	342.4	2360.76	245.65	349.4	2409.03	250.57	356.4	2457.29	255.50	363.4	2505.55
240.80	342.5	2361.45	245.72	349.5	2409.72	250.64	356.5	2457.98	255.57	363.5	2506.24
240.87	342.6	2362.14	245.79	349.6	2410.41	250.71	356.6	2458.67	255.64	363.6	2506.93
240.94	342.7	2362.83	245.86	349.7	2411.10	250.78	356.7	2459.36	255.71	363.7	2507.62
241.01	342.8	2363.52	245.93	349.8	2411.79	250.86	356.8	2460.05	255.78	363.8	2508.31
241.08	342.9	2364.21	246.00	349.9	2412.48	250.93	356.9	2460.74	255.85	363.9	2509.00

Table 4–16 (continued)
STRESS CONVERSION

Metric (KG/mm²)	English (KSI)	SI (MN/m²)	Metric (KG/mm²)	English (KSI)	SI (MN/m²)	Metric (KG/mm²)	English (KSI)	SI (MN/m²)	Metric (KG/mm²)	English (KSI)	SI (MN/m²)
255.92	364.0	2509.69	260.84	371.0	2557.95	265.76	378.0	2606.22	270.68	385.0	2654.48
255.99	364.1	2510.38	260.91	371.1	2558.64	265.83	378.1	2606.91	270.75	385.1	2655.17
256.06	364.2	2511.07	260.98	371.2	2559.33	265.90	378.2	2607.60	270.82	385.2	2655.86
256.13	364.3	2511.76	261.05	371.3	2560.02	265.97	378.3	2608.29	270.89	385.3	2656.55
256.20	364.4	2512.45	261.12	371.4	2560.71	266.04	378.4	2608.98	270.96	385.4	2657.24
256.27	364.5	2513.14	261.19	371.5	2561.40	266.11	378.5	2609.67	271.03	385.5	2657.93
256.34	364.6	2513.83	261.26	371.6	2562.09	266.18	378.6	2610.36	271.10	385.6	2658.62
256.41	364.7	2514.52	261.33	371.7	2562.78	266.25	378.7	2611.04	271.17	385.7	2659.31
256.48	364.8	2515.21	261.40	371.8	2563.47	266.32	378.8	2611.73	271.24	385.8	2660.00
256.55	364.9	2515.90	261.47	371.9	2564.16	266.39	378.9	2612.42	271.31	385.9	2660.69
256.62	365.0	2516.59	261.54	372.0	2564.85	266.46	379.0	2613.11	271.38	386.0	2661.38
256.69	365.1	2517.28	261.61	372.1	2565.54	266.53	379.1	2613.80	271.46	386.1	2662.07
256.76	365.2	2517.97	261.68	372.2	2566.23	266.60	379.2	2614.49	271.53	386.2	2662.76
256.83	365.3	2518.65	261.75	372.3	2566.92	266.67	379.3	2615.18	271.60	386.3	2663.44
256.90	365.4	2519.34	261.82	372.4	2567.61	266.74	379.4	2615.87	271.67	386.4	2664.13
256.97	365.5	2520.03	261.89	372.5	2568.30	266.81	379.5	2616.56	271.74	386.5	2664.82
257.04	365.6	2520.72	261.96	372.6	2568.99	266.89	379.6	2617.25	271.81	386.6	2665.51
257.11	365.7	2521.41	262.03	372.7	2569.68	266.96	379.7	2617.94	271.88	386.7	2666.20
257.18	365.8	2522.10	262.10	372.8	2570.37	267.03	379.8	2618.63	271.95	386.8	2666.89
257.25	365.9	2522.79	262.17	372.9	2571.05	267.10	379.9	2619.32	272.02	386.9	2667.58
257.32	366.0	2523.48	262.24	373.0	2571.74	267.17	380.0	2620.01	272.09	387.0	2668.27
257.39	366.1	2524.17	262.32	373.1	2572.43	267.24	380.1	2620.70	272.16	387.1	2668.96
257.46	366.2	2524.86	262.39	373.2	2573.12	267.31	380.2	2621.39	272.23	387.2	2669.65
257.53	366.3	2525.55	262.46	373.3	2573.81	267.38	380.3	2622.08	272.30	387.3	2670.34
257.60	366.4	2526.24	262.53	373.4	2574.50	267.45	380.4	2622.77	272.37	387.4	2671.03
257.68	366.5	2526.93	262.60	373.5	2575.19	267.52	380.5	2623.46	272.44	387.5	2671.72
257.75	366.6	2527.62	262.67	373.6	2575.88	267.59	380.6	2624.14	272.51	387.6	2672.41
257.82	366.7	2528.31	262.74	373.7	2576.57	267.66	380.7	2624.83	272.58	387.7	2673.10
257.89	366.8	2529.00	262.81	373.8	2577.26	267.73	380.8	2625.52	272.65	387.8	2673.79
257.96	366.9	2529.69	262.88	373.9	2577.95	267.80	380.9	2626.21	272.72	387.9	2674.48
258.03	367.0	2530.38	262.95	374.0	2578.64	267.87	381.0	2626.90	272.79	388.0	2675.17
258.10	367.1	2531.07	263.02	374.1	2579.33	267.94	381.1	2627.59	272.86	388.1	2675.86
258.17	367.2	2531.75	263.09	374.2	2580.02	268.01	381.2	2628.28	272.93	388.2	2676.54
258.24	367.3	2532.44	263.16	374.3	2580.71	268.08	381.3	2628.97	273.00	388.3	2677.23
258.31	367.4	2533.13	263.23	374.4	2581.40	268.15	381.4	2629.66	273.07	388.4	2677.92
258.38	367.5	2533.82	263.30	374.5	2582.09	268.22	381.5	2630.35	273.14	388.5	2678.61
258.45	367.6	2534.51	263.37	374.6	2582.78	268.29	381.6	2631.04	273.21	388.6	2679.30
258.52	367.7	2535.20	263.44	374.7	2583.47	268.36	381.7	2631.73	273.28	388.7	2679.99
258.59	367.8	2535.89	263.51	374.8	2584.15	268.43	381.8	2632.42	273.35	388.8	2680.68
258.66	367.9	2536.58	263.58	374.9	2584.84	268.50	381.9	2633.11	273.42	388.9	2681.37
258.73	368.0	2537.27	263.65	375.0	2585.53	268.57	382.0	2633.80	273.49	389.0	2682.06
258.80	368.1	2537.96	263.72	375.1	2586.22	268.64	382.1	2634.49	273.56	389.1	2682.75
258.87	368.2	2538.65	263.79	375.2	2586.91	268.71	382.2	2635.18	273.63	389.2	2683.44
258.94	368.3	2539.34	263.86	375.3	2587.60	268.78	382.3	2635.87	273.70	389.3	2684.13
259.01	368.4	2540.03	263.93	375.4	2588.29	268.85	382.4	2636.56	273.78	389.4	2684.82
259.08	368.5	2540.72	264.00	375.5	2588.98	268.92	382.5	2637.24	273.85	389.5	2685.51
259.15	368.6	2541.41	264.07	375.6	2589.67	268.99	382.6	2637.93	273.92	389.6	2686.20
259.22	368.7	2542.10	264.14	375.7	2590.36	269.06	382.7	2638.62	273.99	389.7	2686.89
259.29	368.8	2542.79	264.21	375.8	2591.05	269.14	382.8	2639.31	274.06	389.8	2687.58
259.36	368.9	2543.48	264.28	375.9	2591.74	269.21	382.9	2640.00	274.13	389.9	2688.27
259.43	369.0	2544.17	264.35	376.0	2592.43	269.28	383.0	2640.69	274.20	390.0	2688.96
259.50	369.1	2544.85	264.42	376.1	2593.12	269.35	383.1	2641.38	274.27	390.1	2689.64
259.57	369.2	2545.54	264.49	376.2	2593.81	269.42	383.2	2642.07	274.34	390.2	2690.33
259.64	369.3	2546.23	264.57	376.3	2594.50	269.49	383.3	2642.76	274.41	390.3	2691.02
259.71	369.4	2546.92	264.64	376.4	2595.19	269.56	383.4	2643.45	274.48	390.4	2691.71
259.78	369.5	2547.61	264.71	376.5	2595.88	269.63	383.5	2644.14	274.55	390.5	2692.40
259.85	369.6	2548.30	264.78	376.6	2596.57	269.70	383.6	2644.83	274.62	390.6	2693.09
259.92	369.7	2548.99	264.85	376.7	2597.25	269.77	383.7	2645.52	274.69	390.7	2693.78
260.00	369.8	2549.68	264.92	376.8	2597.94	269.84	383.8	2646.21	274.76	390.8	2694.47
260.07	369.9	2550.37	264.99	376.9	2598.63	269.91	383.9	2646.90	274.83	390.9	2695.16
260.14	370.0	2551.06	265.06	377.0	2599.32	269.98	384.0	2647.59	274.90	391.0	2695.85
260.21	370.1	2551.75	265.13	377.1	2600.01	270.05	384.1	2648.28	274.97	391.1	2696.54
260.28	370.2	2552.44	265.20	377.2	2600.70	270.12	384.2	2648.97	275.04	391.2	2697.23
260.35	370.3	2553.13	265.27	377.3	2601.39	270.19	384.3	2649.66	275.11	391.3	2697.92
260.42	370.4	2553.82	265.34	377.4	2602.08	270.26	384.4	2650.34	275.18	391.4	2698.61
260.49	370.5	2554.51	265.41	377.5	2602.77	270.33	384.5	2651.03	275.25	391.5	2699.30
260.56	370.6	2555.20	265.48	377.6	2603.46	270.40	384.6	2651.72	275.32	391.6	2699.99
260.63	370.7	2555.89	265.55	377.7	2604.15	270.47	384.7	2652.41	275.39	391.7	2700.68
260.70	370.8	2556.58	265.62	377.8	2604.84	270.54	384.8	2653.10	275.46	391.8	2701.37
260.77	370.9	2557.27	265.69	377.9	2605.53	270.61	384.9	2653.79	275.53	391.9	2702.06

Table 4–16 (continued)
STRESS CONVERSION

Metric (KG/mm²)	English (KSI)	SI (MN/m²)	Metric (KG/mm²)	English (KSI)	SI (MN/m²)	Metric (KG/mm²)	English (KSI)	SI (MN/m²)	Metric (KG/mm²)	English (KSI)	SI (MN/m²)
275.60	392.0	2702.74	280.52	399.0	2751.01	285.45	406.0	2799.27	290.37	413.0	2847.53
275.67	392.1	2703.43	280.60	399.1	2751.70	285.52	406.1	2799.96	290.44	413.1	2848.22
275.74	392.2	2704.12	280.67	399.2	2752.39	285.59	406.2	2800.65	290.51	413.2	2848.91
275.81	392.3	2704.91	280.74	399.3	2753.08	285.66	406.3	2801.34	290.58	413.3	2849.60
275.88	392.4	2705.50	280.81	399.4	2753.77	285.73	406.4	2802.03	290.65	413.4	2850.29
275.95	392.5	2706.19	280.88	399.5	2754.46	285.80	406.5	2802.72	290.72	413.5	2850.98
276.03	392.6	2706.88	280.95	399.6	2755.14	285.87	406.6	2803.41	290.79	413.6	2851.67
276.10	392.7	2707.57	281.02	399.7	2755.83	285.94	406.7	2804.10	290.86	413.7	2852.36
276.17	392.8	2708.26	281.09	399.8	2756.52	286.01	406.8	2804.79	290.93	413.8	2853.05
276.24	392.9	2708.95	281.16	399.9	2757.21	286.08	406.9	2805.48	291.00	413.9	2853.74
276.31	393.0	2709.64	281.23	400.0	2757.90	286.15	407.0	2806.17	291.07	414.0	2854.43
276.38	393.1	2710.33	281.30	400.1	2758.59	286.22	407.1	2806.86	291.14	414.1	2855.12
276.45	393.2	2711.02	281.37	400.2	2759.28	286.29	407.2	2807.55	291.21	414.2	2855.81
276.52	393.3	2711.71	281.44	400.3	2759.97	286.36	407.3	2808.23	291.28	414.3	2856.50
276.59	393.4	2712.40	281.51	400.4	2760.66	286.43	407.4	2808.92	291.35	414.4	2857.19
276.66	393.5	2713.09	281.58	400.5	2761.35	286.50	407.5	2809.61	291.42	414.5	2857.88
276.73	393.6	2713.78	281.65	400.6	2762.04	286.57	407.6	2810.30	291.49	414.6	2858.57
276.80	393.7	2714.47	281.72	400.7	2762.73	286.64	407.7	2810.99	291.56	414.7	2859.26
276.87	393.8	2715.16	281.79	400.8	2763.42	286.71	407.8	2811.68	291.63	414.8	2859.95
276.94	393.9	2715.84	281.86	400.9	2764.11	286.78	407.9	2812.37	291.70	414.9	2860.63
277.01	394.0	2716.53	281.93	401.0	2764.80	286.85	408.0	2813.06	291.77	415.0	2861.32
277.08	394.1	2717.22	282.00	401.1	2765.49	286.92	408.1	2813.75	291.84	415.1	2862.01
277.15	394.2	2717.91	282.07	401.2	2766.18	286.99	408.2	2814.44	291.91	415.2	2862.70
277.22	394.3	2718.60	282.14	401.3	2766.87	287.06	408.3	2815.13	291.98	415.3	2863.39
277.29	394.4	2719.29	282.21	401.4	2767.56	287.13	408.4	2815.82	292.06	415.4	2864.08
277.36	394.5	2719.98	282.28	401.5	2768.24	287.20	408.5	2816.51	292.13	415.5	2864.77
277.43	394.6	2720.67	282.35	401.6	2768.93	287.27	408.6	2817.20	292.20	415.6	2865.46
277.50	394.7	2721.36	282.42	401.7	2769.62	287.34	408.7	2817.89	292.27	415.7	2866.15
277.57	394.8	2722.05	282.49	401.8	2770.31	287.41	408.8	2818.58	292.34	415.8	2866.84
277.64	394.9	2722.74	282.56	401.9	2771.00	287.49	408.9	2819.27	292.41	415.9	2867.53
277.71	395.0	2723.43	282.63	402.0	2771.69	287.56	409.0	2819.96	292.48	416.0	2868.22
277.78	395.1	2724.12	282.70	402.1	2772.38	287.63	409.1	2820.65	292.55	416.1	2868.91
277.85	395.2	2724.81	282.77	402.2	2773.07	287.70	409.2	2821.33	292.62	416.2	2869.60
277.92	395.3	2725.50	282.84	402.3	2773.76	287.77	409.3	2822.02	292.69	416.3	2870.29
277.99	395.4	2726.19	282.92	402.4	2774.45	287.84	409.4	2822.71	292.76	416.4	2870.98
278.06	395.5	2726.88	282.99	402.5	2775.14	287.91	409.5	2823.40	292.83	416.5	2871.67
278.13	395.6	2727.57	283.06	402.6	2775.83	287.98	409.6	2824.09	292.90	416.6	2872.36
278.20	395.7	2728.26	283.13	402.7	2776.52	288.05	409.7	2824.78	292.97	416.7	2873.05
278.27	395.8	2728.94	283.20	402.8	2777.21	288.12	409.8	2825.47	293.04	416.8	2873.73
278.35	395.9	2729.63	283.27	402.9	2777.90	288.19	409.9	2826.16	293.11	416.9	2874.42
278.42	396.0	2730.32	283.34	403.0	2778.59	288.26	410.0	2826.85	293.18	417.0	2875.11
278.49	396.1	2731.01	283.41	403.1	2779.28	288.33	410.1	2827.54	293.25	417.1	2875.80
278.56	396.2	2731.70	283.48	403.2	2779.97	288.40	410.2	2828.23	293.32	417.2	2876.49
278.63	396.3	2732.39	283.55	403.3	2780.66	288.47	410.3	2828.92	293.39	417.3	2877.18
278.70	396.4	2733.08	283.62	403.4	2781.34	288.54	410.4	2829.61	293.46	417.4	2877.87
278.77	396.5	2733.77	283.69	403.5	2782.03	288.61	410.5	2830.30	293.53	417.5	2878.56
278.84	396.6	2734.46	283.76	403.6	2782.72	288.68	410.6	2830.99	293.60	417.6	2879.25
278.91	396.7	2735.15	283.83	403.7	2783.41	288.75	410.7	2831.68	293.67	417.7	2879.94
278.98	396.8	2735.84	283.90	403.8	2784.10	288.82	410.8	2832.37	293.74	417.8	2880.63
279.05	396.9	2736.53	283.97	403.9	2784.79	288.89	410.9	2833.06	293.81	417.9	2881.32
279.12	397.0	2737.22	284.04	404.0	2785.48	288.96	411.0	2833.75	293.88	418.0	2882.01
279.19	397.1	2737.91	284.11	404.1	2786.17	289.03	411.1	2834.43	293.95	418.1	2882.70
279.26	397.2	2738.60	284.18	404.2	2786.86	289.10	411.2	2835.12	294.02	418.2	2883.39
279.33	397.3	2739.29	284.25	404.3	2787.55	289.17	411.3	2835.81	294.09	418.3	2884.08
279.40	397.4	2739.98	284.32	404.4	2788.24	289.24	411.4	2836.50	294.16	418.4	2884.77
279.47	397.5	2740.67	284.39	404.5	2788.93	289.31	411.5	2837.19	294.23	418.5	2885.46
279.54	397.6	2741.36	284.46	404.6	2789.62	289.38	411.6	2837.88	294.30	418.6	2886.15
279.61	397.7	2742.04	284.53	404.7	2790.31	289.45	411.7	2838.57	294.38	418.7	2886.83
279.68	397.8	2742.73	284.60	404.8	2791.00	289.52	411.8	2839.26	294.45	418.8	2887.52
279.75	397.9	2743.42	284.67	404.9	2791.69	289.59	411.9	2839.95	294.52	418.9	2888.21
279.82	398.0	2744.11	284.74	405.0	2792.38	289.66	412.0	2840.64	294.59	419.0	2888.90
279.89	398.1	2744.80	284.81	405.1	2793.07	289.73	412.1	2841.33	294.66	419.1	2889.59
279.96	398.2	2745.49	284.88	405.2	2793.76	289.81	412.2	2842.02	294.73	419.2	2890.28
280.03	398.3	2746.18	284.95	405.3	2794.45	289.88	412.3	2842.71	294.80	419.3	2890.97
280.10	398.4	2746.87	285.02	405.4	2795.13	289.95	412.4	2843.40	294.87	419.4	2891.66
280.17	398.5	2747.56	285.09	405.5	2795.82	290.02	412.5	2844.09	294.94	419.5	2892.35
280.24	398.6	2748.25	285.17	405.6	2796.51	290.09	412.6	2844.78	295.01	419.6	2893.04
280.31	398.7	2748.94	285.24	405.7	2797.20	290.16	412.7	2845.47	295.08	419.7	2893.73
280.38	398.8	2749.63	285.31	405.8	2797.89	290.23	412.8	2846.16	295.15	419.	2894.42
280.45	398.9	2750.32	285.38	405.9	2798.58	290.30	412.9	2846.85	295.22	419.	2895.11

Table 4–16 (continued)
STRESS CONVERSION

Metric (KG/mm²)	English (KSI)	SI (MN/m²)	Metric (KG/mm²)	English (KSI)	SI (MN/m²)	Metric (KG/mm²)	English (KSI)	SI (MN/m²)	Metric (KG/mm²)	English (KSI)	SI (MN/m²)
295.29	420.0	2895.80	300.21	427.0	2944.06	305.13	434.0	2992.32	310.05	441.0	3040.59
295.36	420.1	2896.49	300.28	427.1	2944.75	305.20	434.1	2993.01	310.12	441.1	3041.28
295.43	420.2	2897.18	300.35	427.2	2945.44	305.27	434.2	2993.70	310.19	441.2	3041.97
295.50	420.3	2897.87	300.42	427.3	2946.13	305.34	434.3	2994.39	310.26	441.3	3042.66
295.57	420.4	2898.56	300.49	427.4	2946.82	305.41	434.4	2995.08	310.33	441.4	3043.35
295.64	420.5	2899.25	300.56	427.5	2947.51	305.48	434.5	2995.77	310.41	441.5	3044.04
295.71	420.6	2899.93	300.63	427.6	2948.20	305.55	434.6	2996.46	310.48	441.6	3044.72
295.78	420.7	2900.62	300.70	427.7	2948.89	305.62	434.7	2997.15	310.55	441.7	3045.41
295.85	420.8	2901.31	300.77	427.8	2949.58	305.69	434.8	2997.84	310.62	441.8	3046.10
295.92	420.9	2902.00	300.84	427.9	2950.27	305.76	434.9	2998.53	310.69	441.9	3046.79
295.99	421.0	2902.69	300.91	428.0	2950.96	305.84	435.0	2999.22	310.76	442.0	3047.48
296.06	421.1	2903.38	300.98	428.1	2951.65	305.91	435.1	2999.91	310.83	442.1	3048.17
296.13	421.2	2904.07	301.05	428.2	2952.33	305.98	435.2	3000.60	310.90	442.2	3048.86
296.20	421.3	2904.76	301.12	428.3	2953.02	306.05	435.3	3001.29	310.97	442.3	3049.55
296.27	421.4	2905.45	301.20	428.4	2953.71	306.12	435.4	3001.98	311.04	442.4	3050.24
296.34	421.5	2906.14	301.27	428.5	2954.40	306.19	435.5	3002.67	311.11	442.5	3050.93
296.41	421.6	2906.83	301.34	428.6	2955.09	306.26	435.6	3003.36	311.18	442.6	3051.62
296.48	421.7	2907.52	301.41	428.7	2955.78	306.33	435.7	3004.05	311.25	442.7	3052.31
296.55	421.8	2908.21	301.48	428.8	2956.47	306.40	435.8	3004.74	311.32	442.8	3053.00
296.63	421.9	2908.90	301.55	428.9	2957.16	306.47	435.9	3005.42	311.39	442.9	3053.69
296.70	422.0	2909.59	301.62	429.0	2957.85	306.54	436.0	3006.11	311.46	443.0	3054.38
296.77	422.1	2910.28	301.69	429.1	2958.54	306.61	436.1	3006.80	311.53	443.1	3055.07
296.84	422.2	2910.97	301.76	429.2	2959.23	306.68	436.2	3007.49	311.60	443.2	3055.76
296.91	422.3	2911.66	301.83	429.3	2959.92	306.75	436.3	3008.18	311.67	443.3	3056.45
296.98	422.4	2912.35	301.90	429.4	2960.61	306.82	436.4	3008.87	311.74	443.4	3057.14
297.05	422.5	2913.03	301.97	429.5	2961.30	306.89	436.5	3009.56	311.81	443.5	3057.82
297.12	422.6	2913.72	302.04	429.6	2961.99	306.96	436.6	3010.25	311.88	443.6	3058.51
297.19	422.7	2914.41	302.11	429.7	2962.68	307.03	436.7	3010.94	311.95	443.7	3059.20
297.26	422.8	2915.10	302.18	429.8	2963.37	307.10	436.8	3011.63	312.02	443.8	3059.89
297.33	422.9	2915.79	302.25	429.9	2964.06	307.17	436.9	3012.32	312.09	443.9	3060.58
297.40	423.0	2916.48	302.32	430.0	2964.75	307.24	437.0	3013.01	312.16	444.0	3061.27
297.47	423.1	2917.17	302.39	430.1	2965.43	307.31	437.1	3013.70	312.23	444.1	3061.96
297.54	423.2	2917.86	302.46	430.2	2966.12	307.38	437.2	3014.39	312.30	444.2	3062.65
297.61	423.3	2918.55	302.53	430.3	2966.81	307.45	437.3	3015.08	312.37	444.3	3063.34
297.68	423.4	2919.24	302.60	430.4	2967.50	307.52	437.4	3015.77	312.44	444.4	3064.03
297.75	423.5	2919.93	302.67	430.5	2968.19	307.59	437.5	3016.46	312.51	444.5	3064.72
297.82	423.6	2920.62	302.74	430.6	2968.88	307.66	437.6	3017.15	312.58	444.6	3065.41
297.89	423.7	2921.31	302.81	430.7	2969.57	307.73	437.7	3017.84	312.66	444.7	3066.10
297.96	423.8	2922.00	302.88	430.8	2970.26	307.80	437.8	3018.52	312.73	444.8	3066.79
298.03	423.9	2922.69	302.95	430.9	2970.95	307.87	437.9	3019.21	312.80	444.9	3067.48
298.10	424.0	2923.38	303.02	431.0	2971.64	307.94	438.0	3019.90	312.87	445.0	3068.17
298.17	424.1	2924.07	303.09	431.1	2972.33	308.01	438.1	3020.59	312.94	445.1	3068.86
298.24	424.2	2924.76	303.16	431.2	2973.02	308.09	438.2	3021.28	313.01	445.2	3069.55
298.31	424.3	2925.45	303.23	431.3	2973.71	308.16	438.3	3021.97	313.08	445.3	3070.24
298.38	424.4	2926.13	303.30	431.4	2974.40	308.23	438.4	3022.66	313.15	445.4	3070.92
298.45	424.5	2926.82	303.37	431.5	2975.09	308.30	438.5	3023.35	313.22	445.5	3071.61
298.52	424.6	2927.51	303.44	431.6	2975.78	308.37	438.6	3024.04	313.29	445.6	3072.30
298.59	424.7	2928.20	303.52	431.7	2976.47	308.44	438.7	3024.73	313.36	445.7	3072.99
298.66	424.8	2928.89	303.59	431.8	2977.16	308.51	438.8	3025.42	313.43	445.8	3073.68
298.73	424.9	2929.58	303.66	431.9	2977.85	308.58	438.9	3026.11	313.50	445.9	3074.37
298.80	425.0	2930.27	303.73	432.0	2978.54	308.65	439.0	3026.80	313.57	446.0	3075.06
298.87	425.1	2930.96	303.80	432.1	2979.22	308.72	439.1	3027.49	313.64	446.1	3075.75
298.95	425.2	2931.65	303.87	432.2	2979.91	308.79	439.2	3028.18	313.71	446.2	3076.44
299.02	425.3	2932.34	303.94	432.3	2980.60	308.86	439.3	3028.87	313.78	446.3	3077.13
299.09	425.4	2933.03	304.01	432.4	2981.29	308.93	439.4	3029.56	313.85	446.4	3077.82
299.16	425.5	2933.72	304.08	432.5	2981.98	309.00	439.5	3030.25	313.92	446.5	3078.51
299.23	425.6	2934.41	304.15	432.6	2982.67	309.07	439.6	3030.94	313.99	446.6	3079.20
299.30	425.7	2935.10	304.22	432.7	2983.36	309.14	439.7	3031.62	314.06	446.7	3079.89
299.37	425.8	2935.79	304.29	432.8	2984.05	309.21	439.8	3032.31	314.13	446.8	3080.58
299.44	425.9	2936.48	304.36	432.9	2984.74	309.28	439.9	3033.00	314.20	446.9	3081.27
299.51	426.0	2937.17	304.43	433.0	2985.43	309.35	440.0	3033.69	314.27	447.0	3081.96
299.58	426.1	2937.86	304.50	433.1	2986.12	309.42	440.1	3034.38	314.34	447.1	3082.65
299.65	426.2	2938.55	304.57	433.2	2986.81	309.49	440.2	3035.07	314.41	447.2	3083.34
299.72	426.3	2939.23	304.64	433.3	2987.50	309.56	440.3	3035.76	314.48	447.3	3084.02
299.79	426.4	2939.92	304.71	433.4	2988.19	309.63	440.4	3036.45	314.55	447.4	3084.71
299.86	426.5	2940.61	304.78	433.5	2988.88	309.70	440.5	3037.14	314.62	447.5	3085.40
299.93	426.6	2941.30	304.85	433.6	2989.57	309.77	440.6	3037.83	314.69	447.6	3086.09
300.00	426.7	2941.99	304.92	433.7	2990.26	309.84	440.7	3038.52	314.76	447.7	3086.78
300.07	426.8	2942.68	304.99	433.8	2990.95	309.91	440.8	3039.21	314.83	447.8	3087.47
300.14	426.9	2943.37	305.06	433.9	2991.64	309.98	440.9	3039.90	314.90	447.9	3088.16

Table 4—16 (continued)
STRESS CONVERSION

Metric (KG/mm²)	English (KSI)	SI (MN/m²)
314.98	448.0	3088.85
315.05	448.1	3089.54
315.12	448.2	3090.23
315.19	448.3	3090.92
315.26	448.4	3091.61
315.33	448.5	3092.30
315.40	448.6	3092.99
315.47	448.7	3093.68
315.54	448.8	3094.37
315.61	448.9	3095.06
315.68	449.0	3095.75
315.75	449.1	3096.44
315.82	449.2	3097.12
315.89	449.3	3097.81
315.96	449.4	3098.50
316.03	449.5	3099.19
316.10	449.6	3099.88
316.17	449.7	3100.57
316.24	449.8	3101.26
316.31	449.9	3101.95
316.38	450.0	3102.64

From *Aerospace Structural Metals Handbook*, AFML 68-115, 1973 ed., Mechanical Properties Data Center, Belfour Stulen, Traverse City, Mich., 1972. With permission.

Table 4–17
MERCURY MANOMETER CONVERSION FACTORS

Equivalent Pressures in psi and Inches of Water per Inch of Mercury

For values in N/m^2, multiply values in psi by 6894.8.

Observed temperature of mercury column			Equivalent values			
				Inches of water		
		psi	Mercury column at 32°F	60°F; 15.56°C	68°F; 20°C	77°F; 25°C
°F	°C					
0	−17.78	0.49275	1.0032	13.652	13.663	13.680
10	−12.22	0.49225	1.0022	13.638	13.649	13.665
20	−6.67	0.49175	1.0012	13.625	13.636	13.652
30	−1.11	0.49126	1.0002	13.611	13.622	13.638
32	0.00	0.49116	1.0000	13.609	13.620	13.635
35	1.67	0.49101	0.9997	13.604	13.615	13.631
40	4.44	0.49076	0.9992	13.598	13.609	13.624
45	7.22	0.49051	0.9987	13.591	13.602	13.618
50	10.00	0.49026	0.9982	13.584	13.595	13.611
55	12.78	0.49002	0.9977	13.577	13.588	13.604
60	15.56	0.48977	0.9972	13.570	13.581	13.597
65	18.33	0.48952	0.9967	13.564	13.575	13.590
68	20.00	0.48938	0.9964	13.560	13.570	13.586
70	21.11	0.48928	0.9962	13.557	13.568	13.584
75	23.89	0.48904	0.9957	13.550	13.561	13.577
77	25.00	0.48894	0.9955	13.547	13.558	13.574
80	26.67	0.48879	0.9952	13.543	13.554	13.570
85	29.44	0.48854	0.9947	13.536	13.547	13.563
90	32.22	0.48830	0.9942	13.530	13.541	13.556
95	35.00	0.48805	0.9937	13.523	13.534	13.549
100	37.78	0.48780	0.9932	13.516	13.527	13.543
110	43.33	0.48732	0.9922	13.502	13.513	13.529
120	48.89	0.48683	0.9912	13.489	13.500	13.515
130	54.44	0.48634	0.9902	13.475	13.486	13.502
140	60.00	0.48585	0.9892	13.462	13.472	13.488

From Bolz, R. E. and Tuve, G. L., Eds., *Handbook of Tables for Applied Engineering Science,* 2nd ed., CRC Press, Cleveland, 1973, 860.

Table 4–18
VISCOSITY CONVERSIONS–LIQUIDS

Absolute viscosity				Kinematic viscosity						
Centi-poises	$\dfrac{lb_m}{ft\ sec}$	$\dfrac{N\text{-}sec}{m^2}$	Mac-Michael	Centi-stokes	$\dfrac{ft^2}{sec}$	$\dfrac{m^2}{sec}$	Saybolt Universal Seconds	Redwood standard seconds	Engler degrees	Ford 3
25	.0168	.025	—	25	.000269	.000025	115	112	3.4	
50	.0336	.050	185	50	.000538	.000050	230	225	6.4	28
75	.0504	.075	240	75	.000807	.000075	346	335	9.4	38
100	.0672	.100	295	100	.001076	.000100	462	448	13	49
125	.0840	.125	350	125	.001346	.000125	577	558	16	61
150	.1008	.150	405	150	.001615	.000150	692	673	19	73
175	.1176	.175	465	175	.001884	.000175	807	786	22	84
200	.1344	.200	520	200	.002153	.000200	923	898	26	95
225	.1512	.225	575	225	.002422	.000225	1038	1010	29	106
250	.1680	.250	625	250	.002691	.000250	1153	1120	32	116
275	.1848	.275	675	275	.002960	.000275	1269	1233	35	125
300	.2016	.300	725	300	.003229	.000300	1384	1345	39	135
325	.2184	.325	780	325	.003498	.000325	1500	1457	43	144
350	.2352	.350	840	350	.003767	.000350	1616	1570	45	152
375	.2520	.375	900	375	.004037	.000375	1731	1683	49	161
400	.2688	.400	960	400	.004306	.000400	1846	1797	53	170
425	.2856	.425	1020	425	.004575	.000425	1960	1910	56	178
450	.3024	.450	1080	450	.004844	.000450	2075	2023	60	188
475	.3192	.475	1160	475	.005113	.000475	2190	2136	63	198
500	.3360	.500	1210	500	.005382	.000500	2307	2240	67	208
525	.3528	.525	1270	525	.005651	.000525	2422	2355	71	217
550	.3696	.550	1330	550	.005920	.000550	2537	2460	75	226
575	.3864	.575	1385	575	.006189	.000575	2652	2575	79	234
600	.4032	.600	1440	600	.006458	.000600	2766	2690	82	242
625	.4200	.625	1500	625	.006728	.000625	2882	2800	85	250
650	.4368	.650	1560	650	.006997	.000650	2999	2915	88	258
675	.4536	.675	1620	675	.007266	.000675	3116	3030	91	266
700	.4704	.700	1680	700	.007535	.000700	3230	3143	94	274
725	.4872	.725	1740	725	.007804	.000725	3345	3256	97	282
750	.5040	.750	1800	750	.008073	.000750	3460	3370	101	290

From Bolz, R. E. and Tuve, G. L., Eds., *Handbook of Tables for Applied Engineering Science,* 2nd ed., CRC Press, Cleveland, 1973, 861.

Notes: In all equations relating viscosity to other physical properties or processes (e.g., fluid flow or heat transfer), the consistent units–fps, MKS, or cgs–should be used (see Table 4–22 for conversion factors). In consistent units kinematic viscosity is equal to absolute viscosity divided by density.

A rotating-cup instrument (MacMichael) gives a reading directly convertible to absolute viscosity. The short-tube viscometers give readings directly convertible to kinematic viscosity.

Values in this table are approximate. The conversion factors vary slightly with temperature; if highly accurate conversions are required, the "ASTM Viscosity Tables" should be used.

The U.S. standard for light liquids is the Saybolt Universal short-tube viscometer. For heavy road oils and fuel oils the Saybolt Furol instrument is used. For the same oil the Saybolt Universal Seconds are approximately ten times the Saybolt Furol values.

The Redwood No. 1 is the English viscometer; the Engler, the German viscometer. Engler degrees represent the ratio of the outflow time of the liquid under test to that for a like volume of water at the same temperature. Engler seconds are sometimes given; this value is about 51 times the Engler degrees.

REFERENCES

For a conversion chart including several other viscometer scales, see *Handbook of Chemistry,* 10th ed., Lange, N. A., Ed., McGraw-Hill, 1961, 1844.

Table 4–19
GRAVITY CONVERSIONS–LIQUIDS

Liquids lighter than water

Sp gr	$\frac{lb}{ft^3}$	$\frac{kg}{m^3}$	$\frac{lb}{gal}$	°Bé	°API
1.00	62.38	999.2	8.34	10.0	10.0
.99	61.76	989.2	8.25	11.4	11.4
.98	61.14	979.2	8.17	12.9	12.9
.97	60.51	969.2	8.09	14.3	14.4
.96	59.89	959.2	8.00	15.8	15.9
.95	59.27	949.2	7.92	17.4	17.5
.94	58.64	939.2	7.84	19.0	19.0
.93	58.02	929.2	7.75	20.5	20.6
.92	57.40	919.3	7.67	22.2	22.3
.91	56.77	909.3	7.59	23.8	24.0
.90	56.15	899.3	7.50	25.6	25.7
.89	55.53	889.3	7.42	27.3	27.5
.88	54.90	879.3	7.34	29.1	29.3
.87	54.28	869.3	7.25	30.9	31.1
.86	53.66	859.3	7.17	32.8	33.0
.85	53.03	849.3	7.09	34.7	35.0
.84	52.40	839.3	7.00	36.7	37.0
.83	51.78	829.4	6.92	38.7	39.0
.82	51.15	819.4	6.84	40.7	41.0
.81	50.52	809.4	6.75	42.8	43.2
.80	49.90	799.4	6.67	45.0	45.4
.79	49.28	789.4	6.59	47.2	47.6
.78	48.66	779.4	6.50	49.5	49.9
.77	48.03	769.4	6.42	51.8	52.2
.76	47.40	759.4	6.34	54.2	54.7
.75	46.77	749.4	6.25	56.7	57.2
.74	46.15	739.4	6.17	59.2	59.7
.73	45.54	729.4	6.09	61.8	62.3
.72	44.91	719.4	6.00	64.4	65.0
.71	44.29	709.4	5.92	67.2	67.8
.70	43.67	699.4	5.83	70.0	70.6
.69	43.04	689.5	5.75	72.9	73.6
.68	42.41	679.5	5.67	75.9	76.6
.67	41.79	669.5	5.59	79.0	79.7
.66	41.17	659.5	5.51	82.1	82.9
.65	40.54	649.5	5.42	85.4	86.2
.64	39.92	639.5	5.34	88.7	89.6
.63	39.29	629.5	5.25	92.2	93.1
.62	38.67	619.5	5.17	95.8	96.7
.61	38.05	609.5	5.09	99.5	100.5
.60	37.43	599.5	5.00	103.3	104.3
.59	36.80	589.5	4.92	107.4	108.5
.58	36.18	579.5	4.83	111.6	112.8
.57	35.56	569.5	4.75	115.9	117.2
.56	34.93	559.6	4.67	120.2	121.6
.55	34.31	549.6	4.59	124.5	126.1

Liquids heavier than water

Sp gr	$\frac{lb}{ft^3}$	$\frac{kg}{m^3}$	$\frac{lb}{gal}$	°Bé	°Twad	°Brix
1.00	62.38	999.2	8.34	0	0	0
1.01	63.00	100.9	8.41	1.44	2	2.5
1.02	63.63	101.9	8.50	2.84	4	5.1
1.03	64.26	102.9	8.58	4.22	6	7.6
1.04	64.88	103.9	8.67	5.58	8	10.0
1.05	65.50	104.9	8.75	6.91	10	12.4
1.06	66.12	105.9	8.83	8.21	12	14.8
1.07	66.75	106.9	8.92	9.49	14	17.1
1.08	67.37	107.9	9.00	10.74	16	19.4
1.09	67.99	108.9	9.08	11.97	18	21.7
1.10	68.62	109.9	9.17	13.18	20	24.0
1.11	69.24	110.9	9.25	14.37	22	26.2
1.12	69.86	111.9	9.33	15.54	24	28.3
1.13	70.49	112.9	9.42	16.68	26	30.4
1.14	71.11	113.9	9.50	17.81	28	32.5
1.15	71.73	114.9	9.58	18.91	30	34.6
1.16	72.35	115.9	9.67	20.00	32	36.6
1.17	72.98	116.9	9.76	21.07	34	38.6
1.18	73.60	117.9	9.84	22.12	36	40.5
1.19	74.22	118.9	9.92	23.15	38	42.4
1.20	74.85	119.9	10.00	24.17	40	44.3
1.21	75.48	120.9	10.09	25.16	42	46.2
1.22	76.10	121.9	10.17	26.15	44	48.0
1.23	76.72	122.9	10.26	27.11	46	49.8
1.24	77.34	123.9	10.34	28.06	48	51.5
1.25	77.97	124.9	10.42	29.00	50	53.3
1.26	78.59	125.9	10.50	29.92	52	55.1
1.27	79.21	126.9	10.59	30.83	54	56.9
1.28	79.83	127.9	10.67	31.72	56	58.6
1.29	80.46	128.9	10.75	32.60	58	60.3
1.30	81.09	129.9	10.84	33.46	60	62.0
1.31	81.71	130.9	10.92	34.31	62	63.7
1.32	82.34	131.9	11.00	35.15	64	65.3
1.33	82.96	132.9	11.09	35.98	66	67.0
1.34	83.58	133.9	11.17	36.79	68	68.8
1.35	84.21	134.9	11.25	37.59	70	70.4
1.36	84.84	135.9	11.34	38.38	72	72.0
1.37	85.46	136.9	11.42	39.16	74	73.6
1.38	86.08	137.9	11.51	39.93	76	75.2
1.39	86.70	138.9	11.59	40.68	78	76.7
1.40	87.33	139.9	11.67	41.43	80	78.2
1.41	87.95	140.9	11.76	42.17	82	79.8
1.42	88.57	141.9	11.84	42.89	84	81.2
1.43	89.19	142.9	11.92	43.60	86	82.6
1.44	89.82	143.9	12.00	44.31	88	84.1
1.45	90.44	145.0	12.09	45.00	90	85.6

Notes: The standard temperature for Baumé and oil gravity data is 60°F (15.6°C); specific gravity is referred to water at the same temperature.

°Bé, degrees Baumé, is actually two hydrometer scales. The NBS equation for liquids lighter than water is sp gr = 140/(130 + °Bé); for liquids heavier than water, sp gr = 145/(145–°Bé). For oils the API equation is 141.5(131.5 + °API).

°Twad, degrees Twaddell, applies only to liquids heavier than water. Each degree represents an increase in specific gravity of 0.005.

°Brix, degrees Brix, is a hydrometer scale for sugar solutions. The reading in °Brix is numerically equal to the percentage of sucrose by weight.

Other gravity scales are used for specific liquids such as salt solutions, acids, bleaches, dyes, milk, urine, paint, and alcoholic liquids.

There is more than one *proof* scale for alcoholic liquids, but the most common one is the U.S. Internal Revenue scale, which defines a "100-proof spirit" as one composed of 50% alcohol by volume. This is 42.5% alcohol by weight, sp gr 0.934 $\frac{60°}{60°}$ F. Since there is a contraction when water and alcohol are mixed, the 100-proof spirit

Table 4–19 (continued)
GRAVITY CONVERSIONS – LIQUIDS

consists of 50 parts by volume of alcohol $\left(\text{sp gr } 0.7939 \dfrac{60^\circ}{60^\circ} \text{F}\right)$ and 53.73 parts water. The "proof hydrometer"

reads 200 in absolute alcohol and zero in water at 60°F. The "over-proof" part of the scale reads above 100; the scale reading at any point is twice the alcohol by volume. Thus "86 proof" is 43% alcohol by volume. The British proof scale is not the same.

From Bolz, R. E. and Tuve, G. L., Eds., *Handbook of Tables for Applied Engineering Science,* 2nd ed., CRC Press, Cleveland, 1973, 862.

Table 4–20
DECIBEL CONVERSION

Ratios of Power, Voltage, Pressure, Current, or Sound Level

The *decibel* is a dimensionless ratio of two values of the same quantity. It is most often applied to a *power ratio* and defined as dB = 10 $\log_{10}$ (actual power level/reference power level), or dB = 10 $\log_{10}$ (W_2/W_1). Since power is proportional to the square of potential or of current (e.g., $W = I^2 R = E^2/R$), the decibel can also be defined as dB = 20 $\log_{10}$ (E_2/E_1), or dB = 20 $\log_{10}$ (I_2/I_1). In the case of sound, the potential is measured as a pressure, but the sound "level" is an energy level: dB = 10 $\log_{10}$ $(p_2/p_1)^2$ = 20 $\log_{10}$ (p_2/p_1).

The *reference levels* $(W_1, p_1$ etc.) are not well standardized. For example, sound power is usually measured above 10^{-12} w/cm^2, but both 10^{-11} and 10^{-16} w/cm^2 are used. Sound pressure in air is usually measured above 0.0002 μbar, which is specified in ISO Standards R131 and R357.[a] The reference level is not important in many cases, since the engineer is usually concerned with the difference in levels, i.e., with a power ratio. A most common ratio is 2, which is a difference of approximately 3 dB.

Decibels	Power ratio	Voltage, current, or pressure ratio	Decibels	Power ratio	Voltage, current, or pressure ratio
0	1.000	1.0000	11	12.589	3.5481
0.5	1.1220	1.0593	12	15.849	3.9811
1	1.2589	1.1220	13	19.953	4.4668
1.5	1.413	1.189	14	25.119	5.0119
2	1.5849	1.2589	15	31.623	5.6234
2.5	1.778	1.334	16	39.811	6.3096
3	1.9953	1.4125	17	50.119	7.0795
3.5	2.239	1.496	18	63.096	7.9433
4	2.5119	1.5849	19	79.433	8.9125
4.5	2.818	1.679	20	100.00	10.0000
5	3.1623	1.7783	30	1,000.0	31.623
5.5	3.548	1.884	40	10^4	10^2
6	3.9811	1.9953	50	10^5	316.23
7	5.0119	2.2387	60	10^6	10^3
8	6.3096	2.5119	70	10^7	3,162.
9	7.9433	2.8184	80	10^8	10^4
10	10.0000	3.1623	90	10^9	31,623
			100	10^{10}	10^5

Notes: If the power ratio is less than unity, invert the fraction and find the decibel *loss* from the table.

Table 4–20 (continued)
DECIBEL CONVERSION

A similar unit to the decibel, and sometimes used in place of it by electrical engineers (as in studying attenuation in long lines), is the neper (Np) (after Napier):

$$1 \text{ dB} = 0.1151 \text{ Np.}$$
$$1 \text{ Np} = 8.686 \text{ dB.}$$

[a]$0.0002 \ \mu$bar $= 2 \times 10^{-5} \ N/m^2$ or 2×10^{-4} dyn/cm^2.

From Bolz, R. E. and Tuve, G. L., Eds., *Handbook of Tables for Applied Engineering Science,* 2nd ed., CRC Press, Cleveland, 1973, 863.

Table 4–21
CONVERSION FACTORS FOR PRESSURE

The following densities were used for the table:
mercury at 0°C (32°F): 13 595.9 kg/m^3, 848.764 lb$_m$/ft^3
mercury at 20°C (68°F): 13 546.6 kg/m^3, 845.687 lb$_m$/ft^3
water at 4°C (39.2°F): 999.973 kg/m^3, 62.426 lb$_m$/ft^3
water at 20°C (68°F): 998.203 kg/m^3, 62.316 lb$_m$/ft^3

Example: To convert 10 in. of mercury at 20°C to its equivalent in N/m^2:

$$10 \text{ (in. Hg, 20°C)} \times 3{,}374.11 \ \frac{(N/m^2)}{(\text{in. Hg, 20°C})} = 33{,}741.1 \ N/m^2$$

For conversions not tabulated: Combine conversion factors from any one column as illustrated by the conversion of 1 ft H$_2$O, 20°C to equivalent in lb$_f$/ft^2:

$$1 \text{ (ft H}_2\text{O, 20°C)} \times \frac{2{,}983.9 \left(\dfrac{N/m^2}{\text{ft H}_2\text{O, 20°C}} \right)}{47.880 \left(\dfrac{N/m^2}{\text{lb}_f/\text{ft}_2} \right)} = 62.320 \ \text{lb}_f/\text{ft}^2$$

To obtain → multiply → by number in table	$\dfrac{N}{m^2}$, SI unit	atm	$\dfrac{\text{lb}_f}{\text{in.}^2}$	$\dfrac{\text{kg}_f}{\text{cm}^2}$
N/m^2 (Pascal)	1	$9.869\ 2 \times 10^{-6}$	$1.450\ 4 \times 10^{-4}$	$1.019\ 7 \times 10^{-4}$
atmosphere (760 mm Hg)	101 325	1	14.696	1.033 2
bar	10^5	0.986 92	14.504	1.019 7
dyne/cm^2 (microbar)	0.1	$9.869\ 2 \times 10^{-7}$	$1.450\ 4 \times 10^{-5}$	$1.019\ 7 \times 10^{-6}$
lb$_f$/in.^{2}, psi	6 894.8	0.068 046	1	0.070 307
lb$_f$/ft^2, psf	47.880	$4.725\ 4 \times 10^{-4}$	$6.944\ 4 \times 10^{-3}$	$4.882\ 4 \times 10^{-4}$
kg$_f$/cm^2 (tech atm)	98 066.5	0.967 84	14.223	1
mm H$_2$O at 4°C (39.2°F)	9.806 3	$9.678\ 7 \times 10^{-5}$	$1.422\ 4 \times 10^{-3}$	$1.000\ 1 \times 10^{-4}$
mm H$_2$O at 20°C (68°F)	9.789 0	$9.661\ 4 \times 10^{-5}$	$1.419\ 9 \times 10^{-3}$	$9.983\ 1 \times 10^{-5}$
mm Hg at 0°C (32°F)	133.32	0.001 316	0.019 337	0.001 359 5
mm Hg at 20°C (68°F)	132.84	0.001 311 0	0.019 267	0.001 354 6
in. H$_2$O at 4°C (39.2°F)	249.08	0.002 458 4	0.036 129	0.002 540 2
in. H$_2$O at 20°C (68°F)	248.64	0.002 454 0	0.036 065	0.002 535 7

Table 4—21 (continued)
CONVERSION FACTORS FOR PRESSURE

To obtain ——————▶ multiply ——┐ by number in table ▼	$\dfrac{N}{m^2}$, SI unit	atm	$\dfrac{lb_f}{in.^2.}$	$\dfrac{kg_f}{cm^2}$
in. Hg at 0°C (32°F)	3 386.4	0.033 421	0.491 15	0.034 532
in. Hg at 20°C (68°F)	3 374.1	0.033 300	0.489 37	0.034 407
ft H$_2$O at 20°C (68°F)	2 983.6	0.029 449	0.432 78	0.030 428

Note: Exact values are in boldface type.

From Bolz, R. E. and Tuve, G. L., Eds., *Handbook of Tables for Applied Engineering Science*, 2nd ed., CRC Press, Cleveland, 1973, 864.

Table 4—22
CONVERSION FACTORS FOR VISCOSITY

Example: To convert 0.01 lbm/ft·s to its equivalent in N·s/m²:

$$0.01 \ (\text{lbm/ft} \cdot \text{s}) \times 1.4882 \ \frac{(\text{N} \cdot \text{s/m}^2)}{(\text{lbm/ft} \cdot \text{s})} = 0.014882 \ (\text{N} \cdot \text{s/m}^2)$$

For conversions not tabulated: Combine conversion factors from any one column as illustrated by the conversion of 1 lbf·s/in.² to its equivalent in kg/m·hr:

$$1 \ (\text{lbf} \cdot \text{s/in.}^2) \times \frac{6,894.7 \left(\dfrac{\text{N} \cdot \text{s/m}^2}{\text{lbf} \cdot \text{s/in.}^2}\right)}{2.7778 \times 10^{-4} \left(\dfrac{\text{N} \cdot \text{s/m}^2}{\text{kg/m} \cdot \text{hr}}\right)} = 2.4821 \times 10^7 \ (\text{kg/m} \cdot \text{hr})$$

Table A. Dynamic or Absolute Viscosity, μ

To obtain → multiply ⌐ by number in table ↓	$\dfrac{\text{N} \cdot \text{s}}{\text{m}^2}$, SI unit $\left(\dfrac{\text{kg}}{\text{m} \cdot \text{s}}\right)$	$\dfrac{\text{g}}{\text{cm} \cdot \text{s}}$ $\left(\dfrac{\text{dyne} \cdot \text{s}}{\text{cm}^2}\right)$	$\dfrac{\text{lbm}}{\text{ft} \cdot \text{s}}$ $\left(\dfrac{\text{poundal} \cdot \text{s}}{\text{ft}^2}\right)$	$\dfrac{\text{lbf} \cdot \text{s}}{\text{ft}^2}$ $\left(\dfrac{\text{slug}}{\text{ft} \cdot \text{s}}\right)$
N·s/m²	**1**	**10**	0.671 97	0.020 885
g/cm·s (poise)	**0.1**	**1**	0.067 197	0.002 088 5
centipoise	**0.001**	**0.01**	$6.719\ 7 \times 10^{-4}$	$2.088\ 5 \times 10^{-5}$
kg/m·hr	$2.777\ 8 \times 10^{-4}$	0.002 777 8	$1.866\ 7 \times 10^{-4}$	$5.801\ 5 \times 10^{-6}$
lbm/ft·s	1.488 2	14.882	**1**	0.031 081
lbm/ft·hr	$4.133\ 8 \times 10^{-4}$	0.004 133 8	$2.777\ 8 \times 10^{-4}$	$8.633\ 6 \times 10^{-6}$
slug/ft·s	47.880	478.80	32.174	**1**
slug/ft·hr	0.013 300	0.133 00	0.008 937 2	$2.777\ 8 \times 10^{-4}$
lbm/in.·s	17.858	178.58	**12**	0.372 97
lbm/in.·hr	0.004 960 5	0.049 605	0.003 333 3	$1.036\ 0 \times 10^{-4}$
slug/in.·hr	0.159 60	1.596 0	0.107 25	0.003 333 3
lbf·s/in.² (Reyn)	6 894.7	68 947	4 633.1	**144**

Table B. Kinematic Viscosity, ν

To obtain → multiply ⌐ by number in table ↓	m²/s, SI unit	cm²/s	ft²/s	in.²/s
m²/s	**1**	**10⁴**	10.764	1 550.0
cm²/s (stoke)	**10⁻⁴**	**1**	0.001 076 4	0.155 00
centistoke	**10⁻⁶**	**0.01**	$1.076\ 4 \times 10^{-5}$	0.001 550 0
m²/hr	$2.777\ 8 \times 10^{-4}$	2.777 8	$2.990\ 0 \times 10^{-3}$	0.430 56
ft²/s	**0.092 903 04**	**929.030 4**	**1**	**144**
ft²/hr	$\mathbf{2.580\ 64 \times 10^{-5}}$	0.258 064	$2.777\ 8 \times 10^{-4}$	**0.04**
in.²/s	$\mathbf{6.451\ 6 \times 10^{-4}}$	**6.451 6**	$6.944\ 4 \times 10^{-3}$	**1**
in.²/hr	$1.792\ 1 \times 10^{-7}$	$1.792\ 1 \times 10^{-3}$	$1.929\ 0 \times 10^{-6}$	$2.777\ 8 \times 10^{-4}$

Note: Exact values are shown in boldface type.

From Bolz, R. E. and Tuve, G. L., Eds., *Handbook of Tables for Applied Engineering Science,* 2nd ed., CRC Press, Cleveland, 1973, 865.

Table 4–23

CONVERSION FACTORS FOR ENERGY

The SI (International System) unit is the joule, J, which is a newton meter, N·m. The thermochemical calorie (4.184 J, exactly) is used unless otherwise noted. The abbreviation *IT* denotes the international steam table calorie (4.186 8 J, exactly). The corresponding thermochemical Btu and international steam table Btu are defined in terms of the corresponding calories by the relationship that 1 Btu/lbm·R = 1 ca/g·K.

Example: To convert 100 Btu (thermochemical) to its equivalent in J:

$$100(\text{Btu, thermochemical}) \times 1{,}054.35\left(\frac{J}{\text{Btu, thermochemical}}\right) = 1.05435 \times 10^5\,J$$

For conversions not tabulated: Combine conversion factors from any one column as illustrated by the conversion of 1 ft-lbf to its equivalent in kgf-m:

$$1(\text{ft} \cdot \text{lbf}) \frac{1.3558\left(\dfrac{J}{\text{ft} \cdot \text{lbf}}\right)}{9.80665\left(\dfrac{J}{\text{kgf} \cdot \text{m}}\right)} = 0.13825\,\text{kgf} \cdot \text{m}$$

To obtain ⟶
multiply ⟶ by number in table

	J	ft·lbf	Btu	cal
J	**1**	0.737 56	$9.484\ 5 \times 10^{-4}$	0.239 01
erg	1×10^{-7}	$7.375\ 6 \times 10^{-8}$	$9.484\ 5 \times 10^{-11}$	$2.390\ 1 \times 10^{-8}$
kgf·m	**9.806 65**	7.233 0	$9.301\ 1 \times 10^{-3}$	2.343 85
ft·lbf	1.355 8	**1**	$1.285\ 9 \times 10^{-3}$	0.324 05
Btu	1 054.35	777.65	**1**	252.00
Btu (IT)	1 055.06	778.17	1.000 669	252.16
cal	**4.184**	3.086 0	$3.968\ 3 \times 10^{-3}$	**1**
cal (IT)	**4.186 8**	3.088 0	$3.971\ 0 \times 10^{-3}$	1.000 669
electron volt	$1.602\ 1 \times 10^{-19}$	$1.181\ 6 \times 10^{-19}$	$1.519\ 5 \times 10^{-22}$	$3.829\ 1 \times 10^{-20}$
kWh	$\mathbf{3.6 \times 10^6}$	$2.655\ 2 \times 10^6$	3 414.4	860 421
hp·hr	$2.684\ 5 \times 10^6$	1.98×10^6	2 546.1	641 616

Note: Exact values are shown in boldface type.

From Bolz, R. E. and Tuve, G. L., Eds., *Handbook of Tables for Applied Engineering Science*, 2nd ed., CRC Press, Cleveland, 1973, 866.

Table 4–24
CONVERSION FACTORS FOR THERMAL CONDUCTIVITY, k

The thermochemical calorie (4.184 J) and the thermochemical Btu (1 054.350 J) are used throughout. The absolute watt (joule/second = newton-meter/second) is used throughout.

Example: To convert 1 000 (Btu·in/hr·ft² ·deg R) to its equivalent in W/m·K:

$$1,000(\text{Btu} \cdot \text{in./hr} \cdot \text{ft}^2 \cdot \text{deg R}) \times 0.14413 \left(\frac{\text{W/m} \cdot \text{K}}{\text{Btu} \cdot \text{in./hr} \cdot \text{ft}^2 \cdot \text{deg R}} \right) = 144.13 \ (\text{W/m} \cdot \text{K})$$

For conversions not tabulated: Combine conversion factors from any one column as illustrated by the conversion of 1 Btu/hr·ft·deg R to its equivalent in kcal/hr·m·K:

$$1(\text{Btu/hr} \cdot \text{ft} \cdot \text{deg R}) \times \frac{1.7296 \left(\frac{\text{W/m} \cdot \text{K}}{\text{Btu/hr} \cdot \text{ft} \cdot \text{deg R}} \right)}{1.1622 \left(\frac{\text{W/m} \cdot \text{K}}{\text{kcal/hr} \cdot \text{m} \cdot \text{K}} \right)} = 1.4882 \ (\text{kcal/hr} \cdot \text{m} \cdot \text{K})$$

To obtain ⟶ multiply ⌐ by number in table

	$\dfrac{W}{m \cdot K}$, SI unit	$\dfrac{cal}{s \cdot cm \cdot K}$	$\dfrac{Btu}{hr \cdot ft \cdot deg\ R}$	$\dfrac{Btu\text{-}in.}{hr \cdot ft^2 \cdot deg\ R}$
W/m·K	1	0.002 390 1	0.578 18	6.938 1
W/cm·K	**100**	0.239 01	57.818	693.81
W/ft·deg R	5.905 5	0.014 114	3.414 4	40.973
W/in.·deg R	70.866	0.169 37	40.973	491.68
cal/s·cm·K	**418.4**	1	241.91	2 902.9
cal/s·in.·deg R	296.50	0.708 66	171.43	2 057.2
kcal/hr·m·K	1.162 2	0.002 777 8	0.671 97	8.063 6
Btu/s·ft·deg R	6 226.5	14.882	**3 600**	**43 200**
Btu/s·in.·deg R	74 717 × 10⁴	178.58	**43 200**	5.184 × 10⁵
Btu·in./s·ft²·deg R	518.87	1.240 1	**300**	**3 600**
Btu/hr·ft·deg R	1.729 6	0.004 133 8	**1**	**12**
Btu/hr·in.·deg R	20.755	0.049 605	**12**	**144**
Btu·in./hr·ft²·deg R	0.144 13	0.000 344 48	0.083 333	**1**
ft·lbf/hr·ft·deg R	0.002 224 1	5.315 7 × 10⁻⁶	0.001 285 9	0.015 431

Note: Exact values are shown in boldface type.

From Bolz, R. E. and Tuve, G. L., Eds., *Handbook of Tables for Applied Engineering Science,* 2nd ed., CRC Press, Cleveland, 1973, 867.

Table 4–25
DECIMAL EQUIVALENTS OF COMMON FRACTIONS

		1/64 = 0.015 625			11/32	22/64 = 0.343 75				43/64 = 0.671 875
	1/32	2/64 = .031 25				23/64 = .359 375	11/16	22/32	44/64 = .687 5	
		3/64 = .046 875	3/8	12/32	24/64 = .375			45/64 = .703 125	
1/16	2/32	4/64 = .062 5			25/64 = .390 625			23/32	46/64 = .718 75	
		5/64 = .078 125		13/32	26/64 = .406 25			47/64 = .734 375	
	3/32	6/64 = .093 75			27/64 = .421 875	3/4	24/32	48/64 = .75	
		7/64 = .109 375	7/16	14/32	28/64 = .437 5			49/64 = .765 625	
1/8	4/32	8/64 = .125			29/64 = .453 125		25/32	50/64 = .781 25	
		9/64 = .140 625		15/32	30/64 = .468 75			51/64 = .796 875	
	5/32	10/64 = .156 25			31/64 = .484 375	13/16	26/32	52/64 = .812 5	
		11/64 = .171 875	1/2	16/32	32/64 = .50			53/64 = .828 125	
3/16	6/32	12/64 = .187 5			33/64 = .515 625		27/32	54/64 = .843 75	
		13/64 = .203 125		17/32	34/64 = .531 25			55/64 = .859 375	
	7/32	14/64 = .218 75			35/64 = .546 875	7/8	28/32	56/64 = .875	
		15/64 = .234 375	9/16	18/32	36/64 = .562 5			57/64 = .890 625	
1/4	8/32	16/64 = .25			37/64 = .578 125		29/32	58/64 = .906 25	
		17/64 = .265 625		19/32	38/64 = .593 75			59/64 = .921 875	
	9/32	18/64 = .281 25			39/64 = .609 375	15/16	30/32	60/64 = .937 5	
		19/64 = .296 875	5/8	20/32	40/64 = .625			61/64 = .953 125	
5/16	10/32	20/64 = .312 5			41/64 = .640 625		31/32	62/64 = .968 75	
		21/64 = .328 125		21/32	42/64 = .656 25			63/64 = .984 375	

From Selby, S. M., Ed., *Standard Mathematical Tables*, 19th ed., Chemical Rubber, Cleveland, 1971, 7.

Table 4–26
PROPERTIES AND USES OF ADSORBENTS

For density in kg/m^3, multiply lb/ft^3 by 16.02.

Adsorbent	Shape of particles[a]	Bulk dry density, lb/ft^3	Surface area sq m/g	Uses and method of regeneration
Active alumina	G	50	250	Drying gases and liquids; catalyst; cat-
	S	55	350	alyst support; defluoridation of alky-lates; neutralization of lube oils. Can be regenerated.
Activated bauxite	C, G	~53	—	Decolorizing petroleum products and dry-ing of gases. Regeneration by heating.
Aluminosilicates (Molecular sieves)	C, S, P	~44	770	Selective adsorption based on molecular size and shape; drying of gases and liquids; catalyst support. Regenera-tion by heating or elution.
Carbon or charcoal:				Water treatment; air and gas purifica-
Bone	G	20–30		tion; gas masks and smoke filters; sol-
Coal	G	20–30	500–1200	vent recovery and purification; sugar
Petroleum	C	28–34	800–1100	refining; decolorizing of solutions;
Shell	G	10–20		decolorizing natural products. Ad-
Wood	G	10–35	625–1400	sorbed gases can be evaporated.
Clay	P	30–45	225–300	Refining petroleum fractions; purifying vegetable oils, juices; catalyst base.
Fuller's earth	G	30–40	130–250	Uses same as for clay. Regeneration by washing and burning adsorbed organic matter.
Silica gel	G, P	~25	320	Drying of gases; adsorption, from solu-
	S	50	650	tions; hydrocarbons; catalyst base. Regeneration by evaporation of adsorbed liquid.

[a]Shape of particles indicated as follows: C, cylindrical pellets; G, granular; P, powder; S, spherical beads.

Notes: Both surface areas and pore sizes are important in adsorption.

Distinction should be made between adsorption of gases by nonporous solids such as smooth metals and by porous

Table 4—26 (continued)
PROPERTIES AND USES OF ADSORBENTS

solids such as the adsorbents listed in this table. Nonporous inorganic block-solids have surface areas in the range below 10 m² /g; their adsorption is correspondingly small but definitely measurable.

Gas and vapor adsorption increases as the temperature is reduced and the pressure of the gas is increased. The usual method for quantitative expression of this relationship is in terms of the adsorption at constant temperature. The adsorption isotherms are plotted with adsorption (e.g., g of adsorbate/g of adsorbent) on the ordinate and pressure (actual pressure/ saturation pressure) on the abscissas. These isotherms have several typical shapes, but in any case the adsorption is much higher if the gas is not highly superheated, i.e., it increases as saturation is approached at the given temperature. The BET classification system (proposed by S. Brunauer, P. H. Emmet, and E. Teller in *J. Am. Chem. Soc.*, 60, 309, 1938, but still widely quoted) for adsorption isotherms recognizes five types of such curves.

REFERENCES

Gregg, S. J. and Sing, K. S. W., *Adsorption, Surface Area, and Porosity,* Academic Press, New York, 1967.
Hassler, J. W., *Activated Carbon,* Chemical Publishing, New York, 1963.
Symposium on Activated Carbon, Atlas Chemical Industries, Wilmington, Del., 1968.

From Bolz, R. E. and Tuve, G. L., Eds., *Handbook of Tables for Applied Engineering Science,* 2nd ed., CRC Press, Cleveland, 1973, 363.

Table 4—27
CLASSIFICATION OF PAINTS AND COATINGS[a]

Materials and practices in the field of paints and coatings are diverse and changing rapidly with the shift from natural to synthetic materials. This table and those following quote data on well-known practices and materials only, using industrial terminology. For more comprehensive data consult the references.

Classifications of coating materials are related to their uses, as indicated in the following table. Typical organic coatings are prepared with pigments and drying oils, with resins and solvents, or with latex or resin emulsions. The pigments are mixed for color, and dyes are sometimes added. Minor components may accelerate or retard drying. Many resin varnishes and lacquers are nearly transparent, but dyes, pigments, and metallic powders may be added for color effects. Plasticizers are used to improve the elastic qualities of the film.

The inexpensive coatings used in large quantities are simple formulations, but special uses may demand highly complex mixtures. Among the recognized and desirable properties are hiding power or opaqueness, light, heat, and chemical resistance, bleeding resistance, and ease of application. Other identified properties are penetration, chalking, gloss or texture, hardness, mold resistance, stain resistance, and package stability.

Class	Description or source	Forms and properties	Typical uses
Oil paint	Vegetable oil and mineral pigment	Linseed (or other) drying oil with opaque but colored (or white) pigment	Exterior and general painting; interior trim; industrial products
Latex paint	Partially polymerized liquid resins in water	Water paint for porous and absorbent materials; alkali resistant	Wood and masonry surfaces; wall board; sheathings and sidings; interiors
Varnish	Drying oil, resin, solvent, and drier	Usually transparent to amber in color; glossy and impervious	Furniture, woodwork, floors, boats, trim, sealers
Lacquer	Nitrocellulose, acrylic, or other resins in volatile solvent	Quick-drying by solvent evaporation	Automobiles; transport equipment; metal products; oven-dried finishes
Whitewash	Lime, water, and additives	Age-old water paint for decorative and reflective purposes	Inexpensive outdoor or barn paint

Table 4–27 (continued)
CLASSIFICATION OF PAINTS AND COATINGS

Class	Description or source	Forms and properties	Typical uses
Cement-water	White cement, lime, and pigment	Water paint for smooth finish on coarse masonry	Concrete or tile finish or grout-coat
Glue or size	Gelatin, skin, bone, starch	Water solutions	Sealer for plaster, etc.
Bituminous	Petroleum; coal tar	Paints; hot mastic; emulsions	Underground; waterproofing; roofs; tanks
Fire retardant	Brominated resins; insulation; glass	Flame resistant and/or insulating or glazing	Prevention of flame spread over combustible surfaces or textiles
Chlorinated rubber	65% chlorine with poly-isoprene and plasticizer	Paints; chemical and water resistant	Traffic, swimming pool, and masonry paints; quick-drying industrial coatings
Strippable coating	Cellulose esters, or other resins, solvent and oil	Spray or hot-dip (oil migrates to interface)	Protection of machine parts and assemblies; "mothballing"
Fluorescent	Added fluorescent dyes	Absorbed violet reemitted as yellow, orange	High visibility for signs, hazard protection, advertising
Wax polish	Hard wax plus resin	Organic solvent, water emulsion, or both	Floors, furniture, automobiles, paneling

[a]For corrosion-resistant coatings, see Table 4–33.

REFERENCES

Roberts, A. G., *Organic Coatings – Properties, Selection, and Use,* National Bureau of Standards, Washington, D.C., 1968.

Morgans, W. M., *Outlines of Paint Technology,* Griffin, London, 1969.

Martens, C. R., *Technology of Paints, Varnishes, and Lacquers,* Van Nostrand Reinhold, New York, 1968.

Myers, R. R. and Long, J. S., Eds., *Treatise on Coatings,* Vol. 1, Marcel Dekker, New York, 1967.

Parker, D. H., *Principles of Surface Coating Technology,* John Wiley & Sons, New York, 1965.

Payne, H. F., *Organic Coating Technology,* John Wiley & Sons, New York, 1954.

Von Fischer, W. and Bobalek, E. G., Eds., *Organic Protective Coatings,* Reinhold, New York, 1953.

Abraham, H., *Asphalts and Allied Substances,* D. Van Nostrand, Princeton, N.J., 1945.

National Paint, Varnish, and Lacquer Association, *Raw Materials Index,* Washington, D.C.

Bennet, H., Bishop, J. L., Wulfinghoff, M. F., *Practical Emulsions,* Vol. 2, Chemical Publishing, New York, 1968.

From Bolz, R. E. and Tuve, G. L., Eds., *Handbook of Tables for Applied Engineering Science,* 2nd ed., CRC Press, Cleveland, 1973, 352.

Table 4–28
ORGANIC COATINGS – DRYING OILS AND DRIERS

Drying oils contain glycerides or esters from a combination of glycerol with unsaturated fatty acids. The drying of a thin oil film is a complex chemical process. Oxygen plays a major part, and the process is retarded at low temperatures and in the absence of light. It is accelerated by the catalytic action compounds of certain metals (e.g., lead).

Typical Properties of Drying Oils

Oil	Specific gravity	Viscosity, cp	Refractive index	Saponification value[a]	Acid value[b]	Iodine value[c]
Linseed (flax)						
Raw	0.93	40	1.48	192	2	177
Boiled	0.95	100	1.48		3	165
Blown	0.97	300	1.48		5	120
Tung	0.94		1.52	193	7	162
Fish	0.93		1.48	191	6	165
Tobacco seed	0.92		1.48	190	5	140
Soybean	0.93		1.47	190	2	135
Tall oil				169	160	

[a]Saponification value is the quantity of potassium hydroxide required to saponify 1 g of oil.
[b]Acid value is a quantitative index of the amount of free organic acid in an oil.
[c]Iodine value is a measure of the degree of unsaturation of the oil and, hence, of its drying properties.

Notes: Raw linseed oil dries slowly. Boiled oil, which has been heated with lead, cobalt, or manganese compounds to reduce drying time, is preferable for interior paints. Blown oil has been thickened by blowing air through hot oil.
Tung-nut oil is fast-drying and has high-water resistance.
Fish oil dries slowly and has a strong odor and darker color unless refined. It is heat resistant and elastic and is used in high-temperature and roofing paints.
Soybean and other seed oils are semidrying but are used in oil blends to give elasticity and durability to the paint film.
Tall oil is a semidrying oil, a by-product of the sulfite papermaking process. It is expensive and is used with various synthetic resins.

From Bolz, R. E. and Tuve, G. L., Eds., *Handbook of Tables for Applied Engineering Science,* 2nd ed., CRC Press, Cleveland, 1973, 353.

Table 4–29
ORGANIC COATINGS – SOLVENTS AND THINNERS

Typical Designations and Properties

Name	Alternative name	Specific gravity	Boiling range, deg C	Flash point, deg C (closed)	ASTM No.
PETROLEUM PRODUCTS					
Mineral spirits	Petroleum ether; ligroin	0.66	60–80	5	D484
Naphtha thinner	White spirits; turpentine substitute		150–190	40	D235
Aromatic naphtha		0.85			
COAL-TAR PRODUCTS					
Toluene	Toluol	0.87	110	5	D362
Xylene (3 isomers)		0.86	137–148	27	D364
Coal-tar naphtha	Heavy naphtha		<190	37	
CONIFEROUS TREE PRODUCTS					
Turpentine (gum, wood)		0.85	150–180	30–36	D13
Dipentene			175–195	54	
Pine oil			200–230		D802
ORGANIC COMPOUNDS					
Acetone		0.79	56	–18	D329
Amyl acetate	Banana oil	0.88	146	24	
Butanol	Butyl alcohol	0.81	118	25	D304
Denatured alcohol	Methylated ethanol	0.79	78	14	
Ethyl acetate	Acetic ester	0.90	77	5	
Isopropyl alcohol	2-Propanol	0.79	82	14	D770
Methanol	Wood alcohol	0.79	65	14	D1152
Propyl alcohol	*n*-Propanol	0.80	98	22	
Trichloroethylene		1.46	87	–	

Compiled from several sources and presented in Bolz, R. E. and Tuve, G. L., Eds., *Handbook of Tables for Applied Engineering Science*, 2nd ed., CRC Press, Cleveland, 1973, 353.

Table 4–30
ORGANIC COATINGS – NATURAL RESINS AND MATERIALS[a]

Natural and Processed Materials Used in Paint, Varnish, and Protective Coatings and Finishes

Common name	Alternate name	Source	Properties	Uses
Rosin	Colophony; gum and wood rosins; tall oil rosin; phenolic blend	U.S.A., France, Portugal, Spain; tapped from coniferous trees	Depends on processing; dispersion, adhesion, gloss	Wide variety of rosin products, e.g., ester gum, limed rosin, zinc resinate, maleic resin; for varnishes, enamels, lacquers
Copal resin	Class name for several resins	Africa, East Indies, Philippines	Hard, glossy film for varnish or enamel	Largely displaced by synthetic resins
Damar resin	Singapore; dipterocarpus	East Indies	Soft, non-yellowing, fume-resistant film	Specialty finishes; paper varnish (not widely used)
Lac	Shellac base; seed lac	India; insect excretion from tree sap	M.p., about 80 deg C; acid value, 70; iodine value, 20; soluble in alcohols; color, orange or darker; tough; fast-drying	Sealer and base coat; foundry patterns; white shellac on floors; largely displaced by synthetics

Table 4–30 (continued)
ORGANIC COATINGS – NATURAL RESINS AND MATERIALS

Common name	Alternate name	Source	Properties	Uses
Rubber	Chlorinated rubber; cyclized rubber	Natural rubber latex	Tough; resistant to abrasion, moisture, chemicals, mildew, petroleum; quick-drying	In blended paints for chemical apparatus, marine uses, traffic areas, swimming pools, rust protection
Asphalt	Bitumen; gilsonite; manjak	Petroleum residues; some natural deposits, as in Trinidad	Softens at 46 deg C; good adhesion; waterproof and chemical-resistant	Roof and waterproof paints; corrosion protection; sound control
Pitch	Coal tar	Residue from coal-tar distillation	Impervious to water; high dielectric strength; heavy coating	Ship-bottom paint; moisture and corrosion protection; chemically resistant paint
Gum	Examples: gum arabic, gum tragacanth	Africa and Asia	Thickening agent and base for water colors	Seldom used except in artists' paints
Glue	Glue size	Skin, bone, animal, and fish scrap	Binder for water paint and ceiling white; adhesion; reduces "suction"	Older types of water paint and wall-size; sealers
Casein	Milk protein	Skimmed milk	Adhesion; thickening	Gel and thickener for water paints, including latex; wallboard base coat
Dextrin	Starch protein	Grain and seeds; potatoes	Adhesion; stiffener	Sizing and coating for fabric, paper, wallboard
Wax	Beeswax, paraffin wax, carnauba wax, etc.	Natural waxes; petroleum	Resistant to moisture, spray, marring, light abrasion; high gloss	Widely used in protective polishes; small amounts added to decorative enamels and varnishes

[a]See also Tables 4–31 and 4–32.

From Bolz, R. E. and Tuve, G. L., Eds., *Handbook of Tables for Applied Engineering Science,* 2nd ed., CRC Press, Cleveland, 1973, 354.

Table 4–31
ORGANIC COATINGS – SYNTHETIC RESINS

Classes, Modifications, Properties, and Applications of Synthetic Resins for Paints and Coatings

Class	Kinds and modifications	Properties or advantages	Usage
Acrylic	Emulsions and water-soluble polymers; thermoplastic or thermosetting, soluble in organic solvents	Thermoplastic; resistant to age, light, water, chemicals, oil; hardness; flexibility	Very widely used in both solution and emulsion; clear coatings; enamels; latex paints; automobile and metal finishes
Alkyd	Alkyd-oil; styrenated; modified; combined with other resins	Solubility; compatibility; durability; hardness; toughness; gloss retention; adhesion	Most widely used of all resins; air-dry or baked; paints, varnishes, enamels, primers
Amino	Urea and melamine; blends	Hardness; color retention; fast baking; durability	Durable finishes for automobiles, appliances, metal surfaces
Cellulose	Cellulose nitrate and others; also combinations with alkyd and amino resins	Flammable; compatibility with oils, plasticizers, and other resins	Quick-drying lacquers, enamels, dopes
Epoxy	Epoxy polyamide; esterified; amine catalyzed	Adhesion; toughness; chemical resistance; quick drying	Baking enamels; marine varnish; can coating; chemical equipment
Fluorocarbon	Polytetrafluoroethylene; polychlorotrifluoroethylene	Resistant to moisture, heat, chemicals, fungi, and abrasion	Chemical-resistant linings for tanks and industrial equipment; electrical insulation; weatherproofing
Phenolic	Bakelite; modified oil-soluble	Resistant to chemicals and solvents; water- and weather-proof	Primer, sealer, varnish; structural and marine paints
Polyamide	Nylon; thermoplastic polyamides	Strength, toughness, abrasion resistance; oil and solvent resistance; good adhesion	Wear-resistant and chemical-resistant coating for metal textiles, leather, paper, industrial equipment
	Glycol esters of unsaturated and saturated dibasic acids, such as with phthalic acid isomers or maleic or fumaric acids. Unsaturated acid types are usually used as copolymers with styrene or acrylic monomers	No volatile thinners need to be expelled in forming thick or thin films that are hard and adhesive	High-build coatings with or without fillers for wood, concrete, and other non-metallic substrates
Polyethylene	Low- and high-density; chlorinated unsaturated	Odorless, tasteless, non-toxic; waterproof; high strength; toughness; flexibility	Coatings for food cartons; wire insulation; flame-sprayed coatings on metals; textile coatings
Polystyrene	Used largely in copolymers and latex	Flexibility; adhesion; toughness; weather resistance	With butadiene or alkyd for latex paints; solution-type outdoor paints
Polyurethane	Urethane oil or alkyd; moisture-cured and two-package compositions	Flexibility; high gloss; abrasion and chemical resistance; toughness; dielectric strength	Varnishes for severe service; enamels; marine paints; textile coatings; wire coatings; concrete finish
Rubbers	Neoprene; butyl; nitrile; SBR, and copolymers of isoprene, butadiene, polypropylene	Variety of properties including chemical resistance, resilience, abrasion resistance, oil resistance	Coatings where resilience, elasticity, and abrasion resistance are needed
Silicone	Copolymer or modified with metal powder or frit	Stable in the 200–500 deg C range with heat-resistant pigments	Heat-stable and high temperature paints; corrosion-resistant paints
Vinyl	Acetate, chloride; copolymers; plastisols	Toughness; flexibility; chemical resistance; wear resistance; dielectric strength	Masonry finish; textured coatings; outdoor uses

From Bolz, R. E. and Tuve, G. L., Eds., *Handbook of Tables for Applied Engineering Science,* 2nd ed., CRC Press, Cleveland, 1973, 355.

Table 4–32
PIGMENTS FOR PAINTS AND COATINGS[a]

Table 4–32
PIGMENTS FOR PAINTS AND COATINGS[a]

The term *pigments* was formerly applied to natural color materials such as those used by artists. Surface coatings today are increasingly dependent on the synthetic resins. The following table covers most of the common classes of paint solids other than the synthetic resins (Table 4–30), the natural resins and gums (Table 4–31), and the water paint powders such as lime and cement.

A paint mixture to be applied as a surface coating is often described in terms of two constituents — the pigment and the vehicle. The liquid vehicle wets the surface and dries or cures into a film, while the pigment, dispersed in the vehicle, functions largely as a radiation absorber and reflector, providing opacity or color. Common insoluble pigments are mixtures of very fine powders or flakes of metal oxides or salts and sometimes the metals themselves. Other constituents are added to the pigment or to the vehicle for special purposes such as controlling the gloss, hardness, adhesion, abrasion resistance, and weather resistance of the film or providing corrosion resistance or increasing the dielectric strength.

For latex paints, and for solution paints such as shellac or asphalt paint, the dual-constituent classification in terms of a film-forming vehicle and a surface-hiding pigment is not applicable. The same is true for such water paints as whitewash, and for many of the complex formulations that often include both dispersions and solutions and even emulsions.

Common Paint Pigments

Name and composition	Typical particle size, μm	Specific gravity	Refractive index	Hiding power	Properties and uses	ASTM No.
WHITE PIGMENTS (OPAQUE)						
Titanium dioxide (anatase or rutile)	0.25	3.5	2.7	Excellent	Brilliant white; anatase form is chalking; non-toxic	D476
White lead (basic carbonate, about 68%), $2PbCO_3 \cdot Pb(OH)_2$; flake white		6.7	2.0	Good	Durable film; tends to darken; weather and water resistant; toxic; primer and undercoat	D81
Zinc oxide, ZnO; Chinese white	0.2–0.3	5.65	2.0	Good	Mildew resistant; highly durable; outdoor oil paints	D79
Zinc sulfide			2.37	Very good	Not widely used	D477
Antimony oxide	0.5–2.0	5.5	2.1	Good	Little used except in fire-retardant paint	
Lithopone (regular), approx 70% $BaSO_4$ and 30% ZnS	0.2–2.0	4.2	1.9	Fair	Interior oil and emulsion paints	D477
WHITE EXTENDER PIGMENTS (LOW REFRACTIVE INDEX)						
Calcium carbonate, $CaCO_3$; precipitated chalk; whiting	2–5	2.7	1.58	Poor	Ceiling white; undercoat	D1199
Calcium sulfate, $CaSO_4 \cdot 2H_2O$; gypsum		2.35	1.53	Poor	Filler; limited use in paints	
Magnesium silicate, $H_2Mg_3(SiO_3)_4$; talc		2.7	1.57	Poor	Flatting; anti-settling; chemically inert	D605
Barium sulfate, $BaSO_4$; barytes; barite		4.4	1.64	Poor	Undercoats; fillers; chemically stable	D602
China clay, $Al_2O_3 \cdot 2SiO_2 \cdot 2H_2O$; kaolin	1.0 (avg)	2.5	1.56	Poor	Undercoats; thickening agent	
Hydrated silica	1–10	2.2			Flatting; consistency control	D604

Table 4–32 (continued)
PIGMENTS FOR PAINTS AND COATINGS

Name and composition	Typical particle size, μm	Specific gravity	Refractive index	Hiding power	Properties and uses	ASTM No.
BLACK, GRAY, BROWN PIGMENTS						
Carbon black; furnace black; impingement black	.02–.09		Opaque	Good	Very widely used; jet-black gloss	
Lampblack, C; oil black		1.8	Opaque	Good	Undercoats; primers; high adhesion; blue undertone; matte	D209
Graphite (amorphous)		2.3	Opaque	Good	Topcoat for steel structures; high coverage; durable	
Iron oxide black, Fe_3O_4; magnetite (synthetic)	0.5		Opaque	Good	Metal paints and fillers	D769
Bone black; ivory drop black	325 mesh (solution)	2.3	Opaque	Good	Black undercoat and filler	D210
Asphaltum; cut-back asphalt; petroleum paint (black to brown)			Opaque	Fair	High adhesion; automotive uses; roof paint; chemical and water resistance; also emulsions	
Blue lead (45% lead sulfate, 30% lead oxide); basic lead sulfate, sublimed			Opaque	Good	Structural steel primer (gray)	D405
Sienna and umber, largely Fe_2O_3; range of brown and gray (to red synthetics)		3.3	1.9	Fair	Widely used tinting colors; low cost	
COLORS						
Prussian blue (ferrocyanides); iron blue		1.8	1.55	Fair	Low cost; non-bleeding; high-temperature baking; light tints may fade	D261
Cobalt blue (cobalt and aluminum oxides)		3.8	1.7	Fair	Expensive; art finishes; highly stable and resistant	
Cobalt green (cobalt and zinc oxides)		4.0	1.95	Fair	Highly stable and resistant; not widely used	
Chrome green (yellow $PbCrO_4$ with $PbSO_4$ blended with iron blue)		5.1	2.5	Excellent	Widely used; wide range of greens; good color retention	D263; D212
Red lead (synthetic), Pb_3O_4 >85%	2.0	8.7	2.4	Excellent	Durable; good adhesion; corrosion protection; primer for steel; fades in sun	D83; D49
Iron oxide, Fe_2O_3; red hematite; red to maroon and brown		5.2	2.5	Excellent	Inexpensive; widely used	
Cadmium red, largely CdS		4.5	2.7	Excellent	Bright tones; non-bleeding high-temperature bake	
Manganese violet			1.7	Good		
Chrome yellow and orange, largely $PbCrO_4$ (with $PbSO_4$ light and PbO orange)		5.9	2.3	Excellent	Bright tints	D211
Zinc-yellow, ZnO and CrO_3; zinc chromate		3.5	1.9	Good	Exterior paints; resists darkening	D478
Yellow ochre, Fe_2O_3; ferrite yellow; natural or synthetic	0.5–1.5	3.5	2.0	Good	Inexpensive; low-temperature bake	D85
Cadmium yellow, largely CdS with CdSe; also lithopone with $BaSO_4$		4.3	2.4	Excellent	Interior oil and water paints	

Table 4–32 (continued)
PIGMENTS FOR PAINTS AND COATINGS

Name and composition	Typical particle size, μm	Specific gravity	Refractive index	Hiding power	Properties and uses	ASTM No.
ORGANIC TONERS	0.01–0.1	Variable	Varies with wavelength	Poor to excellent	Cover spectral range from red to violet; some with hiding power are used in mass tone or solid colors; most with low hiding power give vivid colors and are used in mixtures with titanium dioxide or aluminum flake	
METAL PIGMENTS						
Aluminum (leafing—flakes overlap)			Opaque	Excellent	Brilliant metallic finish; durable	D962
Aluminum (non-leafing)	100–200 mesh		Opaque	Good	Less brilliant; widely used; chemical and heat resistance; durable	D962
Aluminum (extra fine)	400 mesh		Opaque	Excellent	High hiding power; durable	D962
Zinc dust	325 mesh		Opaque	Good	Rust-inhibitive primer for steel and galvanized iron; weatherproof paints	D520
Copper bronze, 2% zinc			Opaque	Good	Decorative copper finish; fungicidal paint	
Gold bronze, Cu, Zn, and Al			Opaque	Good	Decorative color range, pale gold to red gold; exterior and interior	D267
FLUORESCENT PIGMENTS						
Organic-dyed resins (ultraviolet reflectors)			Variable	Fair	Maximum visibility in yellow-orange range; high-visibility coatings; safety paints	

[a]This table has been compiled from several sources. While it includes many common pigments, it is in no sense a complete list. The synthetic organic pigments are entirely omitted, and many common mixtures and blends are not included. A more complete treatment should also give attention to the science of color, to phenomena of spectral absorption and reflectance, and to the various methods for specifying, matching, and measuring colors (see References, Table 4–27, and Table 4–52).

From Bolz, R. E. and Tuve, G. L., Eds., *Handbook of Tables for Applied Engineering Science,* 2nd ed., CRC Press, Cleveland, 1973, 356.

Table 4–33

PAINTS AND COATINGS FOR CORROSION RESISTANCE

Corrosion is a very general term applied to a variety of processes by which metal surfaces are attacked and converted to oxides, sulfides, or other compounds. As much corrosion is electro-chemical, methods of protection are based on isolation of the metal from the environment that contains oxygen, moisture, sulfur, etc., and on minimizing galvanic potentials. The methods by which these steps are accomplished are so diverse that no single summary can adequately cover them.

Galvanic corrosion is prevented by elimination of contact between metals, whether direct or through an electrolyte. A most common case is copper and iron. Surface treatment of steel with a phosphate wash or a primer such as red lead or zinc chrome may form a thin protective layer of stable compound, over which an impervious and durable coating will adhere.

The following table lists many of the common protective coatings, most of which are used on iron and steel, since they present the corrosion problem of greatest economic importance. Asphalt and red lead have long been widely used, but there is an increasing use of protective metals and of synthetic resins. For detailed discussions consult the references listed in Table 4–27.

Name or protective constituent	Description or composition	Advantages	Typical application
Aluminum	Leafing or non-leafing paints, varnishes and primers (hot dipping)	High hiding; fume-resistant; low cost; high coverage; ultraviolet protection	One-coat protection for w use; additive for asphalt and phenolic resin paint
Asphalt	Petroleum residue and solvents; with asbestos, vermiculite, and perlite	Low cost; moisture and chemical resistance; good adhesion	Thinned or hot; pipe lines tanks, undercoat; buried structures
Barium potassium chromate	With barium and chromium oxides	High-strength film; elastic and durable	For steel or light metals; a drying or baking types
Calcium plumbate	Calcium and lead oxides	Quick-drying; smooth; hard; salt-resistant	Primers for structural and galvanized steel
Carbon black	Furnace black or amorphous graphite	High hiding; low cost; high coverage; high chemical resistance	With black iron oxide for shop and foundry paint topcoat varnish
Chlorinated rubber	10–25% in oil or alkyd paints or lacquers	Chemical resistance; low permeability; wide color range	Marine paint; machinery finish; high-build paints chemical equipment
Etch primer	Phosphoric acid or phosphate; often with zinc chromate	Versatile water-solution wash as quick-dry primer	Clean-metal primers for g overcoat adhesion; vari metals, including galvan
Lacquer	Resin lacquers, as nitro-cellulose and acrylic	Transparent or colors; quick drying; cold or hot application	For highly finished metal surfaces, including brass
Lead cyanamide	Yellow $Pb(CN)_2$	Anodic protection	Primer for steel; with linse oil and pigments
Pitch, coal-tar	Coal tar in aromatic solvent	Black, brown, green; glossy; hard; adhesive; chemical resistance	Emulsion topcoat; with ep or other resin for sea-wa immersion
Phenolic resin	Air-dry varnish and baking lacquer	Very low permeability; good baking enamel; good adhesion	Chemically resistant paints electric insulation varni tank linings
Red lead	Pb_3O_4 and PbO with linseed oil	Readily available; weather-resistant; good adhesion	Most widely used where linseed oil is common; outdoor ferrous structu
Red iron oxide	30–50% in oil or chlorinated rubber paints	Low cost; one or two coats without primer; high-hiding; durable	Ferrous metal primer or o coat protection

Table 4–33 (continued)
PAINTS AND COATINGS FOR CORROSION RESISTANCE

Name or protec- tive constituent	Description or composition	Advantages	Typical applications
Stainless steel flake	Barrier coat	Inherently corrosion resistant; good appearance	Metallic barrier for chemical resistance
Zinc chrome	10% or more in non-ferrous primers; with red iron oxide for ferrous metals	High-hiding, sealing, and adhesion; salt-resistant	On light metals; with red iron oxide on ferrous metals; alkyd paints

Notes: Surface preparation is very important prior to coating treatment. Solvent degreasing, alkali washing, sand or shot blasting, pickling, acid or phosphate dipping, flame treatment, wire brushing, and abrasive cleaning are among the methods that should receive consideration for surface preparation.

Dual protective films and even complex mixtures are often used for final-coat or one-coat protection. Examples are pigment with drying oil, pigment and alkyd or phenolic vehicle, powdered or leafing metal with polyvinyl chloride or other resin, pigment and chlorinated rubber, and pigment mixtures with both chlorinated rubber and drying oil. In each case there is at least a dual barrier coat, sometimes over a primer.

High pigment content (40–90%) is characteristic of most protective paints, since these solids act as barrier coats.

Synthetic resins are now being used in many corrosion-protective coatings. Alkyd, PVC, epoxy, vinyl, and certain copolymers are common in paints with various protective pigments.

REFERENCE

Burns, R. M. and Bradley, W. W., *Protective Coatings for Metals,* 3rd ed., Van Nostrand Reinhold, New York, 1967.

From Bolz, R. E. and Tuve, G. L., Eds., *Handbook of Tables for Applied Engineering Science,* 2nd ed., CRC Press, Cleveland, 1973, 359.

Table 4–34
INORGANIC SURFACE COATINGS[a]

Functions, Materials, and Application Methods

Functions — Inorganic coatings are applied to metals and to some other materials to protect and to *modify* the surface properties, as follows:

Mechanical: for resistance to abrasion, erosion, shock, and for control of hardness, texture, smoothness, lubrication.

Chemical: for resistance to oxidation, corrosion, chemical reaction, diffusion.

Thermal: for heat insulation, ablation, control of heat transfer by conduction and by radiation.

Electrical: for control of electrical resistance, magnetic properties, thermionic performance.

Radiation: for control of light and radiative properties including reflection, absorption, diffusion or transmission of radiation (total or spectral); for optical imaging by phosphor.

Materials — Coating materials are used singly and in numerous combinations. Most coatings are chemical compounds of the metallic elements.

Oxides of aluminum, silicon, calcium, magnesium, zirconium, chromium, beryllium, thorium, hafnium, nickel, etc.

Other oxygen compounds: the aluminates, silicates, chromates, zirconates.

Hard metal compounds: the carbides, nitrides, borides, silicides.

Intermetallics: the aluminides, beryllides, stannides.

Combinations of materials: classified under such common class names as enamels, glasses, ceramics, cermets, refractories, composites.

Application methods — These may include more than one of the following steps or processes:

Table 4–34 (continued)
INORGANIC SURFACE COATINGS

Adhesion	Flame spraying	Spreading
Atomized spraying	Fusion or firing	Trowelling
Dipping (hot or cold)	Packed retorts	Vapor deposition (pyrolysis)
Electrophoresis	Painting; brushing	
Electroplating	Slip or slurry (fired)	

Examples – These might include a very long list and a wide variety of uses, but those listed in the following table are illustrative.

Common Inorganic Coatings

Coating name or function	Base or substrate	Coating description, processes, and typical compositions	Typical uses
Enamel (porcelain)	Steel	Surface coated with slip or slurry of powdered glasses and colors. Oxides of Si, Al, Na, B, and Ca predominate	Cookware, signs, tanks, tubs, pails, housewares
Enamel (vitreous)	Cast iron	Similar to above	Sanitary ware, structural decoration
Enamel	Aluminum	Low-melting frit (below 1050 deg F) containing PbO, Li_2O, SrO, TiO_2, etc. Aluminum alloy must be enameling grade	Architectural trim, siding, wall coverings; highway and advertising signs
Anodized	Aluminum	Anodic oxidation and coloring. Processes use phosphoric, sulfuric, or chromic acid	Architectural and vehicle trim; housewares; aircraft materials; handrails
Reflector	Glass, metal, or crystal	Mirror metals or paints for control of light or electromagnetic radiation	Solar reflection, optical instruments, thermal radiation control
Photo converter	Metal or glass	Amplifying, photoemissive, and photo-voltaic surfaces with compounds of Pb, Cd, Ge, and Cs	Solar cells; photomultipliers, amplifiers, detection, and measurement
Phosphor	Glass	Optical imaging, using zinc and cadmium sulfides, silicates	Cathode-ray and television screens; instrument displays, illumination tubes; particle detectors
Ablation coating	Metal	Wound or woven glass fibers with phenolic resin binder	Rockets, nosecones, re-entry vehicles
Lubricant	Metal pair	Lubricating powders and mixtures, usually oxides of Pb, Bi, Cd, or di-sulfides of Mo or W	High-temperature and space applications; journal and ball bearings; extrusion dies
Electrical insulation	Metal wire or surface	Enameled conductors baked from slurry-covered surface; thermionic layers	Large coils and magnets; structural separators
Diffusion coating	Metals	Reactive coating material diffused into substrate by heat treatment. Compounds of Al, Cr, Si, and Ti most common	Oxidation-resistant or hard-surface layers on metals and alloys
Ceramics and cermets	Metals and alloys	Sprayed ceramic mixtures and electro-deposited particles from stirred aqueous bath	High-temperature applications with short-term erosion
Fire retardants	Wood, fiber, or plastic	Compounds of P, Cl, Sb, Br, and B	Fireproofing of combustible surfaces
Thermal insulation	Metals	Thick, porous refractory and fiber coatings and bonded ceramic foams; refractory cements	High-temperature protection for structural materials

[a]For data on organic coatings, see Tables 4–27 through 4–33.

Table 4–34 (continued)
INORGANIC SURFACE COATINGS

REFERENCES

Plunkett, J. D., *NASA Contributions to the Technology of Inorganic Coatings*, NASA SP-5014, National Aeronautics and Space Administration, 1964.

Huminik, J., *High-temperature Inorganic Coatings*, Van Nostrand Reinhold, New York, 1963.

Campbell, I. E. and Sherwood, E. M., Eds., *High-temperature Materials and Technology*, John Wiley & Sons, New York, 1967.

Van Horn, K. R., Ed., *Aluminum, Fabrication and Finishing*, Vol. 3, American Society for Metals, Metals Park, O., 1967.

Andrews, A. I., *Porcelain Enamels*, Garrard, Champaigne, Ill., 1961.

From Bolz, R. E. and Tuve, G. L., Eds., *Handbook of Tables for Applied Engineering Science*, 2nd ed., CRC Press, Cleveland, 1973, 360.

Table 4–35
SUMMARY OF ELECTROPLATING PRACTICE

Average Operating Conditions

Metal	Principal uses	Type of solution	Principal ingredients	Temp, °F	CD ASF	Volts	Cathode efficiency, %	Time to deposit, 0.001 in.
Cadmium	Protection	Cyanide	CdO, NaCN, brighteners	70–95	15–45	1–4	90	20 min
Chromium	Decorative Engineering (hard) Cylinder liners (porous)	Chromic acid	CrO_3, H_2SO_4	120	250	6–8	15	2 hr
Copper	Electroforming	Acid	$CuSO_4 \cdot 5H_2O$, H_2SO_4	75–120	15–40	1–2	100	35 min
	Undercoat for other metals	Cyanide	CuCN, NaCN, Na_2CO_3	75–100	5–15	1.5–3	50	90 min
	Stop-off in case-hardening, etc.	Rochelle	Above + rochelle salts	140–160	20–60	2–3	60	45 min
		Many other types, *e.g.*, fluoborate, pyrophosphate, amine, all-potassium cyanide		—		—		—
Gold	Decorative	Cyanide	$KAu(CN)_2$, K_2CO_3, KCN (Solutions vary considerably, depending on color wanted)	120–160	5–15	2–6	80	—
Indium	Bearing surfaces	Cyanide	$InCl_3$, NaCN, addition agent	Room	10–150	—	40	—
		Sulfate	$In_2(SO_4)_3$, Na_2SO_4	Room	20	—	75	—
		Fluoborate	$In(BF_4)_3$, H_3BO_3, NH_4BF_4	70–90	50–100	—	50	—
Iron	Electroforming	Chloride	$FeCl_2$, $CaCl_2$	190	60	—	95	20 min
	Repair	Sulfate	$FeSO_4(NH_4)_2SO_4$	Room	20	—	95	1 hr
Lead	Protection	Fluoborate	$Pb(BF_4)_2$, HBF_4, glue	Room	10–80	0.5	100	40 min
	Bearing surfaces	Sulfamate	Pb sulfamate, sulfamic acid, addition agents	75–120	5–40	3–8	100	20 min
Nickel	Protection Decorative Electroforming Undercoat for Cr, etc.	Sulfate-chloride	$NiCl_2$, $NiSO_4$, NH_4 ion, H_3BO_3 (Formulations differ widely, depending on purpose)	75–100	Varies greatly	0.5–3	95	30 min
Rhodium	Decorative Optical	Sulfate Phosphate	Prepared salts	110–120	10–80	2.5–5	15	--
Silver	Decorative Protective Bearing surfaces	Cyanide	AgCN, KCN, K_2CO_3, CS_2 (Or Na in place of K)	80	5–15	1	100	—
Tin	Protection Food and dairy Bearings Electrical To enable easy soldering	Sulfate	$SnSO_4$, H_2SO_4, addition agents	Room	40	1–3	90	15 min
		Fluoborate	$Sn(BF_4)_2$, HBF_4, addition agents	75–100	50	—	100	10 min
		Other acid electrolytes		—	—	—	—	—
		Stannate	Na_2- or $K_2Sn(OH)_6$, Na- or KOH	150–190	40	4–8	80	30 min
Zinc	Protection	Sulfate	$ZnSO_4$, NH_4Cl, addition agents	75–100	15–400	—	99	10 min
		Cyanide	$Zn(CN)_2$, NaCN, NaOH, brighteners	100	10–50	—	85	40 min

Table 4—35 (continued)
SUMMARY OF ELECTROPLATING PRACTICE

ALLOYS

Metal	Principal uses	Type of solution	Principal ingredients	Temp, °F	CD ASF	Volts	Cathode efficien- cy, %	Time to deposit, 0.001 in.
Brass	Rubber-bonding Decorative	Cyanide	CuCN, Zn(CN)$_2$, NaCN, Na$_2$CO$_3$	75–100	3–10	2–3	75	—
Bronze	Decorative Undercoat for chromium Stop-off for steel	Cyanide-stannate	CuCN, KCN, KOH, K$_2$Sn(OH)$_6$, rochelle salt	155	20–100	3–6	70	30 min
		Pyrophosphate- cyanide	Sn$_2$P$_2$O$_7$, KCN, CuCN, K$_4$P$_2$O$_7$, addition agents	140–180	20–70	2–5	70	30 min
Lead-tin	Bearings Solderability Electrotyping	Fluoborate	Sn(BF$_4$)$_2$, Pb(BF$_4$)$_2$, HBF$_4$, addition agents	Room	60	1–2	100	—
Tin-zinc	Solderability	Cyanide-stannate	Zn(CN)$_2$, KCN, KOH, K$_2$Sn(OH)$_6$	150	10–75	4–5	80–95	30 min

REFERENCES

Gray, A., Ed., *Modern Electroplating,* John Wiley & Sons, New York, 1953.
Plating and Finishing Guidebook-Directory, Finishing Publications (published yearly).
Bandes, H., *Trans. Electrochem. Soc.,* 88, 263, 1945.
Technical Data Sheet No. 127, Metal and Thermit Corp., New York, 1954.
Safranek, W. H. and Faust, C. L., *Proc. Am. Electroplat. Soc.,* 41, 201, 1954.

From Knowlton, A. E., Ed., *Standard Handbook for Electrical Engineers,* 9th ed., McGraw-Hill, New York, copyright © 1957. With permission.

Table 4—36
FLAME-SPRAYED PROTECTIVE COATINGS

Metal and Nonmetal Coatings by Jet Application

The appearance and life of the functional parts of a structure or a machine depend largely on the properties of their surfaces. An important part of the engineering design is, therefore, involved in the selection and processing of materials for surface qualities such as smoothness, hardness, and resistance to deterioration. Surfaces of base materials may actually be modified in molecular structure and composition, physically processed for smoothness or texture, or actually covered with a more suitable surface material by spraying, welding, cladding, plating, veneering, or coating.

Spray guns — Metal is sprayed with a heater-atomizer gun forming a high-temperature jet. The impingement force produces adhesion and metal build-up by face velocities of 300 to 700 fps. The metal is supplied in wire, powder, or molten form. Spray guns are designed accordingly, ranging from self-contained hand guns to large, mounted equipment under automatic control. Oxygen-fuel gas torches are combined with compressed air for atomizing. For refractory metals a d-c arc plasma jet produces temperatures of 8,000°F or more, heating an inert gas (nitrogen or argon). Complex, proprietary designs are common for nozzles, feeders, and controls.

Surface preparation — Surfaces must first be cleaned by washing, degreasing, solvent cleaning, or abrasive blast cleaning. Metal parts are often roughened or grooved; for heavy sprayed deposits some undercutting, shouldering, or dovetailing may be necessary. The main metal deposition may be preceded by a sprayed bonding layer, as when steel alloys are sprayed over a thin bonding layer of molybdenum. Metal work is usually preheated, for example, 50°F above room temperature for drying, or up to 200 to 400°F to aid bonding and to control shrinkage. Final sprayed metal coatings are often sealed for corrosion protection, surface finish, pressure tightness, or appearance. Sealers range from wax or paint to plastic or even ceramic or glass materials.

Table 4–36 (continued)
FLAME-SPRAYED PROTECTIVE COATINGS

Flame-sprayed Metal and Nonmetal Coatings

Spray-deposited coating							
			Typical thickness	Melting point			
Material	Objective	Raw form	mils	°F	°C	Deposited on	Typical applications
Metals							
Aluminum	Corrosion protection	Wire	3-6	1 220	660	Steel	Industrial atmospheres; salt spray; moisture
Aluminum	Protection; appearance	Wire; powder	2-5	1 220	660	Steel, iron	Structures, piping, tanks, piling, bridges, roofs
Aluminum	High-temperature oxidation	Wire; powder	6-15	1 220	660	Steel	Flues, exhausts, tanks, furnace parts
Aluminum	Appearance; wear; conduction	Wire; powder	1-3	1 220	660	Small objects	Metal parts, wood, cloth, reflectors, ceramics
Brass (Cu-Zn)		Wire		1 800	982	Metals, glass	Glass seals; brass castings
Bronze (Cu-Sn)		Wire				Metals, plastics, concrete	Architectural decoration, casting repairs
Bronze (Cu-Al)	Corrosion; wear resistance	Wire		1 900	1 038	Steel, bronze	Machined bronze parts for wear and good finish
Cadmium		Wire; molten	3-100	610	321	Steel	Small parts; fasteners; nuclear shielding
Cobalt (Co-Cr-W)	Hard facing	Powder		2 725	1 495	Steel	Valves, gages, molds
Copper	Protection; conduction; appearance	Wire; powder	2-10	1 980	1 082	Metals, plastics, wood	Sprayed circuits; brazing; decoration, glass seals, contacts
Bismuth (alloy)		Wire; molten	–125	520	271		Plastic molds
Lead	Corrosion protection	Wire; molten	3-20	620	327	Steel	Acid splash or fumes; acid tanks; bonding metal
Molybdenum	Hard surfacing	Wire; powder		4 750	2 621	Metals	Bonding coat; pulley, brake, and bearing surfaces
Monel (Ni-Cu)	Corrosion protection	Wire		2 400	1 316		Valves, packing glands; pump parts
Nickel	Corrosion protection	Wire; powder	15-30	2 645	1 452	Steel, iron	Paper rolls; dye vats; pump rods
Ni-Cr (80-20)	Protection, appearance	Powder	1-10			Steel, iron	Bright, high-finish; heat resistant
Ni-Cr-B	Hard facing	Powder	5-100	1 900	1 038	Steel	Self-fluxing, as-fused, hard, wearing surfaces; valves; glass molds
Silver	Conduction; appearance	Wire		1 760	960	Metals	Conductors; decorative effects
Solder	Bond, protection, appearance	Wire; molten		420	216	Metals	Heat-transfer fins; filling dents and seams; models
Steel		Wire		2 300	1 260	Steel	Machinable build-up on journals, housings
Steel, stainless		Wire; powder		2 600	1 427	Steel	Shafts, glands, impellers, valves, friction surfaces
Steel, superalloy		Powder		2 700	1 482	Steel	Grind-finished repairs
Tin		Molten; wire		450	232	Metals	Food-processing equipment; glass-lined tank repairs
Tungsten		Powder		6 150	3 400		
Zinc		Wire; powder	2-5	787	419	Steel	Structural steel, tanks, fasteners, window frames

Table 4—36 (continued)
FLAME-SPRAYED PROTECTIVE COATINGS

Flame-Sprayed Metal and Nonmetal Coatings

Spray-deposited coating			Typical thickness mils	Melting point		Deposited on	Typical applications
Material	Objective	Raw form		°F	°C		
Zinc		Wire; powder	8-15	787	419		Water and seawater contact; chemical solutions
Ni-Al (80-20)							
NON-METALS							
Alumina	Heat and wear resistance	Powder		3 700	2 038		High-temperature erosion protection; insulation
Cr₃C₂				3,435	1,890		High-temperature seals
Plastic	Insulation; appearance	Powder				Steel, brass, concrete	
WC				5 200	2 870	Alloy steel	
WC-Co		Powder					
Zirconia	High-temperature insulation	Powder		4 700	2 593		Thermal barriers and linings

REFERENCES

Ballard, W. E., *Metal Spraying Technology,* 4th ed., Griffin, London, 1963.
Burns, R. M. and Bradley, W. W., *Protective Coatings for Metals,* 3rd ed., Van Nostrand Reinhold, New York, 1967.

From Bolz, R. E. and Tuve, G. L., Eds., *Handbook of Tables for Applied Engineering Science,* 2nd ed., CRC Press, Cleveland, 1973, 1051.

Table 4-37
PROPERTIES OF PLASTER OF PARIS

The following table gives variations in mechanical properties of pure gypsum (hydrous calcium sulfate) plaster with proportions of water in mix.[a]
For density in kg/m^3, multiply values in lbm/ft^3 by 16.018. For Young's modulus and strength in MN/m^2, multiply values in psi by 0.0068948. The unit grams per liter is identical to kilograms per cubic meter.

Density, lb/ft$_3$	Proportion by mass, %			Retarder,[b] grams per liter of water	Young's modulus, psi	Compressive strength, psi	Tensile strength, psi	Set time, minutes
	Water	Plaster	Diato-maceous earth					
75	60	100	0	0.50	1,000,000	2000	—	—
71	65	100	0	0.25	870,000	1700	274	7
67	70	100	0	0.25	730,000	1500	230	7
62	80	100	0	0.20	600,000	1100	—	11
58	90	100	0	0	500,000	800	145	6.5
53	100	100	0	0	440,000	700	—	10
48	110	100	0	0	368,000	470	103	7
46	120	100	0	0	316,000	400	82	9
43	130	100	0	0	273,000	330	72	9.5
40	140	100	0	0	225,000	280	69	9.5
38	150	100	0	0	190,000	240	—	10
37	160	100	8	0	164,000	200	51	12
36	170	100	8	0	131,000	170	—	12

[a]Poisson's ratio 0.24 and independent of mix.
[b]Sodium citrate.

From Barron, K. and Larocque, G. E., The development of a model for a mine structure, in *Proc. Rock. Mech. Symp.*, Mines Branch of the Department of Energy, Mines, and Resources, Ontario, Canada. With permission.

Table 4–38
CONCRETE FOR VARIOUS TYPES OF CONSTRUCTION

For cement factor in kilograms per cubic meter, multiply the tabulated value in sacks (94 lb) per cubic yard by 55.767.

For aggregate size in millimeters, multiply values in inches by 25.4.

Type of construction	Typical structures	Consistency	Cement factor sacks per cubic yard	Maximum size of aggregate inches
Massive	Dams, heavy piers, large open foundations	Stiff	$2\frac{1}{2}$–5	3–6
Semimassive	Piers, heavy walls, foundations, heavy arches, girders	Stiff; medium	4–6	2–3
Heavy building	Large structural members, small piers, medium footings; wide to moderately wide spacing of reinforcement	Medium wet	5–7	1–2
Light	Small structural members, thin slabs, small columns, heavily reinforced sections, closely spaced reinforcement	Wet	$5\frac{1}{2}$–7	$\frac{1}{2}$–1

From Troxell, G. E. and Davis, H. E., *Composition and Properties of Concrete,* 2nd ed., McGraw-Hill, New York, copyright © 1968. With permission.

Table 4–39

TYPES OF PORTLAND CEMENT AND INCREASE IN STRENGTH OF CONCRETE WITH AGE

Composition and Properties

Portland cement is a mixture consisting mainly of calcium and silicon oxides (as calcium silicates) in powder form with particle sizes largely in the range 10 to 50 μm. While the composition is approximately 65% CaO, 21% SiO_2, and 7% Al_2O_3 (balance from other oxides), the actual compounds are $3CaO \cdot SiO_2$, $2CaO \cdot SiO_2$, $3CaO \cdot Al_2O_3$, and $4CaO \cdot Al_2O_3 \cdot Fe_2O_3$. The proportions of these actual compounds will vary with the type of cement. When water is added and the mixture hardens, the heat of hydration is approximately 100 cal/g cement (180 Btu/lb). The heat is released gradually as hydration proceeds, with a shrinkage of about 8% compared with the total volume of cement-plus-water in the original mix. Shrinkage is minimal if original water content is low. To accelerate the setting of cement, especially in cold weather, calcium chloride (2% or less of weight of cement) often is added. In very hot weather retarders are sometimes used to delay hardening.

Standard ASTM classification	Character and use	Compressive strength,[a] psi/1,000, at age of				
		7 days	28 days	3 months	1 year	5 years
I	Common; general use	3.0	4.3	5.1	5.5	5.7
II	General use; moderate resistance to sulfate attack; lower heat of hydration	2.6	4.2	5.2	5.9	6.4
III	High early strength; shorter curing time	3.8	4.7	5.1	5.4	5.5
IV	Develops less heat; slow curing; resists cracking; high resistance to sulfate attack	1.5	3.5	5.2	6.0	6.5
V	Highest resistance to sulfate attack	2.5	4.1	5.3	6.1	6.7

[a]For strength in MN/m^2, multiply tabulated values in thousands of psi by 6.8948.

Note: Although strengths will vary greatly with materials, proportions, and curing conditions, these are typical.

From Bolz, R. E. and Tuve, G. L., Eds., *Handbook of Tables for Applied Engineering Science,* 2nd ed., CRC Press, Cleveland, 1973, 644.

Table 4–40
RELATIONSHIPS BETWEEN
WATER-CEMENT RATIO AND
COMPRESSIVE STRENGTH OF CONCRETE

	Water-cement ratio, by weight	
Compressive strength at 28 days, psi[a]	Non-air-entrained concrete	Air-entrained concrete
6000	0.41	–
5000	0.48	0.40
4000	0.57	0.48
3000	0.68	0.59
2000	0.82	0.74

[a]Values are estimated average strengths for concrete containing not more than the percentage of air shown in Table 5.3.3 of the original source. For a constant water-cement ratio, the strength of concrete is reduced as the air content is increased.

Strength is based on 6 × 12 in. cylinders moist-cured 28 days at 73.4 ± 3°F (23 ± 1.7°C) in accordance with Section 9(b) of ASTM C31 for Making and Curing Concrete Compression and Flexure Test Specimens in the Field.

Relationship assumes maximum size of aggregate about ¾ to 1 in.; for a given source, strength produced for a given water-cement ratio will increase as maximum size of aggregate decreases.

From *Recommended Practice for Selecting Proportions for Normal and Heavyweight Concrete (ACI 211.1-74)*, American Concrete Institute, Detroit. With permission.

Table 4–41
PROPERTIES OF SPECIAL CONCRETES

A great many varieties of aggregates have been used for concrete, dependent largely on the materials available. In general, high density results in high strength and high thermal conductivity, and vice versa, although such variables as water/cement ratio, percentage of fines, and curing conditions may result in wide differences in properties with the same materials. The following table gives typical examples.

	Approximate density, lb/ft³		Compressive strength, lb/in.²	Thermal conductivity, Btu/hr·ft·deg F
Description; type of aggregate	Aggregate	Concrete		
Frost resisting; 1% CaCl₂ ; normal aggregates	110	140	4,500	1.0
Frost-resisting porous; 6% air entrainment	100	130	3,500	0.85
Lightweight, with expanded shale or clay	50	75	1,500	0.25
Lightweight, with foamed slag	40	75	1,000	0.20
Cinder concrete, fine and coarse	50	80	1,000	0.25
Pulverized fuel ash	60	85	1,200	0.25
Lightweight refractory concrete with aluminous cement	35	65	3,500	0.20
Lightweight, insulating, with perlite	10	35	250	0.15
Lightweight, insulating, with expanded vermiculite	8	30	150	0.10

From Bolz, R. E. and Tuve, G. L., Eds., *Handbook of Tables for Applied Engineering Science*, 2nd ed., CRC Press, Cleveland, 1973, 645.

Table 4–42
EFFECTS OF VARIOUS SUBSTANCES ON UNPROTECTED CONCRETE

In addition to attack by corrosion (e.g., by acids) and other chemical combination (with sulfates and chlorides), there are many types of erosion of concrete. Concrete is highly susceptible to erosion by cavitation occurring in water-flow structures.

Substance	Effect on unprotected concrete
Petroleum oils: heavy, light, and volatile	None
Coal-tar distillates	None or very slight
Inorganic acids	Disintegration
Organic materials	
Acetic acid	Slow disintegration
Oxalic and dry carbonic acids	None
Carbonic acid in water	Slow attack
Lactic and tannic acids	Slow attack
Vegetable oils	Slight or very slight attack
Inorganic salts	
Sulfates of calcium, sodium, magnesium, potassium, aluminum, iron	Active attack
Chlorides of sodium, potassium	None
Chlorides of magnesium, calcium	Slight attack
Miscellaneous	
Milk	Slow attack
Silage juices	Slow attack
Molasses, corn syrup, and glucose	Slight attack

From *Concrete Manual,* 7th ed., U.S. Bureau of Reclamation, 1966.

Table 4–43
STRENGTHS OF METALS AT CRYOGENIC TEMPERATURES

Test Data on 25 Alloys

Key:

UTS = ultimate tensile strength, kpsi
YS = yield strength, kpsi
Elong = percentage elongation in 2 in.
Notch ratio = strength ratio: notched/unnotched
Joint eff = percentage strength ratio welded/clear specimen

Specimens were mostly $\frac{1}{16}$-in. sheet, 2-in. gage length, cut longitudinal to rolling direction. Tests were made in triplicate. The test temperatures were as follows: 297 to 300 K = room temperature; 200 K = –100 deg F; 144 K = –200 deg F; 77 K = –320 deg F; 20 K = –423 deg F; 5 K = –450 deg F.

For MN/m^2, multiply kpsi by 6.8948.

Temp, K	UTS, kspi	YS, kpsi	Elong, %, 2 in.	Notch ratio	Joint eff, %	Temp, K	UTS, kpsi	YS, kpsi	Elong, %, 2 in.	Notch ratio	Joint eff, %
ALUMINUM ALLOYS											
2014: Al + Cu 4.5%, Mn 1%, Si 1%, Mg 0.5%						**5086: Al + Mg 4.0%, Fe 0.5%, Mn 0.45%**					
300	70	63	9.7	0.99	66	300	47	37	10.4	1.00	90
200	73	68	9.5	1.00	60	200	48	38	12.0	0.99	86
144	76	71	9.3	0.99	59	144	52	39	15.4	0.98	89
77	84	76	11.7	0.93	63	77	64	44	25.0	0.89	93
20	96	79	13.6	0.88	82	20	85	47	20.2	0.76	80
5	97	82	10.4	0.83	74	5	80	49	23.4	0.74	86
2020: Al + Cu 4.5%, Li 1.1%, Mn 0.5%						**5456: Al + Mg 5.0%, Mn 0.75%, Zr <0.25%**					
300	79	75	8.0	0.67	—	300	57	45	8.7	0.92	84
200	82	76	6.3	0.65	—	200	57	45	9.3	0.90	84
144	86	83	3.0	0.60	—	144	62	47	11.7	0.91	86
77	95	88	4.0	0.52	—	77	74	53	13.0	0.79	85
20	101	93	2.3	0.50	—	20	87	57	8.7	0.75	74
5	104	95	3.6	0.50	—	5	86	57	9.5	0.73	74
2119: Al + Cu 5.9%, Fe 0.15%, Ti 0.15%						**7002: Al + Zn 3.35%, Mg 2.07%, Cu 0.88%**					
300	60	43	9.0	0.93	—	300	67	57	16.7	1.05	74
200	63	44	9.5	0.92	—	200	71	61	18.0	1.07	73
144	66	47	10.2	0.90	—	144	74	64	18.8	1.08	73
77	76	53	12.2	0.86	—	77	83	70	19.8	1.03	68
20	88	43	16.5	0.62	—	20	104	77	18.9	0.86	56
2219: Al + Cu 6.0%, Mn 0.33%, Fe 0.16%						**7075: Al + Zn 5.6%, Mg 2.5%, Cu 1.6%, Cr 0.3%**					
300	65	52	9.8	0.92	66	300	79	74	9.2	0.90	—
200	69	55	9.3	0.92	67	200	83	77	8.7	0.82	—
144	73	58	10.0	0.92	70	144	85	80	6.7	0.78	—
77	83	64	12.1	0.90	75	77	94	88	5.2	0.68	—
20	96	79	15.3	0.81	77	20	101	95	3.2	0.56	—
5	94	69	12.0	0.68	80						
5052: Al + Mg 2.5%, Fe 0.45%, Cr 0.25%						**7079: Al + Zn 4.5%, Mg 3.3%, Cu 0.6%**					
300	34	25	10.6	0.97	95	300	76	67	9.0	1.00	—
200	36	26	15.1	1.01	93	200	80	68	9.0	0.84	—
144	39	27	20.1	0.96	98	144	86	75	7.0	0.78	—
77	52	29	30.0	0.93	98	77	93	84	4.0	0.68	—
20	73	37	26.5	0.88	92	20	101	94	3.0	0.56	—
5	72	33	27.0	0.83	95	5	102	93	2.5	0.53	—

Table 4–43 (continued)
STRENGTHS OF METALS AT CRYOGENIC TEMPERATURES

Temp, K	UTS, kpsi	YS, kpsi	Elong, %, 2 in.	Notch ratio	Joint eff, %

ALUMINUM ALLOYS (continued)

7178: Al + Zn 7%, Mg 3%, Cu 2%, Cr 0.3%

Temp, K	UTS, kpsi	YS, kpsi	Elong, %, 2 in.	Notch ratio	Joint eff, %
300	94	88	7.5	0.67	—
200	96	93	4.0	0.65	—
144	100	96	4.0	0.60	—
77	109	104	1.2	0.41	—
20	117	113	1.0	0.32	—

MAGNESIUM ALLOYS
LA-91: Mg + Li 9.0%, Al 1.0%

Temp, K	UTS, kpsi	YS, kpsi	Elong, %, 2 in.	Notch ratio	Joint eff, %
297	23	20	36.8	0.95	100
200	32	23	11.2	0.82	—
144	34	23	14.3	0.77	—
77	36	25	15.5	0.77	85
20	46	36	20.5	0.75	88
5	48	41	11.5	—	—

LA-141: Mg + Li 14.5%, Al 1.5%

Temp, K	UTS, kpsi	YS, kpsi	Elong, %, 2 in.	Notch ratio	Joint eff, %
297	20	18	23.7	1.06	95
200	28	21	10.8	0.94	88
144	30	23	13.7	0.94	87
77	33	28	13.8	0.95	87
20	43	39	14.3	0.86	98

ALLOY STEELS
20-Cb, Carpenter: Fe + Ni 25%, Cr 20%, Cu 3.5%, Mo 2.5%

Temp, K	UTS, kpsi	YS, kpsi	Elong, %, 2 in.	Notch ratio	Joint eff, %
300	95	55	33.3	0.92	101
200	109	63	36.2	0.93	102
144	120	71	35.7	0.92	105
77	154	87	64.0	0.82	101
20	163	104	30.1	0.89	117

A286-N:[a] Fe + Ni 26%, Cr 16%, Ti 2%

Temp, K	UTS, kpsi	YS, kpsi	Elong, %, 2 in.	Notch ratio	Joint eff, %
300	93	42	37.3	0.86	100
200	104	48	38.8	0.88	101
144	115	57	43.0	0.87	101
77	144	68	71.0	0.80	99
20	161	81	47.3	0.82	96

A286-H:[a] (Same as above)

Temp, K	UTS, kpsi	YS, kpsi	Elong, %, 2 in.	Notch ratio	Joint eff, %
300	140	94	22.0	0.94	71
200	153	101	25.7	0.92	74
144	162	110	28.2	0.90	79
77	191	122	40.7	0.82	72
20	218	137	28.5	0.83	71

AISI 202-N:[a] Fe + Cr 18%, Mn 8%, Ni 5%

Temp, K	UTS, kpsi	YS, kpsi	Elong, %, 2 in.	Notch ratio	Joint eff, %
300	101	47	56.8	—	106
200	156	70	40.7	—	101
144	176	78	43.7	—	100
77	231	88	51.7	—	100
20	—	—	—	—	—
5	206	111	25.0	—	91

Maraging H:[a] Fe + Ni 18%, Co 8%, Mo 5%

Temp, K	UTS, kpsi	YS, kpsi	Elong, %, 2 in.	Notch ratio	Joint eff, %
300	254	245	2.8	1.09	94
200	275	266	2.8	1.08	100
144	288	274	3.0	1.06	102
77	321	309	2.5	0.90	99
20	365	355	3.2	0.41	86

NICKEL ALLOYS
Inconel X-H:[a] Ni + Cr 15%, Fe 7%, Ti 2.5%, Mn 1.0%, Al 0.7%

Temp, K	UTS, kpsi	YS, kpsi	Elong, %, 2 in.	Notch ratio	Joint eff, %
300	180	125	25.3	0.90	67
200	194	132	26.7	0.87	69
144	203	136	27.2	0.84	71
77	220	139	32.0	0.79	72
20	224	140	28.0	0.82	79

Waspaloy-H:[a] Ni + Cr 19%, Co 14%, Mo 4.3%, Ti 3.0%, Al 1.3%

Temp, K	UTS, kpsi	YS, kpsi	Elong, %, 2 in.	Notch ratio	Joint eff, %
300	178	116	26.3	0.81	79
200	193	128	20.5	0.81	86
144	203	139	18.8	0.79	83
77	205	142	15.0	0.80	92
20	197	154	10.2	0.85	97

K Monel-H:[a] Ni + Cu 29%, Al 3%, Fe 1.5%, Mn 1.0%

Temp, K	UTS, kpsi	YS, kpsi	Elong, %, 2 in.	Notch ratio	Joint eff, %
300	148	106	22.7	0.92	67
200	156	111	24.0	0.96	70
144	165	119	24.7	0.93	72
77	177	128	30.7	0.92	76
20	192	137	28.3	0.90	83

René 41-H: Ni + Cr 20%, Co 10%, Mo 10%, Ti 3.0%, Fe 3.0%

Temp, K	UTS, kpsi	YS, kpsi	Elong, %, 2 in.	Notch ratio	Joint eff, %
300	199	147	16.0	—	77
200	201	152	14.0	—	86
144	210	167	12.0	—	81
77	229	179	12.0	—	85
20	239	199	9.0	—	85

TITANIUM ALLOYS
TiAlV: Ti + Al 5.9%, V 4%, Fe 0.12%

Temp, K	UTS, kpsi	YS, kpsi	Elong, %, 2 in.	Notch ratio	Joint eff, %
297	140	133	11.0	1.02	100
200	161	158	9.3	1.00	99
144	178	177	6.5	0.99	100
77	218	214	13.0	0.82	102
20	240	240	1.7	0.61	96
5	242	—	0.2	0.62	100

TiAlSn: Ti + Al 5.2%, Sn 2.4%, Fe 0.32%

Temp, K	UTS, kpsi	YS, kpsi	Elong, %, 2 in.	Notch ratio	Joint eff, %
297	134	128	12.8	1.20	102
200	157	152	11.8	1.10	100
144	172	169	9.3	1.10	100
77	213	207	14.0	0.81	101
20	234	234	5.0	0.66	104
5	235	—	1.3	0.62	100

TiVCr: Ti + V 13.4%, Cr 11.3%, Al 2.8%, Fe 0.18%

Temp, K	UTS, kpsi	YS, kpsi	Elong, %, 2 in.	Notch ratio	Joint eff, %
297	137	137	13.3	1.20	106
200	182	182	6.0	1.10	106
144	218	215	4.5	0.96	106
77	285	282	2.5	0.54	60
20	289	—	0.7	0.40	34
5	301	—	0.0	—	—

[a]N: annealed; H; thermally age hardened, welded before aging.

Table 4–43 (continued)
STRENGTHS OF METALS AT CRYOGENIC TEMPERATURES

Notes: Only the major alloying elements are listed herewith; for minor alloying elements, see original data.

Data on specimens cut transverse to rolling direction are given in original source, as are data on aged and cold-worked alloys.

Note that 5, 20, and 77 K the approximately the atmospheric boiling points of liquid helium, hydrogen, and nitrogen, respectively; 200 K is just above the evaporation temperature of dry ice.

Notch ratios of less than one indicate that the material is weakened by the presence of stress-raising defects.

From *Effects of Low Temperature on Structural Metals*, NASA SP-5012, National Aeronautics and Space Administration, 1964 (revised edition, 1968).

Table 4–44
CRYOGENIC THERMAL INSULATION[a]

Description and Advantages

Classes

1. **Liquid and vapor shields** — Very low-temperature, valuable, or dangerous liquids such as helium or fluorine are often shielded by an intermediate cryogenic liquid or vapor container that must in turn be insulated by one of the methods described below.

2. **Multilayer reflecting shields** — Foil or aluminized plastic alternated with paper-thin glass- or plastic-fiber sheets; lowest conductivity, low density, and heat storage; good stability; minimum support structure.

3. **Opacified evacuated powders** — Contain metallic flakes to reduce radiation; conform to irregular shapes.

4. **Evacuated dielectric powders** — Very fine powders of low-conductivity adsorbent; moderate vacuum requirement; minimum fire hazard in oxygen.

5. **Vacuum flasks (Dewar)** — Tight shield-space with highly reflecting walls and high vacuum; minimum heat capacity; rugged; small thickness.

6. **Gas-filled powders** — Same powders as Class 4 but with air or inert gas; low cost; easy application; no vacuum requirement.

7. **Expanded foams** — Very light foamed plastic; inexpensive; minimum weight but bulky; self supporting.

8. **Porous fiber blankets** — Blanket material of fine fibers, usually glass; minimum cost and easy installation but not an adequate insulation for most cryogenic applications.

Insulation Properties

Class	Descriptive name	Approximate density $\frac{lbm}{ft^3}$	$\frac{kg}{m^3}$	Approximate specific heat $\frac{Btu}{lbm \cdot deg\ F}$	$\frac{kJ}{kg \cdot K}$	Range of mean conductivities $\frac{Btu}{hr \cdot ft \cdot deg\ F}$	$\frac{mW}{m \cdot K}$	Interspace pressure, mm Hg[b]
2	Multilayer	5	80	.22	0.92	.000023–.00012	0.04–0.2	10^{-4}
3	Opacified powder	7	110	.23	0.96	.00015–.0004	0.26–0.7	10^{-4}
4	Evacuated powder	6	100	.25	1.05	.00057–.00115	1.0–2.0	10^{-4}
5	Vacuum flask	–	–	–	–	.0029	5.0	10^{-6}
6	Gas-filled powder	6	100	.25	1.05	.001–.004	1.7–7.0	760
7	Expanded foam	2	30	0.4	1.67	.0029–.020	5.0–35	760
8	Fiber blanket	8	130	0.5	2.09	.02–.026	35–45	760

Table 4-44 (continued)
CRYOGENIC THERMAL INSULATION

Structural Support

For those insulating materials and constructions requiring structural support, the relative strengths, weights, heat capacities, and conductivities of the supporting materials are important.

Material	Tensile yield strength S, 1000's psi[c]	Density, ρ		Specific heat, c_p		Mean thermal conductivity k,[d] 20–300°K		Relative		
		$\dfrac{lbm}{ft^3}$	$\dfrac{kg}{m^3}$	$\dfrac{Btu}{lbm \cdot deg\, F}$	$\dfrac{kJ}{kg \cdot K}$	$\dfrac{Btu}{hr \cdot ft \cdot deg\, F}$	$\dfrac{W}{m \cdot K}$	$\dfrac{S}{k}$	$\dfrac{\rho}{k}$	$\dfrac{c_p\,\rho}{S}$
Aluminum alloy	50	170	2720	.22	0.92	50	86	1	3	.75
"K" Monel®	100	520	8330	.13	0.54	10	17	10	52	.68
Stainless steel	100	500	8010	.12	0.50	5.4	9.3	18	93	.60
Titanium alloy	100	625	10010	.06	0.25	3.5	6.1	29	180	.37
Nylon	15	70	1120	.4	1.67	.17	0.29	88	41	1.9
Teflon	2	120	1920	.25	1.05	.14	0.24	14	86	15.0

[a]For other thermal conductivities, see Tables 4–51 and 4–71.

[b]For N/m² multiply by 133.32.

[c]For MN/m² multiply tabulated values in 1000's psi by 6.8948.

[d]For solid members; perforation and lamination used to reduce condition.

REFERENCE

Thermal Insulation Systems, NASA SP-5027, National Aeronautics and Space Administration, 1967.

From Bolz, R. E. and Tuve, G. L., Eds., *Handbook of Tables for Applied Engineering Science,* 2nd ed., CRC Press, Cleveland, 1973, 529.

Table 4-45
LOW-TEMPERATURE COOLING BATHS

Cryogenic and Refrigerating Fluids – Atmospheric Pressure

Liquid	Boiling point			Liquid	Boiling point		
	K	°C	°F		K	°C	°F
Helium	4.2	−268.95	−452.1	Refrigerant 14 (CF_4)	145.	−128.	−198.
Hydrogen	20.4	−252.7	−422.8	Refrigerant 13 ($CClF_3$)	192.	−81.	−114.
Neon	27.1	−246.	−410.8	Carbon dioxide (CO_2)[a]	195.	−78.	−108.5
Nitrogen	77.4	−195.8	−320.4	Propylene (C_3H_6)	225.	−48.	−54.
Air	79.	−194.	−317.	Refrigerant 502 (Azeotrope)	227.	−46.	−50.
Argon	87.3	−185.9	−302.6	Propane (C_3H_8)	225.	−48.	−54.
Oxygen	90.2	−183.0	−297.5	Refrigerant 22 ($CHClF_2$)	232.	−41	−41.
Methane (CH_4)	111.	−162.	−259.	Refrigerant 12 (CCl_2F_2)	243.	−30	−22.

[a]Solid; sublimes.

Notes: Low-temperature baths are conveniently prepared by adding dry ice or liquid nitrogen to acetone, ether, chloroform, or one of the alcohols, stirring the mixture until a slush is formed. Bath

Table 4–45 (continued)
LOW-TEMPERATURE COOLING BATHS

temperatures will then be fixed by the freezing point and will range from about −82°F for chloroform to −197°F for propyl alcohol. The lowest bath temperature with dry ice is −108°F. Minimum temperatures attainable (with difficulty) using water-ice and salt mixtures are as follows: NaCl (23.3%), −6°F; $CaCl_2$ (30%), −67°F.

Warning: Adequate safety precautions are necessary when using low-temperature baths. In addition to skin "burns," toxic and explosion hazards may be present. In the initial cooling of a bath from room temperature, using dry ice or liquid nitrogen, the evolution of gas may be so rapid as to approach an explosion. Adequate venting and room ventilation are necessary. Liquid nitrogen absorbs and condenses oxygen (shown by bluish color) and may also produce "liquid air" in a heat-transfer device such as a vacuum cold trap. Later reevaporation of the air calls for adequate venting, or explosive forces are produced. Bath liquids and vapors may be toxic or irritant to skin or eyes (e.g., acetone or ketone) and contact should be avoided.

From Bolz, R. E. and Tuve, G. L., Eds., *Handbook of Tables for Applied Engineering Science*, 2nd ed., CRC Press, Cleveland, 1973, 587.

Table 4–46
APPLICATIONS OF CRYOGENICS

Cryogenic applications are based largely on the use of liquefied "permanent" gases. Following the long period of development of gas-liquefaction systems (1875–1935) and the improvement of cryogenic insulation, the uses of the liquefied gases have expanded rapidly. Included in the partial list of applications given below are several uses of liquefied gases that are as yet in the early stages of development. The future rate or extent of each of these developments is not predictable, but the prospects are sufficient to warrant their inclusion.

Table A. Applications at Cryogenic Temperatures

Type of use	Brief description	Gases	Minimum temperature K	°F
Space simulation	Pressures of 10^{-10} to 10^{-14} torr by cryopumping. Plasma engines	He	5	−451
High-capacity electromagnets	High-intensity electromagnets with low-power consumption at cryogenic temperatures; compact, high-strength electromagnets with superconducting coils of special alloys (e.g., niobium–tin)	He	5	−451
Superconducting power cables	Long-distance and high-capacity underground power transmission with negligible I^2R losses (or superconductors)	He	5	−451·
Cryotrons	Thin-film (or wire-wound) superconducting switching elements used as memory, logic element, etc. in computers	He	5	−451
Frictionless bearings	Superconductors at cryogenic temperatures support bearing loads by magnetic repulsion	He	5	−451
Electronic "noise" suppression	Thermal noise in amplifiers suppressed at cryogenic temperatures; cryogenically cooled masers	He H₂	5 20	−451 −423

Table 4–46 (continued)
APPLICATIONS OF CRYOGENICS

Table A. Applications at Cryogenic Temperatures (continued)

Type of use	Brief description	Gases	Minimum temperature	
			K	°F
High-capacity electric motors and generators	Increased capacity of conventional designs at low temperatures; compact motors for liquid-cryogen pumps by special design using superconducting motor and ac or pulsed dc in stator windings	He C_3H_8	5 225	−451 −54
Cryogenic infrared detectors	Photoconductor sensors highly sensitive to wavelengths above 10 μm at cryogenic temperatures	H_2 He	20 5	−423 −451
Gas liquefaction separation, rectification, refining	Large-scale production of liquid natural gas; natural-gas separations	N_2 He	77 5	−320 −451
Modification of "radiation damage" to materials	Alleviation of damage to materials resulting from neutron bombardment	He H_2	5 20	−451 −423
Bubble chambers	Experiments involving elementary particles and nucleation in slightly superheated cryogenic liquid	H_2	20	−423
Rocket propulsion	Liquid hydrogen fuel. Liquid oxygen oxidizer for liquid or solid fuels	H_2 O_2	20 90	−423 −297
Neutron moderator	D_2O ice, cooled by liquid hydrogen, used for slowing down of fast neutrons in a thermal reactor	D_2O-H_2	20	(−423)[a]
Food freezing	Food frozen by liquid nitrogen spray or food cartons conveyed through liquid nitrogen. N_2 also used in frozen food transport	N_2	77	−320
Cryobiology	Cryogenic long-time preservation of blood, sperm, and other bio-materials	N_2	77	−320
Cryosurgery	Used for cryogenic brain surgery for treatment of Parkinson's disease; eye surgery	CO_2	200	−100
High vacuum and "cold traps"	Outgassing of high-vacuum systems, removal of oil vapors and residuals. Deposition of thin metal films under high vacuum; semiconductor manufacture	CO_2 N_2 H_2	195 77 20	−108 −320 −423
Reduction of chemical reaction rate	Excessive rates of chemical reaction at usual temperature may be slowed as desired by very low temperatures	CO_2 (bath) N_2	195 77	−108 −320
Refrigerated transport	Dry ice in trucks, railway cars, ships, planes, and in insulated cartons (liquid nitrogen for some transport)	CO_2 N_2	195[b] 77	−108[b] −320
Cutting-tool coolant	Low-temperature heat removal at cutting-tool face when machining superalloys	CO_2 (bath) N_2	195 77	−108 −320
Metal working and fabrication	Stretching, forming, and rolling of metals at cryogenic temperatures. Precipitation hardening and work hardening of austenitic steels	CO_2 (bath) N_2	195 77	−108 −320

[a]Heavy-water ice cooled to 20 K (36° R).
[b]Solid CO_2 sublimes at 195 K, −108.5 °F.

Table 4—46 (continued)
APPLICATIONS OF CRYOGENICS

Table B. Reevaporated Gases

Type of use	Typical application	Gases
Oxygen steel processes	Oxygen lances in open-hearth, converter, or electric-furnace operation; oxygen-enriched blast in blast furnace	O_2
Chemical industries	Nitrogen for raw material in chemical manufacture. Hydrogenation processes	N_2 H_2
Combustion processes	Fuel gases with air, or oxygen-enriched combustion processes	H_2 CH_4 C_3H_8 C_4H_{10}
Flame cutting and welding	Gas welding and cutting processes, metals, and alloys	O_2 H_2 C_2H_2

REFERENCE

Timmerhaus, K. D., Ed., *Advances in Cryogenic Engineering* Series, Plenum, New York.

From Bolz, R. E. and Tuve, G. L,, Eds., *Handbook of Tables for Applied Engineering Science,* 2nd ed., CRC Press, Cleveland, 1973, 588.

Table 4–47
CRYOGENIC AND REFRIGERATING LIQUIDS—SI UNITS

| Fluid | Boiling | | | | Density, kg/m³ | Volume ratio (to room temp), gas/liq | Latent heat of vaporization, kJ/kg | Specific heat, c_p | | Viscosity | | Thermal conductivity | | Dielectric constant |
	K	°R	°C	°F				Liquid, kJ/kg·K	Gas, kJ/kg·K	Liquid, mN·s/m²	Gas, mN·s/m²	Liquid, mW/m·K	Gas, mW/m·K	
Air	79	142	−194	−318	875	740:1	205	1.97	1.02	0.080 6	0.006 61		7.44	1.52
Argon	87	157	−186	−303	1 400	840:1	162	1.14	0.531	0.252	0.008 27	123	6.05	
Carbon dioxide[a]	195	350	−79	−110	1 560	730:1	572	1.33	0.795				14.7[c]	1.59[c]
Ethane	185	334	−88	−126	548	420:1	488	2.51	1.05				25.1[d]	
Fluorine	85	154	−188	−306	1 500	880:1	166	1.55	0.812	0.245	0.007 36	135	7.20	1.43
Helium 3	3.20	5.76	−270	−454	58.9	600:1	8.48	4.60		0.001 62		17.1		
Helium 4	4.215	7.59	−269	−452	125	600:1	20.7	4.56	6.82	0.003 57		27.0	9.69	1.049 2
Hydrogen	20	36.7	−253	−423	71.0	800:1	446	9.79	11.9	0.013 1	0.001 28	118	15.6	1.226
Methane	111	201	−162	−259	424	550:1	509	3.45	1.66	0.119	0.001 05	111		1.68
Neon	27	48.8	−246	−411	1 200	1 400:1	86.7	1.84	1.17	0.124	0.004 55	130	9.86	
Nitrogen	77.4	139.2	−196	−320	810	700:1	198	2.04	1.08	0.158	0.005 54	139	7.18	1.434
Oxygen[e]	90.2	162.	−183	−297	1 140	860:1	213	1.90	1.40					1.51
Propane	231	416	−42	−44	581	310:1	425	2.20	1.21					1.27
Propylene	225	406	−48	−54	614	330:1	437	2.59	1.52					
Refrigerant 12	243	438	−30	−22	1 490	280:1	165	0.891	0.456	0.371	0.010 9			2.13
Refrigerant 13	192	346	−81	−114	1 520	350:1	149	0.895	0.413			60.5		
Refrigerant 13B1	215	388	−58	−72	1 990	310:1	119	1.69	0.849					
Refrigerant 22	232	419	−41	−41	1 410	380:1	234	1.05	0.464	0.351	0.010 5	110[b]	7.78	2.44

[a]Sublimes.
[b]At 253.15 K.
[c]At 273.15 K.
[d]At 324.82 K.
[e]Solidifies at −218° C, 55 K.

From Bolz, R. E. and Tuve, G. L., Eds., *Handbook of Tables for Applied Engineering Science*, 2nd ed., CRC Press, Cleveland, 1973, 590.

Table 4–48
DIFFUSION OF SOLUTES INTO WATER

Dilute Solutions at 20°C

Diffusion constant, D, English: for square feet per hour multiply by 10^{-5}. Diffusion constant, D, metric: for square meters per second multiply by 10^{-9}

Diffusion constant, D[a]

Substance	English	Metric	Schmidt number, $\left(\frac{\mu}{\rho D}\right)$[b]
H_2	19.8	5.13	196
O_2	6.97	1.80	558
CO_2	6.85	1.77	568
NH_3	6.81	1.76	571
N_2	6.35	1.64	613
Acetylene	6.04	1.56	644
Cl_2	4.72	1.22	824
HCl	10.2	2.64	381
HNO_3	10.1	2.6	390
H_2SO_4	6.70	1.73	581
NaOH	5.84	1.51	666
NaCl	5.22	1.35	744
Ethyl alcohol	3.87	1.00	1005
Acetic acid	3.41	0.88	1140
Phenol	3.25	0.84	1200
Glycerol	2.79	0.72	1400
Sucrose	1.74	0.45	2230

[a]The following relationship may be used to estimate the effect of temperature of the diffusion constant

$$\frac{D_1}{D_2} = \frac{T_1}{T_2}\frac{\mu_2}{\mu_1}$$

where T = temperature, °K and μ = solution viscosity, centipoises. The diffusion constant varies with concentration because of the changes in viscosity and the degree of ideality of the solution.

[b]Based on $\mu/\rho = 0.01005$ cm^2/sec for water at 20°C. Applies only for dilute solutions.

From Bolz, R. E. and Tuve, G. L., Eds., *Handbook of Tables for Applied Engineering Science*, 2nd ed., CRC Press, Cleveland, 1973, 545.

Table 4–49
DIFFUSION OF WATER VAPOR INTO AIR

Values of Diffusion Constant and Schmidt Number

Temp, °C	Diffusion constant, D		$\left(\frac{\mu}{\rho D}\right)$[a]
	ft^2/hr	cm^2/sec	
0	0.844	0.218	0.608
10	0.898	.232	.610
20	0.952	.246	.612
30	1.01	.260	.614
40	1.06	.275	.615
50	1.12	.290	.616
60	1.18	.305	.618
70	1.24	.321	.619
80	1.30	.337	.619

[a]The values of $\left(\frac{\mu}{\rho D}\right)$ were calculated using the viscosity and density of dry air. Thus, the values apply only when the diffusing water vapor is very dilute.

REFERENCES

Perry, R. H., Chilton, C. H., and Kirkpatrick, S. D., Eds., *The Chemical Engineers' Handbook*, McGraw-Hill Book Company, New York, 1963.

National Research Council, *International Critical Tables of Numerical Data*, Vol. 5, McGraw-Hill, New York, 1929.

Landolt-Börnstein Physikalisch-Chemische Tabellen, Springer, Berlin, 1923.

Foust, A. S., Wenzel, L. A., Clump, C. W., Maus, L., and Anderson, L. B., *Principles of Unit Operations*, John Wiley & Sons, New York, 1960.

From Bolz, R. E. and Tuve, G. L., Eds., *Handbook of Tables for Applied Engineering Science*, 2nd ed., CRC Press, Cleveland, 1973, 545.

Table 4–50
DIFFUSION OF GASES AND VAPORS INTO AIR

Values of Diffusion Constant and Schmidt Number

At 1 atm Pressure

Substance	Diffusion constant, D, ft² /hr		Diffusion constant, D, cm²/sec		$\left(\dfrac{\mu}{\rho D}\right)$ [a]	
	0°C	25°C	0°C	25°C	0°C	25°C
H₂	2.37	2.76	0.611	0.712	0.217	0.216
NH₃	0.766	0.886	0.198	0.229	0.669	0.673
N₂	0.691		0.178		0.744	
O₂	0.689	0.80	0.178	0.206	0.744	0.748
CO₂	0.550	0.635	0.142	0.164	0.933	0.940
CS₂	0.36	0.414	0.094	0.107	1.41	1.44
Methyl alcohol	0.513	0.615	0.132	0.159	1.00	0.969
Formic acid	0.509	0.615	0.131	0.159	1.01	0.969
Acetic acid	0.411	0.515	0.106	0.133	1.25	1.16
Ethyl alcohol	0.394	0.461	0.102	0.119	1.30	1.29
Chloroform	0.352		0.091		1.46	
Diethylamine	0.342	0.406	0.0884	0.105	1.50	1.47
n-Propyl alcohol	0.329	0.387	0.085	0.100	1.56	1.54
Propionic acid	0.328	0.383	0.0846	0.099	1.57	1.56
Methyl acetate	0.325	0.387	0.0840	0.100	1.58	1.54
Butylamine	0.318	0.391	0.0821	0.101	1.61	1.53
Ethyl ether	0.304	0.360	0.0786	0.093	1.69	1.66
Benzene	0.291	0.341	0.0751	0.088	1.76	1.75
Ethyl acetate	0.277	0.330	0.0715	0.085	1.85	1.81
Toluene	0.274	0.325	0.0709	0.084	1.87	1.83
n-Butyl alcohol	0.272	0.348	0.0703	0.090	1.88	1.71
i-Butyric acid	0.263	0.313	0.0679	0.081	1.95	1.90
Chlorobenzene		0.283		0.073		2.11
Aniline	0.236	0.279	0.0610	0.072	2.17	2.14
Xylene	0.228	0.275	0.059	0.071	2.25	2.17
Amyl alcohol	0.228	0.271	0.0589	0.070	2.25	2.20
n-Octane	0.195	0.232	0.0505	0.060	2.62	2.57
Naphthalene	0.199	0.20	0.0513	0.052	2.58	2.96

[a]Based on $\dfrac{\mu}{\rho}$ = 0.1325 cm²/sec for air at 0°C and 0.1541 cm²/sec for air at 25°C; applies only when the diffusing gas or vapor is very dilute.

From Bolz, R. E. and Tuve, G. L., Eds., *Handbook of Tables for Applied Engineering Science,* 2nd ed., CRC Press, Cleveland, 1973, 546.

Table 4—51

HEAT TRANSMISSION THROUGH BUILDING STRUCTURES[a]

Temperature control within structures requires the use of accurate and consistent data on heat transfer coefficients. Extensive tables have been compiled by the American Society of Heating, Refrigerating and Air-conditioning Engineers (*ASHRAE Handbook of Fundamentals*, chap. 26), and these are accepted as standard. The following tables give some of the basic coefficients (Table A) and examples of overall coefficients for common wall and roof constructions (Table B) from this source.

Table A. Basic Coefficients

Emissivities — Aluminum foil, 1 surface, 0.05; 2 surfaces, 0.03. Aluminum paint, 0.50. Nonmetallic surface, 0.90. (Blackbody = 1.0.)

Coefficients in Btu/hr·ft² ·deg F (W/m² ·K) — Indoor nonmetallic surfaces, still air; vertical, 1.46 (8.29); horizontal ceiling, winter, 1.63 (9.26); summer, 1.08 (6.13). Vertical air spaces: non-metallic, 1.04 (5.91); foil on both surfaces, 0.34 (1.93). Horizontal air space: nonmetallic surfaces, winter, 1.18 (6.70); summer, 1.01 (5.74). Outdoor surface coefficients, winter, 6.0 (34) for 15 m/h (6.7 m/s) wind; summer, 4.0 (23) for 7.5 m/h (3.35 m/s) wind.

For density in kg/m^3, multiply values in lbm/ft^3 by 16.018.

Conductivity, k, Btu in./ft² ·hr·°F (at 75°F)

Material	Density, lbm ft³	Thermal conductivity Btu·in. hr·ft² ·°F	W m·K
Asbestos cement board	120	4.0	0.58
Blanket of batt insulation	3	0.27	0.039
Brick, common	120	5.0	0.72
Brick, face	130	9.0	1.30
Cement mortar	116	5.0	0.72
Concrete, gypsum fiber	51	1.66	0.24
Concrete, lightweight	40	1.15	0.17
Corkboard	7	0.27	0.039
Hardboard, wood fiber	65	1.40	0.20
Mineral fiberboard	18	0.35	0.05
Mineral wool fill	3	0.28	0.04
Plaster, cement sand	116	5.0	0.72
Plaster, gypsum perlite	45	1.5	0.22
Plaster, vermiculite	45	1.7	0.25
Plywood	34	0.80	0.115
Redwood bark fill	4	0.27	0.039
Sheathing, wood fiber	22	0.41	0.059
Shredded wood, cemented	22	0.60	0.086
Vermiculite	8	0.47	0.068
Wood, oak, maple	45	1.10	0.159
Wood or cane fiberboard	15	0.35	0.050
Wood, pine, fir	32	0.80	0.115

Table 4–51 (continued)
HEAT TRANSMISSION THROUGH BUILDING STRUCTURES

Table B. Overall Coefficients (Transmittance) Air to Air

Structure	U, $W/m^2 \cdot K$	U, $Btu/hr \cdot ft^2 \cdot °F$
Single-glass window	6.2	1.1
Double-insulating glass with ½-in. air space, or storm window	3.2	0.56
Frame wall: wood sheathing, lath, and sand plaster	1.5	0.26
Frame wall: insulating sheathing, lath, and lightweight plaster	1.1	0.19
Frame wall: full fibrous insulating between studs	0.40	0.07
Brick veneer, insulating sheathing, lightweight plaster	1.2	0.21
Brick wall, 8-in. solid, sand plaster	2.6	0.45
Brick wall, 8-in., gypsum lath, furred, lightweight plaster	1.5	0.27
Clay tile wall, 8-in. hollow, gypsum lath, furred, lightweight plaster	1.3	0.23
Masonry cavity wall: 4-in. face brick, air space, 4-in. cinder block, gypsum lath, furred, lightweight plaster, two aluminum foil surfaces or ¾-in. insulation	0.74	0.13
Roof, pitched, shingle, unventilated rafter space, lath and plaster ceiling	1.7	0.30
Roof, pitched, asbestos cement, slate or tile shingles on wood sheathing, plastered ceiling, 3-in. insulation between joists	0.40	0.07
Roof, flat, metal deck, plaster ceiling on gypsum board	1.9	0.33
Roof, flat, preformed 2-in. insulating slab deck, acoustical ceiling on gypsum board	0.74	0.13

[a]For other data on thermal conductivities, see Tables 4–44 and 4–71.

From Bolz R. E and Tuve. G. L , Eds., *Handbook of Tables for Applied Engineering Science,* 2nd ed., CRC Press, Cleveland, 1973, 682.

Table 4–52
REFLECTANCE AND APPEARANCE OF COLORS

The following table gives the percentage reflectance in daylight of typical glossy- or smooth-surface finishes on wood, paper, metal, and other materials. So-called 'fluorescent' colors are not included.

Descriptive name	Light reflected, percent	Federal color No.[a]
White	85	–
Light colors		
Light ivory	75	13711
Soft yellow	75	13695
Cream; off-white	75	–
Light buff; light gray	70	–
Peach	64	12648
Suntan	60	13613
Light green	55	14516
Light blue	50	15526
Medium colors		
Brilliant yellow	58	13538
Highlight buff	55	13578
Clear green	50	–
Pearl gray	46	16492
Wood finish, maple	40	–
Dark colors		
Light navy gray	28	16251
Medium green	25	14277
Vivid orange	23	12246
Clear blue	19	15177
Radiation purple	15	17142
Wood finish, oak or walnut	15	–
Medium navy gray	14	16187
Light red	13	–
Medium brown	10	–
Wood finish, red mahogany	9	–
Fire red	7	11105
Passive green	7	14077
Maroon; dark green	7	–
Deep navy gray	7	16081
Marine Corps green	4	14052
Dark brown	4	–

[a]Federal Color Standards No. 595. Color numbers beginning with digit 1 are gloss finish.

Appearance of colors

Intense or saturated hues protrude.

Incandescent lighting tends to dull the green, blue, lavender, and purple shades. Olive green appears brown.

"Cool white" fluorescent lamps make reds appear less bright, and they somewhat dull yellow and blue shades. Illuminated by "warm white" fluorescent lighting, the shades of gray and olive green are changed, and the blues and purples appear dull.

Colored light sources change the hue of most pigment colors, except those that are nearly the same color as the light.

From Bolz, R. E. and Tuve, G. L., Eds., *Handbook of Tables for Applied Engineering Science*, 2nd ed., CRC Press, Cleveland, 1973, 682.

Table 4–53
CHARACTERISTICS OF PARTICLES AND PARTICLE DISPERSOIDS

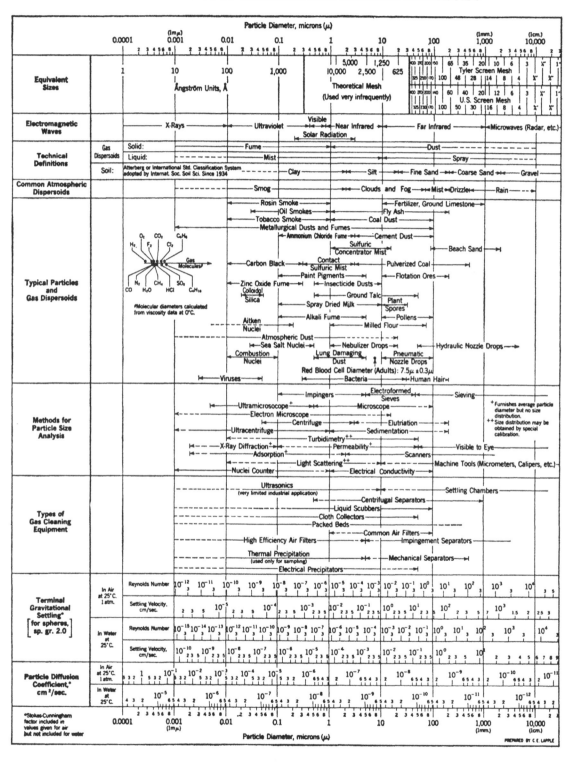

From Lapple, C. E., *Stanford Res. Inst. J.*, 5, 95, 1961. With permission.

Table 4–54

DUST EXPLOSION CHARACTERISTICS

The following table is based on laboratory test results by the U.S. Bureau of Mines on dried samples of fine dusts (passing 200-mesh sieve).[a] The values below probably represent "the most hazardous conditions" for these materials.

Type of dust	Ignition temperature of dust cloud, °C	Minimum igniting energy, J	Minimum explosive concentration, oz/ft³	Maximum explosion pressure, psig	Maximum rate of pressure rise, psi/sec	Terminal oxygen concentration, %[b]	Relative explosion hazard
Agricultural							
Alfalfa	530	0.320	0.105	92	2,200	—	Moderate
Cereal grass	550	0.800	0.250	52	500	—	Weak
Coffee	720	0.160	0.085	53	300	—	Weak
Corn	400	0.040	0.055	95	6,000	—	Strong
Corncob	480	0.080	0.040	110	3,100	—	Strong
Cornstarch	390	0.030	0.040	115	9,000	—	Severe
Cotton linters	520	1.920	0.500	48	150	—	Moderate
Cottonseed	530	0.120	0.055	96	3,000	—	Moderate
Grain, mixed	430	0.030	0.055	115	5,500	—	Strong
Grass seed	490	0.260	0.290	34	400	—	Weak
Malt, brewers	400	0.035	0.055	92	4,400	—	Strong
Milk, skim	490	0.050	0.050	83	2,100	—	Strong
Peanut hull	460	0.050	0.045	82	4,700	—	Strong
Peat, sphagnum	460	0.050	0.045	84	2,200	—	Strong
Pecan nutshell	440	0.050	0.030	106	4,400	—	Strong
Potato starch	440	0.025	0.045	97	8,000	—	Severe
Rice	440	0.050	0.050	93	2,600	—	Strong
Soy flour	550	0.100	0.060	111	1,600	15	Moderate
Sugar, powdered	370	0.030	0.045	91	1,700	—	Strong
Wheat flour	440	0.060	0.050	104	4,400	—	Strong
Wheat straw	470	0.050	0.055	99	6,000	—	Strong
Carbonaceous							
Charcoal, hardwood mix, volatile content 27.1%	530	0.020	0.140	100	1,800	18	Strong
Coal, Ill., No. 7, volatile content 48.6%	600	0.050	0.040	84	1,800	15	Strong

Table 4–54 (continued)
DUST EXPLOSION CHARACTERISTICS

Type of dust	Ignition temperature of dust cloud, °C	Minimum igniting energy, J	Minimum explosive concentration, oz/ft³	Maximum explosion pressure, psig	Maximum rate of pressure rise, psi/sec	Terminal oxygen concentration, %[b]	Relative explosion hazard
Coal, Pa. (Pittsburgh), volatile content 37.0%	610	0.060	0.055	83	2,300	17	Strong
Gilsonite, Utah, volatile content 86.5%	580	0.025	0.020	78	3,700	–	Severe
Lignite, Cal., volatile content 60.4%	450	0.030	0.030	90	8,000	–	Severe
Pitch, coal tar, volatile content 58.1%	710	0.020	0.035	88	6,000	–	Severe
Metals							
Aluminum	650	0.015	0.045	100	10,000[b]	2	Severe
Copper	900	–	–	–	–	–	Fire
Iron	420	0.020	0.100	46	6,000[b]	10	Strong
Magnesium	520	0.020	0.020	94	10,000[b]	0	Severe
Tin	630	0.080	0.190	37	1,300	15	Moderate
Titanium	460	0.010	0.045	80	10,000[b]	0	Severe
Uranium	20	0.045	0.060	53	3,400	0	Severe
Zinc	600	0.640	0.480	48	1,800	9	Weak
Plastics							
Acetal resin (polyformaldehyde)	440	0.020	0.035	89	4,100	11	Severe
Acrylic polymer resin	480	0.010	0.030	85	6,000	11	Severe
Methyl methacrylate-ethyl acrylate							
Alkyd resin	500	0.120	0.155	15	150	15	Weak
Alkyd molding compound							
Amino resin	450	0.080	0.075	89	3,600	17	Strong
Urea-formaldehyde molding compound							
Cellulose fillers	430	0.020	0.035	110	5,500	17	Severe
Wood flour							
Cellulose resin	320	0.010	0.025	102	6,000	11	Severe
Ethyl cellulose molding compound							

Table 4—54 (continued)
DUST EXPLOSION CHARACTERISTICS

Type of dust	Ignition temperature of dust cloud, °C	Minimum igniting energy, J	Minimum explosive concentration, oz/ft³	Maximum explosion pressure, psig	Maximum rate of pressure rise, psi/sec	Terminal oxygen concentration, %[b]	Relative explosion hazard
Epoxy resin	530	0.020	0.020	86	6,000	12	Severe
Phenolic resin	500	0.020	0.030	92	10,000[b]	14	Severe
Phenol-formaldehyde molding compound							
Rayon (viscose) flock	520	0.240	0.055	88	1,700	–	Moderate
Rubber, synthetic	320	0.030	0.030	93	3,100	15	Severe

[a]See U.S. Bureau of Mines Reports of Investigations No. 5624, 5753, 5971, 6516.
[b]The terminal oxygen concentration is the limiting oxygen concentration in air-CO_2 atmosphere required to prevent ignition of dust clouds by electric spark.

From Bolz, R. E. and Tuve G L., Eds., *Handbook of Tables for Applied Engineering Science*, 2nd ed., CRC Press, Cleveland, 1973, 784.

Table 4-55

TYPICAL FIRE-RESISTANCE RATINGS FOR REINFORCED CONCRETE CONSTRUCTIONS

Ratings are dependent on protective cover of concrete over steel. For length in millimeters, multiply values in inches by 25.4.

Fire-resistance rating, hours[a]	Structural members	Cover over steel, inches
1	Beams, medium size	3/4–7/8
	Slabs, unrestrained	1
2	Beams, medium size	1–1 1/4
	Slabs, unrestrained	1 1/2
3	Beams, medium size	1 1/2
	Columns, 12–14 in.	1 1/2
	Slabs, unrestrained	2
	(Solid concrete walls, 6 in.)	–
4	Columns, 12–14 in.	2
	Columns, > 16 in.	1 1/2
	Slabs, unrestrained	2

[a]The standard test for fire resistance rating for a large wall specifies measurement of the time required for a temperature rise of 250° F on the unexposed face when the other face is subjected to a "standard fire."

Note: Concrete with aggregates containing more than 30% quartz, chert, flint, or granite has an inferior fire rating unless mesh is used.
High cement content or light aggregates improve the fire rating.
Cover is defined as the minimum distance from steel to exposed surface.
Prestressed concrete members require more cover for a given rating.

From Bolz, R. E. and Tuve, G. L., Eds., *Handbook of Tables for Applied Engineering Science,* 2nd ed., CRC Press, Cleveland, 1973, 782.

Table 4-56
COMBUSTION RETARDANTS FOR SOLID MATERIALS

Fire and Flame Retardants for Combustible, Structural Materials, Sheets, Fabrics, and Plastics

Variables — Conditions for the combustion of solids in air are so diverse that most data and tests provide only relative values. Among the variables, in addition to the chemical composition, are size, shape, and orientation of the materials and the heat source, initial and ambient temperatures, humidity and moisture content, intensity of radiation, air motion, and duration of the exposure. In addition, the material itself may agglomerate, melt, vaporize, decompose, distort, or intumesce (swell), thus greatly altering the access of oxygen to support combustion.

Fire retardants — In most cases a fire-retardant treatment also will raise the maximum service temperature, and both objectives deserve attention. Oxygen supply is largely determined by surface/volume ratio; three classes are quickly recognized: (1) bulk solids and shapes, (2) sheets and fabrics, and (3) particulates and dusts. The common combustible solids — wood, paper, fabrics, and plastics — each present their special problems. The methods of application also are distinctive, e.g., compounding (filling), coating, and impregnating.

The chemistry of fire retardants deals largely with six elements — phosphorus, chlorine, bromine, boron, antimony, and nitrogen. As many treatments involve two or more retardants, a great variety of compounds of these six elements have been used. Inert and refractory coatings of oxides

Table 4–56 (continued)
COMBUSTION RETARDANTS FOR SOLID MATERIALS

Fire and Flame Retardants for Combustible, Structural Materials, Sheets, Fabrics, and Plastics (continued)

or mineral powders comprise another class of fire retardants, merging into the insulating protective coverings as the thickness and porosity are increased. Many kinds of fillers are used in plastics and rubbers; these usually reduce the flammability, but residues of plasticizers and solvents have opposite effects.

Standard tests – Arbitrary methods of testing for flammability, fire resistance, or fire endurance have been prescribed by the ASTM,[a] the Underwriters' Laboratories, and similar agencies in other countries.

Tabular summary – The following table presents a partial list of the fire-retardant treatments for wood, paper, textiles, and plastics. Many of these same materials are used in fire-retardant treatment or compounding for rubbers, asphalts, and bitumens. Not included in this table are several of the incombustible, or reflective, coatings

that may be applied to the surfaces of combustible materials.

A single table hardly can suggest the many complexities of the methods for fire retardation. Flammability is reduced by stabilizing a material to reduce the decomposition into volatile, combustible products. Some treatments increase the residual char, which then acts as a barrier against propagation of the flame. Certain additives melt and form a hard skin; others decompose and give a protective blanket of inert gas. These are complex processes that are not easily analyzed, but there is a vast literature reporting the efforts and the specific results. It should be mentioned that when certain fire-retarding compounds are used together, their effectiveness increases. Examples are combinations of phosphorus compounds with chlorine compounds and antimony in combination with halogens.

Chemical Fire Retardants for Solids

Key for methods of treatment:

I = immersion or impregnation;
M = mixture compounded;
S = surface application

Class of compound	Examples	Fire retardant for			
		Wood[b]	Paper[c]	Cellulose fabrics	Plastics
Phosphorus compounds					
Phosphoric acid compounds	TCP,DCP,TPP,THPC	I, S		I	M
Ammonium phosphate	$(NH_4)_2 PO_4$;$NH_4 H_2 PO_4$	M	I, M	I	M
p-Halogen compounds	PCl_3	I			M
Bromine compounds					
Organic	CHBr aromatics		M, S	I	M
Bromides	$MgBr_2$;$ZnBr_2$;$NH_4 Br$	I			
Chlorine compounds					
Organic	Chlorinated paraffin	M	M, S	I	M
Zinc chloride	$ZnCl_2$	I			
Boron compounds					
Borax; boric acid	$Na_2 B_2 O_7$;$H_3 BO_3$	S	M, S	I	
Nitrogen compounds					
Ammonium compounds	$(NH_4)_2 SO_4$;$(NH_4)_2 HPO_4$	S	M		M
Miscellaneous materials					
Silicates	Sodium silicate	I, S	S, M		
Fire-retardant paints		S			
Titanium salts	$(TiCl_4)$; $Ti_2 (SO_4)_5$				M

Table 4-56 (continued)
COMBUSTION RETARDANTS FOR SOLID MATERIALS

Chemical Fire Retardants for Solids (continued)

Class of compound	Examples	Fire retardant for			
		Wood[b]	Paper[c]	Cellulose fabrics	Plastics
Antimony compounds (often with halogens)					
Oxide	Sb_4O_6 (Sb_2O_3)	M	M, S	I	M
Chloride	$SbCl_3$	I			
Organic		M	M	I	

[a]See the ASTM Index for flammability tests such as No. D568, D635, D777, D1230, D2859, E162, and E286. See also fire tests such as D1360-61, E84, etc.
[b]Including fiber wallboards.
[c]Including cover stock, cardboards, and boxboards.

REFERENCES

Lyons, J. W., *The Chemistry and Uses of Fire Retardants,* John Wiley & Sons, New York, 1970.
Goundry, J. H., *Fireproofing,* American Elsevier, New York, 1970.

From Bolz, R. E. and Tuve, G. L., Eds., *Handbook of Tables for Applied Engineering Science,* 2nd ed., CRC Press, Cleveland, 1973, 781.

Table 4-57
FLAMMABILITY LIMITS FOR GASES AND VAPORS IN AIR

At Atmospheric Pressure and Approximately Room Temperature[a]

Compound	Formula	Flammability limits in air, % of total volume		Compound	Formula	Flammability limits in air, % of total volume	
		Lower (lean)	Upper (rich)			Lower (lean)	Upper (rich)
Acetaldehyde	C_2H_4O	4.	57.	Ethyl bromide	C_2H_5Br	6.7	11.3
Acetone	C_3H_6O	2.6	12.8	(Bromoethane)			
(2-Propanone)				Ethyl chloride	C_2H_5Cl	3.6	15.4
Acetonitrile	C_2H_3N	4.5	16.	(Chloroethane)			
Acrolein	C_3H_4O	2.7	40.	Ethyl ether	$C_4H_{10}O$	1.7	48.
Acrylonitrile	C_3H_3N	3.	17.	(Diethyl ether)			
Allyl alcohol	C_3H_6O	2.5	18.	Ethyl formate	$C_3H_6O_2$	2.7	16.5
Allyl bromide	C_3H_5Br	4.35	7.3	Ethyl nitrite	$C_2H_5NO_2$	3.	50.
(3-Bromopropene)				Ethylene dichloride	$C_2H_4Cl_2$	6.2	16.
Allyl chloride	C_3H_5Cl	3.3	11.2	(1,2-Dichloroethane)			
(3-Chloropropene)				Ethylene oxide	C_2H_4O	3.	90.
Ammonia	NH_3	15.5	27.	(Oxirane)			
Amyl chloride	$C_5H_{11}Cl$	1.5	8.6	Hydrocyanic acid	HCN	5.6	41.
(1-Chloropentane)				Hydrogen	H_2	4.	75.
1,3-Butadiene	C_4H_6	2.	12.	Hydrogen sulfide	H_2S	4.3	45.5
2-Butoxyethanol	$C_6H_{14}O_2$	1.1	12.7	Isopropyl acetate	$C_5H_{10}O_2$	1.8	7.8
Butyl acetate	$C_6H_{12}O_2$	1.7	10.	Methyl acetate	$C_3H_6O_2$	3.1	16.
Butyl alcohol	$C_4H_{10}O$	1.4	18.	Methyl alcohol	CH_4O	7.3	40.
(1-Butanol)				Methylamine	CH_5N	4.9	20.8

Table 4–57 (continued)
FLAMMABILITY LIMITS FOR GASES AND VAPORS IN AIR

At Atmospheric Pressure and Approximately Room Temperature (continued)

Compound	Formula	Flammability limits in air, % of total volume Lower (lean)	Upper (rich)	Compound	Formula	Flammability limits in air, % of total volume Lower (lean)	Upper (rich)
Butylamine	$C_4H_{11}N$	1.7	9.8	Methyl bromide	CH_3Br	10.	16.
Butyl chloride	C_4H_9Cl	1.8	10.1	(Bromomethane)			
(1-Chlorobutane)				Methyl butyl ketone	$C_6H_{12}O$	1.25	8.
Butyl ether	$C_8H_{18}O$	1.5	7.6	Methyl chloride	CH_3Cl	8.1	19.5
Carbon disulfide	CS_2	1.	50.	Methyl ether	C_2H_6O	2.	20.
Carbon monoxide	CO	12.	74.	Methyl ethyl ketone	C_4H_8O	1.8	11.
Chloroprene	C_4H_5Cl	4.	20.	Methyl formate	$C_2H_4O_2$	5.	22.7
Crotonaldehyde	C_4H_6O	2.1	15.5	n-Propyl acetate	$C_5H_{10}O_2$	2.0	8.
Cyanogen	C_2N_2	6.	42.	n-Propyl alcohol	C_3H_8O	2.1	13.5
Cyclohexane	C_6H_{12}	1.3	8.3	Propylamine	C_3H_9N	2.	10.4
Cyclopropane	C_3H_6	2.4	10.4	Propyl chloride	C_3H_7Cl	2.6	11.1
n-Decane	$C_{10}H_{22}$	.7	5.4	(1-Chloropropane)			
Deuterium	D_2	5.	75.	Propylene dichloride	$C_3H_6Cl_2$	3.4	14.5
Diborane	B_2H_6	.9	98.	(1,2-Dichloropropane)			
Dichlorobenzene	$C_6H_4Cl_2$	2.2	9.2	Propylene oxide	C_3H_6O	2.	22.
Diethylamine	$C_4H_{11}N$	1.8	10.1	Pyridine	C_5H_5N	1.8	12.4
Dimethylamine	C_2H_7N	2.8	14.4	Triethylamine	$C_6H_{15}N$	1.2	8.
Dioxane	$C_4H_8O_2$	2.	22.2	Trimethylamine	C_3H_9N	2.	11.6
2-Ethoxyethanol	$C_4H_{10}O_2$	1.7	15.6	Vinyl acetate	$C_6H_6O_2$	2.6	21.7
Ethyl acetate	$C_4H_8O_2$	2.5	11.5	Vinyl chloride	C_2H_3Cl	4.	22.
Ethyl alcohol	C_2H_6O	4.3	19.	(Chloroethylene)			
Ethylamine	C_2H_7N	3.5	14.	Xylene	C_8H_{10}	1.	7.

[a]Flammable or explosive limits in pure air will differ greatly at other temperatures or pressures. In general, the effect of increasing temperature or pressure is to widen the flammable range.

REFERENCE

Williams-Steiger Occupational Safety and Health Act of 1970, Chapter XVII, Title 29, *Code of Federal Regulations*, April 13, 1971 (amended August 13, 1971), Part 1910 (Occupational Safety and Health Standards); published in *Fed. Register*, 36 (105), May 29, 1971.

From Bolz, R. E. and Tuve, G. L., Eds., *Handbook of Tables for Applied Engineering Science*, 2nd ed., CRC Press, Cleveland, 1973, 783.

Table 4–58
SURFACE FLAMMABILITY OF WOODS

Tests on Bare Wood, Plywood, Wallboards, and Painted Wood Surfaces
Basis of Test Data

A report titled "Surface Flammability of Various Wood-Base Building Materials" was issued by the Forest Products Laboratory in 1959.[a] A similar report on painted wood surfaces appeared in 1963 in the *Official Digest* of the FSPT,[b] and this material was reprinted in the *NFPA Quarterly* in 1964.[c] All tests were made in a tunnel furnace by a uniform method.[c] From these tests three index numbers were computed and tabulated. In each case an index number of 100 was assigned to the performance of a standard red oak specimen; the comparative index numbers were determined as follows:

1. Flame-spread index. If the flame spread was faster than for the standard red-oak specimen, the index was determined by the length of time for the flames to reach the end of the test specimen, as compared with the standard time (18.4 min) for the flame to reach the end of the red-oak specimen. For a flame spread slower than that for red oak, the index was based on the ratio of the distances reached by the flames on the two specimens in the standard 18.4 min period (87 in. for red oak).

2. Index of heat contributed. This index was based on the readings of thermocouples in the furnace stack.

3. Smoke density index. This index was obtained from the readings of photoelectric smoke meter in the furnace stack.

A zero reference for both the smoke-density and heat-contributed indices was established by using a test specimen of asbestos millboard.

It is apparent that this is an arbitrary test procedure that gives approximate comparative results only. Tests of identical specimens often varied ten points or more on the index scales (sometimes with one above 100 and the other below 100). It must be concluded that the various woods and finishes cannot be ranked in absolute order and that the tests serve rather to establish classes or categories of sufrace-flammability performance. This also is to be expected from the fact that the specimens within a class will vary, i.e., no two trees are exactly alike and no two manufacturers of wallboard of the same description will produce identical specimens.

Although specific index numbers are quoted in the following table, they must be considered as approximate test results rather than absolute ratings.

Description of test specimens	lb/ft³	Moisture %	Index numbers			Evaluation of flammability
			Flame spread	Heat contributed	Smoke density	
Tests on 1-in. lumber						
Alder	29.7	6.5	121	121	112	8
Aspen	27.3	6.1	121	124	72	7
Bald cypress	29.6	6.9	112	109	389	14
Basswood	28.0	5.4	128	128	79	3
Beech	46.8	5.8	101	140	132	24
Birch	39.4	5.6	96	94	86	26
Cedar, red	33.8	8.9	109	94	224	16
Chestnut	29.0	6.1	120	92	17	10
Cottonwood	27.3	5.0	134	135	125	1
Elm, slippery	38.5	6.1	89	108	151	29
Fir, Douglas	27.1	5.9	116	77	81	11
Fir, white	29.9	6.7	115	96	109	12
Hemlock	29.0	7.4	108	95	78	17

Table 4–58 (continued)
SURFACE FLAMMABILITY OF WOODS

Tests on Bare Wood, Plywood, Wallboards, and Painted Wood Surfaces (continued)

Description of test specimens	lb/ft³	Moisture %	Index numbers Flame spread	Heat contributed	Smoke density	Evaluation of flammability
Tests on 1-in. lumber (continued)						
Larch	34.9	6.8	106	98	56	19
Mahogany	29.7	6.6	104	70	34	22
Mahogany, Philippine	35.3	5.4	106	81	55	20
Maple, sugar	41.7	6.2	95	83	76	28
Oak, red	39.0	5.0	100	100	100	25
Oak, white	40.8	6.4	95	91	40	27
Pine, Northern white	23.9	5.7	132	104	193	2
Pine, ponderosa	27.7	6.5	114	102	230	13
Pine, Southern yellow	29.2	6.8	102	115	158	23
Pine, sugar	24.0	6.2	125	98	262	4
Pine, western white	25.9	6.0	123	113	274	6
Poplar, yellow	31.0	5.7	124	125	155	5
Redwood	25.6	6.4	121	68	188	9
Spruce, sitka	26.8	6.8	112	80	57	15
Sweet gum	32.3	6.8	105	105	68	21
Walnut, black	37.4	5.1	107	114	85	18
Tests on plywood						
¼-in., 3 ply, fir, interior, protein glue	30.7	6.3	123	119	136	
¼-in., 3 ply, fir, interior, resin glue	34.6	5.0	121	122	81	
⅜-in., 3 ply, fir, exterior, resin glue	33.6	4.8	114	112	96	
⅜-in., 3 ply, fir, exterior, paint A, 1 coat	34.8	4.8	83	58	477	
⅜-in., 3 ply, fir, exterior, paint A, 2 coats	33.9	5.1	53	19	968	
⅜-in., 3 ply, fir, exterior, paint B, 1 coat	34.1	5.0	65	21	747	
⅜-in., 3 ply, fir, exterior, paint B, 2 coats	34.1	5.0	34	10	1 143	
⅜-in., 3 ply, fir, exterior, paint C, 2 coats	34.1	5.0	81	38	1 000	
⅜-in., 3 ply, fir, exterior, paper, plastic overlay	36.	5.0	105	112	141	
Tests on fiberboard						
Four different ½-in. insulating fiberboards	19.	5.0	125	91	105	
Three heavier fiberboards	32.	4.0	122	161	—	
Tests on hardboard						
Willow, oak, wax, resin, 0.22 in.	51.4	3.2	119	177	131	
Fir, pine, wax, resin (4 makes)	61.4	4.3	96	169	407	
Fir, redwood, dense, oil-tempered	70.8	3.1	90	165	657	
Tests on particle boards						
Nine wood-chip boards	36–67	3.8–6.6	85–104	80–150	55–650	

Table 4–58 (continued)
SURFACE FLAMMABILITY OF WOODS

Conclusions From Test Results

Index numbers obtained by flame-spread tests of 29 species of structural lumber by the tunnel-furnace method gave a graduated scale from the 130 range for flammable softwoods, such as white pine, cottonwood, basswood, and poplar, to the 90 range for the dense hardwoods, such as white oak, birch, elm, and sugar maple. Although the "heat-contributed" index followed the flame-spread index roughly, the smoke density was dependent on other factors. The flame-spread index of 1/4- and 3/8-in. fir plywood was somewhat higher than for a 1-in. fir board. The flame-spread index for either structural or insulating fiberboard differed little from those for the softwoods.

The flammability of pressed hardboards varied from the softwood range for the lighter boards (50 lb/ft^3) to the hardwood range for oil-tempered board of high density (65 lb/ft^3).

The particle boards tested were no more flammable than hardwood; however, the range was rather wide, as there is a great variety of such boards.

Many tests were made on painted specimens, most of them with 3/8-in. Douglas fir plywood as a base. For some coatings such as interior oil-base paints and enamels, varnish, shellac, lacquer, and asphalt paint the flammability was changed little from that of bare plywood, but the smoke and heat were increased. The effectiveness of fire-retardant paints was clearly demonstrated, especially if two coats were used or an intumescent (foaming) composition was employed. Most of the fire-retardant paints reduced the flammability of the fir-plywood specimen to that of the least-flammable hardwood. The best of the fire-retardant paints reduced the flammability index to the 30 range, on a scale of red oak = 100 and asbestos millboard = 0.

[a]Bruce, H. D. and Downs, L. E., Surface Flammability of Various Wood-base Building Materials, Forest Products Laboratory Report No. 2140, Forest Service, U.S. Department of Agriculture, 1959.
[b]*Official Digest*, Federation of Societies for Paint Technology, August 1963.
[c]Eickner, W. C. and Peters, C. C., *Natl. Fire Protection Q.*, April 1964.

From Bolz, R. E. and Tuve, G. L., Eds., *Handbook of Tables for Applied Engineering Science*, 2nd ed., CRC Press, Cleveland, 1973, 778.

Table 4–59
ION-EXCHANGE RESINS

SYMBOLS:

Manufacturers:
1. Dow Chemical Co., Midland, Mich. 48640
2. Diamond Shamrock Chemical Co., Redwood City, Cal. 94063
3. Ionac Chemical Corp., Birmingham, N.J. 08011
4. Nalco Chemical Co., Chicago, Ill. 60638
5. Rohm and Haas Co., Philadelphia, Pa. 19105

Characters:
S = strong
W = weak

Physical form:
b = beads
g = granules

Anion-exchange Resins

Trade name	Manufacturer	Character	Active group	Matrix	Effective pH	Selectivity	Order of selectivity
Dowex® 1	1	S	Trimethyl benzyl ammonium	Polystyrene	0–14	Cl/H approx. 25	I, NO_3, Br, Cl, acetate, OH, F
Dowex 21 K	1	S	Trimethyl benzyl ammonium	Polystyrene	0–14	Cl/H approx. 1.5	I, NO_3, Br, Cl, acetate, OH, F
Duolite® A-101 D	2	S	Trimethyl benzyl ammonium	Polystyrene	0–14	—	—
Ionac® A-540	3	S	Trimethyl benzyl ammonium	Polystyrene	0–14	—	—
Amberlite® IRA-400	5	S	Trimethyl benzyl ammonium	Polystyrene	0–14	—	—
Dowex 2	1	S	Dimethyl ethanol benzyl ammonium	Polystyrene	0–14	Cl/H approx. 1.5	I, NO_3, Br, Cl, acetate, OH, F
Duolite A-102 D	2	S	Dimethyl ethanol benzyl ammonium	Polystyrene	0–14	—	—
Ionac A-550	3	S	Dimethyl ethanol benzyl ammonium	Polystyrene	0–14	—	—
Amberlite IRA-410	5	S	Dimethyl ethanol benzyl ammonium	Polystyrene	0–14	—	—
Duolite A-30 B	2	W	Tertiary amine; quaternary ammonium	Epoxy polyamine	0–9	—	—
Ionac A-300	3	W	Tertiary amine; quaternary ammonium	Epoxy polyamine	0–12	—	—

Table 4–59 (continued)

Anion-exchange Resins

Trade name	Manufacturer	Character	Active group	Matrix	Effective pH	Selectivity	Order of selectivity
Duolite A-6	2	W	Tertiary amine	Phenolic	0–5	—	—
Duolite A-7	2	W	Secondary amine	Phenolic	0–4	—	—
Amberlite IR-45	5	W	Primary, secondary, and tertiary amine	Polystyrene	0–9	—	—

Trade name	Manufacturer	Total exchange capacity, meq/ml[a]	Total exchange capacity, meq/g[a]	Maximum thermal stability, °C	Physical form	Standard mesh range	Ionic form as shipped[b]	Shipping density, lb/ft^3
Dowex 1	1	1.33	3.5	50 (OH^-) / 150 (Cl^-)	b	20–50 (wet)	Cl^-	44
Dowex 21 K	1	1.25	4.5	50 (OH^-) / 150 (Cl^-)	b	20–50 (wet)	Cl^-	43
Duolite A-101 D	2	1.4	4.2	60 (OH^-) / 100 (Cl^-)	b	16–50	Cl^-	—
Ionac A-540	3	1.0	3.6	60 (OH^-) / 100 (Salt)	b	16–50	Salt	43–66
Amberlite IRA-400	5	1.4	3.8	60 (OH^-)	b	—	Cl^-	—
Dowex 2	1	1.33	3.5	30 (OH^-) / 150 (Cl^-)	b	20–50 (wet)	Cl^-	44
Duolite A-102 D	2	1.4	4.2	40 (OH^-) / 100 (Cl^-)	b	16–50	Cl^-	—
Ionac A-550	3	1.3	3.5	40 (OH^-) / 100 (Salt)	b	16–50	Salt	43–46
Amberlite IRA-410	5	1.4	3.3	40 (OH^-) / 75 (Cl^-)	b	—	Cl^-	—
Duolite A-30 B	2	2.6	8.7	80	b	16–50	Salt	—
Ionac A-300	3	1.8	5.5	40	g	16–50	Salt	19–21
Duolite A-6	2	2.4	7.6	60	g	16–50	Salt	—
Duolite A-7	2	2.4	9.1	40	g	16–50	Salt	—
Amberlite IR-45	5	1.9	5.2	100	b	—	Free base	—

Table 4—59 (continued)
Cation-exchange Resins

Trade name	Manufacturer	Character	Active group	Matrix	Effective pH	Selectivity	Order of selectivity
Dowex 50	1	S	Nuclear sulfonic acid	Polystyrene	0–14	Na/H approx. 1.2	Ag, Cs, Rb, K, NH$_4$, Na, H, Li, Ba, Sr, Ca, Mg, Be
Dowex MPC-1	4	S	Nuclear sulfonic acid	Polystyrene	0–14	—	—
Duolite C-20	2	S	Nuclear sulfonic acid	Polystyrene	0–14	—	—
Ionac C-240	3	S	Nuclear sulfonic acid	Polystyrene	0–14	—	—
Amberlite IR-120	5	S	Nuclear sulfonic acid	Polystyrene	0–14	—	—
Duolite C-3	2	S	Methylene sulfonic	Phenolic	0–9	—	—
Dowex CCR-1	4	W	Carboxylic	Phenolic	0–9	—	—
Duolite ES-63	2	W	Phosphonic	Polystyrene	4–14	—	—
Duolite ES-80	2	W	Carboxylic	Acrylic	6–14	—	—
Amberlite IRC-84	5	W	Carboxylic	Acrylic	4–14	—	—

Trade name	Manufacturer	Total exchange capacity, meq/ml[a]	Total exchange capacity, meq/g[a]	Maximum thermal stability, °C	Physical form	Standard mesh range	Ionic form as shipped[b]	Shipping density, lb/ft^3
Dowex 50	1	Na$^+$ 1.9 / H$^+$ 1.7 H$^+$ form	Na$^+$ 4.8 / H$^+$ 5.0 H$^+$ form	150	b	20–50 (wet)	H$^+$ or Na$^+$	H$^+$ 50 / Na$^+$ 53
Dowex MPC-1	4	1.6–1.8 H$^+$ form	4.5–4.9 H$^+$ form	150	b	20–40 (wet)	Na$^+$	50
Duolite C-20	2	2.2	5.1	150	b	16–50	Na$^+$	—
Ionac C-240	3	1.9	4.6	140 (Na$^+$) 130 (H$^+$)	b	16–50	Na$^+$	50–55
Amberlite IR-120	5	1.9	4.4	120	b	—	Na$^+$ or H$^+$	—
Duolite C-3	2	1.1	2.9	60	g	16–50	H$^+$	—
Dowex CCR-1	4	—	—	38	g	20–50	H$^+$ (dry)	21
Duolite ES-63	2	3.3	6.6	100	b	16–50	H$^+$	—
Duolite ES-80	2	3.5	10.2	100	b	16–50	H$^+$	—
Amberlite IRC-84	5	3.5	10.3	120	b	—	H$^+$	—

Table 4–59 (continued)

Applications of Ion-exchange Resins

1. Replacement of deleterious by unobjectionable ions, as in water softening.

2. Concentration or recovery of a valuable substance, as in the extraction of uranium and other metals from ores, and the concentration and purification of streptomycin.

3. Removal of harmful substances from waste effluent; for example, cyanide from plating baths, thiocyanate from gas-works liquors, and radioactive contaminants from wastes of nuclear power plants.

4. Removal of acids by OH-resins and alkalis by H-resins, as in the removal of formic acid from commercial formaldehyde solutions and the removal of acid catalyst from esterification mixtures.

5. Removal of both anions and cations, as in water deionization, sugar purification, and glycerol refining.

6. Other applications include analytical separations; treatment of wine, fruit juices, and milk; use in the manufacture of pulp and paper, pharmaceuticals, and petroleum products; in hydrometallurgy; in the isolation and purification of biochemical products; and as catalysts in chemical reactions.

Regeneration of Ion-exchange Resins

Ion exchange resins must be regenerated at intervals to restore their capacity to exchange ions. In water softening the resin is regenerated with a salt brine solution containing 10 to 20% NaCl. In formaldehyde purification the weak-base resins can be regenerated with dilute solutions of caustic soda, soda ash, or ammonia. In sugar purification sulfuric acid and caustic soda or ammonia are usually the reagents involved. Streptomycin is eluted from the resin with aqueous HCl. In the sulfuric acid leach process uranium is eluted with a solution of ammonium nitrate and nitric acid or sodium chloride and sulfuric acid.

REFERENCES

^{a}Meq = milligram equivalents. The equivalent weight of an element or ion is its atomic or formula weight divided by its valence. The exchange capacity of a resin is the number of milligram equivalents of ions in the solution that 1 g (dry) or 1 ml (wet) of the resin will exchange.
^{b}Changes in ionic form will cause resins to swell 5 to 25% or more.

Abrams, I. M. and Benezra, L., Ion exchange polymers, in *Encyclopedia of Polymer Science and Technology*, Vol. 7, John Wiley & Sons, New York, 1967, 692.

Condensed from Weast, R. C., Ed., *Handbook of Chemistry and Physics*, 53rd ed., CRC Press, Cleveland, 1972.

Table 4—60

USEFUL RANGE OF SOLIDS FOR DRY LUBRICATION

For load in MN/m^2, multiply values in psi by
6.8948×10^{-3}.

Lubricant	Load,[a] psi	Temperature, °F
Molybdenum disulfide	2,000–yield point of metal	–300–+750[b]
Graphite powder	2,000–100,000	–300–+1,200
Tungsten disulfide powder	2,000–yield point of metal	Low not investigated, max. 850[b]
Polytetrafluoroethylene (PTFE) powder and sintered shapes	150–3,000	–300–+500
Polytetrafluoroethylene and other plastics containing fillers	50–4,000	Max. 500
Organic binder coatings, graphite–MoS_2 type	2,000–yield point of metal	–100–500
Inorganic binder coatings, graphite–MoS_2 type	2,000–yield point of metal	–300–1,000

[a]Speed affects the load capacity to a marked degree, but solid lubricants should only be considered at low speeds. As speed increases, load-carrying capacity decreases.
[b] In inert and reducing atmospheres these solids are used at temperatures above 2,000° F.

From DiSapio, A. and Gerstung, H. S., Solid and bonded-film lubricants, *Machine Design* (The Bearings Reference Issue), 38(6), 8, 1966. With permission.

Table 4—61
AMERICAN WOODS—PROPERTIES AND USES

For weight-density in kg/m^3, multiply value in
lb/ft^3 by 16.02.

Species	Specific gravity Green	Dry	Characteristics	Uses	Weight lb/ft³, green	lb/ft³, air-dry 12%	lb/1,000 board ft, air-dry 12%
Alder, red	0.37	0.41	Low shrinkage; moderate in strength, shock resistance, hardness, and weight[a]	Furniture; sash; doors; millwork	46	28	2,330
Ash, black	0.45	0.49	Light in weight[a]	Cabinets; veneer; cooperage, containers	52	34	2,830
Ash, Oregon	0.50	0.55	Similar to but lighter than white ash[a]	Similar to white ash	46	38	3,160
Ash, white	0.54	0.58	Heavy; hard; stiff; strong; high shock resistance[a]	Handles; ladder rungs; baseball bats; farm implements; car parts	48	41	3,420
Bald cypress (Southern cypress)			Moderate in strength, weight, hardness, and shrinkage[c]	Building construction; beams; posts; ties; tanks; ships; paneling	51	32	2,670
Beech, American	0.56	0.64	Heavy; high strength, shock resistance, and shrinkage; uniform texture[a]	Flooring; furniture; handles; kitchenware; ties (treated)	54	45	3,750

Table 4–61 (continued)
AMERICAN WOODS–PROPERTIES AND USES

Species	Specific gravity		Characteristics	Uses	Weight		
	Green	Dry			lb/ft³, green	lb/ft³, air-dry 12%	lb/1,000 board ft, air-dry 12%
Birch	0.57	0.63	Heavy; high strength, shock resistance, and shrinkage; uniform texture[a]	Interior finish; dowels; ties (treated); veneer; musical instruments	57	44	3,670
Cottonwood	0.37	0.40	Uniform texture; does not split readily; moderate in weight, strength, hardness, and shrinkage	Crates; trunks; car parts; farm implements	49	28	2,330
Douglas fir	0.41	0.44	Moderate in strength, weight, shock resistance, and shrinkage[b]	Building and construction; poles; veneer; plywood; ships; furniture; boxes	38	34	2,830
Elm	0.57	0.63	Moderate in strength, weight, and hardness; high in shock resistance and shrinkage; good in bending[a]	Cooperage; baskets; crates; veneer; vehicle parts	54	34	2,920
Hemlock, Eastern	0.38	0.40	Moderate in weight, strength, and hardness[a]	Building and construction; boxes	50	28	2,330
Hemlock, Western	0.38	0.42	Moderate in weight, strength, and hardness[a]	Sash; doors; posts; piles; building and construction	41	29	2,420
Hickory, true	0.65	0.73	High toughness, hardness, shock resistance, strength, and shrinkage[a]	Dowels; spokes; poles; shafts; gymnasium equipment	63	51	4,250
Incense cedar	0.35		Uniform texture; easy to season; low shrinkage, shock resistance, weight, and stiffness[c]	Lumber; fence posts; ties; poles; shingles	45		
Larch, Western	0.48	0.52	Moderate in strength, weight, shock resistance, hardness, and shrinkage[b]	Doors; sash; posts; pilings; building and construction	48	36	3,000
Locust, black	0.66	0.69	High in shock resistance, weight, and hardness; very high strength; moderate shrinkage[c]	Mine timbers; posts; poles; ties	58	48	4,000
Maple	0.44	0.48	High in hardness, weight, strength, shock resistance, and shrinkage; uniform texture[a]	Flooring; furniture; trim; spools; farm implements	54	40	3,330
Oak, red and white	0.57	0.63	High in hardness, weight, strength, shock resistance, and shrinkage; red,[a] white[b]	Trim; ships; flooring; ties; furniture; cooperage; piles	64	44	3,670
Pine, jack			Coarse texture; low strength, stiffness, shock resistance, and shrinkage	Box lumber; fuel; mine timber; ties; poles; posts			
Pine, lodgepole	0.38	0.41	Moderate in weight, hardness, strength, shock resistance, and shrinkage; easy to work[b]	Poles; mine timber; ties; construction	39	29	2,420
Pine			High shrinkage; moderate strength, stiffness, hardness, and shock resistance	General construction; ties; poles; posts			

Table 4–61 (continued)
AMERICAN WOODS–PROPERTIES AND USES

Species	Specific gravity		Characteristics	Uses	lb/ft³, green	lb/ft³, air-dry 12%	lb/1,000 board ft, air-dry 12%
	Green	Dry					
Pine, Ponderosa	0.38	0.40	Moderate in weight, shock resistance, shrinkage, and hardness; easy to work[a]	Building; paneling; sash; frames	45	28	2,330
Pine, S. yellow	0.47	0.51	Moderate in shock resistance, shrinkage, and hardness; high in strength[b]	Building and construction; poles; pilings; boxes	55	41	3,420
Pine, sugar	0.35	0.36	Low shock resistance; easy to work; moderate strength[a]	Sash; counters; blinds; patterns	52	25	2,080
Pine, Western white	0.36	0.38	Moderate in strength, shock resistance, shrinkage, and hardness; easy to work[a]	Building and construction; patterns; boxes	35	27	2,250
Red cedar, Eastern and Western	0.44	0.47	High shock resistance; low stiffness and shrinkage; moderate in strength and hardness[c]	Fence posts; closet liners; chests; flooring	37	37	2,750
Redwood	0.38	0.40	Low shrinkage; medium in weight, strength, hardness, and shock resistance[c]	Posts; doors; interiors; cooling towers	50	28	2,330
Spruce, Eastern	0.38	0.40	Moderate in hardness, shock resistance, weight, shrinkage, and strength[a]	Building; millwork; boxes; ladders	34	28	2,330
Spruce, Engelmann	0.31	0.33	Generally straight grained; light in weight; low strength as a beam or post; low shock resistance; moderate shrinkage	Mine timber; ties; poles; flooring; studding; paper	39	23	1,920
Spruce, Sitka	0.37	0.40	Moderate in weight, hardness, strength, shock resistance, and shrinkage[a]	Important in boat and plane construction; sash; doors; boxes; siding	33	28	2,330
Sycamore	0.46	0.49	High shrinkage; moderate in weight, strength, hardness, and shock resistance[a]	Boxes; ties; posts; veneer; flooring; butcher blocks	52	34	2,830
Tamarack	0.49	0.53	Coarse texture; moderate in strength, hardness, shrinkage, and shock resistance	Ties; mine timber; posts; poles; tanks; scaffolding	47	37	3,080
Tupelo			Uniform texture; moderate in strength, hardness, shock resistance; high shrinkage; interlocked grain makes splitting difficult[a]	Flooring; planking; crates; furniture			
Walnut, black	0.51	0.55	Moderate shrinkage; high weight, strength, hardness, and shock resistance; easily worked and glued[c]	Gun stocks; cabinets; plywood; furniture; veneer	58	38	3,170

Table 4–61 (continued)
AMERICAN WOODS–PROPERTIES AND USES

| Species | Specific gravity | | Characteristics | Uses | Weight | | |
	Green	Dry			lb/ft³, green	lb/ft³, air-dry 12%	lb/1,000 board ft, air-dry 12%
White cedar	0.31	0.32	Low shrinkage, weight, shock resistance, and strength; soft; easily worked[c]	Poles; posts; ties; tanks; ships	24	23	1,920
Willow, black			High strength and shock resistance; low beam strength and weight; interlocked grain	Lumber; veneer; charcoal; furniture; subflooring; studding			

[a]Decay resistance low.
[b]Decay resistance medium.
[c]Decay resistance high.

From Parker, E. R., *Materials Data Book*, McGraw-Hill, New York, copyright © 1967, 252. With permission.

Table 4–62
ALLOWABLE UNIT STRESSES FOR LUMBER

Grading and Specifications of Softwood Lumber

American Softwood Lumber Standard — A voluntary standard for softwood lumber has been developing since 1922. Five editions of Simplified Practice Recommendation R16 were issued from 1924–53 by the Department of Commerce; the present NBS Voluntary Product Standard PS 20-70, *American Softwood Lumber Standard*, was issued in 1970. It was supported by the American Lumber Standards Committee, which functions through a widely representative National Grading Rule Committee.

The *American Softwood Lumber Standard*, PS 20-70, gives the size and grade provisions for American Standard lumber and describes the organization and procedures for compliance enforcement and review. It lists commercial name classifications and complete definitions of terms and abbreviations.

PS 20-70 lists 11 softwood species, viz., cedar, cypress, fir, hemlock, juniper, larch, pine, redwood, spruce, tamarack, and yew. Five dimensional tables show the standard dressed (surface planed) sizes for almost all types of lumber, including matched tongue-and-grooved and shiplapped flooring, decking, siding, etc. Dry or seasoned lumber must have 19% or less moisture content, with an allowance for shrinkage of 0.7 to 1.0% for each four points of moisture content below this maximum. Green lumber has more than 19% moisture. Table A illustrates the relation between nominal size and dressed or green sizes.

Table 4−62 (continued)
ALLOWABLE UNIT STRESSES FOR LUMBER

Table A. Nominal and Minimum-dressed Sizes

This table applies to boards, dimensional lumber, and timbers. The thicknesses apply to all widths and all widths to all thicknesses.

Item	Thicknesses			Face widths		
		Minimum dressed			Minimum dressed	
	Nominal	Dry,[a] inches	Green, inches	Nominal	Dry,[a] inches	Green, inches
Boards[b]				2	1 1/2	1 9/16
				3	2 1/2	2 9/16
				4	3 1/2	3 9/16
				5	4 1/2	4 5/8
	1	3/4	25/32	6	5 1/2	5 5/8
				7	6 1/2	6 5/8
	1 1/4	1	1 1/32	8	7 1/4	7 1/2
				9	8 1/4	8 1/2
	1 1/2	1 1/4	1 9/32	10	9 1/4	9 1/2
				11	10 1/4	10 1/2
				12	11 1/4	11 1/2
				14	13 1/4	13 1/2
				16	15 1/4	15 1/2
Dimension				2	1 1/2	1 9/16
				3	2 1/2	2 9/16
				4	3 1/2	3 9/16
	2	1 1/2	1 9/16	5	4 1/2	4 5/8
	2 1/2	2	2 1/16	6	5 1/2	5 5/8
	3	2 1/2	2 9/16	8	7 1/4	7 1/2
	3 1/2	3	3 1/16	10	9 1/4	9 1/2
				12	11 1/4	11 1/2
				14	13 1/4	13 1/2
				16	15 1/4	15 1/2
Dimension				2	1 1/2	1 9/16
				3	2 1/2	2 9/16
				4	3 1/2	3 9/16
				5	4 1/2	4 5/8
	4	3 1/2	3 9/16	6	5 1/2	5 5/8
	4 1/2	4	4 1/16	8	7 1/4	7 1/2
				10	9 1/4	9 1/2
				12	11 1/4	11 1/2
				14		13 1/2
				16		15 1/2
Timbers	5 and thicker		1/2 off	5 and wider		1/2 off

[a]Maximum moisture content of 19% or less.

[b]Boards less than the minimum thickness for 1-in. nominal but 5/8-in. or greater thickness dry (11/16 in. green) may be regarded as American Standard Lumber, but such boards shall be marked to show the size and condition of seasoning at the time of dressing. They shall also be distinguished from 1-in. boards on invoices and certificates.

From *American Softwood Lumber Standard,* NBS 20-70, National Bureau of Standards, Washington, D.C., 1970 (available from Superintendent of Documents).

Table 4–62 (continued)
ALLOWABLE UNIT STRESSES FOR LUMBER

National Design Specification – Table B is condensed from the 1971 edition of *National Design Specification for Stress-grade Lumber and Its Fastenings,* as recommended and published by the National Forest Products Association, Washington, D.C. This specification was first issued by the National Lumber Manufacturers Association in 1944; subsequent editions have been issued as recommended by the Technical Advisory Committee. The 1971 edition is a 65-page bulletin with a 20-page supplement giving "Allowable Unit Stresses, Structural Lumber," from which Table B has been condensed. The data on working stresses in this Supplement have been determined in accordance with the corresponding ASTM Standards, D245-70 and D2555-70.

Table B. Species, Sizes, Allowable Stresses, and Modulus of Elasticity

Normal loading conditions: moisture content not over 19%, No. 1 grade, visual grading. To convert psi to N/m^2, multiply by 6,895.

Species[a]	Sizes, nominal	Typical grading agency, 1971[b]	Extreme fiber in bending[c]	Tension parallel to grain	Compression perpendicular	Compression parallel	Modulus of elasticity, psi
Cedar							
Northern white	2 × 4	NL, NH	1,100	600	205	675	800,000
	2 or 4 × 6+	NL, NH	1,000	575	205	675	800,000
Western	2 × 4	NC	1,450	725	285	975	1,100,000
	2 or 4 × 6+	NC, WW	1,250	725	285	975	1,100,000
Fir							
Balsam	2 × 4	NL, NH	1,300	675	170	825	1,200,000
	2 or 4 × 6+	NL, NH	1,150	650	170	825	1,200,000
Douglas (larch)	2 × 4	WC, NC	2,400	1,200	385	1,250	1,800,000
	2 or 4 × 6+	WC, NC	1,750	1,000	385	1,250	1,800,000
Hemlock							
Eastern (tamarack)	2 × 4	NL, NH	1,750	900	365	1,050	1,300,000
	2 or 4 × 6+	NL, NH	1,500	875	365	1,050	1,300,000
Hem-fir	2 × 4	WC, NC	1,600	825	245	1,000	1,500,000
	2 or 4 × 6+	WC, NC	1,400	800	245	1,000	1,500,000
Mountain	2 × 4	WC, WW	1,700	850	370	1,000	1,300,000
	2 or 4 × 6+	WC, WW	1,450	850	370	1,000	1,300,000
Pine							
Idaho white	2 × 4	WW	1,400	725	240	925	1,400,000
	2 or 4 × 6+	WW	1,200	700	240	925	1,400,000
Lodgepole	2 × 4	WW	1,500	750	250	900	1,300,000
	2 or 4 × 6+	WW	1,300	750	250	900	1,300,000
Northern	2 × 4	NL, NH	1,600	825	280	975	1,400,000
	2 or 4 × 6+	NL, NH	1,400	800	280	975	1,400,000
Ponderosa (sugar)	2 × 4	WW, NC	1,400	700	250	850	1,200,000
	2 or 4 × 6+	WW, NC	1,200	700	250	850	1,200,000
Red	2 × 4	NC	1,350	700	280	825	1,300,000
	2 or 4 × 6+	NC	1,150	675	280	825	1,300,000
Southern	2 × 4	SP	2,000	1,000	405	1,250	1,800,000
	2 or 4 × 6+	SP	1,750	1,000	405	1,250	1,800,000

Table 4–62 (continued)
ALLOWABLE UNIT STRESSES FOR LUMBER

Table B. Species, Sizes, Allowable Stresses, and Modulus of Elasticity (continued)

Species[a]	Sizes, nominal	Typical grading agency, 1971[b]	Allowable unit stresses, psi[d]				Modulus of elasticity, psi
			Extreme fiber in bending[c]	Tension parallel to grain	Compression perpendicular	Compression parallel	
Redwood							
California	2 or 4 × 2 or 4	RI	1,950	1,000	425	1,250	1,400,000
	2 or 4 × 6 to 12	RI	1,700	1,000	425	1,250	1,400,000
Spruce							
Eastern	2 × 4	NL, NH	1,500	750	255	900	1,400,000
	2 or 4 × 6+	NL, NH	1,250	750	255	900	1,400,000
Engelmann	2 × 4	WW	1,300	675	195	725	1,200,000
	2 or 4 × 6+	WW	1,150	650	195	725	1,200,000
Sitka	2 × 4	WC	1,550	775	280	925	1,500,000
	2 or 4 × 6+	WC	1,300	775	280	925	1,500,000

[a]Grade designations are not entirely uniform. Values in the table apply approximately to "No. 1." There is seldom more than one better grade than No. 1, and this may be designated as select, select structural, dense, or heavy. In addition to lower grades 2 and 3, there may be other lower grades, designated as construction, standard, stud, and utility. In bending and tension the allowable unit stresses in the lowest recognized grade (utility) are of the order of one eighth to one sixth of the allowable stresses for grade No. 1. The tabular values for allowable bending stress are for the extreme fiber in "repetitive member uses" and edgewise use. The original tables give correction factors, which are less than unity for moist locations and for short-time loading; they are greater than unity if the moisture content of the wood in service is 15% or less. In general, all data apply to uses within covered structures. From the extensive tables, only the No. 1 grade in nominal 2 × 4 size and 2-in. or 4-in. planks, 6 in., and wider have been selected for illustration.

In a few cases the allowable stresses specified for the Canadian products will vary slightly from those given here for the same species by the U.S. agencies.

[b]Grading agencies represented by letters in this column are as follows:

NC = National Lumber Grades Authority (a Canadian agency)
NH = Northern Hardwood and Pine Manufacturers Association
NL = Northeastern Lumber Manufacturers Association
RI = Redwood Inspection Service
SP = Southern Pine Inspection Bureau
WC = West Coast Lumber Inspection Bureau
WW = Western Wood Products Association

[c]It is assumed that all members are so framed, anchored, tied, and braced that they have the necessary rigidity.

[d]For short term loads, these values may be increased: add 15% for 2-month snow load; add 33% for wind or earthquake; add 100% for impact load.

Note: Allowable unit stresses in horizontal shear are in the range 60 to 100 psi for No. 1 grade.

REFERENCES

Wood Handbook, Handbook No. 72, U.S. Department of Agriculture, Washington, D.C., 1955.

Timber Construction Manual, American Institute of Timber Construction, John Wiley & Sons, New York, 1966.

National Design Specification for Stress-grade Lumber and Its Fastenings, National Forest Products Association, Washington, D.C., 1971.

Table 4–63
STRESS-GRADE LUMBER–MAXIMUM END LOADS

Allowable Unit Stresses of Wood for End Grain in Bearing Parallel to Grain

These allowable unit stresses apply to the net area in bearing. When the stress in end-grain bearing exceeds 75% of the adjusted allowable unit stresses, bearing shall be on a metal plate, strap, or other durable, rigid, homogeneous material of adequate strength.

To convert stresses to N/m^2, multiply by 6,895.

Species	Unseasoned, psi	Seasoned, psi
Ash, commercial white	1,510	2,060
Balsam fir	980	1,330
Beech	1,310	1,790
Birch, sweet and yellow	1,260	1,720
California redwood (close grain)	1,720	2,340
California redwood (open grain)	1,270	1,730
Cottonwood, Eastern	840	1,150
Douglas fir–larch (dense)	1,730	2,360
Douglas fir–larch	1,480	2,020
Douglas fir, South	1,340	1,820
Eastern hemlock–tamarack	1,270	1,730
Eastern spruce	1,060	1,450
Eastern white pine	990	1,360
Engelmann spruce	860	1,170
Hem-fir	1,230	1,680
Hickory and pecan	1,510	2,050
Idaho white pine	1,080	1,470
Lodgepole pine	1,060	1,450
Maple, black and sugar	1,260	1,710
Mountain hemlock	1,170	1,600
Northern pine	1,150	1,570
Northern white cedar	810	1,110
Oak, red and white	1,160	1,590
Ponderosa pine–sugar pine	1,000	1,360
Red pine and northern species	970	1,320
Sitka spruce	1,090	1,480
Southern cypress	1,460	1,990
Southern pine (dense)	1,730	2,360
Southern pine (med. grain)	1,480	2,020
Southern pine (open grain)	1,260	1,720
Spruce–pine–fir	1,040	1,410
Subalpine fir	840	1,150
Sweetgum and tupelo	1,120	1,530
Western cedars	1,140	1,560
Western white pine	1,030	1,400
Yellow poplar	980	1,340

From *National Design Specification for Stress-grade Lumber and Its Fastenings*, National Forest Products Association, Washington, D.C., 1973, II-3. With permission.

Table 4–64
ELECTRICAL RESISTANCE OF VARIOUS SPECIES OF WOOD

This table gives the average of measurements made along the grain between two pairs of needle electrodes 1 1/4 in. apart and driven to a depth of 5/16 in., measured at 80°F.

Species	Electrical resistance, megohms, for various moisture contents						
	7%	8%	9%	10%	12%	16%	20%
SOFTWOODS							
Cypress, Southern	12,600	3,980	1,410	630	120	11.2	1.78
Douglas fir (coast region)	22,400	4,780	1,660	630	120	11.2	2.14
Fir, white	57,600	15,850	3,980	1,120	180	16.6	3.02
Hemlock, Western	22,900	5,620	2,040	850	185	16.2	2.52
Larch, Western	39,800	11,200	3,980	1,445	250	19.9	3.39
Pine							
Eastern white	20,900	5,620	2,090	850	200	19.9	3.31
Ponderosa	39,800	8,910	3,310	1,410	300	25.1	3.55
Southern longleaf	25,000	8,700	3,160	1,320	270	24.0	3.72
Southern shortleaf	43,600	11,750	3,720	1,350	255	22.4	3.80
Sugar	22,900	5,250	1,660	645	140	15.9	3.02
Redwood	22,400	4,680	1,550	615	100	7.2	1.74
Spruce, Sitka	22,400	5,890	2,140	830	165	15.5	3.02
HARDWOODS							
Ash, commercial white	12,000	2,190	690	250	55	5.0	0.89
Birch	87,000	19,950	4,470	1,290	200	18.2	3.55
Gum, red	38,000	6,460	2,090	815	160	15.1	2.63
Hickory, true		31,600	2,190	340	50	3.7	0.71
Maple, sugar	72,400	13,800	3,160	690	105	10.2	2.24
Oak							
Commercial red	14,400	4,790	1,590	630	125	11.3	2.09
Commercial white	17,400	3,550	1,100	415	80	7.2	1.15

From *Wood Handbook,* Handbook No. 72, U.S. Department of Agriculture, Washington, D.C., 1955.

Table 4–65
LOADS FOR NAILS AND SCREWS IN WOOD

Table A. Grouping of Wood Species

The holding power of wood fastenings depends largely on the density of the wood, hard to soft. Loads increase with the specific gravity of wood.

Group No.	Typical species	Specific gravity
1	Hickory, hard maple, oak, beech, birch	.62–.75
2	Douglas fir (larch), Southern pine, gum	.51–.55
3	Hemlock, spruce, Northern pine, cypress, redwood	.42–.48
4	Balsam fir, cedar, Engelmann spruce, soft white pine	.31–.41

Table 4–65 (continued)
LOADS FOR NAILS AND SCREWS IN WOOD

Table B. Sizes and Allowable Design Loads

The following table gives the normal range of loading in pounds for joints secured by nails, spikes, wood screws, lag screws, or bolts. For load in kilograms, multiply tabular values by 0.453 6.

Symbols:

L = length; D = diameter

Fastener description (seasoned wood only)	Nominal gage	Length and diameter, inches	Wood species, group			
			1	2	3	4
Nails,[a] allowable withdrawal load per individual nail, per in. penetration into side grain	6 d	2 × .113	47–76	29–34	18–25	9–17
	8 d	2.5 × .131	55–87	34–39	21–29	10–20
	10 d	3 × .148	62–78	38–44	23–33	12–22
	12 d	3.25 × .148	62–78	38–44	23–33	12–22
	16 d	3.5 × .162	68–85	42–49	25–36	13–24
Spikes, allowable withdrawal load per individual spike per in. penetration into side grain	12	3.25 × .192	80–127	49–57	30–42	15–29
	16	3.5 × .207	86–138	53–61	33–46	16–31
	20	4 × .225	94–150	58–67	35–50	18–33
	40	5 × .263	110–174	68–79	41–58	21–39
Nails and spikes, allowable lateral load in total lb shear for penetration from 10 diam for Group 1 to 14 diam for Group 4, into final member	6 d	2 × .113	78	63	51	41
	8 d	2.5 × .131	97	78	64	51
	10 d	3 × .148	116	94	77	62
	12 d	3.25 × .148	116	94	77	62
	16	3.5 × .207	191	155	126	101
	20	4 × .225	218	176	144	116
	40	5 × .623	276	223	182	146
Wood screws,[b] allowable withdrawal load per in. penetration to full length of threaded section	6	.67 L × .138	151–222	102–118	69–91	38–66
	8	.67 L × .164	180–263	121–141	82–108	45–79
	10	.67 L × .190	208–306	141–164	95–125	53–91
	12	.67 L × .216	237–347	160–186	109–142	59–103
	16	.67 L × .268	294–430	199–231	135–176	73–128
Wood screws,[c] allowable lateral load for penetration 7 times the shank diameter into final member	8	7 D × .164	129	106	87	68
	10	7 D × .190	173	143	117	91
	12	7 D × .216	224	185	151	118
	16	7 D × .268	345	284	233	181
Lag screws,[d] allowable withdrawal load lb per in. penetration into side grain	3/8	10 D × .375	421–561	313–356	235–287	145–226
	1/2	10 D × .500	523–697	389–443	291–357	180–281
	5/8	10 D × .625	618–824	460–524	344–421	213–332
	3/4	10 D × .75	708–944	528–601	395–483	244–380
Lag bolts, allowable lateral load using ½-in. metal side pieces; penetration into side grain, single shear; load parallel to grain[e]	5/16	4 × .312	410	355	290	235
	3/8	6 × .375	630	545	490	430
	1/2	8 × .5	1 140	985	880	775
	3/4	10 × .75	2 540	2 190	1 970	1 625
	1	12 × 1	4 520	3 900	3 290	2 630
Bolted joints, double shear (joint consisting of 3 members; 2 side members, each ½ the thickness of main member. Bolt length given for main-member thickness only.)	1/2	1.5 × .5		820–1 120		
	5/8	3 × .625		1 700–2 340		
	3/4	4.5 × .75		2 500–3 440		
	7/8	7.5 × .875		3 390–4 670		
	1	11.5 × 1		4 460–6 140		

Table 4–65 (continued)
LOADS FOR NAILS AND SCREWS IN WOOD

[a]Loads for threaded, hardened steel nails, in 6 d to 20 d sizes, are the same as for common nails.

[b]Wood screws shall not be driven with a hammer. Soap or other lubricant may be used, and lead holes are permitted, usually 70% of root diameter of thread. Spacing of screws shall be such as to prevent splitting.

[c]Tabular values are for screws inserted into side grain. With metal side plates loads may be increased 25%. If screw is inserted in end grain, allowable loads are to be reduced 33%.

[d]Penetration of threaded portion about 10 D, but withdrawal resistance approximates tensile (root) strength for penetrations as follows: species group 1 = 7 D; group 2 = 8 D; group 3 = 10 D; group 4 = 11 D. For penetration into end grain, allowable loads are reduced 25%. Lead holes for threaded section about 75% of shank diameter. Lag screws shall not be driven with a hammer.

[e]For load perpendicular to grain, the allowable loads are much less.

Note: The examples in Table 4–65 indicate the ranges of the allowable loads specified in the extensive tables in National Design Specification for Stress-grade Lumber and Its Fastenings, National Forest Products Association, Washington, D.C., 1971.

From Bolz, R. E. and Tuve, G. L., Eds., *Handbook of Tables for Applied Engineering Science,* 2nd ed., CRC Press, Cleveland, 1973, 632.

Table 4-66
UNIT STRESSES FOR LAMINATED TIMBERS

The following table gives the range of allowable unit stresses for "structural glued-laminated timber," in which the grain of all laminations is approximately parallel, net finished widths 2 1/4 to 14 1/2 in. These values are for dry conditions of use (moisture content < 16%) and normal loading duration. Lower unit stresses are specified for wet conditions of use, for curved members, and for deep and slender beams (see References). Ranges indicate various grades and also numbers of laminations.

For stress and modulus of elasticity in MN/m^2, multiply values in psi by 0.0068948.

Allowable unit stresses, psi

Laminations of	Modulus of elasticity, millions of psi	Bending, load parallel[a]	Bending, load perpendicular	Tension, parallel to grain	Compression, parallel to grain	Compression, perpendicular to grain	Horizontal shear
Douglas fir or larch	1.82	900–2,300	1,500–2,600	1,800–2,400	1,200–1,800	385–450	195
Douglas fir, Coast region	1.80	1,200–2,600	2,200–2,600	1,600	1,500	450	165
Southern pine	1.80		1,800–2,600	2,200–2,600	1,800–2,000	385–450	200
California redwood	1.30	1,000–2,200	1,400–2,200	1,800–2,200	1,800–2,200	325	125
Ash, commercial white	1.60	1,100–2,300	—	1,400–2,300	1,600–2,200	610	230
Birch, sweet or yellow	1.80	1,200–2,450	—	1,500–2,450	1,800–2,400	610	230
Cottonwood, Eastern	1.10	600–1,250	—	750–1,250	900–1,200	180	110
Hickory	2.00	1,500–3,100	—	1,900–3,100	2,200–3,000	730	260
Maple, hard	1.80	1,200–2,450	—	1,500–2,450	1,800–2,400	610	230
Oak, red or white	1.60	1,100–2,300	—	1,350–2,300	1,500–2,000	610	230

[a]Parallel or perpendicular with respect to the wide face of the laminations.

REFERENCES

National Design Specification for Stress-grade Lumber and Its Fastenings, National Forest Productions Association, Washington, D.C., 1971.
Standards for Structural Glued-laminated Members Assembled with WWPA Grades of Douglas Fir and Larch Lumber, Western Wood Products Association, 1966.
Standard Specifications for Structural Glued-laminated Douglas Fir Timber, West Coast Lumber Inspection Bureau, 1963.
Standard Specifications, Structural Glued-laminated California Redwood Timber, California Redwood Association, 1965.

Table 4–66 (continued)
UNIT STRESSES FOR LAMINATED TIMBERS

Standard Specification for the Design and Fabrication of Hardwood Glued-laminated Lumber for Structural, Marine, and Vehicular Uses, Southern Hardwood Producers, Inc., Appalachian Hardwood Manufacturers, Inc., and Northern Hardwood and Pine Manufacturers Association.

Standard Specifications for Structural Glued-laminated Timber, American Institute of Timber Construction, 1970.

From Bolz, R. E. and Tuve, G. L., Eds., *Handbook of Tables for Applied Engineering Science,* 2nd ed., CRC Press, Cleveland, 1973, 634.

Table 4–67
SAFE LOADS FOR PIPE COLUMNS

Following are approximate maximum axial loads for "Standard Pipe" (Schedule 40) used as columns, plain or filled with concrete. For load in kN, multiply tabulated values in thousands of pounds by 4.4482.

Maximum load in thousands of pounds for column lengths, feet

Description	6	7	8	9	10	11	12
3-in. pipe, OD 3.5 in., wall thickness 0.216 in.							
Plain	25	24	23	21	20	18	16
3½-in. pipe, OD 4.0 in., wall thickness 0.226 in.							
Plain	31	30	29	28	26	24	23
Filled	38	37	36	35	33	31	30
4-in. pipe, OD 4.5 in., wall thickness 0.237 in.							
Plain	38	37	36	34	33	31	30
Filled	46	45	44	43	41	40	38
5-in. pipe, OD 5.563 in., wall thickness 0.258 in.							
Plain	52	51	50	49	48	47	46
Filled	66	65	64	63	62	60	59
6-in. pipe, OD 6⅝ in., wall thickness 0.280 in.							
Plain	69	68	67	66	65	64	63
Filled	88	87	86	85	84	83	82

Table 4–67 (continued)
SAFE LOADS FOR PIPE COLUMNS

Maximum load in thousands of pounds for column lengths, feet

Description	6	7	8	9	10	11	12
8-in. pipe, OD 8⅝ in., wall thickness 0.322 in.							
Plain	105	104	103	102	101	100	99
Filled	126	125	124	123	122	121	120
10-in. pipe, OD 10¾ in., wall thickness 0.365 in.							
Plain	150	149	148	147	146	145	144
Filled	203	202	201	200	199	198	197

From Bolz, R. E. and Tuve, G. L., Eds., *Handbook of Tables for Applied Engineering Science*, 2nd ed., CRC Press, Cleveland, 1973, 632.

Table 4–68
PROPERTIES OF MOLTEN SALTS

For density in kg/m³, multiply values in g/cm³ by 1,000. For viscosity in N·s/m², multiply values in centipoise by 0.001.

Salts	Melting point, °C	Near melting point				+ 100° (approx.)			
		Temperature, °C	Density, g/cm³	Electrical conductivity per ohm-cm	Viscosity, centipoise	Temperature, °C	Density, g/cm³	Electrical conductivity per ohm-cm	Viscosity, centipoise
$AgBr$	430	447	5.562	2.896	3.30	547	5.458	3.113	2.53
$AgCl$	455	467	4.861	3.868	2.24	567	4.774	4.225	1.78
AgI	558	607	5.526	2.43	2.95	707	5.425	2.60	2.28
$AgNO_3$	210	217	3.954	0.692	—	317	3.852	1.122	2.55
$AlCl_3$	192	207	1.209	—	0.320	307	0.938	—	—
B_2O_3	450	1137	1.508	—	5020.	1237	1.499	—	3580.
BeF	540	697	—	0.61×10^{-5}	2.62	797	—	13.77×10^{-5}	0.154
Bu_4NBF_4	162	167	0.935	—	9.00	267	0.877	—	1.91
$CaBr_2$	730	747	3.108	1.409	—	847	3.058	1.716	—
$CaCl_2$	782	787	2.078	2.059	3.34	887	2.036	2.501	1.96
$CdBr_2$	568	587	4.054	1.097	2.73	687	3.946	1.301	—
$CdCl_2$	568	577	3.381	1.884	—	677	3.299	2.150	1.91
$CsCl$	645	667	2.768	1.167	1.28	767	2.662	1.474	0.94
CsI	621	657	3.140	0.707	1.72	757	3.022	0.916	1.26
$CuCl$	422	527	3.618	3.703	2.54	627	3.542	3.841	1.80
$HgCl_2$	277	287	4.336	3.49×10^{-5}	1.74	387	4.050	7.447×10^{-5}	—
HgI_2	257	277	5.164	0.0282	2.53	377	4.841	0.0180	—
KBr	735	747	2.116	1.639	1.18	847	2.034	1.886	0.92
KCl	770	787	1.518	2.203	1.15	887	1.460	2.439	0.86
K_2CO_3	896	907	1.892	2.053	—	1007	1.848	2.342	—
$K_2Cr_2O_7$	398	417	2.273	0.247	11.87	517	2.204	0.507	—
KI	685	727	2.404	1.369	1.45	827	2.308	1.569	1.13
KNO_3	337	357	1.856	0.691	2.63	457	1.783	0.986	1.61
KOH	360	407	1.714	2.56	2.21	507	1.670	3.14	1.25
$LaCl_3$	870	877	3.196	1.521	—	977	3.118	1.810	3.94
$LiBr$	550	597	2.499	4.951	1.52	697	2.433	5.470	1.10
$LiCl$	610	637	1.490	5.864	1.59	737	1.447	6.354	1.09
Li_2CO_3	618	737	1.826	4.097	—	837	1.789	4.892	2 91
LiI	449	487	3.093,	3.967	2.17	587	3.001	4.427	1.53
$LiNO_3$	254	277	1.768'	0.928	5.85	377	1.713	—	2.95
$NaBr$	750	787	2.309	3.018	1.28	887	2.227	3.319	1.02
$NaCl$	800	817	1.547	3.629	1.38	917	1.493	3.926	0.95
Na_2CO_3	854	867	1.968	2.900	—	967	1.923	3.288	1.63
NaF	980	997	1.944	4.937	—	1097	1.888	5.211	—
NaI	662	677	2.726	2.292	1.45	777	2.631	2.603	1.09
Na_2MoO_4	687	747	2.765	1.243	—	847	2.703	1.575	—
$NaNO_2$	285	297	1.801	1.329	3.04	397	1.726	1.872	—
$NaNO_3$	310	317	1.898	1.015	2.86	417	1.827	1.453	1.77
$NaOH$	318	357	1.767	2.44	3.79	457	1.719	3.34	2.11
Na_2WO_4	698	777	3.792	1.145	—	877	3.712	1.35	—
$PbCl_2$	498	507	4.942	1.486	4.41	607	4.792	1.957	2.69
$RbBr$	680	697	2.699	1.125	1.45	797	2.592	1.359	1.13
$RbCl$	715	737	2.229	1.549	1.29	837	2.141	1.818	0.98
RbI	640	657	2.886	0.879	1.39	757	2.772	1.056	1.06
$RbNO_3$	316	327	2.466	0.457	3.68	427	2.369	0.673	—

Table 4–68 (continued)
PROPERTIES OF MOLTEN SALTS

		Near melting point				+100° (approx.)			
Salts	Melting point, °C	Temperature, °C	Density, g/cm³	Electrical conductivity per ohm-cm	Viscosity, centipoise	Temperature, °C	Density, g/cm³	Electrical conductivity per ohm-cm	Viscosity, centipoise
SnCl₂	245	267	3.339	0.884	—	367	3.214	1.493	—
SnCl₄	−33	37	2.186	—	0.73	137	1.917	—	0.34
SrCl₂	875	897	2.713	2.082	3.18	997	2.655	2.466	—
TlCl	429	447	5.597	1.154	—	547	5.417	1.515	—
ZnCl₂	275	327	2.514	0.00268	2900.	427	2.469	0.0323	—

Note: Tables in the original source are in 10-degree increments, and several other salts are included. "Best equations" for each property are listed, and over 250 keyed references are included.

From Janz, G. J., Dampier, F. W., Lakshiminarayanan, G. R., Lorenz, P. K., and Tomkins, R. P. T., *Molten Salts: Electrical Conductance, Density, and Viscosity Data,* Vol. 1, National Standard Reference Data Series–NBS 15, National Bureau of Standards, Washington, D. C., October 1968.

Table 4–69
AVERAGE AMOUNTS OF THE ELEMENTS IN THE EARTH'S CRUST

In Grams Per Metric Ton or Parts Per Million

Element	Quantity	Element	Quantity	Element	Quantity
O	466,000	N	46	Br	1.6
Si	277,200	Ce	46	Ho	1.2
Al	81,300	Sn	40	Eu	1.1
Fe	50,000	Y	28	Sb	1?
Ca	36,300	Nd	24	Tb	0.9
Na	28,300	Nb	24	Lu	0.8
K	25,900	Co	23	Tl	0.6
Mg	20,900	La	18	Hg	0.5
Ti	4,400	Pb	16	I	0.3
H	1,400	Ga	15	Bi	0.2
P	1,180	Mo	15	Tm	0.2
Mn	1,000	Th	12	Cd	0.15
S	520	Cs	7	Ag	0.1
C	320	Ge	7	In	0.1
Cl	314	Sm	6.5	Se	0.09
Rb	310	Gd	6.4	Ar	0.04
F	300	Be	6	Pd	0.01
Sr	300	Pr	5.5	Pt	0.005
Ba	250	Sc	5	Au	0.005
Zr	220	As	5	He	0.003
Cr	200	Hf	4.5	Te	0.002?
V	150	Dy	4.5	Rh	0.001
Zn	132	U	4	Re	0.001
Ni	80	B	3	Ir	0.001
Cu	70	Yb	2.7	Os	0.001?

Table 4—69 (continued)
AVERAGE AMOUNTS OF THE ELEMENTS IN THE EARTH'S CRUST

In Grams Per Metric Ton or Parts Per Million (continued)

Element	Quantity	Element	Quantity	Element	Quantity
W	69	Er	2.5	Ru	0.001?
Li	65	Ta	2.1		

From Mason, B., *Principles of Geochemistry,* John Wiley & Sons, New York, 1952. With permission.

Table 4—70
COMPOSITION OF SEA WATER

Table A. Elements in Solution

Excluding Dissolved Gases

Element	Concentration, parts per million (approximate)	Percent by weight
Oxygen	857,000.	85.7000
Hydrogen	108,000.	10.8000
Chlorine	19,000.	1.9000
Sodium	10,500.	1.0500
Magnesium	1,275.	0.1275
Sulfur	885.	0.0885
Calcium	400.	0.0400
Potassium	380.	0.0380
Bromine	65.	0.0065
Carbon	30.	0.0030
Strontium	13.	0.0013
Boron	4.6	0.00046
Silicon	2.	0.0002
Fluorine	1.3	0.00013
Aluminum	1.	0.0001

Table B. Ionic Constituents

Anions in Sea Water of 34.4 Salinity per Mil, or 3.44% by Weight

Tabular values are given in percent by weight.

Chloride	1.897	Bromide	0.0065
Sulfate	0.265	Borate	0.0027
Bicarbonate	0.014		

Table 4–70 (continued)
COMPOSITION OF SEA WATER

Table C. Artificial Sea Water

To simulate the physical properties, a 3.4% solution of sodium chloride or natural sea salt may be used.

For a more exact chemical reproduction the following is an average.

Salt	Grams
NaCl	25.
MgCl$_2$	3.
MgSO$_4$ (or NaSO$_4$)	4.
CaCl$_2$	1.
KCl	0.7
NaHCO$_3$	0.2
NaBr (or KBr)	0.1
TOTAL	34.0

Note: Add water to make 1 kg.

USN Specification 44T27b–1940 is as follows:

NaCl	23 g
Na$_2$SO$_4$ · 10H$_2$O	8 g
Stock solution	20 ml
Sterile distilled water to 1 liter.	

The stock solution is as follows:

Magnesium chloride	550 g
Calcium chloride	110 g
Potassium bromide	45 g
Potassium chloride	10 g
Sterile distilled water to 1 liter.	

From Bolz, R. E. and Tuve, G. L., Eds., *Handbook of Tables for Applied Engineering Science,* 2nd ed., CRC Press, Cleveland, 1973, 659.

Table 4–71
PROPERTIES OF COMMON SOLID MATERIALS

Material	Specific gravity	Specific heat $\frac{Btu}{lbm \cdot deg\,R} = \frac{cal}{g \cdot K}$	$\frac{kJ}{kg \cdot K}$	Thermal conductivity $\frac{Btu}{hr \cdot ft \cdot deg\,F}$	$\frac{cal}{sec \cdot cm \cdot deg\,C}$	$\frac{W}{m \cdot K}$
Asbestos cement board	1.4	0.2	.837	0.35	.001 45	0.607
Asbestos millboard	1.0	0.2	.837	0.08	.000 33	0.14
Asphalt	1.1	0.4	1.67			
Beeswax	0.95	0.82	3.43			
Brick, common	1.75	0.22	.920	0.42	.001 7	0.71
Brick, hard	2.0	0.24	1.00	0.75	.003 1	1.3
Chalk	2.0	0.215	.900	0.48	.002 0	0.84
Charcoal, wood	0.4	0.24	1.00	0.05	.000 21	0.088
Coal, anthracite	1.5	0.3	1.26			
Coal, bituminous	1.2	0.33	1.38			
Concrete, light	1.4	0.23	.962	0.25	.001 0	0.42
Concrete, stone	2.2	0.18	.753	1.0	.004 1	1.7
Corkboard	0.2	0.45	1.88	0.025	.000 1	0.04
Earth, dry	1.4	0.3	1.26	0.85	.003 5	1.5
Fiberboard, light	0.24	0.6	2.51	0.035	.000 14	0.058
Fiber hardboard	1.1	0.5	2.09	0.12	.000 5	0.2
Firebrick	2.1	0.25	1.05	0.8	.003 3	1.4
Glass, window	2.5	0.2	.837	0.55	.002 3	0.96

Table 4—71 (continued)
PROPERTIES OF COMMON SOLID MATERIALS

Material	Specific gravity	Specific heat $\dfrac{Btu}{lbm \cdot deg\ R} = \dfrac{cal}{g \cdot K}$	$\dfrac{kJ}{kg \cdot K}$	Thermal conductivity $\dfrac{Btu}{hr \cdot ft \cdot deg\ F}$	$\dfrac{cal}{sec \cdot cm \cdot deg\ C}$	$\dfrac{W}{m \cdot K}$
Gypsum board	0.8	0.26	1.09	0.1	.000 41	0.17
Hairfelt	0.1	0.5	2.09	0.03	.000 12	0.050
Ice (32°)	0.9	0.5	2.09	1.25	.005 2	2.2
Leather, dry	0.9	0.36	1.51	0.09	.000 4	0.2
Limestone	2.5	0.217	.908	1.1	.004 5	1.9
Magnesia (85%)	0.25	0.2	.837	0.04	.000 17	0.071
Marble	2.6	0.21	.879	1.5	.006 2	2.6
Mica	2.7	0.12	.502	0.4	.001 7	0.71
Mineral wool blanket	0.1	0.2	.837	0.025	.000 1	0.04
Paper	0.9	0.33	1.38	0.07	.000 3	0.1
Paraffin wax	0.9	0.69	2.89	0.15	.000 6	0.2
Plaster, light	0.7	0.24	1.00	0.15	.000 6	0.2
Plaster, sand	1.8	0.22	.920	0.42	.001 7	0.71
Plastics, foamed	0.2	0.3	1.26	0.02	.000 08	0.03
Plastics, solid	1.2	0.4	1.67	0.11	.000 45	0.19
Porcelain	2.5	0.22	.920	0.9	.003 7	1.5
Sandstone	2.3	0.22	.920	1.0	.004 1	1.7
Sawdust	0.15	0.21	.879	0.05	.000 2	0.08
Silica aerogel	0.11	0.2	.837	0.015	.000 06	0.02
Vermiculite	0.13	0.2	.837	0.035	.000 14	0.058
Wood, balsa	0.16	0.7	2.93	0.03	.000 12	0.050
Wood, oak	0.7	0.5	2.09	0.10	.000 41	0.17
Wood, white pine	0.5	0.6	2.51	0.07	.000 29	0.12
Wool, felt	0.3	0.33	1.38	0.04	.000 17	0.071
Wool, loose	0.1	0.3	1.26	0.02	.000 8	0.3

From Bolz, R. E. and Tuve, G. L., Eds., *Handbook of Tables for Applied Engineering Science,* 2nd ed., CRC Press, Cleveland, 1973, 177.

Table 4–72
GENERAL CLASSIFICATION OF COMMON ROCKS

Table A. Igneous Rocks

Solidified from a Molten State

Coarse-grained crystalline	Fine-grained crystalline (or crystals and glass)	Fragmental (crystalline or glassy)
Origin: deep intrusion, slowly cooled	Origin: quickly cooled, volcanic or shallow intrusive	Origin: explosive volcanic fragments deposited as sediments
Granite	Rhyolite	Ash and pumice (volcanic dust or cinders)
Diorite	Andesite	Tuff (consolidated volcanic ash)
Gabbro	Basalt	Agglomerate (coarse and fine volcanic debris)
	Obsidian and pitchstone (essentially glass—suddenly chilled, few or no crystals)	

Table B. Mineral Constituents of Igneous Rocks

Minerals Key:

Q = quartz (SiO_2): hard, shiny, no true cleavage

O = orthoclase feldspar: silicates; regular cleavage

P = plagioclase feldspar: nearly white, good cleavage

A = amphibole (magnesium, iron, calcium) and/or biotite (black mica)

B = pyroxene; nearly black

Coarsely crystalline	Principal constituent minerals	Finely crystalline or porphyritic
Granite	Q+O+A(+P)[a]	Rhyolite
Syenite	O+A(+P)[a]	Trachyte
Quartz monzonite	Q+O+P+A	Dellenite
Monzonite	O+P+A	Latite
Quartz diorite	Q+P+A or B (+O)[a]	Dacite
Diorite	P+A or B (+O)[a]	Andesite
Gabbro	P+B	Basalt

[a]Small amount.

Table 4—72 (continued)
GENERAL CLASSIFICATION OF COMMON ROCKS

Table C. Sedimentary Rocks

Sediments Transported by Water, Air, Ice, Gravity

Mechanically deposited	Chemically or biochemically deposited
A–Unconsolidated	A–Calcareous
Clay ⎫	Limestone ($CaCO_3$)
Silt ⎪ According to	Dolomite ($CaCO_3 \cdot MgCO_3$)
Sand ⎬ particle size	Marl (calcareous shale)
Gravel ⎪	Caliche (calcareous soil)
Cobbles ⎭	Coquina (shell limestone)
B–Consolidated	B–Siliceous
Shale (consolidated clay)	Chert
Siltstone (consolidated silt)	Flint
Sandstone (consolidated sand)	Agate ⎫ Spring deposit,
Conglomerate (consolidated gravel	Opal ⎬ vein or
or cobbles–rounded)	Chalcedony ⎭ cavity filling
Breccia (angular fragments)	
	C–Others
	Coal, phosphate, salines, etc.

Table D. Metamorphic Rocks

Igneous or Sedimentary Rocks
Changed by Heat, Pressure

A–Foliated
  Slate: dense, dark, splits into thin plates
    (metamorphosed shale)
  Schist: predominantly micaceous, semiparallel
    lamellae
  Gneiss: granular, banded, subordinately micaceous

B–Massive
  Marble: coarsely crystalline, calcareous
    (metamorphosed limestone)
  Quartzite: dense, very hard, quartzose (metamor-
    phosed sandstone)

From *Concrete Manual,* 7th ed., U.S. Bureau of
Reclamation, 1966, Tables 10 and 11.

Table 4–73
CHEMICAL COMPOSITION OF AVERAGE ROCKS

Constituent	Igneous rock	Shale	Sandstone	Limestone	Sediment[a]
SiO_2	59.14	58.10	78.33	5.19	57.95
TiO_2	1.05	0.65	0.25	0.06	0.57
Al_2O_3	15.34	15.40	4.77	0.81	13.39
Fe_2O_3	3.08	4.02	1.07	0.54	3.47
FeO	3.80	2.45	0.30		2.08
MgO	3.49	2.44	1.16	7.89	2.65
CaO	5.08	3.11	5.50	42.57	5.89
Na_2O	3.84	1.30	0.45	0.05	1.13
K_2O	3.13	3.24	1.31	0.33	2.86
H_2O	1.15	5.00	1.63	0.77	3.23
P_2O_5	0.30	0.17	0.08	0.04	0.13
CO_2	0.10	2.63	5.03	41.54	5.38
SO_3		0.64	0.07	0.05	0.54
BaO	0.06	0.05	0.05		
C		0.80			0.66
	99.56	100.00	100.00	99.84	99.93

[a]Shale, 82%; sandstone, 12%; limestone, 6% (after Leith and Mead, 1915).

After Clarke, from Pettijohn, F. J., *Sedimentary Rocks,* 2nd ed., Harper & Row, New York, 1957, 106. With permission.

Table 4–74
PROPERTIES OF CLEAR FUSED QUARTZ

Property	English or metric system value	International system of units (SI) value	Property	English or metric system value	International system of units (SI) value
Density	2.2 g/cm³	2.2×10^3 kg/m³	Annealing point	1140°C	1410°K
Hardness	4.9 Mohs' scale		Strain point	1070°C	1340°K
Tensile strength	7,000 psi	4.8×10^7 N/m²	Electrical resistivity	$10^{9.5}$ ohm cm (350°C)	
Compressive strength	>160,000 psi	$>1.1 \times 10^9$ N/m²	Dielectric properties Constant	(20°C and 1 Mc) 3.75	(293°K and 1 Mhz) 3.75
Bulk modulus	5.3×10^6 psi	3.7×10^{10} N/m²	Strength	410 volts/mil	1.6×10^7 V/m
Rigidity modulus	4.5×10^6 psi	3.1×10^{10} N/m²	Loss factor	$<4 \times 10^{-4}$	$<4 \times 10^{-4}$
Young's modulus	10.4×10^6 psi	7.17×10^{10} N/m²	Dissipation factor	$<1 \times 10^{-4}$	$<1 \times 10^{-4}$
Poisson's ratio	.16	.16	Index of refraction	1.4585	1.4585
Coefficient of thermal expansion	5.5×10^{-7} cm/cm °C (20°C–320°C)	5.5×10^{-7} m/m °K (293°K–593°K)	Velocity of sound-shear wave	3.75×10^5 cm/sec	3.75×10^3 m/s
Thermal conductivity	3.3×10^{-3} g cal cm/cm² sec °C	1.4 W/m² °K	Velocity of sound-compressive wave	5.90×10^5 cm/sec	5.90×10^3 m/s
Specific heat	.18 g cal/g	750 J/kg	Sonic attenuation	<.033 db/ft Mc	<.11 db/m Mhz
Fusion temperature	1800°C	2070°K			
Softening point	1670°C	1940°K			

Note: These data apply to a specific grade of commercially available material. The term "fused silica" is often used to include the entire group of materials made by fusing of silica (SiO_2). All such products contain small amounts of impurities, but the clear varieties can have a purity of 99.98 Alumina (Al_2O_3) is the major impurity. Fused silica has an extremely low coefficient of expansion and does not react with most acids, metals, chlorine, or bromide at ordinary temperatures. It has good mechanical and electrical properties and is almost perfectly elastic. Its radiant transmission is high in the ultraviolet as well as in the visible region. Fused quartz or silica products are available in rod, ribbon, and other solid forms such as tubing, chemical glassware, and other fabricated quartzware.

From Bolz, R. E. and Tuve, G. L., Eds., *Handbook of Tables for Applied Engineering Science*, 2nd ed., CRC Press, Cleveland, 1973, 187. Properties data from *Fused Quartz Catalog*, courtesy of The General Electric Company.

Table 4–75
SOIL MECHANICS—CLASSES OF SOILS

Comparison of Classification Systems

Classification System							
American Society for Testing and Materials	Colloids[a]	Clay	Silt	Fine sand	Coarse sand	Gravel	Boulders
American Association of State Highway Officials Soil Classification	Colloids[a]	Clay	Silt	Fine sand	Coarse sand	Fine gravel / Medium gravel / Coarse gravel	Cobbles
U.S. Department of Agriculture Soil Classification	Clay	Silt	Very fine sand / Fine sand / Medium sand / Coarse sand / Very coarse sand			Fine gravel	Coarse gravel
Federal Aviation Agency Soil Classification	Clay	Silt	Fine sand	Coarse sand		Gravel	
Unified Soil Classification (Corps of Engineers, Department of the Army, and Bureau of Reclamation)	Fines (silt or clay)		Fine sand	Medium sand	Coarse sand	Fine gravel / Coarse gravel	Cobbles
Massachusetts Institute of Technology	Clay	Silt (Fine / Medium / Coarse)	Sand (Fine / Medium / Coarse)			Gravel	

[a]Colloids are included in clay fraction in test reports.

From Bolz, R. E. and Tuve, G. L., Eds., *Handbook of Tables for Applied Engineering Science*, 2nd ed., CRC Press, Cleveland, 1973, 635.

Table 4–76
SOIL MECHANICS – VOLUME AND WEIGHT RELATIONSHIPS

Density, Porosity, and Moisture

Table A. Representative Specific Gravities of Some Dense Soils

Soil	Specific gravity
Normal inorganic clay	2.70
Silt	2.70
Loess	2.70
Sand	2.65
Diatomaceous earth	2.65
Bentonite clay	2.34

From Bolz, R. E. and Tuve, G. L., Eds., *Handbook of Tables for Applied Engineering Science,* 2nd ed., CRC Press, Cleveland, 1973, 636.

Table B. Relative Density of Sands

Very dense	85–100%
Dense	65–85%
Medium	35–65%
Loose	15–35%

From Bolz, R. E. and Tuve, G. L., Eds., *Handbook of Table. for Applied Engineering Science,* 2nd ed., CR ` Press, Cleveland, 1973, 636.

Table C. Porosity of Soils in Natural State

Porosity is the percentage ratio, volume of voids to total volume. *Void ratio* is the ratio of volume of voids to volume of moist solids. For unit weights in kilograms per cubic meter, multiply values in grams per cubic centimeters.

				Unit weights			
			Water	g/cm^3		lb/ft^3	
Description	Porosity, %	Void ratio	content, %	Dry	Sat.	Dry	Sat.
Uniform sand, loose	46	0.85	32	1.43	1.89	90	118
Uniform sand, dense	34	0.51	19	1.75	2.09	109	130
Mixed-grained sand, loose	40	0.67	25	1.59	1.99	99	124
Mixed-grained sand, dense	30	0.43	16	1.86	2.16	116	135
Glacial till, very mixed grained	20	0.25	9	2.12	2.32	132	145
Soft glacial clay	55	1.2	45	–	1.77	–	110
Stiff glacial clay	37	0.6	22	–	2.07	–	129

Table 4–76 (continued)
SOIL MECHANICS – VOLUME AND WEIGHT RELATIONSHIPS

Density, Porosity, and Moisture (continued)

Table C. Porosity of Soils in Natural State (continued)

Description	Porosity, %	Void ratio	Water content, %	Unit weights g/cm³ Dry	Sat.	lb/ft³ Dry	Sat.
Soft slightly organic clay	66	1.9	70	–	1.58	–	98
Soft very organic clay	75	3.0	110	–	1.43	–	89
Soft bentonite	84	5.2	194	–	1.27	–	80

From Terzaghi, K. and Peck, R. B., *Soil Mechanics in Engineering Practice,* 2nd ed., John Wiley & Sons, New York, 1967, 28 (Table 6.3). With permission.

Table D. Degree of Saturation, Decimal

Description	Degree of saturation
Dry soil	0.0
Humid soil	<0.25
Damp soil	0.26–0.50
Moist soil	0.51–0.75
Wet soil	0.76–0.99
Saturated	1.00

From Bolz, R. E. and Tuve, G. L., Eds., *Handbook of Tables for Applied Engineering Science,* 2nd ed., CRC Press, Cleveland, 1973, 636.

Table 4–77
SOIL MECHANICS–PLASTICITY AND WATER CONTENT OF SOILS

Moisture, Consistency, and Atterberg Limits

Definitions:

Plastic limit: the percentage of water content by weight at which soil crumbles when rolled into thin threads.

Liquid limit: the percentage of water content by weight at which soil has a very small but measurable shearing strength.

Plasticity index: the arithmetical difference between the liquid limit and the plastic limit.

Terminology for Degree of Plasticity

Description	Range of plasticity index
Nonplastic	0–5
Moderately plastic	5–15
Plastic	15–40
Highly plastic	>40

Atterberg Limits for Clays

Clay	Adsorbed ion	Plastic limit	Liquid limit
Kaolinite	Sodium	26	52
	Calcium	36	73
Illite	Sodium	34	61
	Calcium	40	90
Montmorillonite	Sodium	97	900
	Calcium	63	177

From Bolz, R. E. and Tuve, G. L., Eds., *Handbook of Tables for Applied Engineering Science*, 2nd ed., CRC Press, Cleveland, 1973, 637.

Table 4–78
SOIL MECHANICS–COMPARATIVE PROPERTIES OF CLAY SOILS

Consistency, Sensitivity, Brittleness, and Activity

Definitions:

Consistency: the relative compressive strength of the unconfined clay.

Sensitivity: the ratio of unconfined compressive strengths (undisturbed material to remolded material).

Brittleness: the percentage of strain at failure in an unconfined compression test.

Activity: the ratio of the plasticity index to the clay fraction, i.e., (liquid limit – plastic limit)/(percentage by weight of particles smaller than 2 μm).

Consistency		Sensitivity		Brittleness		Activity	
Description	Range	Description	Range	Description	Range	Description	Range
Very soft	<.25	Insensitive	<2	Brittle	3–8	Inactive	<.75
Soft	.25–.50	Moderately		Semibrittle	8–14	Normal	.75–1.25
Medium	.50–1.0	sensitive	2–4	Plastic	14–20	Active	>1.25
Stiff	1.0–2.0	Sensitive	4–8				
Very stiff	2.0–4.0	Very sensitive	8–16				
Hard	>4.0	Slightly quick	16–32				
		Medium quick	32–64				
		Quick	>64				

From Bolz, R. E. and Tuve, G. L., Eds., *Handbook of Tables for Applied Engineering Science,* 2nd ed., CRC Press, Cleveland, 1973, 637.

Table 4–79
SOIL MECHANICS–DENSITY AND PENETRATION

Density and consistency of soil may be determined during boring and sampling operations by the standard penetration test. In this test the number of blows required per foot of penetration of a standard sampling spoon is observed. A 140-lb hammer is used, with a 30-in. drop; the sampling spoon is 2.0 in. OD and 1.375 in. ID. The density and consistency are characterized in the following table.

Relative density of sand		Relative consistency of clay	
Number of blows per foot	Description	Number of blows per foot	Description
0–4	Very loose	0–2	Very soft
4–10	Loose	2–4	Soft
10–30	Medium	4–8	Medium
30–50	Dense	8–15	Stiff
>50	Very dense	15–30	Very stiff
		>30	Hard

Note: For sands and silts a correction for depth (and overburden pressure) or a correction for location of the water table may be required.

From Bolz, R. E. and Tuve, G. L., Eds., *Handbook of Tables for Applied Engineering Science,* 2nd ed., CRC Press, Cleveland, 1973, 638.

Table 4–80
SOIL MECHANICS–PERMEABILITY OF SOILS

The *coefficient of permeability, k,* is a simple index of the water conductivity or drainage characteristics of a soil. It is determined by using a *permeameter,* which is an arrangement for measuring the flow through a soil sample under a given hydraulic head.

It is assumed that the rate of flow is directly proportional to the head h (laminar-type flow). The coefficient is based on the gross area of the flow conduit. The coefficient of permeability is a measure of the velocity of flow per unit of head, or the velocity, based on gross area, at which the head loss is unity. In metric units the dimensions are

$$q = VA. \qquad \frac{cm^3}{sec} = \frac{cm}{sec} \times cm^2$$

$$q = khA. \qquad \frac{cm^3}{sec} = \left(\frac{cm}{sec \times cm}\right) (cm) (cm^2).$$

In the field the permeability is determined by a pumping test in which the hydraulic gradients at fixed distances from the pumping point are measured while a constant and known rate of flow is being maintained.

Natural soil deposits often consist of layers, each of which has a different permeability. The flow resistances are additive (Ohm's law), and the individual coefficients of permeability are analogous to individual conductivities. An average coefficient of permeability can thus be computed.

The following table gives a classification of soils in terms of the numerical values of the coefficient of permeability. The units of k are cm/sec for a head of 1 cm.

Descriptive Ranges of Permeability

Degree of permeability	Values of k, cm/sec	Degree of permeability	Values of k, cm/sec
High	> .1	Very low	$10^{-5}-10^{-7}$
Medium	.1–.001	Practically	
Low	.001–.00001	impervious	$< 10^{-7}$

The value of $k = 0.1$ is typical for a medium to coarse, clean sand or a sand and gravel mixture. Homogeneous clays are often nearly impervious. The three intermediate ranges characterize very fine sands, silts, and mixtures, including those with organic materials.

For soils with a coefficient greater than 0.001, the values of k can be readily determined, using either the constant-head permeameter or a pumping test. The falling-head permeameter can be used both for these and for finer soils, but as the fineness increases so does the importance of a large amount of experience in making the tests.

From Bolz, R. E. and Tuve, G. L., Eds., *Handbook of Tables for Applied Engineering Science,* 2nd ed., CRC Press, Cleveland, 1973, 638.

Table 4–81
BUILDING FOUNDATIONS AND PILES

Typical Building Code Requirements for Soil Bearing
Pressures and Design of Pile Foundations

Table A. Allowable Bearing Pressures

Soil descriptions[a]	Maximum bearing pressure[b]	
	tons/ft²	MN/m²
Rock, bedrock, solid, massive, sound	100	1,380
Rock, sound foliated (limestone, slate)	40	550
Rock, hard sedimentary (hard shale, sand-stone, conglomerate)	25	245
Rock, broken bedrock	10	138
Hardpan	10	138
Gravel, very compact	10	138
Gravel, compact	6	83
Gravel, loose	4	55
Sand or silty gravel	1–3	14–40

[a]Descriptive terminology used in building codes is far from uniform, but the terms here given occur in several codes.

[b]Each of these values is to be found in several of the building codes of large U.S. cities or government authorities. Modifications or corrections are recognized for conditions such as depth of stratum, depth of embedment, and extent of overburden. For gravel and sand the allowable bearing pressures are sometimes stated in terms of on-site soil test results.

Table B. Allowable Design Stresses for Loaded Piles

Widely Used Specifications in Building Codes

Key:

BS = basic stress for clear timber as per ASTM D25 and D2555
CS = compressive strength of concrete
YS = yield strength of steel

Type of pile	Number of components	Allowable unit stress, psi[a]		
		Core	Reinforcing steel	Steel shell
Steel H-section	1	35% YS	–	–
Timber	1	60% BS	–	–
Precast concrete	2	22.5% CS	35% YS	–
Cast-in-place concrete	3	22.5% CS	35% YS	35% YS
Steel pipe (filled)	3	22.5% CS	35% YS	35% YS

[a]Many city codes specifying numerical values for allowable stresses and variations among cities are occasionally 2 to 1 or even more. A few codes use percentages other than those here quoted. A limit of 1,000 psi (6,895 MN/m²) is often specified for timber piling.

Table 4–81 (continued)
BUILDING FOUNDATIONS AND PILES

Typical Building Code Requirements for Soil Bearing
Pressures and Design of Pile Foundations (continued)

Notes: Most piles are designed as "short" columns, in compression only; for long
piles in soft soil, a "long-column" treatment may be necessary.

Estimation of axial-load distribution along the pile may be included in a
building code, but there is little agreement on the basis for estimating. The
percentage of the total load being carried at the tip of the pile is usually
assumed to be less than one third; for shorter piles in coarse material on
bedrock, it may even be 100%.

Possible pile damage, due to overdriving, obstructions, or defects, should
be considered in the design decisions.

Pile spacing is often specified in the building codes, with 24- or 30-in.
(0.61–0.76 m) as minimum for cylindrical piles.

Success of a pile foundation depends on the ability of the soil to support
the pile, which, in turn, depends on the adhesion and shear properties of
the soil. The use of tapered piles and rigid-body caps is common. In any
case the limitations on eventual settling should be examined, especially in
locations where settling has been a problem, as in Mexico City.

For soil conditions producing very heavy corrosion, it may be necessary to
avoid steel piles.

For extreme loads the caisson-type pile, terminating in a recess in bedrock,
is recommended.

REFERENCES

Building Code Requirements for Reinforced Concrete, ACl 318, American
Concrete Institute.

Design Manual, DM7, Bureau of Yards and Docks, U.S. Navy.

Foundation Piling: A Survey of Practice, Building Research Advisory Board,
Report No. 4, National Academy of Sciences-National Research Council.

Southern Standard Building Code and *Uniform Building Code*.

*Specification for the Design Fabrication and Erection of Structural Steel for
Buildings*, American Institute of Steel Construction.

Consult also (1) *Journal of the Soil Mechanics and Foundations Division,
Proceedings of the ASCE*, (2) *Proceedings of the International Conferences on
Soil Mechanics and Foundation Engineering*, and (3) city building codes.

From Bolz, R. E. and Tuve, G. L., Eds., *Handbook of Tables for Applied
Engineering Science*, 2nd ed., CRC Press, Cleveland, 1973, 639.

Table 4–82
SOIL CHARACTERISTICS PERTINENT TO ROADS AND AIRFIELDS

The following table is based on the "Unified Soil Classification System" used by the military services and the Bureau of Reclamation. Other classification systems include those of the Highway Research Board and the Civil Aeronautics Administration. See also Table 4–75 for comparisons.

Key:

A—excellent	F—poor	SR—steel-wheeled rollers
B—good to excellent	G—poor to very poor	SF—sheepsfoot roller
C—good	NS—not suitable	RT—rubber-tired equipment
D—fair to good	CT—crawler-type tractor	M—close control of moisture required
E—fair to poor		

Soil class and symbol	Value as foundation[a] X	Value as foundation[a] Y	Compressibility and expansion	Compaction equipment[b]	Dry weight,[c] lb/ft^3	Field CBR[d]	Subgrade modulus,[e] k, psi/in.
GW—well graded gravels or gravel-sand mixtures, little or no fines	A	C	Almost none	CT, RT, SR	125–140	60–80	>300
GP—poorly graded gravels or gravel-sand mixtures, little or no fines	B	E	Almost none	CT, RT, SR	110–130	25–60	>300
GM—silty gravels, gravel-sand-silt mixtures	B	D	Very slight	RT, SF, M	130–145	40–80	>300
	C	F	Slight	RT, SF	120–140	20–40	200–300
GC—clayey gravels, gravel-sand-clay mixtures	C	F	Slight	RT, SF	120–140	20–40	200–300
SW—well graded sands or gravelly sands, little or no fines	C	F	Almost none	CT, RT	110–130	20–40	200–300
SP—poorly graded sands or gravelly sands, little or no fines	D	F–NS	Almost none	CT, RT	100–120	10–25	200–300
SM—silty sands, sand-silt mixtures	C	F	Very slight	RT, SF, M	120–135	20–40	200–300
	D	NS	Slight to medium	RT, SF	105–130	10–20	200–300
SC—clay-like sands, sand-clay mixtures	D	NS	Slight to medium	RT, SF	105–130	10–20	200–300
ML—inorganic silts and very fine sands, rock flour, silty, or clay-like fine sands, or clay-like silts with slight plasticity	E	NS	Slight to medium	RT, SF, M	100–125	5–15	100–200
CL—inorganic clays of low to medium plasticity, gravelly clays, sandy clays, silty clays, lean clays	E	NS	Medium	RT, SF	100–125	5–15	100–200
OL—organic silts and organic silt-clays of low plasticity	F	NS	Medium to high	RT, SF	90–105	4–8	100–200
MH—inorganic silts, micaceous or diatomaceous, fine, sandy or silty soils, elastic silts	F	NS	High	SF	80–100	4–8	100–200
CH—inorganic clays of high plasticity, fat clays	G	NS	High	SF	90–110	3–5	50–100
OH—organic clays of medium to high plasticity, organic silts	G	NS	High	SF	80–105	3–5	50–100
PT—peat and other highly organic soils	NS	NS	Very high	None			

Table 4—82 (continued)
SOIL CHARACTERISTICS PERTINENT TO ROADS AND AIRFIELDS

[a]Values in column X are for subgrades and base courses not subject to frost action, while those in column Y apply to base directly under bituminous pavement. The A rating in column Y has been reserved for base materials consisting of high-quality processed crushed stone.

[b]The equipment listed will usually produce the required densities with a reasonable number of passes when moisture conditions and thickness of lift are properly controlled. In some instances several types of equipment are listed, because variable soil characteristics within a given soil group may require different equipment. A combination of two types may be necessary.

Processed base materials and other angular materials: Steel-wheeled rollers are recommended for hard angular materials with limited fines or screenings. Rubber-tired equipment is recommended for softer materials subject to degradation.

Finishing: Rubber-tired equipment is recommended for rolling during final shaping operations for most soils and processed materials.

Equipment size: The following sizes of equipment are necessary to assure the high densities required for airfield construction:

Crawler-type tractor—total weight in excess of 30,000 lb.

Rubber-tired equipment—wheel load in excess of 15,000 lb. Wheel loads as high as 40,000 lb may be necessary to obtain the required densities for some materials (based on contact pressure of approximately 65–150 psi).

Sheepsfoot roller—unit pressure (on 6–12 sq in. foot) to be in excess of 250 psi and unit pressures as high as 650 psi may be necessary to obtain the required densities for some materials. The area of the feet should be at least 5% of the total peripheral area of the drum, using the diameter measured to the faces of the feet.

[c]Unit dry weights are for compacted soil at optimum moisture content for modified AASHO compactive effort.

[d]The percentage ratio of the resistance to penetration developed by a subgrade soil to that developed by a specimen of standard crushed rock base material is called the California bearing ratio (CBR). It was developed by the California Division of Highways and is applied in connection with the design of flexible pavements.

[e]The subgrade modulus is a test ratio representing the slope of the stress-strain (or pressure-settlement) diagram. It is actually the ratio of the pressure on the 30-in. test plate in psi to the settlement in inches when the latter reaches a specified value (either 0.05 or 0.10 in.).

Condensed from Appendix B: Characteristics of soil groups pertaining to roads and airfields, *Unified Soil Classification System,* U.S. Corp of Engineers, Technical Memorandum No. 3-357, Vol. III, USAE Waterways Experiment Station, March 1953.

Table 4–83
ALLOWABLE SOIL PRESSURES FOR COLUMN FOOTINGS

Allowable bearing capacities or foundation soil loadings are often specified in local building codes. These are applicable to the design of ordinary extended footings, and the loads and capacities are those at footing grade. There are certain assumptions — that no weaker layer exists within the stress zone and that the stress zones do not overlap. Mats or continuous footings cannot be loaded as heavily. Even for preliminary estimates a relation between allowable bearing capacity and standard penetration resistance by test (ASTM D1586) is a much better basis than mere soil-texture classification. The following table gives typical ranges for the bearing loads as related to the penetration resistance determined in a soil test. The required blows per foot for a 2-in. OD sampling spoon represent the test data.

Estimating Bearing Loads From Results of Penetration Tests

Numbers in this table represent the minimum numbers of hammer blows per linear foot by standard penetration test ASTM D1586.

Description of soil below footing grade	Maximum soil load, thousands of psi								
	2	3	4	5	6	8	10	12	14
Gravel, sand, clay—well graded mixture		3	4	6	8	14	22	30	40
Gravel, sand, clay—poorly graded mixture		6	8	12	15	24	36	50	65
Gravel, sand—well graded		3	5	8	10	18	30	55	
Sandy clay	4	7	9	12	16	25	38	50	
Inorganic clay, no silt	8	12	15	20	25	40	60		
Fine sand, uniform	6	10	15	22	30	50			
Organic clay	16	30	65						
Fine silt	15	25	45						
Rough guides									
Sand, gravel, clay—stiff mixture									50
Compact gravel mixture								55	
Sandy clay							45		
Firm gravel						20			
Hard clay					35				
Firm, coarse sand				20					
Loose gravel				10					
Medium sandy clay			10						
Medium stiff clay		15							
Firm silt	15								

From Bolz, R. E. and Tuve, G. L., Eds., *Handbook of Tables for Applied Engineering Science,* 2nd ed., CRC Press, Cleveland, 1973, 641.

Table 4–84
TYPICAL PROPERTIES OF FOUNDATION ROCKS

For strength in MN/m^2, multiply values in psi
by 0.0068948.

Name of rock	Compressive strength, psi[a]	Shear strength, psi	Tangent of angle of internal friction
Andesite	18,900	4,060	5.7
Basalt	24,600	4,500	7.4
Diorite	12,200	2,000	9.2
Gneiss	16,100	2,500	10.0
Granite	22,300	3,250	11.8
Graywacke	8,700	1,700	6.5
Limestone	15,300	2,150	12.6
Sandstone	13,000	2,450	7.0
Schist	10,000	1,260	15.7
Shale	10,300	1,160	19.7
Siltstone	4,000	720	7.7
Tuff	430	100	4.9

[a]Uniaxial compressive strength for a specimen having a length-to-diameter ratio of 2. The product of the principal stresses is assumed to be linear.

From U.S. Bureau of Reclamation Report SP-39, August 1953.

Table 4–85
STATIC MECHANICAL PROPERTIES OF ROCK

For strength in MN/m², multiply tabulated values in thousands of psi by 6.8948. For modulus of elasticity in GN/m², multiply tabulated values in millions of psi by 6.8948.

| Rock type and number of test groups[a] | Compressive strength of test group, thousands of psi | | | Modulus of rupture of test group, thousands of psi | | | Modulus of elasticity of test group, millions of psi | | | | |
| | | | | | | | Static[b] | | Dynamic[c] | | |
	Max.	Min.	>50% of data within	Max.	Min.	>50% of data within	Max.	Min.	Max.	Min.	>50% of data within
Amphibolite (13)	74.9	30.4	—	7.4	4.0	—	—	—	15.1	6.7	—
Basalt (9)	52.0	11.8	—	6.6	2.1	—	—	—	12.4	5.9	—
Dibase (10)	51.8	23.2	40–50	8.0	4.5	5.0–5.5	—	—	13.9	10.2	—
Diorite (11)	48.3	22.5	25–35	7.3	2.0	—	—	—	6.12	3.6	4.0–5.0
Dolomite (10)	52.0	9.0	—	3.8	2.5	—	—	—	12.3	3.2	—
Gneiss (15)	36.4	22.2	25–35	3.1	1.2	2.0–3.0	—	—	15.1	3.5	7.0–10
Granite (17)	42.6	23.0	25–35	3.9	1.2	2.0–3.0	11.0	2.5	11.9	1.5	—
Greenstone (11)	45.5	16.6	—	6.7	1.7	—	9.0	6.9	15.2	3.4	10–13
Limestone[d] (46)	37.6	5.3	20–30	5.2	0.4	—	12.0	4.2	14.1	1.2	—
Marble (8)	34.5	6.7	30–35	3.3	1.7	—	—	—	—	—	—
Marlstone (15)	28.2	10.4	10–20	4.8	0.4	—	4.8	0.6	7.0	1.5	—
Quartzite (11)	91.2	21.1	—	6.4	1.2	—	—	—	—	—	—
Sandstone (48)	34.1	4.8	—	3.6	0.6	0.6–1.0	7.3	1.4	8.0	0.8	—
Shale (18)	33.5	10.9	10–20	4.2	0.3	—	7.6	1.7	9.9	1.5	1.5–2.5
Siltstone (8)	45.8	5.0	—	5.0	1.1	—	—	—	9.3	1.0	—

[a]The number following rock type indicates the number of petrographically distinct groups tested in uniaxial compression. A smaller number of groups were tested for the other listed mechanical properties.
[b]Tangent modulus of elasticity at the midstrength point.
[c]Determined by the resonant bar velocity method.
[d]Excluding chalk and coral rock.

From U.S. Bureau of Mines Reports of Investigations 3891, 4459, 5130, and 5244.

Table 4–86
CLASSIFICATION OF NATURAL FIBERS

Name	Source	Principal producers	Approximate production, 10^6 lb/year	Typical uses
ANIMAL ORIGIN				
Wool (pure, dry)	Sheep	Australia, Argentina	3 000	Warm clothing, carpet, felt
Silk	Silkworm caterpillar	Japan, China	70	Fine fabrics
Cashmere	Goat	India, Tibet	70	Quality clothes
Mohair	Goat	U.S.A., Turkey	30	Upholstery, linings, suitings, rugs
Camel hair	Camel	China, Mongolia	2	Overcoats, knits
VEGETABLE ORIGIN				
Cotton	Seed	U.S.A., U.S.S.R.	25 000	Almost all textile uses
Jute	Bast	India, Pakistan	4 500	Sacking, bale wrapping, curtains, bags, oakum
Sisal	Leaf	Mexico, Brazil	1 300	Rope, twine, rugs
Flax	Bast	U.S.S.R., Poland	1 200	Strong fabrics, paper
Kenaf	Bast	India, Pakistan	1 100	Rope, twine, carpet, canvas, bags
Hemp	Bast	U.S.S.R., Yugoslavia	500	Rope, sacking, canvas
Henequen	Leaf	Mexico, Cuba	300	Rope, twine, canvas, bags
Abaca (Manila)	Leaf	Phillipines, Guatemala	200	Rope, marine cable
Sunn	Bast	India, Pakistan	150	Rope, twine, carpet, paper
Ramie	Bast	China, Japan	25	Canvas
MINERAL ORIGIN				
Asbestos	Ore	Canada, U.S.S.R.	2 000	Brakes and clutches, building materials, packings, fireproofing
Glass[a]	Sand	U.S.A.	—	Composites, insulation, draperies, tire cord, filters
Aluminum silicate[a]	Ore	U.S.A.	—	Packings and insulation for high temperatures

[a]Here classified as natural fibers for convenience, although they are man-made by processing.

REFERENCE

Mauersberger, H. R., Ed., *Matthews' Textile Fibers,* John Wiley & Sons, New York, 1954.

From Bolz, R. E. and Tuve, G. L., Eds., *Handbook of Tables for Applied Engineering Science,* 2nd ed., CRC Press, Cleveland, 1973, 171.

Table 4–87
PROPERTIES OF NATURAL FIBERS

Because there are great variations within a given fiber class, average properties may be misleading. The following typical values are only a rough comparative guide.

Name	Specific gravity	Tenacity, g/denier	Tensile strength, 10³psi	Elongation at break (dry), %	Standard regain, % of dry[b]	Fiber diameter, microns	Fiber length, inches	Fiber shape and kind	Resistant to
ANIMAL ORIGIN									
Wool	1.32	1.0–1.7	17–29	23–35	15–18	17–40	1.5–5	Oval, crimped, scales	Age, weak acids, solvents
Silk	1.25	3.5–5	90	20–25	10	10–13		Flexible, soft, smooth	Heat, solvents, weak acids, wear
Cashmere						15–16	1–4	Round, scales, soft	
Mohair	1.32	1.2–1.5		30	13	24–50	6–12	Round, silky	Wear, age, solvents, weak acids
Camel hair	1.32	1.8		40	13	10–40	1–6	Oval, striated	Age, solvents
VEGETABLE ORIGIN									
Cotton	1.54	2–5	30–120	5–11	7.5–8.5	10–20	0.5–2	Flat, convoluted, ribbon	Age, heat, washing, wear, solvents, alkalies, insects
Jute (bast)	1.5		50	1–1.5	14	15–20		Woody, rough, polygon	
Sisal (leaf)	1.49	2.2	75	2–2.5	13	10–30	Strand 30–40	Stiff, straight	
Flax (bast)	1.52	4–7		2–3	12	15–18	Strand 40–50	Soft, fine	Age, solvents, washing, insects, weak acids, and alkalies
Kenaf (bast)			45			15–30		Polygon or oval	
Hemp (bast)	1.48			2		18–25	Strand 30–70	Polygon or oval, irregular	
Henequen (leaf)			60				Strand 30–60	Finer than sisal	
Abaca (leaf) (Manila)	1.48	2.3–2.9	100	2–3	13		Strand 30–120		
MINERAL ORIGIN									
Asbestos	2.5		40–200			Various	0.5–10	Smooth, straight	Heat to 400 deg C, acids, chemicals, organisms
Glass[a]	2.5	7–12	200–500	3–4.5	0	Various		Circular, smooth	Chemicals, insects
Silicate[a] (Ca, Al, Mg)	2.85				0				Heat to 900 deg C, most chemicals, insects, rot

[a]Here classified as natural fibers for convenience, although they are man-made by processing.
[b]Expected equilibrium moisture regain of dry fiber, in percent of dry weight, when exposed in air at 70 deg F, 65% relative humidity.

Note: Wide variations may be expected, especially for different grades of cotton. Wet strength is lower (for rayon, very much lower), but it depends on the duration of soaking. The strength of yarn is only a fraction of the cumulative strength of all individual fibers.

Most fibers exhibit relaxation of stress at constant strain and also increase in elongation at constant load (creep). The stress-strain curve is greatly affected by the rate of extension. When the stress is removed, there is a quick elastic recovery, a delayed recovery, and a permanent set. Hence the elastic behavior of any fiber depends on its

Table 4—87 (continued)
PROPERTIES OF NATURAL FIBERS

stress-strain history. The elastic recoveries of nylon and wool are high; those of cotton, flax, and rayon are much lower.

The heat capacity (specific heat) of most fibers is about one third that of water.

Other fibers: Fur hair is slightly coarser than silk fibers. Camel and llama hairs are almost as coarse as wool but only about one third the size of human hair. Horse hair is over 100 μm; hog bristles, over 200 μm. Jute, sisal, and hemp are intermediate between cotton and wool. These are rough average sizes, and many natural fibers range 50% above or below such averages.

From Bolz, R. E. and Tuve, G. L., Eds., *Handbook of Tables for Applied Engineering Science,* 2nd ed., CRC Press, Cleveland, 1973, 172.

Table 4–88

CLASSIFICATION OF MAN-MADE FIBERS AND FABRICS

Key to U.S. Manufacturers:

A – American Viscose Corp.

B – Beaunit Fibers, Div. of Beaunit Corp.

C – Celanese Fibers Co., Div. of Celanese Corp.

E – Tennessee Eastman Co., Div. of Eastman Kodak Co.

F – Fiber Industries, Inc.

G – W. R. Grace & Co., Dawbarn Div.

H – Hercules Powder Co.

K – American Enka Corp.

M – Monsanto Co., Textiles Div. (Chemstrand)

N – Vectra Co., Div. of National Plastics Products Co., Inc.

P – E. I. du Pont de Nemours & Co., Inc.

S – Firestone Synthetic Fibers Co.

U – Uniroyal, Inc., Textile Div.

UC – Union Carbide Corp.

V – FMC Corp., American Viscose Div.

Chemical class; common name (sources)	Typical proprietary names (and manufacturer)	Resistant to	Damaged by[a]	Typical uses
Cellulose fibers (natural)				
Acetate	Esteron®, etc. (E) Celacloud®, etc. (C)	Petroleum chemicals, dilute acids, weak alkalies, mildew, moths	Oxidizers, many solvents, strong alkalies, heat (above 140°C)	Clothing, satins, drapes, linings, knits
Triacetate	Arnel® (C)	Petroleum solvents, bleaches, insects, mildew, heat	Strong acids, strong alkalies, most solvents	Pleated garments, knits, drip-dry wear, table covers
Viscose rayon	Enka®, etc. (K) Avisco®, etc. (A)	Dilute alkalies, insects, solvents	Acids, strong alkalies, heat (above 150°C), moisture, mildew	Clothing, carpets, curtains, upholstery, linings
High-tenacity viscose	Tenasco® (V)	Moisture, solvents	Strong acids, strong alkalies, heat (above 150°C)	Tire cord, belting, hose
Polynosic viscose	Avril® (A)	Moisture, solvents, insects	Strong acids, strong alkalies, heat (above 150°C)	Dress fabrics, knits, drapes
Cuprammonium rayon (cupro)	Bemberg®, etc. (B)	Bleaches, weak alkalies	Strong oxidizers, mildews, some insects, strong acids, heat (above 230°C)	Sheer fabrics, drapes, upholstery, satins
Protein fibers (natural)				
Animal: casein (milk)	Now little used	Solvents	Strong acids, alkalies, mildew, heat (above 100°C)	Largely for blending with wool, cotton, rayon, etc.
Vegetable–seed: soybeans, peanuts, corn	Now little used	Acids, solvents, moths, mildew	Alkalies, heat above 150°C	Blends with wool, cotton, etc.
Vegetable–latex: rubber (vulcanized)	Lastex® (U)	Moisture, insects	Heat (above 110°C), oxidation, ozone, oils, hydrocarbons, fats, solvents	Corsetry, swimwear, footwear, supports

Table 4–88 (continued)
CLASSIFICATION OF MAN-MADE FIBERS AND FABRICS

Chemical class; common name (sources)	Typical proprietary names (and manufacturer)	Resistant to	Damaged by[a]	Typical uses
Synthetic fibers				
Polyacrylonitrile (acrylic)	Orlon® (P) Acrilan® (M) Creslan® (AC) Cantrece® (P) Verel®-copolymer (E) Dynel®-copolymer (P)	Dilute acids and alkalies, solvents, insects, mildew, weather	Strong alkalies and acids, heat (above 180°C), acetone, ketones	Outdoor fabrics, carpets, knits, furlike fabrics, blankets
Polyamide	Nylon[b] (P) 6, 6.6, etc. Chemstrand nylon® (M)	Alkalies, molds, solvents, moths	Strong acids, phenol, bleaches, heat (above 170°C)	Tire cord, carpet, upholstery, apparel, belting, hose, tents
Polyester	Dacron® (P) Kodel® (E) Fortrel (F)	Weak acids and alkalies, solvents, oils, mildew, moths	Phenol; heat (above 200°C)	Apparel, curtains, rope, twine, sailcloth, belting, fiberfill
Polyethylene (olefin, low density)	DLP® (G)	Alkalies, acids (except nitric), insects, mildew	Oil and grease; heat (above 90°C) oxidizers	Outdoor fabrics; filter fabrics; decorative coverings
Polyethylene (olefin, high density)	DLP (G)	Alkalies, acids (except nitric), insects, mildew	Oil and grease; heat (above 100°C) oxidizers	Rope, twine, fishnets
Polypropylene (olefin)	Herculon® (H) Polycrest® (U)	Alkalies, acids, solvents, insects, mildew	Heat (above 110°C)	Rope, twine, outdoor fabrics, carpets, upholstery
Polyurethane; spandex[b]	Lycra® (P) Spandelle® (S)	Solvents, oils, alkalies, insects, oxidation	Heat (above 140°C); strong acids	Elastic garments, swimwear, hosiery, tricot fabrics, knits
Polyvinyl chloride (PVC)	Vinyon[b] HH (V) Dynel®-copolymer (U)	Acids and alkalies, insects, mildew, alcohol, oils	Ethers, esters, aromatic hydrocarbons, ketones; hot acids; heat (above 70°C)	Nonwoven materials; felts; filters; blends with other fibers
Polyvinyl alcohol (PVA)	(Foreign manufacture)	Acids, alkalies, insects, mildew, oils	Heat (above 160°C); phenol, cresol, formic acid	Wide range of industrial and apparel uses; rope, work clothes; fish nets
Polyvinylidene chloride	Saran[b] by Vectra (N)	Acids, most alkalies, alcohol, bleaches, insects, mildew, weather	Heat (above 90°C); many solvents	Outdoor fabrics; insect screen; curtains, upholstery; carpet; work clothes
Polytetrafluoroethylene (PTFE)	Teflon® (P)	Almost all chemicals, solvents, insects, mildew	Heat (above 250°C); fluorine at high temperature	Corrosion-resistant packings, etc., tapes, filters, bearings

Table 4–88 (continued)
CLASSIFICATION OF MAN-MADE FIBERS AND FABRICS

aFabrics are often damaged by heat in ironing. The synthetics will not withstand the usual 400°F that is used for cotton and linen. For acetate and olefins the ironing temperature should be below 250°F, as for silk and wool. For most other synthetics a 300°F ironing temperature is recommended, but triacetate will tolerate higher temperature without damage.

bThe names *nylon, vinyon, saran,* and *spandex* are recognized as generic terms rather than proprietary names.

From Bolz, R. E. and Tuve, G. L., Eds., *Handbook of Tables for Applied Engineering Science*, 2nd ed., CRC Press, Cleveland, 1973, 173.

Table 4–89
PROPERTIES OF MAN-MADE FIBERS[a]

Chemical class; common name (sources)	Specific gravity	Tenacity g/denier	Tensile strength, 10³ psi	Elongation at break, %	Regain (standard)	Softening point, °C	Melting point, °C	Flammability	Brittleness temp, °C
CELLULOSE FIBERS (NATURAL)									
Acetate	1.30	1.–1.3	18–25	20–30	6.5	140	230	Melts and burns	
Triacetate	1.32	1.2–1.4	20–28	25–30	3–4.5	225	300	Melts and burns	
Viscose rayon	1.51	2–2.6	30–46	17–25	13.		200[b]	Burns readily	
High-tenacity viscose	1.53	3–5	60–80	10–12	10		200[b]	Burns readily	< –114
Polynosic viscose	1.53	3–5	60–80	8–20	7		200[b]	Burns readily	
Cuprammonium rayon (cupro)	11.52	1.7–2.3	30–45	10–17	12.5		250[b]	Burns readily	
PROTEIN FIBERS (NATURAL)									
Animal: casein (milk)	1.3	1.0	15	60–70	14	100	150	Slow	
Vegetable—seed: soybeans, peanuts, corn	1.3	0.7–0.9	11–14	40–60	11–15	150	250	Slow	
Vegetable—latex: rubber (vulcanized)	1.0	0.4–0.6	4–7	700–900	0	300		Burns	–60

Table 4–89 (continued)
PROPERTIES OF MAN-MADE FIBERS

Chemical class; common name (sources)	Specific gravity	Tenacity, g/denier	Tensile strength, 10^3 psi	Elonga- tion at break, %	Regain (standard)	Soften- ing point, °C	Melting point, °C	Flamma- bility	Brittleness temp, °C
SYNTHETIC FIBERS									
Polyacrylonitrile (acrylic)	1.17	2–5	50–75	25–40	2	190	260	Burns	
Polyamide (nylon)	1.14	4–9	70–120	20–40	4	200	215–250	Slow	< –100
Polyester (PET dacron)	1.38	4–8	70–120	10–50	0.4	225	250–290	Low	
Polyethylene (olefin, low density)	0.92	3–6	40–70	25–40	0.15	90–120	120	Slow	–114
Polyethylene (olefin, high density)	0.95	5–7	60–80	10–20	0.01	120–130	140	Slow	–114
Polypropylene (olefin)	0.91	4.5–8	45–80	15–30	0–0.5	145	160–170	Self-ext. low	–70
Polyurethane (spandex)	1.1	0.5–1.0	7–16	500–700	1.0	190	250	Burns	
Polyvinyl chloride (PVC)	1.38	0.7–2	12–17	100–125	0.1	70	140[b]	No; chars	< –100
Polyvinyl alcohol (PVA)	1.3	3–7	60–90	15–28	5	230	240	Slow	
Polyvinylidene chloride (saran)	1.7	2	40	20–30	0.1	115–135	170	No	
Polytetrafluoroethylene (PTFE)	2.1	1.2–1.4	33	15–30	0	225	300[b]	No	

[a]For additional properties of fibers, see Table 4–87.
[b]Decomposition; does not melt.

Note: Mechanical properties are for room temperature and humidity and based on cross section.

From Bolz, R. E. and Tuve, G. L., Eds., *Handbook of Tables for Applied Engineering Science*, 2nd ed., CRC Press, Cleveland, 1973, 175.

Table 4–90
FIBERS FOR SPECIAL USES

Desired characteristics	Fibers that are superior for these requirements	Desired characteristics	Fibers that are superior for these requirements
Moisture resistant Nonabsorbent, fast drying, high wet strength, non-swelling, nonshrinking	Glass, Teflon®, saran, PVC, rubber, polyethylene, polypropylene saran, spandex, acrylic polyester	**Fire resistant** Nonflammable or very slow burning; flame resistant	Asbestos, Teflon, PVC, saran polypropylene, polyvinyl alcohol
Climate resistant Minimum deterioration from sun, rain, sea water, insects, mildew, and other environmental factors	Glass, Teflon, saran, PVC, rubber, acrylic, poly-propylene, spandex, poly-ester, polyvinyl alcohol	**Lightweight** Low density and high strength-weight ratio	Polypropylene, polyethylene, polyurethane, nylon
Chemical resistant Unharmed by acids, alkalies, salts, oils, common solvents	Asbestos, Teflon, saran, polypropylene, polyvinyl alcohol, polyester	**High tenacity** High breaking strength for given diameter; high ulti-mate tensile strength	Glass, nylon, polyester, ramie, polypropylene, poly-vinyl alcohol, flax, silk, high-tensile viscose
Temperature resistant Retains properties at high and low temperatures; heat resistant	Asbestos, Teflon, nylon, poly-ester, cupro, triacetate	**Hard wearing** High resistance to friction and abrasion	Flax, silk, cotton, kenef, sunn, PVC, polyvinyl alcohol, polyester, polypropylene, nylon, acrylic
		Elastic and resilient No permanent set after large deformation; springy	Rubber, spandex, polyester, wool, PVC, nylon, acrylic, silk

From Bolz, R. E. and Tuve, G. L., Eds., *Handbook of Tables for Applied Engineering Science*, 2nd ed., CRC Press, Cleveland, 1973, 176.

Table 4—91

TEXTILE YARN AND FIBER SIZES

The *fineness* of silk, cotton, rayon, and other man-made yarns is usually expressed in terms of the length of yarn per unit weight, or the weight of a given length. The *denier* is the weight in grams of 9,000 m of yarn. (Sometimes the *international denier* is used; it is defined as the weight in grams of 10,000 m of yarn.) A similar unit is the *count,* which is the number of hanks or skeins per pound. The hank, for silk or cotton, is 840 yd, but for wool, worsted, linen, or man-made fiber, it may be different. In the metric system the *count* is the number of meters per gram of yarn.

Conversion Table

To convert from	To	Multiply by
Gram/meter	Denier	.0001111
Milligram/kilometer	Denier	111.11
Gram/meter	International denier	.0001
Denier	Ounce/1,000 yd	0.003584
Denier	Pound/10,000 yd	0.002240
Count (cotton, silk)	Yard/pound	840

Note: There is no simple conversion from these units of fineness to the diameter in microns, nor between breaking strength per denier (tenacity) and tensile strength per unit area.

From Bolz, R. E. and Tuve, G. L., Eds., *Handbook of Tables for Applied Engineering Science,* 2nd ed., CRC Press, Cleveland, 1973, 176.

Table 4–92
INDUSTRIAL WOOL FELT SPECIFICATIONS AND DATA

Applications:

Vibration isolation
Liquid and gas filtration
Sealing

Fluid transfer and retention (wicking and lubrication)
Spacing, padding, and shock absorption
Polishing

Cushioning and packaging
Sound absorption and attenuation
Thermal insulation
Frictional material

NTA class	Corresponding to		Specifications						Typical design data		
	SAE number	CF 206 ASTM CS 185	Wool content fiber basis, minimum %	Other fibers, maximum %	Wool content chemical basis, minimum %	Tensile strength, minimum psi[a]	Density, lb/ft³	Color	Compressional resistance, %[b]	Tear strength[c]	Durometer hardness[d]
14R1	–	–	45	55	40	–	8.7	White	35	–	8
17R1	–	–	50	50	45	–	10.6	White	45	–	12
17R2	F26	8R5	45	55	40	–	10.6	Gray	40	–	12
18R1	F10	9R1	100	–	95	240	11.2	White	50	15	15
18R2	F11	9R2	90	10	87	225	11.2	Gray	50	10	15
18R3	F13	9R4	80	20	75	150	11.2	Gray	50	10	15
26R1	F5	12R1	95	5	90	425	16.2	White	65	20	23
26R2	F6	12R2	90	10	87	300	16.2	Gray	65	18	23
26R3	F7 and F55	12R3	85	15	80	275	16.2	Gray	65	15	23
34R1	F1 and F50	16R1	100	–	95	600	21.2	White	75	25	27
34R2	F2	16R2	95	5	90	500	21.2	Black/gray	75	20	27
34R3	F3 and F51	16R3	90	10	85	400	21.2	Gray	75	15	27
38R1	–	18R1	100	–	95	650	23.7	White	80	25	37
38R2	–	–	100	–	95	550	23.7	Gray	80	20	37
43S1	–	20S1	100	–	95	700	26.8	White	85	35	45
56S1	–	26S1	100	–	95	800	35.0	White	90	40	62
68S1	–	32S1	100	–	95	900	42.5	White	92	45	73

[a] ASTM D-461 method.
[b] NTA test method.
[c] ASTM D-2262 method.
[d] Shore durometer.
From Northern Textile Association – Felt Manufacturers Council. With permission.

Table 4–93

PERMEABILITY OF MATERIALS TO WATER VAPOR

For Comparative Estimates of Water Vapor Transmission through Walls

Values in this table give the permeance of each material for typical thickness. The equation $W = nA\Delta P$ is applicable, where W = rate of vapor transmission, n = permeance, A = area of flow path, and ΔP = vapor-pressure difference, from one side to the other, i.e., corresponding to the dewpoints on the two sides.

Values in the table are results of laboratory tests either by the dry-cup method of ASTM E96 or some other test method. The permeance is in grain/hr·ft^2·in. Hg (= perms); corresponding units should be used for area and vapor-pressure difference to obtain W in grain/hr. For permeance in SI units, kg/s·N, multiply the tabulated values by 57.213×10^{-12}

Material	Permeance Dry-cup method	Permeance Other methods	Material	Permeance Dry-cup method	Permeance Other methods
STRUCTURAL MATERIALS			FELTS AND BUILDING AND ROOFING PAPERS		
Acrylic, glass-fiber reinforced sheet, 56 mil	0.12	—	Blanket thermal insulation back-up paper, asphalt-coated (31)	0.4	0.6–4.2
Asbestos-cement board, 0.2 in. thick	0.54	—	Duplex sheet, asphalt-laminated, aluminum foil one side (43)	0.002	0.176
Brick masonry, 4 in. thick	—	0.8	Felt, 15 lb asphalt (70)	1.0	5.6
Concrete, 1:2:4 mix, per inch	—	3.2	Felt, 15 lb tar (70)	4.0	18.2
Concrete block, 8 in. cored, limestone aggregate	—	2.4	Kraft paper and asphalt-laminated, reinforced 30–120–30 (34)	0.3	1.8
Gypsum sheathing, $\frac{1}{2}$ in. asphalt impregnated	20	—	Roll roofing, saturated and coated (326)	0.05	0.24
Hardboard, $\frac{1}{8}$ in. standard	—	11	Sheathing paper, asphalt-saturated, uncoated (22)	3.3	20.2
Hardboard, $\frac{1}{8}$ in. tempered	—	5	Single-kraft, double-infused (16)	31	42
Plaster on plain gypsum lath, with studs	—	20	Vapor-barrier paper, asphalt-saturated, coated (43)	0.2–0.3	0.6
Plaster on metal lath, $\frac{3}{4}$ in.	—	15	LIQUID-APPLIED COATING MATERIALS		
Plywood, Douglas-fir, exterior, $\frac{1}{4}$ in. thick	—	0.7	*Paint—2 coats*		
Plywood, Douglas-fir, interior $\frac{1}{4}$ in. thick	—	1.9	Aluminum varnish on wood	0.3–0.5	—
Polyester, glass-fiber reinforced sheet, 48 mil	0.05	—	Asphalt paint on plywood	—	0.4
Tile masonry, glazed, 4 in. thick	—	0.12	Enamels on smooth plaster	—	0.5–1.5
			Sealers or flat paint on interior-insulation board	—	0.9–2.1
THERMAL INSULATIONS, PER INCH THICKNESS			Various primers plus 1 coat flat oil paint on plaster	—	1.6–3.0
Corkboard	2.1–2.6	—	*Paint—3 coats*		
Expanded polystyrene, bead	2.0–5.8	—	Asphalt cut-back mastic, $\frac{1}{16}$ in. dry	0.14	--
Expanded polystyrene, extruded	1.2	—	Asphalt cut-back mastic, $\frac{3}{16}$ in. dry	0.0	—
Expanded polyurethane (R–11 blown)	0.4–1.6	—	Chloro-sulfonated polyethylene mastic, 3.5 oz/sq ft	1.7	—
PLASTIC AND METAL FOILS AND FILMS			Chloro-sulfonated polyethylene mastic, 7.0 oz/sq ft	0.06	—
Aluminum foil, 0.35 mil	0.05	—	Exterior paint, white lead-zinc oxide and oil on wood	0.9	—
Aluminum foil, 1 mil	0.0	—	Hot melt asphalt, 2 oz/sq ft	0.5	—
Cellulose acetate, 10 mil	0.46	—	Hot melt asphalt, 3.5 oz/sq ft	0.1	—
Polyester, 1 mil	0.7	—			
Polyethylene, 2 mil	0.16	—			
Polyethylene, 6 mil	0.06	—			
Polyethylene, 10 mil	0.03	—			
Polyvinylchloride, plasticized, 4 mil	0.8–1.4	—			
Polyvinylchloride, unplasticized, 2 mil	0.68	—			

Table 4–93 (continued)
PERMEABILITY OF MATERIALS TO WATER VAPOR

For Comparative Estimates of Water Vapor Transmission through Walls (continued)

Material	Permeance		Material	Permeance	
	Dry-cup method	Other methods		Dry-cup method	Other methods
			Polyvinyl-acetate latex coating, 4 oz/sq ft	5.5	—
			Styrene-butadiene latex coating, 2 oz/sq ft	11	—

Based on *ASHRAE Handbook of Fundamentals,* American Society of Heating, Refrigerating and Air-Conditioning Engineers, New York, 1972. With permission.

Table 4—94

SOUND FREQUENCIES AND SCALES

A pure tone or musical note is produced if the vibrations consist of a single frequency within the audible range. A "harmonic" is a partial tone, the frequency of which is an integral multiple of the fundamental or lowest frequency. A harmonic series consists of integral multiples.

If the frequency is doubled, the tone rises one octave on the musical scale. The common musical scale in Western countries has 12 half tones, with an interval or frequency ratio between successive tones of $\sqrt[12]{2}$. In this equally tempered scale the intervals are given the following names.

Name	Frequency ratio	Name	Frequency ratio
Semitone	1.05946	Perfect fifth	1.49831
Whole tone or major second	1.12246	Minor sixth	1.58740
Minor third	1.18921	Major sixth	1.68179
Major third	1.25992	Minor seventh	1.78180
Perfect fourth	1.33484	Major seventh	1.88775
Augmented fourth[a]	1.41421	Octave	2.00000

In a complex tone the ear judges the pitch by the lowest frequency, or fundamental, and interprets the tone quality in terms of the accompanying higher frequencies, or overtones. The audible range, which is about 16 to 20,000 hz, varies among individuals and among animals.

Very complex sounds may be given special names such as white noise. A "noise" is defined as an unwanted sound, and it may or may not have a prevailing frequency. Electronic generators are available for generating sound combinations of almost any desired pattern.

Engineering specifications and measurements are most likely to deal with sound intensity or energy, but the direct measurement is that of sound pressure in decibels expressed on a logarithmic scale.[b] Loudness is the human subjective interpretation of sound intensity and is expressed in phons (or sones). Sounds are analyzed with respect to frequency through the use of electronic band-pass filters, e.g., octave bands or 1/3-octave bands, with special attention to five octaves in the auditory range, 125 to 4,000 hz.

The following diagram provides a comparison of ranges in the acoustic spectrum, as represented by average human voices and by the usual string and wind instruments of the orchestra. Frequencies for the C-scale on the piano are also given, based on the tuning of this instrument to American Standard Pitch, A = 440 hz (formerly C at 256 hz).

[a]Also called diminished fifth.

[b]Sound measurement techniques are covered by standard codes; see ANSI-S1.2 and entire ANSI S-series of over 30 standards.

605

Table 4–94 (continued)

The Acoustic Spectrum in Hertz

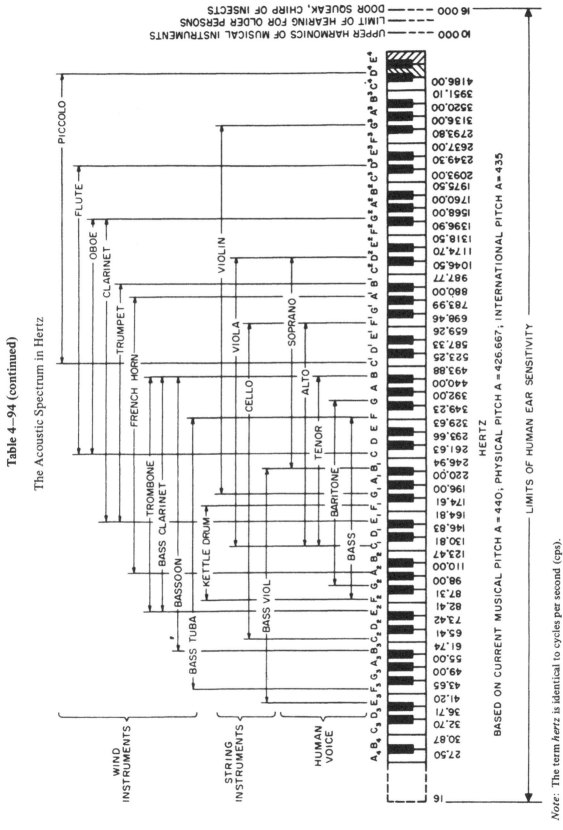

Note: The term *hertz* is identical to cycles per second (cps).

From *Reference Data for Radio Engineers*, 5th ed., Howard W., Sams, Indianapolis, 1968. With permission.

Table 4–95
WAXES AND RELATED MATERIALS

Most waxes are commercially available in several grades, depending on source, refinement, bleaching, or processing. Thus, the properties of a single type of wax may vary over a rather wide range, and the values given in this table should be considered only as typical.

Name, source, type, color	Specific gravity	Melting point, °C	Flash point, °C	Acid value[a]	Iodine value[b]	Saponification value[a]	Dielectric constant, at 10^6 Hz
VEGETABLE WAXES							
Bayberry (myrtle)	.93	50		3.5	3	212	
Candelilla (brown) from Mexico	.98	68	241	16	24	57	2.5
Carnauba (palm) from Brazil	1.00	85	300	6	12	83	2.9
Castor oil (hydrogenated)	.98	86	315	2	4	180	
Cotton (yellow)	.96	80		30	24	70	
Esparto (grass)	.99	74	255	25	16	68	
Japan (sumac)	.98	51	200	10	9	220	3.1
Ouricuri (palm) from Brazil	1.02	82	280	12	8	80	
Sugar cane (brown)	.97	80		15	20	60	2.9
ANIMAL WAXES							
Beeswax, yellow	.96	64	245	20	10	93	2.8
Chinese insect	.96	82		10	2	90	3.7
Lanolin (sheepswool) refined	.94	40		8	25	107	
Shellac (insect)	.97	77		15	5	110	
Spermaceti (whale)	.93	45	245	2	4	123	8.
MINERAL, SYNTHETIC, WAXLIKE							
Ceresin	.90	70		0	8	0	
Microcrystalline[c]	.93	70	260				2.3
Moutan (lignite)	1.02	85		40		100	
Ozocerite	.90	75		0	8	0	2.4
Paraffin[c]	.92	60	205				2.2
Polyethylene	.93	100	315	15		25	
Polyethylene glycol	1.1	45	250				

[a]*Acid value* is milligrams of KOH to neutralize the free fatty acids. *Saponification value* is milligrams of KOH for complete saponification.

[b]*Iodine value* is a measure of unsaturated linkages present and is expressed as grams iodine absorbed per 100 g of sample.

[c]There are so many grades of paraffin wax and microcrystalline petroleum wax that the properties of any specific wax should be checked.

Note: The *index of refraction* of most waxes is between 1.4 and 1.5

REFERENCES

Bennett, H., *Industrial Waxes,* Vol. 1, Chemical Publishing, New York, 1963.
Weast, R. C., Ed., *Handbook of Chemistry and Physics,* 52nd ed., Chemical Rubber, Cleveland, 1971.
Lange, N. A., Ed., *Handbook of Chemistry,* 9th ed., Handbook Publishers, Sandusky, Oh., 1956.
Davidsohn, A. and Milwidsky, B. M., *Polishes,* 4th ed., CRC Press, Cleveland, 1968.
Knaggs, N. S., *Adventures in Man's First Plastic,* Reinhold, New York, 1947.

From Bolz, R. E. and Tuve, G. L., Eds., *Handbook of Tables for Applied Engineering Science,* 2nd ed., CRC Press, Cleveland, 1973, 159.

Table 4–96

SOUND ABSORPTION – GENERAL BUILDING MATERIALS AND FURNISHINGS

The following table gives the sound absorption coefficients of various materials. The reverberation time is defined as the time required for the reverberant sound intensity to decrease to one millionth of its initial intensity, or 60 dB. The time is given in seconds by the relation, $T_\tau = 0.049$ $\frac{V}{\alpha}$, where V is the volume of the room in cubic feet and α is the absorption. The absorption α in sabins is computed by multiplying the area in square feet of each surface by its absorption coefficient and taking the sum of the products. The absorption of seats and audience is also computed on an area basis using the coefficients given below.

Complete tables of coefficients of the various materials that normally constitute the interior finish of rooms may be found in the various books on architectural acoustics. The following short list will be useful in making simple calculations of the reverberation in rooms.

Materials	Coefficients					
	125 cps	250 cps	500 cps	1,000 cps	2,000 cps	4,000 cps
Brick, unglazed	.03	.03	.03	.04	.05	.07
Brick, unglazed, painted	.01	.01	.02	.02	.02	.03
Carpet, heavy, on concrete	.02	.06	.14	.37	.60	.65
Same, on 40 oz hairfelt or foam rubber	.08	.24	.57	.69	.71	.73
Same, with impermeable latex backing on 40 oz hairfelt or foam rubber	.08	.27	.39	.34	.48	.63
Concrete block, coarse	.36	.44	.31	.29	.39	.25
Concrete block, painted	.10	.05	.06	.07	.09	.08
Fabrics						
Light velour, 10 oz/yd^2, hung straight, in contact with wall	.03	.04	.11	.17	.24	.35
Medium velour, 14 oz/yd^2, draped to half area	.07	.31	.49	.75	.70	.60
Heavy velour, 18 oz/yd^2, draped to half area	.14	.35	.55	.72	.70	.65
Floors						
Concrete or terrazzo	.01	.01	.015	.02	.02	.02
Linoleum, asphalt, rubber, or cork tile on concrete	.02	.03	.03	.03	.03	.02
Wood	.15	.11	.10	.07	.06	.07
Wood parquet in asphalt on concrete	.04	.04	.07	.06	.06	.07
Glass						
Large panes of heavy plate glass	.18	.06	.04	.03	.02	.02
Ordinary window glass	.35	.25	.18	.12	.07	.04
Gypsum board, 1/2 in. nailed to 2 × 4's 16 in. o.c.	.29	.10	.05	.04	.07	.09
Marble or glazed tile	.01	.01	.01	.01	.02	.02
Openings						
Stage, depending on furnishings			.25– .75			
Deep balcony, upholstered seats			.50–1.00			
Grills, ventilating			.15– .50			
Plaster, gypsum or lime, smooth finish on tile or brick	.013	.015	.02	.03	.04	.05
Plaster, gypsum or lime, rough finish on lath	.02	.03	.04	.05	.04	.03
Same, with smooth finish	.02	.02	.03	.04	.04	.03
Plywood paneling, 3/8 in. thick	.28	.22	.17	.09	.10	.11
Water surface, as in a swimming pool	.008	.008	.013	.015	.020	.025
Air, sabins/1,000 ft^3					2.3	7.2

Table 4—96 (continued)

SOUND ABSORPTION — GENERAL BUILDING MATERIALS AND FURNISHINGS

Absorption of Seats and Audience

Values given are in sabins per square foot of seating area or per unit.

Materials	Coefficients					
	125 cps	250 cps	500 cps	1,000 cps	2,000 cps	4,000 cps
Audience, seated in upholstered seats, per square foot of floor area	.60	.74	.88	.96	.93	.85
Unoccupied cloth-covered upholstered seats, per square foot of floor area	.49	.66	.80	.88	.82	.70
Unoccupied leather-covered upholstered seats, per square foot of floor area	.44	.54	.60	.62	.58	.50
Wooden pews, occupied, per square foot of floor area	.57	.61	.75	.86	.91	.86
Chairs, metal or wood seats, each, unoccupied	.15	.19	.22	.39	.38	.30

From *Sound Absorption Coefficients of Architectural Acoustical Materials,* Bulletin 22, Acoustical Materials Association, New York, 1962, 68. With permission.

Table 4–97
SOUND ABSORPTION – SPECIAL ACOUSTICAL MATERIALS

Armstrong Cork Company

Material	Thickness	Flame resistance*	Surface	Mounting	Coefficients					
					125 cps	250 cps	500 cps	1,000 cps	2,000 cps	4,000 cps
Acoustical Fire Guard[g] Full Random	5/8 in.	A	Perforated,[b] painted	X-2	.43	.54	.67	.96	.79	.59
	3/4 in.	A	Same as above	X-2 1/4	.44	.60	.83	.99	.66	.56
Acoustical Fire Guard Fissured	5/8 in.	A	Fissured, painted	X-2 1/2	.48	.60	.67	.71	.75	.73
	3/4 in.	A	Same as above	X-2 3/4	.42	.68	.74	.74	.71	.64
Travertone Fire Guard[h]	3/4 in.	A	Fissured, painted	7	.75	.68	.69	.85	.93	.90
Travertone	3/4 in.	A	Fissured, painted	1	.08	.32	.79	.93	.87	.80
				X-2 1/4	.36	.62	.81	.84	.87	.91
Ventilating Travertone	3/4 in.	A	Fissured, painted	Write to manufacturer for sound absorption values.						
Travertone Embossed	3/4 in.	A	Embossed, painted	1	.08	.23	.79	.93	.88	.86
				X-10 3/4	.65	.56	.63	.76	.88	.91
Travertone, Golden and Silver	3/4 in.	A	Fissured, painted; Inlaid silver or gold colored metallic flecks	1	.13	.41	.86	.84	.76	.62
				X-2 1/4	.37	.68	.77	.68	.71	.64
Minatone Classic	1/2 in.	A	Perforated,[f] painted	1	.11	.14	.68	.87	.68	.45
	5/8 in.	A	Same as above	1	.12	.25	.83	.87	.64	.52
				X-10 1/2	.46	.44	.68	.80	.72	.56
Minaboard® Fissured	5/8 in.	A	Fissured, painted	7	.46	.44	.58	.78	.80	.74
Minaboard Classic	5/8 in.	A	Perforated,[f] painted	X-2 1/2	.39	.58	.72	.90	.72	.59
Crestone®	5/8 in.	A	Striated, painted	1	.11	.37	.61	.63	.68	.70
				X-2	.39	.63	.57	.56	.65	.72
Cushiontone® Straight Row	1/2 in.	C, D	Perforated,[a] painted	1	.07	.17	.65	.73	.73	.67
				2	.16	.57	.58	.69	.71	.70
	3/4 in.	C, D	Same as above	1	.12	.31	.83	.93	.75	.65
				2	.18	.66	.69	.94	.77	.63
				X-2 1/4	.32	.65	.75	.89	.72	.64

*The Federal specification establishes specific criteria by which materials may be classified from "A" to "D," depending on their performance in the test. No specific terms are given to describe these classes but materials classified "A" are usually considered as "incombustible" and those classified "D" as "combustible." Classes "B" and "C" represent materials of intermediate flame resistance.

For the flame resistance tests for which ratings are shown in this table, materials were mounted by bolting them directly to an asbestos cement board panel.

Table 4–97 (continued)
SOUND ABSORPTION – SPECIAL ACOUSTICAL MATERIALS

Armstrong Cork Company (continued)

Material	Thickness	Flame resistance	Surface	Mounting	Coefficients					
					125 cps	250 cps	500 cps	1,000 cps	2,000 cps	4,000 cps
Cushiontone Textured	9/16 in.	C, D	Fissured, perforated, painted	1	.14	.23	.60	.73	.83	.64
			Same as above	2	.17	.60	.52	.71	.81	.71
	3/4 in.	C, D		1	.21	.33	.77	.78	.82	.70
				2	.16	.67	.60	.76	.81	.69
				X-2 1/4	.39	.64	.64	.72	.79	.68
Cushiontone Classic	1/2 in.	C, D	Perforated,[f] painted	1	.10	.19	.64	.78	.72	.52
				2	.11	.57	.54	.66	.71	.54
	3/4 in.	C, D	Same as above	1	.20	.32	.84	.81	.74	.55
				2	.22	.69	.72	.76	.71	.54
				X-2 1/4	.35	.62	.71	.71	.68	.52
Cushiontone Silver (classic pattern)	1/2 in.	D	Perforated,[f] painted Inlaid silver colored metallic flecks	1	.20	.25	.57	.62	.56	.39
				2	.18	.59	.45	.53	.51	.40
Cushiontone Golden (textured pattern)	9/16 in.	D	Fissured, perforated painted. Inlaid gold colored metallic flecks	1	.14	.33	.55	.52	.56	.53
				2	.21	.64	.39	.51	.56	.52
Cushiontone Centennial Rhapsody	1/2 in.	D	Perforated,[f] painted with design, tongue and groove	1	.22	.30	.62	.54	.49	.43
				2	.28	.67	.46	.51	.54	.44
Tahiti	1/2 in.	D	Same as above	1	.22	.32	.62	.52	.48	.36
				2	.22	.67	.46	.50	.46	.39
Autumn Leaves	1/2 in.	D	Same as above	1	.20	.33	.64	.52	.51	.49
				2	.27	.67	.48	.52	.52	.45
Arrestone® Diagonal	1 9/16 in.	A	Perforated, enameled metal[c]	3	.27	.57	.99	.99	.75	.64

Table 4–97 (continued)
SOUND ABSORPTION – SPECIAL ACOUSTICAL MATERIALS

Armstrong Cork Company (continued)

Material	Thickness	Flame resistance	Surface	Mounting	Coefficients					
					125 cps	250 cps	500 cps	1,000 cps	2,000 cps	4,000 cps
Full Random		A	Perforated, enameled metal[d]	3	.25	.58	.99	.99	.76	.61
Gridtone®	3/4 in.	A	Perforated, embossed, painted metal[e]	X-8 3/4	.64	.84	.85	.78	.68	.40

Material	Thickness	Ceiling detail at partition	Attenuation factors–decibels				
			125 cps	250 cps	500 cps	1,000 cps	4,000 cps
Acoustical Fire Guard, Full Random	5/8 in.	Interrupted	31	29	36	44	59
Acoustical Fire Guard, Classic	3/4 in.	Interrupted	27	29	39	50	54
Acoustical Fire Guard, Lay-in Units Classic	5/8 in.	Interrupted	29	32	37	44	61
Travertone	3/4 in.	Interrupted	29	25	28	32	55
Travertone Vinyl Face	3/4 in.	Continuous	26	32	40	46	57
Minaboard Classic	5/8 in.	Interrupted	24	28	35	43	56
Cushiontone Classic	3/4 in.	Interrupted	29	29	40	50	49
Ventilating Travertone	3/4 in.	Interrupted	21	21	23	26	52

[a]Straight row perforated 529 holes/ft², 3/16 in. diameter.
[b]Full random perforated 353 holes/ft²; 250 of 3/16 in. and 103 of 1/4 in. diameters.
[c]Diagonal perforated, enameled metal pan with mineral wool sound-absorbing pad. Pans perforated 1,105 holes/ft², 3/32 in. diameter; bevels and flanges unperforated. Thickness includes tee bar.
[d]Full random perforated, enameled metal pan with mineral wool sound-absorbing pad. Pan perforated 717 holes/ft², 498 of 7/64 in. and 219 of 5/32 in. diameters. Thickness includes tee bar.
[e]Random perforations in two sizes and a field of directional bars set in relief, painted metal facing unit backed with 3/4 in. glass fiber sound-absorbing pad.

Table 4–97 (continued)
SOUND ABSORPTION – SPECIAL ACOUSTICAL MATERIALS

Armstrong Cork Company (continued)

f Hundreds of small perforations scattered in lacelike fashion.
g Acoustical ceiling constructions classified by Underwriters' Laboratories, Inc.

No. 4177-3–5/8 in. Acoustical Fire Guard–Design 8–2 hr
No. 4177-4–5/8 in. Acoustical Fire Guard–Design 9–1 hr
No. 4177-5–3/4 in. Acoustical Fire Guard–Design 31–4 hr
No. 4177-6–3/4 in. Acoustical Fire Guard–Design 21–4 hr

h Acoustical ceiling constructions classified by Underwriters' Laboratories, Inc.

No. 4177-9–3/4 in. Travertone Fire Guard–Design 7–1 1/2 hr

Baldwin-Ehret-Hill, Incorporated

Material	Thickness	Flame resistance	Surface	Mounting	Coefficients					
					125 cps	250 cps	500 cps	1,000 cps	2,000 cps	4,000 cps
Hiltonia®, regular pattern, perforated mineral tile	5/8 in.	A	Perforated,[a] painted	1	.03	.23	.71	.92	.69	.58
				2	.11	.32	.73	.99	.91	.70
				X-10 1/2	.68	.67	.74	.96	.94	.75
Hiltonia, random pattern, perforated mineral tile	1/2 in.	A	Perforated,[b] painted	1	.05	.16	.48	.99	.86	.58
				2	.10	.40	.68	.97	.86	.62
	5/8 in.	A	Same as above	1	.06	.17	.61	.99	.78	.47
				2	.11	.36	.84	.99	.85	.56
				7	.61	.57	.69	.99	.83	.51
	3/4 in.	A	Same as above	1	.08	.21	.70	.99	.77	.52
				7	.70	.69	.80	.99	.82	.56
Hiltonia, pin point pattern, perforated mineral tile	1/2 in.	A	Perforated,[c] painted	1	.05	.13	.50	.99	.84	.58
	5/8 in.	A	Same as above	1	.03	.16	.71	.99	.85	.74
				7	.63	.52	.69	.92	.82	.65
	3/4 in.	A	Same as above	1	.10	.21	.90	.99	.85	.61
				X-10 3/4	.53	.62	.73	.96	.91	.74
Panatone® with B-E-H mineral fiber pads	1 9/16 in.	A	Perforated, enameled metal[d]	3	.26	.60	.99	.98	.78	.66

Table 4–97 (continued)
SOUND ABSORPTION – SPECIAL ACOUSTICAL MATERIALS

Baldwin-Ehret-Hill, Incorporated (continued)

Material	Thickness	Flame resistance	Surface	Mounting	Coefficients 125 cps	250 cps	500 cps	1,000 cps	2,000 cps	4,000 cps
Claritone® regular drilled	1/2 in.	C, D	Perforated,[a] painted[e]	1	.03	.16	.58	.82	.87	.71
				2	.13	.57	.56	.74	.84	.71
	3/4 in.	C, D	Same as above	1	.04	.29	.77	.88	.81	.55
				2	.08	.65	.65	.87	.87	.65
				X-10 3/4	.51	.37	.59	.86	.90	.64
B-E-H Ceiling Board	3/4 in.	A	Textured, painted	7	.85	.67	.74	.91	.93	.92

Material	Thickness	Surface	Ceiling detail at partition	Attenuation factors–decibels 125 cps	250 cps	500 cps	1,000 cps	4,000 cps
Styltone® fissured mineral tile	3/4 in.	See preceding table	Interrupted	28	33	33	37	58

[a] Perforated 484 holes/ft², 3/16 in. diameter, uniformly spaced.
[b] Perforated 320 holes/ft², 111 of 1/4 in. and 209 of 3/16 in. diameters, randomly spaced.
[c] Perforated 1,596 holes/ft², 3/32, 5/64, 1/16, 3/64 in. diameters, randomly spaced.
[d] Perforated, enameled metal pan backed with sound-absorbing mineral fiber pad. Pan perforated with 1,024 holes/ft², 0.109 in. diameter. Thickness includes tee bar.
[e] Available with standard, factory-applied paint finish or with factory-applied flame-resistant paint finish which gives flame resistance rating of Class "C" and light reflection of .80.

Table 4–97 (continued)

SOUND ABSORPTION – SPECIAL ACOUSTICAL MATERIALS

The Celotex Corporation

Material	Thickness	Flame resistance	Surface	Mounting	Coefficients					
					125 cps	250 cps	500 cps	1,000 cps	2,000 cps	4,000 cps
Acousti-Celotex® cane tile, standard pattern	1/2 in.	C, D	Perforated,[a] painted[b]	1	.03	.18	.64	.82	.79	.71
				2	.15	.55	.55	.69	.75	.73
				X-2 1/2	.39	.62	.59	.69	.73	.71
	3/4 in.	C, D	Perforated,[a] painted[b]	1	.12	.21	.80	.99	.80	.60
				2	.15	.53	.80	.92	.77	.51
				X-2 3/4	.31	.71	.80	.90	.76	.63
Hush-Tone® tile, standard pattern	1/2 in.	D	Perforated,[a] painted	2	.15	.55	.55	.69	.75	.73
	3/4 in.	C, D	Same as above	1	.18	.41	.79	.80	.69	.51
				2	.26	.61	.64	.76	.69	.57
				X-2 3/4	.33	.68	.71	.67	.72	.55
Celotone® standard	3/4 in.	A	Fissured, painted	1	.08	.23	.79	.98	.87	.80
				X-2 3/4	.28	.58	.96	.84	.89	.78
Acousteel®	1 9/16 in.	A	Perforated, enameled metal[c]	3	.32	.67	.99	.96	.72	.56
Steelacoustic® Standard	1 1/4 in.	A	Slotted, enameled metal[d]	X-9 1/4	.25	.63	.78	.70	.81	.62
Random		A	Same as above	X-9 1/4	.54	.75	.72	.69	.63	.38
Supracoustic®	3/4 in.	A	Fissured, painted	7	.89	.87	.75	.95	.86	.60
	1 in.	A	Same as above	7	.93	.88	.81	.99	.89	.68

Table 4–97 (continued)
SOUND ABSORPTION – SPECIAL ACOUSTICAL MATERIALS

The Celotex Corporation (continued)

Material	Thickness	Ceiling detail at partition	Attenuation factors—decibels					
			125 cps	250 cps	500 cps	1,000 cps	4,000 cps	
Acousti-Celotex cane tile, standard pattern	3/4 in.	Interrupted	23	25	33	39	50	
Perforated mineral fiber tile								
MP-1, standard	5/8 in.	Interrupted	27	29	32	34	57	
MP-6, standard	1 in.	Interrupted	28	32	39	49	59	
Celotone standard	3/4 in.	Interrupted	23	24	27	30	56	
Chase Celotone	3/4 in.	Interrupted	28	33	36	44	64	
Protectone® mineral fiber tile								
PSP-1 Serene	5/8 in.	Interrupted	30	33	34	39	60	
PF-4 Natural fissured	3/4 in.	Interrupted	27	30	29	34	59	

[a]Perforated 484 holes/ft², 3/16 in. diameter, 17/32 in. on centers.

[b]Available either with standard No. 232 two-coat, hot-rolled finish, or with No. 9 flame-resisting factory-applied finish which provides a Class "C" flame resistance rating according to Federal Specification SS-A-118b.

[c]Perforated, enameled metal pan with mineral wool sound-absorbing pad. Perforated in diagonal pattern, 1,105 holes/ft², 3/32 in. diameter. Flanges and bevels unperforated. Thickness includes tee bar. Also available in standard and random perforating patterns.

[d]Slotted enameled metal units with 1 1/4 in. mineral fiber absorbent pads laminated.

Table 4–97 (continued)
SOUND ABSORPTION – SPECIAL ACOUSTICAL MATERIALS

Elof Hansson, Inc.

Material	Thickness	Flame resistance	Surface	Mounting	Coefficients					
					125 cps	250 cps	500 cps	1,000 cps	2,000 cps	4,000 cps
Hansonite regular perforated acoustical tile	1/2 in.	C, D	Perforated[a], painted[b]	1	.09	.19	.58	.73	.78	.73
				2	.13	.59	.49	.66	.76	.75
	3/4 in.	C, D	Same as above	1	.21	.32	.68	.87	.86	.71
				2	.29	.65	.59	.80	.91	.71
Hansonite random perforated mineral tile	1/2 in.	A	Perforated[c], painted[f]	1	.05	.16	.48	.99	.86	.58
				2	.10	.40	.68	.97	.86	.62
	5/8 in.	A	Same as above	1	.06	.17	.61	.99	.78	.47
				2	.11	.36	.84	.99	.85	.56
				7	.61	.57	.69	.99	.83	.51
	3/4 in.	A	Same as above	1	.08	.21	.70	.99	.77	.52
				7	.70	.69	.80	.99	.82	.56
Hansoguard®[g], Hansostar® perforated fire protective acoustical tile	5/8 in.	A	Perforated[d], painted[f]	7	.41	.38	.66	.89	.77	.60
Hansoguard[g] fissured fire protective acoustical tile	5/8 in.	A	Fissured, painted	7	.53	.43	.62	.73	.72	.68
Hansoboard® Gold Mist mineral ceiling board	5/8 in.	A	Perforated[d], painted[e]	7	.56	.51	.69	.89	.76	.52
Hansonite perforated asbestos board	3 3/16 in.	A	Perforated[h], unpainted	X-13 1/4	.83	.93	.88	.87	.69	.48

Table 4–97 (continued)

SOUND ABSORPTION – SPECIAL ACOUSTICAL MATERIALS

Elof Hansson, Inc. (continued)

Material	Thickness	Ceiling detail at partition	Attenuation factors–decibels					
			125 cps	250 cps	500 cps	1,000 cps	4,000 cps	
Hansonite fissured mineral tile Note[i]	3/4 in.	Interrupted	23	26	26	29	52	
		Interrupted	25	36	46	54	61	
Hansoguard random perforated fire protective acoustical tile	5/8 in.	Interrupted	26	30	35	40	61	
Hansopan® regular perforated metal pan with 1 1/4 in. paper-wrapped pad	1 9/16 in.	Interrupted	23	23	25	29	52	
Hansoboard perforated mineral ceiling board	5/8 in.	Interrupted	26	31	31	37	61	

[a]Perforated 484 holes/ft², 3/16 in. diameter, uniformly spaced.

[b]Factory painted white. Also available factory painted white with flame resistant finish which meets Federal Specification SS-A-118b Class "C" requirements.

[c]Perforated 320 holes/ft², 111 of 1/4 in. and 209 of 3/16 in. diameters, randomly spaced.

[d]Perforated 1,596 holes/ft², 3/64, 1/16, 5/64, and 3/32 in. diameters, randomly spaced.

[e]Factory painted white base with textured gold overlay; flame resistant finish meets Federal Specification SS-A-118b Class "C" requirements.

[f]Factory painted white, face and bevels.

[g]Acoustical ceiling constructions classified by Underwriters' Laboratories, Inc. No. 4355-1–5/8 in. Hansoguard–Design 18–2 hr.

[h]Perforated 3/16 in. diameter holes on 1/2 in. straight centers.

[i]Tile cemented to 5/8 in. gypsum board suspended ceiling.

Table 4–97 (continued)
SOUND ABSORPTION – SPECIAL ACOUSTICAL MATERIALS

E. F. Hauserman Company

Material	Thickness	Flame resistance	Surface	Mounting	Coefficients					
					125 cps	250 cps	500 cps	1,000 cps	2,000 cps	4,000 cps
Acoustic ceiling	3 in.	A	Perforated metal[a,b]	4	.71	.60	.81	.99	.69	.50
Acousti-Wall	2 3/4 in.	A	Perforated metal[a,c]	4	.70	.47	.77	.99	.69	.44
Signature wall panel	2 1/4 in.	A	Perforated metal[a,c]	4	.41	.49	.78	.99	.67	.42

[a]Color as selected, factory baked enamel.
[b]Perforated steel face, airspace with 1 in. glass fiber board, 3/8 in. gypsum board, rock wool and unperforated back.
[c]Construction similar to Note b except 1/4 in. gypsum board.

Johns-Manville Sales Corporation

Material	Thickness	Flame resistance	Surface	Mounting	Coefficients					
					125 cps	250 cps	500 cps	1,000 cps	2,000 cps	4,000 cps
Sanacoustic® "W," J-M mineral wool pad, paper wrapped, 1 1/4 in. thick	1 9/16 in.	A	Perforated,[a] enameled metal	3 / 7	.39 / .71	.73 / .68	.99 / .87	.99 / .98	.90 / .83	.72 / .70
Sanacoustic 50/50 pattern, J-M mineral wool pad, paper wrapped, 1 1/4 in. thick	1 9/16 in.	A	Perforated and unperforated, enameled metal[b]	3	.50	.59	.66	.71	.56	.32
Sanacoustic "W," J-M microlite blanket; 1 1/2 in. thick, 0.75 lb. density	3 in.	A	Perforated,[a] enameled metal[c]	X-12 1/2	.70	.77	.79	.84	.74	.54
Perforated Transite panels, J-M microlite blanket; 1 in. thick, 1.0 lb. density	1 3/16 in.	A	Perforated[d] painted	5	.20	.30	.69	.98	.68	.25

Table 4–97 (continued)
SOUND ABSORPTION – SPECIAL ACOUSTICAL MATERIALS

Johns-Manville Sales Corporation (continued)

Material	Thickness	Flame resistance	Surface	Mounting	Coefficients 125 cps	250 cps	500 cps	1,000 cps	2,000 cps	4,000 cps
Perforated Transite panels, J-M microlite blanket; 2 in. thick, 1.0 lb. density	2 3/16 in.	A	Perforated,[d] painted	8	.32	.62	.99	.86	.63	.37
Spintone® tile, uniform drilled mineral tile	1/2 in.	A	Perforated,[e] painted	1	.09	.23	.62	.75	.77	.77
	5/8 in.	A	Same as above	1	.18	.28	.68	.95	.84	.66
				7	.44	.50	.65	.90	.92	.67
	3/4 in.	A	Same as above	1	.20	.31	.78	.95	.78	.67
				7	.54	.55	.70	.92	.85	.71
Spintone tile, random drilled mineral tile	1/2 in.	A	Perforated,[f] painted	1	.14	.26	.64	.65	.64	.58
	5/8 in.	A	Same as above	1	.20	.35	.68	.71	.80	.78
				7	.58	.50	.67	.92	.91	.72
	3/4 in.	A	Same as above	7	.45	.46	.64	.89	.85	.65
Firedike® tile,[h] uniform drilled mineral tile	3/4 in.	A	Perforated,[e] painted	X-10 3/4	.44	.47	.59	.83	.87	.68
				7	.54	.55	.70	.92	.85	.71
Spanglas, random pebbled	1 1/4 in.	A	Random pebbled, painted	X-11 1/4	.81	.85	.87	.96	.99	.86
				7	.84	.87	.89	.99	.96	.86
Acousti-Shell®	2 in.		Dome shaped,[g] glass fiber fabric facing	7	.78	.76	.68	.79	.81	.81
Fibertone®, uniform drilled	1/2 in.	C, D	Perforated,[e] painted	1	.03	.16	.58	.82	.87	.71
				2	.13	.57	.56	.74	.84	.71
	3/4 in.	C, D	Same as above	1	.04	.29	.77	.88	.81	.55
				2	.08	.65	.65	.87	.87	.65
Solo tile, random perforated	7/8 in.		Perforated,[i] painted	X-10 3/4	.51	.37	.59	.86	.90	.64
				1	.21	.45	.59	.58	.58	.35
Airacoustic®	1/2 in.	A	Unpainted	6	.35	.43	.40	.74	.85	.78
	1 in.	A	Same as above	6	.60	.56	.69	.75	.82	.82

Table 4–97 (continued)
SOUND ABSORPTION – SPECIAL ACOUSTICAL MATERIALS

Johns-Manville Sales Corporation (continued)

Material	Thickness	Ceiling detail at partition	Attenuation factors—decibels				
			125 cps	250 cps	500 cps	1,000 cps	4,000 cps
Firedike tile, uniform drilled mineral tile	3/4 in.	Interrupted	33	34	39	46	65
Sanacoustic "W," J-M mineral wool pad, paper wrapped, 1 1/4 in. thick	1 9/16 in.	Interrupted	23	23	25	29	52
Sanacoustic "W," J-M mineral wool pad, paper wrapped, 1 1/4 in. thick; backed with 1/8 in. J-M Flexboard® attenuation baffles	1 9/16 in.	Interrupted	30	34	48	54	58
Spintone tile, random drilled mineral tile	5/8 in.	Interrupted	27	31	36	42	59

[a] Perforated enameled metal pan with sound-absorptive element as stated under Material. Pan perforated 925 holes/ft², 0.108 in. diameter. Thickness includes tee bar.

[b] One half perforated metal pan backed with sound-absorptive element as stated under Material; one half unperforated metal pan unbacked. Thickness includes tee bar. Pan perforated, 1,625 holes/ft², 0.10 in. diameter.

[c] Blanket supported by tee bars about 1 1/2 in. above face of metal pan.

[d] Transite perforated 600 holes/ft², 3/16 in. diameter.

[e] Perforated 484 holes/ft², 3/16 in. diameter.

[f] Perforated 312 holes/ft², 132 of 1/4 in. and 180 of 3/16 in. diameters.

[g] Three-dimensional glass fiber lay-in panel, approximately 5/32 in. thick, with glass fiber fabric finish available in beige, green, and blue.

[h] Acoustical ceiling constructions classified by Underwriters' Laboratories, Inc.

No. 4400-1–3/4 in. Firedike Mineral Tile–Design 29–2 hr.

[i] Perforated 350 hexagonal shaped holes/ft², 86 holes 3/16 in. diameter across the flats, and 264 holes 5/32 in. diameter across the flats.

Table 4–97 (continued)
SOUND ABSORPTION – SPECIAL ACOUSTICAL MATERIALS

Kaiser Gypsum Company, Incorporated

Material	Thickness	Flame resistance	Surface	Mounting	Coefficients					
					125 cps	250 cps	500 cps	1,000 cps	2,000 cps	4,000 cps
Kaiser Fir-Tex® perforated acoustical tile, regular drilled	1/2 in.	C, D	Perforated,[a] painted[b]	1	.12	.18	.58	.74	.84	.78
				2	.14	.60	.64	.68	.83	.77
				7	.51	.38	.48	.79	.94	.91
	3/4 in.	C, D	Same as above	1	.19	.24	.71	.97	.88	.65
				2	.27	.58	.63	.94	.92	.71
				X-10 3/4	.61	.39	.57	.90	.90	.72

[a]Perforated 529 holes/ft², 13/64 in. diameter.
[b]Factory-painted two coats face and bevels. Also available factory-painted two coats face and bevels with slow-burning finish which meets Federal Specification SS-A-118b, Class "C" flame resistance requirements.

National Gypsum Company

Material	Thickness	Flame resistance	Surface	Mounting	Coefficients					
					125 cps	250 cps	500 cps	1,000 cps	2,000 cps	4,000 cps
Acoustifibre®, uniform pattern	1/2 in.	C, D	Perforated,[a] painted[b]	1	.07	.22	.61	.83	.80	.72
				2	.12	.53	.54	.77	.85	.78
				7	.35	.40	.48	.75	.89	.84
	3/4 in.	C, D	Same as above	1	.14	.32	.78	.92	.88	.66
				2	.15	.71	.63	.91	.88	.72
				7	.44	.42	.58	.94	.89	.67
Acoustiroc® fissured pattern	3/4 in.	A	Fissured, painted	1	.11	.32	.77	.88	.87	.84
				7	.62	.58	.61	.79	.90	.83
Travacoustic®	3/4 in.	A	Fissured, painted	1	.09	.23	.67	.99	.92	.89
				7	.69	.71	.73	.89	.88	.90
Travacoustic sculptured pattern	3/4 in.	A	Striated, painted	1	.04	.16	.62	.98	.95	.91
				X-10 3/4	.71	.76	.78	.82	.93	.90
Gold Bond® perforated asbestos panels	3/16 in.	A	Perforated, auto-claved[d]	7	.82	.88	.66	.73	.66	.51

Table 4–97 (continued)
SOUND ABSORPTION – SPECIAL ACOUSTICAL MATERIALS

National Gypsum Company (continued)

Material	Thickness	Flame resistance	Surface	Mounting	Coefficients					
					125 cps	250 cps	500 cps	1,000 cps	2,000 cps	4,000 cps
	15/16 in.	A	Perforated, auto-claved^c	7	.77	.64	.60	.78	.68	.42
	1 3/16 in.	A	Perforated, auto-claved^c,e	5	.08	.20	.57	.96	.69	.24
			Perforated, auto-claved^c	5	.09	.31	.56	.93	.68	.23
				7	.75	.66	.62	.75	.65	.44
	2 3/16 in.	A	Same as above	8	.18	.55	.98	.98	.58	.44
				7	.93	.81	.86	.96	.65	.45

Material	Thickness	Ceiling detail at partition	Attenuation factors—decibels				
			125 cps	250 cps	500 cps	1,000 cps	4,000 cps
Acoustifibre, full random pattern^f	3/4 in.	Interrupted	22	25	37	45	46
		Interrupted	*31*	*34*	*48*	*53*	*61*
Fire-Shield Solitude®							
Full random pattern	3/4 in.	Interrupted	33	35	43	51	65
Needlepoint pattern	5/8 in.	Interrupted	*31*	*34*	*42*	*50*	*61*
Fire-Shield Acoustiroc®							
Textured pattern	3/4 in.	Interrupted	28	34	37	45	63
Full random pattern	3/4 in.	Interrupted	25	26	24	26	53
Fissured pattern	3/4 in.	Interrupted	27	31	31	36	60
Travacoustic^f	3/4 in.	Interrupted	24	25	31	37	50
		Interrupted	*31*	*34*	*47*	*53*	*58*
Acoustimetal® uniform pattern with RIS blanket^g	3 5/8 in.	Interrupted	22	26	33	41	49
Fire-Shield Solitude grid panels, fissured	5/8 in.	Interrupted	29	33	41	49	55

Table 4–97 (continued)
SOUND ABSORPTION – SPECIAL ACOUSTICAL MATERIALS

National Gypsum Company (continued)

[a] Perforated 484 holes/ft², 3/16 in. diameter, uniformly spaced.
[b] Factory painted two coats, which includes flame resisting paint giving Class "C" flame resistance rating. Also available with Class "D" finish.
[c] Asbestos board perforated 550 holes/ft², 3/16 in. diameter; autoclaved white finish.
[d] Asbestos board backed with flame resistant paper sound-absorbing element, perforated 550 holes/ft², 3/16 in. diameter; autoclaved white finish.
[e] Pad thickness 3/4 in.
[f] Tile cemented to 1/2 in. gypsum board suspended ceiling.
[g] Semithick Gold Bond RIS blankets installed on top of pans and pads, vapor barrier side toward the pan.

Owens-Corning Fiberglas Corporation

Material	Thickness	Flame resistance	Surface	Mounting	Coefficients					
					125 cps	250 cps	500 cps	1,000 cps	2,000 cps	4,000 cps
Fiberglas® acoustical tile	1/2 in.	A	Textured, painted	1	.14	.19	.66	.78	.80	.64
Type TXW	3/4 in.	A	Same as above	X-10 3/4	.13	.33	.78	.86	.78	.69
			Same as above	7	.68	.74	.74	.81	.87	.76
Type TXW-TL	3/4 in.	A	Same as above	7	.80	.83	.69	.85	.84	.72
Fiberglas Frescor acoustical tile	3/4 in.	A	Stippled,[a] painted	1	.70	.54	.68	.81	.81	.59
					.17	.29	.76	.91	.92	.76
			Same as above	X-10 3/4	.70	.81	.77	.83	.88	.77
			Same as above	X-10 3/4	.63	.74	.77	.82	.87	.73
Type TL	3/4 in.	A	Same as above	7	.60	.49	.72	.89	.85	.63
Fiberglas ceiling board	3/4 in.	A	Textured, painted	X-10 3/4	.69	.76	.92	.88	.88	.88
	1 in.	A	Same as above	X-11	.65	.89	.93	.98	.92	.91
	1 1/4 in.	A	Same as above	X-11 1/4	.65	.78	.91	.95	.97	.93
Type TL	3/4 in.	A	Same as above	7	.51	.53	.75	.95	.97	.80
Fiberglas Sonocor® ceiling board	1 in.	C	Embossed membrane faced[b]	7	.77	.57	.84	.93	.75	.51
	2 in.		Same as above	7	.53	.86	.94	.91	.63	.39
	3 in.		Same as above	7	.84	.87	.99	.85	.58	.30

Table 4–97 (continued)
SOUND ABSORPTION – SPECIAL ACOUSTICAL MATERIALS

Owens-Corning Fiberglas Corporation (continued)

Material	Thickness	Ceiling detail at partition	Attenuation factors—decibels				
			125 cps	250 cps	500 cps	1,000 cps	4,000 cps
Fiberglas acoustical tile Type TXW-TL	3/4 in.	Interrupted	29	26	33	38	50
Fiberglas Frescor acoustical tile Type TL	3/4 in.	Interrupted	29	23	31	36	49
Fiberglas ceiling board, Type TL	3/4 in.	Interrupted	24	23	31	38	52
Fiberglas Frescor ceiling board, Type TL	3/4 in.	Interrupted	26	24	32	39	54
Fiberglas cloth faced ceiling board, Type TL	3/4 in.	Interrupted	23	28	32	31	48

[a] Stippled, random white on white.
[b] Thin embossed plastic membrane cemented to board face.

Simpson Timber Company

Material	Thickness	Flame resistance	Surface	Mounting	Coefficients					
					125 cps	250 cps	500 cps	1,000 cps	2,000 cps	4,000 cps
Simpson perforated acoustical tile PCP Random Perforated	1/2 in.		Perforated,[a] painted	1	.14	.27	.63	.76	.71	.49
				2	.08	.66	.55	.68	.62	.50
	3/4 in.		Same as above	1	.15	.42	.77	.78	.62	.45
				2	.19	.65	.67	.76	.58	.42
				7	.39	.46	.65	.77	.61	.49
Simpson perforated acoustical tile, standard drilled	1/2 in.	C, D	Perforated,[b] painted[c]	1	.13	.19	.61	.80	.83	.74
				2	.07	.65	.52	.62	.75	.77
	3/4 in.	C, D	Same as above	1	.29	.37	.69	.81	.85	.77
				2	.22	.62	.66	.88	.86	.58
				X-10 3/4	.34	.31	.61	.83	.81	.61
Simpson Forestone® ceiling board	1 in.	C	Fissured, painted	X-10 3/4	.39	.37	.48	.62	.77	.80

Table 4–97 (continued)
SOUND ABSORPTION – SPECIAL ACOUSTICAL MATERIALS

Simpson Timber Company (continued)

Material	Thickness	Flame resistance	Surface	Mounting	Coefficients					
					125 cps	250 cps	500 cps	1,000 cps	2,000 cps	4,000 cps
Simpson fissured mineral tile	3/4 in.	A	Fissured, painted	1	.06	.19	.67	.99	.91	.90
				X-10 3/4	.81	.74	.70	.80	.94	.89
Simpson metal acoustical units	1 9/16 in.	A	Perforated, enameled metal[d]	3	.31	.68	.99	.98	.79	.62

[a]Random drilled 457 holes/ft², 63 holes 13/16 in. and 92 holes 11/16 in. diameters, 302 holes punch perforated, approximately 5/16 in. diameter.
[b]Perforated 484 holes/ft², 3/16 in. diameter.
[c]Painted by manufacturer two coats face and bevels. Also available with factory-applied finish which meets Federal Specification SS-A-118b, Class "C" flame resistance.
[d]Perforated, enameled metal pan with sound-absorbing pad. Pan perforated 1,013 holes/ft², 0.109 in. diameter. Thickness includes tee bar.

United States Gypsum Company

Material	Thickness	Flame resistance	Surface	Mounting	Coefficients					
					125 cps	250 cps	500 cps	1,000 cps	2,000 cps	4,000 cps
Acoustone® "F"	3/4 in.	A	Fissured, painted	1	.03	.27	.83	.99	.82	.71
				7	.68	.67	.65	.84	.87	.74
Acoustone "db"	3/4 in.	A	Fissured, painted	7	.58	.48	.67	.99	.92	.85
Auditone® regular perforated	1/2 in.	C, D	Perforated,[a] painted[b]	1	.14	.23	.66	.68	.78	.76
				2	.18	.61	.51	.62	.78	.74
	3/4 in.	C, D	Same as above	1	.24	.42	.64	.69	.81	.79
				2	.23	.62	.56	.72	.83	.85
Perfatone®	1 9/16 in.	A	Perforated,[c] enameled metal	7	.36	.35	.62	.88	.83	.74
				7	.74	.79	.88	.99	.77	.56
Ceiling board	3/4 in.	A	Fissured, painted	7	.89	.87	.75	.95	.86	.60
	1 in.	A	Same as above	7	.93	.88	.81	.99	.89	.68

Table 4–97 (continued)

SOUND ABSORPTION – SPECIAL ACOUSTICAL MATERIALS

United States Gypsum Company (continued)

Material	Thickness	Ceiling detail at partition	Attenuation factors—decibels					
			125 cps	250 cps	500 cps	1,000 cps	4,000 cps	
Acoustone "F"d	3/4 in.	Interrupted	23	26	26	29	52	
		Interrupted	25	36	46	54	61	
Acoustone "db"	3/4 in.	Interrupted	31	33	41	49	59	
Auditone random perforated	3/4 in.	Interrupted	26	33	43	50	49	
Perfatone	1 9/16 in.	Interrupted	21	21	21	24	44	
Motif'd® Acoustone "db," Galaxy design No. 33	3/4 in.	Interrupted	27	32	35	41	61	
Acoustone "120"	3/4 in.	Interrupted	29	32	36	41	60	

aPerforated 529 holes/ft², 3/16 in. diameter.
bFactory painted face and bevels. Also furnished, factory painted, with special paint finish giving Class "C" flame resistance rating.
cPerforated enameled metal, 1,105 holes/ft², 3/32 in. diameter. Thickness includes tee bar.
dTile cemented to 5/8 in. Sheetrock Firecode® gypsum board suspended ceiling.

Table 4–97 (continued)
SOUND ABSORPTION – SPECIAL ACOUSTICAL MATERIALS

Wood Conversion Company

Material	Thickness	Flame resistance	Surface	Mounting	Coefficients 125 cps	250 cps	500 cps	1,000 cps	2,000 cps	4,000 cps
Nu-Wood®, regular pattern, regularly perforated cellulose fiber tile	1/2 in.	C, D	Perforated,[a] painted[b]	1	.09	.19	.58	.73	.78	.73
	3/4 in.	C, D	Same as above	2	.13	.59	.49	.66	.76	.75
Lo-Tone®, regular pattern, regularly perforated mineral tile	5/8 in.	A	Perforated,[a] painted	1	.21	.32	.68	.87	.86	.71
				2	.29	.65	.59	.80	.91	.71
				X-10 3/4	.03	.23	.71	.92	.69	.58
Lo-Tone FR[c] random pattern mineral tile	5/8 in.	A	Perforated,[d] painted	7	.11	.32	.73	.99	.91	.70
					.68	.67	.74	.96	.94	.75
					.48	.41	.57	.96	.89	.59
Lo-Tone metal pan regular pattern	1 9/16 in.	A	Perforated,[e] enameled metal	3	.39	.73	.99	.99	.90	.72
random pattern		A	Perforated,[f] enameled metal	3	.39	.76	.99	.92	.61	.38
Lo-Tone asbestos board, regularly perforated	1 3/16 in.	A	Perforated,[g] painted	5	.20	.32	.62	.99	.70	.30
	2 3/16 in.	A	Same as above	8	.36	.51	.99	.88	.63	.46

Table 4–97 (continued)
SOUND ABSORPTION – SPECIAL ACOUSTICAL MATERIALS

Wood Conversion Company (continued)

Material	Thickness	Ceiling detail at partition	Attenuation factors–decibels				
			125 cps	250 cps	500 cps	1,000 cps	4,000 cps
Lo-Tone, random pattern, random perforated mineral tile	5/8 in.	Interrupted	26	24	26	28	54
Lo-Tone FR, random pattern mineral tile	5/8 in.	Interrupted	26	30	35	40	61
Lo-Tone, metal pan, regular pattern	1 9/16 in.	Interrupted	23	23	25	29	52

[a]Perforated 484 holes/ft², 3/16 in. diameter, uniformly spaced.

[b]Factory-painted face and bevels. Also available with factory-applied paint finish providing Class "C" flame resistance rating.

[c]Acoustical ceiling constructions classified by Underwriters' Laboratories, Inc.

    No. 4355-1–5/8 in. Lo-Tone FR–Design 18–2 hr
    No. 4355-3–5/8 in. Lo-Tone FR–Design 24–2 hr

[d]Perforated 320 holes/ft², 1/4 and 3/16 in. diameters, randomly spaced.

[e]Perforated, enameled metal pan with sound-absorbing mineral pad. Pan random perforated 925 holes/ft², 0.108 in. diameter. Thickness includes tee bar.

[f]Perforated, enameled metal pan with sound-absorbing mineral pad. Pan random perforated 293 holes/ft², 166 of 9/64 and 127 of 3/16 in. diameters. Thickness includes tee bar.

[g]Perforated 550 holes/ft², 3/16 in. diameter.

From *Sound Absorption Coefficients of Architectural Acoustical Materials*, Bulletin 22, Acoustical Materials Association, New York, 1962, 29. With permission.

Table 4–98
WIRE TABLES

Comparison of Wire Gauges

A. Diameter of Wire in Inches

Gauge No.	Brown and Sharpe	Birmingham or Stubs'	Washburn and Moen	Imperial or British standard	Stubs' steel	U.S. standard plate
00000000	–	–	–	–	–	–
0000000	–	–	–	.500	–	–
000000	–	–	–	.464	–	.46875
00000	–	–	–	.432	–	.4375
0000	.4600	.454	.3938	.400	–	.40625
000	.4096	.425	.3625	.372	–	.375
00	.3648	.380	.3310	.348	–	.34375
0	.3249	.340	.3065	.324	–	.3125
1	.2893	.300	.2830	.300	.227	.28125
2	.2576	.284	.2625	.276	.219	.265625
3	.2294	.259	.2437	.252	.212	.25
4	.2043	.238	.2253	.232	.207	.234375
5	.1819	.220	.2070	.212	.204	.21875
6	.1620	.203	.1920	.192	.201	.203125
7	.1443	.180	.1770	.176	.199	.1875
8	.1285	.165	.1620	.160	.197	.171875
9	.1144	.148	.1483	.144	.194	.15625
10	.1019	.134	.1350	.128	.191	.140625
11	.09074	.120	.1205	.116	.188	.125
12	.08081	.109	.1055	.104	.185	.109375
13	.07196	.095	.0915	.092	.182	.09375
14	.06408	.083	.0800	.080	.180	.078125
15	.05707	.072	.0720	.072	.178	.0703125
16	.05082	.065	.0625	.064	.175	.0625
17	.04526	.058	.0540	.056	.172	.05625
18	.04030	.049	.0475	.048	.168	.05
19	.03589	.042	.0410	.040	.164	.04375
20	.03196	.035	.0348	.036	.161	.0375
21	.02846	.032	.0318	.032	.157	.034375
22	.02535	.028	.0286	.028	.155	.03125
23	.02257	.025	.0258	.024	.153	.028125
24	.02010	.022	.0230	.022	.151	.025
25	0.01790	0.020	0.0204	0.020	0.148	0.021875
26	0.01594	0.018	0.0181	0.018	0.146	0.01875
27	0.01419	0.016	0.0173	0.0164	0.143	0.0171875
28	0.01264	0.014	0.0162	0.0149	0.139	0.015625
29	0.01126	0.013	0.0150	0.0136	0.134	0.0140625
30	0.01003	0.012	0.0140	0.0124	0.127	0.0125
31	0.008928	0.010	0.0132	0.0116	0.120	0.0109375
32	0.007950	0.009	0.0128	0.0108	0.115	0.01015625
33	0.007080	0.008	0.0118	0.0100	0.112	0.009375
34	0.006304	0.007	0.0104	0.0092	0.110	0.00859375

Table 4–98 (continued)
WIRE TABLES

Comparison of Wire Gauges (continued)

A. Diameter of Wire in Inches (continued)

Gauge No.	Brown and Sharpe	Birmingham or Stubs'	Washburn and Moen	Imperial or British standard	Stubs' steel	U.S. standard plate
35	0.005614	0.005	0.0095	0.0084	0.108	0.0078125
36	0.005000	0.004	0.0090	0.0076	0.106	0.00703125
37	0.004453	–	0.0085	0.0068	0.103	0.006640625
38	0.003965	–	0.0080	0.0060	0.101	0.00625
39	0.003531	–	0.0075	0.0052	0.099	–
40	0.003145	–	0.0070	0.0048	0.097	–
41	–	–	0.0066	0.0044	0.095	–
42	–	–	0.0062	0.0040	0.092	–
43	–	–	0.0060	0.0036	0.088	–
44	–	–	0.0058	0.0032	0.085	–
45	–	–	0.0055	0.0028	0.081	–
46	–	–	0.0052	0.0024	0.079	–
47	–	–	0.0050	0.0020	0.077	–
48	–	–	0.0048	0.0016	0.075	–
49	–	–	0.0046	0.0012	0.072	–
50	–	–	0.0044	0.0010	0.069	–

B. Diameter of Wire in Centimeters

Gauge No.	Brown and Sharpe	Birmingham or Stubs'	Washburn and Moen	Imperial or British standard	Stubs' steel	U.S. standard plate
00000000	–	–	–	–	–	–
0000000	–	–	1.245	1.27	–	1.27
000000	–	–	1.172	1.18	–	1.191
00000	–	–	1.093	1.10	–	1.111
0000	1.168	1.15	1.000	1.02	–	1.032
000	1.040	1.08	0.9208	0.945	–	0.9525
00	0.9266	0.965	0.8407	0.884	–	0.8731
0	0.8252	0.864	0.7785	0.823	–	0.7938
1	0.7348	0.762	0.7188	0.762	0.577	0.7144
2	0.6543	0.721	0.6668	0.701	0.556	0.6747
3	0.5827	0.658	0.6190	0.640	0.538	0.6350
4	0.5189	0.605	0.5723	0.589	0.526	0.5953
5	0.4620	0.559	0.5258	0.538	0.518	0.5556
6	0.4115	0.516	0.4877	0.488	0.511	0.5159
7	0.3665	0.457	0.4496	0.447	0.505	0.4763
8	0.3264	0.419	0.4115	0.406	0.500	0.4366
9	0.2906	0.376	0.3767	0.366	0.493	0.3969
10	0.2588	0.340	0.3429	0.325	0.485	0.3572
11	0.2305	0.305	0.3061	0.295	0.478	0.3175
12	0.2053	0.277	0.2680	0.264	0.470	0.2778
13	0.1828	0.241	0.232	0.234	0.462	0.2381
14	0.1628	0.211	0.203	0.203	0.457	0.1984
15	0.1450	0.183	0.183	0.183	0.452	0.1786

Table 4–98 (continued)
WIRE TABLES

Comparison of Wire Gauges (continued)

B. Diameter of Wire in Centimeters (continued)

Gauge No.	Brown and Sharpe	Birmingham or Stubs'	Washburn and Moen	Imperial or British standard	Stubs' steel	U.S. standard plate
16	0.1291	0.165	0.159	0.163	0.445	0.1588
17	0.1150	0.147	0.137	0.142	0.437	0.1429
18	0.1024	0.124	0.121	0.122	0.427	0.1270
19	0.09116	0.107	0.104	0.102	0.417	0.1111
20	0.08118	0.089	0.0884	0.0914	0.409	0.09525
21	0.07229	0.081	0.0808	0.0813	0.399	0.08731
22	0.06439	0.071	0.0726	0.0711	0.394	0.07938
23	0.05733	0.064	0.0655	0.0610	0.389	0.07144
24	0.05105	0.056	0.0584	0.0559	0.384	0.06350
25	0.04547	0.051	0.0518	0.0508	0.376	0.05556
26	0.04049	0.046	0.0460	0.0457	0.371	0.04763
27	0.03604	0.041	0.0439	0.0417	0.363	0.04366
28	0.03211	0.036	0.0411	0.0378	0.353	0.03969
29	0.02860	0.033	0.0381	0.0345	0.340	0.03572
30	0.02548	0.030	0.0356	0.0315	0.323	0.03175
31	0.02268	0.025	0.0335	0.0295	0.305	0.02778
32	0.02019	0.023	0.0325	0.0274	0.292	0.02580
33	0.01798	0.020	0.0300	0.0254	0.284	0.02381
34	0.01601	0.018	0.0264	0.0234	0.279	0.02183
35	0.01426	0.013	0.024	0.0213	0.274	0.01984
36	0.01270	0.010	0.023	0.0193	0.269	0.01786
37	0.01131	–	0.022	0.0173	0.262	0.01687
38	0.01007	–	0.020	0.0152	0.257	0.01588
39	0.008969	–	0.019	0.0132	0.251	–
40	0.007988	–	0.018	0.0122	0.246	–
41	–	–	0.017	0.0112	0.241	–
42	–	–	0.016	0.0102	0.234	–
43	–	–	0.015	0.0091	0.224	–
44	–	–	0.015	0.0081	0.216	–
45	–	–	0.014	0.0071	0.206	–
46	–	–	0.013	0.0061	0.201	–
47	–	–	0.013	0.0051	0.196	–
48	–	–	0.012	0.0041	0.191	–
49	–	–	0.012	0.0030	0.183	–
50	–	–	0.011	0.0025	0.175	–

From Weast, R. C., Ed., *Handbook of Chemistry and Physics,* 55th ed., CRC Press, Cleveland, 1974.

Table 4—99
TWIST DRILL AND STEEL WIRE GAUGE

Inches

No.	Size	No.	Size	No.	Size	No.	Size	No.	Size
1	0.2280	17	0.1730	33	0.1130	49	0.0730	65	0.0350
2	0.2210	18	0.1695	34	0.1110	50	0.0700	66	0.0330
3	0.2130	19	0.1660	35	0.1100	51	0.0670	67	0.0320
4	0.2090	20	0.1610	36	0.1065	52	0.0635	68	0.0310
5	0.2055	21	0.1590	37	0.1040	53	0.0595	69	0.02925
6	0.2040	22	0.1570	38	0.1015	54	0.0550	70	0.0280
7	0.2010	23	0.1540	39	0.0995	55	0.0520	71	0.0260
8	0.1990	24	0.1520	40	0.0980	56	0.0465	72	0.0250
9	0.1960	25	0.1495	41	0.0960	57	0.0430	73	0.0240
10	0.1935	26	0.1470	42	0.0935	58	0.0420	74	0.0225
11	0.1910	27	0.1440	43	0.0890	59	0.0410	75	0.0210
12	0.1890	28	0.1405	44	0.0860	60	0.0400	76	0.0200
13	0.1850	29	0.1360	45	0.0820	61	0.0390	77	0.0180
14	0.1820	30	0.1285	46	0.0810	62	0.0380	78	0.0160
15	0.1800	31	0.1200	47	0.0785	63	0.0370	79	0.0145
16	0.1770	32	0.1160	48	0.0760	64	0.0360	80	0.0135

Centimeters

No.	Size	No.	Size	No.	Size	No.	Size	No.	Size
1	0.5791	17	0.4394	33	0.2870	49	0.1854	65	0.0889
2	0.5613	18	0.4305	34	0.2819	50	0.1778	66	0.0838
3	0.5410	19	0.4216	35	0.2794	51	0.1702	67	0.0813
4	0.5309	20	0.4089	36	0.2705	52	0.1613	68	0.0787
5	0.5220	21	0.4039	37	0.2642	53	0.1511	69	0.0743
6	0.5182	22	0.3988	38	0.2578	54	0.1397	70	0.0711
7	0.5105	23	0.3912	39	0.2527	55	0.1321	71	0.0660
8	0.5055	24	0.3861	40	0.2489	56	0.1181	72	0.0635
9	0.4978	25	0.3797	41	0.2438	57	0.1092	73	0.0610
10	0.4915	26	0.3734	42	0.2375	58	0.1067	74	0.0572
11	0.4851	27	0.3658	43	0.2261	59	0.1041	75	0.0533
12	0.4801	28	0.3569	44	0.2184	60	0.1016	76	0.0508
13	0.4699	29	0.3454	45	0.2083	61	0.0991	77	0.0457
14	0.4623	30	0.3264	46	0.2057	62	0.0965	78	0.0406
15	0.4572	31	0.3048	47	0.1994	63	0.0940	79	0.0368
16	0.4496	32	0.2946	48	0.1930	64	0.0914	80	0.0343

From Weast, R. C., Ed., *Handbook of Chemistry and Physics,* 55th ed., CRC Press, Cleveland, 1974.

Table 4-100
DIMENSIONS OF WIRE

Stubs' Gauge

Diameter and cross-section are given in English and metric units for the Birmingham or Stubs' gauge.

Gauge No.	Diameter, inches	Section, square inches	Diameter, centimeters	Section, square centimeters
0000	0.454	0.16188	1.1532	1.0444
000	.425	.14186	1.0795	0.9152
0	.380	.11341	0.9652	.7317
0	0.340	0.09079	0.8636	0.5858
1	.300	.07069	.7620	.4560
2	.284	.06335	.7214	.4087
3	.259	.05269	.6579	.3399
4	.238	.04449	.6045	.2870
5	0.220	0.03801	0.5588	0.2452
6	.203	.03237	.5156	.20881
7	.180	.02545	.4572	.16147
8	.165	.02138	.4191	.13795
9	.148	.01720	.3759	.11099
10	0.134	0.01410	0.3404	0.09098
11	.120	.011310	.3048	.07297
12	.109	.009331	.2769	.06160
13	.095	.007088	.2413	.04573
14	.083	.005411	.2108	.03491
15	0.072	0.004072	0.1829	0.02627
16	.065	.0033183	.16510	.021409
17	.058	.0026421	.14732	.017046
18	.049	.0018857	.12446	.012166
19	.042	.0013854	.10668	.008938
20	0.035	0.0009621	0.08890	0.006207
21	.032	.0008042	.08128	.005189
22	.028	.0006158	.07112	.003973
23	.025	.0004909	.06350	.003167
24	.022	.0003801	.05588	.002452
25	0.020	0.0003142	0.05080	0.002027
26	.018	.0002545	.04572	.0016417
27	.016	.0002011	.04064	.0012972
28	.014	.0001539	.03556	.0009932
29	.013	.0001327	.03302	.0008563
30	0.012	0.0001181	0.03048	0.0007297
31	.010	.00007854	.02540	.0005067
32	.009	.00006362	.02286	.0004104
33	.008	.00005027	.02032	.0003243
34	.007	.00003848	.01778	.0002483
35	0.005	0.00001963	0.01270	0.0001267
36	.004	.00001257	.01016	.0000811

Table 4–100 (continued)
DIMENSIONS OF WIRE

British Standard Gauge

Diameter and cross-section are given in English and metric units for the British Standard Gauge.

Gauge No.	Diameter, inches	Section, square inches	Diameter, centimeters	Section, square centimeters
0000000	0.500	0.1963	1.2700	1.267
000000	.464	.1691	1.1786	1.091
00000	0.432	0.1466	1.0973	0.9456
0000	.400	.1257	1.0160	.8107
000	.372	.1087	0.9449	.7012
00	.348	.0951	.8839	.6136
0	0.324	0.0825	0.8230	0.5319
1	.300	.07069	.7620	.4560
2	.276	.05983	.7010	.3858
3	.252	.04988	.6401	.3218
4	.232	.04227	.5893	.2727
5	0.212	0.03530	0.5385	0.2277
6	.192	.02895	.4877	.18679
7	.176	.02433	.4470	.15696
8	.160	.02010	.4064	.12973
9	.144	.01629	.3658	.10507
10	0.128	0.01287	0.3251	0.08302
11	.116	.010568	.2946	.06818
12	.104	.008495	.2642	.05480
13	.092	.006648	.2337	.04289
14	.080	.005027	.2032	.03243
15	0.072	0.004071	0.1829	0.02627
16	.064	.003217	.16256	.020755
17	.056	.002463	.14224	.015890
18	.048	.001810	.12192	.011675
19	.040	.001257	.10160	.008107
20	0.036	0.001018	0.09144	0.006567
21	.032	.0008042	.08128	.005189
22	.028	.0006158	.07112	.003973
23	.024	.0004524	.06096	.002922
24	.022	.0003801	.05588	.002452
25	0.020	0.0003142	0.05080	0.002027
26	.0180	.0002545	.04572	.0016417
27	.0164	.0002112	.04166	.0013628
28	.0148	.0001728	.03759	.0011099
29	.0136	.0001453	.03454	.0009363
30	0.0124	0.0001208	0.03150	0.0007791
31	.0116	.00010568	.02946	.0006818
32	.0108	.00009161	.02743	.0005910
33	.0100	.00007854	.02540	.0005067
34	.0092	.00006648	.02337	.0004289
35	0.0084	0.00005542	0.02134	0.0003575
36	.0076	.00004536	.01930	.0002927
37	.0068	.00003632	.01727	.0002343

Table 4–100 (continued)
DIMENSIONS OF WIRE

British Standard Gauge (continued)

Gauge No.	Diameter, inches	Section, square inches	Diameter, centimeters	Section, square centimeters
38	.0060	.00002827	.01524	.0001824
39	.0052	.00002124	.01321	.0001370
40	0.0048	0.00001810	0.01219	0.0001167
41	.0044	.00001521	.01118	.0000982
42	.0040	.00001257	.01016	.0000811
43	.0036	.00001018	.00914	.0000656
44	.0032	.00000804	.00813	.0000519
45	0.0028	0.00000616	0.00711	0.0000397
46	.0024	.00000452	.00610	.0000212
47	.0020	.00000314	.00508	.0000203
48	.0016	.00000201	.00406	.0000129
49	.0012	.00000113	.00305	.0000073
50	0.0010	0.00000079	0.00254	0.0000051

From Weast, R. C., Ed., *Handbook of Chemistry and Physics,* 55th ed., CRC Press, Cleveland, 1974.

Table 4–101
PLATINUM WIRE

Mass in Grams Per Foot

B. and S. gauge	Diameter, inches	Mass, grams per foot	B. and S. gauge	Diameter, inches	Mass, grams per foot
10	.1019	37.5	23	.02257	1.8
11	.09074	28.0	24	.02010	1.4
12	.08081	22.0	25	.01790	1.1
13	.07196	17.5	26	.01594	0.9
14	.06408	14.0	27	.01420	0.7
15	.05707	11.0	28	.01264	0.6
16	.05082	9.0	29	.01126	0.45
17	.04526	7.0	30	.01003	0.35
18	.04030	5.7	31	.008928	0.28
19	.03589	4.4	32	.007950	0.22
20	.03196	3.4	33	.007080	0.17
21	.02846	2.9	34	.006305	0.15
22	.02535	2.3	35	.005615	0.11

From Weast, R. C., Ed., *Handbook of Chemistry and Physics,* 55th ed., CRC Press, Cleveland, 1974.

Table 4–102

ALLOWABLE CARRYING CAPACITIES OF CONDUCTORS

(National Electrical Code)

The ratings in the following tabulation are those permitted by the National Electrical Code for flexible cords and for interior wiring of houses, hotels, office buildings, industrial plants, and other buildings.

The values are for copper wire. For aluminum wire the allowable carrying capacities shall be taken as 84% of those given in the table for the respective sizes of copper wire with the same kind of covering.

Size, AWG	Area, circular mils[a]	Diameter of solid wires, mils	Rubber insulation, amperes	Varnished cambric insulation, amperes	Other insulations and bare conductors, amperes
18	1,624.	40.3	3[b]	–	6[c]
16	2,583.	50.8	6[b]	–	10[c]
14	4,107.	64.1	15	18	20
12	6,530.	80.8	20	25	30
10	10,380.	101.9	25	30	35
8	16,510.	128.5	35	40	50
6	26,250.	162.0	50	60	70
5	33,100.	181.9	55	65	80
4	41,740.	204.3	70	85	90
3	52,630.	229.4	80	95	100
2	66,370.	257.6	90	110	125
1	83,690.	289.3	100	120	150
0	105,500.	325.0	125	150	200
00	135,100.	364.8	150	180	225
000	167,800.	409.6	175	210	275
0000	211,600.	460	225	270	325

[a]1 mil = 0.001 in.
[b]The allowable carrying capacities of No. 18 and 16 are 5 and 7 A, respectively, when in flexible cords.
[c]The allowable carrying capacities of No. 18 and 16 are 10 and 15 A, respectively, when in cords for portable heaters (types AFS, AFSJ, HC, HPD, and HSJ).

From Weast, R. C., Ed., *Handbook of Chemistry and Physics,* 55th ed., CRC Press, Cleveland, 1974.

Table 4–103
WIRE TABLE, STANDARD ANNEALED COPPER

American Wire Gauge (B. and S.), Metric Units

Gauge No.	Diameter in mils at 20°C	Cross-section at 20°C		Ohms per 1,000 ft[a]			
		Circular mils	Square inches	0°C (32°F)	20°C (68°F)	50°C (122°F)	75°C (167°F)
0000	460.0	211,600	0.1662	0.04516	0.04901	0.05479	0.05961
000	409.6	167,800	.1318	.05695	.06180	.06909	.07516
00	364.8	133,100	.1045	.07181	.07793	.08712	.09478
0	324.9	105,500	.08289	.09055	.09827	.1099	.1195
1	289.3	83,690	.06573	.1142	.1239	.1385	.1507
2	257.6	66,370	.05213	.1440	.1563	.1747	.1900
3	229.4	52,640	.04134	.1816	.1970	.2203	.2396
4	204.3	41,740	.03278	.2289	.2485	.2778	.3022
5	181.9	33,100	.02600	.2887	.3133	.3502	.3810
6	162.0	26,250	.02062	.3640	.3951	.4416	.4805
7	144.3	20,820	.01635	.4590	.4982	.5569	.6059
8	128.5	16,510	.01297	.5788	.6282	.7023	.7640
9	114.4	13,090	.01028	.7299	.7921	.8855	.9633
10	101.9	10,380	.008155	.9203	.9989	1.117	1.215
11	90.74	8,234	.006467	1.161	1.260	1.408	1.532
12	80.81	6,530	.005129	1.463	1.588	1.775	1.931
13	71.96	5,178	.004067	1.845	2.003	2.239	2.436
14	64.08	4,107	.003225	2.327	2.525	2.823	3.071
15	57.07	3,257	.002558	2.934	3.184	3.560	3.873
16	50.82	2,583	.002028	3.700	4.016	4.489	4.884
17	45.26	2,048	.001609	4.666	5.064	5.660	6.158
18	40.30	1,624	.001276	5.883	6.385	7.138	7.765
19	35.89	1,288	.001012	7.418	8.051	9.001	9.792
20	31.96	1,022	.0008023	9.355	10.15	11.35	12.35
21	28.45	810.1	.0006363	11.80	12.80	14.31	15.57
22	25.35	642.4	.0005046	14.87	16.14	18.05	19.63
23	22.57	509.5	.0004002	18.76	20.36	22.76	24.76
24	20.10	404.0	.0003173	23.65	25.67	28.70	31.22
25	17.90	320.4	.0002517	29.82	32.37	36.18	39.36
26	15.94	254.1	.0001996	37.61	40.81	45.63	49.64
27	14.20	201.5	.0001583	47.42	51.47	57.53	62.59
28	12.64	159.8	.0001255	59.80	64.90	72.55	78.93
29	11.26	126.7	.00009953	75.40	81.83	91.48	99.52
30	10.03	100.5	.00007894	95.08	103.2	115.4	125.5
31	8.928	79.70	.00006260	119.9	130.1	145.5	158.2
32	7.950	63.21	.00004964	151.2	164.1	183.4	199.5
33	7.080	50.13	.00003937	190.6	206.9	231.3	251.6
34	6.305	39.75	.00003122	240.4	260.9	291.7	317.3
35	5.015	31.52	.00002476	303.1	329.0	367.8	400.1
36	5.000	25.00	.00001964	382.3	414.8	463.7	504.5
37	4.453	19.83	.00001557	482.0	523.1	584.8	636.2
38	3.965	15.72	.00001235	607.8	659.6	737.4	802.2
39	3.531	12.47	.000009793	766.4	831.8	929.8	1,012
40	3.145	9.888	.000007766	966.5	1,049	1,173	1,276

Table 4–103 (continued)
WIRE TABLE, STANDARD ANNEALED COPPER

American Wire Gauge (B. and S.), Metric Units (continued)

Gauge No.	Pounds per 1,000 ft	Feet per pound	Feet per ohm[b]			
			0°C (32°F)	20°C (68°F)	50°C (122°F)	75°C (167°F)
0000	640.5	1.561	22,140	20,400	18,250	16,780
000	507.9	1.968	17,560	16,180	14,470	13,300
00	402.8	2.482	13,930	12,830	11,480	10,550
0	319.5	3.130	11,040	10,180	9,103	8,367
1	253.3	3.947	8,758	8,070	7,219	6,636
2	200.9	4.977	6,946	6,400	5,725	5,262
3	159.3	6.276	5,508	5,075	4,540	4,173
4	126.4	7.914	4,368	4,025	3,600	3,309
5	100.2	9.980	3,464	3,192	2,855	2,625
6	79.46	12.58	2,747	2,531	2,264	2,081
7	63.02	15.87	2,179	2,007	1,796	1,651
8	49.98	20.01	1,728	1,592	1,424	1,309
9	39.63	25.23	1,370	1,262	1,129	1,038
10	31.43	31.82	1,087	1,001	895.6	823.2
11	24.92	40.12	861.7	794.0	710.2	652.8
12	19.77	50.59	683.3	629.6	563.2	517.7
13	15.68	63.80	541.9	499.3	446.7	410.6
14	12.43	80.44	429.8	396.0	354.2	325.6
15	9.858	101.4	340.8	314.0	280.9	258.2
16	7.818	127.9	270.3	249.0	222.8	204.8
17	6.200	161.3	214.3	197.5	176.7	162.4
18	4.917	203.4	170.0	156.6	140.1	128.8
19	3.899	256.5	134.8	124.2	111.1	102.1
20	3.092	323.4	106.9	98.50	88.11	80.99
21	2.452	407.8	84.78	78.11	69.87	64.23
22	1.945	514.2	67.23	61.95	55.41	50.94
23	1.542	648.4	53.32	49.13	43.94	40.39
24	1.223	817.7	42.28	38.96	34.85	32.03
25	0.9699	1,031	33.53	30.90	27.64	25.40
26	.7692	1,300	26.59	24.50	21.92	20.15
27	.6100	1,639	21.09	19.43	17.38	15.98
28	.4837	2,067	16.72	15.41	13.78	12.67
29	.3836	2,607	13.26	12.22	10.93	10.05
30	.3042	3,287	10.52	9.691	8.669	7.968
31	.2413	4,145	8.341	7.685	6.875	6.319
32	.1913	5,227	6.614	6.095	5.452	5.011
33	.1517	6,591	5.245	4.833	4.323	3.974
34	.1203	8,310	4.160	3.833	3.429	3.152
35	.09542	10,480	3.299	3.040	2.719	2.499
36	.07568	13,210	2.616	2.411	2.156	1.982
37	.06001	16,660	2.075	1.912	1.710	1.572
38	.04759	21,010	1.645	1.516	1.356	1.247
39	.03774	26,500	1.305	1.202	1.075	0.9886
40	.02993	33,410	1.035	0.9534	0.8529	.7840

Table 4–103 (continued)
WIRE TABLE, STANDARD ANNEALED COPPER

American Wire Gauge (B. and S.), Metric Units (continued)

Gauge No.	Diameter in mils at 20° C	Ohms per pound			Pounds per ohm
		0° C (32° F)	20° C (68° F)	50° C (122° F)	20° C (68° F)
0000	460.0	0.00007051	0.00007652	0.00008554	13,070
000	409.6	.0001121	.0001217	.0001360	8,219
00	364.8	.0001783	.0001935	.0002163	5,169
0	324.9	.0002835	.0003076	.0003439	3,251
1	289.3	.0004507	.0004891	.0005468	2,044
2	257.6	.0007166	.0007778	.0008695	1,286
3	229.4	.001140	.001237	.001383	808.6
4	204.3	.001812	.001966	.002198	508.5
5	181.9	.002881	.003127	.003495	319.8
6	162.0	.004581	.004972	.005558	201.1
7	144.3	.007284	.007905	.008838	126.5
8	128.5	.01158	.01257	.01405	79.55
9	114.4	.01842	.01999	.02234	50.03
10	101.9	.02928	.03178	.03553	31.47
11	90.74	.04656	.05053	.05649	19.79
12	80.81	.07404	.08035	.08983	12.45
13	71.96	.1177	.1278	.1428	7.827
14	64.08	.1872	.2032	.2271	4.922
15	57.07	.2976	.3230	.3611	3.096
16	50.82	.4733	.5136	.5742	1.947
17	45.26	.7525	.8167	.9130	1.224
18	40.30	1.197	1.299	1.452	0.7700
19	35.89	1.903	2.065	2.308	.4843
20	31.96	3.025	3.283	3.670	.3046
21	28.46	4.810	5.221	5.836	.1915
22	25.35	7.649	8.301	9.280	.1205
23	22.57	12.16	13.20	14.76	.07576
24	20.10	19.34	20.99	23.46	.04765
25	17.90	30.75	33.37	37.31	.02997
26	15.94	48.89	53.06	59.32	.01885
27	14.20	77.74	84.37	94.32	.01185
28	12.64	123.6	134.2	150.0	.007454
29	11.26	196.6	213.3	238.5	.004688
30	10.03	312.5	339.2	379.2	.002948
31	8.928	497.0	539.3	602.9	.001854
32	7.950	790.2	857.6	958.7	.001166
33	7.080	1,256	1,364	1,524	.0007333
34	6.305	1,998	2,168	2,424	.0004612
35	5.615	3,177	3,448	3,854	.0002901
36	5.000	5,051	5,482	6,128	.0001824
37	4.453	8,032	8,717	9,744	.0001147
38	3.965	12,770	13,860	15,490	.00007215
39	3.531	20,310	22,040	24,640	.00004538
40	3.145	32,290	35,040	39,170	.00002854

Table 4–103 (continued)
WIRE TABLE, STANDARD ANNEALED COPPER

American Wire Gauge (B. and S.), Metric Units (continued)

Gauge No.	Diameter in millimeters at 20°C	Cross-section in square millimeters at 20°C	Ohms per kilometer[c]			
			0°C	20°C	50°C	75°C
0000	11.68	107.2	0.1482	0.1608	0.1798	0.1956
000	10.40	85.03	.1868	.2028	.2267	.2466
00	9.266	67.43	.2356	.2557	.2858	.3110
0	8.252	53.48	.2971	.3224	.3604	.3921
1	7.348	42.41	.3746	.4066	.4545	.4944
2	6.544	33.63	.4724	.5127	.5731	.6235
3	5.827	26.67	.5956	.6465	.7227	.7862
4	5.189	21.15	.7511	.8152	.9113	.9914
5	4.621	16.77	.9471	1.028	1.149	1.250
6	4.115	13.30	1.194	1.296	1.449	1.576
7	3.665	10.55	1.506	1.634	1.827	1.988
8	3.264	8.366	1.899	2.061	2.304	2.506
9	2.906	6.634	2.395	2.599	2.905	3.161
10	2.588	5.261	3.020	3.277	3.663	3.985
11	2.305	4.172	3.807	4.132	4.619	5.025
12	2.053	3.309	4.801	5.211	5.825	6.337
13	1.828	2.624	6.054	6.571	7.345	7.991
14	1.628	2.081	7.634	8.285	9.262	10.08
15	1.450	1.650	9.627	10.45	11.68	12.71
16	1.291	1.309	12.14	13.17	14.73	16.02
17	1.150	1.038	15.31	16.61	18.57	20.20
18	1.024	0.8231	19.30	20.95	23.42	25.48
19	0.9116	.6527	24.34	26.42	29.53	32.12
20	.8118	.5176	30.69	33.31	37.24	40.51
21	.7230	.4105	38.70	42.00	46.95	51.08
22	.6438	.3255	48.80	52.96	59.21	64.41
23	.5733	.2582	61.54	66.79	74.66	81.22
24	.5106	.2047	77.60	84.21	94.14	102.4
25	.4547	.1624	97.85	106.2	118.7	129.1
26	.4049	.1288	123.4	133.9	149.7	162.9
27	.3606	.1021	155.6	168.9	188.8	205.4
28	.3211	.08098	196.2	212.9	238.0	258.9
29	.2859	.06422	247.4	268.5	300.1	326.5
30	.2546	.05093	311.9	338.6	378.5	411.7
31	.2268	.04039	393.4	426.9	477.2	519.2
32	.2019	.03203	496.0	538.3	601.8	654.7
33	.1798	.02540	625.5	678.8	758.8	825.5
34	.1601	.02014	788.7	856.0	956.9	1,041
35	.1426	.01597	994.5	1,079	1,207	1,313
36	.1270	.01267	1,254	1,361	1,522	1,655
37	.1131	.01005	1,581	1,716	1,919	2,087
38	.1007	.007967	1,994	2,164	2,419	2,632
39	.08969	.006318	2,514	2,729	3,051	3,319
40	.07987	.005010	3,171	3,441	3,847	4,185

Table 4–103 (continued)
WIRE TABLE, STANDARD ANNEALED COPPER

American Wire Gauge (B. and S.), Metric Units (continued)

Gauge No.	Diameter in millimeters at 20°C	Kilograms per kilometer	Meters per gram	Meters per ohm[d]			
				0°C	20°C	50°C	75°C
0000	11.68	953.2	0.001049	6,749	6,219	5,563	5,113
000	10.40	755.9	.001323	5,352	4,932	4,412	4,055
00	9.266	599.5	.001668	4,245	3,911	3,499	3,216
0	8.252	475.4	.002103	3,366	3,102	2,774	2,550
1	7.348	377.0	.002652	2,669	2,460	2,200	2,022
2	6.544	299.0	.003345	2,117	1,951	1,745	1,604
3	5.827	237.1	.004217	1,679	1,547	1,384	1,272
4	5.189	188.0	.005318	1,331	1,227	1,097	1,009
5	4.621	149.1	.006706	1,056	972.9	870.2	799.9
6	4.115	118.2	.008457	837.3	771.5	690.1	634.4
7	3.665	93.78	.01066	664.0	611.8	547.3	503.1
8	3.264	74.37	.01345	526.6	485.2	434.0	399.0
9	2.906	58.98	.01696	417.6	384.8	344.2	316.4
10	2.588	46.77	.02138	331.2	305.1	273.0	250.9
11	2.305	37.09	.02696	262.2	242.0	216.5	199.0
12	2.053	29.42	.03400	208.3	191.9	171.7	157.8
13	1.828	23.33	.04287	165.2	152.2	136.1	125.1
14	1.628	18.50	.05406	131.0	120.7	108.0	99.24
15	1.450	14.67	.06816	103.9	95.71	85.62	78.70
16	1.291	11.63	.08595	82.38	75.90	67.90	62.41
17	1.150	9.226	.1084	65.33	60.20	53.85	49.50
18	1.024	7.317	.1367	51.81	47.74	42.70	39.25
19	0.9116	5.803	.1723	41.09	37.86	33.86	31.13
20	.8118	4.602	.2173	32.58	30.02	26.86	25.69
21	.7230	3.649	.2740	25.84	23.81	21.30	19.58
22	.6438	2.894	.3455	20.49	18.88	16.89	15.53
23	.5733	2.295	.4357	16.25	14.97	13.39	12.31
24	.5106	1.820	.5494	12.89	11.87	10.62	9.764
25	.4547	1.443	.6928	10.22	9.417	8.424	7.743
26	.4049	1.145	.8736	8.105	7.468	6.680	6.141
27	.3606	0.9078	1.102	6.428	5.922	5.298	4.870
28	.3211	.7199	1.389	5.097	4.697	4.201	3.862
29	.2859	.5709	1.752	4.042	3.725	3.332	3.063
30	.2546	.4527	2.209	3.206	2.954	2.642	2.429
31	.2268	.3590	2,785	2.542	2.342	2.095	1.926
32	.2019	.2847	3.512	2.016	1.858	1.662	1.527
33	.1798	.2258	4.429	1.599	1.473	1.318	1.211
34	.1601	.1791	5.584	1.268	1.168	1.045	0.9606
35	.1426	.1420	7.042	1.006	0.9265	0.8288	.7618
36	.1270	.1126	8.879	0.7974	.7347	.6572	.6041
37	.1131	.08931	11.20	.6324	.5827	.5212	.4791
38	.1007	.07083	14.12	.5015	.4621	.4133	.3799
39	.08969	.05617	17.80	.3977	.3664	.3278	.3013
40	.07987	.04454	22.45	.3154	.2906	.2600	.2390

Table 4–103 (continued)
WIRE TABLE, STANDARD ANNEALED COPPER

American Wire Gauge (B. and S.), Metric Units (continued)

Gauge No.	Ohms per kilogram			Grams per ohm
	0°C	20°C	50°C	20°C
0000	0.0001554	0.0001687	0.0001886	5,928,000
000	.0002472	.0002682	.0002999	3,728,000
00	.0003930	.0004265	.0004768	2,344,000
0	.0006249	.0006782	.0007582	1,474,000
1	.0009936	.001078	.001206	927,300
2	.001580	.001715	.001917	583,200
3	.002512	.002726	.003048	366,800
4	.003995	.004335	.004846	230,700
5	.006352	.006893	.007706	145,100
6	.01010	.01096	.01225	91,230
7	.01606	.01743	.01948	57,380
8	.02553	.02771	.03098	36,080
9	.04060	.04406	.04926	22,690
10	.06456	.07007	.07833	14,270
11	.1026	.1114	.1245	8,976
12	.1632	.1771	.1980	5,645
13	.2595	.2817	.3149	3,550
14	.4127	.4479	.5007	2,233
15	.6562	.7122	.7961	1,404
16	1.043	1.132	1.266	883.1
17	1.659	1.801	2.013	555.4
18	2.638	2.863	3.201	349.3
19	4.194	4.552	5.089	219.7
20	6.670	7.238	8.092	138.2
21	10.60	11.51	12.87	86.88
22	16.86	18.30	20.46	54.64
23	26.81	29.10	32.53	34.36
24	42.63	46.27	51.73	21.61
25	67.79	73.57	82.25	13.59
26	107.8	117.0	130.8	8.548
27	171.4	186.0	207.9	5.376
28	272.5	295.8	330.6	3.381
29	433.3	470.3	525.7	2.126
30	689.0	747.8	836.0	1.337
31	1,096	1,189	1,329	0.8410
32	1,742	1,891	2,114	.5289
33	2,770	3,006	3,361	.3326
34	4,404	4,780	5,344	.2092
35	7,003	7,601	8,497	.1316
36	11,140	12,090	13,510	.08274
37	17,710	19,220	21,480	.05204
38	28,150	30,560	34,160	.03273
39	44,770	48,590	54,310	.02058
40	71,180	77,260	86,360	.01294

Table 4–103 (continued)
WIRE TABLE, STANDARD ANNEALED COPPER

American Wire Gauge (B. and S.), Metric Units (continued)

[a]Resistance at the stated temperatures of a wire with a length of 1,000 ft at 20°C.
[b]Length at 20°C of a wire with a resistance of 1 Ω at the stated temperature.
[c]Resistance at the stated temperatures of a wire with a length of 1 km at 20°C.
[d]Length at 20°C of a wire with a resistance of 1 Ω at the stated temperatures.

From Weast, R. C., Ed., *Handbook of Chemistry and Physics*, 55th ed., CRC Press, Cleveland, 1974.

Table 4–104
ALUMINUM WIRE TABLE

Hard-drawn Aluminum Wire at 20°C (or 68°F) American Wire Gauge (B. and S.), English Units

Gauge No.	Diameter in mils	Cross-section		Ohms per 1,000 ft	Pounds per 1,000 ft	Pounds per ohm	Feet per ohm
		Circular mils	Square inches				
0000	460	212,000	0.166	0.0804	195	2,420	12,400
000	410	168,000	.132	.101	154	1,520	9,860
00	365	133,000	.105	.128	122	957	7,820
0	325	106,000	.0829	.161	97.0	602	6,200
1	289	83,700	.0657	.203	76.9	379	4,920
2	258	66,400	.0521	.256	61.0	238	3,900
3	229	52,600	.0413	.323	48.4	150	3,090
4	204	41,700	.0328	.408	38.4	94.2	2,450
5	182	33,100	.0260	.514	30.4	59.2	1,950
6	162	26,300	.0206	.648	24.1	37.2	1,540
7	144	20,800	.0164	.817	19.1	23.4	1,220
8	128	16,500	.0130	1.03	15.2	14.7	970
9	114	13,100	.0103	1.30	12.0	9.26	770
10	102	10,400	.00815	1.64	9.55	5.83	610
11	91	8,230	.00647	2.07	7.57	3.66	484
12	81	6,530	.00513	2.61	6.00	2.30	384
13	72	5,180	.00407	3.29	4.76	1.45	304
14	64	4,110	.00323	4.14	3.78	0.911	241
15	57	3,260	.00256	5.22	2.99	.573	191
16	51	2,580	.00203	6.59	2.37	.360	152
17	45	2,050	.00161	8.31	1.88	.227	120
18	40	1,620	.00128	10.5	1.49	.143	95.5
19	36	1,290	.00101	13.2	1.18	.0897	75.7
20	32	1,020	.000802	16.7	0.939	.0564	60.0
21	28.5	810	.000636	21.0	.745	.0355	47.6
22	25.3	642	.000505	26.5	.591	.0223	37.8
23	22.6	509	.000400	33.4	.468	.0140	29.9
24	20.1	404	.000317	42.1	.371	.00882	23.7
25	17.9	320	.000252	53.1	.295	.00555	18.8
26	15.9	254	.000200	67.0	.234	.00349	14.9
27	14.2	202	.000158	84.4	.185	.00219	11.8

Table 4–104 (continued)
ALUMINUM WIRE TABLE

Hard-drawn Aluminum Wire at 20°C (or 68°F) American Wire Gauge (B. and S.), English Units (continued)

Gauge No.	Diameter in mils	Cross-section Circular mils	Cross-section Square inches	Ohms per 1,000 ft	Pounds per 1,000 ft	Pounds per ohm	Feet per ohm
28	12.6	160	.000126	106.	.147	.00138	9.39
29	11.3	127	.0000995	134.	.117	.000868	7.45
30	10.0	101	.0000789	169.	.0924	.000546	5.91
31	8.9	79.7	.0000626	213.	.0733	.000343	4.68
32	8.0	63.2	.0000496	269.	.0581	.000216	3.72
33	7.1	50.1	.0000394	339.	.0461	.000136	2.95
34	6.3	39.8	.0000312	428.	.0365	.0000854	2.34
35	5.6	31.5	.0000248	540.	.0290	.0000537	1.85
36	5.0	25.0	.0000196	681.	.0230	.0000338	1.47
37	4.5	19.8	.0000156	858.	.0182	.0000212	1.17
38	4.0	15.7	.0000123	1080.	.0145	.0000134	0.924
39	3.5	12.5	.00000979	1360.	.0115	.00000840	.733
40	3.1	9.9	.00000777	1720.	.0091	.00000528	.581

Gauge No.	Diameter in millimeters	Cross-section in square millimeters	Ohms per kilometer	Kilograms per kilometer	Grams per ohm	Meters per ohm
0000	11.7	107	0.264	289	1,100,000	3,790
000	10.4	85.0	.333	230	690,000	3,010
00	9.3	67.4	.419	182	434,000	2,380
0	8.3	53.5	.529	144	273,000	1,890
1	7.3	42.4	.667	114.	172,000	1,500
2	6.5	33.6	.841	90.8	108,000	1,190
3	5.8	26.7	1.06	72.0	67,900	943
4	5.2	21.2	1.34	57.1	42,700	748
5	4.6	16.8	1.69	45.3	26,900	593
6	4.1	13.3	2.13	35.9	16,900	470
7	3.7	10.5	2.68	28.5	10,600	373
8	3.3	8.37	3.38	22.6	6,680	296
9	2.91	6.63	4.26	17.9	4,200	235
10	2.59	5.26	5.38	14.2	2,640	186
11	2.30	4.17	6.78	11.3	1,660	148
12	2.05	3.31	8.55	8.93	1,050	117
13	1.83	2.62	10.8	7.08	657	92.8
14	1.63	2.08	13.6	5.62	413	73.6
15	1.45	1.65	17.1	4.46	260	58.4
16	1.29	1.31	21.6	3.53	164	46.3
17	1.15	1.04	27.3	2.80	103	36.7
18	1.02	0.823	34.4	2.22	64.7	29.1
19	0.91	.653	43.3	1.76	40.7	23.1
20	.81	.518	54.6	1.40	25.6	18.3
21	.72	.411	68.9	1.11	16.1	14.5
22	.64	.326	86.9	0.879	10.1	11.5

Table 4–104 (continued)
ALUMINUM WIRE TABLE

Hard-drawn Aluminum Wire at 20°C (or 68°F) American Wire Gauge (B. and S.), English Units (continued)

Gauge No.	Diameter in millimeters	Cross-section in square millimeters	Ohms per kilometer	Kilograms per kilometer	Grams per ohm	Meters per ohm
23	.57	.258	110	.697	6.36	9.13
24	.51	.205	138	.553	4.00	7.24
25	.45	.162	174	.438	2.52	5.74
26	.40	.129	220	.348	1.58	4.55
27	.36	.102	277	.276	0.995	3.61
28	.32	.0810	349	.219	.626	2.86
29	.29	.0642	440	.173	.394	2.27
30	.25	.0509	555	.138	.248	1.80
31	.227	.0404	700	.109	.156	1.43
32	.202	.0320	883	.0865	.0979	1.13
33	.180	.0254	1110	.0686	.0616	0.899
34	.160	.0201	1400	.0544	.0387	.712
35	.143	.0160	1770	.0431	.0244	.565
36	.127	.0127	2230	.0342	.0153	.448
37	.113	.0100	2820	.0271	.00963	.355
38	.101	.0080	3550	.0215	.00606	.282
39	.090	.0063	4480	.0171	.00381	.223
40	.080	.0050	5640	.0135	.00240	.177

From Weast, R. C., Ed., *Handbook of Chemistry and Physics,* 55th ed., CRC Press, Cleveland, 1974.

Table 4–105
CROSS-SECTION AND MASS OF WIRES
U.S. Measure

Diameters are given in mils (1 mil = .001 in.), and area in square mils (1 mil^2 = .000001 in.^{2}). For sections and masses for one tenth the diameters given, divide by 100; for sections and masses ten times the diameter multiply by 100.

Diameter in mils	Cross-section in square mils	Pounds per foot			
		Copper, density 8.90	Iron, density 7.80	Brass, density 8.56	Aluminum, density 2.67
10	78.54	0.000303	0.0002656	0.0002915	0.0000909
11	95.03	0367	03214	03527	01100
12	113.10	0436	03825	04197	01309
13	132.73	0512	04488	04926	01536
14	153.94	0594	05206	05713	01782
15	176.71	0.000682	0.0005976	0.0006558	0.0002045
16	201.06	0776	06799	07461	02327
17	226.98	0876	07675	08423	02627

Table 4–105 (continued)
CROSS-SECTION AND MASS OF WIRES

U.S. Measure (continued)

Pounds per foot

Diameter in mils	Cross-section in square mils	Copper, density 8.90	Iron, density 7.80	Brass, density 8.56	Aluminum, density 2.67
18	254.47	0982	08605	09443	02946
19	283.53	1094	09588	10522	03282
20	314.16	0.001212	0.001062	0.01166	0.0003636
21	346.36	1336	1171	1285	04009
22	380.13	1467	1286	1411	04400
23	415.48	1603	1405	1542	04809
24	452.39	1746	1530	1679	05237
25	490.87	0.001894	0.001660	0.001822	0.0005682
26	530.93	2046	1795	1970	06147
27	572.56	2209	1936	2125	06628
28	615.75	2376	2082	2285	07127
29	660.52	2549	2234	2451	07646
30	706.86	0.002727	0.002390	0.002623	0.0008182
31	754.77	2912	2552	2801	08737
32	804.25	3103	2720	2985	09309
33	855.30	3300	2892	3174	09900
34	907.92	3503	3070	3369	10509
35	962.11	0.003712	0.003253	0.003570	0.001114
36	1,017.88	3927	3442	3777	1178
37	1,075.21	4149	3636	3990	1245
38	1,134.11	4376	3844	4218	1316
39	1,194.59	4609	4040	4433	1383
40	1,256.64	0.004849	0.004249	0.004664	0.001455
41	1,320.25	5094	4465	4900	1528
42	1,385.44	5346	4685	5141	1604
43	1,452.20	5603	4911	5389	1681
44	1,520.53	5867	5142	5643	1760
45	1,590.43	0.006137	0.005378	0.005902	0.001841
46	1,661.90	6412	5620	6167	1924
47	1,734.94	6694	5867	6438	2008
48	1,809.56	6982	6119	6715	2095
49	1,885.74	7276	6377	6998	2183
50	1,963.50	0.007576	0.006640	0.007287	0.002273
51	2,042.82	7882	6098	7581	2365
52	2,123.72	8194	7181	7881	2458
53	2,206.18	8512	7460	8187	2554
54	2,290.22	8837	7744	8499	2651
55	2,375.83	0.009167	0.008034	0.008817	0.002750
56	2,463.01	09504	08329	09140	2851
57	2,551.76	09846	08629	09470	2954
58	2,642.08	10195	08934	09805	3058
59	2,733.97	10549	09245	10146	3165

Table 4–105 (continued)
CROSS-SECTION AND MASS OF WIRES

U.S. Measure (continued)

		Pounds per foot			
Diameter in mils	Cross-section in square mils	Copper, density 8.90	Iron, density 7.80	Brass, density 8.56	Aluminum, density 2.67
60	2,827.43	0.01091	0.00956	0.01049	0.003273
61	2,922.47	1128	0988	1085	3383
62	3,019.07	1165	1021	1120	3495
63	3,117.25	1203	1054	1157	3608
64	3,216.99	1241	1088	1194	3724
65	3,318.31	0.01280	0.01122	0.01231	0.003841
66	3,421.19	1320	1157	1270	3960
67	3,525.65	1360	1192	1308	4081
68	3,631.68	1401	1228	1348	4204
69	3,739.28	1443	1264	1388	4328
70	3,848.45	0.01485	0.01302	0.01429	0.004456
71	3,959.19	1528	1339	1469	4583
72	4,071.50	1571	1377	1511	4713
73	4,185.39	1615	1415	1553	4845
74	4,300.84	1660	1454	1596	4978
75	4,417.86	0.01705	0.01494	0.01639	0.005114
76	4,536.46	1751	1534	1684	5251
77	4,656.63	1797	1575	1728	5390
78	4,778.36	1844	1616	1773	5531
79	4,901.67	1892	1658	1819	5674
80	5,026.55	0.01939	0.01700	0.01865	0.005818
81	5,153.00	1988	1743	1912	5965
82	5,281.02	2038	1786	1960	6113
83	5,410.61	2088	1830	2008	6263
84	5,541.77	2138	1874	2057	6415
85	5,674.50	0.02189	0.01919	0.02106	0.006568
86	5,808.80	2241	1964	2156	6724
87	5,944.68	2294	2010	2206	6881
88	6,082.12	2347	2057	2257	7040
89	6,221.14	2400	2104	2309	7201
90	6,361.73	0.02455	0.02151	0.02360	0.007364
91	6,503.88	2509	2199	2414	7528
92	6,647.61	2565	2248	2467	7695
93	6,792.91	2621	2297	2521	7863
94	6,939.78	2678	2347	2575	8033
95	7,088.22	0.02735	0.02397	0.02630	0.008205
96	7,238.23	2793	2448	2686	8378
97	7,389.81	2851	2499	2742	8554
98	7,542.96	2910	2551	2799	8731
99	7,697.69	2970	2603	2857	8910
100	7,853.98	0.03030	0.02656	0.02915	0.009091

Table 4–105 (continued)
CROSS-SECTION AND MASS OF WIRES

Metric Measure

Diameters are given in thousandths of a centimeter and area of section in square thousandths of a centimeter; 1 $(cm/1000)^2$ = .000001 cm^2. For sections and masses for diameters one tenth or ten times those of the table, divide or multiply by 100.

Diameter in thousandths of a centimeter	Cross-section in square thousandths of a centimeter	Grams per meter			
		Copper, density 8.90	Iron, density 7.80	Brass, density 8.56	Aluminum, density 2.67
10	78.54	0.06990	0.06126	0.06723	0.02097
11	95.03	.08458	.07412	.08135	.02537
12	113.10	.10065	.08822	.09681	.03020
13	132.73	.11813	.10353	.11362	.03544
14	153.94	.13701	.12008	.13177	.04110
15	176.71	0.1573	0.1378	0.1513	0.04718
16	201.06	.1789	.1568	.1721	.05368
17	226.98	.2020	.1770	.1943	.06060
18	254.47	.2265	.1985	.2178	.06794
19	283.53	.2523	.2212	.2427	.07570
20	314.16	0.2796	0.2450	0.2689	0.08388
21	346.36	.3083	.2702	.2965	.09248
22	380.13	.3383	.2965	.3254	.10149
23	415.48	.3698	.3241	.3557	.11093
24	452.39	.4026	.3529	.3872	.12079
25	490.87	0.4369	0.3829	0.4202	0.1311
26	530.93	.4725	.4141	.4545	.1418
27	572.56	.5096	.4466	.4901	.1529
28	615.75	.5480	.4803	.5271	.1644
29	660.52	.5879	.5152	.5654	.1764
30	706.86	0.6291	0.5514	0.6051	0.1887
31	754.77	.6717	.5887	.6461	.2015
32	804.25	.7158	.6273	.6884	.2147
33	855.30	.7612	.6671	.7321	.2284
34	907.92	.8081	.7082	.7772	.2424
35	962.11	0.856	0.7504	0.8236	0.2569
36	1,017.88	.906	.7939	.8713	.2718
37	1,075.21	.957	.8387	.9204	.2871
38	1,134.11	1.012	.8866	.9730	.3035
39	1,194.59	.063	.9318	1.0230	.3190
40	1,256.64	1.118	0.980	1.076	0.3355
41	1,320.25	.175	1.030	.130	.3525
42	1,385.44	.233	.081	.186	.3699
43	1,452.20	.292	.133	.243	.3877
44	1,520.53	.353	.186	.302	.4060
45	1,590.43	1.415	1.241	1.361	0.4246
46	1,661.90	.479	.296	.423	.4437
47	1,734.94	.544	.353	.485	.4632
48	1,809.56	.611	.411	.549	.4832

Table 4-105 (continued)
CROSS-SECTION AND MASS OF WIRES

Metric Measure (continued)

Diameter in thousandths of a centimeter	Cross-section in square thousandths of a centimeter	Grams per meter			
		Copper, density 8.90	Iron, density 7.80	Brass, density 8.56	Aluminum, density 2.67
49	1,885.74	.678	.471	.614	.5035
50	1,963.50	1.748	1.532	1.681	.5243
51	2,042.82	.818	.593	.753	.5454
52	2,123.72	.890	.657	.818	.5670
53	2,206.18	.964	.721	.888	.5891
54	2,290.22	2.038	.786	.960	.6115
55	2,375.83	2.114	1.853	2.034	0.6343
56	2,463.01	.192	.921	.108	.6576
57	2,551.76	.271	.990	.184	.6813
58	2,642.08	.351	2.061	.262	.7054
59	2,733.97	.433	.132	.340	.7300
60	2,827.43	2.516	2.205	2.420	0.7549
61	2,922.47	.601	.280	.502	.7803
62	3,019.07	.687	.355	.584	.8061
63	3,117.25	.774	.431	.668	.8323
64	3,216.90	.863	.509	.760	.8589
65	3,318.31	2.953	2.588	2.840	0.8860
66	3,421.19	3.045	.669	.929	.9135
67	3,525.65	.138	.750	3.018	.9413
68	3,631.68	.232	.833	.109	.9697
69	3,739.28	.328	.917	.201	.9984
70	3,848.45	3.426	3.003	3.295	1.028
71	3,959.19	.524	.088	.389	.057
72	4,071.50	.624	.176	.485	.087
73	4,185.39	.725	.265	.583	.117
74	4,300.84	.828	.355	.682	.148
75	4,417.86	3.932	3.446	3.782	1.180
76	4,536.46	4.037	.538	.883	.211
77	4,656.63	.144	.632	.986	.243
78	4,778.36	.253	.727	4.090	.276
79	4,901.67	.362	.823	.177	.309
80	5,026.55	4.474	3.921	4.303	1.342
81	5,153.00	.586	4.019	.411	.376
82	5,281.02	.700	.119	.521	.410
83	5,410.61	.815	.220	.631	.445
84	5,541.77	.932	.323	.744	.480
85	5,674.50	5.050	4.426	4.857	1.515
86	5,808.80	.170	.531	.972	.551
87	5,944.68	.291	.637	5.089	.587
88	6,082.12	.413	.744	.206	.624
89	6,221.14	.537	.852	.325	.661

Table 4–105 (continued)
CROSS-SECTION AND MASS OF WIRES

Metric Measure (continued)

		Grams per meter			
Diameter in thousandths of a centimeter	Cross-section in square thousandths of a centimeter	Copper, density 8.90	Iron, density 7.80	Brass, density 8.56	Aluminum, density 2.67
90	6,361.73	5.662	4.962	5.446	1.699
91	6,503.88	.788	5.073	.567	.737
92	6,647.61	.916	.185	.690	.775
93	6,792.91	6.046	.298	.815	.814
94	6,939.78	.176	.413	.940	.853
95	7,088.22	6.309	5.529	6.068	1.893
96	7,238.23	.442	.646	.196	.933
97	7,389.81	.577	.764	.326	.973
98	7,542.96	.713	.884	.457	2.014
99	7,697.69	.851	6.004	.589	.055
100	7,853.98	6.990	6.126	6.723	2.097

From Weast, R. C., Ed., *Handbook of Chemistry and Physics*, 55th ed., CRC Press, Cleveland, 1974.

Table 4–106
RESISTANCE OF WIRES

The following table gives the approximate resistance of various metallic conductors. The values have been computed from the resistivities at 20°C, except as otherwise stated, and for the dimensions of wire indicated. Owing to differences in purity in the case of elements and of composition in alloys, the values can be considered only as approximations.

The following dimensions have been adopted in the computations.

	Diameter			Diameter	
B. and S. gauge	Millimeters	Mils (1 mil = .001 in.)	B. and S. gauge	Millimeters	Mils (1 mil = .001 in.)
10	2.588	101.9	26	0.4049	15.94
12	2.053	80.81	27	0.3606	14.20
14	1.628	64.08	28	0.3211	12.64
16	1.291	50.82	30	0.2546	10.03
18	1.024	40.30	32	0.2019	7.950
20	0.8118	31.96	34	0.1601	6.305
22	0.6438	25.35	36	0.1270	5.000
24	0.5106	20.10	40	0.07987	3.145

Table 4–106 (continued)
RESISTANCE OF WIRES

B. and S. No.	Ohms per centimeter	Ohms per foot	B. and S. No.	Ohms per centimeter	Ohms per foot
Advance® (0°C) ρ = 48. × 10^{-6} ohm-cm			Aluminum ρ = 2.828 × 10^{-6} ohm-cm		
10	.000912	.0278	10	.0000538	.00164
12	.00145	.0442	12	.0000855	.00260
14	.00231	.0703	14	.000136	.00414
16	.00367	.112	16	.000216	.00658
18	.00583	.178	18	.000344	.0105
20	.00927	.283	20	.000546	.0167
22	.0147	.449	22	.000869	.0265
24	.0234	.715	24	.00138	.0421
26	.0373	1.14	26	.00220	.0669
27	.0470	1.43	27	.00277	.0844
28	.0593	1.81	28	.00349	.106
30	.0942	2.87	30	.00555	.169
32	.150	4.57	32	.00883	.269
34	.238	7.26	34	.0140	.428
36	.379	11.5	36	.0223	.680
40	.958	29.2	40	.0564	1.72
Brass ρ = 7.00 × 10^{-6} ohm-cm			Climax® ρ = 87. × 10^{-6} ohm-cm		
10	.000133	.00406	10	.00165	.0504
12	.000212	.00645	12	.00263	.0801
14	.000336	.0103	14	.00418	.127
16	.000535	.0163	16	.00665	.203
18	.000850	.0259	18	.0106	.322
20	.00135	.0412	20	.0168	.512
22	.00215	.0655	22	.0267	.815
24	.00342	.104	24	.0425	1.30
26	.00543	.166	26	.0675	2.06
27	.00686	.209	27	.0852	2.60
28	.00864	.263	28	.107	3.27
30	.0137	.419	30	.171	5.21
32	.0219	.666	32	.272	8.28
34	.0348	1.06	34	.432	13.2
36	.0552	1.68	36	.687	20.9
40	.140	4.26	40	1.74	52.9
Constantan (0°C) ρ = 44.1 × 10^{-6} ohm-cm			Copper, annealed ρ = 1.724 × 10^{-6} ohm-cm		
10	.000838	.0255	10	.0000328	.000999
12	.00133	.0406	12	.0000521	.00159
14	.00212	.0646	14	.0000828	.00253
16	.00337	.103	16	.000132	.00401
18	.00536	.163	18	.000209	.00638
20	.00852	.260	20	.000333	.0102
22	.0135	.413	22	.000530	.0161
24	.0215	.657	24	.000842	.0257

Table 4–106 (continued)
RESISTANCE OF WIRES

B. and S. No.	Ohms per centimeter	Ohms per foot	B. and S. No.	Ohms per centimeter	Ohms per foot
Constantan (0°C) $\rho = 44.1 \times 10^{-6}$ ohm-cm			Copper, annealed $\rho = 1.724 \times 10^{-6}$ ohm-cm		
26	.0342	1.04	26	.00134	.0408
27	.0432	1.32	27	.00169	.0515
28	.0545	1.66	28	.00213	.0649
30	.0866	2.64	30	.00339	.103
32	.138	4.20	32	.00538	.164
34	.219	6.67	34	.00856	.261
36	.348	10.6	36	.0136	.415
40	.880	26.8	40	.0344	1.05
Eureka® (0°C) $\rho = 47. \times 10^{-6}$ ohm-cm			Excello $\rho = 92. \times 10^{-6}$ ohm-cm		
10	.000893	.0272	10	.00175	.0533
12	.00142	.0433	12	.00278	.0847
14	.00226	.0688	14	.00442	.135
16	.00359	.109	16	.00703	.214
18	.00571	.174	18	.0112	.341
20	.00908	.277	20	.0178	.542
22	.0144	.440	22	.0283	.861
24	.0230	.700	24	.0449	1.37
26	.0365	1.11	26	.0714	2.18
27	.0460	1.40	27	.0901	2.75
28	.0580	1.77	28	.114	3.46
30	.0923	2.81	30	.181	5.51
32	.147	4.47	32	.287	8.75
34	.233	7.11	34	.457	13.9
36	.371	11.3	36	.726	22.1
40	.938	28.6	40	1.84	56.0
German silver $\rho = 33. \times 10^{-6}$ ohm-cm			Gold $\rho = 2.44 \times 10^{-6}$ ohm-cm		
10	.000627	.0191	10	.0000464	.00141
12	.000997	.0304	12	.0000737	.00225
14	.00159	.0483	14	.000117	.00357
16	.00252	.0768	16	.000186	.00568
18	.00401	.122	18	.000296	.00904
20	.00638	.194	20	.000471	.0144
22	.0101	.309	22	.000750	.0228
24	.0161	.491	24	.00119	.0363
26	.0256	.781	26	.00189	.0577
27	.0323	.985	27	.00239	.0728
28	.0408	1.24	28	.00301	.0918
30	.0648	1.97	30	.00479	.146
32	.103	3.14	32	.00762	.232
34	.164	4.99	34	.0121	.369

Table 4–106 (continued)
RESISTANCE OF WIRES

B. and S. No.	Ohms per centimeter	Ohms per foot	B. and S. No.	Ohms per centimeter	Ohms per foot
German silver ρ = 33. $\times$ 10^{-6} ohm-cm			Gold ρ = 2.44 $\times$ 10^{-6} ohm-cm		
36	.260	7.94	36	.0193	.587
40	.659	20.1	40	.0487	1.48
Iron ρ = 10. $\times$ 10^{-6} ohm-cm			Lead ρ = 22. $\times$ 10^{-6} ohm-cm		
10	.000190	.00579	10	.000418	.0127
12	.000302	.00921	12	.000665	.0203
14	.000481	.0146	14	.00106	.0322
16	.000764	.0233	16	.00168	.0512
18	.00121	.0370	18	.00267	.0815
20	.00193	.0589	20	.00425	.130
22	.00307	.0936	22	.00676	.206
24	.00489	.149	24	.0107	.328
26	.00776	.237	26	.0171	.521
27	.00979	.299	27	.0215	.657
28	.0123	.376	28	.0272	.828
30	.0196	.598	30	.0432	1.32
32	.0312	.952	32	.0687	2.09
34	.0497	1.51	34	.109	3.33
36	.0789	2.41	36	.174	5.29
40	.200	6.08	40	.439	13.4
Magnesium ρ = 4.6 $\times$ 10^{-6} ohm-cm			Manganin ρ = 44. $\times$ 10^{-6} ohm-cm		
10	.0000874	.00267	10	.000836	.0255
12	.000139	.00424	12	.00133	.0405
14	.000221	.00674	14	.00211	.0644
16	.000351	.0107	16	.00336	.102
18	.000559	.0170	18	.00535	.163
20	.000889	.0271	20	.00850	.259
22	.00141	.0431	22	.0135	.412
24	.00225	.0685	24	.0215	.655
26	.00357	.109	26	.0342	1.04
27	.00451	.137	27	.0431	1.31
28	.00568	.173	28	.0543	1.66
30	.00903	.275	30	.0864	2.63
32	.0144	.438	32	.137	4.19
34	.0228	.696	34	.218	6.66
36	.0363	1.11	36	.347	10.6
40	.0918	2.80	40	.878	26.8
Molybdenum ρ = 5.7 $\times$ 10^{-6} ohm-cm			Monel® metal ρ = 42. $\times$ 10^{-6} ohm-cm		
10	.000108	.00330	10	.000798	.0243
12	.000172	.00525	12	.00127	.0387

Table 4–106 (continued)
RESISTANCE OF WIRES

B. and S. No.	Ohms per centimeter	Ohms per foot	B. and S. No.	Ohms per centimeter	Ohms per foot
Molybdenum			Monel		

B. and S. No.	Ohms per centimeter	Ohms per foot	B. and S. No.	Ohms per centimeter	Ohms per foot
Molybdenum $\rho = 5.7 \times 10^{-6}$ ohm-cm			Monel® metal $\rho = 42. \times 10^{-6}$ ohm-cm		
14	.000274	.00835	14	.00202	.0615
16	.000435	.0133	16	.00321	.0978
18	.000693	.0211	18	.00510	.156
20	.00110	.0336	20	.00811	.247
22	.00175	.0534	22	.0129	.393
24	.00278	.0849	24	.0205	.625
26	.00443	.135	26	.0326	.994
27	.00558	.170	27	.0411	1.25
28	.00704	.215	28	.0519	1.58
30	.0112	.341	30	.0825	2.51
32	.0178	.542	32	.131	4.00
34	.0283	.863	34	.209	6.36
36	.0450	1.37	36	.331	10.1
40	.114	3.47	40	.838	25.6
Nichrome $\rho = 150. \times 10^{-6}$ ohm-cm			Nickel $\rho = 7.8 \times 10^{-6}$ ohm-cm		
10	.0021281	.06488	10	.000148	.00452
12	.0033751	.1029	12	.000236	.00718
14	.0054054	.1648	14	.000375	.0114
16	.0085116	.2595	16	.000596	.0182
18	.0138383	.4219	18	.000948	.0289
20	.0216218	.6592	20	.00151	.0459
22	.0346040	1.055	22	.00240	.0730
24	.0548088	1.671	24	.00381	.116
26	.0875760	2.670	26	.00606	.185
28	.1394328	4.251	27	.00764	.233
30	.2214000	6.750	28	.00963	.294
32	.346040	10.55	30	.0153	.467
34	.557600	17.00	32	.0244	.742
36	.885600	27.00	34	.0387	1.18
38	1.383832	42.19	36	.0616	1.88
40	2.303872	70.24	40	.156	4.75
Platinum $\rho = 10. \times 10^{-6}$ ohm-cm			Silver (18°C) $\rho = 1.629 \times 10^{-6}$ ohm-cm		
10	.000190	.00579	10	.0000310	.000944
12	.000302	.00921	12	.0000492	.00150
14	.000481	.0146	14	.0000783	.00239
16	.000764	.0233	16	.000124	.00379
18	.00121	.0370	18	.000198	.00603
20	.00193	.0589	20	.000315	.00959
22	.00307	.0936	22	.000500	.0153

Table 4–106 (continued)
RESISTANCE OF WIRES

B. and S. No.	Ohms per centimeter	Ohms per foot	B. and S. No.	Ohms per centimeter	Ohms per foot
Platinum $\rho = 10. \times 10^{-6}$ ohm-cm			Silver (18°C) $\rho = 1.629 \times 10^{-6}$ ohm-cm		
24	.00489	.149	24	.000796	.0243
26	.00776	.237	26	.00126	.0386
27	.00979	.299	27	.00160	.0486
28	.0123	.376	28	.00201	.0613
30	.0196	.598	30	.00320	.0975
32	.0312	.952	32	.00509	.155
34	.0497	1.51	34	.00809	.247
36	.0789	2.41	36	.0129	.392
40	.200	6.08	40	.0325	.991
Steel, piano wire (0°C) $\rho = 11.8 \times 10^{-6}$ ohm-cm			Steel, invar. (35% Ni) $\rho = 81. \times 10^{-6}$ ohm-cm		
10	.000224	.00684	10	.00154	.0469
12	.000357	.0109	12	.00245	.0746
14	.000567	.0173	14	.00389	.119
16	.000901	.0275	16	.00619	.189
18	.00143	.0437	18	.00984	.300
20	.00228	.0695	20	.0156	.477
22	.00363	.110	22	.0249	.758
24	.00576	.176	24	.0396	1.21
26	.00916	.279	26	.0629	1.92
27	.0116	.352	27	.0793	2.42
28	.0146	.444	28	.100	3.05
30	.0232	.706	30	.159	4.85
32	.0368	1.12	32	.253	7.71
34	.0586	1.79	34	.402	12.3
36	.0931	2.84	36	.639	19.5
40	.236	7.18	40	1.62	49.3
Tantalum $\rho = 15.5 \times 10^{-6}$ ohm-cm			Tin $\rho = 11.5 \times 10^{-6}$ ohm-cm		
10	.000295	.00898	10	.000219	.00666
12	.000468	.0143	12	.000348	.0106
14	.000745	.0227	14	.000553	.0168
16	.00118	.0361	16	.000879	.0268
18	.00188	.0574	18	.00140	.0426
20	.00299	.0913	20	.00222	.0677
22	.00476	.145	22	.00353	.108
24	.00757	.231	24	.00562	.171
26	.0120	.367	26	.00893	.272
27	.0152	.463	27	.0113	.343
28	.0191	.583	28	.0142	.433
30	.0304	.928	30	.0226	.688
32	.0484	1.47	32	.0359	1.09

Table 4—106 (continued)
RESISTANCE OF WIRES

B. and S. No.	Ohms per centimeter	Ohms per foot	B. and S. No.	Ohms per centimeter	Ohms per foot
Tantalum $\rho = 15.5 \times 10^{-6}$ ohm-cm			Tin $\rho = 11.5 \times 10^{-6}$ ohm-cm		
34	.0770	2.35	34	.0571	1.74
36	.122	3.73	36	.0908	2.77
40	.309	9.43	40	.230	7.00
Tungsten $\rho = 5.51 \times 10^{-6}$ ohm-cm			Zinc $(0°C)$ $\rho = 5.75 \times 10^{-6}$ ohm-cm		
10	.000105	.00319	10	.000109	.00333
12	.000167	.00508	12	.000174	.00530
14	.000265	.00807	14	.000276	.00842
16	.000421	.0128	16	.000439	.0134
18	.000669	.0204	18	.000699	.0213
20	.00106	.0324	20	.00111	.0339
22	.00169	.0516	22	.00177	.0538
24	.00269	.0820	24	.00281	.0856
26	.00428	.130	26	.00446	.136
27	.00540	.164	27	.00563	.172
28	.00680	.207	28	.00710	.216
30	.0108	.330	30	.0113	.344
32	.0172	.524	32	.0180	.547
34	.0274	.834	34	.0286	.870
36	.0435	1.33	36	.0454	1.38
40	.110	3.35	40	.115	3.50

From Weast, R. C., Ed., *Handbook of Chemistry and Physics,* 55th ed., CRC Press, Cleveland, 1974.

Table 4–107

PHYSICAL PROPERTIES OF GLASS SEALING AND LEAD WIRE MATERIALS

	Copper		Glass sealing materials									
	OFC	0.02P	Dumet	42 Ni	Gas free 42% Ni	42-6 NiCr	46 Ni	52 Ni	27 Cr	Kovar®	W	Mo
Analysis: Carbon				0.10	0.05	0.10	0.10	0.10	0.15	0.02		
Manganese				0.50	0.50	0.50	0.50	0.50	0.60	0.30		
Silicon			Copper Clad	0.25	0.25	0.25	0.25	0.25	0.40	0.20		
Chromium			42% Nickel	–	–	5.75	–	–	28.00	–		
Nickel			Iron	42	42	42.5	46	51	0.50	29		
Copper	99.95	99.90	See GE Spec. DS 8311-01					–	–			
Other		P 0.02		Bal.Fe	Ti0.4 Bal.Fe	Bal.Fe	Ti0.4 Bal.Fe	Bal.Fe	Bal.Fe	Co17 Bal.Fe		
Density: g/cm³	8.94	8.94	8.26–8.32	8.12	8.12	8.12	8.17	8.30	7.60	8.36	19.3	10.2
lb/in.³	0.323	0.323	0.298–0.301	0.293	0.293	0.293	0.295	0.300	0.274	0.302	0.697	0.369
Thermal conductivity, 20–100°C												
cal/cm/sec/cm²/°C	0.948	0.8	0.2–0.3	0.025	0.025	0.029	0.028	0.032	0.054	0.04	0.31	0.34
Btu/in./hr/ft$_v$/°F	2,750	2,320	580–870	74	74	84	81	93	158	116	900	1,000
Electrical resistivity (20°C)												
μΩ-cm	1.71	2.03	7.3–12.0	72	72	95	46	43	63	49	5.5	5.2
Ω/cir mil ft	10.3	12.2	44–72	430	430	570	275	258	380	294	33	31
Elec. Cond. % IACS	101	85	23–14	2.3	2.3	1.8	3.6	3.9	2.8	3.4	31	33
Curie temperature °C	–	–	380	380	380	295	460	530	610	435		
Melting temperature °C	1,083	1,083	–	1,425	1,425	1,425	1,425	1,425	1,425	1,450	3,410	2,610
°F	1,981	1,981	–	2,597	2,597	2,597	2,597	2,597	2,597	2,642	6,170	4,730
Specific heat, cal/g	0.092	0.092	0.11	0.12	0.12	0.12	0.12	0.12	0.14	0.11	0.033	0.066
Thermal expansion in./in./°C $\times 10^7$			Radial									
25–100°C	168	168	60–80	50.1	43.4	65.5	71.0	99.5	94.6	58.6	45	51
25–200°C	172	172		47.1	44.1	70.8	73.7	101.0	100.5	52.0	46	
25–300°C	177	177	60–80	47.6	46.1	82.6	75.0	101.0	105.3	51.3	46	
25–350°C	178	178	65–85	50.5	64.1	90.4	74.4	100.0	107.0	48.9	46	
25–400°C	181	181	80–100	62.5	85.6	100.0	74.3	100.0	107.8	50.6	46	
25–500°C	183	183	100–140	83.2	100.1	115.0	86.8	102.1	111.2	61.5	46	
25–600°C	188	188		99.0		125.8	100.2	110.0	112.6	78.0		57

Table 4-107 (continued)

PHYSICAL PROPERTIES OF GLASS SEALING AND LEAD WIRE MATERIALS

	Copper		Glass sealing materials									
	OFC	0.02P	Dumet	42 Ni	Gas free 42% Ni	42-6 NiCr	46 Ni	52 Ni	27 Cr	Kovar®	W	Mo
Mechanical properties (annealed)												
Ultimate str. (1,000 psi)	35	35	74	82	80	80	82	80	85	75	490	120
Yield str. (1,000 psi)			50	34	34	40	34	40	55	50	360	110
% Elong. (2 in.)			30	30	30	30	27	35	25	30	8	30
Rockwell hardness				B76	B76	B80	B76	B83	B85	B68	C25	B88
Elastic modulus (10⁶ psi)	16	16		21.5	21.0	23	23	24	30	20	50	47

	Nickel 200	Nickel 211	AISI Type 302	AISI Type 316	AISI Type 430	400 Monel®	600 Inconel®	Advance®	CDA 752	70-30 Brass
Metal alloy lead wires										
Analysis: Carbon	0.06	0.10	0.15 max	0.08 max	0.12 max	0.12	0.04			
Manganese	0.25	4.75	2.00 max	2.00 max	1.00 max	0.90	0.20		0.50 max	
Silicon	0.05	0.05	1.00 max	1.00 max	1.00 max	0.15				
Chromium			17-19	16-18	14-18		16			
Nickel	99.5	95.0	8-10	10-14	0.50 max	66	76	43	18	
Copper	0.05					31.5		Bal.	65	70
Other				Mo 2-3			Fe 7.20		Bal. Zn	Bal. Zn
Density: g/cm³	8.89	8.72	7.9	7.9	7.7	8.84	8.41	8.9	8.73	8.53
lb/in.³	0.321	0.315	0.285	0.285	0.278	0.319	0.304	0.322	0.316	0.308
Thermal conductivity, 20-100°C										
cal/cm/sec/cm²/°C	0.15	0.12	0.04	0.04	0.05	0.062	0.036	0.051	0.08	0.29
Btu/in./hr/ft²/°F	435	350	116	116	145	180	104	148	232	845
Electrical resistivity (20°C)										
μΩ-cm	9.5	18.3	72	74	60	48.2	98.1	49	28.7	6.16
Ω/cir mil ft	57	110	433	445	361	290	590	294	173	37
Elec. Cond. % IACS	18	9	2.3	2.3	2.9	3.6	1.7	3.5	6	28
Curie temperature °C	360	352			610	43/60	-125	-	-	-

Table 4–107 (continued)

PHYSICAL PROPERTIES OF GLASS SEALING AND LEAD WIRE MATERIALS

	Nickel 200	Nickel 211	AISI Type 302	AISI Type 316	AISI Type 430	400 Monel®	600 Inconel®	Advance®	CDA 752	70–30 Brass
Melting temperature °C	1,455	1,427	1,421	1,399	1,510	1,349	1,427	1,210	1,110	954
°F	2,651	2,600	2,590	2,550	2,750	2,460	2,600	2,210	2,030	1,750
Specific heat, cal/g	0.13	0.13	0.12	0.12	0.11	0.102	0.109	0.094	0.09	0.09
Thermal expansion, in./in./°C × 10⁷										
25–100°C	133	133	166	166	101	140	115	149		
25–200°C	139	139				145				
25–300°C	144	144	171	175	110	150			160	199
25–350°C	146	146				155				
25–400°C	148	148				160				
25–500°C	172	153	180	180	120	165		163		
25–600°C										
Mechanical properties (annealed)										
Ultimate str. (1,000 psi)	70		90	85	75	85	100	60	60	50
Yield str. (1,000 psi)	25		37	35	45	40	45		30	–
% Elong. (2 in.)	45		55	55	30	40	40		40	60
Rockwell hardness	B62		B82	B80	B82	B68	B74		–	–
Elastic modulus (10⁶ psi)	30	30	29	29	29	26	31	18	18	16

Metal alloy lead wires

Table 4–107 (continued)
PHYSICAL PROPERTIES OF GLASS SEALING AND LEAD WIRE MATERIAL

	Kul-grid®	Clad materials (Also see Dumet)		Pure metals (Also see W and Mo glass scaling materials)						
		40% CCFe	30% CCFe	Au	Ag	Al	Pt	Ta	Fe	Ti
Analysis: Carbon										
Manganese										
Silicon										
Chromium										
Nickel	27	37.5	26							
Copper										
Other.	Cu core	Fe core	Fe core							
Density: g/cm³	8.89	8.15	8.15	19.3	10.5	2.69	21.45	16.6	7.87	4.51
lb/in.³	0.321	0.294	0.294	0.698	0.379	0.097	0.775	0.600	0.284	0.163
Thermal conductivity, 20–100°C										
cal/cm/sec/cm²/°C		0.46	0.38	0.71	1.00	0.57	0.165	0.13	0.18	0.43
Btu/in./hr/ft²/°F		1,330	1,100	2,050	2,900	1,650	480	380	520	1250
Electrical resistivity (20°C)										
μΩ-cm	2.3	4.4	5.9	2.19	1.629	2.65	9.83	12.45	9.71	7.0
Ω/cir mil ft	14	26.4	35.3	13	9.8	16	59	75	58	42
Elec. Cond. % IACS	70	40	30	78	105	65	16	14	18	25
Curie temperature °C		770	770	–	–	–	–	–	770	–
Melting temperature °C				1,063	960	660	1,769	3,000	1,536	1,668
°F				1,945	1,760	1,220	3,217	5,425	2,797	3,035
Specific heat, cal/g		0.10	0.10	0.031	0.056	0.225	0.031	0.034	0.11	0.124
Thermal expansion, in./in./°C × 10⁷										
25–100°C				142	196	239	91	65	122	88
25–200°C						243			129	91
25–300°C						253				
25–350°C										
25–400°C									138	94
25–500°C				152	206		96	66		
25–600°C						287			145	97

Table 4–107 (continued)

PHYSICAL PROPERTIES OF GLASS SEALING AND LEAD WIRE MATERIALS

	Clad materials (Also see Dumet)			Pure metals (Also see W and Mo glass scaling materials)							
	Kul-grid ®	40% CCFe	30% CCFe	Au	Ag	Al	Pt	Ta	Fe	Ti
Mechanical properties (annealed)										
Ultimate str. (1,000 psi)			20	22	13	40	55	40		
Yield str. (1,000 psi)				8	5	12		20		
% Elong. (2 in.)			45	48	40	30	30	45		
Rockwell hardness										
Elastic modulus (10⁶ psi)	24	24	11.6	11	9	21.3	27	30	16.8	

Courtesy of Lamp Metals and Components Department, General Electric Company, Cleveland.

Table 4–108
BINARY PHASE DIAGRAM CHART

J. E. Selle
Monsanto Research Corp.

The chart, attached to the inside back cover of this book, provides a general description of the binary phase diagrams existing between two elements and summarizes some of the more important details of the phase diagrams. Fifty-five elements are considered and are arranged alphabetically. When it is desired to know the type of diagram existing between two elements, A and B (also arranged alphabetically), locate the intercept between element A along the bottom of the diagram and element B along the left leg of the triangle. The color code indicates the general type of diagram, complete solid solubility, eutectic, peritectic, monotectic, or complete immiscibility. Diagrams that are not of one of these simple generalized types are left uncolored. All systems with a common element can be summarized by locating that element along the bottom of the diagram and comparing the various systems to the upper left and the upper right.

The total number of intermediate phases in the system is indicated in the box along with the number of congruent melting compounds or incongruent melting compounds in the system with no range of solid solubility. The number of eutectics present in the system is indicated along with the composition and temperature of the lowest melting eutectic. For each system, the maximum terminal solid solubility, if any, is also indicated, along with the temperature of this maximum. This temperature may or may not correspond to other critical temperatures in the system and is given to indicate the temperature of maximum solid solubility. No correlation with room temperature solubility is implied. When more than one allotropic modification is present, maximum available solubility data are given for each phase.

Monotectics and syntectics are indicated in yellow. Liquid immiscibility and partial liquid immiscibility are indicated in green. When complete immiscibility has been observed the word "Immisc." appears. When some liquid solubility has been observed, usually less than 0.1 atomic percent, the box is blank except for the color. This case is also a variation of a eutectic, at about 0 atomic percent, of the higher melting constituent, except that the amount of liquid solubility is lower.

At the top of each square, the numbers 1, 2, or 3 are listed to indicate the reference in which more detailed information on each system can be found. An underlined number indicates that a diagram is given in that reference. The references are

1. **Hansen, M. and Anderko, K.**, *Constitution of Binary Alloys*, 2nd ed., McGraw-Hill, New York, 1958.
2. **Elliott, R. P.**, *Constitution of Binary Alloys: First Supplement*, McGraw-Hill, New York, 1965.
3. **Shunk, F. A.**, *Constitution of Binary Alloys: Second Supplement*, McGraw-Hill, New York, 1969.

Thus, a designation 1 2 3 indicates that some information is available in all three references, and that a diagram is given in Reference 2. Very little attempt has been made to update the date beyond these references.

All temperatures are listed in centigrade degrees, and compositions are given in atomic percent except where indicated as w/o (weight percent).

In this chart columbium (Cb) is listed as niobium (Nb).

Key: Color Coding System

Blue

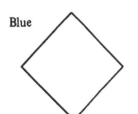

Solid solubility existing in one phase completely across the diagram. The presence of a maximum or minimum in the liquidus is indicated along with the temperature and composition in percent of element B.

663

Table 4–108 (continued)
BINARY PHASE DIAGRAM CHART

Key: Color Coding System (continued)

Yellow

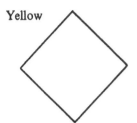

Indicates a known monotectic or syntectic. The temperature and composition of the monotectic or syntectic are given.

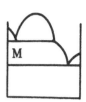

Orange

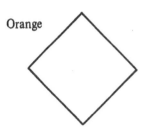

Indicates a peritectic type diagram. Several incongruent compounds may occur in succession but no congruent compounds are present. A eutectic may be found. The temperature and composition of the eutectic and limits of terminal solid solubility are given when appropriate.

Green

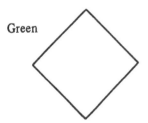

Indicates either complete immiscibility or partial liquid immiscibility. In the latter case the liquid solubility is less than 0.1% at the temperature reported. Complete immiscibility is indicated by "Immisc."

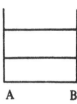

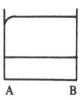

Table 4–108 (continued)
BINARY PHASE DIAGRAM CHART

Key: Color Coding System (continued)

Red

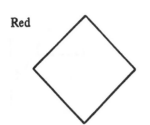

Indicates a simple eutectic diagram with no inter-metallic compounds forming directly from the melt. The presence of terminal solid solutions is indicated by the temperature and composition of the maximum limit of solid solubility.

A   B

Section 5

Materials Standards

5.1 ANALYTICAL STANDARDS

G. W. Latimer, Jr.
Mead Johnson Company

Modern instrumental methods of analysis depend upon the use of standard materials for calibration. Many of these new methods (e.g., electron probe, solids mass spectrometer) have placed stringent requirements on sample homogeneity.

The materials listed in the following tables are believed suitable for use as calibration standards in many methods of analysis. Inclusion in the listing, however, should not be construed as an endorsement of a particular firm's product, of the suitability of a material for a given purpose, nor of the homogeneity of the material. As always, the standard will have to be judged on the basis of the certificates provided and the user's personal knowledge and experience.

The list is not meant to be exhaustive.

Since the composition of these standards changes, the list of elements or materials presented are simply "typical." Specific details should be sought from a firm's catalogue.

Table 5.1–1
ALLOYS AND HIGH-PURITY METALS

A. Alloys

Matrix	Source(s)	Alloy type	Elements for which analytical values are available
Aluminum	JMC, NBS, Alcoa, WW, SPEX, BBAS, ALA, Apex, MK, JMC, PS		Cu, Cr, Fe, Ni, Mg, Pb, Si, Sn, Ti, Zn
Copper	JA, NBS,	Brass	As, Be, Cu, Fe, Ni, Pb, Sn, Zn
	SPEX, ARL,	Bronze	Cu, Fe, Ni, P, Pb, S, Sn, Zn
	CAAS, WW, JMC	Gilding	Cu, Fe, Ni, P, Pb, Sn, Zn
Cobalt	NBS		C, Co, Cr, Cu, Fe, Mn, Mo, Nb, Ni, P, S, Si, Ta, Ti, V, W
Iron	BBAS, NBS	Mild steel	Al, B, C, Co, Cr, Cu, Mg, Mn, Mo, Ni, Pb, Si, Sn, Ti, V, W, Zr
	BBAS, NBS	Low alloy steel	Cr, Cu, Mn, Mo, Ni, Si, V
	BBAS	Low tungsten steel	C, Cr, Mn, Mo, Ni, P, S, Si, W
	NBS	Tool steel	Cr, Co, Cu, Mn, Mo, Si, V, W
	NBS	Maraging steel	Al, B, C, Cu, Cr, Co, Mn, Mo, Ni, O_2 (g), P, S, Si, Ti, Zr
		High temp steel	
	BBAS, NBS	Carbon steel	C, Mn, Nb, P, S, Si

Table 5.1–1 (continued)
ALLOYS AND HIGH-PURITY METALS

A. Alloys

Matrix	Source(s)	Alloy type	Elements for which analytical values are available
	BBAS	Austenitic stainless	C, Cr, Co, Mn, Mo, Nb, Ni, P, Pb, S, Si, Ti
	BBAS	Ferritic stainless	C, Cr, Mn, Mo, Ni, P, S, Si
Lead	Alfa, MK,	Babbit solder	As, Bi, Cd, Cu, Fe, Sb, Sn
	NBS		As, Ag, Bi, Cu, Sb, Sn
Magnesium	Alcoa, Dow		
Nickel	NBS, INCO		C, Co, Cr, Cu, Fe, Mn, Mo, Ni, P, S, Si
Nobel metals	NBS,[a] SPEX	Gold	Au, Ag, Hf, Ga, In, Ir, Pd, Pt, Re, Rh, Ru
		Platinum	
		Silver	Ag, Cu
Rare earths	SPEX		Ce, Dy, Er, Eu, Gd, Ho, La, Lu, Nd, Pr, Sc, Sm, Tb, Tm, Y, Yb
Tin	NBS, MK, Alfa		Ag, As, Bi, Cd, Co, Cu, Ni, Pb, Sb, Zn
Titanium	NBS		Al, Cr, Fe, H_2 (g), Mn, Mo, N_2 (g), O_2 (g), V
Tungsten			
Zinc	NBS, Apex, MK, SZA		Al, Cd, Cr, Cu, Fe, Mg, Mn, Ni, Pb, Si, Sn
Zirconium	NBS		Al, B, Cr, Cu, Fe, Mn, N_2 (g), Ni, Ti, U, W

B. Metals
(99–99.9999%)

LEICO, JMC, NBS, MK

C. Metals in Solution[b]
Alfa, NSL, CONOSTAN, SPEX

[a]Suitable for microprobe standardization.

[b]Suitable for AA standardizations. Some materials are oil soluble.

Table compiled by G. W. Latimer, Jr.

Table 5.1–2
MINERALS, ORES, AND STANDARDS

A. Minerals and Ores

Sources	Types
SOMAR,[a] NBS, GFS, USAEC, USGS, CNSC, CNRS, SASE	Amphiboles, borates, carbides, carbonates, clays, epidotes, feldspars, feldspathoids, garnets, halides, humites, iron-contg., limestones, lithium-contg., nitrites, nickel oxide-contg., manganese-contg., micas, molybdates, olivines, oxides (including hydrous and multiple), phosphates, pyroxenes, pyrosenoids, scapolites, silicates (including disilicates, metasilicates, pyrosilicates, and subsilicates), sodalites, sulfates, sulfides, sulfosalts, thorium-contg., tin-contg., tungstates, uranium-contg., vanadates, zeolites.

B. Standard Ores and Minerals

Description	Source
Bauxite	NBS
Limestone, dolomitic argillaceous	SPEX, NBS
Chrome ore, Grecian	BBAS, SPEX
Iron ores, various	BA, SPEX
Sibley	NBS
Norite	NBS
Spodumene	
Petalite	NBS
Lepidolite	
Manganese ore	NBS
Tin, NEI concentrate	NBS
Feldspar, soda and potash	SPEX, NBS
Clay, flint and plastic	NBS
Zinc, tristate concentrate	NBS
Sulfide	CAAS

[a]Suitable for diffraction identification.

Table compiled by G. W. Latimer, Jr.

Table 5.1–3
INDEX FOR MATERIALS STANDARDS
SOURCES IN TABLES 5.1–1 AND 5.1–2

Symbol	Address
Alcoa	Aluminum Co. of America Research Lab., New Kensington, Pa.
ALA	Aluminum Laboratory Ltd. Analytical Div., P.O. Box 645, Arvida, Quebec, Canada
Alfa	Alfa Products, 8 Congress Ave., Beverly, Mass. 01915
Apex	Apex Smelting, 2537 W. Taylor St., Chicago, Ill. 60612

Table 5.1–3 (continued)
INDEX FOR MATERIALS STANDARDS
SOURCES IN TABLES 5.1–1 AND 5.1–2

Symbol	Address
ARL	Applied Research Laboratories, Inc., Box 1710, Glendale, Cal. 92609
BBAS	British Bureau of Analytical Samples, Newham Hall, Middlesbourgh, Yorkshire, England
CAAS	Canadian Association for Applied Spectroscopy, Copper Standards Committee, Dept. of Mines and Technical Surveys, 555 Booth St., Ottawa, Ontario, Canada
CNRS	Centre National de la Recherche Scientifique, B.P. 682, Nancy, France
CNSC	Canadian Association for Applied Spectroscopy, Nonmetallic Standards, Dept. of Geological Sciences, McGill University, Montreal, Quebec, Canada
CONOSTAN	Continental Oil Co., P.O. Box 1267, Ponca City, Okla. 74601
Dow	Dow Chemical Company, Midland, Mich. 48640
GFS	G. Frederick Smith Chemical, P.O. Box 23344, Columbus, O. 43023
INCO	International Nickel, Huntington, W. Va.
JMC	Johnson, Matthey, 74 Hatton Garden, London, E.C. 1, England
LEICO	LEICO Industries, 250 West 57th St., New York, N.Y. 10019
MK	Morris P. Kirk, 2717 So. Indiana St., Los Angeles, Cal. 90023
NBS	Office of Standard Reference Materials, National Bureau of Standards, Washington, D.C. 20234
NSL	National Spectrographic Laboratories, 6300 Euclid Ave., Cleveland, O. 44103
PS	Pechiny-Service, D.I./CA-M, 23 Rue Balzac, Paris, France
SASE	Schweitzerische Arbeitsgemeinschaft für Steine & Erden Mineralogisch-Petrographische Institut, Sahlistrasse 6, Bern, Switzerland
SOMAR	Somar Laboratories, 54 E. 11th St., New York, N.Y. 10003
SPEX	SPEX Industries, P.O. Box 798, Metuchen, N.J. 08840
SZA	Societe Zinc et Alliages, 32 Rue Collange, Lavollois-Perret, Seine, France.
USGS	United States Geological Survey Analytical Laboratories, Washington, D.C. 20234
USAEC	Atomic Energy Commission, New Brunswick Laboratory, P.O. Box 150, New Brunswick, N.J. 07102
WW	Wieland Werke A.G., Ulm Donau Postfache 636, German Federal Republic

Table compiled by G. W. Latimer, Jr.

Table 5.1−4
POLYMERS, RUBBERS, AND FIBERS

Material	Description	Source
Rubbers	Natural	NBS
	Styrene-butadiene, types 1500 and 1503	NBS
	Butyl	NBS
	Acrylonitrile-butadiene	NBS
Polystyrene	Molecular weight distribution: narrow	NBS
	wide	NBS
	$7.2 \pm 0.5 \times 10^6$ g/mol	Duke
	Suspensions .088−19.5: varying RSD	Duke
	Beads 3.5−100.00: varying RSD	Duke
Asbestos	Canadian and Rhodesian chrysotile fibers	Duke
	Crocidolite fibers	Duke
	Amosite fibers	Duke
	Anthophyllite fibers	Duke

Table compiled by G. W. Latimer, Jr.

Table 5.1−5
BIOCHEMICAL STANDARDS

Type	Material	Source
Mixture or std. soln.	Bile acids	AS
	Brain cerebroside hydolyzates	AS
	Cholesterol	Analabs
	Fat and oils, AOCS references	AS
	Fatty acids, methyl esters	AS
	NIH methyl ester mixtures	AS, Analabs
	Glycerides, mono-	AS, Analabs
	di-	
	tri-	
	Hydrocarbons	AS, Analabs, NBS
	Oils, animal and vegetable	Analabs
	Pesticides	AS, Analabs, City Chemical
	Polar lipids	AS, Analabs
	Amino acids	Analabs
High purity (99%)	Amino acids	Biorad
	Alcohols, fatty saturated and unsaturated	Analabs, AS
	Aldehydes, fatty	AS
	Amphetamines	Analabs
	Barbiturates	Analabs, AS
	Bile acids	Suplco, AS
	Carbohydrates	AS, Senn, NBS
	Cholesterol	NBS, AS
	Compesterol	AS
	Creatinine	NBS
	Fatty acids, st. chain even and odd	Analabs, AS
	branched chain iso and anteiso	Analabs
	unsaturated	Analabs
	dibasic	Analabs, AS

Table 5.1–5 (continued)
BIOCHEMICAL STANDARDS

Type	Material	Source
	Glycerides, mono-, di-, tri-	AS
	Glyceryl ethers	AS, Analabs
	Lipids, glyco, shingo, sulfo, phospho	Analabs, AS, Suplco
	α-Hydroxyacids	Analabs
	Methyl esters, of fatty acids	AS, Analabs, Varian
	of hydroxy acids	AS
	of dibasic acids	AS
	Nicotinic acid	NBS
	Pesticides	Analabs, City Chemical
	Herbicides	Varian
	Insecticides	AS, Varian
	Miticides	Varian
	Nematocides	Varian
	Phytanic acid and metabolites	Analabs
	Stigmasteryl palmitate	AS
	Urea	NBS
	Uric acid	NBS
	Cholesteryl esters	AS

Index to Suppliers

Symbol	Firm
Analabs	Analabs, Inc., P.O. Box 501, North Haven, Conn. 06473
AS	Applied Science Lab., P.O. Box 440, State College, Pa. 16801
Biorad	Bio-Rad Lab., 32nd and Griffin, Richmond, Cal. 94804
City Chemical	City Chemical, Inc., 132 W. 22nd St., New York, N.Y. 10011
NBS	National Bureau of Standards, Office of Standard Reference Materials, Washington, D.C. 20234
Suplco	Sopelco, Inc., Sopelco Park, Bellefonte, Pa. 18823
Varian	Varian Aerograph, 2700 Mitchell Dr., Walnut Creek, Cal. 94598.

Table compiled by G. W. Latimer, Jr.

Table 5.1−6
RADIOCHEMICAL STANDARDS

Element	Isotopes	Firms
Actinium	227	GN, ICNC, NENC
Aluminum	26	GN, ICNC
Americium	241	A/S, AAEC, BIO, CN, CEN, EIC, GN, HVE, IA, ICNC, IPL, JEN, NENC, NME, OR, TRLAB
Antimony	122, 124, 125	A/S, RC
Argon	37	GN
Arsenic	76, 77	CEA
Barium	131, 133, 137m, 140	GN
Beryllium	7, 10	GN
Bismuth	206, 207, 210	GN
	207	GN, HVE, ICNC, IPL
	210	HVE, IPL, NCA
Bromine	82	AAEC, CN, CEN, CEA, GN, IA, IPL, NENC, RC
Cadmium	109, 115m	GN, NENC
Calcium	45, 47	A/S, GN
Carbon	14	A/S, BA, CN, CEN, CEA, GN, HVE, HAS, ICNC, IPL, NBS, NENC
Cerium	139, 141, 144	A/S, RC
Cesium	131, 134, 137	A/S, CEA, NENC, RC
Chlorine	36	A/S, BA, CN, CEN, GN, HVE, ICNC, IPL, NBS, NENC, TRLAB
Chromium	51	A/S, CN, CEA, GN, HAS, IA, ICNC, RC
Cobalt	56, 57, 58, 60	GN
Copper	64	A/S, AAEC, IA, NENC, RC
Curium	242, 244	IPL
	244	JEN
Dysprosium	159	GN, NENC
Erbium	169	NENC
Europium	154, 155	GN
Gadolinium	153	GN, ICNC
Gallium	68	GN
Germanium	68, 71	ICNC
Gold	195, 198, 199	A/S, GN, RC
Hafnium	175, 181	GN
Holmium	166m	NENC
Hydrogen	3	A/S, BI, CN, CEN, CEA, CN, HAS, IPL, NENC, NBS, RC, TRLAB
Indium	114m, 114	GN, NENC
	114	A/S, RC
Iodine	125, 129	GN, ICNC
	131, 132	RC
Iridium	192, 194	GN
Iron	55, 59	A/S, CN, CEN, CEA, GN, ICNC, IPL, UR
Krypton	85m	ICNC, IPL
	85	CEN, GN, NBS, TRLAB
Lanthanum	140	A/S, GN, IA, ICNC, RC
Lead	210	A/S, BA, GN, HVE, ICNC, IPL
Lutetium	171	CN, CEN, CEA, GN

Table 5.1–6 (continued)
RADIOCHEMICAL STANDARDS

Element	Isotopes	Firms
Manganese	52, 54	GN
Mercury	197, 203	A/S, CN, CEN, ICNC, RC
Molybdenum	99	A/S, AAEC, GN, NENC, RC
Neodymium	147, 149	GN
Neptunium	237	CN, CEN, GN, IPL, JEN, RC
Nickel	59, 63	GN, ICNC, NENC
Niobium	94, 95	GN
Osmium	185	GN
Phosphorus	32	AAEC, CN, CEN, CEA, GN, HAS, IA, RC
Plutonium	238, 239	GN, IPL
Polonium	208, 210	GN
Potassium	40, 42	ICNC
Praseodymium	142, 143	GN
Promethium	145	NENC
	147	A/S, CN, CEN, GN, ICNC, NBS
Protactinium	231, 234	IPL
	234m	NENC
Radium	226	A/S, GN, HVE, ICNC, NVPD, UM
	228	IPL
Radon	222	GN
Rhenium	186	GN
Rubidium	83, 84, 86	GN
Ruthenium	103, 106	A/S, CN, GN, RC
Samarium	151, 153	GN
Scandium	46, 47	A/S, RC
Selenium	75	A/S, CN, CEN, CEA, GN, NENC, RC, UR
Silver	110m, 110, 111	GN
Sodium	22, 24	A/S, CN, CEN, CEA, GN, IA, ICNC, NENC, RC
Strontium	85, 89, 90	CN, CEN, CEA, GN, NENC, RC, UR
Sulfur	35	A/S, BA, CN, CEN, CEA, GN, ICNC, IPL, NENC, RC
Tantalum	182	A/S, GN, NENC, RC, UR
Technetium	99m	AAEC, CN, CEN, GN, IA, ICNC, SQ
	99	A/S, BA, HVE, NENC, TRLAB, UR
Tellurium	125m, 127m	ICNC, NENC
	129m	NENC
	129	ICNC
Terbium	160, 161	A/S, GN, ICNC, NENC, RC
Thallium	204	A/S
Thorium	228, 230, 232	GN, IPL, NENC
Thulium	170, 171	ICNC, NENC
Tin	113, 119m	GN, HVE, IPL, NENC
Titanium	44	NENC
Tungsten	181, 185	GN, NENC
	187	A/S, RC
Uranium	233	JEN, IPL
Vanadium	49	ICNC
Xenon	133	GN, NENC

Table 5.1–6 (continued)
Table 5.1–6 (continued)
RADIOCHEMICAL STANDARDS

Element	Isotopes	Firms
Ytterbium	169	GN, NENC
Yttrium	87, 90, 91	A/S, RC
Zinc	65	A/S, AECAN, CN, CEN, CEA, GN, IA, ICNC, IPL, NENC, RC
Zirconium	95, 97	A/S, RC

Table compiled by G. W. Latimer, Jr.

Table 5.1–7
INDEX TO SOURCES IN TABLE 5.1–6

Symbol	Address
A/S	Amersham/Searle, 2626 S. Clearbrook Dr., Arlington Heights, Ill. 60005
AAEC	Australian AEC Research Establishment, Private Mail Bag, Post Office Sutherland, Sydney, NSW, Australia 2232
AECAN	Atomic Energy Commission of Canada, Commercial Products, P.O. Box 93, Ottawa, Ontario, Canada
BA	Baird-Atomic, 125 Middlesex Tpk., Bedford, Mass. 01730
BI	Beckman Instruments, 2500 Harbor Blvd., Fullerton Cal. 92634
BIO	Bionuclear, P.O. Box 12634, Houston, Tex. 77017
CEA	Commissariat a L'Energie Atomique, Dept. des Radioelements, C.E.N. Saclay BP 2, 91 Gif-sur-Yvette, France
CEN	C.E.N.-S.C.K., Dept. des Radioisotopes, Boeretang 200, 2400 MOL, Belgium, B24
CN	Capintec Nuclear, 63 E. Sandford Blvd., Mount Vernon, N.Y. 10550
EIC	Eberline Instrument Corp., P.O. Box 2108, Santa Fe, N.M. 87501
GN	General Nuclear Inc., 9320 Travenor, Houston, Texas 77034
HAS	Hungarian Academy of Sciences, Institute of Isotopes, Budapest 114, POB 77, Hungary
HVE	High Voltage Engineering Corp., South Bedford St., Burlington, Mass. 01803
IA	Institutt for Atomenergi, Isotope Laboratories; P.O. Box 40, 2007 Kjeller, Norway
ICNC	International Chemical and Nuclear Corp., Chemical and Radioisotope Div., 2727 Campus Dr., Irvine, Cal. 92664
IPL	Isotope Products Laboratories, 404 South Lake St., Burbank, Cal. 91502
JEN	Junta de Energia Nuclear, Direccion de Quimica e Isotopes, Avenida Complutens, Madrid 3, Spain
NBS	Office of Standard Reference Materials, National Bureau of Standards, Washington, D.C. 20234
NENC	New England Nuclear Corp., 575 Albany St., Boston, Mass. 02118
OR	Ortec, Inc., P.O. Box C, 100 Midland Road, Oak Ridge, Tenn. 37830

Table 5.1–7 (continued)
INDEX TO SOURCES IN TABLE 5.1–6

Symbol	Address
RC	Radio Chemical Center, White Lion Road, Amersham Bucks, England
SQ	Squibb and Sons, Inc., Georges Road, New Brunswick, N.J. 08903
TRLAB	Tracerlab, 1601 Trapelo Road, Waltham, Mass. 02154
UM	Union Miniere, Rue de la Chancellerie, 11, 1000-Brussels, Belgium
UR	Universal Radioisotopes, 635 S. 31st St., Richmond, Cal. 94804

Table compiled by G. W. Latimer, Jr.

Table 5.1–8
GLASSES AND REFRACTORY MATERIALS

Type	Elements determined	Source
Chrome refractory	Si, Al, Fe, Ti, Zr, Mn, P, Cr, Ca, Mg	NBS
Silica refractory	Al, Fe, Ti, Mn, P, Ca, Mg, Li, Na, K	NBS
Magnesite refractory	Si, Al, Fe, Ti, Mn, P, Cr, Ca, Mg, Li, Na, K	NBS
Pb-Ba glass	Si, Al, Fe, Ti, Zr, Mn, Ca, Mg, Na, K, Ba, Pb, P, As, SO_3, Cl	NBS
Opal glass	Si, Al, Fe, Ti, Zr, Mn, Ca, Mg, Na, Pb, P, As, Zn, Cl, F	NBS
High boron glass	Si, Al, Fe, Ti, Zr, Mg, Na, K, As, SO_3, Cl, B	NBS

Source Index

Code	Firm
Duke	Duke Standards Company, 445 Sherman Ave., Palo Alto, Calif. 94306
NBS	Office of Standard Reference Materials, National Bureau of Standards, Washington, D.C. 20234

Table compiled by G. W. Latimer, Jr.

Table 5.1–9

BORIDES, CARBIDES, NITRIDES, OXIDES, SELENIDES, AND TELLURIDES

A. Analysis Provided

Material	Elements determined	Source
Nickel oxide	Mn, Si, Cu, Cr, Co, Ti, Al, Fe, Mg	NBS
Silicon carbide	Ti, Al, Fe, Mg, Zr, Ca, Si, C	NBS

B. Purity Greater Than 99.5%

Element	Boride	Carbide	Nitride	Oxide	Selenide	Sulfide	Telluride
Al				Alfa ROC			Alfa ROC
Sb				Alfa ROC PCR	ROC		Alfa ROC
Ba				PCR	Alfa		Alfa
Be				ROC			
Bi				Alfa ROC	ROC	ROC	Alfa, ROC
B			ROC	ROC			
Cd				Alfa ROC	Alfa ROC	ROC	Alfa, ROC
Ca				Alfa ROC			Alfa
Ce	REO	Alfa		REO			REO
Cr	REO, ROC				Alfa		Alfa
Cu				Alfa REO	ROC	ROC	ROC
Co			Alfa		Alfa		Alfa
Dy	REO, ROC		REO	REO	ROC		Alfa, ROC
Er	REO, ROC		REO	REO, ROC	Alfa, ROC	ROC	Alfa, ROC
Eu	REO, ROC		REO	REO			REO
Fe				Alfa ROC	ROC	ROC	Alfa, ROC
Gd	REO, ROC		REO	REO, ROC	ROC	ROC	ROC
Ga	Alfa		Alfa ROC	Alfa ROC	Alfa, ROC	ROC	Alfa, ROC
Ge			Alfa ROC	Alfa ROC	ROC	ROC	Alfa
Au				Alfa ROC		ROC	
Hf	Alfa			Alfa ROC			Alfa
Ho	REO, ROC		REO	REO, ROC	ROC	ROC	REO
In			Alfa	REO, ROC	Alfa, ROC	ROC	Alfa, ROC
Ir				Alfa ROC			
La	ROC	REO	Alfa ROC	REO ROC	ROC	ROC	REO, REO
Pb				Alfa ROC	Alfa ROC	ROC	Alfa, ROC
Lu	REO		Alfa	REO, ROC			Alfa
Mg		Alfa	Alfa ROC	ROC	Alfa		Alfa
Mn		Alfa	Alfa ROC	ROC	Alfa		Alfa

Table 5.1–9 (continued)

BORIDES, CARBIDES, NITRIDES, OXIDES, SELENIDES, AND TELLURIDES

B. Purity Greater Than 99.5%

Element	Boride	Carbide	Nitride	Oxide	Selenide	Sulfide	Telluride
Mo	ROC	Alfa	Alfa	Alfa ROC, PCR	Alfa, ROC	ROC	Alfa
Nd	REO, ROC		REO	REO, ROC			REO, ROC
Ni				ROC	ROC		Alfa, ROC
Nb	Alfa, ROC		Alfa	REO, ROC PCR	Alfa		
Os				PCR			
Pd				Alfa, ROC			
P			Alfa	Alfa, ROC			
Pt				Alfa, ROC	ROC	ROC	
Pr	REO, ROC		REO	REO, ROC			REO
Re	Alfa			Alfa, ROC PCR			Alfa
Ru				ROC, Alfa			
Rh				REO			
Sc	Alfa		REO	ROC			Alfa
Sm	Alfa, ROC		REO	REO		ROC	REO
Si			Alfa	Alfa			
Ag							Alfa
Sn				ROC	ROC	ROC	ROC
Sr	Alfa, ROC			Alfa			Alfa
Ta	Alfa, ROC	Alfa	ROC	Alfa, ROC			Alfa, ROC
Te				Alfa			
Tb	ROC		REO	REO, ROC			REO
Tl				Alfa			Alfa
Th	Alfa	Alfa		Alfa, ROC			Alfa
Tm	Alfa		REO	REO		ROC	REO, ROC
Ti				Alfa, ROC	Alfa		Alfa
W	Alfa, ROC	Alfa		Alfa, ROC	Alfa		Alfa
U	Alfa, ROC	Alfa		Alfa, NBS			Alfa
V				Alfa			Alfa
Yb	REO, ROC		REO	REO			REO
Y	REO, ROC	Alfa	REO	REO, ROC		ROC	REO, ROC
Zn				Alfa, PCR	Alfa		Alfa
Zr	Alfa, ROC			Alfa, ROC			Alfa, ROC

C. Source Index

Code	Firm
Alfa	Ventron Corporation, Beverly, Mass. 01915
GFS	G. Frederick Smith Chemical Co., 867 McKinley Ave., Columbus, O. 43222
NBS	Office of Standard Reference Materials, National Bureau of Standards, Washington, D.C. 20234
PCR	PCR, Inc., P.O. Box 1466, Gainesville, Fla. 32601
REO	
ROC	Research Organic/Inorganic Chemical Corp., 507-519 Main St., Belleville, N.J. 07109

Table compiled by G. W. Latimer, Jr.

Table 5.1−10
AVAILABILITY[a] AND PRICE LIST

A. Standard Reference Materials

SRM	Type	Unit	Price		SRM	Type	Unit	Price
1b	Limestone, argillaceous	50 g	$ 36		132b	Steel, tool	150 g	$ 37
3b	Iron, white	110 g	37		133a	Steel, stainless (Cr13-Mo0.3-S0.3)	150 g	37
4j	Iron, cast	150 g	37		134a	Steel, Mo8-W2-Cr4-V1	150 g	37
5L	Iron, cast	150 g	45					
6g	Iron, cast	150 g	40		136c	Potassium dichromate, oxidimetric	60 g	36
					139a	Steel, Cr-Ni-Mo (AISI 8640)	150 g	37
7g	Iron, cast (high phosphorus)	150 g	37		140b	Benzoic acid	2 g	32
8j	Steel, bessemer (simulated), 0.1C	150 g	37		141b	Acetanilide	2 g	32
10g	Steel, bessemer, 0.2C	150 g	37		142	Anisic acid	2 g	30
11h	Steel, B.O.H. 0.2C	150 g	37					
12h	Steel, B.O.H. 0.4C	150 g	37		143b	Cystine	2 g	33
					147	Triphenyl phosphate	2 g	32
13e	Steel, B.O.H. 0.6C	150 g	37		148	Nicotinic acid	2 g	28
14e	Steel, B.O.H. 0.8C	150 g	37		152a	Steel, B.O.H. 0.5C, 0.03 Sn	150 g	37
15g	Steel, B.O.H. 0.1C	150 g	37		153a	Steel, Co8-Mo9-W2-Cr4-V2	150 g	37
16e	Steel, B.O.H. 1.1C	150 g	37					
17	Sucrose (cane sugar)	60 g	30		154b	Titanium Dioxide	90 g	54
					155	Steel, CrO.5-W0.5	150 g	37
19g	Steel, A.O.H. 0.2C	150 g	37		157a	Nickel silver (Cu58-Ni12-Zn29)	135 g	37
20g	Steel, AISI 1045	150 g	37		158a	Bronze, silicon	150 g	37
25c	Ore, manganese	100 g	31		160b	Steel, stainless, Cr19-Ni14-Mo3		
27e	Ore, iron, Sibley	100 g	32			(SAE 316)	150 g	37
30f	Steel, Cr-V (SAE 6150)	150 g	37					
					162a	Monel-type (Ni64-Cu31)	150 g	37
32e	Steel, Ni-Cr (SAE 3140)	150 g	37		163	Steel, 0.9C, 0.9Mn, 1.0Cr	100 g	44
33d	Steel, Ni-Mo (SAE 4820)	150 g	37		166c	Steel, stainless, low-carbon	100 g	29
36b	Steel, Cr2-Mo1	150 g	37		168	Cobalt-base alloy, Co41-Mo4-Nb3-Ta1-W4	150 g	37
37e	Brass, sheet	150 g	37		171	Magnesium-base alloy	100 g	37
39i	Benzoic acid, calorimetric	30 g	36					
					173a	Titanium alloy 6Al-4V	100 g	37
40h	Sodium oxalate, oxidimetric	60 g	36		174	Titanium alloy 4Al-4Mn	100 g	37
41a	Dextrose (glucose)	70 g	30		176	Titanium alloy 5Al-2.5Sn	100 g	37
42g	Tin, freezing-point std.	350 g	60		178	Steel, basic oxygen 0.4C	150 g	37
43h	Zinc, freezing-point	350 g	50		180	Fluorspar, high-grade	120 g	44
44f	Aluminum, freezing-point std.	200 g	75					
					181	Ore, lithium (Spodumene)	45 g	31
45d	Copper, freezing-point std.	450 g	50		182	Ore, lithium (Petalite)	45 g	31
49e	Lead, freezing-point std.	600 g	50		183	Ore, lithium (Lepidolite)	45 g	31
50c	Steel, W18-Cr4-V1	150 g	37		184	Bronze, leaded-tin	150 g	37
51b	Steel, electric furnace 1.2C	150 g	37		185e	Potassium hydrogen phthalate, pH	60 g	39
53e	Bearing metal, lead-base	150 g	37					
					186Ic	Potassium dihydrogen phosphate, pH	30 g	39
54d	Bearing metal, tin-base	170 g	37		186IIc	Disodium hydrogen phosphate, pH	30 g	34
55e	Iron, ingot	150 g	37		187b	Borax	30 g	34
57	Silicon, refined	60 g	33		188	Potassium hydrogen tartrate, pH	60 g	34
58a	Ferrosilicon (Si 75%)	75 g	50		189	Potassium tetroxalate, pH	65 g	34
59a	Ferrosilicon (Si 50%)	50 g	44					
					191	Sodium bicarbonate, pH	30 g	37
64b	Ferrochromium (high carbon)	100 g	35		192	Sodium carbonate, pH	30 g	37
65d	Steel, basic electric, 0.3C	150 g	37		193	Potassium Nitrate, Fertilizer	90 g	49
69a	Bauxite	50 g	31		194	Ammonium dihydrogen phosphate, Fertilizer	90 g	49
70a	Feldspar, potash	40 g	36		195	Ferrosilicon (75% Si, High Purity)	75 g	50
71	Calcium molybdate	60 g	33					
					196	Ferrochromium (low carbon)	100 g	49
72f	Steel, Cr-Mo (SAE X4130)	150 g	37		198	Silica refractory (0.2% Al$_2$O$_3$)	45 g	31
73c	Steel, stainless Cr13 (SAE 420)	150 g	37		199	Silica refractory (0.5% Al$_2$O$_3$)	45 g	31
76a	*Burned Refractory (Al$_2$O$_3$, 40%)*				217b-5	2,2,4-Trimethylpentane	5 ml	62
77a	*Burned Refractory (Al$_2$O$_3$, 60%)*				217b-8S	2,2,4-Trimethylpentane	8 ml	69
78a	*Burned Refractory (Al$_2$O$_3$, 70%)*							
					217b-25	2,2,4-Trimethylpentane	25 ml	184
79a	Fluorspar	120 g	44		217b-50	2,2,4-Trimethylpentane	50 ml	334
82b	Iron, nickel-chromium cast	150 g	37		300	Toluidine red toner	40 g	30
83c	Arsenic trioxide, oxidimetric	75 g	36		301	Yellow ocher	45 g	30
84h	Potassium phthalate, acid, acidimetric	60 g	30		302	Raw sienna	45 g	30
85b	Aluminum alloy, wrought	75 g	37					
					303	Burnt sienna	50 g	30
87a	Aluminum-silicon alloy	75 g	37		304	Raw umber	45 g	30
88a	Limestone, dolomitic	50 g	36		305	Burnt umber	50 g	30
89	Glass, lead-barium	45 g	31		306	Venetian red	60 g	30
90	Ferrophosphorus	75 g	33		307	Metallic brown	60 g	30
91	Glass, opal	45 g	31					
					308	Indian red	50 g	30
92	Glass, low boron	45 g	31		309	Mineral red	65 g	30
93a	Glass, high boron	ea	54		310	Bright red oxide	50 g	30
94c	Zinc-base die-casting alloy	150 g	37		311	Carbon black (high color)	10 g	30
97a	Clay, flint	60 g	86		312	Carbon black (all purpose)	20 g	30
98a	Clay, plastic	60 g	86					
					313	Black iron oxide	42 g	30
99a	Feldspar, soda	40 g	36		314	Yellow iron oxide, light lemon	20 g	30
100b	Steel, manganese (SAE T1340)	150 g	37		315	Yellow iron oxide, lemon	20 g	30
101f	Steel, stainless, Cr18-Ni9 (SAE 304)	100 g	37		316	Yellow iron oxide, orange	25 g	30
103a	Chrome refractory	60 g	31		317	Yellow iron oxide, dark orange	40 g	30
104	Magnesite, burned	60 g	31					
					318	Lampblack	15 g	30
105	Steel, high-sulfur 0.2C carbon only	150 g	29		319	Primrose chrome yellow	65 g	30
106b	Steel, Cr-Mo-Al (Nitralloy G)	150 g	37		320	Lemon chrome yellow	60 g	30
107b	Iron, cast, Ni-Cr-Mo	150 g	37		321	Medium chrome yellow	65 g	30
112	Silicon carbide	85 g	31		322	Light chrome orange	100 g	30
113a	*Zinc Concentrate*							
					323	Dark chrome orange	100 g	30
114L	Cement, turbidimetric and fineness std.	set(20)	57		324	Ultramarine blue	37 g	30
115a	Iron, cast, Cu-Ni-Cr	150 g	37		325	Iron blue	25 g	30
120b	Phosphate Rock (Florida)	90 g	49		326	Light chrome green	60 g	30
121d	Steel, Cr17-Ni11-Ti0.3, AISI 321	150 g	37		327	Medium chrome green	50 g	30
122e	Iron, cast, (car-wheel)	150 g	37					
					328	Dark chrome green	45 g	30
123c	Steel, Cr17-Ni11-Nb0.7, AISI 348	150 g	37		*329*	*Zinc concentrate*		
124d	Bronze (Cu85-Pb5-Sn5-Zn5) ounce metal	150 g	37		330	Copper, millheads	100 g	50
125b	Steel, high silicon	150 g	37		331	Copper, milltails	100 g	50
126c	Steel, high-nickel (36% Ni)	150 g	37		332	Copper, concentrate	50 g	50
127b	Solder (Sn40-Pb60)	150 g	37					
					333	Molybdenum, concentrate	35 g	50
129c	Steel, high-sulfur	150 g	37		335	Steel, B.O.H. 0.1C (carbon only)	300 g	31
131b	Steel, low-carbon silicon	100 g	31		336	Steel, Cr-V (carbon only), 1-g pins	75 g	35
					337	Steel, B.O.H. 1.1C (carbon only)	300 g	31
					339	Steel, stainless, Cr17-Ni9-0.2Se (SAE 303Se)	150 g	44

Table 5.1–10 (continued)
AVAILABILITY AND PRICE LIST

SRM	Type	Unit	Price	SRM	Type	Unit	Price
340	Ferroniobium	100 g	$ 49	481	Microprobe, Gold-silver wires	set	$ 13
341	Iron, ductile	150 g	37	482	Microprobe, Gold-copper wires	set	13
342	Iron, nodular	150 g	37	483	Microprobe, Iron-3% silicon	ea	5
342a	Iron, nodular	150 g	39	485	Austenite in ferrite	ea	8
343	Steel, stainless, Cr16-Ni2 (SAE 431)	150 g	37	493	Iron carbide in ferrite	ea	8
344	Steel, stainless, Cr15-Ni7-Mo2-Al1	150 g	37	592	Hydrocarbon blends - Blend No. 1	set	
345	Steel, stainless, Cr16-Ni4-Cu3	150 g	37	593	Hydrocarbon blends - Blend No. 2	set	
346	Steel, valve (Cr22-Ni4-Mn9)	150 g	44	594	Hydrocarbon blends - Blend No. 3	set	
348	Steel, Ni26-Cr15 (A286)	150 g	37	595	Hydrocarbon blends - Blend No. 4	set	
349	Nickel-base alloy (Ni57-Co14-Cr20)	150 g	37	596	Hydrocarbon blends - Blend No. 5	set	
350	Benzoic acid, acidimetric	30 g	36	597	Hydrocarbon blends - Blend No. 6	set	
352	Titanium, unalloyed, for hydrogen	20 g	39	598	Hydrocarbon blends - Blend No. 7	set	
353	Titanium, unalloyed, for hydrogen	20 g	39	599	Hydrocarbon blends - Blend No. 8	set	
354	Titanium, unalloyed, for hydrogen	20 g	39	607	Potassium Feldspar, Trace Rubidium and Strontium	5 g	4
355	Titanium, unalloyed, for oxygen	20 g	44	608	Glass, trace elements, set 1 each 614 and 616	set	20
356	Titanium alloy, 6Al-4V	20 g	44	609	Glass, trace elements, set 1 each 615 and 617	set	20
360a	Zircaloy-2	100 g	59	610	Glass, trace elements 500 ppm, 3 mm	ea	6
361	Steel, AISI 4340, chip	150 g	37	611	Glass, trace elements 500 ppm, 1 mm	ea	6
362	Steel, AISI 94B17 (modified), chip	150 g	37	612	Glass, trace elements 50 ppm, 3 mm	ea	6
363	Steel, Cr-V (modified), chip	150 g	37	613	Glass, trace elements 50 ppm, 1 mm	ea	6
364	Steel, high carbon (modified), chip	150 g	37	614	Glass, trace elements 1 ppm, 3 mm	ea	6
365	Iron, electrolytic, chip	150 g	37	615	Glass, trace elements 1 ppm, 1 mm	ea	
366	Set 1 ea of 361, 362, 363, 364 and 365	set	104	616	Glass, trace elements .02 ppm, 3 mm	ea	
370d	Zinc oxide (Set of 4)	8 kg	38	617	Glass, trace elements .02 ppm, 1 mm	ea	
371f	Sulfur (Set of 4)	6 kg	42	618	Glass, trace elements 3 mm	set	20
372g	Stearic acid (Set of 4)	3.2 kg	35	619	Glass, trace elements, 1 mm	set	
373f	Benzothiazyl disulfide (Set of 4)	2 kg	44	620	Glass plate, soda lime	pkg(3)	
374c	Tetramethylthiuram disulfide	2 kg	44	625	Zinc-base A	ea	
375f	Channel black (Set of 4)	28 kg	71	626	Zinc-base B	ea	
376a	Light magnesia	450 g	29	627	Zinc-base C	ea	
377	Phenyl-beta-naphthylamine	600 g	31	628	Zinc-base D	ea	
378a	Oil furnace black (Set of 4)	28 kg	40	629	Zinc-base E	ea	
379	Conducting black	5.5 kg	30	630	Zinc-base F	ea	
380	Calcium carbonate	6 kg	29	631	Zinc spelter (Modified)	ea	
381	Calcium silicate	4 kg	29	633	Cement, Portland B (red)		
382a	Gas furnace black (Set of 4)	32 kg	56	634	Cement, Portland C (gold)		
383	Mercaptobenzothiazole (Set of 4)	3.2 kg	37	635	Cement, Portland D (blue)		
384a	N-tertiary-Butyl-2-benzo-thiazolesulfenamide (Set of 4)	4.5 kg	63	636	Cement, Portland F (yellow)		
385b	Natural rubber	31.4 kg	109	637	Cement, Portland G (pink)		
386g	Styrene-butadiene type 1500	34 kg	71	638	Cement, Portland I (green)		
388f	Butyl rubber	37 kg	109	639	Cement, Portland J (clear)		
389	Styrene-butadiene, type 1503	34 kg	58	641	Titanium alloy 8Mn(A)	ea	
391	Acrylonitrile-butadiene rubber	25 kg	109	642	Titanium alloy 8Mn(B)	ea	
404a	Steel, basic electric	ea	34	643	Titanium alloy 8Mn(C)	ea	
405a	Steel, medium manganese	ea	34	644	Titanium alloy 2Cr-2Fe-2Mo(A)	ea	
407a	Steel, chromium-vanadium	ea	34	645	Titanium alloy 2Cr-2Fe-2Mo(B)	ea	
408a	Steel, chromium-nickel	ea	34	646	Titanium alloy 2Cr-2Fe-2Mo(C)	ea	
409b	Steel, nickel	ea	34	654a	Titanium alloy, 6Al-4V	ea	
413	Steel, A.O.H. 0.4C	ea	34	661	Steel, AISI 4340, rod	ea	
414	Steel, Cr-Mo (SAE 4140)	ea	34	662	Steel, AISI 94B17 (modified), rod	ea	
417a	Steel, B.O.H. 0.4C	ea	34	663	Steel, Cr-V (modified), rod	ea	
418	Steel, Cr-Mo (SAE X4130)	ea	34	664	Steel, high carbon (modified), rod	ea	
418a	Steel, Cr-Mo (SAE X4130)	ea	34	665	Iron, electrolytic, rod	ea	
420a	Iron, ingot	ea	34	666	Set of one each (661 & 665)	set	
427	Steel, Cr-Mo (boron only) (SAE 4150)	ea	34	667	Set of one each (662 & 663)	set	
431	Tin A	ea	39	668	Set of one each (661, 662, 663, 664 and 665)	set	
432	Tin B	ea	39	671	Nickel oxide 1	25 g	
433	Tin C	ea	39	672	Nickel oxide 2	25 g	
434	Tin D	ea	39	673	Nickel oxide 3	25 g	
435	Tin E	ea	39	680 L-1	Platinum, high-purity	ea	
436	Steel, special Cr6-Mo3-W10	ea	39	680 L-2	Platinum, high-purity	ea	
437	Steel, special Cr8-Mo2-W3-Co3	ea	39	681 L-1	Platinum, doped	ea	
438	Steel, Mo high speed (AISI-SAE-M30)	ea	39	681 L-2	Platinum, doped	ea	1.
439	Steel, Mo high speed (AISI-SAE-M36)	ea	39	682	Zinc, high-purity	ea	
440	Steel, special W high speed Cr2-W13-Co12	ea	39	683	Zinc metal	ea	
441	Steel, W high speed (AISI-SAE-TI)	ea	39	685-R	Gold, high-purity (rod)	ea	
442	Steel, stainless, Cr16-Ni10	ea	39	685-W	Gold, high-purity (wire)	ea	
443	Steel, stainless, Cr18.5-Ni9.5	ea	39	700c	Paper, light-sensitive	pkg	
444	Steel, stainless, Cr20.5-Ni10	ea	39	701c	Paper, standard faded strips	bklt	1
445	Steel, stainless, Cr13-Mo0.9 (Modified AISI 410)	ea	39	702	Plastic chips, light-sensitive	pkg	
446	Steel, stainless, Cr18-Ni9 (Modified AISI 321)	ea	39	703	Plastic chips, light-sensitive	pkg	
447	Steel, stainless, Cr24-Ni13 (Modified AISI 309)	ea	39	704a	Paper, internal tearing resistance	set(4)	
448	Steel, stainless, Cr9-Mo0.3 (Modified AISI 403)	ea	39	705	Polystyrene, narrow molecular weight	5 g	
449	Steel, stainless, Cr5.5-Ni6.5	ea	39	706	Polystyrene, broad molecular weight	18 g	
450	Steel, stainless, Cr3-Ni25	ea	39	707	Water vapor permeance, 12 sheets	pkg	
461	Steel, low-alloy A	ea	39	708	Glass, relative stress optical coefficient		
462	Steel, low-alloy B	ea	39	709	Glass, extra dense lead, 4 x 4 x 5 cm	500 g	
463	Steel, low-alloy C	ea	39	710	Glass, soda-lime silica	900 g	
464	Steel, low-alloy D	ea	39	711	Glass, lead-silica	1.3kg	
465	Iron, ingot E	ea	39	712	Glass, mixed alkali lead silicate	225 g	
466	Iron, ingot F	ea	39	713	Glass, dense barium crown	225 g	
467	Steel, low-alloy G	ea	39	714	Glass, alkaline earth alumina silicate	225 g	
468	Steel, low-alloy H	ea	39	715	Glass, alkali-free aluminosilicate	200 g	
479	Microprobe, Fe-Cr-Ni Alloy	ea	54	716	Glass, neutral (borosilicate)	250 g	
480	Microprobe, Tungsten - 20% Molybdenum alloy	ea	129	717	Glass, standard, borosilicate	450 g	
				718	Polycrystalline alumina, Elasticity	ea	

Table 5.1–10 (continued)

AVAILABILITY AND PRICE LIST

SRM	Type	Unit	Price
720	Sapphire, synthetic (Al_2O_3)	15 g	$ 60
723	Tris(hydroxymethyl)aminomethane, basimetric	50 g	55
724	Tris(hydroxymethyl)aminomethane, calorimetric	50 g	44
725	Mossbauer Differential Chemical Shift	ea	159
726	Selenium	450 g	49
728	Zinc	450 g	47
731L1	Borosilicate glass, thermal expansion, 2 in.	ea	75
731L2	Borosilicate glass, thermal expansion, 4 in.	ea	123
731L3	Borosilicate glass, thermal expansion, 6 in.	ea	171
733	Thermocouple wire, Silver - 28% Gold, 32 AWG (0.2019 mm dia.) and 3 meters long	ea	89
734S	Iron, electrolytic, thermal conductivity, rod 6.4 mm dia., 305 mm long	ea	79
734L1	Iron, electrolytic, thermal conductivity, rod, 31.8 mm dia., 152 mm long	ea	89
734L2	Iron, electrolytic, thermal conductivity, rod 31.8 mm dia., 305 mm long	ea	154
735S	*Stainless steel, thermal conductivity, rod 0.65 cm dia., 30 cm long*		
735M1	Stainless steel, thermal conductivity, rod 1.25 cm dia., 15 cm long	ea	104
735M2	Stainless steel, thermal conductivity, rod 1.25 cm dia., 30 cm long	ea	154
735L1	*Stainless steel, thermal conductivity, rod 3.5 cm dia., 5 cm long*		
735L2	*Stainless steel, thermal conductivity, rod 3.5 cm dia., 10 cm long*		
736L1	Copper, thermal expansion, 2 in.	ea	75
736L2	Copper, thermal expansion, 4 in.	ea	123
736L3	Copper, thermal expansion, 6 in.	ea	171
737L1	*Tungsten, thermal expansion*		
737L2	*Tungsten, thermal expansion*		
737L3	*Tungsten, thermal expansion*		
739L1	Fused-silica, thermal expansion, 2 in.	ea	75
739L2	Fused-silica, thermal expansion, 4 in.	ea	123
739L3	Fused-silica, thermal expansion, 6 in.	ea	171
740	Zinc, primary freezing-point std.	350 g	100
741	Tin, primary freezing-point std.	350 g	125
742	Alumina, high temperature melting point	10 g	67
745	Gold, vapor pressure std.	ea	89
746	Cadmium, vapor pressure std.	ea	69
747	*Platinum, vapor pressure std.*		
748	Silver, vapor pressure std.	ea	79
749	*Tungsten, vapor pressure*		
755	Quartz, SiO_2	2 g	39
756	Potassium nitrate	5 g	39
758	DTA temperature std. (125-435 °C)	set(5)	49
759	DTA temperature std. (295-675 °C)	set(5)	49
760	DTA temperature std. (570-940 °C)	set(5)	49
763-1	*Aluminum, magnetic susceptibility, cylinder*		
763-2	*Aluminum, magnetic susceptibility, wire*		
763-3	*Aluminum, magnetic susceptibility, (GOUY) rod*		
764-1	*Platinum, magnetic susceptibility, cylinder*		
764-2	*Platinum, magnetic susceptibility, wire*		
765-1	*Palladium, magnetic susceptibility, cylinder*		
765-2	*Palladium, magnetic susceptibility, wire*		
765-3	*Palladium, magnetic susceptibility, sponge*		
766-1	*Manganese Fluoride, magnetic susceptibility, cube*		
767	Superconducting fixed point	ea	250
803a	Steel, A.O.H. 0.6C	ea	34
D803a	Steel, A.O.H. 0.6C	ea	39
804a	Steel, basic electric	ea	34
805a	Steel, medium manganese	ea	34
D805a	Steel, medium manganese	ea	39
807a	Steel, chromium-vanadium	ea	34
D807a	Steel, chromium-vanadium	ea	39
808a	Steel, chromium-nickel	ea	34
809b	Steel, nickel	ca	34
D809b	Steel, nickel	ea	39
810a	Steel, Cr2-Mo1	ea	34
817a	Steel, B.O.H. 0.4C	ea	34
820a	Iron, ingot	ea	34
D820a	Iron, ingot	ea	39
821	Steel, Cr-W, 0.9C	ea	34
827	Steel, Cr-Mo (boron only) (SAE 4150)	ea	34
D836	Steel, special (Cr6-Mo3-W10)	ea	54
837	Steel, special (Cr8-Mo2-W3-Co3)	ea	47
D837	Steel, special (Cr8-Mo2-W3-Co3)	ea	54
838	Steel, Mo high speed (AISI-SAE-M30)	ea	47
D838	Steel, Mo high speed (AISI-SAE-M30)	ea	$ 54
839	Steel, Mo high speed (AISI-SAE-M36)	ea	47
D839	Steel, Mo high speed (AISI-SAE-M36)	ea	54
840	Steel, special W high speed (Cr2-W13-Co12)	ea	47
D840	Steel, special W high speed (Cr2-W13-Co12)	ea	54
841	Steel, W high speed (AISI-SAE-TI)	ea	47
D841	Steel, W high speed (AISI-SAE-TI)	ea	54
845	Steel, Cr13-Mo0.9 (Modified AISI 410)	ea	47
D845	Steel, Cr13-Mo0.9 (Modified AISI 410)	ea	54
846	Steel, Cr18-Ni9 (Modified AISI 321)	ea	54
D847	Steel, Cr24-Ni13 (Modified AISI 309)	ea	54
849	Steel, Cr5.5-Ni6.5	ea	47
D849	Steel, Cr5.5-Ni6.5	ea	54
850	Steel, Cr3-Ni25	ea	47
D850	Steel, Cr3-Ni25	ea	54
911	Cholesterol, clinical	0.5 g	34
912	Urea, clinical	25 g	40
913	Uric acid, clinical	10 g	34
914	Creatinine, clinical	10 g	40
915	Calcium carbonate, clinical	20 g	34
916	Bilirubin, clinical	100 mg	96
917	D-Glucose, clinical	25 g	47
918	Potassium chloride, clinical	30 g	44
919	Sodium chloride, clinical	30 g	44
920	D-Mannitol, clinical	50 g	61
921	*Cortisol*		
922	Tris(hydroxymethyl)aminomethane clinical	25 g	44
923	Tris(hydroxymethyl)aminomethane hydrochloride, clinical	35 g	44
924	Lithium carbonate, clinical	30 g	54
925	*VMA (4-Hydroxy-3-methoxymandelic acid) clinical*		
930a	Glass filters for spectrophotometry, clinical	set(3)	304
931	Liquid filters for spectrophotometry, clinical, 3 sets of 4	set	69
944	Plutonium sulfate tetrahydrate assay	0.5 g	80
945	Plutonium metal, std matrix	5 g	504
946	Plutonium, 12% isotopic	0.25 g	154
947	Plutonium, 18% isotopic	0.25 g	154
948	Plutonium sulfate hydrate	0.25 g	71
949d	Plutonium metal assay	0.5 g	154
950a	Uranium oxide (U_3O_8)	25 g	32
951	Boric acid	100 g	59
952	Boric acid, 95% enriched [10]B	0.25 g	44
953	Neutron density monitor wire, 1 meter long	ea	43
953-L1	Neutron density monitor wire, 5 meters long	ea	100
953-L2	Neutron density monitor wire, 10 meters long	ea	171
953-L3	Neutron density monitor wire, 25 meters long	ea	385
960	Uranium metal, assay	26 g	54
975	Sodium chloride - isotopic	0.25 g	44
976	Copper metal - isotopic	0.25 g	44
977	Sodium bromide - isotopic	0.25 g	44
978	Silver nitrate - isotopic	0.25 g	44
979	Chromium nitrate - isotopic	0.25 g	44
980	Magnesium metal - isotopic	0.25 g	44
981-3	Lead - isotopic	set	109
984	Rubidium chloride, isotopic	1 g	47
987	Strontium carbonate, isotopic	1 g	44
988	Strontium-84 spike, isotopic	1 mg	154
999	Potassium chloride, primary	60 g	57
1000	Enameled iron plaques	set(3)	29
1002b	Hardboard sheet, 4 specimens	set	39
1003	Glass spheres (5-30 μm)	40 g	37
1004	Glass beads	63 g	52
1006	Smoke density std., non-flaming	pkg(3)	36
1007	Smoke density std., flaming	pkg(3)	34
1008	Photographic step tablet, 0-4	ea	72
1009	Photographic step tablet 0-3	ea	58
1010a	Microcopy test chart	set	14
1011	Cement, Portland	set	32
1013	Cement, Portland	set	32
1014	Cement, Portland	set	32
1015	Cement, Portland	set	32
1016	Cement, Portland	set	32
1017a	Glass beads (sieve nos. 50-140)	84 g	44
1018a	Glass beads (sieve nos. 25-60)	74 g	44
1019	Glass spheres (sieves No.8-18)	100 g	35
1020	Zinc sulfide phosphor	14 g	28
1021	Zinc silicate phosphor	28 g	28
1022	Zinc sulfide phosphor	14 g	28
1023	Zinc-cadmium sulfide phosphor (Ag activator)	14 g	30
1024	Zinc-cadmium sulfide phosphor (Cu activator)	14 g	28
1025	Zinc phosphate phosphor	28 g	28
1026	Calcium tungstate phosphor	28 g	28
1027	Magnesium tungstate phosphor	28 g	28
1028	Zinc silicate phosphor	28 g	28
1029	Calcium silicate phosphor	14 g	28

Table 5.1–10 (continued)
AVAILABILITY AND PRICE LIST

SRM	Type	Unit	Price
1030	Magnesium arsenate phosphor	28 g	$ 28
1031	Calcium halophosphate phosphor	28 g	28
1032	Barium silicate phosphor	28 g	28
1033	Calcium phosphate phosphor	28 g	28
1051b	Barium cyclohexanebutyrate	5 g	35
1052b	Bis(1-phenyl-1,3-butanediono) oxovanadium (IV)	5 g	35
1053a	Cadmium cyclohexanebutyrate	5 g	35
1055b	Cobalt cyclohexanebutyrate	5 g	35
1057b	Dibutyltin bis(2-ethylhexanoate)	5 g	35
1059b	Lead cyclohexanebutyrate	5 g	35
1060a	Lithium cyclohexanebutyrate	5 g	35
1061c	Magnesium cyclohexanebutyrate	5 g	35
1062a	Manganous cyclohexanebutyrate	5 g	35
1063a	Methyl borate	5 g	35
1064	Mercuric cyclohexanebutyrate	5 g	35
1065b	Nickel cyclohexanebutyrate	5 g	35
1066a	Octaphenylcyclotetrasiloxane	5 g	35
1069b	Sodium cyclohexanebutyrate	5 g	35
1070a	Strontium cyclohexanebutyrate	5 g	35
1071a	Triphenyl phosphate	5 g	35
1073b	Zinc cyclohexanebutyrate	5 g	35
1074a	Calcium 2-ethylhexanoate	5 g	35
1075a	Aluminum 2-ethylhexanoate	5 g	35
1076	Potassium erucate	5 g	35
1077a	Silver 2-ethylhexanoate	5 g	35
1078b	Tris(1-phenyl-1,3-butanediono) chromium (III)	5 g	37
1079b	Tris(1-phenyl-1,3-butanediono) iron (III)	5 g	35
1080	Bis(1-phenyl-1,3-butanediono) copper (II)	5 g	35
1089	Gasometric, Set: 1 ea of 1095, 1096, 1097, 1098, and 1099	set(5)	79
1090	Gasometric, Iron, ingot	ea	59
1091	Gasometric, Steel, stainless (AISI 431)	ea	59
1092	Gasometric, Steel, vacuum-melted	ea	59
1093	Gasometric, Steel, valve	ea	59
1094	Gasometric, Steel, maraging	ea	59
1095	Gasometric, Steel, AISI 4340, rod	ea	37
1096	Gasometric, Steel, AISI 94B17 (modified), rod	ea	37
1097	Gasometric, Steel, Cr-V (modified), rod	ea	37
1098	Gasometric, Steel, high-carbon (modified), rod	ea	37
1099	Gasometric, Iron, electrolytic, rod	ea	37
1101	Brass, cartridge B	ea	69
C 1101	Brass, cartridge B	ea	69
1102	Brass, cartridge C	ea	69
C 1102	Brass, cartridge C	ea	69
1103	Brass, free-cutting A	ea	69
C 1103	Brass, free-cutting A	ea	69
1104	Brass, free-cutting B	ea	69
C 1104	Brass, free-cutting B	ea	69
1105	Brass, free-cutting C	ea	69
C 1105	Brass, free-cutting C	ea	69
1106	Brass, naval A	ea	69
C 1106	Brass, naval A	ea	69
1107	Brass, naval B	ea	69
C 1107	Brass, naval B	ea	69
1108	Brass, naval C	ea	69
C 1108	Brass, naval C	ea	69
1109	Brass, red A	ea	69
C 1109	Brass, red A	ea	69
1110	Brass, red B	ea	69
C 1110	Brass, red B	ea	69
1111	Brass, red C	ea	69
C 1111	Brass, red C	ea	69
1112	Gilding metal A	ea	69
C 1112	Gilding metal A	ea	69
1113	Gilding metal B	ea	69
C 1113	Gilding metal B	ea	69
1114	Gilding metal C	ea	69
C 1114	Gilding metal C	ea	69
1115	Bronze, commercial A	ea	69
C 1115	Bronze, commercial A	ea	69
1116	Bronze, commercial B	ea	69
C 1116	Bronze, commercial B	ea	69
1117	Bronze, commercial C	ea	69
C 1117	Bronze, commercial C	ea	69
1118	Brass, aluminum A	ea	69
C 1118	Brass, aluminum A	ea	69
1119	Brass, aluminum B	ea	69
C 1119	Brass, aluminum B	ea	69
1120	Brass, aluminum C	ea	69
C 1120	Brass, aluminum C	ea	69
1121	Beryllium copper CA-172	ea	69
C 1121	Beryllium copper CA-172	ea	69

SRM	Type	Unit	Price
1122	Beryllium copper CA-170	ea	$ 69
C 1122	Beryllium copper CA-170	ea	69
1123	Beryllium copper CA-175	ca	69
C 1123	Beryllium copper CA-175	ea	69
1131	Solder (Sn40-Pb60)	ea	54
1132	Bearing metal, lead-base	ea	54
1134	Steel, high silicon	ea	54
1135	Steel, high-silicon	ea	54
1136	Steel, high-sulfur	ea	54
1138	Steel, cast 1	ea	69
1139	Steel, cast 2	ea	69
1140	Iron, ductile 1	ea	69
1141	Iron, ductile 2	ea	69
1142	Iron, ductile 3	ea	69
1143	Iron, blast furnace 1	ea	69
1144	Iron, blast furnace 2	ea	69
1147	Iron, white cast	ea	69
1148	Iron, white	ea	69
1149	Iron, white	ea	69
1152	Steel, stainless B (Cr18-Ni10)	ea	69
1154	Steel, stainless D (Cr19-Ni10)	ea	69
1155	Steel, stainless, Cr18-Ni12-Mo2	ea	69
1156	Steel, maraging (disk form)	ea	69
1157	Steel, tool	ea	54
1158	Steel, high nickel (36% Ni)	ea	54
1159	Nickel-base alloy, 49% Ni, balance Fe	ea	69
1160	Nickel-base alloy, 80% Ni, 4% Mo, balance Fe	ea	69
1165	Iron, ingot E	ea	69
1166	Iron, ingot F	ea	69
1167	Steel, low-alloy G	ea	69
1171	Steel, Cr17-Ni11-Ti0.3, AISI 321, disk	ea	54
1172	Steel, Cr17-Ni11-Nb0.7, AISI 348, disk	ea	54
1185	Steel, stainless, AMS 5360A, AISI 316 alloy	ea	69
1197	High-temperature alloy, M308		
1198	High-temperature alloy, Incaloy 901		
1199	High-temperature alloy, L605		
1200	High-temperature alloy, S816		
1201	High-temperature alloy, Hastaloy X		
1206-2	High temperature alloy, Rene 41	ea	54
1207-1	High temperature alloy, Waspaloy (No. 1)	ea	54
1207-2	High temperature alloy, Waspaloy (No. 2)	ea	54
1208-1	High temperature alloy, Inco 718 (No. 1)	ea	54
1208-2	High temperature alloy, Inco 718 (No. 2)	ea	54
1209	High temperature alloy, Set, 1 ea of 1206-2, 1207-1, 1207-2, 1208-1, and 1208-2	set	189
1210	Zirconium metal A	ea	94
1261	Steel, AISI 4340, disk	ea	49
1262	Steel, AISI 94B17 (modified), disk	ea	49
1263	Steel, Cr-V (modified), disk	ea	49
1264	Steel, high carbon (modified), disk	ea	49
1265	Iron, electrolytic, disk	ea	49
1266	Set, 1 ea of 1261, 1262, 1263, 1264, and 1265	set	179
1301	Metal coating, nonmagnetic, 0.00010 in thick	ea	39
1302	Metal coating, nonmagnetic, 0.00025 in thick	ea	39
1303	Metal coating, nonmagnetic, 0.00050 in thick	ea	39
1304	Metal coating, nonmagnetic, 0.00075 in thick	ea	39
1305	Metal coating, nonmagnetic, 0.0010 in thick	ea	39
1306	Metal coating, nonmagnetic, 0.0015 in thick	ea	39
1307	Metal coating, nonmagnetic, 0.0020 in thick	ea	39
1308	Metal coating, nonmagnetic, 0.0025 in thick	ea	39
1309	Metal coating, nonmagnetic, 0.0027 in thick	ea	39
1310	Metal coating, nonmagnetic, 0.0032 in thick	ea	39
1311	Metal coating, nonmagnetic, 0.0055 in thick	ea	39
1312	Metal coating, nonmagnetic, 0.0080 in thick	ea	39
1313	Metal coating, nonmagnetic, 0.010 in thick	ea	39
1314	Metal coating, nonmagnetic, 0.015 in thick	ea	39
1315	Metal coating, nonmagnetic, 0.020 in thick	ea	39
1316	Metal coating, nonmagnetic, 0.025 in thick	ea	39
1317	Metal coating, nonmagnetic, 0.03 in thick	ea	39
1318	Metal coating, nonmagnetic, 0.04 in thick	ea	39
1319	Metal coating, nonmagnetic, 0.06 in thick	ea	39
1320	Metal coating, nonmagnetic, 0.08 in thick	ea	39
1331	Metal coating, magnetic, 0.00012 in thick	ea	39
1332	Metal coating, magnetic, 0.00035 in thick	ea	39
1333	Metal coating, magnetic, 0.00055 in thick	ea	39
1334	Metal coating, magnetic, 0.00075 in thick	ea	39
1335	Metal coating, magnetic, 0.0010 in thick	ea	39
1336	Metal coating, magnetic, 0.0013 in thick	ea	39
1337	Metal coating, magnetic, 0.0016 in thick	ea	39
1338	Metal coating, magnetic, 0.0020 in thick	ea	39
1339	Metal coating, magnetic, 0.0025 in thick	ea	39
1341	Metal coating, magnetic, 0.00012 in thick	ea	39
1342	Metal coating, magnetic, 0.00035 in thick	ea	39
1343	Metal coating, magnetic, 0.00065 in thick	ea	39
1344	Metal coating, magnetic, 0.0010 in thick	ea	39
1345	Metal coating, magnetic, 0.0015 in thick	ea	39
1346	Metal coating, magnetic, 0.0020 in thick	ea	5
1351	Set of one each 1307 and 1311	set(2)	5
1352	Set of one each 1332 and 1334	set(2)	5
1353	Set of one each 1335 and 1339	set(2)	5

Table 5.1–10 (continued)
AVAILABILITY AND PRICE LIST

SRM	Type	Unit	Price	SRM	Type	Unit	Price
1361	Set of one each 1302, 1303, 1305, and 1307	set(4)	$ 75	1627	Sulfur dioxide permeation tube 2 cm	ea	$ 65
1362	Set of one each 1306, 1310, 1311, and 1312	set(4)	75	1630	Trace mercury in coal	50 g	49
				1631	Sulfur in coal, three concentrations, 5 sets of 3	set	57
1363	Set of one each 1313, 1314, 1315, and 1316	set(4)	75	1651	Zirconium-barium chromate heat source powder (ca 350 cal/g)	50 g	59
1364	Set of one each 1317, 1318, 1319, and 1320	set(4)	75	1652	Zirconium-barium chromate heat source powder (ca 390 cal/g)	50 g	59
1365	Set of one each 1331, 1332, 1333, and 1334	set(4)	75	1653	Zirconium-barium chromate heat source powder (ca 425 cal/g)	50 g	59
1366	Set of one each 1335, 1336, 1337, and 1338	set(4)	75	1654	α-Quartz for hydrofluoric acid solution calorimetry	25 g	179
1367	Set of each 1341, 1342, 1343, and 1344	set(4)	75	1810	Linerboard for tape test	pkg	37
1368	Set of one each 1312, 1313, 1314, and 1315	set(4)	75	2001	Aluminum on glass, specular spectral reflectance	ea	279
1369	Set of one each 1316, 1317, 1318, and 1319	set(4)	75	2002	Aluminum on glass, specular spectral reflectance	ea	279
1370	Set of one each 1312, 1313, 1314, 1315, 1316, 1317, 1318, and 1319	set(8)	146	2003	Aluminum on glass, specular spectral reflectance	ea	279
1371	Gold coating (Fe-Ni-Co) 30 microinches	ea	70	2005	Gold on glass, specular spectral reflectance	ea	279
1372	Gold coating (Fe-Ni-Co) 60 microinches	ea	70	2006	Gold on glass, specular spectral reflectance	ea	279
1373	Gold coating (Fe-Ni-Co) 120 microinches	ea	70	2007	Gold on glass, specular spectral reflectance	ea	279
1374	Gold coating (Fe-Ni-Co) 280 microinches	ea	70	2008	Gold on glass, specular spectral reflectance	ea	279
1375	Gold coating (Nickel) 30 microinches	ea	70	2101-5	Color std.	set	379
1376	Gold coating (Nickel) 60 microinches	ea	70	2106	ISCC-NBS color charts	set	9
1377	Gold coating (Nickel) 120 microinches	ea	70	2141	Urea	2 g	37
1378	Gold coating (Nickel) 350 microinches	ea	70	2142	o-Bromobenzoic acid	2 g	37
1381	Set of one each 1371 and 1372	set(2)	113	2143	p-fluorobenzoic acid		
1382	Set of each 1372 and 1373	set(2)	113	2144	m-chlorobenzoic acid		
1383	Set of each 1373 and 1374	set(2)	113	2186-I	Potassium dihydrogen phosphate, pD	30 g	45
1384	Set of one each 1375 and 1376	set(2)	113	2186-II	Disodium hydrogen phosphate, pD	30 g	45
1385	Set of one each 1376 and 1377	set(2)	113	2191	Sodium bicarbonate, pD	30 g	45
1386	Set of one each 1377 and 1378	set(2)	113	2192	Sodium carbonate, pD	30 g	45
1398	Set of one each 1371, 1372, 1373, and 1374	set(4)	186	2201	Sodium chloride ion-selective electrode	125 g	38
				2202	Potassium chloride ion-selective electrode	160 g	38
1399	Set of one each 1375, 1376, 1377, and 1378	set(4)	186	2301	Gold coating (epoxy) 30 microinches	ea	70
1402	Emittance std., 1/2 in. disk	ea	184	2302	Gold coating (epoxy) 60 microinches	ea	70
1403	Emittance std., 7/8 in. disk	ea	194	2303	Gold coating (epoxy) 120 microinches	ea	70
1404	Emittance std., 1 in. disk	ea	209	2304	Gold coating (epoxy) 280 microinches	ea	70
1405	Emittance std., 1 1/8 in. disk	ea	244	2305	Set of one each 2301 and 2302	set(2)	113
1406	Emittance std., 1 1/4 in. disk	ea	259	2306	Set of one each 2302 and 2303	set(2)	113
1407	Emittance std., 2 in. × 2 in.	ea	394	2307	Set of one each 2303 and 2304	set(2)	113
1408	Emittance std., 1 in. × 10 in.	ea	759	2308	Set of one each 2301, 2302, 2303, and 2304	set(4)	186
1409	Emittance std., 3/4 in. × 10 in.	ea	609	2311	Gold coating (copper) 30 microinches	ea	70
1420	Emittance std., 1/2 in. disk	ea	184	2312	Gold coating (copper) 60 microinches	ea	70
1421	Emittance std., 7/8 in. disk	ea	184	2313	Gold coating (copper) 120 microinches	ea	70
1422	Emittance std., 1 in. disk	ea	184	2314	Gold coating (copper) 280 microinches	ea	70
1423	Emittance std., 1 1/8 in. disk	ea	184	2315	Set of one each 2311 and 2312	set(2)	113
1424	Emittance std., 1 1/4 in. disk	ea	184	2316	Set of one each 2312 and 2313	set(2)	113
1425	Emittance std., 2 in. × 2 in.	ea	184	2317	Set of one each 2313 and 2314	set(2)	113
1427	Emittance std., 3/4 in. × 10 in.	ea	184	2318	Set of one each 2311, 2312, 2313, and 2314	set(4)	186
1428	Emittance std., 1/4 in. × 8 in.	ea	184	2331	Tin coating 60 microinches	ea	70
1440	Emittance std., 1/2 in. disk	ea	184	2332	Tin coating 110 microinches	ea	70
1441	Emittance std., 7/8 in. disk	ea	184	2533	Tin coating 160 microinches	ea	70
1442	Emittance std., 1 in. disk	ea	184	2334	Tin coating 275 microinches	ea	70
1443	Emittance std., 1 1/8 in. disk	ea	184	2335	Tin coating 650 microinches	ea	70
1444	Emittance std., 1 1/4 in. disk	ea	184	2336	Tin coating 750 microinches	ea	70
1445	Emittance std., 2 in. × 2 in.	ea	184	2338	Set of one each 2332 and 2335	set(2)	113
1475	Polyethylene, linear	50 g	104	2339	Set of one each 2331, 2333, 2334, and 2336	set(4)	186
1476	Polyethylene, branched	50 g	79	2340	Set of one each 2331, 2332, 2333, 2334, 2335, and 2336	set(6)	265
1511	Cyclohexane - dielectric	400 ml	129	3200	Tape, magnetic, secondary std.	ea	699
1512	1,2 Dichloroethane dielectric	400 ml	124	4200-B	Cesium-137, gamma-ray point source	ea	64
1513	Nitrobenzene dielectric	400 ml	124	4201-B	Niobium-94, gamma-ray point source	ea	156
1516	Permittivity Std., 38 mm × 2.5 mm	ea	197	4202	Cadmium-109, gamma-ray point source	ea	97
1517	Permittivity Std., 38 mm × 5 mm	ea	197	4203	Cobalt-60, gamma-ray point source		
1518	Permittivity Std., 51 mm × 2.5 mm	ea	197	4205	Thorium-228, gamma-ray point source	ea	102
1519	Permittivity Std., 51 mm × 5 mm	ea	197	4206	Thorium-228, gamma-ray point source	ea	102
1541	Mossbauer, iron foil	ea	154	4207	Cesium-137, gamma-ray point source	ea	64
1571	Botanical, orchard leaves, trace element	75 g	72	4210	Cobalt-60, gamma-ray point source	ea	90
1573	Botanical, tomato leaves			4211	Americium-241, gamma-ray point source	ea	132
1577	Biological, Liver, bovine	50 g	92	4212	Krypton-85, gamma-ray point source	ea	164
1578	Biological, Tuna, albacore			4213	Americium-241, gamma-ray point source	ea	132
1579	Powdered lead-base paint	35 g	35	4214	Cobalt-57, gamma-ray point source		
1601	Carbon dioxide in nitrogen, 308 ppm	cyl	154	4215	Mixed radionuclides, gamma-ray point source		
1602	Carbon dioxide in nitrogen, 346 ppm	cyl	154	4216	Mixed radionuclides, gamma-ray point source		
1603	Carbon dioxide in nitrogen, 384 ppm	cyl	154				
1604a	Oxygen in nitrogen, 1.5 ppm	cyl	114	4222	Carbon-14(n-hexadecane) soln.std.	3 g	59
1605	Oxygen in nitrogen, 10 ppm	cyl	114	4223	Carbon-14(n-hexadecane) soln.std.	3 g	59
1606	Oxygen in nitrogen, 112 ppm	cyl	114	4224	Carbon-14(n-hexadecane) soln.std.	3 g	59
1607	Oxygen in nitrogen, 211 ppm	cyl	114	4226	Nickel-63, soln.std.	4 g	153
1608	Oxygen in nitrogen, 978 ppm	cyl	114	4228	Selenium-75, soln.std.	4.6 g	122
1609	Oxygen in nitrogen, 20.98 mole percent	cyl	114	4229	Aluminum-26, soln. std.	4.6 g	204
1610	Hydrocarbon in air, 0.103 mole percent	cyl	178	4230	Chromium-51, soln. std.		
1611	Hydrocarbon in air, 0.0107 mole percent	cyl	178	4231	Cobalt-56, soln. std.		
1613	Hydrocarbon in air, 0.000102 mole percent	cyl	178	4232	Silver-110m, soln. std.		
1621	Sulfur in residual fuel oil, 1.05 wt percent	100 ml	34	4233	Cesium-137-Barium-137m, soln. std.		
1622	Sulfur in residual fuel oil, 2.14 wt percent	100 ml	34	4234	Barium-140-Lanthanum-140, soln. std.		
1623	Sulfur in residual fuel oil, 0.268 wt percent	100 ml	34	4235	Krypton-85, gamma-ray gas std.	ea	104
1624	Sulfur in distillate fuel oil, 0.211 wt percent	100 ml	34				
1625	Sulfur dioxide permeation tube 10 cm	ea	65				
1626	Sulfur dioxide permeation tube 5 cm	ea	65				

Table 5.1–10 (continued)
AVAILABILITY AND PRICE LIST

SRM	Type	Unit	Price $
4236	*Xenon-133, gas std.*		
4240	*Bismuth-207, gamma-ray point source*		
4242-B	Mixed radionuclides	450 ml	54
4243-B	Mixed radionuclides	50 ml	54
4244-B	Mixed radionuclides	15 ml	54
4245	*Carbon-14 (Na_2CO_3 in H_2O)*		
4246	*Carbon-14 (Na_2CO_3 in H_2O)*		
4247	*Carbon-14 (Na_2CO_3 in H_2O)*		
4252	Mixed radionuclides, test std.	450 ml	50
4253	Mixed radionuclides, test std.	50 ml	50
4300	Argon-37, gas std.	10 ml	72
4301	Argon-37, gas std.	10 ml	72
4302	*Argon-39, gas std.*		
4303	*Argon-39, gas std.*		
4304	*Xenon-131m, gas std.*		
4305	*Xenon-131m, gas std.*		
4306	*Xenon-133, gas std.*		
4307	*Xenon-133, gas std.*		
4900	*Polonium-210, alpha-particle source*		
	On Request		
4901	*Polonium-210, alpha-particle source*		
	On Request		
4902	*Polonium-210, alpha-particle source*		
	On Request		
4904-D	Americium-241, alpha-particle source	ea	128
4906	Plutonium-238, alpha-particle source	ea	162
4907	*Gadolinium-148*		
4921-C	Sodium-22, soln. std.	3 g	46
4922-E	Sodium-22, soln. std.	5 g	65
4925	Carbon-14 (benzoic acid in toluene)	3 g	52
4926	Hydrogen-3 (water)	25 g	52
4927	Hydrogen-3 (water)	3 g	52
4929-C	Iron-55, soln. std.	4 g	119
4935-C	Krypton-85, beta-particle gas std.	10 ml	104
4940-B	Promethium-147, soln. std.	3 g	64
4941-C	Cobalt-57, soln. std.	5 g	112
4943	Chlorine-36, soln. std.	3 g	47
4947	Hydrogen-3 (tritiated toluene)	4 g	50
4949	*Iodine-129*		
4950-B	Radium solution std., 10^{-9} g (Rd analysis)	20 g	85
4951	Radium solution std., 10^{-11} g (Rd analysis) ...	100 g	52
4953	Radium solution std., 10^{-8} g (Rd analysis)	20 g	85
4955	Radium solution std., 0.1 µg Ra	5 g	67
4956	Radium solution std., 0.2 µg Ra	5 g	67
4957	Radium solution std., 0.5 µg Ra	5 g	67
4958	Radium solution std., 1 µg Ra	5 g	67

SRM	Type	Unit	Price $
4959	Radium solution std., 2 µg Ra	5 g	
4960	Radium solution std., 5 µg Ra	5 g	
4961	Radium solution std., 10 µg Ra	5 g	
4962	Radium solution std., 20 µg Ra	5 g	
4963	Radium solution std., 50 µg Ra	5 g	
4964-B	Radium solution std., 102 µg Ra	5 g	
4990-B	Carbon-14, contemporary std. for dating ...	1 lb	
4991-C	Sodium-22, gamma-ray point source	ea	
4996-B	Sodium-22, gamma-ray point source	ea	
U-0002	Uranium oxide - depleted (U-235)	1 g	
U-005	Uranium oxide - depleted (U-235)	1 g	
U-010	Uranium oxide - enriched (U-235)	1 g	
U-015	Uranium oxide - enriched (U-235)	1 g	
U-020	Uranium oxide - enriched (U-235)	1 g	
U-030	Uranium oxide - enriched (U-235)	1 g	
U-050	Uranium oxide - enriched (U-235)	1 g	
U-100	Uranium oxide - enriched (U-235)	1 g	
U-150	Uranium oxide - enriched (U-235)	1 g	
U-200	Uranium oxide - enriched (U-235)	1 g	
U-350	Uranium oxide - enriched (U-235)	1 g	
U-500	Uranium oxide - enriched (U-235)	1 g	
U-750	Uranium oxide - enriched (U-235)	1 g	
U-800	Uranium oxide - enriched (U-235)	1 g	
U-850	Uranium oxide - enriched (U-235)	1 g	
U-900	Uranium oxide - enriched (U-235)	1 g	
U-930	Uranium oxide - enriched (U-235)	1 g	
U-970	Uranium oxide - enriched (U-235)	1 g	

B. Research Materials

RM	Type	Unit	Price $
RM-1C	Ultra-purity aluminum, single crystal cube	ea	
RM-1R	Ultra-purity aluminum, polycrystaline rod	ea	

C. General Materials

GM	Type	Unit	Price $
GM-1	Hydrogen in steel	set	
GM-2	Hydrogen in steel	set	
GM-5	Nickel and Vanadium in Residual Oil	500 ml	
GM-2007	Clay, Attapulgus	18 kg	

[a]SRM's listed in italics are in preparation.

From *Supplement on Standard Reference Materials,* NBS Special Publication 260, U.S. Government Printing Office, Washington, D.C., 1973.

5.2 GENERAL STANDARDS

INTERNATIONAL ORGANIZATIONS PRODUCING NUCLEAR STANDARDS

Early in 1963 Section Committee N6, Reactor Safety Standards of the American Standards Association (now the American National Standards Institute) established Subcommittee N6.9 to maintain an indexed catalog of standards of interest to the nuclear industry for the U.S.A. and other countries. In order to fulfill its designated functions, Subcommittee N6.9 first compiled and published the compilation of U.S.A. nuclear standards, and next undertook a survey of nuclear standards of other countries. This table reprints part of the fifth edition of the Compilation of National and International Nuclear Standards (ORNL-NSIC-63); the fourth edition of the Compilation of U.S.A. Nuclear Standards was published in 1967 as ORNL-NSIC-43.

Report ORNL-NSIC-63 contains a description of the organizations involved. Readers interested in additional descriptive information are referred to the bibliography of this report and to the *Selected Bibliography of Radiation Protection Organizations* published in 1963 by the American Conference of Governmental Industrial Hygienists. An additional reference is the *Atomic Handbook*, edited by John W. Shurtall, published by Morgan Brothers, Ltd., London.

While committee members have undertaken to contact all organizations and committees actively engaged in the generation of nuclear standards, the nature of the work is such that information is soon outdated. Therefore, the subcommittee cannot guarantee the complete accuracy of the report.

Table 5.2–1

International organization	Standard organization	Informant, title, address
1. Bureau Veritas	Maritime Technical Committee; Subcommittee for Nuclear Energy	P. Weiss, Secretary Bureau Veritas 31 rue Henri Rochefort Paris 17, France
2. CERN (European Organization for Nuclear Research)		Prof. Victor F. Weisskopf M.I.T. Cambridge, Massachusetts 02139 United States of America
		Prof. B. Gregory Director General CERN 1211 Meyrin Geneva 23, Switzerland
3. EURATOM (European Atomic Energy Community)		M. Pierre Chatenet, President European Atomic Energy Community 51 rue Belliard Brussels, Belgium
		M. Jean Rey, President Commission of the European Communities 24, av. de la Joyeuse Entree Brussels 4, Belgium

Table 5.2–1 (continued)
INTERNATIONAL ORGANIZATIONS PRODUCING NUCLEAR STANDARDS

International organization	Standard organization	Informant, title, address
4. IMCO (Inter-governmental Maritime Consultative Organization)		T. Busha, Head External Relations and Legal Matters Section Intergovernmental Maritime Consultative Organization 22 Berners Street London, W.1, England
5. IAEA (Inter-national Atomic Energy Agency)		Hon. Sigvard Eklund, Director General International Atomic Energy Agency Kärntnerring 11 Vienna 1, Austria
6. ICRP (International Commission on Radiological Protection)		E. Eric Pochin, Chairman International Commission on Radiological Protection University College Hosp. Med. School University Street London, W.C.1, England
		F. D. Sowby, Scientific Secretary International Commission on Radiological Protection Clifton Avenue Sutton, Surrey, England
	A. Committee 1. Radiation Effects	H. G. Newcombe, Head Biology Branch Atomic Energy of Canada Ltd. Chalk River, Ontario, Canada
	B. Committee 2. Internal Exposure	K. Z. Morgan, Director Health Physics Division Oak Ridge National Laboratory P. O. Box X Oak Ridge, Tennessee 37830
	C. Committee 3. External Exposure	Bo Lindell, Director Statens Strålskyddsinstitut Karolinska ajukhuset Stockholm 60, Sweden
	D. Committee 4. Application Recommendations	H. Jammet, Chef du Department de la Protection Sanitaire au Commissariat a l'Energie Atomique B.P. No. 6 Fontenay-aux-Roses (Seine), France
	E. Committee 5. Radioactive Waste Handling and Disposal	
	F. Committee 6. Committee on RBE	
	G. Joint ICRP and ICRU Study	

Table 5.2–1 (continued)
INTERNATIONAL ORGANIZATIONS PRODUCING NUCLEAR STANDARDS

International organization	Standard organization	Informant, title, address
7. ICRU (International Commission on Radiation Units and Measurements)		Dr. L. S. Taylor, Chairman International Commission on Radiation Units and Measurements Suite 402 4201 Connecticut Avenue, N.W. Washington, D.C. 20008
	IA. Fundamental Physical Parameters and Measurement Techniques	Dr. F. W. Spiers, Chairman
	IB. Medical and Biological Applications	Dr. H. Vetter, Chairman Austria
	IIB. X-rays, Gamma rays and Electrons	Dr. W. Pohlit, Chairman Germany
	IIC. Heavy Particles	Dr. W. K. Sinclair, Chairman
	IID. Medical and Biological Applications (Therapy)	Dr. W. J. Meredith, Chairman
	IIE. Medical and Biological Applications (Diagnosis)	Dr. Olle Olsson, Chairman
	IIF. Neutron Fluence and Kerma	Dr. A. H. W. Aten, Jr., Chairman
	IIIA. Protection Instrumentation and Its Application	Dr. L. Larsson, Chairman
8. IEC (International Electrotechnical Commission)		L. Ruppert, General Secretary International Electrotechnical Commission 1 rue de Verembe Geneva, Switzerland
	A. Technical Committee No. 45, Measuring Instruments Used in Connection with Ionizing Radiation	A. Rys, Chairman IEC Technical Committee No. 45 Inspection Generale, CEA 29 rue de la Federation Paris 15, France
	Working Group No. 1, Classification and Terminology	A. Rys (as above)
	Working Group No. 2, Safety of Radiation Instruments	S. J. Dagg, Chairman Central Electricity Generating Board 20 Newgate Street London, E.C.1, England
	Working Group No. 3, Interchangeability	W. Böhme, Chairman Siemens A.G. Rheinbrückenstrasse 50 75 Karlsruhe, Germany

Table 5.2–1 (continued)
INTERNATIONAL ORGANIZATIONS PRODUCING NUCLEAR STANDARDS

International organization	Standard organization	Informant, title, address
	Working Group No. 4, Reactor Instrumentation	
	Working Group No. 5, Measuring Instruments Used in Prospecting and Mining	P. Fabre, D.P.R.M. Direction of Productions, CEA B.P. No. 4 92-Chatillon sous Bagneux France
	Working Group No. 6, Electrical Measuring Instruments Using Sealed Radiation Sources	M. A. Goldmann, Chairman Comitato Nazionale Energie Nucleare C.N.E.N. Via Belisario, 15 Roma, Italy
	Working Group No. 7, Testing Methods	M. L. Strackee Hoofd. Afd. Radioactiviteit Rijksinsituut voor de Volksgezondheid Sterrenbos 1 Utrecht, Netherlands
	Working Group No. 8, Health Physics Instruments	
	Working Group No. 9, Radiation Detectors	L. Costrell, Chairman National Bureau of Standards Washington, D.C. 20234
	1. Subcommittee No. 45A Reactor Instrumentation	S. H. Hanauer, Chairman University of Tennessee Knoxville, Tennessee 37916
	Working Group A1, Reactor Instrumentation Principles	J. L. Petrie, Secretariat Walden House 24 Cathedral Place London, E.C.4, England
	Working Group A2, Reactor Instruments	J. Furet, D.E.G. CEA, Saclay–BP. No. 2 91 Gif-sur-Yvette, France
	2. Subcommittee No. 45B, Health Physics Instrumentation	Prof. Brunello Rispoli, Chairman Via Belisario 15 Roma, Italy
	B. Technical Committee No. 62, Medical X-ray Equipment	Secretariat: Germany
9. ISO (International Standards Organization)		Olle Sturen, Secretary-General International Standards Organization 1 rue de Verembe 1211 Geneva 20, Switzerland

Table 5.2–1 (continued)
INTERNATIONAL ORGANIZATIONS PRODUCING NUCLEAR STANDARDS

International organization	Standard organization	Informant, title, address
	A. Technical Committee No. 85, Nuclear Energy	Secretariat United States of America Standards Institute 10 East 40th Street New York, N.Y. 10016
	Subcommittee No. 1, Terminology, Definitions, Units, and Symbols	Secretariat United States of America Standards Institute 10 East 40th Street New York, N.Y. 10016
	Subcommittee No. 2, Radiation Protection	M. R. Frontard, Director General Association Francaise de Normalisation 23 rue Notre-Dame des Victoires 75, Paris 2e, France
	Subcommittee No. 3, Reactor Safety	H. A. R. Binney, C. B., Director British Standards Institution 2 Park Street, London, W.1, England
	Working Group No. 1, Siting	
	Working Group No. 2, Meteorological Aspects	
	Working Group No. 3, Containment Structures	L. P. Zick Chicago Bridge and Iron Co. 901 West 22nd Street Oak Brook, Ill. 60521
	Working Group No. 4, Effects of Nuclear Radiation in Steels	
	Working Group No. 5, Criticality Safety	Secretariat: United Kingdom
	Working Group No. 6, Nuclear Reactor Pressure Vessels	G. Wiesenach, Chairman Germany
	Subcommittee No. 4, Radioisotopes	l'Ing. J. Wodzicki, President Polski Komitet Normalizacyjny Ul, Swietokrzyska 14 Warsaw 51, Poland
	B. Technical Committee No. 12, Quantities, Units, Symbols, Conversion Factors, and Conversion Proposals	Secretariat: Denmark Dansk Standardiseringsraad Aurehøjvej 12 Kobenhavn Hellerup, Denmark
10. Lloyds		The Secretary Lloyds Register of Shipping 71 Fenchurch Street London, E.C.3, England

Table 5.2–1 (continued)
INTERNATIONAL ORGANIZATIONS PRODUCING NUCLEAR STANDARDS

International organization	Standard organization	Informant, title, address
11. OECD (Organization for Economic Co-operation and Development); European Nuclear Energy Agency (ENEA)	Committee on Reactor Safety Technology	Mr. F. R. Farmer, Chairman Safety Division United Kingdom Atomic Energy Authority Risley, Warrington, Lancs. England
		Henri B. Smets, Secretariat Committee on Reactor Safety Technology 38 Boulevard Suchet Paris 16, France
12. United Nations		Kurt Waldheim, Secretary-General United Nations United Nations Building New York, New York 10017
	A. Food and Agriculture Organization (FAO)	G. Wortley, Head Pesticide Residues and Food Protection Section Joint FAO-IAEA Division of Atomic Energy in Agriculture IAEA Kärtner Ring 11 Vienna, Austria
	B. International Labor Organization (ILO)	Dr. Luigi Parmeggiani, Chief Occupational Safety and Health Branch Conditions of Work and Life Department International Labor Organization CH 1211 Geneva 22, Switzerland
	C. United Nations Scientific Committee on the Effects of Atomic Radiation (UNSCEAR)	Francesco Sella, Secretary United Nations United Nations Building New York, New York 10017
	D. World Health Organization (WHO)	Dr. W. H. P. Seelentag Chief Medical Officer Radiation Health World Health Organization Avenue Appia, 1211 Geneva 10, Switzerland
	E. World Meteorological Organization (WMO)	Dr. K. Langlo, Chief Technical Division World Meteorological Organization Palais des Nations Geneva, Switzerland

Table 5.2–1 (continued)
INTERNATIONAL ORGANIZATIONS PRODUCING NUCLEAR STANDARDS

ACKNOWLEDGMENT

This chapter is reprinted from "Compilation of National and International Nuclear Standards" by William B. Cottrell, Chairman of U.S.A. Standards Institute Subcommittee N6.9, which includes W. B. Allred, J. P. Blakely, D. Davis, J. B. Godel, J. E. McEwen, Jr., M. Novick, C. Roderick, and J. C. Russ.

LEGAL NOTICE

This report was prepared as an account of Government sponsored work. Neither the United States, nor the Commission, nor any person acting on behalf of the Commission:

A. Makes any warranty or representation, expressed or implied, with respect to the accuracy, completeness, or usefulness of the information contained in this report, or that the use of any information, apparatus, method, or process disclosed in this report may not infringe privately owned rights; or

B. Assumes any liabilities with respect to the use of, or for damages resulting from the use of any information, apparatus, method, or process disclosed in this report.

As used in the above, "person acting on behalf of the Commission" includes any employee or contractor of the Commission, or employee of such contractor, to the extent that such employee or contractor of the Commission, or employee of such contractor prepares, disseminates, or provides access to, any information pursuant to his employment or contract with the Commission, or his employment with such contractor.

From Cottrell, W. B., in *Handbook of Laboratory Safety*, 2nd ed., Steere, N. V., Ed., Chemical Rubber, Cleveland, 1971, 503.

Table 5.2-2
RADIATION PROTECTION STANDARDS

Important Dates in the Development of Radiation Protection Standards

1915	British Roentgen Society proposals for radiation protection
1921	British adopt radiation protection recommendations
1922	American Roentgen Ray Society adopts radiation protection rules
1928	Unit of X-ray intensity proposed by Second International Congress of Radiology
1928	International Committee on X-ray and Radium Protection established
1928	First international recommendations on radiation protection adopted by Second International Congress of Radiology
1929	Advisory Committee on X-ray and Radium Protection established (United States of America)
1931	The roentgen adopted as a unit of X-radiation
1934	Tolerance dose of 0.1 roentgen/day recommended by Advisory Committee on X-ray and Radium Protection (March)
1934	Tolerance dose of 0.2 roentgen/day recommended by International Committee on X-ray and Radium Protection (July)
1941	Advisory Committee on X-ray and Radium Protection recommends 0.1 μCi permissible body burden for radium
1946	Advisory Committee on X-ray and Radium Protection reorganized as the National Committee on Radiation Protection
1949	National Committee on Radiation Protection lowers basic maximum permissible dose (MPD) for radiation workers to 0.3 rem/week. Risk-benefit philosophy introduced
1950	International Commission on Radiological Protection and International Commission on Radiological Units reorganized from pre-war committees
1950	International Commission on Radiological Protection adopts basic MPD of 0.3 rem/week for radiation workers
1953	International Commission on Radiological Units introduces concept of *absorbed dose*
1956	National Academy of Sciences and International Commission on Radiological Protection recommend lower basic permissible dose for radiation workers of 5 rem/year
1957	National Committee on Radiation Protection and Measurements introduces age proration concept for occupational exposure and 0.5 rem/year for individuals in population
1959	International Commission on Radiological Protection recommends limitation of genetically significant dose to population of 5 rem in 30 years

Table 5.2–2 (continued)
RADIATION PROTECTION STANDARDS

1964   Federal Radiation Council introduces concept of protective action guides
1971   National Council on Radiation Protection and Measurements recommends same value of 15 rem/year for all
        noncritical organs

From Taylor, L. S., *Radiation Protection Standards*, Chemical Rubber, Cleveland, 1971, 101. (See this source for an extensive bibliography.)

Radiation Protection Publications

Since so many agencies have been involved in the development of U.S. radiation protection standards, the following list is offered to indicate some of the contributions. The role of the NCRP is discussed first, and the participating agencies are then listed alphabetically.

NCRP (National Council on Radiation Protection and Measurements): Formed as a committee in 1946, succeeding the ACRP (see below), and was chartered by Congress in 1964 as a Council and a cooperating organization with ICRP, FRC, and others. By 1971 there were 28 organizations collaborating with the Council, and it had issued over 40 reports and major publications on radiation protection. The Council functions through 34 scientific committees; its published results and recommendations are widely accepted and form the basis for most of the rules and standards on radiation protection now current in the United States. The groups collaborating with NCRP include most of the concerned divisions of the U.S. government, as well as the scientific societies and trade associations that have some direct concern with radiation protection.

ABCC (Atomic Bomb Casualty Commission): Initiated by AEC, through NAS, the ABCC combined the efforts of U.S. and Japanese scientists in a continuing study of atomic-bomb survivors.

ACRP (Advisory Committee on X-ray and Radium Protection): Formed in 1929, it was a small group with representation from AMA, NBS, and manufacturers. It issued a major report in 1931 and issued several more reports throughout the 1930's. In 1946 ACRP became the National Committee on Radiation Protection.

AEC (Atomic Energy Commission): A postwar development that took over the protection activities of the Manhattan District, the AEC has been very active in the field and has published reports and also regulations governing the operations under its jurisdiction.

ARRS (American Roentgen Ray Society): Published a set of recommendations for radiation protection in 1920. Active committee work and issuance of reports on the subject have continued since.

BMRC (British Medical Research Council): An active group that published an independent report in 1956, *The Hazards to Man of Nuclear and Allied Radiations* (HMS-Cmd 9780, 1956).

BRS (British Roentgen Society): Adopted a pioneering resolution in 1915 that steps be taken toward the adoption of stringent rules for the personal safety of operators conducting roentgen-ray examinations.

FRC (Federal Radiation Council): Established by executive order of the President in 1959 to coordinate the work of federal agencies and advise the President. It has worked in close consultation with the NAS and the HCRP. In 1960 the FRC began publishing standards and reports (available from Superintendent of Documents).

IAEA (International Atomic Energy Agency): A large organization within the family of the United Nations. It has issued hundreds of papers and reports, largely on nuclear science and its peaceful uses, but also on safety and on legal aspects.

ICRP (International Committee on X-ray and Radium Protection): Organized in 1928 and did major work in the 10 years following. Its recommendations covered working conditions, shielding, and permissible dose levels (0.2 R/day) for radiation workers. Beginning in 1950, there have been at least 15 reports on later ICRP activities (published largely by Pergamon Press, London).

ICRU (International Commission on Radia-

Table 5.2–2 (continued)
RADIATION PROTECTION STANDARDS

tion Units and Measurements): A large international committee, formed in the 1920's.

JCAE (Joint Committee on Atomic Energy): A U.S. Congress committee. It has held several hearings dealing with fallout, waste disposal, and radiation safety and standards; at least 25 documents on these activities are available (from the Superintendent of Documents).

NACOR (National Advisory Committee on Radiation): Under the U.S. Public Health Service, NACOR was organized in 1958, largely as an outgrowth of the ICAE hearings. It functioned effectively in the public health field and issued reports up to the late 1960's.

NAS–NRC (National Academy of Sciences-National Research Council): Government-chartered organizations. On invitation they have taken an active part in radiation protection research, especially since 1956, and have published a dozen or more reports.

NBS (National Bureau of Standards): Participated in much of the radiation protection activity; over the years of 1931–64 it has published more than 30 NBS handbooks in the field.

These publications matched closely the recommendations promulgated by the NCRP.

TPC (Tri-partite Conferences): A series of three meetings (1949, 1950, 1953) with representation from the United States, Canada, and England. The proceedings were held classified.

UNSCEAR (United Nations Scientific Committee on the Effects of Atomic Radiation): Especially active through 1958–1966, when several reports were published by the United Nations. These dealt largely with fallout and its possible biological effects, and with the effects of nuclear explosions.

Miscellaneous organizations: Other organizations have functioned effectively in radiation protection, including several in the medical, biological, and standardization fields. Many similar activities are carried on in other nations and within states and provinces. Of particular interest are the codes and standards of the ASA and its successors (the USASI and the ANSI), the ISO (International Standards Organization), the IEC (International Electrotechnical Commission), and the ASTM (see latest ANSI and ASTM lists of U.S. standards).

From Bolz, R. E. and Tuve, G. L., Eds., *Handbook of Tables for Applied Engineering Science*, 2nd ed., CRC Press, Cleveland, 1973, 754.

Table 5.2–3
COMPONENTS OF THE ATMOSPHERE[a]

Average Composition of Dry Air

For most engineering applications the following accepted values for the "average" composition of the atmosphere are adequate. These values are for sea level or any land elevation. Proportions remain essentially constant to 50,000 ft (15,240 m) altitude.

Gas	Molecular weight	Percentage by volume, mol fraction	Percentage by weight
Nitrogen	N_2 = 28.016	78.09	75.55
Oxygen	O_2 = 32.000	20.95	23.13
Argon	Ar = 39.944	0.93	1.27
Carbon dioxide	CO_2 = 44.010	0.03	0.05
		100.00	100.00

Table 5.2–3 (continued)
COMPONENTS OF THE ATMOSPHERE

For many engineering purposes the percentages 79% N_2–21% O_2 by volume and 77% N_2–23% O_2 by weight are sufficiently accurate, the argon being considered as nitrogen with an adjustment of molecular weight to 28.16.

Other gases in the atmosphere constitute less than 0.003% (actually 27.99 ppm by volume), as given in the following table.

Minor Constituents of Dry Air

Gas	Molecular weight	Parts per million	
		By volume	By weight
Neon	Ne = 20.183	18.	12.9
Helium	He = 4.003	5.2	0.74
Methane	CH_4 = 16.04	2.2	1.3
Krypton	Kr = 83.8	1.	3.0
Nitrous oxide	N_2O = 44.01	1.	1.6
Hydrogen	H_2 = 2.0160	0.5	0.03
Xenon	Xe = 131.3	0.08	0.37
Ozone	O_3 = 48.000	0.01	0.02
Radon	Rn = 222.	(0.06×10^{-12})	

Minor constituents may also include dust, pollen, bacteria, spores, smoke particles, SO_2, H_2S, hydrocarbons, and larger amounts of CO_2 and ozone, depending on weather, volcanic activity, local industrial activity, and concentration of human, animal, and vehicle population. In certain enclosed spaces the minor constituents will vary considerably with industrial operations and with occupancy by humans, plants, or animals.

The above data do not include water vapor, which is an important constituent in all normal atmospheres.

[a]For atmospheric conditions at high altitudes, see Tables 5.2–6 and 5.2–7.

Compiled from several sources; presented in Bolz, R. E. and Tuve, G. L., Eds., *Handbook of Tables for Applied Engineering Science,* 2nd ed., CRC Press, Cleveland, 1973, 649.

Table 5.2—4
U.S. STANDARD ATMOSPHERE, TO 300,000 ft[a]

45° North Latitude, July

Symbols:

Z, ft = geometric altitude, feet
Z, m = geometric altitude, meters
H, ft = geopotential altitude, feet
t, °F = temperature, degrees Fahrenheit
t, °C = temperature, degrees Celsius
P, in. Hg = pressure, inches of mercury. For atmospheres multiply by 0.0334210. For psia multiply by 0.491154. For kN/m² multiply by 3.3864.

ρ, English = density. For lb_m/ft^3 multiply by 10^{-3}.
ρ, metric = density. For kg/m^3 multiply by 10^{-3}.
V_s, fps = speed of sound, ft/sec. For m/sec multiply by 0.3048.
μ = viscosity. For lb_m/ft sec multiply by 10^{-5}. For $N \cdot s/m^2$ (= kg/m · s) multiply by 10^{-6} and by 14.882.
k = thermal conductivity. For Btu/sec ft °R multiply by 10^{-5}. For W/m · K multiply by 622.65.

Z, ft	H, ft	Z, m	t, °F	t, °C	P, in. Hg	ρ, English	ρ, metric	V_s, fps	μ	k
0	0	0	73.5	23.1	29.93	74.4	1,192.	1,132	1.228	.417
1,000	1,000	305	70.7	21.4	28.89	72.2	1,157	1,129	1.223	.415
2,000	2,000	610	68.0	20.0	27.89	70.1	1,123	1,126	1.218	.413
3,000	3,000	915	65.2	18.4	26.91	68.0	1,090	1,123	1.213	.411
4,000	3,999	1,220	62.4	16.9	25.96	65.9	1,057	1,120	1.209	.409
5,000	4,999	1,525	59.7	15.4	25.05	63.9	1,025	1,117	1.204	.407
6,000	5,998	1,830	57.0	13.9	24.16	62.0	994	1,114	1.199	.405
7,000	6,998	2,135	53.9	12.2	23.30	60.1	964	1,111	1.193	.403
8,000	7,997	2,440	50.4	10.2	22.46	58.4	936	1,107	1.187	.401
9,000	8,996	2,745	46.9	8.3	21.65	56.6	908	1,103	1.180	.398
10,000	9,995	3,050	43.4	6.3	20.8ȯ	55.0	881	1,100	1.174	.396
11,000	10,994	3,355	40.0	4.4	20.09	53.3	855	1,096	1.168	.393
12,000	11,993	3,660	36.6	2.6	19.35	51.7	829	1,092	1.162	.391
13,000	12,992	3,965	33.2	.7	18.63	50.1	803	1,088	1.155	.389
14,000	13,991	4,270	29.8	−1.2	17.94	48.6	779	1,085	1.149	.386
15,000	14,989	4,575	26.5	−3.1	17.26	47.1	755	1,081	1.143	.384
16,000	15,988	4,880	23.1	−4.9	16.61	45.6	731	1,077	1.137	.381
17,000	16,986	5,185	19.7	−6.8	15.97	44.2	708	1,073	1.130	.379
18,000	17,984	5,490	16.4	−8.7	15.36	42.8	686	1,070	1.124	.377
19,000	18,983	5,795	13.0	−10.6	14.77	41.4	664	1,066	1.118	.374
20,000	19,981	6,100	9.61	−12.5	14.19	40.1	643	1,062	1.111	.372
21,000	20,979	6,405	6.02	−14.4	13.63	38.8	622	1,058	1.105	.369
22,000	21,977	6,710	2.44	−16.4	13.10	37.6	602	1,054	1.098	.367
23,000	22,975	7,015	−1.14	−18.4	12.57	36.4	583	1,050	1.091	.364
24,000	23,972	7,320	−4.72	−20.4	12.07	35.2	564	1,046	1.084	.361
25,000	24,970	7,625	−8.31	−22.4	11.58	34.0	545	1,042	1.077	.359
26,000	25,968	7,930	−11.9	−24.4	11.11	32.9	527	1,037	1.070	.356
27,000	26,965	8,235	−15.5	−26.4	10.65	31.8	510	1,033	1.063	.353
28,000	27,962	8,540	−19.0	−28.3	10.21	30.7	492	1,029	1.056	.351
29,000	28,960	8,845	−22.6	−30.4	9.79	29.7	476	1,025	1.049	.348
30,000	29,957	9,150	−26.2	−32.3	9.38	28.7	460	1,021	1.043	.346
32,000	31,951	9,760	−33.3	−36.3	8.60	26.7	428	1,012	1.029	.340
34,000	33,945	10,370	−40.4	−40.2	7.87	24.9	399	1,004	1.014	.335
36,000	35,938	10,980	−47.6	−44.2	7.19	23.1	371	995	1.000	.330
38,000	37,931	11,590	−54.7	−48.2	6.56	21.5	344	987	.986	.325
40,000	39,923	12,200	−61.8	−52.1	5.98	19.9	319	978	.972	.319
45,000	44,903	13,725	−71.5	−57.5	4.71	16.1	258	966	.952	.312
50,000	49,880	15,250	−71.5	−57.5	3.70	12.7	203	966	.952	.312
55,000	54,855	16,775	−71.5	−57.5	2.91	9.95	159	966	.952	.312
60,000	59,828	18,300	−68.8	−56.0	2.29	7.78	125	969	.957	.314
65,000	64,798	19,825	−65.6	−54.2	1.81	6.08	97.5	973	.964	.316
70,000	69,766	21,350	−62.3	−54.4	1.43	4.77	76.4	977	.970	.319
75,000	74,731	22,875	−59.0	−50.6	1.13	3.75	60.3	981	.977	.321
80,000	79,694	24,400	−55.8	−48.8	.898	2.95	47.2	985	.984	.324
85,000	84,655	25,925	−52.5	−46.9	.714	2.33	37.3	989	.990	.326
90,000	89,613	27,450	−48.7	−44.8	.569	1.84	29.4	994	.998	.329

Table 5.2–4 (continued)
U.S. STANDARD ATMOSPHERE, TO 300,000 ft

Z, ft	H, ft	Z, m	t, °F	t, °C	P, in. Hg	ρ, English	ρ, metric	V_s, fps	μ	k
0	0	0	73.5	23.1	29.93	74.4	1,192.	1,132	1.228	.417
100,000	99,523	30,500	−37.3	−38.5	.364	1.14	18.3	1,008	1.021	.337
125,000	124,255	38,125	−4.6	−20.3	.126	.368	5.90	1,046	1.084	.361
150,000	148,929	45,750	29.3	−1.50	.047	.129	2.06	1,084	1.148	.386
175,000	173,544	53,375	32.5	.28	.019	.0503	.807	1,088	1.154	.388
200,000	198,100	61,000	−1.2	−18.4	.0071	.0205	.329	1,050	1.091	.364
250,000	247,039	76,250	−116.2	−82.3	.0007	.0028	.0444	909	.857	.278
300,000	295,746	91,500	−156.6	−104.8	.00004	.0002	.0025	855	.770	.247

[a]For variations of high altitude atmosphere with latitude and season, see Table 5.2–6.

Condensed from *U.S. Standard Atmosphere Supplements,* U.S. Government Printing Office, Washington, D.C., 1966.

Table 5.2–5
U.S. STANDARD ATMOSPHERE, 1962[a]

Middle-latitude, Year-round Mean Conditions

The "U.S. Standard Atmosphere, 1962" was developed to serve the aerospace community as a mean basis for design and operation of vehicles and for general scientific considerations. For all practical purposes it is in agreement with the ICAO Standard Atmosphere over the altitude range that they have in common.[b]

Several earlier standards existed (see below); for realistic tables accounting for departures from the mean conditions due to geography, season, time of day, and solar activity, the "U.S. Standard Atmosphere, 1966" should be consulted.

The 1962 standard and the 1966 supplement (almost 300 pages each) were both cosponsored by the National Aeronautics and Space Administration, the U.S. Air Force, the former U.S. Weather Bureau, and COESA.[c]

The "U.S. Standard Atmosphere, 1962" agrees in general but differs in detail from the COSPAR International Reference Atmosphere, 1961.[d] Among the earlier standards were the COESA standard, "U.S. Extension to the ICAO Standard Atmosphere, 1958"; "ICAO Standard, 1952," adopted by NACA and published as NACA Report 1235, 1955; "ARDC Model Atmosphere, 1959";

and "U.S.S.R. Standard Atmosphere, 1960." A later standard is "COSPAR International Reference Atmosphere, 1965."

Basis of the 1962 Tables[e]

These tables depict idealized year-round mean conditions at 45°N latitude, for the range of solar activity that occurs between sunspot minimum and sunspot maximum. In this model the atmosphere is defined in terms of temperature. The low-altitude temperature patterns are shown with abrupt changes at the following geometric altitudes (in kilometers): 11, 20, 32, 47, 52, 61, and 79. Variations with season, weather, latitude, and solar activity are not included in this mean table (see other tables and their references).

The "U.S. Standard Atmosphere, 1962" treats air as a clean, dry, perfect-gas mixture ($C_p/C_v = 1.40$), having a molecular weight to 90 km of 28.9644 (C-12 scale). The principal sea level constituents are assumed to be N_2−78.084%, O_2−20.9476%, Ar−0.934%, CO_2−0.0314%, Ne−0.001818%, He−0.000524%, and methane−0.0002%.

Assigned mean conditions at sea level are as follows:

Table 5.2–5 (continued)
U.S. STANDARD ATMOSPHERE, 1962

P = 0.1013250 MN/m² = 2,116.22 psf = 14.696 psi

T = 288.15 K = 15 deg C = 59 deg F

ρ = 1.2250 kg/m³ = 0.076474 lb/ft³

g = 980.655 m/s² = 32.1741 ft/s²

R = 8.31432 J/mol · K = 1,545.31 ft lb/lb · mol · deg R.

The viscosities as tabulated are based on the equation:

$\mu = 1.458 \times 10^{-6} T^{1.5}/T + 110.4,$

where temperature is in Kelvin. The thermal conductivities as tabulated are based on the equation:

$k = 6.325 \times 10^{-7} T^{1.5}/T + 245.4 \times 10^{-(12/T)},$

where temperature is in Kelvin. The velocity of sound is based on the equation:

$V_s = 1.4 [c_p/c_v (RT/M_O)].$

[a]For variations with geography and season, see Tables 5.2-4, 5.2-6, and 5.2-7.

[b]ICAO – International Civil Aviation Organization.

[c]COESA – Committee on Extension to the Standard Atmosphere, a group of 30 U.S. scientific and engineering organizations, including APL-Johns Hopkins, Battelle, Boeing, FAA, Smithsonian Institution, JPL-Cal. Tech., Lockheed, NBS, and NRL.

[d]COSPAR – Committee on Space Research.

[e]For a 30-page discussion, see the original source.

Table A. U.S. Standard Atmosphere – English Units

Altitude Geometric, Z, ft	Altitude Geopotential, H, ft	Temperature, t deg F	Temperature, t deg R	Pressure, P in. Hg	Pressure, P psia	Density, ρ, lb/ft³ × 10⁵	Gravity, g, ft/s²	Molecular weight, M	Velocity of sound, V_s, fps	Viscosity, μ, lb/ft·s × 10⁵	Conductivity, k, kcal × 10⁵ m·s·K
0	0	59.00	518.7	29.92	14.7	7 647.	32.174	28.96	1 116.	1.202	.406 7
500	500	57.22	516.9	29.38	14.4	7 536.	32.173	28.96	1 115.	1.199	.405 5
1 000	1 000	55.43	515.1	28.86	14.2	7 426.	32.171	28.96	1 113.	1.196	.404 2
1 500	1 500	53.65	513.3	28.33	13.9	7 317.	32.169	28.96	1 111.	1.193	.403 0
2 000	2 000	51.87	511.5	27.82	13.7	7 210.	32.168	28.96	1 109.	1.190	.401 7
2 500	2 500	50.09	509.8	27.32	13.4	7 104.	32.166	28.96	1 107.	1.186	.400 2
3 000	3 000	48.30	508.0	26.82	13.2	6 998.	32.165	28.96	1 105.	1.183	.399 2
3 500	3 499	46.52	506.2	26.33	12.9	6 895.	32.163	28.96	1 103.	1.180	.398 0
4 000	3 999	44.74	504.4	25.84	12.7	6 792.	32.162	28.96	1 101.	1.177	.396 7
4 500	4 499	42.96	502.6	25.37	12.5	6 690.	32.160	28.96	1 099.	1.173	.395 4
5 000	4 999	41.17	500.8	24.90	12.2	6 590.	32.159	28.96	1 097.	1.170	.394 2
5 500	5 499	39.39	499.1	24.43	12.0	6 491.	32.157	28.96	1 095.	1.167	.392 9
6 000	5 998	37.61	497.3	23.98	11.8	6 393.	32.156	28.96	1 093.	1.164	.391 7
6 500	6 498	35.83	495.5	23.53	11.6	6 296.	32.154	28.96	1 091.	1.160	.390 4
7 000	6 998	34.05	493.7	23.09	11.3	6 200.	32.152	28.96	1 089.	1 157	.389 1
7 500	7 497	32.26	491.9	22.66	11.1	6 105.	32.151	28.96	1 087.	1.154	.387 9
8 000	7 997	30.48	490.2	22.23	10.9	6 012.	32.149	28.96	1 085.	1.150	.386 6
8 500	8 497	28.70	488.4	21.81	10.7	5 919.	32.148	28.96	1 083.	1.147	.385 3
9 000	8 996	26.92	486.6	21.39	10.5	5 828.	32.146	28.96	1 081.	1.144	.384 0
9 500	9 496	25.14	484.8	20.98	10.3	5 738.	32.145	28.96	1 079.	1.140	.382 8

Table 5.2–5 (continued)
U.S. STANDARD ATMOSPHERE, 1962

Table A. U.S. Standard Atmosphere – English Units (continued)

Altitude		Temperature, t		Pressure, P		Density, ρ, lb/ft³ $\times 10^5$	Gravity, g, ft/s²	Molecular weight, M	Velocity of sound, V_g, fps	Viscosity, μ, lb/ft·s $\times 10^5$	Conductivity, k, kcal $\times 10^5$ m·s·K
Geometric, Z, ft	Geopotential, H, ft	deg F	deg R	in. Hg	psia						
l0 000	9 995	23.36	483.0	20.58	10.1	5 648.	32.145	28.96	1 077.	1.137	.381 5
11 000	10 994	19.79	479.5	19.80	9.72	5 473.	32.140	28.96	1 073.	1.130	.379 0
12 000	11 993	16.23	475.9	19.03	9.34	5 302.	32.137	28.96	1 069.	1.124	.376 4
13 000	12 992	12.67	472.3	18.30	8.99	5 135.	32.134	28.96	1 065.	1.117	.373 8
14 000	13 991	9.12	468.8	17.58	8.63	4 973.	32.131	28.96	1 061.	1.110	.371 3
15 000	14 989	5.55	465.2	16.89	8.29	4 814.	32.128	28.96	1 057.	1.104	.368 7
16 000	15 988	+1.99	461.6	16.23	7.97	4 659.	32.125	28.96	1 053.	1.097	.366 1
17 000	16 986	−1.58	458.1	15.58	7.65	4 508.	32.122	28.96	1 049.	1.090	.363 6
18 000	17 984	−5.14	454.5	14.95	7.34	4 361.	32.119	28.96	1 045.	1.083	.361 0
19 000	18 983	−8.70	451.0	14.35	7.05	4 217.	32.115	28.96	1 041.	1.076	.358 4
20 000	19 981	−12.2	447.4	13.76	6.76	4 077.	32.112	28.96	1 037.	1.070	.355 8
21 000	20 979	−15.8	443.9	13.20	6.48	3 941.	32.109	28.96	1 033.	1.063	.353 2
22 000	21 977	−19.4	440.3	12.65	6.21	3 808.	32.106	28.96	1 029.	1.056	.350 6
23 000	22 975	−22.9	436.7	12.12	5.95	3 679.	32.103	28.96	1 024.	1.049	.348 0
24 000	23 972	−26.5	433.2	11.61	5.70	3 553.	32.100	28.96	1 020.	1.042	.345 4
25 000	24 970	−30.0	429.6	11.09	5.44	3 431.	32.097	28.96	1 016.	1.035	.342 8
26 000	25 968	−33.6	426.1	10.64	5.22	3 311.	32.094	28.96	1 012.	1.028	.340 1
27 000	26 965	−37.2	422.5	10.18	5.00	3 195.	32.091	28.96	1 008.	1.021	.337 5
28 000	27 962	−40.7	418.6	9.74	4.78	3 082.	32.088	28.96	1 003.	1.014	.334 9
29 000	28 960	−44.3	415.4	9.31	4.57	2 973.	32.085	28.96	999.	1.007	.332 2
30 000	29 957	−47.8	411.8	8.90	4.37	2 866.	32.082	28.96	995.	1.000	.329 6
32 000	31 951	−54.9	404.7	8.12	3.99	2 661.	32.08	28.96	987.	0.986	.324 6
34 000	33 945	−62.0	397.6	7.40	3.63	2 468.	32.07	28.96	978.	0.971	.319 0
36 000	35 938	−69.2	390.5	6.73	3.30	2 285.	32.06	28.96	969.	0.956	.313 7
38 000	37 931	−69.7	390.0	6.12	3.05	2 079.	32.06	28.96	968.	0.955	.313 3
40 000	29 923	−69.7	390.0	5.56	2.73	1 890.	32.05	28.96	968.	0.955	.313 3
45 000	44 903	−69.7	390.0	4.375	2.148	1 487.	32.04	28.96	968.	0.955	.313 3
50 000	49 880	−69.7	390.0	3.444	1.691	1 171.	32.02	28.96	968.	0.955	.313 3
55 000	54 855	−69.7	390.0	2.712	1.332	922.	32.00	28.96	968.	0.955	.313 3
60 000	59 828	−69.7	390.0	2.135	1.048	726.	31.99	28.96	968.	0.955	.313 3
65 000	64 798	−69.7	390.0	1.681	0.825	572.	31.97	28.96	968.	0.955	.313 3
70 000	69 766	−67.4	392.2	1.325	0.651	447.9	31.96	28.96	971.	0.960	.315 0
75 000	74 731	−64.7	394.8	1.046	0.514	351.1	31.94	28.96	974.	0.966	.317 0
80 000	79 694	−62.0	397.7	0.827	0.406	273.1	31.93	28.96	978.	0.971	.319 1
85 000	84 655	−59.2	400.4	0.655	0.322	217.0	31.91	28.96	981.	0.977	.321 1
90 000	89 613	−56.5	403.1	0.520	0.255	171.0	31.90	28.96	984.	0.982	.323 1
100 000	99 523	−51.1	408.6	0.329	0.162	106.8	31.87	28.96	991.	0.993	.327 2
200 000	198 100	−2.6	457.0	0.006	0.003	1.696	31.57	28.96	1 048.	1.088	.362 8
300 000	295 745	−126.8	332.9	3.7×10^{-5}	1.8×10^{-5}	1.5×10^{-2}	31.27	28.96	895.	0.834	.269 8
400 000	392 471	+233.9	693.6	6.3×10^{-7}	3.1×10^{-7}	1.2×10^{-4}	30.97	27.97	—	—	—
500 000	488 291	1 203.	1 663.	1.4×10^{-7}	0.7×10^{-7}	1.0×10^{-5}	30.68	26.86	—	—	—
1 000 000	954 232	2 125.	2 584.	5.1×10^{-9}	2.5×10^{-9}	2.0×10^{-7}	29.30	22.52	—	—	—
1 500 000	1 399 317	2 221.	2 681.	5.5×10^{-10}	2.7×10^{-10}	1.8×10^{-8}	28.00	18.68	—	—	—
2 000 000	1 824 911	2 251.	2 710.	9.2×10^{-11}	4.5×10^{-11}	2.6×10^{-9}	26.79	16.77	—	—	—

Table 5.2–5 (continued)
U.S. STANDARD ATMOSPHERE, 1962
Table B. U.S. Standard Atmosphere – Metric Units

Altitude Geometric, Z, m	Altitude Geopotential, H, m	Temperature, t deg C	Temperature, t K	Pressure, P mbar	Pressure, P mm Hg	Density, ρ, kg/m³	Gravity, g, m/s²	Molecular weight, M	Velocity of sound, V_s, m/s	Viscosity, μ, kg/m·s × 10⁵	Conductivity, k, kcal × 10⁵ m·s·K
0	0	15.00	288.15	1 013.3	760.	1.225	980.7	28.96	340.3	1.789	.605 3
200	200	13.70	286.85	989.5	742.	1.202	980.6	28.96	339.5	1.783	.602 9
400	400	12.40	285.55	966.1	724.	1.179	980.5	28.96	338.8	1.777	.600 4
600	600	11.10	284.25	943.2	707.	1.156	980.5	28.96	338.0	1.771	.598 0
800	800	9.80	282.95	920.8	691.	1.134	980.4	28.96	337.2	1.764	.595 5
1 000	1 000	8.50	281.65	898.8	674.	1.112	980.4	28.96	336.4	1.758	.593 1
1 200	1 200	7.20	280.35	877.2	658.	1.090	980.3	28.96	335.7	1.752	.590 6
1 400	1 400	5.90	279.05	856.0	642.	1.069	980.2	28.96	334.9	1.745	.588 1
1 600	1 600	4.60	277.75	835.3	627.	1.048	980.2	28.96	334.1	1.739	.585 7
1 800	1 799	3.30	276.45	814.9	611.	1.027	980.1	28.96	333.3	1.732	.583 2
2 000	1 999	2.00	275.15	795.0	596.	1.007	980.0	28.96	332.5	1.726	.580 7
2 200	2 199	0.70	273.86	775.5	582.	0.987	980.0	28.96	331.7	1.720	.578 4
2 400	2 399	−0.59	272.56	756.3	567.	0.967	979.9	28.96	331.0	1.713	.575 9
2 600	2 599	−1.89	271.26	737.6	553.	0.947	979.9	28.96	330.2	1.707	.573 3
2 800	2 799	−3.19	269.96	719.2	539.	0.928	979.8	28.96	329.4	1.700	.570 8
3 000	2 999	−4.49	268.66	701.2	526.	0.909	979.7	28.96	328.6	1.694	.568 3
3 200	3 198	−5.79	267.36	683.6	513.	0.891	979.7	28.96	327.8	1.687	.565 8
3 400	3 398	−7.09	266.06	666.3	500.	0.872	979.6	28.96	327.0	1.681	.563 4
3 600	3 598	−8.39	264.76	649.4	487.	0.854	979.6	28.96	326.2	1.674	.560 9
3 800	3 798	−9.69	263.47	632.8	475.	0.837	979.5	28.96	325.4	1.668	.558 4
4 000	3 997	−10.98	262.17	616.6	462.	0.819	979.4	28.96	324.6	1.661	.555 9
4 200	4 197	−12.3	260.87	600.7	451.	0.802	979.4	28.96	323.8	1.655	.553 4
4 400	4 397	−13.6	259.57	585.2	439.	0.785	979.3	28.96	323.0	1.648	.550 8
4 600	4 597	−14.9	258.27	570.0	428.	0.769	979.3	28.96	322.2	1.642	.548 3
4 800	4 796	−16.2	256.97	555.1	416.	0.752	979.2	28.96	321.4	1.635	.545 8
5 000	4 996	−17.5	255.68	540.5	405.	0.736	979.1	28.96	320.5	1.628	.543 3
5 200	5 196	−18.8	254.38	526.2	395.	0.721	979.1	28.96	319.7	1.622	.540 8
5 400	5 395	−20.1	253.08	512.3	384.	0.705	979.0	28.96	318.9	1.615	.538 3
5 600	5 595	−21.4	251.78	498.6	374.	0.690	978.9	28.96	318.1	1.608	.535 7
5 800	5 795	−22.7	250.48	485.2	364.	0.675	978.9	28.96	317.3	1.602	.533 2
6 000	5 994	−24.0	249.19	472.2	354.	0.660	978.8	28.96	316.5	1.595	.530 7
6 200	6 194	−25.3	247.89	459.4	345.	0.646	978.8	28.96	315.6	1.588	.528 2
6 400	6 394	−26.6	246.59	446.9	335.	0.631	978.7	28.96	314.8	1.582	.525 6
6 600	6 593	−27.9	245.29	434.7	326.	0.617	978.6	28.96	314.0	1.575	.523 1
6 800	6 793	−29.2	244.00	422.7	317.	0.604	978.6	28.96	313.1	1.568	.520 5
7 000	6 992	−30.5	242.70	411.1	308.	0.590	978.5	28.96	312.3	1.561	.518 0
7 500	7 491	−33.7	239.46	383.0	287.	0.572	978.4	28.96	310.2	1.544	.511 6
8 000	7 990	−36.9	236.22	356.5	267.	0.526	978.2	28.96	308.1	1.527	.505 2
8 500	8 489	−40.2	232.97	331.6	249.	0.496	978.1	28.96	306.0	1.510	.498 8
9 000	8 987	−43.4	229.73	308.0	231.	0.467	977.9	28.96	303.8	1.493	.492 4
9 500	9 486	−46.7	226.49	285.8	214.	0.440	977.7	28.96	301.7	1.475	.485 9
10 000	9 984	−49.9	223.25	265.0	199.	0.414	977.6	28.96	299.5	1.458	.479 4
11 000	10 981	−56.4	216.77	227.0	170.	0.365	977.3	28.96	295.2	1.422	.466 4
12 000	11 977	−56.5	216.65	194.0	146.	0.312	977.0	28.96	295.1	1.422	.466 2
13 000	12 973	−56.5	216.65	165.8	124.	0.267	976.7	28.96	295.1	1.422	.466 2
14 000	13 969	−56.5	216.65	141.7	106.	0.228	976.4	28.96	295.1	1.422	.466 2
15 000	14 965	−56.5	216.65	121.1	90.8	0.195	976.1	28.96	295.1	1.422	.466 2
16 000	15 960	−56.5	216.65	103.5	77.7	0.166	975.8	28.96	295.1	1.422	.466 2
17 000	16 955	−56.5	216.65	88.5	66.4	0.142	975.4	28.96	295.1	1.422	.466 2
18 000	17 949	−56.5	216.65	75.7	56.7	0.122	975.1	28.96	295.1	1.422	.466 2

Table 5.2–5 (continued)
U.S. STANDARD ATMOSPHERE, 1962

Table B. U.S. Standard Atmosphere – Metric Units (continued)

Altitude		Temperature, t		Pressure, P		Density, ρ, kg/m³	Gravity, g, m/s²	Molecular weight, M	Velocity of sound, V_s, m/s	Viscosity, μ, kg/m·s × 10⁵	Conductivity, k, kcal × 10⁶ m·s·K
Geometric, Z, m	Geopotential, H, m	deg C	K	mbar	mm Hg						
19 000	18 943	−56.5	216.65	64.7	48.5	0.104	974.8	28.96	295.1	1.422	.466 2
20 000	19 937	−56.5	216.65	55.3	41.4	0.088 9	974.5	28.96	295.1	1.422	.466 2
22 000	21 924	−54.6	218.57	40.5	30.4	0.064 5	973.9	28.96	296.4	1.433	.470 2
24 000	23 910	−52.6	220.56	29.7	22.3	0.046 9	973.3	28.96	297.8	1.444	.474 2
26 000	25 894	−50.6	222.54	21.9	16.4	0.034 3	972.7	28.96	299.1	1.454	.478 2
28 000	27 877	−48.6	224.53	16.2	12.1	0.025 1	972.1	28.96	300.4	1.465	.482 0
30 000	29 859	−46.6	226.51	12.0	9.0	0.018 4	971.5	28.96	301.7	1.475	.485 9
35 000	34 808	−36.6	236.51	5.75	4.31	0.008 5	970.0	28.96	308.3	1.529	.505 8
40 000	39 750	−22.8	250.35	2.87	2.15	0.004 0	968.4	28.96	317.2	1.601	.533 0
50 000	49,610	−2.5	270.65	0.798	0.598	0.001 0	965.4	28.96	329.8	1.704	.572 1
100 000	98 451	−63.1	210.02	3 × 10⁻⁴	2.6 × 10⁻⁴	5 × 10⁻⁷	950.1	28.88	—	—	—
200 000	193 898	+962.8	1 236.	1.3 × 10⁻⁶	1 × 10⁻⁶	3.3 × 10⁻¹⁰	921.7	25.56	—	—	—
400 000	376 312	1 214.	1 487.	4 × 10⁻⁸	3 × 10⁻⁸	6.5 × 10⁻¹²	867.9	19.94	—	—	—
700 000	630 530	1 234.	1 508.	1.2 × 10⁻⁹	8.9 × 10⁻¹⁰	1.5 × 10⁻¹³	795.6	16.17	—	—	—

Condensed and computed from *U.S. Standard Atmosphere, 1962*, U.S. Government Printing Office, Washington, D.C., 1962.

Table 5.2–6
VARIATIONS OF HIGH-ALTITUDE ATMOSPHERE

Values for 20,000 ft Geometric Altitude (6,100 m)

Symbols:

H, ft = geopotential altitude, feet. For meters multiply by 0.3048.

t, °F = temperature, degrees Fahrenheit.

t, °C = temperature, degrees Celsius.

P, psia = atmospheric pressure, psia. For atmospheres multiply by 0.068046. For kN/m² multiply by 6.8948.

ρ = density. For lb_m/ft³ multiply by 10⁻³. For kg/m³ multiply by 10⁻³ and by 16.018.

V_s, fps = speed of sound, ft/sec. For m/sec multiply by 0.3048.

μ = viscosity. For lb_m/ft sec multiply by 10⁻⁵. For N · s/m² (= kg/m · s) multiply by 10⁻⁵ and by 1.4882.

k = thermal conductivity. For Btu/sec ft°R multiply by 10⁻⁵. For W/m · K multiply by 0.62265.

Table 5.2–6 (continued)
VARIATIONS OF HIGH-ALTITUDE ATMOSPHERE

Values for 20,000 ft Geometric Altitude (6,100 m)

Latitude, °N	Season	H, ft	t, °F	t, °C	P, psia	ρ	V_s, fps	μ	k
30	Winter	19,953	− 1.20	−18.4	6.92	40.8	1,050	1.091	.364
30	Summer	19,953	18.89	− 7.9	7.06	39.8	1,072	1.129	.378
45	Winter	19,981	−21.99	−30.0	6.62	40.8	1,026	1.051	.349
45	Summer	19,981	9.61	−12.4	6.97	40.1	1,062	1.111	.372
60	Winter (cold)	20,006	−39.36	−39.7	6.39	41.0	1,005	1.016	.336
60	Winter (warm)	20,006	−39.36	−39.7	6.39	41.0	1,005	1.016	.336
60	Summer	20,006	− 5.01	−20.6	6.79	40.3	1,045	1.083	.361
75	Winter (cold)	20,026	−48.66	−44.8	6.24	41.0	994	.998	.329
75	Winter (warm)	20,026	−48.66	−44.8	6.24	41.0	994	.998	.329
75	Summer	20,026	−12.73	−24.9	6.71	40.5	1,036	1.069	.355

Condensed from *U.S. Standard Atmosphere Supplements*, U.S. Government Printing Office, Washington, D.C., 1966.

Table 5.2–7
EXTREMELY HIGH-ALTITUDE CONDITIONS

The following table gives temperature and molecular weight of atmosphere above 390,000 ft (120 km) geometric altitude (summer model).[a]

For pressure in N/m^2, multiply values in millibars by 100.

Altitudes, geometric		Exospheric temperature,[b] 600°K			Exospheric temperature,[b] 2,100°K		
km	ft	Temp, °K	Molecular weight	Pressure, millibars	Temp, °K	Molecular weight	Pressure, millibars
120	394,000	380.0	26.77	2.30×10^{-5}	379.6	26.76	2.27×10^{-5}
140	459,000	483.9	25.40	5.92×10^{-6}	819.9	25.78	8.16×10^{-6}
160	525,000	535.0	23.90	2.00×10^{-6}	1,142.6	25.07	4.55×10^{-6}
180	590,000	561.1	22.32	7.80×10^{-7}	1,375.3	24.47	2.93×10^{-6}
200	656,000	574.1	20.79	3.36×10^{-7}	1,544.8	23.92	2.03×10^{-6}
250	820,000	593.7	17.85	5.53×10^{-8}	1,796.5	22.65	9.49×10^{-7}
300	984,000	598.4	16.06	1.19×10^{-8}	1,942.0	21.50	5.01×10^{-7}
400	1,312,000	599.9	11.30	9.91×10^{-10}	2,058.4	19.55	1.70×10^{-7}
1,000	3,280,000	600.0	1.28	5.44×10^{-11}	2,100.0	13.29	1.94×10^{-9}

[a]Atmospheric density varies from 9.05×10^{-14} to 1.93×10^{-8} kg/m^3.
[b]Exospheric temperature depends on solar activity and season.

REFERENCES

U.S. Standard Atmosphere Supplements, U.S. Government Printing Office, Washington, D.C., 1966.
Minzer, R. A., Extension of Tables for the U.S. Standard Atmosphere, CRL-67-0335, USAF, 1967.

From Bolz, R. E. and Tuve, G. L., Eds., *Handbook of Tables for Applied Engineering Science*, 2nd ed., CRC Press, Cleveland, 1973, 655.

Table 5.2–8
MEASUREMENT AND TEST STANDARDS

Standards may be established for a variety of objectives: for the control of quantity or quality, for establishing capacity ratings, for improving interchangeable manufacture, or for the protection of users or consumers. Standards have a long history, as do many standards organizations. In fact, there are so many changes in standards that many organizations revise their lists annually. An attempt is made in the following list to give current information, but this list is in no sense exhaustive. If its currency is to be verified, the user should contact the issuing organization.

Although standards for measurement and testing are of first concern here, the list of organizations and the standards issued or promoted by them cover a much wider field, including standards for materials, processes, and performance. The fact that all standards specify numerical values that must be established by measurement is a sufficient justification for treating all types of standards in this list.

While trade practices and industrial standards set up by trade associations are only mentioned herewith, it is recognized that many national and even international standards originate from trade standards. The "engineering standards" issued by the following organizations are established by authoritative but impartial groups that have no direct interest in the manufacture, marketing, or procurement of the materials or equipment to which the standards apply. Although U.S. Government and military standards fail to meet this criterion, these standards may be fully as important to many industrial organizations as the recognized "national standards." They represent a class of their own, however, and their literature is voluminous.[a]

Standards set up by the national engineering societies are intended for nationwide use and are therefore classified as national standards.

National Engineering Standards

NBS: The U.S. National Bureau of Standards is a government agency within the U.S. Department of Commerce, established in 1901 and charged with the "development and maintenance of the national standards of measurement." One of its divisions is the Institute of Basic Standards. An important NBS function is the development of standard practices, codes, and specifications, but these are developed largely for Government departments and do not necessarily come into nationwide use in engineering.

NBS regularly broadcasts frequency and time standards and maintains a Radio Standards Laboratory in Boulder, Col. (see Table 5.2-11).

In keeping with its function to "provide the means and methods for making measurements consistent with the national standards of measurement," NBS conducts testing, evaluation, and calibration services, provides standard reference materials, and cooperates in the setting of international standards. NBS publications are extensive and have been the center of activity of the "National Standards Reference Data System," which undertakes to compile critically evaluated, quantitative data on substances and materials. As far as engineers in industry are concerned, NBS codes for testing and measurement are usually considered advisory only, but on basic units they are the highest authority.

ANSI: The American National Standards Institute is an independent cooperative association that promulgates and publishes national standards, reprints and sells international standards, lists certain other technical standards, and advertises the standards of various industry groups. It issues an annual catalogue offering copies of some 4,000 standards. Many of these are joint standards, approved and adopted by two or more agencies, but now designated as American National Standards (formerly called "ASA Standards" and "USA Standards"). The date of each standard or of its latest revision is indicated in each case.

ASTM: The American Society for Testing and Materials is a membership society that has become the foremost U.S. source of information on the specification and testing of materials. Its membership includes individuals, corporations, libraries, and over 1,000 government agencies. It publishes annual editions of ASTM Standards covering 32 fields, as listed in the following table. The annual editions include 4,500 standards, comprising 30,000 pages and a 250-page index.

Table 5.2–8 (continued)
MEASUREMENT AND TEST STANDARDS

ASTM part No.	Subject	Month of new issue
1	Steel-piping materials	April
2	Ferrous castings	April
3	Steel sheet, strip, bar, rod, wire; metallic-coated products	April
4	Structural and boiler steel; forgings; ferro-alloys; filler metal	April
5	Copper and copper alloys	July
6	Light metals and alloys	July
7	General nonferrous metals; electrodeposited coatings; metal powders	July
8	Magnetic properties; materials for electron tubes, thermostats, electrical heating	November
9	Cement; lime; gypsum	November
10	Concrete and mineral aggregates	November
11	Bituminous materials; soils	April
12	Mortars; clay and concrete pipe and tile; masonry units; asbestos-cement products	April
13	Refractories; glass; ceramic materials	April
14	Thermal insulation; acoustical materials; fire tests; building constructions	November
15	Paper; packaging; cellulose; casein; flexible barrier materials; leather	April
16	Structural sandwich construction; wood; adhesives	July
17	Petroleum products – motor fuels; solvents; fuel oils; lubricating oils; cutting oils; grease	November
18	Petroleum products – LPG; light pure hydrocarbons; wax petrolatum	November
19	Gaseous fuels; coal and coke	November
20	Paint – materials specifications and tests; naval stores; aromatic hydrocarbons	April
21	Paint – tests for formulated products and applied coatings	April
22	Soap; antifreezes; wax polishes; halogenated organic solvents	July
23	Industrial water; atmospheric analysis	November
24	Textiles – general methods and definitions	November
25	Textiles – fibers and products	November
26	Plastics – specifications	July
27	Plastics – methods of testing	July
28	Rubber; carbon black; gaskets	July
29	Electrical insulating materials	July
30	General testing methods; quality control; appearance tests; temperature measurement; radiation; spectroscopy	July
31	Metallography; nondestructive tests; fatigue; corrosion	July
32	Chemical analysis of metals; metal-bearing ores	April
33	Index	September

NFPA: The National Fire Protection Association publishes a ten-volume set of "National Fire Codes" (7,000 pages), which recommend standards and practices for fire prevention, protection, and safety.

UL: The Underwriters' Laboratories is an approval and testing agency sponsored by the American Insurance Association. UL issues test instructions and approval rules. It grants approvals of electrical and structural materials and of equipment and methods relative to fire and casualty hazards and safety. It publishes standards, specifications, and annual lists of approved materials and equipment with names of manufacturers. Located at 207 E. Ohio St., Chicago, Ill., UL also maintains testing laboratories at Chicago and Northbrook, Ill., Melville, Long Island, N.Y., and Santa Clara, Cal.

Engineering Society Standards. Most of the national engineering societies are active in promoting standards relating to their particular field of engineering. Several of the larger societies have publication programs in this field and also cooperate actively with the ANSI.[b]

Trade-association Standards. Trade-association standards vary from simple protective agreements to large public service operations of benefit to the technical community and to users. Because many trade standards are of great importance to engineers, their existence should not be overlooked. This is particularly true of those associations that issue "approval requirements."[c]

TABLE 5.2–8   (continued)
MEASUREMENT AND TEST STANDARDS

Safety Standards. This group of standard recommendations to promote safety and health is important to engineers as well as to the medical profession, industrial hygienists, toxicologists, and others.[d]

International Standards Organizations

ISO: The International Organization for Standardization is a federation of 56 national member bodies similar to the ANSI. It issues three annual publications: the *Memento,* giving names of member organizations and lists of committees and projects, the *Catalog,* listing adopted standards and draft recommendations, and the *Journal,* which includes the calendar of ISO meetings. The ANSI catalogue lists and offers copies of approximately 700 ISO standards.

IEC: The International Electro-technical Commission promotes about 400 international standards and publishes many of them in French and Russian as well as English. It publishes an annual *IEC Central Office Report*, giving information on its technical work in the electrical field.

COPANT: The Pan-American Standards Commission issues 50 or more standards in Spanish. It coordinates the national standards programs of its member bodies in the Western hemisphere.

CEE: The International Commission on Rules for the Approval of Electrical Equipment issues specifications that are widely followed in Europe. While its membership is all European, the United States has observer status. It is somewhat similar in function to the Underwriters' Laboratories, Inc. (UL approval) in the United States.

[a] The Global Engineering Documentation Services, Inc., Newport Beach, Calif., lists 250,000 documents. See also DODISS in Sweets Microfilms, published by the McGraw-Hill Information Systems Co.

[b] The American National Standards Institute, formerly the United States of America Standards Institute (USASI).

[c] For information on trade associations, see *Directory of National Trade and Professional Associations of the United States,* Columbia Publishers, Washington, D.C., 1969, and *Gale Encyclopedia of Trade Associations,* Gale Research Co., Book Tower, Detroit, Mich. 48226, 1968.

[d] For additional information consult the following handbooks:

Sunshine, I., Ed., *Handbook of Analytical Toxicology,* Chemical Rubber, Cleveland, 1969.
Faulkner, W. R., King, J. W., and Damm, H. C., Eds., *Handbook of Clinical Laboratory Data,* 2nd ed., Chemical Rubber, Cleveland, 1968.
Steere, N. V., Ed., *Handbook of Laboratory Safety,* 2nd ed., Chemical Rubber, Cleveland, 1971.
Wang, Y., Ed., *Handbook of Radioactive Nuclides,* Chemical Rubber, Cleveland, 1969.

From Bolz, R. E. and Tuve, G. L., Eds., *Handbook of Tables for Applied Engineering Science,* 2nd ed., CRC Press, Cleveland, 1973, 1010.

Table 5.2—9

FIXED POINTS FOR CALIBRATION OF TEMPERATURE-MEASURING INSTRUMENTS

Substance	Phase change at standard atmosphere pressure	Temperature °C	Temperature °F	Substance	Phase change at standard atmosphere pressure	Temperature °C	Temperature °F
Helium	Melts	-270.98	-455.76	Benzophenone	Boils	305.9[b]	582.6
Helium	Boils	-268.93	-452.07	Cadmium	Melts	320.9[b]	609.6
Hydrogen (normal)	Boils	-252.87	-423.17	Lead	Melts	327.3[b]	621.1
Oxygen	Melts	-277.	-375.6	Potassium dichromate	Melts	397.5	747.5
Nitrogen	Boils	-195.85	-320.43	Zinc	Melts	419.58[b]	787.24
Oxygen	Boils	-182.96[a]	-297.33	Sulfur	Boils	444.6[a]	832.3
Isopentane	Melts	-160.	-256.	Lead chloride	Melts	501.0	933.8
Methyl cyclohexane	Melts	-126.	-194.8	Calcium nitrate	Melts	561.0	1,041.8
Carbon disulfide	Melts	-112.	-169.6	Antimony	Melts	630.5[b]	1,166.9
Toluene	Melts	-95.	-139.	Aluminum	Melts	660.1[b]	1,220.2
Carbon dioxide	Sublimes	-78.5[b]	-109.3	Potassium chloride	Melts	770.3	1,418.5
Chloroform	Melts	-63.5	-82.	Sodium chloride	Melts	801.4	1,474.5
Mercury	Melts	-38.87[b]	-37.97	Sodium carbonate	Melts	852.0	1,565.6
Carbon tetrachloride	Melts	-22.9	-9.22	Sodium sulfate	Melts	884.7	1,624.5
Water ice	Melts	0.0[a]	32.	Silver	Melts	961.93[a]	1,763.47
Sodium sulfate	Melts	+32.38	90.28	Gold	Melts	1,064.43[a]	1,947.97
Acetylene dichloride	Boils	55.0	131.	Copper	Melts	1,083.[b]	1,981.4
Ethyl alcohol	Boils	78.3	172.94	Lithium silicate	Melts	1,201.	2,193.8
Water	Boils	100.0[a]	212.	Barium fluoride	Melts	1,280.	2,336.
Toluene	Boils	110.0	230.	Nickel	Melts	1,453.	2,647.4
Benzoic acid	Boils	122.36[b]	252.25	Cobalt	Melts	1,480.	2,696.
Bromobenzene	Boils	156.6	313.88	Iron	Melts	1,530.	2,786.
Aniline	Boils	184.5	364.1	Palladium	Melts	1,552.[b]	2,826.
Nitrobenzene	Boils	209.0	408.2	Platinum	Melts	1,769.[b]	3,216.
Naphthalene	Boils	218.0[b]	424.4	Rhodium	Melts	1,960.	3,560.

Table 5.2—9 (continued)

FIXED POINTS FOR CALIBRATION OF TEMPERATURE-MEASURING INSTRUMENTS

Substance	Phase change at standard atmosphere pressure	Temperature	
		°C	°F
Tin	Melts	231.9[b]	449.4
Diphenyl	Boils	254.6	490.2

Substance	Phase change at standard atmosphere pressure	Temperature	
		°C	°F
Iridium	Melts	2,443.	4,429.4
Tungsten	Melts	3,380.	6,116.

[a] Primary standards
[b] Secondary standards

From Bolz, R. E. and Tuve, G. L., Eds., *Handbook of Tables for Applied Engineering Science*, 2nd ed., CRC Press, Cleveland, 1973, 1012.

Table 5.2–10
STANDARD AIR AND ATMOSPHERE[a]

Past and Present Standards

The composition and properties of air and of the atmospheric envelope surrounding the earth have been variously defined and standardized for engineering purposes. Older standards cannot be readily changed or abandoned. Various assumed standards are convenient for different purposes and locations. There are large accumulations of equipment-performance data based on "standard" test codes. Each code uses some accepted but arbitrary definition of air, often neglecting the minor constituents or even the water vapor. Great numbers of tests of materials for aeronautical or space uses have been made in atmospheres stimulated according to early "standard atmosphere" patterns.

Since there is no universally accepted "standard air," the following are offered for reference and comparison. Slight differences in the assumed composition of air may be included in the variations from one standard to another.

Various Standards for Dry Air

Density		Absolute pressure				Temperature			Molecular weight	Reference
lb/ft³	kg/m³	in. Hg (32°F)	psi	mm Hg (0°C)	MN/m² (bars/10)	°F	°C	K		
.080722	1.29304	29.921	14.696	760.0	.101325	32.0	0.0	273.15	28.966	1
.076474	1.2250	29.921	14.696	760.0	.101325	59.0	15.0	288.15	28.9644	2
.0754	1.21	29.921	14.696	760.0	.101325	68.0	20.0	293.15	–	3
.075	1.2014	29.921	14.696	760.0	.101325	69.5	20.8	293.95	–	4
.0705	1.13	29.00	14.25	738.	.0984	85.0	29.4	302.55	–	5
.074	1.1847	29.921	14.696	760.	.101325	77.0	25.0	298.15	–	6

REFERENCES

1. **Hilsenrath, J. et al.,** *Tables of the Thermal Properties of Gases,* NBS Circular 564, National Bureau of Standards, Washington, D.C., 1955.
2. *U.S. Standard Atmosphere, 1962,* U.S. Government Printing Office, Washington, D.C., 1962; this is the standard of NASA, USAF, USWB, USN, NBS, and ICAO.
3. ASME Test Codes. The intermediate temperature of 20°C is used in many practical gas tables.
4. ASHRAE Standard (dry air at 69.5°F; saturated air at 60°F; air at 68°F, 50% relative humidity). This is approximately equal to the ISO "Standard Reference Atmosphere," 20°C, 65% relative humidity, 1,013 mbar.
5. SAE Engine and Gas Turbine Test Codes.
6. "Room Conditions." Widely used as a typical indoor condition, i.e., dry air at 25°C.

[a]See also Tables 5.2–3 and 5.2–5.

From Bolz, R. E. and Tuve, G. L., Eds., *Handbook of Tables for Applied Engineering Science,* 2nd ed., CRC Press, Cleveland, 1973, 1013.

Table 5.2–11
TIME AND FREQUENCY STANDARDS

International *standard time*, or Greenwich Mean Time (GMT), is scientifically known as Coordinated Universal Time (UTC). It is one of four values on the Universal Time scale. The basic value UTO is derived from mean solarrotation data. Corrected for regular periodic variations, this becomes UT1. Irregular variations also occur; these are reported from all parts of the world to the Bureau International de l'Heure (BIH) in Paris. BIH in turn issues the corrections to produce the astronomer's time, UT2. Coordinated Universal Time (UTC) formerly operated with a frequency offset from the atomic scale of -300×10^{-10} to agree approximately with the rotation of the earth. Occasional step adjustments in time of 0.10 s were made to compensate for unpredictable variations in the earth's rate of rotation. On January 1, 1972, the UTC second was made equal to the SI second; to maintain approximate agreement with UT1, it will be necessary to add or subtract a leap-second every 6 to 18 months (preferably December 31 and June 30).

International Atomic Time (IAT) as maintained by the BIH now differs from UTC by an integral number of seconds.

In the United States the time corrections are made and sent out by radio broadcast by the National Bureau of Standards through the operation of four radio stations — WWVB, WWV, WWVH, and WWVL. The total annual correction to maintain UTC in close agreement with UT2 is approximately 200 parts in 10^{10}; the NBS time signals are stable to one part in 10^{12} of frequency. The UTC (NBS) signals are maintained within 5 μs of the time signals of the U.S. Naval Observatory, UTC (USNO).

Time signals from the Bureau's radio station WWV can be heard on the telephone. By dialing (303)499-7111, listeners can hear the accurate shortwave signals from Fort Collins, Colorado, as received at the Bureau in Boulder, Colorado. These signals are a national service provided by the U.S. Department of Commerce.

The signals include a voice announcement of Greenwich Mean Time (GMT) every minute, plus standard audio-frequency tones and special announcements of interest to geophysicists and navigators. The time and frequency signals are the most accurate in the U.S. available to telephone users — callers from "the lower 48" should receive time signals accurate to within 30-40 ms — and are controlled ultimately by the NBS Atomic Frequency Standard in Boulder.

The eight NBS services are as follows: standard radio frequencies, standard audio frequencies, standard musical pitch, standard time intervals, time signals, UT2 corrections, radio propagation forecasts, and geophysical alerts. The broadcast services of the NBS stations are summarized in Table A.

The low frequency radio station WWVB (Fort Collins, Col.) will continue to broadcast seconds pulses without offset (as before January 1, 1972). However, the designation of its time scale, SAT (Stepped Atomic Time), will be changed to UTC (NBS), and its adjustments will be the same as for the UTC scale discussed above. (Previously, SAT incorporated step adjustments of 200 ms made on the first of the month, as necessary.)

UTC (NBS) as broadcast by WWV, WWVH, and WWVB will be maintained within ±0.7 s of UT1 and with ± 1 ms of UTC (BIH). For navigators and others who may need to know UT1 to better than 0.7 s, a special code will be broadcast to indicate the difference between UT1 and UTC with a resolution of 0.1 s.

Standard time and frequency broadcasts are also transmitted from at least 15 other countries, with accuracy of 20 parts in 10^9 or better. Most of these stations are operating under the UTC system.

Basic Time and Frequency Standards

Ephemeris time is determined by observations of the moon. In 1956 the International Committee on Weights and Measures defined the constant second in terms of ephemeris time at noon January 1, 1900, as 1/31,566,925.9747 of the tropical year. Actually, the precision indicated by 12 significant figures cannot be justified by present experimental observations of the solar system.

Sidereal time is determined from observation of the stars; it is not perfectly uniform because of variations in the rotational speed of the earth.

Table 5.2–11 (continued)
TIME AND FREQUENCY STANDARDS

Atomic time is the term usually applied to the most accurate known methods for measuring time *intervals.* Earlier standards were based on induced vibrations of tuning forks, then later of quartz crystals. Vibrations of the molecules by electrical excitation produced the "ammonia clock," in which high-frequency electronic circuits, synchronized with the ammonia-molecule vibration, controlled a standard clock with an accuracy of one or two parts in a billion. More recently the gas cell and atomic-beam devices have provided even much greater accuracy, perhaps 1 sec in 300 years.

The cesium-beam-controlled oscillator was designated as the official frequency standard by the 12th General Conference of Weights and Measures, 1964 (see Table B).

Table A. Radio-broadcast Services of NBS Stations*

Services	WWV	WWVH	WWVB	WWVL
Standard radio frequencies:				
20 khz				X
60 khz			X	
2.5 Mhz	X	X		
5 Mhz	X	X		
10 Mhz	X	X		
15 Mhz	X	X		
20 Mhz	X			
25 Mhz	X			
Standard audio frequencies:				
440 hz	X	X		
600 hz	X	X		
1,000 hz	X			
Standard time intervals	X	X	X	
Time signals	X	X	X	
Time code	X		X	
UT2 corrections	X	X	X	
Radio propagation forecasts	X			
Geophysical alerts	X	X		

Note: The station locations and radiated powers are as follows:
    At Fort Collins, Colorado: WWV–(10 kW for 5, 10, and 15 Mhz; 2.5 kW for 2.5, 20, and 25 Mhz)
           WWVB–(12 kW)
           WWVL–(1.8 kW)
    At Puunene, Maui, Hawaii: WWVH–(2 kW for 5, 10, and 15 Mhz; 1 kW for 2.5 Mhz)

*Data from *Reference Data for Radio Engineers,* 5th ed., Howard W. Sams, Indianapolis, 1968.

Table 5.2–11
TIME AND FREQUENCY STANDARDS

Table B. Standard Frequency Sources*

Typical Reported Values

Characteristic	Rubidium gas cell	Cesium beam	Hydrogen maser
Stability (rms deviation, 1 hr)	1×10^{12}	8×10^{13}	3×10^{14}
Resonance events per second	10^{12}	10^6	10^{12}
Nominal resonance frequency, Mhz	6,834.682608	9,192.631770	1,420.405751
Detectable drift	Yes	Yes	No

*Data from *Reference Data for Radio Engineers,* 5th ed., Howard W. Sams, Indianapolis, 1968.

REFERENCES

Viezbicke, P. P., Ed., *NBS Frequency and Time Broadcast Services,* NBS Special Publication 236, National Bureau of Standards, Washington, D. C., 1971.
NBS Tech. News Bull., 53, 5, May 1969.
NBS Tech. News Bull., 55, 303, December 1971.
Frequency-Time Broadcast Services, National Bureau of Standards, Boulder, Colorado 80302.

From Bolz. R. E. and Tuve, G. L., Eds., *Handbook of Tables for Applied Engineering Science,* 2nd ed., CRC Press, Cleveland, 1973, 1013.

Table 5.2 – 12

ENGLISH-METRIC EQUIVALENTS FOR STANDARD TENSILE TESTS

Dimensions of Standard Round Tensile-test Specimens

One-half Inch Round, Two-inch Gage Length

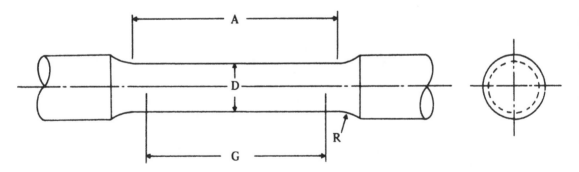

In view of the increasing use of the international system (SI) of units, the dimensions for standard test specimens are given in both English and metric units.

CONVERSION FACTORS: 1 in. = 25.4 mm (exactly)
1 lb_f = 0.45359237 kg_f (exactly)

Table 5.2–12 (continued)
ENGLISH-METRIC EQUIVALENTS FOR STANDARD TENSILE TESTS

Dimensions of Standard Round Tensile-test Specimens

One-half Inch Round, Two-inch Gage Length

	Standard specimen		Small-size specimen proportional to standard	
	Inches 0.500	Millimeters 12.5	Inches 0.350	Millimeters 8.75
Nominal diameter				
G–Gage length	2.000 ± 0.005	50.80 ± 0.13	1.400 ± 0.005	35.56 ± 0.13
D–Diameter	0.500 ± 0.010	12.70 ± 0.25	0.350 ± 0.007	8.89 ± 0.18
R–Radius of fillet, min	3/8	approx. 9.5	1/4	approx. 6.5
A–Length of reduced section, min	2 1/4	approx. 57	1 3/4	approx. 44.5

From *ASTM Metric Practice Guide*, National Bureau of Standards Handbook 102, U.S. Government Printing Office, Washington, D.C., 1967.

Table 5.2 – 13
BAROMETER CORRECTIONS

Barometer Corrections for Temperature

To correct the observed reading of a mercury barometer or U tube to the 32°F standard, add or subtract the following values in inches of mercury.

Observed reading of mercury column, inches Hg

Actual temperature of mercury column, °F	20	22	24	26	28	30	32
				Add the correction			
–20	0.09	0.10	0.11	0.11	0.12	0.13	0.14
0	0.05	0.06	0.06	0.07	0.07	0.08	0.08
20	0.02	0.02	0.02	0.02	0.02	0.02	0.02
				Subtract the correction			
40	0.02	0.02	0.02	0.03	0.03	0.03	0.03
60	0.06	0.06	0.07	0.07	0.08	0.08	0.09
80	0.09	0.10	0.11	0.12	0.13	0.14	0.15
100	0.13	0.14	0.15	0.17	0.18	0.19	0.20

Table 5.2–13 (continued)
BAROMETER CORRECTIONS

Barometer Corrections for Elevation

To correct the observed reading of a mercury barometer or U tube to the equivalent reading at a higher elevation, subtract the following values in inches of mercury for each 100-ft difference in elevation (add for lower elevation).

Mean elevation, ft	Mean atmospheric temperature, °F						
	−20	0	20	40	60	80	100
0	0.13	0.12	0.12	0.11	0.11	0.10	0.10
1,000	0.12	0.12	0.11	0.11	0.10	0.10	0.10
2,000	0.12	0.11	0.11	0.10	0.10	0.10	0.09
3,000	0.11	0.11	0.10	0.10	0.10	0.09	0.09
4,000	0.11	0.10	0.10	0.10	0.09	0.08	0.08
5,000	0.10	0.10	0.10	0.09	0.09	0.08	0.08
6,000	0.10	0.10	0.09	0.09	0.08	0.08	0.08
7,000	0.10	0.09	0.09	0.09	0.08	0.08	0.08

Barometer Corrections for Gravity

To correct the observed reading of a mercury barometer or U tube to the equivalent reading at standard gravity, add or subtract the following values in inches of mercury.

North latitude, degree	Elevation, feet							
	0	0	2,000	2,000	4,000	4,000	6,000	6,000
	Height of column, inches Hg							
	30	28	28	26	26	24	24	22
	Subtract the correction							
25	0.05	0.05	0.05	0.05	0.05	0.05	0.06	0.05
30	0.04	0.04	0.04	0.04	0.05	0.04	0.05	0.04
35	0.03	0.03	0.03	0.03	0.03	0.03	0.04	0.03
40	0.02	0.01	0.02	0.02	0.02	0.02	0.03	0.02
45	0.00	0.00	0.01	0.01	0.01	0.01	0.01	0.01
	Add the correction							
50	0.01	0.01	0.01	0.01	0.00	0.00	0.00	0.00

From Bolz, R. E. and Tuve, G. L., Eds., *Handbook of Tables for Applied Engineering Science,* 2nd ed., CRC Press, Cleveland, 1973, 1016.

Table 5.2 – 14

EMERGENT STEM CORRECTION FOR LIQUID-IN-GLASS THERMOMETERS

Accurate thermometers are calibrated with the entire stem immersed in the bath that determines the temperature of the thermometer bulb. However, for reasons of convenience, it is common practice when using a thermometer to permit its stem to extend out of the apparatus. Under these conditions both the stem and the mercury in the exposed stem are at a temperature different from that of the bulb. This introduces an error into the observed temperature. Since the coefficient of thermal expansion of glass is less than that of mercury, the observed temperature will be less than the true temperature if the bulb is hotter than the stem and greater than the true temperature, providing the thermal gradient is reversed. For exact work the magnitude of this error can only be determined by experiment. However, for most purposes it is sufficiently accurate to apply the following equation, which takes into account the difference of the thermal expansion of glass and mercury:

$$T_c = T_o + F \times L(T_o - T_m)$$

where

T_c = corrected temperature

T_o = observed temperature

T_m = mean temperature of exposed stem. The mean temperature of the exposed stem may be determined by fastening the bulb of a second thermometer against the midpoint of the exposed liquid column.

L = the length of the exposed column in degrees above the surface of the substance whose temperature is being determined.

F = correction factor. For approximate work and when the liquid in the thermometer is mercury, a value for F of 0.00016 is generally used. For more accurate work with mercury-filled thermometers, values as given in the following table are used. For thermometers filled with organic liquids, it is customary to use 0.001 for the value of F.

Values of F for Various Glasses

T_m °C	Corning 0041	Corning 8800	Corning 8810	Jena 16 III	Jena 59 III
50	0.000157	0.000166	0.000156	0.000158	0.000164
150	0.000159	0.000167	0.000157	0.000158	0.000165
250	0.000163	0.000168	0.000161	0.000161	0.000170
350	0.000168	0.000173	0.000166	–	0.000177

From Weast, R. C., Ed., *Handbook of Chemistry and Physics*, 53rd ed., Chemical Rubber, Cleveland, 1972, D-147.

5.3 TABLES OF SPECIFIC
COMPOSITIONAL STANDARDS

The following tables are those of selected compositional standards, indicating form and nominal composition, for which accurate certified analyses are available in the Standard Reference Materials (SRM numbers) program of the National Bureau of Standards. Specific information on these as well as other Standard Reference Materials may be obtained from:

> Office of Standard Reference Materials
> Room B311, Chemistry Building
> National Bureau of Standards
> Washington, D.C. 20234

Table 5.1—10 comprises the price and availability listing of these standards in 1973. More detailed information is available in *Catalogue of Standard Reference Materials*, NBS Special Publication 260, July 1970, and a series of 260-1, -2, etc. publications on the preparation and analysis of specific standards. These are available from the Superintendent of Documents, U.S. Government Printing Office, Washington, D.C. 20402.

Table 5.3—1

STEELS (GRANULAR FORM)

SRM No.	Name	Wt/unit (grams)	C	Mn	P	S	Si	Cu	Ni
163	Low alloy, 1.0 Cr	100	0.933	0.897	0.007	0.027	0.488	0.087	0.081
101f	Stainless, (AISI 304L)	100	.014	.087	.008	.008	.876	.030	9.96

SRM No.	Cr	V	Mo	W	Co	N	As	Sb	Ga
163	0.982	—	0.029	—	—	0.007	—	—	—
101f	18.49	0.034	.007	(0.0002)	0.088	—	(0.003)	(0.0009)	(0.004)

From *Catalogue of Standard Reference Materials*, NBS Special Publication 260, U.S. Government Printing Office, Washington, D.C., July 1970.

Table 5.3–2
STEELS (CHIP FORM)

SRM No.	Name	Wt/unit (grams)	C	Mn	P	S Grav	S Comb	Si
8i	Bessemer, 0.1 C	150	0.077	0.511	0.080	0.063	0.063	0.020
10g	Bessemer, 0.2 C	150	.240	.850	.086	.109	.109	.020
15g	Basic open hearth, 0.1 C	150	.097	.485	.005	–	.026	.095
335	Basic open hearth, 0.1 C (C only)	300	.092	–	–	–	–	–
11h	Basic open hearth, 0.2 C	150	.200	.510	.010	–	.026	.211
12h	Basic open hearth, 0.4 C	150	.407	.842	.018	–	.027	.235
152a	Basic open hearth, 0.5 C, 0.03 Sn	150	.486	.717	.012	–	.030	.202
13g	Basic open hearth, 0.6 C	150	.61	.85	.006	–	.030	.355
14e	Basic open hearth, 0.8 C	150	.753	.404	.008	–	.039	.177
16e	Basic open hearth, 1.1 C	150	1.09	.381	.021	–	.029	.20
337	Basic open hearth, 1.1 C (C only)	300	1.07	–	–	–	–	–
178	Basic oxygen, 0.4 C	150	0.395	.824	.012	.032	.014	.163
19g	Acid open hearth, 0.2 C	150	.223	.554	.046	.014	.033	.186
51b	Electric furnace, 1.2 C	150	1.21	.573	.013	.014	.014	.246
65d	Basic electric, 0.3 C	150	0.264	.730	.015	.010	.010	.370
100b	Manganese (SAE T1340)	150	.397	1.89	.023	.029	.028	.210
105	High-sulfur, 0.2 C (C only)	150	.193	–	–	–	–	–
30f	Cr-V (SAE 6150)	150	.49	0.79	.010	–	.010	.28
32e	Ni-Cr (SAE 3140)	150	.409	.798	.008	.022	.021	.278
33d	Ni-Mo (SAE 4820)	150	.173	.537	.006	.010	.011	.253
72f	Cr-Mo (SAE X4130)	150	.301	.545	.014	.024	.024	.256
111b	Ni-Mo (SAE 4620)	150	.193	.706	.012	.015	.015	.302
106b	Cr-Mo-Al (nitralloy G)	150	.326	.506	.008	.016	.017	.274
139a	Cr-Ni-Mo (AISI 8640)	150	.404	.780	.013	.019	.019	.241
50c	W18-Cr4-V1 (tool)	150	.719	.342	.022	.010	.009	.311
132a	Mo5-W6-Cr4-V2 (tool)	150	.825	.268	.029	.005	.006	.190
134a	Mo8-W2-Cr4-V1 (tool)	150	.808	.218	.018	.007	.007	.323
153a	Co8-Mo9-W2-Cr4-V2 (tool)	150	.902	.192	.023	.007	.007	.270
155	Cr0.5-W0.5 (low alloy)	150	.905	1.24	.015	.010	.011	.322
73c	Stainless (Cr13) (SAE 420)	150	.310	0.330	.018	–	.036	.181

Table 5.3–2 (continued)
STEELS (CHIP FORM)

SRM No.	Name	Wt/unit (grams)	C	Mn	P	S		Si
						Grav	Comb	
133a	Stainless (Cr13–Mo0.3–S0.3)	150	.120	1.03	0.26	3.26	.330	.412
121c	Cr18-Ni10-Ti0.4 (SAE 32a)	150	.038	1.31	.028	—	.009	.64
160b	Cr18-Ni12-Mo2.4 (AISI 316)	150	.046	1.64	.020	—	.018	.509
339	Cr17-Ni9-0.2Se (SAE 303Se)	150	.052	0.738	.129	—	.013	.654
343	Stainless (SAE 431)	150	.150	—	—	—	—	—
346	Valve (Cr22-Ni4-Mn9)	150	.541	9.15	.018	—	.063	.239
126b	Ni36 (high nickel)	150	.090	0.380	—	—	—	.200
36b	Cr2-Mo1 (low alloy)	150	.114	.404	.007	—	.019	.258
131b	Low-carbon silicon (C only)	100	.0018	—	—	—	—	—
344	Cr15-Ni7-Mo2-Al1	150	.069	.57	.018	—	.019	.395
345	Cr16-Ni4-Cu3	150	.048	.224	.018	0.012	.012	.610
348	Ni26-Cr14.5 (A 286)	150	.044	1.48	.015	—	.002	.54

SRM No.	Cu	Ni	Cr	V	Mo	W	Co	Ti	Sn	Al (Total)	Nb	N	Other
8i	0.016	0.009	0.009	0.012	0.003	—	—	—	—	—	—	0.018	—
10g	.008	.005	.008	.007	.002	—	—	—	—	—	—	.015	—
15g	—	—	—	—	—	—	—	—	—	—	—	—	—
335	—	—	—	—	—	—	—	—	—	—	—	—	—
11h	—	—	—	—	—	—	—	—	—	—	—	—	—
12h	.073	.032	.074	.003	.006	—	—	—	—	—	—	.006	—
152a	.023	.056	.046	.001	.036	—	—	—	0.032	—	—	—	—
13g	—	—	—	—	—	—	—	—	—	—	—	—	—
14e	.072	.053	.071	.002	.013	—	—	—	—	.060	—	—	—
16e	—	—	—	—	—	—	—	—	—	—	—	—	—
337	.032	.010	.016	.001	.003	—	—	—	—	—	—	—	—
178	—	—	—	—	—	—	—	—	—	—	—	—	—
19g	.093	.066	.374	.012	.013	—	0.012	.027	.008	.031	.026	—	—

Table 5.3–2 (continued)
STEELS (CHIP FORM)

SRM No.	Cu	Ni	Cr	V	Mo	W	Co	Ti	Sn	Al (Total)	Nb	N	Other
51b	.071	.053	.455	.002	.014	—	—	—	.008	—	—	.011	—
65d	.051	.060	.049	.002	.025	—	—	—	.004	.059	Al_2O_3 .009	.013	—
100b	.064	.030	.063	.003	.237	—	—	—	—	—	—	.004	—
105	—	—	—	—	—	—	—	—	—	—	—	—	—
30f	.076	.071	.05	.18	—	—	—	—	—	—	—	—	—
32e	.127	1.19	.678	.002	.023	—	—	—	.011	—	—	.009	—
33d	.123	3.58	.143	.002	.246	—	—	—	—	—	—	(.011)	—
72f	.062	0.055	.891	.005	.184	—	—	—	—	—	—	.009	—
111b	.028	1.81	.070	.003	.255	—	—	—	—	.043	—	—	—
106b	.117	0.217	1.18	.003	.199	—	—	—	—	1.07	—	—	—
139a	.096	.510	0.486	.003	.183	—	—	—	—	—	—	.008	—
50c	.079	.069	4.13	1.16	.082	18.44	—	—	.018	—	—	.012	As 0.022
132a	.120	.137	4.21	1.94	4.51	6.20	—	—	—	—	—	—	—
134a	.101	.088	3.67	1.25	8.35	2.00	8.47	—	—	—	—	—	—
153a	.094	.168	3.72	2.06	8.85	1.76	—	—	—	—	—	.024	—
155	.083	.100	0.485	0.014	0.039	0.517	—	—	—	—	—	—	—
73c	.080	.246	12.82	.030	.091	—	—	—	—	—	—	.037	—
133a	.118	.241	12.89	.026	.294	—	—	—	—	—	—	.032	—
121c	.14	10.51	17.58	.048	.16	—	—	.42	—	—	—	—	—
160b	.172	12.26	18.45	.047	2.38	—	0.101	—	—	—	—	.039	Pb 0.001
339	.199	8.89	17.42	.058	0.248	—	.096	—	—	—	—	—	Se 0.247
343	—	2.14	15.76	.036	—	—	—	—	—	—	—	.074	—
346	—	3.94	21.61	.058	—	—	—	—	—	—	—	.441	—
126b	.082	35.99	0.066	(.001)	(.006)	—	.032	—	—	—	—	—	—
36b	.179	0.203	2.18	.004	.996	—	—	—	—	—	—	—	—
131b	—	—	—	—	—	—	—	—	—	—	—	—	—
344	.106	7.28	14.95	.040	2.40	—	—	.076	—	1.16	—	—	—
345	3.44	4.24	16.04	.041	0.122	—	.089	—	—	—	.231	—	Ta .002
348	0.22	25.8	14.54	.25	1.3	—	—	2.24	—	0.23	—	B .0031	Fe 53.3

From *Catalogue of Standard Reference Materials*, NBS Special Publication 260, U.S. Government Printing Office, Washington, D.C., July 1970.

Table 5.3–3
INGOT IRON AND LOW-ALLOY STEELS

SRM No.				Chemical composition (nominal weight percent)				
7/32 in. D × 4 in. long	1/2 in. D × 2 in. long	1 1/4 in. D × 1/4 in. disk	Name	Mn	Si	Cu	Ni	Cr
–	803a	D803a	Acid open hearth, 0.6C	1.04	0.34	0.096	0.190	0.101
404a	804a	–	Basic electric	0.88	.44	.050	.040	.025
405a	805a	D805a	Medium manganese	1.90	.27	.032	.065	.037
407a	807a	D807a	Chromium-vanadium	0.76	.29	.132	.169	.92
408a	808a	–	Chromium-nickel	.76	.28	.10	1.20	.655
409b	809b	D809b	Nickel	.46	.27	.104	3.29	.072
–	810a	–	Cr2-Mo1	–	.36	.11	0.24	2.39
413	–	–	Acid open hearth, 0.4C	.67	.22	.25	.18	0.055
414	–	–	Cr-Mo (SAE 4140)	.67	.26	.11	.080	.99
417a	817a	–	Basic open hearth, 0.4C	.78	–	.13	.062	.050
418	–	–	Cr-Mo (SAE X4130)	.52	.28	–	.11	.96
418a	–	–	Cr-Mo (SAE X4130)	.52	.27	.040	.125	1.02
420a	820a	D820a	Ingot iron	.017	–	.027	.0092	0.0032
–	821	–	Cr-W, 0.9C	1.24	–	.080	.10	.49
427	827	–	Cr-Mo (SAE 4150) (B only)	–	–	–	–	–

SRM No.			Chemical composition (nominal weight percent)						
7/32 in. D × 4 in. long	1/2 in. D × 2 in. long	1 1/4 in. D × 1/4 in. disk	V	Mo	W	Co	Sn	Al Total	B
–	803a	D803a	0.005	0.033	–	–	–	–	–
404a	804a	–	.002	.007	–	–	–	–	–
405a	805a	D805a	–	.005	–	–	–	0.056	–
407a	807a	D807a	.146	–	–	–	–	–	–
408a	808a	–	.002	.065	–	–	–	–	–
409a	809a	D809a	.002	.009	–	0.025	0.012	–	–
–	810a	–	–	.91	–	–	–	–	–
413	–	–	.007	.006	–	–	–	–	–
414	–	–	.003	.32	–	–	.014	.020	–
417a	817a	–	–	.013	–	–	.036	–	–
418	–	–	–	.22	–	–	–	–	–
418a	–	–	–	.21	–	–	–	–	–
420a	820a	D820a	–	.0013	–	.006	.0017	.003	–
–	821	–	.012	.040	0.52	–	–	–	–
427	827	–	–	–	–	–	–	–	0.0027

From *Catalogue of Standard Reference Materials*, NBS Special Publication 260, U.S. Government Printing Office, Washington, D.C., July 1970.

Table 5.3—4
SPECIAL INGOT IRONS AND LOW-ALLOY STEELS

SRM No. 7/32 in. D × 4 in. long	SRM No. 1 1/4 in. D × 3/4 in. disk	Name	Chemical composition (nominal weight percent)								
			C	Mn	P	S	Si	Cu	Ni	Cr	V
461	—	Low alloy A	0.15	0.36	0.053	(0.02)	0.047	0.34	1.73	0.13	0.024
462	—	Low alloy B	.40	.94	.045	(.02)	.28	.20	0.70	.74	.058
463	1163	Low alloy C	.19	1.15	.031	(.02)	.41	.47	.39	.26	.10
464	—	Low alloy D	.54	1.32	.017	(.02)	.48	.094	.135	.078	.295
465	1165	Ingot iron E	.037	0.032	.008	(.01)	.029	.019	.026	.004	.002
466	1166	Ingot iron F	.065	.113	.012	(.01)	.025	.033	.051	.011	.007
467	1167	Low alloy G	.11	.275	.033	(.01)	.26	.067	.088	.036	.041
468	1168	Low alloy H	.26	.47	.023	(.02)	.075	.26	1.03	.54	.17
—	1170	Selenium (0.3 Se)	.089	.79	.109	.207	.163	Se 0.29	—	—	—

SRM No. 7/32 in. D × 4 in. long	SRM No. 1 1/4 in. D × 3/4 in. disk	Chemical composition (nominal weight percent)											
		Mo	W	Co	Ti	As	Sn	Al (Total)	Nb	Ta	B	Pb	Zr
461	—	0.30	0.012	0.26	(0.01)	0.028	0.022	(0.005)	0.011	0.002	0.0002	(0.003)	(<0.005)
462	—	.080	.053	.11	.037	.046	.066	.023	.096	.036	.0005	.006	.063
463	1163	.12	.105	.013	.010	.10	.013	.027	.195	.15	.0012	.012	.20
464	—	.029	.022	.028	.004	.018	.043	.005	.037	.069	.005	.020	.010
465	1165	.005	(.001)	.008	.20	.010	.001	.19	(.001)	.001	.0001	(<.0005)	(.002)
466	1166	.011	(.006)	.046	.057	.014	.005	.015	.005	.002	(.0002)	(.0013)	(<.005)
467	1167	.021	.20	.074	.26	.14	.10	.16	.29	.23	(.0002)	.0006	.094
468	1168	.20	.077	.16	.011	.008	.009	.042	.006	.005	.009	(<.0005)	(<.005)
—	1170	—	—	—	—	—	—	—	—	—	—	—	—

From *Catalogue of Standard Reference Materials*, NBS Special Publication 260, U.S. Government Printing Office, Washington, D.C., July 1970.

Table 5.3—5
STAINLESS STEELS

A. Group I

SRM No. 7/32 in. D × 4 in. long	Name	Chemical composition (nominal weight percent)								
		Mn	Si	Cu	Ni	Cr	V	Mo	W	Co
442	Cr16-Ni10	2.88	(0.09)	0.11	9.9	16.1	0.032	0.12	(0.08)	0.13
443	Cr18.5-Ni9.5	3.38	(.15)	.14	9.4	18.5	.064	.12	(.09)	.12
444	Cr20.5-Ni10	4.62	(.65)	.24	10.1	20.5	.12	.23	(.17)	.22

SRM No. 7/32 in. D × 4 in. long	Chemical composition (nominal weight percent)							
	Ti	Sn	Nb	Ta	B	Pb	Zr	Zn
442	0.002	0.0035	0.032	(0.0006)	0.0005	0.0017	(0.004)	(.003)
443	.003	.006	.056	(.0008)	.0012	.0025	–	(.005)
444	.019	.014	.20	(.004)	.0033	.0037	(.011)	(.004)

B. Group II

SRM No. 7/32 in. D × 4 in. long	1/2 in. D × 2 in. long	1 1/4 in. D × 1/4 in. disks	Name	Chemical composition (nominal weight percent)					
				Mn	Si	Cu	Ni	Cr	V
445	845	D845	Cr13-Mo0.9 (mod. AISI 410)	0.77	0.52	0.065	0.28	13.31	(0.05)
446	846	D846	Cr18-Ni9 (mod. AISI 321)	.53	1.19	.19	9.11	18.35	(.03)
447	847	D847	Cr24-Ni13 (mod. AISI 309)	.23	0.37	.19	13.26	23.72	(.03)
448	–	D848	Cr9-Mo0.3 (mod. AISI 403)	2.13	1.25	.16	.52	9.09	(.02)
449	849	D849	Cr5.5-Ni6.5	1.63	0.68	.21	6.62	5.48	(.01)
450	850	D850	Cr3-Ni25	–	.12	.36	24.8	2.99	(.006)

SRM No. 7/32 in. D × 4 in. long	1/2 in. D × 2 in. long	1 1/4 in. D × 1/4 in. disks	Chemical composition (nominal weight percent)					
			Mo	W	Ti	Sn	Nb	Ta
445	845	D845	0.92	(0.42)	(0.03)	–	0.11	(0.002)
446	846	D846	.43	(.04)	(.34)	(0.02)	.60	(.030)
447	847	D847	.059	(.06)	(.02)	–	.03	(.002)
448	–	D848	.33	(.14)	(.23)	(.05)	.49	(.026)
449	849	D849	.15	(.19)	(.11)	(.07)	.31	(.021)
450	850	D850	–	(.21)	(.05)	(.09)	.05	(.002)

Table 5.3–5 (continued)
STAINLESS STEELS

C. Group III

SRM No. 1 1/4 in. D × 3/4 in. disks	Name	Chemical composition (nominal weight percent)								
		C	Mn	P	S	Si	Cu	Ni	Cr	V
1152	(Cr18-Ni10)	0.163	1.19	0.017	0.017	0.654	0.497	10.21	18.49	0.044
1154	(Cr19-Ni10)	.094	1.74	.038	.033	1.09	.560	10.25	19.58	.061
1155	(AISI 316)	.046	1.63	.020	.018	0.50	.169	12.18	18.45	.047
1185	(AISI 316)	.11	1.22	.019	.016	.40	.067	13.18	17.09	–

SRM No. 1 1/4 in. D × 3/4 in. disks	Chemical composition (nominal weight percent)									
	Mo	Co	Ti	As	Sn	Al	Nb	Ta	B	Pb	Zr
1152	0.366	(0.095)	(0.12)	(0.01)	(0.004)	(0.003)	(0.20)	(0.085)	(0.005)	(0.001)	(0.03)
1154	.463	(.12)	(.48)	(.03)	(.023)	(.035)	(.26)	(.045)	(.0006)	(.012)	(.022)
1155	2.38	.101	–	–	–	–	–	–	–	.001	–
1185	2.01	–	<.001	–	–	–	<.001	<.001	–	–	–

From *Catalogue of Standard Reference Materials*, NBS Special Publication 260, U.S. Government Printing Office, Washington, D.C., July 1970.

Table 5.3–6
TOOL STEELS

SRM No.			Name	Chemical composition (nominal weight percent)							
7/32 in. D × 4 in. long	1/2 in. D × 2 in. long	1 1/4 in. D × 1/4 in. disk		Mn	Si	Cu	Cr	V	Mo	W	Co
436	836	D836	Special (Cr6-Mo3-W10)	0.21	0.32	0.075	6.02	0.63	2.80	9.7	–
437	837	D837	Special (Cr8-Mo2-W3-Co3)	.48	.53	–	7.79	3.04	1.50	2.8	2.9
438	838	D838	Mo high speed (AISI-SAE-M30)	.20	.17	.17	4.66	1.17	8.26	1.7	4.9
439	839	D839	Mo high speed (AISI-SAE-M36)	.18	.21	.12	2.72	1.50	4.61	5.7	7.8
440	840	D840	Special W high speed (Cr2-W13-Col 12)	.15	.14	.059	2.12	2.11	0.070	13.0	11.8
441	841	D841	W high speed (AISI-SAE-T1)	.27	.16	.072	4.20	1.13	.84	18.5	–

From *Catalogue of Standard Reference Materials*, NBS Special Publication 260, U.S. Government Printing Office, Washington, D.C., July 1970.

Table 5.3-7
MARAGING STEEL

SRM No. 1 1/4 in. D × 3/4 in. disk	Name	C	Mn	P	S	Si	Cu
1156	Maraging, (Ni 19)	0.023	0.21	0.011	0.012	0.184	0.025

SRM No. 1 1/4 in. D × 3/4 in. disk	Ni	Cr	Mo	Co	Ti	Al	Zr	B	Ca
1156	19.0	0.20	3.1	7.3	0.21	0.047	0.004	0.003	<0.001

From *Catalogue of Standard Reference Materials*, NBS Special Publication 260, U.S. Government Printing Office, Washington, D.C., July 1970.

Table 5.3-8
HIGH-TEMPERATURE ALLOYS

SRM No. 1 1/4 in. D × 3/4 in. disk	Name	C	Mn	P	S	Si	Cu	Ni	Al
1194	A 286	0.81	0.67	0.011	0.008	0.71	0.047	24.06	1.45
1185	Cr17-Ni13 (AISI 316, AMS 5360A)	.11	1.22	.019	.016	.40	.067	13.18	–
1155	Cr18-Ni12-Mo2 (AISI 316)	.046	1.63	.020	.018	.502	.169	12.18	–

SRM No. 1 1/4 in. D × 3/4 in. disk	Fe	Cr	V	Mo	Co	Ti	Nb	Ta	Pb	Zr	B
1194	51.3	16.35	0.32	1.27	2.77	1.45	–	–	–	0.026	0.0090
1185	–	17.09	–	2.01	–	<0.001	<0.001	<0.001	–	–	–
1155	–	18.45	.047	2.38	0.101	–	–	–	0.001	–	–

From *Catalogue of Standard Reference Materials*, NBS Special Publication 260, U.S. Government Printing Office, Washington, D.C., July 1970.

Table 5.3-9
OXYGEN STANDARDS

SRM No.	Name	Unit	Oxygen (ppm)	Nitrogen (ppm)
1090	Ingot iron	Rods 1/4 in. D × 4 in. long	491	(60)
1091	Stainless steel (AISI 431)	Rods 5/16 in. D × 4 in. long	131	(945)
1092	Vacuum-melted steel	Rods 1/4 in. D × 4 in. long	28	(4)
1093	Valve steel	Rods 1/4 in. D × 4 in. long	60	(4807)
1094	Maraging steel	Rods 1/4 in. D × 4 in. long	4.5	(71)

From *Catalogue of Standard Reference Materials*, NBS Special Publication 260, U.S. Government Printing Office, Washington, D.C., July 1970.

Table 5.3—10
CAST IRONS (CHIP FORM)

Chemical composition (nominal weight percent)

SRM No.	Name	Wt/unit (grams	C Total	C Graphitic	Mn	P	S Grav.	S Comb.	Si	Cu
3b	White	110	2.44	–	0.353	0.086	–	0.088	1.04	0.050
4j	Cast	150	2.99	2.38	.79	.17	–	.062	1.31	.24
5L	Cast	150	2.59	1.99	.68	.280	–	.123	1.83	1.01
6g	Cast	150	2.84	2.00	1.06	.56	–	.123	1.06	0.50
7g	Cast (high phosphorus)	150	2.69	2.59	0.612	.794	0.061	0.060	2.41	.128
55e	Ingot	150	0.0112	–	.035	.003	.012	.011	0.001	.065
82b	Cast (Ni-Cr)	150	2.85	2.37	.745	.025	–	.007	2.10	.038
107b	Cast (Ni-Cr-Mo)	150	2.75	1.87	.510	.058	.067	.067	1.35	.235
115a	Cast (Cu-Ni-Cr)	150	2.62	1.96	1.00	.086	.064	.065	2.13	5.52
122e	Cast (car wheel)	150	3.51	2.78	0.528	.349	–	.074	0.510	0.033
341	Ductile	150	1.81	1.23	.92	.024	.007	.007	2.44	.152
342	Nodular	150	2.45	2.14	.369	.020	.014	.014	2.85	.14
342a	Nodular	150	1.86	1.38	.275	.018	–	.006	2.73	.14

Chemical composition (nominal weight percent)

SRM No.	Ni	Cr	V	Mo	Co	Ti	As	Sn	Al (Total)	Mg	N
3b	0.010	0.052	0.006	0.002	–	–	–	–	–	–	–
4j	.068	.09	.03	.080	–	0.05	0.03	–	–	–	–
5L	.086	.15	.036	.020	–	.05	<.005	–	–	–	0.006
6g	.136	.37	.06	.035	–	.06	.04	–	–	–	.006
7g	.120	.048	.010	.012	–	.044	.014	–	–	–	.004
55e	.038	.006	<.001	.011	0.007	–	.007	0.007	0.002	–	.004
82b	1.22	.333	.027	.002	–	.027	–	–	–	–	–
107b	2.12	.560	.008	.750	–	.016	–	–	–	–	(.008)
115a	14.49	1.98	.014	.050	–	.020	–	–	–	–	–
122e	0.080	(0.038)	(.032)	(.001)	–	(.026)	(.018)	–	–	–	(.009)
341	20.32	1.98	.012	.010	–	.018	–	–	–	0.068	–
342	0.023	0.032	.005	.009	–	.019	–	–	–	.053	–
342a	.06	.034	–	–	–	.020	–	–	–	.069	–

From *Catalogue of Standard Reference Materials*, NBS Special Publication 260, U.S. Government Printing Office, Washington, D.C., July 1970.

Table 5.3–11

CAST STEELS, WHITE CAST IRONS, DUCTILE IRONS AND BLAST FURNACE IRONS (SOLID FORM)

SRM No. 1 1/4 in. thick / 1/2 in. disk	Name	Chemical composition (nominal weight percent)									
		C	Mn	P	S	SI	Cu	Ni	Cr	V	Mo
1174a	White (special 1)	3.45	0.180	0.168	0.168	0.283	0.170	0.035	0.018	0.008	0.008
1175a	White (special 2)	1.98	1.62	.648	.018	3.47	1.50	2.99	2.41	.222	1.49
1147	White (4i)	3.60	0.78	.160	.059	1.31	0.23	0.070	0.093	.032	0.078
1148	White (5L)	2.89	.66	.300	(.11)	1.82	.99	.091	.146	.036	.022
1149	White (6g)	3.28	1.05	.564	.127	1.04	.49	.138	.363	.055	.036
1140	Ductile (No. 1)	3.18	0.725	.0070	.010	1.92	.10	.028	.030	.030	.090
1141	Ductile (No. 2)	3.64	.480	.072	.020	1.11	.21	.54	.145	.0090	.05
1142	Ductile (No. 3)	2.94	.18	.20	.015	3.33	1.02	1.65	.053	.006	.022
1138	Cast steel (No. 1)	0.120	.43	.053	.053	0.34	0.09	0.10	.12	.020	.05
1139	Cast steel (No. 2)	.792	.98	.011	.013	.85	.40	.93	1.96	.24	.51
1143	Blast furnace (No. 1)	3.91	.414	.158	.028	1.68	.144	.115	0.145	.008	(.005)
1144	Blast furnace (No. 2)	4.27	1.33	.112	.021	0.276	.090	.021	.019	.004	.007

SRM No. 1 1/4 in. thick / 1/2 in. disk	Chemical composition (nominal weight percent)													
	Co	Ti	As	Sb	Sn	Al	Te	Zr	B	Bi	Ce	Y	Pb	Mg
1174a	0.009	0.011	0.024	0.17	0.23	(0.001)	0.072	(0.02)	0.040	(0.008)	—	—	(0.01)	—
1175a	.11	.35	.19	.022	.025	(.03)	.009	(.03)	.005	(.017)	—	—	.006	—
1147	—	.049	.022	—	—	—	.016	—	—	—	—	—	—	—
1148	—	.050	(.022)	—	—	—	.015	—	—	—	—	—	—	—
1149	—	.062	.036	—	—	—	.013	—	—	—	—	—	—	—
1140	—	.10	(.07)	—	—	(.01)	—	—	—	—	(0.09)	(<0.002)	—	0.019
1141	—	.013	(.04)	—	—	(.005)	—	—	—	—	(.05)	.040	—	.044
1142	—	.008	(.015)	—	—	(.09)	—	—	—	—	(.015)	.01	—	.10
1138	—	—	—	—	—	—	—	—	—	—	—	—	—	—
1139	—	—	—	—	—	—	—	—	—	—	—	—	—	—
1143	—	.17	(.004)	—	—	—	.020	—	—	—	—	—	—	—
1144	—	.44	(.004)	—	—	—	.020	—	—	—	—	—	—	—

From *Catalogue of Standard Reference Materials*, NBS Special Publication 260, U.S. Government Printing Office, Washington, D.C., July 1970.

725

Table 5.3–12
STEELMAKING ALLOYS

SRM No.	Name	Wt/unit (grams)	Chemical composition (nominal weight percent)						
			C	Mn	P	S	Si	Cu	Ni
57	Refined silicon	60	0.087	0.034	0.008	0.005	96.80	0.02	0.002
59a	Ferrosilicon-50%	50	.04	.76	.016	–	48.2	.05	.03
64b	Ferrochromium (high carbon)	100	4.30	.208	.012	.062	1.42	–	–
196	Ferrochromium (low carbon)	100	0.035	.28	–	–	.38	–	–
71	Calcium molybdate	60	–	–	–	–	–	–	–
90	Ferrophosphorus	75	–	–	26.2	–	–	–	–
340	Ferroniobium	100	.060	1.71	.035	–	4.39	–	–

SRM No.	Chemical composition (nominal weight percent)											
	Cr	V	Mo	Ti	Al	Nb	Zr	Ca	Mg	Fe	B	N
57	0.025	–	–	0.10	0.67	–	0.025	0.73	0.01	0.65	–	–
59a	.08	–	–	–	.35	–	–	.04	–	50.0	0.06	–
64b	68.03	0.15	–	–	–	–	–	–	–	–	–	0.033
196	70.87	.12	–	–	–	–	–	–	–	–	–	–
71	–	–	35.3	.06	–	–	–	–	–	1.92	–	–
90	–	–	–	–	–	–	–	–	–	–	–	–
340	–	–	–	.89	–	57.51	Ta 3.73	–	–	–	–	–

From *Catalogue of Standard Reference Materials*, NBS Special Publication 260, U.S. Government Printing Office, Washington, D.C., July 1970.

Table 5.3–13
ALUMINUM-BASE ALLOYS

SRM No.	Name	Wt/unit (grams)	Chemical composition (nominal weight percent)												
			Mn	Si	Cu	Ni	Cr	V	Ti	Sn	Ga	Fe	Pb	Mg	Zn
85b	Wrought	75	0.61	0.18	3.99	0.084	0.211	0.006	0.022	—	0.019	0.24	0.021	1.49	0.030
86c	Casting	75	.041	.68	7.92	.030	.029	—	.035	—	—	.90	.031	0.002	1.50
87a	Al-Si	75	.26	6.24	0.30	.57	.11	<.01	.18	0.05	.02	.61	.10	.37	0.16

From *Catalogue of Standard Reference Materials*, NBS Special Publication 260, U.S. Government Printing Office, Washington, D.C., July 1970.

Table 5.3—14
COPPER-BASE ALLOYS

SRM No.	Name	Wt/unit (grams)	Chemical composition (nominal weight percent)					
			Mn	P	S	Si	Cu	Ni
37e	Brass, sheet	150	—	—	—	—	69.61	0.53
52c	Bronze, cast	150	—	0.001	0.002	—	89.25	.76
124d	Bronze, ounce metal	150	—	.02	.093	—	83.60	.99
157a	Nickel silver	135	0.174	.009	—	—	58.61	11.82
158a	Bronze, silicon	150	1.11	.026	—	3.03	90.93	0.001
184	Bronze, leaded tin	150	—	.009	—	—	88.96	.50

SRM No.	Chemical composition (nominal weight percent)								
	Co	As	Sn	Fe	Al	Pb	Sb	Ag	Zn
37e	—	—	1.00	0.004	—	1.00	—	—	27.85
52c	—	—	7.85	.004	—	0.011	—	—	2.12
124e	—	0.02	4.56	.18	—	5.20	0.17	0.02	5.06
157a	0.022	—	0.021	.174	—	0.034	—	—	29.09
158a	—	—	.96	1.23	0.46	.097	—	—	2.08
184	—	—	6.38	0.005	--	1.44	—	—	2.69

From *Catalogue of Standard Reference Materials,* NBS Special Publication 260, U.S. Government Printing Office, Washington, D.C., July 1970.

Table 5.3—15
NICKEL OXIDE

SRM No.	Name	Wt/unit (grams)	Chemical composition (nominal weight percent)								
			Mn	Si	Cu	Cr	Co	Ti	Al	Fe	Mg
671	Oxide 1	25	0.13	0.047	0.20	0.025	0.31	0.024	0.009	0.39	0.030
672	Oxide 2	25	.095	.11	.018	.003	.55	.009	.004	.079	.020
673	Oxide 3	25	.0037	.006	.002	.0003	.016	.003	.001	.029	.003

From *Catalogue of Standard Reference Materials,* NBS Special Publication 260, U.S. Government Printing Office, Washington, D.C., July 1970.

Table 5.3—16
TITANIUM-BASE ALLOYS

SRM No.	Name	Wt/unit (grams)	Chemical composition (nominal weight percent)									
			C	Mn	Si	Cu	V	Mo	Sn	Al	Fe	N
173a	6Al-4V	100	0.025	—	0.037	0.002	4.06	0.005	—	6.47	0.15	0.018
174	4Al-4Mn	100	—	4.57	.015	—	—	—	—	4.27	.175	.012
176	5Al-2.5Sn	100	.015	0.0008	—	.003	—	.0003	2.47	5.16	.070	.010

From *Catalogue of Standard Reference Materials,* NBS Special Publication 260, U.S. Government Printing Office, Washington, D.C., July 1970.

Table 5.3–17
COPPER-BASE ALLOYS FOR INSTRUMENTAL ANALYSIS

Chemical composition (nominal weight percent)

SRM No.		Name	Cu	Zn	Pb	Fe	Sn	Ni	Al	Sb	As
–	C1100	Cartridge brass A	67.43	32.20	0.106	0.072	0.055	0.052	0.008	0.018	0.019
1101	C1101	Cartridge brass B	69.50	30.30	.05	.037	.016	.013	.0006	.012	.009
1102	C1102	Cartridge brass C	72.85	27.10	.020	.011	.006	.005	.0007	.005	.004
1103	C1103	Free-cutting brass A	59.23	35.7	3.73	.26	.88	.16	–	–	–
1104	C1104	Free-cutting brass B	61.33	35.3	2.77	.088	.43	.070	–	–	–
1105	C1105	Free-cutting brass C	63.7	34.0	2.0	.044	.21	.043	–	–	–
1106	C1106	Naval brass A	59.08	40.08	0.032	.004	.74	.025	–	–	–
1107	C1107	Naval brass B	61.21	37.34	.18	.037	1.04	.098	–	–	–
1108	C1108	Naval brass C	64.95	34.42	.063	.050	.39	.033	–	–	–
1109	C1109	Red brass A	82.2	17.4	.075	.053	.10	.10	–	–	–
1110	C1110	Red brass B	84.59	15.20	.033	.033	.051	.053	–	–	–
1111	C1111	Red brass C	87.14	12.81	.013	.010	.019	.022	–	–	–
1112	C1112	Gilding metal A	93.38	6.30	.057	.070	.12	.100	–	–	–
1113	C1113	Gilding metal B	95.03	4.80	.026	.043	.064	.057	–	–	–
1114	C1114	Gilding metal C	96.45	3.47	.012	.017	.027	.021	–	–	–
1115	C1115	Commercial bronze A	87.96	11.73	.013	.13	.10	.074	–	–	–
1116	C1116	Commercial bronze B	90.37	9.44	.042	.046	.044	.048	–	–	–
1117	C1117	Commercial bronze C	93.01	6.87	.069	.014	.021	.020	–	–	–
1118	C1118	Aluminum brass A	75.1	21.9	.025	.065	–	–	2.80	.010	.007
1119	C1119	Aluminum brass B	77.1	20.5	.050	.030	–	–	2.14	.050	.040
1120	C1120	Aluminum brass C	80.1	18.1	.105	.015	–	–	1.46	.100	.090
1121	C1121	Beryllium copper CABRA alloy 165-170	97.49	(0.01)	(.002)	.085	.01	.012	.07	–	–
1122	C1122	Beryllium copper CABRA alloy 25-172	97.45	(.01)	(.003)	.16	(.01)	(.01)	.17	–	–
1123	C1123	Beryllium copper CABRA alloy 10-175	97.10	(.01)	(.001)	.04	(.01)	(.01)	.02	–	–

Table 5.3—17 (continued)
COPPER-BASE ALLOYS FOR INSTRUMENTAL ANALYSIS

SRM No.		Chemical composition (nominal weight percent)									
		Be	Bi	Cd	Mn	P	Si	Ag	Te	Co	Cr
–	C1100	0.0015	0.0010	0.013	0.003	0.010	(0.010)	0.019	0.0035	–	–
1101	C1101	.00055	.0004	.0055	.0055	.0020	(.005)	.003	.0015	–	–
1102	C1102	.00003	.0005	.0045	.0045	.0048	(.002)	.0010	.0003	–	–
1103	C1103	–	–	–	–	.003	–	–	–	–	–
1104	C1104	–	–	–	–	.005	–	–	–	–	–
1105	C1105	–	–	–	–	.003	–	–	–	–	–
1106	C1106	–	–	–	.005	–	–	–	–	–	–
1107	C1107	–	–	–	–	–	–	–	–	–	–
1108	C1108	–	–	–	.025	–	–	–	–	–	–
1109	C1109	–	–	–	–	.006	–	–	–	–	–
1110	C1110	–	–	–	–	–	–	–	–	–	–
1111	C1111	–	–	–	–	–	–	–	–	–	–
1112	C1112	–	–	–	–	.009	–	–	–	–	–
1113	C1113	–	–	–	–	.008	–	–	–	–	–
1114	C1114	–	–	–	–	.009	–	–	–	–	–
1115	C1115	–	–	–	–	.005	–	–	–	–	–
1116	C1116	–	–	–	–	.008	–	–	–	–	–
1117	C1117	–	–	–	–	.002	–	–	–	–	–
1118	C1118	–	–	–	–	.13	.0021	–	–	–	–
1119	C1119	–	–	–	–	.070	.0015	–	–	–	–
1120	C1120	–	–	–	–	.018	.0011	–	–	–	–
1121	C1121	1.90	–	–	(.004)	(.005)	.11	(.005)	–	0.295	(0.002)
1122	C1122	1.75	–	–	(.004)	(.004)	.17	(.005)	–	.220	(.002)
1123	C1123	0.46	–	–	(.002)	(.002)	.03	(.009)	–	2.35	(.002)

From *Catalogue of Standard Reference Materials*, NBS Special Publication 260, U.S. Government Printing Office, Washington, D.C., July 1970.

Table 5.3–18
TITANIUM-BASE ALLOYS FOR AEROSPACE INDUSTRY

SRM No.	Name	Unit size	Chemical composition (nominal weight percent)					
			Mn	Cr	Fe	Mo	Al	V
641	8Mn (A)	1 1/4 in. D × 3/4 in. disks	6.68	–	–	–	–	–
642	8Mn (B)	1 1/4 in. D × 3/4 in. disks	9.08	–	–	–	–	–
643	8Mn (C)	1 1/4 in. D × 3/4 in. disks	11.68	–	–	–	–	–
644	2Cr-2Fe-2Mo (A)	1 1/4 in. D × 3/4 in. disks	–	1.03	1.36	3.61	–	–
645	2Cr-2Fe-2Mo (B)	1 1/4 in. D × 3/4 in. disks	–	1.96	2.07	2.38	–	–
646	2Cr-2Fe-2Mo (C)	1 1/4 in. D × 3/4 in. disks	–	3.43	2.14	1.11	–	–
654	6Al-4V (B)	1 1/4 in. D × 3/4 in. disks	–	–	–	–	6.03	3.83

From *Catalogue of Standard Reference Materials*, NBS Special Publication 260, U.S. Government Printing Office, Washington, D.C., July 1970.

Table 5.3–19
TITANIUM-BASE ALLOYS – OXYGEN AND HYDROGEN ONLY

SRM No.	Name	Unit size	Wt/unit (grams)	Oxygen (ppm)	Hydrogen (Wt %)
352	Unalloyed titanium for hydrogen	1/4 in. square × 0.05 in. thick	20	–	0.0032
353	Unalloyed titanium for hydrogen	1/2 in. square × 0.05 in. thick	20	–	.0098
354	Unalloyed titanium for hydrogen	1/2 in. square × 0.05 in. thick	20	–	.0215
355	Unalloyed	Rod–1/2 in. D × 2 in. long	–	3031	–
356	Alloy, 6Al-4V	Rod–.425 in. D × 1 3/4 in. long	–	1332	–

From *Catalogue of Standard Reference Materials*, NBS Special Publication 260, U.S. Government Printing Office, Washington, D.C., July 1970.

Table 5.3—20
DIE CASTING ALLOYS AND SPELTER

SRM No.	Name	Unit size	Chemical composition (nominal weight percent)					
			Cu	Al	Mg	Fe	Pb	Cd
625	Zinc-base A-ASTM AG 40A	1¾ in. square × ¾ in. thick	0.034	3.06	0.070	0.036	0.0014	0.0007
626	Zinc-base B-ASTM AG 40A	1¾ in. square × ¾ in. thick	.056	3.56	.020	.103	.0022	.0016
627	Zinc-base C-ASTM AG 40A	1¾ in. square × ¾ in. thick	.132	3.88	.030	.023	.0082	.0051
628	Zinc-base D-ASTM AC 41A	1¾ in. square × ¾ in. thick	.611	4.59	.0094	.066	.0045	.0040
629	Zinc-base E-ASTM AC 41A	1¾ in. square × ¾ in. thick	1.50	5.15	.094	.017	.0135	.0155
630	Zinc-base F-ASTM AC 41A	1¾ in. square × ¾ in. thick	0.976	4.30	.030	.023	.0083	.0048
631	Zinc spelter (modified)	1¾ in. square × ¾ in. thick	.0013	0.50	(<.001)	.005	(.001)	.0002

SRM No.	Chemical composition (nominal weight percent)									
	Sn	Cr	Mn	Ni	Si	In	Ga	Ca	Ag	Ge
625	0.0006	0.0128	0.031	0.0184	0.017	—	—	—	—	—
626	.0012	.0395	.048	.047	.042	—	—	—	—	—
627	.0042	.0038	.014	.0029	.021	—	—	—	—	—
628	.0017	.0087	.0091	.030	.009	—	—	—	—	—
629	.012	.0008	.0017	.0075	.078	—	—	—	—	—
630	.0040	.0031	.0106	.0027	.022	—	—	—	—	—
631	.0001	.0001	.0015	(<.0005)	<.002	(0.0023)	(0.002)	<0.001	(<0.0005)	(0.0002)

From *Catalogue of Standard Reference Materials*, NBS Special Publication 260, U.S. Government Printing Office, Washington, D.C., July 1970.

Table 5.3–21
HIGH-PURITY METALS

SRM No.	Name	Unit size	Chemical compositions (nominal parts per million by weight)				
			Cu	Ni	Sn	Pb	Zr
685W	High-purity gold (wire)	1.4 mm D × 102 mm long	0.1	–	–	–	–
685R	High-purity gold (rod)	5.9 mm D × 25 mm long	.1	–	–	–	–
680L1	High-purity platinum (wire)	0.51 mm D × 102 mm long	.1	<1	–	<1	<0.1
680L2	High-purity platinum (wire)	0.51 mm D × 1.0 m long					
681L1	Doped-platinum (wire)	0.51 mm D × 102 mm long	5.1	0.5	–	12	11
681L2	Doped-platinum (wire)	0.51 mm D × 1.0 m long					
682	High-purity zinc	Semicircular segments 57 mm D × 19 mm long	0.042	–	(0.02)	–	–
683	Zinc metal	Semicircular segments 57 mm D × 19 mm long	5.9	–	(.02)	11.1	–

Chemical compositions (nominal parts per million by weight)

SRM No.	Ag	Mg	In	Fe	O	Pd	Au	Rh	Ir	Cd	Tl
685W[a]	[0.1]	–	0.007	0.3	[2]	–	–	–	–	–	–
685R[a]	[.1]	–	.007	2	[<2]	–	–	–	–	–	–
680L1	.1	<1	–	0.7	4	0.2	<1	<0.2	<0.01	–	–
680L2											
681L1	2.0	12	–	5	7	6	9	9	11	–	–
681L2											
682[a]	(0.02)	–	–	(0.1)	–	–	–	–	–	(0.1)	–
683[a]	1.3	–	–	2.2	–	–	–	–	–	1.1	(0.2)

[a]Certificate gives upper limits for other elements found to be present.

From *Catalogue of Standard Reference Materials*, NBS Special Publication 260, U.S. Government Printing Office, Washington, D.C., July 1970.

Table 5.3 − 22

PRIMARY, WORKING, AND SECONDARY STANDARD CHEMICALS

These SRMs are high-purity chemicals defined as primary, working, and secondary standards in accordance with recommendations of the Analytical Chemistry Section of the International Union of Pure and Applied Chemistry (reference: *Analyst*, 90, 251, 1965). These definitions are as follows:

Primary Standard: a commercially available substance of purity 100 ± 0.02% (Purity 99.98+ %).

Working Standard: a commercially available substance of purity 100 ± 0.05% (Purity 99.95+ %).

Secondary Standard: a substance of lower purity which can be standardized against a primary grade standard.

SRM No.	Name	Wt/unit (grams)	Certified use	Purity on basis of titration
17	Sucrose	60	Polarimetric value	a
40h	Sodium oxalate	.60	Reductometric value	99.95
41a	Dextrose (D-glucose)	70	Reductometric value	b
83c	Arsenic trioxide	75	Reductometric value	99.99
84h	Acid potassium phthalate	60	Acidimetric value	99.99
136c	Potassium dichromate	60	Oxidimetric value	99.98
350	Benzoic acid	30	Acidimetric value	99.98
950a	Uranium oxide (U_3O_8)	25	Uranium oxide standard value	99.94
951	Boric acid	100	Acidimetric and boron isotopic value	100.00

aSucrose—moisture <0.01%, reducing substances <0.02%, ash 0.003%.
bDextrose—moisture <0.2%, ash <0.01%.

From *Catalogue of Standard Reference Materials*, NBS Special Publication 260, U.S. Government Printing Office, Washington, D.C., July 1970.

Table 5.3−23
METALLO-ORGANIC COMPOUNDS

SRM No.	Constituent certified		Wt/unit (grams)	Name
	Element	Wt. percent		
1075a	Al	8.1	5	Aluminum 2-ethylhexanoate
1051b	Ba	28.7	5	Barium cyclohexanebutyrate
1063a	B	2.4	5	Menthyl borate
1053	Cd	24.0	5	Cadmium cyclohexanebutyrate
1074a	Ca	12.5	5	Calcium 2-ethylhexanoate
1078a	Cr	9.7	5	Tris(1-phenyl-1,3-butanediono)chromium(III)
1055b	Co	14.8	5	Cobalt cyclohexanebutyrate
1080	Cu	16.5	5	Bis(1-phenyl-1,3-butanediono)copper(II)
1079b	Fe	10.3	5	Tris(1-phenyl-1,3-butanediono)iron(III)
1059b	Pb	36.7	5	Lead cyclohexanebutyrate
1060a	Li	4.1	5	Lithium cyclohexanebutyrate
1061b	Mg	6.5	5	Magnesium cyclohexanebutyrate
1062a	Mn	13.8	5	Manganous cyclohexanebutyrate
1064	Hg	36.2	5	Mercuric cyclohexanebutyrate
1065b	Ni	13.9	5	Nickel cyclohexanebutyrate

Table 5.3−23 (continued)
METALLO-ORGANIC COMPOUNDS

SRM No.	Constituent certified		Wt/unit (grams)	Name
	Element	Wt. percent		
1071a	P	9.5	5	Triphenyl phosphate
1066a	Si	14.1	5	Octaphenylcyclotetrasiloxane
1076	K	10.1	5	Potassium erucate
1077a	Ag	42.6	5	Silver 2-ethylhexanoate
1069b	Na	12.0	5	Sodium cyclohexanebutyrate
1070a	Sr	20.7	5	Strontium cyclohexanebutyrate
1057b	Sn	23.0	5	Dibutyltin bis(2-ethylhexanoate)
1052b	V	13.0	5	Bis(1-phenyl-1,3-butanediono)oxovanadium(IV)
1073b	Zn	16.7	6	Zinc cyclohexanebutyrate

From *Catalogue of Standard Reference Materials,* NBS Special Publication 260, U.S. Government Printing Office, Washington, D.C., July 1970.

Table 5.3−24
ANALYZED GASES

SRM No.	Name	Vol/unit (liters as STP)	Constituents certified
1601	Carbon dioxide in nitrogen	68	CO_2, 308 ± 3 ppm
1602	Carbon dioxide in nitrogen	68	CO_2, 346 ± 3 ppm
1603	Carbon dioxide in nitrogen	68	CO_2, 384 ± 4 ppm
1604	Oxygen in nitrogen	68	O_2, 3 ppm
1605	Oxygen in nitrogen	68	O_2, 10 ppm
1606	Oxygen in nitrogen	68	O_2, 112 ppm
1607	Oxygen in nitrogen	68	O_2, 212 ppm
1608	Oxygen in nitrogen	68	O_2, 978 ppm
1609	Oxygen in nitrogen	68	O_2, 20.95 mol %

From *Catalogue of Standard Reference Materials,* NBS Special Publication 260, U.S. Government Printing Office, Washington, D.C., July 1970.

Table 5.3—25
ORES

SRM No.	Name	Wt/unit (grams)	Chemical composition (nominal weight percent)								
			Fe	Mn	P	P_2O_5	SiO_2	Li_2O	Sn	Zn	Available oxygen
27e	Iron (Sibley)	100	66.58	—	0.042	—	3.65	—	—	—	—
28a	Iron (Norrie)	50	—	0.435	—	—	—	—	—	—	—
181	Lithium (Spodumene)	45	—	—	—	—	—	6.4	—	—	—
182	Lithium (Petalite)	45	—	—	—	—	—	4.3	—	—	—
183	Lithium (Lepidolite)	45	—	—	—	—	—	4.1	—	—	16.7
25c	Manganese	100	—	57.85	—	0.22	2.36	—	—	—	—
138	Tin (N.E.I. concentrate)	50	—	—	—	—	—	—	74.8	—	—
113	Zinc (tri-state concentrate)	50	—	—	—	—	—	—	—	61.1	—

SRM No.	Name	Wt/unit (grams)	Chemical composition (nominal weight percent as the oxide)						
			Al_2O_3	Fe_2O_3	TiO_2	P_2O_5	ZrO_2	V_2O_5	SiO_2
69a	Bauxite	50	55.0	5.8	2.78	0.08	0.18	0.03	6.01

SRM No.	Chemical composition (nominal weight percent as the oxide)								
	Cr_2O_3	CaO	BaO	MgO	MnO	Na_2O	K_2O	SO_3	Loss on ignition
69a	0.05	0.29	0.01	0.02	<0.01	<0.01	<0.01	0.04	29.55

From *Catalogue of Standard Reference Materials*, NBS Special Publication 260, U.S. Government Printing Office, Washington, D.C., July 1970.

Table 5.3–26
CEMENTS

SRM No.	Name	Wt/unit (grams)	Chemical composition (nominal weight percent as the oxide)				
			SiO_2	Al_2O_3	Fe_2O_3	TiO_2	P_2O_5
1011	Portland	15	21.03	5.38	2.07	0.25	0.33
1013	Portland	15	24.17	3.30	3.07	.20	.20
1014	Portland	15	19.49	6.38	2.50	.25	.32
1015	Portland	15	20.65	5.04	3.27	.26	.05
1016	Portland	15	21.05	4.97	3.71	.34 .	.13

Chemical composition (nominal weight percent as the oxide)

SRM No.	CaO (+SrO)	SrO	MgO	SO_3	Mn_2O_3	Na_2O	K_2O	Li_2O	Rb_2O	Loss on ignition
1011	66.60	0.11	1.12	1.75	0.03	0.08	0.26	(0.002)	(0.001)	1.13
1013	64.34	.08	1.39	1.80	.05	.20	.32	(.001)	(.004)	0.99
1014	63.36	.26	2.80	2.70	.07	.24	.99	(.005)	(.007)	.81
1015	61.48	.11	4.25	2.28	.06	.16	.87	(.004)	(.005)	1.70
1016	65.26	.25	0.42	2.27	.04	.55	.04	(.012)	(<.001)	1.20

From *Catalogue of Standard Reference Materials*, NBS Special Publication 260, U.S. Government Printing Office, Washington, D.C., July 1970.

Table 5.3–27
MINERALS

SRM No.	Name	Wt/unit (grams)	Chemical composition (nominal weight percent as the oxide)					
			SiO_2	Fe_2O_3	Al_2O_3	TiO_2	MnO	CaO
1b	Limestone, argillaceous	50	4.92	0.75	1.12	0.046	0.20	50.9
88a	Limestone, dolomitic	50	1.20	.28	0.19	.02	.03	30.1
70a	Feldspar, potash	40	67.1	.075	17.9	.01	–	0.11
99a	Feldspar, soda	40	65.2	.065	20.5	.007	–	2.14
97a	Clay, flint	60	43.7	.45	38.8	1.90	–	0.11
98a	Clay, plastic	60	48.9	1.34	33.2	1.61	–	.31

Chemical composition (nominal weight percent as the oxide)

SRM No.	SrO	MgO	Cr_2O_3	Na_2O	K_2O	Li_2O	ZrO_2	BaO	Rb_2O	P_2O_5	CO_2	Loss on ignition
1b	0.14	0.36	–	0.04	0.25	–	–	–	–	0.08	40.4	41.1
88a	.01	21.3	–	.01	.12	–	–	–	–	.01	46.6	46.7
70a	–	–	–	2.55	11.8	–	–	0.02	0.06	–	–	0.40
99a	–	0.02	–	6.2	5.2	–	–	.26	–	0.02	–	0.25
97a	.18	.15	0.03	0.037	0.50	0.11	0.063	.078	–	.36	–	13.32
98a	.039	.42	.03	.082	1.04	.070	0.042	.03	–	.11	–	12.44

From *Catalogue of Standard Reference Materials*, NBS Special Publication 260, U.S. Government Printing Office, Washington, D.C., July 1970.

Table 5.3—28
REFRACTORIES

SRM No.	Name	Wt/unit (grams)	Chemical composition (nominal weight percent as the oxide)				
			SiO_2	Al_2O_3	Total as Fe_2O_3	FeO	TiO_2
103a	Chrome refractory	60	4.6	29.96	–	12.43	0.22
198	Silica refractory	45	–	0.16	0.66	–	.02
199	Silica refractory	45	–	.48	.74	–	.06
104	Burned magnesite	60	2.54	.84	7.07	–	.03

SRM No.	Chemical composition (nominal weight percent as the oxide)									Loss on ignition
	ZrO_2	MnO	P_2O_5	Cr_2O_3	CaO	MgO	Li_2O	Na_2O	K_2O	
103a	0.01	0.11	0.01	32.06	0.69	18.54	–	–	–	–
198	<.01	.008	.022	–	2.71	0.07	0.001	0.012	0.017	0.21
199	.01	.007	.015	–	2.41	.13	.002	.015	.094	.17
104	–	.43	.057	0.026	3.35	85.67	.001	.015	.015	–

From *Catalogue of Standard Reference Materials,* NBS Special Publication 260, U.S. Government Printing Office, Washington, D.C., July 1970.

Table 5.3—29
CARBIDES

SRM No.	Name	Wt/unit (grams)	Carbon		Silicon		Chemical composition (nominal weight percent)					
			Total	Free	Total	SiC	Fe	Al	Ti	Zr	Ca	Mg
112	Silicon carbide	85	29.10	0.09	69.11	96.85	0.45	0.23	0.025	0.027	0.03	0.02

From *Catalogue of Standard Reference Materials,* NBS Special Publication 260, U.S. Government Printing Office, Washington, D.C., July 1970.

Table 5.3–30
GLASSES

SRM No.	Name	Wt/unit (grams)	Chemical composition (nominal weight percent as the oxide)									
			SiO$_2$	PbO	Al$_2$O$_3$	Fe$_2$O$_3$	ZnO	MnO	TiO$_2$	ZrO$_2$	CaO	BaO
89	Lead-barium	45	65.35	17.50	0.18	0.049	–	0.088	0.01	0.005	0.21	1.40
91	Opal	45	67.53	0.097	6.01	.081	0.08	.008	.019	.010	10.48	–
92	Low-boron	45	–	–	–	–	–	–	–	–	–	–
93	High-boron	45	80.60	–	1.94	.076	–	–	.027	.013	–	–

SRM No.	Chemical composition (nominal weight percent as the oxide)										Loss on ignition
	MgO	K$_2$O	Na$_2$O	B$_2$O$_3$	P$_2$O$_5$	As$_2$O$_5$	As$_2$O$_3$	SO$_3$	Cl	F	
89	0.03	8.40	5.70	–	0.23	0.36	0.03	0.03	0.05	–	0.32
91	.008	3.25	8.48	–	.022	.10	.091	–	.014	5.72	–
92	–	–	–	0.70	–	–	–	–	–	–	–
93	.026	0.16	4.16	12.76	–	.14	.085	.009	.036	–	–

From *Catalogue of Standard Reference Materials*, NBS Special Publication 260, U.S. Government Printing Office, Washington, D.C., July 1970.

Table 5.3–31A
HYDROGEN ION STANDARDS, pH

SRM No.	Name	pH(S) (at 25°C)	Wt/unit (grams)
185d	Acid potassium phthalate	4.004	60
186Ic	Potassium dihydrogen phosphate	See above	30
186IIb	Disodium hydrogen phosphate	See above	30
187a	Borax	9.180	30
188	Potassium hydrogen tartrate	3.557	60
189	Potassium tetroxalate	1.679	65
191	Sodium bicarbonate		30
192	Sodium carbonate	10.01	30

From *Catalogue of Standard Reference Materials,* NBS Special Publication 260, U.S. Government Printing Office, Washington, D.C., July 1970.

Table 5.3–31B
DEUTERIUM ION STANDARDS, pD

SRM No.	Name	pD(S) values	Wt/unit (grams)
2186-I	Potassium dihydrogen phosphate		30
2186-II	Disodium hydrogen phosphate	7.43	30
2191	Sodium bicarbonate		30
2192	Sodium carbonate	10.74	30

From *Catalogue of Standard Reference Materials,* NBS Special Publication 260, U.S. Government Printing Office, Washington, D.C., July 1970.

Table 5.3–32A
STANDARD RUBBERS

SRM No.	Name	Wt/unit (grams)
385b	Natural	34,000
386g	Styrene-butadiene, type 1500	34,000
388e	Butyl	34,000
389	Styrene-butadiene, type 1503	34,000
391	Acrylonitrile-butadiene	25,000

From *Catalogue of Standard Reference Materials,* NBS Special Publication 260, U.S. Government Printing Office, Washington, D.C., July 1970.

Table 5.3–32B
RUBBER COMPOUNDING MATERIALS

SRM No.	Name	Wt/unit (grams)
370d	Zinc oxide	8,000
371f	Sulfur	6,000
372g	Stearic acid	3,200
373e	Benzothiazyl disulfide	2,000
375f	Channel Black	28,000
376a	Light magnesia	450
377	Phenyl-beta-naphthylamine	600
378a	Oil furnace black	28,000
379	Conducting black	5,500
380	Calcium carbonate	6,000
381	Calcium silicate	4,000
382a	Gas furnace black	40,000
383	Mercaptobenzothiazole	3,200
384	N-tertiary-butyl-2-benzo-thiazolesulfenamide	3,200

From *Catalogue of Standard Reference Materials,* NBS Special Publication 260, U.S. Government Printing Office, Washington, D.C., July 1970.

Index

INDEX

C

Printed and bound by CPI Group (UK) Ltd, Croydon, CR0 4YY

23/10/2024

01778254-0014